Automotive Simulation World Congress 2013 (ASWC 2013)

Frankfurt am Main, Germany
29 - 30 October 2013

ISBN: 978-1-5108-4102-4

Printed from e-media with permission by:

Curran Associates, Inc.
57 Morehouse Lane
Red Hook, NY 12571

Some format issues inherent in the e-media version may also appear in this print version.

Printed by Curran Associates, Inc. (2017)

For permission requests, please contact ANSYS Inc.
at the address below.

ANSYS Inc.
Southpointe
2600 ANSYS Drive
Canonsburg, PA 15317
USA

Phone: 844.Go.ANSYS
844.462.6797
Fax: 724.514.9494

www.ansys.com

Additional copies of this publication are available from:

Curran Associates, Inc.
57 Morehouse Lane
Red Hook, NY 12571 USA
Phone: 845-758-0400
Fax: 845-758-2633
Email: curran@proceedings.com
Web: www.proceedings.com

AGENDA DAY 1

Time	October 29, 2013
8:00	**Registration and Welcome Coffee**
09:00	**Welcome and Introduction** – Sandeep Sovani, ANSYS, Inc. 1
09:15	**Simulation-Driven Product Development in the Automotive Industry - Status & Progress** – Swaminathan Subbiah, ANSYS, Inc. 6
09:45	**High-Performance CFD Analysis - Enabling Success at the Track** – Alan Peasland, Red Bull Technology .....N/A
10:15	**Enterprise High-Performance and Cloud Computing** – Barbara Hutchings, ANSYS, Inc. 13
10:45	**Break**
11:15	**Empty Test Benches - A Discussion About the Five Dilemmas of Virtual Product Clearance** – Christoph Schöttl, MAN Truck & Bus, München 20
11:45	**Model-Based Systems Engineering: Bringing Safety-Critical Embedded Software Together with Full Systems Simulation** – Eric Bantegnie, Esterel Technologies 27
12:15	**Diversity of Drivetrains and Fuels Define Combustion Engine Development** – Roland Baar, Technische Universität Berlin 47
12:45	**Lunch Break**

Salon 9+10
Sandeep Sovani (Chairperson)

	Salon 10	Salon 8	Salon 7	Salon 9	Salon 6
Breakout Sessions	Engines	Powertrain	Body, Chassis & Interior	Aerodynamics & Underhood	Electrification, Electronics & Embedded Systems
Chair-persons	Laz Foley, ANSYS, Inc., Lebanon	Wolfgang Bauer, ANSYS Germany, Otterfing	Pierre Louat, ANSYS France, Lyon	Matteo Aroni, ANSYS Italy, Milano	Alain Michel, ANSYS France, Montigny
14:00	**Simulation of In-Cylinder Flow Phenomena in Turbocharged Direct Injection Spark Ignition Engines** D. Linse, BMW 61	**Throttle Valve for Switchable Water Pump Application - Focus on Calculation & Simulation Activities** E. Dattoli, Pierburg Pump Technology France 235	**Optimisation of the Defroster Ducts of a Coach Bus** B. Tavukçu, Anadolu Isuzu 396	**Numerical Computation of the Flow around a Racing Car** B. Schoeneberg, CFX Berlin Software 541	**Total Battery Simulation** S. Stanton, ANSYS, Inc. 729
14:30	**Cooling 760 Horses - Fluid-Structure-Interaction Simulations of Water-Cooled Diesel Cylinder Heads** ..... 74 W. Haslinger, ANSYS Germany	**Modelling of the Fuel Supply Module and its Components** O. Kurenkov, Continental Automotive 246	**Sunburned Products: Numerical Aging Caused by Sun Exposition** A. Müller, HTCO ...... N/A	**ANSYS Solutions in Competition: Formula Student** T. Grundmann, ANSYS Germany 554	**Active Force Feedback Pedal - The Smart Way to Save Fuel** J. Heydenreich, PhilonNet 749
15:00	**Numerical Simulation of the Biofuel Blend Engine In-Cylinder Combustion Process** N. Bellato, Magneti Marelli Powertrain ..... N/A	**Fluent Thermal and Fluidic Modelling to Robustify and Optimize the Design of Power Electronics Modules Carried on El. Vehicles** I. Mohand-Kaci, Renault .... N/A	**Use of the Discrete Ordinate Model for Temperature Calculation in Light Systems at Volkswagen** M. Kleimeyer, Volkswagen ..408	**A Revolution in Solar Vehicle Design** R. Stephens, Cambridge University 568	**Structural Integrity Assessment of a Li-Ion Battery Pack using Simulations** S. Kottalgi, ANSYS India 767
15:30	**Break**				
Breakout Sessions	Engines	Powertrain	Body, Chassis & Interior	Aerodynamics & Underhood	Electrification, Electronics & Embedded Systems
Chair-persons	Martin Kunz, ANSYS Germany, Otterfing	Frank Kaufmann, ANSYS Germany, Darmstadt	Jacques Depont, ANSYS France, Lyon	Anders Jonson, ANSYS Sweden, Goteborg	Gerd Prillwitz, ANSYS Germany, Otterfing
16:00	**ANSYS FLUENT Use of GPUs for Automotive CFD Applications** S. Posey, NVIDIA 30	**Development of SCR Exhaust Aftertreatment Mixing System with CFD** C. Chauvin, Donaldson Europe 258	**Design and Analysis Tools for Complex Composite Applications in the Automotive Industry** ..... N/A T. Hirche, ANSYS Switzerland	**Advanced Mesh Morphing for Automotive Applications using RBF Morph** M. Biancolini, University of Rome 586	**Collaborative Co-Simulation Cloud for Functional Mockup Design Assessments** M. Nagasawa, Cybernet Systems 780
16:30	**Diesel Particulate Filter Simulations in Exhaust After-Treatment** R. Reinelt, ANSYS Germany 107	**Multiphase CFD Simulation of Urea Injection on Off-Road Vehicles as Part of a Tier-4 B Aftertreatment System and Validation of Optimised Mixer Performance** F. Grassi, Officine Metallurgiche Cornaglia ..... N/A	**Heat-Up Simulation of Car Bodies in Paint Shop Ovens** Ch. Knüsel, Dürr Systems ... 424	**Advanced Mesh Morphing Applications in Motorsport** S. Invernizzi, Dallara Engineering ..... N/A	**FEA-Based Efficiency Map Generation of Different Electric Traction Machines** O. Winter, AIT Austrian Institute of Technology ..... N/A
17:00	**Efficient Product Development using ANSYS Software** L. Berger, CFD Schuck Ingenieurgesellschaft 123	**Aero-Thermal Modelling of High Pressure Hydrogen Storage Systems for Automotive Applications** A. Mack-Gardner, Adam Opel ..... N/A	**Quality meets Efficiency - Developing Best Practice Guidelines for Cooling Applications in Automotive Manufacturing Tool Design** M.Lanfrit, ANSYS Germany . 433	**Aerodynamics of Trucks – From CFD to Full Scale Prototypes** M. Karlsson, Linköping University ..... N/A	**Multiphysics Power Connector Simulations** M. Hanke, CADFEM 787
17:30	**Design Optimisation using Adjoint Solver and Mesh Morphing & Optimisation** N. Kulkarni, Intelligent Energy ..... N/A	**Acoustic Transmission Loss Prediction for Muffler Performance Improvement** S. Kottalgi, ANSYS India 268	**Shock-Absorber: A Study on the Dynamic Performance of a Solenoid Hydraulic Valve for Automotive Semi-Active Suspensions Coupling ANSYS Fluent with Lumped Parameter Modelling** M. Pelosi, Öhlins Racing 451	**Numerical Study of Aerodynamic Characteristics of Sports Car with Movable Flaps and Deformable Airbags** J. Piechna, Warsaw University of Technology 608	**Design of a FSPM Bridged Motor Rotor using ANSYS Simulation Tools** X. Wu, TriVista Engineering 797
18:30	Shuttles to Evening Event				

Status of Agenda, October 16, 2013

Follow us on Twitter: #ASWC_2013

Time	October 30, 2013				
8:00	Registration and Welcome Coffee				
	Salon 10	Salon 8	Salon 7	Salon 9	Salon 6
Breakout Sessions	Engines	Powertrain	Body, Chassis & Interior	Aerodynamics & Underhood	Electrification, Electronics & Embedded Systems
Chair-persons	Tammo Grundmann, ANSYS Germany, Otterfing	Marco Lanfrit, ANSYS Germany, Darmstadt	Laurent Chignier, ANSYS France, Lyon	Frederic Meslay, ANSYS France, Montegny	Leon Voss, ANSYS Germany, Hannover
8:30	**Validation and Verification of ANSYS Internal Combustion Engine Software** M. Kuntz, ANSYS Germany 129	**CFD-Based Evaluation of an Automotive Exhaust System,** R. Löffler, ANSYS Germany 286	**Significance of Simulation in the Journey of Tire from Rubber to Road** C. Patil, ANSYS India 465	**The Design of Formula SAE Car Body Aerodynamics Using ANSYS Fluent Optimisation Tools** M. Gugala, Queen Mary University of London 623	**From Radar Traps to Navigation Systems - How Simulations Can Help Reduce Cost and Improve Efficiency** S. Muff, ANSYS Germany 816
9:00	**Diesel Engine Combustion Modelling with Fluent and Tabkin** F. Tap, Dacolt ..... N/A	**Numerical Simulation of Jet Noise of Automotive Exhaust Systems using Scale-Resolving Turbulence Models and an Acoustic Analogy** J. Hillenbrand, BMW 297	**High Frequency Structural Vibrations on ANSYS Meshes - The DEA Tool for Modelling Vibrations and Acoustics in Automotive Engineering Structures** G. Tanner, The University Nottingham 481	**Continuous Improvements on the Wheel Hub of a Formula Student Race Car** H. Rönnebeck, Hochschule Amberg-Weiden 641	**Integrated Carbon Fiber Antennas for Vehicular Communication** C. Römelsberger, CADFEM 838
9:30	**Lagrangian Spark-Ignition Model Incorporating Chemical Kinetics for Turbo-Charged Direct-Injection SI Engines** L. Schäfer, BMW 153	**Air-Cooling System Study of the FP7 EUNICE In-Wheel Drive Concept using CFD** M. Rudolph, AIT Austrian Institute of Technology	**Fracture Mechanical Approach for Fatigue of Adhesive Joints** S. Vervoort, Hottinger Baldwin Meßtechnik 491	**Aerodynamic Design of an Active Rim for the EUNICE In-Wheel Electrical Drive Concept** A. Zanon, AIT Austrian Institute of Technology ..... N/A	**Microwave Measurement Portable Laboratories** P. Gareau, Microwave Vision
10:00	**Scale-Resolving Simulation of In-Cylinder Flow of an IC-Engine and Comparison to Experimental Data** S. Buhl, TU Bergakademie Freiberg 162	**An Automated Routine for Postprocessing and Report Generation in CFD-Post** F. Hahn, MANN+HUMMEL 315	**Modelling Aspects of Fatigue Analyses of Rubber Components** M. Alvelid, Alvelid Engineering ..... N/A	**The Benefit of SpaceClaim for Ansys Users** M. Hübner, SpaceClaim 648	**Signal, power integrity and EMI analysis of a high voltage lithium battery charger for full electric vehicle** P. Montanari, Meta System. 847
10:30	Break				
Breakout Sessions	Engines	Powertrain	Body, Chassis & Interior	Aerodynamics & Underhood	Electrification, Electronics & Embedded Systems
Chair-persons	Karl Kühlert, ANSYS Inc, Lebanon	Frank Kaufmann, ANSYS Germany, Darmstadt	Thomas Lehnhäuser, ANSYS Germany, Darmstadt	Marco Oswald, ANSYS Germany, Darmstadt	Simon Muff, ANSYS Germany, Otterfing
11:00	**Design Chain Simulation Supports SME Competitiveness: A Case of Piston Rod** L. Bellati, Fonderie Abor ... 176	**Simulation-Driven Design of EGR Coolers** P. Mandloi, ANSYS India ... 333	**Vehicle Brake Squeal: Morphing Allows a Robust Solution** S. Weiland, TRW Automotive ..... N/A	**Parameter-Free CFD Optimisation with Adjoint Methods for Automotive** O. Zühlke, ANSYS Germany ..652	**Towards Virtual Prototyping: Integrated Simulation of Software and Physics** C. Schrader, Esterel Technologies 858
11:30	**Quality Assurance & Best Practice Development for Internal Combustion Engine Flow Simulations** W. Bauer, ANSYS Germany 188	**Leveraging SGI's UV and VizServer Platform to Enable Multiple CAE Users to Run ANSYS Software with Remote Visualization in a Data Center** T. DeVarco, SGI 356	**Suitable Simulation Tools for the Development of a CFRP Sports Car** M. Hörmann, CADFEM 499	**Numerical Aerodynamic Investigations of the DrivAer Car Model using ANSYS Fluent** A. Joshi, TATA Consultancy Services	**Effective Automotive System Design and Validation Through the Adoption of Standardised Modelling Interfaces and Models Such as the VDA FAT-AK30 Library** ..... 867 G.Prillwitz, ANSYS, Germany
12:00	**BLDC Motor for Automotive Cooling Fan Assembly: Heat Sink Optimisation** D. Parodi, Johnson Electric 205	**A Complete Loop of Numerical Analyses about a Variable Oil Pump** R. Maccherini, Pierburg Pump Technology ..... N/A	**CAE-based Robustness Evaluation using ANSYS and optiSLang for Automotive Applications** J. Will, Dynardo ..... N/A	**DrivAer - Aerodynamic Investigations for a New Realistic Generic Car Model using ANSYS CFD** ..... 689 T. Frank, ANSYS Germany	**High-Performance Computing for a Full Vehicle Electro-Magnetic Simulation** A. Michel, ANSYS France 892
12:30	**Advanced Simulation of Filter Elements - A Practitioner's View on the Evolution from GAMBIT TO ANSYS Workbench 14.5** E. Nißler, MANN+HUMMEL . 225	**Faster Simulations using Smart Initial Solutions** G.Endicott, Honda Research Institute EU 384	**CAE Process and Crash Simulation of Fiber-Reinforced Composite Vessels** M. Chatiri, CADFEM 523	**The Influence of Mesh Characteristics on CFD Simulations for Automotive Applications** V. Skaperdas, BETA CAE Systems 710	**Modelling Inhomogeneous Heat Generation for Lithium-Ion Battery Cells** ..... 907 L. Kostetzer, CADFEM
13:00	Lunch Break				
14:00	**The Future of Integrated Simulation: Where We Come From, Where We Have To Go** – Peter Pirro, John Deere ..... 918 (Salon 9+10, Georg Scheuerer (Chairperson))				
14:30	**Cycle-to-Cycle Variations in IC Engines: Is LES an Academic Tool Only or Can It Solve Technical Problems?** – Christian Hasse, TU Bergakademie Freiberg				
15:00	Break				
15:30	**System Simulation of a Hybrid Electric Vehicle** – Scott Stanton, ANSYS, Inc. 939				
16:00	**Intel and the Future of HPC for Robust Design Simulations** – Stephen Wheat, Intel, Phoenix ..... N/A				
16:30	**Award Ceremony & Farewell** – Georg Scheuerer, ANSYS Germany 952				
17:00	End				

ADDITIONAL PAPERS

2013 Automotive Simulation World Congress

Welcome to the ASWC 2013!

AGENDA

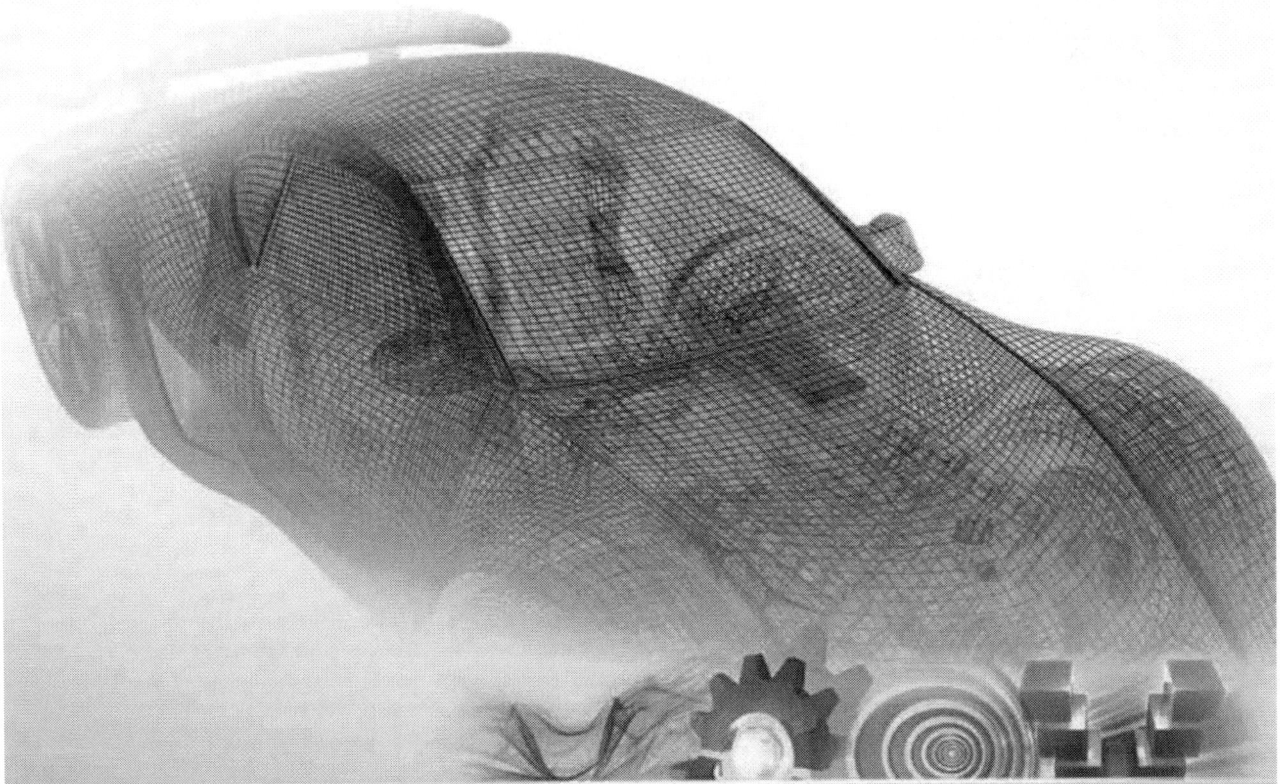

October 29 - 30 • Frankfurt am Main, Germany

Premium Sponsors

2013 Automotive Simulation World Congress

Welcome

The **World Automotive Simulation Congress** provides an unique opportunity to get an update on the latest technologies in Computer Aided Engineering and Simulation in a comprehensive multi-physics environment.
We thank you for coming and are happy to present our agenda. We look forward to spending time with you sharing information and ideas for the next two days.

Evening Event

Join us on Tuesday, October 29, 2013 for an evening at a very special location: Klassikstadt Frankfurt, where you can look forward to strolling through a collection of vintage cars, supported by experienced guides, and accompanied by live music and a buffet dinner.
For more information please visit: **www.klassikstadt.de**

KLASSIKSTADT

Shuttle Bus Service

A shuttle bus service between the Congress Venue and Klassikstadt will be offered. Buses will leave at **18:30 h** at the main entrance of the Steigenberger Airport Hotel. Between **22:00 h** and **22:30 h** shuttle buses will be available to take you back to Steigenberger Airport Hotel, to Meininger Hotel, to Park Inn by Radisson and to InterCity Hotel (all nearby the Airport).

WiFi Acess

WiFi is available within the Congress & Exhibition Area.
Login: **conf-ansys**
Password: **ASWC2013**

Follow us on Twitter: #ASWC_2013

Awards

The ASWC will close with a session in which awards will be presented in the following four categories:

- Best Paper
- Best Use of HPC
- Best Application of Multi-Disciplinary or Multi-Scale Simulation
- Most Innovative Use of Simulation

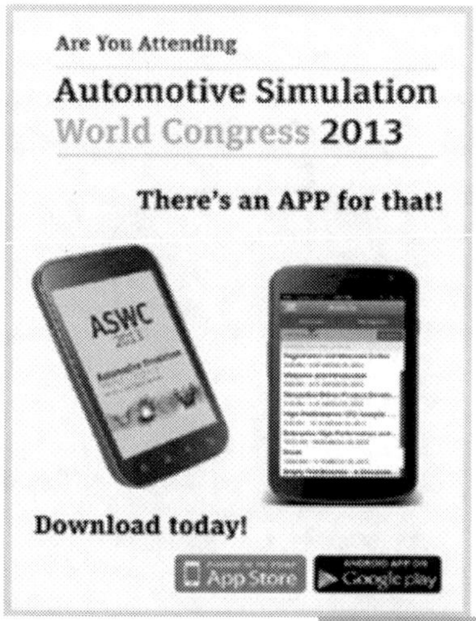

LOG IN NOW

Conference Proceedings Download

This year's conference proceedings will be distributed by download using ANSYS Engineering Knowledge Manager (ANSYS EKM). Access will be password-protected. You will receive the access information by email as soon as the proceedings are available.

You will have access to all available presentations. These will be structured on the server as in the sessions and topics at the ASWC 2013. A full-text search supports your browsing for presentations, which you may have missed in Frankfurt. Please note that it is not necessary to install ANSYS EKM or any other program to download presentations.

Keynote Speaker at ASWC 2013

Dr. Christoph Schöttl

Vice President "Engineering Simulation"

MAN Bus & Truck

Dr. Schöttl has more than 20 years simulation experience with MAN. Starting as technical staff member in the department of technical calculations in 1992 and later as Head of "Simulation Bus Structures" and Head of "Technical Calculation Truck & Bus". From 2009 to 2012 he was responsible for all kinds of virtual and real testing of complete trucks and busses as well as components in terms of MBS, crash, fatigue and stiffness. Since the beginning of this year, Dr. Christoph Schöttl is Head of the central department "Engineering Simulation" of MAN Bus & Truck AG in Munich. Before joining MAN in 1992, Dr. Schöttl was employed as Assistant Professor at the University of the Armed Forces in Munich.

Alan Peasland

Head of Technical Partnerships

Infiniti Red Bull Racing

Alan Peasland is the Head of Technical Partnerships for Infiniti Red Bull Racing and has over 20 years of experience with 3D CAD and CAM tools and their respective data management systems. He transitioned from a design engineer into a PLM administrator and manager, ultimately leading to the challenging world of Formula One technical management. As head of CAD/PLM he has been driving the PLM solution forwards and increasing the use of advanced design and management technologies at Infiniti Red Bull Racing. Recently, moving into technical partnership management where he will ensure Infiniti Red Bull Racing will maximize their benefits from fostering relationships with the very best technical and innovation partners.

Dr. Peter J.M. Pirro

Manager Global Mid Tractor Product Verification & Validation

John Deere

Dr. Pirro has more than 35 years simulation experience, mostly inside John Deere. He started his Deere career in 1986 as Finite Element Specialist at the John Deere Combine Product Engineering Department Zweibrücken. After assisting with the implementation of the CAD system CADAM, he moved to the position of Manager Product Verification & Validation for Combines and Hay & Forage Equipment, covering both simulation and hardware testing. He was then appointed Engineering Manager for Self Propelled Forage Harvesters, before moving to his current role in 2006, where he has responsibilities for all types of simulation, laboratory testing of components, full vehicle testing, in all areas of the world for the Mid Tractor Product Line.

Prof. Dr. Christian Hasse

Professor for Numerical Thermo-Fluid Dynamics

University of Technology Freiberg

Prof. Dr. Christian Hasse has more than 15 years of experience in modelling of reactive and non-reactive flows. He obtained his Ph.D degree from RWTH Aachen University (supervisor Prof. Dr. N. Peters) in 2004. The topic of this dissertation was modelling of diesel engine combustion. From 2000-2004 he worked as an dependent software consultant for Cummins Engine Company (USA). In 2004, he joined BMW engine development in Munich. From 2006-2010, he assumed responsibility for emissions and OBD for a turbo-charged four-cylinder engine and supervised 3 Ph.D students at BMW. Since 2010, he is professor for Numerical Thermo-Fluid Dynamics at the University of Technology Freiberg, where he continues his research in modelling and HPC simulation of reactive and non-reactive flows. Special engine related topics are biofuel combustion and pollutant formation, spray modelling and LES of in-cylinder flows. Currently 20 researchers are employed at his chair.

Dr. Stephen Wheat

General Manager of Intel's HPC Business Unit

Dr. Wheat is the General Manager of Intel's HPC Business Unit responsible for the development of Intel's HPC business strategy and the pursuit of that strategy through platform architecture, eco-system development and collaborations. In this role, Dr. Wheat has influenced the deployment of several Top10 systems and even more Top500 HPC systems. He possesses a depth of knowledge and breadth of experience that gives him a unique perspective in understanding large-scale HPC deployments, with management and software architecture experience from numerous divisions at Intel. Awards recognizing Dr. Wheat's work include the Gordon Bell Prize in 1994 and multiple selections as a top pick for HPCwire's People to Watch List. He also has patent in Dynamic Load Balancing in HPC systems.

Prof. Dr. Roland Baar

Professor for Combustion Enginges

Technical Universiät Berlin

Prof. Dr. Roland Baar started his professional career at Volkswagen in 1997, where he soon became manager within Diesel engine development, responsible for pre-development of engine calibration. In 2003 he also added responsibilities for engine component development, for turbochargers and other components. He joined Voith Turbo as the responsible director for turbocharger development Voith Turbo in 2006, and also started as lecturer for combustion engines and turbochargers at Leibniz University of Hannover as an honorary professor in 2008. Since 2011 he is a full professor for combustion engines at Technical University of Berlin, focused on thermodynamics and fluid mechanics at the engine level (Gasoline and Diesel) and component level (turbochargers and injection systems), both experimental and simulation.

"Thank You for Your support " to all our Sponsors of ASWC 2013!

Premium Sponsor

Intel Corporation's Technical Computing Group is your ally in helping you explore more of your simulation ideas in less time than ever before. We know your robust design modeling and simulation work will benefit from powerful Intel® Xeon® E5-2600v2 processors found in:

- High performance workstations
- Compact and easy to access simulation appliances and
- Intel Cluster Ready supercomputers with advanced Intel® TrueScale high speed fabrics.

Combined, these technologies will help you to create, test and modify your ideas faster; enabling you to embrace new designs that surpasses product expectations. **www.intel.com**

Co-Premium Sponsor

HP brings the depth of high-performance computing expertise, and breadth of software, systems and services to enable the highest levels of agility, efficiency, and certainty:

Performance – Speed advancements with a converged infrastructure purpose-built for scale. Adapt easily with a unified, modular cluster portfolio that utilizes leading-edge Intel® Xeon® processors.

Agility – Quickly deploy an optimally configured system for your application, using our in-depth HPC and industry expertise.

Efficiency – Optimize to meet any workload with lightweight, powerful management that spans local to cloud models.

www.hp.com/go/hpc

Gold Sponsors

Global supercomputing leader Cray provides innovative systems that enable scientists and engineers in government, industry and academia to meet existing and future computational challenges. Building on expertise in developing, marketing and servicing the world's most advanced supercomputers, Cray offers a comprehensive portfolio of high performance computing (HPC) systems delivering unrivaled sustained performance on a wide range of challenging applications. **www.cray.com**

The power to do more

Michael Dell founded our company in 1984 in Austin, Texas, with an unprecedented idea — by selling computer systems directly to customers, we could deliver the most effective computing solutions to meet their needs. Today 96,000 team members around the world are committed to making technology work harder for customers and communities. Our business is aligned to address the unique needs of large enterprises, public institutions (healthcare, education and government), small and medium businesses, and consumers. We ship more than 110,000 systems every day to customers in 180 countries and make technology more accessible to people and organizations around the world.

www.dell.com

All-in-one Solution for Automotive Simulation

High Performance Computing (HPC) solutions from Fujitsu deliver the performance to meet today's HPC challenges, while scaling for tomorrow's growing needs. With integrated and certified "Ready-to-Go" HPC cluster solutions a special focus is put on optimized HPC solutions for ANSYS automotive workloads. SIMPLIFIED HPC means capturing knowledge and expertise within a solution that broadens HPC access and eases work, even for experienced users. From our comprehensive experience of real models and customer needs, we have architected this PRIMERGY HPC cluster solution to deliver the outcomes you need for better designs –faster. **www.ts.fujitsu.com/hpc**

Gompute delivers comprehensive solutions for High Performance Computing, in-house, in-the-cloud or both. With over 10 years' of experience with the ANSYS product portfolio, providing solutions and services to the IT and Engineering communities, we can deliver a collaborative and highly productive work environment for geographically distributed engineering teams. Combined with Gompute On-Demand, we ensure that you have the option to either scale out or completely host your HPC environment, ranging from 1 -> 1000's of cores in a Gompute owned Super Computing Centre. To learn more on how we do HPC, visit **www.gompute.com.**

SGI and its server portfolio deliver a unified compute and storage solution that reduces overall system management requirements and costs, and simplifies data management and archiving. SGI UV 2 products provide the only single node, cache-coherent shared memory platform that can start small, and grow seamlessly as your needs develop. Based on the Intel® Xeon® E5, SGI UV 2000 can consolidate the entire CAE workflow onto a single platform. SGI® ICE X and SGI® Rackable servers provide best of breed cluster computing. And finally, SGI® Modular Infinite Storage manages the vast amount of engineering data created by CAE bandwidth-intensive applications. **www.sgi.com/**

CE I - Creator of EnSight

EnSight is an advanced 3D visualization and post-processing system for CFD, CSM, and other simulation disciplines. It can handle extremely large datasets very efficiently and supports remote visu-alization. It directly reads binary .rst, .rth, .rmq, .rfl files from ANSYS mechanical. From ANSYS Fluent and ANSYS CFX you can export data in EnSight format. Alternatively, the direct readers in EnSight will import your native data. EnSight can post-process multiple cases from various sources at the same time for result comparison. All available functions can be processed in batch. Together with included Python scripting your post-processing can thus be completely automated. **www.ceisoftware.com/**

Silver Sponsors

CPU 24/7 GmbH & Co. KG, based in Potsdam/Berlin, is an innovative company specialised in providing High Performance Computing (HPC) power 'on demand' for industry and universities, for applications in development and research. Due to many years of experience and expertise in the provision of HPC power, e.g. in the field of computational fluid dynamics, CPU 24/7 offers these HPC On Demand resources in a timely manner with a secure ready-to-work workplace environment – either as flexibly and online bookable computing capacities for a limited period of time (Resource Area) or also as continuous state-of-the-art HPC system geared to one's individual needs (Tailored Configurations). **www.cpu-24-7.com/**

IBM High Performance Computing. Data-intensive. Energy-efficient. Intuitive.

Success in Technical Computing is about more than compute power.That is why IBM brings together enterprise-class, leading edge technologies for compute, storage, and systems networking, accompanied by a powerful, intuitive software portfolio to manage clusters, workloads and data. Organizations can improve productivity by leveraging a powerful portfolio of performance optimized systems and software. Learn more about the IBM Application Ready Solution for ANSYS. This is a pre-architected, pre-tested, pre-configured system, bringing together the best of IBM and ANSYS, reducing time to insight whilst minimizing risk. **www.ibm.com**

Sharc is the developer of Harpoon and Periscope, the revolutionary meshing and visualisation software.

Harpoon is a fully automatic body fitted Hex-Dominant and Polyhedral mesher. Turn-round times from CAD to CFD solutions are reduced by orders of magnitude for complex geometry. Periscope is the industry leading visualisation package for CFD in terms of speed and ease of use. Advanced features include parallel-processing, flow-feature extraction and movie creation through an intuitive GUI or in batch via extremely simple scripting. Periscope is complimented by our free to use viewer Periview, allowing users to interact with even their largest datasets on the most basic laptops. This is enabled through use of revolutionary data formats and rendering engines. **www.sharc.co.uk**

Your Tour Guide through the Exhibition

The partner road map which you received in your delegate pack provides you with information of our sponsoring partners participating in the Automotive Simulation World Congress, as well as their joint solutions with ANSYS. Throughout the road map you will find all of our sponsoring partners logos and company descriptions. Upon visiting each booth, you can present your road map to each sponsoring partner and they will place a sticker on the designated sponsor section. You must obtain stickers of the participating sponsors to be eligible for the prize raffle. When you have obtained all stickers, please write your name and company name on the last page and take it to the registration area.

The draw will be held during the Award Session on the second day of the event, Wednesday, October 30.

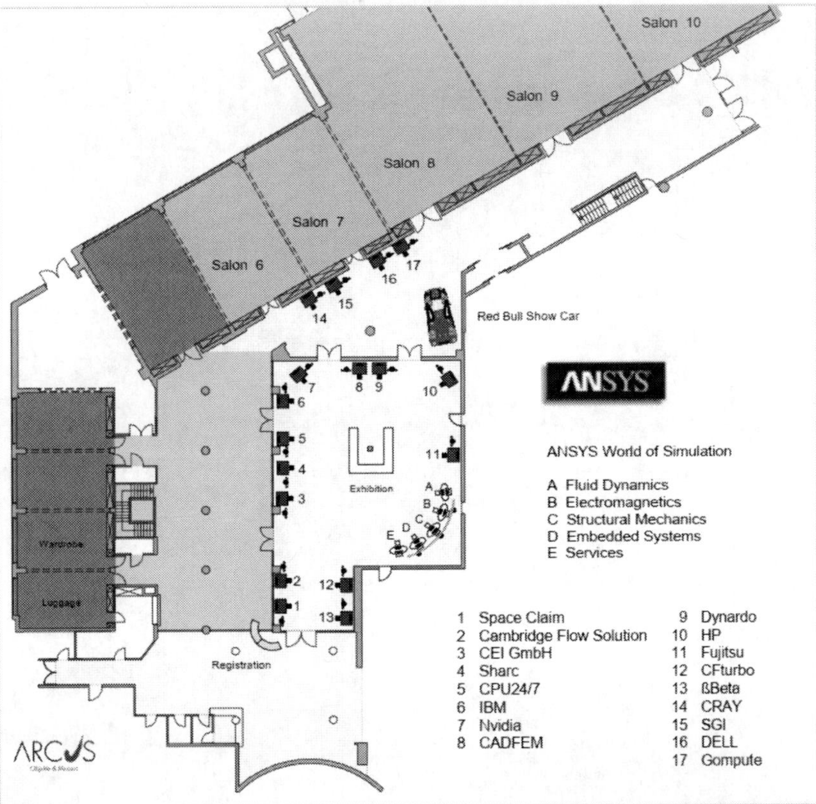

Exhibition Plan

Congress Venue:

Steigenberger Airport Hotel Frankfurt
Unterschweinstiege 16
60549 Frankfurt/Germany
T: +49 69 6975-0
F: +49 69 69 75-2505
www.steigenberger.com

Organizer / Contact:

ANSYS Germany GmbH
Birkenweg 14 a
64295 Darmstadt / Germany
T: +49 6151 3644-0
F: +49 6151 3644-44
www.ansys-germany.com
workshops-germany@ansys.com

Exhibitors

Mediapartners & Associations

2013 **Automotive Simulation** World Congress

Welcome and Introduction

Sandeep Sovani, Ph.D.
Director, Global Automotive Industry

Evolution of Automotive Simulation

1970s	1980s	1990s	2000s	2010s
What is simulation ?	Does it work ?	Does it work for my application ?	How can I use it in production ?	How can I use simulation for the whole system?

Evolution of the ASWC

2013 Automotive Simulation World Congress
October 29 - 30 • Frankfurt am Main, Germany

ASWC in now an annual conference rotating across Americas, Europe, Asia

- 2003 eacc — EUROPEAN AUTOMOTIVE CFD CONFERENCE
- 2005 eacc — EUROPEAN AUTOMOTIVE CFD CONFERENCE — FRANKFURT
- 2007 eacc — EUROPEAN AUTOMOTIVE CFD CONFERENCE — FRANKFURT
- 2009 easc — EUROPEAN AUTOMOTIVE SIMULATION CONFERENCE — MUNICH

- 2012 Automotive Simulation World Congress — DETROIT
- 2013 Automotive Simulation World Congress — FRANKFURT
- 2014 Automotive Simulation World Congress — TOKYO
- 2015 Automotive Simulation World Congress — AMERICAS
- 2016 Automotive Simulation World Congress — EUROPE

2014 ASWC:
Email announcement will be sent with details

Agenda

2013 Automotive Simulation World Congress
October 29 - 30 • Frankfurt am Main, Germany

DAY 1 – Oct 29

Keynotes:
ANSYS
Infiniti – Red Bull Racing
ANSYS
MAN Truck and Bus
Esterel Technologies
Technical University of Berlin

Lunch

Engines	Powertrain	Body, Chassis, Interior	Aerody-namics and Underhood	Electri-fication, Electronics Embedded Systems

Evening Event

DAY 2 – Oct 30

Engines	Powertrain	Body, Chassis, Interior	Aerody-namics and Underhood	Electri-fication, Electronics Embedded Systems

Lunch

Keynotes:
John Deere
Technical University of Freiberg
ANSYS
Intel
Awards Ceremony

2013 **Automotive Simulation** World Congress
October 29 - 30 • Frankfurt am Main, Germany

Evening Event

Klassikstadt

Departure: Shuttle Buses will leave at 6:30 PM outside main lobby of Steinberger Airport Hotel

Return: Shuttle Buses will leave from 10:00 PM to 10:30PM and will go to Steigenberger Airport Hotel, Meininger Hotel, Park Inn by Radisson, and InterCity Hotel

2013 **Automotive Simulation** World Congress
October 29 - 30 • Frankfurt am Main, Germany

Are You Attending

Automotive Simulation World Congress 2013

There's an APP for that!

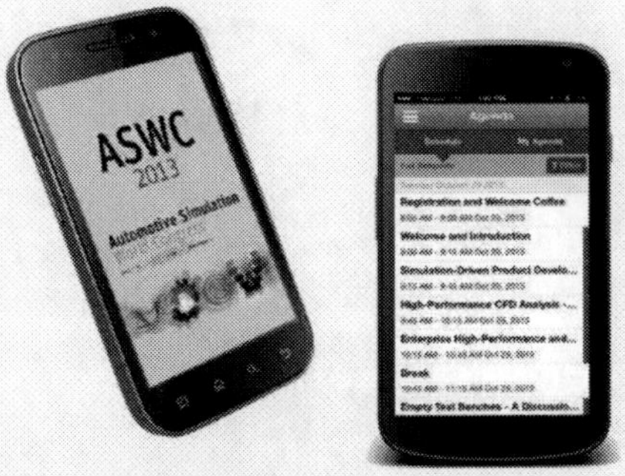

Download today!

Available on the iPhone App Store | ANDROID APP ON Google play

Twitter @ANSYS_Inc

LinkedIn ANSYSInc

Facebook ANSYSInc

Follow our Blog
ansys-blog.com

Conference Proceedings

2013 Automotive Simulation World Congress
October 29 - 30 • Frankfurt am Main, Germany

- Proceedings will be made available after the conference
- Download from ANSYS Knowledge Management Server
- Password needed for access. You will receive an email with instructions

---

2013 Automotive Simulation World Congress

Thank you to our sponsors

intel, hp, FUJITSU, GOMPUTE, CRAY, DELL (The power to do more), sgi, sharc, IBM, CADFEM, CPU 24|7, CEI

2013 Automotive Simulation
World Congress
Thank you to our exhibitors
dynardo®
dynamic software & engineering
CFturbo®
BETA
CAE Systems SA
NVIDIA.
CAMBRIDGE
FLOW SOLUTIONS
Proprietary and customised CFD
SPACECLAIM
CORPORATION

2013 Automotive Simulation World Congress

Simulation-Driven Product Development in the Automotive Industry: Status & Progress

S. Subbiah, Ph.D.
Vice President, Corporate Product Strategy
ANSYS, Inc.

2013 Automotive Simulation World Congress
October 29 - 30 • Frankfurt am Main, Germany

Agenda

Our observations of changes in the engineering landscape . . .

. . . their implications on simulation practices . . .

. . . and ANSYS's recent technological advances to help customers meet the new challenges.

2013 Automotive Simulation World Congress
October 29 - 30 • Frankfurt am Main, Germany

Status & Progress

- **16 papers** on Engines
- **15 papers** on rest of the Powertrain
- **15 papers** on Body, Chassis and Interior
- **16 papers** on Aerodynamics & Underhood
- **17 papers** on Electrification, Electronics and Embedded Systems

"Science can amuse and fascinate us all, but it is engineering that changes the world."

ISAAC ASIMOV

Thursday, October 31, 2013 2013 Automotive Simulation World Congress 3

2013 Automotive Simulation World Congress
October 29 - 30 • Frankfurt am Main, Germany

New Frontiers in Engineering Simulation

- Looking ahead to the next 3-5 years..
  - *What issues will be important in Engineering Simulation?*
- Major trends in engineering and product development
  - Implications for Automotive Industry

2013 Automotive Simulation World Congress
October 29 - 30 • Frankfurt am Main, Germany

Trend 1: Products are Increasingly More *Instrumented* and *Interconnected*

Cars That Communicate With Each Other

Wireless Replaces Wire Harnesses

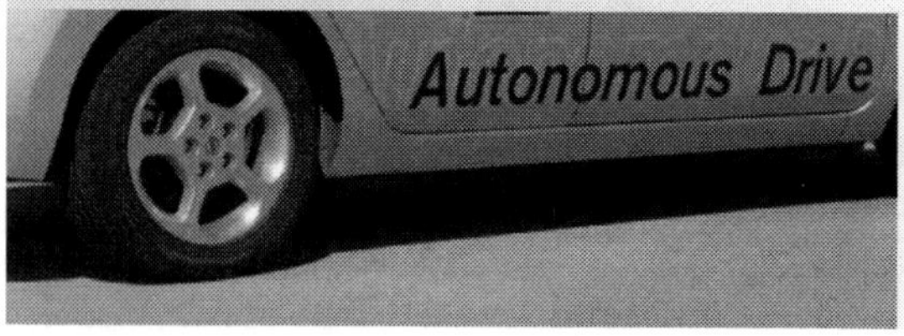

2013 Automotive Simulation World Congress
October 29 - 30 • Frankfurt am Main, Germany

Trend 1: Smart Products Software, Hardware And Electronics Require A Systems-Level Approach

•Best Practices:
- Requirements allocation and traceability
- Behavioral modeling
- Model-based controls design, automatic code generation and certification
- 3D Multphysics simulation
- Behavioral model /3D model co-simulation
- Software/Hardware in the loop

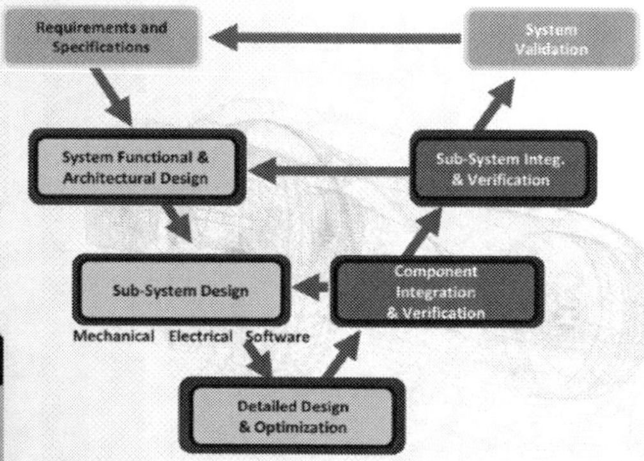

Key Capabilities Required:

Complete virtual Prototype

- Comprehensive Multiphysics
- 3D and/or behavior models
- Electronics, Hardware, software
- Model based software simulation & certified code generation

Trend 2: Selecting the *Optimal* design is *not enough*

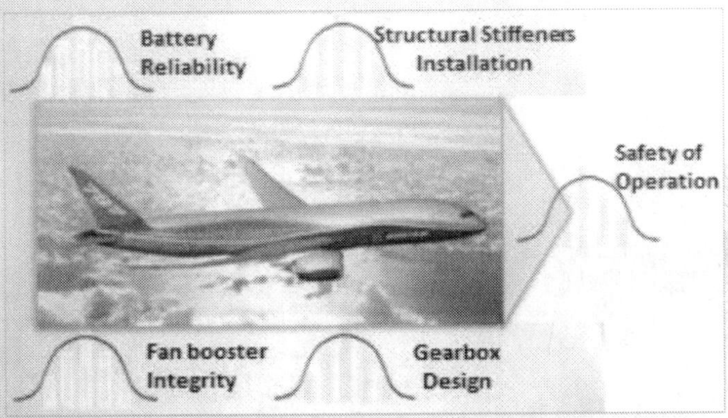

Variations in operating conditions, manufacturing processes and material properties ***create uncertainty*** in the overall success of a product design.

2013 Automotive Simulation World Congress
October 29 - 30 • Frankfurt am Main, Germany

Trend 2: Warranty, Liability, And Competition Drive "Product Integrity Through Robust Design"

Variations in operating conditions, manufacturing processes and material properties aggregate to create uncertainty in the overall success of a product design

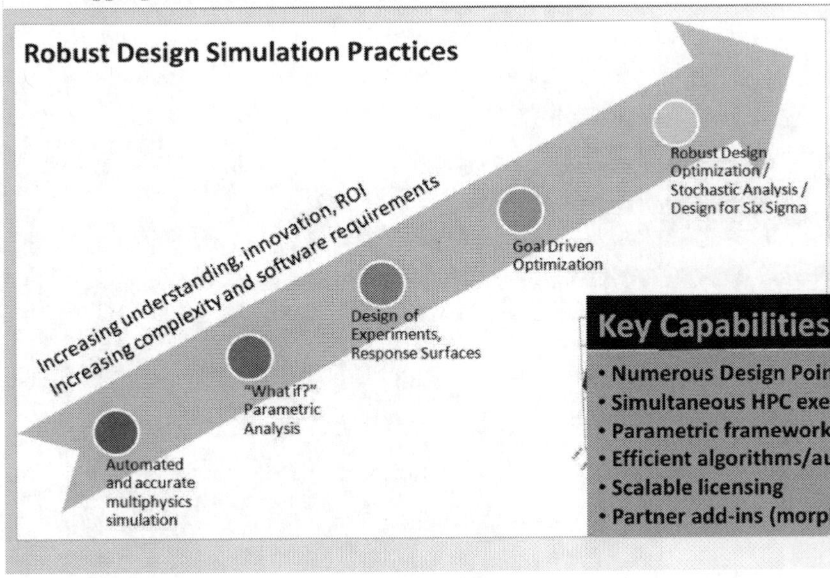

Key Capabilities Required:

- Numerous Design Points evaluated
- Simultaneous HPC execution
- Parametric framework
- Efficient algorithms/automation (Adjoints)
- Scalable licensing
- Partner add-ins (morphing, optimizers)

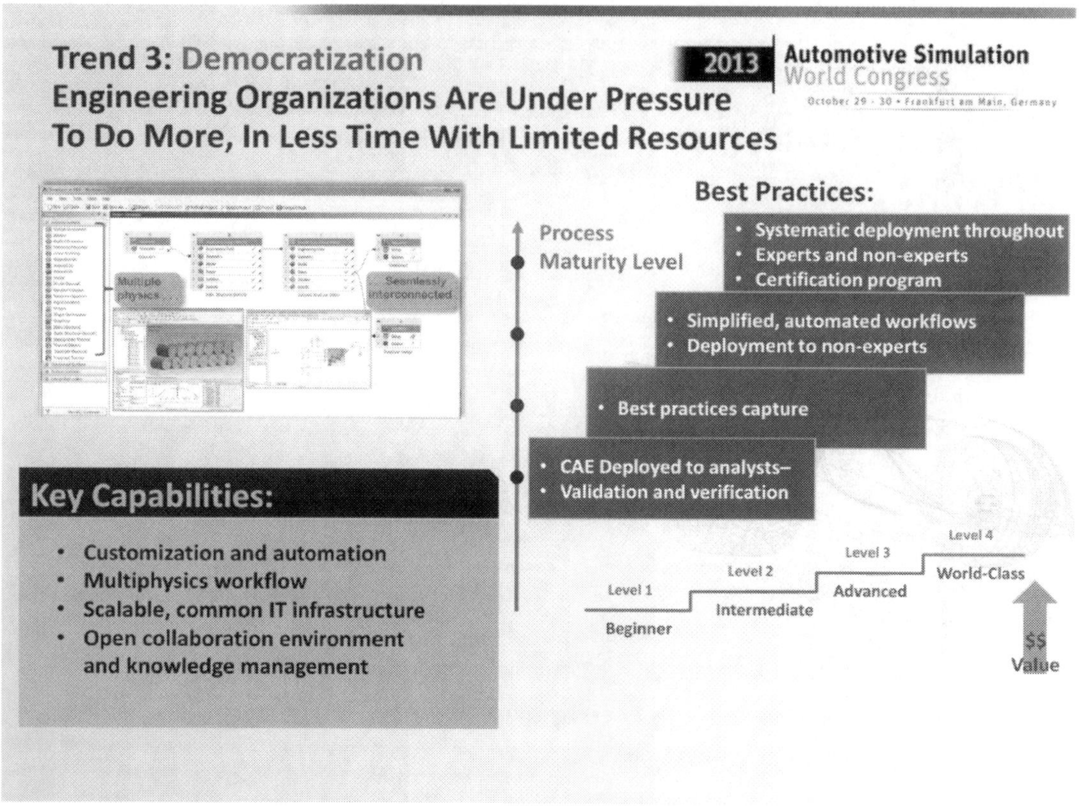
Trend 3: Democratization
Engineering Organizations Are Under Pressure
To Do More, In Less Time With Limited Resources
2013
Automotive Simulation
World Congress
October 29 - 30 • Frankfurt am Main, Germany
Multiple physics...
Seamlessly interconnected
Best Practices:
Process
Maturity Level
Systematic deployment throughout
Experts and non-experts
Certification program
Simplified, automated workflows
Deployment to non-experts
Best practices capture
CAE Deployed to analysts–
Validation and verification
Level 1
Beginner
Level 2
Intermediate
Level 3
Advanced
Level 4
World-Class
$$
Value
Key Capabilities:
Customization and automation
Multiphysics workflow
Scalable, common IT infrastructure
Open collaboration environment and knowledge management

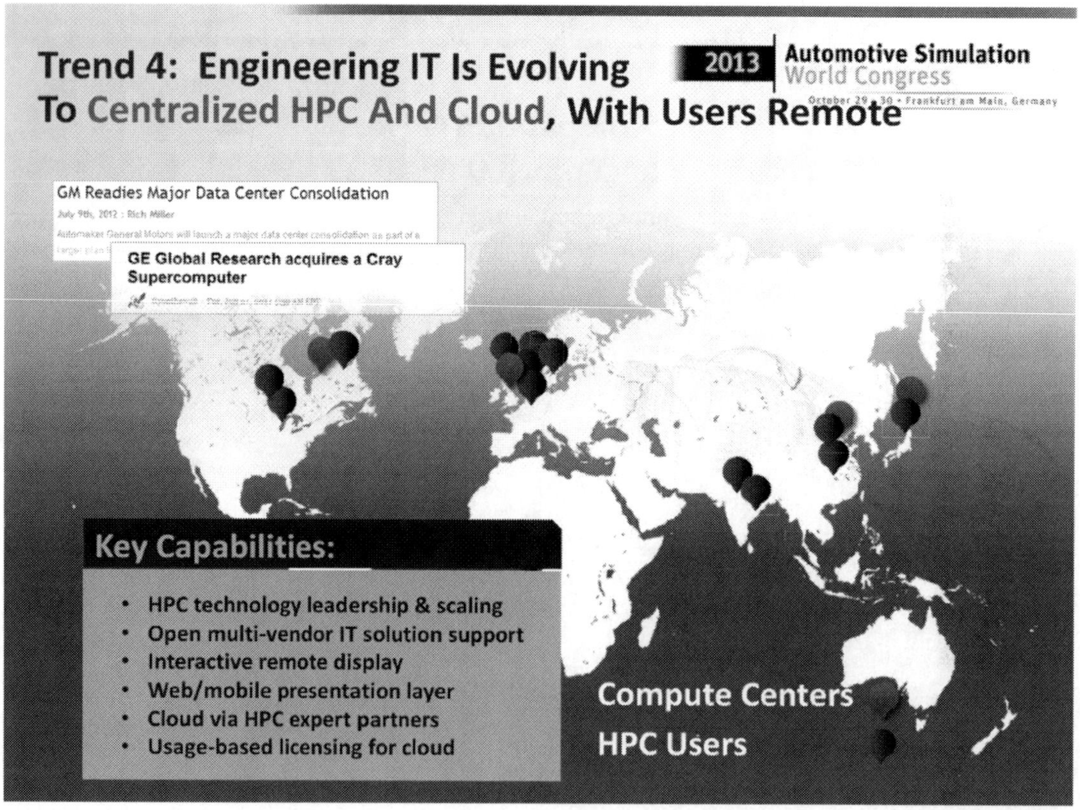
Trend 4: Engineering IT Is Evolving
To Centralized HPC And Cloud, With Users Remote
2013
Automotive Simulation
World Congress
October 29 - 30 • Frankfurt am Main, Germany
GM Readies Major Data Center Consolidation
GE Global Research acquires a Cray Supercomputer
Key Capabilities:
HPC technology leadership & scaling
Open multi-vendor IT solution support
Interactive remote display
Web/mobile presentation layer
Cloud via HPC expert partners
Usage-based licensing for cloud
Compute Centers
HPC Users

Trend 5 : Simulation IP Management Call For Better Management Of Simulation IP due to Globalization of R&D

2013 **Automotive Simulation** World Congress
October 29 - 30 • Frankfurt am Main, Germany

Best Practices:

- **Global collaborative simulation to enable** data management, knowledge capture & HPC
- **Simulation decision support to enable** audit trails & traceability

No Development Is Complete Without Simulation

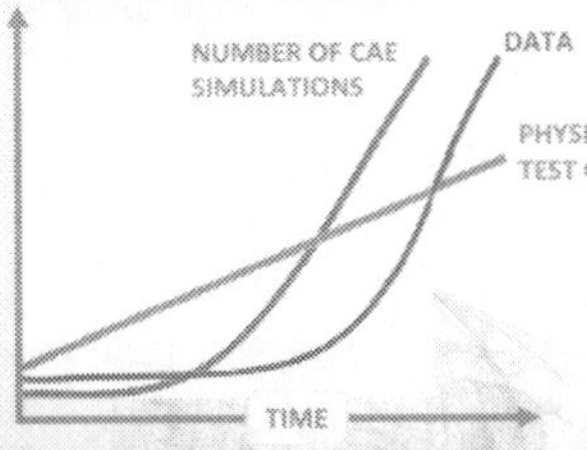

Simulation Knowledge Retention & Reuse - A Key Value Driver For Design Success

Workbench and EKM
Key Capabilities

- Extension of ANSYS Workbench platform to define, manage and execute simulations
- Manage CAE assets (ANSYS & Non-ANSYS)
- Integrate local/distributed simulation teams.

HIGH SDM MATURITY

	Level 1: ISOLATED ISLANDS	Level 2: ARCHIPELAGO	Level 3: CENTRALIZED / CLOUD
PEOPLE	• Engineers	• Overlap of Engineering & IT	• Engineering & IT Separated
PROCESS	• Manual / Adhoc	• Flexible / Agile • Governance	• Digital Library • Automation
TECH-NOLOGY	• Desperate Data Repositories	• Workgroup Storage • Replicated	• Centralized Data • Links to PLM

LOW SDM MATURITY

Global Megatrends

2013 **Automotive Simulation** World Congress
October 29 - 30 • Frankfurt am Main, Germany

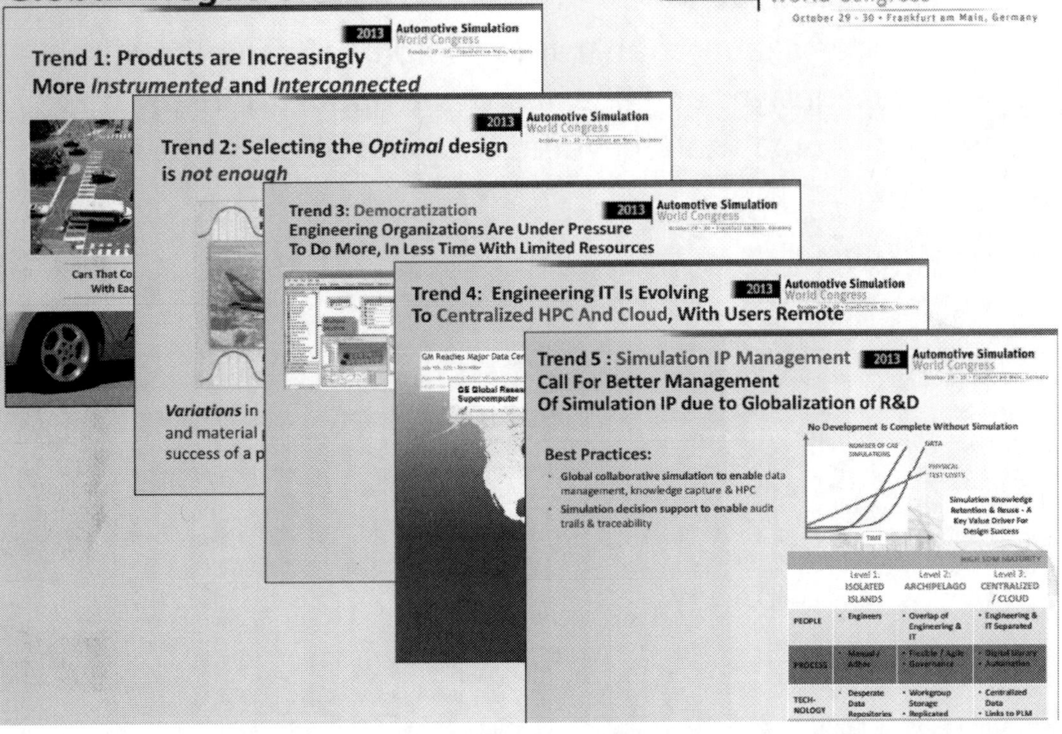

2013 Automotive Simulation World Congress
October 29 - 30 • Frankfurt am Main, Germany

Closing Thoughts…

- Successful Product Design will require ***recognizing sophisticated interactions*** (multiphysics, multi-scale, multi-domain, multi-disciplinary) to ensure product performance

- Safety and Reliability of Products cannot be ensured with a few design point evaluations, - ***comprehensive and intelligent design exploration*** will be needed to assure ***Product Integrity***

- Worldwide collaboration, ***capturing global know-how***, re-using past designs and IP will provide ***opportunity for step changes*** in organizational efficiency and effectiveness

2013 **Automotive Simulation** World Congress

Enterprise HPC & Cloud Computing for Engineering Simulation

Barbara Hutchings
Director, Strategic Partnerships
ANSYS, Inc.

Historical Perspective
Evolution of Computing for Simulation

- Pendulum swing:
  - Centralized to distributed to centralized
  - Remote to local to remote
- Consistent trends:
  - Scale up (cores/processors, workloads)
  - Globalization / collaboration
  - IT complexity (applications/users)

Mainframes | Client/Server | Workstation/PCs | Local Clusters | Shared HPC Grid | Cloud, Mobile

1980 | 1990 | 2000

2013 Automotive Simulation World Congress
October 29 - 30 • Frankfurt am Main, Germany

Rising Role of IT
Focus on Innovation *and* Efficiency

- "Looking ahead, executives expect IT to create new platforms to support innovation and growth..."

"...the highest current priorities for IT: improving the effectiveness and efficiency of business processes and reducing IT costs."

Source: A rising role for IT: McKinsey Global Survey Results, McKinsey Quarterly, December 2011

2013 Automotive Simulation World Congress
October 29 - 30 • Frankfurt am Main, Germany

Current Trends
IT Environment for Engineering Simulation

- **Scale-up of high performance computing**
  - new/ emerging architectures
  - more and bigger workloads
- **Efficient centralized infrastructure (or cloud)**
  - with effective remote access
- **Collaboration hubs with secure and scalable data access**
- **Support for mobile platforms**

GM Readies Major Data Center Consolidation
By Data Center Knowledge

GE Global Research Acquires a Cray Supercomputer
Press Release: Cray Inc – Tue, Jun 21, 2011 3:05 AM EDT

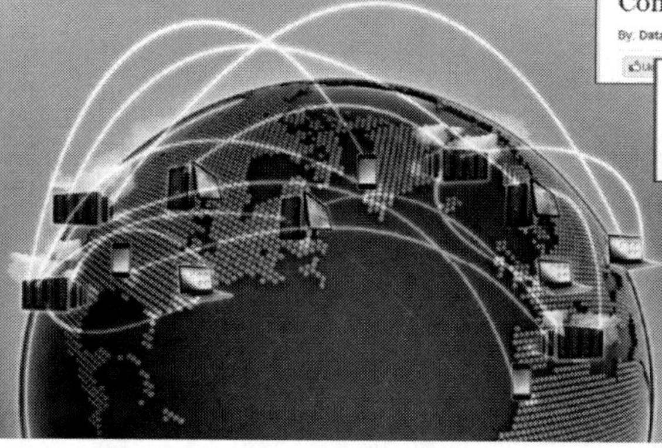

2013 Automotive Simulation World Congress
October 29 - 30 • Frankfurt am Main, Germany

Current Trends / ANSYS Strategy
Spectrum of Private, Hosted and Public Cloud

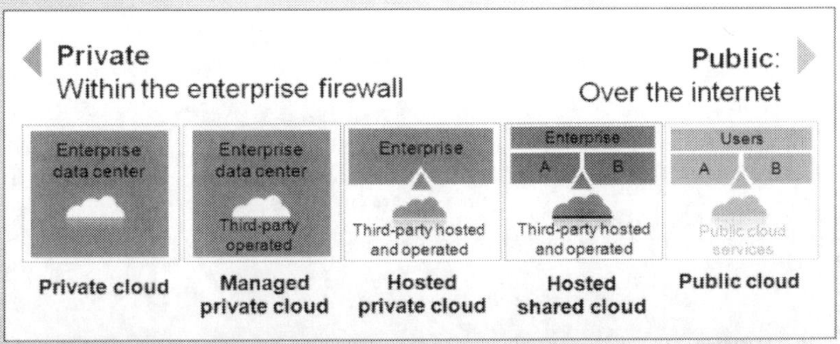

Software & User Environment: Seamless usability and remote access

Hardware: Private cloud or hosted HPC (w/ data security) → later

Business: Evolution / Add flexibility of "usage" to traditional license model

Graphic courtesy of IBM

2013 Automotive Simulation World Congress
October 29 - 30 • Frankfurt am Main, Germany

ANSYS Cloud & HPC – Product Strategy
Optimizing Remote Simulation Workflow

•Local computing infrastructure for simulation is transitioning to centralized HPC resources, shared by a globally distributed workforce

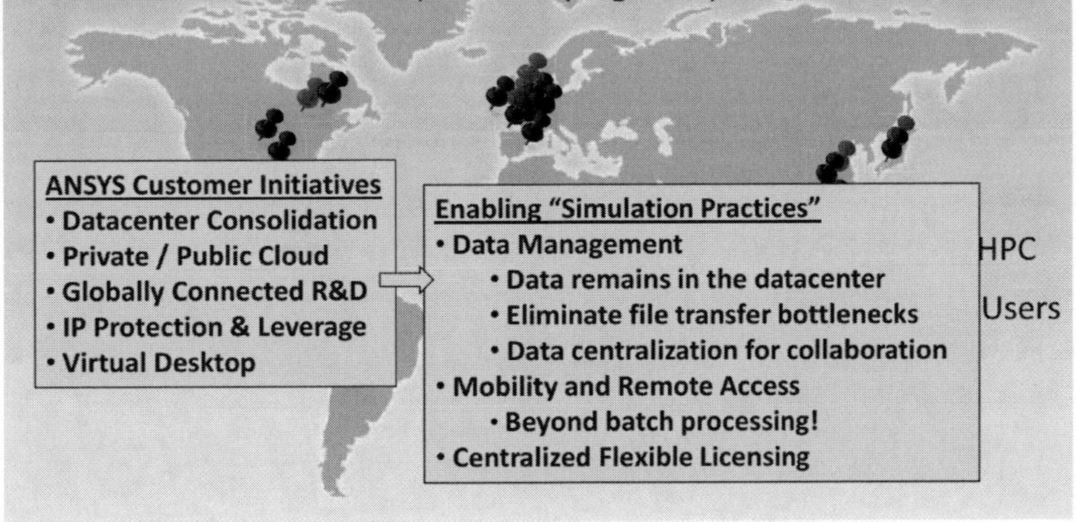

ANSYS Customer Initiatives
- Datacenter Consolidation
- Private / Public Cloud
- Globally Connected R&D
- IP Protection & Leverage
- Virtual Desktop

Enabling "Simulation Practices"
- Data Management
  - Data remains in the datacenter
  - Eliminate file transfer bottlenecks
  - Data centralization for collaboration
- Mobility and Remote Access
  - Beyond batch processing!
- Centralized Flexible Licensing

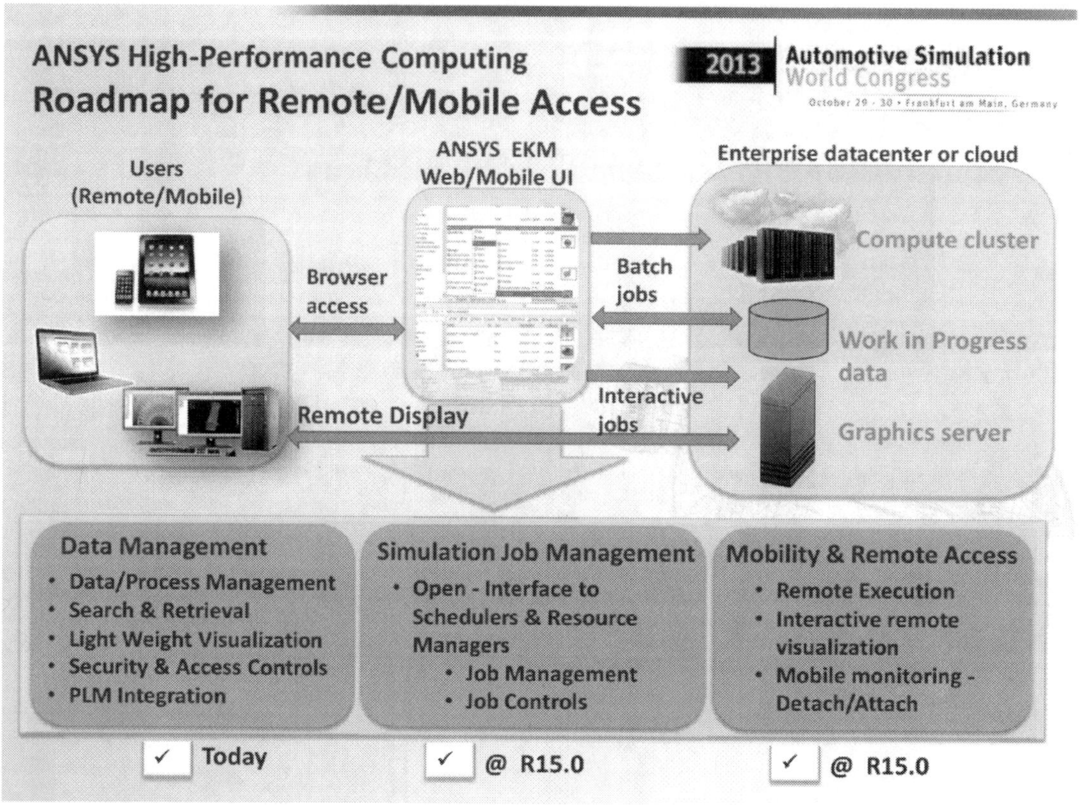
ANSYS High-Performance Computing
Roadmap for Remote/Mobile Access
2013 Automotive Simulation World Congress
October 29 - 30 • Frankfurt am Main, Germany
Users (Remote/Mobile)
ANSYS EKM Web/Mobile UI
Enterprise datacenter or cloud
Browser access
Batch jobs
Compute cluster
Work in Progress data
Interactive jobs
Remote Display
Graphics server
Data Management
• Data/Process Management
• Search & Retrieval
• Light Weight Visualization
• Security & Access Controls
• PLM Integration
Simulation Job Management
• Open - Interface to Schedulers & Resource Managers
• Job Management
• Job Controls
Mobility & Remote Access
• Remote Execution
• Interactive remote visualization
• Mobile monitoring - Detach/Attach
✓ Today
✓ @ R15.0
✓ @ R15.0

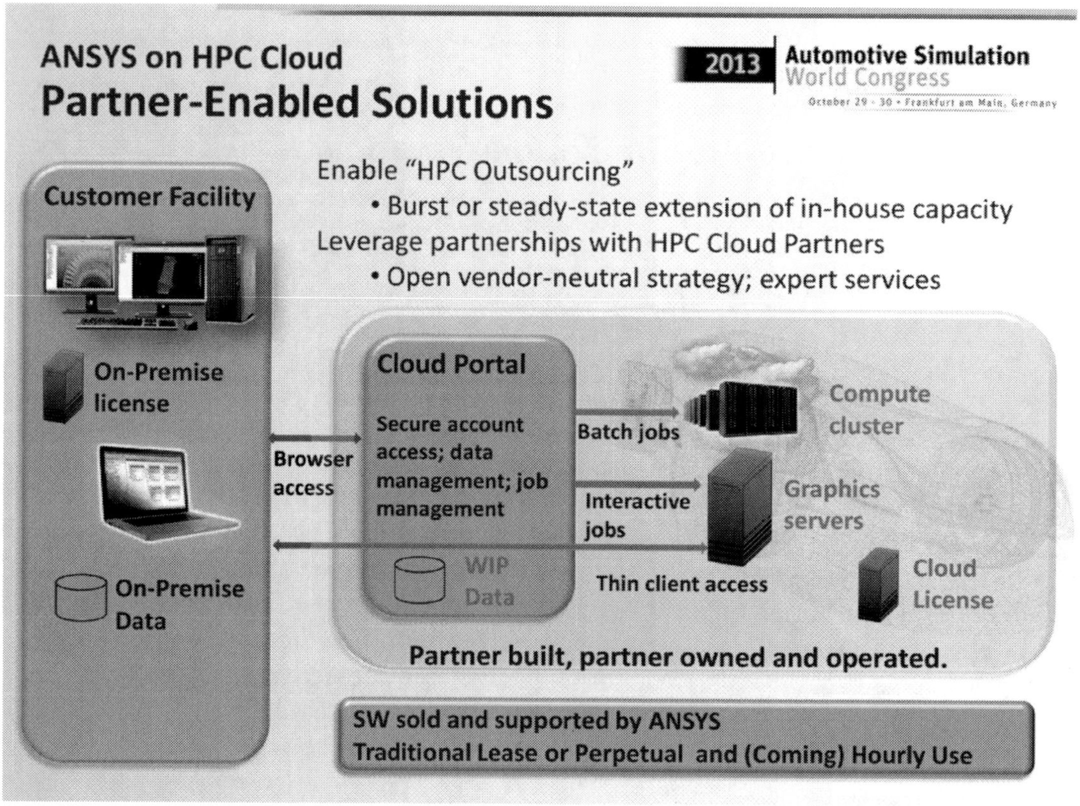
ANSYS on HPC Cloud
Partner-Enabled Solutions
2013 Automotive Simulation World Congress
October 29 - 30 • Frankfurt am Main, Germany
Enable "HPC Outsourcing"
• Burst or steady-state extension of in-house capacity
Leverage partnerships with HPC Cloud Partners
• Open vendor-neutral strategy; expert services
Customer Facility
On-Premise license
On-Premise Data
Browser access
Cloud Portal
Secure account access; data management; job management
WIP Data
Batch jobs
Compute cluster
Interactive jobs
Graphics servers
Thin client access
Cloud License
Partner built, partner owned and operated.
SW sold and supported by ANSYS
Traditional Lease or Perpetual and (Coming) Hourly Use

Cloud Hosting
ANSYS Cloud Partner Program

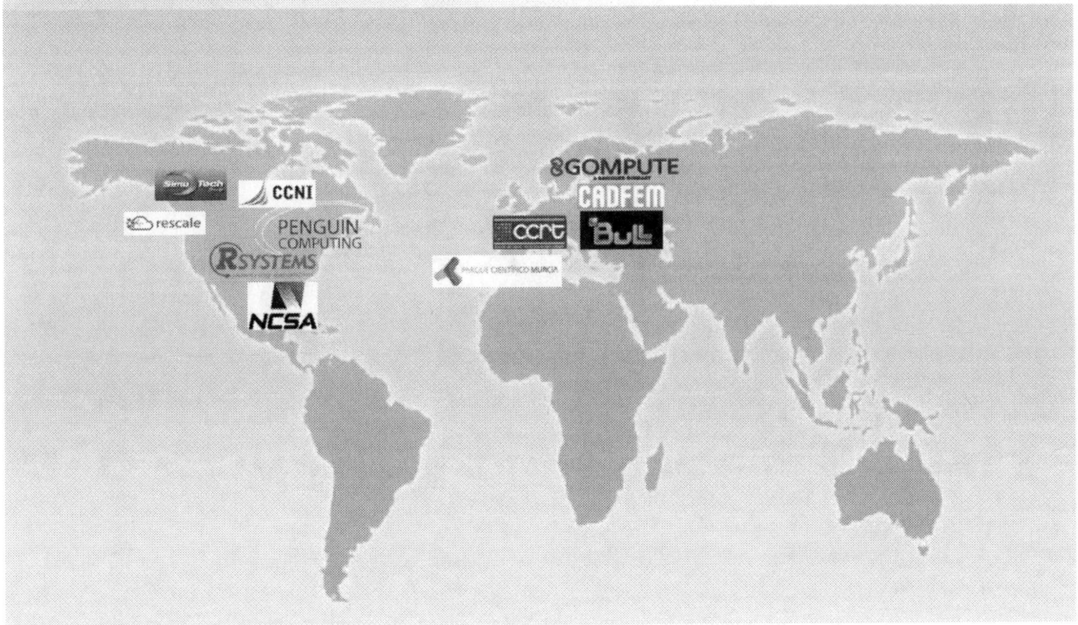

2013 Automotive Simulation World Congress
October 29 - 30 • Frankfurt am Main, Germany

ANSYS Cloud Partners
Customer Experience – Gompute HPC Cloud

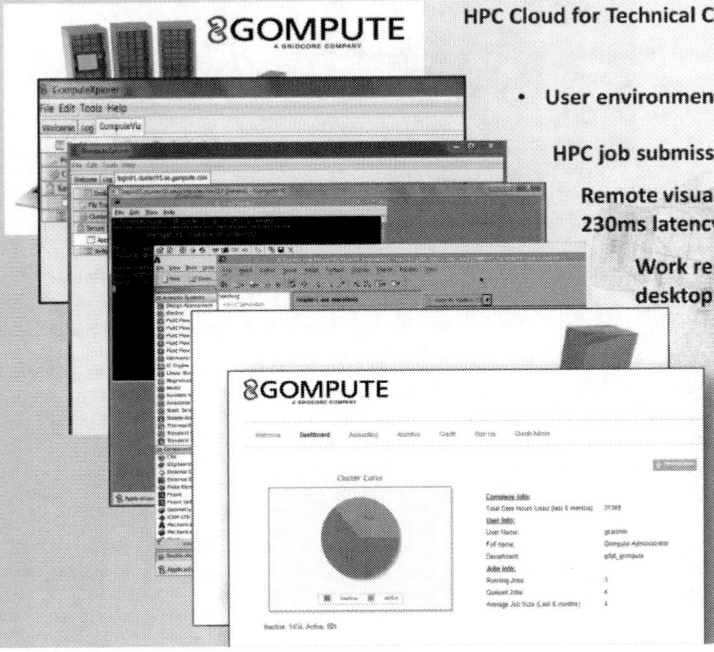

HPC Cloud for Technical Computing

- User environment (GomputeXplorer)
  - HPC job submission
  - Remote visualization (working well up to 230ms latency links)
  - Work remotely, with same GUI as local desktop
  - Scale up your simulations to 1000´s of cores, or run multiple jobs at same time
  - Accounting control with jobs, credits and license usage

2013 Automotive Simulation World Congress
October 29 - 30 • Frankfurt am Main, Germany

ANSYS Cloud Partners
Customer Experience – CADFEM Engineering Cloud

Lo[Start your ANSYS in Cloud

Secure VPN Connection

Customers use cloud in exactly the same manner as their own equipment

- Combined license usage: on premise + cloud-based
- Portal = pure Login
- Keep it simple

2013 Automotive Simulation World Congress
October 29 - 30 • Frankfurt am Main, Germany

ANSYS IT Industry Partnerships
HPC Scale-Up and Performance

•Partnerships with IT industry leaders, ensuring optimized HPC performance, a roadmap to the future, and wrap-around support

- **ANSYS and Intel – 60% speed-up on Sandy Bridge/ Ivybridge; R&D focused on Intel's Many Integrated Core (Xeon Phi)**
- **ANSYS and NVIDIA – GPU acceleration of ANSYS Mechanical; very active R&D engagement with NVIDIA across full portfolio**
- **ANSYS and Cray – Support for extreme scalability of ANSYS CFD, up to 1000's of cores**

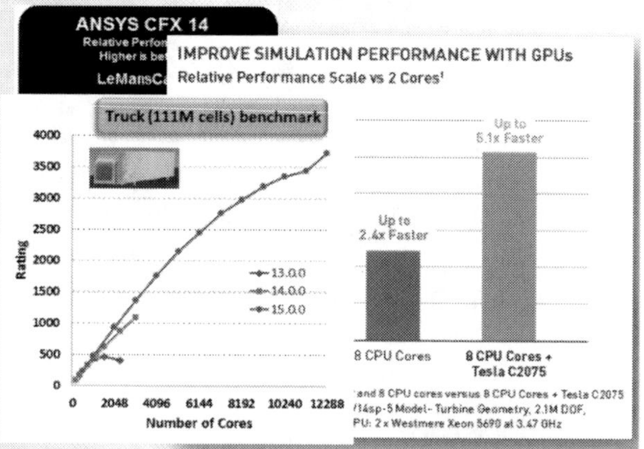

ANSYS HPC performance optimizes the utilization of licenses, hardware, and people

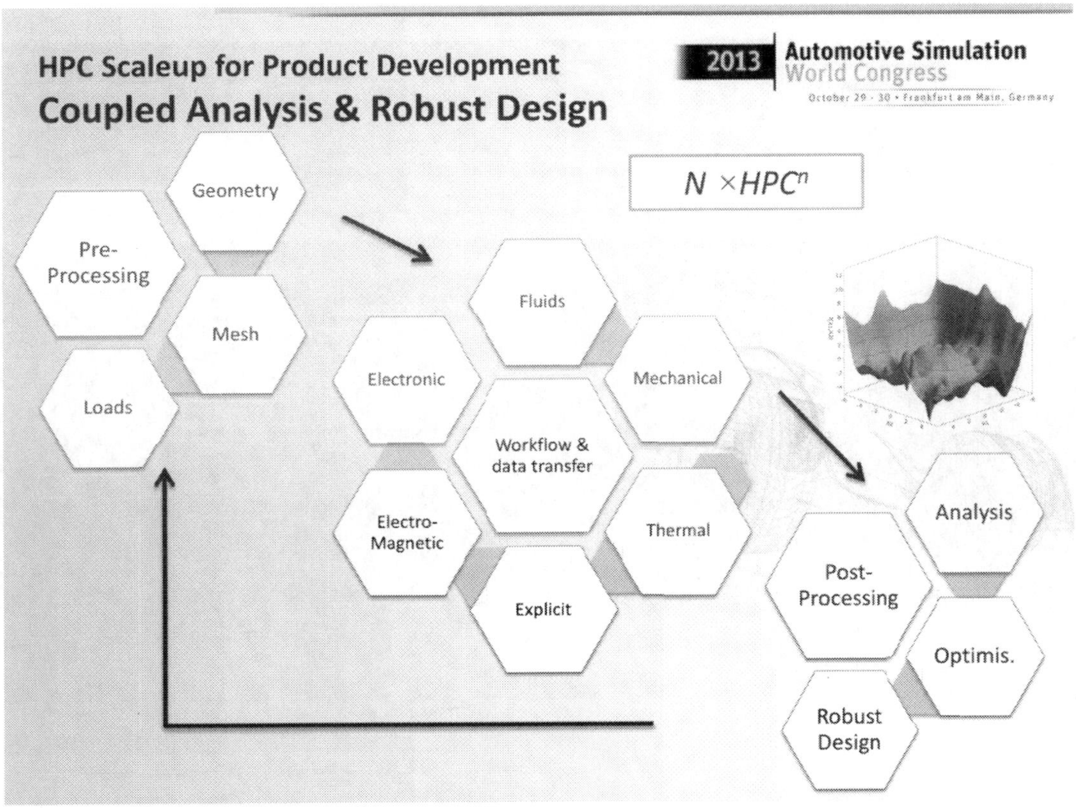
HPC Scaleup for Product Development
Coupled Analysis & Robust Design
2013
Automotive Simulation
World Congress
October 29 - 30 • Frankfurt am Main, Germany
N ×HPC^n
Geometry
Pre-Processing
Mesh
Loads
Fluids
Electronic
Mechanical
Workflow & data transfer
Electro-Magnetic
Thermal
Explicit
Analysis
Post-Processing
Optimis.
Robust Design

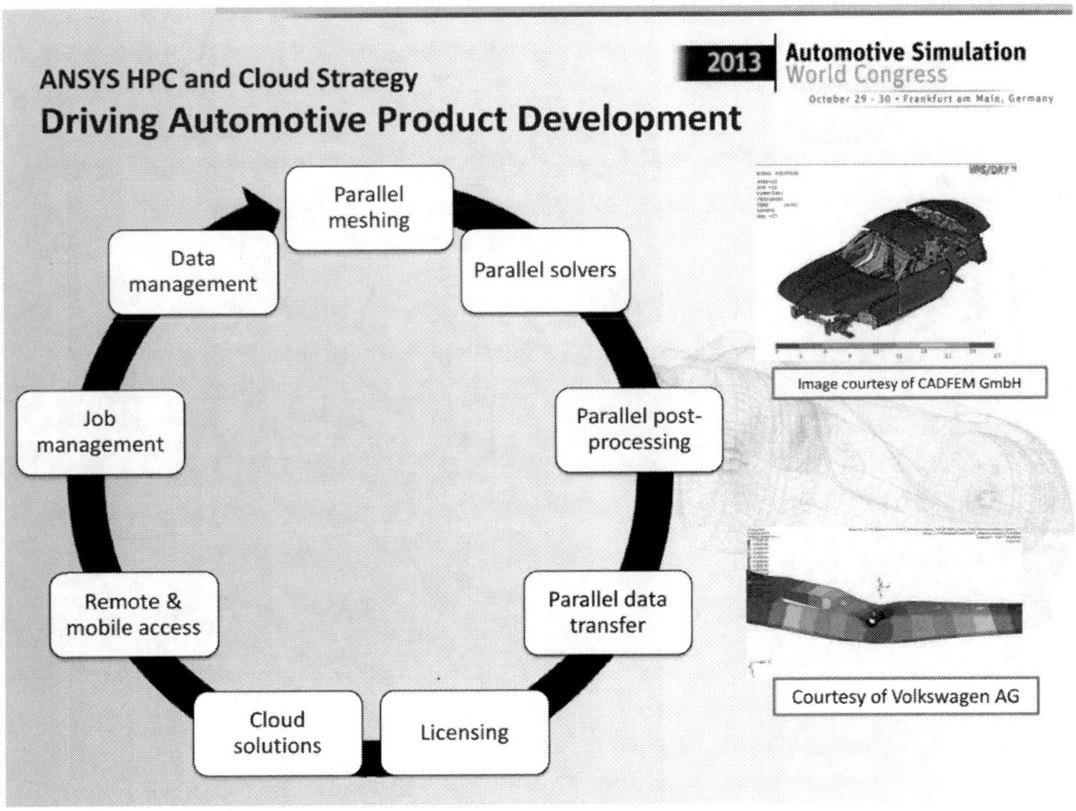
ANSYS HPC and Cloud Strategy
Driving Automotive Product Development
2013
Automotive Simulation
World Congress
October 29 - 30 • Frankfurt am Main, Germany
Parallel meshing
Parallel solvers
Parallel post-processing
Parallel data transfer
Licensing
Cloud solutions
Remote & mobile access
Job management
Data management
Image courtesy of CADFEM GmbH
Courtesy of Volkswagen AG

MAN Truck & Bus AG

„Empty test benches“

„A discussion about the five dilemmas of virtual product clearance“

Dr. Christoph Schöttl

October 2013

… empty test benches

Agenda

MAN

1	Planning of scenarios
2	Dilemma (#1): "Optimization of resources"
3	Dilemma (#2): "Evaluation criterion"
4	Dilemma (#3): "Comparison of faults"
5	Dilemma (#4): "Acceptance of alternative clearance"
6	Dilemma (#5): "Formation of opinion and communication"
7	Summary

Planning of scenarios

MAN

Forecast:

A forecast is a projection of the past or the present into the future

Mathematics: A forecast is comparable to the integration of an initial value problem

A scenario allows - consciously - a break with the past

Mathematics: Scenarios show analogy to a boundary value problem

Source: *„Scenario thinking and planning"*, Dr. Angela Wilkinson, SAID Business school, University of Oxford, 14. GMP

Planning of scenarios

Think Again!

Saïd Business School

Analysis of 80 cases of 'failed strategies:

- "Misleading prejudgments" in 82% of cases
- "Misleading experiences" (64%)
- "Inappropriate self-interest", including wanting to be popular (69%)
- "Inappropriate attachments" (43%)

Source: Think Again" by S Finkelstein, J Whitehead, and A Campbell, Harvard Business Press.

Planning of scenarios

Scenario-based planning

Saïd Business School

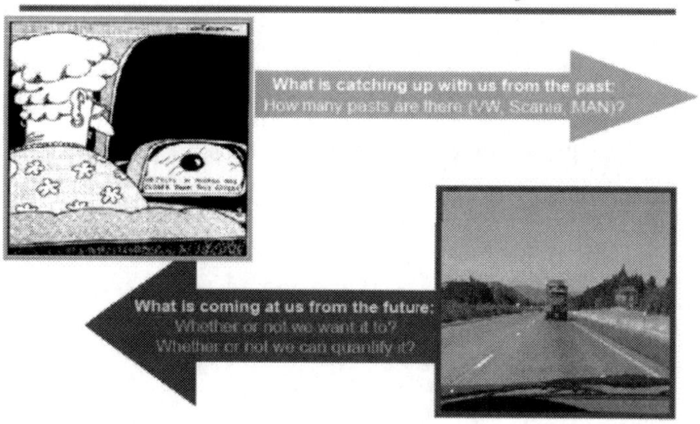

Source: Oxford Scenarios Programme

Optimization of resources
Dilemma (#1)

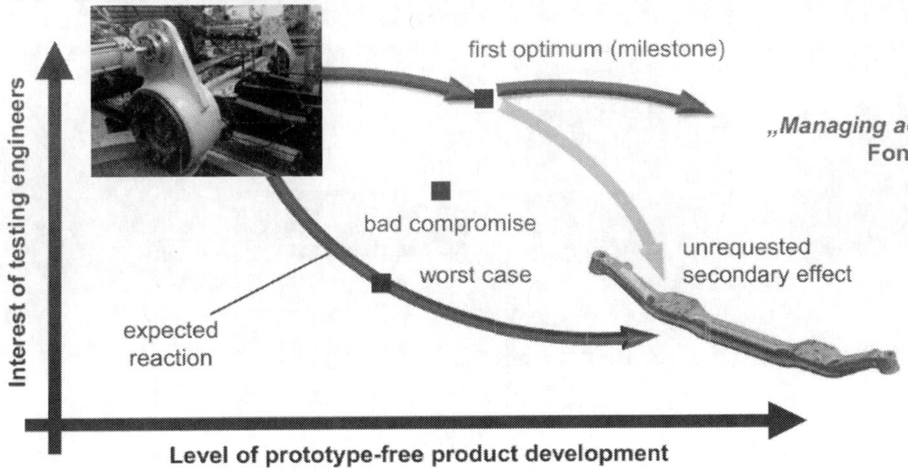

Source:
„Managing across cultures"
Fons Trompenaars
WHU-Seminar

Evaluation criterion
Dilemma (#2)

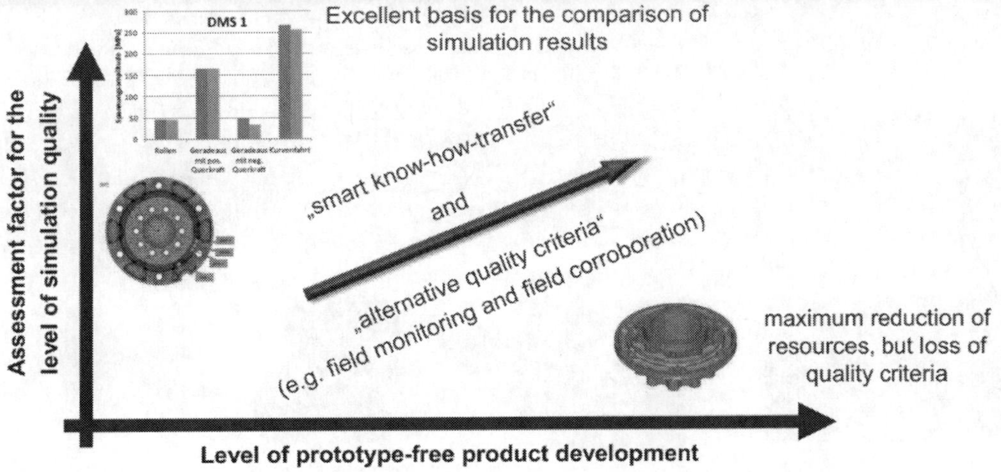

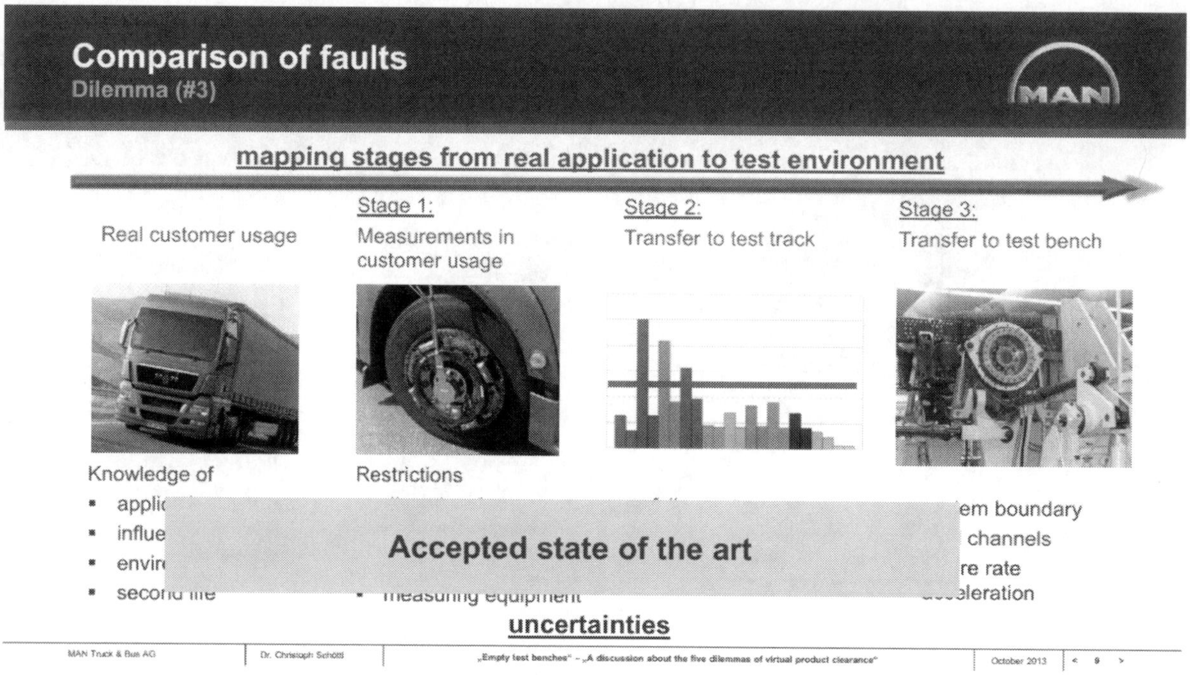
Comparison of faults
Dilemma (#3)
MAN
mapping stages from real application to test environment
Real customer usage
Stage 1:
Measurements in customer usage
Stage 2:
Transfer to test track
Stage 3:
Transfer to test bench
Knowledge of
Restrictions
Accepted state of the art
uncertainties
MAN Truck & Bus AG
Dr. Christoph Schött
„Empty test benches" – „A discussion about the five dilemmas of virtual product clearance"
October 2013
9

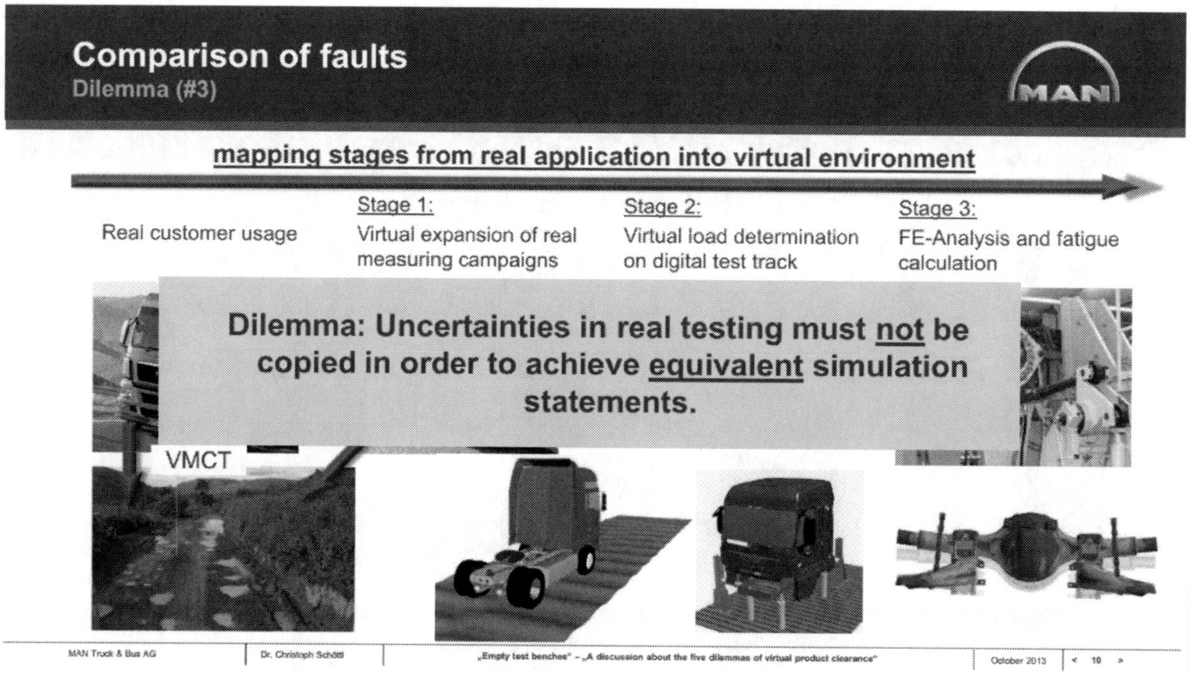
Comparison of faults
Dilemma (#3)
MAN
mapping stages from real application into virtual environment
Real customer usage
Stage 1:
Virtual expansion of real measuring campaigns
Stage 2:
Virtual load determination on digital test track
Stage 3:
FE-Analysis and fatigue calculation
Dilemma: Uncertainties in real testing must not be copied in order to achieve equivalent simulation statements.
VMCT
MAN Truck & Bus AG
Dr. Christoph Schött
„Empty test benches" – „A discussion about the five dilemmas of virtual product clearance"
October 2013
10

Acceptance of alternative clearance
Dilemma (#4)

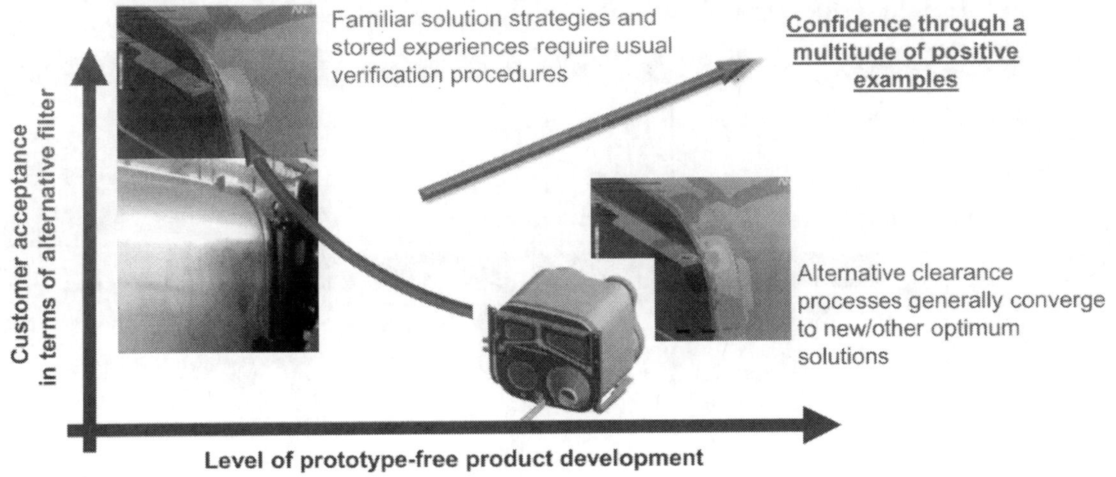

Formation of opinion and communication
Dilemma (#5)

Case study: Project manager requests FE-Analysis from supplier

Project manager: „… *We requested from our supplier a FEM-analysis of the <component>*

- Avoid third person communication about simulation processes
- Take corrective action in these cases
- Guide the simulation task back into the regular process

critically and can not be confirmed. An external FEM-Analysis is rarely comparable (… to an internal FEM …) and therefore, can not be used for a clearance as well. …“

Summary

Do the 5 described dilemmas correspond to your own experiences?

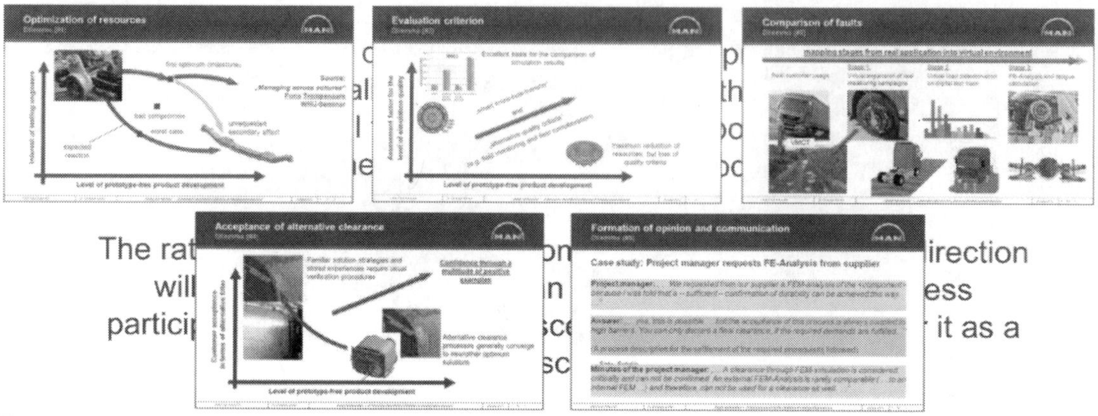

The rat … irection
will … ess
partici … it as a

MAN Truck & Bus AG

Thank you for your attention!

2013 **Automotive Simulation** World Congress

Model Based Systems Engineering: Bringing Together Safety Critical Embedded Software and Full System Simulation

Eric BANTEGNIE
President & CEO, Esterel Technologies
VP Systems Business Unit, ANSYS

When software & electronics become dominant in a product architecture

- Larger solution space – more design degrees of freedom
- Multiplicity of architectural interfaces have to be designed, tested and maintained
- Electronics & software are increasing the risk of catastrophic failure modes and the cost of safety certification

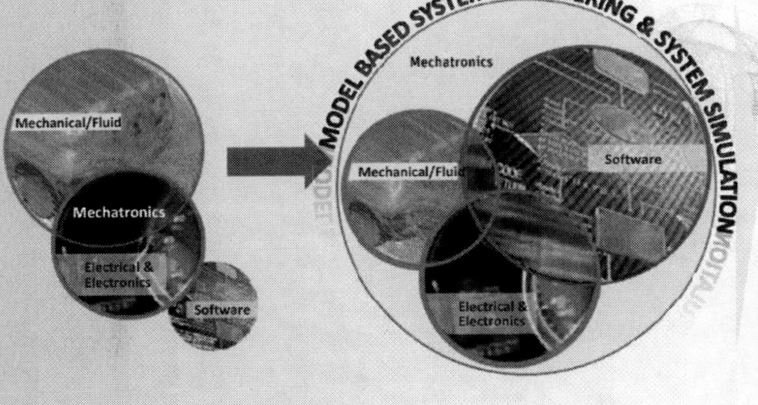

Manage Complexity

to design innovative, market leading products

Coordinate Interdisciplinary Engineering

to reduce design changes and development costs

Early & Reliable Verification

to deliver high quality, safe, and reliable products to the market faster

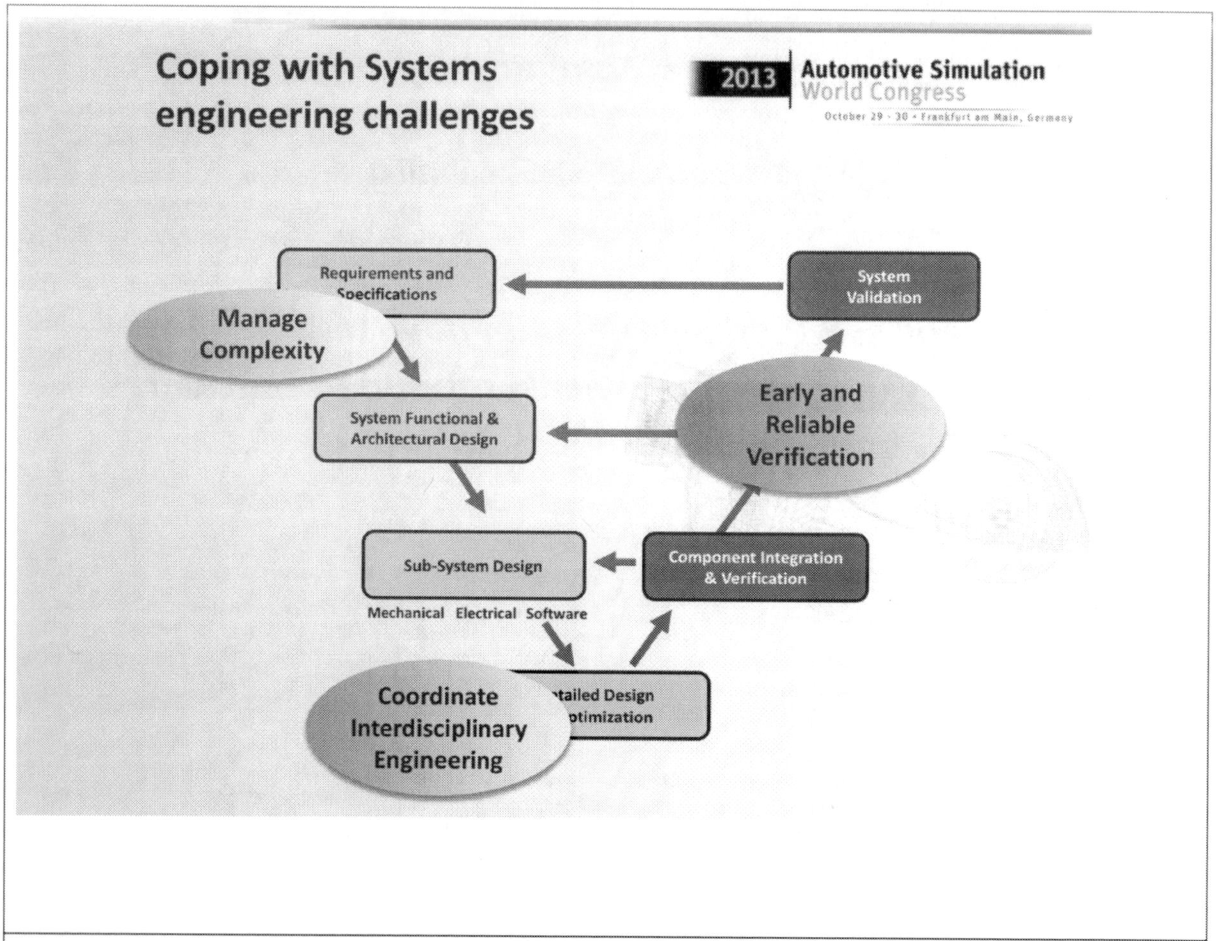
Coping with Systems engineering challenges
2013 Automotive Simulation World Congress
October 29 - 30 • Frankfurt am Main, Germany
Requirements and Specifications
System Validation
Manage Complexity
System Functional & Architectural Design
Early and Reliable Verification
Sub-System Design
Component Integration & Verification
Mechanical Electrical Software
Coordinate Interdisciplinary Engineering

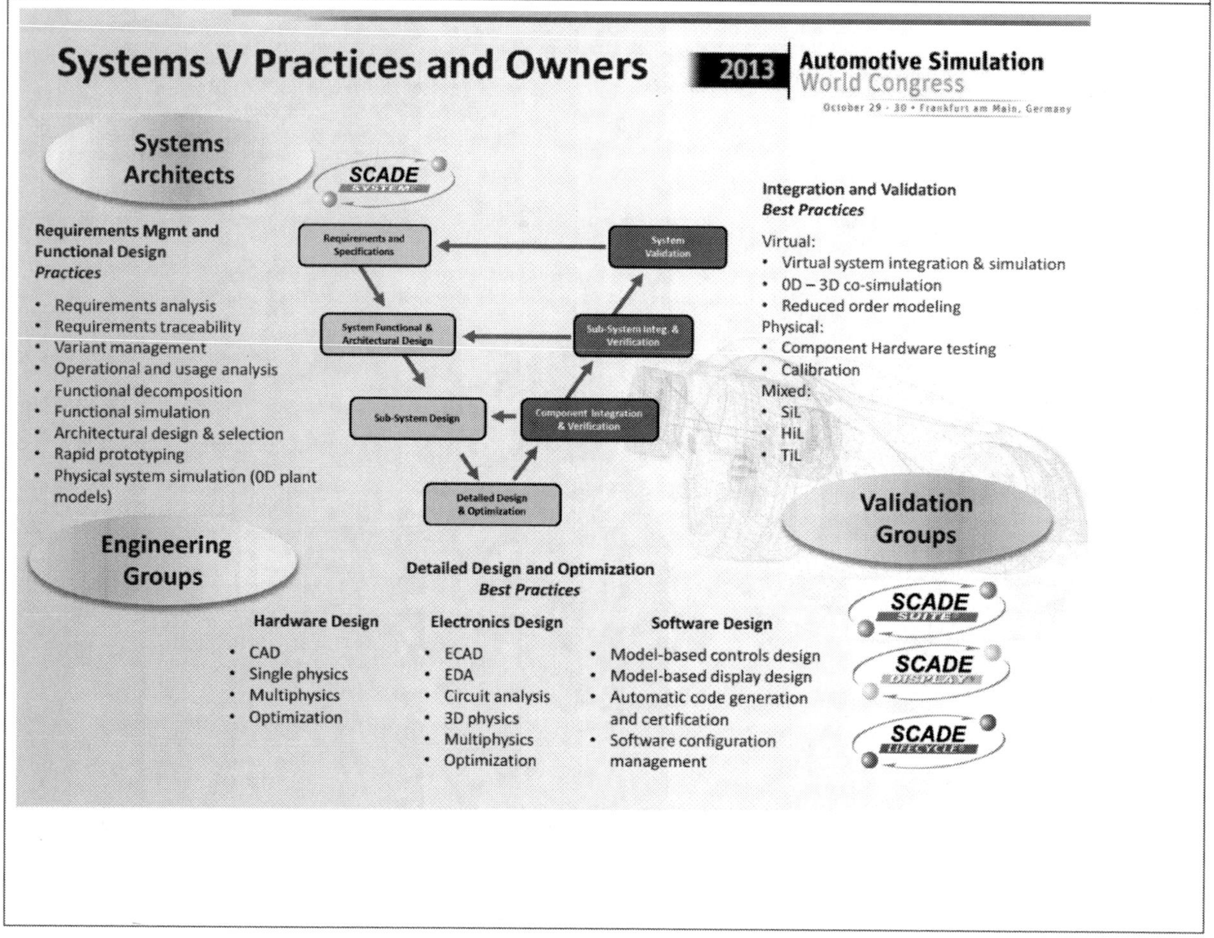
Systems V Practices and Owners
2013 Automotive Simulation World Congress
October 29 - 30 • Frankfurt am Main, Germany
Systems Architects
SCADE SYSTEM
Requirements Mgmt and Functional Design
Practices
• Requirements analysis
• Requirements traceability
• Variant management
• Operational and usage analysis
• Functional decomposition
• Functional simulation
• Architectural design & selection
• Rapid prototyping
• Physical system simulation (0D plant models)
Requirements and Specifications
System Validation
System Functional & Architectural Design
Sub-System Integ. & Verification
Sub-System Design
Component Integration & Verification
Detailed Design & Optimization
Integration and Validation
Best Practices
Virtual:
• Virtual system integration & simulation
• 0D – 3D co-simulation
• Reduced order modeling
Physical:
• Component Hardware testing
• Calibration
Mixed:
• SiL
• HiL
• TiL
Validation Groups
Engineering Groups
Detailed Design and Optimization
Best Practices
Hardware Design
• CAD
• Single physics
• Multiphysics
• Optimization
Electronics Design
• ECAD
• EDA
• Circuit analysis
• 3D physics
• Multiphysics
• Optimization
Software Design
• Model-based controls design
• Model-based display design
• Automatic code generation and certification
• Software configuration management
SCADE SUITE
SCADE DISPLAY
SCADE LIFECYCLE

2013 Automotive Simulation World Congress

System Functional and Architectural Design

SCADE SYSTEM®

2013 Automotive Simulation World Congress

October 29 - 30 • Frankfurt am Main, Germany

SCADE SYSTEM®

Model Based System Engineering

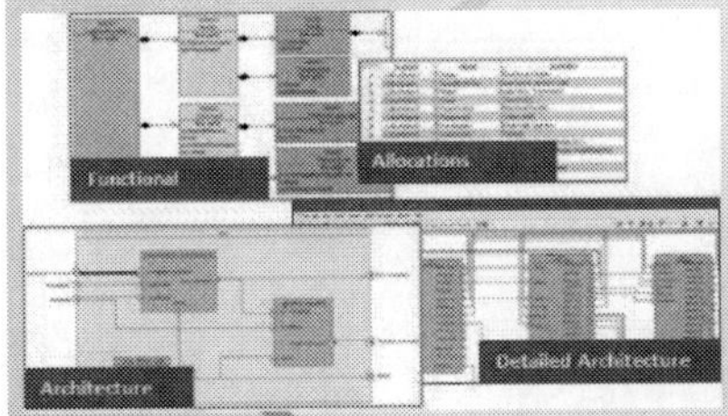

Model Checking

Model Diff

System/Software Sync Up (including I/Os)

DESIGN

VERIFY

INTEGRATE

SCADE System delivers powerful functional and architecture modeling capabilities

2013 **Automotive Simulation** World Congress
October 29 - 30 • Frankfurt am Main, Germany

Systems Engineering

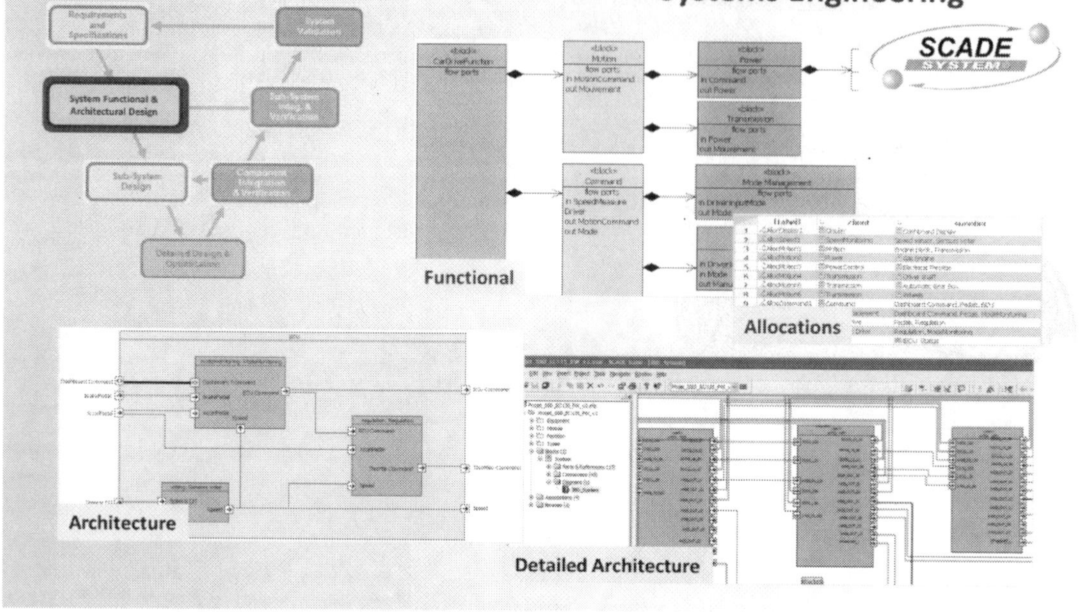

Step 1: Define what the system shall do

2013 **Automotive Simulation** World Congress
October 29 - 30 • Frankfurt am Main, Germany

- Draw the functions and interfaces with SCADE System

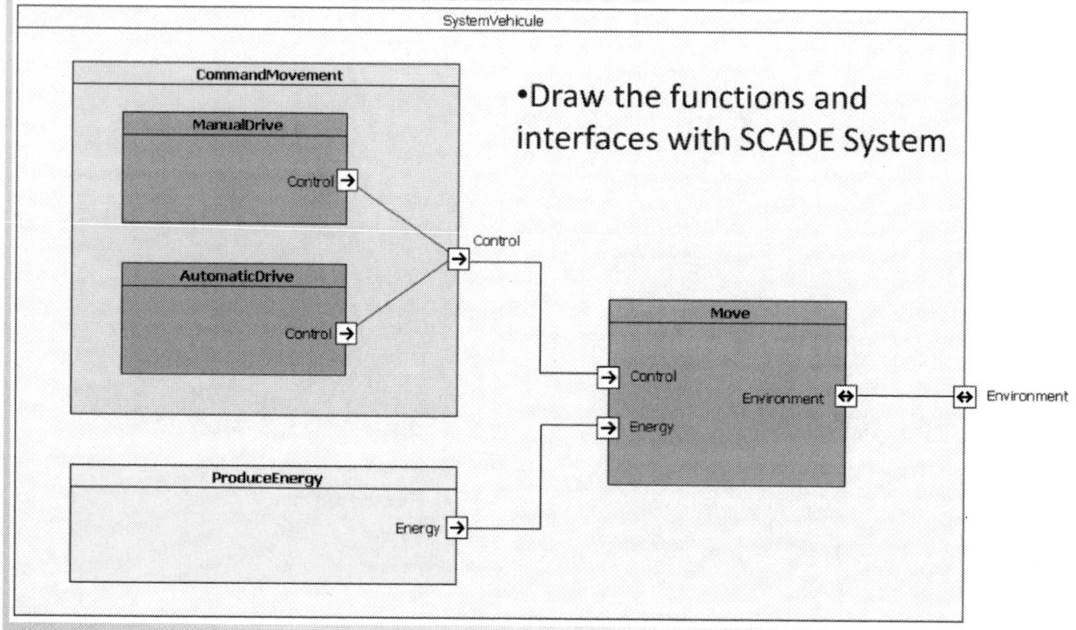

2013 Automotive Simulation World Congress
October 29 - 30 • Frankfurt am Main, Germany

Step 2: Define how the system is built

Draw the system architecture in SCADE System

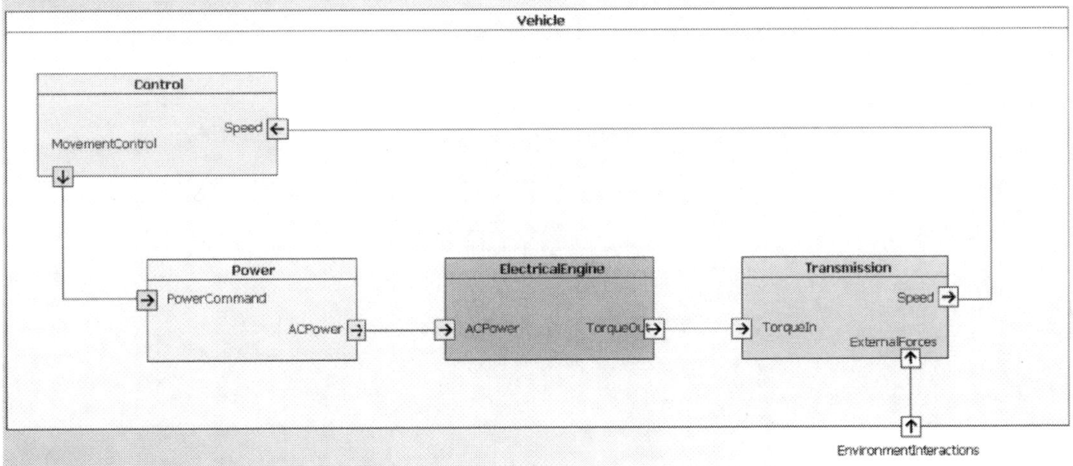

2013 Automotive Simulation World Congress
October 29 - 30 • Frankfurt am Main, Germany

Step 3: Bring further details

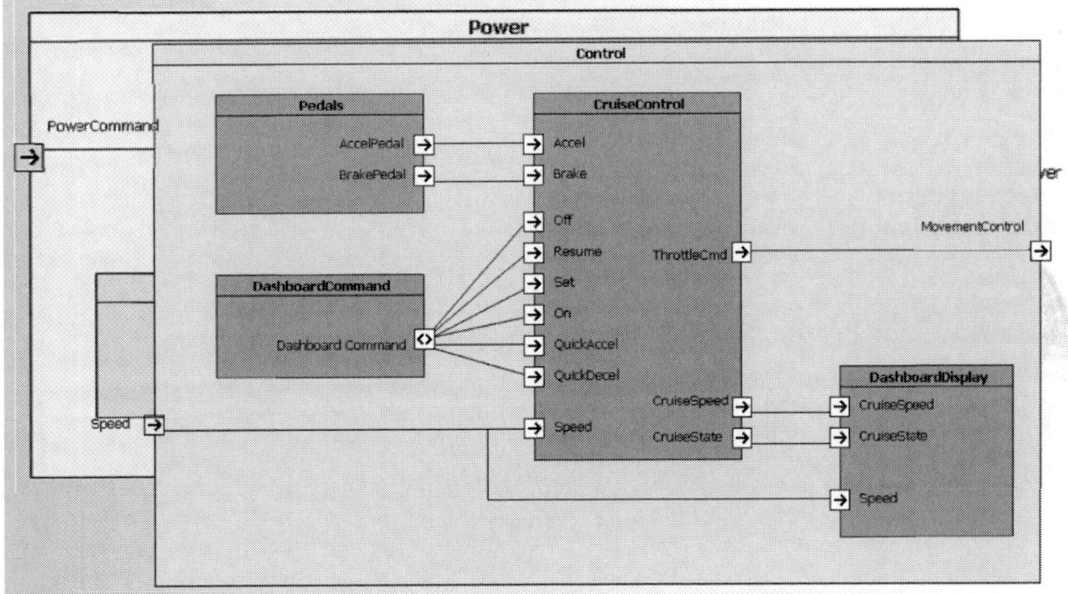

Step 4: Verify design consistency

2013 Automotive Simulation World Congress
October 29 - 30 • Frankfurt am Main, Germany

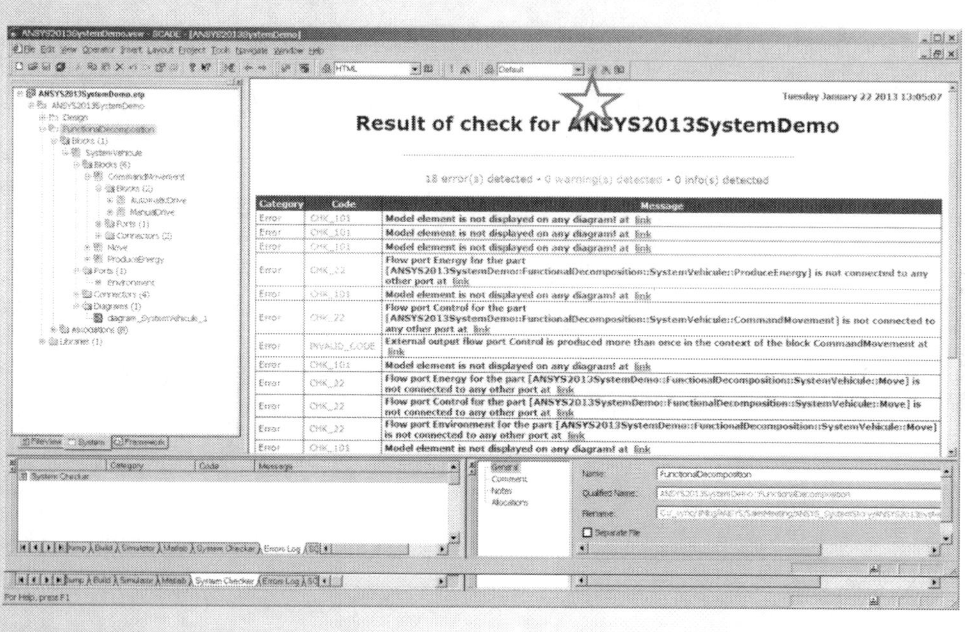

System – Software Synchronization

2013 Automotive Simulation World Congress
October 29 - 30 • Frankfurt am Main, Germany

- Avoid duplication of efforts and inconsistencies between system structural models and software behavioral models
- System design and Software components evolve independently
- On-demand re-synchronization of interfaces

Interfaces described in SCADE System model

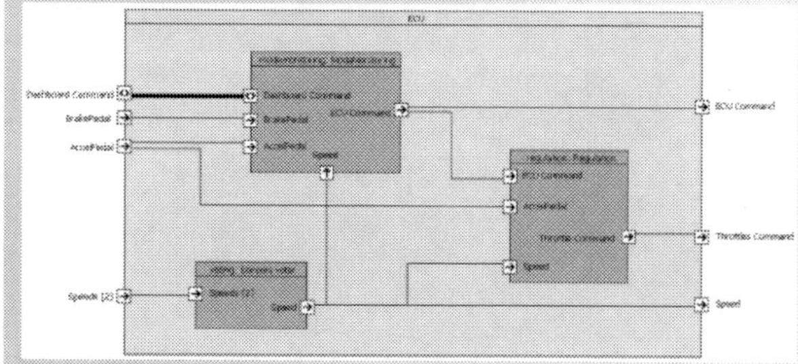

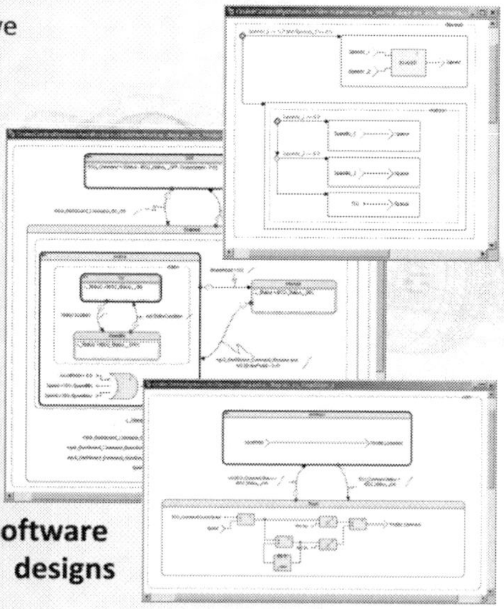

Software designs

2013 Automotive Simulation
World Congress
Safety Critical
Embedded Software Design
SCADE SUITE®
SCADE DISPLAY®
SCADE LIFECYCLE®

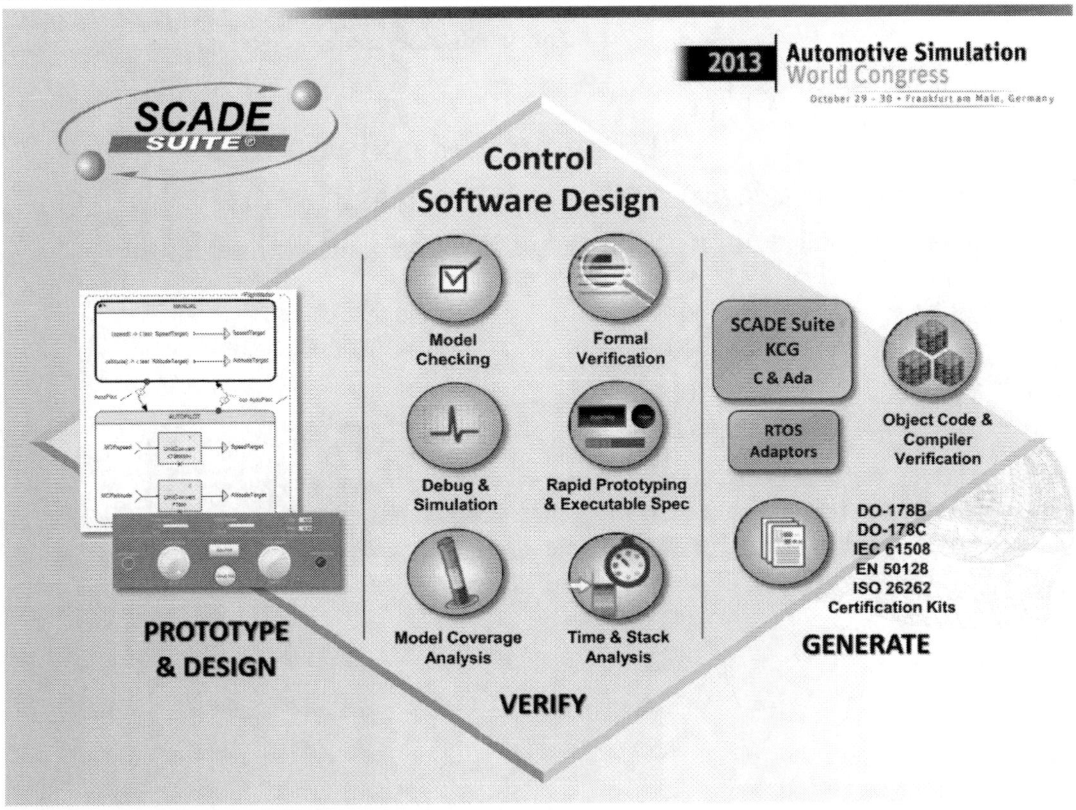
2013 Automotive Simulation
World Congress
October 29 - 30 • Frankfurt am Main, Germany
SCADE SUITE®
Control
Software Design
Model
Checking
Formal
Verification
Debug &
Simulation
Rapid Prototyping
& Executable Spec
Model Coverage
Analysis
Time & Stack
Analysis
SCADE Suite
KCG
C & Ada
RTOS
Adaptors
Object Code &
Compiler
Verification
DO-178B
DO-178C
IEC 61508
EN 50128
ISO 26262
Certification Kits
PROTOTYPE
& DESIGN
VERIFY
GENERATE

SCADE Suite-based Cruise Control

2013 Automotive Simulation World Congress
October 29 - 30 • Frankfurt am Main, Germany

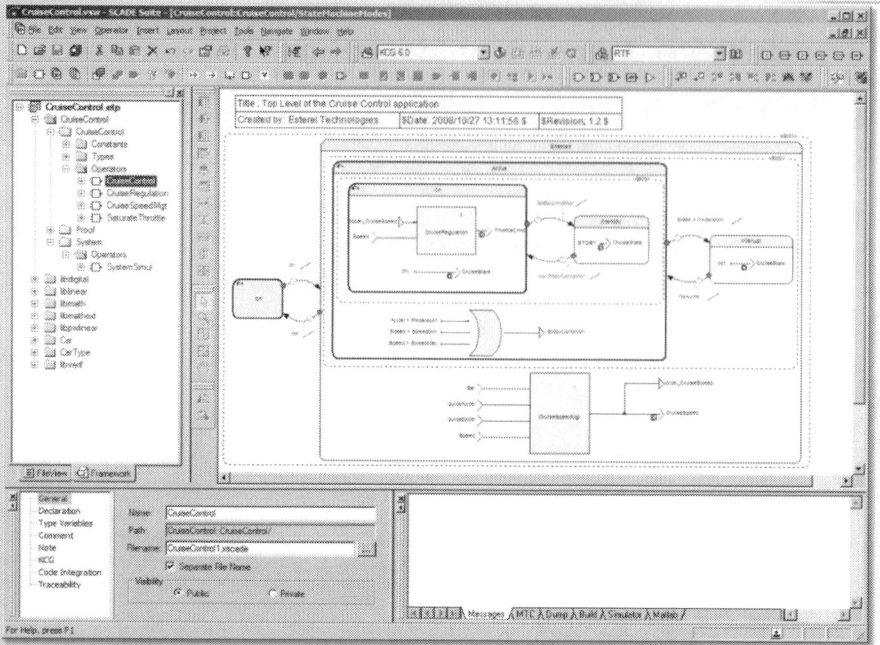

2013 Automotive Simulation World Congress
October 29 - 30 • Frankfurt am Main, Germany

HIL Simulation Support

- Download & run SCADE generated code on to a real physical system and/or a HIL infrastructure

Ex: National Instruments HIL (PXI, Compact RIO) with real sensors & actuators

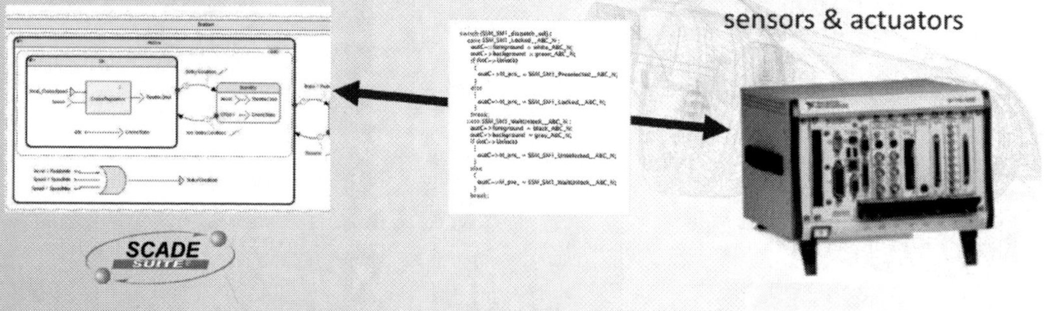

BACK

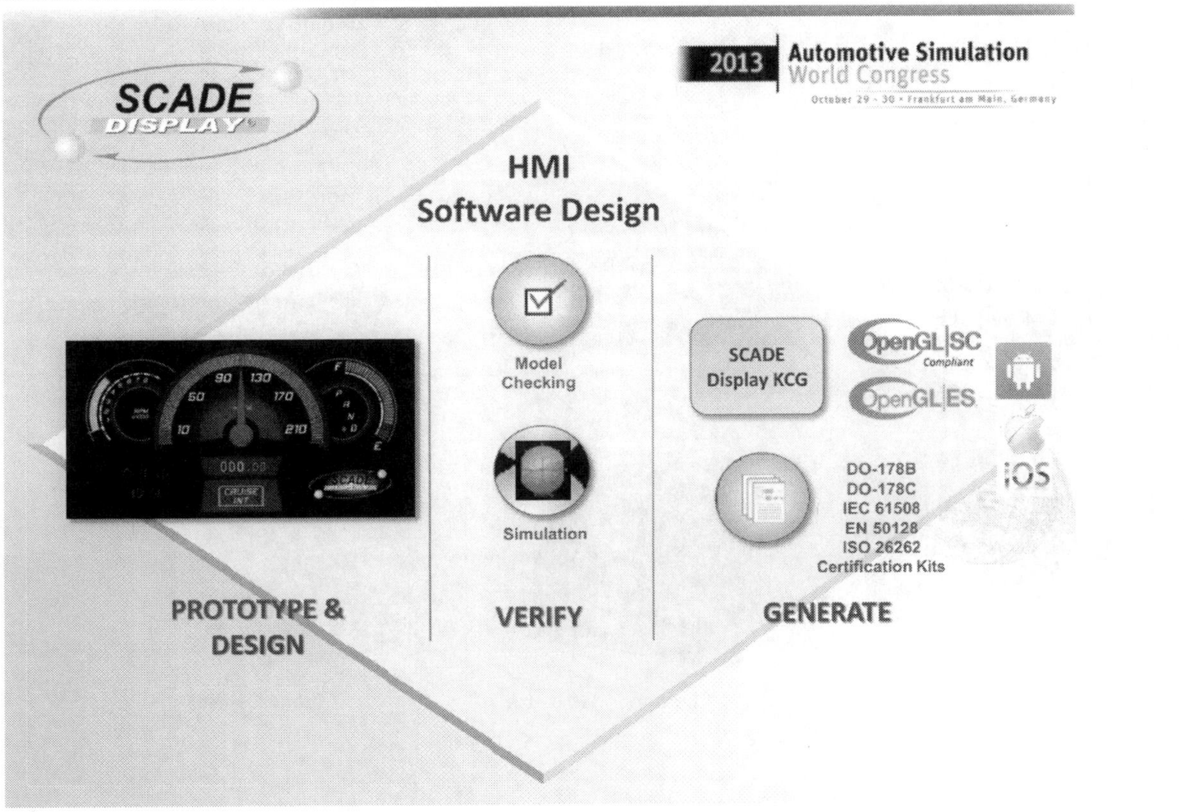
2013
Automotive Simulation
World Congress
October 29 - 30 • Frankfurt am Main, Germany
SCADE
DISPLAY
HMI
Software Design
Model
Checking
Simulation
SCADE
Display KCG
OpenGL SC
Compliant
OpenGL ES
iOS
DO-178B
DO-178C
IEC 61508
EN 50128
ISO 26262
Certification Kits
PROTOTYPE &
DESIGN
VERIFY
GENERATE

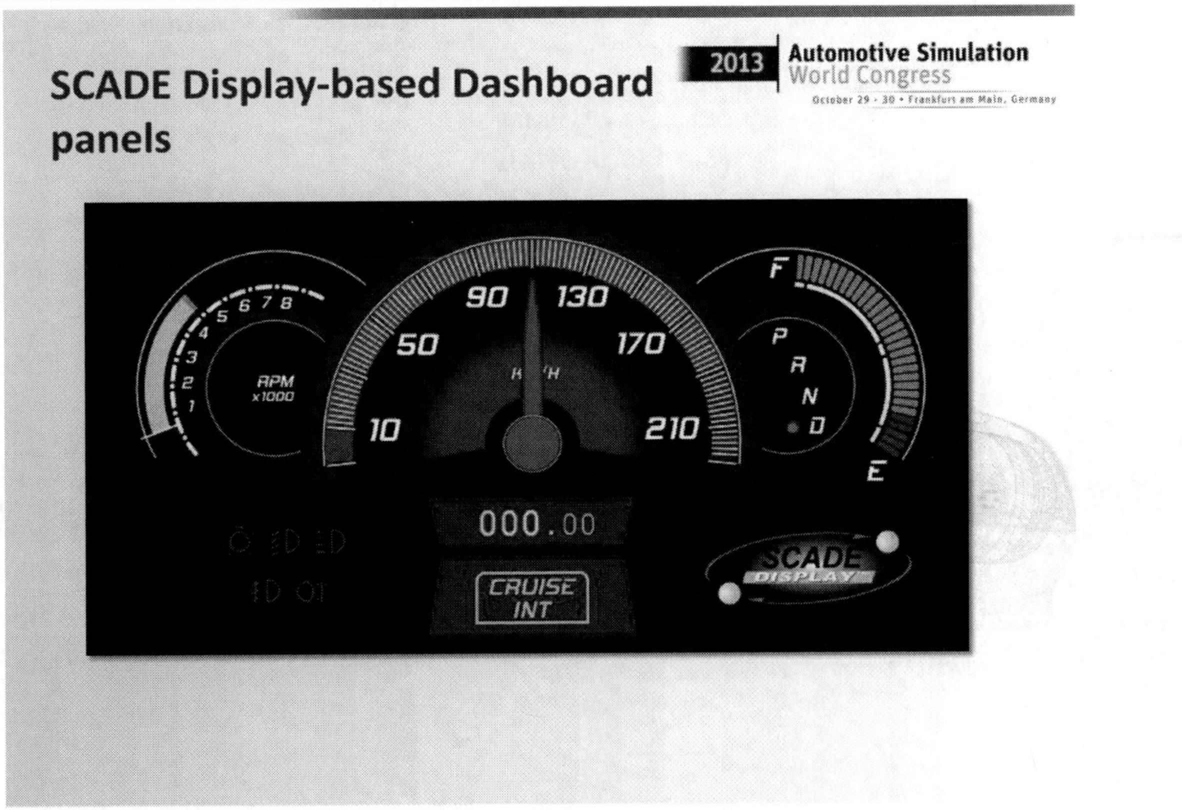
2013
Automotive Simulation
World Congress
October 29 - 30 • Frankfurt am Main, Germany
SCADE Display-based Dashboard
panels
RPM
x1000
90
130
50
170
10
210
F
P
R
N
D
E
000.00
CRUISE
INT
SCADE
DISPLAY

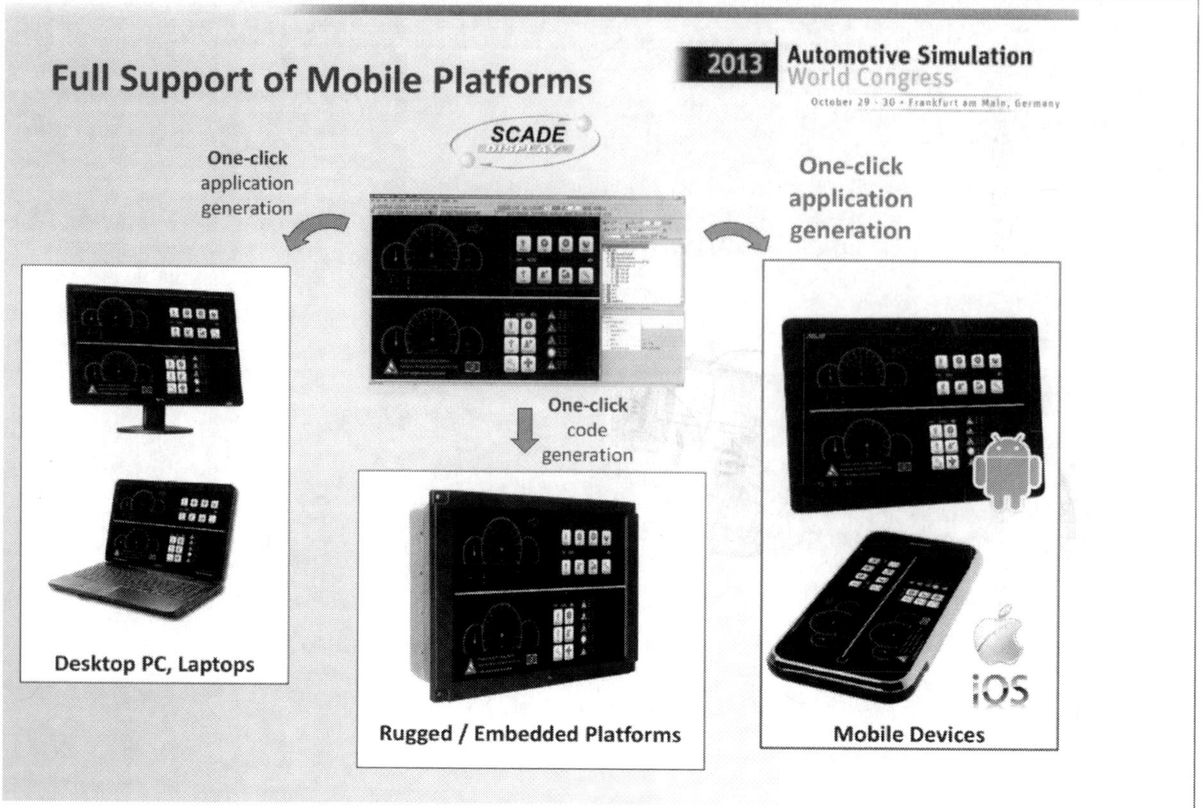
Full Support of Mobile Platforms
2013 Automotive Simulation World Congress
October 29 - 30 • Frankfurt am Main, Germany
SCADE
One-click application generation
One-click application generation
One-click code generation
Desktop PC, Laptops
Rugged / Embedded Platforms
iOS
Mobile Devices

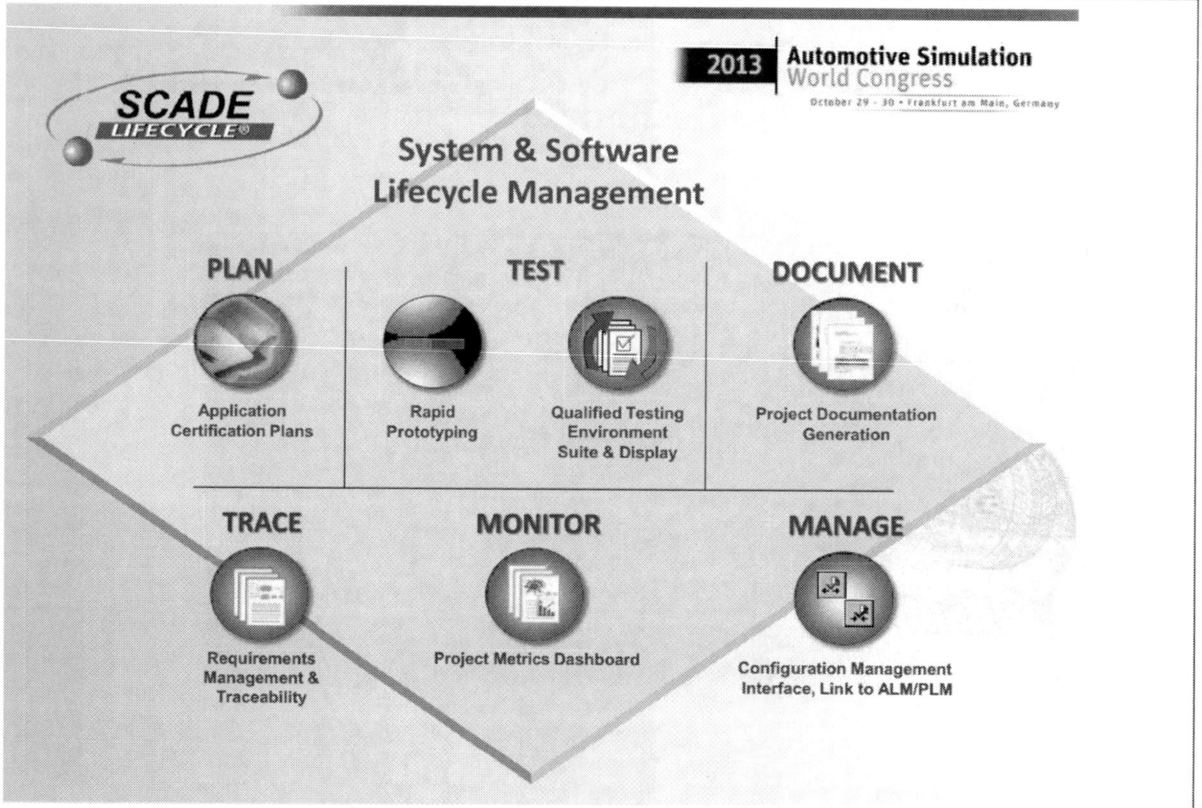
2013 Automotive Simulation World Congress
October 29 - 30 • Frankfurt am Main, Germany
SCADE LIFECYCLE®
System & Software Lifecycle Management
PLAN
TEST
DOCUMENT
Application Certification Plans
Rapid Prototyping
Qualified Testing Environment Suite & Display
Project Documentation Generation
TRACE
MONITOR
MANAGE
Requirements Management & Traceability
Project Metrics Dashboard
Configuration Management Interface, Link to ALM/PLM

Verify requirements traceability with SCADE LifeCycle

2013 Automotive Simulation World Congress
October 29 - 30 • Frankfurt am Main, Germany

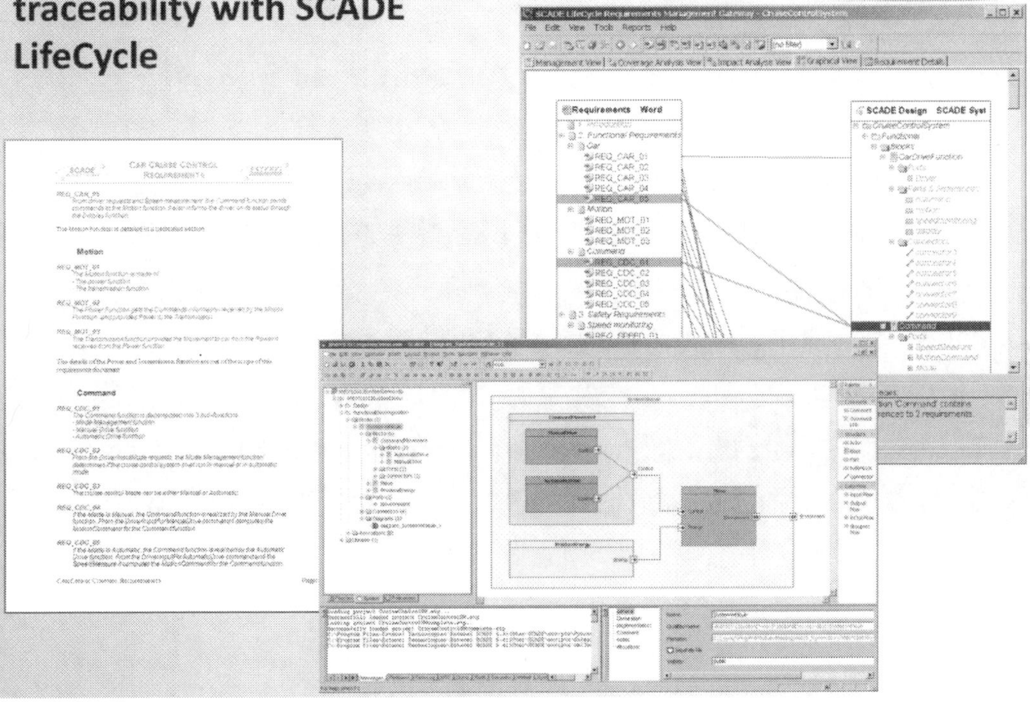

What is unique about SCADE ?

2013 Automotive Simulation World Congress
October 29 - 30 • Frankfurt am Main, Germany

- SCADE is developed specifically to be able to address critical system and software applications
- SCADE Suite and Display Code Generators are certified/qualified according to the following international safety standards:
  - **ISO 26262 certification up to ASIL D – Automotive (TUV Sud)**
  - IEC 61508 certification up to SIL 3 – Industrial & Energy
    - IEC 60880 full compliance – Nuclear Instrumentation & Control
    - IEC 62304 full compliance – Medical Systems
    - EN 13849 full compliance – Industrial Machines Safety
  - DO-178B / DO-178C qualification up to Level A – Aerospace & Defense
  - EN 50128 certification up to SIL 3/4 – Rail Transportation
- Same products qualified at the highest level of safety across 5 market segments by 10 safety authorities, worldwide

2013 Automotive Simulation World Congress
October 29 - 30 • Frankfurt am Main, Germany

What costs reduction does this bring ?

- Application software can be modelled, simulated and debugged on host using a notation that has been designed for safety critical systems
- Generated code will behave like the Applications Models
- All unit testing to verify the coding process are eliminated
- Time spent on HIL, testing the correctness of application software decreases by more than 80% versus manual methods

2013 Automotive Simulation World Congress
October 29 - 30 • Frankfurt am Main, Germany

Comparative Software Development Effort per Level including Testing

ISO 26262 Level	No cert Effort	ASIL A Effort	ASIL B Effort	ASIL C Effort	ASIL D Effort
Comparative Effort	Baseline	No cert +10%	ASIL A +36%	ASIL B +80%	ASIL C +30%
Effort	100	110	150	270	350+

ISO 26262 Methodology Handbook

- Contents:
  - Development and verification steps
    - Model-based development with SCADE
    - Simulation and Model Test Coverage
    - Formal verification
    - Automatic code generation with KCG
    - C compiler verification activities
  - Set of guidelines for developing efficient models, generating efficient code, etc.

ESTEREL

SCADE Suite at Subaru Electric Vehicles

- Subaru chose SCADE Suite for the design of its electric vehicle engine controls:
  - Vehicle dynamics
  - Engine functions
  - Vehicle energy consumption
    - » heating & air conditioning
    - » breaking
    - » body controls
  - Battery load management

SCADE Suite at LIEBHERR Construction

2013 Automotive Simulation World Congress
October 29 - 30 • Frankfurt am Main, Germany

- Liebherr chose SCADE Suite as its corporate standard embedded software development environment
- For all its design centers for earth moving machines, cranes and any heavy duty product

2013 Automotive Simulation World Congress

Full Sytem Simulation and Virtual Prototyping

E/E System Physical Simulation with Simplorer

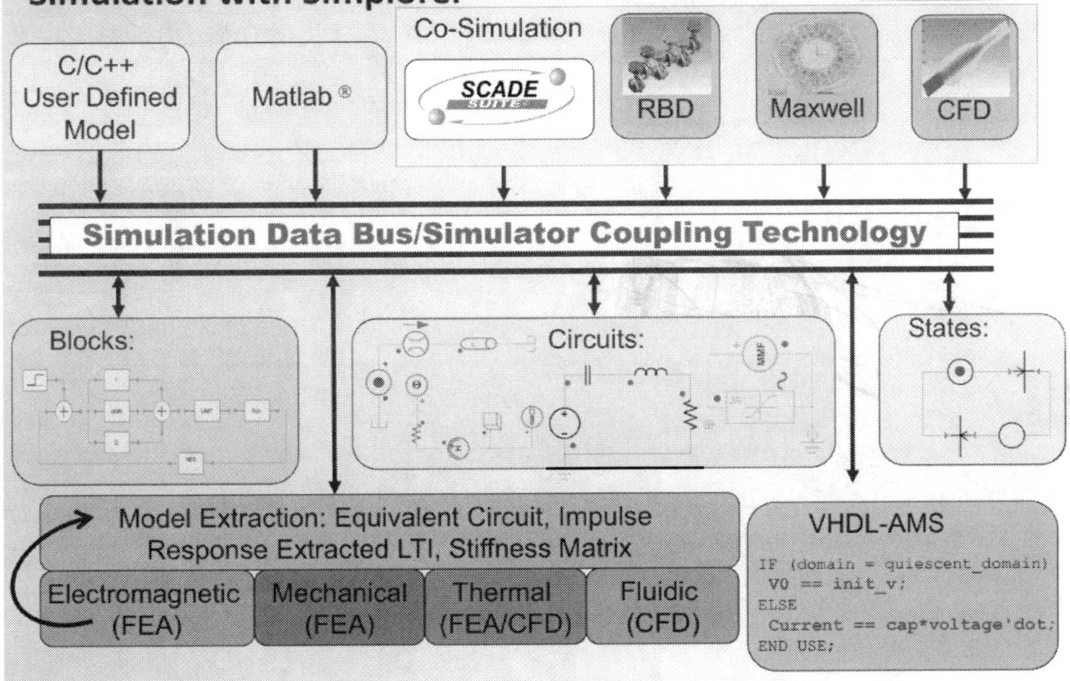

2013 Automotive Simulation World Congress
October 29 - 30 • Frankfurt am Main, Germany

SIMPLORER VHDL-AMS

VHDL-AMS:

- Mixed signal modeling language standardized by the IEEE/IEC
- IEEE standard 1076.1, IEC Standard 61691-2
- Rich Electronics library development: Electrical, Power electronics, HEV, etc..

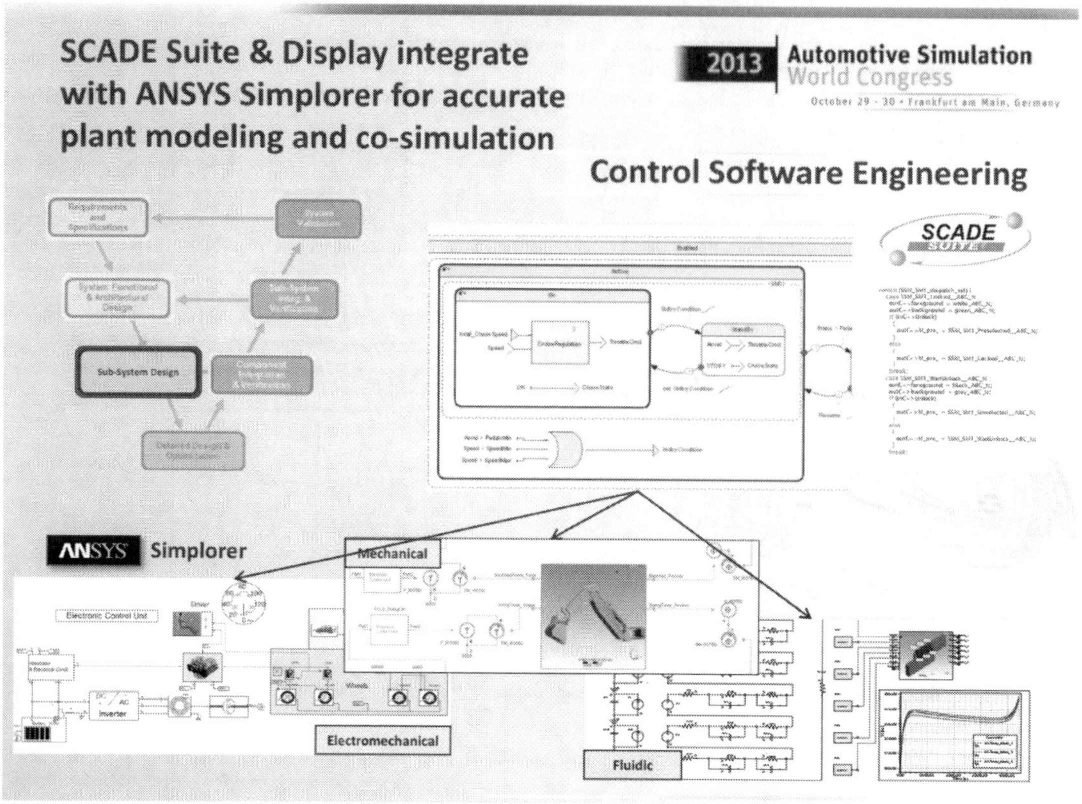
SCADE Suite & Display integrate with ANSYS Simplorer for accurate plant modeling and co-simulation
2013 Automotive Simulation World Congress
October 29 - 30 • Frankfurt am Main, Germany
Control Software Engineering
SCADE
Sub-System Design
ANSYS Simplorer
Mechanical
Electronic Control Unit
Electromechanical
Fluidic

Battery Physics – Thermal
ANSYS CFD Viewer
ANSYS
Temperature
TemperatureOnCells
319
317
316
315
313
312
311
309
308
307
305
304
302
301
300
298
297
296
294
293
[K]

2013 Automotive Simulation World Congress
October 29 - 30 • Frankfurt am Main, Germany

Reduced Order Model for CFD

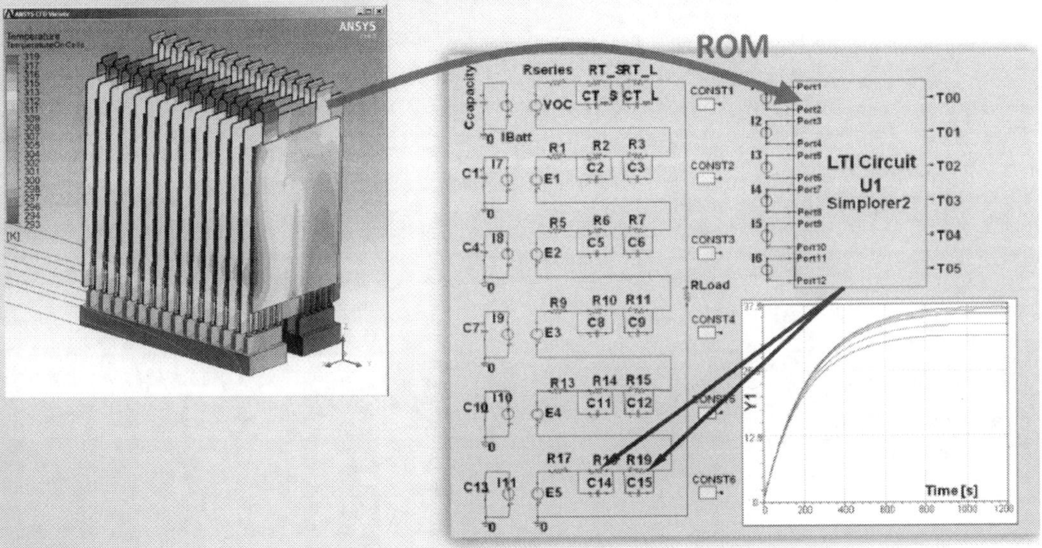

2013 Automotive Simulation World Congress
October 29 - 30 • Frankfurt am Main, Germany

Sensor Physics and Behavior via Reduced Order Model (ROM)

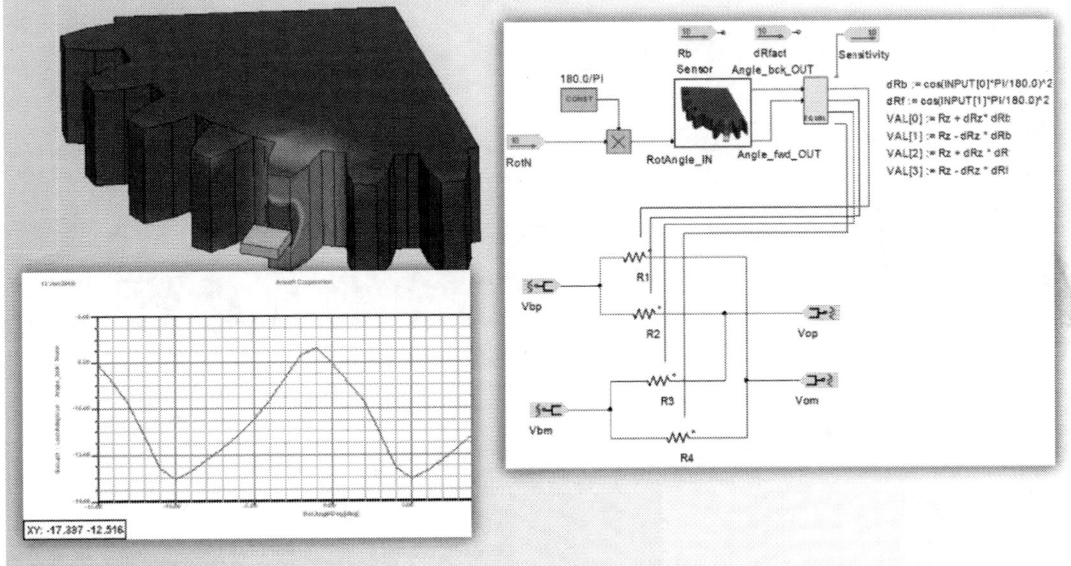

Electric Machine Physics and Behavioral ROM

2013 **Automotive Simulation** World Congress
October 29 - 30 • Frankfurt am Main, Germany

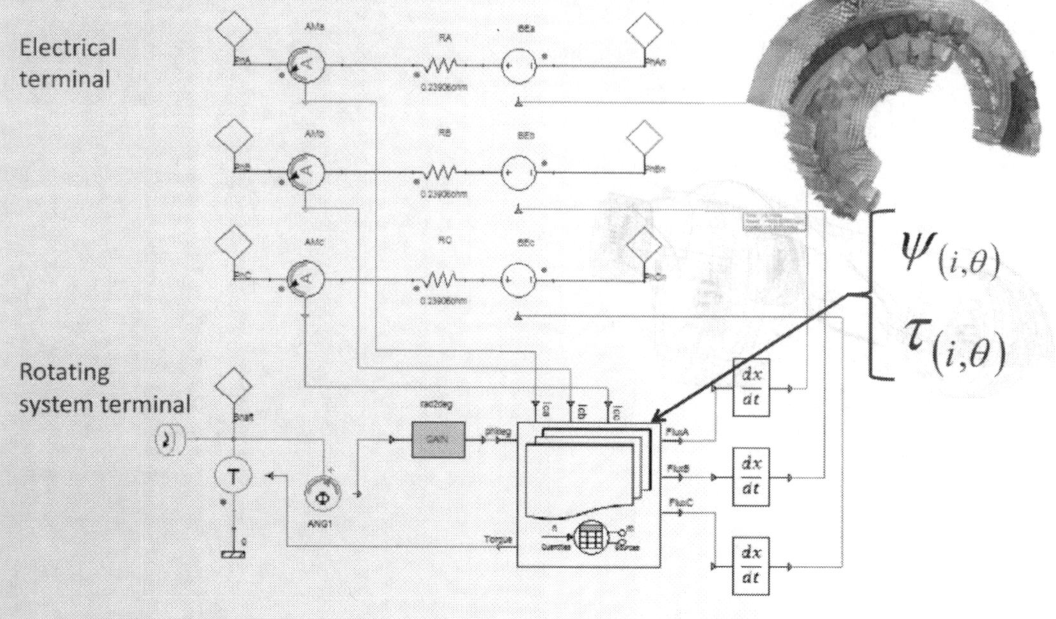

The System Simulation Model

2013 **Automotive Simulation** World Congress
October 29 - 30 • Frankfurt am Main, Germany

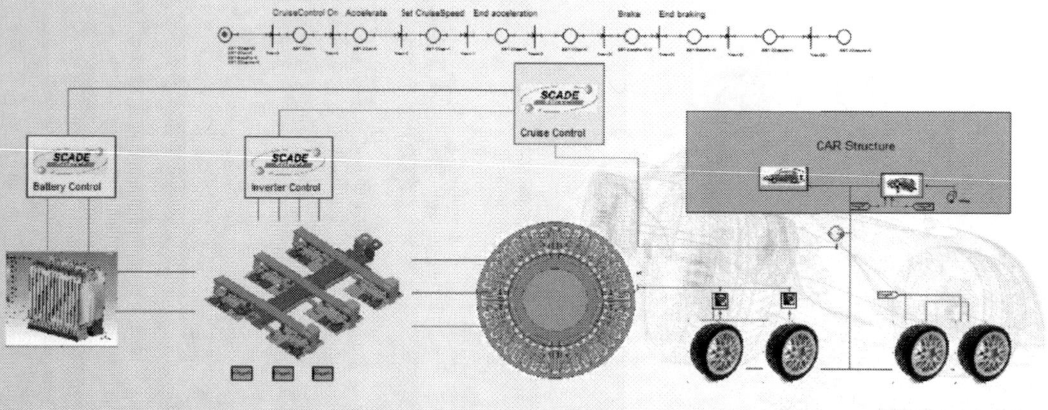

Build the Virtual Prototype

2013 **Automotive Simulation** World Congress
October 29 - 30 • Frankfurt am Main, Germany

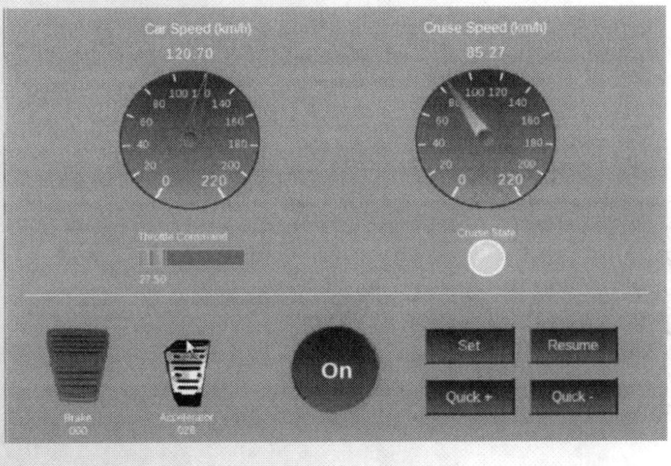

SCADE Rapid Prototyper Simulation Interface

A Complete Virtual Prototype !

2013 **Automotive Simulation** World Congress
October 29 - 30 • Frankfurt am Main, Germany

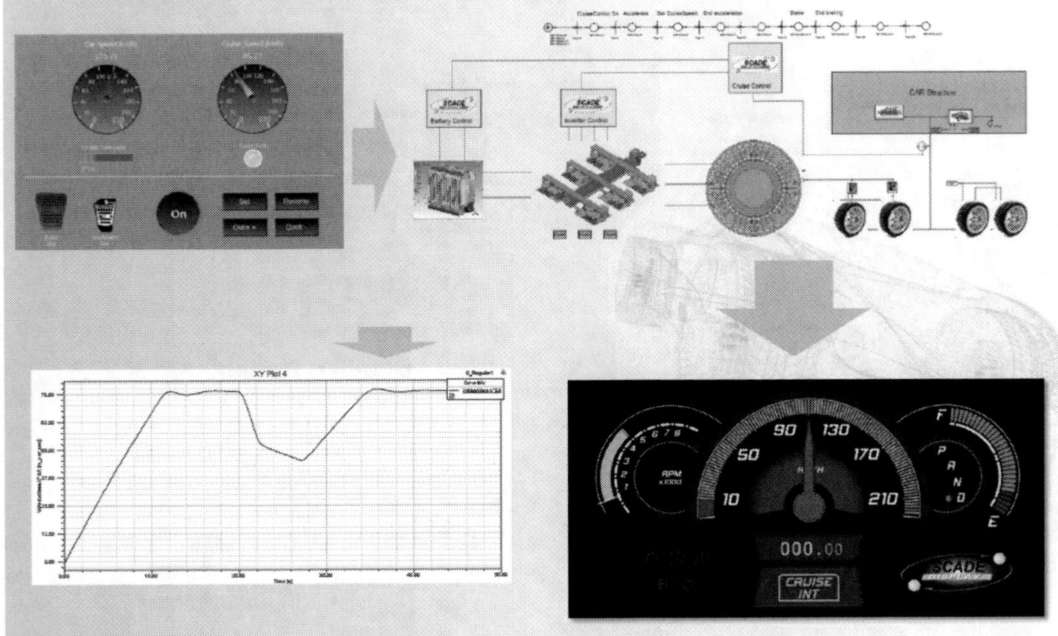

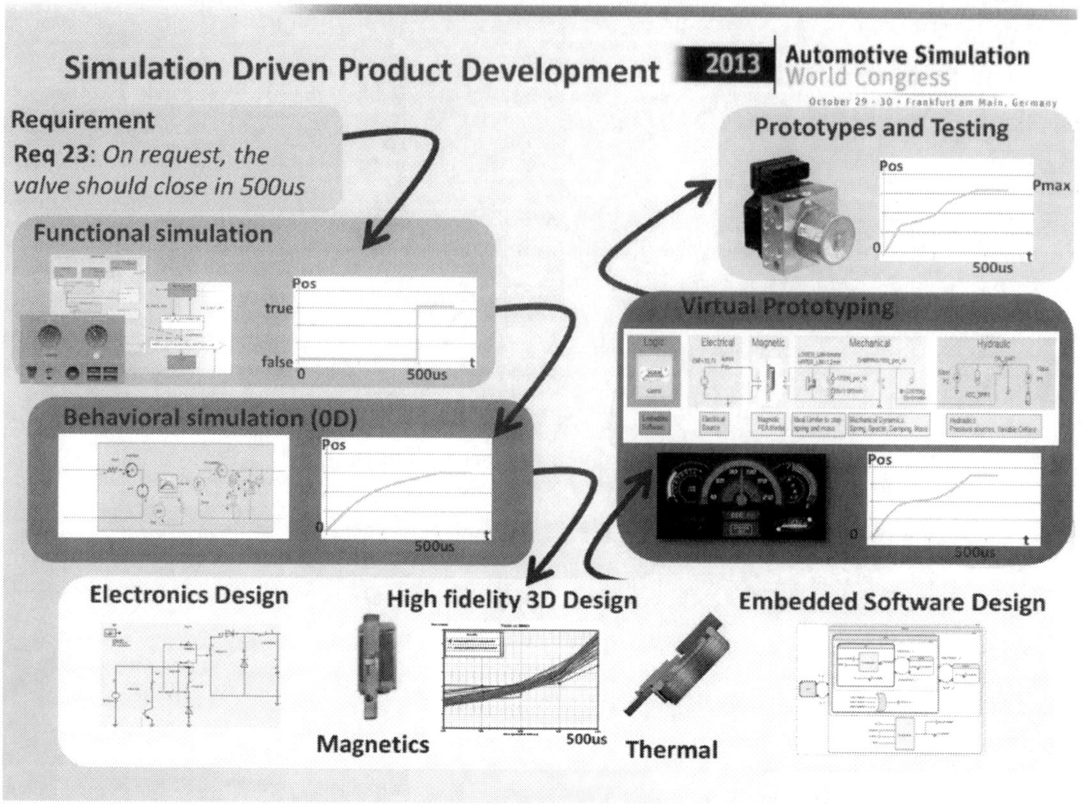
Simulation Driven Product Development
2013 Automotive Simulation World Congress
October 29 - 30 • Frankfurt am Main, Germany
Requirement
Req 23: On request, the valve should close in 500us
Functional simulation
Pos
true
false
0
500us
t
Behavioral simulation (0D)
Pos
0
500us
t
Prototypes and Testing
Pos
Pmax
0
500us
t
Virtual Prototyping
Pos
0
500us
t
Electronics Design
High fidelity 3D Design
Embedded Software Design
Magnetics
500us
Thermal

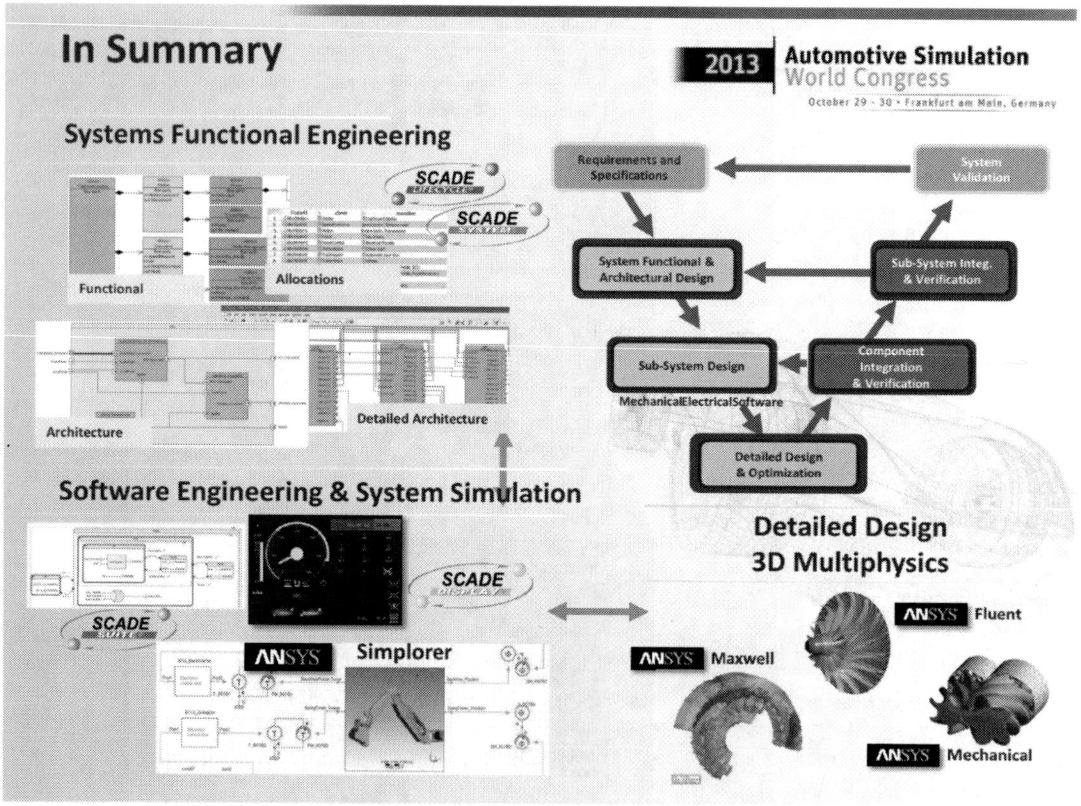
In Summary
2013 Automotive Simulation World Congress
October 29 - 30 • Frankfurt am Main, Germany
Systems Functional Engineering
SCADE
SCADE
Functional
Allocations
Architecture
Detailed Architecture
Requirements and Specifications
System Validation
System Functional & Architectural Design
Sub-System Integ. & Verification
Sub-System Design
Component Integration & Verification
MechanicalElectricalSoftware
Detailed Design & Optimization
Software Engineering & System Simulation
SCADE
SCADE
ANSYS Simplorer
Detailed Design 3D Multiphysics
ANSYS Fluent
ANSYS Maxwell
ANSYS Mechanical

Technische Universität Berlin
FG Verbrennungskraftmaschinen

Diversity of Drivetrains and Fuels Define Combustion Engine Development

Prof. Dr.-Ing. Roland Baar
Frankfurt, 29.10.2013

... from specs definition to market introduction ...

"I call my invention 'wheel', but still needs to develop the production stage and select a supplier."

5 phases during product life cycle

- o Product planing
- o Concept development
- o Series development
- o Production preparation
- o Continuous improvement

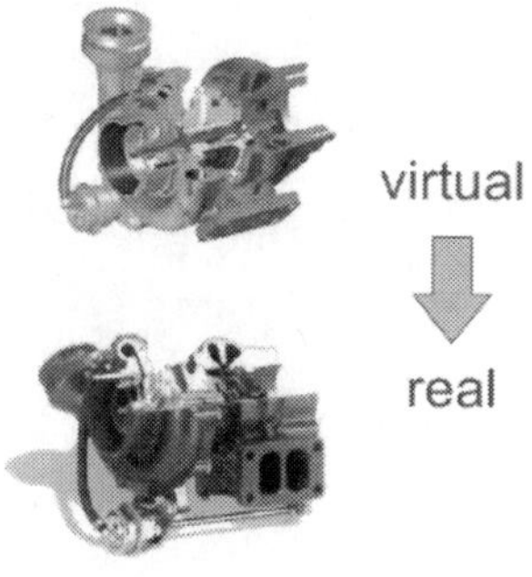

⇛ Added value is driven by complexity of processes
⇛ Virtual work is getting more important

Product development: who & when

▼ milestone

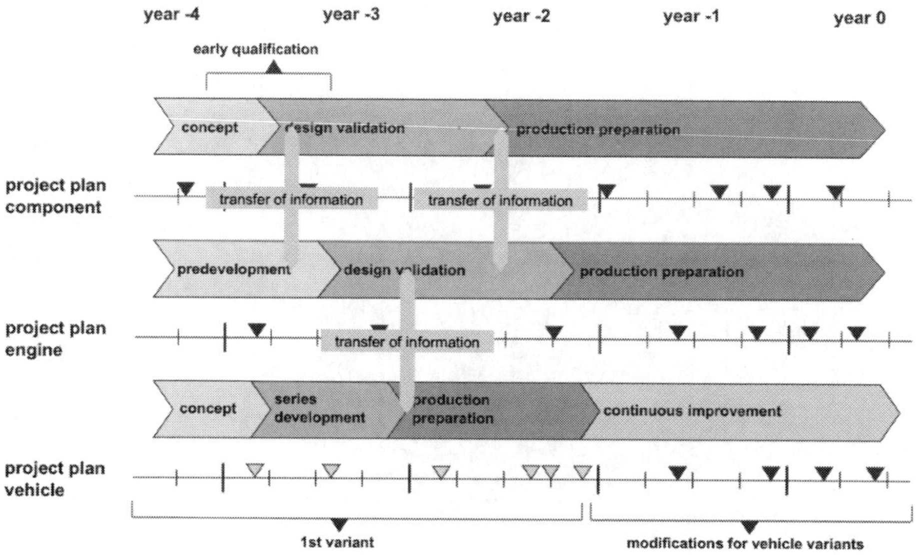

Complexity in powertrains

Design and calibration

Testing and simulation

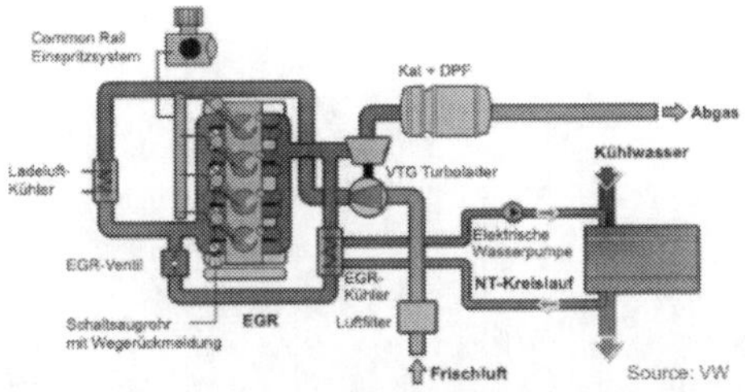

Cycle assessment

Models in different quality

engine | models | basic science

Gas and heat exchange

Injection hydraulics

Air fuel mixing

Exhaust after treatment.

Controll

Boosting

Combustion

Research roadmap boosting and gas exchange

competence

acoustics

pulsa-tion

friction losses

CFD

surge

2012 | 2013 | 2014 | +5 years

ball bearing

axial turbine

two stage

advanced boosting system

electri-fication

air bearing

small turbo

matching

technology

Technology to improve turbo transients
How to compare?

Ball bearing

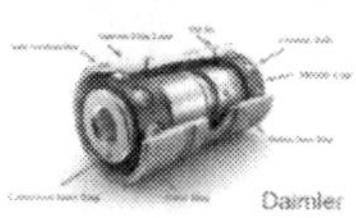

Titanium-Aluminid

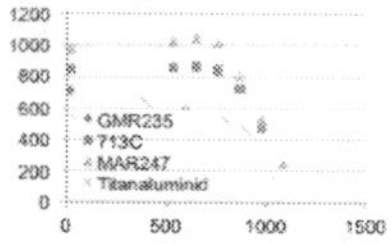

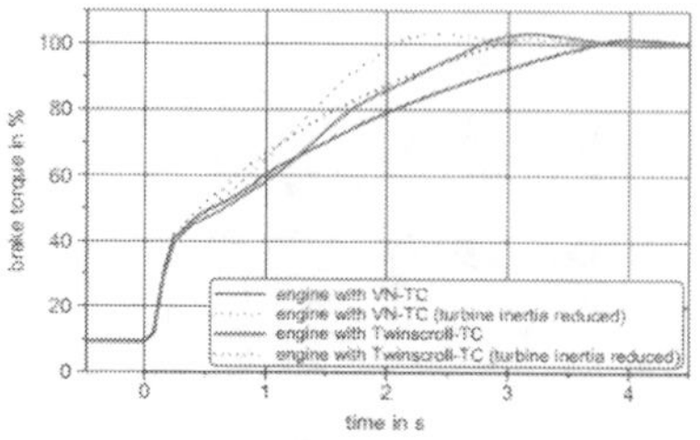

Controll

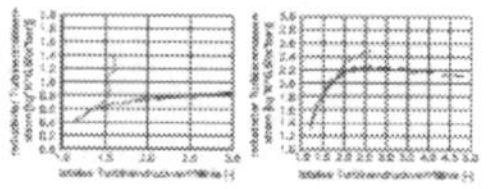

E-Turbo

Mixed flow-/
axialturbine

Scavenging /
LP-EGR

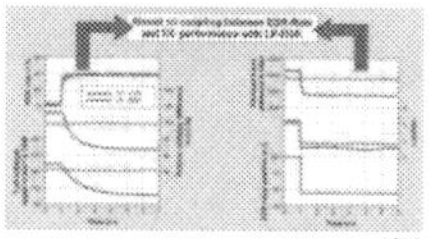

twostage
turbo

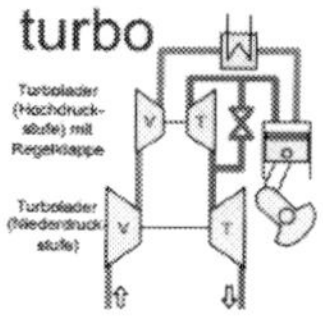

Isolation /
heat manag.

Research roadmap
injection, mixing and combustion

pressure chamber

fuel lab

injection analyzer

CFD

optical engine

competence

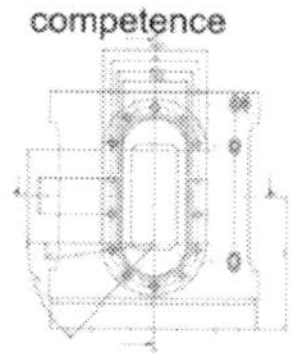

2012 | 2013 | 2014 | +5 years

EGR

blends

VVT

control

thermo management

advanced combustion system

technology

Combustion

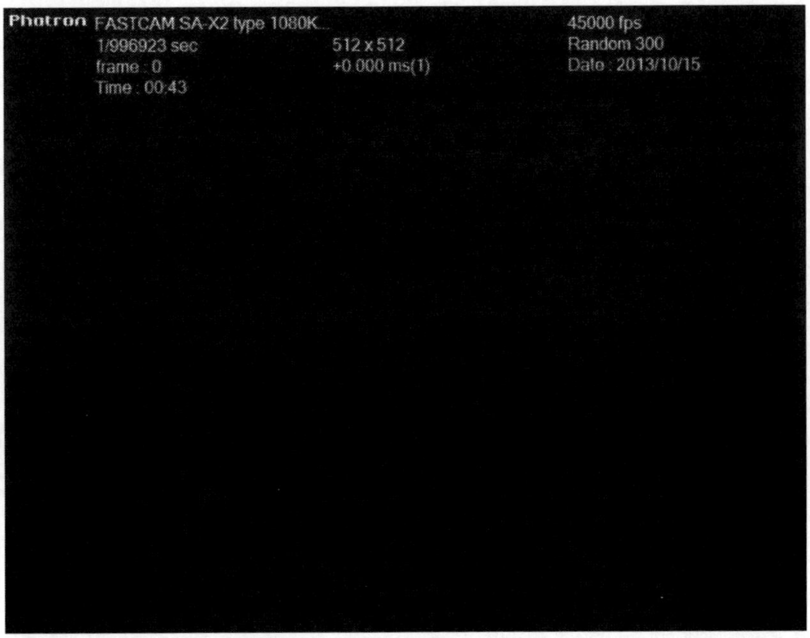

Technology to improve combustion Where are the needs and options?

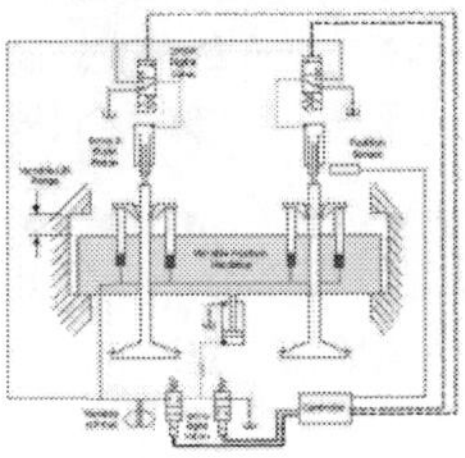

- Variable valve
  - Timing
  - Duration
  - Lift
- Secondary lift for each valve
- Valve deactivation
- Valve opening at high cylinder pressure

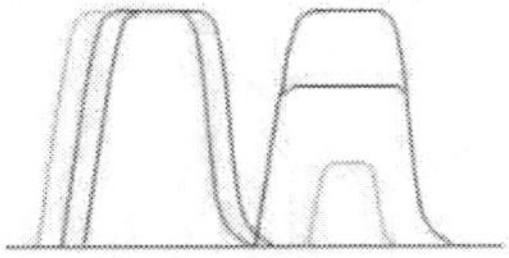

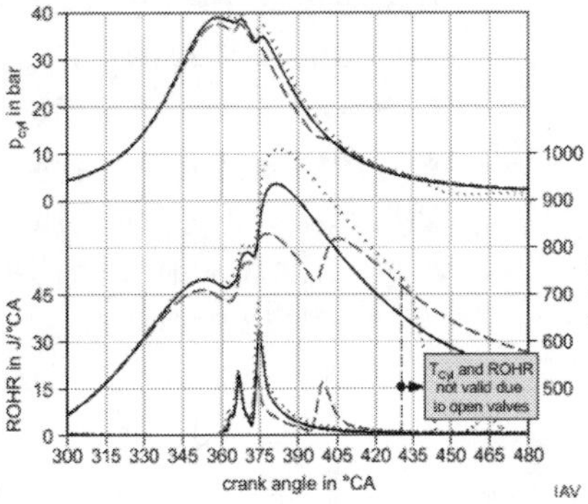

The Model – What is that?

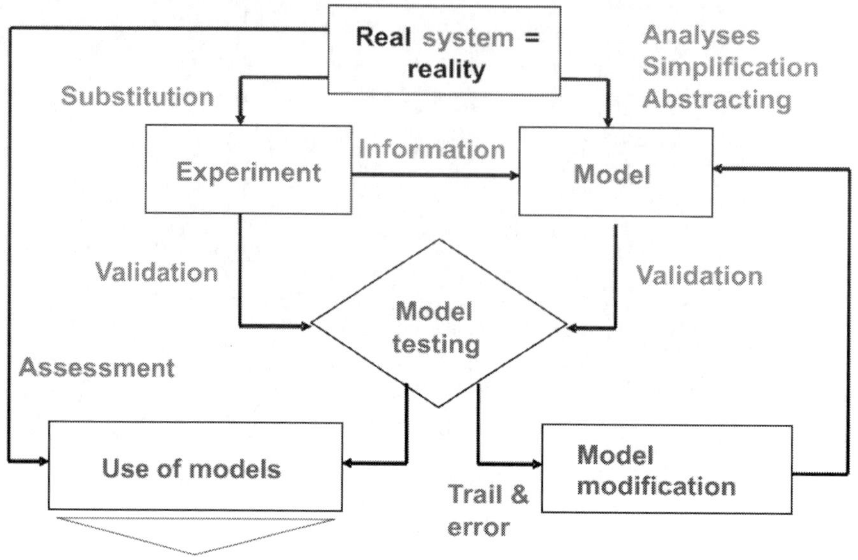

ANSYS Keynote - Frankfurt 2013 | Baar | 29.10.2013
FG Verbrennungskraftmaschinen

11

Problem statement

- Both experimental and simulated data is defective.
- CFD uses simplified models of the physics a validation with experimental data is essential.
- CFD does not provide the true value it represents a correlation with a certain range of tolerance.

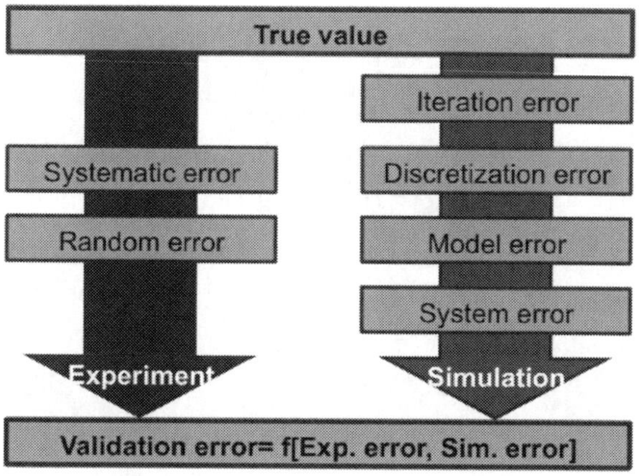

ANSYS Keynote - Frankfurt 2013 | Baar | 29.10.2013
FG Verbrennungskraftmaschinen

12

Turbocharger performance map

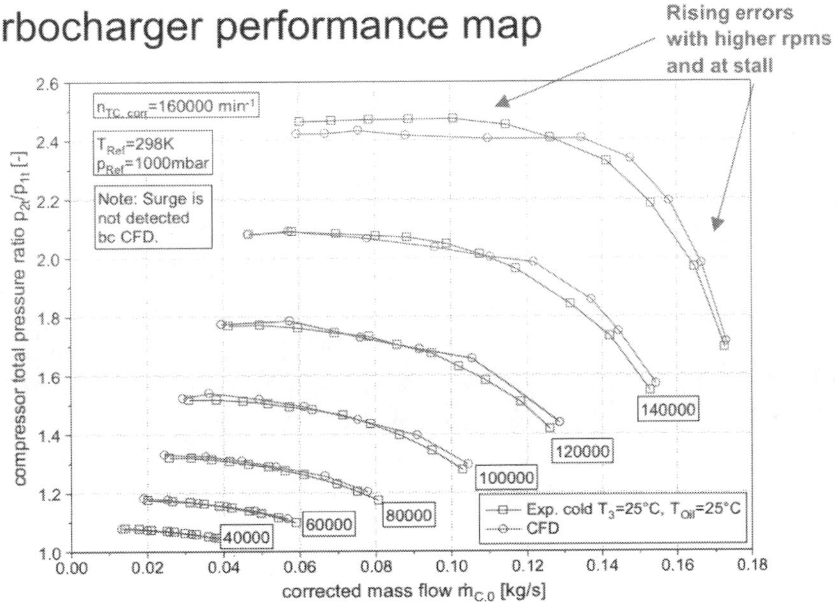

ANSYS Keynote - Frankfurt 2013 | Baar | 29.10.2013
FG Verbrennungskraftmaschinen

13

Problem I – heat
Diabatic effects in a turbocharger

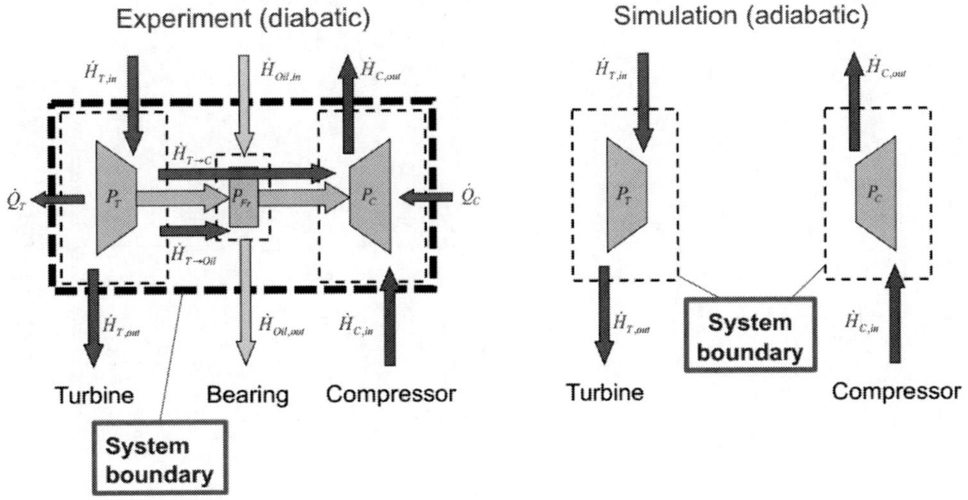

ANSYS Keynote - Frankfurt 2013 | Baar | 29.10.2013
FG Verbrennungskraftmaschinen

14

Turbocharger performance map

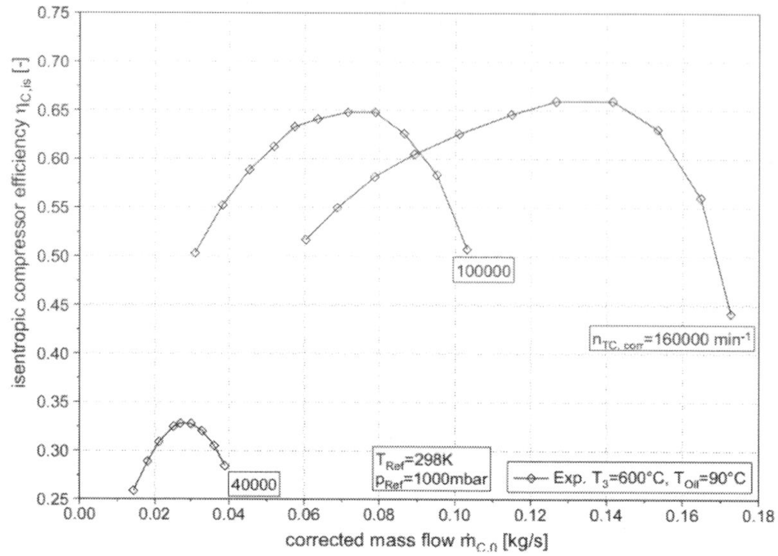

ANSYS Keynote - Frankfurt 2013 | Baar | 29.10.2013
FG Verbrennungskraftmaschinen

15

Turbocharger performance map

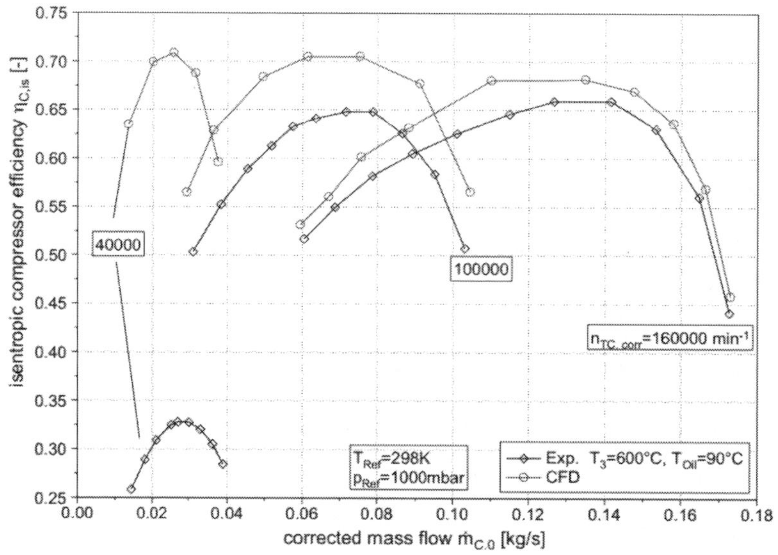

ANSYS Keynote - Frankfurt 2013 | Baar | 29.10.2013
FG Verbrennungskraftmaschinen

16

Turbocharger performance map

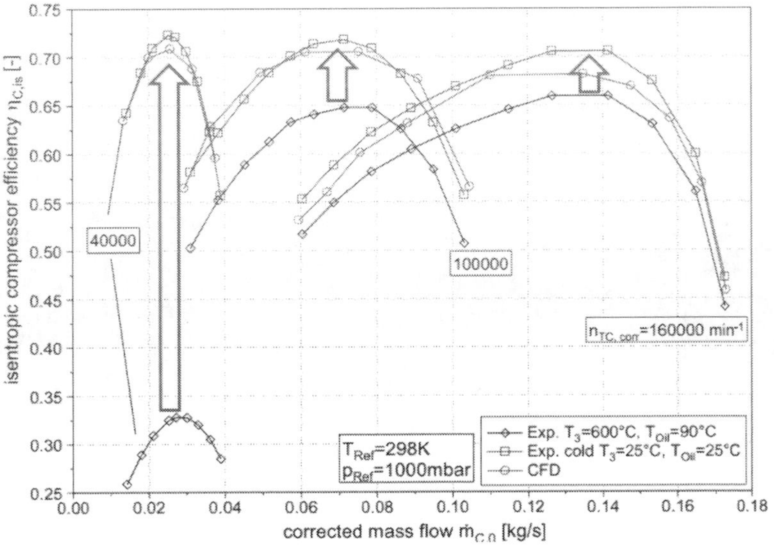

Turbocharger performance map

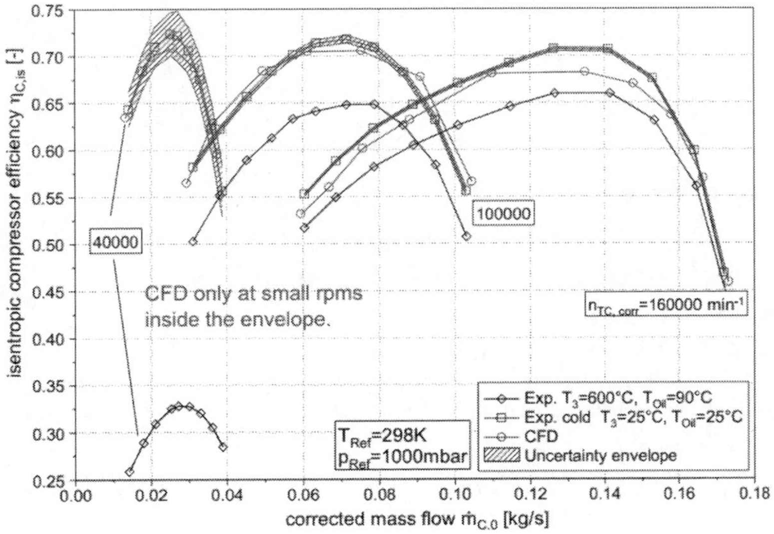

Problem II – measurement technology
Turbine outlet temperature

- Heavy gradients in temperature distribution
- Gap flow
- Complex flow (swirl)
- Depending on operation point

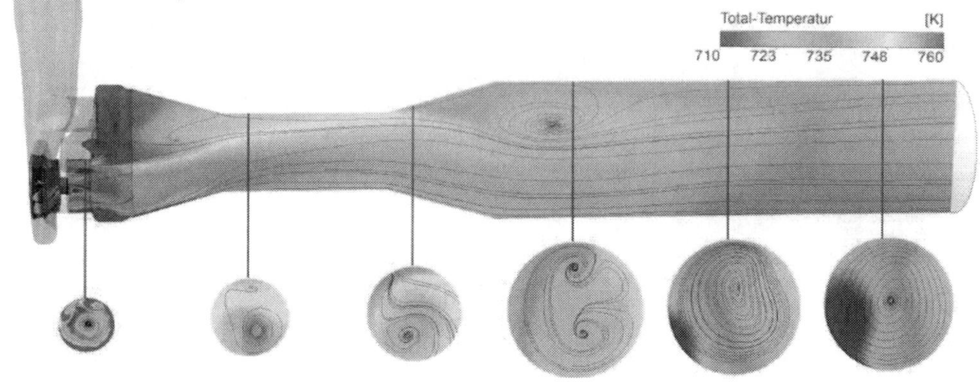

Turbine efficiency measurement

Temperature measurement according turbine (T_4) is unreliable and complex

Thermodynamic reasons:

- high exhaust gas temperatures cause driving temperature difference
- heat loss by convection and radiation,

Aerodynamic causes:

- swirl character of the hot gas flow,
- local Mach number and local α the tube wall,
- temperature gradient from center to wall due to velocity gradients across the flow cross-sectional

- Indirect determination of $\eta_{T.is}$ about friction models for mechanical efficiency
- Direct determination on T_4 and prevention of heat loss

Mixing of turbine outlet temperature

ANSYS CFX 14.5 (SST)		Standard	V-element mixer
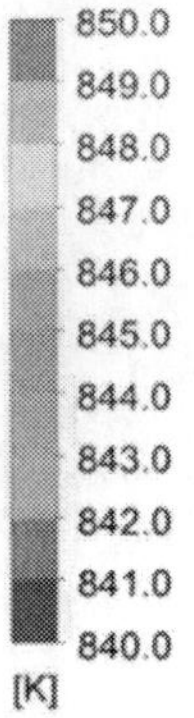	**CAD**		
	Cross section		
	Outlet	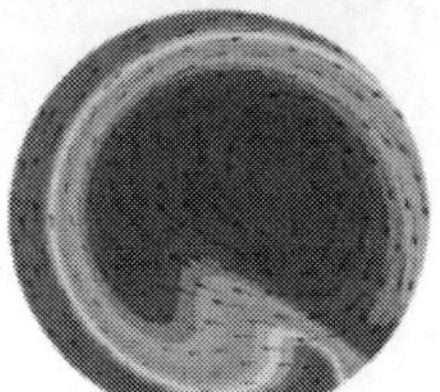	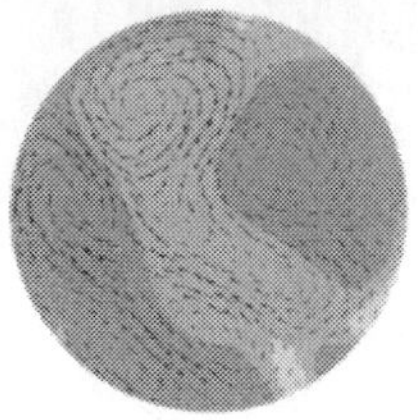

Measurement improvement through isolation and mixer

- Significant influence of heat losses to temperature measurement
- Positive effect of mixer especially at low mass flow
- Positive effect of isolation in the complete range of operation

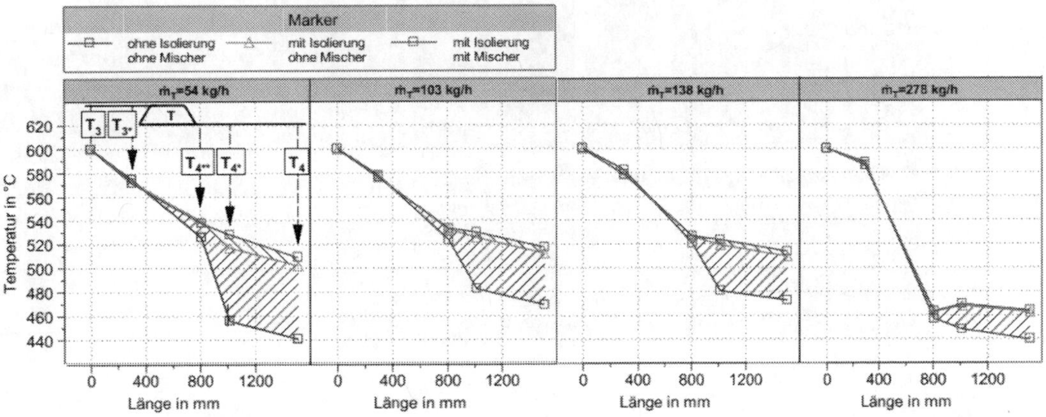

Problem III – system interaction
Compressor flow instabilities

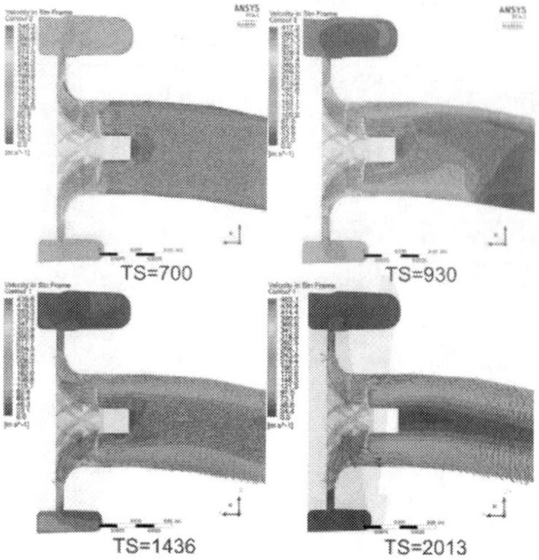

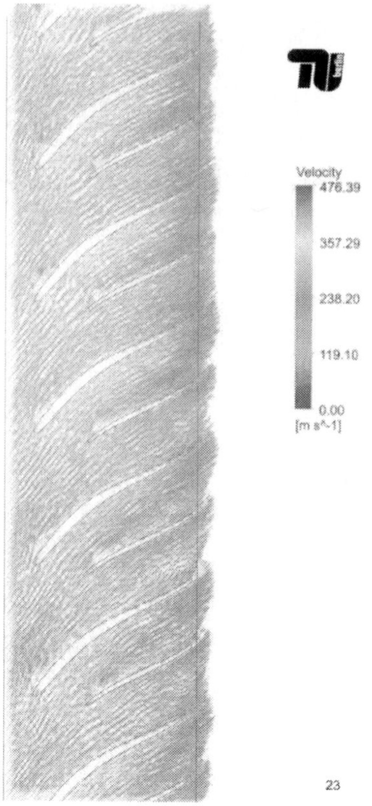

Wall pressure during instability
Selection of measurement principle

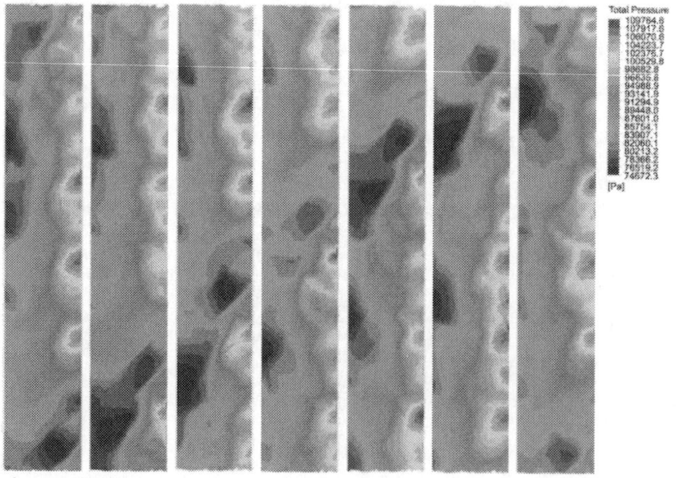

1 revolution

Position of pressure sensors

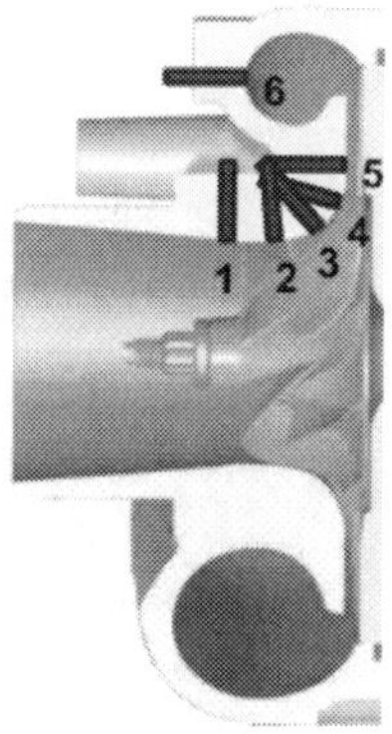

Effects of technology trends

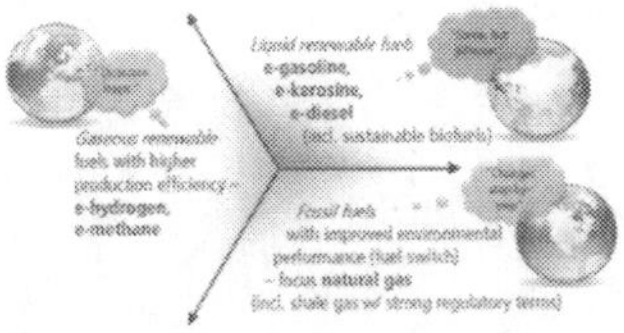

Source: FVV2013 / ludwig bölkow systemtechnik

Diversity of fuels
Fossil limits and GHG
... different costs
... different CO2-potential
... unknown technology
... distribution
... public pressure

Diversity of drivetrains
More individual solutions
... VVT
... LP-EGR
... Miller&Atkinson
... Exhaust temperature
... Engine control

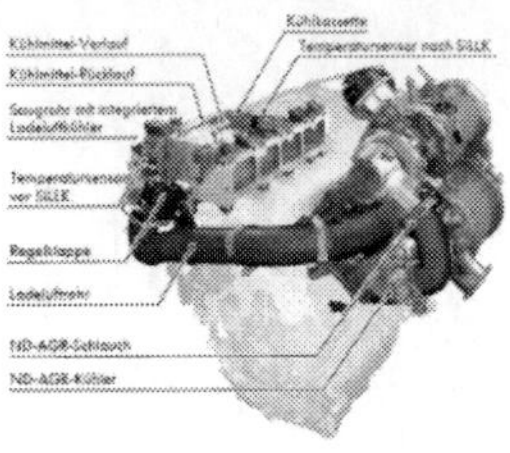

Source: Volkswagen

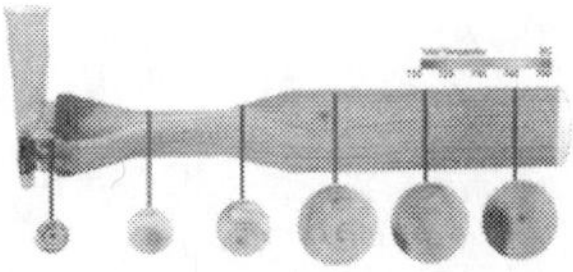

Development tools
Experiment & simulation
... what is true?
... integration of models
....balance within DVP&R
... efficiency of use
... more research needed

Antoine de Saint-Exupery ("Der kleine Prinz"):

„If you want to build a ship, do not gather all your men in order to cut down trees, but rather prepare the instruments and distribute work, but teach them passion for the distant and infinite sea “

Thank you!

Contact:
Prof. Dr.-Ing. Roland Baar

TECHNISCHE UNIVERSITÄT BERLIN
Fachgebiet Verbrennungskraftmaschinen
Carnotstr. 1A
10587 Berlin

phone: +49 30 314 26946
fax: +49 30 314 26105
email: roland.baar@tu-berlin.de
internet: www.vkm.tu-berlin.de

ASWC, October 29-30, 2013

SIMULATION OF IN-CYLINDER FLOW PHENOMENA IN TURBOCHARGED DISI ENGINES.

Dirk Linse, Andreas Kleemann

BMW GROUP

GASOLINE ENGINE DEVELOPMENT PROCESS. SIMULATION OF IN-CYLINDER FLOW PHENOMENA.

- The development of advanced combustion engines to cope with the increasing demands with respect to emissions, performance, and fuel consumption is nowadays only possible with a combination of experimental and simulation methods.
- Computational fluid dynamics (CFD) has been established as an efficient tool for the description and analysis of the flow inside IC engines.
- The 3D simulation of in-cylinder flow phenomena is fully integrated in the engine development process at BMW.
- It is extensively used in order to improve and optimize the spray targeting, combustion chamber geometry, intake port design etc. .

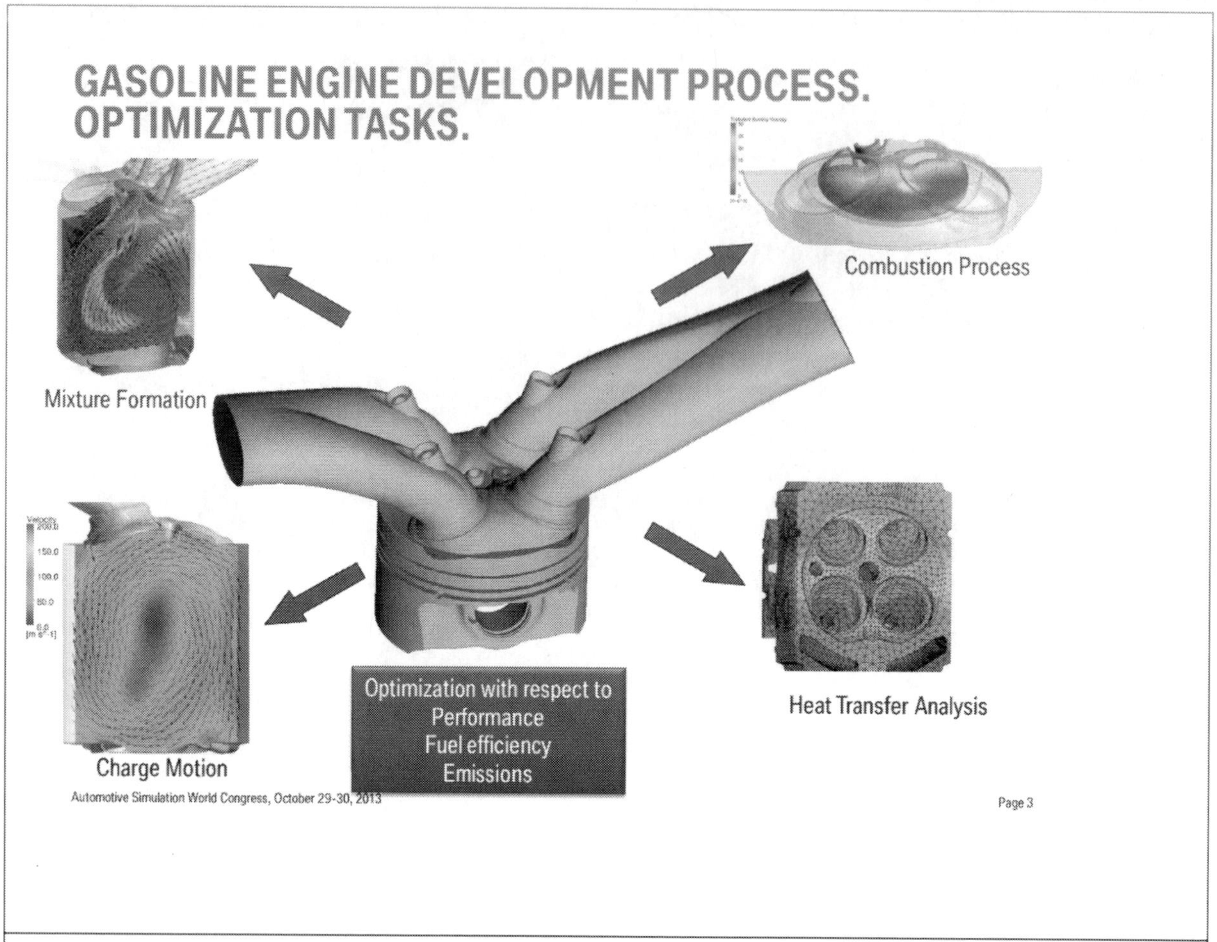
GASOLINE ENGINE DEVELOPMENT PROCESS.
OPTIMIZATION TASKS.
Mixture Formation
Combustion Process
Charge Motion
Heat Transfer Analysis
Optimization with respect to
Performance
Fuel efficiency
Emissions
Automotive Simulation World Congress, October 29-30, 2013
Page 3

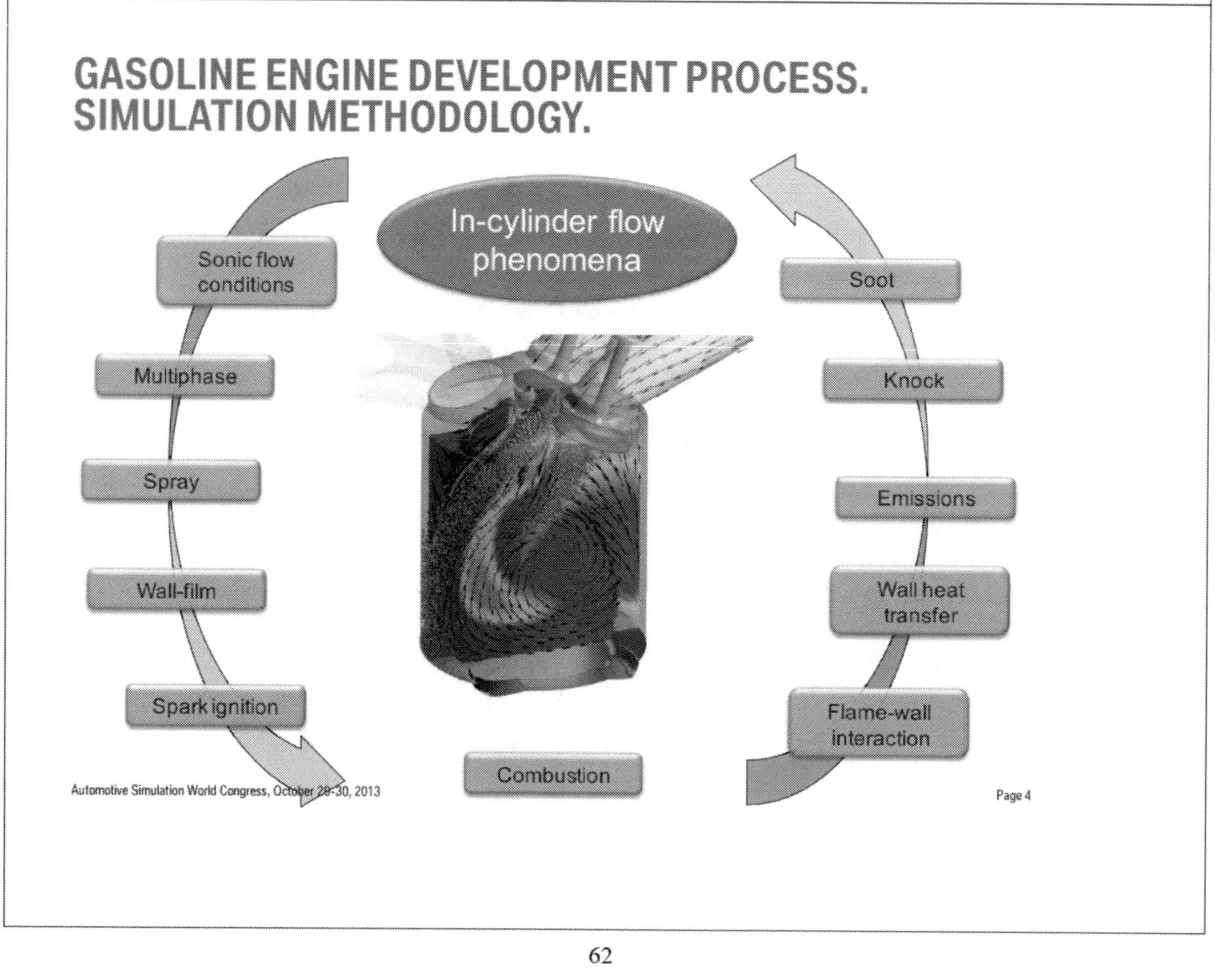
GASOLINE ENGINE DEVELOPMENT PROCESS.
SIMULATION METHODOLOGY.
In-cylinder flow phenomena
Sonic flow conditions
Multiphase
Spray
Wall-film
Spark ignition
Combustion
Flame-wall interaction
Wall heat transfer
Emissions
Knock
Soot
Automotive Simulation World Congress, October 29-30, 2013
Page 4

GASOLINE ENGINE DEVELOPMENT PROCESS. SIMULATION METHODOLOGY.

- These phenomena have a direct impact on fuel efficiency, emissions levels, and performance of internal combustion (IC) engines.
- For the improvement of the combustion process predictive, robust and accurate physical models are needed in the framework of a comprehensive simulation methodology.
- Unfortunately, not all models in CFX used for turbulent reactive flows meet these requirements.

New physical models have been implemented into ANSYS CFX in order to increase the accuracy of the simulation results.

GASOLINE ENGINE DEVELOPMENT PROCESS. SIMULATION METHODOLOGY.

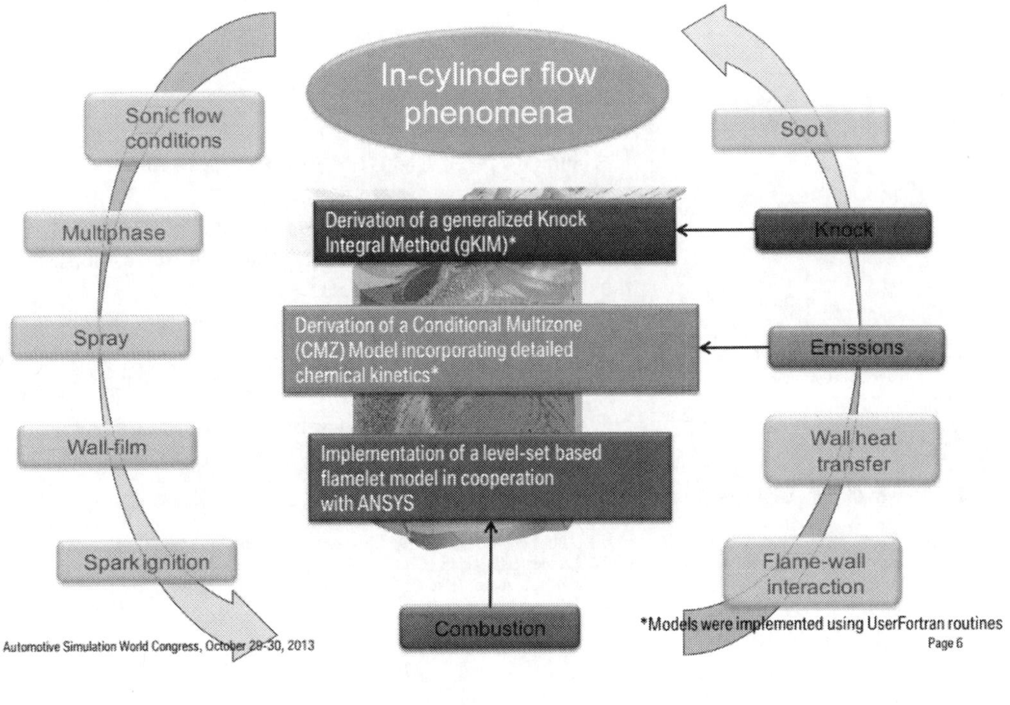

*Models were implemented using UserFortran routines

GASOLINE ENGINE DEVELOPMENT PROCESS. SIMULATION METHODOLOGY.

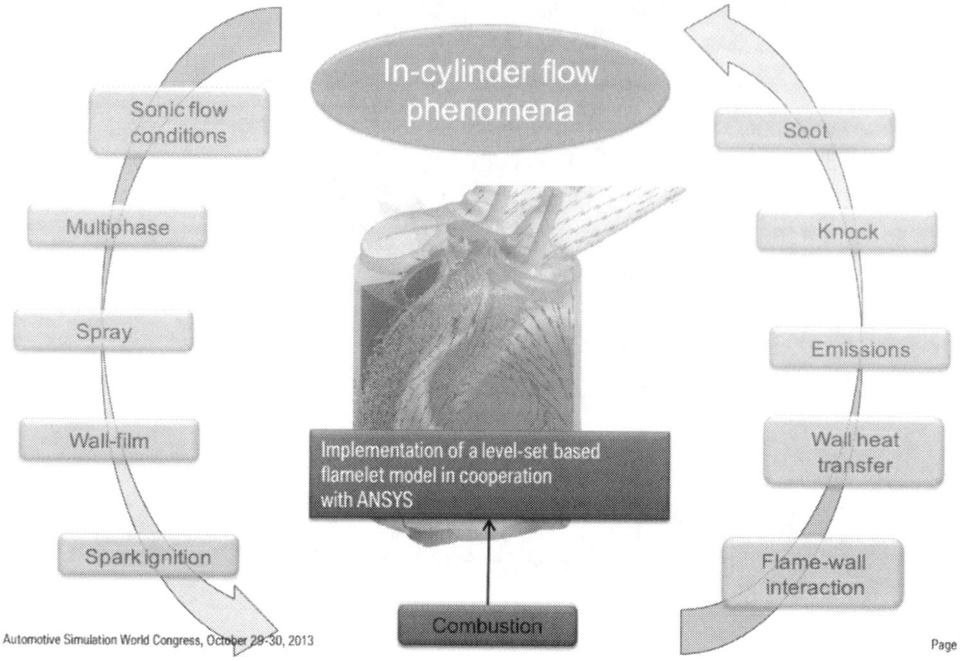

COMBUSTION PROCESS. LEVEL-SET BASED FLAMELET MODEL.

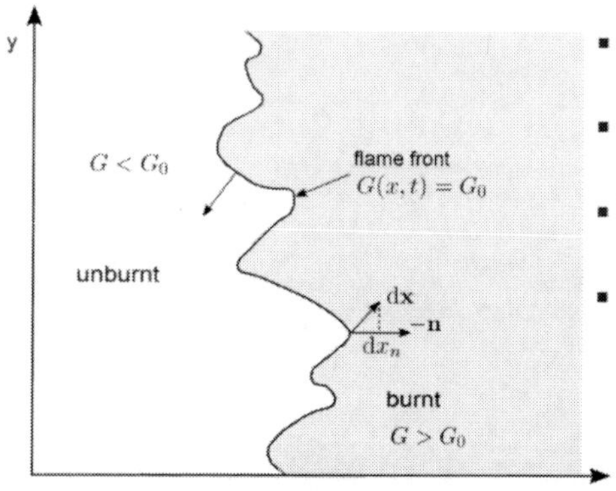

- Geometrical description of the flame front.
- Flame front is represented by the iso-surface $G(\mathbf{x}, t) = G_0$
- The flame front is tracked using the level-set approach.
- It has proven in recent years to be successful in predicting the combustion process in spark ignition engines without the use of "tuning" parameters.

The level-based flamelet model was implemented into ANSYS CFX [1,2] and it is now available as a beta feature in CFX 14.0 and higher versions.

[1] Toninel, S., Forkel, H., Frank, T., Durst, B., Hasse, C., Linse, D. (2009): Implementation and Validation of the G-equation Model Coupled with Flamelet Libraries for Simulation Premixed Combustion in I.C. Engines. SAE Technical Paper, 2009-01-0709.
[2] Linse, D. (2013): Modeling of Knock and Nitric Oxide Emissions in Turbocharged Direct Injection Spark Ignition Engines. Ph.D. Thesis.

COMBUSTION PROCESS. LEVEL-SET BASED FLAMELET MODEL.

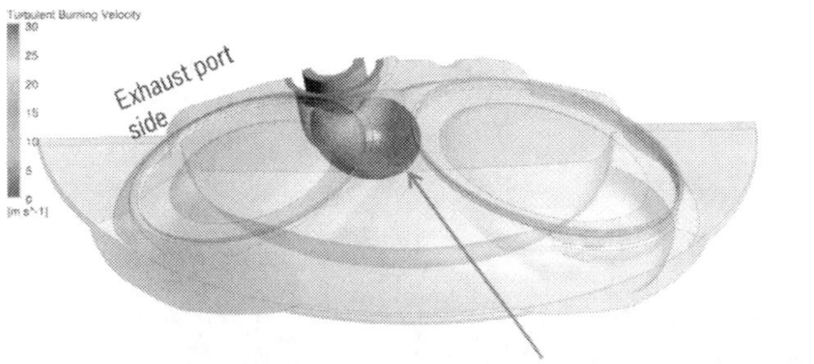

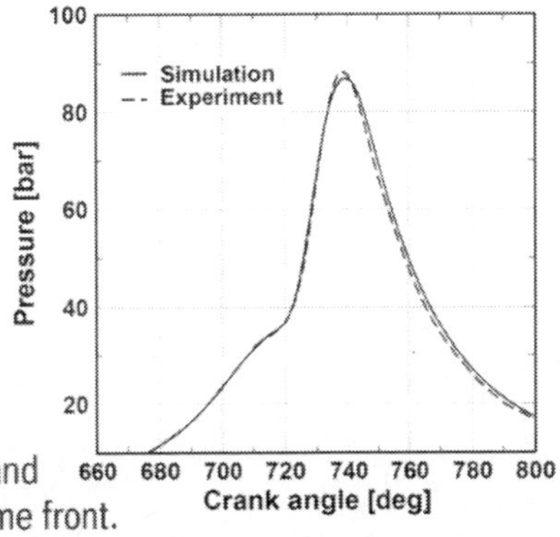

Local thermodynamical properties, turbulence level, and velocity field directly affect the propagation of the flame front.

Analyzing the impact of geometry modifications as well as different spray targetings on the overall combustion process and flame propagation.

GASOLINE ENGINE DEVELOPMENT PROCESS. SIMULATION METHODOLOGY.

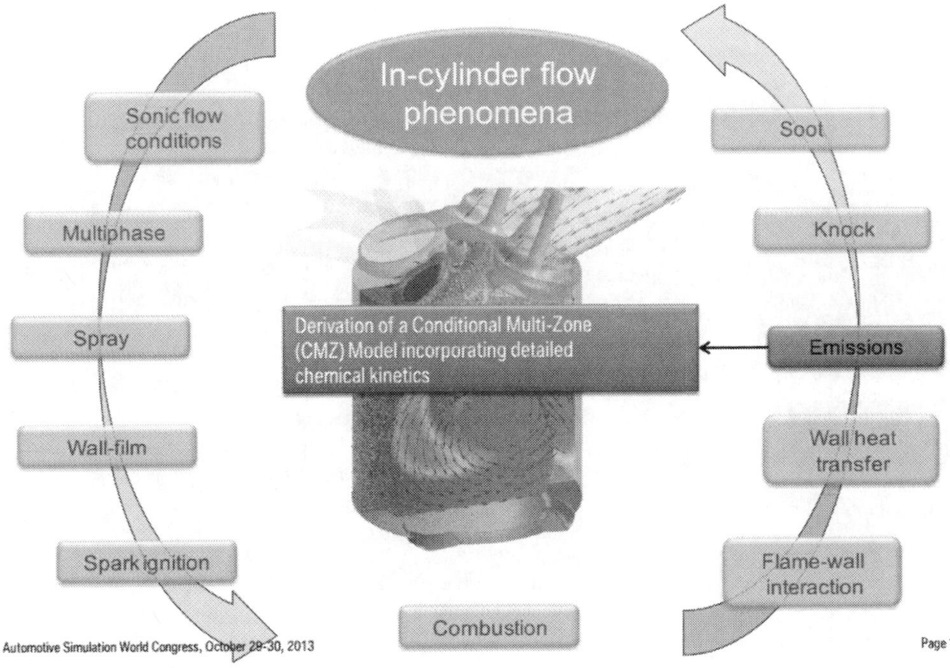

NITRIC OXIDE EMISSIONS. PHENOMENOLOGICAL DESCRIPTION.

- Thermal NO is generally the most important formation mechanism in SI engines.
- The Zeldovich mechanism describes the formation of nitric oxide at high temperatures in close to stoichiometric fuel-air mixtures.
- NO is mainly formed in the post-flame region.
- The reactions involved in the formation of NO are slow.

$$N_2 + O \Leftrightarrow NO + N$$
$$N + O_2 \Leftrightarrow NO + O$$
$$N + OH \Leftrightarrow NO + H$$

Development of a Conditional Multizone (CMZ) Model in order to simulate the slow process of nitric oxide emissions using detailed chemistry.

MODELING OF NITRIC OXIDE EMISSIONS. CONDITIONAL MULTIZONE MODEL.

- Simulation of flow dynamics and mixture formation.
- Modeling of the flame front propagation and heat release.

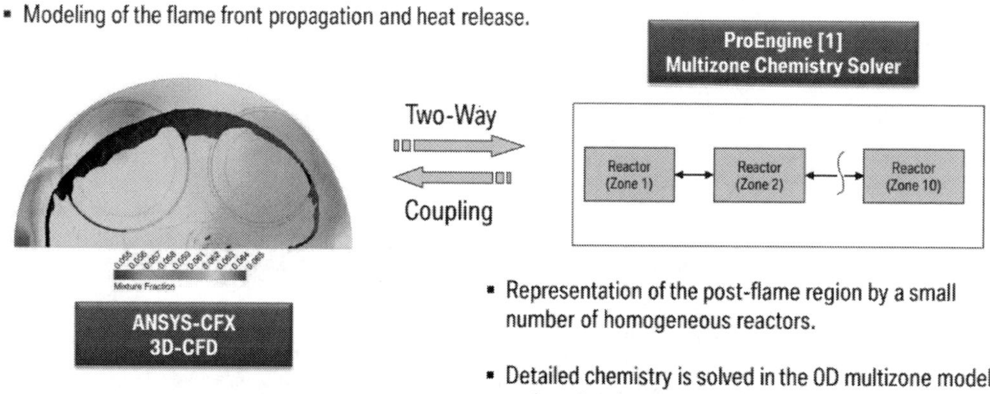

- Representation of the post-flame region by a small number of homogeneous reactors.
- Detailed chemistry is solved in the 0D multizone model and not in every computational cell.

[1] Linse, D. (2013): Modeling of Knock and Nitric Oxide Emissions in Turbocharged Direct Injection Spark Ignition Engines. Ph.D. Thesis.

MODELING OF NITRIC OXIDE EMISSIONS. CONDITIONAL MULTIZONE MODEL.

Representation of zones in the physical domain.

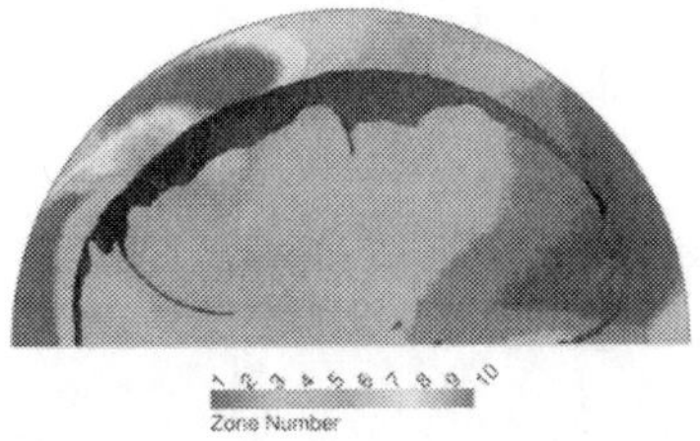

Each zone is subdivided in a burnt and unburnt part.

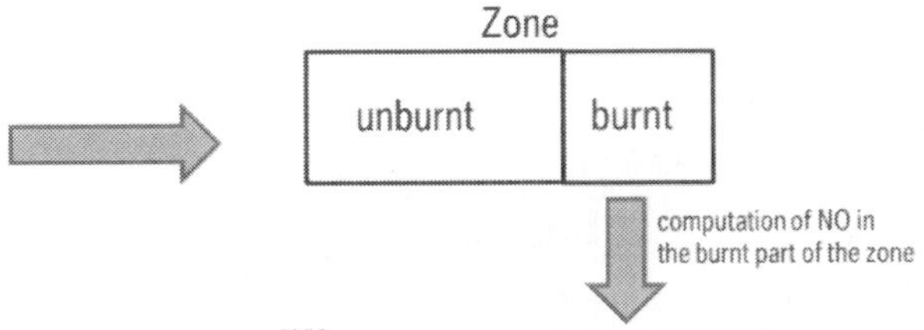

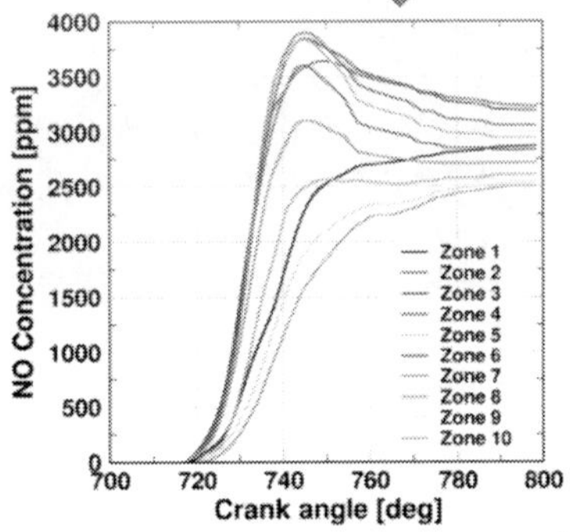

Distribution of nitric oxide concentration.

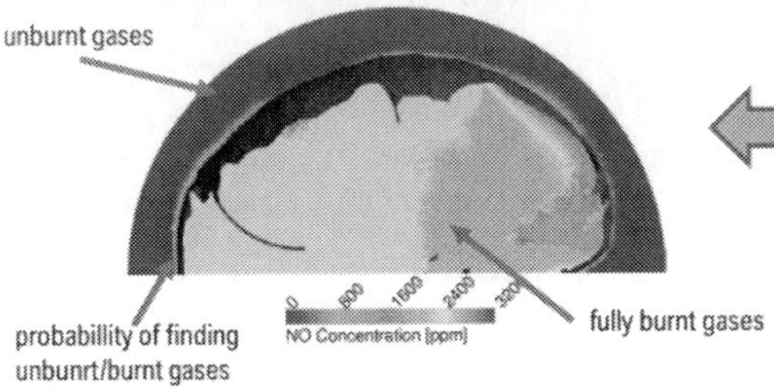

GASOLINE ENGINE DEVELOPMENT PROCESS. SIMULATION METHODOLOGY.

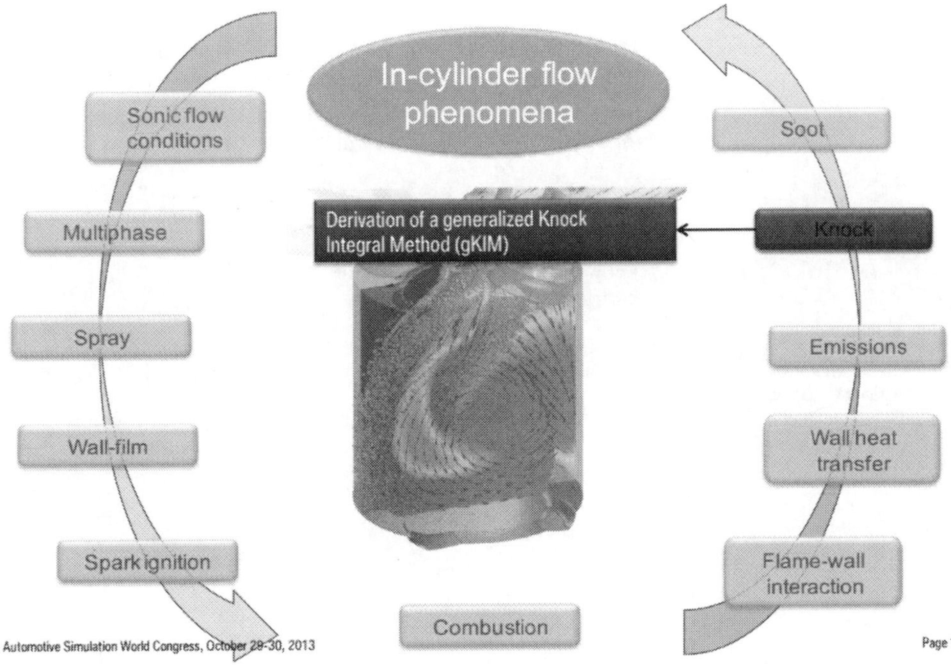

ON THE NATURE OF ENGINE KNOCK. PHENOMENOLOGICAL DESCRIPTION.

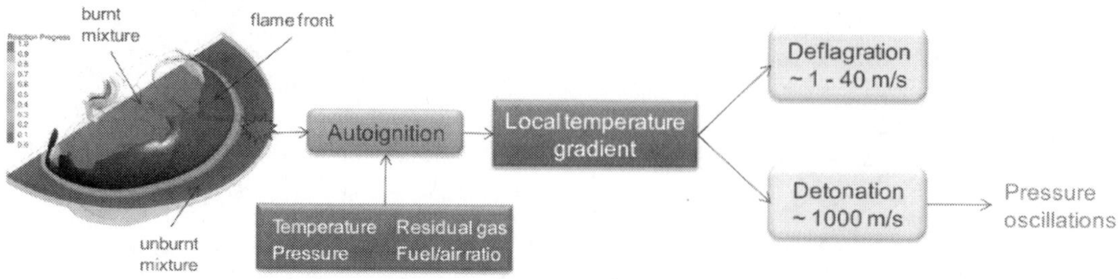

- Autoignition is promoted by chemical processes in the unburnt fuel-air mixture.
- It is triggered by compression of the unburnt fuel-air mixture caused by the pressure rise of both the moving piston and the expanding flame front.
- The local temperature gradient determines the mode of combustion (deflagration/detonation).
- Engine knock can lead to severe damage of the engine. It also limits the efficiency and performance of the engine.

ON THE NATURE OF ENGINE KNOCK. ANALYSIS OF KNOCKING CYCLES.

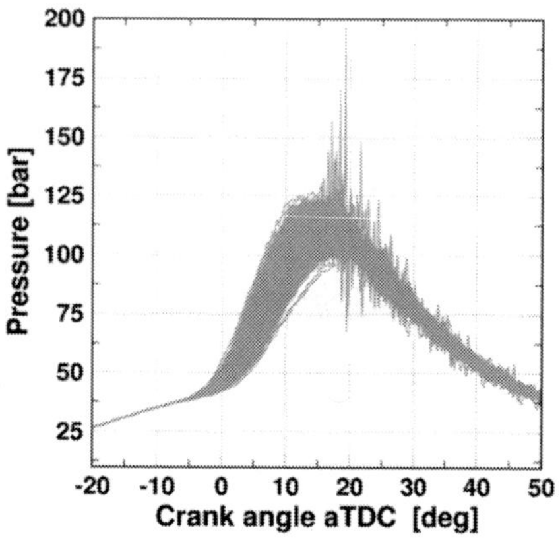

- Measurement of 1000 consecutive cycles for a full-load operating point at 5000 RPM.
- 270 cycles were identified as knocking cycles.
- Cycle-by-cycle variations are clearly visible.
- Variation of peak pressure, knock intensity and knock onset.

ON THE NATURE OF ENGINE KNOCK. ANALYSIS OF KNOCKING CYCLES.

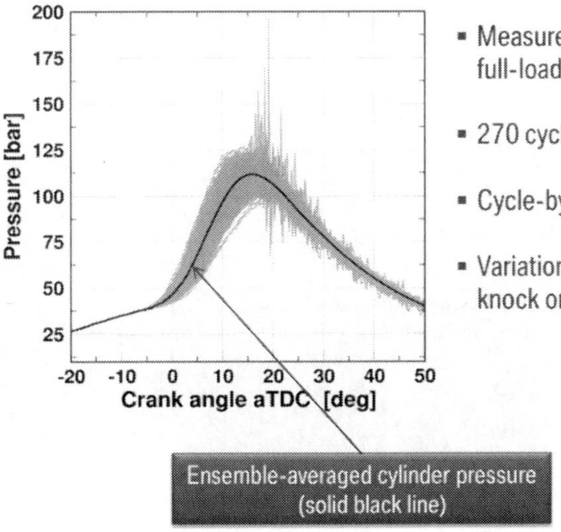

- Measurement of 1000 consecutive cycles for a full-load operating point at 5000 RPM.
- 270 cycles were identified as knocking cycles.
- Cycle-by-cycle variations are clearly visible.
- Variation of peak pressure, knock intensity and knock onset.

ON THE NATURE OF ENGINE KNOCK. PHENOMENOLOGICAL DESCRIPTION.

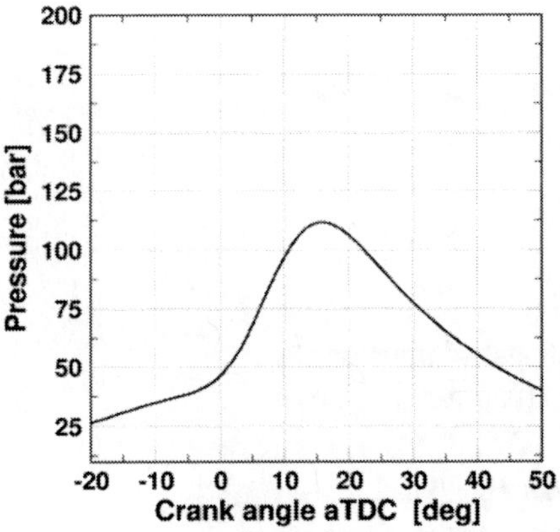

- Smooth pressure trace – no pressure oscillations.
- All statistical information is lost.
- It is not possible to determine the mean knock onset or knock intensity solely from the averaged cylinder pressure history.

→ Direct consequences for the modeling of engine knock in the context of RANS-based CFD simulations.

A comprehensive knock model needs to provide additional statistical information regarding the probability of autoignition.

MODELING OF ENGINE KNOCK.
DEFINITION OF THE IGNITION PROGRESS VARIABLE.

Introduction of an ignition progress variable c_I

$c_I = 0$ ➔ no chemical reactions in the unburnt mixture
$c_I = 1$ ➔ ignition of the unburnt mixture

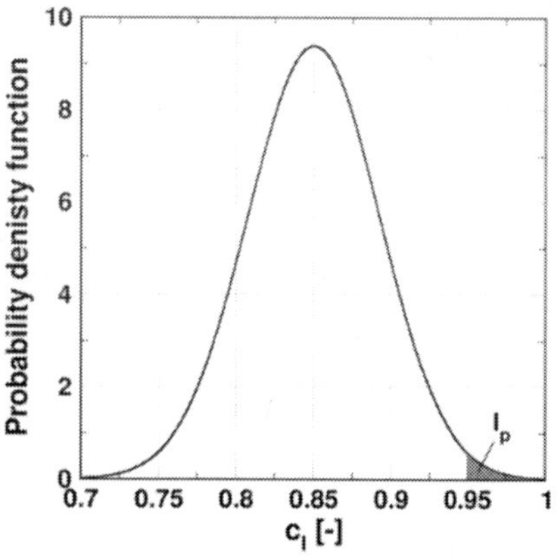

- The new knock model [1,2] computes the statistical distribution of the ignition progress variable.
- Detailed chemistry and turbulence chemistry interaction are accounted for.
- A sound definition of the ignition probability can be derived according to:

$$I_p = prob(c_I > 0.95) = \int_{0.95}^{1} \widetilde{P}(c_I; \mathbf{x}, t)\, \mathrm{d}c_I$$

[1] Linse, D., Kleemann, A., Hasse, C. (2013): Probability Density Function Approach Coupled with Detailed Chemical Kinetics for the Prediction of Knock in Turbocharged Direct Injection Spark Ignition Engines. Accepted by Combustion and Flame.
[2] Linse, D. (2013): Modeling of Knock and Nitric Oxide Emissions in Turbocharged Direct Injection Spark Ignition Engines. Ph.D. Thesis.

SIMULATION OF ENGINE KNOCK.
IGNITION PROGRESS AND IGNITION PROBABILITY.

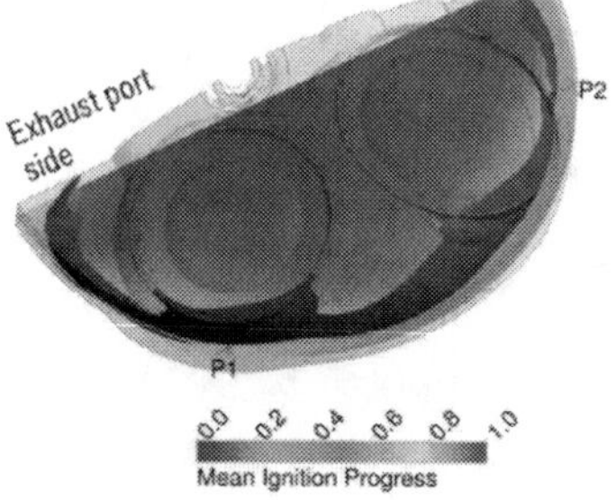

At location P2 ➔ $\widetilde{c}_I \approx 0.25$

At location P1 ➔ $\widetilde{c}_I \approx 0.7$

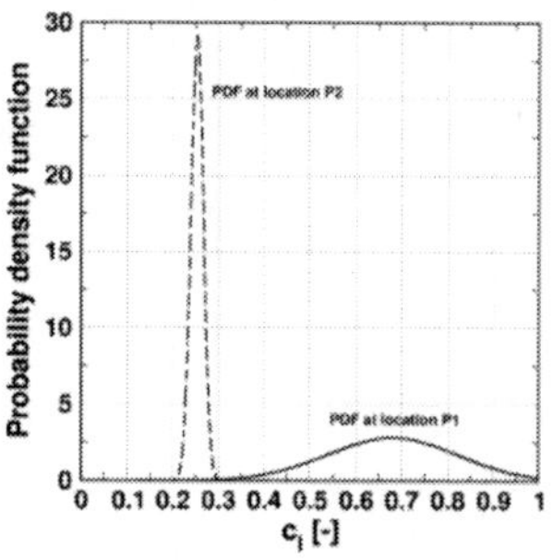

- The new knock model provides additional information about the probability distribution of the ignition progress variable.
- The ignition probability I_p at P1 is approximately 4%.

$$I_p = prob(c_I > 0.95) = \int_{0.95}^{1} \widetilde{P}(c_I; \mathbf{x}, t)\, \mathrm{d}c_I$$

SIMULATION OF ENGINE KNOCK. EVALUATION OF THE IGNITION PROBABILITY.

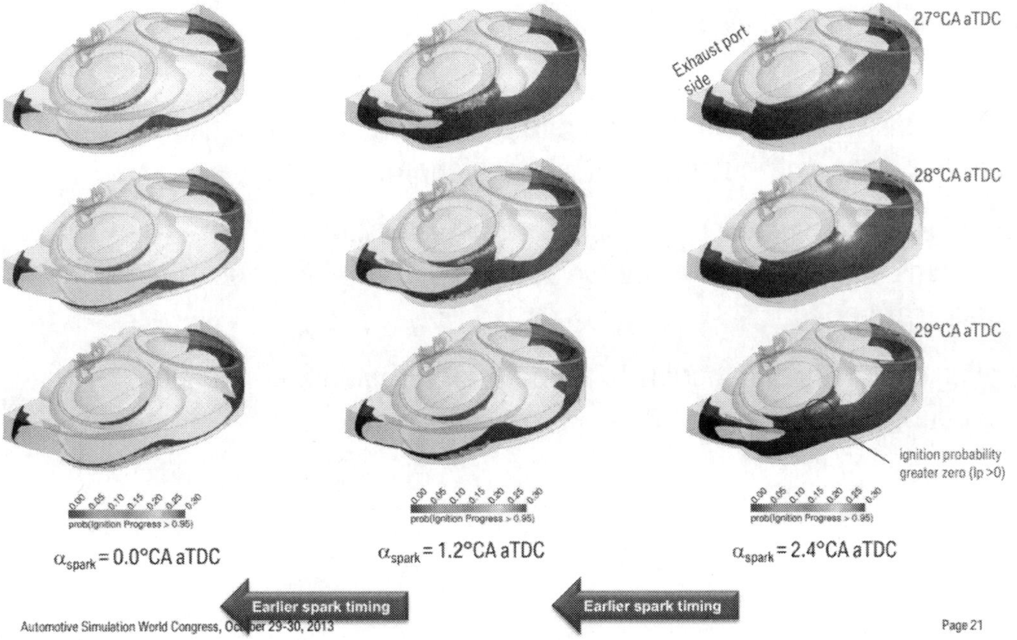

GASOLINE ENGINE DEVELOPMENT PROCESS. SIMULATION METHODOLOGY.

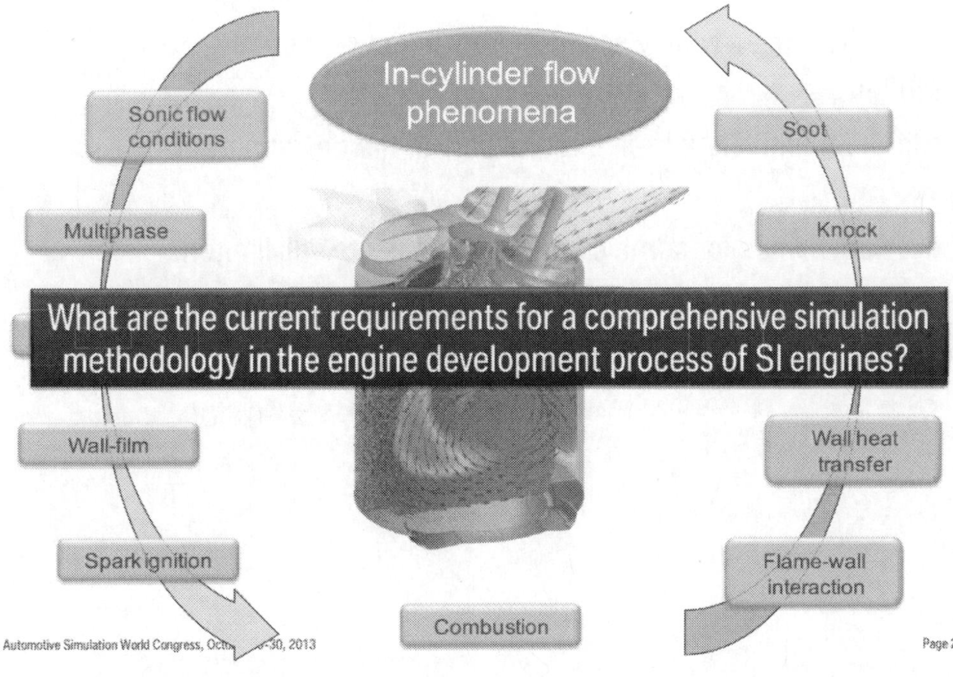

SIMULATION OF IN-CYLINDER FLOW PHENOMENA. ENGINE DEVELOPMENT REQUIREMENTS.

- Multicomponent fuel: evaporation and combustion. Modeling of surrogate fuels.
- Comprehensive soot model to compute particle number densities and particle size distribution (EURO 6 Regulation).
- In order to further increase the accuracy of the combustion simulations, improved spark ignition and flame-wall interaction models are needed.
- Detailed chemistry and models to account for turbulence-chemistry interaction.

SIMULATION OF IN-CYLINDER FLOW PHENOMENA. CONCLUSIONS.

- 3D-CFD has become a widely accepted tool in powertrain development.
- In-cylinder flow simulation is an integral part of the development process.
- Predictive capabilities for some simulation cases are still limited.
- Enhancement of CFX is necessary in order to cope with current and future trends in powertrain development.

SIMULATION OF IN-CYLINDER FLOW PHENOMENA. THANK YOU VERY MUCH FOR YOUR ATTENTION.

Automotive Simulation World Congress, October 29-30, 2013

Page 25

ANSYS CADFEM

Cooling 760 Horses – FSI-Simulation of Water-Cooled Diesel Cylinder Head

Realize Your Product Promise™

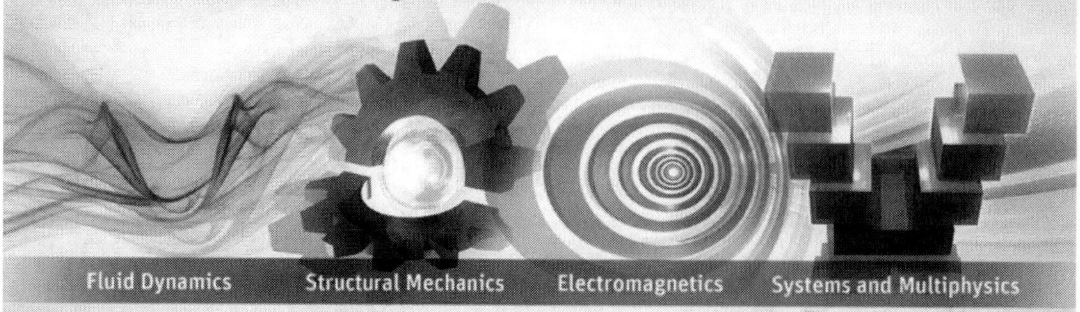

Wolfgang Haslinger
ANSYS Continental Europe
Wolfgang.Haslinger@ansys.com

ANSYS

Acknowledgements

Andreas Linke (MAN Diesel & Turbo SE)

Matthias Alberts (CADFEM)

Marold Moosrainer (CADFEM)

ANSYS

Cooling 760 Horses …

ANSYS

MAN Diesel & Turbo – 32/44CR Diesel Engine

Medium speed, four-stroke engine:

- 560kW/cyl.
- Marine: Ferries, Tankers, Special purpose vessels, …
- Stationary applications

http://www.mandieselturbo.de/1006145/Presse/Pressemeldungen/Fachpressemeldungen/Marine-Power/Marine-GenSets/32%2F44CR.html
http://www.pferdeheimat.net/wp-content/uploads//2012/06/Pferd-191.jpg

ANSYS

Overview

- Motivation
- Cooling concept
- Simulation strategy
- Fluid simulation
- Mechanical simulation
- Summary

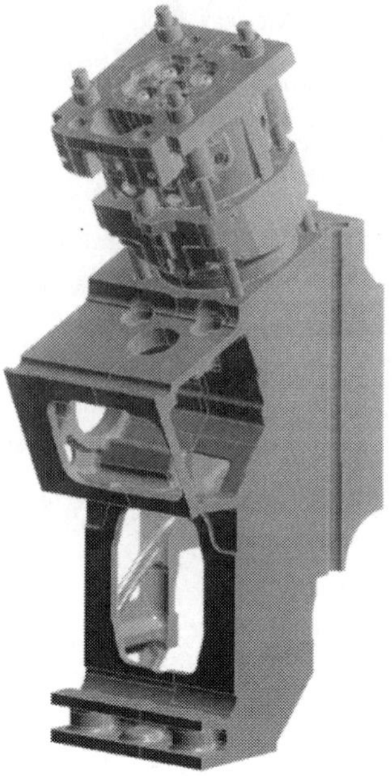

5 November 11, 2013

ANSYS

Overview

- Motivation
- Cooling concept
- Simulation strategy
- Fluid simulation
- Mechanical simulation
- Summary

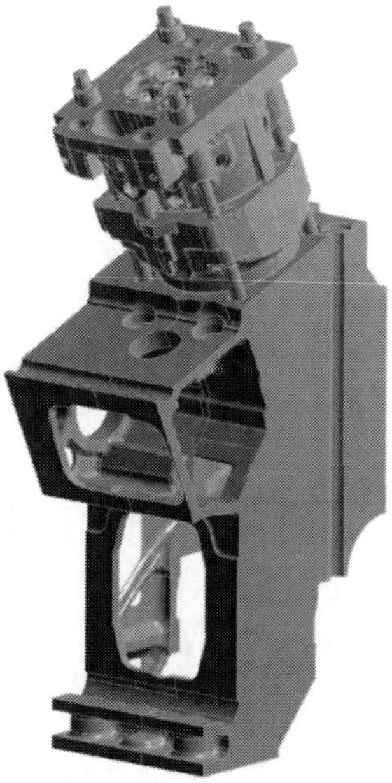

6 November 11, 2013

ANSYS Motivation

Engine development objectives
- High specific power output
- High fuel efficiency
- Low exhaust emissions
- High reliability
- Long maintenance intervals

Critical temperature & stress limits
- Critical temperatures
- Critical stresses
  - Plastification
  - Crack building & propagation
- Critical deformation: Leakages

Numerical simulation techniques
- Internal combustion
- Water cooling
- Solid temperatures
- Stresses & displacements

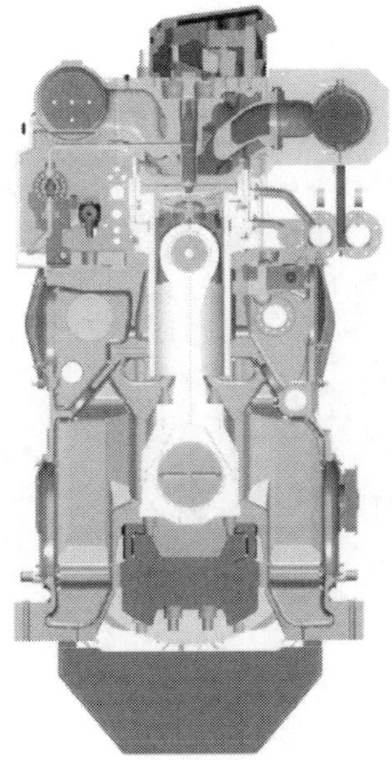

http://www.mandieselturbo.de/1006145/Presse/Pressemeldungen/Fachpressemeldungen/Marine-Power/Marine-GenSets/32%2F44CR.html

ANSYS Overview

- Motivation
- Cooling concept
- Simulation strategy
- Fluid simulation
- Mechanical simulation
- Summary

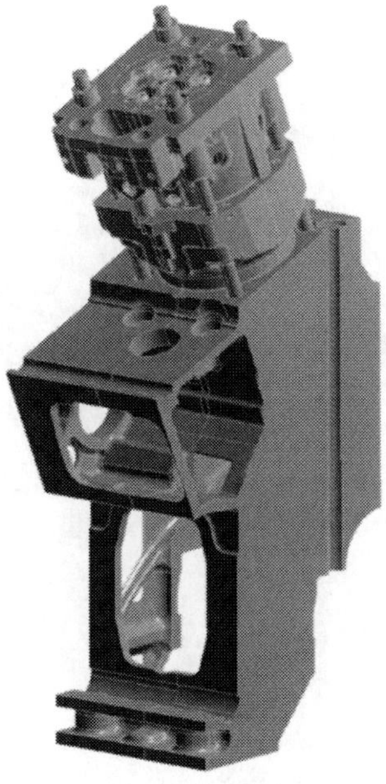

ΛNSYS Cylinder Head Cooling Concept

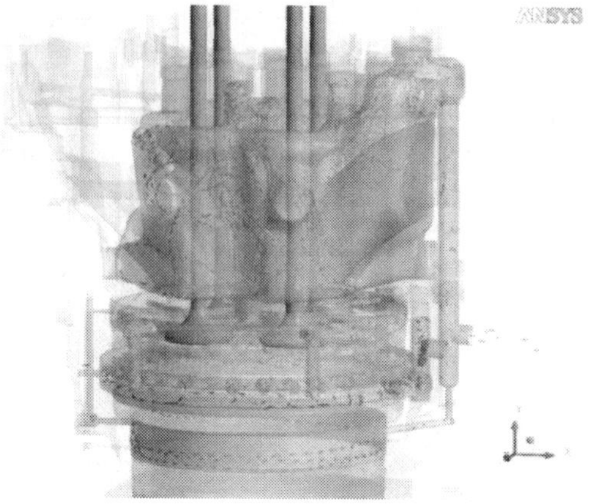

Very different and complex flow topologies
→ Strongly coupled coolant flow and solid temperatures
→ "Fluid-Structure-Interaction"

ΛNSYS Overview

- Motivation
- Cooling concept
- Simulation strategy
- Fluid simulation
- Mechanical simulation
- Summary

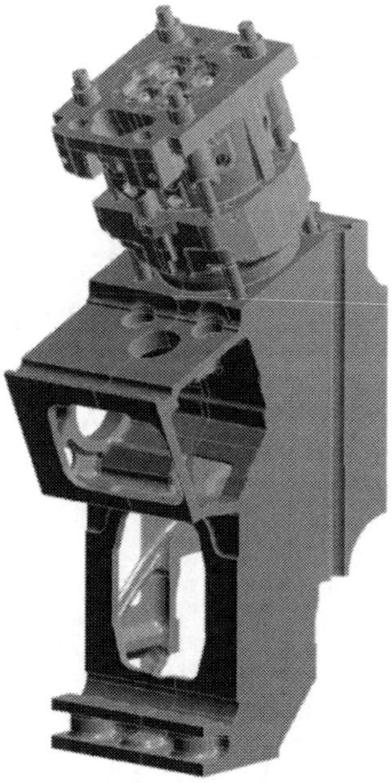

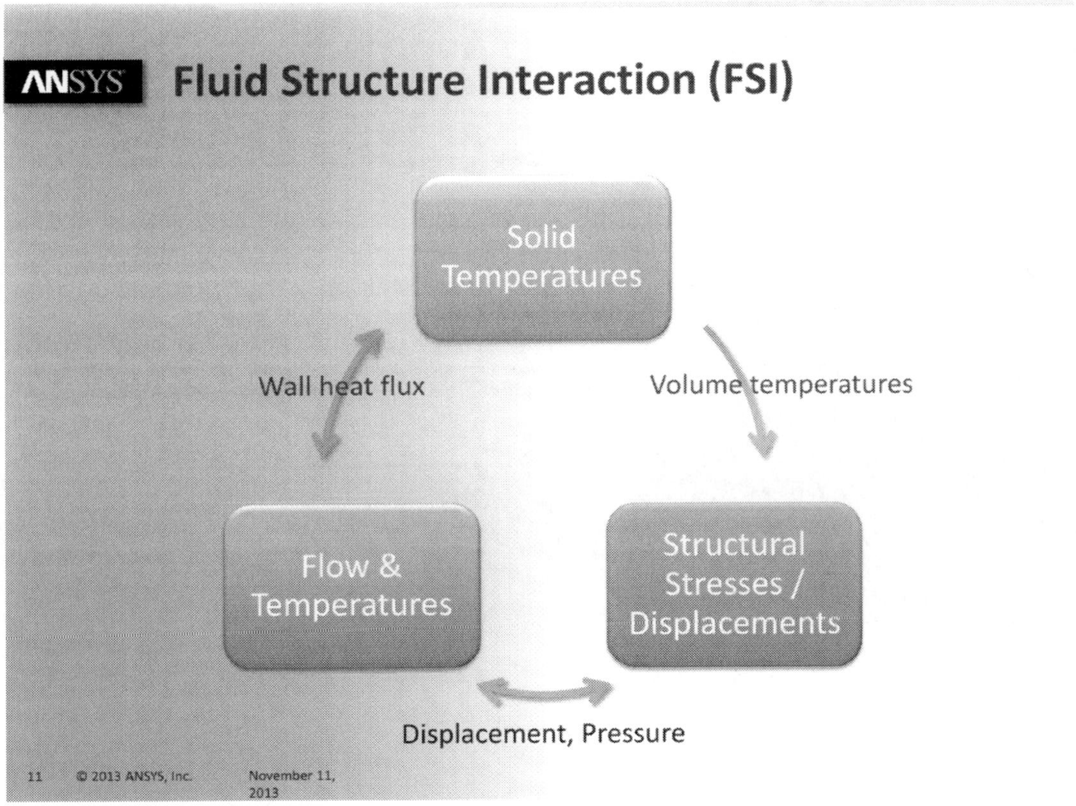
ANSYS
Fluid Structure Interaction (FSI)
Solid Temperatures
Wall heat flux
Volume temperatures
Flow & Temperatures
Structural Stresses / Displacements
Displacement, Pressure
11
© 2013 ANSYS, Inc.
November 11, 2013

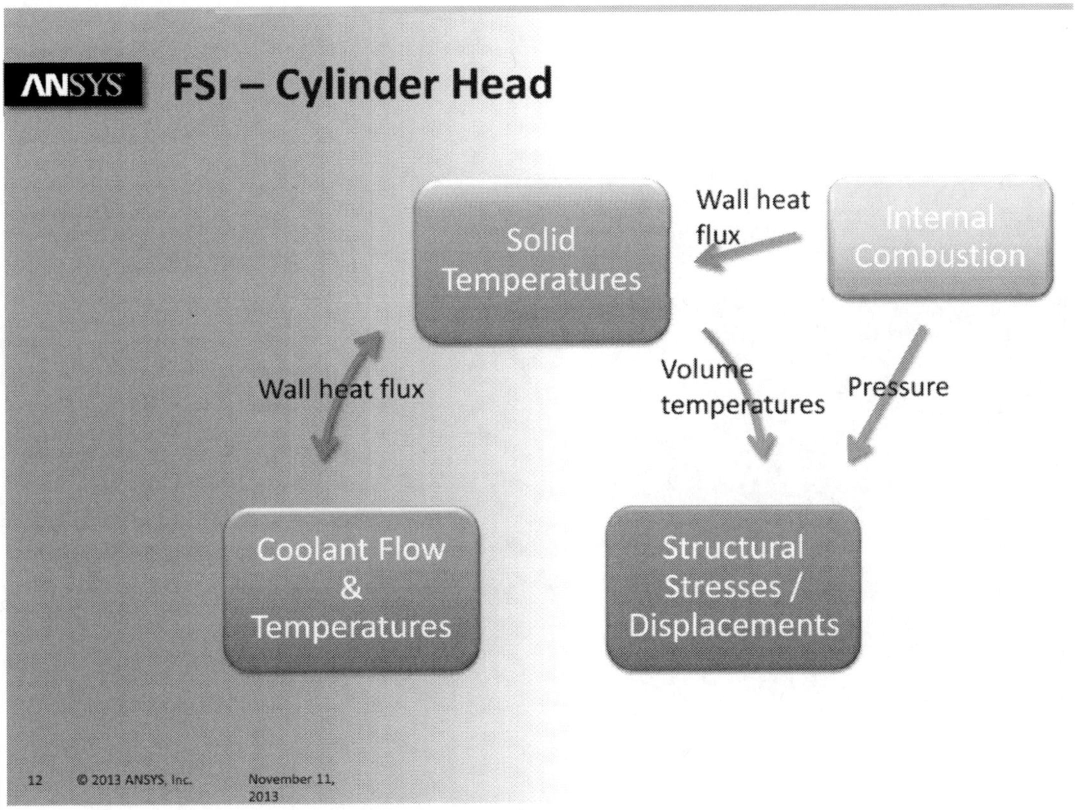
ANSYS
FSI – Cylinder Head
Solid Temperatures
Wall heat flux
Internal Combustion
Wall heat flux
Volume temperatures
Pressure
Coolant Flow & Temperatures
Structural Stresses / Displacements
12
© 2013 ANSYS, Inc.
November 11, 2013

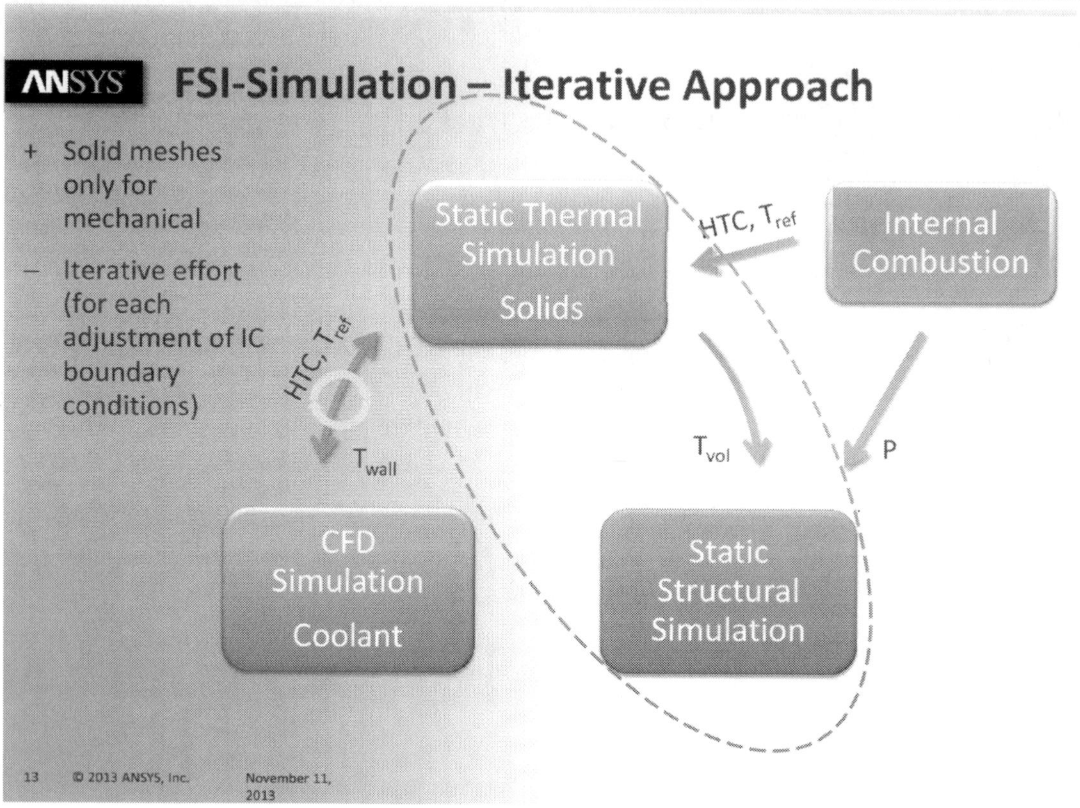
ANSYS
FSI-Simulation – Iterative Approach
+ Solid meshes only for mechanical
– Iterative effort (for each adjustment of IC boundary conditions)
Static Thermal Simulation
Solids
HTC, Tref
Internal Combustion
HTC, Tref
Twall
Tvol
P
CFD Simulation
Coolant
Static Structural Simulation
13 © 2013 ANSYS, Inc. November 11, 2013

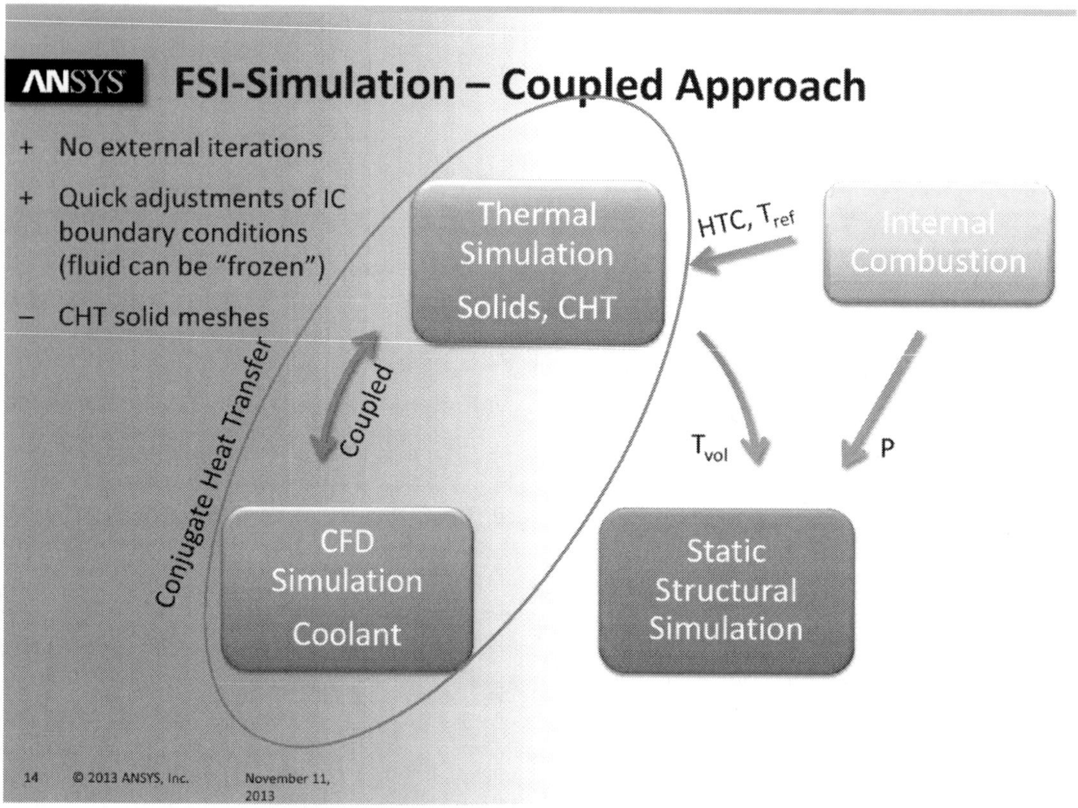
ANSYS
FSI-Simulation – Coupled Approach
+ No external iterations
+ Quick adjustments of IC boundary conditions (fluid can be "frozen")
– CHT solid meshes
Thermal Simulation
Solids, CHT
HTC, Tref
Internal Combustion
Conjugate Heat Transfer
Coupled
Tvol
P
CFD Simulation
Coolant
Static Structural Simulation
14 © 2013 ANSYS, Inc. November 11, 2013

ANSYS Overview

- Motivation
- Cooling concept
- Simulation strategy
- **Fluid simulation**
- Mechanical simulation
- Summary

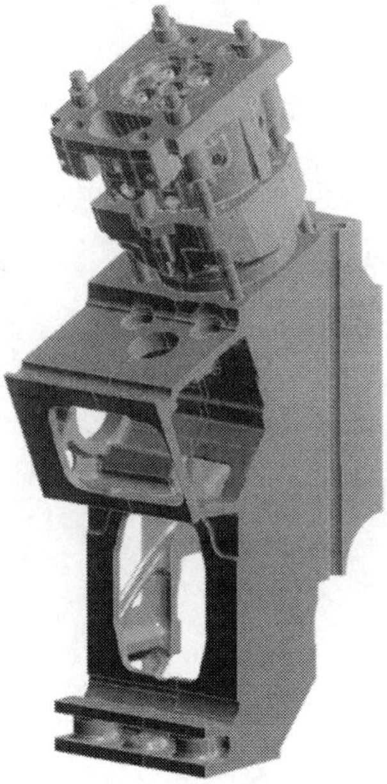

ANSYS Engine Parts

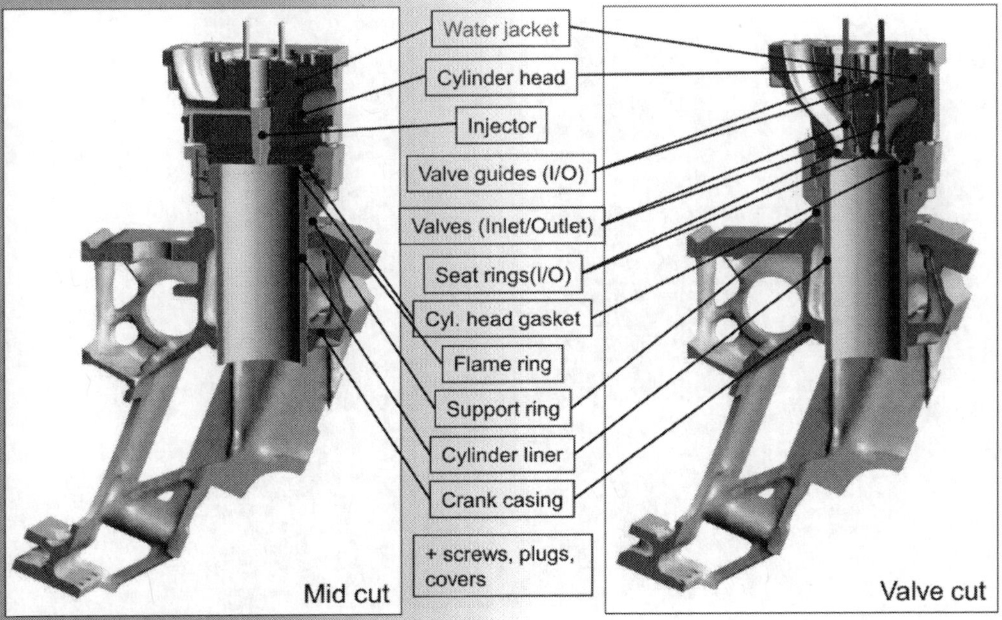

Overall ~ 40 different parts, 16 different materials

ANSYS Numerical Grids for CHT-Simulation

ANSYS ICEM CFD Hexa, Tetra, Prism

	Nodes	Elements
Fluid	11 Mio.	14 Mio.
Solids	2,5 Mio	9 Mio.

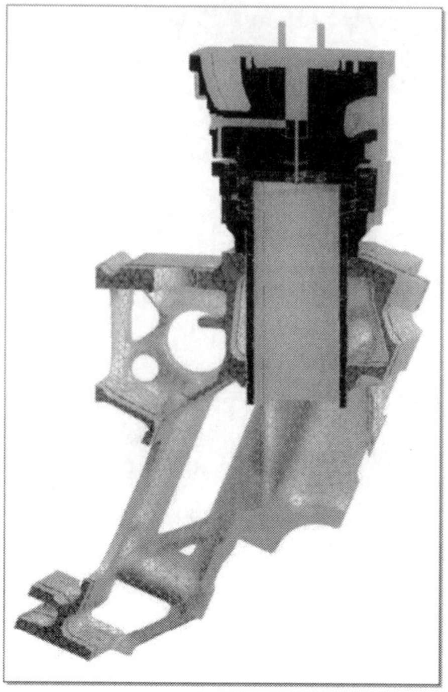

ANSYS Numerical Grids CHT: Zoom

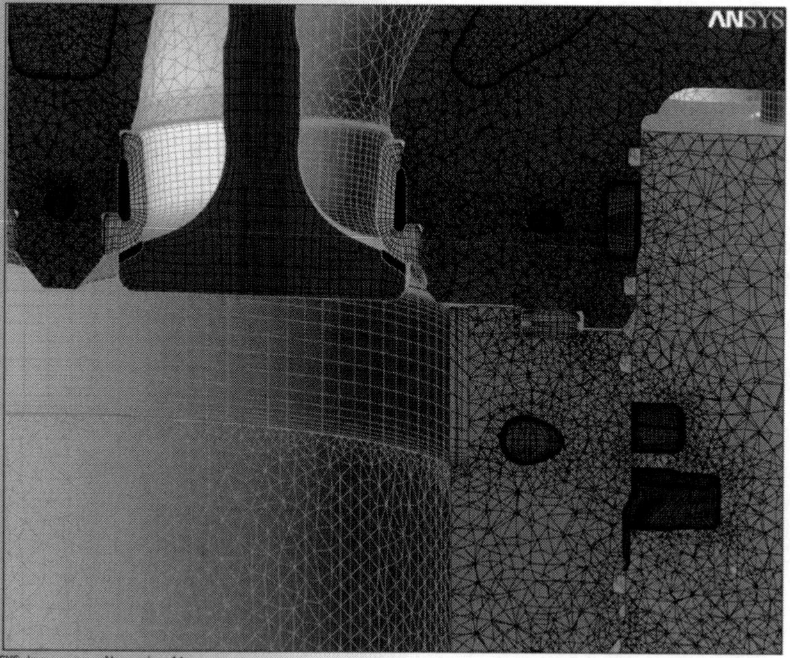

ANSYS Numerical & Physical Setup CHT

Solver: ANSYS CFX

Conjugate heat transfer

Steady-state

Turbulence model: SST

Materials:

- Fluid: Water
- Solids: 16 different materials (temperature dependent properties)

Contact heat resistance

- Valve guide ↔ valve shaft
- Valve head ↔ valve seat
- Valve seat ↔ cylinder head

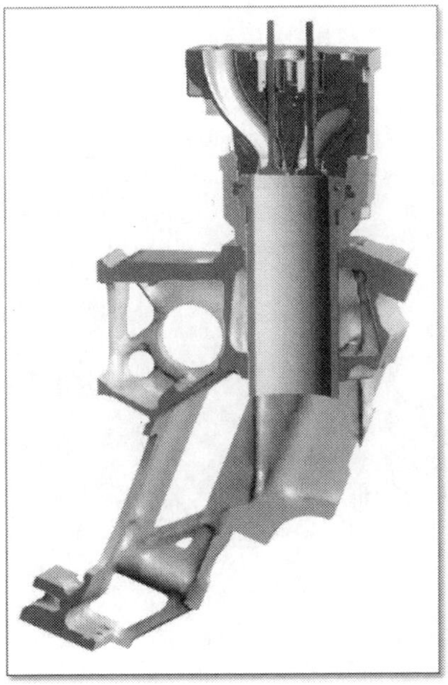

ANSYS Flow Boundary Conditions

Inlet:

- Total pressure
- Total temperature

Outlet:

- Water mass flow

Water-jacket walls:

- Zero slip
- Hydraulically smooth
- Fluid-solid heat transfer

ANSYS Solids Thermal Boundary Conditions

	Heat transfer coefficient	Reference temperature
Cylinder roof	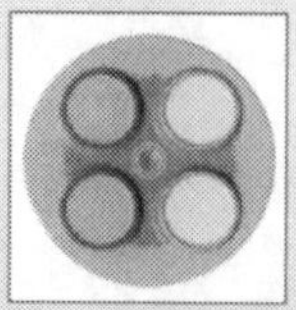	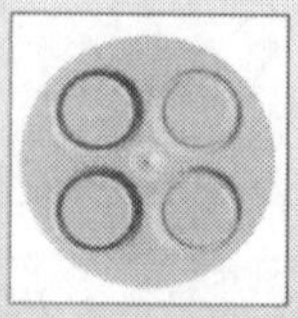
Cylinder liner		
Inflow channel Exhaust channel Oil channels Injector cooling External walls	HTC	T_{ref}

ANSYS CHT-Results: Water-Jacket Walls

Pressure

Temperature

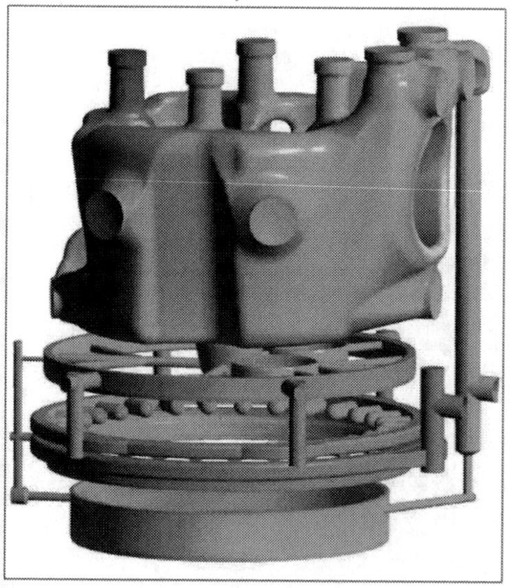

CHT-Results: Solid Temperatures

Mid cut

Valve cut

All temperatures below critical limits
→ Export to mechanical simulation

Overview

- Motivation
- Cooling concept
- Simulation strategy
- Fluid simulation
- **Mechanical simulation**
- Summary

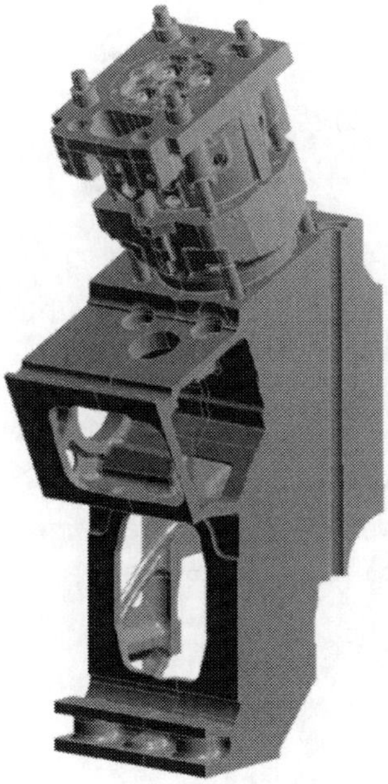

Numerical Grids (Mechanical)

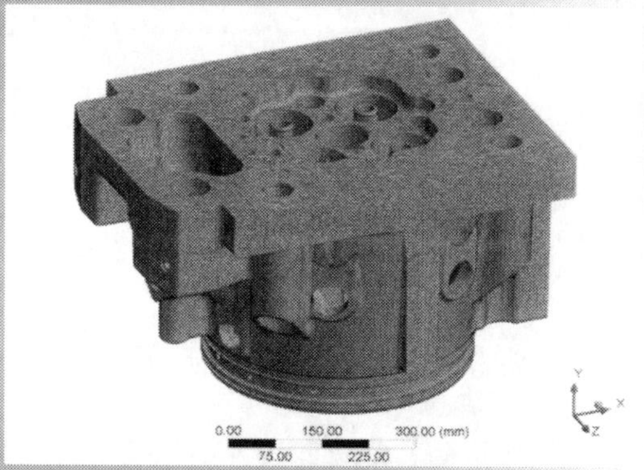

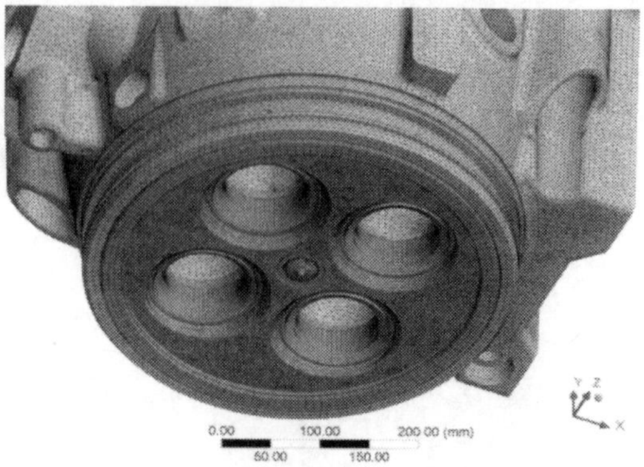

Quadratic Tetrahedrons & Hexahedrons

	Nodes	Elements
Solids	3,2 Mio	1.73 Mio.

Mechanical Contacts

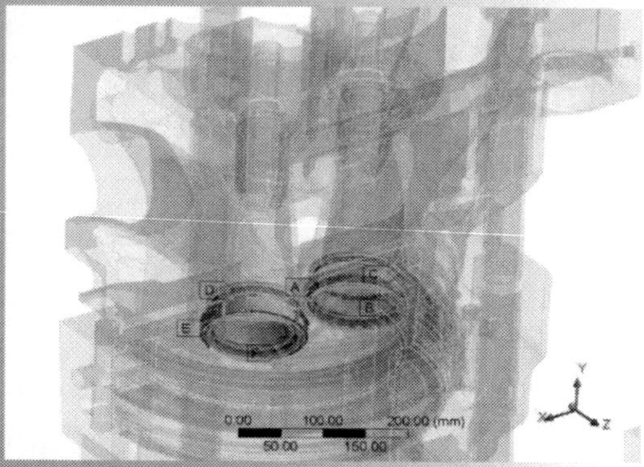

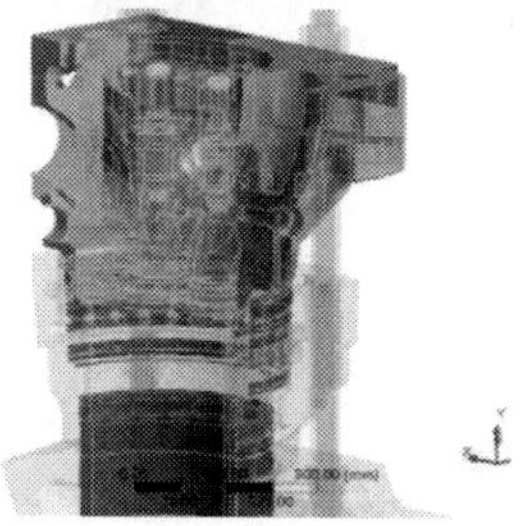

- More than 60 non-linear frictional contacts
- Oversize contacts to account for interference fits

CADFEM
ANSYS

Mechanical Load Cases

LC1: Force fitting of seat rings and valve guides

LC2: Cold assembly with full bolt pretension

LC3: Hot assembly (Temperatures from CFD)

LC4: Hot assembly + max. gas pressure

LC5: Assembly after cool-down

LC6: Assembly after cool-down + max. gas pressure

CADFEM
ANSYS

Results LC3: Deformation

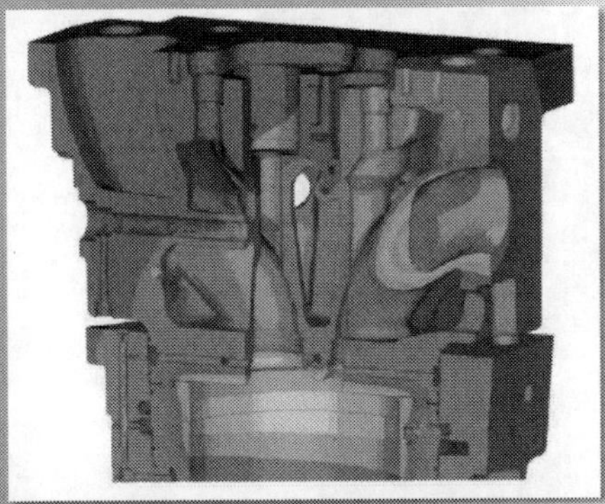

Interpolated temperatures from CFD

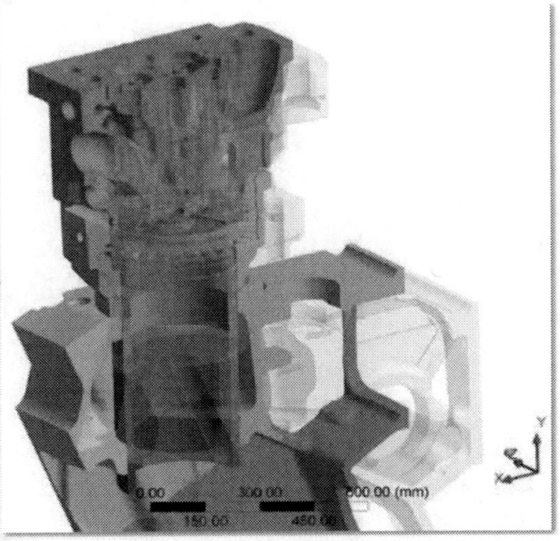

Deformation due to thermal expansion

Results LC4: Mechanical Stresses

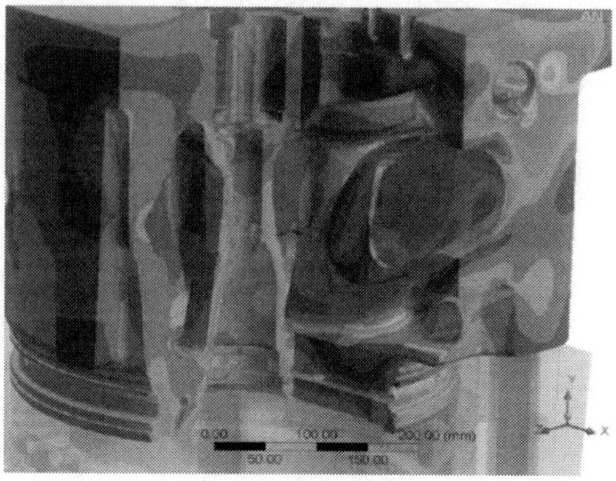

- Mechanical stresses in cylinder head due to temperature field and maximum gas pressure
- Stress level mainly induced by thermal expansion
- No critical yield stresses reached

  November 11, 2013

Overview

- Motivation
- Cooling concept
- Simulation strategy
- Fluid simulation
- Mechanical simulation
- **Summary**

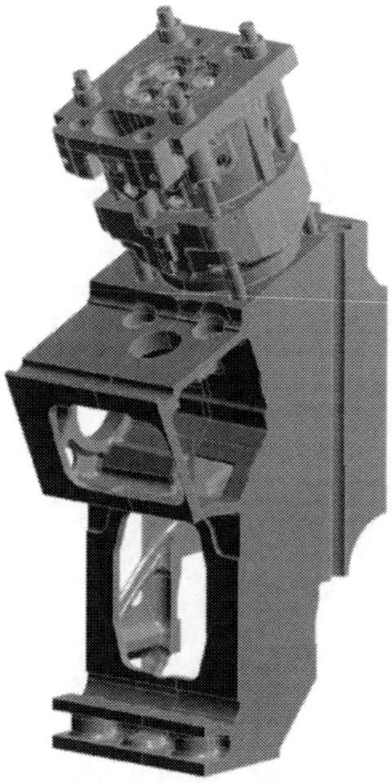

  November 11, 2013

Summary

Diesel engine 32/44CR: Demand for high reliability and long durability

Numerical simulation for evaluation of thermal and mechanical stresses

Simulation strategy: Coupled Fluid-Structure-Interaction

Proof of cooling concept efficiency: maximum temperatures & max. stresses below critical limits

Interesting to investigate cooling of 760 horses but...

31 © 2013 ANSYS, Inc. November 11, 2013

... Easier to Cool one Single Horse ...

32 © 2013 ANSYS, Inc. November 11, 2013

http://timpearson.co.uk/wp-content/uploads/2013/08/horse-shower.jpg

ANSYS FLUENT Use of GPUs for Automotive CFD Applications

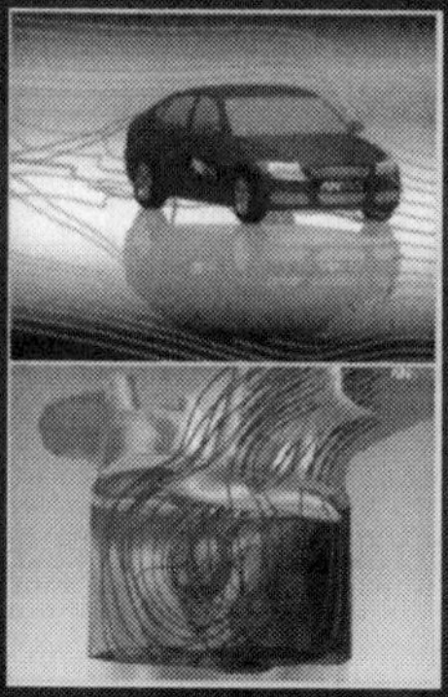

Stan Posey
NVIDIA, Santa Clara, CA, USA; sposey@nvidia.com

GPUs Now Mainstream HPC Technology

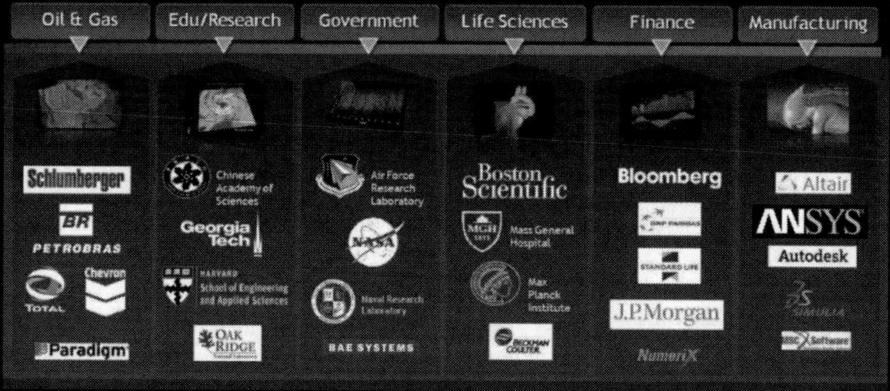

The buyer plans for including accelerators in their next technical computing server purchase has more than doubled from 29% to over 65% in last 20 months.

IDC Market Research
April, 2013

2

GPUs for Servers: Now as Common as CPUs
NVIDIA
Bull
hp
IBM
CRAY
APPRO
sgi
SUPERMICRO
DELL
TYAN
3

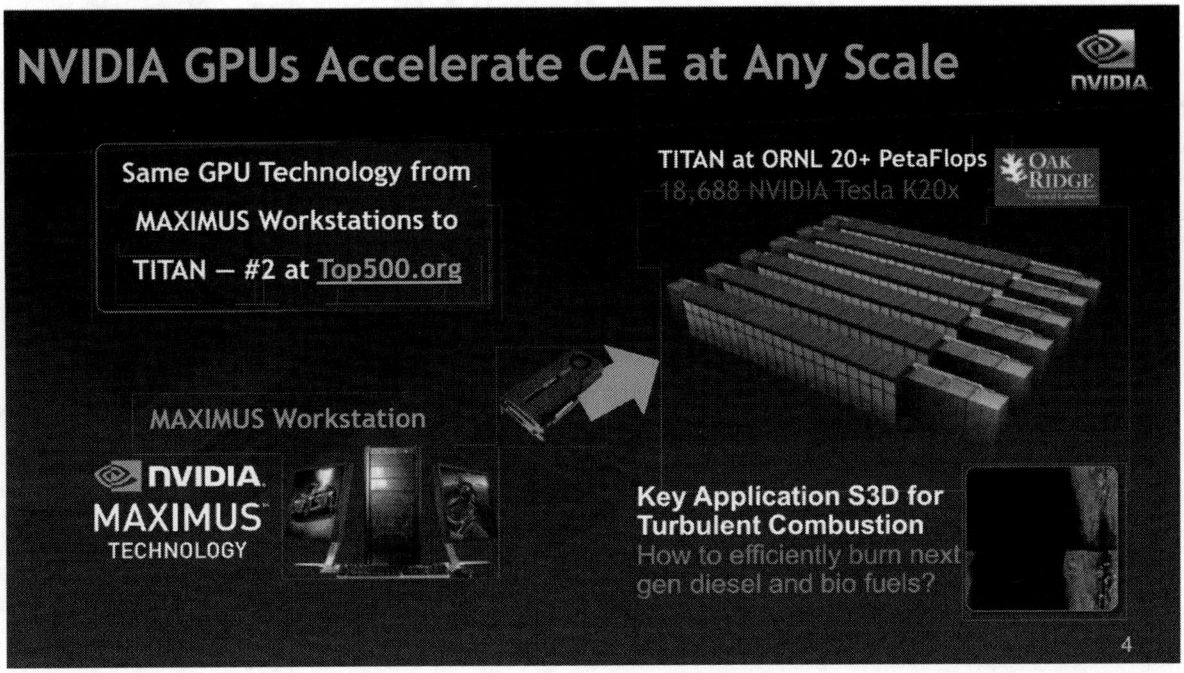

NVIDIA GPUs Accelerate CAE at Any Scale
NVIDIA
Same GPU Technology from MAXIMUS Workstations to TITAN — #2 at Top500.org
TITAN at ORNL 20+ PetaFlops
18,688 NVIDIA Tesla K20x
OAK RIDGE
MAXIMUS Workstation
NVIDIA MAXIMUS TECHNOLOGY
Key Application S3D for Turbulent Combustion
How to efficiently burn next gen diesel and bio fuels?
4

NVIDIA HPC Technology and CAE Strategy

NVIDIA

Technology

- Development of professional GPUs as co-processing accelerators for x86 CPUs

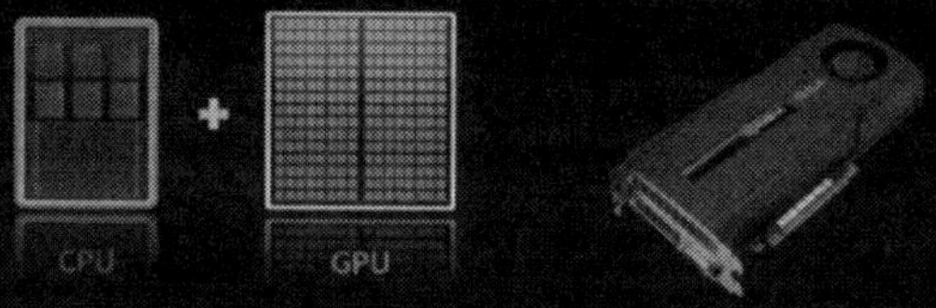

Strategy

- Strategic Alliances
  - Business and technical collaboration with ISVs; Industry customers; Research organizations
- Applications Engineering
  - Technical collaboration with ISVs like ANSYS for development of GPU-accelerated solvers
- Software Development
  - NVIDIA linear solver toolkit (implicit iterative solvers) , CUDA libraries, GPU compilers
- GPU System Integration
  - HP, Dell, IBM, Cray, SGI, Fujitsu, others; Kepler K20 based-systems available since 2012

5

NVIDIA Leadership in Remote Visualization

NVIDIA

6

GPU Motivation: CAE Cost Trends Over 20 Years

NVIDIA

Cost Trends : **Hardware is Cheap, People and Software Continue Cost Increase**

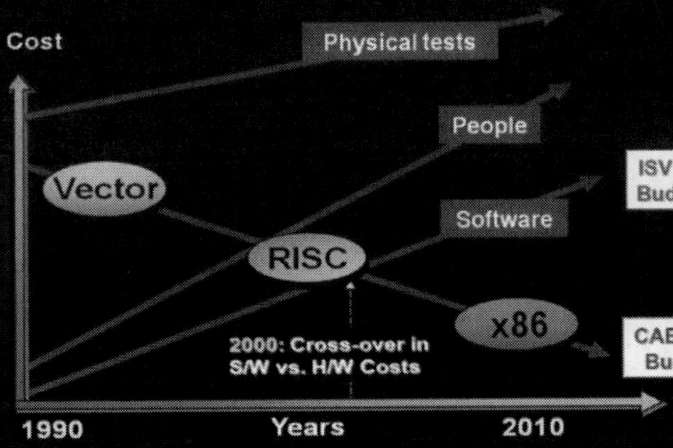

- Historically hardware very expensive vs. ISV software and people
- ISV software budgets are now 4x vs. hardware
- Increasingly important that hardware choices drive cost-performance efficiency in people and ISV software

7

NVIDIA Uses ANSYS CAE in Product Engineering

NVIDIA

ANSYS Icepak – active and passive cooling of IC packages

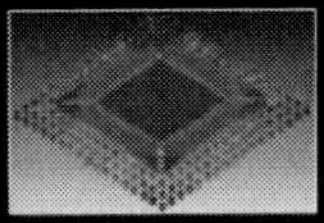

ANSYS Mechanical – large deflection bending of PCBs

ANSYS Mechanical – comfort and fit of 3D emitter glasses

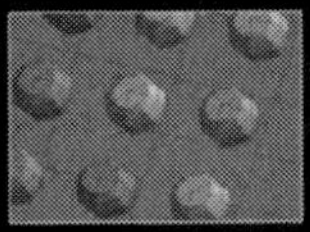

ANSYS Mechanical – shock & vib of solder ball assemblies

8

Progress Summary for GPU-Parallel CAE (I)

NVIDIA

- Strong GPU investments by commercial CAE vendors (ISVs)
  - GPU adoption led by implicit FEA and CEM, followed by CFD
    - Recent CFD breakthroughs in linear solvers (AMG) and preconditioners
- GPUs now production-HPC for leading CAE end-user sites
  - Led by automotive, electronics, and aerospace industries
- GPUs contributing to fast growth in emerging CAE applications
  - New developments in particle-based CFD (LBM, SPH, DEM, etc.)
  - Rapid growth for range of CEM applications and GPU adoption

9

Progress Summary for GPU-Parallel CAE (II)

NVIDIA

- Every ISV has GPU-based products available or undergoing evaluation
- The 4 largest ISVs have products based on GPUs, some at 3rd generation
  - #1 ANSYS, #2 DS SIMULIA, #3 MSC Software, and #4 Altair
- ANSYS 15.0 will have multiphysics capability on GPUs from 3 domains
  - ANSYS Mechanical – 4th gen; ANSYS Fluent – 2nd gen; ANSYS HFSS (transient) – 1st gen
- The top 4 out of 5 ISV applications are available on GPUs today
  - ANSYS Fluent, ANSYS Mechanical, Abaqus/Standard, MSC Nastran, *LS-DYNA implicit only*
- Several new ISVs were founded with GPUs as a competitive strategy
  - Prometech (JP), FluiDyna (DE), Vratis (PL), IMPETUS (SE), Turbostream (UK)

10

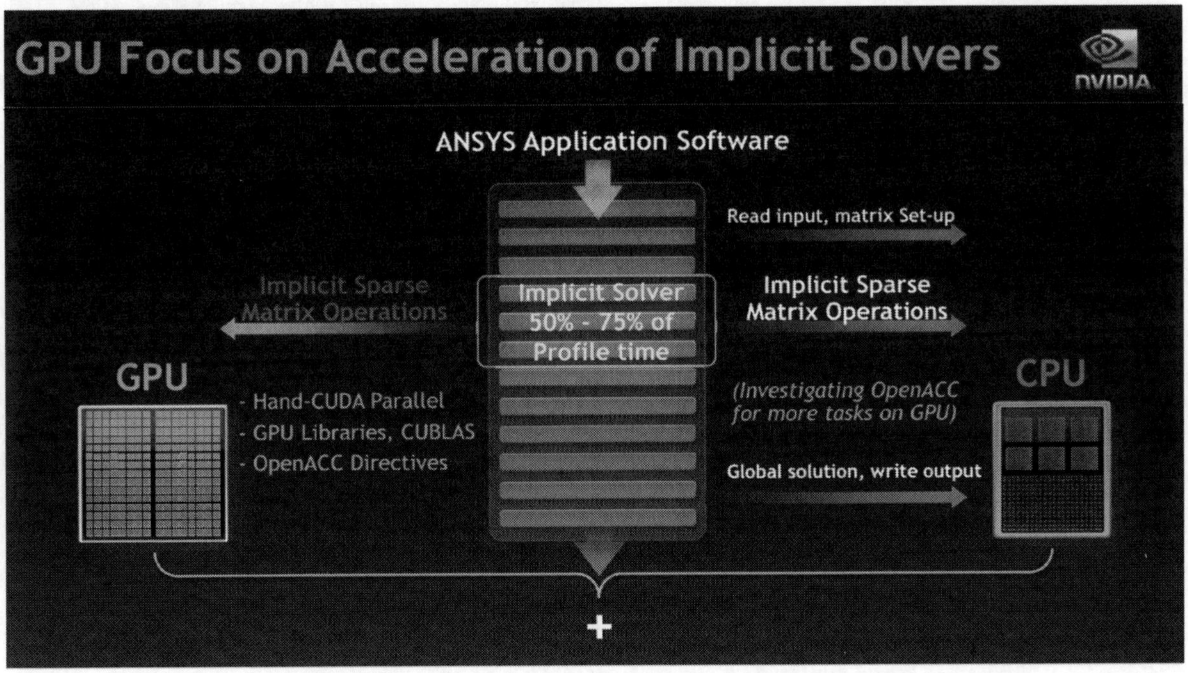
GPU Focus on Acceleration of Implicit Solvers
NVIDIA
ANSYS Application Software
Read input, matrix Set-up
Implicit Sparse Matrix Operations
Implicit Solver 50% - 75% of Profile time
Implicit Sparse Matrix Operations
GPU
- Hand-CUDA Parallel
- GPU Libraries, CUBLAS
- OpenACC Directives
(Investigating OpenACC for more tasks on GPU)
CPU
Global solution, write output
+

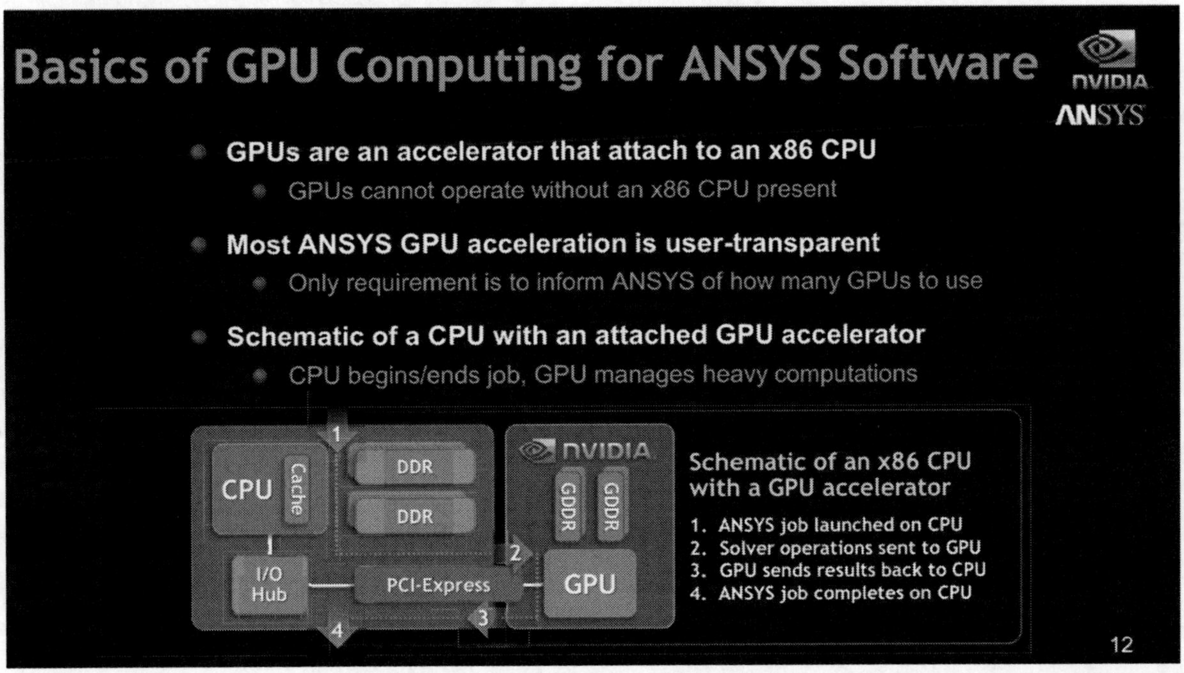
Basics of GPU Computing for ANSYS Software
NVIDIA
ANSYS
GPUs are an accelerator that attach to an x86 CPU
GPUs cannot operate without an x86 CPU present
Most ANSYS GPU acceleration is user-transparent
Only requirement is to inform ANSYS of how many GPUs to use
Schematic of a CPU with an attached GPU accelerator
CPU begins/ends job, GPU manages heavy computations
CPU
Cache
DDR
DDR
I/O Hub
PCI-Express
NVIDIA
GDDR
GDDR
GPU
1
2
3
4
Schematic of an x86 CPU with a GPU accelerator
1. ANSYS job launched on CPU
2. Solver operations sent to GPU
3. GPU sends results back to CPU
4. ANSYS job completes on CPU
12

ANSYS and NVIDIA Collaboration Roadmap

Release	ANSYS Mechanical	ANSYS Fluent	ANSYS EM
13.0 Dec 2010	SMP, Single GPU, Sparse and PCG/JCG Solvers		ANSYS Nexxim
14.0 Dec 2011	+ Distributed ANSYS; + Multi-node Support	Radiation Heat Transfer (beta)	ANSYS Nexxim
14.5 Nov 2012	+ Multi-GPU Support; + Hybrid PCG; + Kepler GPU Support	+ Radiation HT; + GPU AMG Solver (beta), Single GPU	ANSYS Nexxim
15.0 Q4-2013	+ CUDA 5 Kepler Tuning	+ Multi-GPU AMG Solver; + CUDA 5 Kepler Tuning	ANSYS Nexxim ANSYS HFSS (Transient)

13

ANSYS 15.0 License Scheme for GPUs

One HPC Task Required to Unlocked One GPU

Examples:

1 x ANSYS HPC Pack
Total 8 HPC Tasks (4 GPUs Max)

2 x ANSYS HPC Pack
Total 32 HPC Tasks (16 GPUs Max)

(Applies to all schemes: HPC, HPC Pack, HPC Workgroup, HPC Enterprise, etc.)

Valid Configurations:

6 CPU Cores + 2 GPUs

4 CPU Cores + 4 GPUs

Total Use of 2 Servers:

24 CPU Cores + 8 GPUs (3:1 Ratio)

14

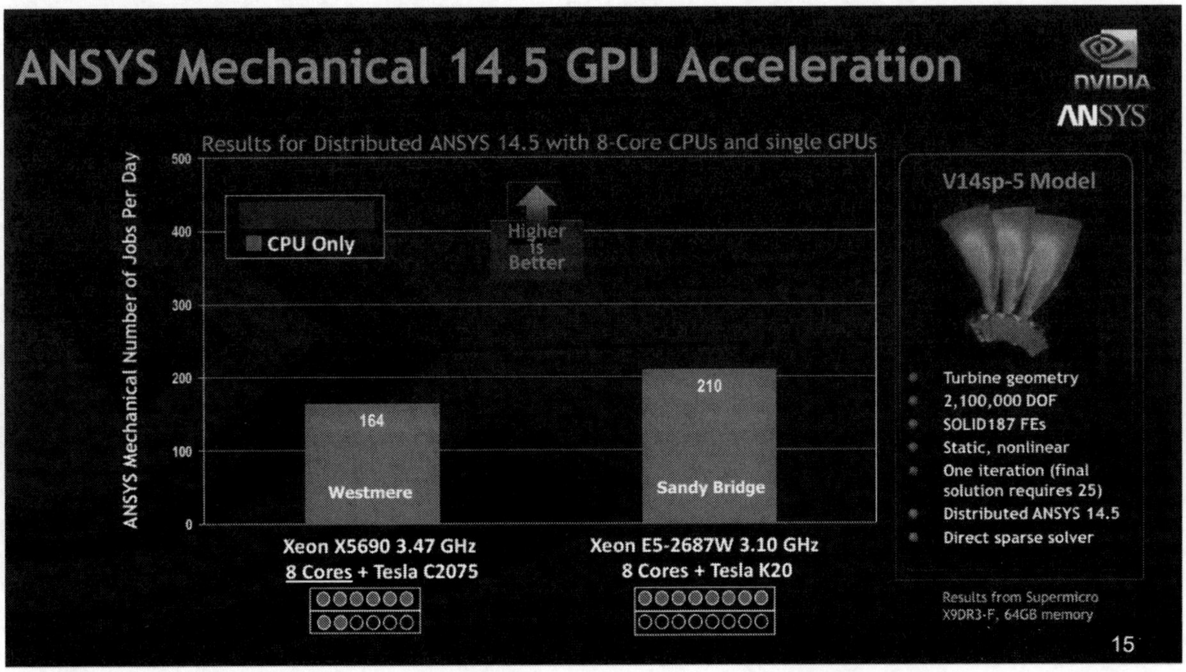
ANSYS Mechanical 14.5 GPU Acceleration
NVIDIA
ANSYS
Results for Distributed ANSYS 14.5 with 8-Core CPUs and single GPUs
ANSYS Mechanical Number of Jobs Per Day
500
400
300
200
100
0
CPU Only
Higher is Better
164
Westmere
210
Sandy Bridge
Xeon X5690 3.47 GHz
8 Cores + Tesla C2075
Xeon E5-2687W 3.10 GHz
8 Cores + Tesla K20
V14sp-5 Model
Turbine geometry
2,100,000 DOF
SOLID187 FEs
Static, nonlinear
One iteration (final solution requires 25)
Distributed ANSYS 14.5
Direct sparse solver
Results from Supermicro X9DR3-F, 64GB memory
15

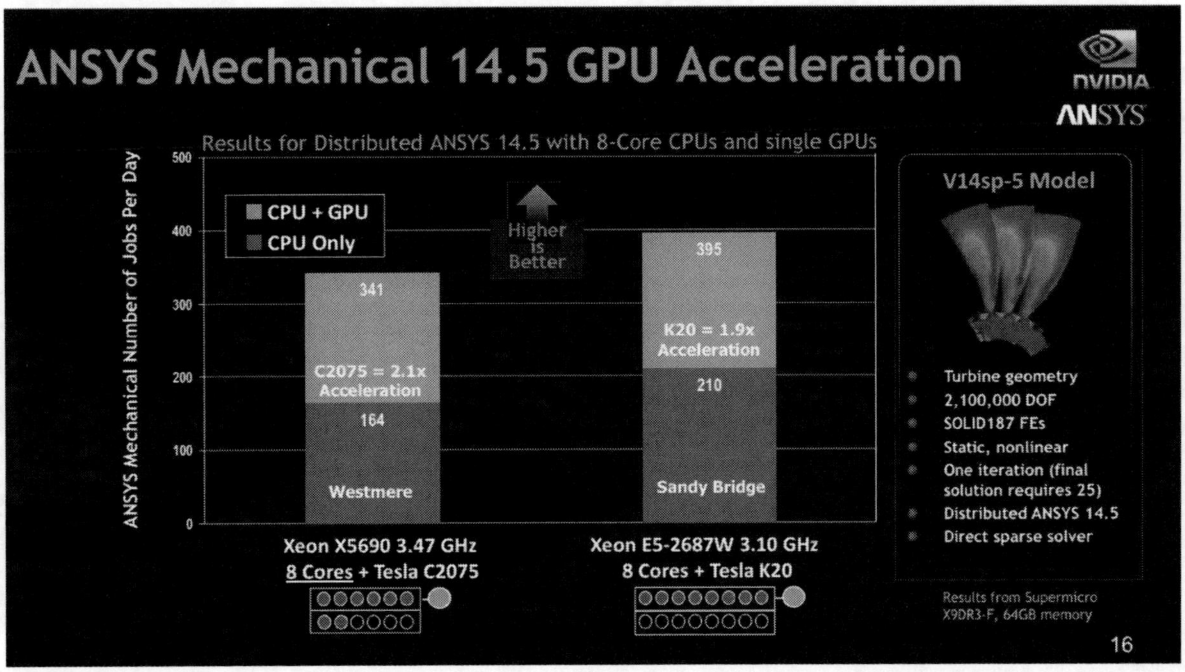
ANSYS Mechanical 14.5 GPU Acceleration
NVIDIA
ANSYS
Results for Distributed ANSYS 14.5 with 8-Core CPUs and single GPUs
ANSYS Mechanical Number of Jobs Per Day
500
400
300
200
100
0
CPU + GPU
CPU Only
Higher is Better
341
C2075 = 2.1x Acceleration
164
Westmere
395
K20 = 1.9x Acceleration
210
Sandy Bridge
Xeon X5690 3.47 GHz
8 Cores + Tesla C2075
Xeon E5-2687W 3.10 GHz
8 Cores + Tesla K20
V14sp-5 Model
Turbine geometry
2,100,000 DOF
SOLID187 FEs
Static, nonlinear
One iteration (final solution requires 25)
Distributed ANSYS 14.5
Direct sparse solver
Results from Supermicro X9DR3-F, 64GB memory
16

ANSYS and NVIDIA Collaboration Roadmap

NVIDIA
ANSYS

Release	ANSYS Mechanical	ANSYS Fluent	ANSYS EM
13.0 Dec 2010	SMP, Single GPU, Sparse and PCG/JCG Solvers		ANSYS Nexxim
14.0 Dec 2011	+ Distributed ANSYS; + Multi-node Support	Radiation Heat Transfer (beta)	ANSYS Nexxim
14.5 Nov 2012	+ Multi-GPU Support; + Hybrid PCG; + Kepler GPU Support	+ Radiation HT; + GPU AMG Solver (beta), Single GPU	ANSYS Nexxim
15.0 Q4-2013	+ CUDA 5 Kepler Tuning	+ Multi-GPU AMG Solver; + CUDA 5 Kepler Tuning	ANSYS Nexxim ANSYS HFSS (Transient)

17

GPU Acceleration in ANSYS Fluent

NVIDIA
ANSYS

Beta Release in14.5, Full Product Support in15.0 (Dec 2013)

- GPU-based Model: Radiation Heat Transfer using OptiX, Product in 14.5
- GPU-based Solver: Coupled Algebraic Multigrid (AMG) PBNS linear solver
- Operating Systems: Both Linux and Win64 for workstations and servers
- Parallel Methods: Shared memory in 14.5; Distributed ANSYS in 15.0
- Multi-GPU Support: Single GPU for 14.5; full multi-GPU, multi-node 15.0
- Model Suitability: Size of 3M cells or less in 14.5, unlimited in 15.0

18

ANSYS Fluent 14.5 and Radiation HT on GPU

NVIDIA
ANSYS

VIEWFAC Utility:
Use on CPUs, GPUs or both ~2x speedup

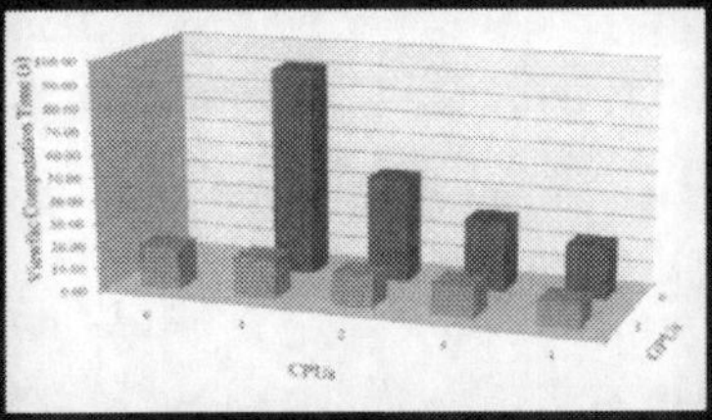

RAY TRACING Utility:
Uses OptiX library from NVIDIA with up to ~15x speedup
(Use on GPU only)

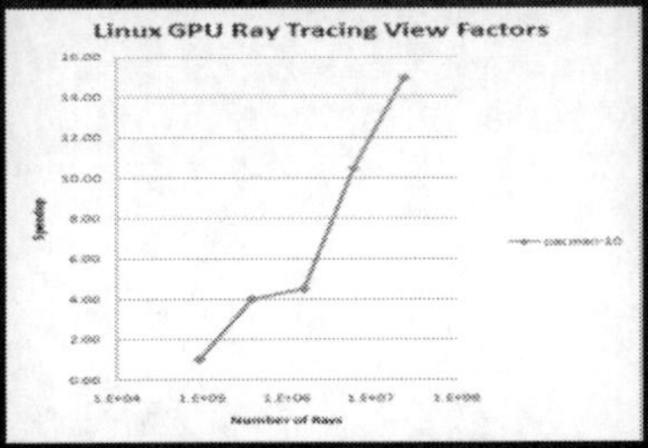

Radiation HT Applications:

- Underhood cooling
- Cabin comfort HVAC
- Furnace simulations
- Solar loads on buildings
- Combustor in turbine
- Electronics passive cooling

19

GPU-based AMG Solver for ANSYS Fluent 15.0

NVIDIA
ANSYS

- **New ANSYS Fluent AMG based on NVIDIA-developed solver toolkit**
  - Developed with support for MPI across multiple nodes and multiple GPUs
  - Solver collaboration on coupled pressure-based Navier-Stokes, others to follow
- **Early results published at Parallel CFD 2013, 20-24 May, Changsha, CN**
  - *GPU-Accelerated Algebraic Multigrid for Applied CFD*

Parallel CFD 2013 :
May 20-24
National Supercomputing Center :
Changsha, Hunan, China
Parallel Computational Fluid Dynamics

ELSEVIER
Available online at www.sciencedirect.com
SciVerse ScienceDirect
Procedia Engineering 00 (2012) 000–000
Procedia Engineering
www.elsevier.com/locate/procedia

Parallel Computational Fluid Dynamics Conference (ParCFD2013)

GPU-Accelerated Algebraic Multigrid for Applied CFD

Robert Strzodka[a], Jonathan Cohen[a], Stan Posey[a] *
NVIDIA Corporation, 2701 San Tomas Expressway, Santa Clara, CA 95050, USA

20

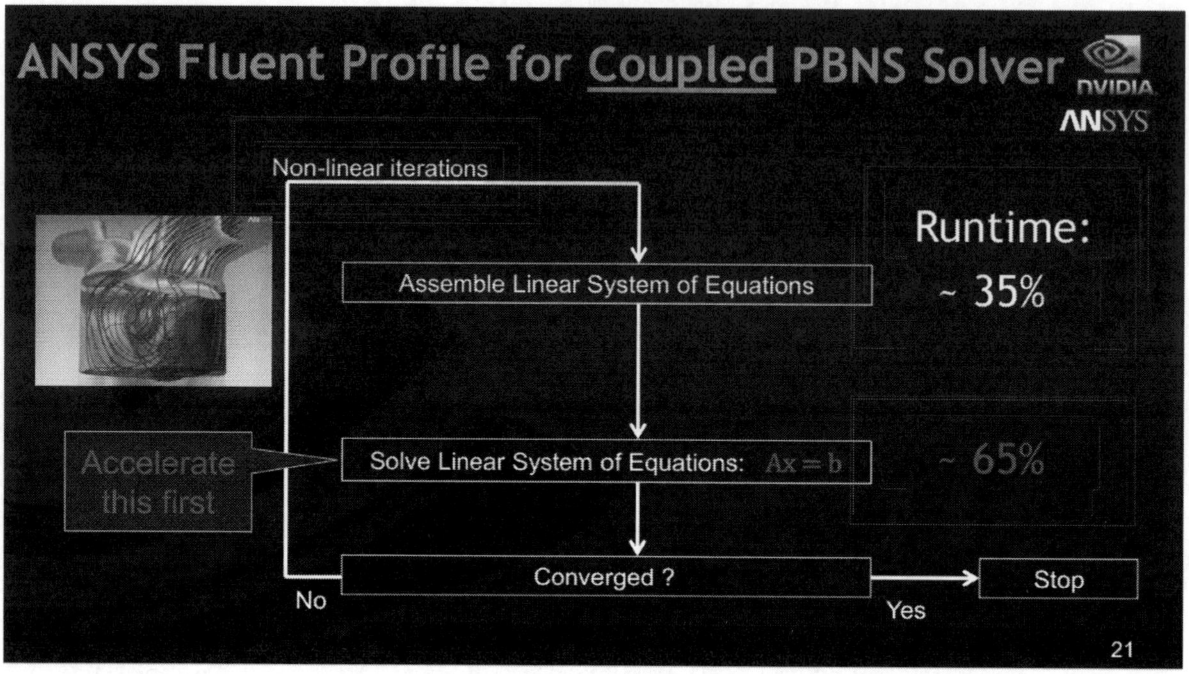
ANSYS Fluent Profile for Coupled PBNS Solver
NVIDIA
ANSYS
Non-linear iterations
Assemble Linear System of Equations
Runtime:
~ 35%
Accelerate this first
Solve Linear System of Equations: Ax = b
~ 65%
Converged ?
No
Yes
Stop
21

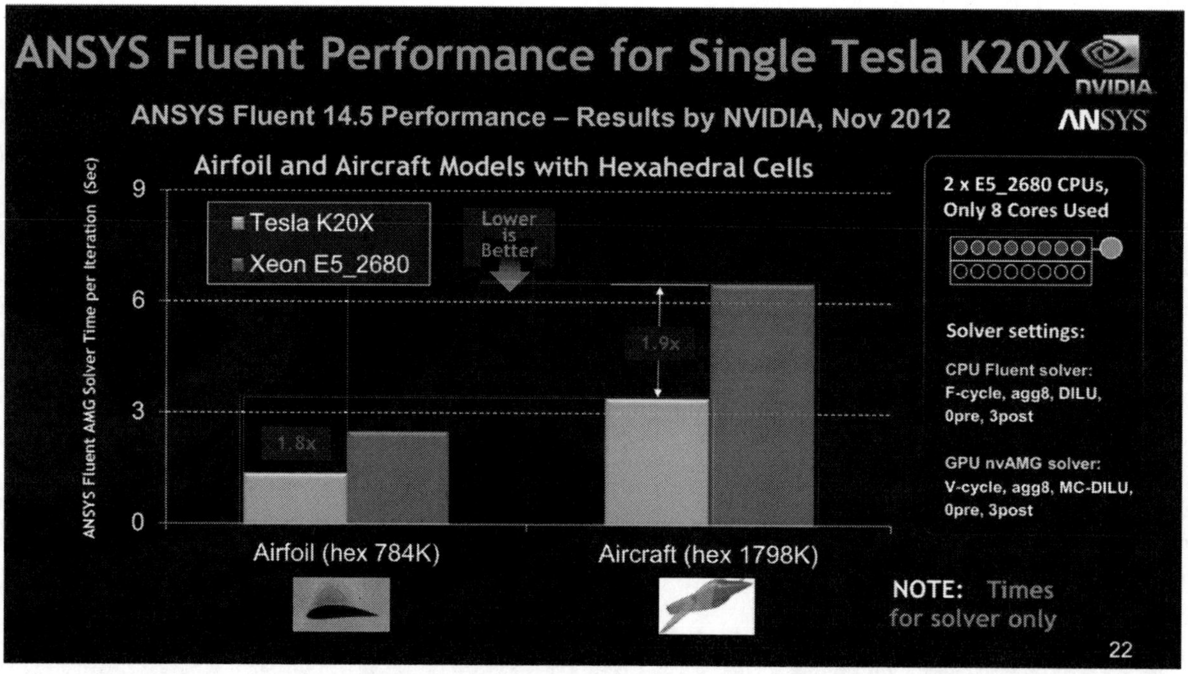
ANSYS Fluent Performance for Single Tesla K20X
NVIDIA
ANSYS
ANSYS Fluent 14.5 Performance – Results by NVIDIA, Nov 2012
Airfoil and Aircraft Models with Hexahedral Cells
ANSYS Fluent AMG Solver Time per Iteration (Sec)
Tesla K20X
Xeon E5_2680
Lower is Better
9
6
3
0
1.8x
1.9x
Airfoil (hex 784K)
Aircraft (hex 1798K)
2 x E5_2680 CPUs, Only 8 Cores Used
Solver settings:
CPU Fluent solver: F-cycle, agg8, DILU, 0pre, 3post
GPU nvAMG solver: V-cycle, agg8, MC-DILU, 0pre, 3post
NOTE: Times for solver only
22

Multi-GPU Preview

ANSYS Fluent 15.0

Available Late 2013

(Preview 3 from Aug 2013)

23

GPUs and Distributed Cluster Computing

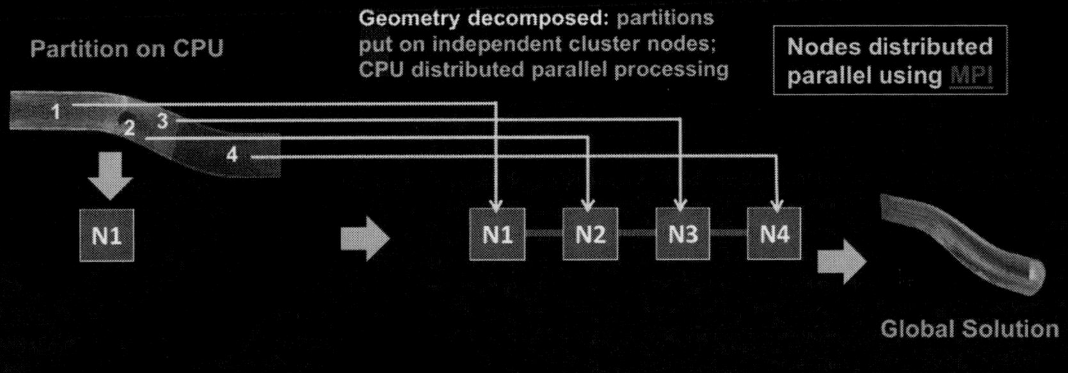

24

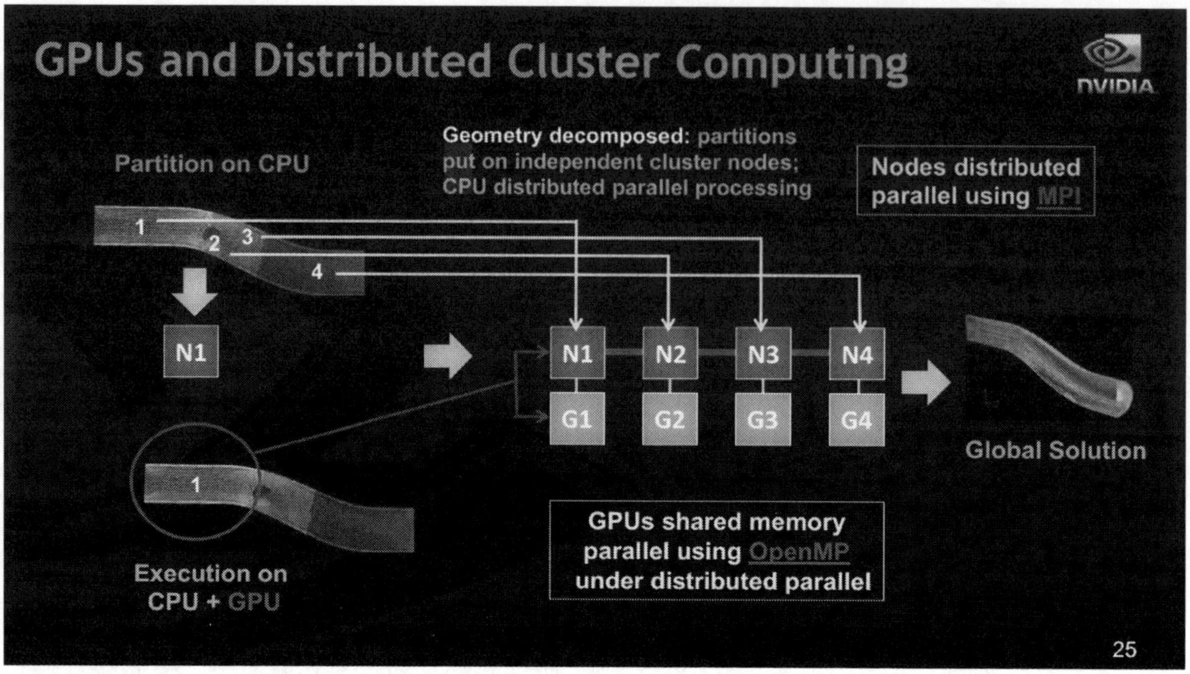
GPUs and Distributed Cluster Computing
NVIDIA
Partition on CPU
Geometry decomposed: partitions put on independent cluster nodes; CPU distributed parallel processing
Nodes distributed parallel using MPI
1
2
3
4
N1
N1
N2
N3
N4
G1
G2
G3
G4
Global Solution
1
Execution on CPU + GPU
GPUs shared memory parallel using OpenMP under distributed parallel
25

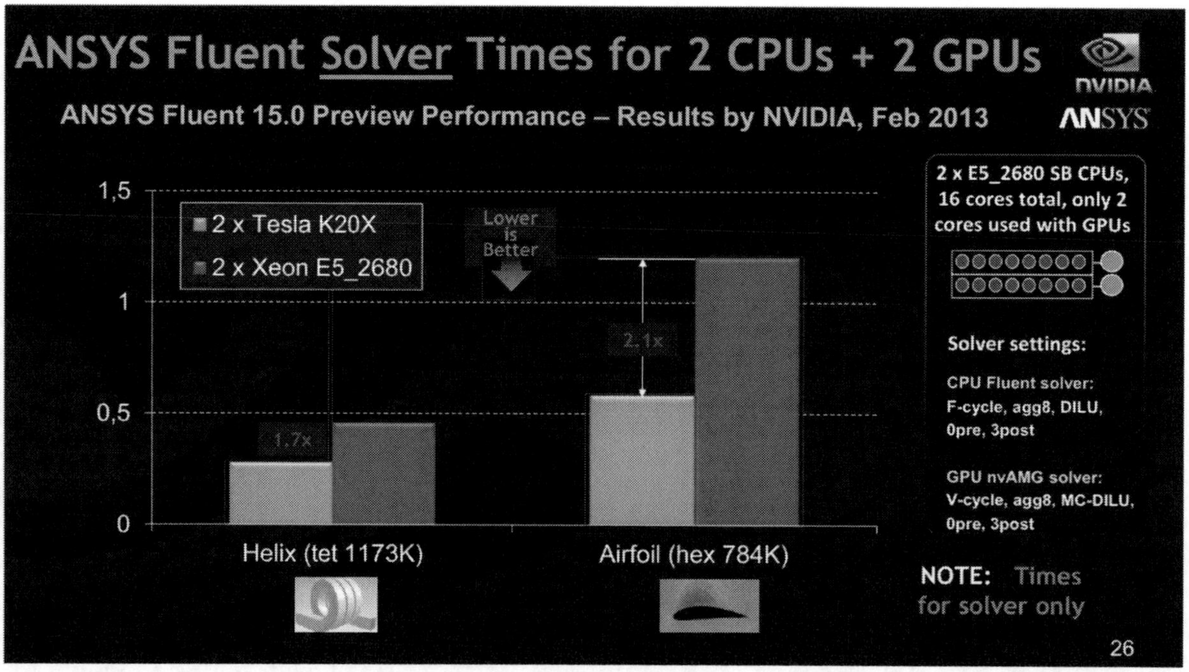
ANSYS Fluent Solver Times for 2 CPUs + 2 GPUs
NVIDIA
ANSYS Fluent 15.0 Preview Performance – Results by NVIDIA, Feb 2013
ANSYS
1,5
1
0,5
0
2 x Tesla K20X
2 x Xeon E5_2680
Lower is Better
2.1x
1.7x
Helix (tet 1173K)
Airfoil (hex 784K)
2 x E5_2680 SB CPUs, 16 cores total, only 2 cores used with GPUs
Solver settings:
CPU Fluent solver: F-cycle, agg8, DILU, 0pre, 3post
GPU nvAMG solver: V-cycle, agg8, MC-DILU, 0pre, 3post
NOTE: Times for solver only
26

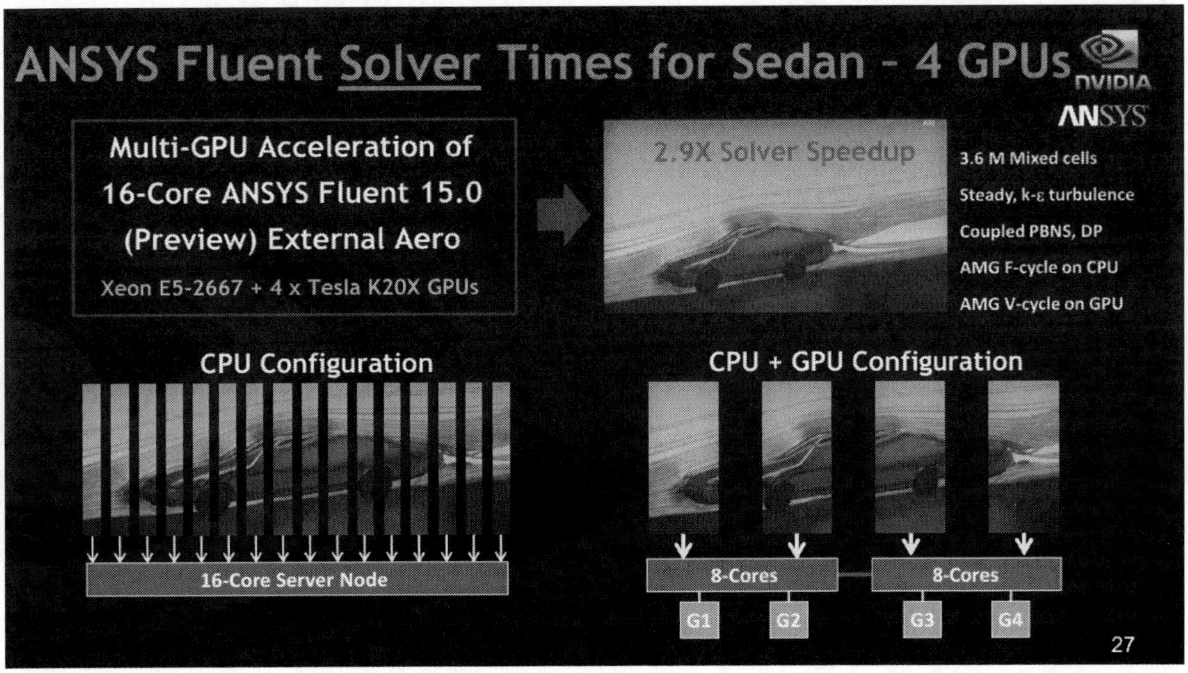
ANSYS Fluent Solver Times for Sedan - 4 GPUs
NVIDIA
ANSYS
Multi-GPU Acceleration of
16-Core ANSYS Fluent 15.0
(Preview) External Aero
Xeon E5-2667 + 4 x Tesla K20X GPUs
2.9X Solver Speedup
3.6 M Mixed cells
Steady, k-ε turbulence
Coupled PBNS, DP
AMG F-cycle on CPU
AMG V-cycle on GPU
CPU Configuration
CPU + GPU Configuration
16-Core Server Node
8-Cores
8-Cores
G1
G2
G3
G4
27

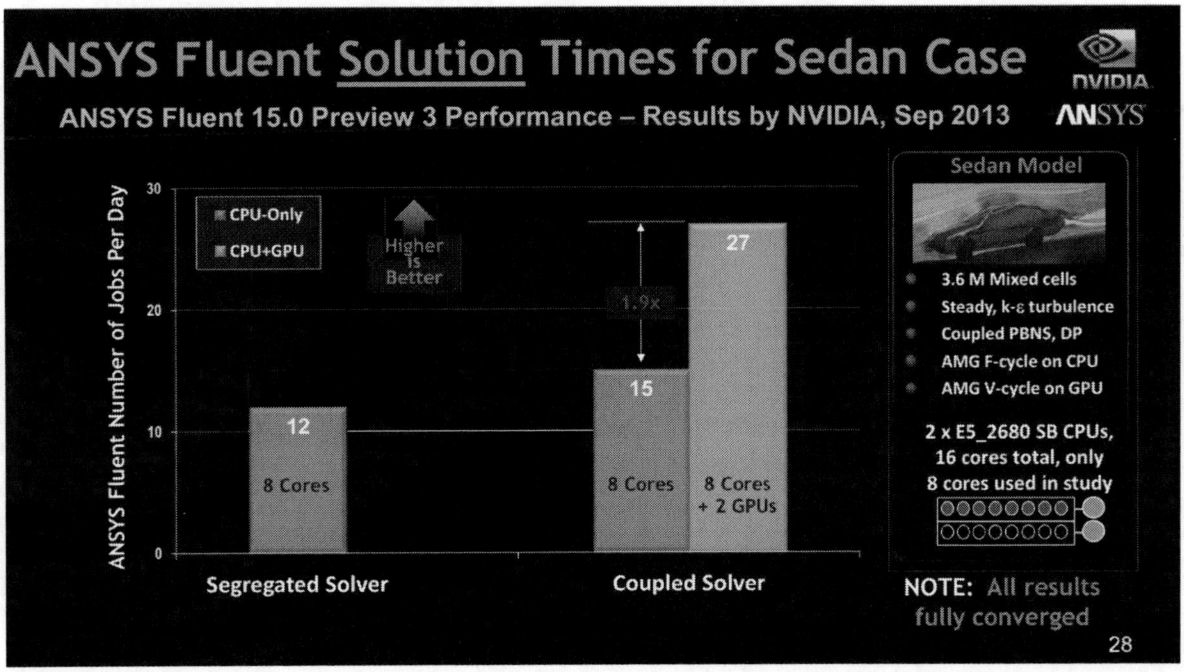
ANSYS Fluent Solution Times for Sedan Case
NVIDIA
ANSYS Fluent 15.0 Preview 3 Performance – Results by NVIDIA, Sep 2013
ANSYS
ANSYS Fluent Number of Jobs Per Day
30
20
10
0
CPU-Only
CPU+GPU
Higher is Better
1.9x
12
8 Cores
15
8 Cores
27
8 Cores + 2 GPUs
Segregated Solver
Coupled Solver
Sedan Model
3.6 M Mixed cells
Steady, k-ε turbulence
Coupled PBNS, DP
AMG F-cycle on CPU
AMG V-cycle on GPU
2 x E5_2680 SB CPUs, 16 cores total, only 8 cores used in study
NOTE: All results fully converged
28

ANSYS Fluent Convergence for Truck Case

NVIDIA

Coupled PBNS and stable convergence for drag coefficient at ~500 iterations

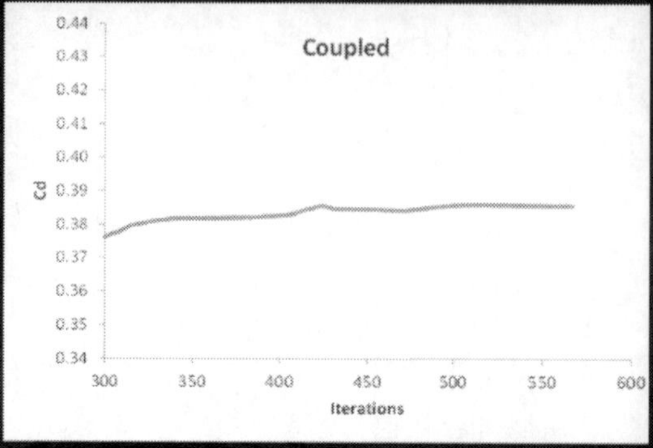

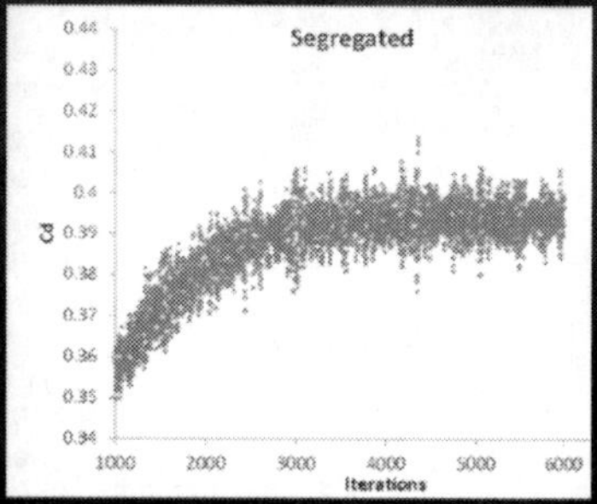

Segregated PBNS and oscillating behavior for drag coefficient, not converged after ~6000 iterations

Truck Body Model

- 14 M Mixed cells
- Steady, k-ε turbulence
- PBNS, double precision
- Default URF's for each
- CPU: AMG F-cycle
- GPU: FGMRES with AMG Preconditioner

29

ANSYS Fluent <u>Solution</u> Times for Truck Case

NVIDIA

ANSYS Fluent 15.0 Preview 3 Performance – Results by NVIDIA, Sep 2013 ANSYS

- Same solution times: 64 cores vs. 32 cores + 8 GPUs
- Frees up 32 CPUs and HPC licenses for additional job(s)
- Approximate 56% increase in overall productivity for 25% increase in cost

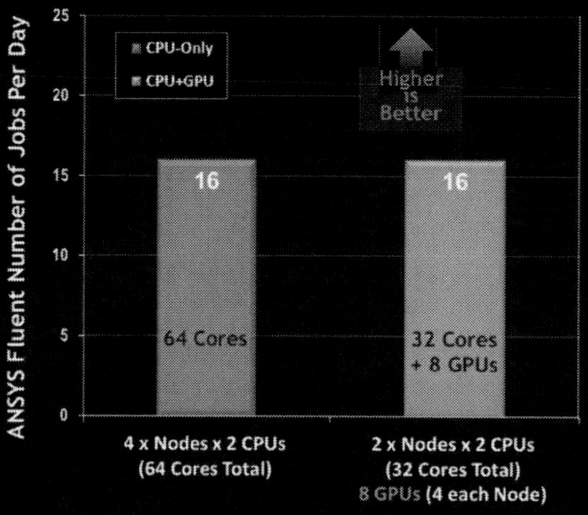

Truck Body Model

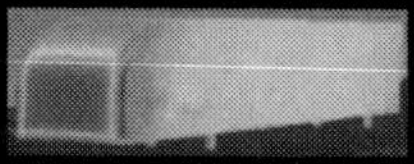

- 14 M Mixed cells
- Steady, k-ε turbulence
- Coupled PBNS, DP
- Total solution times
- CPU: AMG F-cycle
- GPU: FGMRES with AMG Preconditioner

NOTE: All results fully converged

30

Additional Information

NVIDIA

Configuration Details on Workstation or Server

- www.nvidia.com/teslawtb
- www.nvidia.com/workstationwtb

Test-drive GPU Computing with Your ANSYS Simulations

- Contact ANSYS or email cae@nvidia.com

Read More About ANSYS and GPU Computing

- ANSYS Unveils GPU Computing for Accelerated Engineering Simulations
- Speed Up Simulations with a GPU, Article in ANSYS Advantage Magazine
- Speeding to a Solution, Article in ANSYS Advantage Magazine
- HPC Delivers a 3-D View, Article in ANSYS Advantage Magazine

For More Information on NVIDIA and ANSYS Solutions

- www.nvidia.com/ansys

31

Acknowledgements

NVIDIA

ANSYS

ANSYS
www.ansys.com

- Mr. Jeff Beisheim, ANSYS Mechanical Parallel Solver Development
- Dr. Sunil Sathe, ANSYS Fluent Parallel Solver Development
- Dr. Prasad Alavilli, Manager, ANSYS Fluent Parallel HPC Development

NVIDIA

NVIDIA
www.nvidia.com

- Mr. Jon Cohen, Manager, NVIDIA Computational Library Development
- Dr. Joe Eaton, Manager, NVIDIA Linear Solver Tool Kit Development
- Dr. Steve Rennich, Applications Engineer, Developer Technology Group
- Dr. Bhushan Desam, ANSYS Alliances and CFD Market Development
- Mr. Vijay Sellappan, Applications Engineer, CAE Technology

32

Thank You and Questions?

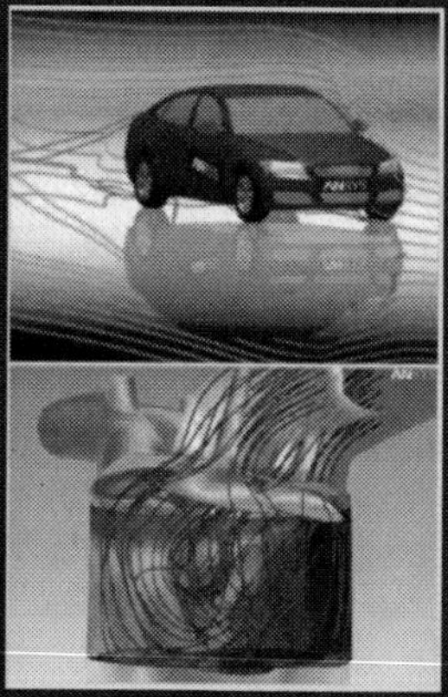

Stan Posey
NVIDIA, Santa Clara, CA, USA; sposey@nvidia.com

2013 Automotive Simulation World Congress

Advantages of Diesel Particulate Filter Simulation in Exhaust Aftertreatment Systems

Dr. Rolf Reinelt, ANSYS Germany

Agenda

- Introduction
- Theory
- Grid generation
- Results
  - Eberspächer test case
  - SAE paper by Turner et al.
- Conclusions

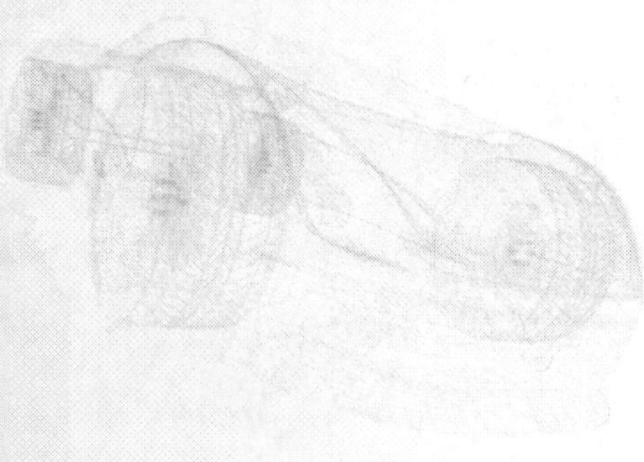

Introduction

- Diesel Particulate Filters (DPF) nowadays are applied in the majority of Diesel cars and trucks
- For the development main goals are
  - costs => size of filter
  - efficient removal of particulate matter (PM) =>pore size
  - low back pressure => uniform flow, regeneration
  - mileage => volume for ash deposit
  - durability => failure due to overheating

2013 Automotive Simulation World Congress
October 29 - 30 • Frankfurt am Main, Germany

Introduction

- A DPF involves very different scales of time and length
  - engine cycle $\Delta\phi=1°$ CA
  - Driving cycle $\Delta t=1s$
  - Ash deposit (200000 km)

  - Filter diameter (500000 cells)
  - Channel width (240 million cells)
  - Porous holes in ceramic

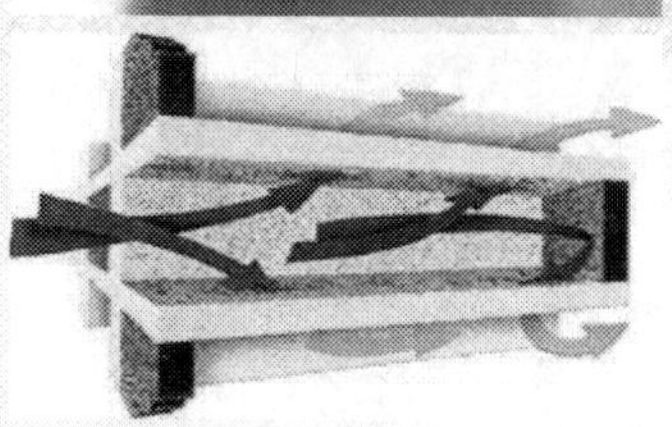

2013 Automotive Simulation World Congress
October 29 - 30 • Frankfurt am Main, Germany

Introduction

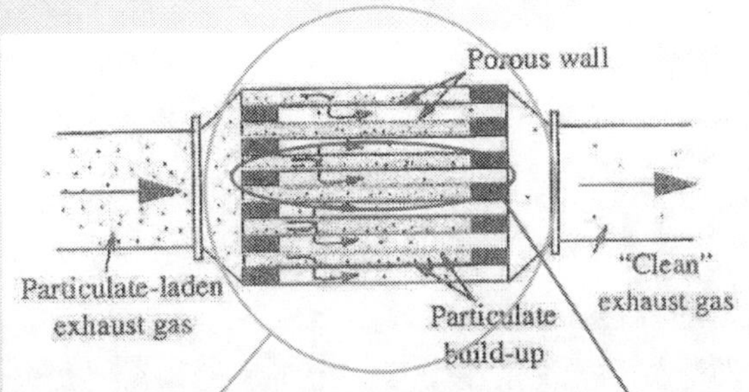

Macroscopic approach
(Porous media)

Microscopic approach
(Single channels)

2013 Automotive Simulation World Congress
October 29 - 30 • Frankfurt am Main, Germany

Introduction

- A CFD simulation can give the following answers:
  - soot distribution in the filter
  - pressure field => pressure loss
  - peak temperatures and temperature gradients during regeneration
  - regeneration efficiency

Agenda

- Introduction
- Theory
- Grid generation
- Results
  - Eberspächer test case
  - SAE paper by Turner et al.
- Conclusions

2013 Automotive Simulation World Congress
October 29 - 30 • Frankfurt am Main, Germany

1D theory for DPF modeling

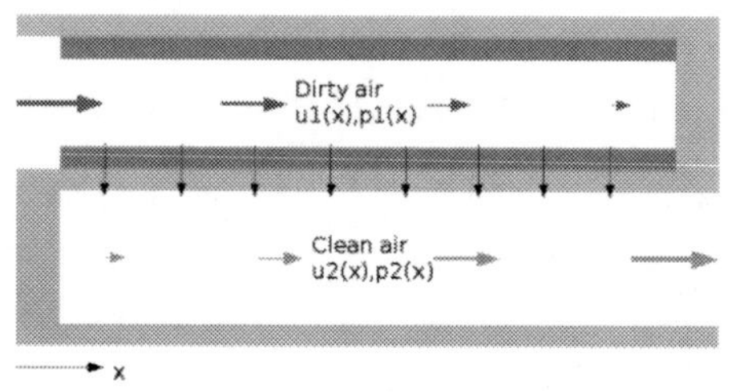

$$\frac{d^2u_2}{dx^2} - 2A_1\frac{du_2}{dx} - (A_2 + A_2^*)u_2 + A_2 = 0$$

- ordinary differential eqn.
- easy to solve with finite-difference method
- small C-program

- mass conservation
- momentum conservation
- pressure drop across the wall
- constant density
- isothermal

1D theory for DPF modeling

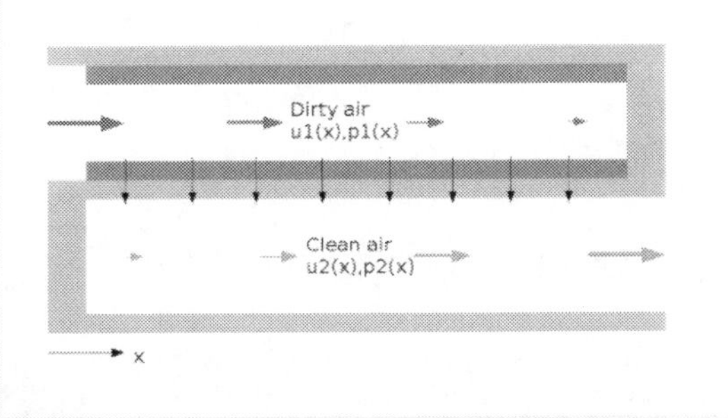

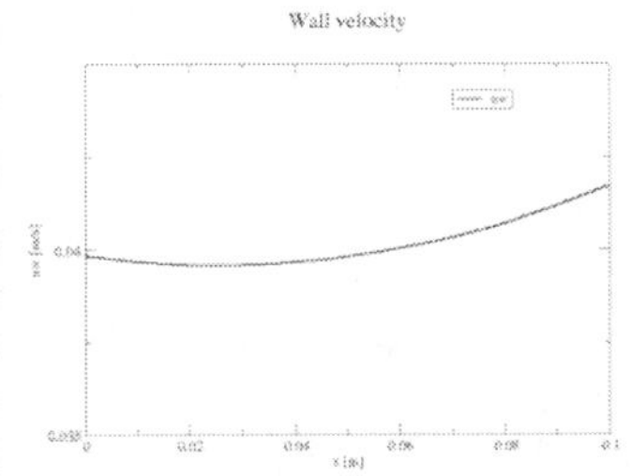

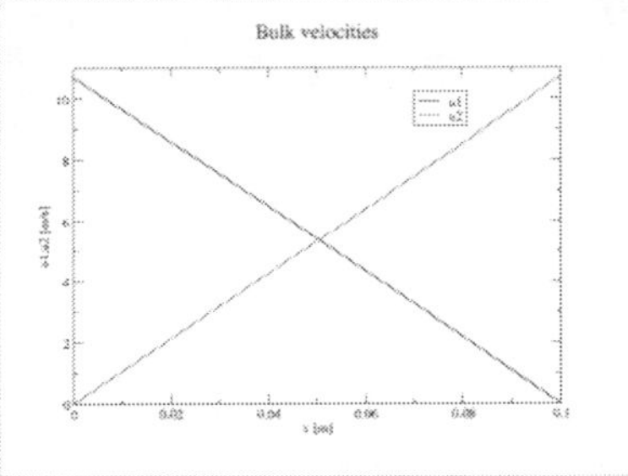

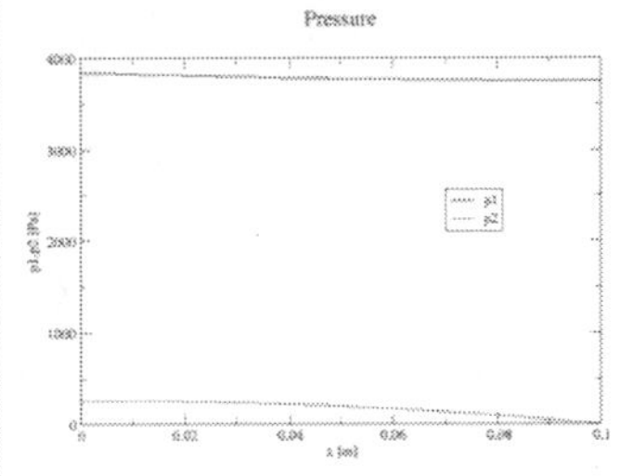

Theory for DPF modeling

- Limitation of 1D theory
  - 3D flow before and after the filter
  - 3D heat conduction in the filter

Use 1D theory in a porous media triple zone approach

1. Bulk velocity and pressure and temperature in the dirty air channels.
2. Bulk velocity and pressure and temperature in the clean air channels.
3. Temperature in the solid.

Theory for DPF modeling

- Align flow with filter axis using porous media
- Mass source terms are driven by pressure difference of the dirty and clean air zone
- Energy source terms are driven by temperature differences.

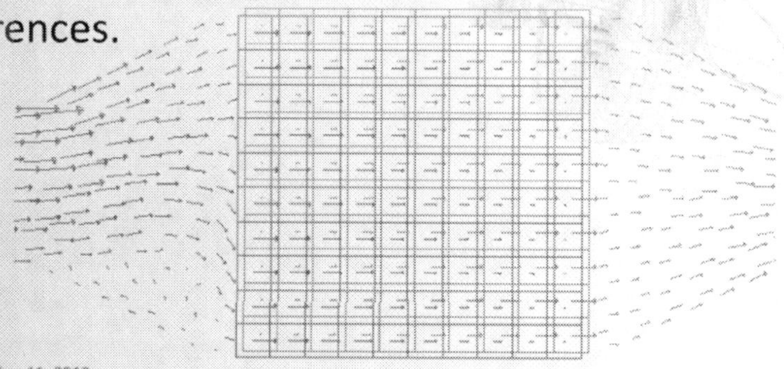

2013 Automotive Simulation World Congress
October 29 - 30 • Frankfurt am Main, Germany

Theory for DPF modeling

- Wall flow velocity uw is calculated from pressure difference

$$\Delta p = u_w \mu \left[\frac{d_p}{k_p} - \frac{a}{2k_s} \ln\left(1 - \frac{2d_s}{a}\right) \right]$$

- Mass source is proportional to wall flow velocity

$$Q_m = \frac{u_w \rho_{air} a N}{2(a + d_p)^2}$$

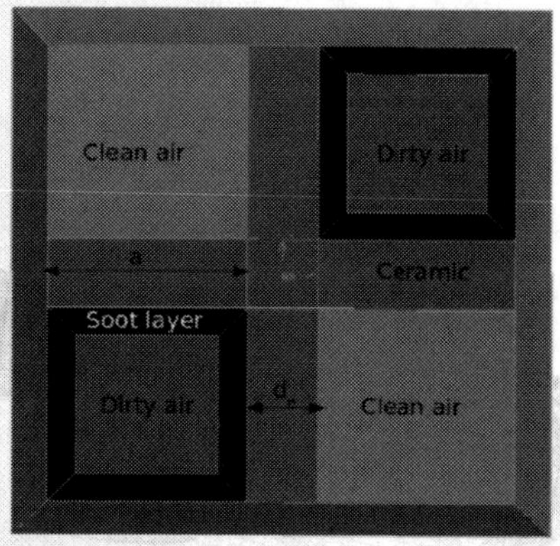

- Exhaust air flowing through the wall is in equilibrium with wall temperature

$$(T_1' - T_1)c_{p,air}Q_m + \frac{T_1' - T_3}{\Delta t} c_{p,3}\rho_3 \frac{V_3}{V_t} = Q_{reg}$$

2013 Automotive Simulation World Congress
October 29 - 30 • Frankfurt am Main, Germany

Theory for DPF modeling

- Heat exchange for flow along a wall

$$Q_{ch,i} = h_i \left(\frac{A}{V} \right)_i (T_i - T_3)$$

$$\left(\frac{A}{V} \right)_i = \frac{2(a - 2d_s)}{(a + d_p)^2}$$

- Conductivities in the solid (Orthotropic thermal conductivity)

$$\lambda_x = \frac{d_p}{d_p + a} \qquad \lambda_z = \frac{d_p^2 + 2ad_p}{(d_p + a)^2}$$

2013 Automotive Simulation World Congress
October 29 - 30 • Frankfurt am Main, Germany

Chemical reaction model

- Take into account burning of soot by O2 and NO2
- Do not solve for species in Fluent
- O2 and NO2 are constant from inflow to 1 to 2 and from 3 to 4 to 5 to outflow
- Reaction rate is of Arrhenius type $R = K \cdot \rho_s \cdot \rho_{NO_2/O_2} \cdot e^{\frac{-E}{RT}}$

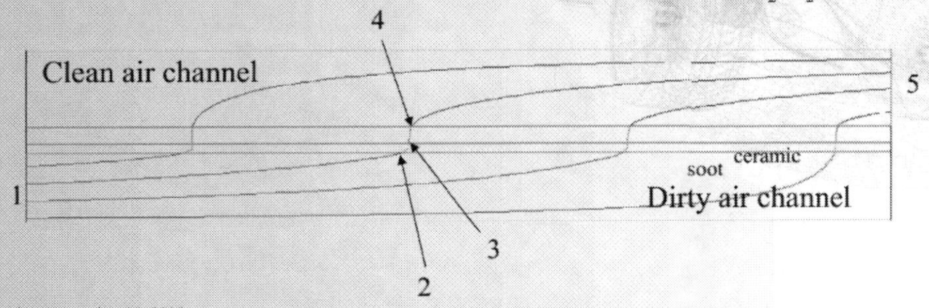

Agenda

- Introduction
- Theory
- Grid generation
- Results
  - Eberspächer test case
  - SAE paper by Turner et al.
- Conclusions

Grid generation

- Do we have to generate grids taking into account the glue layers?

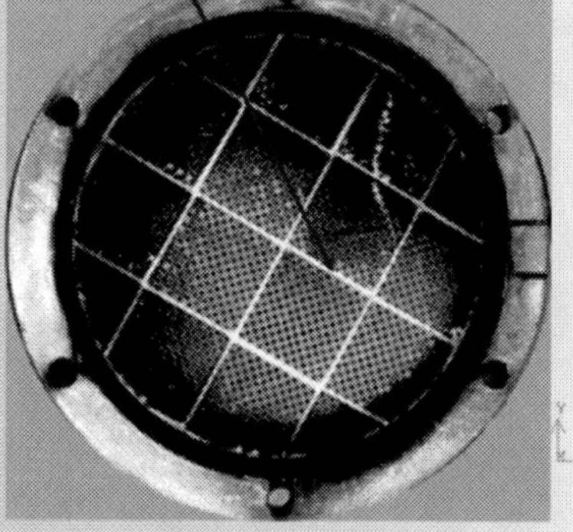

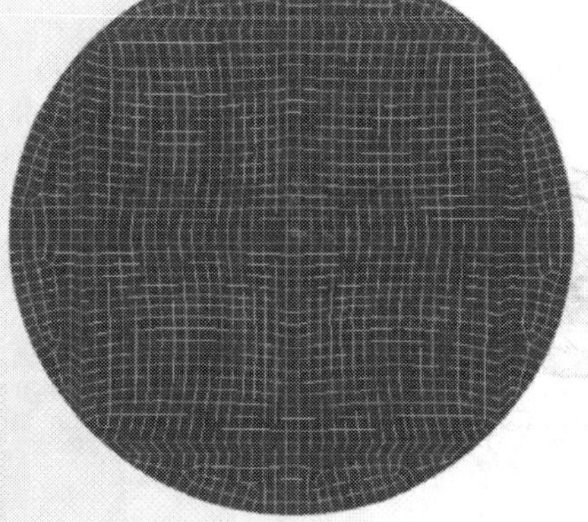

Yes, if we want to see the effects of the glue layer.

Grid generation

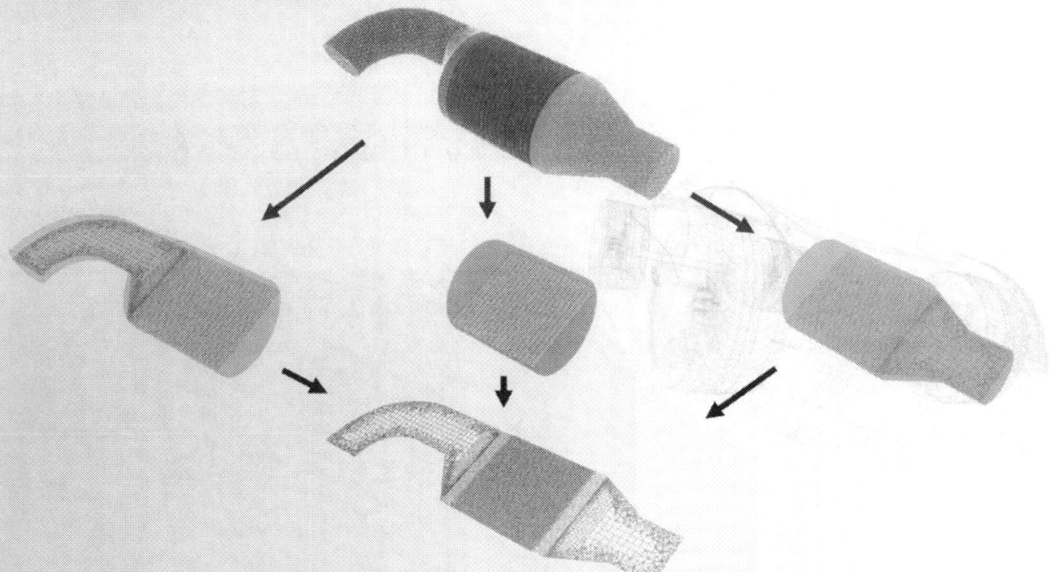

Agenda

- Introduction
- Theory
- Grid generation
- Results
  - Eberspächer test case
  - SAE paper by Turner et al.
- Conclusions

Regeneration test case of Eberspächer

- Realistic geometry (5.66"x7", 300/12)
- Measurements include
  - Film of unsteady temperature distribution at the back of the filter (infrared camera)
  - Temperature measurement at different points in the filter
  - CFD mesh was provided
  - Loading at 303kg/h, 270°C
  - Regeneration at 909kg/h, 630°C

Early soot loading

0.13/0.0925=1.41

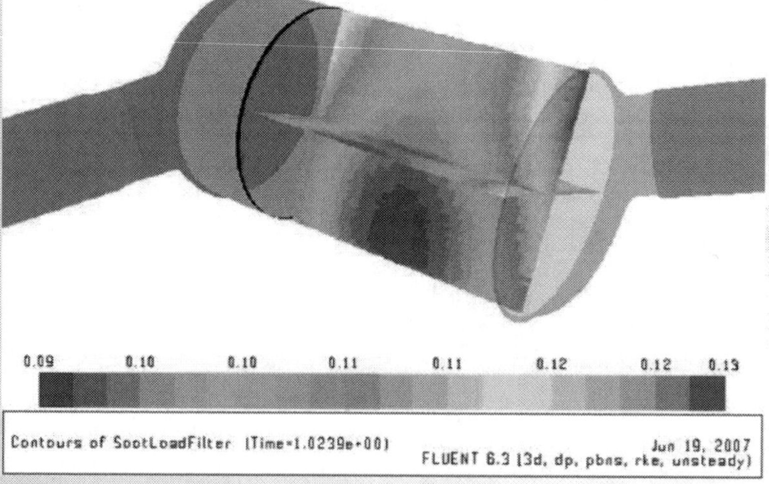

2013 Automotive Simulation World Congress
October 29 - 30 • Frankfurt am Main, Germany

Late soot loading

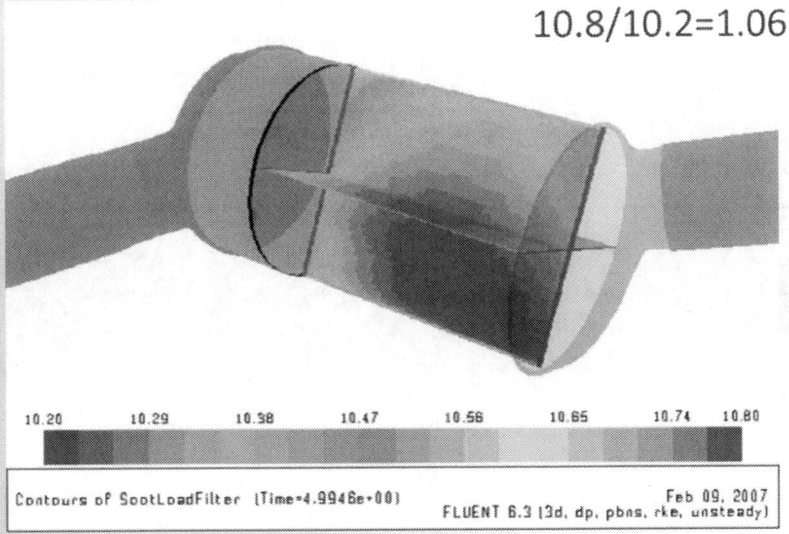

2013 Automotive Simulation World Congress
October 29 - 30 • Frankfurt am Main, Germany

Pressure loss (loading and regeneration)

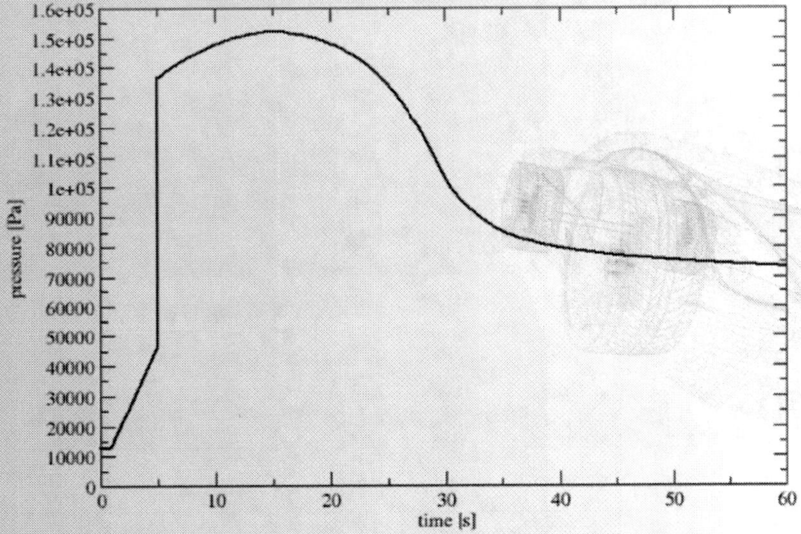

Temperature history in the middle of the filter

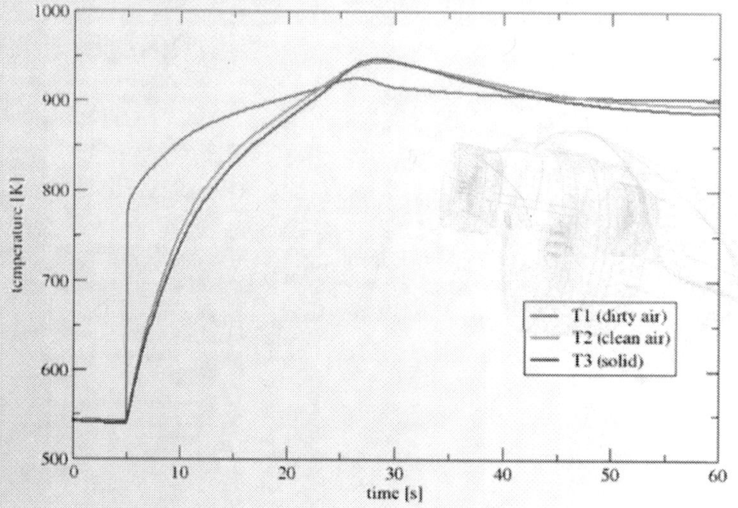

Solid temperature after 10s

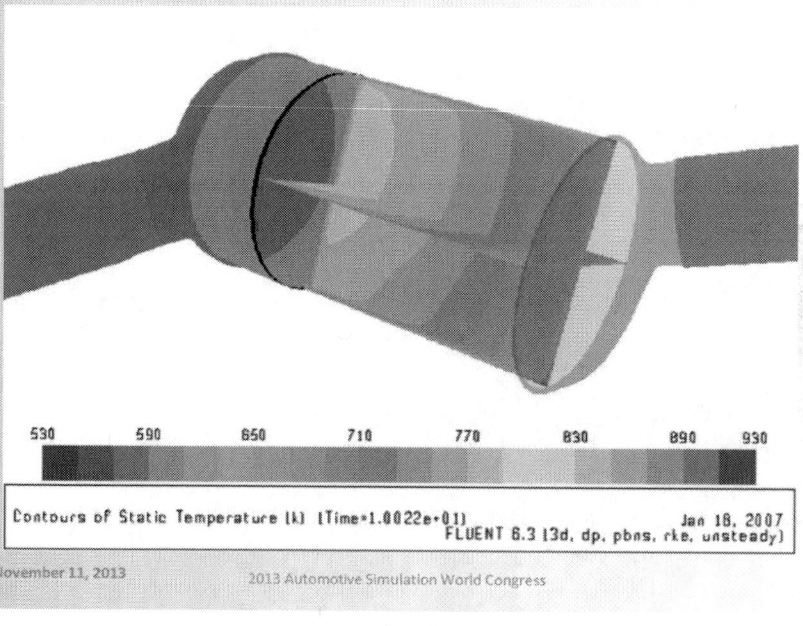

2013 Automotive Simulation World Congress
October 29 - 30 • Frankfurt am Main, Germany

Solid temperature after 30s

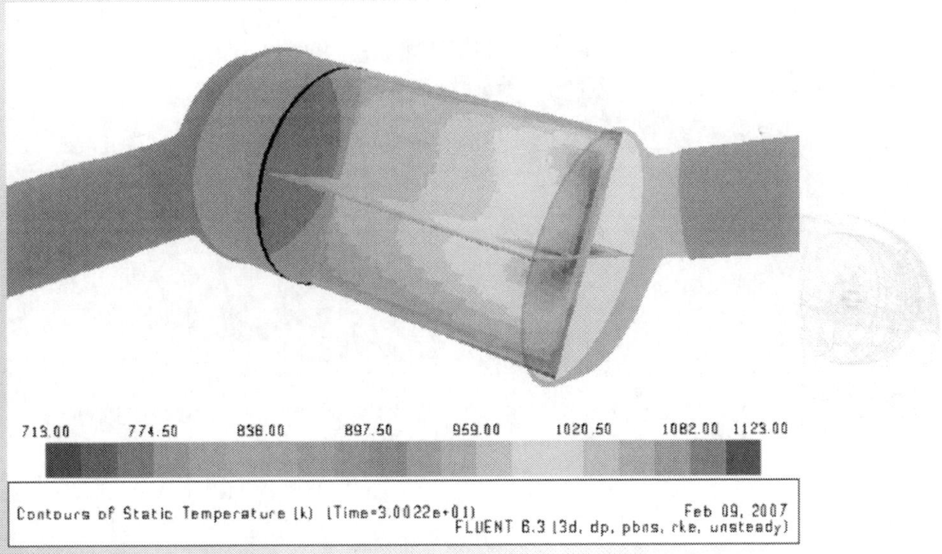

2013 Automotive Simulation World Congress
October 29 - 30 • Frankfurt am Main, Germany

Reaction rate after 30s

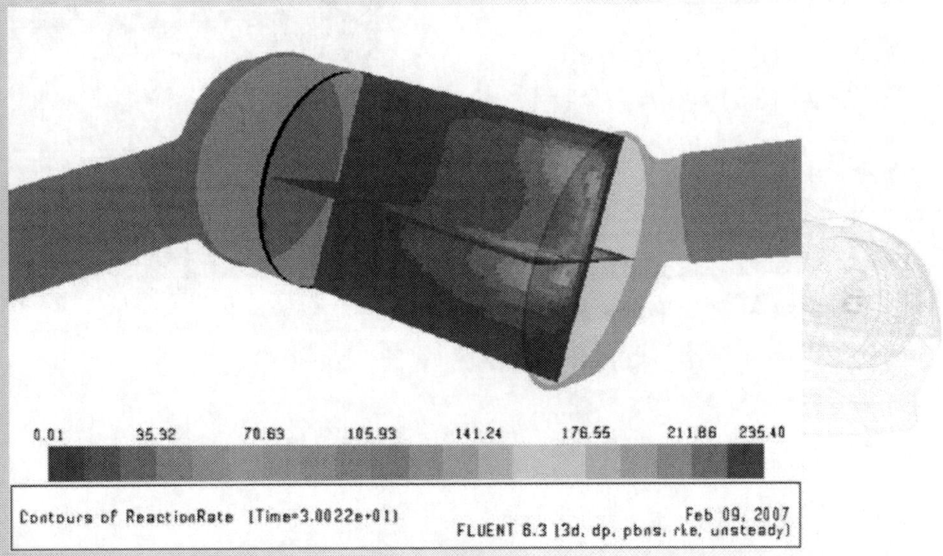

2013 Automotive Simulation World Congress
October 29 - 30 • Frankfurt am Main, Germany

Soot loading after 30s

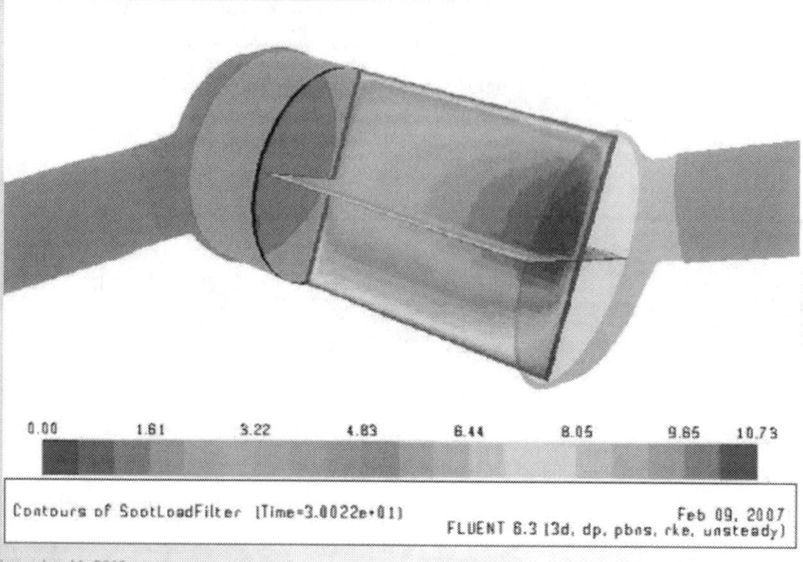

2013 Automotive Simulation World Congress
October 29 - 30 • Frankfurt am Main, Germany

SAE paper by Turner et al.

- SAE 2011-01-1241
- Turner, Thornhill, McCullough, Patel
- Comparison Fluent and PIV measurements
  - Good agreement
  - Showing here results for medium mass flow rate

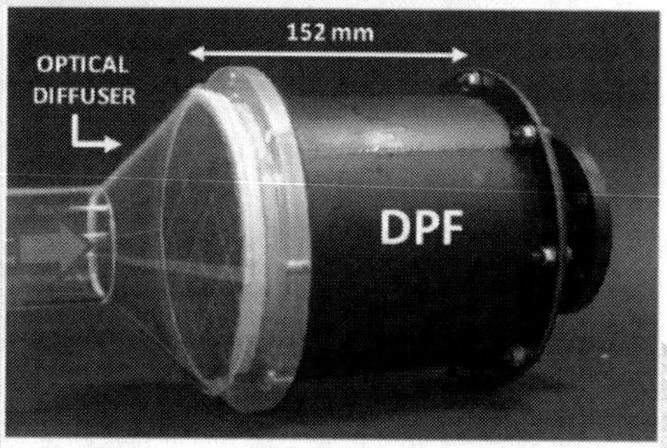

Figure 2. DPF with Optical Cone in Position

SAE paper by Turner et al.

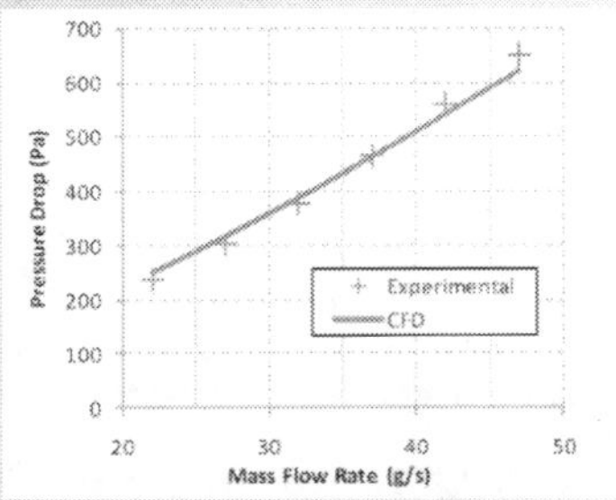

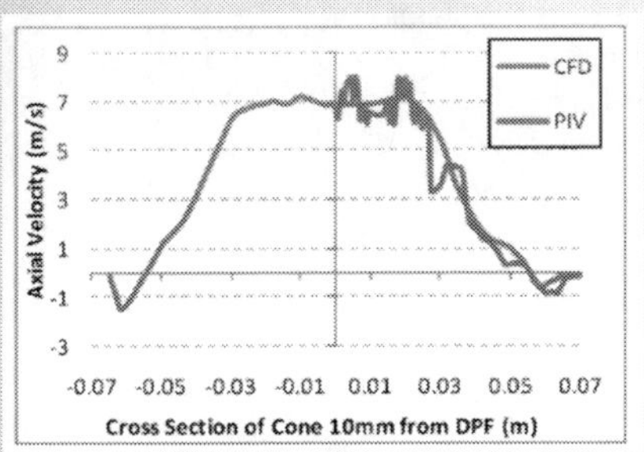

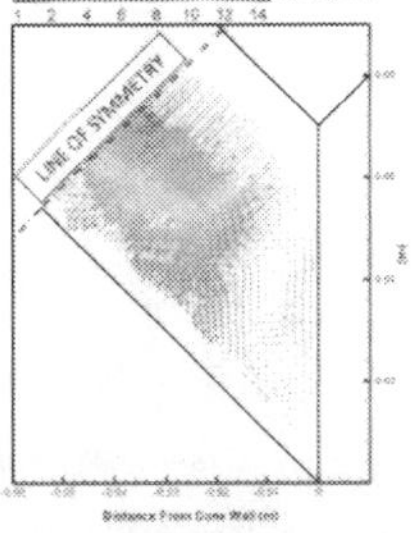

Figure 14. PIV Vector Plot of Cone, MFR 0.032kg/s

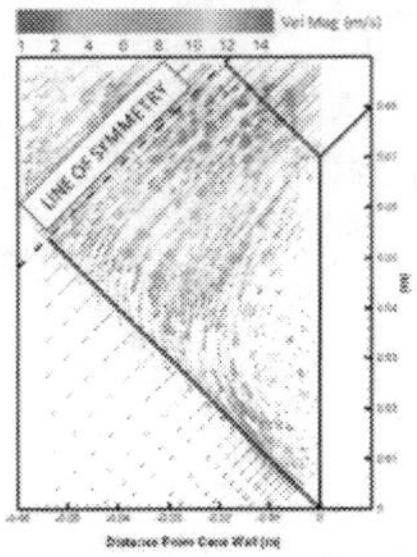

Figure 15. CFD Vector Plot of Cone, MFR 0.032kg/s

2013 Automotive Simulation World Congress
October 29 - 30 • Frankfurt am Main, Germany

Agenda

- Introduction
- Theory
- Grid generation
- Results
  - Eberspächer test case
  - SAE paper by Turner et al.
- Conclusions

Conclusions

- Regeneration is programmed with know-how from Eberspächer
- First results of regeneration are plausible
- Comparison with experimental data needs to be done
- A version without Eberspächer know-how is available

CFD SCHUCK
Ingenieurgesellschaft mbH

Efficient product development using ANSYS software

Automotive Simulation World Congress 2013
Dipl.-Ing. Ludwig Berger
Email: lb@cfd-schuck.de
www.cfd-schuck.de

CFD SCHUCK
Ingenieurgesellschaft mbH

Outline

- CFD-Schuck: Company profile
- ANSYS Software at CFD Schuck
- Overview of activities in the fields of automotive
- Investigation of an intake port with CFX-Ice

CFD SCHUCK
Ingenieurgesellschaft mbH

Company profile

- Provider of CFD services
  - 35 employees distributed across 2 locations
- more than 20 years of experience
  - Foundation in Heidenheim in 1990
  - Opening of a branch in Munich in 1999
- all key commercial CFD software packages
- Company philosophy: Using CFD as development tool

CFD SCHUCK
Ingenieurgesellschaft mbH

Company profile

ANSYS Software at CFD Schuck

- CAD data preparation and mesh generation
  - IcemCFD: Tetra, Hexa
- CFD-Solver
  - Fluent
  - CFX
- Workflow
  - Workbench

CFD SCHUCK
Ingenieurgesellschaft mbH

Automotive activities

- Exhaust technology, turbocharger, compressor
- Air intake, charge air cooler, manifolds, water jackets
- Intake ports, in-cylinder flows
- Refuelling (liquid fuels, hydrogen)

- Complete vehicle, engine compartment, under hood flows
- Brakes (heat management)

- HVAC
- Lights

CFD SCHUCK
Ingenieurgesellschaft mbH

Intake Port investigation

- Design and optimisation of engine intake ports
  => Flow coefficient, swirl number
- Methodologies
  - Steady-state flow rigs
  - Numerical simulation
    - „step by step procedure" for different valve lifts
      - Mesh generation
      - CFD simulation and post processing
    - Software controlled procedure based on CFX-ICE
      - Mesh generation needed only for once
      - Transient simulation of all valve lifts

CFD SCHUCK
Ingenieurgesellschaft mbH

Intake Port investigation

CFX-Ice Setup via GUI

- Motion Setup
  - Crank radius, rod length
  - Geometry parameters,
  - Valve lift curve
  - Valve angle

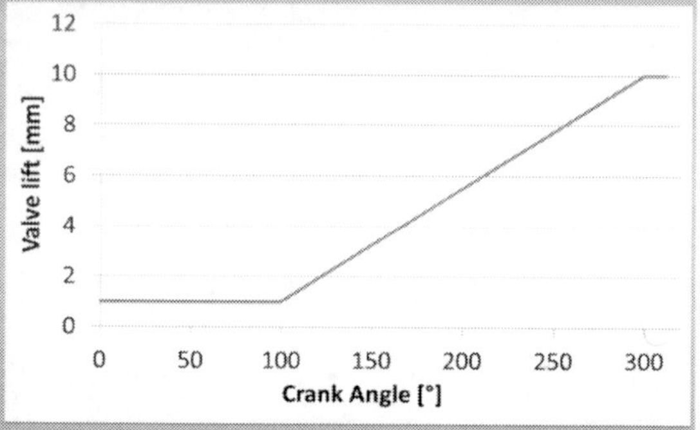

- Meshing Setup
  - 7 Layer
  - static / dynamic mesh parts
  - Mesh quality criteria

2013 ASWC: Efficient product development using ANSYS Software 29/10/2013 7

CFD SCHUCK
Ingenieurgesellschaft mbH

Intake Port investigation

CFX-Ice Setup via GUI

- Flow Setup
  - Transient, isotherm
  - Turbulence: k-ω-sst
  - CA min, CA max, CA steps,
  - Boundary conditions
    - Inlet: Total pressure 1 bar
    - Outlet: Opening p_{stat}=-50 mbar
- Simulation setup
  - Mesh run, simulation, restart

Firmenion Januar 2013 8

CFD SCHUCK
Ingenieurgesellschaft mbH

Intake Port investigation

Results: Isentropic Area

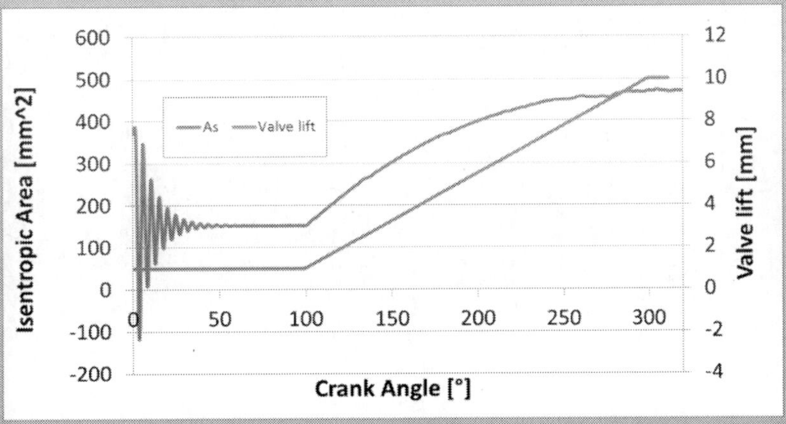

CFD SCHUCK
Ingenieurgesellschaft mbH

Intake Port investigation

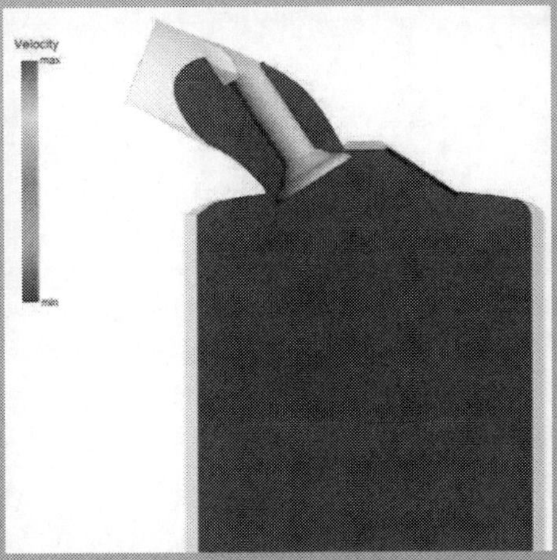

CFD SCHUCK
Ingenieurgesellschaft mbH

Intake Port investigation

- Benefits
  - Application of existing Setups of In-cylinder flows
  - Flexibility
    - Continuously or stepwise representation of valve lift
  - Minimising costs for design optimisation
    - Small efforts for design studies

CFD SCHUCK
Ingenieurgesellschaft mbH

Conclusion

- CFD Schuck:
  - Service provider for automotive product development
  - Need for reliable CFD-software
  - Workflow-optimised use of ANSYS products shown by typical applications

⇨ ANSYS products = important part of CFD-software used at CFD Schuck

2013 Automotive Simulation World Congress

Validation and Verification of ANSYS Internal Combustion Engine Software

Martin Kuntz, ANSYS Germany
Nagendra Kumar, ANSYS India

Introduction

- Validation and verification for Internal Combustion Engines
  - Procedure
  - Applications
  - Simulation environment
  - Examples

2013 Automotive Simulation World Congress
October 29 - 30 • Frankfurt am Main, Germany

Procedure

- Demonstration
  - Illustrate application of software to generic case
  - Realistic geometry, real physics, no validation data
- Regression
  - Prove functionality of software
  - Complete workflow, individual features
- Verification
  - Verify, that model is implemented correctly
  - Simplified geometry, simple physics
- Validation
  - Demonstrate simulation accuracy
  - Realistic geometry, real physics, with experimental data

2013 Automotive Simulation World Congress
October 29 - 30 • Frankfurt am Main, Germany

IC Engine Applications

- Component simulations
  - Intake port, intake manifold, water jackets, fuel injectors
  - Spray bomb
- IC engine simulations
  - Cold flow
    - Charge motion
  - Combustion
    - Thermal management
    - Emissions

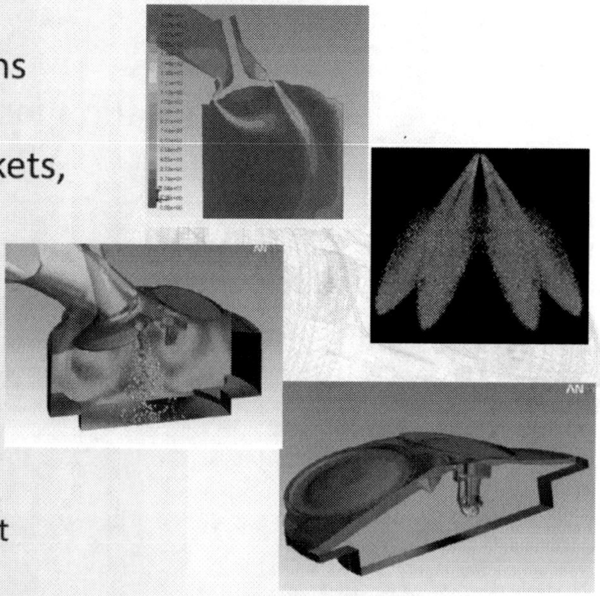

4

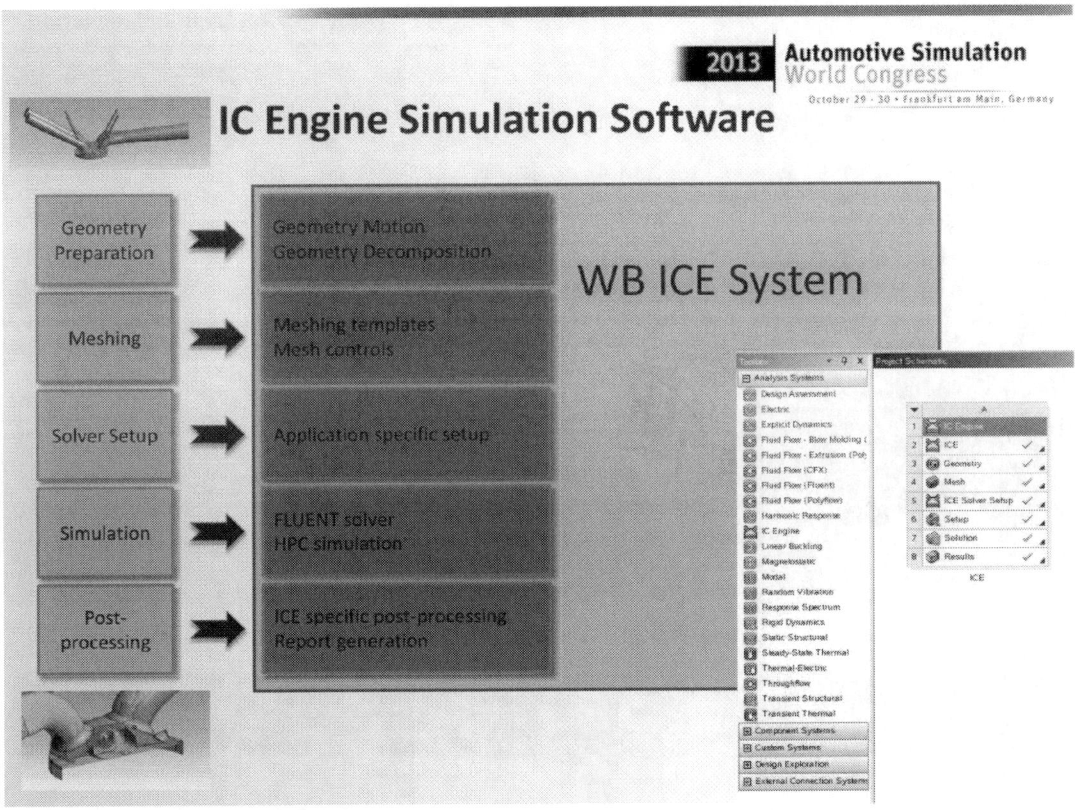
2013
Automotive Simulation
World Congress
October 29 - 30 • Frankfurt am Main, Germany
IC Engine Simulation Software
Geometry Preparation
Meshing
Solver Setup
Simulation
Post-processing
Geometry Motion
Geometry Decomposition
Meshing templates
Mesh controls
Application specific setup
FLUENT solver
HPC simulation
ICE specific post-processing
Report generation
WB ICE System

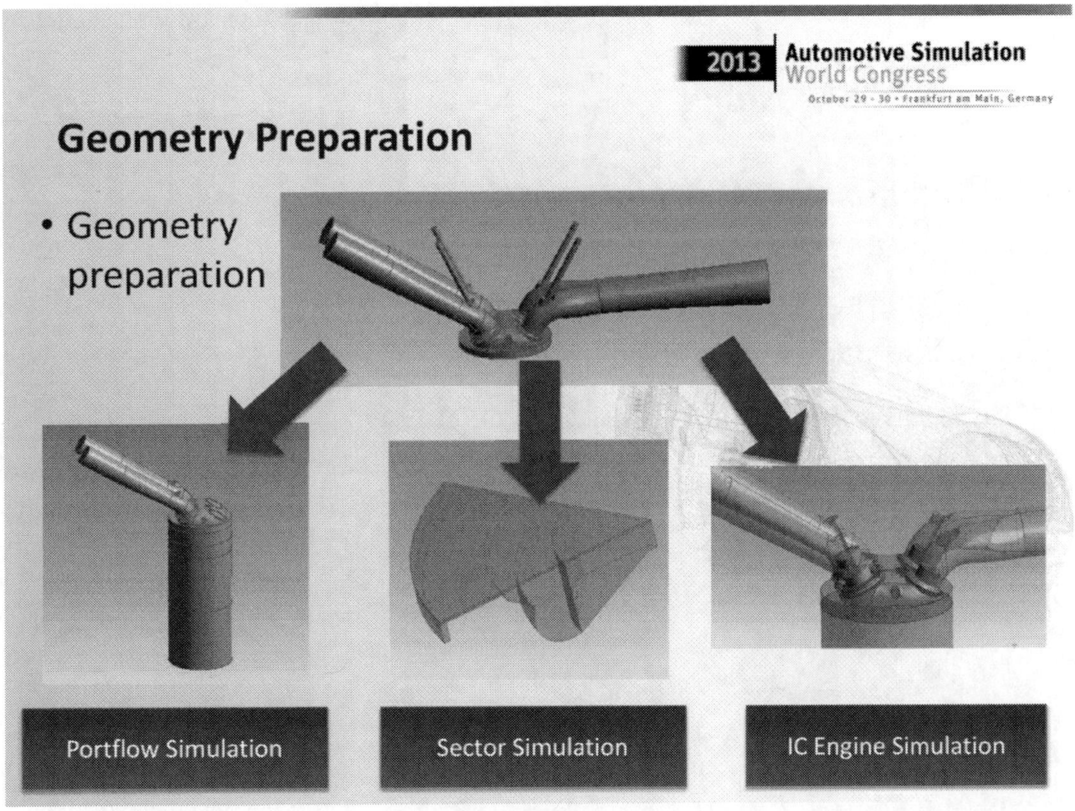
2013
Automotive Simulation
World Congress
October 29 - 30 • Frankfurt am Main, Germany
Geometry Preparation
• Geometry preparation
Portflow Simulation
Sector Simulation
IC Engine Simulation

2013 Automotive Simulation World Congress
October 29 - 30 • Frankfurt am Main, Germany

Meshing

- Initial meshing
  - Predefined meshing templates
  - Application specific mesh topology, mesh controls

Portflow Simulation

Sector Simulation

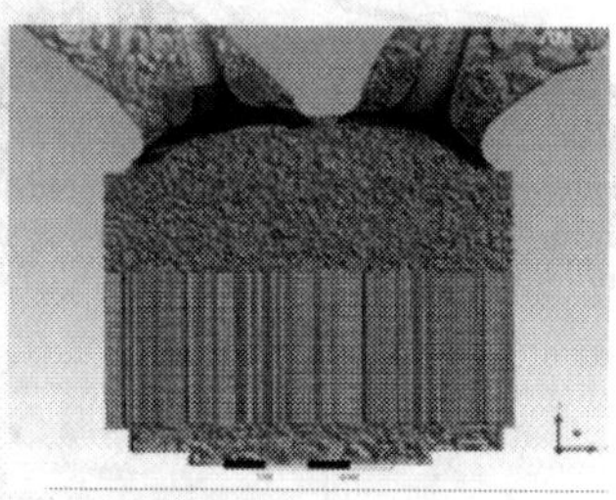

IC Engine Simulation

2013 Automotive Simulation World Congress
October 29 - 30 • Frankfurt am Main, Germany

Key Grid Meshing

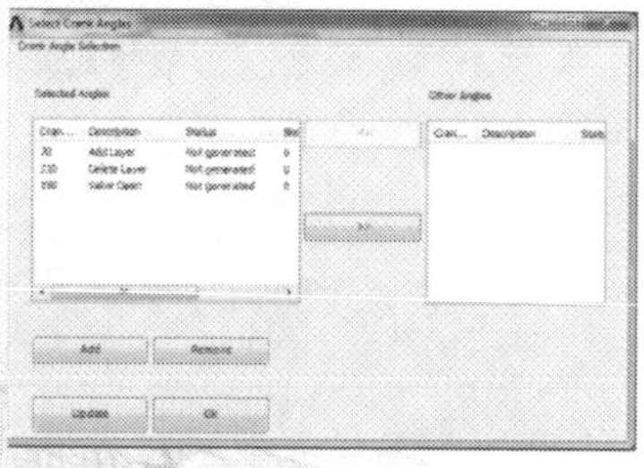

- Additional meshes at events during the run
  - Increased mesh control and quality during the simulation
  - Physics specific adaption of mesh

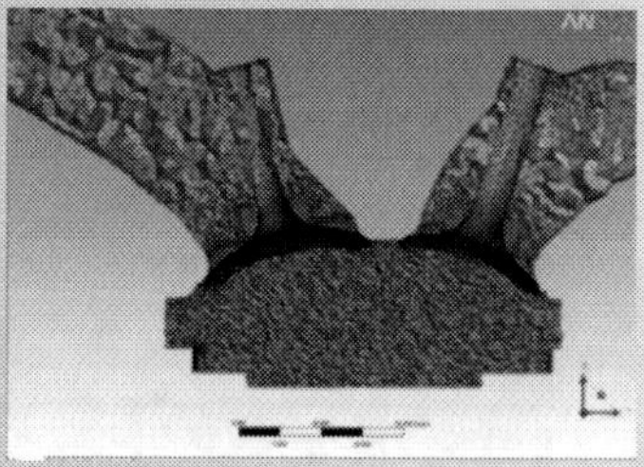

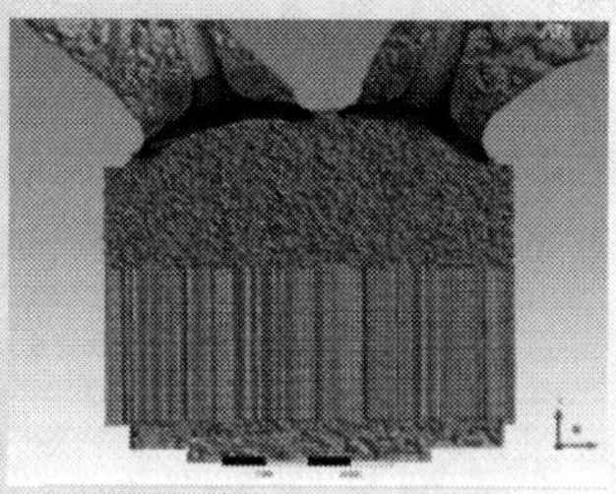

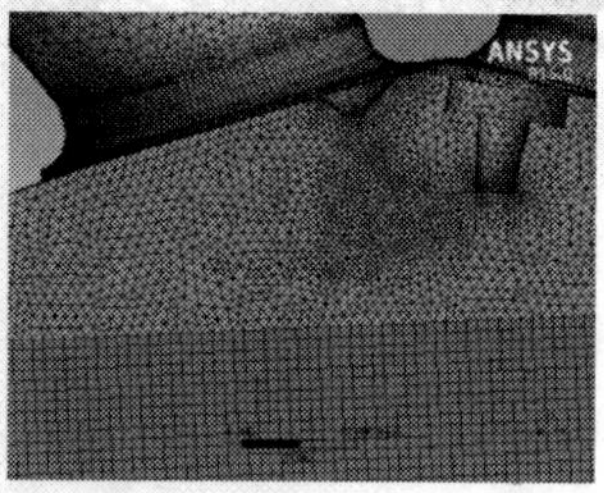

Solver Setup & Simulation

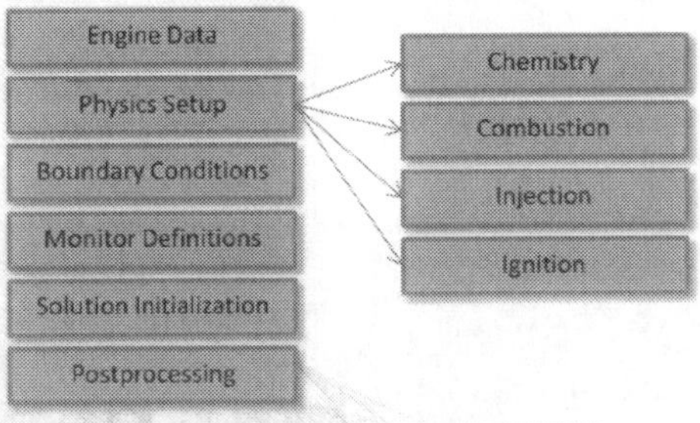

- Single User Interface with customized panels to define complete IC engine setup
- Default parameter settings for different applications based on best practise
- Simulation launch

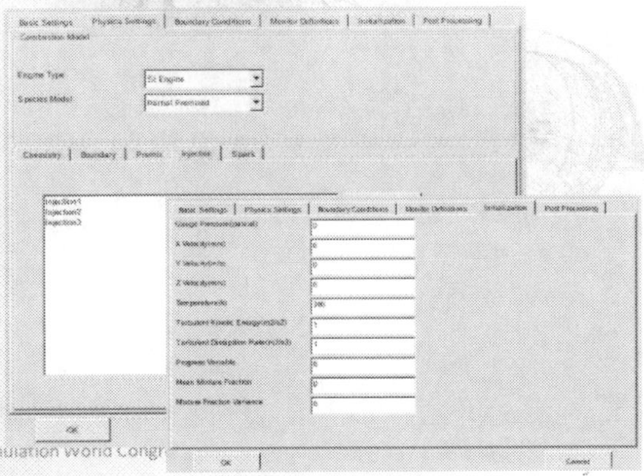

2013 Automotive Simulation World Congress
October 29 - 30 • Frankfurt am Main, Germany

Post-processing

- Automatic report generation
- Manual post-processing

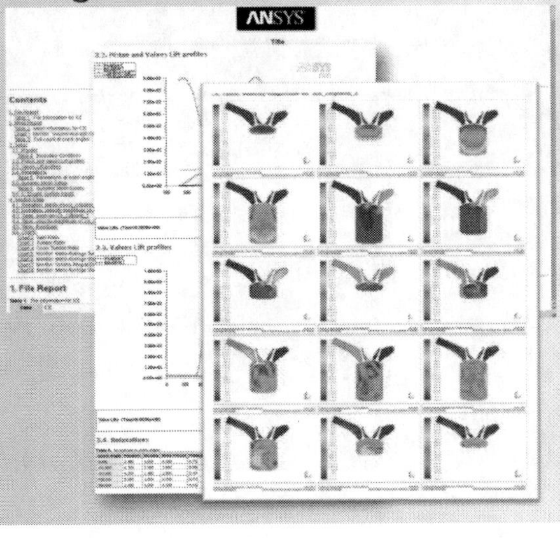

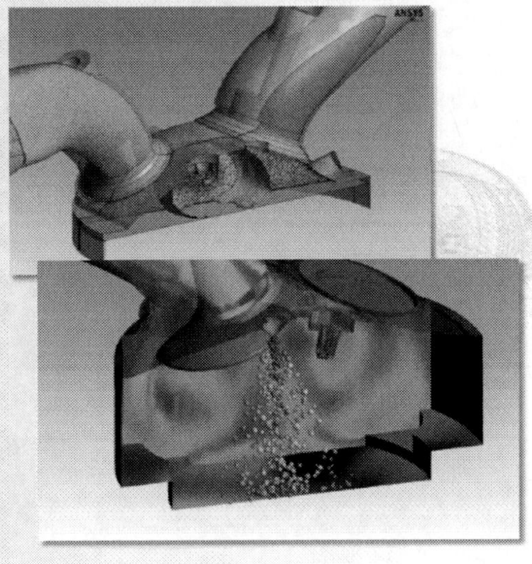

Introduction

- Validation and verification for Internal Combustion Engines
  - Procedure
  - Applications
  - Simulation environment
  - Examples

2013 Automotive Simulation World Congress
October 29 - 30 • Frankfurt am Main, Germany

Procedure

- Demonstration
  - Illustrate application of software to generic case
  - Realistic geometry, real physics, no validation data
- Regression
  - Prove functionality of software
  - Complete workflow, individual features
- Verification
  - Verify, that model is implemented correctly
  - Simplified geometry, simple physics
- Validation
  - Demonstrate simulation accuracy
  - Realistic geometry, real physics, with experimental data

Demonstration Examples

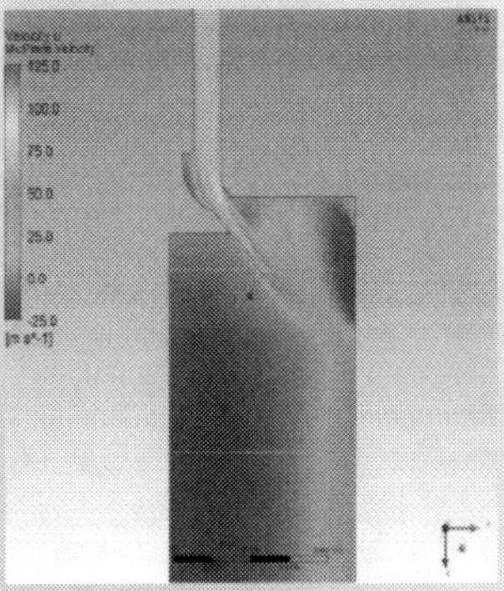

Portflow Simulation

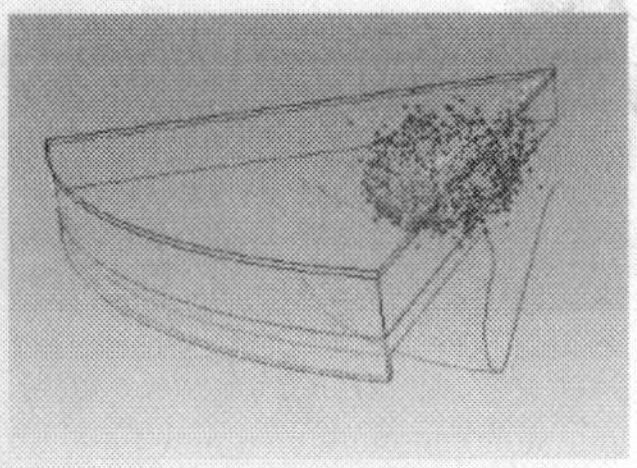

Sector Simulation

IC Engine Simulation

2013 Automotive Simulation World Congress
October 29 - 30 • Frankfurt am Main, Germany

Demonstration: Port Flow Simulation

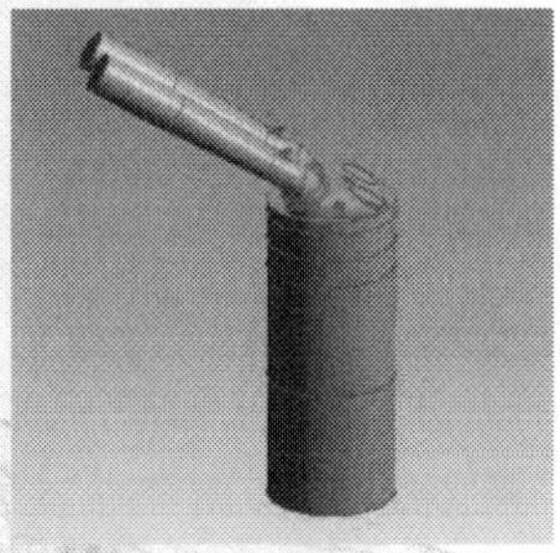

- Design point study
- Workbench parameter manager
  - Input: valve lift, Output: mass flow

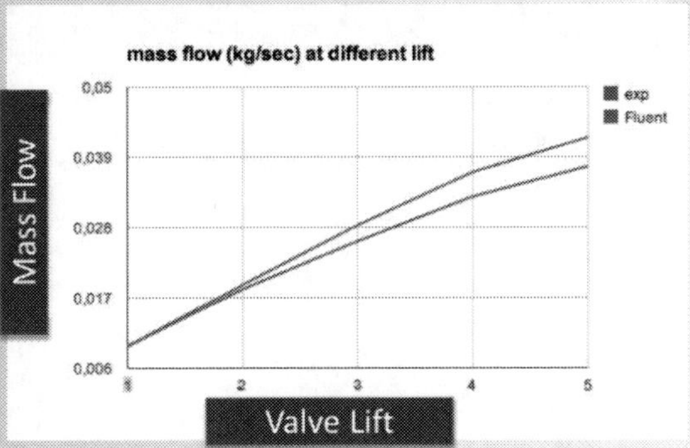

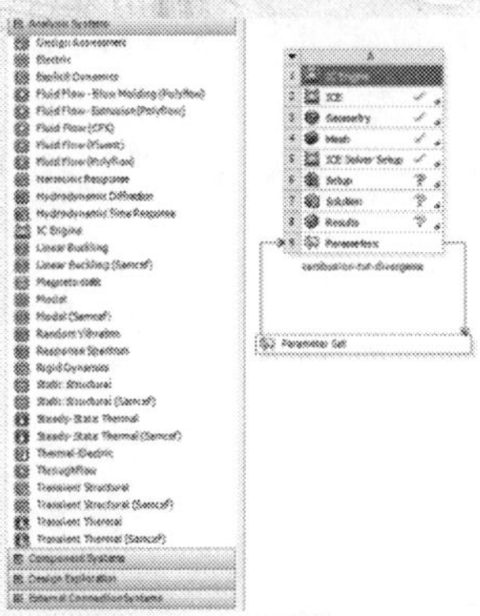

2013 Automotive Simulation World Congress

Demonstration: Combustion on Sector Geometry

- Combustion setup from geometry to report for diesel engine of a sector geometry

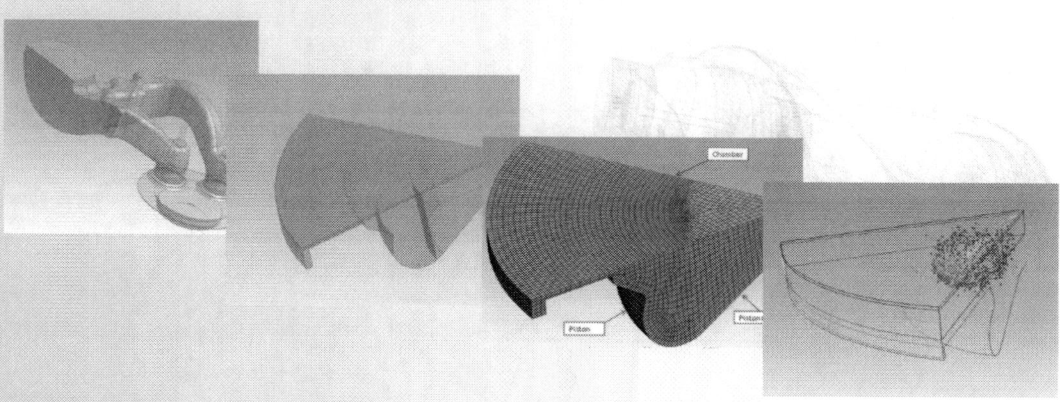

2013 Automotive Simulation World Congress
October 29 - 30 • Frankfurt am Main, Germany

Demonstration: Combustion on Sector Geometry

- Report

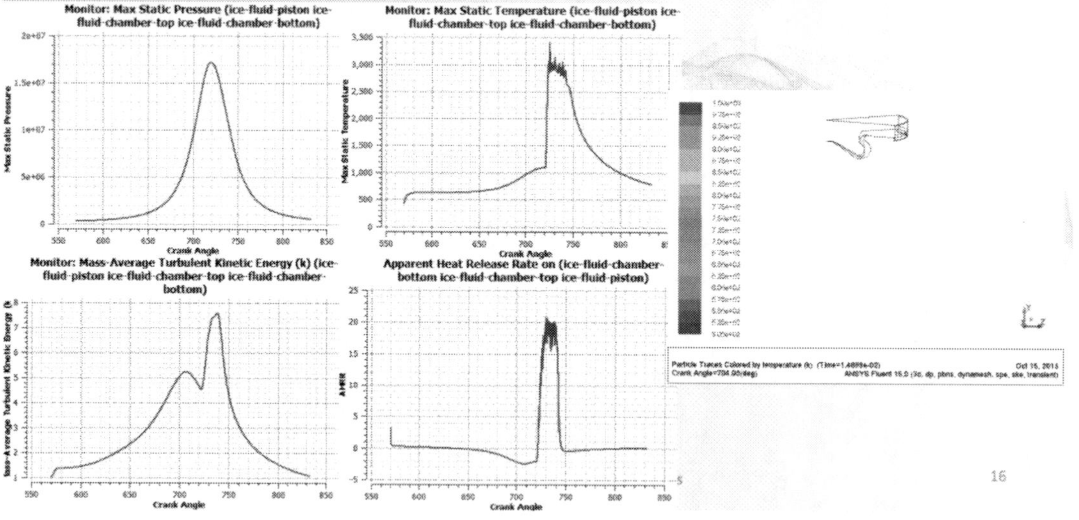

Demonstration : Combustion Flow

- Direction-injection spark-ignition combustion setup
  - Complete cycle setup
    - Material Iso-octane
    - Spray injection
    - Liquid evaporation model
    - Spark ignition
    - G equation combustion

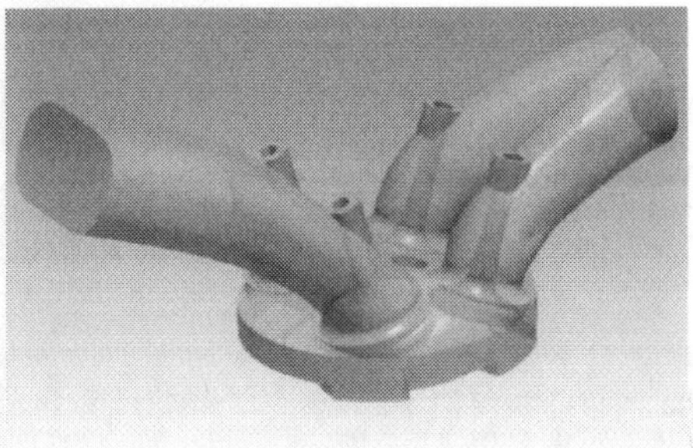

Courtesy: BMW

Demonstration : Combustion Flow

- Direction-injection spark-ignition combustion setup

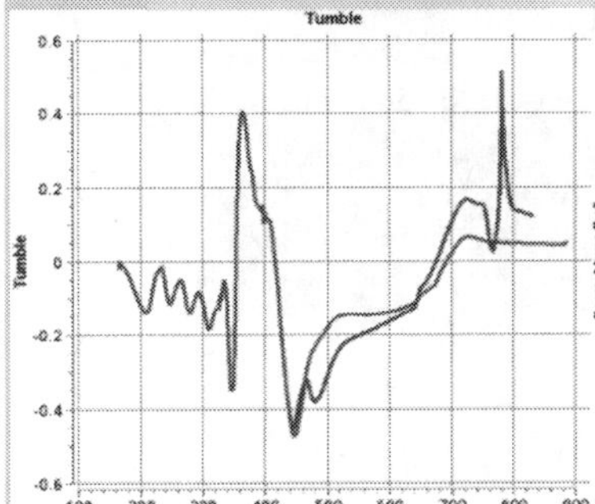

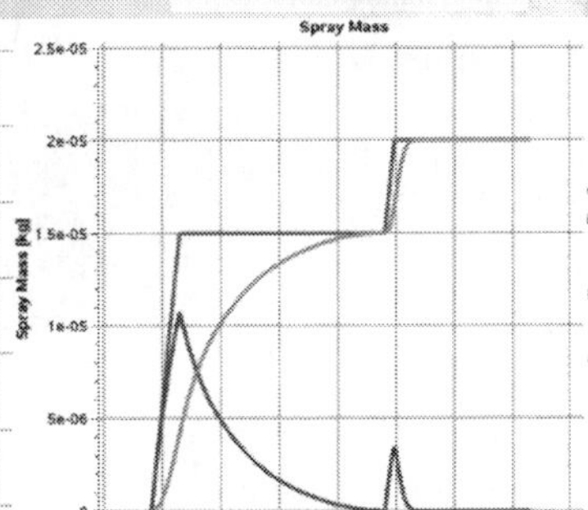

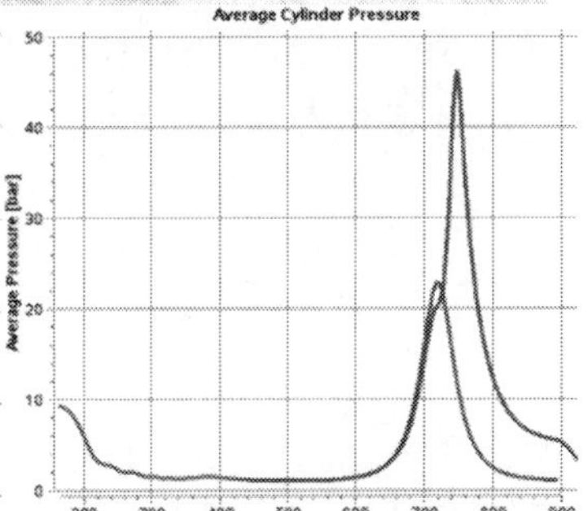

Regression Testing

- Testing and maintaining stability of software
  - 15 different engines cases
- Automatic test decks
  - Daily testing of pre-processing → decomposition, meshing
  - Weekly testing of solver → mesh motion, cold flow
- Manual compatibility study for different engine types to test complete workflow
- Tutorials
- Dedicated team for testing and validation

2013 Automotive Simulation World Congress
October 29 - 30 • Frankfurt am Main, Germany

Validation & Verification

- Applications
  - Cases with data from public data bases
  - Benchmark cases / customer cases
- Process
  - Reduced complexity, validate individual models
  - Validate real engine application
- Requirements
  - Valuable, accurate data
  - No confidentiality, no restrictions to publish

- Testcases
  - Isolated physics
    - Spraybox
    - Premixed combustion
  - Portflow
    - Thobois
  - IC Engines
    - Direct-injection spark-ignition engine

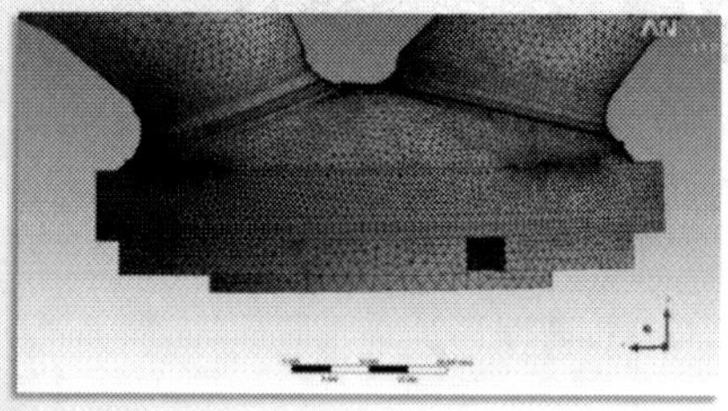

Validation: Hiroyasu Spray Box

- Spray box test
- Non-evaporating spray

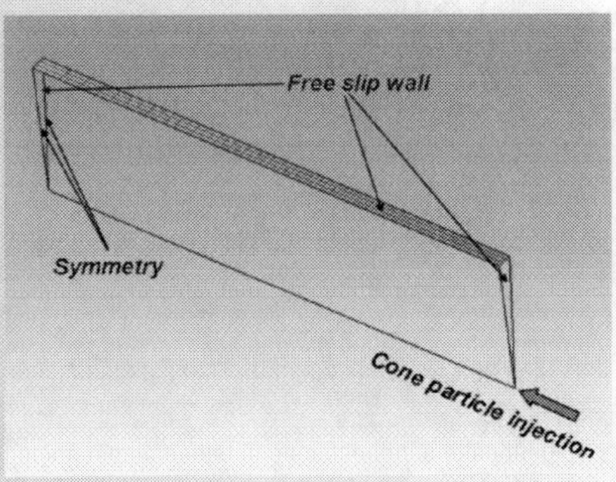

	Case 1	Case 2	Case 3
Gas parameters			
Gas type	N_2		
Temperature [K]	300	300	300
Pressure [MPa]	1.1	3.0	5.0
Fuel Properties			
Fuel type	$C_{12}H_{26}$		
Density [kg/m³]	840		
Surface tension [kg/s²]	0.0205		
Spray parameters			
Initial Temperature [K]	300	300	300
Injection Velocity [m/s]	102	90	86
Particle Mass Flow Rate [g/s]	6.05	5.36	5.13
Nozzle diameter [mm]	0.3	0.3	0.3
Injection rate, single pulse, ms	2.5	4	4

Validation: Hiroyasu Spray Box

- Mass penetration vs. time
- Case 1, 2 and 3
- SSD break-up model

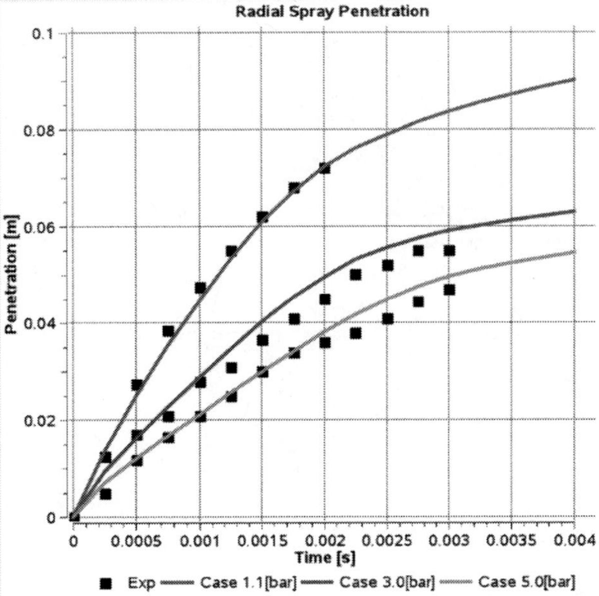

2013 Automotive Simulation World Congress
October 29 - 30 • Frankfurt am Main, Germany

Validation: Hamamoto Testcase

- Spark-ignition + premixed combustion in a closed vessel with fixed walls.
- Propane/air mixture:
  - Equivalence ratio ϕ=1.0
- Measured data
  - Optical access
  - Pressure transducer
  - Described in
    - Hamamoto et al. (1988)
    - Ewald (2005)

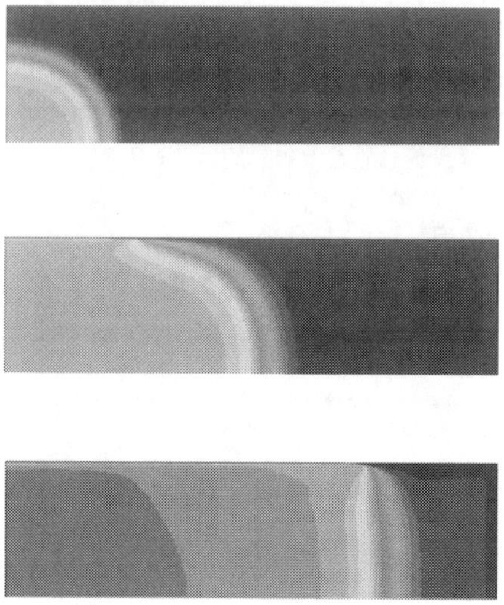

2013 Automotive Simulation World Congress
October 29 - 30 • Frankfurt am Main, Germany

Validation: Hamamoto Testcase

- Meshes
  - 2D meshes
  - 700, 2800, 11000, 45000 cells
- Models
  - SST
- Combustion model
  - G-Equation
  - Flame speed: Peters

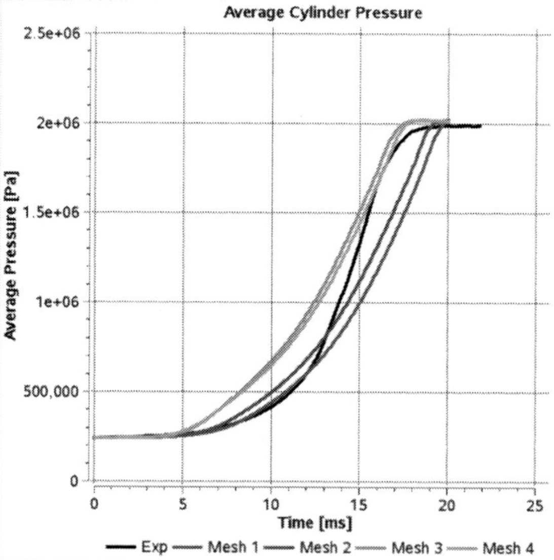

Validation: Thobois Port Flow

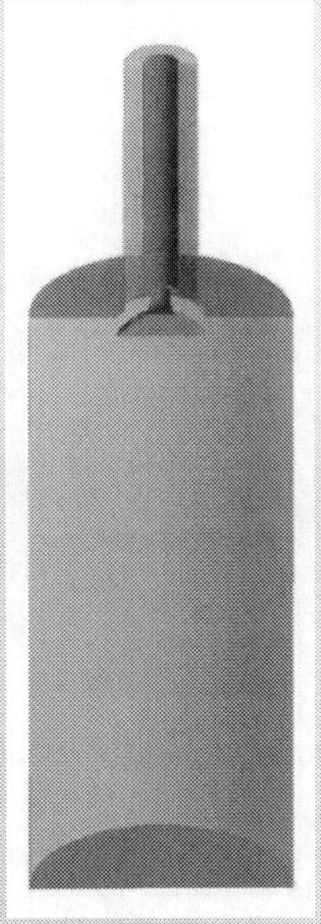

- Steady intake flow conditions
- Accuracy analysis
  - Influence of mesh size, mesh type
  - Influence of turbulence model
- Testcase defined in
  - Large Eddy Simulations in IC Engine Geometries
    - Thobois, Rymer, Souleres, Poinsot
    - SAE Paper, 2004-01-1854

Validation: Thobois Port Flow

- Meshes
  - 2D Hex mesh: 7.000 → 430.000 cells
  - 3D Hybrid mesh: 0.77M → 5.2M cells
  - 3D Cutcell mesh: 1.1M cells

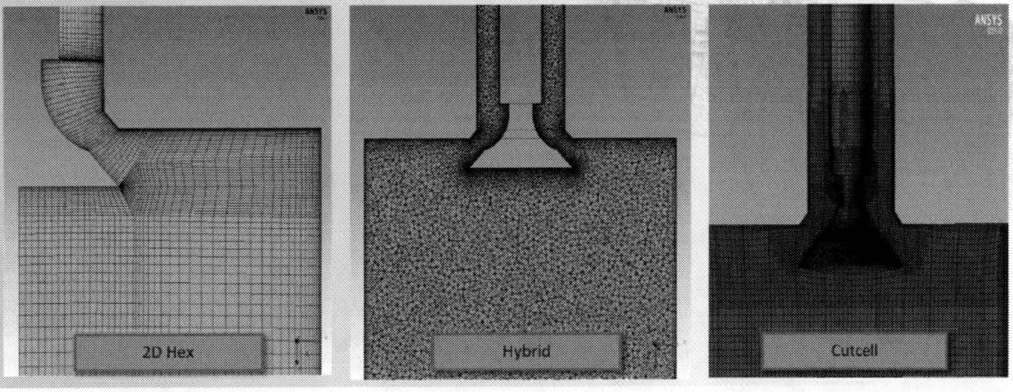

2013 Automotive Simulation World Congress
October 29 - 30 • Frankfurt am Main, Germany

Validation: Thobois Port Flow

- Inflow: 65 m/s, RANS turbulence model
- Steady state geometry, transient flow effects
- Transient simulation with solution averaging

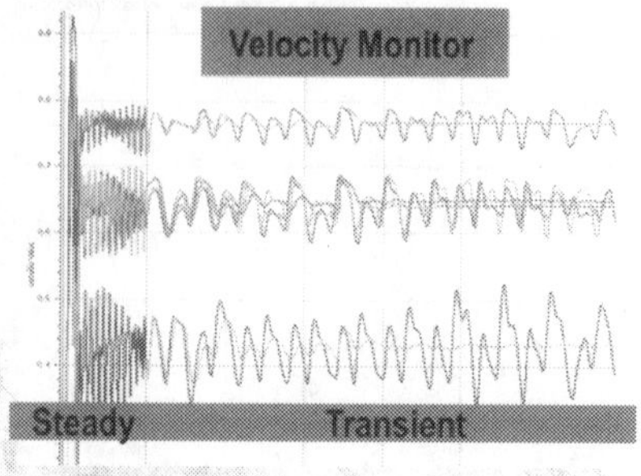

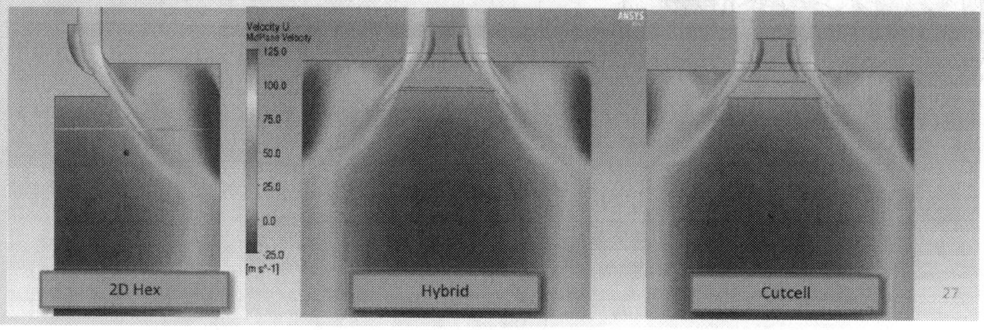

27

2013 Automotive Simulation World Congress
October 29 - 30 • Frankfurt am Main, Germany

Validation: Thobois Port Flow

- Mesh Study on 2D Hex Meshes

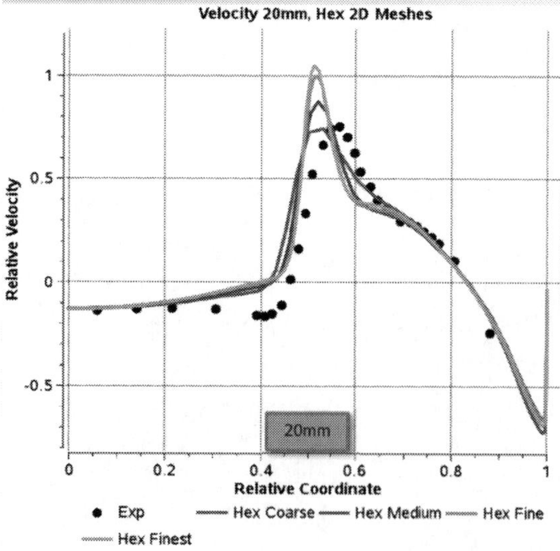

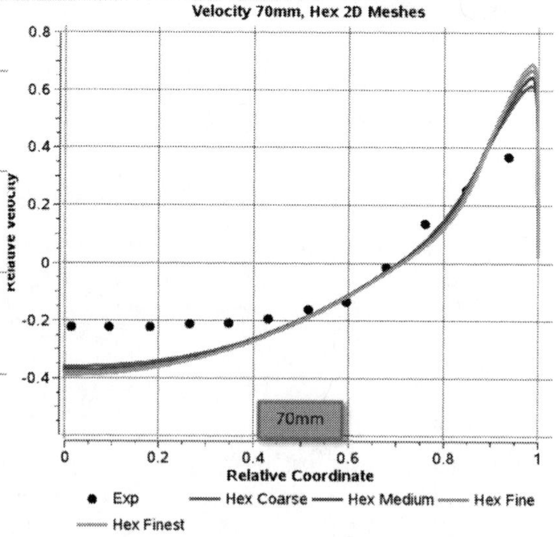

Validation: Thobois Port Flow

- Mesh Study on 3D meshes (hybrid / cutcell)

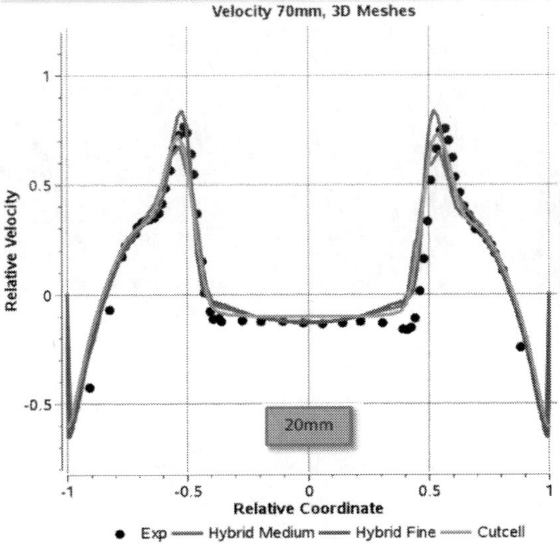

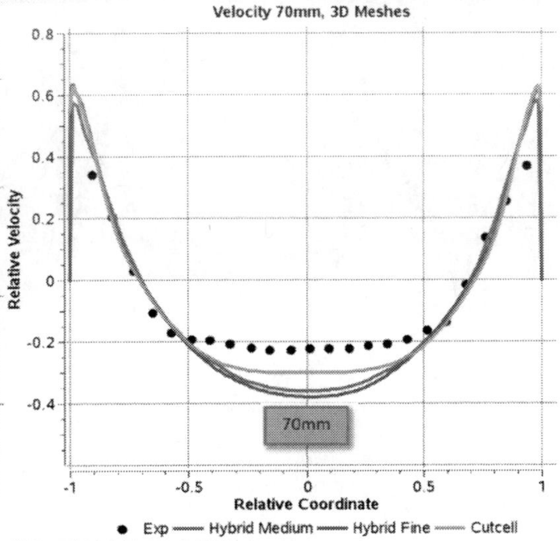

2013 Automotive Simulation World Congress
October 29 - 30 • Frankfurt am Main, Germany

Validation: Thobois Port Flow

- Turbulence model variation

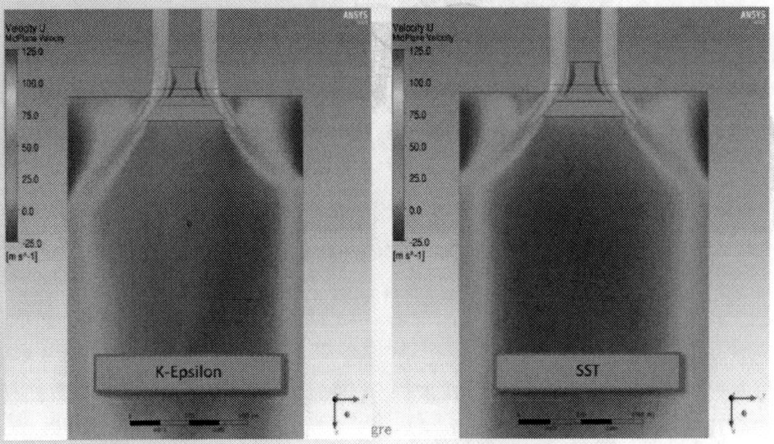

2013 **Automotive Simulation** World Congress
October 29 - 30 • Frankfurt am Main, Germany

Validation: Thobois Port Flow

- Turbulence Study

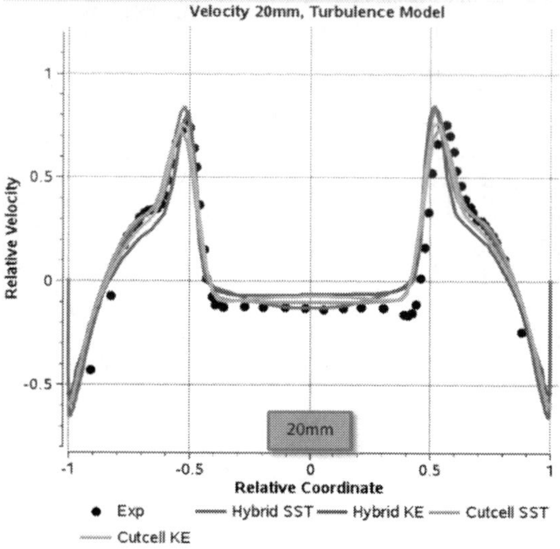

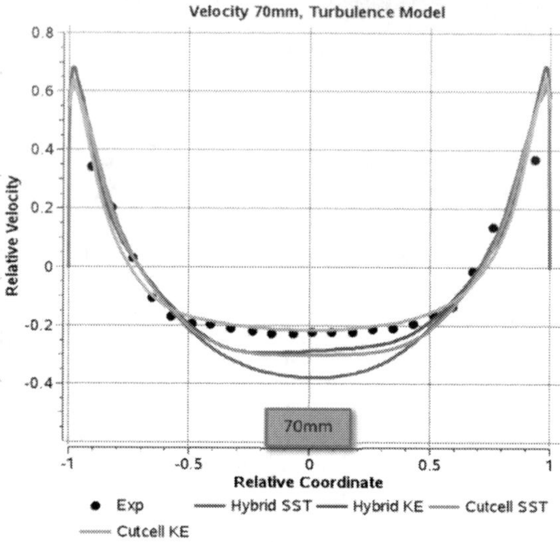

2013 **Automotive Simulation** World Congress
October 29 - 30 • Frankfurt am Main, Germany

Wisconsin Engine

- Single-cylinder direct-injection spark-ignition engine
- Research project conducted at University of Wisconsin
- Measurement campaigns
  - High speed in-cylinder measurements
    - Motored engine conditions
    - Homogeneous conditions
    - Direct-Injection conditions
  - Injector characterization

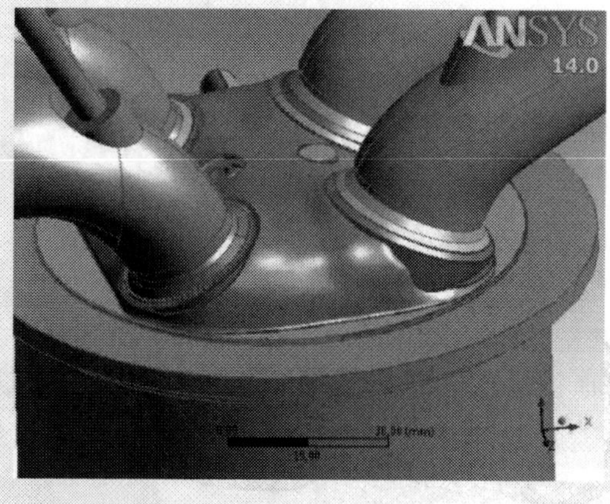

Engine Research Center
University of Wisconsin-Madison

THE UNIVERSITY WISCONSIN MADISON

2013 Automotive Simulation World Congress
October 29 - 30 • Frankfurt am Main, Germany

Validation: Wisconsin Engine

Engine Specifications	
Engine Type	4-Stroke, 4-Valve, SI
Chamber Geometry	Pentroof
Fueling type	Spray-guided direct-injection
Displacement	696 cm^3
Compression Ratio	12:1
Injection Pressure	11 MPa (typical)

- Measurement data
  - High-speed in-cylinder measurements
    - Pressure, temperature
    - Cylinder head &liner temperature
  - Injector characterization
    - Spray penetration
    - Diameter and velocity distribution
- Measurements
  - Motored operation
  - Fired homogeneous (fully vaporized) spark-ignition operation with premixed air/fuel mixtures
  - Direct-injection spark-ignition operation
  - Injector characterization

2013 Automotive Simulation World Congress
October 29 - 30 • Frankfurt am Main, Germany

Validation: Wisconsin Engine

- Motored Engine Conditions
- Experimental data
  - In-cylinder pressure
- Repeatability of measurements

Motored Engine Conditions	
Engine Speeds (RPM)	1000, 2000
Intake Air Pressures (kPa)	40, 80
Intake Air Temperatures (°C)	40
Coolant Temperatures (°C)	80

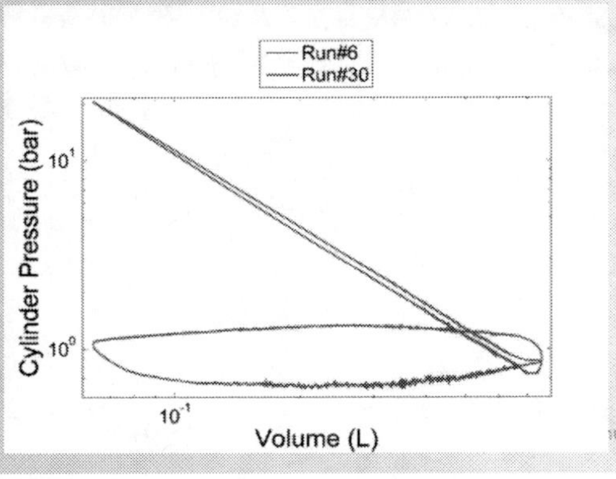

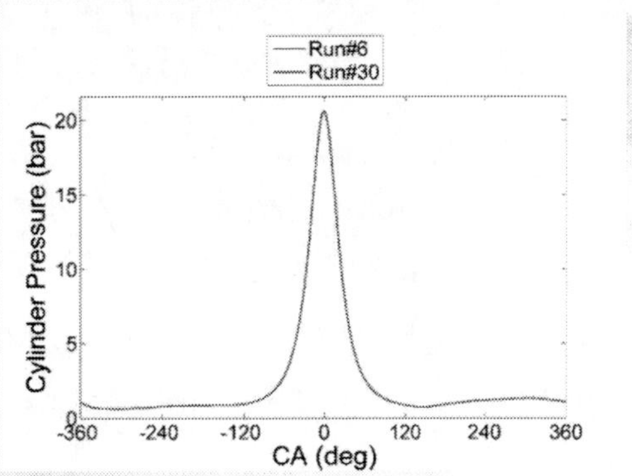

2013 Automotive Simulation World Congress
October 29 - 30 • Frankfurt am Main, Germany

Validation: Wisconsin Engine

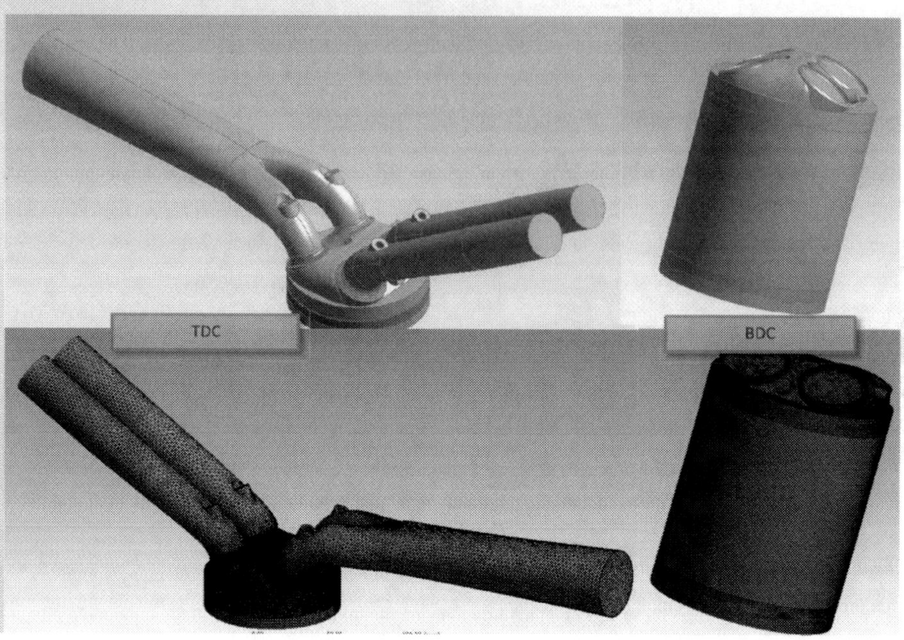

35

2013 Automotive Simulation World Congress
October 29 - 30 • Frankfurt am Main, Germany

Validation: Wisconsin Engine

- Validation runs

CF#	6	18	19	22
Engine RPM	2000	2000	2000	1000
Intake Pressure (*PROFILE*) [kPa]	80	80	80	80
Post Head [C]	78.8	39.7	38.5	39.7
Coolant T [C]	80	40	40	40
Intake Manifold [C]	40	80	80	40
Exhaust Manifold [C]	80.3	64.7	55.3	46.8
Peak cylinder P (bar)	20.62	19.4	20.0	16.4

2013 Automotive Simulation World Congress
October 29 - 30 • Frankfurt am Main, Germany

Validation: Wisconsin Engine

Mesh Information	Coarse	Medium	Fine
TDC (Million elements)	0.57	1.9	5.7
BDC (Million elements)	0.83	10	13.3

- Validation runs

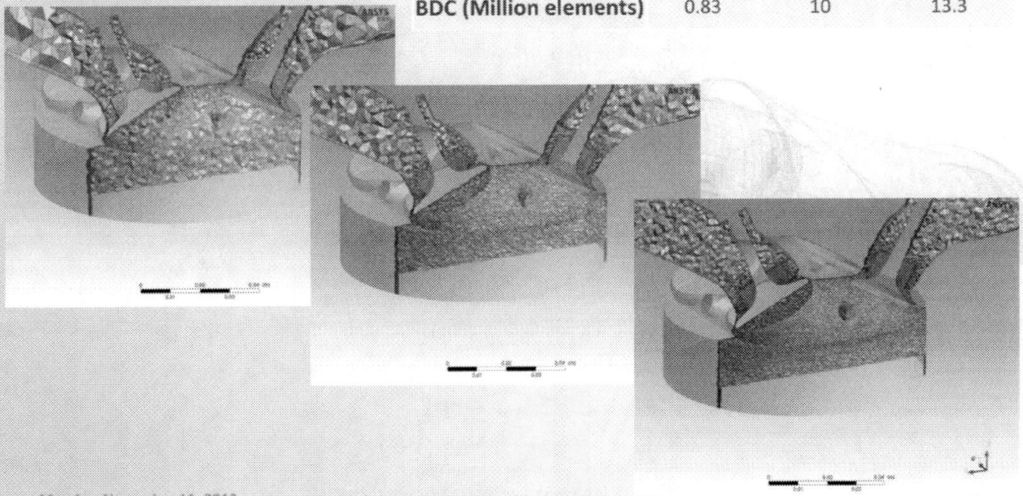

2013 Automotive Simulation World Congress
October 29 - 30 • Frankfurt am Main, Germany

Validation: Wisconsin Engine

- Mesh Study CF#19
- Average cylinder pressure

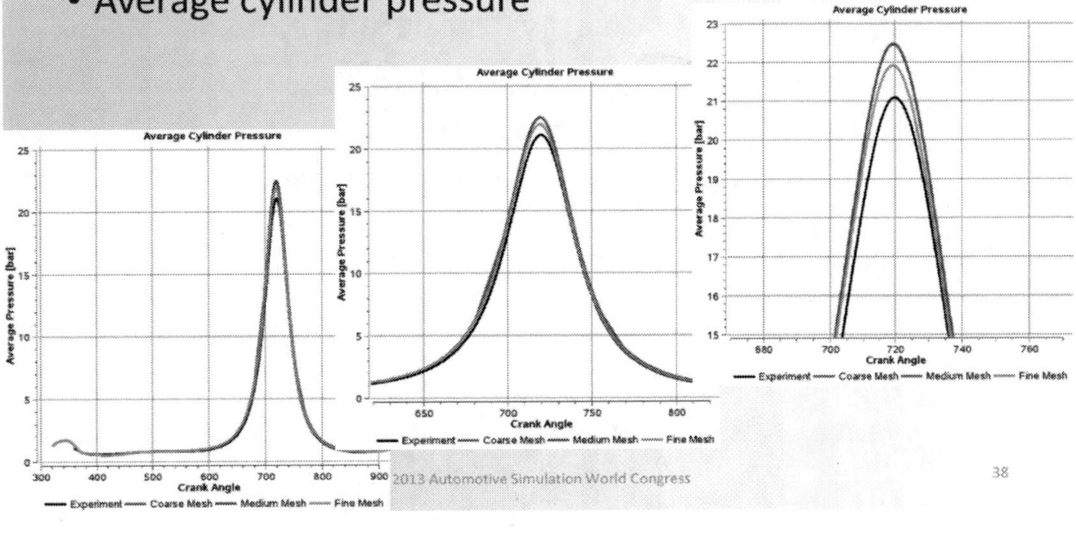

2013 Automotive Simulation World Congress
October 29 - 30 • Frankfurt am Main, Germany

Validation: Wisconsin Engine

- Mesh study CF#19
- Cylinder pressure vs. volume

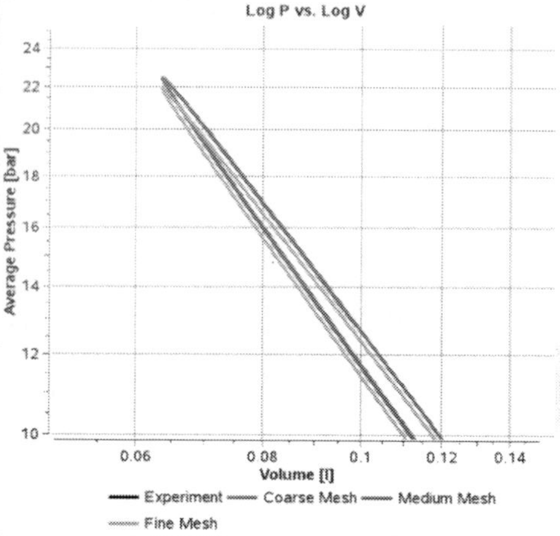

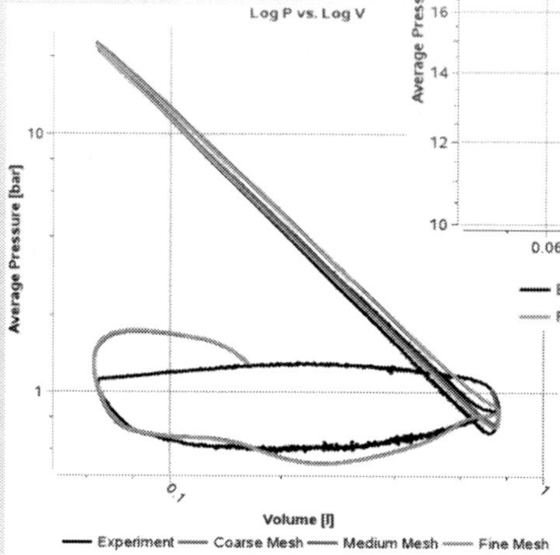

39

2013 Automotive Simulation World Congress
October 29 - 30 • Frankfurt am Main, Germany

Validation: Wisconsin Engine

- Mesh study CF#19
- Average cylinder data
  - Turbulent kinetic energy, temperature, mass

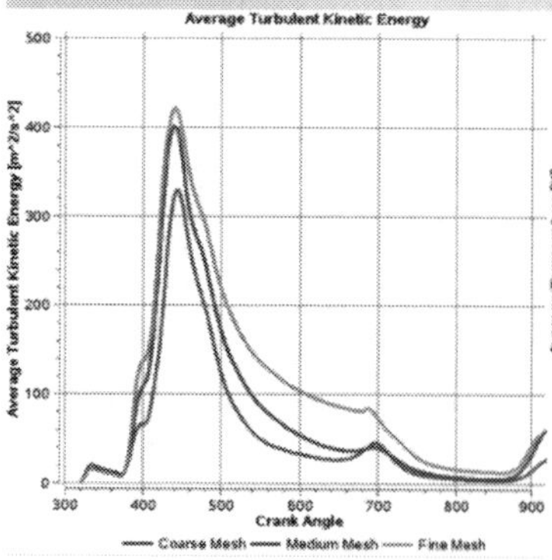

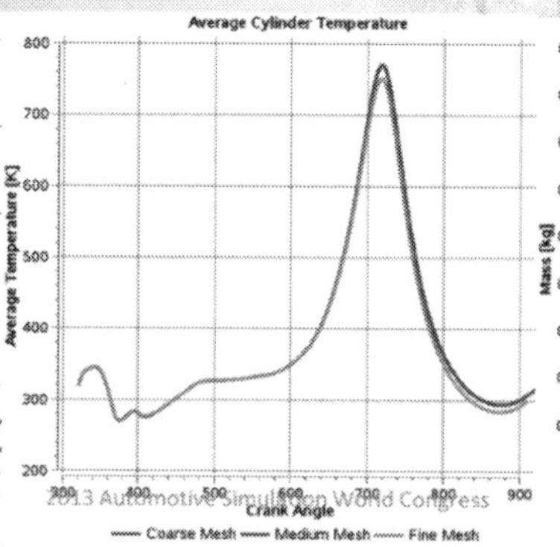

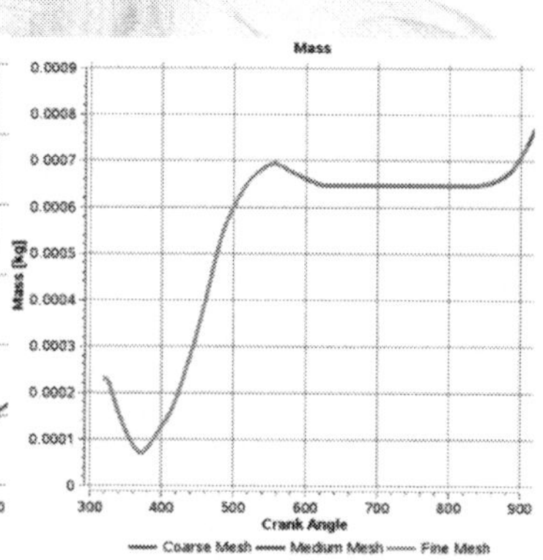

2013 Automotive Simulation World Congress

2013 Automotive Simulation World Congress
October 29 - 30 • Frankfurt am Main, Germany

Validation: Wisconsin Engine

- Operating conditions

CF#	6	19	22
Engine RPM	2000	2000	1000
Intake Pressure [kPa]	80	80	80
Coolant T [C]	80	40	40
Intake Manifold [C]	40	80	40

- Average cylinder pressure

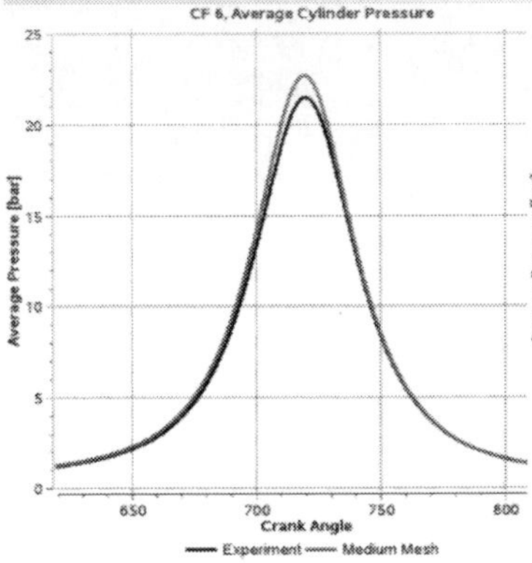

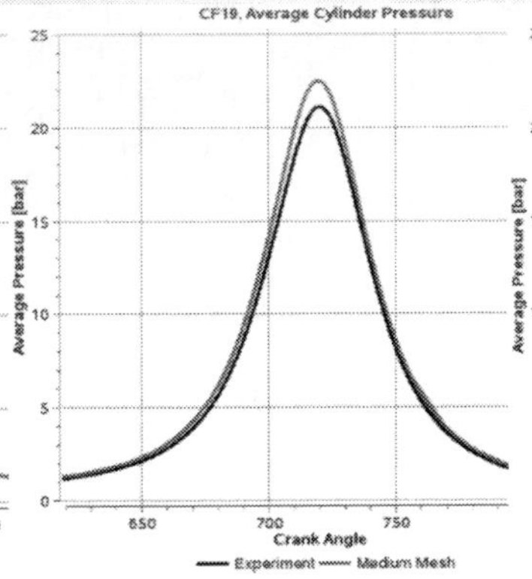

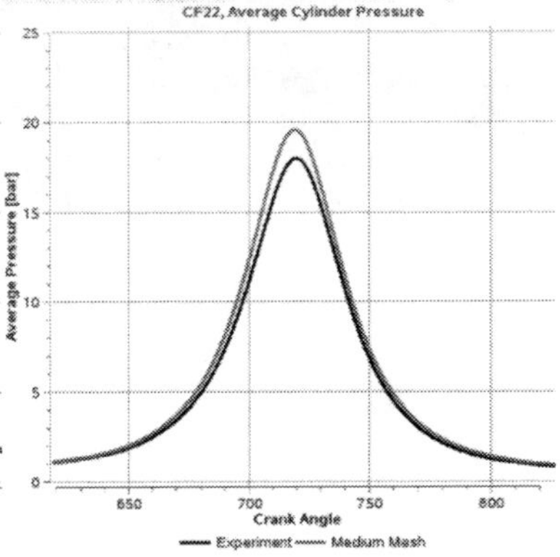

2013 Automotive Simulation World Congress
October 29 - 30 • Frankfurt am Main, Germany

Validation: Wisconsin Engine

- Turbulence model variation: SST vs. k-epsilon
- Operating condition CF#18
- Average cylinder pressure

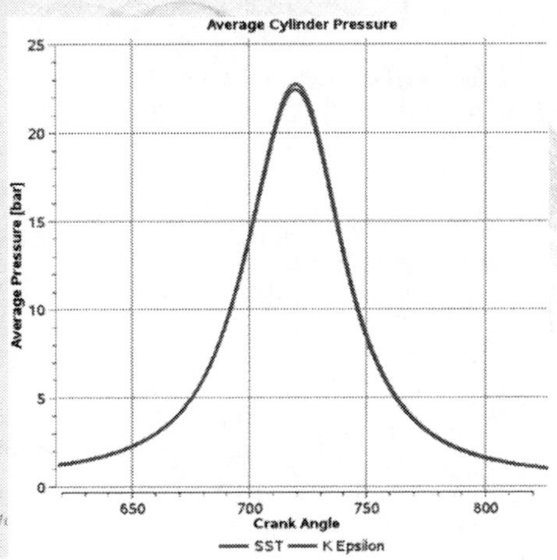

2013 Automotive Simulation World Congress
October 29 - 30 • Frankfurt am Main, Germany

Validation: Wisconsin Engine

- Homogeneous Combustion
- Measurement data
  - High-speed data
    - Cylinder pressure
  - Heat release rate

Homogeneous Combustion Conditions	
Engine Speeds (RPM)	1000, 2000
Intake Air Pressures (kPa)	40, 80
Intake Air Temperatures (˚C)	100
Coolant Temperatures (˚C)	80
Spark Timing	Sweep
Equivalence Ratios	.87, .97, 1.1
Injection Pressure (MPa)	10

2013 Automotive Simulation World Congress
October 29 - 30 • Frankfurt am Main, Germany

Validation: Wisconsin Engine

- Homogeneous Combustion
- Repeatability of measurements

HC#	2
Engine RPM	1000
Intake Pressure [kPa]	60
Coolant T [C]	80
Intake Manifold [C]	100
Spark Timing ˚bTDC	12.5
Equivalence Ratio	0.97

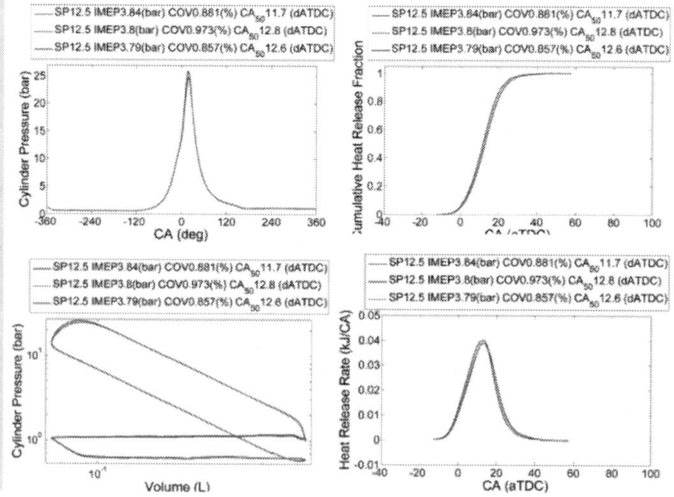

Validation: Wisconsin Engine

- Homogeneous Combustion
  - Cylinder pressure, heat release rate

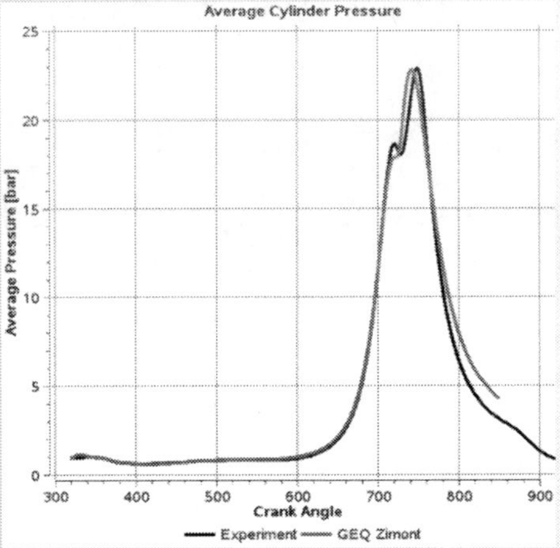

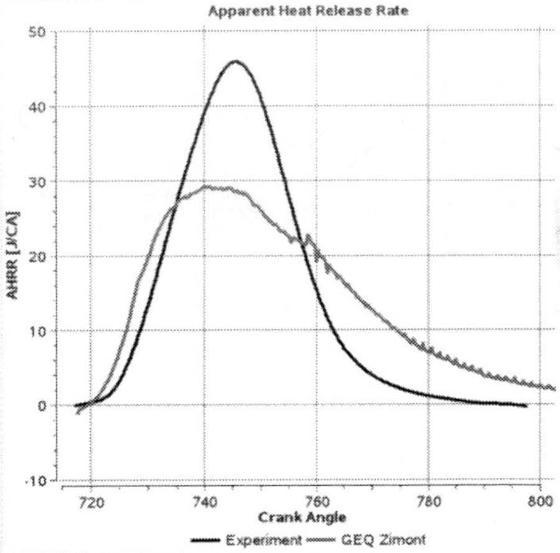

Next Steps

- Extension of IC Engine functionality
  - Improved performance and robustness
- Continuation of validation work
  - More sensitivity analysis
  - Fired engine conditions
- New customer focused examples

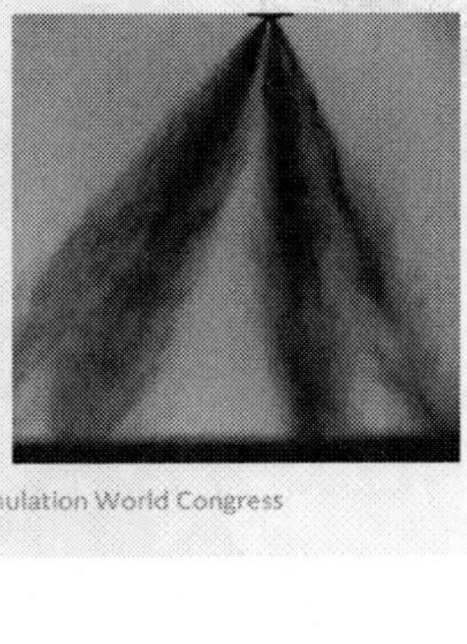

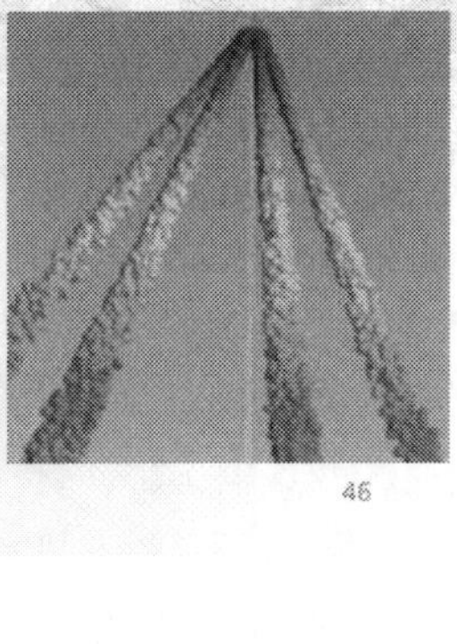

Automotive Simulation World Congress
October 29 - 30 • Frankfurt am Main, Germany

Summary

- Workflow for simulating IC Engine related flows
  - Automation with option for customization
- Strong regression suite and good coverage of manual testing
- Demonstration cases using the IC Engine workflow
- Validation of workflow for engines at different operating conditions
  - Good agreement for validation results with experimental data

2013 Automotive Simulation World Congress 47

Automotive Simulation World Congress
October 29 - 30 • Frankfurt am Main, Germany

Any Questions ?

2013 Automotive Simulation World Congress 48

ASWC, October 29-30, 2013

LAGRANGIAN SPARK-IGNITION MODEL INCORPORATING CHEMICAL KINETICS FOR TURBOCHARGED DIRECT-INJECTION SI ENGINES.

Lukas Schäfer, Dr. Dirk Linse, Dr. Oliver Hausner, BMW Group, München

Prof. Dr. Christian Hasse, Technische Universität Bergakademie, Freiberg

BMW
GROUP

LAGRANGIAN SPARK-IGNITION MODEL. INTRODUCTION.

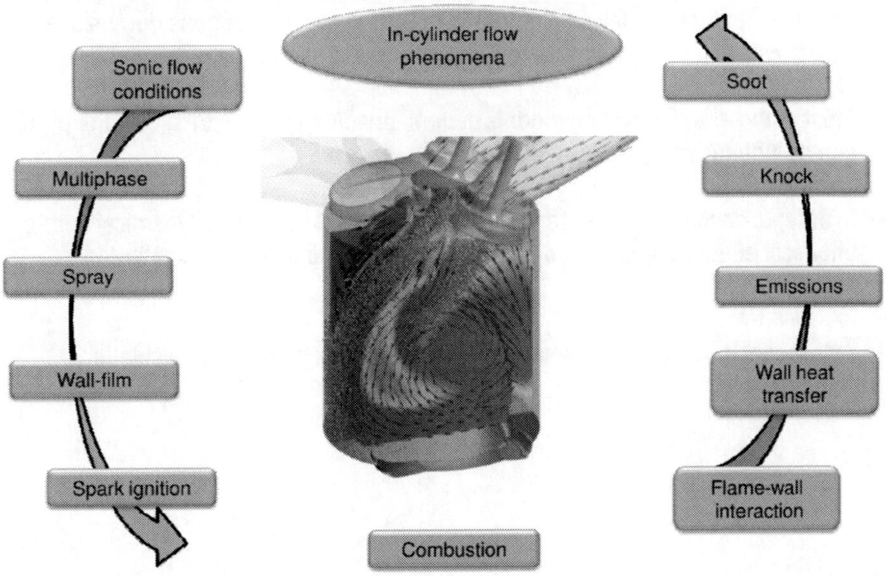

LAGRANGIAN SPARK-IGNITION MODEL. SPARK IGNITION PHENOMENA.

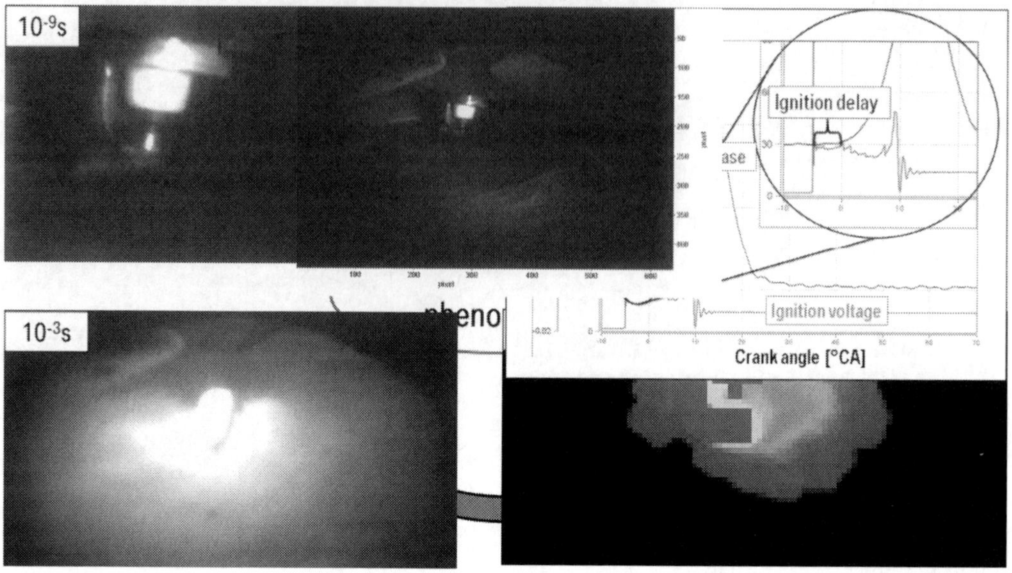

LAGRANGIAN SPARK-IGNITION MODEL. MOTIVATION.

- The most ignition models implemented in CFD-codes do not consider the spark dynamics and the thermodynamical and chemical effects.
- Most of the existing ignition models depend on tuning parameters, thus the physics can be influenced by the user.
- In order to account for the spark dynamics and the local thermodynamical and chemical effects a mesh independent ignition submodel was implemented in ANSYS-CFX.
- This model incorporates chemical kinetics for the ignition delay calculation.

LAGRANGIAN SPARK-IGNITION MODEL. MODELING APPROACH.

Modeling approach

Spark dynamics:
modeled by Langrangian particles

Ignition delay:
computation of the ignition delay with homogeneous reactor approach

Flame kernel propagation:
transition from ignition model to combustion model (G-equation)

Initial flame kernel:
the small flame surface is represented by marker particles

LAGRANGIAN SPARK-IGNITION MODEL. MODELING THE SPARK DYNAMICS.

- In the new igr ticles.
- These particl les are convected by

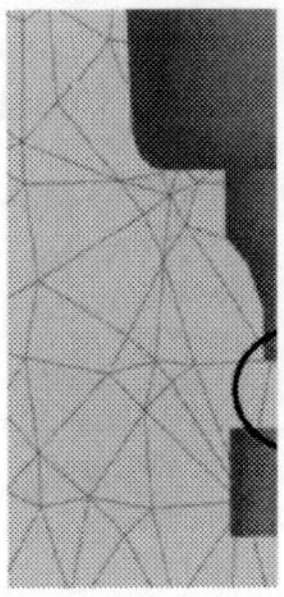
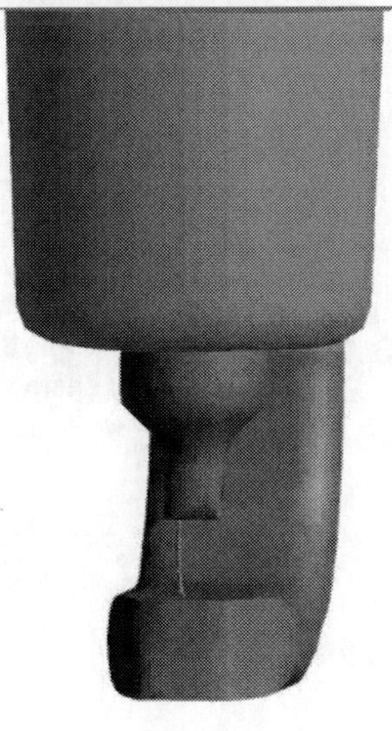
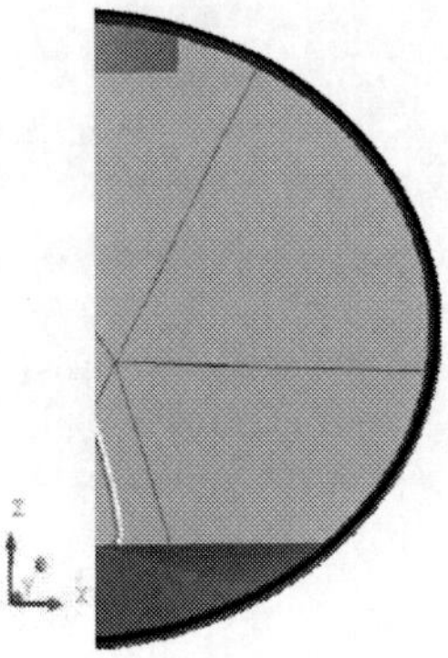

LAGRANGIAN SPARK-IGNITION MODEL. MODELING APPROACH.

Spark dynamics:
modeled by Langrangian particles

Ignition delay:
computation of the ignition delay with homogeneous reactor approach

Modeling approach

Flame kernel propagation:
transition from ignition model to combustion model (G-equation)

Initial flame kernel:
the small flame surface is represented by marker particles

LAGRANGIAN SPARK-IGNITION MODEL. MODELING THE IGNITION DELAY.

- **Ignition delay**: time between spark-timing and the first flame kernel development.

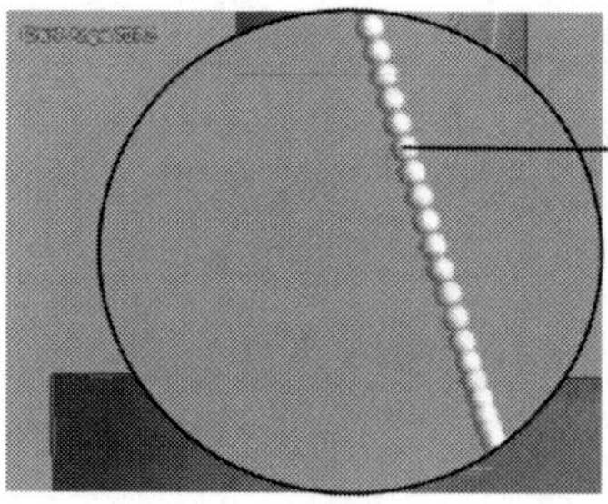

Spark energy

initializing with:
T, p, Z, V

Homogeneous reactor:
solving the equations for energy, mass, species and pressure.

Heat losses

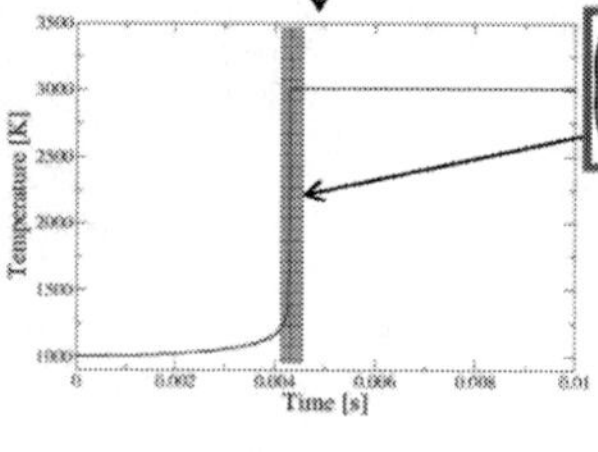

$$\left(\frac{dT}{dt}\right)_{max} = ignition$$

LAGRANGIAN SPARK-IGNITION MODEL. MODELING THE IGNITION DELAY.

- The local flow field leads to convection and deformation of the spark channel.
- Local spark curvature leads to regions with higher energy release.

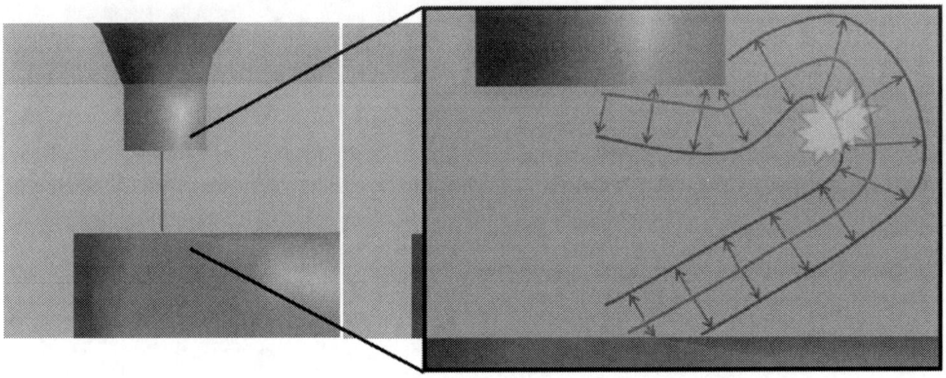

LAGRANGIAN SPARK-IGNITION MODEL. MODELING THE IGNITION DELAY.

- Depending on the local curvature a different amount of spark energy is inserted into the spark/ homogeneous reactor (HR).

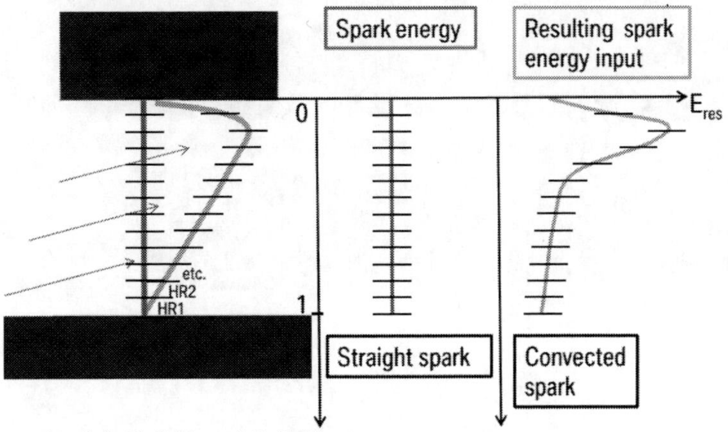

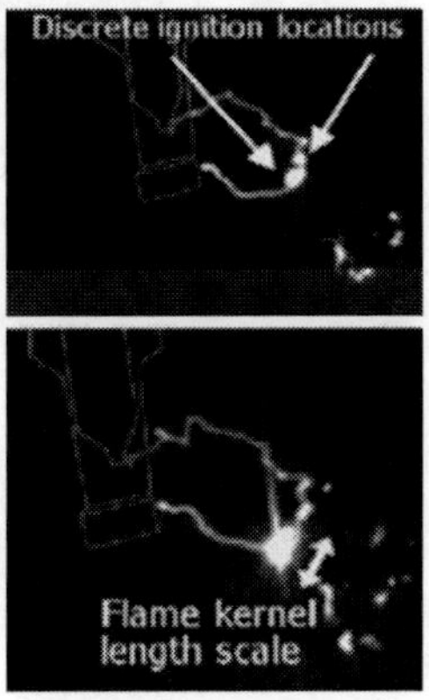

Propulsion Systems Research Laboratory, General Motors Global Research & Development, Warren, MI, USA

LAGRANGIAN SPARK-IGNITION MODEL. MODELING APPROACH.

Spark dynamics:
modeled by Langrangian particles

Ignition delay:
computation of the ignition delay with homogeneous reactor approach

Modeling approach

Flame kernel propagation:
transition from ignition model to combustion model (G-equation)

Initial flame kernel:
the small flame surface is represented by marker particles

LAGRANGIAN SPARK-IGNITION MODEL. MODELING THE FLAME KERNEL DEVELOPMENT.

- The ignition time is represented by the maximum temperature gradient:

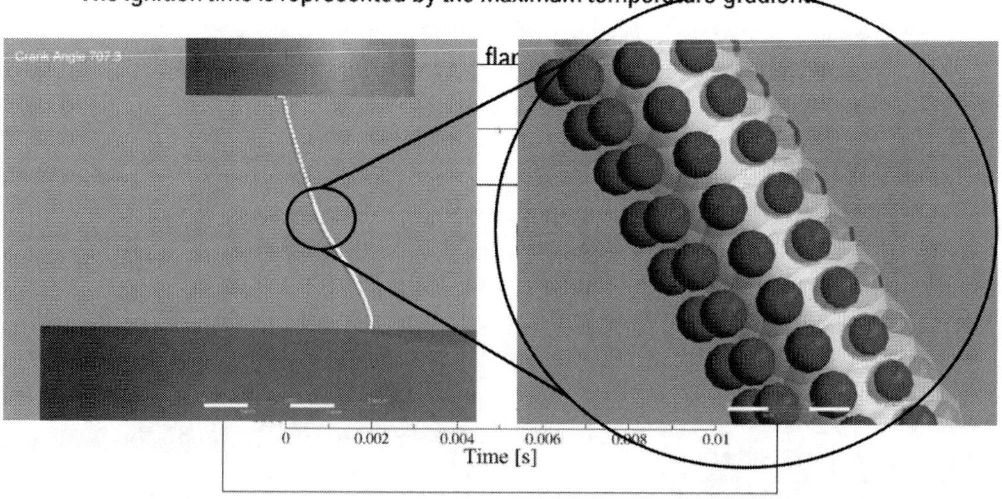

LAGRANGIAN SPARK-IGNITION MODEL. MODELING THE FLAME KERNEL DEVELOPMENT.

- The initial flame kernel propagation is calculated by:

$$s_K = \max\left(s_L, s_T - \frac{2}{r_K} D_t\right)$$

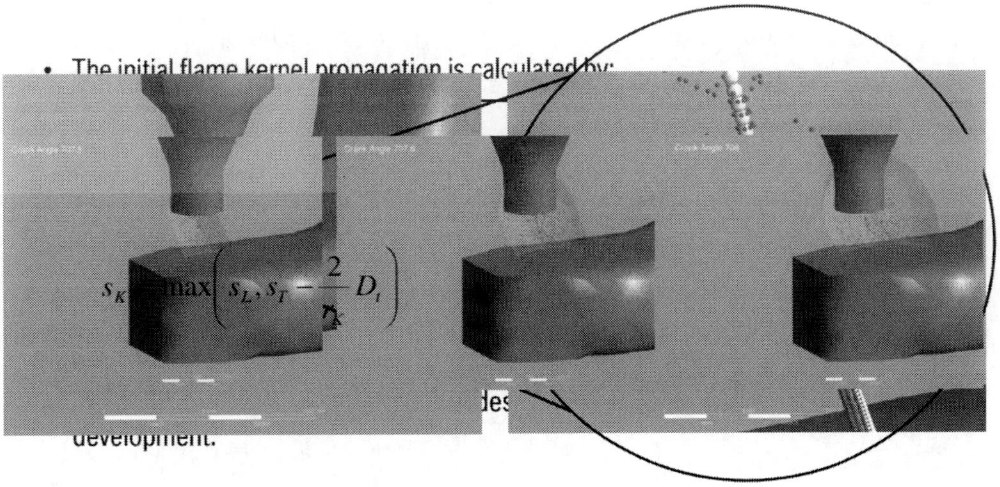

...des... development.

LAGRANGIAN SPARK-IGNITION MODEL. MODELING APPROACH.

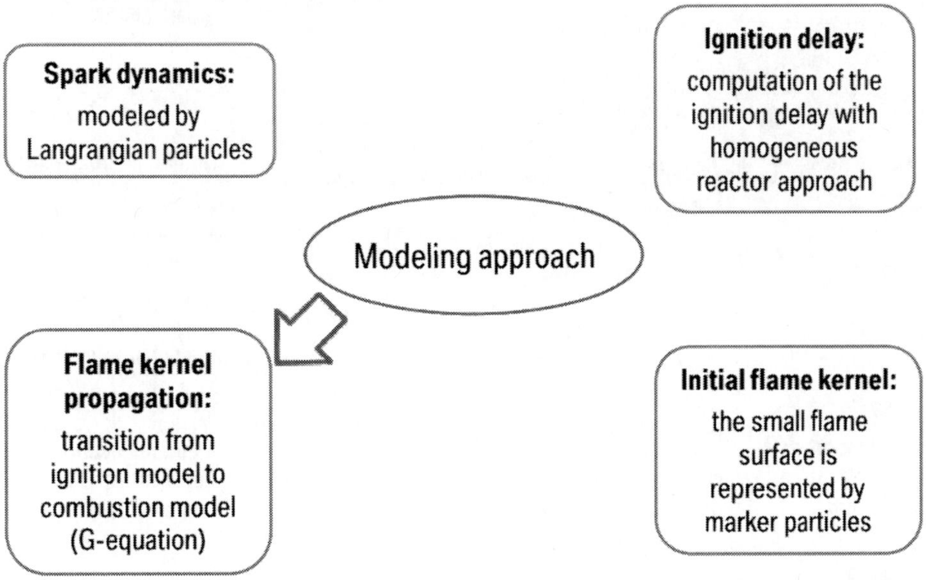

LAGRANGIAN SPARK-IGNITION MODEL. TRANSITION FROM IGNITION MODEL TO COMBUSTION MODEL.

- As the single flame kernels exceed a radius of 3mm the flame surface can be described by the mesh.
- At this time the flame kernel surfaces describe the flame surface in the G-Equation model.
- From this timestep the G-Equation and the expansion of the flame is calculated by the G-Equation in the CFX-Solver.

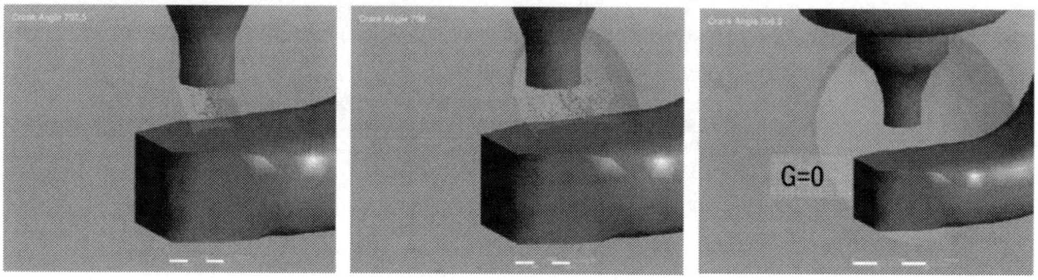

LAGRANGIAN SPARK-IGNITION MODEL. OVERVIEW.

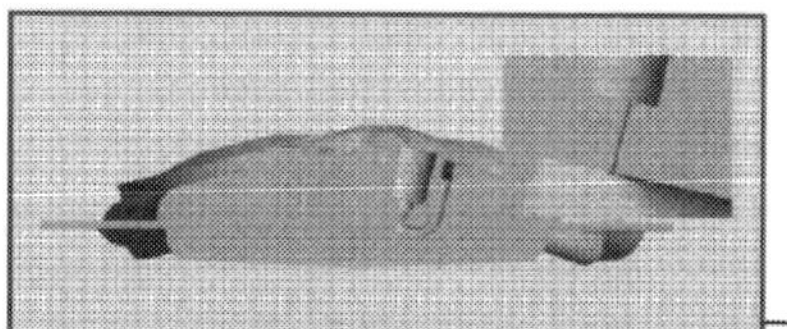

3-D CFD-Simulation

At ignition time: initialising with T, p, Z, V → Homogeneous reactor

Consideration of: wall heat loss and spark energy → Homogeneous reactor

Homogeneous reactor → At any time step; transfer of: T,p, Z, ignition information → If ignition occurs: Initialising of the flame kernels

If ignition occurs: Initialising of the flame kernels → Modeling the flame kernel development → 3-D CFD-Simulation

Consideration of: flame-wall-interaction → Modeling the flame kernel development

LAGRANGIAN SPARK-IGNITION MODEL. MODELING APPROACH - SUMMARY.

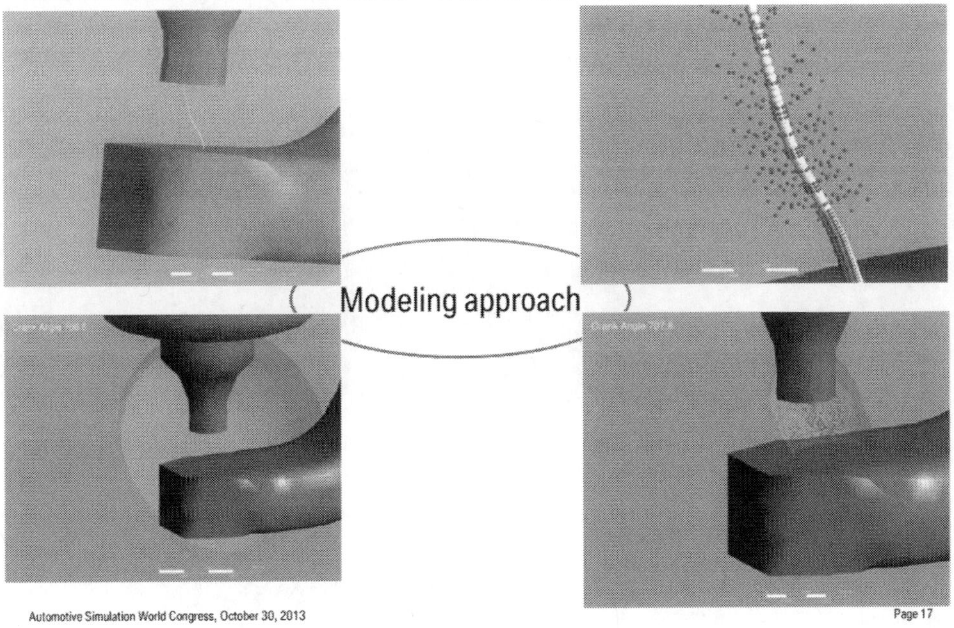

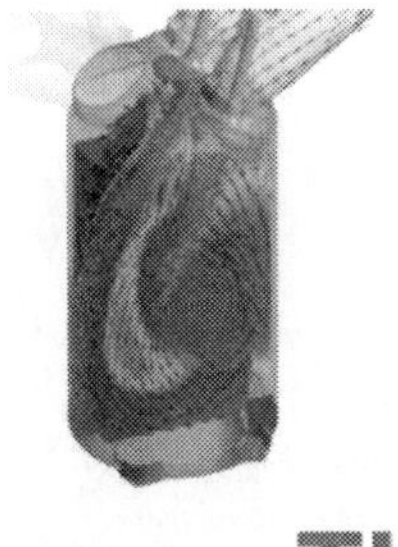

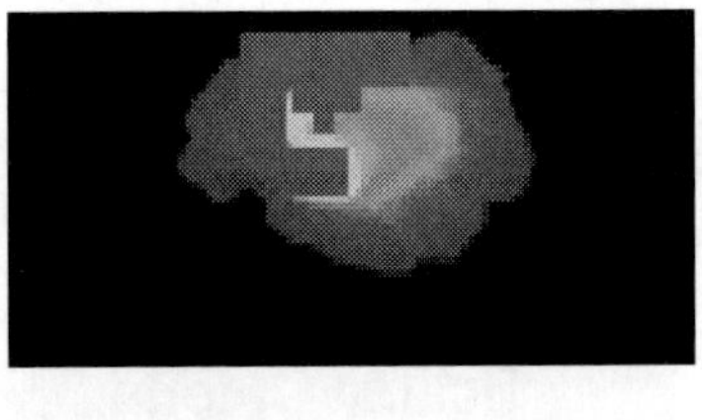

THANK YOU FOR YOUR ATTENTION

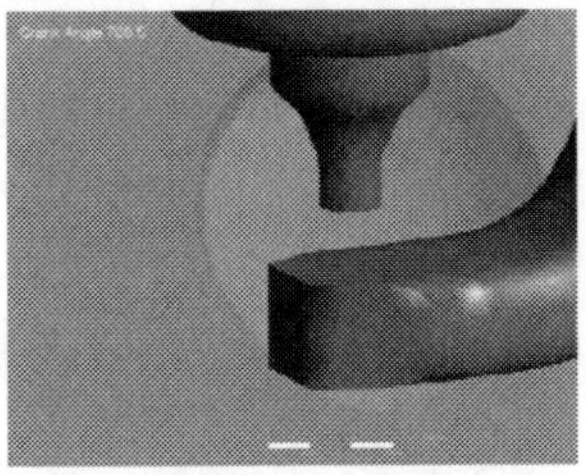

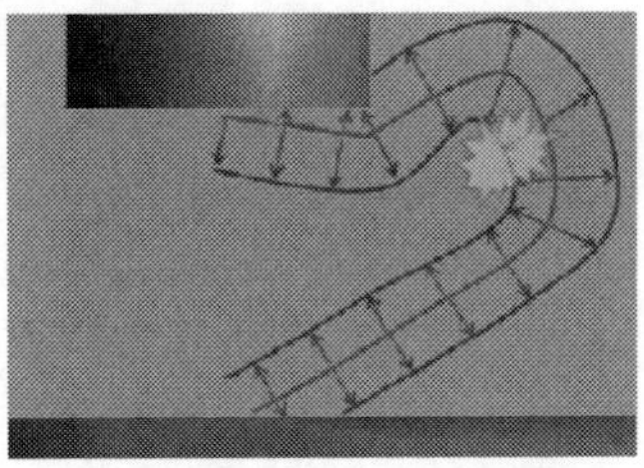

TECHNISCHE UNIVERSITÄT
BERGAKADEMIE FREIBERG
Die Ressourcenuniversität. Seit 1765.

Scale-Resolving Simulations of IC-Engine In-Cylinder Flows and Comparison to Experimental Data

Stefan Buhl

CHAIR OF NUMERICAL THERMO-FLUID DYNAMICS

Chair of Numerical Thermo-Fluid Dynamics | TU Bergakademie Freiberg | 09599 Freiberg
S. Buhl| +49 3737 394814 | stefan.buhl@iec.tu-freiberg.de | www.ntfd.tu-freiberg.de

Overview

- Motivation
- General Information
- Numerical Setup
- Quality Criteria
- Results
- Summary

2

Motivation

"Cycle to Cycle Variation" (CCV) is an actual topic in scientific research

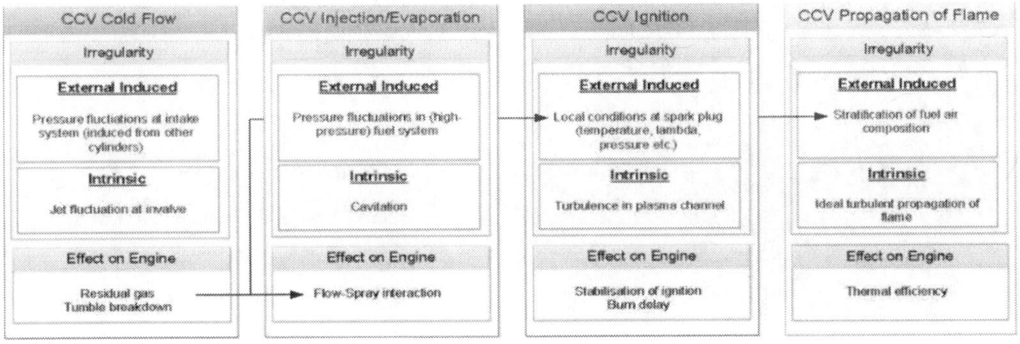

Brief excerpt of chain of effects at CCV

- Focus on cold flow
- Scale Resolving Simulations (SRS) offer the possibility for investigations of CCV

3

Overview

- Motivation
- General Information
- Numerical Setup
- Quality Criteria
- Results
- Summary

4

General Information - Engine

Picture:
http://www.sandia.gov/ecn/engines/engineFlows/TCCEngine/engGeo.php

- Research engine of Sandia International Laboratories
- Transparent piston and cylinder
- Technical data:
  - Engine speed = 800 rpm
  - Bore = 92 mm
  - Stroke = 86 mm
  - Rod length = 234.95 mm
  - Max. valve lift = 8.9 mm
  - One invalve and one exvalve

5

General Information - PIV

Picture:
http://www.sandia.gov/ecn/engines/engineFlows/TCCEngine/engGeo.php

Further information on PIV:
Abraham, Reuss, Sick: High-Speed Particle Image Velocimetry Study of In-Cylinder Flows with Improved Dynamic Range

2d-2c PIV data available (transparent cylinder/piston)

- Silicon droplets as tracer particle
- Optimized pulse interval for current velocity range
- Multi frame to increase the accuracy and dynamic range of velocity
  - Low res.: 1.46 mm (between valves)
  - High res.: 0.8 mm (combustion chamber)
- 70 cycles
- Delta ca: 5 deg for full cycle

6

Overview

- Motivation
- General Information
- Numerical Setup
    - Boundary Conditions
    - Workflow for CCV
    - Mesh
- Quality Criteria
- Results
- Summary

7

Numerical Setup – Boundary Conditions

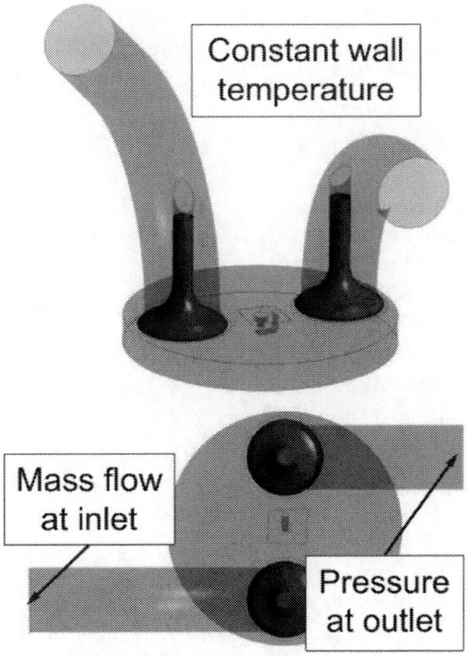

- Turbulence model:
  SAS-Forced
- Turbulent wall function for heat transfer:
  Han-Reitz
- Time step width:
  0.05 deg (intake stroke)
  0.25 deg

8

Numerical Setup – Workflow for CCV

C1

RANS

9

Numerical Setup – Workflow for CCV

C1

C2

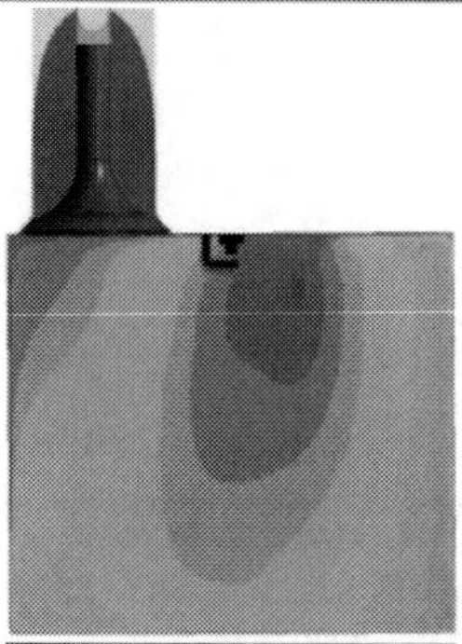

Initial flow field for cycle 2

RANS

↓

SRS

10

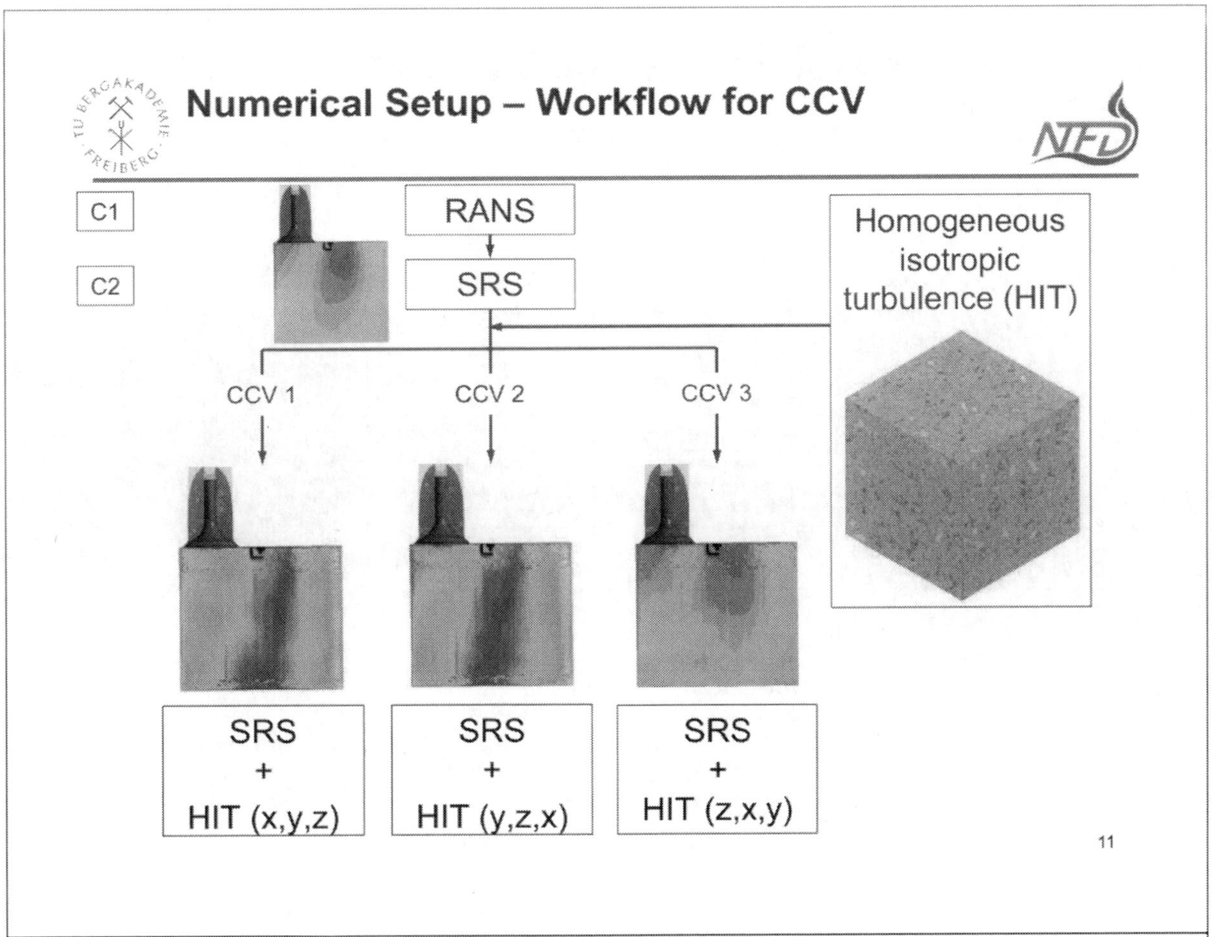
Numerical Setup – Workflow for CCV
C1
C2
RANS
SRS
Homogeneous isotropic turbulence (HIT)
CCV 1
CCV 2
CCV 3
SRS + HIT (x,y,z)
SRS + HIT (y,z,x)
SRS + HIT (z,x,y)
11

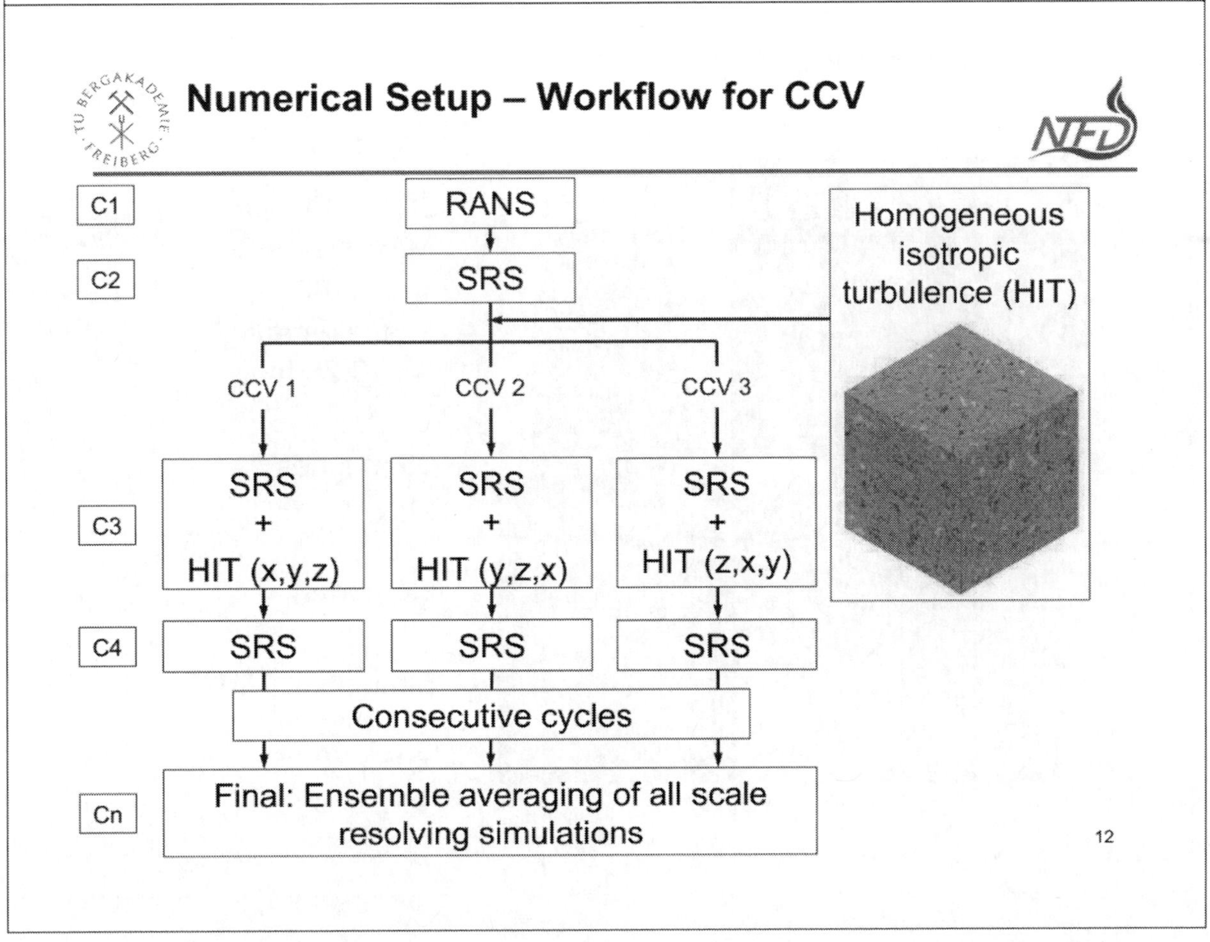
Numerical Setup – Workflow for CCV
C1
C2
RANS
SRS
Homogeneous isotropic turbulence (HIT)
CCV 1
CCV 2
CCV 3
C3
SRS + HIT (x,y,z)
SRS + HIT (y,z,x)
SRS + HIT (z,x,y)
C4
SRS
SRS
SRS
Consecutive cycles
Cn
Final: Ensemble averaging of all scale resolving simulations
12

Numerical Setup – Workflow for CCV

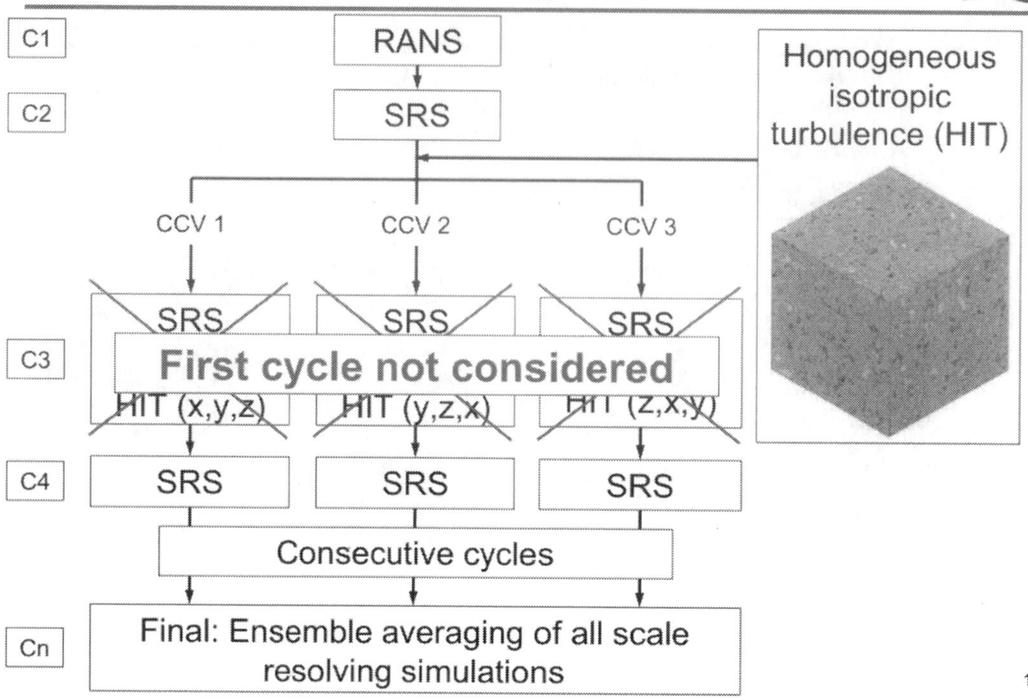

13

Numerical Setup – Mesh

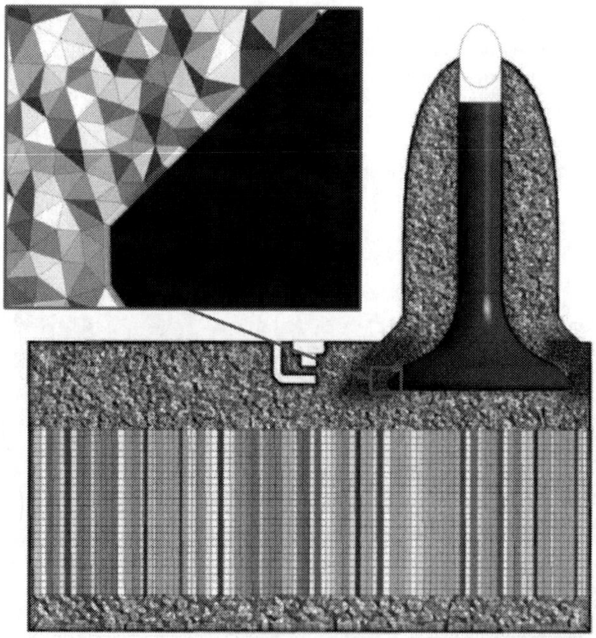

- 10 Prism-Layer
- Max. element size: 1 mm
- Min. element size: 0.25 mm
- 65 meshes
- Max. 4.1 million nodes (13.2 million elements)
- Simulation Time: ~ 8 days on 48 cores → ~ 9200 CPUh

14

Overview

- Motivation
- General Information
- Numerical Setup
- Quality Criteria
    - CFL & $Tke_{resolved}$
    - y^+ & Boundary Layer Resolution
- Results
- Summary

15

Quality Criteria – Spatial/Temporal Resolution (ca = 452 deg)

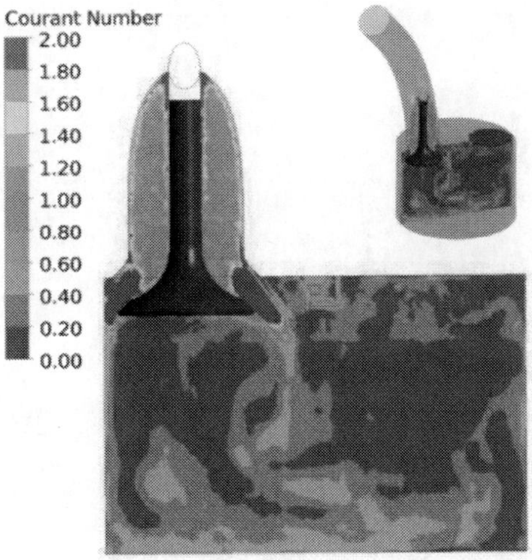

Volume averaged CFL ≈ 0.4

Max. CFL ≈ 90 (at Sparkplug)

→ Adequate temporal resolution

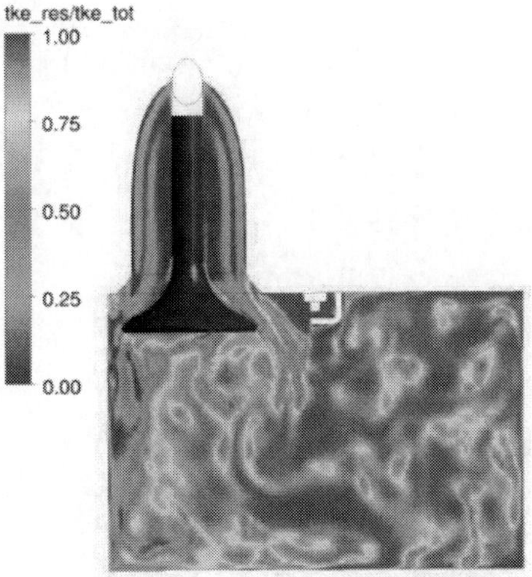

Volume averaged in chamber

$k_{res}/k_{tot} \approx 0.72$

→ Adequate spatial resolution

16

Quality Criteria – Resolution at Boundary Layer (ca = 452 deg)

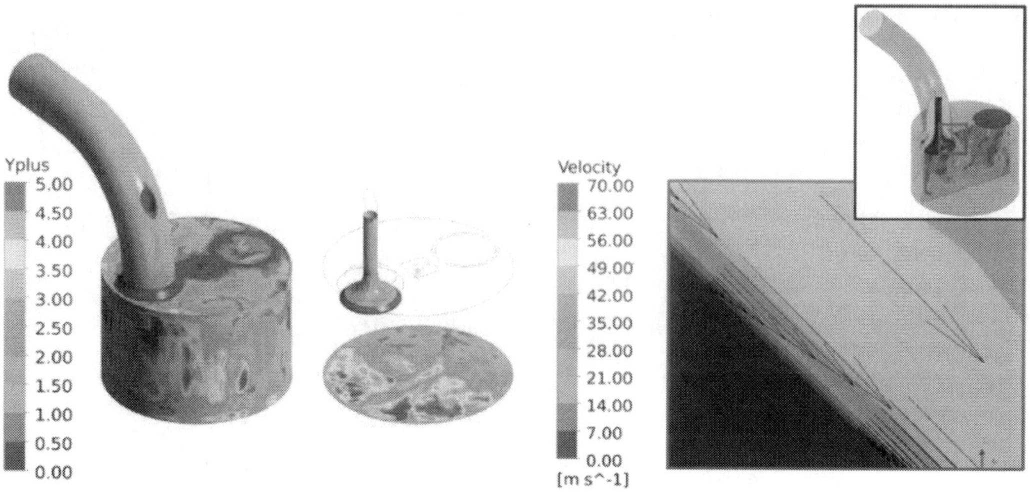

Average $y^+ \approx 1.3$

Max. $y^+ \approx 4.9$

→ Adequate resolution at boundary layer

17

Overview

- Motivation
- General Information
- Numerical Setup
- Quality Criteria
- Results
- Summary

18

Results – Cylinder Pressure & Swirl Ratio (PIV Plane) Experiment vs. SAS

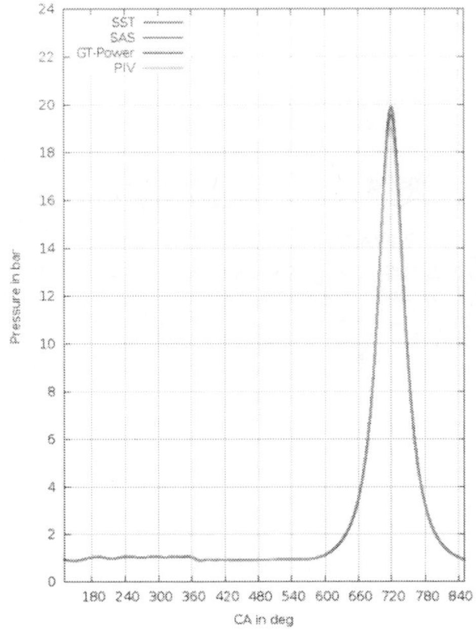

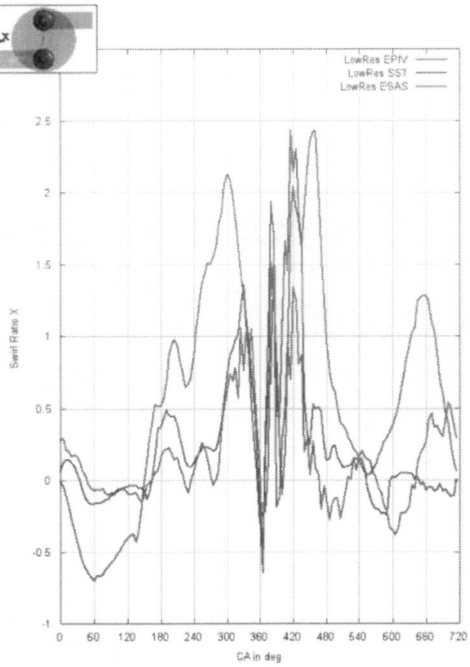

19

Results – Comparison Experiment and Simulation

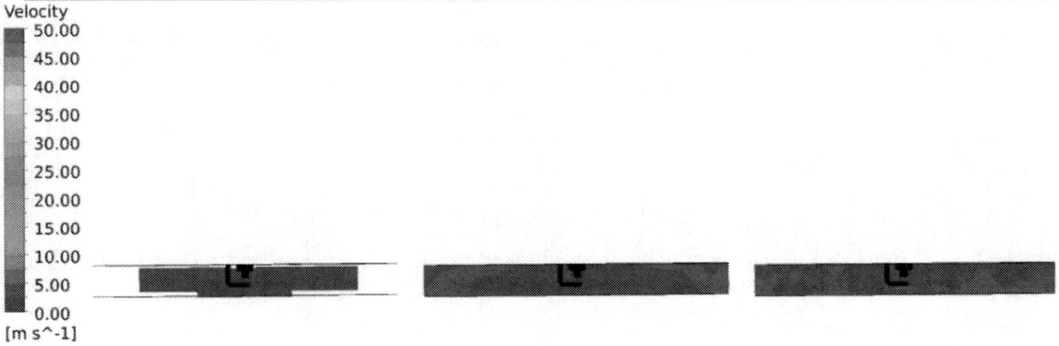

Ensemble averaged PIV RANS (SST) Ensemble averaged SAS

20

Results – Comparison Cycle 4

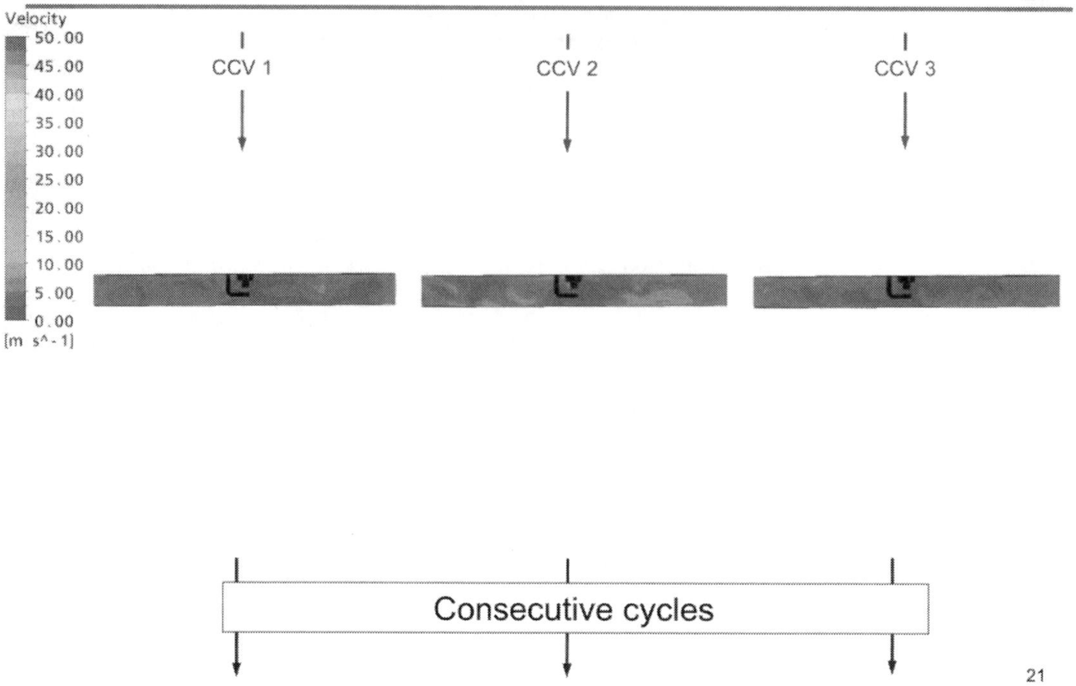

21

Results – Swirl Ratio (chamber) SST vs. SAS

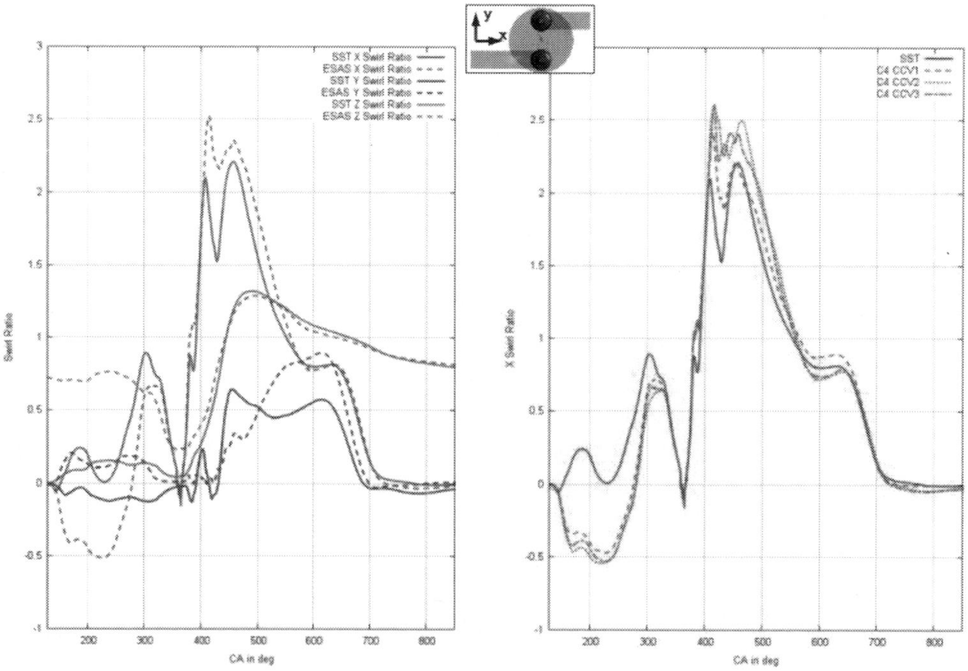

22

Results – TKE_{SGS} SST vs. SAS

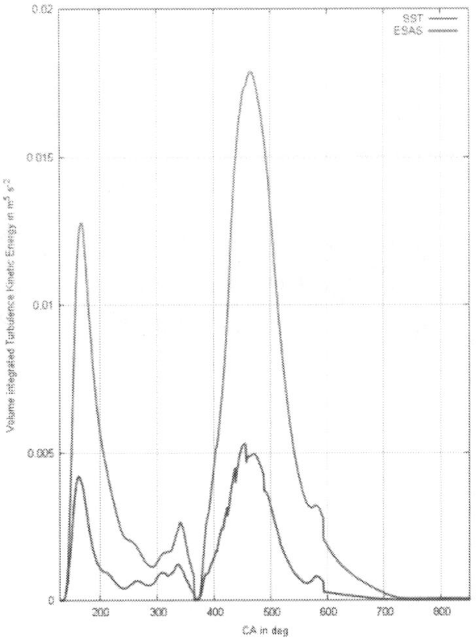

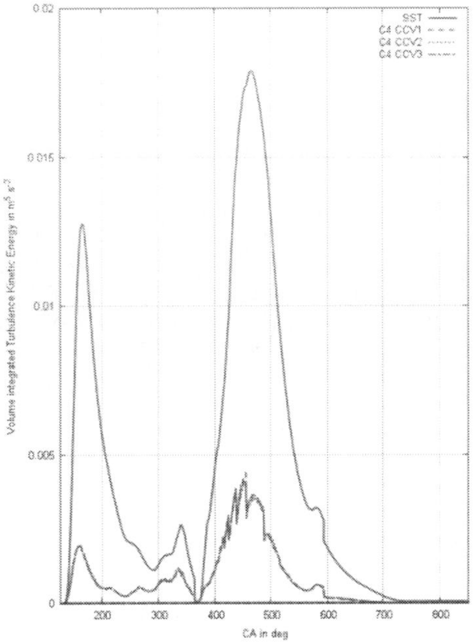

23

Results – Resolved Fluctuations

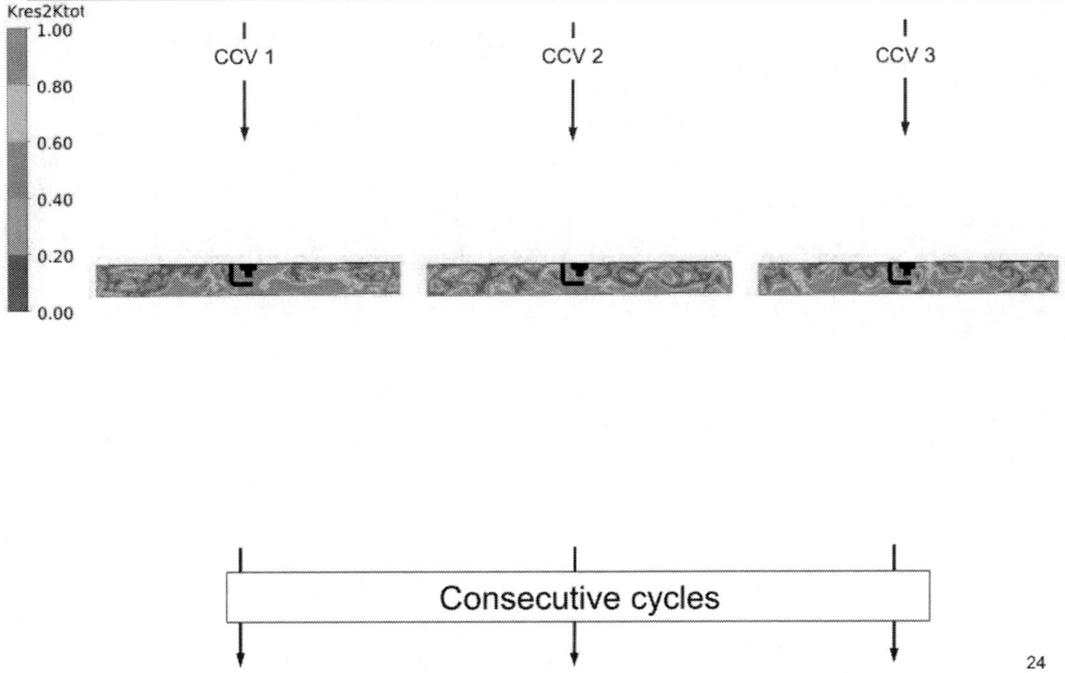

24

Overview

- Motivation
- General Information
- Numerical Setup
- Quality Criteria
- Results
- Summary

25

Summary

- SAS-Forced turbulence model successfully used in ICE simulations
- Conditioning (e.g. filtering) of numerical data necessary
- Further steps
  - Calculation of more cycles for better comparison to experimental data
  - Statistical evaluation
  - Detailed POD analysis

26

Thanks to:

- Frank Hartmann
- ANSYS Germany

27

ENGIN SOFT

Key partner in Design Process Innovation

DESIGN CHAIN SIMULATION SUPPORTS SME COMPETITIVENESS: A CASE OF PISTON ROD

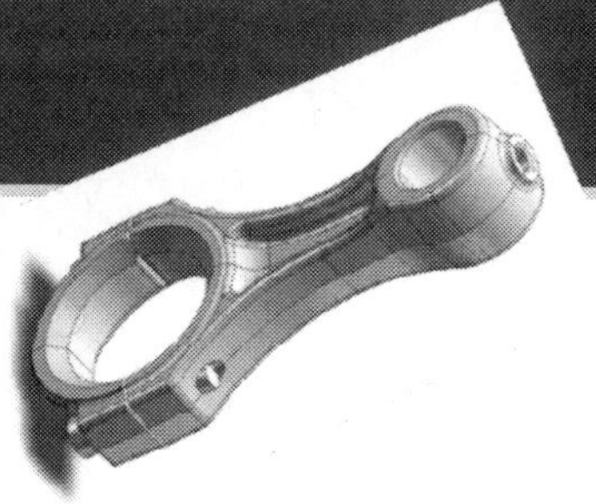

L. Bellati – A.B.O.R. S.p.A., Italy

G. Scarpa, N. Gramegna, F. Lago, P. Bortolato - Enginsoft S.p.A., Italy

Content

Acknowledgement

The optimization work was carried out in close collaboration with the technicians of the A.B.O.R Foundry,
particular acknowledgment to Mr. Luca Bellati.

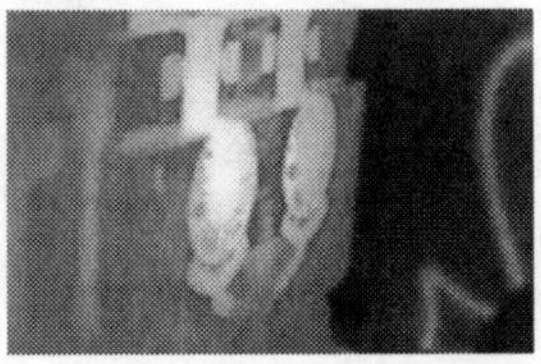

The ABOR company has a manufacturing facility that covers an area of 3800 square meters and an open area of about 1200 square meters. The die casting production lines are equipped with different presses from 200 to 800 tons.

ENGINSOFT company profile

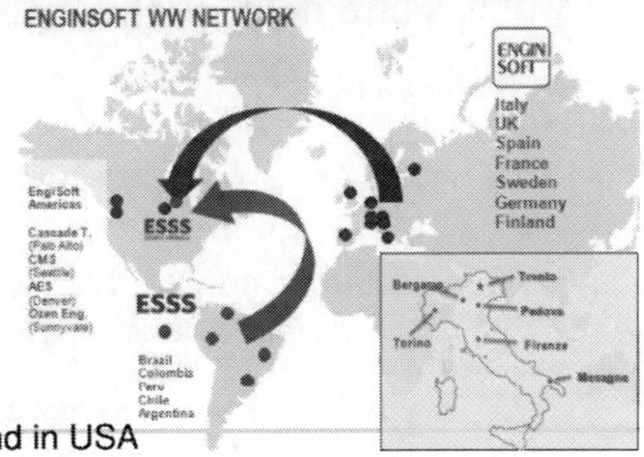

- TYPE: private company
- HISTORY: founded in 1984 on the basis of other activities/structures active since 1973

ACTIVITIES:

- Leading group (in Italy) in CAE/iDP/PIDO
- Subsidiaries/partnerships all over Europe and in USA
- Supply of software, services, consultancy, training
- Development of software technologies
- Participation in industrial research projects (EU or national funding)
- MIUR-recognized research centre for CAE/iDP technology transfer

Current approach to numerical simulation

- **Simulation in product development:**
  - Structural behavior (strength/durability/fatigue)
  - CFD problems
  - Crash events
  - Vibration and dynamic (multi-body)
  - Manufacturing process

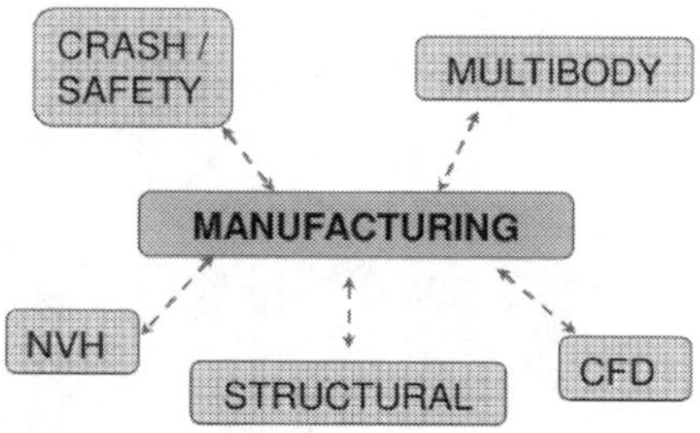

- Manufacturing process simulations are usually considered only to evaluate the producibility of a part.
- All the other simulations are performed only to ensure the fulfill of required performances in terms of strength, on-service behavior, durability and safety.

➢ **An innovative design methodology is necessary to link these two unconnected worlds!**

Design chain approach - Automotive

- Market's needs:
  - lighter and safer vehicles
  - reduction on environmental footprint
- Manufacturers' needs:
  - use of innovative materials (e.g. light alloys)
  - new developments methodologies
  - novel manufacturing processes
  - shorter time-to-market
  - reduction on development costs

- Applicability of Design Chain approach…
  - final local safety factor (crash / structural)
  - layout definition (size)
  - fuel consumption (weight)
  - fatigue prediction
  - …
- … as function of material selection
- … as function of production process parameters
- … as function of management approach
- … linked to optimization techniques

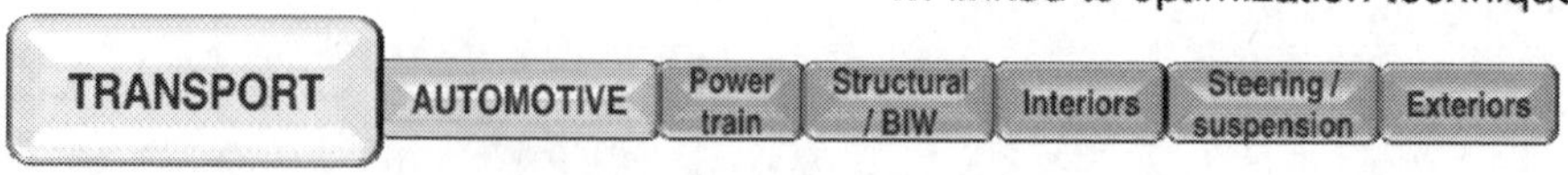

The case study of Piston Rod

Automotive components require:

- Acceptable manufacturing cost vs forging process
- Low weight: light alloy vs steel
- High mechanical properties in the crucial areas
- Structural integrity maximize fatigue life
- High productivity

The PIDO (Process Integration and Design Optimization) methodology can be divided in sub-tasks :

1. **Piston Rod** traditional structural analysis: evaluation of critical zones in terms of maximum compressive and tensile stresses in function of the applied loads
2. Optimization of the HPDC process: Automatic Optimization technology to maximize the **Piston Rod quality**
3. Advanced structural analysis exploiting the **Piston Rod mechanical properties**: to verify the component and to validate the HPDC process.

Customer requirements

The case study was commissioned to Enginsoft in order to design the process production (equipment and parameters) to get

Piston Rod produced by High Pressure Die Casting process

with high performance under high loads and high speed 'rotation for automotive equipment

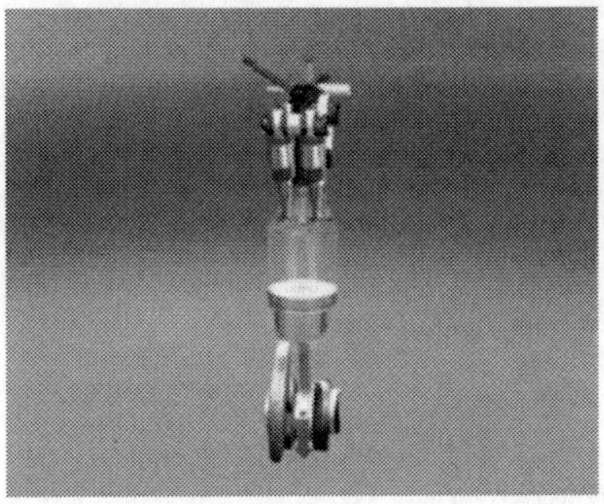
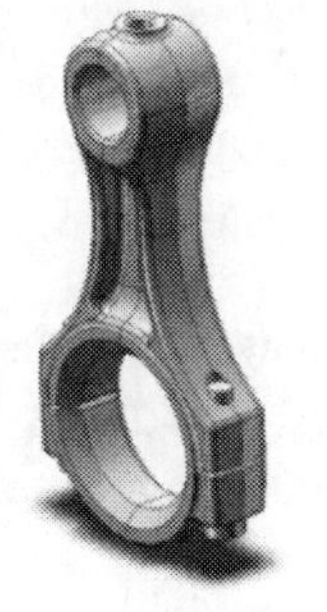
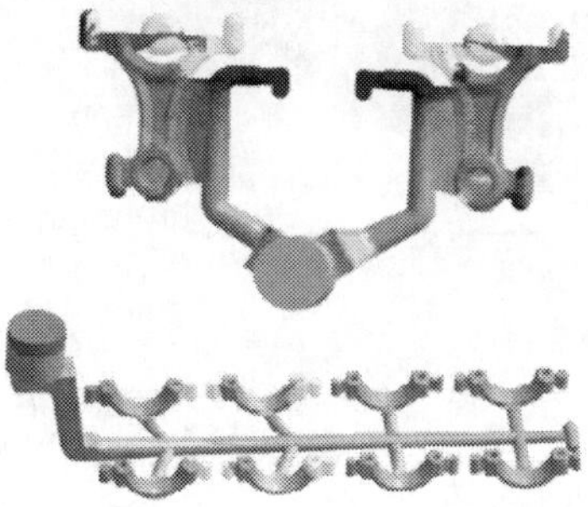

Piston Rod traditional structural analysis

Simulation Type: Elasto-Plastic Structural Analysis (ANSYS) using the nominal and homogeneous mechanical properties of the selected Al alloy (EN 46100) and no defects in the Piston Rod

Aim:

- The design of the Piston Rod
- Design of manufacturing process to ensure the highest quality with particular reference to the identified areas with high stresses.

The work regards the virtual structural analysis of the Piston Rod coupled with foot cap to which are applied the constraints, loads and boundary conditions

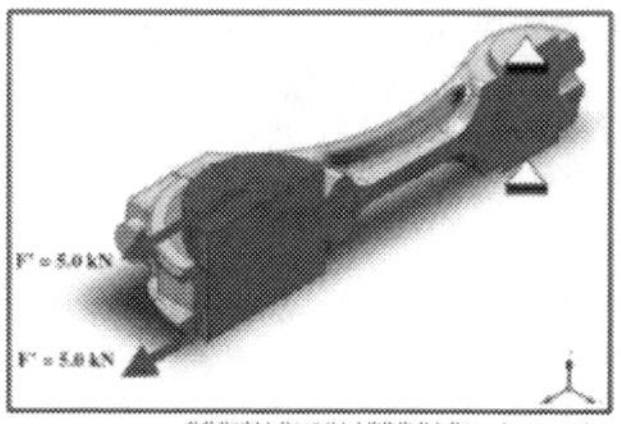

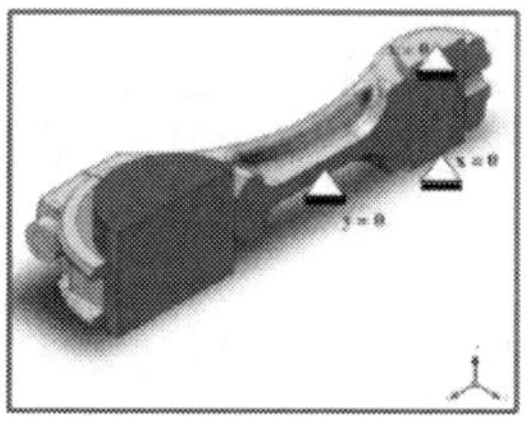

ENGIN SOFT Key partner in Design Process Innovation

Piston Rod - Traditional Structural Analysis

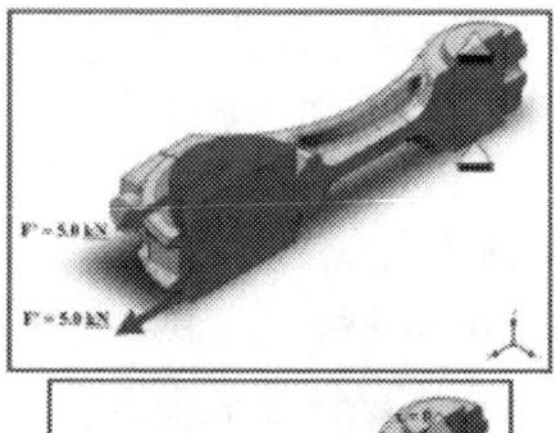

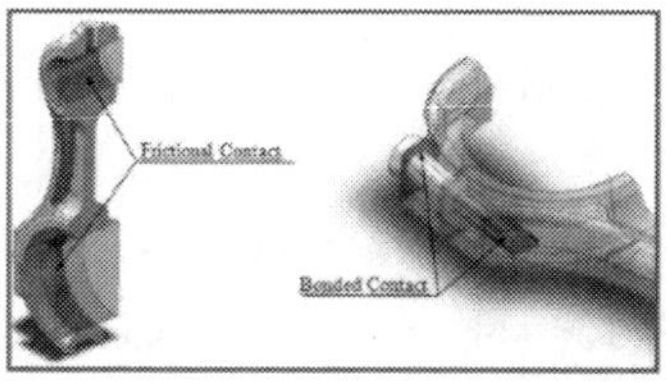

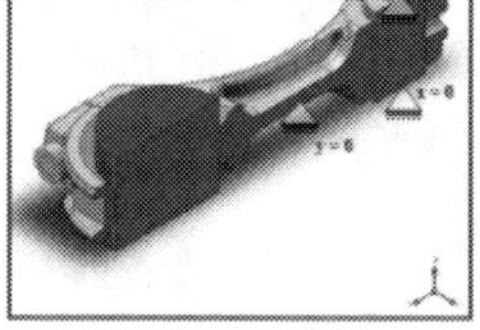

Numerical Model Description

- **Friction Coefficient = 0,3**
- **Tightening Preload Screw = 13,3 kN**
- **Traction axial load = 10 kN**
- **Twork = 100°C**

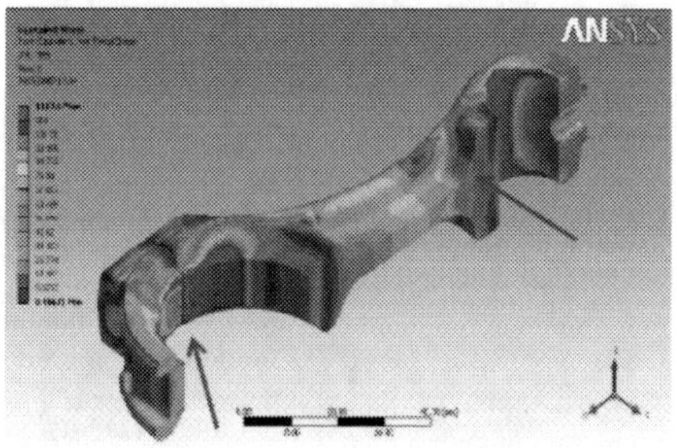

Analysis Results

- **Presence of maximum stress concentration in the head and foot part**

New Piston Rod requirements

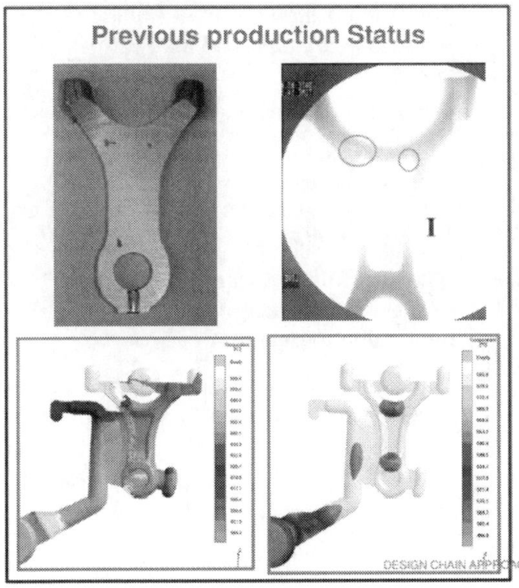

New Piston Rod production requirements:

- Technological Charateristics of the Piston Rod (Mechanical Properties)
- Speed and competitiveness in terms of quality and costs

The designer have to ensure the maximum reduction of waste and the best quality of the cast.

Casting process Optimization to find out the best design that improve the quality of the new Piston Rod.

The aim is pursued using MAGMAfrontier that change the relevant geometric and process parameters to simulate a large number of variants.

Optimization Settings (Step1): Objectives

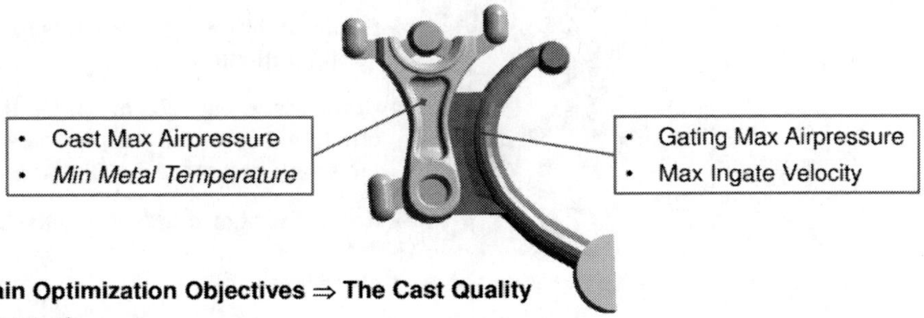

Main Optimization Objectives ⇒ The Cast Quality

In terms of:

- Reduction of air entrapment as a function of the maximum value air-pressure achieved (expressed in mbar) for the Cast
- Reduction of air entrapment as a function of the maximum value air-pressure achieved (expressed in mbar) for the Gating System

Constrains:

- Minimum filling temperature higher than the critical temperature of undercooling (Tliq +15°C)
- Maximum ingate velocity less than the critical value indicated by the foundry (40 m/s)

ENGIN SOFT

Key partner in Design Process Innovation

Optimization Settings (Step2): Variables

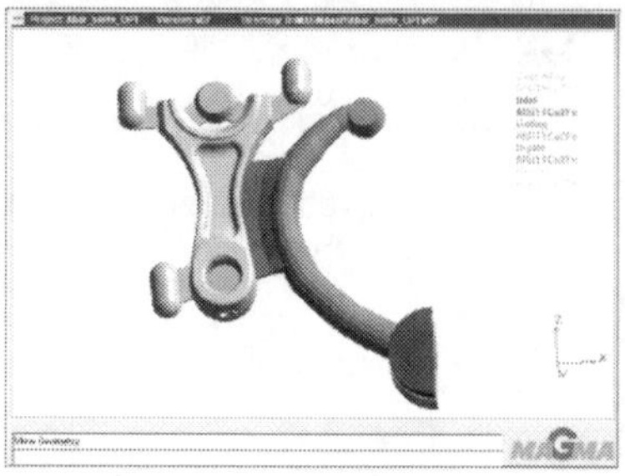

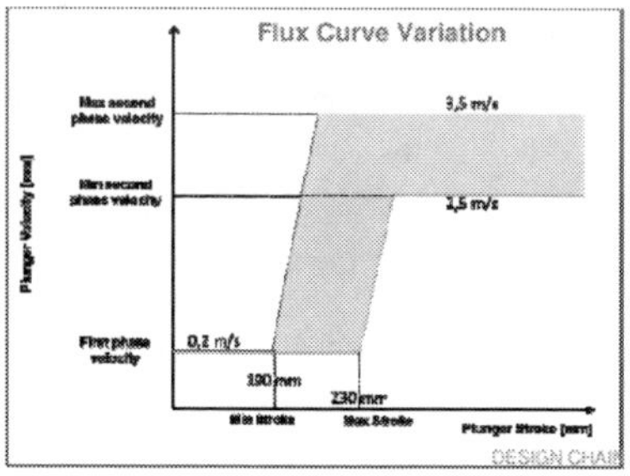

- **The Optimization setting establishes the following variables and their ranges of variation:**
  - geometry parameters to define the gating system
  - process parameters to control the plunger movement during the filling phase (injection curve).
- **The computation start from the DoE (representative samples of the entire population of designs)**
- **The subsequent generation are created by MOGA (Multi Objective Genetic Algorithm)**

ENGIN SOFT

Key partner in Design Process Innovation

Optimization Results Analisys

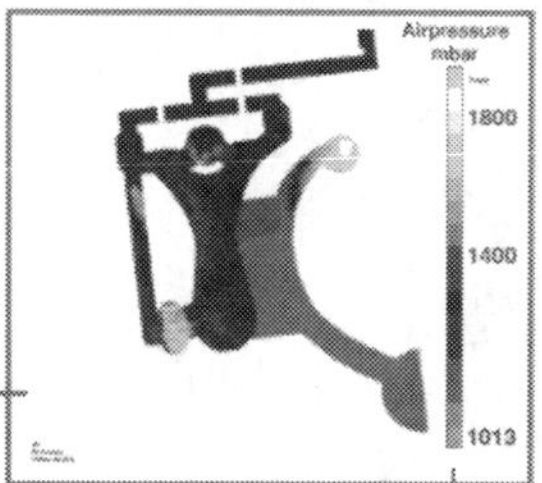

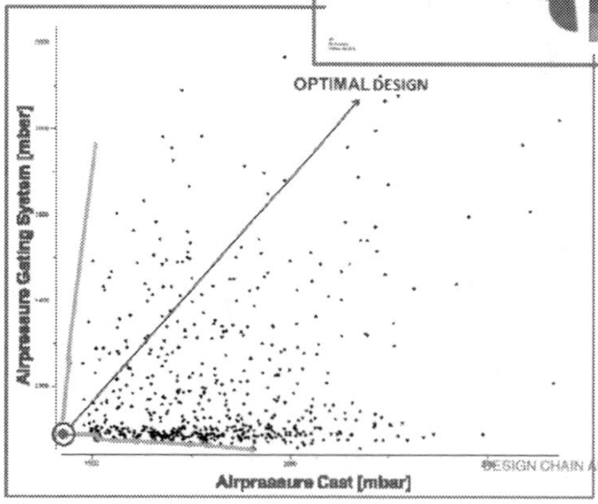

- **Optimization results: 2840 computed configurations**
- **Comparison between the pressure of entrapped air of the Piston Rod with the entrapped air of the Gating System**
- **Pareto Frontier representing the best set of solutions that minimize the two objectives**
- **The comparison between the different configurations in a two-dimensional diagram allows to identify the best design, in terms of principal quality indicators.**

ENGIN SOFT

Key partner in Design Process Innovation

Thermal Steady State Simulation

Filling Temperature

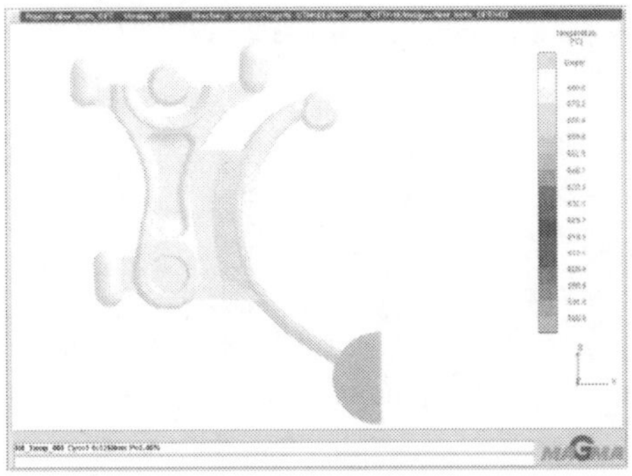

Cast Solidification

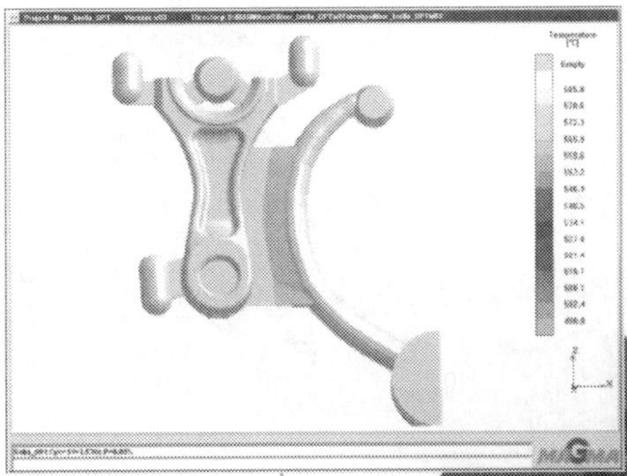

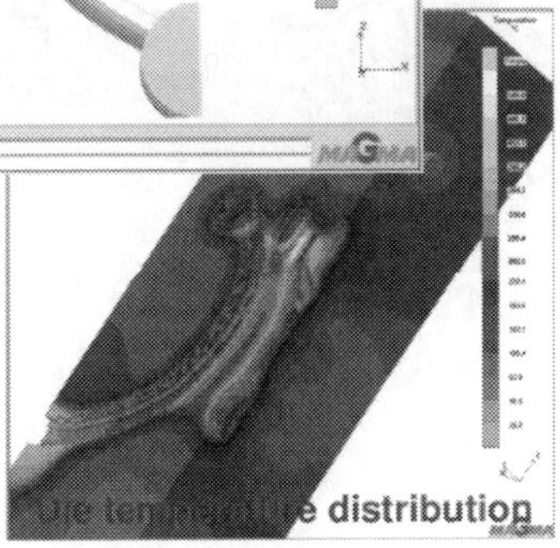

The Thermal steady state simulation allows to verify:

- The melt temperature during filling to evaluate finally the presence of cold shots
- The solidification behavior to evaluate the shrinkage problems
- The dies temperature distribution to evaluate the cooling channels behaviour and the durability of the dies (Dies Life)

ENGIN SOFT

Key partner in Design Process Innovation

Die casting mechanical properties

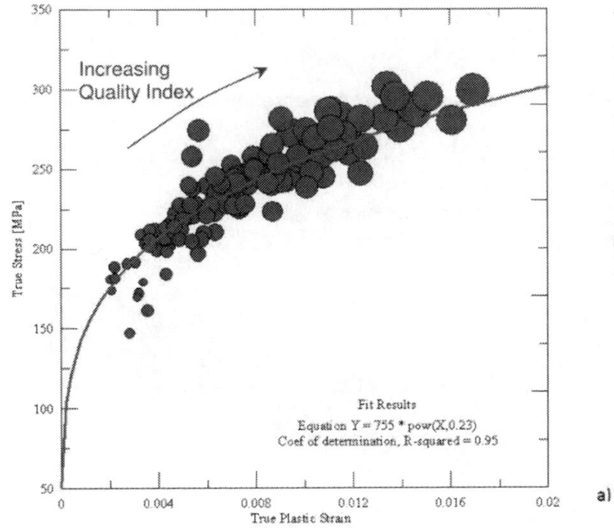

The dimension of the porosity reduces the quality, so the machanical properties (UTS).

Yield Strength (YS) depends on microstructure – constant for HPDC components

FP6 NADIA Project – www.nadiaproject.org

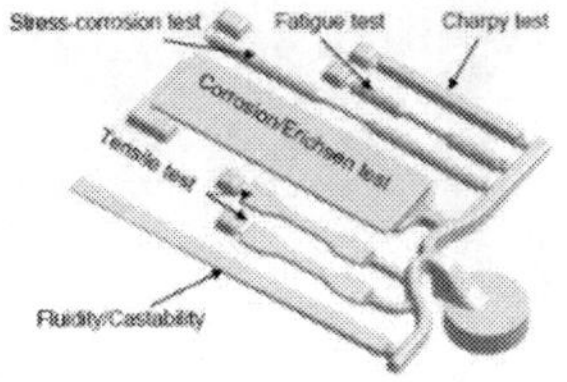

ENGIN SOFT

Key partner in Design Process Innovation

Mechanical Properties Prediction

Local Mechanical Properties ⇒ correlated to cast defects combining defects criteria in a polynomial function.

The implementation of the prediction model of the local **Ultimate Tensile Stress (UTS)** and the relative Elongation to fracture is made possible by exploiting the curve 'Stress-Strain' of the material used for the Piston Rod.

$$\mathrm{UTSfinal} = \mathrm{UTSnominal} - a{*}\mathrm{Airpressure}^{A} - b{*}\mathrm{FlowLength}^{B} - c{*}\mathrm{Aircontact}^{C} - d{*}\mathrm{Coolrate}^{D}$$

Stress Strain curve (EN 46100)

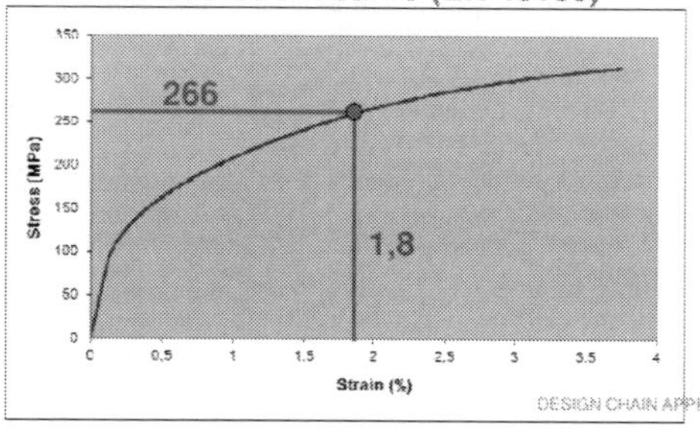

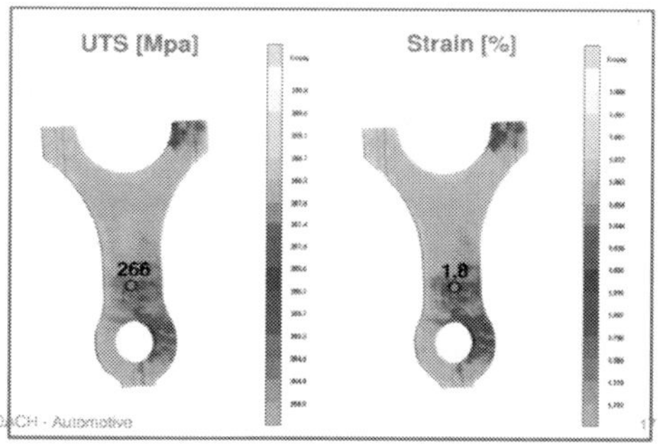

ENGIN SOFT

Key partner in Design Process Innovation

Residual Stress Analysis

- Calculation of residual stresses generated at the end of the production process on the Piston Rod after the machining of the gating system and the overflows.
- Measured values are particularly low

Local Mechanical Properties and Residual Stress represent the realistic material status for the final Structural Analysis

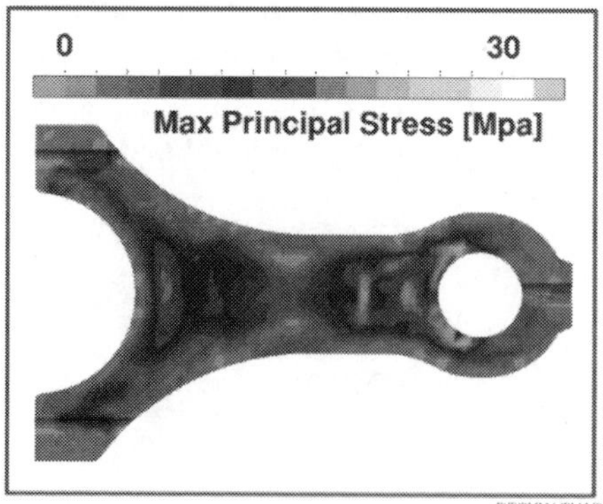

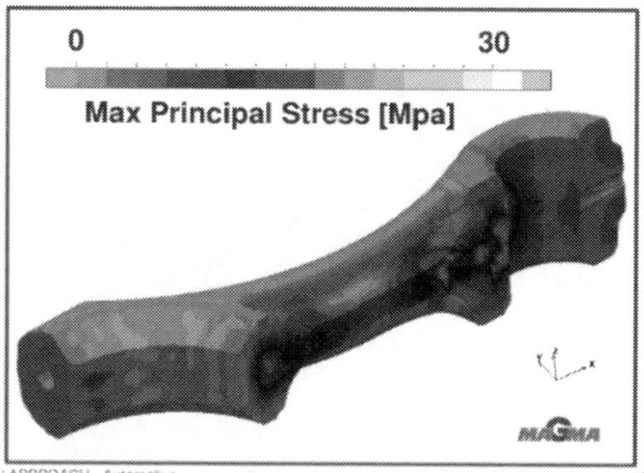

ENGIN SOFT

Key partner in Design Process Innovation

Result transfer

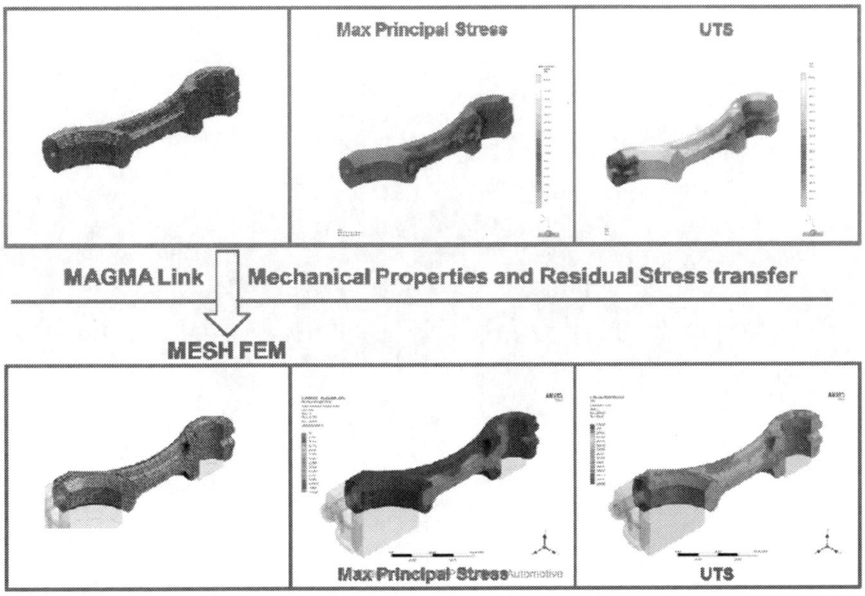

19

ENGIN SOFT

Key partner in Design Process Innovation

Advanced structural analysis (Design Chain)

- **The final mechanical simulation, considering the properties deriving from the production process, evaluates the realistic performance of the Piston Rod.**
- **The previous critical tested areas (foot and head of the piston rod) does not reach the values for breaking.**
- **The innovative process-product approach provides the possibility to reduce the security factor and of course the weight**

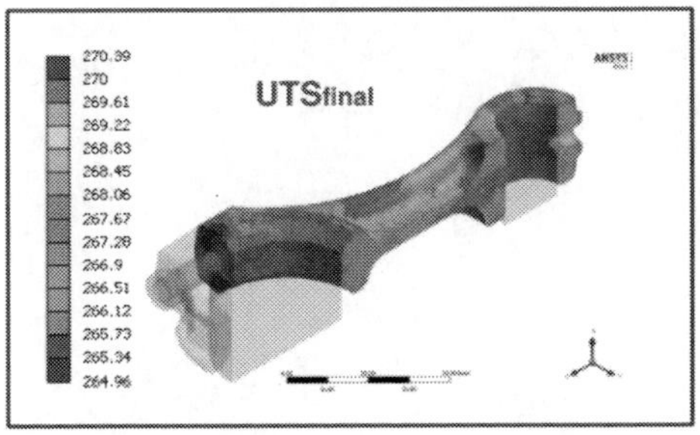

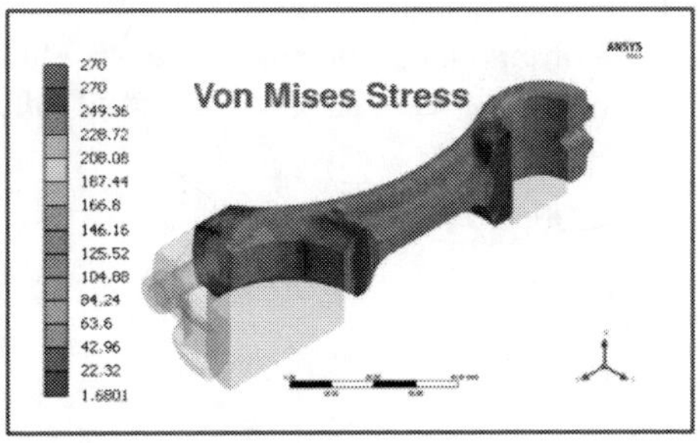

DESIGN CHAIN APPROACH - Automotive 20

Key partner in Design Process Innovation

Design chain approach

- This chain methodology links manufacturing and product design :

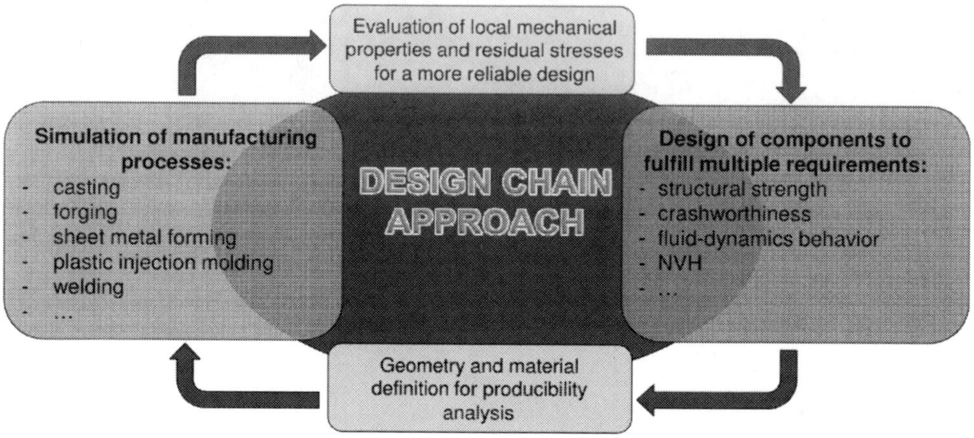

DESIGN CHAIN APPROACH - Automotive 21

Key partner in Design Process Innovation

Design chain approach – Main advantages

- This methodology raises the main advantages of classical numerical simulations to an upper level:
  - reliable and accurate simulations;
  - realistic material modelling;
  - possible design with limited safety factors;
  - weight and material use reduction;
  - product performances improvements (in terms of strength and safety);
  - market appeal;
  - shorter time to market, due to reduced or absent iterative tests;
  - competitiveness;
  - ... and many others.

- Numerical optimization tools can help designers to combine parameters afferent to different areas, evaluate their influence on the final product

- ... in all application fields

DESIGN CHAIN APPROACH - Automotive 22

ENGIN SOFT
Key partner in Design Process Innovation

Conclusions

- The described PIDO methodology is addressed to the High Pressure Die Casting and it is aimed at the product design (HPDC foundry and industrial end-users)
- The innovative approach offers the ability to view the real mechanical behavior of the casting, and consequently the ability to validate and to predict the right response of the component.
- The Design Chain method allows to meet the continuing needs and requirements of:
  - Acceptable manufacturing cost vs forging process
  - Low weight: light alloy vs steel
  - High mechanical properties in the crucial areas
  - Structural integrity maximize fatigue life
  - High productivity

DESIGN CHAIN APPROACH - Automotive 23

ENGIN SOFT
Key partner in Design Process Innovation

Nicola Gramegna

- n.gramegna@enginsoft.it
- Via Giambellino 7, 35129 Padova, Italy
- Phone +39 049 7705311
- www.enginsoft.it
- www.caeconference.it

Thank you

DESIGN CHAIN APPROACH - Automotive 24

2013 Automotive Simulation World Congress

Quality Assurance & Best Practice Development for Internal Combustion Engine Flow Simulations

Wolfgang Bauer, ANSYS Germany GmbH

Outline

Motivation

Quality Assurance & Best Practise
- **Numerical Uncertainty**
- **Model Uncertainty**

Summary and Outlook

Motivation

2013 Automotive Simulation World Congress
October 29 - 30 • Frankfurt am Main, Germany

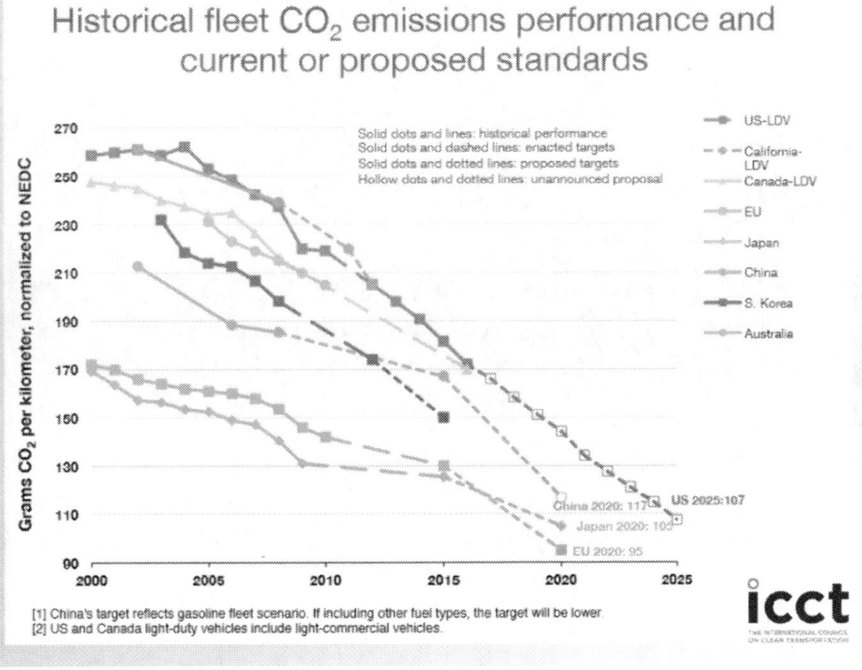

Multi-Physics System

2013 Automotive Simulation World Congress
October 29 - 30 • Frankfurt am Main, Germany

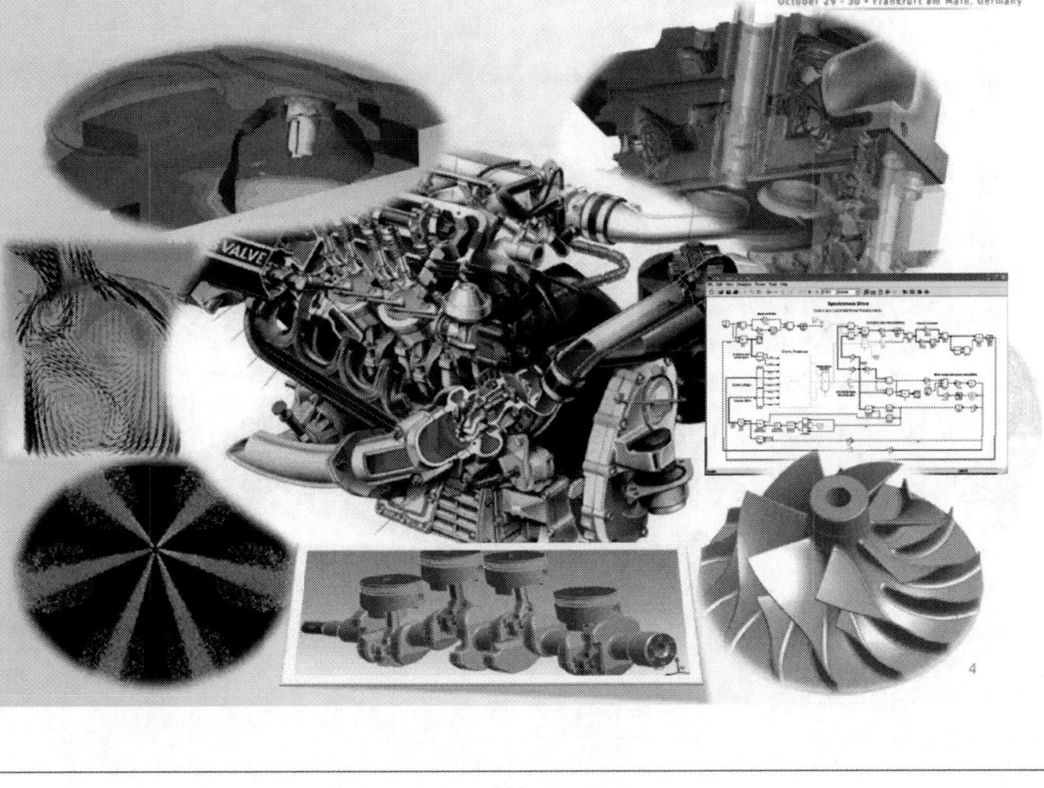

4

Internal Combustion Engines

2013 Automotive Simulation World Congress
October 29 - 30 • Frankfurt am Main, Germany

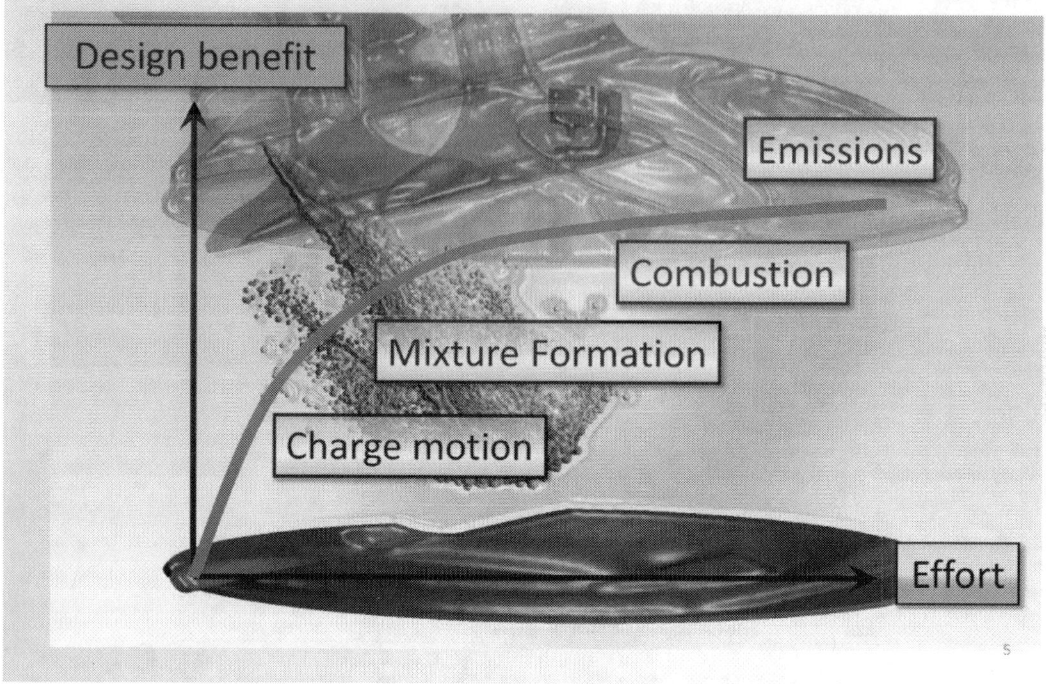

5

Quality Assurance

2013 Automotive Simulation World Congress
October 29 - 30 • Frankfurt am Main, Germany

- Iteration errors
  - Difference between 'converged' solution and solution at iteration 'n'
- Solution errors
  - Difference between converged solution on current grid and 'exact' solution of model equations
  - 'Exact' solution → Solution on infinitely fine grid
- Model errors
  - Difference between 'exact' solution of model equations and reality (data or analytic solution)
- Systematic errors
  - Discrepancies remain, even if numerical and model errors are insignificant

6

Iteration Error, 1

2013 Automotive Simulation World Congress
October 29 - 30 • Frankfurt am Main, Germany

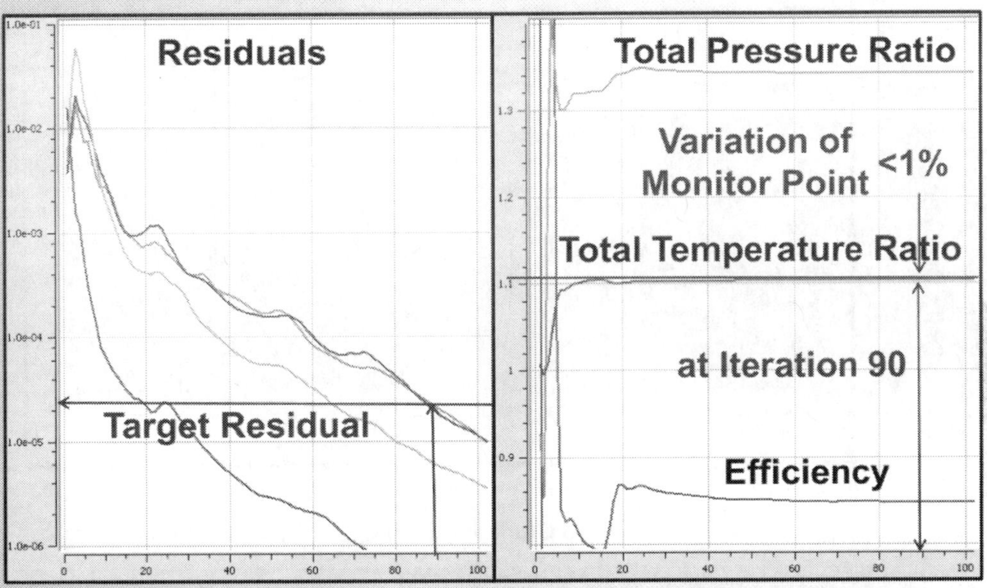

Convergence: steady state

7

Iteration Error, 2

2013 Automotive Simulation World Congress
October 29 - 30 • Frankfurt am Main, Germany

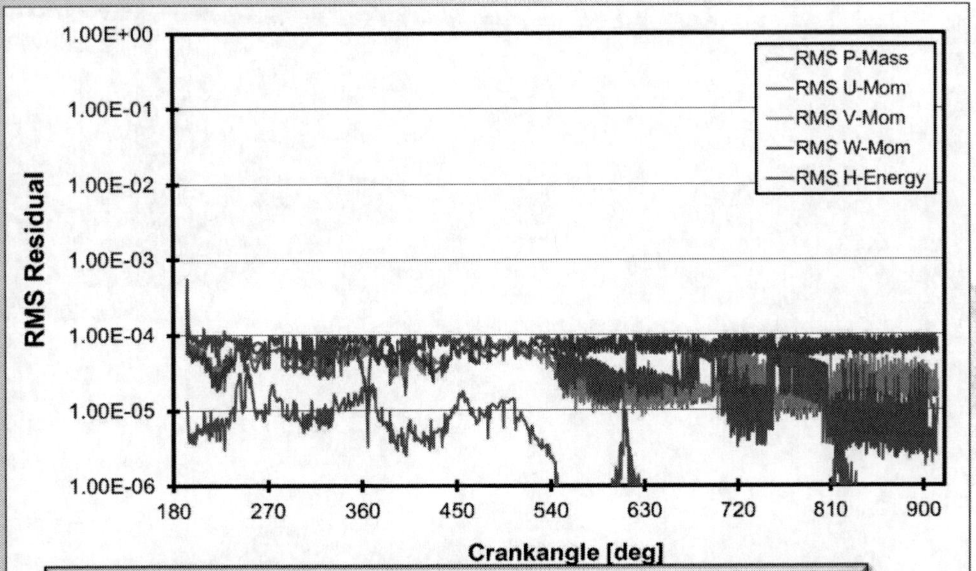

Transient: Each timestep has to be converged!

8

Iteration Error, 3

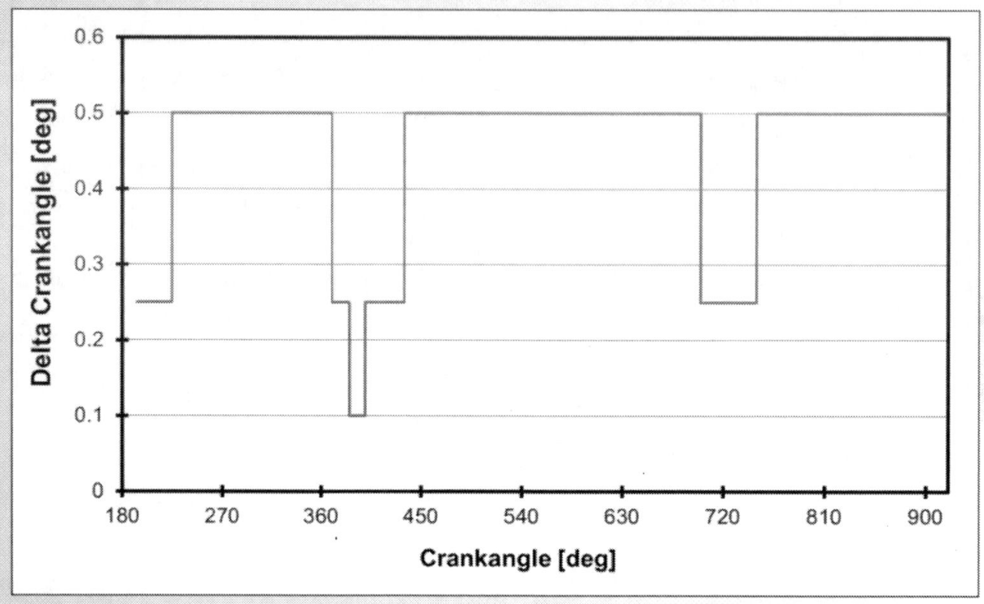

9

Grid resolution, 1

2013 Automotive Simulation World Congress
October 29 - 30 • Frankfurt am Main, Germany

Grid resolution, 2

2013 Automotive Simulation World Congress
October 29 - 30 • Frankfurt am Main, Germany

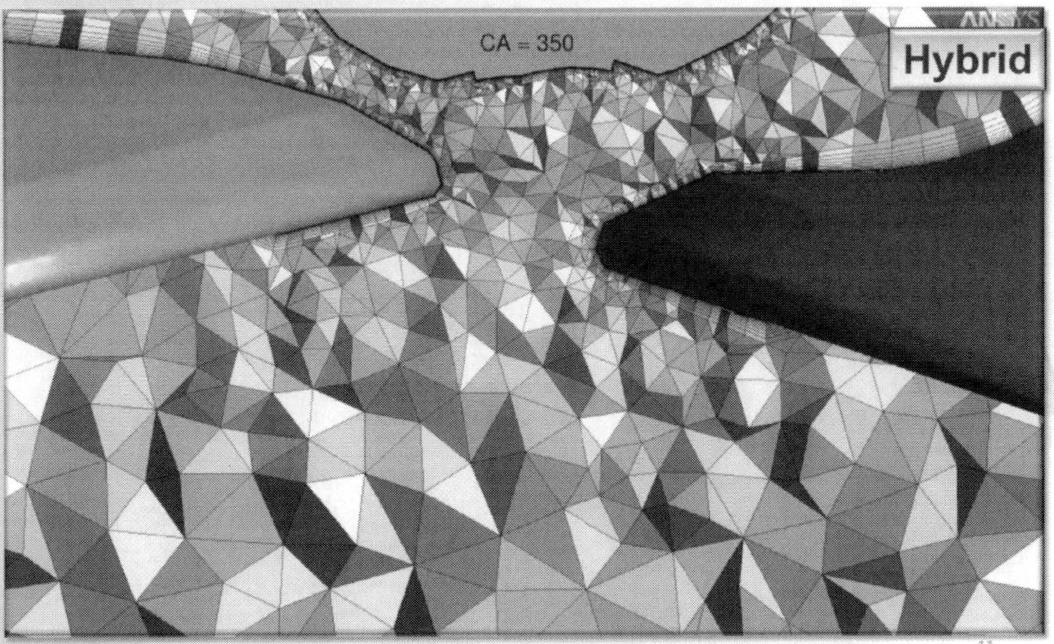

11

Grid resolution, 3

2013 Automotive Simulation World Congress
October 29 - 30 • Frankfurt am Main, Germany

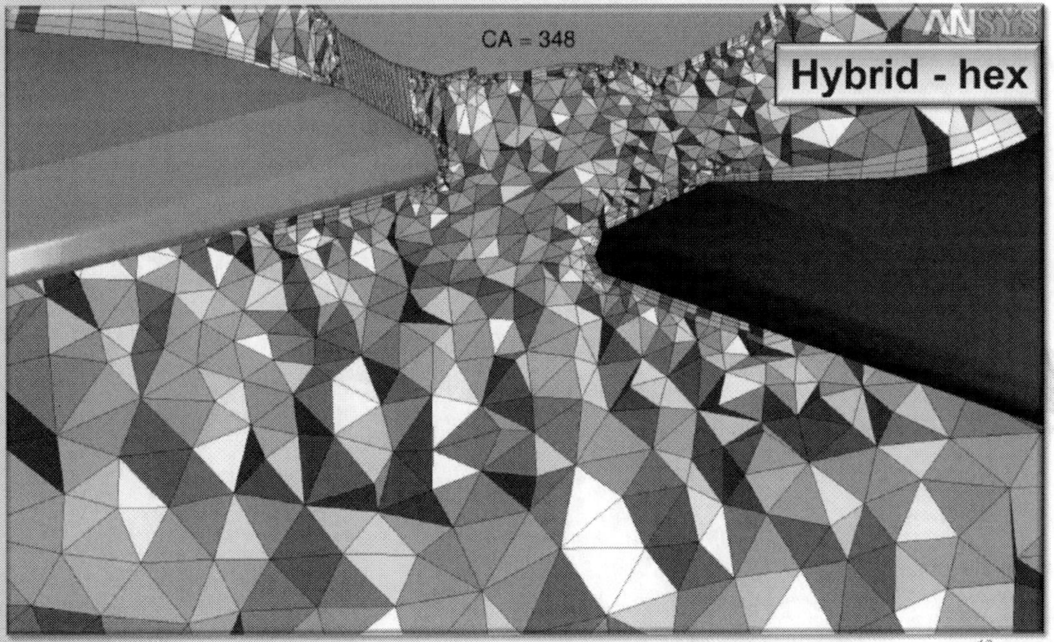

12

Grid resolution, 4

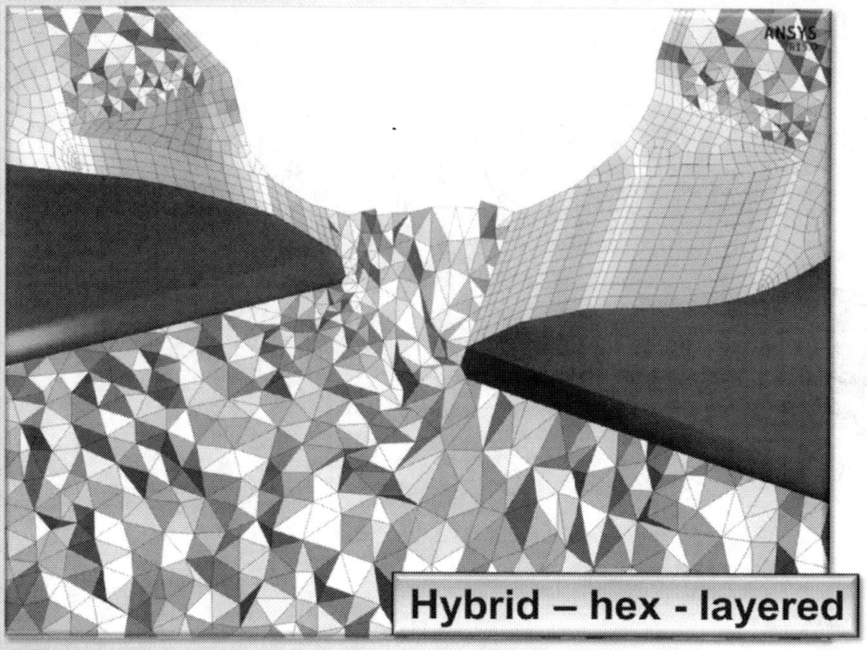

Discretization Error, 2

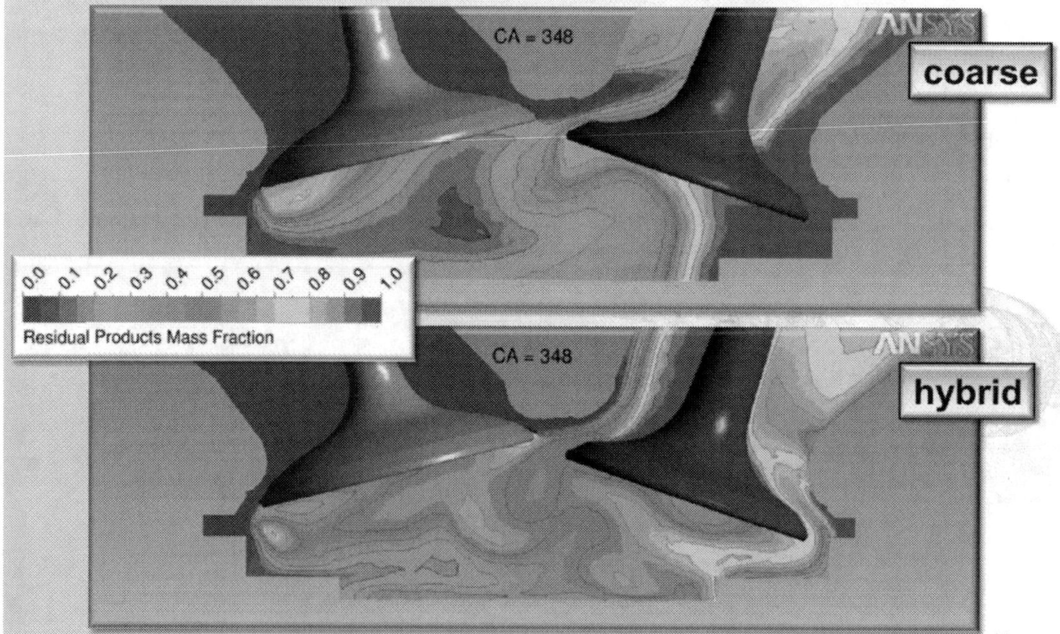

Discretization error

2013 Automotive Simulation World Congress
October 29 - 30 • Frankfurt am Main, Germany

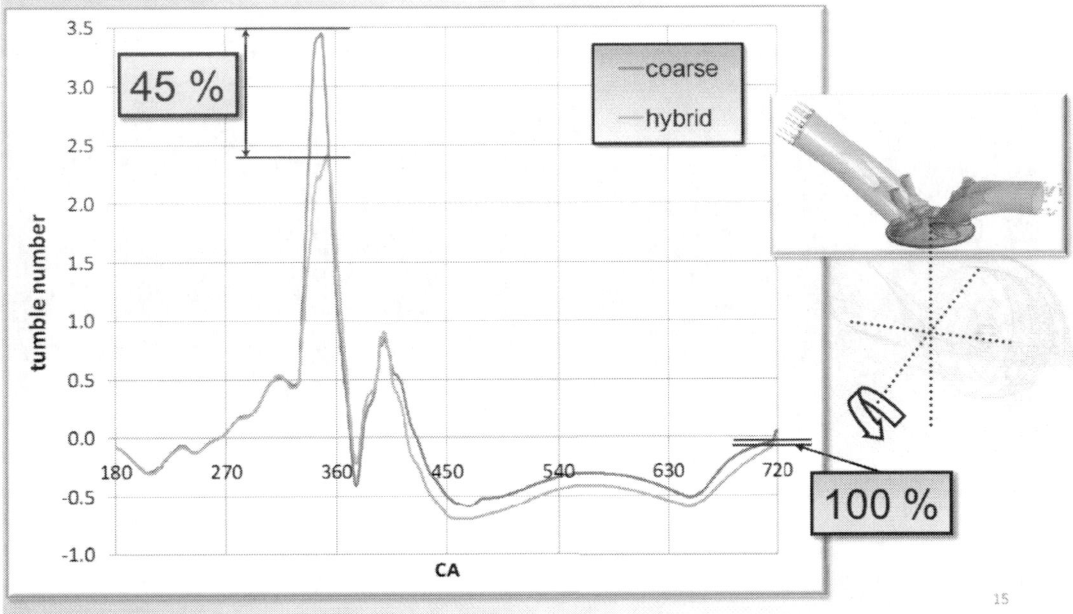

15

Grid: Spray refinement

2013 Automotive Simulation World Congress
October 29 - 30 • Frankfurt am Main, Germany

16

Fuel injection

2013 Automotive Simulation World Congress
October 29 - 30 • Frankfurt am Main, Germany

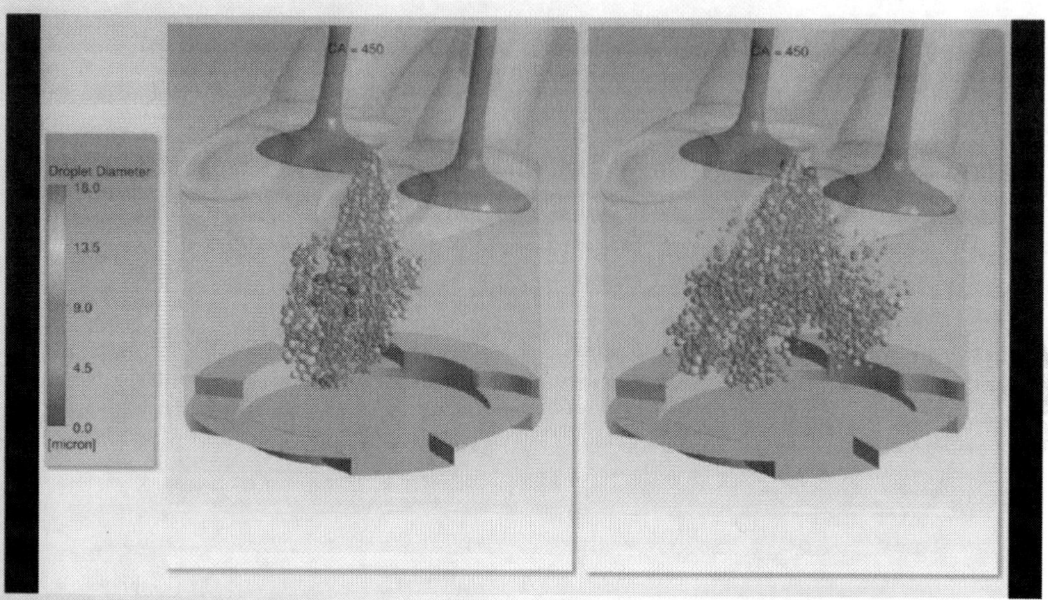

Without refinement

With refinement

17

Grid Size for GDI Engine

2013 Automotive Simulation World Congress
October 29 - 30 • Frankfurt am Main, Germany

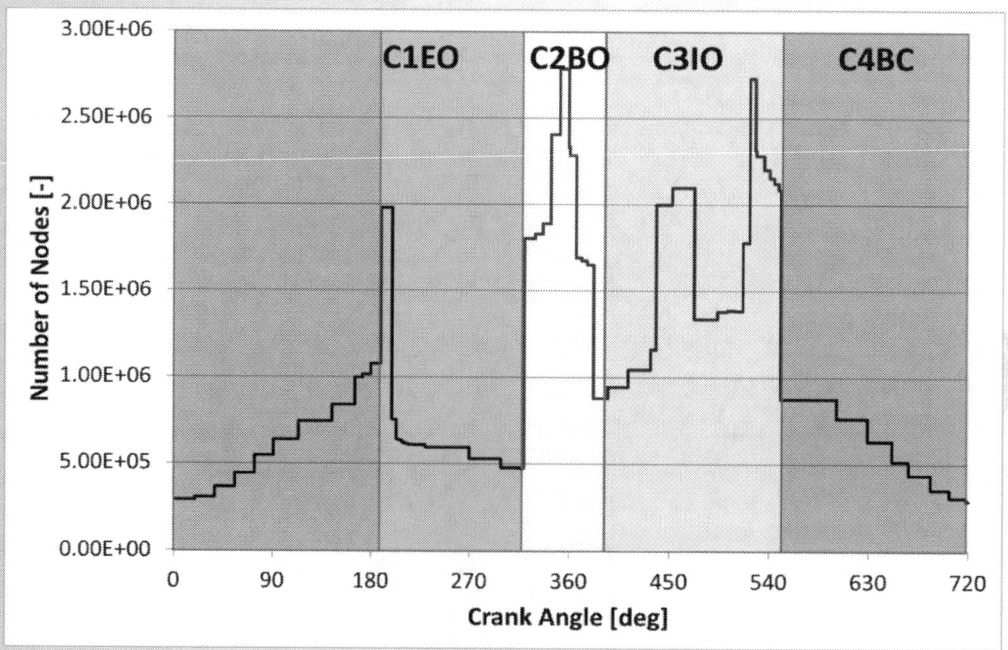

Multi-hole: Cold

2013 Automotive Simulation World Congress
October 29 - 30 • Frankfurt am Main, Germany

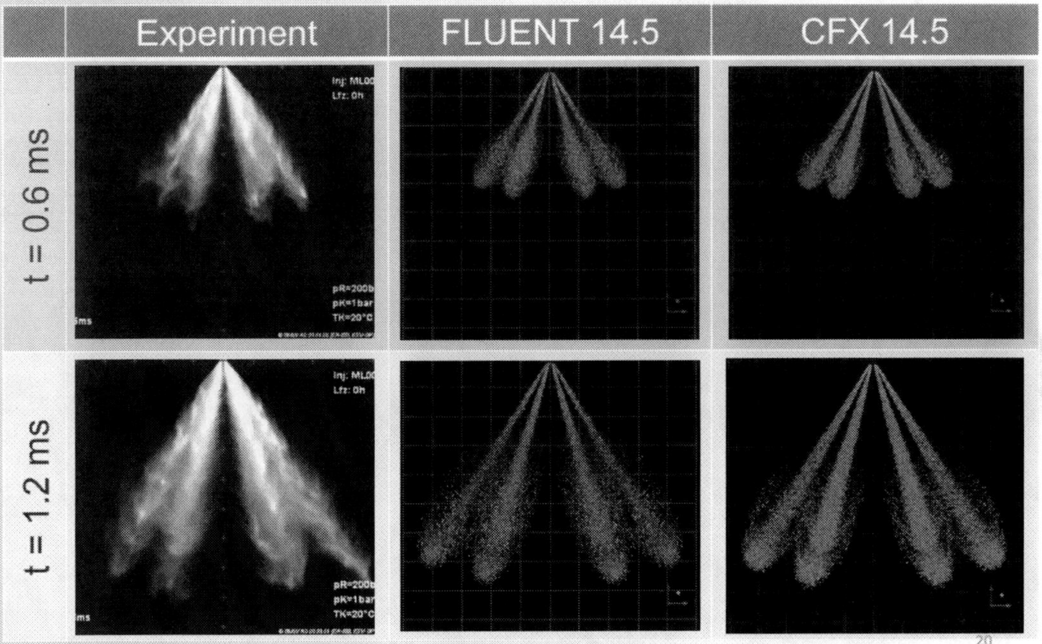

20

Advanced Numerics, 1

2013 Automotive Simulation World Congress
October 29 - 30 • Frankfurt am Main, Germany

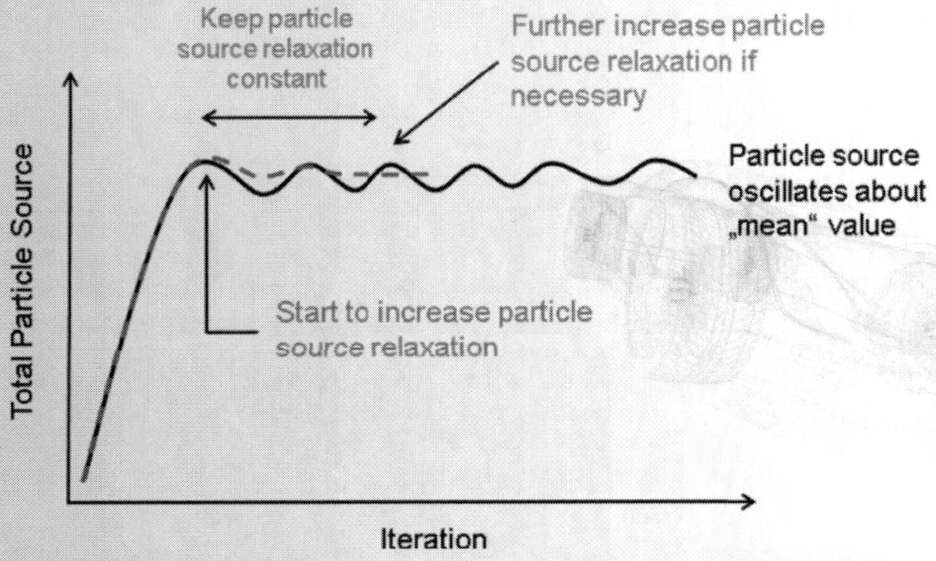

21

Advanced Numerics, 2

2013 Automotive Simulation World Congress
October 29 - 30 • Frankfurt am Main, Germany

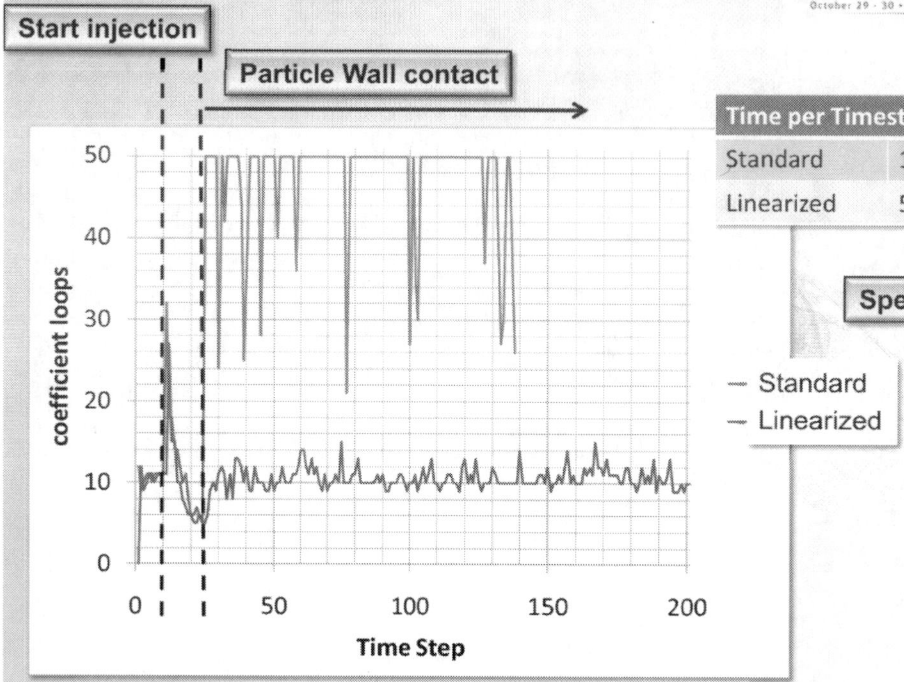

Time per Timestep	
Standard	1831
Linearized	513.2

Speed up 3.6!!

22

Convergence at each time step!

2013 Automotive Simulation World Congress
October 29 - 30 • Frankfurt am Main, Germany

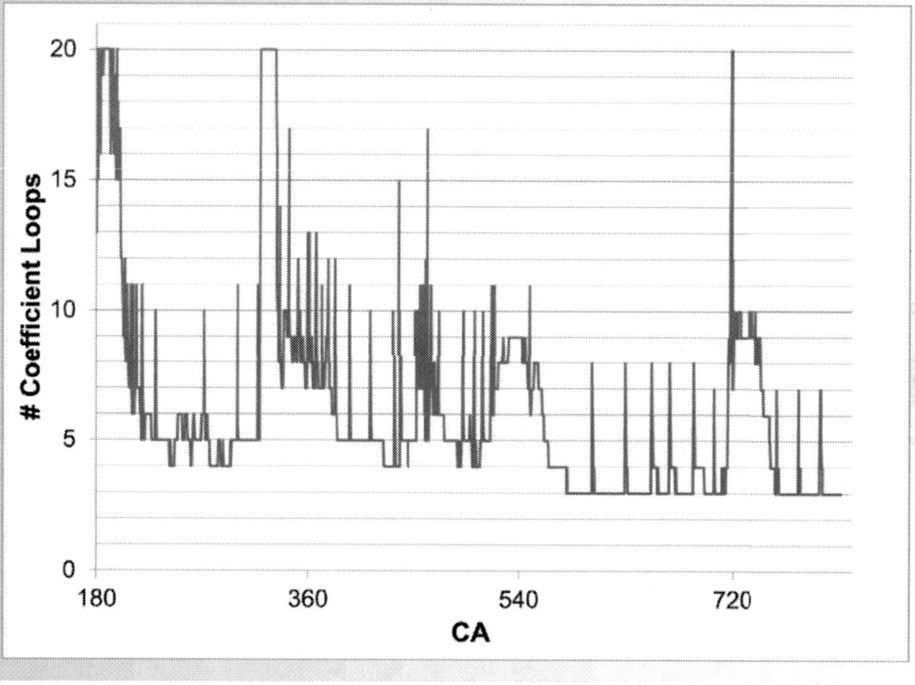

23

Discretzation Error

2013 Automotive Simulation World Congress
October 29 - 30 • Frankfurt am Main, Germany

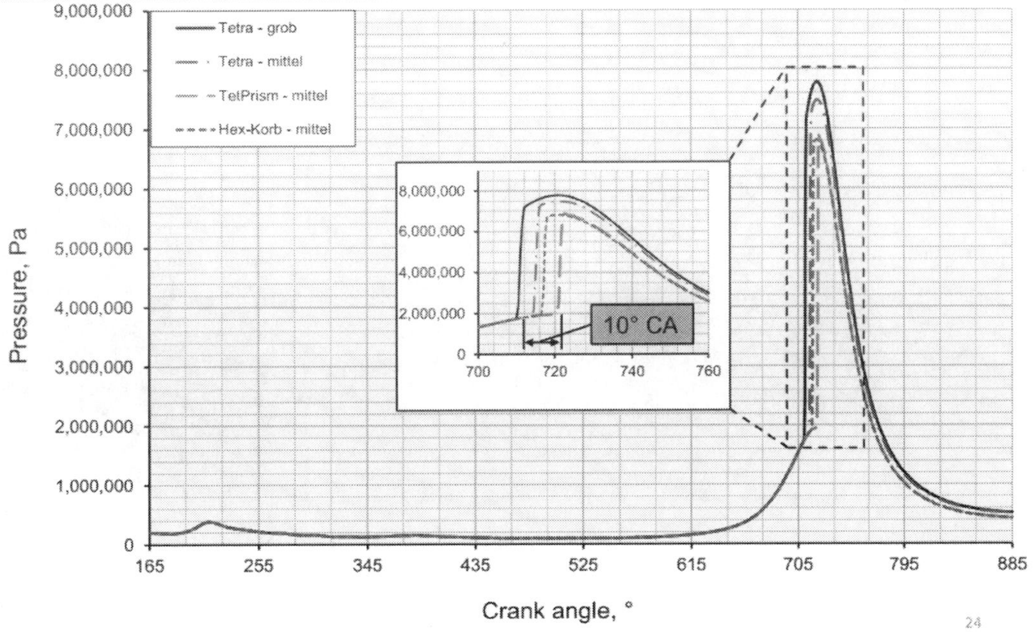

24

Spatial Resolution: TKE

2013 Automotive Simulation World Congress
October 29 - 30 • Frankfurt am Main, Germany

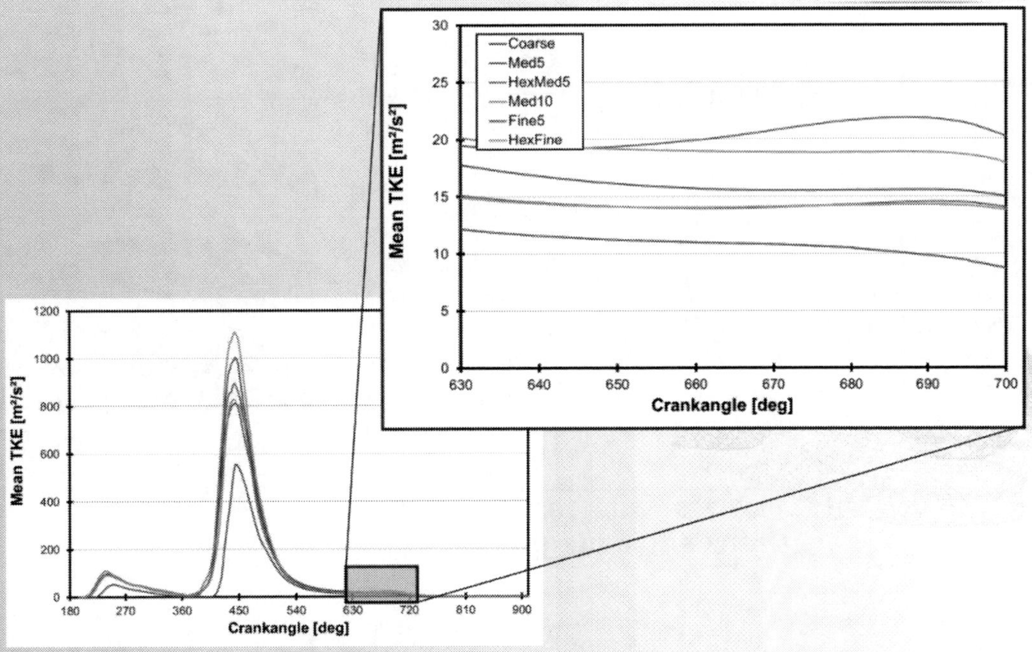

25

Flame Kernel Propagation

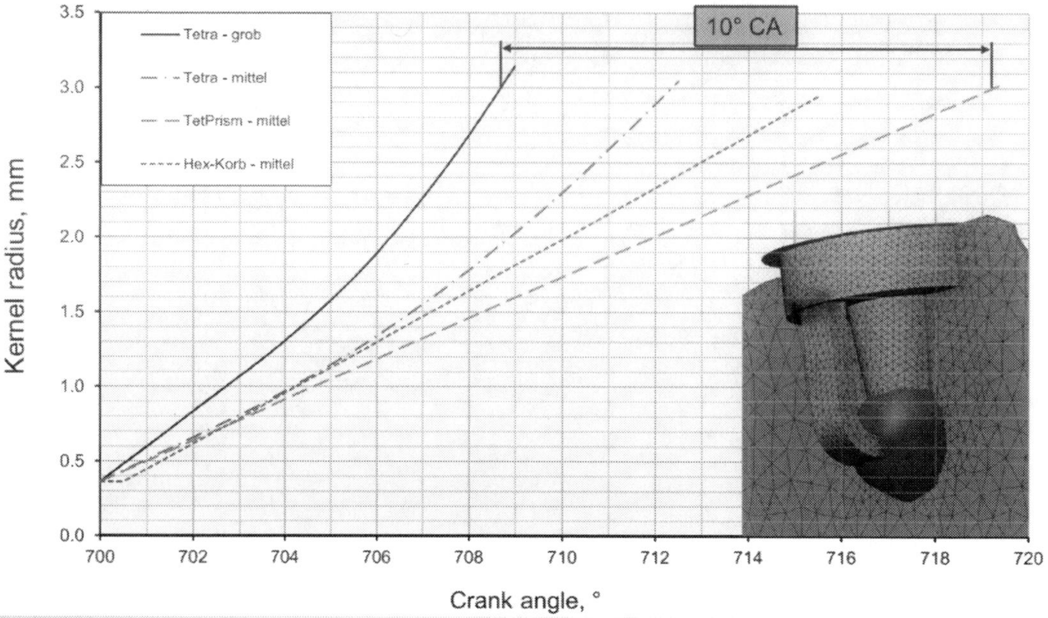

Validation

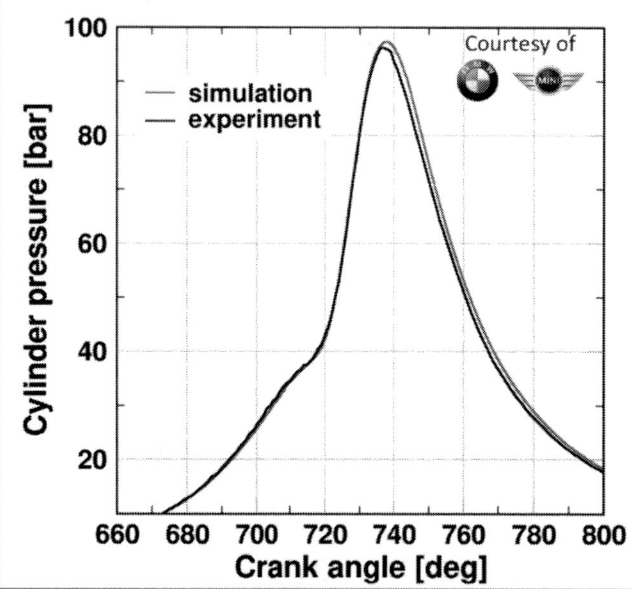

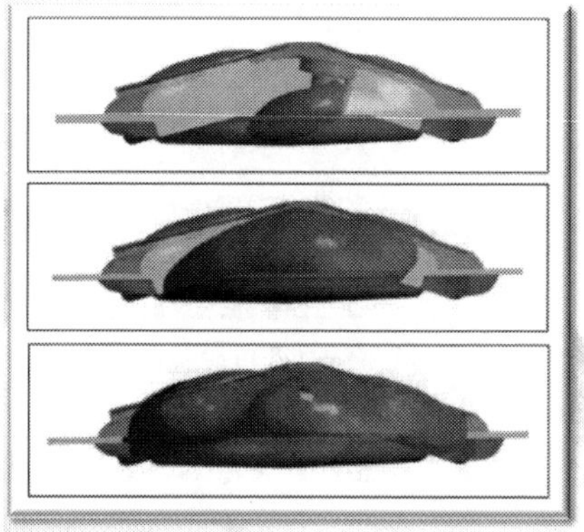

In-Cylinder Flow in a 2V Diesel Engine

2013 **Automotive Simulation** World Congress
October 29 - 30 • Frankfurt am Main, Germany

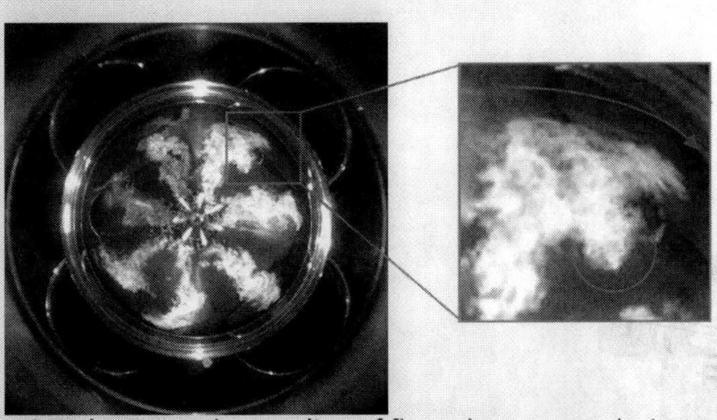

- Instantaneous flow structures (Vortices)
- Cyclic variations
- Turbulence

- Gain better understanding of flow phenomena during gas exchange
  - Experiment
    - Flow investigation using Particle Image Velocimetry (PIV)
    - Generation of a comprehensive database for simulation
  - Simulation

BOSCH

Boundary Conditions

2013 **Automotive Simulation** World Congress
October 29 - 30 • Frankfurt am Main, Germany

- **Special emphasis on simulation needs**

→ **Engine specifications**
- Operating point (1500 rpm, p_{inlet}=1.2 bar)
- Piston and valve lift curves

→ **Geometry**

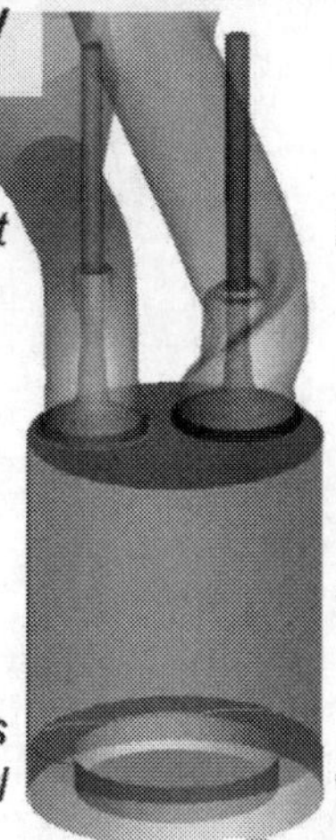

BOSCH

SRS and RANS Simulation

2013 Automotive Simulation World Congress
October 29 - 30 • Frankfurt am Main, Germany

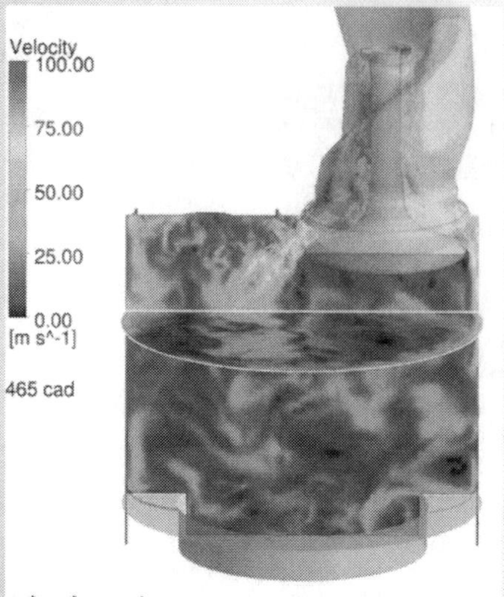

1.2–6.8 mio elements
Δt=0.03 – 0.3 cad, 50 CPU
→ Run time ~ 30 d

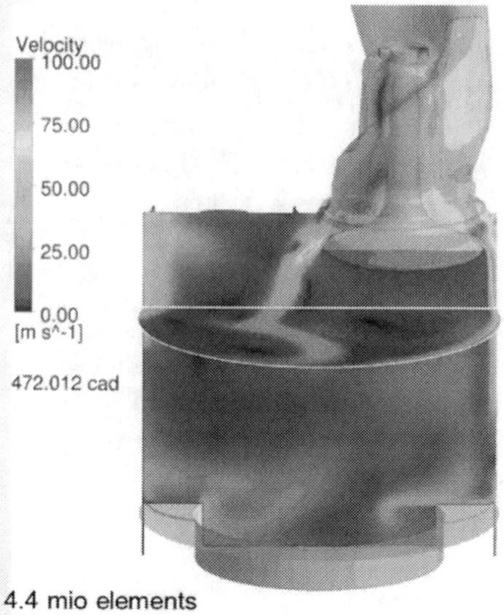

1.2 – 4.4 mio elements
Δt = 0.5 cad, 50 CPU
→ Run time ~ 2.5 d

SRS

URANS

BOSCH

Velocity Profiles

2013 Automotive Simulation World Congress

10 mm below cylinder head
466 cad - Inlet valve max open
631 cad - Mid compression

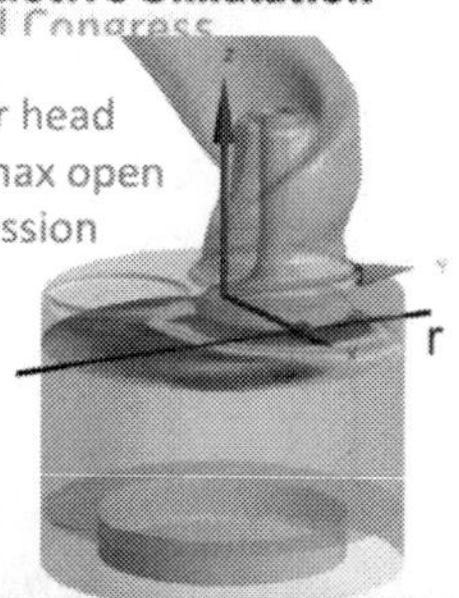

- Comparison of LES to PIV and RANS
  - **Average** for LES, PIV and RANS
  - **Deviation** (±2σ) for LES and PIV
    LES: 11 cycles
    PIV: 200 cycles

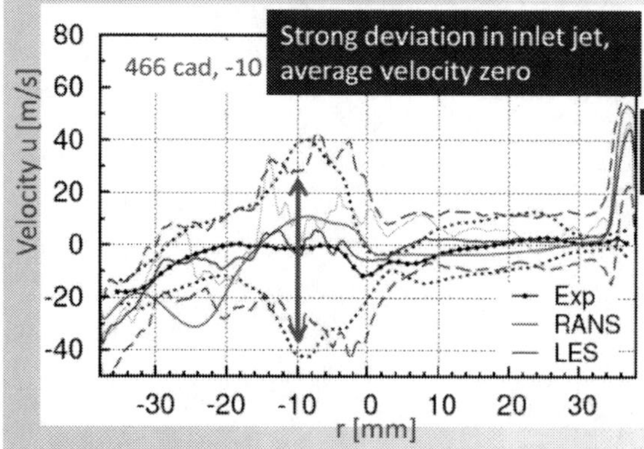

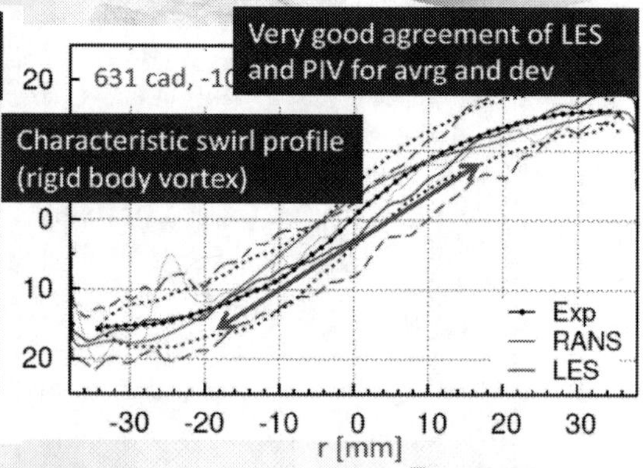

BOSCH

Velocity Field and Swirl Centre

2013 Automotive Simulation World Congress
October 29 - 30 • Frankfurt am Main, Germany

→ Variation of large scale flow characteristics

$$J(\boldsymbol{u}) = \left|\nabla \times \left(\frac{\boldsymbol{u}}{|\boldsymbol{u}|}\right)\right| \longrightarrow \boldsymbol{x}_D : \max_{\boldsymbol{x}} J(\boldsymbol{u}(\boldsymbol{x}))$$

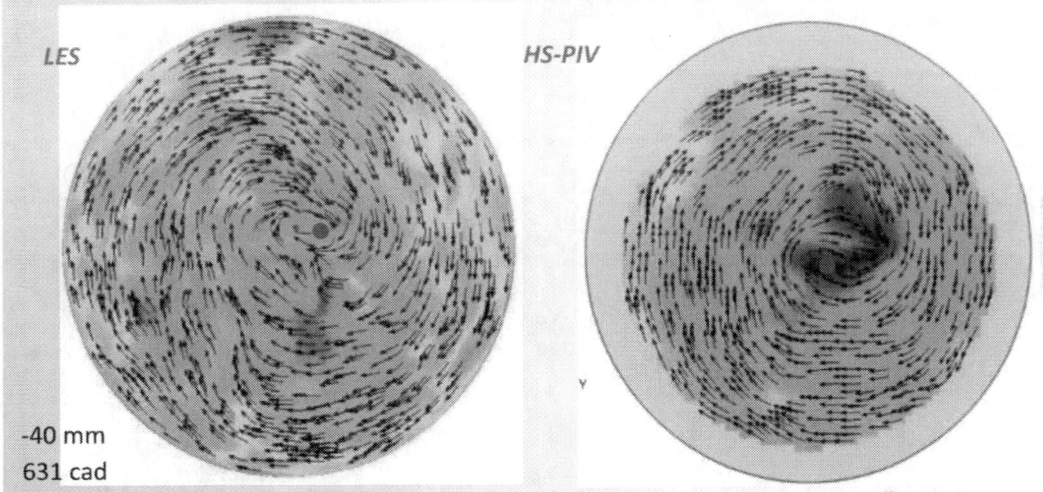

BOSCH

Swirl Centre

2013 Automotive Simulation World Congress
October 29 - 30 • Frankfurt am Main, Germany

- Variation of large scale flow characteristics → cycle-to-cycle variation

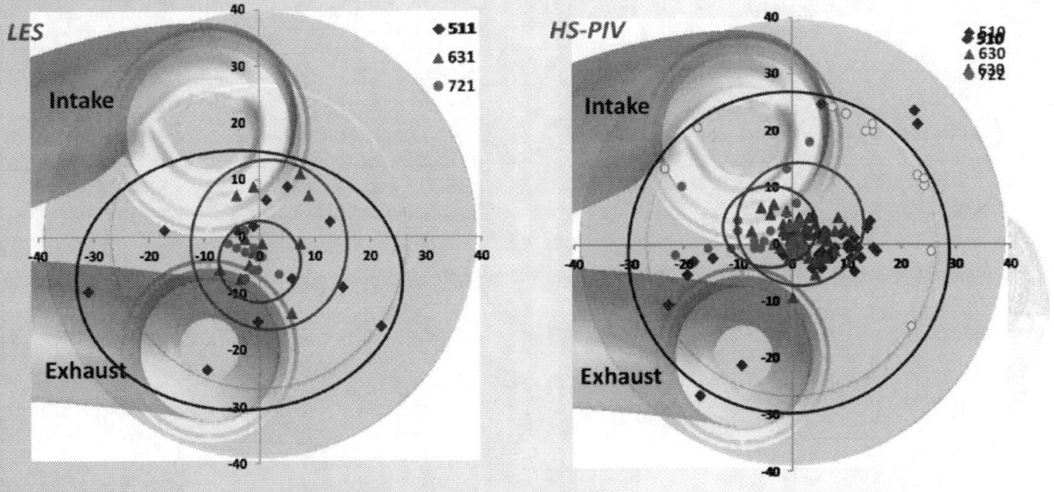

BOSCH

Summary and Outlook

2013 Automotive Simulation World Congress
October 29 - 30 • Frankfurt am Main, Germany

- Simulation process: "backbone"
  - Fully automated
  - Scalable and robust
  - Extensible
- Continuous development
  →Future requirements
- Validation is key to Industry Use
  - Co-operation with OEM's and research institutions

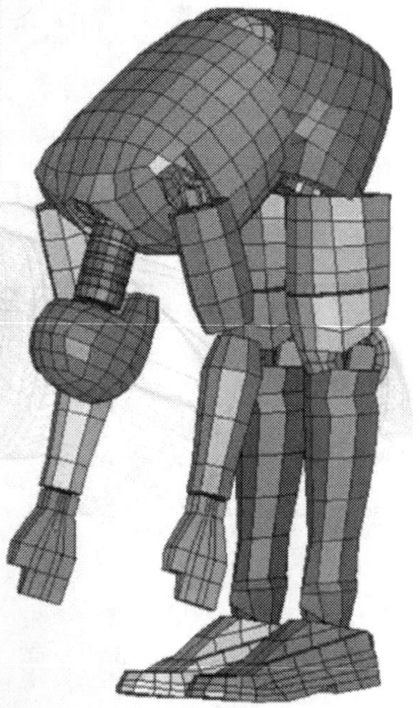

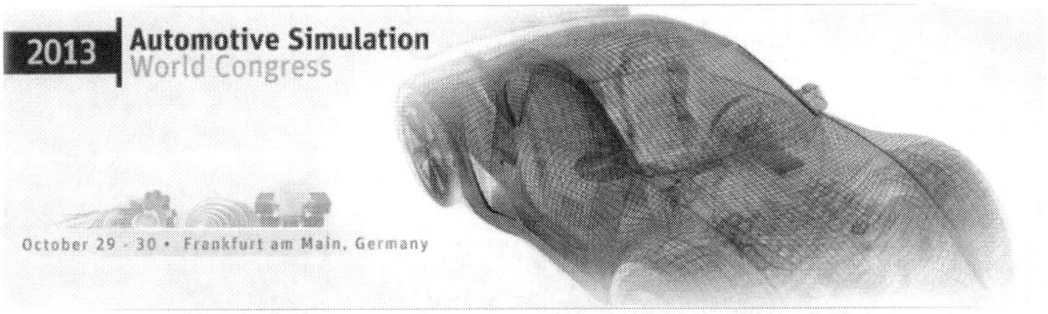

BLDC Motor for Automotive Cooling Fan Assembly: Heat Sink Optimization

Davide Parodi

Fluid Dynamic and Aeroacoustic Engineer, Automotive Product Group, Johnson Electric Asti S.r.l.
Asti, Italy
Email: davide.parodi@johnsonelectric.com
Web: www.johnsonelectric.com

Johnson Electric Asti - GATE

Powertrain cooling

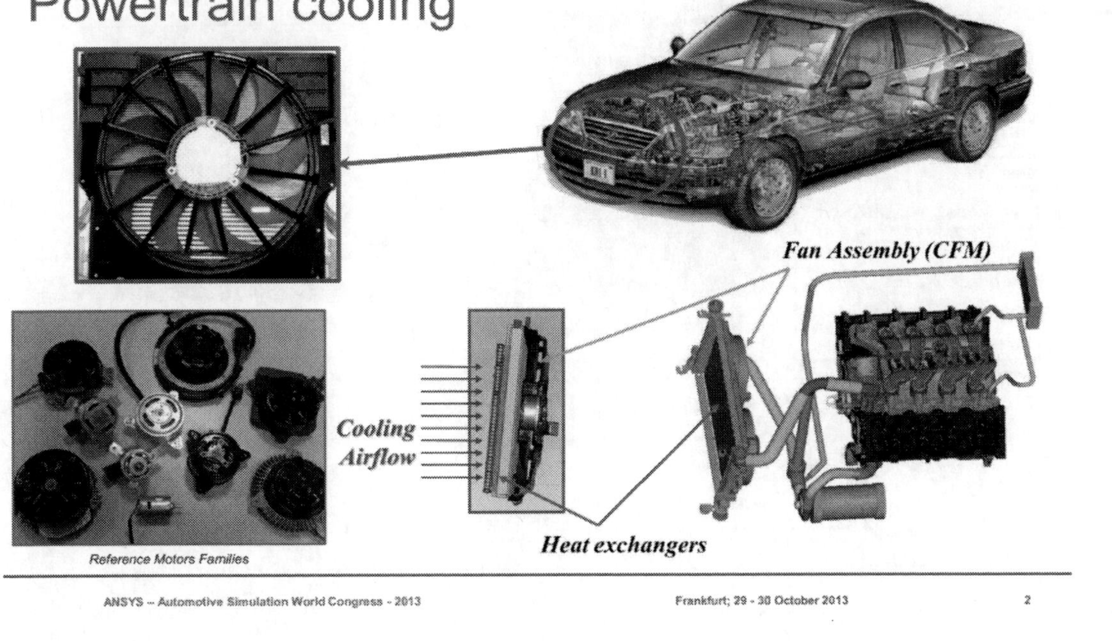

CFD Simulations in JE - Asti

- Commercial software (ANSYS)
- Part of design procedure
- Component optimization
- Data exchange with customers
- Data exchange with suppliers
- Research activities

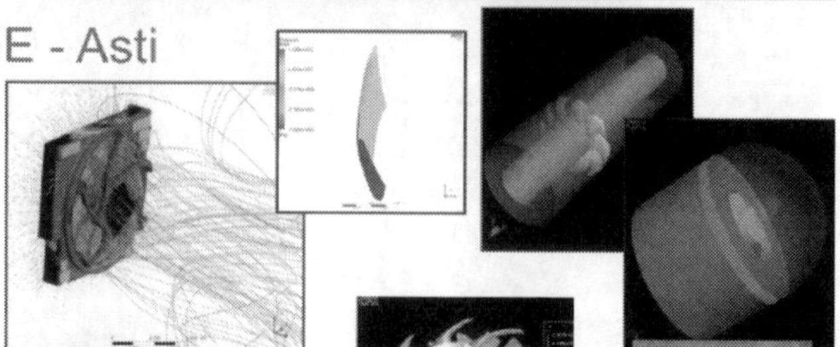

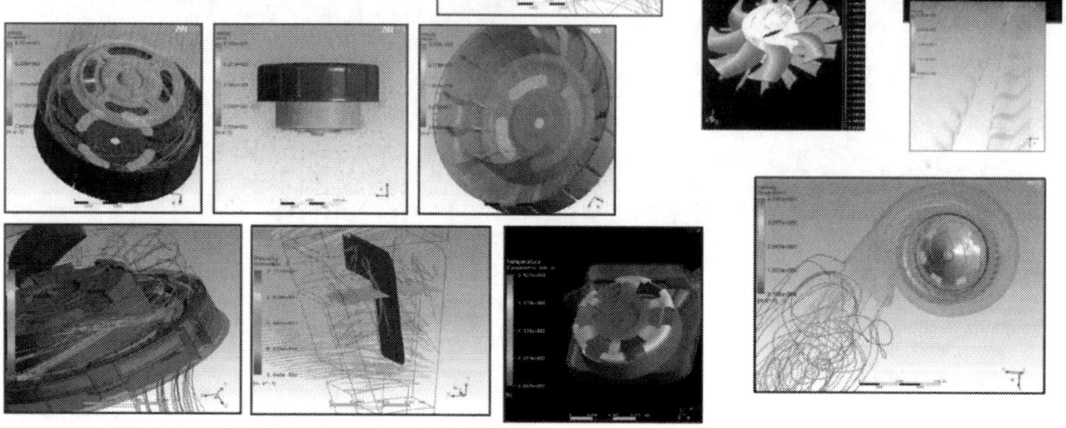

ANSYS – Automotive Simulation World Congress - 2013

Recent investigation activities

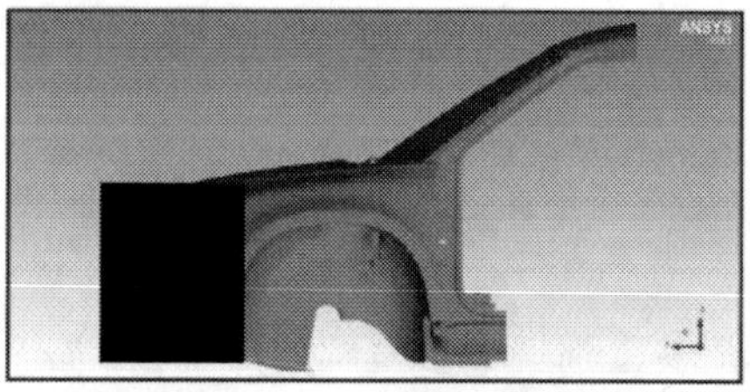

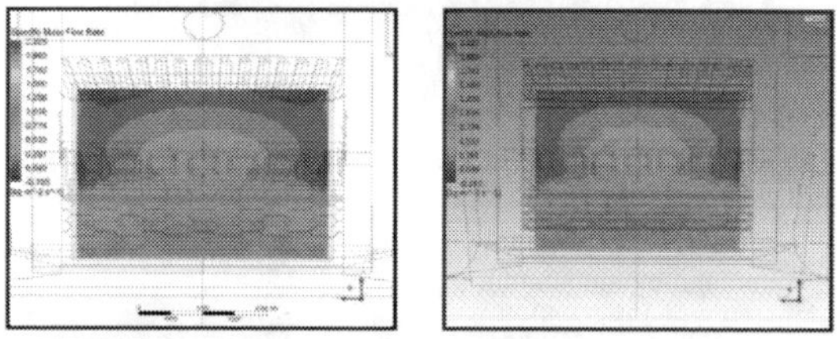

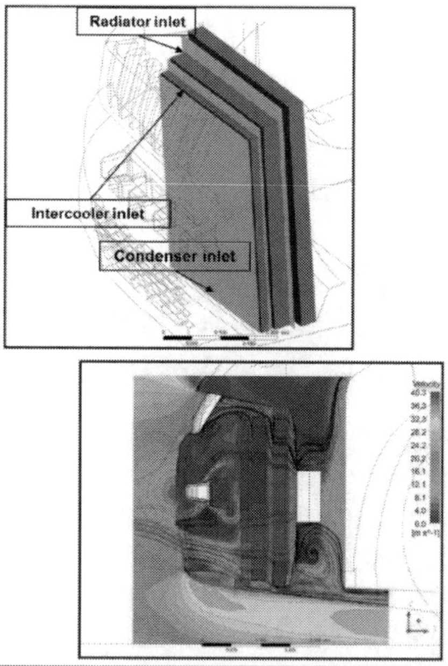

ENGIN SOFT

JOHNSON ELECTRIC
innovating motion

Brushless Electric DC Motor

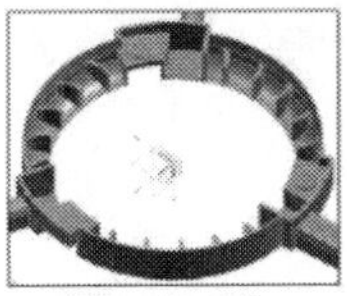

Deflector ring view

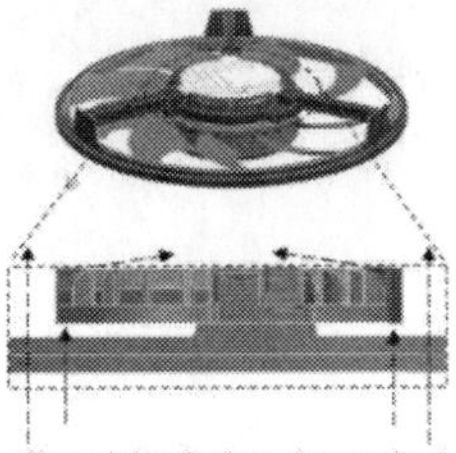

Shroud ring (bottom view section)

The thermal stabilization test

Aluminum Heat Sink with Gap Pad		T Amb 30°C		T Amb 90°C		T Amb 110°C	
		DutyCycle%	86%	DutyCycle%	86%	DutyCycle%	86%
Watt		415		417		416	
Rpm		2670	ΔT	2750	ΔT	2770	ΔT
T AMB	°C	[redacted]	[redacted]	[redacted]	[redacted]	111.1	
[redacted]	°C	[redacted]	[redacted]	[redacted]	[redacted]	[redacted]	[redacted]
[redacted]	°C	[redacted]	[redacted]	[redacted]	[redacted]	[redacted]	[redacted]
[redacted]	°C	[redacted]	[redacted]	[redacted]	[redacted]	[redacted]	[redacted]
[redacted]	°C	[redacted]	[redacted]	[redacted]	[redacted]	[redacted]	[redacted]
[redacted]	°C	[redacted]	[redacted]	[redacted]	[redacted]	[redacted]	[redacted]
[redacted]	°C	[redacted]	[redacted]	[redacted]	[redacted]	[redacted]	[redacted]
[redacted]	°C	[redacted]	[redacted]	[redacted]	[redacted]	[redacted]	[redacted]
L1 ind	°C	[redacted]	[redacted]	[redacted]	[redacted]	126.3	[illegible]
[redacted]	°C	[redacted]	[redacted]	[redacted]	[redacted]	[redacted]	[redacted]
[redacted]	°C	[redacted]	[redacted]	[redacted]	[redacted]	[redacted]	[redacted]
[redacted]	°C	[redacted]	[redacted]	[redacted]	[redacted]	[redacted]	[redacted]
[redacted]	V	[redacted]	[redacted]	[redacted]	[redacted]	[redacted]	[redacted]
[redacted]	A	[redacted]	[redacted]	[redacted]	[redacted]	[redacted]	[redacted]
[redacted]	°C	[redacted]	[redacted]	[redacted]	[redacted]	[redacted]	[redacted]
[redacted]	°C	[redacted]	[redacted]	[redacted]	[redacted]	[redacted]	[redacted]
[redacted]	°C	[redacted]	[redacted]	[redacted]	[redacted]	[redacted]	[redacted]
Q1	°C	[redacted]	[redacted]	[redacted]	[redacted]	132.9	[illegible]
Q2	°C	[redacted]	[redacted]	[redacted]	[redacted]	134.2	[illegible]
Q3	°C	[redacted]	[redacted]	[redacted]	[redacted]	131.8	[illegible]
Q4	°C	[redacted]	[redacted]	[redacted]	[redacted]	132.8	[illegible]
Q5	°C	[redacted]	[redacted]	[redacted]	[redacted]	132.5	[illegible]
Q6	°C	[redacted]	[redacted]	[redacted]	[redacted]	132.3	[illegible]

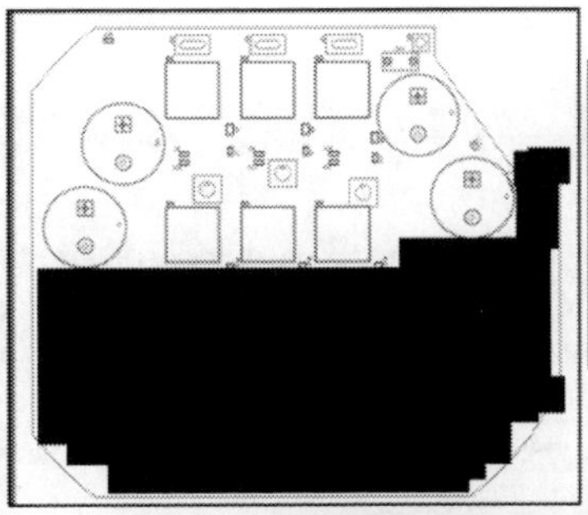

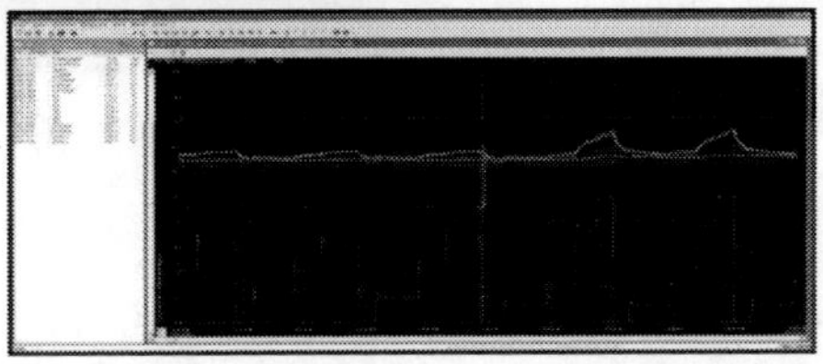

Thermal Stabilization in Oven:

- *Required by some Customers for Product validation*
- *Regularly performed internally for design validation*

ANSYS – Automotive Simulation World Congress - 2013

Purpose of analyses

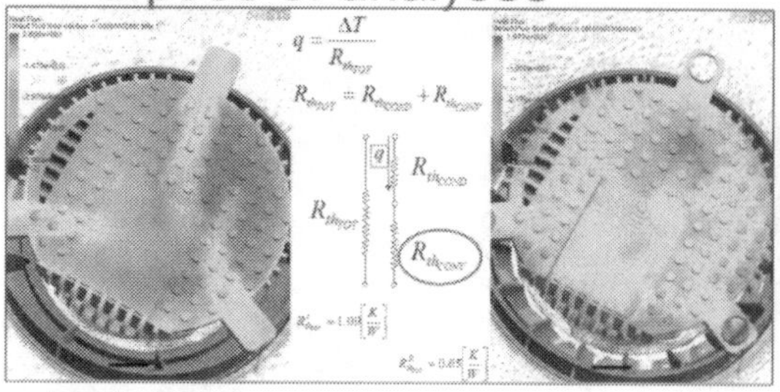

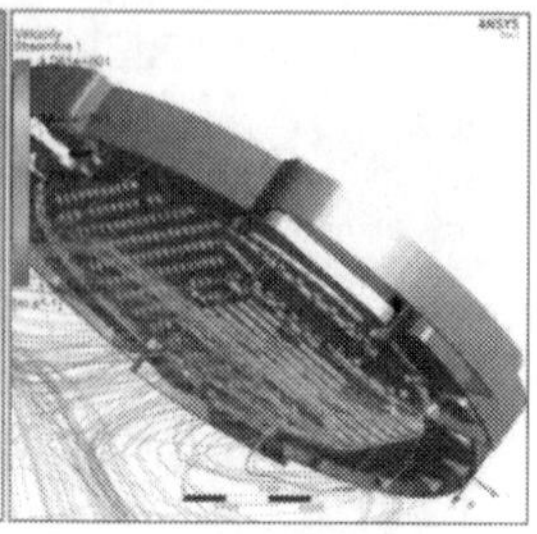

Starting point:

- Technically optimized configuration

Issues:

- Not completely profitable for lower power classes
- Heavy for lower power classes

Ways:

- Investigation on new heat sink design and materials
- Investigation about influence of different gap pad materials

Constraints:

- Electronics temperatures under prescribed safety limits: T_{safe} = 155 [°C]
- Final design: lighter, cheaper, reliable, feasible, etc.

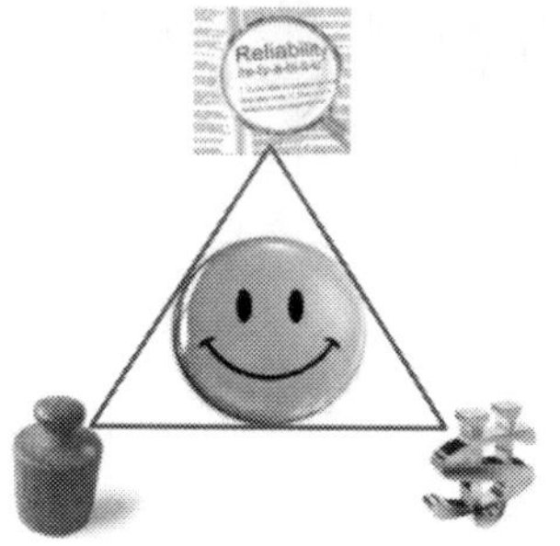

Preliminary options

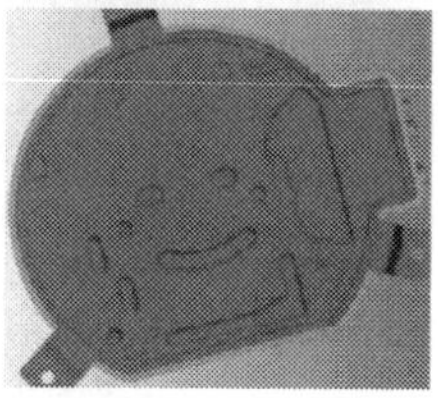

❖ ***Heat sink concept modification***

- From aluminum mold heat sink to metal stamped lamination
- Preliminary metal options: iron, steel

- Materials thermo-physics characteristics strongly different (4 times)
- Material for traditional aluminum heat sink: Thermal conductivity: ~ 240 [W m^{-1} K^{-1}]
- Material for new steel heat sink: Thermal conductivity: ~ 60 [W m^{-1} K^{-1}]

❖ ***Points for differentiation on production process***

- Material thickness reduction necessary
- Significant exchange surface reduction
- Planarity issues

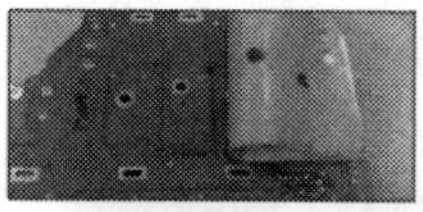

❖ ***Internal coupling material modification***

- From traditional gap – pad to gap filler paste
- Heat transfer conductivity typically reduced

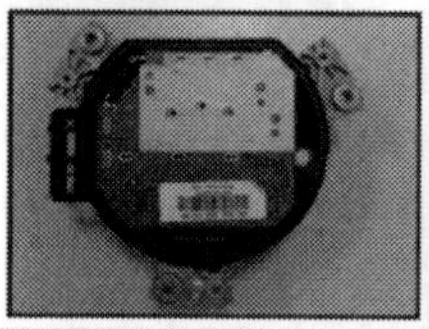

Heat transfer mechanisms modeled

Heat transfer problem: balance between contributions of conduction, convection and radiation.

Radiation: present at every temperature value but important for high temperature values

Temperature range for the present problem: < 180 [°C] → Radiation negligible !

Heat transfer mechanisms solved: ***conduction, convection***

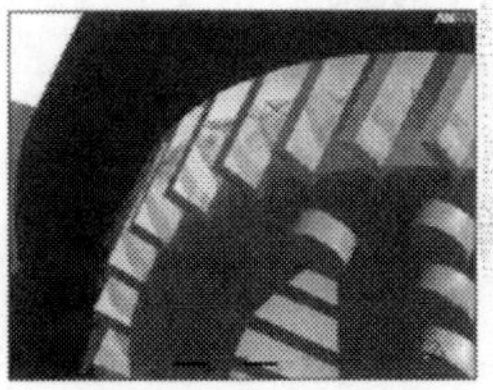

Key parameters on heat transfer analysis:

1. Geometry (thickness, exchange surface)
2. Material properties (specifically: thermal conductivity)
3. Flow field characteristics (local air speed, temperatures, etc.)

- Complex fluid path and geometry
- Problem focused on heat rejection
- Realistic and reliable models should consider all parts affecting fluid motion

Complete fluid-dynamic and thermal analysis

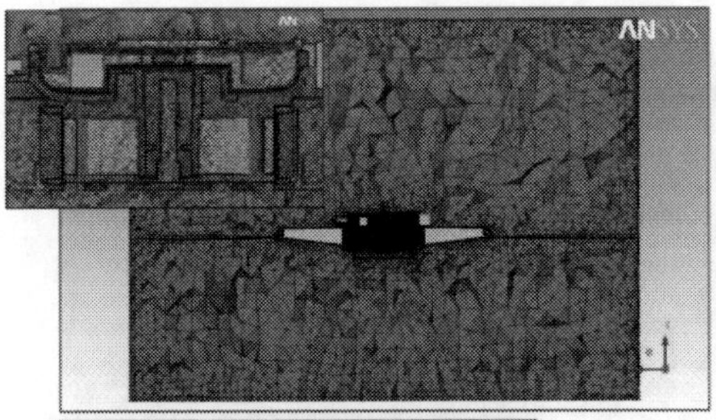

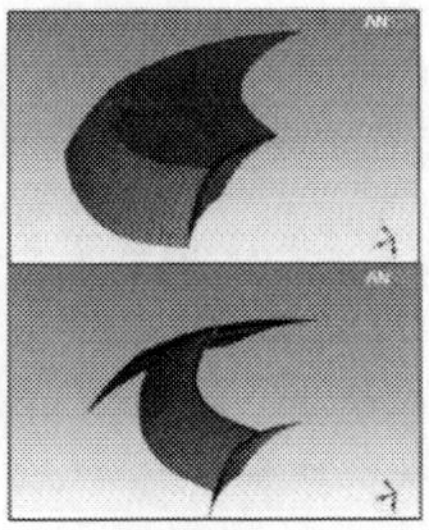

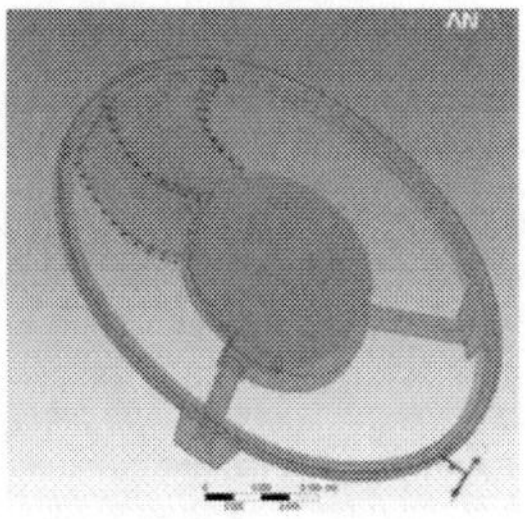

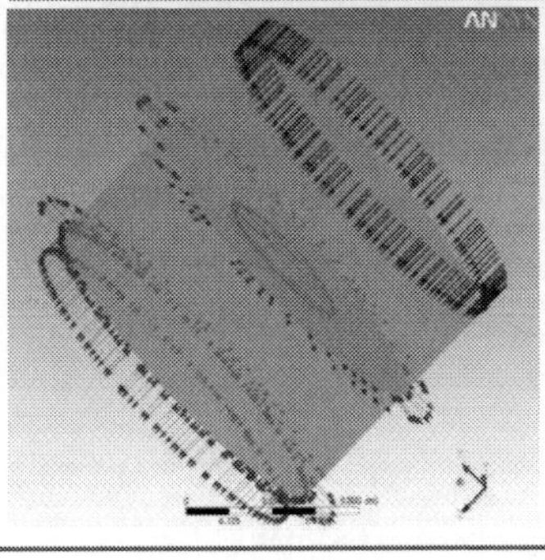

Detailed thermal-fluid dynamic model needed

- **Previous models exploited: fluid domain; solid domain; fan fluid domain**
- **Original Model Size: up to 15 x 10^6 of elements (TETRA + PRISM; HEXA), even exploiting symmetric properties of turbo-machinery for fan**

Fluid domain model simplification without affecting input flow field by means of a cutting plane

ANSYS – Automotive Simulation World Congress - 2013

MOTOR ASSEMBLY – Solid interfaces

- Aluminum heat sink to be compared with metal cover
- PCB Electronic components
- Plastic cover (FRAME)

All of them have been modeled in the current CFD model…

From CFD Post-Processing, their impact can be simplified as surface input temperature, as per the following description…

ANSYS – Automotive Simulation World Congress - 2013

Complete model reduction – Fluid dynamic Inputs (1)

- **Previous CFD analysis used for fluid dynamic inputs to simplify models by means of user functions**
- **95% similar geometry and same upstream thermal-fluid-dynamic field (Electric Power: 400 ÷ 450 [W])**

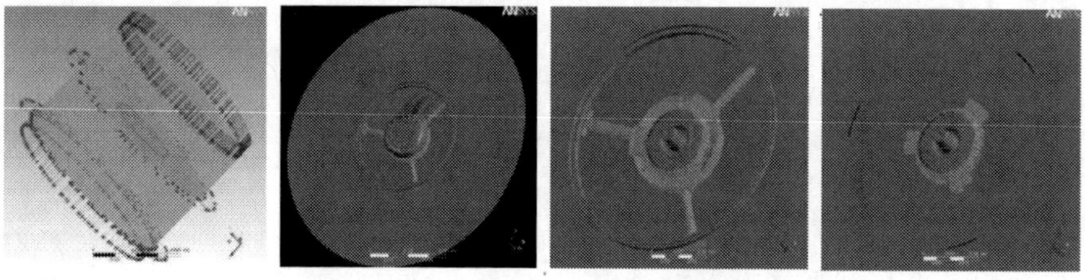

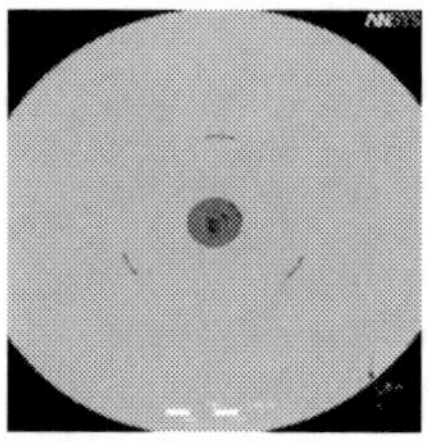

New input surface

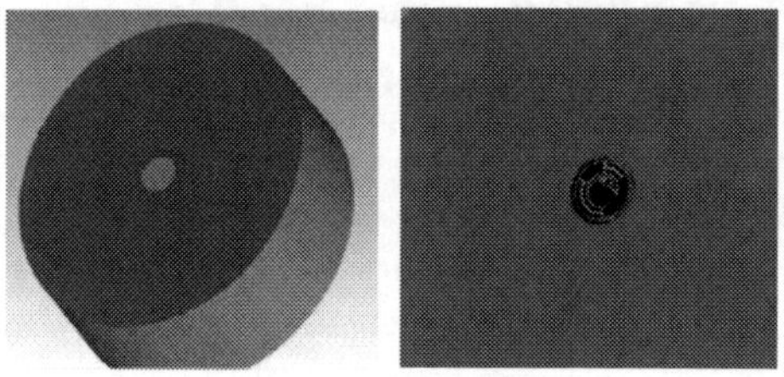

New geometrical model

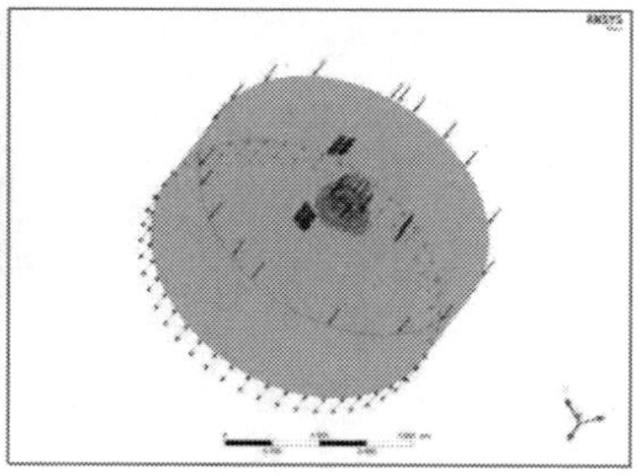

Complete model reduction – Fluid dynamic Inputs (2)

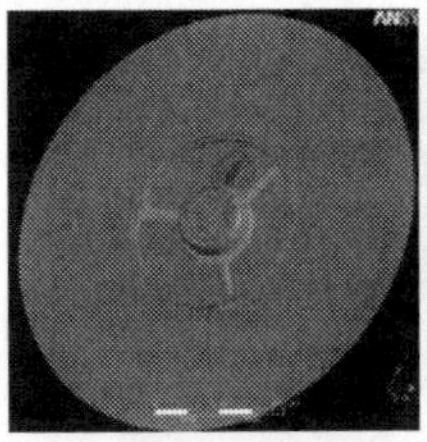

Original CFD model

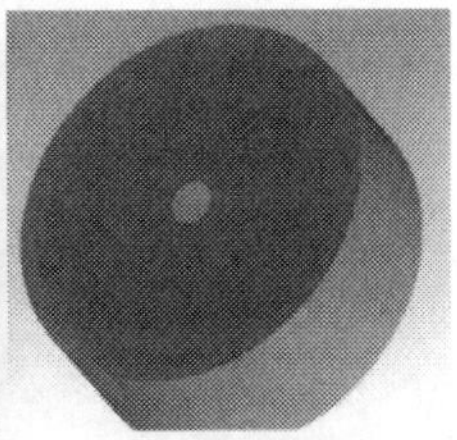

New geometrical model

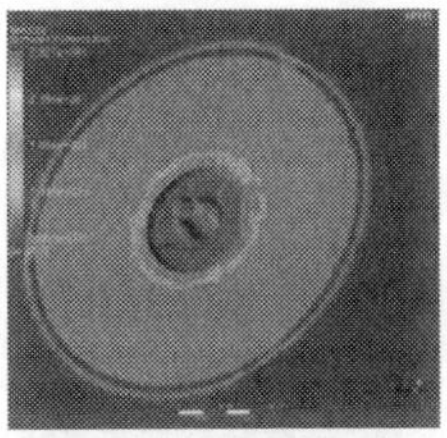

Old velocity profile re-used for the new model

Old velocity profile splitted on its three components

v_x, v_y, v_z

...plus T, TKE, TurbEddyDiss

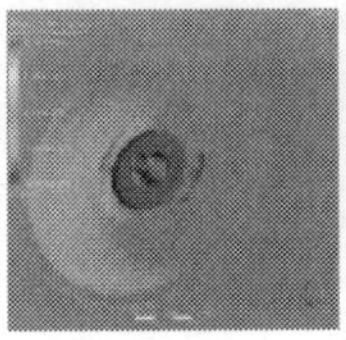

v_x

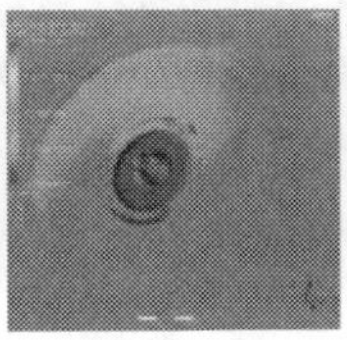

v_y

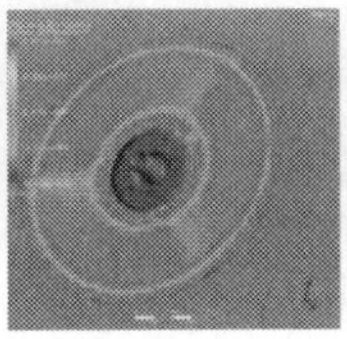

v_z

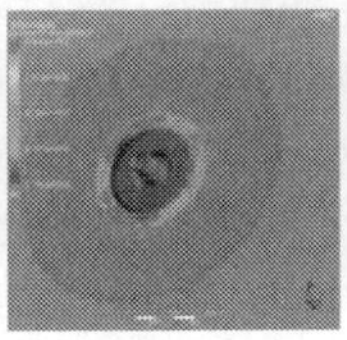

T_{air}

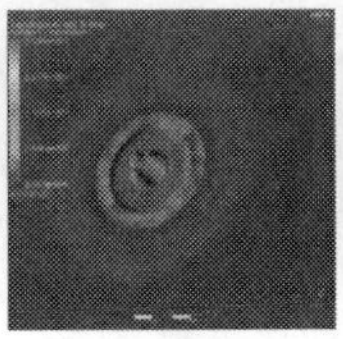

TKE

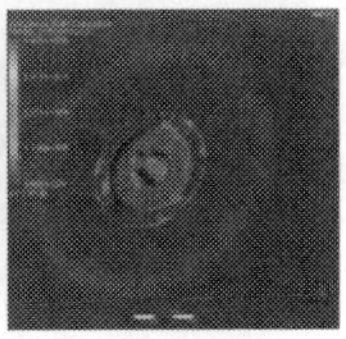

Turbulence Eddy Dissipation

Plastic Frame-Motor Support Interface & Internal Air Material

Aluminum motor support – plastic frame interface: constant wall temperature

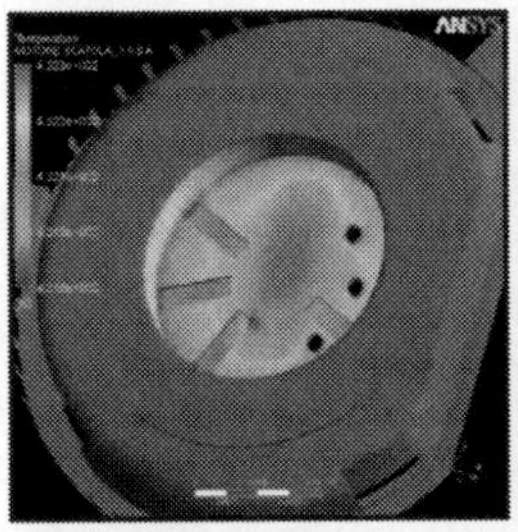

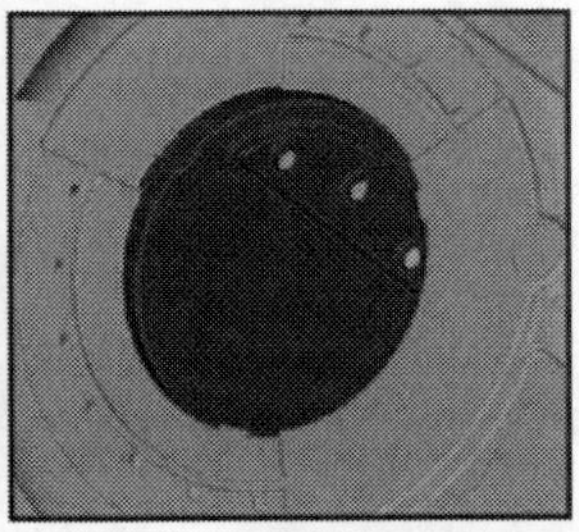

- **Specific heat flux: 19 [W m^{-2}]**
- **Exchange surface: 9.72 x 10^{-9} [m^2]**
- **Heat exchange: 1.847 x 10^{-7} [W]**

➔ Negligible with respect to others

➔ Constant T_{wall}

Internal frame air domain: 'solid air'

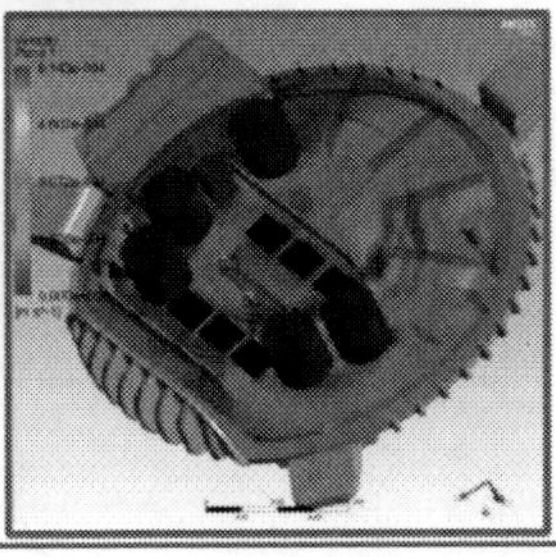

- **Physics: natural convection !**
- **Max internal speed: 0.6 [mm s^-1]**
- **Small volume for fluid motion**

➔ Air modeled as solid material

PCB Modeling

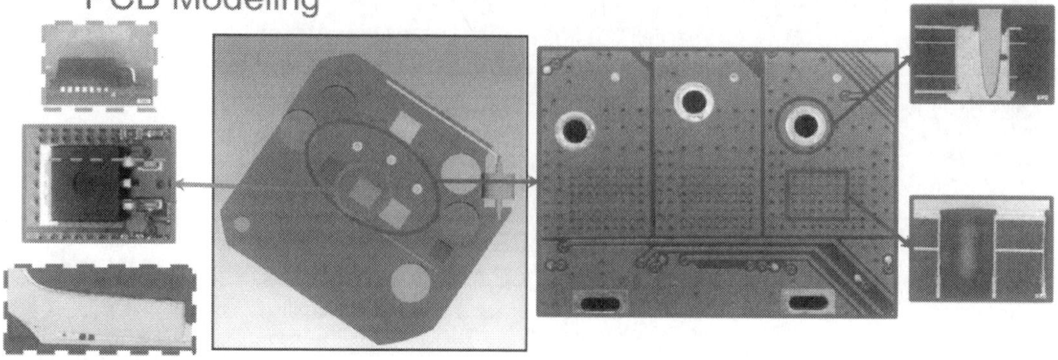

- Complex geometry and internal heat exchange
- Extreme anisotropy (composite materials, miniaturized, multilayer)

➔ Necessity to define resultant/corresponding material

➔ Proper materials thermo-physic characteristics modeled using both analytical approaches and fine tuning CFD analyses

➔ Input thermal powers derived from measured conditions

Thermal Inputs – Main Materials

DOMAIN	MATERIAL	Thermal Conductivity $\left[\frac{W}{mK}\right]$	Source Notes
Gap Pad Gap Filler Paste	Soft filled-polymer material Bi-Component resin + Spacer Beads	2.4 1.8	Supplier Data Sheet
Board	FR4	0.2	Supplier Data Sheet
Coil	Copper	401	ANSYS CFX Library
Coil Core	Ferrite	73	Supplier Data Sheet
Capacitors	Aluminum + PA	237	Modeled
Driver Capacitors	Aluminum + PA	237	Modeled
MOS Fet	Aluminum + Si + PA	237	Modeled
Thermal VIAS	Hybrid	8.94	Modeled
Bus Bars	Copper	401	ANSYS CFX Library
Plastic Frame	PBT	0.3	Supplier Data Sheet
Aluminum Heat Sink	Aluminum	237	ANSYS CFX Library
Steel Heat Sink Iron Heat Sink	Steel Iron	60 75	ANSYS CFX Library

Thermal Inputs – Input Powers

DOMAIN	Input Power	Specificic Dissipated Power $\left[\frac{W}{m^3}\right]$	Source Notes
MOS Fet	2.77 [W] @ 150 [°C] @ 13 [V] @ 500 [W_{el}] 2.55 [W] @ 25 [°C] @ 13 [V] @ 500 [W_{el}] **3.24 [W] @ 150 [°C] @ 12 [V] @ 500 [W_{el}]** **3.02 [W] @ 25 [°C] @ 12 [V] @ 500 [W_{el}]**	9.066×10^6	Calculated from real tests
Coil	0.844 [W] @ 25 [°C] @ 13 [V] @ 500 [W_{el}] **0.987 [W] @ 25 [°C] @ 12 [V] @ 500 [W_{el}]**	1.83×10^6	Calculated from real tests
Capacitors	0.2 [W]	Neglected	Calculated from real tests
Positive Bus Bar	0.17 [W] @ 25 [°C] @ 13 [V] @ 500 [W_{el}] **0.2 [W] @ 25 [°C] @ 12 [V] @ 500 [W_{el}]**	5.824×10^5	Calculated from real tests
Negative Bus Bar	0.2 [W] @ 25 [°C] @ 13 [V] @ 500 [W_{el}] **0.24 [W] @ 25 [°C] @ 12 [V] @ 500 [W_{el}]**	8.097×10^5	Calculated from real tests

Boundary Condition	Averaged Input Temperature	Source Notes
External Environment	110	Thermal Stabilization Condition
Motor Support/Frame Interface	130	Previous CFD Analysis

Aluminum Heat Sink simulation

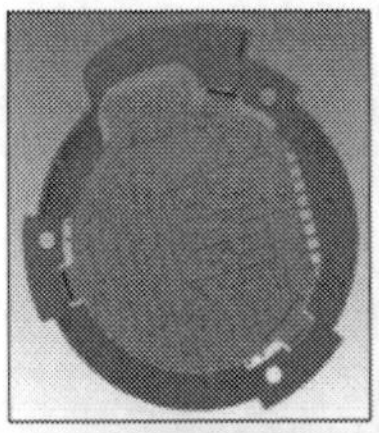

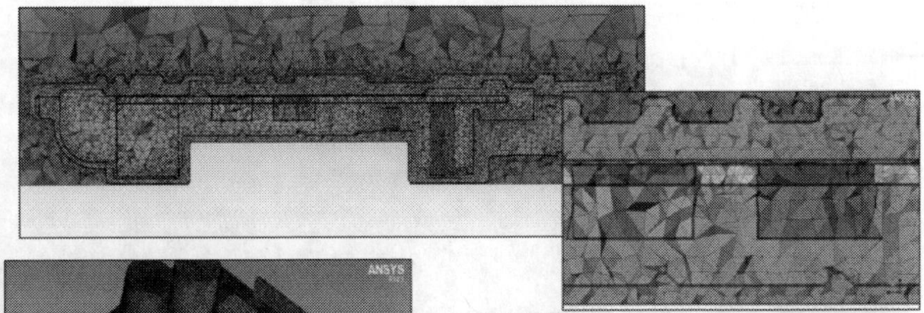

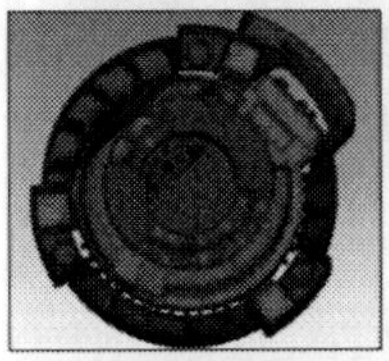

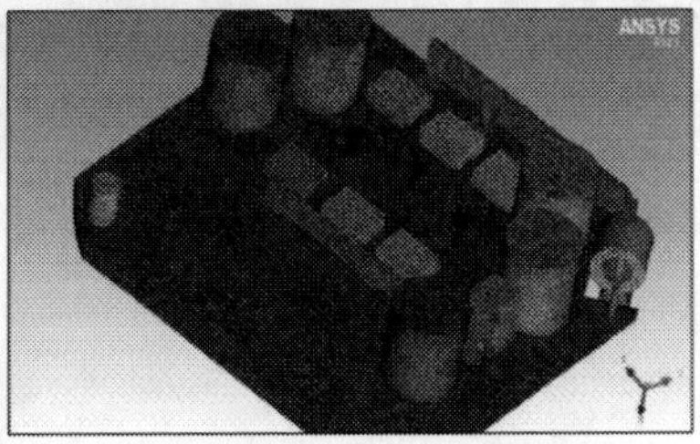

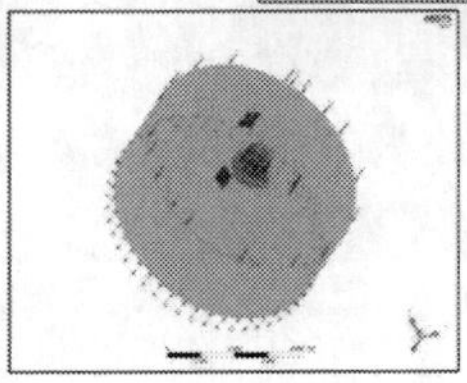

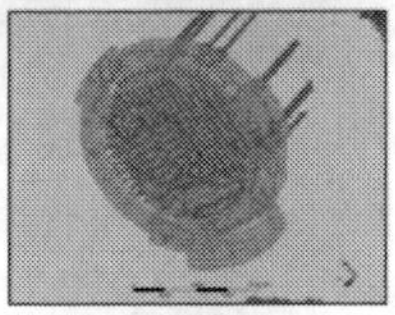

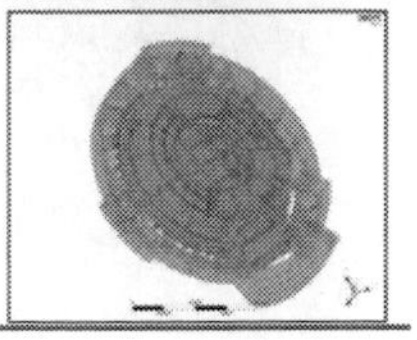

PRESENT CASE CHARACTERISTICS:

Cells Numbers (TETRA+PRISM): ~ 10×10^6

Steady state analysis

Turbulence Model (chosen after sensitivity studies): K - ε with Wall Function

Aluminum Heat Sink – Post Processing

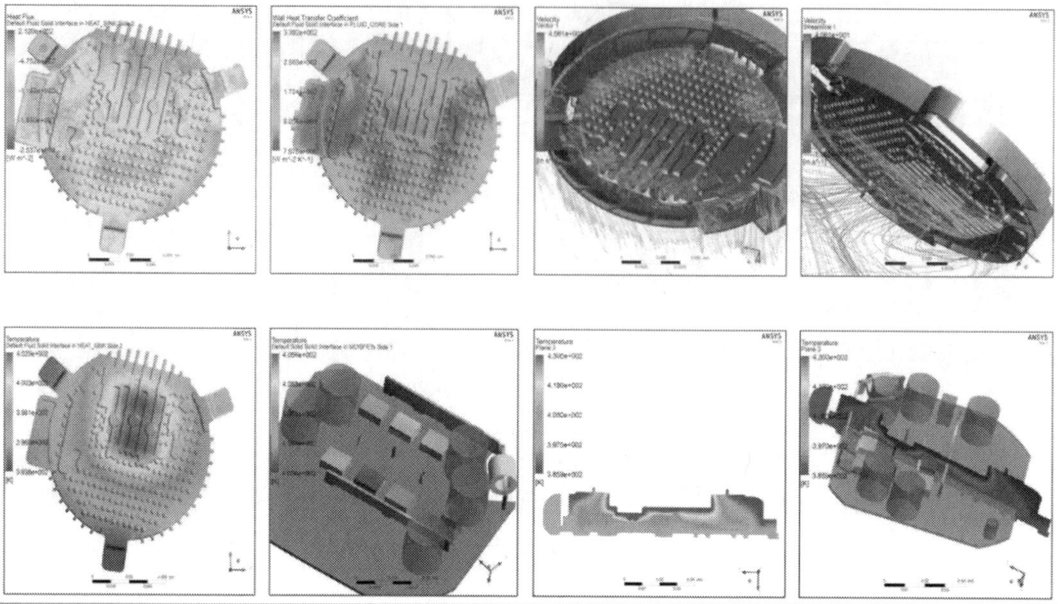

Aluminum Heat Sink - Thermal Stabilization Test

Aluminum Heat Sink with Gap Pad		T Amb 30°C		T Amb 90°C		T Amb 110°C	
		DutyCycle%	86%	DutyCycle%	86%	DutyCycle%	86%
Watt		415		417		415	
Rpm		2670	ΔT	2750	ΔT	2770	ΔT
T AMB	°C	[redacted]	[redacted]	[redacted]	[redacted]	111.1	
[redacted]	°C	[redacted]	[redacted]	[redacted]	[redacted]	[redacted]	[redacted]
[redacted]	°C	[redacted]	[redacted]	[redacted]	[redacted]	[redacted]	[redacted]
[redacted]	°C	[redacted]	[redacted]	[redacted]	[redacted]	[redacted]	[redacted]
[redacted]	°C	[redacted]	[redacted]	[redacted]	[redacted]	[redacted]	[redacted]
[redacted]	°C	[redacted]	[redacted]	[redacted]	[redacted]	[redacted]	[redacted]
[redacted]	°C	[redacted]	[redacted]	[redacted]	[redacted]	[redacted]	[redacted]
[redacted]	°C	[redacted]	[redacted]	[redacted]	[redacted]	[redacted]	[redacted]
L1 ind	°C	[redacted]	[redacted]	[redacted]	[redacted]	126.3	15.2
[redacted]	°C	[redacted]	[redacted]	[redacted]	[redacted]	[redacted]	[redacted]
[redacted]	°C	[redacted]	[redacted]	[redacted]	[redacted]	[redacted]	[redacted]
[redacted]	°C	[redacted]	[redacted]	[redacted]	[redacted]	[redacted]	[redacted]
[redacted]	V	[redacted]	[redacted]	[redacted]	[redacted]	[redacted]	[redacted]
[redacted]	A	[redacted]	[redacted]	[redacted]	[redacted]	[redacted]	[redacted]
[redacted]	°C	[redacted]	[redacted]	[redacted]	[redacted]	[redacted]	[redacted]
[redacted]	°C	[redacted]	[redacted]	[redacted]	[redacted]	[redacted]	[redacted]
[redacted]	°C	[redacted]	[redacted]	[redacted]	[redacted]	[redacted]	[redacted]
Q1	°C	[redacted]	[redacted]	[redacted]	[redacted]	132.9	21.8
Q2	°C	[redacted]	[redacted]	[redacted]	[redacted]	134.2	23.1
Q3	°C	[redacted]	[redacted]	[redacted]	[redacted]	131.8	20.7
Q4	°C	[redacted]	[redacted]	[redacted]	[redacted]	132.0	20.8
Q5	°C	[redacted]	[redacted]	[redacted]	[redacted]	132.5	21.4
Q6	°C	[redacted]	[redacted]	[redacted]	[redacted]	132.3	21.2

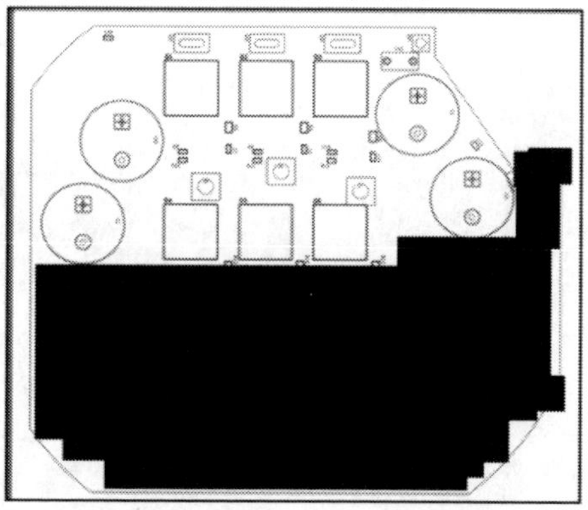

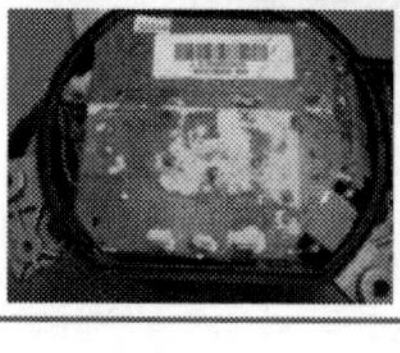

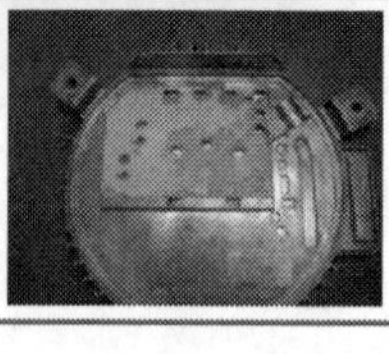

ANSYS – Automotive Simulation World Congress - 2013

Aluminum Heat Sink simulation – Post Processing

- **Generally good agreement between real test and CFD analysis**
- **CFD Model satisfactorily representing real geometry**

CFD Analysis: Aluminum Heat Sink

Q1	133.2	[°C]
Q2	133.6	[°C]
Q3	133.28	[°C]
Q4	133.18	[°C]
Q5	133.76	[°C]
Q6	133.27	[°C]
L1 ind	140	[°C]

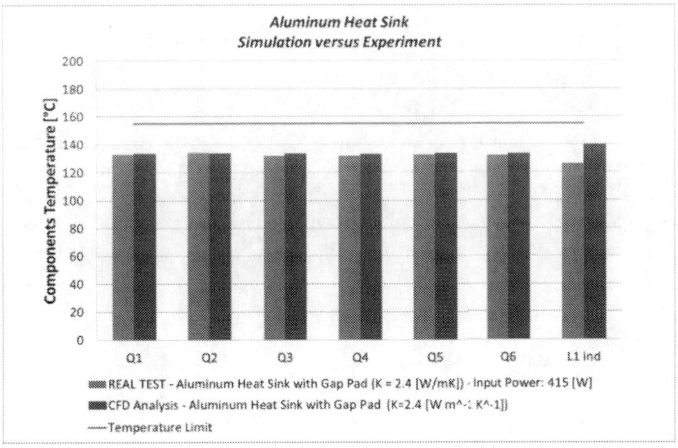

- **Coil temperature difference**

1. **Coil thermal power dependent on current**
2. **Conservativeness reason: current value simulated for coil corresponding to 500 [W], instead of 400 [W]**

ANSYS – Automotive Simulation World Congress - 2013

Metal Heat Sinks simulations (Steel and Iron)

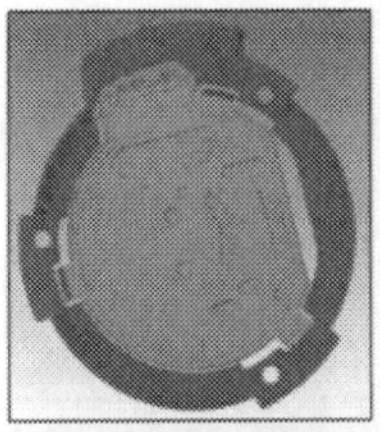
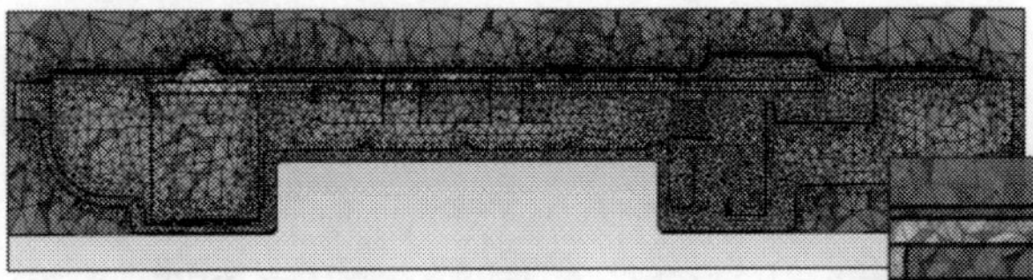
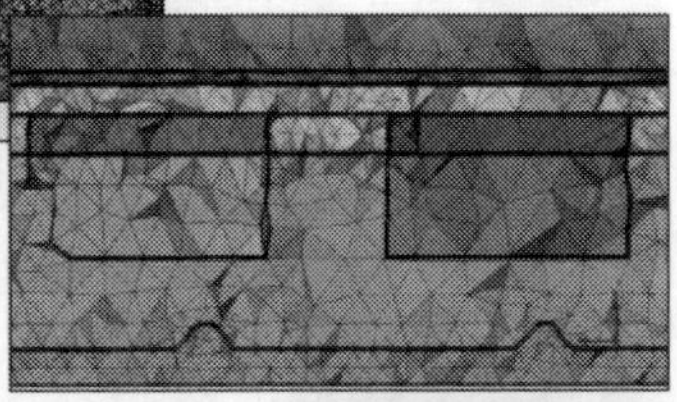
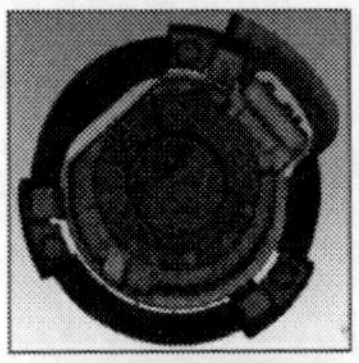
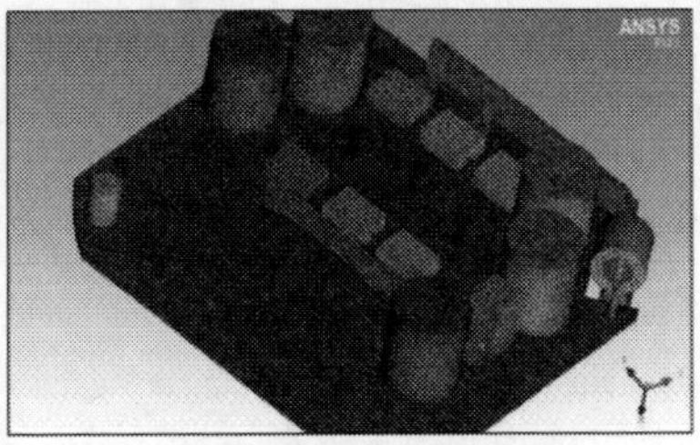
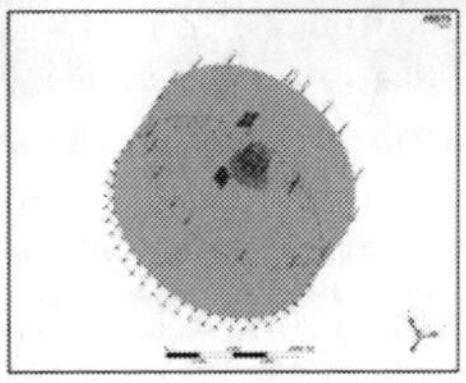
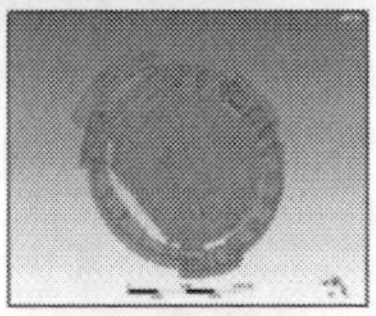

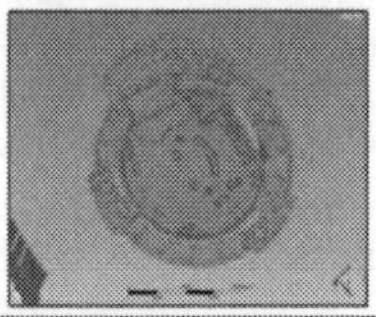

PRESENT CASE CHARACTERISTICS:

Cells Numbers (TETRA+PRISM): < 10 x 10^6

Steady state analysis

Turbulence Model: K - ε with Wall Function

Metal Heat Sinks – Post-Processing (Steel and Iron)

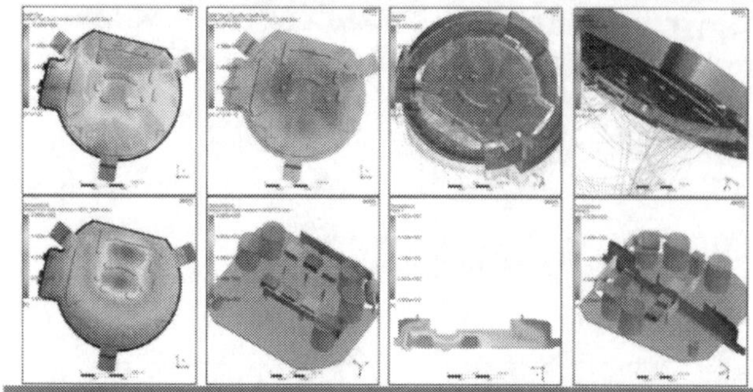

$K_{Steel} = 60 \left[\frac{W}{mK}\right]$

CFD Analysis: Steel Heat Sink

Q1	183.66	[°C]
Q2	189.54	[°C]
Q3	185.52	[°C]
Q4	183.96	[°C]
Q5	189.06	[°C]
Q6	182.21	[°C]
L1 ind	176.25	[°C]

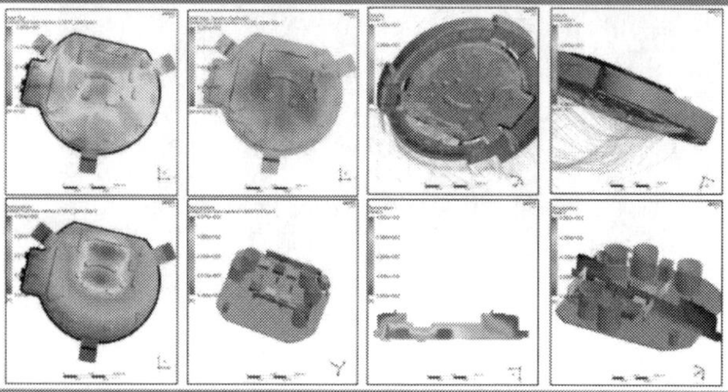

$K_{Fe} = 75 \left[\frac{W}{mK}\right]$

CFD Analysis: Iron Heat Sink

Q1	178.88	[°C]
Q2	184.36	[°C]
Q3	180.7	[°C]
Q4	178.39	[°C]
Q5	183.07	[°C]
Q6	176.92	[°C]
L1 ind	172.28	[°C]

Post Processing Results – Aluminum, Iron and Steel

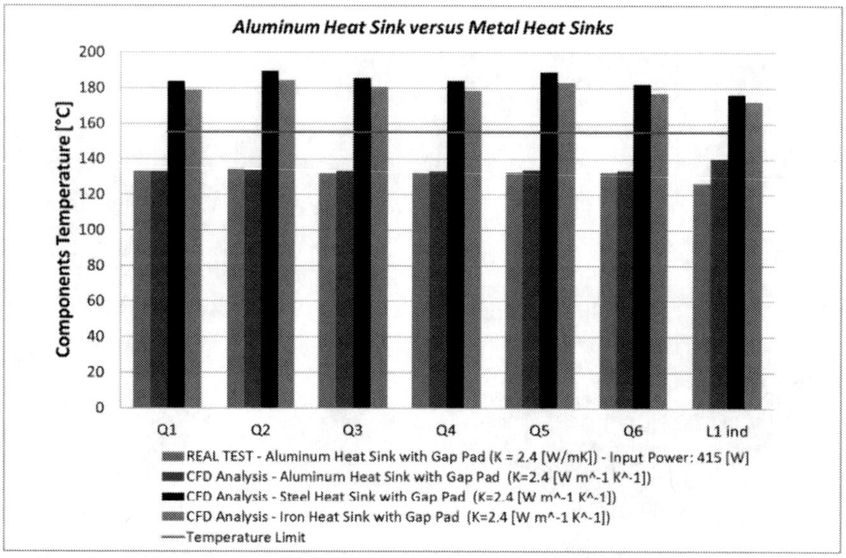

$K_{Al} \sim 240 \left[\frac{W}{mK}\right]$

$K_{Al} = 237 \left[\frac{W}{mK}\right]$

$K_{Steel} = 60 \left[\frac{W}{mK}\right]$

$K_{Fe} = 75 \left[\frac{W}{mK}\right]$

- **Comparison trends in accordance with different material properties**
- **Necessity to improve heat rejection for components temperature reduction !**

ANSYS – Automotive Simulation World Congress - 2013

Reason of heat rejection decrease (Heat sink material)

- **General conduction theory described FOURIER's EQUATION:**

$$q_i = KS\frac{\partial T}{\partial x_i} \rightarrow \boxed{Simplified\ 1D\ form} \rightarrow q_l = KS\frac{\Delta T}{\Delta l} \rightarrow R_{th} = \frac{\Delta T}{q} = \frac{\Delta l}{KS}$$

- q = Heat Flux [W]
- K = Thermal Conductivity of material [W m^-1 K^-1]
- S = Thermal Exchange Surface [m²]
- $\frac{\partial T}{\partial x_i}$ = Temperature Gradient along the i-direction [K m^-1]
- Δi = Material Thickness [m]
- R_{th} = Specific Thermal Resistance [K W^-1]

$$\rightarrow \boxed{R_{th} \cdot S = \frac{\Delta l}{K}}$$

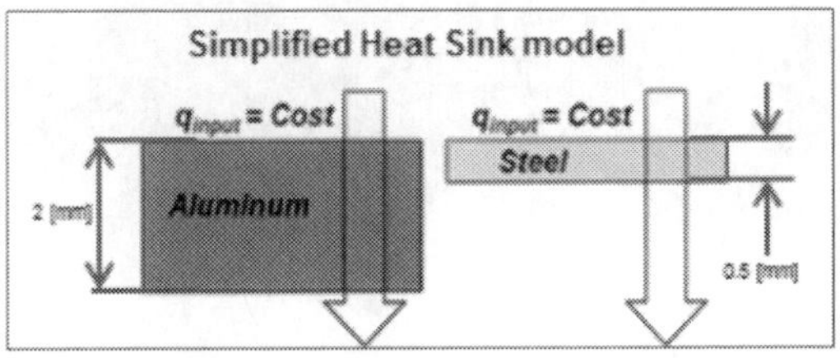

$$R_{th_{Al}} \cdot S = \frac{2.0 \cdot 10^{-3}}{240} \sim 8.33 \cdot 10^{-6} \text{ [K W^-1]}$$

$$R_{th_{St}} \cdot S = \frac{0.5 \cdot 10^{-3}}{60} \sim 8.33 \cdot 10^{-6} \text{ [K W^-1]}$$

Heat sink Material change alone not responsible of heat rejection decrease → "compensation effect" in solid thermal resistance (reduced conductivity with reduced material thickness)

ANSYS – Automotive Simulation World Congress - 2013

Reason of heat rejection decrease (Convection) (1)

- **Convection mechanism:** $q = h \cdot A_{Heat\ Exchange} \cdot (T_{\text{wall}} - T_{air})$
- **Heat path → Simplified multilayer model**
- **T_j = Junction temperature to be reduced**

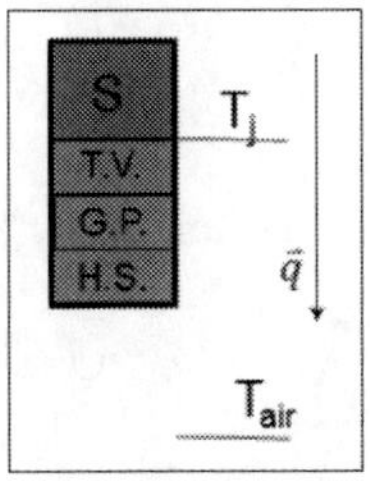

$$q = \frac{T_j - T_{air}}{R}$$

$$R = \frac{l_{T.V.}}{K_{T.V.} \cdot A_{T.V.}} + \frac{l_{G.P.}}{K_{G.P.} \cdot A_{G.P.}} + \frac{l_{H.S.}}{K_{H.S.} \cdot A_{H.S.}} + \frac{1}{h_W \cdot A_{Heat\ Exchange}}$$

- S = Heat Source (MOSFet)
- T_j = Junction Temperature [K]
- T.V. = Thermal Vias (conductivity K, Thickness l, Surface A)
- G.P. = Gap Pad (conductivity K, Thickness l, Surface A)
- H.S. = Heat Sink (with proper conductivity K, Thickness l, Surface A)
- T_{air} = External Air Temperature
- q = Heat Flux moving from Source to air
- R = Global Thermal Resistance
- h = Convection coefficient

$$T_j \sim T_{air} + \frac{q}{A}\left(\frac{l_{T.V.}}{K_{T.V.}} + \frac{l_{G.P.}}{K_{G.P.}} + \frac{l_{H.S.}}{K_{H.S.}} + \frac{1}{h_W}\right)$$

Almost constant for convection

ANSYS – Automotive Simulation World Congress - 2013

Reason of heat rejection decrease (Convection) (2)

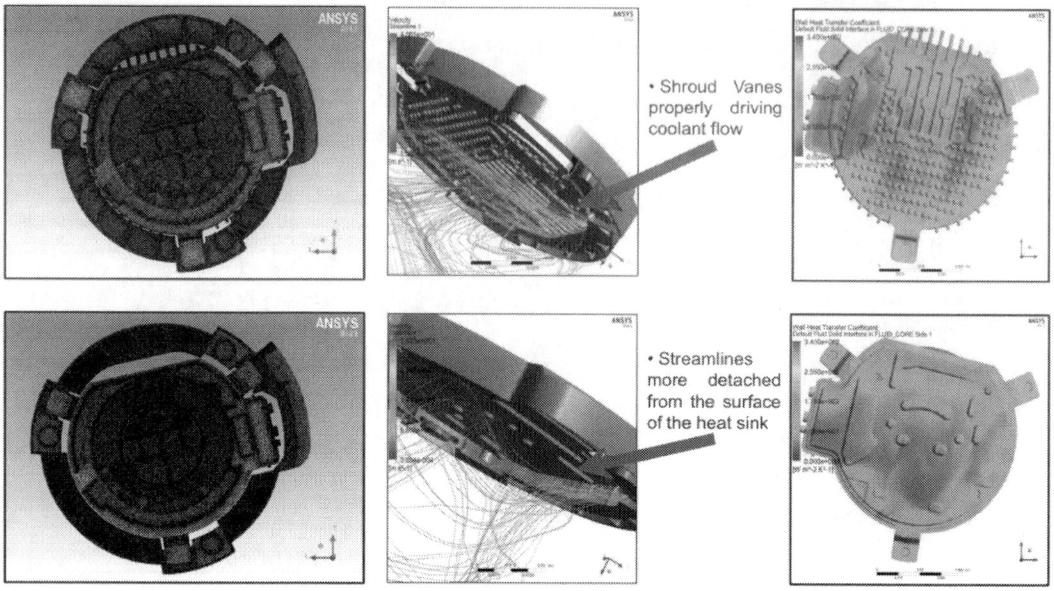

ANSYS – Automotive Simulation World Congress - 2013

Considerations to Optimize

CONDUCTION

- **Different heat sink material →**
  - **Geometry → thickness reduction**
  - **Material → higher conduction, cost compatibility**
  - **Production → feasible and easy solution**

$$T_j \sim T_{air} + \frac{q}{A} \cdot \left(\ldots + \frac{l_{H.S.}}{K_{H.S.}} + \cdots \right)$$

CONVECTION

- **Heat sink ventilation increase →**
  - **Geometry → Heat sink ventilation increase ("flow driver")**
  - **Material → Impact to be defined (costs, etc.)**
  - **Production → feasible and easy solution**

$$T_j \sim T_{air} + \frac{q}{A} \cdot \left(\ldots\ldots\ldots + \frac{1}{h_W} \right)$$

THERMAL EXCHANGE AREA

- **Higher heat exchange surface →**
  - **Geometry → Highest feasible increase**
  - **Material → Choice affecting surface increasing**
  - **Production → feasible and easy solution**
  - **Aluminum: 0.030402 [m^2]**
  - **Steel/Iron: 0.0169583 [m^2]**
  - **Resulting reduction: ≈ -44 [%]**

$$T_j \sim T_{air} + \frac{q}{A} \cdot (\ldots\ldots\ldots\ldots\ldots)$$

Convection improvement: first considerations

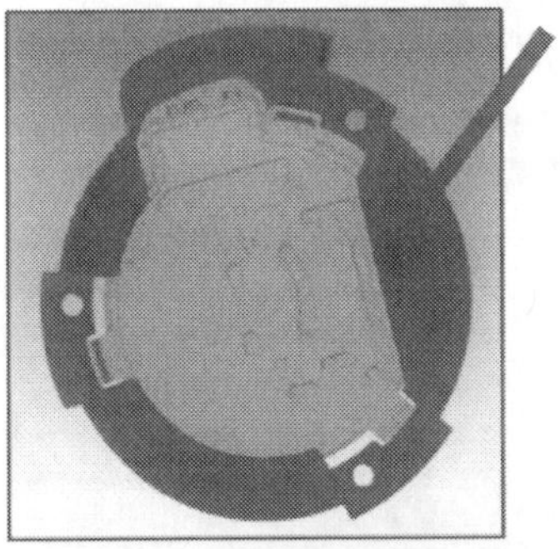

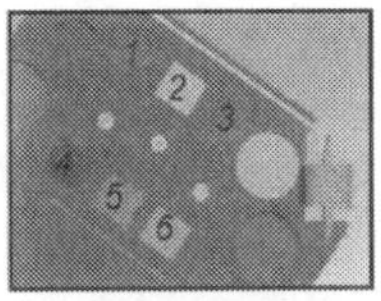

- Averagely 15 [°C] on external MOSFet (1, 2, 3)
- Averagely 10 [°C] on internal MOSFet (4, 5, 6)
- Around 12 [°C] on the coil

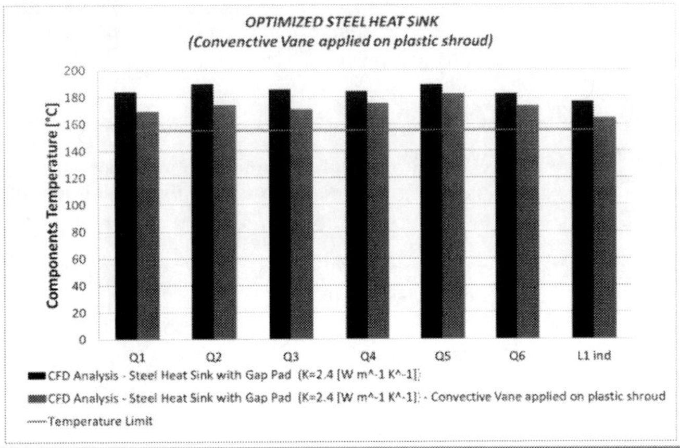

CFD Analysis: Plastic Shroud Convective Vane

Q1	169	[°C]
Q2	173.83	[°C]
Q3	170.72	[°C]
Q4	174.75	[°C]
Q5	182.03	[°C]
Q6	172.96	[°C]
L1 ind	163.77	[°C]

Important note on optimized shroud for Steel Heat Sink

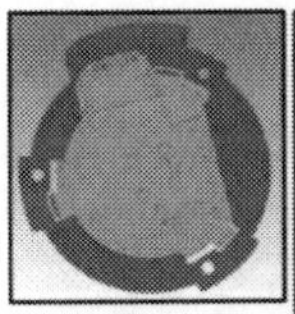

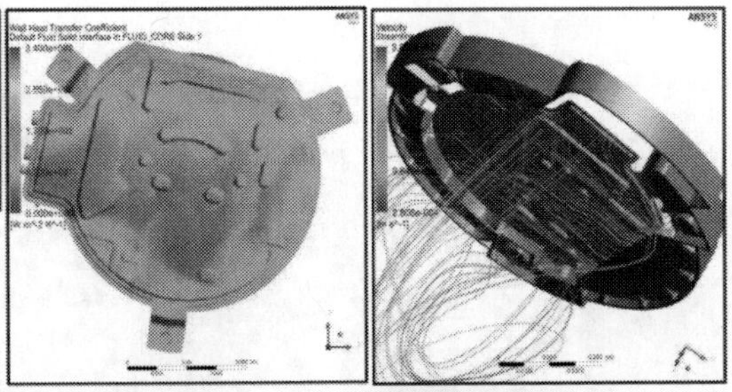

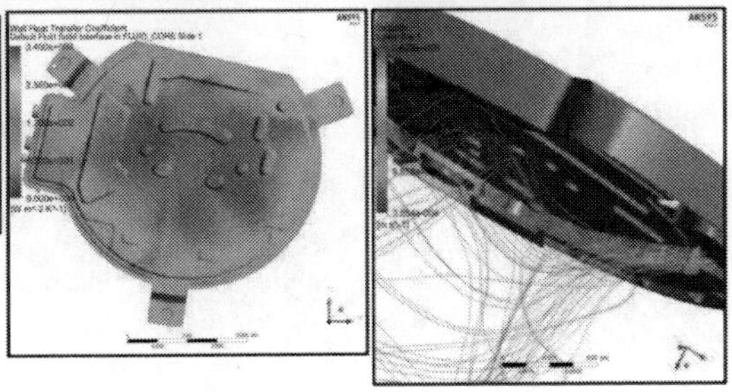

Improved flow driving → High positive impact

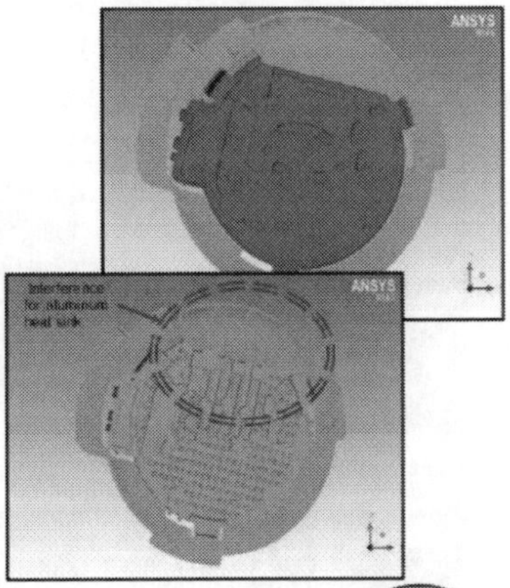

1. **Not completely enough**
2. **Not interchangeability**

Counter measurements

Optimization: needs and constraints

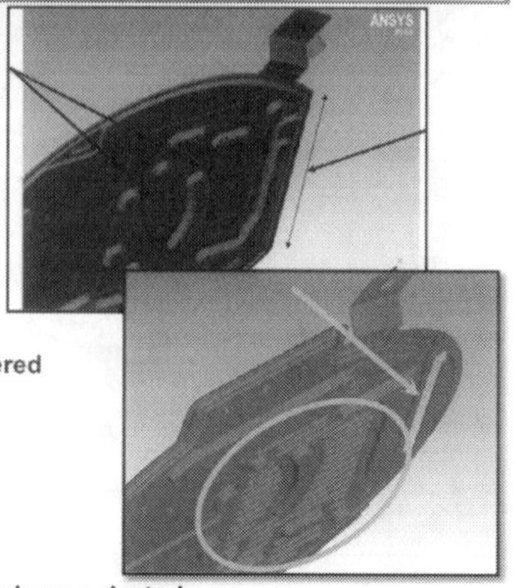

- CONVECTION INCREASE

- **Ventilation:**
  - Approaching to the correct flow driving effect
  - Interchangeability

- THERMAL EXCHANGE AREA

- **Area increase:**
  - "Lost surface" difficult to be completely recovered
  - Limitation due to process
  - Increase with direct stamping on "hot areas"

- CONDUCTION

- **Different material:**
  - **Higher conduction material: aluminum has been selected**
  - **Parallel studies to find best trade-off between stamping process and minimum thickness feasibility**
  - **Parallel studies to warranty reliable contact between PCB and heat sink**

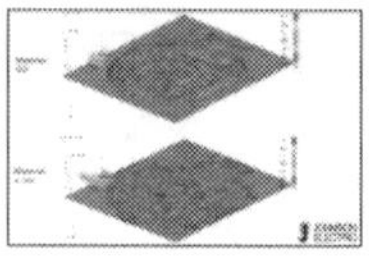

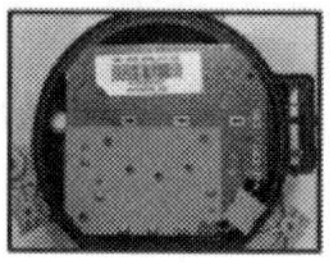

Sensitivity study for "Heat Sink – PCB interface" geometry

- Gap pad and gap filler paste differently adapt to heat sink geometry
- Geometrical sensitivity: different filler shape, same filler material (**$K_{G.P.}$ = $K_{G.F.}$ = 2.4 [W/m K]**)
- Averaged Differences: < 0.18 % (Aluminum Heat Sink); < 0.75 % (Iron Heat Sink)

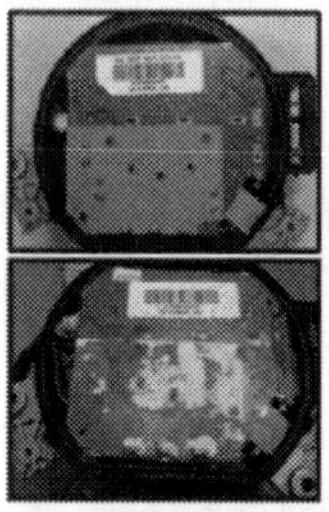

Aluminum Heat Sink with Hybrid Model

Q1	133.2	[°C]
Q2	133.6	[°C]
Q3	133.28	[°C]
Q4	133.18	[°C]
Q5	133.76	[°C]
Q6	133.27	[°C]
L1 ind	140	[°C]

Aluminum Heat Sink with Gap Pad Model

Q1	132.82	[°C]
Q2	133.42	[°C]
Q3	133.11	[°C]
Q4	132.92	[°C]
Q5	133.52	[°C]
Q6	133.2	[°C]
L1 ind	139.53	[°C]

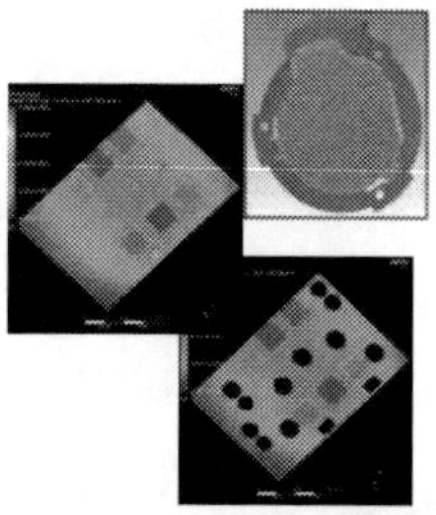

Iron Heat Sink with Hybrid Model

Q1	178.88	[°C]
Q2	184.36	[°C]
Q3	180.7	[°C]
Q4	178.39	[°C]
Q5	183.07	[°C]
Q6	176.92	[°C]
L1 ind	172.28	[°C]

Iron Heat Sink with Gap Pad Model

Q1	178.25	[°C]
Q2	183.73	[°C]
Q3	179.7	[°C]
Q4	175.85	[°C]
Q5	181.64	[°C]
Q6	175.14	[°C]
L1 ind	170.95	[°C]

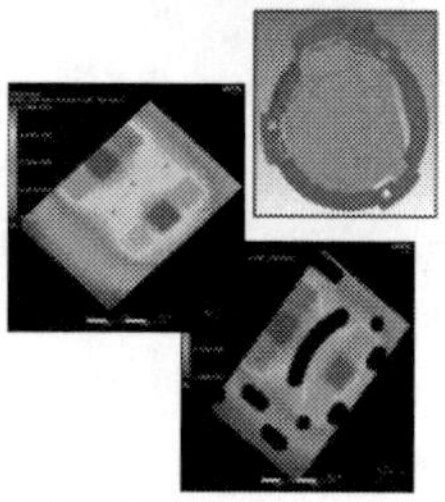

Sensitivity study for “Heat Sink – PCB interface”: filler material

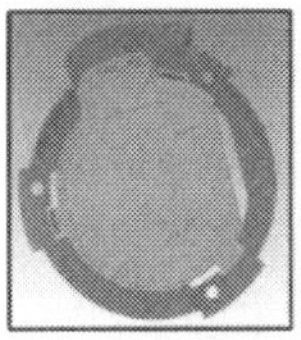

$K_{Fe} = 75 \left[\frac{W}{mK}\right]$

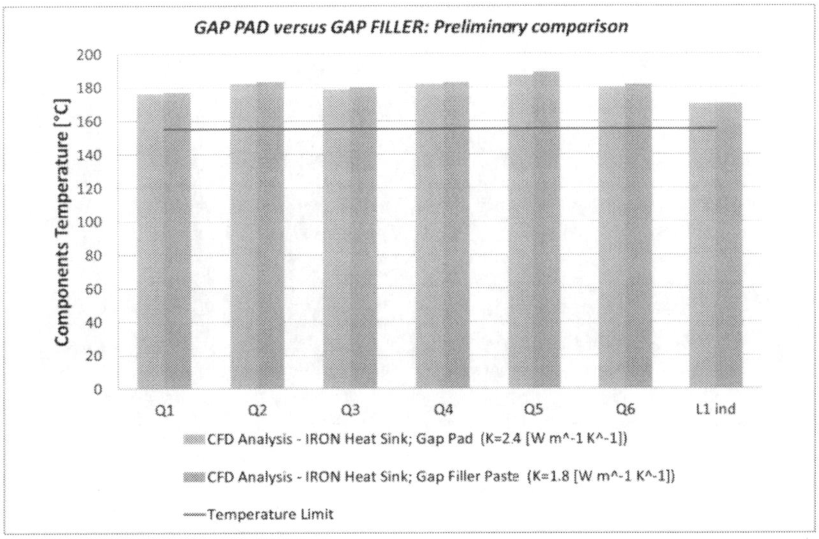

Averaged Temperature difference: < 1 [°C]
Maximum Temperature difference: ≈ 1.5 ÷ 2 [°C]

Gap Filler Paste: Less conductive
Gap Filler Paste: Thinner

Thinned Aluminum Heat Sink with integrated deflector

- **Case 1: Minimum Thickness (0.5 [mm]) with Gap Pad ($K_{G.P.}$ = 2.4 [W m-1 K^-1])**
- **Case 2: Higher Thickness (1.0 [mm]) with Gap Pad ($K_{G.P.}$ = 2.4 [W m-1 K^-1])**

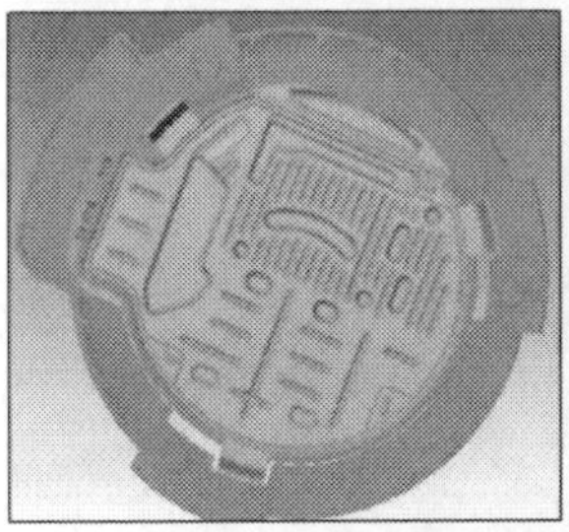

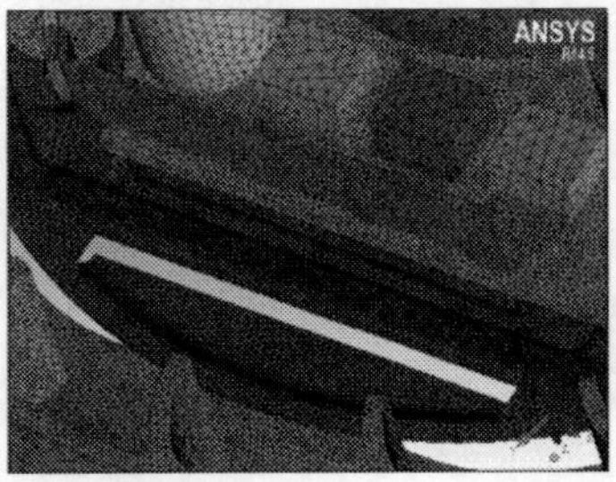

Heat Sink Material: STAMPED (0.5mm) ALUMINUM

Q1	152.98	[°C]
Q2	156.13	[°C]
Q3	154.38	[°C]
Q4	152.25	[°C]
Q5	155.71	[°C]
Q6	152.75	[°C]
L1 ind	151.86	[°C]

Heat Sink Material: STAMPED (1mm) ALUMINUM

Q1	163.42	[°C]
Q2	168.05	[°C]
Q3	165.51	[°C]
Q4	165.7	[°C]
Q5	169.56	[°C]
Q6	164.7	[°C]
L1 ind	159.41	[°C]

Main CFD cases results comparison

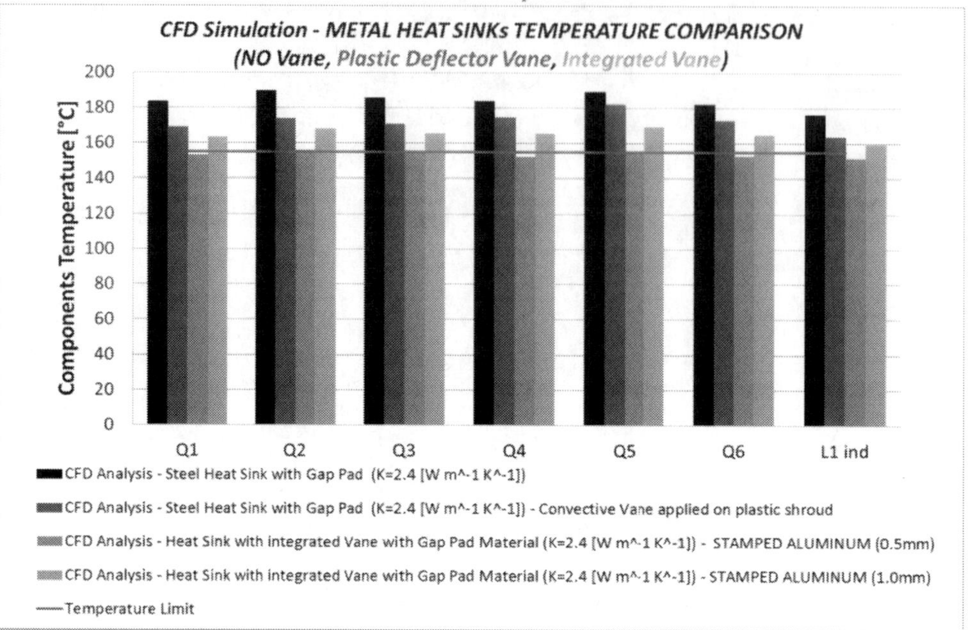

ANSYS – Automotive Simulation World Congress - 2013

Thinned Aluminum Heat Sink with integrated deflector: *convection*

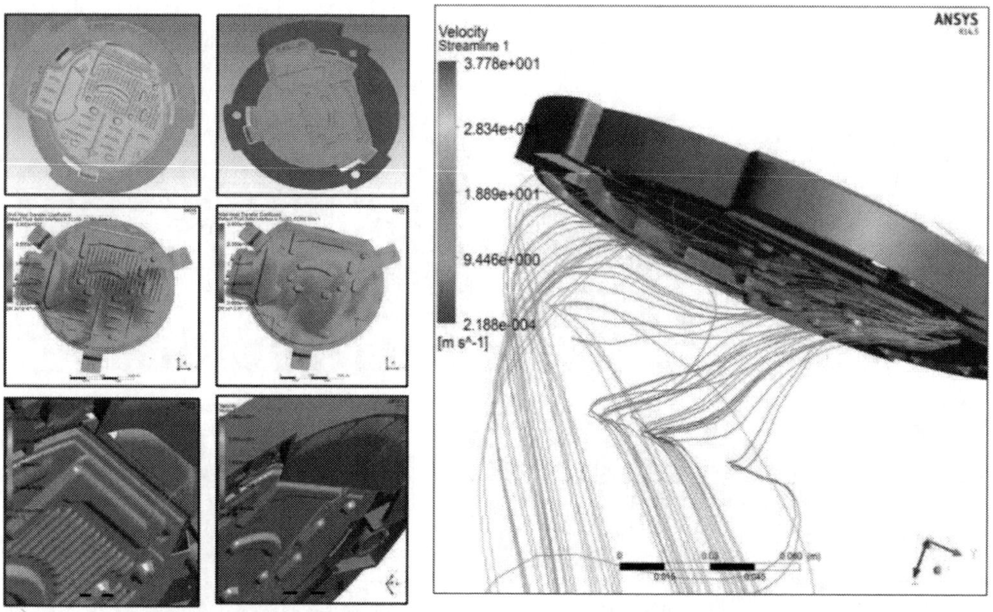

Thinned Aluminum Heat Sink with integrated deflector: R2 (1)

- **Bigger deflection vane**
- **"Cleaner" air passage**

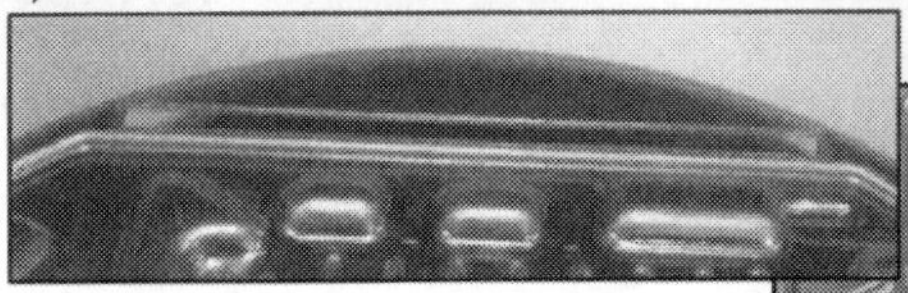

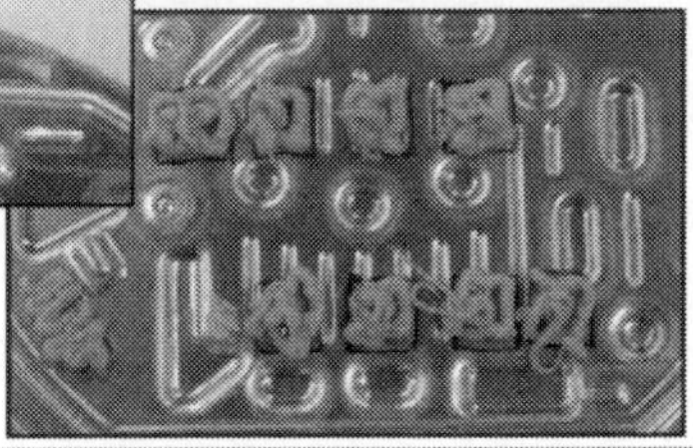

Cross-checked for feasibility

- **Necessity of gap filler paste (cohesion)**
- **Thickness: 0.8 [mm] (cracking)**
- **Reduced surface (cracking)**

Thinned Aluminum Heat Sink with integrated deflector: R2 (2)

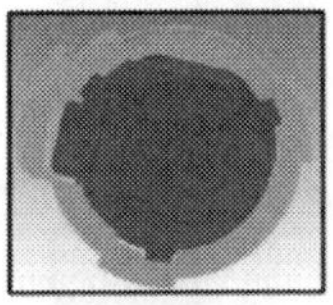

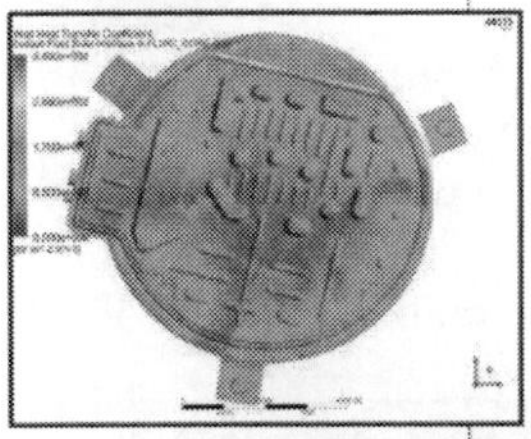

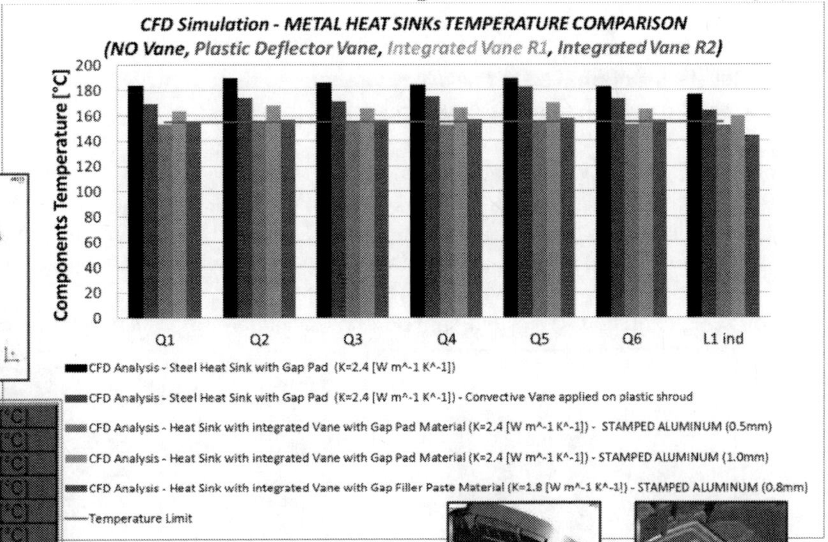

Q1	155.2	[°C]
Q2	156.76	[°C]
Q3	156.13	[°C]
Q4	156.39	[°C]
Q5	157.24	[°C]
Q6	156.12	[°C]
L1 ind	144.05	[°C]

CFD Analysis: Heat Sink, Integrated Vane (R2)

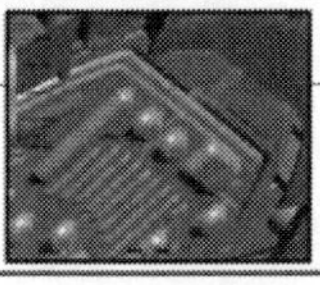

The new heat sink design (R2): real test

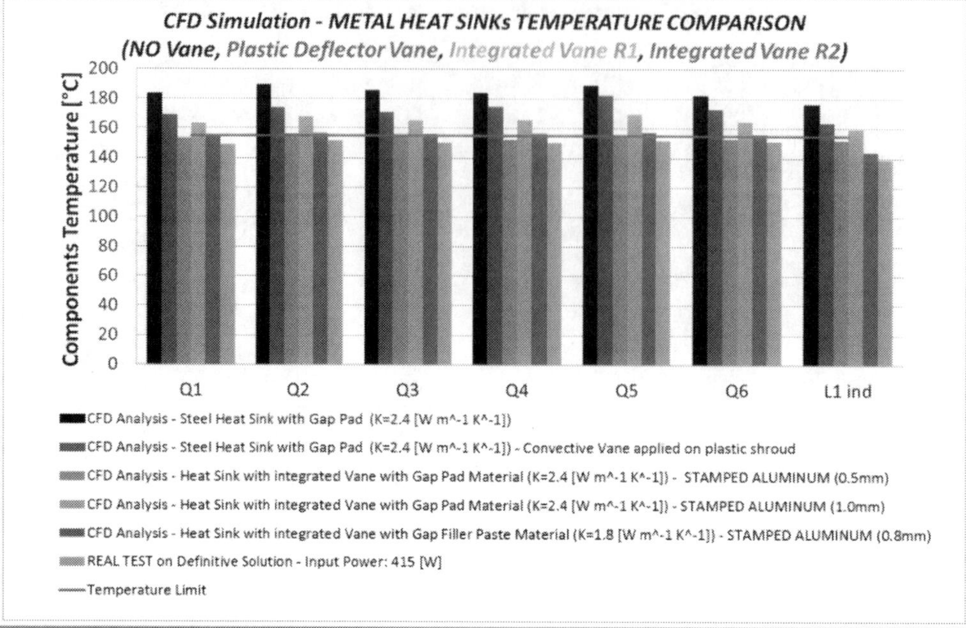

General conclusions

Thanks to the present work a new evolution of a real motor cooling system has been designed with the following optimization:

1. Weight reduction
2. Reliability (internal electronics temperatures under safety limits)
3. Cost reduction (materials usage optimization)

The use of CFD modeling has confirmed :

- Good reliability as simulation tool even with some simplifications due to the high complexity of the analyzed system
- Extreme utility in understanding main thermal exchange mechanisms importance
- Good tool to drive effective design decisions

...allowing to achieve a system configuration working and providing the searched requirements !

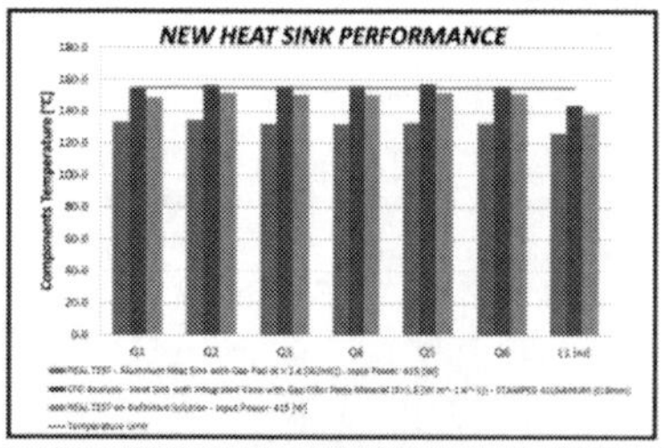

Advanced Simulation of Filter Elements
A Practicioner's View on the Evolution from GAMBIT/FLUENT to ANSYS Workbench 14.5

MANN+HUMMEL

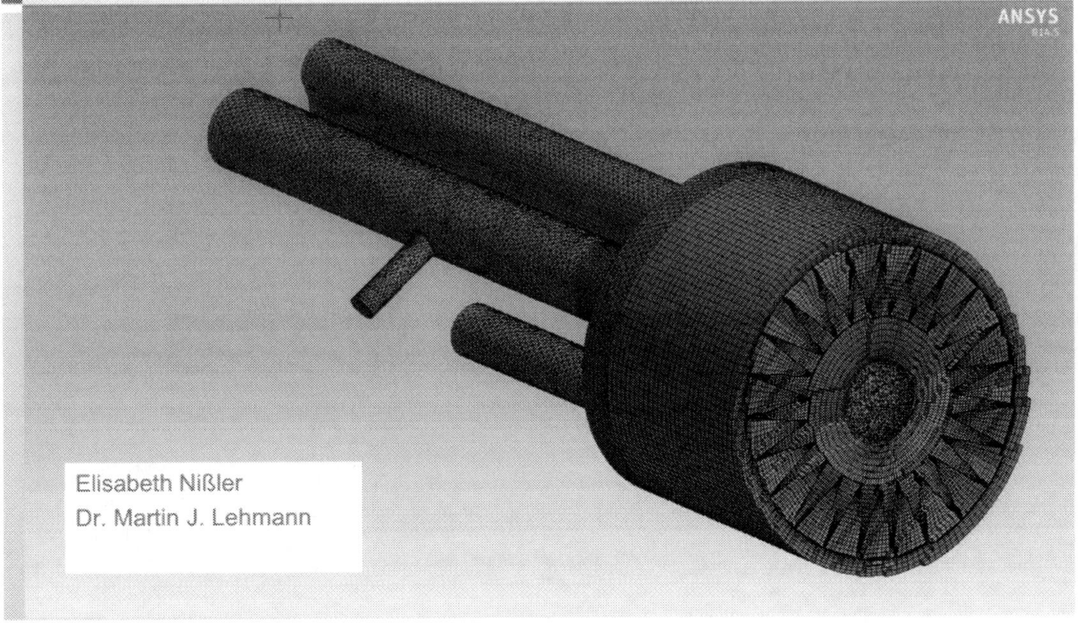

Elisabeth Nißler
Dr. Martin J. Lehmann

MANN+HUMMEL
Welcome to Our World

MANN+HUMMEL

▶▶ Cars – Would You Have Thought About Filters?

2

MANN+HUMMEL
Welcome to Our World

▶▶ > 16 Filter Elements Produced per Second

MANN+HUMMEL
One Purpose: Satisfied Customers.

▶▶ MANN+HUMMEL: Development Partner and Original Equipment Supplier

Automotive Original Equipment	Automotive Aftermarket	Industrial Filtration	Water Filtration

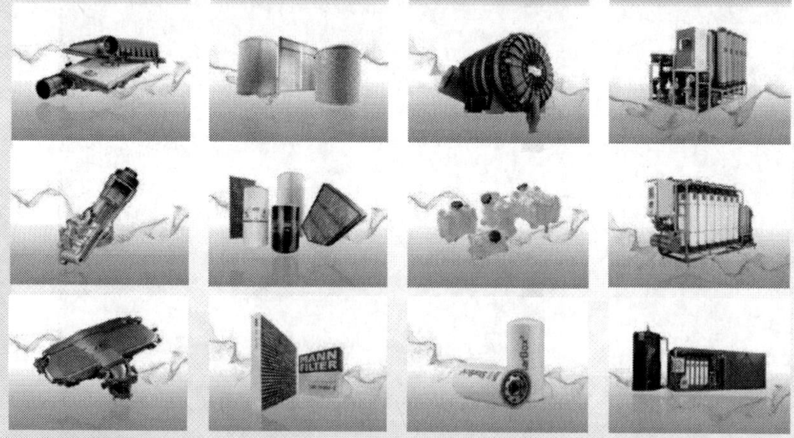

The Future Poses Many Questions. Our Specialists Are Working On The Answers.

MANN+HUMMEL

▶▶ Virtual Development as a Key For Successful Product Development

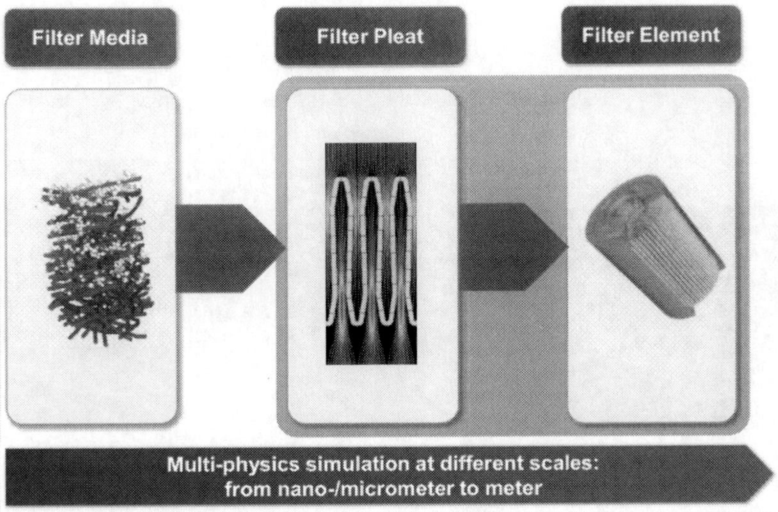

Basic Topic: CFD for Spin-Ons

MANN+HUMMEL

▶▶ Standard product – sophisticated interior

To be found in each car …

… but detailed know-how inside

Reasons for Automation

MANN+HUMMEL

▶▶ Reducing repetitive work – fast answers in projects

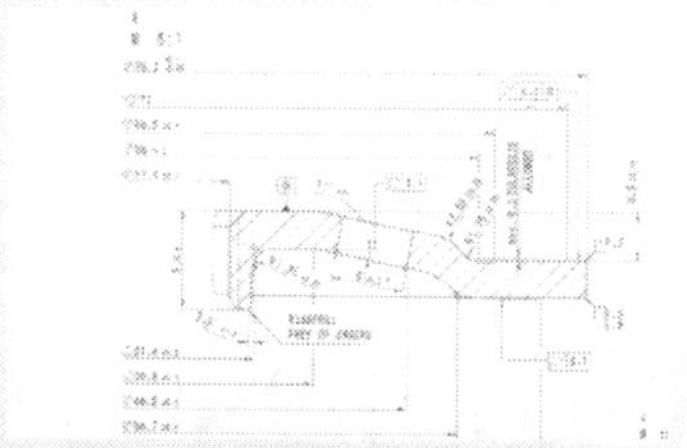

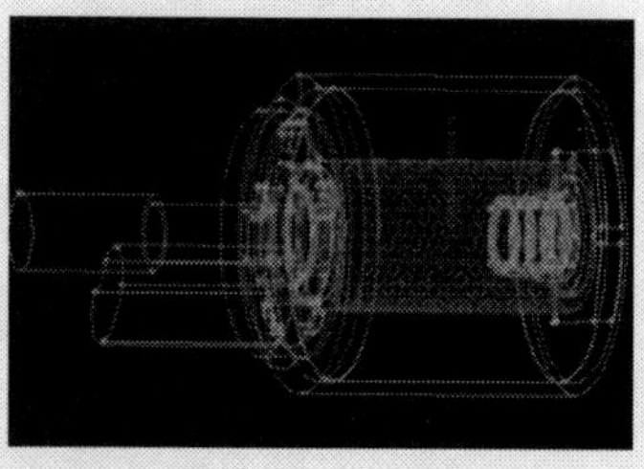

- Complex geometries need a lot of manual work
- Parameters are available from design
- Similar problems – small changes
- Need for DoEs
- Opportunity to "store" know how on resolution, best meshing strategies

Our Tool c

MANN+HUMMEL

▶▶ Workflow

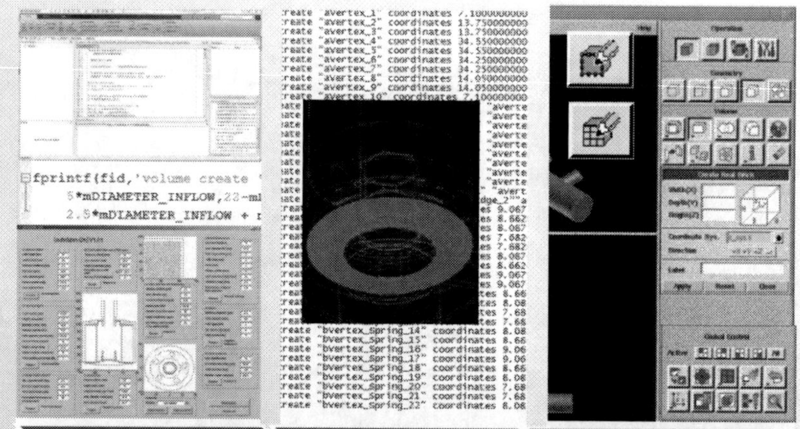

Run customized Matlab GUI → Run GAMBIT journal → Mesh manually

3

Automation Strategy with GAMBIT

MANN+HUMMEL

▶▶ Matlab GUI: compute base points of geometry…

```
    %compute center points of the polygon:
bangle = 2*pi/bNO_DISCRETIZATION_POINTS;
bz_min = bOFFSET_Z_VALVE + bHEIGHT_VALVE + 0.5 * bDIAMETER_SPRING_WIRE * cos(bangle/2);
bz_max = bOFFSET_Z_VALUE + bHEIGHT_VALVE + bHEIGHT_SPRING - ...
                                          0.5 * bDIAMETER_SPRING_WIRE * cos(bangle/2);
%bz_step = (bz_max-bz_min)/(bNUMBER_OF_COILS - 1);
bz_step = (bz_max-bz_min-1*cos(bangle/2)*bDIAMETER_SPRING_WIRE)/(bNUMBER_OF_COILS - 1);
```

Code properties: 5000 lines for element, 5000 lines for housing, extra effort for graphics and GUI

▶▶ … and write GAMBIT commands to a journal file

```
fprintf(fid,'volume create "mSPLITTER_1" height %6.10f radius1 %6.10f radius3 %6.10f offse
    5*mDIAMETER_INFLOW,22-mDIAMETER_INFLOW/2, 22-mDIAMETER_INFLOW/2,...
    2.5*mDIAMETER_INFLOW + mZ_OFFSET + mHEIGHT_INFLOW_AREA_ISO);
fprintf(fid,'volume create "mSPLITTER_2" height %6.10f radius1 %6.10f radius3 %6.10f offse
    5*mDIAMETER_INFLOW,22+mDIAMETER_INFLOW/2, 22+mDIAMETER_INFLOW/2,...
    2.5*mDIAMETER_INFLOW + mZ_OFFSET + mHEIGHT_INFLOW_AREA_ISO);
```

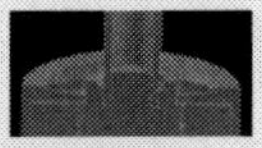

9

Details

MANN+HUMMEL

▶▶ Properties and Advantages

- GUI with preview
- Fully automated generation
- No import issues, fewer crashes
- Geometry optimized for CFD (no gaps, sharp angles…)
- Standardized geometry

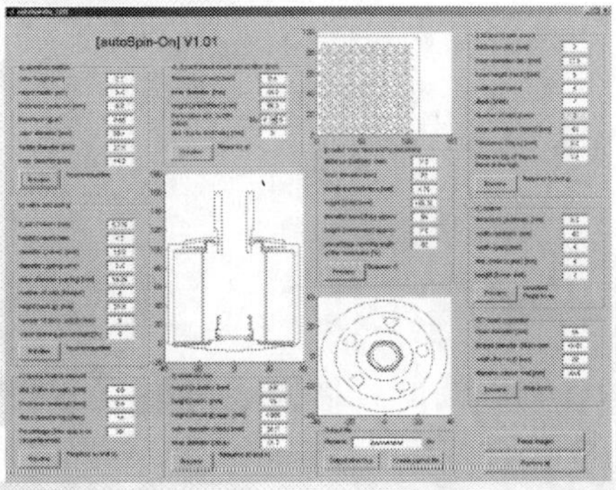

▶▶ Limitations

- Pleat generation is complicated
- Only possible because of a student being a small genius
- No automated meshing
- No automated FLUENT setup

10

From GAMBIT to Workbench 11

MANN+HUMMEL

- GAMBIT discontinued – migration challenging, because:
  - GAMBIT designed for a CFD specialist
  - Mature program for complex meshing
- WB 11 with its advantages and problems:
  - CAD kernel better for complex CAD geometries
  - Entire workflow in a single WB project
  - Not designed for CFD meshes
    (no direct meshing yet, few possibilities of preview)
  - Crashes
- Training needed:
  - Before migrating, training was given and absolutely necessary
  - Helpful: 1-to-1 training at specific problems in house
  - Support needed:
    - ideas – how to reproduce existing solutions?
    - explanations – why do our old strategies not work anymore?

11

Effect on Workflow

MANN+HUMMEL

- Different idea and strategy of CFD procedure

	FLUENT in WB	FLUENT/GAMBIT
Mesh update	👍	👎
Automation for geometry	👍👎	😐👍
Parameters	👍	😐
Handling of boundary conditions	😐👍	👍😐
Features for big geometries	👍👎	👎👍
Automation for FLUENT	👍	👎
Automation for postprocessing	👍	😐
File management	👍😐	😐

12

[autoSpin-On]: based on Workbench

▸▸ Working with standard FLUID workflow

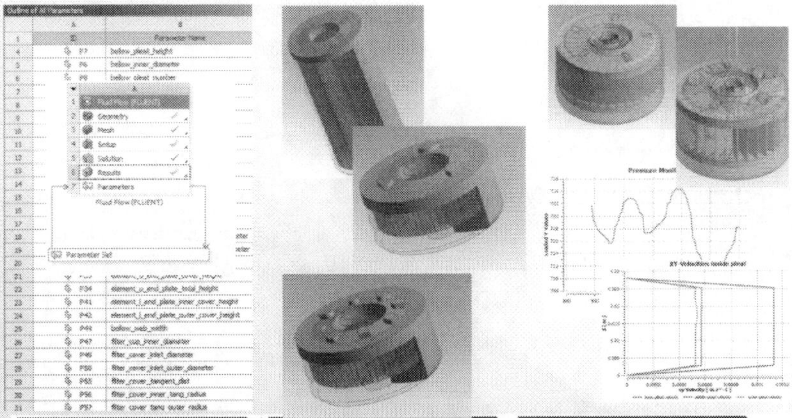

Open project template → Change parameters → Get report

13

Structure of Automated Workflow: Preprocessing

▸▸ Fully parametric model (geometry as wall as meshing)

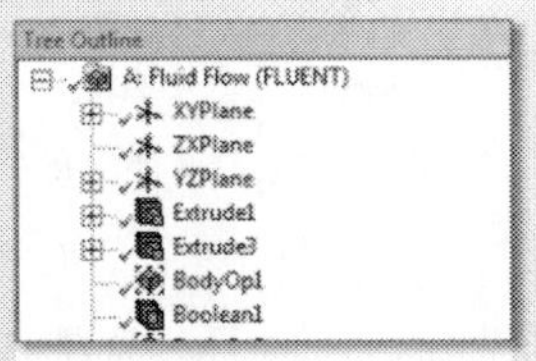

Geometry setup based on DM case ...

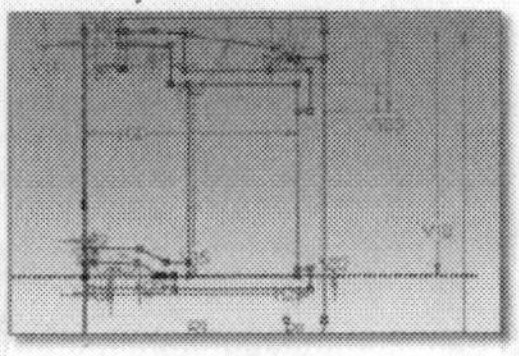

... sketches ...

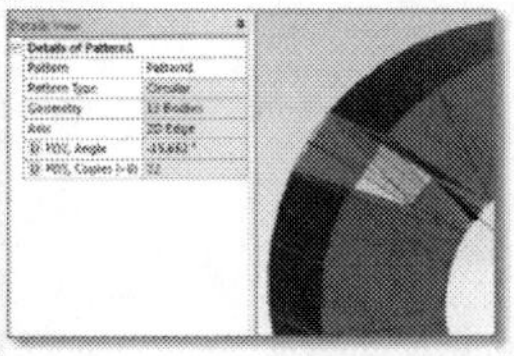

... and patterns

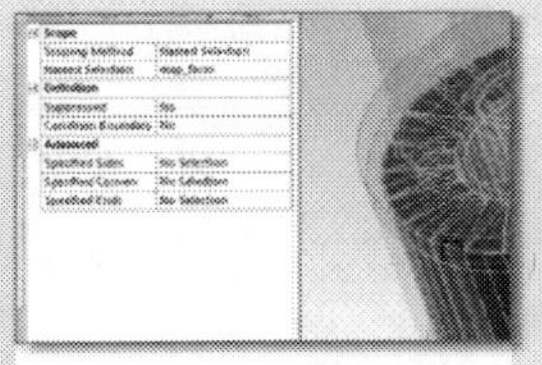

Mesh setup based on named selections

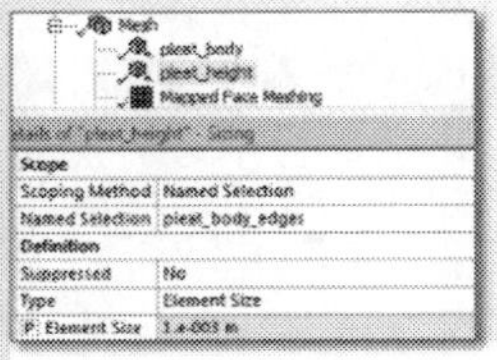

Cell size/number can be changed by user

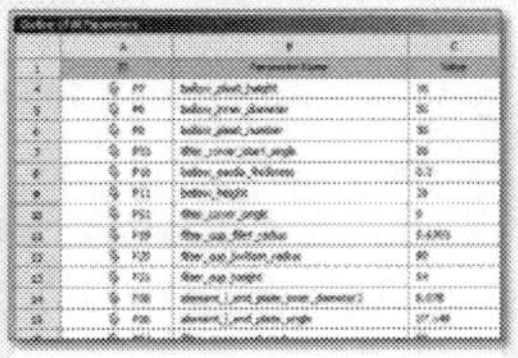

More than 50 parameters

14

Structure of Automated Workflow: CFD Simulation in FLUENT

MANN+HUMMEL

▶▶ Key advantage of the Workbench: automatic update!

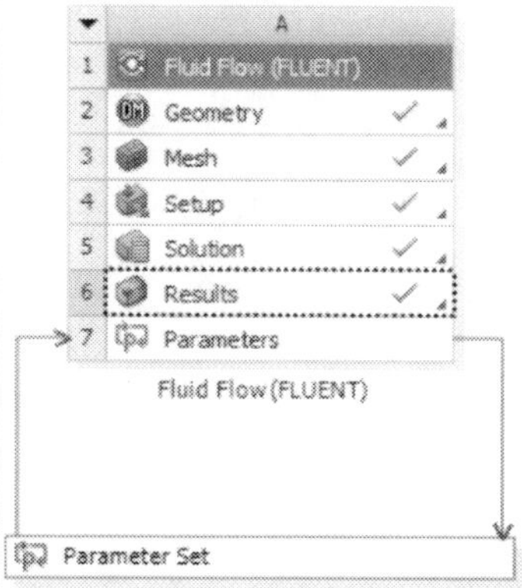

FLUENT set up for desired flow regime

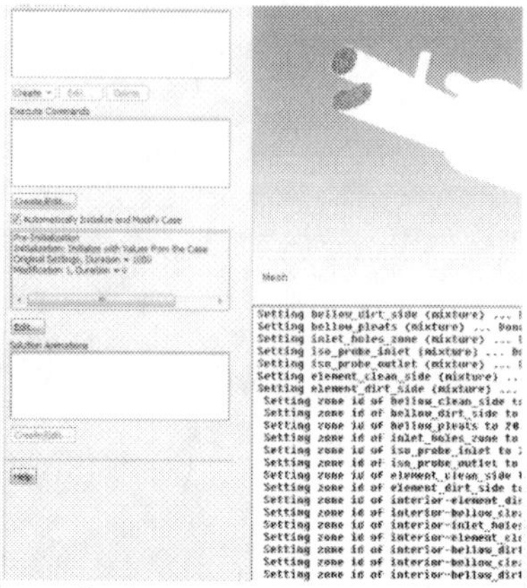

Settings saved automatically

Nice feature: calculation activities

15

Structure of Automated Workflow: Postprocessing in CFD-Post

MANN+HUMMEL

▶▶ Once again: on board tools

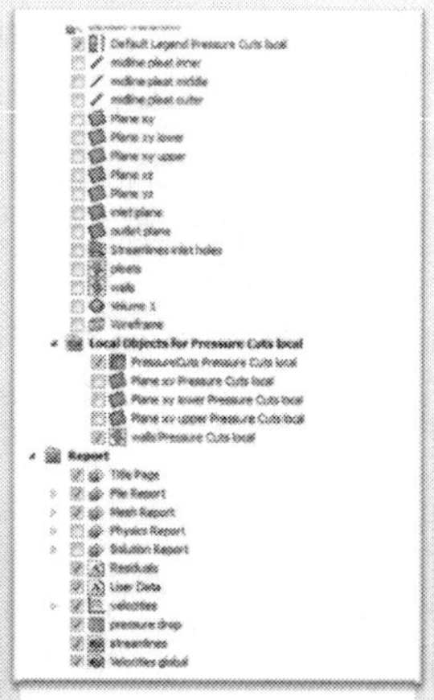

CFD-Post setup

plus additional macros to carry out analysis

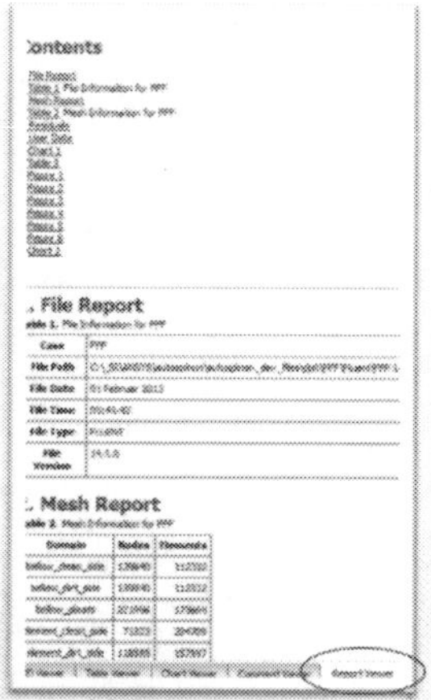

→html report with FLUENT pictures

16

Summary [autoSpin-On]

MANN+HUMMEL

▶▶ Workflow streamlined, usability improved

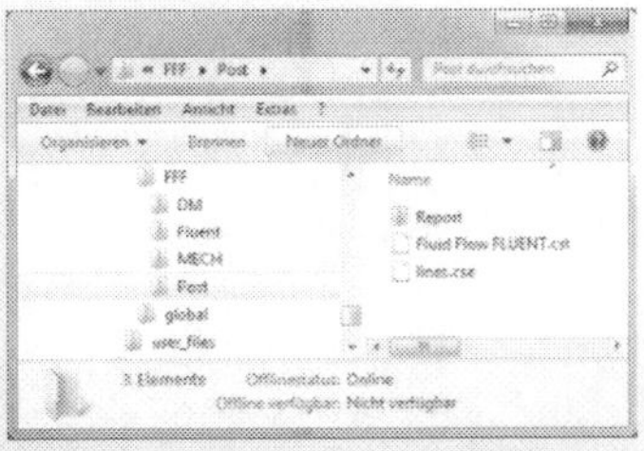

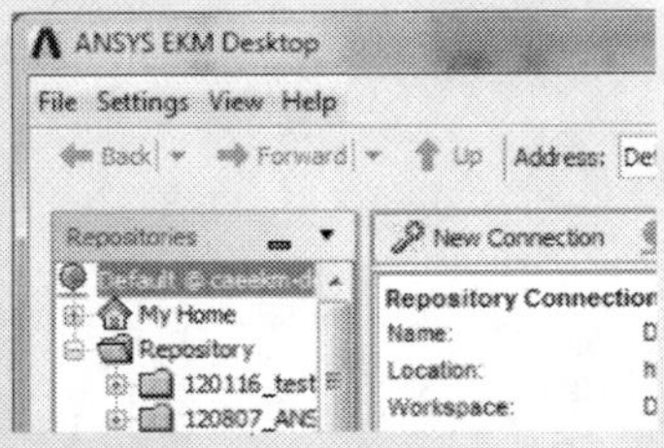

WB, FLUENT set-files and cst-files (automatically) do most of work, just some smaller scripts needed to get parameters and pictures into CFD-Post

Outlook:

- EKM-based: connecting parameters and having the possibility to "forbid" values, which are not possible
- Allow parametric runs

17

Where there is still improvement needed

MANN+HUMMEL

▶▶ Funny, strange, entertaining

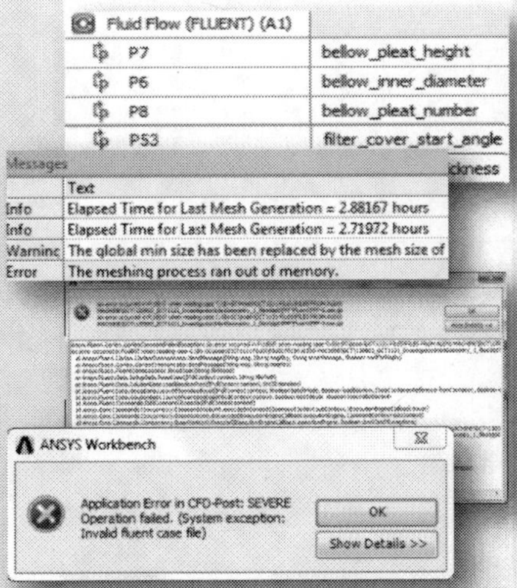

- Error messages often not helpful or too late
- Extremely time-consuming, when patterns are used after slicing a geometry (seems to be a bug)
- Options are often not working for too complex geometries
- No documentation on customization
- CFD-Post: no cell values, language mix often confusing
- Parameters not consequently handled

HARASS SUPPORT!!

18

Conclusions

MANN+HUMMEL

▶▶ Proprietary MANN+HUMMEl tool [autoSpin-on] in WB 14.5.7 offers

- Fast and even accurate solutions
- Standardized results, pictures and values
- Tool of choice for quick design evaluation
- Easy to enhance with additional features
- Simple maintenance
- We don't miss GAMBIT

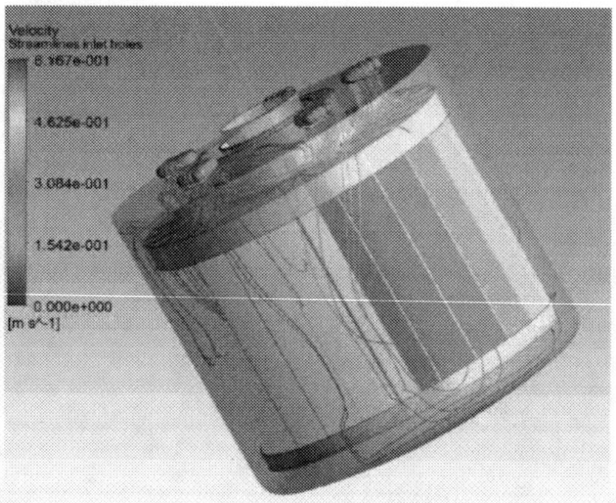

BUT

- Support needed!
- Waiting for WB 14.5.7.1(!) instead of WB 15

19

Throttle Valve for Variable Water Pump applications

Focus on Calculation and Simulation activities

E. DATTOLI, Calculation & Simulation dept. Manager
PTT France
Thionville, September 30th 2013, issue6

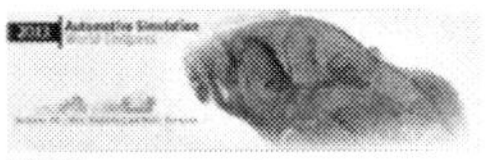

PIERBURG PUMP TECHNOLOGY

➢ **Pierburg Pump Technology**

Rheinmetall AG
Sales: € 4,445 million - Employees: 21,516

KSPG AG

Automotive supplier of engine components and systems

Sales: € 2,313 million
Employees: 11,548

Rheinmetall Defence

European Defence company for ground forces technology

Sales: € 2,141 million
Employees: 9,833

KSPG AG

Divisions

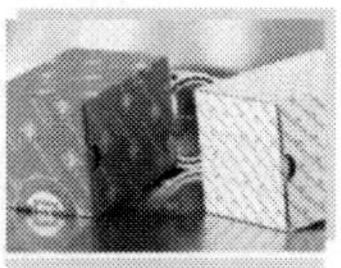

	Hardparts	Mechatronics	Motorservice
Business Units	Pistons	Pierburg	International
	Aluminium-Technology	Pierburg Pump Technology	Domestic
	Plain bearings		
	Large-Bore Pistons		

➢ Water pump for automotive application

Function : cooling liquid circulation

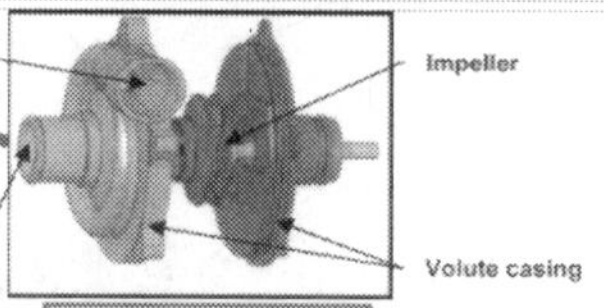

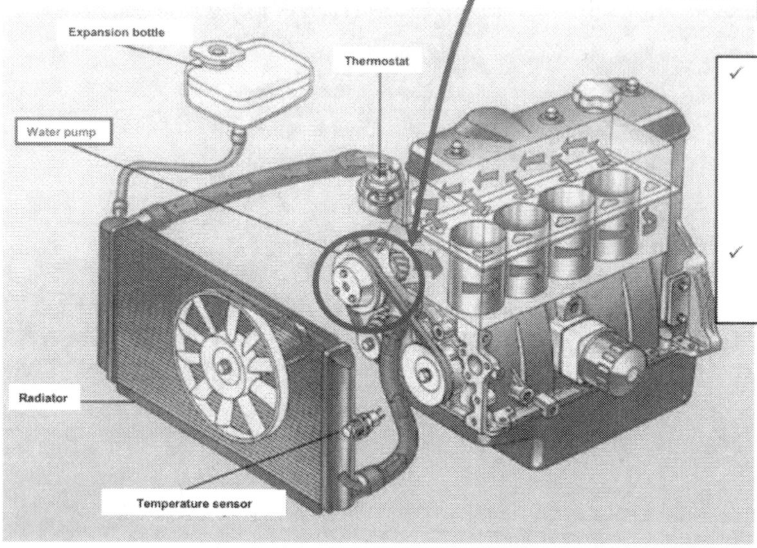

- ✓ The water pump insures the circulation of the cooling liquid between the hot regions of the engine and the radiator so as to evacuate calories generated by fuel combustion.
- ✓ The water pump is generally driven either by timing belt or by accessory belt

➢ Evolution of regulations relevant to cars emission

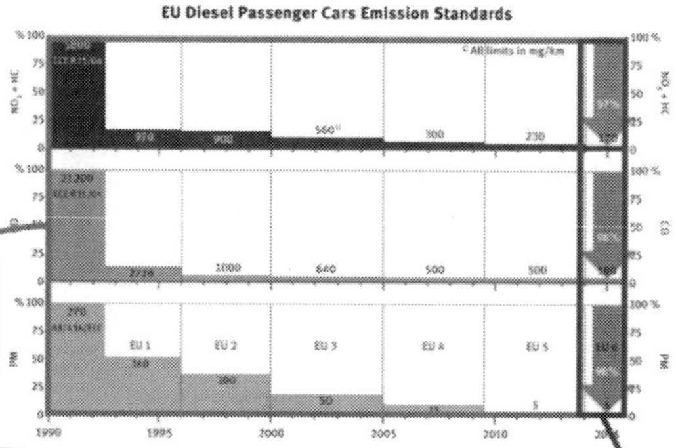

Usual way of dimensioning a coolant pump until Eu6
- Conventional water pumps -

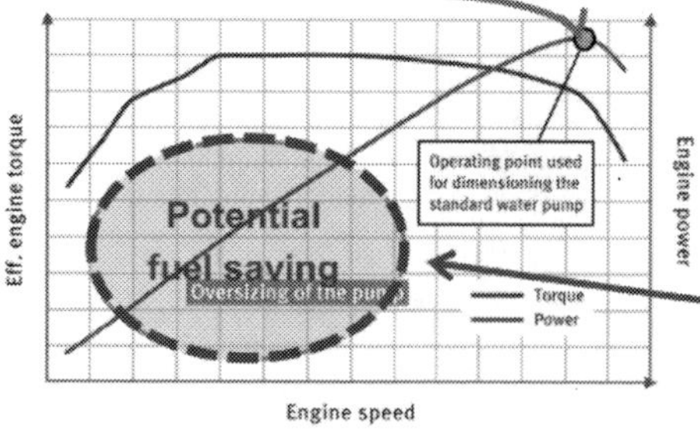

Necessity to find new technical solutions to meet the new emissions standards
- Variable water pumps -

➢ PPT technical responses to the challenge

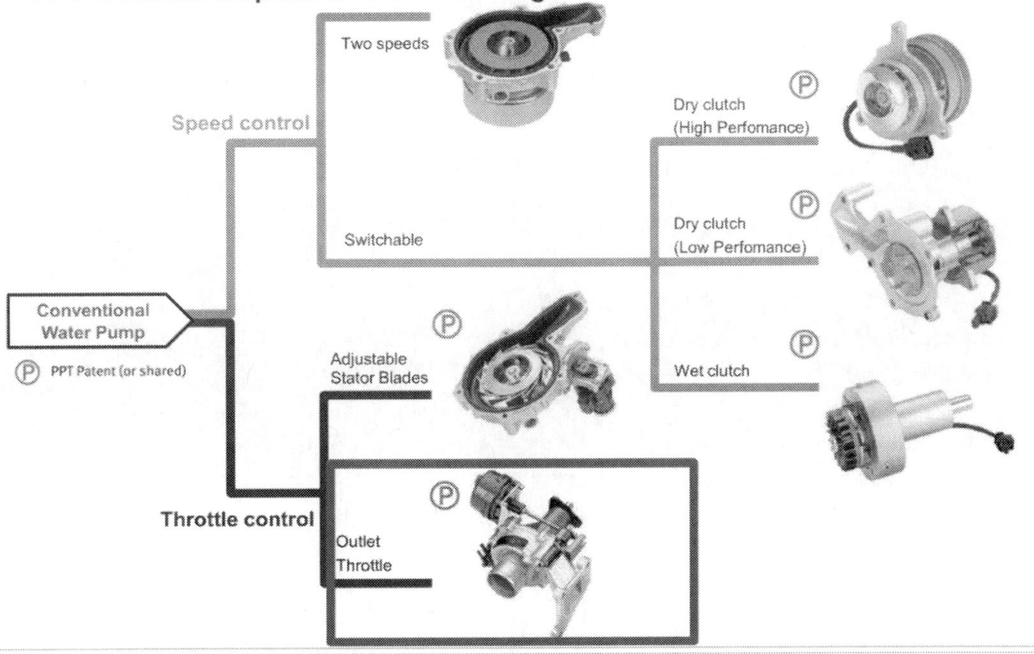

➢ Overview of a throttle valve for variable water pump application

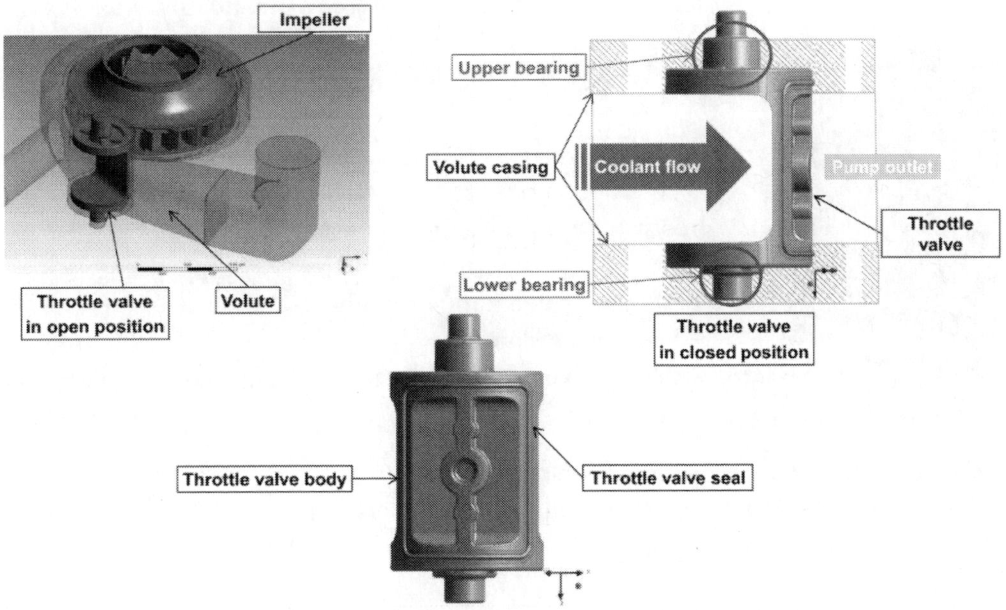

➢ Throttle valve water pump: interest of the concept

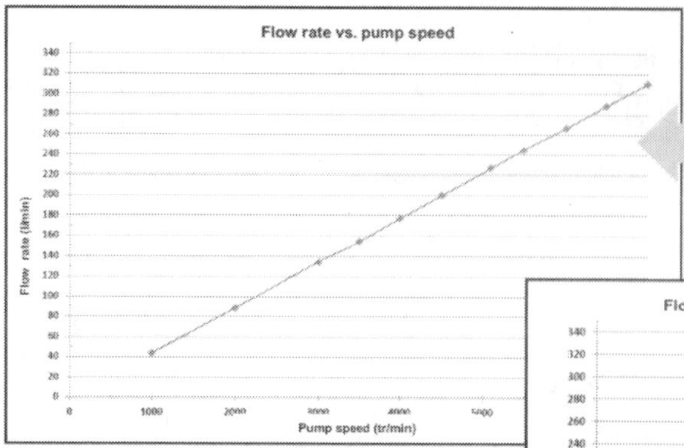

Conventional water pump
→ only one performance curve

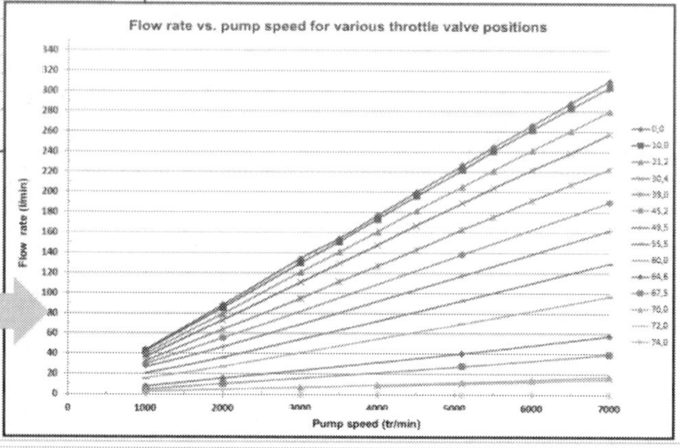

Throttle valve water pump
→'Unlimited' number of performance curves
from zero flow to maximum flow

➢ Design critical points and relevant simulation challenge

➢Throttle valve body: reinforced plastic material
→behaviour under water pressure ??
(robustness, deflection at any valve positions)

➢Throttle valve seal: elastomeric material
→behaviour under compression load ??
(resisting torque , water-tightness, endurance)

➢ Throttle valve actuation load
→maximum reduction !!
(limited packaging, limited power supply & consumption)

Request to Calculation & Simulation department:
→To provide a realistic model of the throttle valve system.
→This model must be reliable enough to orientate and validate any design choices.

➢ Simulation process surrounding throttle valve design

Throttle valve fluid dynamic analysis

↓ Hydraulic optimization → Throttle valve actuator definition

↓ Hydraulic loading → Throttle valve structural analysis

Throttle valve structural analysis → Throttle valve robustness; Throttle valve stiffness behaviour

Throttle valve stiffness behaviour → Throttle valve seal in service compression → Throttle valve water-tightness evaluation; Throttle valve actuator definition

Throttle valve actuator definition

- Throttle valve actuator driving force (throttle valve closing phase)
- Throttle valve actuator fail safe spring driving force (throttle valve opening phase)

Throttle valve for Variable Water Pump applications – Calculation and Simulation activities
PIERBURG PUMP TECHNOLOGY

➢ Throttle valve FE Model description

✓ **Involved geometrical components**

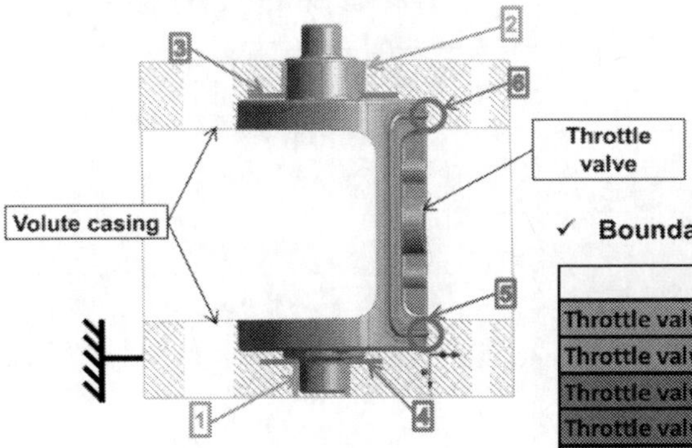

✓ **Boundary conditions / interfaces idealization**

Interfaces		Contact type	
Throttle valve body upper bearing	Volute casing	radial / frictional	1
Throttle valve body lower bearing	Volute casing	radial / frictional	2
Throttle valve body upper flange	Volute casing	normal / frictional	3
Throttle valve body lower flange	Volute casing	normal / frictional	4
Throttle valve seal	Volute casing	radial / frictional	5 6

Friction coefficients values???

✓ **Loads**

- **Step1: Seal pre-compression (assembly phase)**
- **Step2: Hydraulic pressure (in service phase)**

➢ Elastomeric and plastic materials tribological characterization (by external labs)

✓ Determination of friction coefficients for various couples of materials in specific conditions

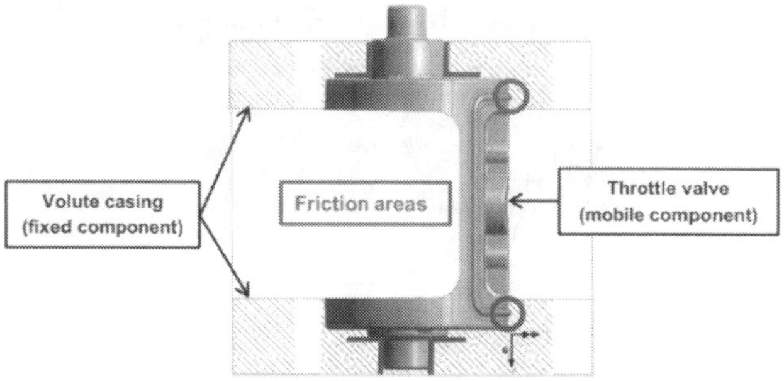

- Considered parameters
  - Temperature
  - Contact pressure
  - Velocity
  - Media (water coolant)
- Considered material couples
  - Elastomer – aluminium
  - Elastomer – plastic
  - Plastic – plastic
  - Plastic – aluminium

cetim
Centre technique des industries mécaniques
www.cetim.fr

InS

➢ Elastomeric and plastic materials tribological characterization (by external labs)

✓ Determination of friction coefficients for various couples of materials

- Test rigs

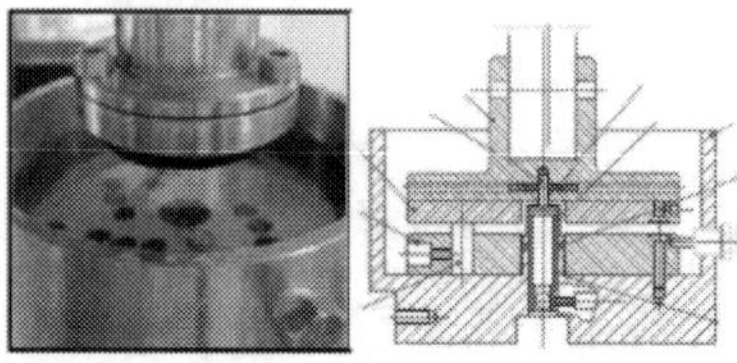

Pin on disk test rig

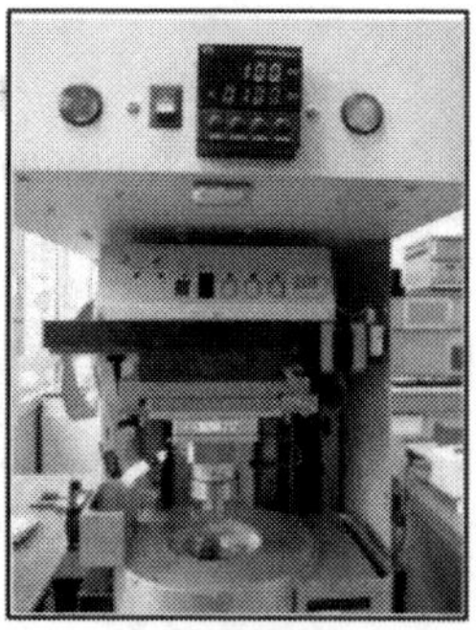

Tribometer

Ring on ring test rig

- Typical output graphs

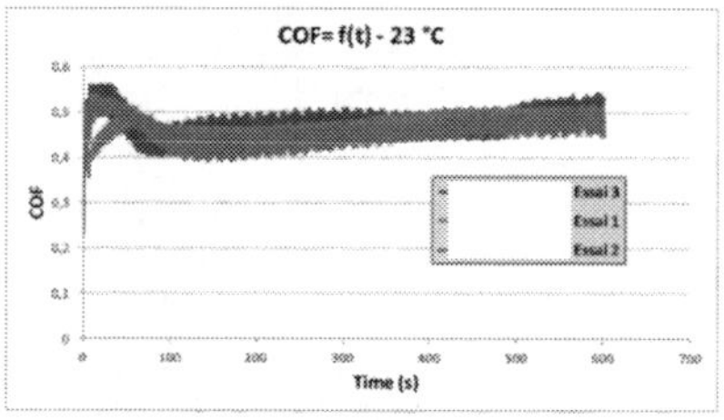

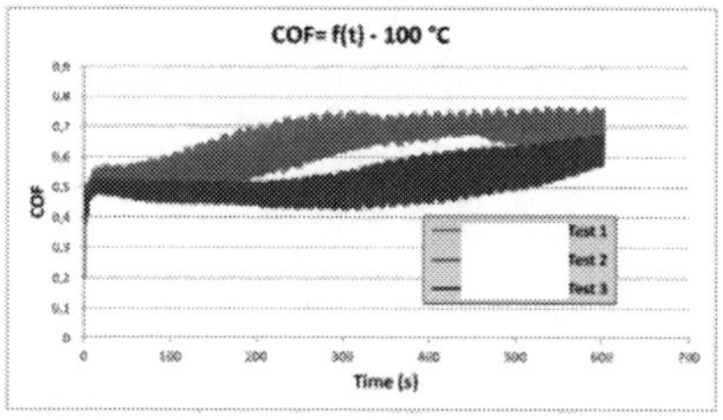

➢ Throttle valve fluid dynamic analysis

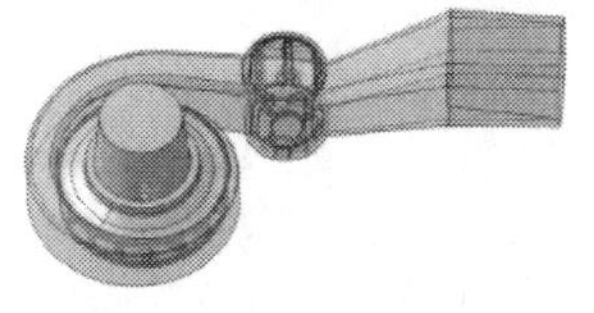

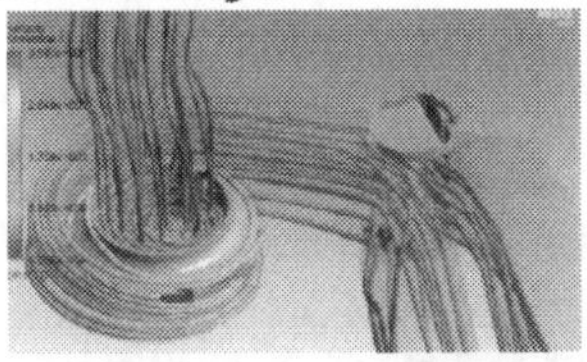

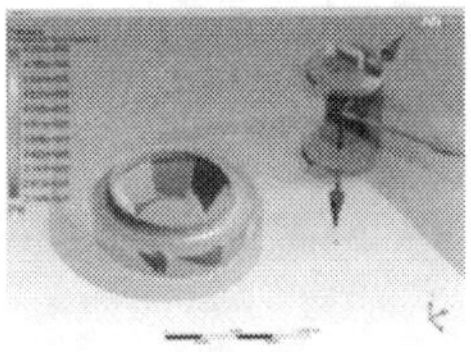

Hydraulic optimization

	Hydraulic force [N]			Hydraulic moment [N*m]		
	x	y	z	x	y	z
Opened	-21,48	-0,67	1,17	0,040	-1,676	-0,033
75 opened	-21,19	-2,02	8,98	0,118	-1,793	-0,054
50 opened	-21,63	3,32	18,51	-0,324	-2,005	0,044
25 opened	-23,67	38,82	76,70	-3,243	-3,236	0,677
Closed	-21,045	127,89	125,88	-10,241	-4,018	2,3978

Hydraulic loading

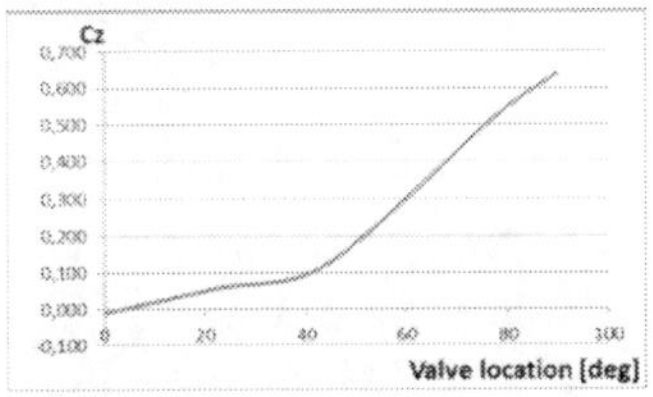

Fluid – Structure interaction

Direct link between hydraulic analysis and structural analysis

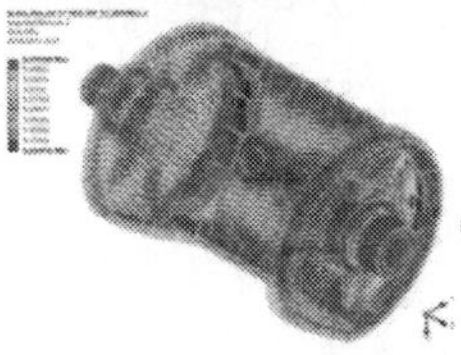

Realistic hydraulic loading acting on the valve at any positions

Stiffness behaviour???

➢ Throttle valve structural analysis: reinforced plastic body idealization

✓ **Rheo-mechanical advanced methodology (for non isotropic & non homogeneous materials)**

- Injection process simulation (rheology)

Process parameters of the Throttle manufacturing		
Clamping force	kN	100
Injection speed	mm/s	30
Holding pressure	bar	850
Holding pressure time	sec.	6
Cooling time	sec.	40
Cylinder temperature	°C	320
Mould temperature	°C	145
Injection time	sec.	1,66

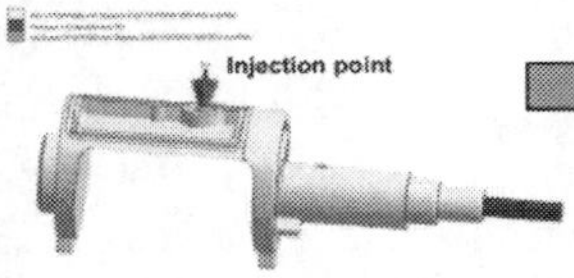

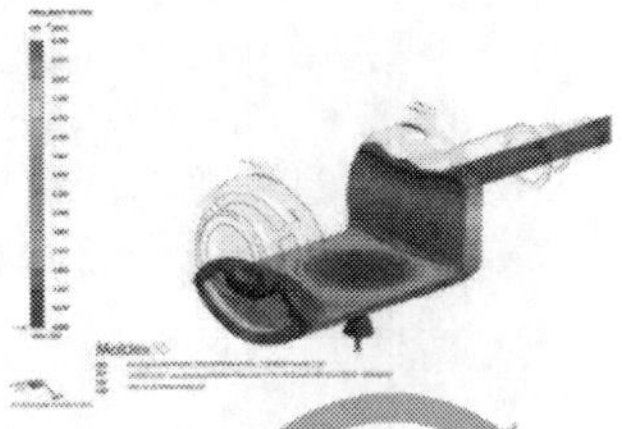

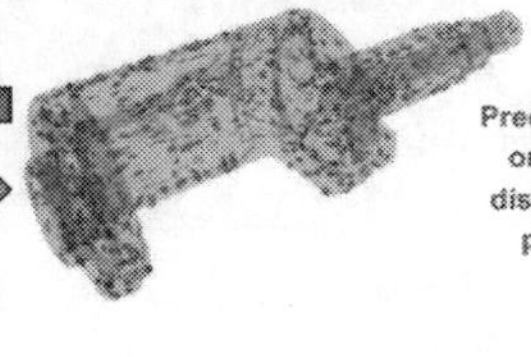

Prediction of fibres orientation and distribution in the plastic matrix

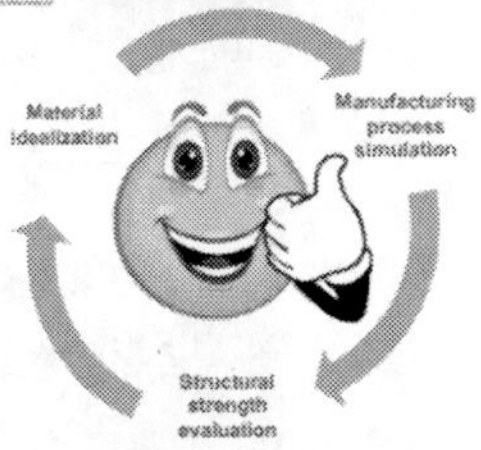

Optimization circle

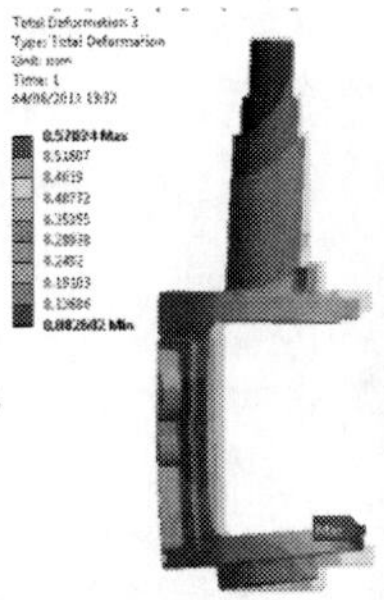

- Advanced composite material idealization

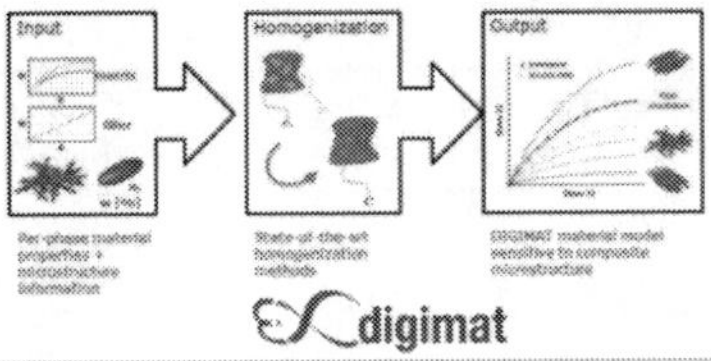

Realistic mechanical behaviour of reinforced plastic body
- Stiffness
- Strength

➢ Simulation-test correlation #1

✓ **Throttle valve stiffness and strength evaluation on a bending test case**

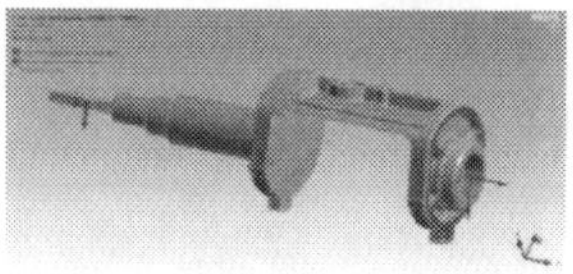

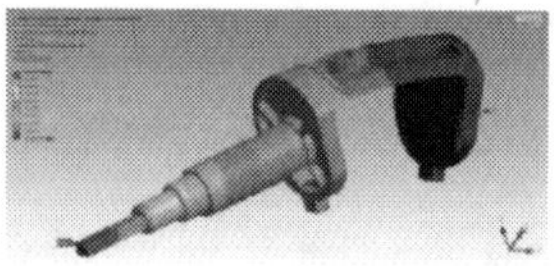

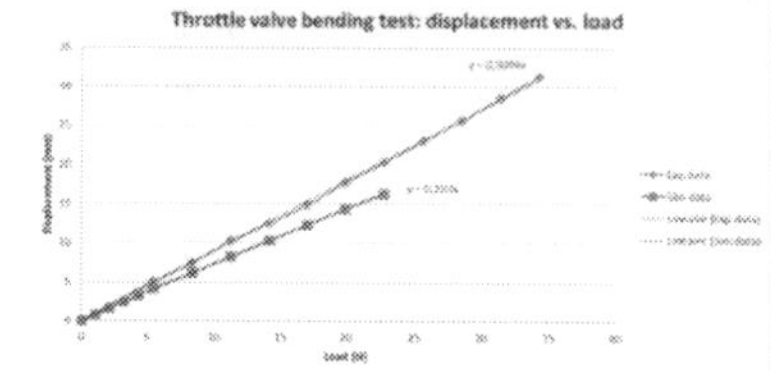

Correlation level on stiffness and strength → less than 20% (under enhancement)

➢ Throttle valve structural analysis: elastomeric seal idealization

✓ **Seal FEM: full 3D geometrical approach**

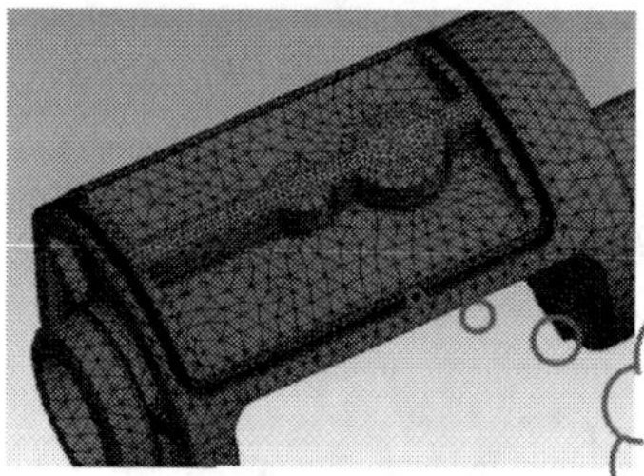

Seal full 3D mesh

Elastomeric materials behaviour laws???

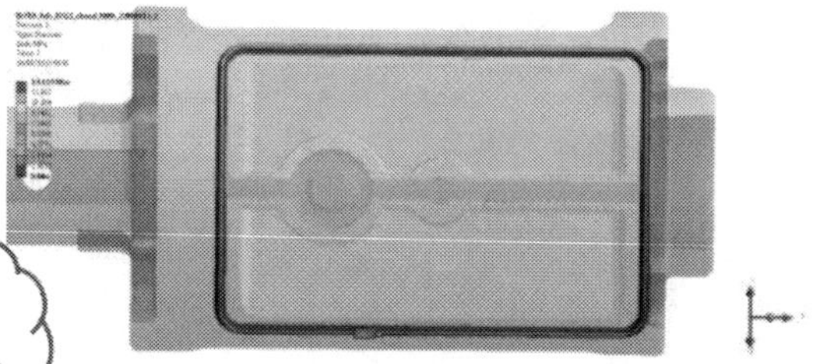

Contact pressure

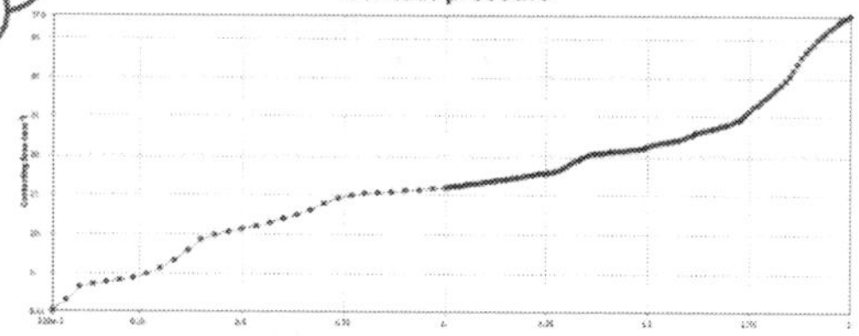

Contact surface

Lip seal deformation

Normal force generated by compression (→ water tightness, resisting torque)

Throttle valve for Variable Water Pump applications – Calculation and Simulation activities

PIERBURG PUMP TECHNOLOGY

➢ Elastomers characterization (by external lab)

✓ **Hyper-elastic behaviour laws** (stress vs. strain)

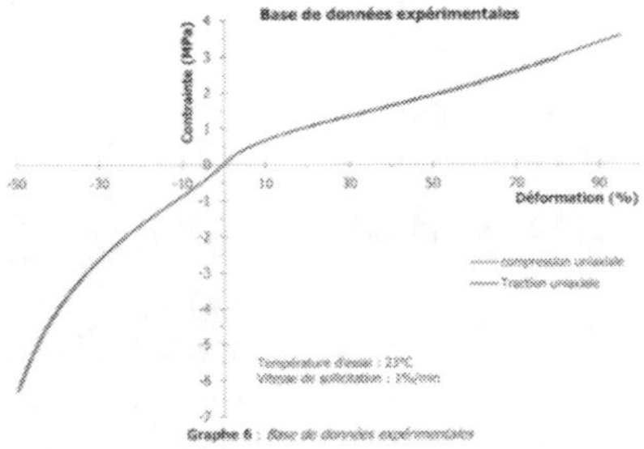

Experimental data

Laboratoire de Recherches et de Contrôle du Caoutchouc et des Plastiques

INSTITUT CARNOT

✓ **Numerical idealization of elastomers hyper-elastic experimental laws**

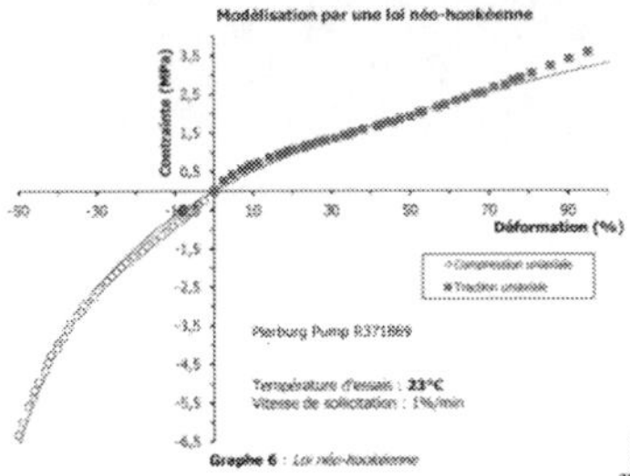

Loi néo-hookéenne

$W = C_{10}(I_1 - 3)$

avec $C_{10} = 9,4303\ 10^{-1}$ MPa

Loi de Rivlin du second ordre complet

$W = C_{10}(I_1 - 3) + C_{01}(I_2 - 3) + C_{20}(I_1 - 3)^2 + C_{11}(I_1 - 3)(I_2 - 3) + C_{02}(I_2 - 3)^2$

avec
$C_{10} = 8,3417\ 10^{-1}$ MPa
$C_{01} = 2,5201\ 10^{-1}$ MPa
$C_{20} = 6,7839\ 10^{-1}$ MPa
$C_{11} = -1,4211$ MPa
$C_{02} = 5,3767\ 10^{-3}$ MPa

Elastomeric numerical behavior laws

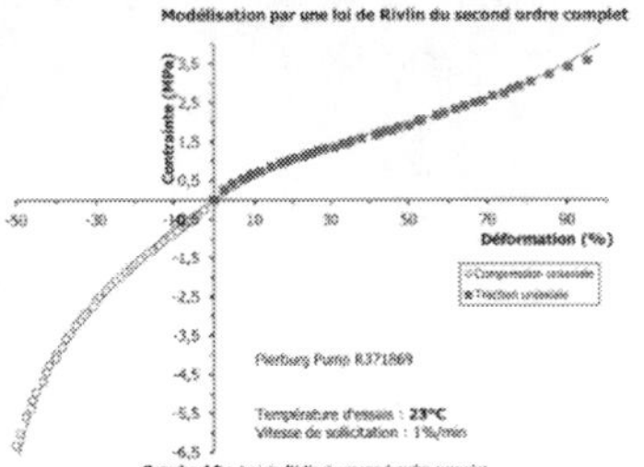

Laboratoire de Recherches et de Contrôle du Caoutchouc et des Plastiques

INSTITUT CARNOT

➢ Throttle valve actuation load : actuator sizing

✓ **Determination of actuator and fail safe spring driving force: analytical approach**

Hydraulic loads

Seal resisting torque

Actuator internal friction

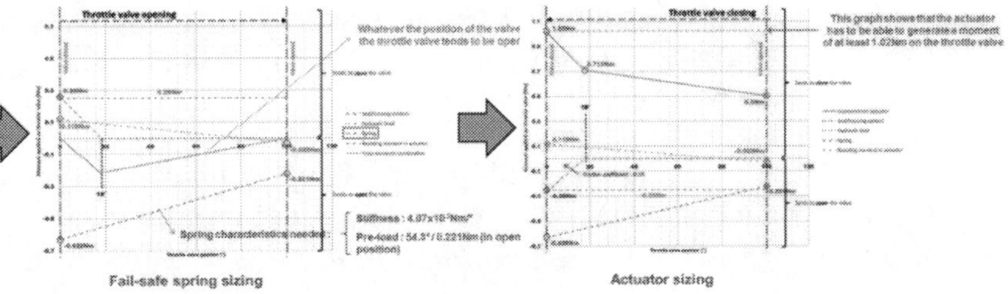

Fail-safe spring sizing

Actuator sizing

✓ **Determination of actuation parameters all along the motion: multi-body simulation approach**

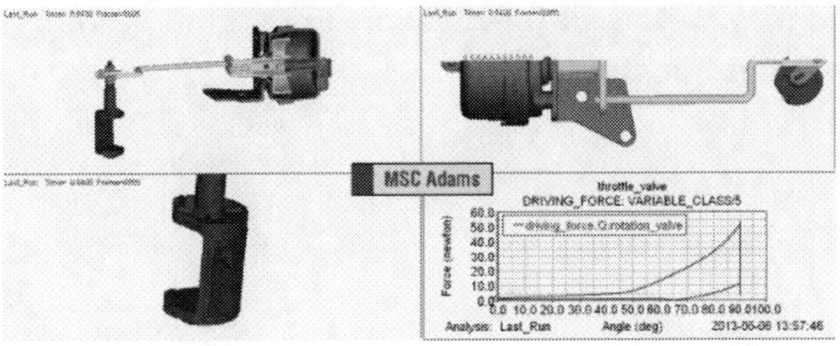

➢ **Simulation-test correlation #2**

✓ **Fail safe spring sizing**

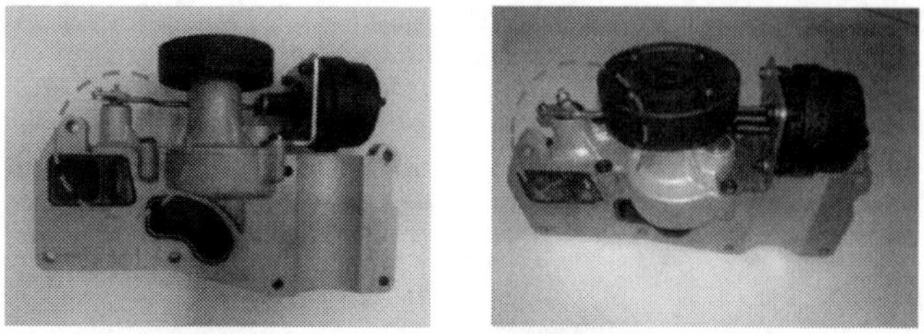

Prototype of a variable water pump fitted with a throttle valve

Water pump on hydraulic test bench

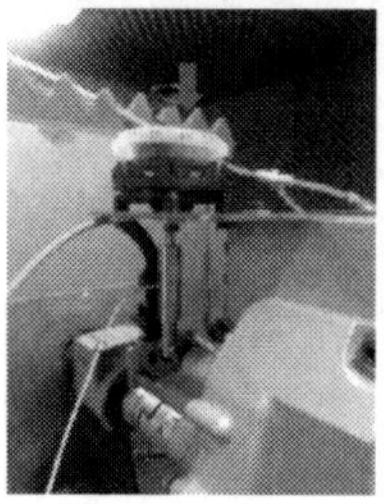

Throttle valve in closed position

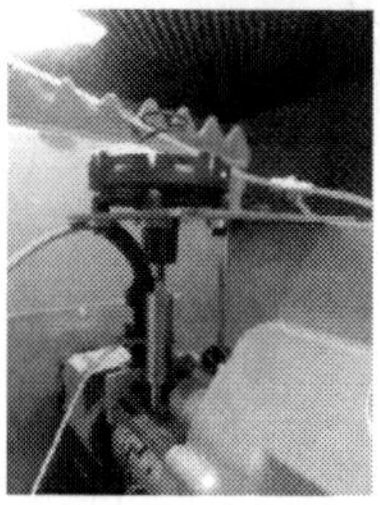

Throttle valve in open position

✓ Fail-safe force measured: 43 N

✓ Fail-safe force calculated: 47 N

✓ Correlation level: 8%

➢ **General conclusion**

✓ The variable water pump operated by means of a throttle valve is part of the range of innovative products proposed by Pierburg Pump Technology so as to help cars manufacturers to meet the more and more stringent regulations in term of gas emissions.

✓ Thanks to a judicious combination of simulation tools supplied with relevant experimental data, PPT France Calculation & Simulation department was able to derive a realistic and reliable model of the throttle valve system.

✓ This model is now used at PPT France in a very standard way to orientate and validate design parameters of throttle valve systems in the frame of answers to Customers Request For Quotations.

Thank you for your attention!

Many thanks to those of my team who have participated to the successful development of this calculation model.

Any questions?

PIERBURG PUMP TECHNOLOGY

Disclaimer

This presentation contains "forward-looking statements". Forward-looking statements are sometimes, but not always, identified by their use of a date in the future or such words as "will", "anticipates", "aims", "could", "may", "should", "expects", "believes", "intends", "plans" or "targets". By their nature, forward-looking statements are inherently predictive, speculative and involve risk and uncertainty because they relate to events and depend on circumstances that will occur in the future. There are a number of factors that could cause actual results and developments to differ materially from those expressed or implied by these forward-looking statements. In particular, such factors may have a material adverse effect on the costs and revenue development of the KSPG group.

All written or oral forward-looking statements attributable to KSPG AG or any group company of KSPG AG or any persons acting on their behalf contained in or made in connection with this presentation are expressly qualified in their entirety by factors of the kind referred to above. No assurances can be given that the forward-looking statements in this presentation will be realized. Except as otherwise stated herein and as may be required to comply with applicable law and regulations, KSPG AG does not intend to update these forward-looking statements and does not undertake any obligation to do so.

Modeling of the fuel supply units and its components

Dr. Oleksiy Kurenkov,

Continental, Division Powertrain,
BU Fuel Supply, Schwalbach

Fuel Supply Unit
Key products

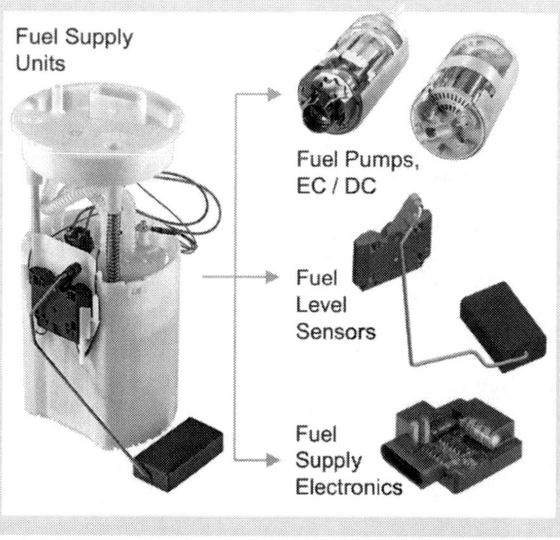

- Fuel delivery systems for gasoline, diesel, alcohol, and LPG applications
- Multiple functions (fuel pump, fuel level measurement, pressure regulator, lifetime fuel filter, electronics)
- Electronic pump speed regulation (pump speed on-demand controlled)

Fuel Supply Unit

Tank integration

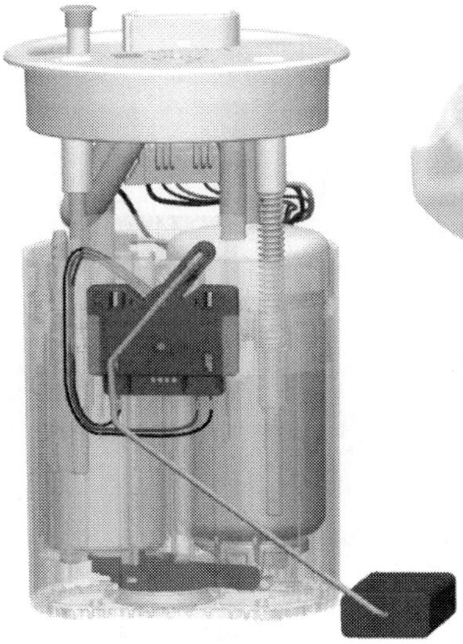

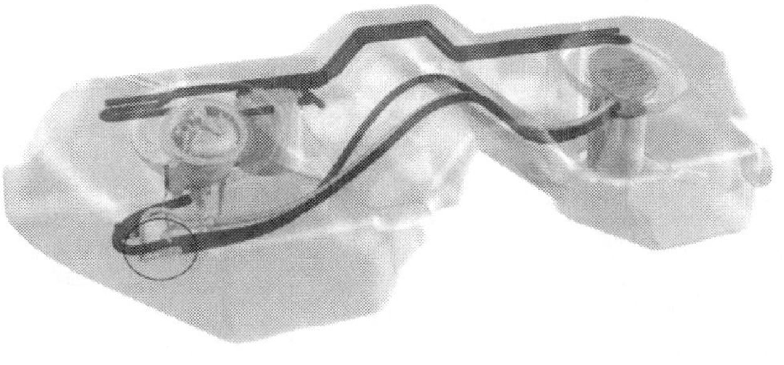

- Modern multi-chamber tanks require many *additional* transfer pumps and complex piping
- fuel level measurement is performed by many sensors

Continental

Major Challenges in the Development of FSU

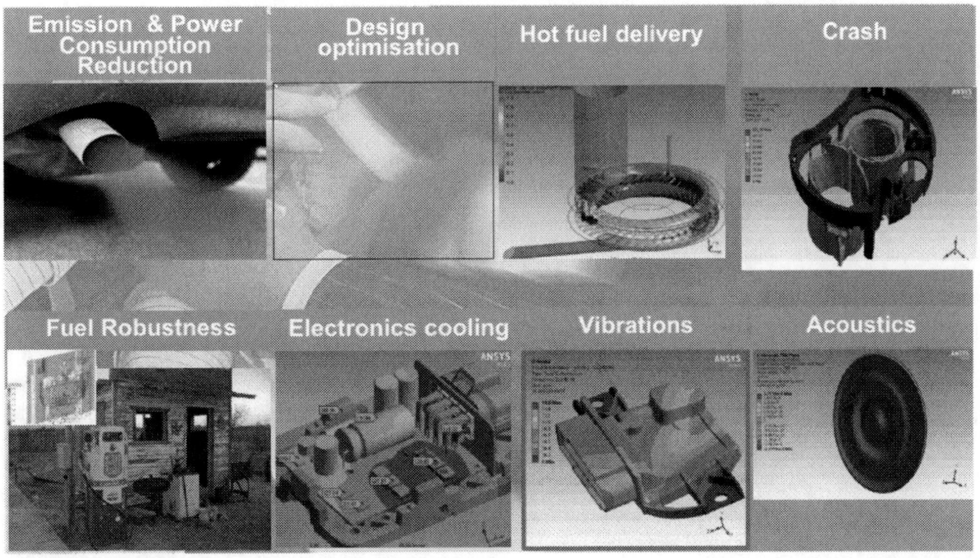

Continental

System Modeling
Process

Components measurement → Component modeling → FSU modeling → load cycle NEDC, full load) → FSU control → Results

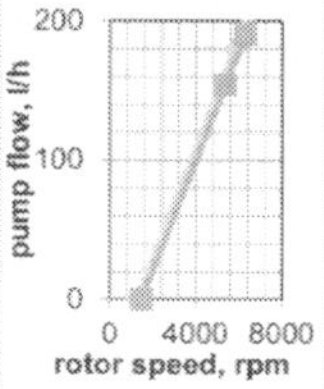

pump characteristic

Ansys components simulation

Amesim 0D/1D simulation tool

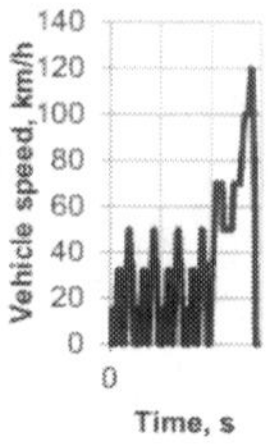

Full load, e.g. 6.1 bar, 260l/h

On-demand regulated FSU

Action	Savings of electrical power, W
Register pumps	14.64
E-jet pumps	7.84
Contact soldering	2.37
Total	24.85

Energy savings

Continental

System Modeling
Modeling approach

- Reference car
  - 1.6 l, 4 cylinders
  - 1360kg
- On demand regulated DC pump (fuel consumption: full-car model)
- Two-channel pump
- One transfer pump

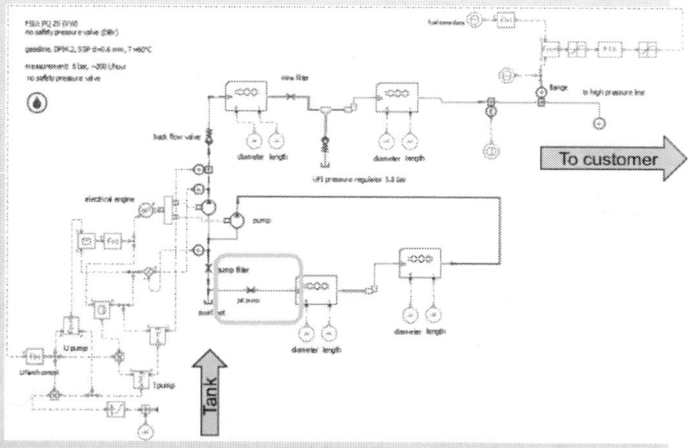

Continental

System Modeling
Results gasoline pumps

Pump	Aver. Power in NEDC, W	Total fuel consumption in NEDC, g	CO2, g/km	CO2 reduction, g/km
G43T/C unregulated Gasoline, DC43, unreg.	119.8	508.38	145.49	*reference*
G43T/C Gasoline, DC43	30.29	503.71	144.15	1.34
G38T/C Gasoline, DC38	29.34	503.65	144.14	1.35
G38T/EC Gasoline, EC38	28.11	503.6	144.12	1.37
G38T/EC part load Gasoline, EC38, low speed opt	12.79	502.75	143.89	1.6

~1.1% gain on CO_2 emissions with regulated pumps

Design optimisation
Parametric optimisation approach

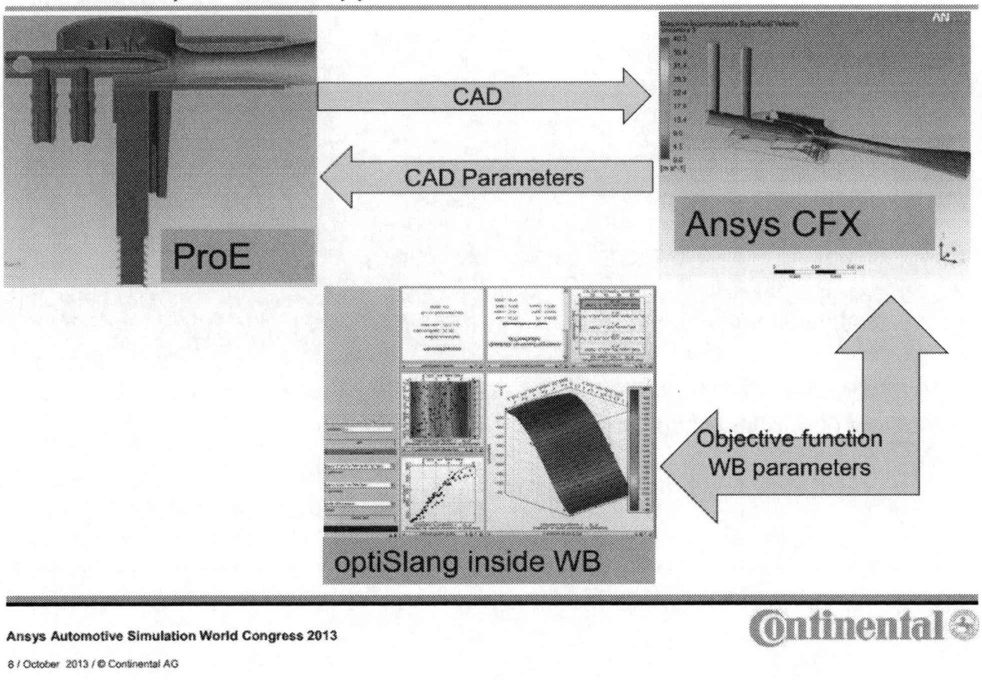

Design optimisation

Pump hydraulics optimisation

- ✓ Inlet/outlet ports improvements
- ✓ reduction of pressure pulsations (acoustics improvement)

THE GRAPHIC ON THIS PAGE WAS NOT AVAILABLE ON THE E-MEDIA VERSION OF THIS PAPER.

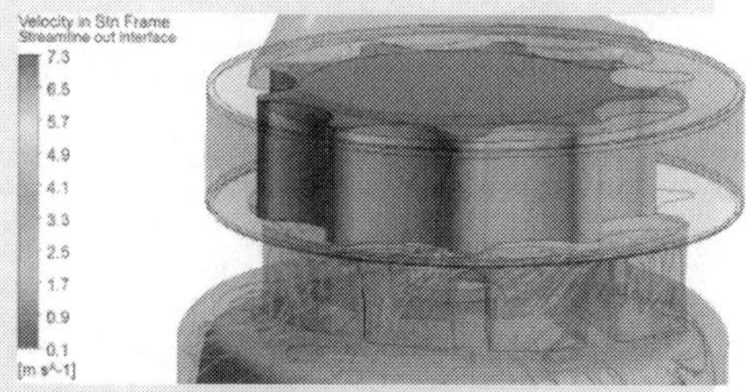

Pump stage structural analysis

Channel heigth change during operation

- Goal: to model change of the channel height during pump operation
- Reducing/increasing of channel heigth drastically influences pump leakages and performance
- Operational conditions
  - Press fit is modeled accordingly to CAD model
  - Hydraulic pressure is applied using CFD simulation data (s. pict.)
  - Temperature +70° C
  - Swelling in fuel included

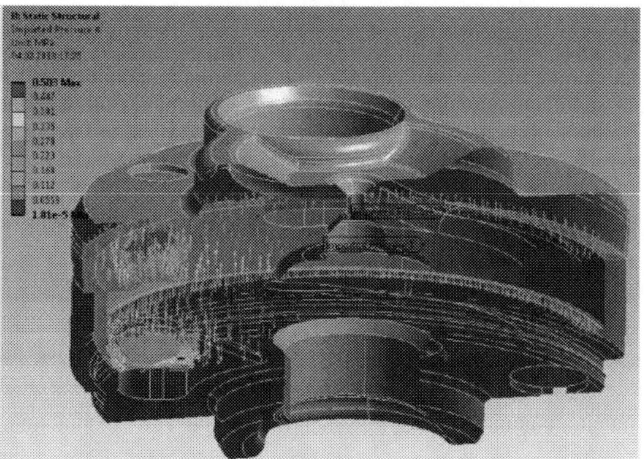

Pump stage structural analysis
Results

- Evaluation of channel height change in the pump
  - Find the minimal distance between deformed parts, housing and cover
- Final result takes into account wheel height change

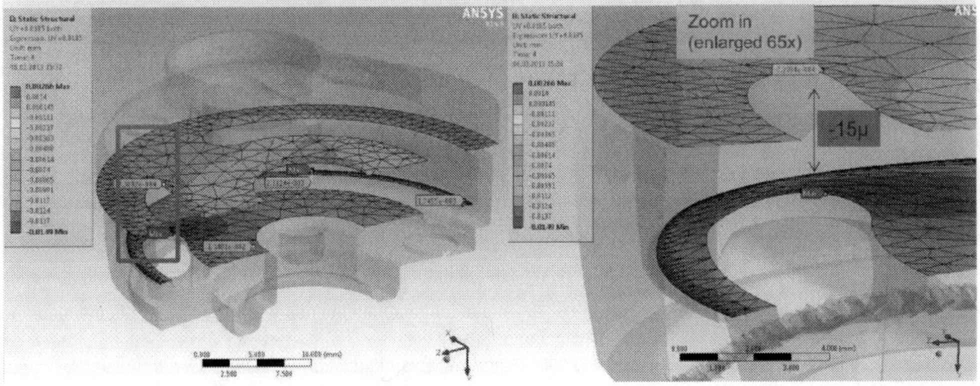

Hot fuel test
Setup

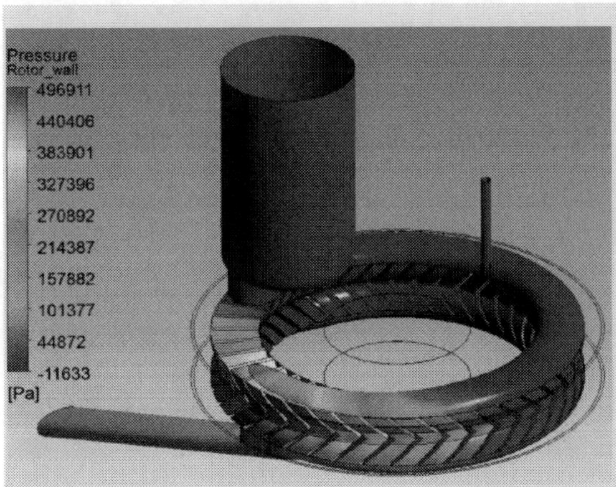

- Due to relatively high vapour pressure of gasoline the gasoline pumps tend to cavitate on inlet
- Increasing of temperature & degaussing of dissolved air even increase the process ->

  As result the delivery rate fails
- Modeling approach
  - cavitation model Rayleigh-Plesset
  - Viscosity, desity, surface tension = func(T)
- Very sensitive to b.c., convergency difficulties

Results hot fuel test

Gasoline vapour buble

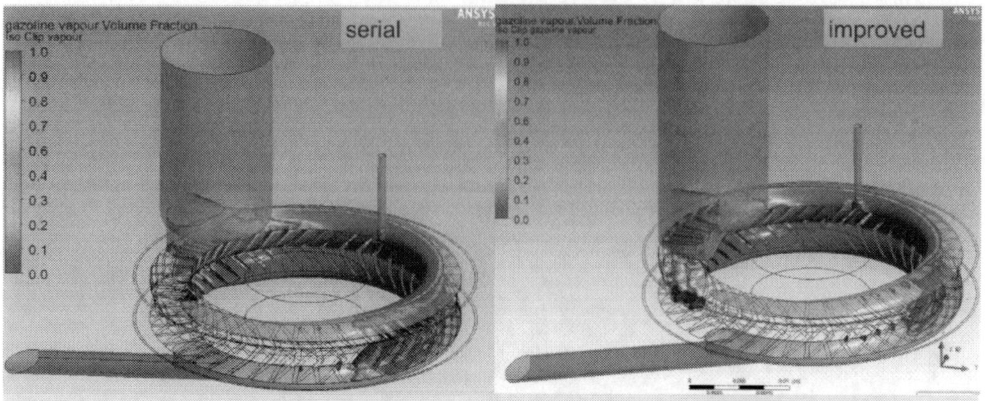

- Gasoline bubble becomes smaller -> more pump flow at high temperatures

Crash & explicit simulations

Ansys LS-Dyna

THE GRAPHIC ON THIS PAGE WAS NOT AVAILABLE ON THE E-MEDIA VERSION OF THIS PAPER.

THE GRAPHIC ON THIS PAGE WAS NOT AVAILABLE ON THE E-MEDIA VERSION OF THIS PAPER.

Ansys implicit

Fuel robustness

Fluid flow

Design 01 **Design 02**

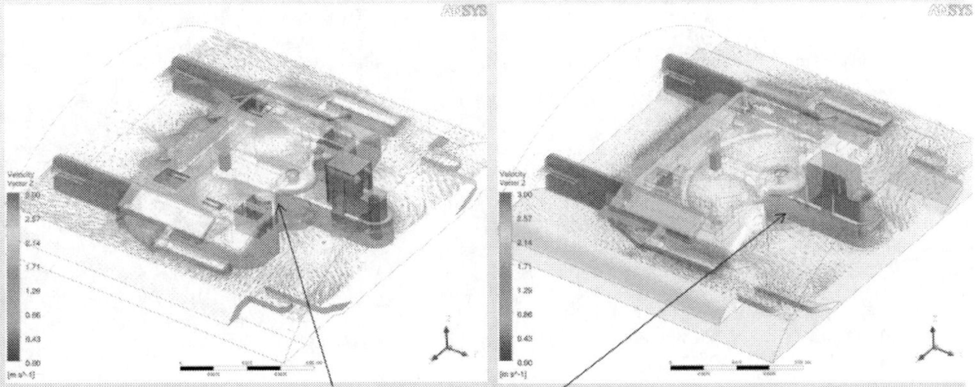

✓Significant improvement in the gap region (flow decreases in the new design)

Fuel robustness

Particles collected in MAPPS after 10 flow cycles

Design 01 **Design 02**

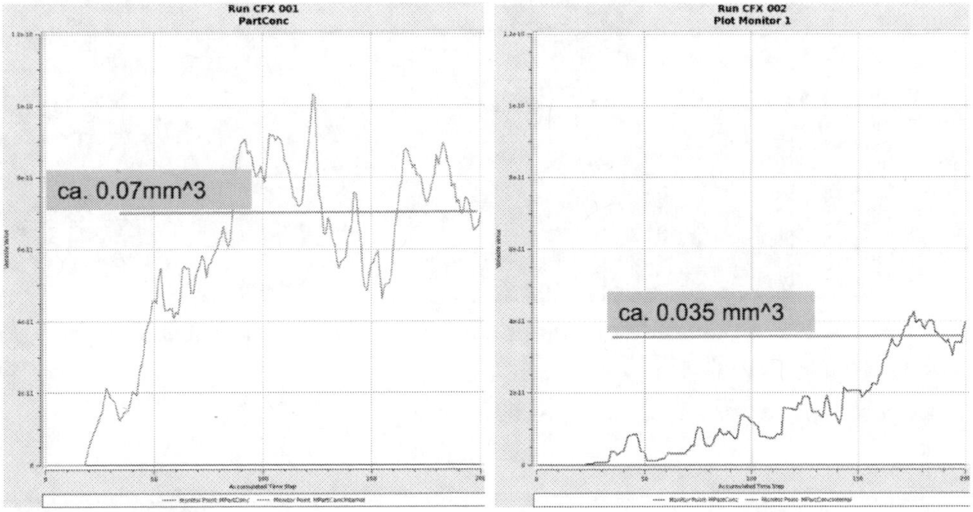

Electronics cooling & thermal shock simulations

Modeling aproach

- Thermal power ~10W (electrical power ~200W)
- Goals:
  - Temperature on components
  - Stresses on solders due to temperature increasing
  - Stress on potting mass due to thermal shock (-40° .. +80° C)
- Isssues
  - To define the thermal power on every electronic component
  - To describe the external heat transfer

Electronics cooling

Results

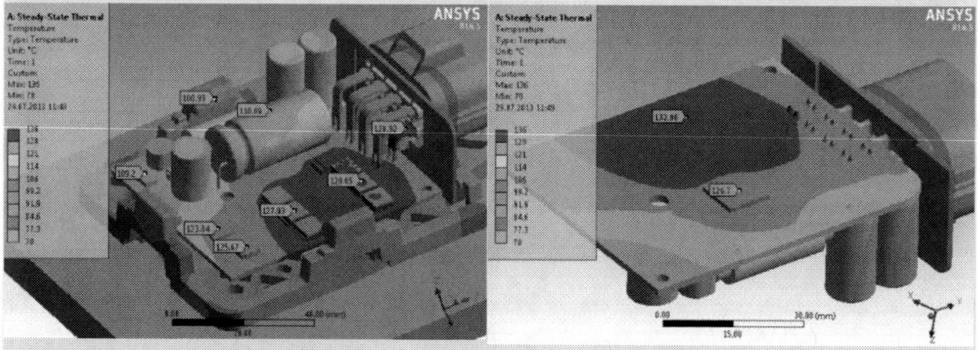

- ✓ the simulated temperature on components fits the experimental values < 10° (typically ~3° C)
- ✓ Stresses on solders & housing are evaluated too

Noise & Vibration

Gerotor dynamics

- gerotor pump modeled in order to optimise gerotor profile (smooth running, less noise)
- pressure changes accordingly to the teeth position
- applying the rotating speed on shaft, we look for the rotating torque oscillations
- using FFT of torque we estimate acoustic excitation

THE GRAPHIC ON THIS PAGE WAS NOT AVAILABLE ON THE E-MEDIA VERSION OF THIS PAPER.

Continental

Acoustics

Echo simulation

- Visualisation of acoustics wawes propagation helps to optimize wawe quide design
- Coupled piezoelectric-structural-acoustics simulation
  - SOLID 226 for piezoelectric elements
  - FLUID 220 for 3D acoustics elements
- Technology is based on Ansys 14.5 with APDL snippets
  - Ansys Acoustics extension
  - FSI surfaces on contacts

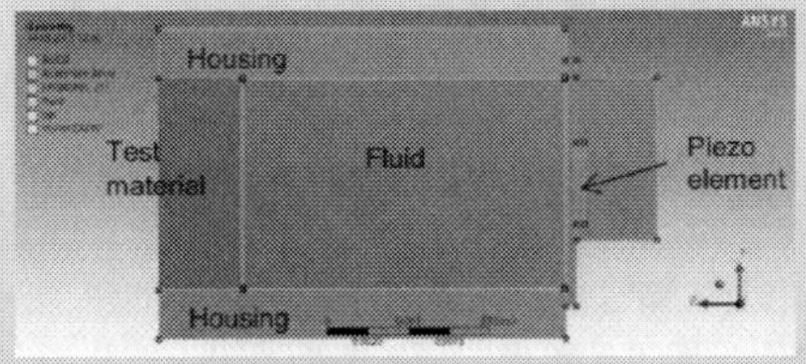

Continental

Acoustics
Model verification

- Voltage response on piezolement validated
- both, reflected sygnal amplitude & echo travelling time are in good agreement with experiment

Voltage response in piezoelectric element

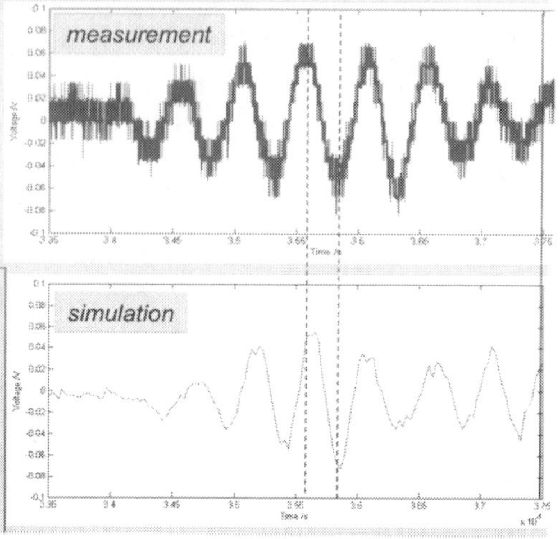

THE GRAPHIC ON THIS PAGE WAS NOT AVAILABLE ON THE E-MEDIA VERSION OF THIS PAPER.

Acoustics
Sensor model

- Quasi 3D Simulation (1 element in vertical direction)
- Perturbation on piezo
  - sinusoidal ΔV=10V, f=2MHz
- Voltage response on piezo in good agreement with measurement (~2%)
- computational cost ~ 3days on 8 Processors
  - RAM demand is a bottleneck
- Full 3D needs much more ressources
  - >16Processors, comp. time ~ weeks

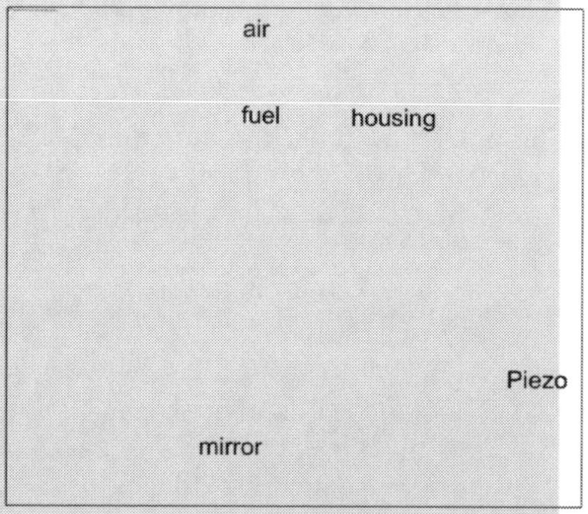

Summary

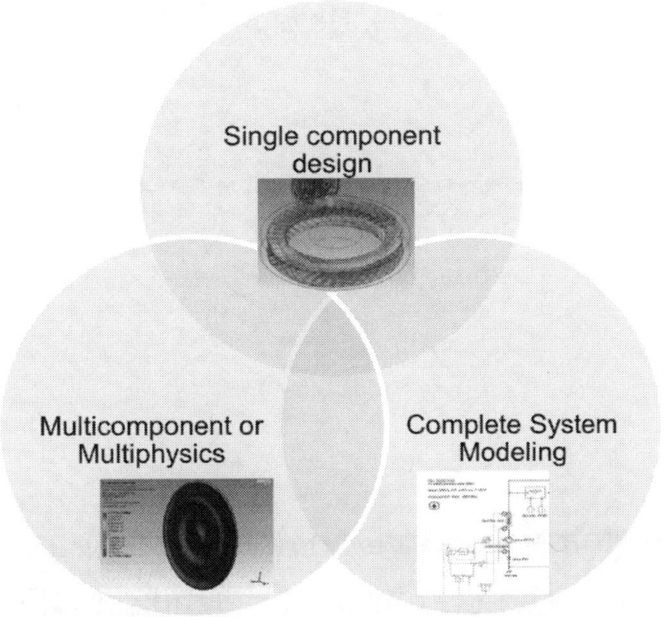
Single component design
Multicomponent or Multiphysics
Complete System Modeling

DEVELOPMENT OF SCR EXHAUST AFTER-TREATMENT MIXING SYSTEM WITH CFD

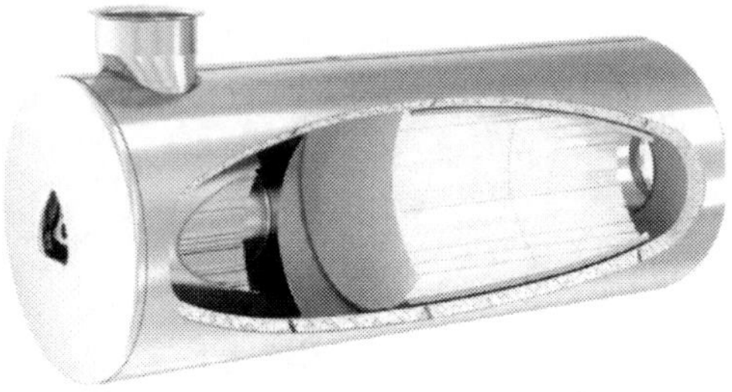

Dr. Corine Chauvin
Donaldson Europe, Belgium

Donaldson FILTRATION SOLUTIONS

ANSYS *Automotive Simulation World Congress 2013* - 29 October 2013

- Introduction Donaldson company
- closed coupled DOC+mixer+SCR system for non-road
- System development with CFD
- CFD/Experimental results
- Conclusion

Donaldson FILTRATION SOLUTIONS

ANSYS *Automotive Simulation World Congress 2013* - 29 October 2013

The Company at a Glance

- Leading WW provider of filtration systems & replacement parts
- Founded in 1915
- Technology-driven company
- Customers served in the industrial and engine markets
- 13 000+ employees worldwide
- Locations in 37+ countries worldwide

Donaldson : *exhaust system integrator*

- Technology selection
- Integration of aftertreatment with engine
- Optimizing trade-off between engine and aftertreatment

- Customize packaging/design
- Vehicle integration
- Modular systems

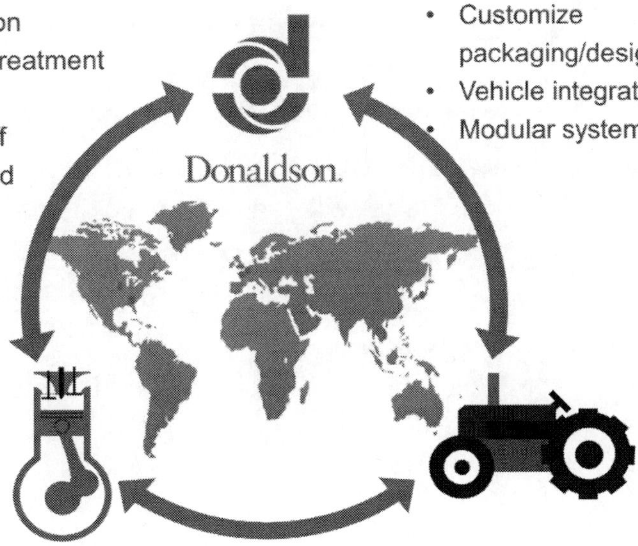

Comparison

Typical

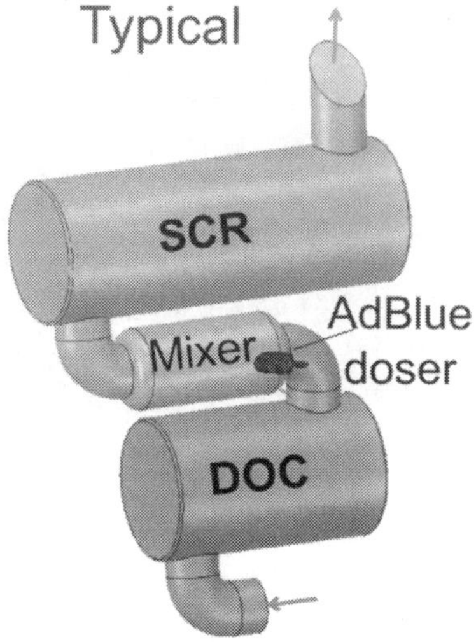

Combined
DOC-mixer-SCR

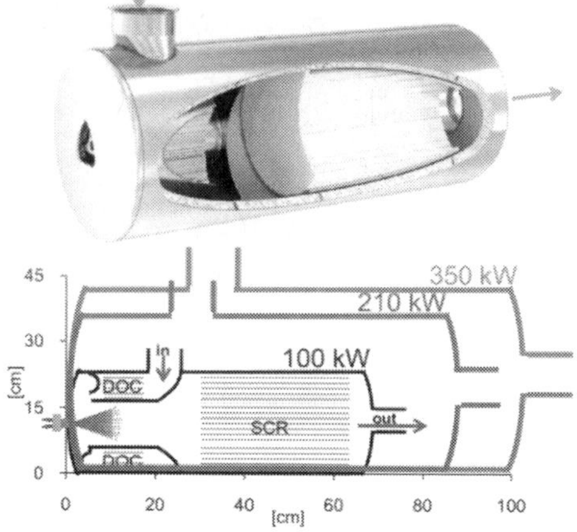

Donaldson.

AdBlue Mixer challenges

- Good mixing & *avoiding deposit formation*
- Densely packaged machines
  - No space for aftertreatment
  - No space for mixer + downstream tubing
- Low/high volume platforms
  - Need for flexible and robust solutions
- SCR-only & no EGR
  - High raw NOx + high efficiency → high dosing rate
  - No DPF regeneration to burn off deposits

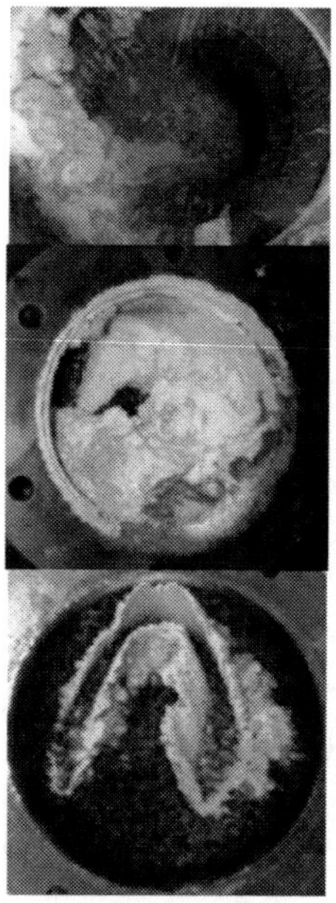

Donaldson.

ANSYS ***Automotive Simulation World Congress 2013*** - *29 October 2013*

Donut mixing system

1. Emitec donut shaped metallic DOC
2. Tangential Swirl baffle
3. Spraying in high turbulence zone with converging flow
4. Well mixed and evaporated AdBlue at SCR inlet face

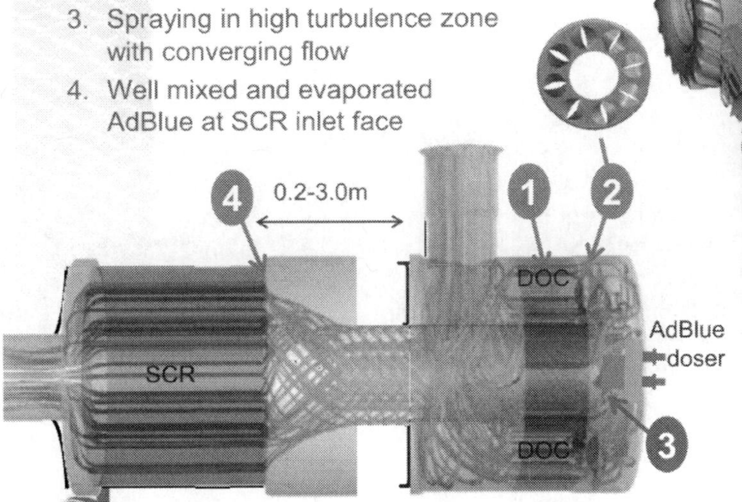

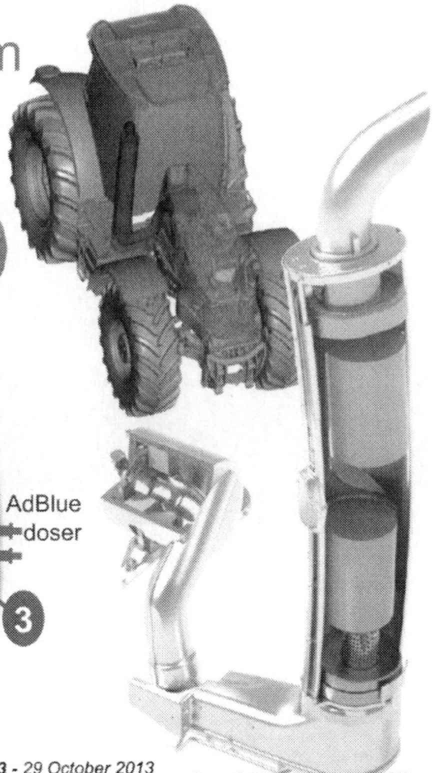

See SAE 2012-01-1087

Close coupled DOC-mixer-SCR

Combines all exhaust components:

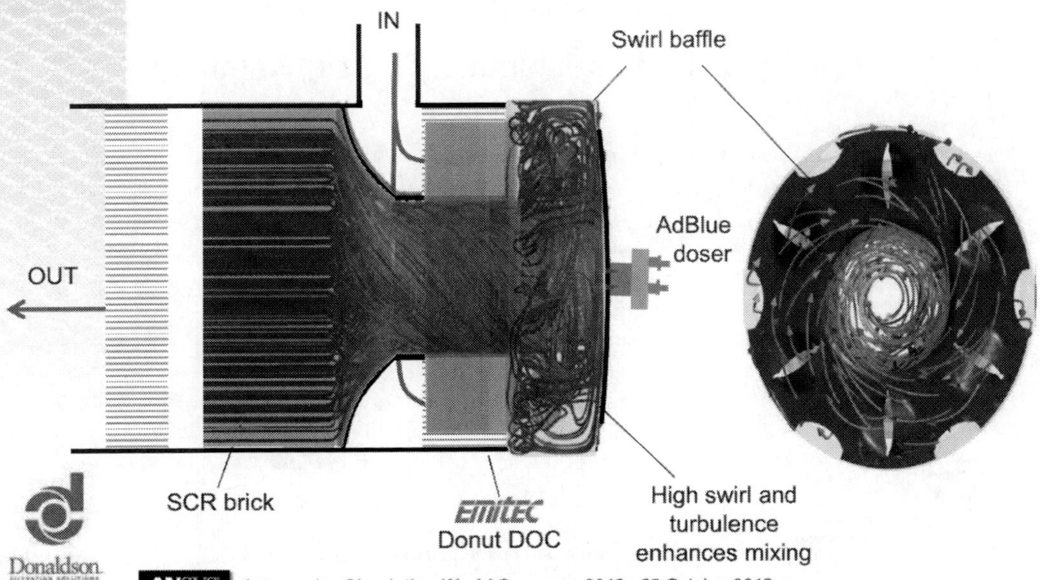

CFD setup

- Fluent 3D 14.5 steady for flow, transient for DPM
- Tet and Hex mesh ≈3 000 000 cells
- Realizable k-eps model
- Air: Ideal Gas, Phys Prop = f (T)
- Working point: engine 4.5L
  - High Flow: 635kg/h@445°C
  - Low Flow: 160kg/h@230°C
- DOC Substrate :
  - Porous media:
    - Viscous resistance = 2.92e07 (1/m2)
    - Inertial resistance = 139.6 (1/m)
- SCR Substrate :
  - Porous media:
    - Viscous resistance = 2.54e07 (1/m2)
    - Inertial resistance = 16.6 (1/m)

Donaldson.

ANSYS *Automotive Simulation World Congress 2013* - 29 October 2013

CFD setup water spray

- Injection of H2O doser 3 solid cones
  - Cone angle α = 20°
  - Spray angle δ=8.5°
  - Single cone angle β = 14°
- Spray diameter distribution: Rosin-Rammler
  - Sauter Mean Diameter ≈ 80μm
  - mean-Ø=25μm
- Low Flow :
  - 160kg/h@230°C
  - Mass_flow_H2O=176g/h

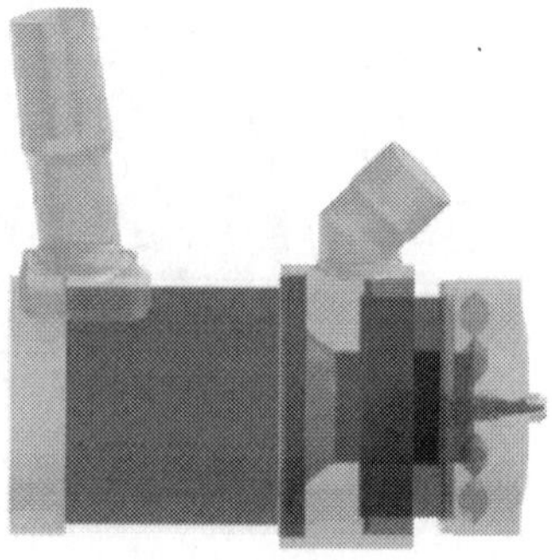

Donaldson.

ANSYS *Automotive Simulation World Congress 2013* - 29 October 2013

CFD results for exhaust line efficiency

- evaluation of exhaust line total back-pressure at HF
- Calculation of flow uniformity γ for DOC and SCR substrates : should be > target value for both flow conditions

$$\gamma = 1 - \sum_{i=1}^{n} \frac{\left|V_i - \overline{V}\right|}{2\overline{V}.S}.S_i$$

- Calculation of urea/water) uniformity γ for SCR substrate: should be > target value value for both flow conditions

$$\gamma_{H_2O} = 1 - \sum_{i=1}^{n} \frac{\left|C_i - \overline{C}\right|}{2\overline{C}.S}.S_i$$

Donaldson.

ANSYS *Automotive Simulation World Congress 2013 - 29 October 2013*

CFD results at LF

In pipe injection system

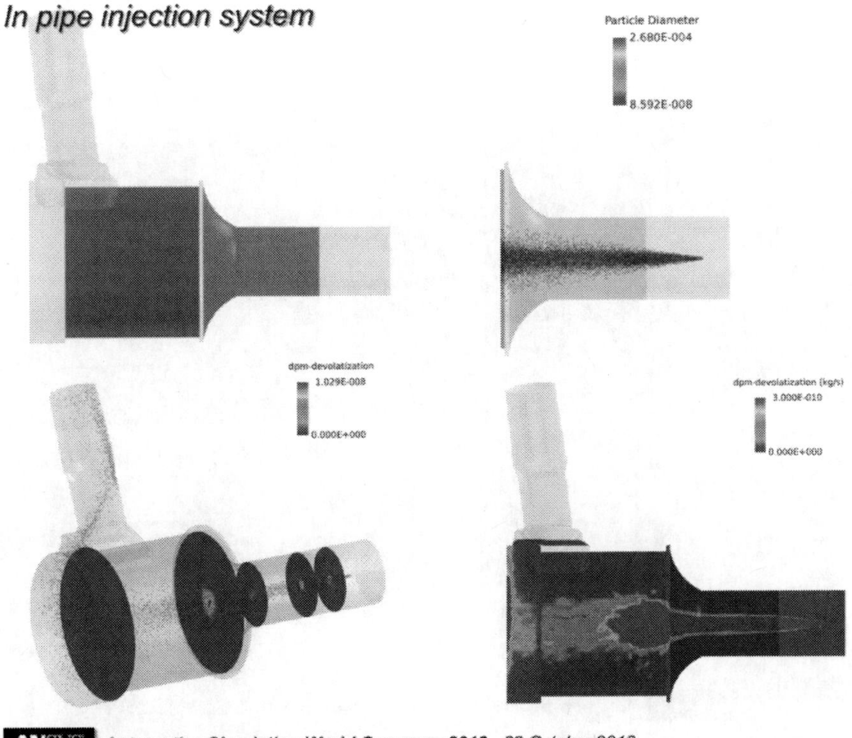

Donaldson.

ANSYS *Automotive Simulation World Congress 2013 - 29 October 2013*

CFD results at LF

Donut injection system

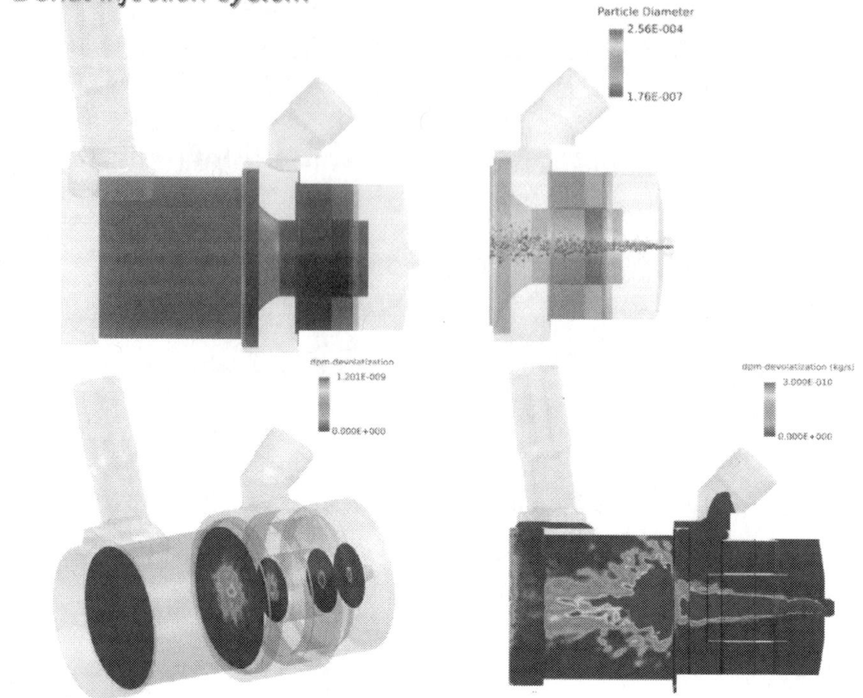

Donaldson.

ANSYS **Automotive Simulation World Congress 2013** - *29 October 2013*

CFD results at LF

Donut + mixer injection system

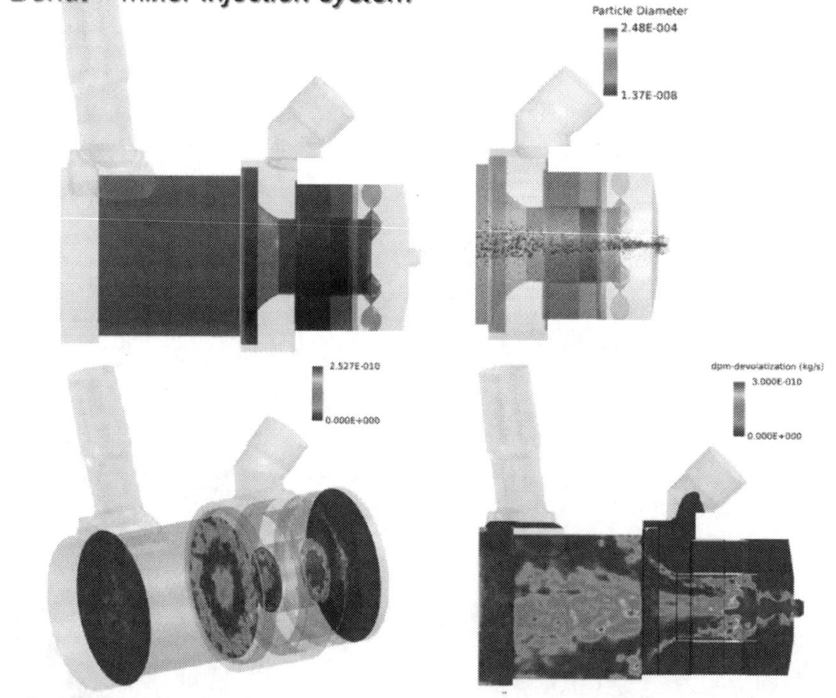

Donaldson.

ANSYS **Automotive Simulation World Congress 2013** - *29 October 2013*

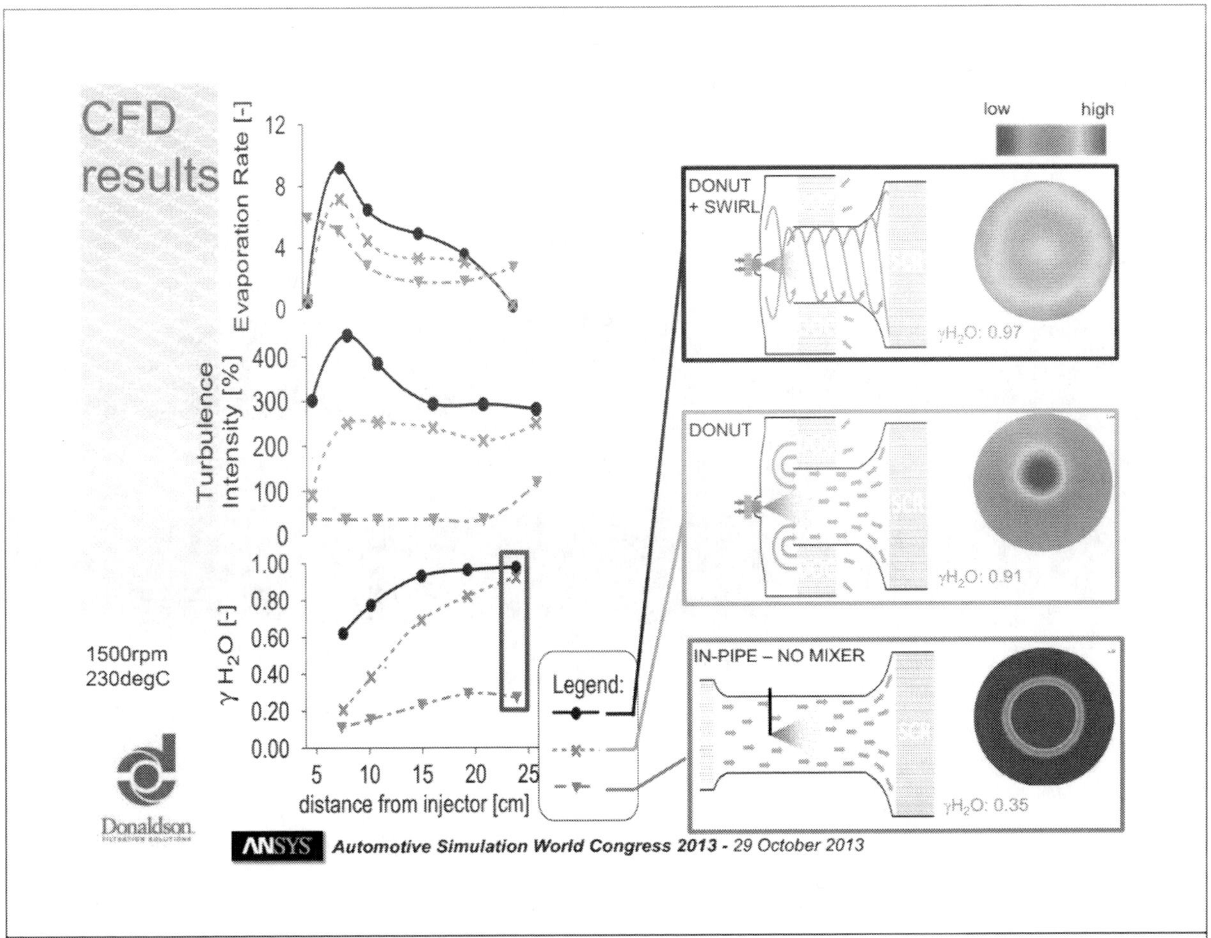
CFD results
Evaporation Rate [-]
Turbulence Intensity [%]
γ H2O [-]
distance from injector [cm]
1500rpm
230degC
Legend:
low
high
DONUT + SWIRL
γH2O: 0.97
DONUT
γH2O: 0.91
IN-PIPE – NO MIXER
γH2O: 0.35
SCR
Donaldson.
ANSYS
Automotive Simulation World Congress 2013 - 29 October 2013

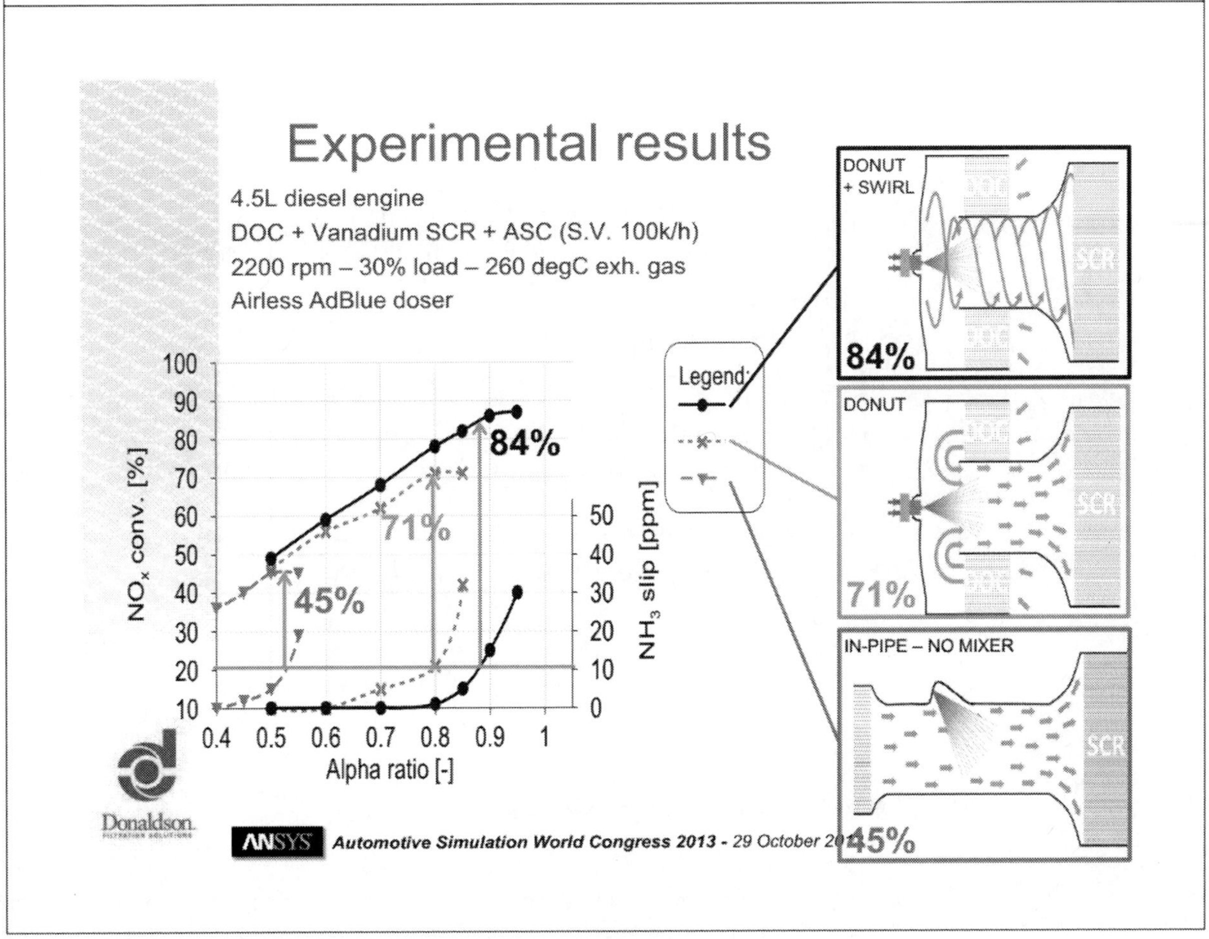
Experimental results
4.5L diesel engine
DOC + Vanadium SCR + ASC (S.V. 100k/h)
2200 rpm – 30% load – 260 degC exh. gas
Airless AdBlue doser
NOx conv. [%]
NH3 slip [ppm]
Alpha ratio [-]
84%
71%
45%
Legend:
DONUT + SWIRL
DONUT
IN-PIPE – NO MIXER
SCR
Donaldson.
ANSYS
Automotive Simulation World Congress 2013 - 29 October 2013

Experimental results: AdBlue deposits

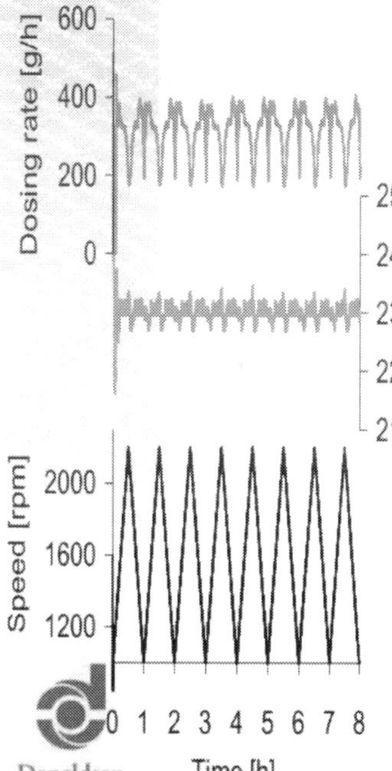

- In pipe injection

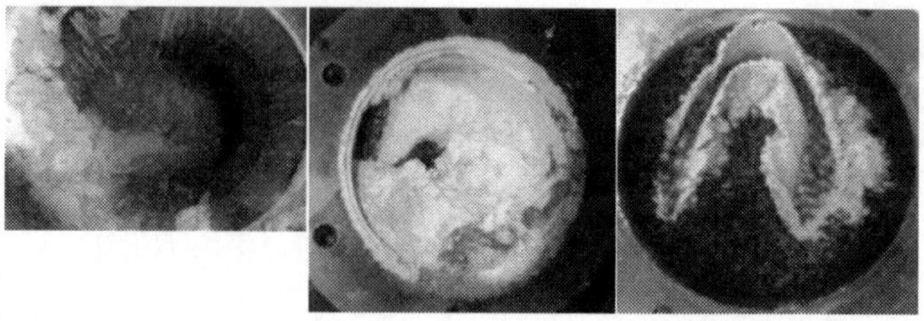

- Swirl mixer

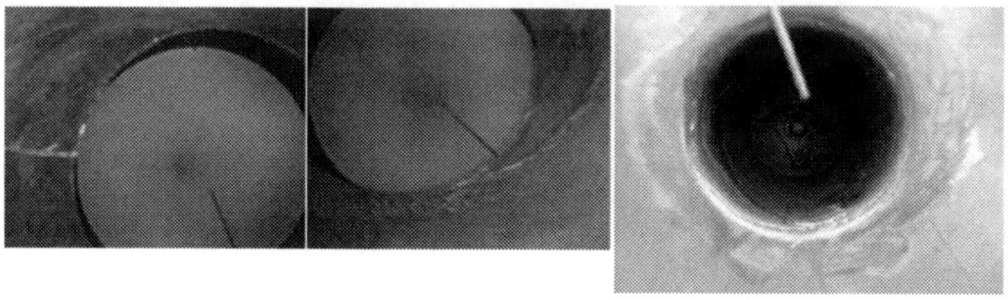

Donaldson.

Conclusion

- Demonstrate how ANSYS-Fluent helps into develop new mixing concept for exhaust after-treatment line.
- Success factors:
  - Creative packaging of DOC and SCR substrate volume
  - High quality AdBlue mixing and evaporation
  - Deposit-free exhaust system
- Next step : evaluation of urea deposit with SCR model including urea decomposition.

Donaldson.

references

- De Rudder K., "Tier 4 high efficiency SCR for agricultural applications", SAE Technical Paper 2012-01-1087
- De Rudder K., Chauvin C., "Close coupled DOC-mixer-SCR for Tier 4 final ", AVL International Commercial Powertrain Conference -Mai 2013.

Any question?

2013 Automotive Simulation World Congress

Acoustic Transmission Loss Prediction for Muffler Performance Improvement

Santosh Kottalgi, Ashok Khondge, David Roche, Peiran Ding

ANSYS

Outline

2013 Automotive Simulation World Congress
October 29 - 30 • Frankfurt am Main, Germany

- **Muffler Acoustics Overview**
- **Acoustics Performance Criteria**
- **ANSYS Advantage**
- **Validation Cases Using ANSYS Acoustics ACT**
- **Summary**

Why Muffler is used?

- **Muffler minimizes noise from Engine exhaust to an acceptable level as per noise regulations**
  - » **Muffler, Resonator and Silencer all have similar Acoustic Function and are used at different noise level attenuation.**

Muffler In Automotive Exhaust System

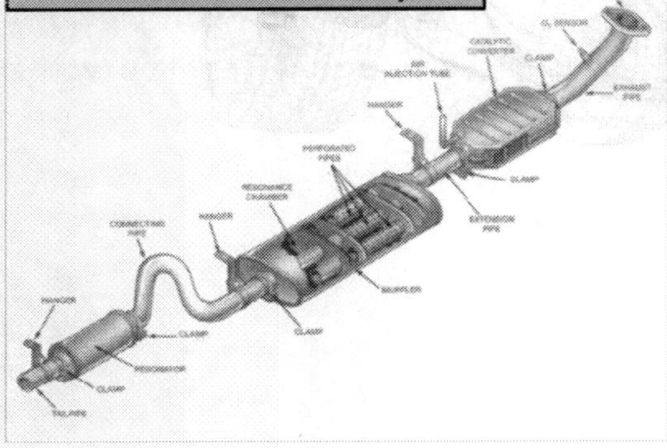

Muffler Design Drivers

2013 Automotive Simulation World Congress
October 29 - 30 • Frankfurt am Main, Germany

Optimum Muffler Design

Trade off between Noise attenuation & Engine Efficiency	Maximum allowable muffler volume and shape	Material Durability Considerations
• Have high noise reduction but lesser back pressure	• Fit it in limited space under vehicle body	• Durable under high temperature and extreme weather conditions

Classification of Mufflers

1. Reactive Mufflers

- Acoustic Energy is reduced through multiple reflections at discontinuities
- Consist of a number of tubular elements of different transverse dimensions joined together so as to cause impedance mismatch at every junction

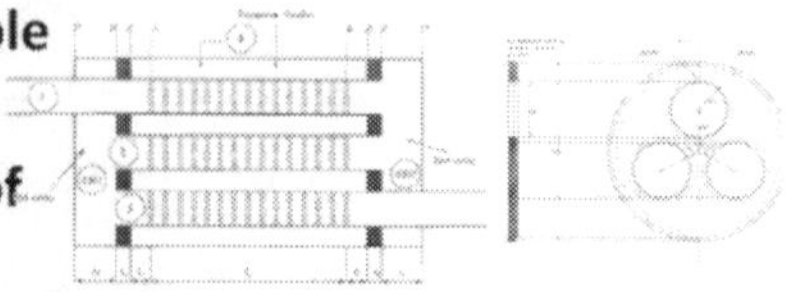

2. Dissipative Mufflers

- Acoustic Energy is dissipated into heat
- ducts are lined or filled with acoustically absorptive material

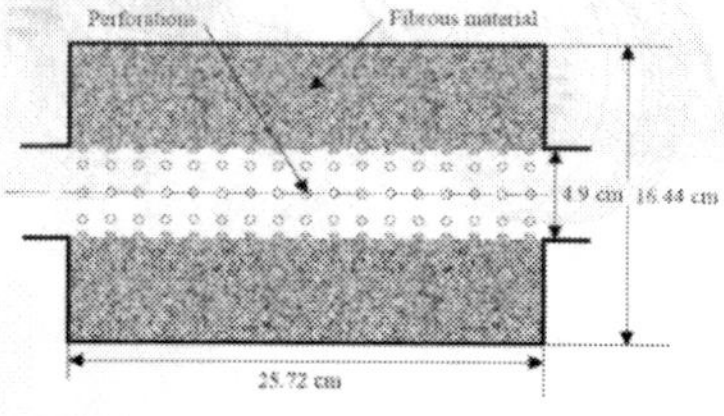

2013 Automotive Simulation World Congress

Muffler Acoustics Performance Criteria

Transmission Loss (TL)

2013 Automotive Simulation World Congress
October 29 - 30 • Frankfurt am Main, Germany

- Transmission loss (TL) is defined as the ratio between the sound power incident (*Win*) to the transmitted sound power (*Wtr*) with reflection free termination at outlet
- In later slides, Transmission loss validation study using ANSYS FEM solver is discussed in details.

$$TL = 10 \cdot \log_{10}\left(\frac{W_{in}}{W_{tr}}\right)$$

Insertion loss (IL)

2013 Automotive Simulation World Congress
October 29 - 30 • Frankfurt am Main, Germany

- Insertion loss (IL) is change in Sound Pressure Level at tail pipe outlet with and without insertion of a muffler

$$IL = 10 \cdot \log_{10}\left(\frac{W_{ref}}{W_{muffler}}\right)$$

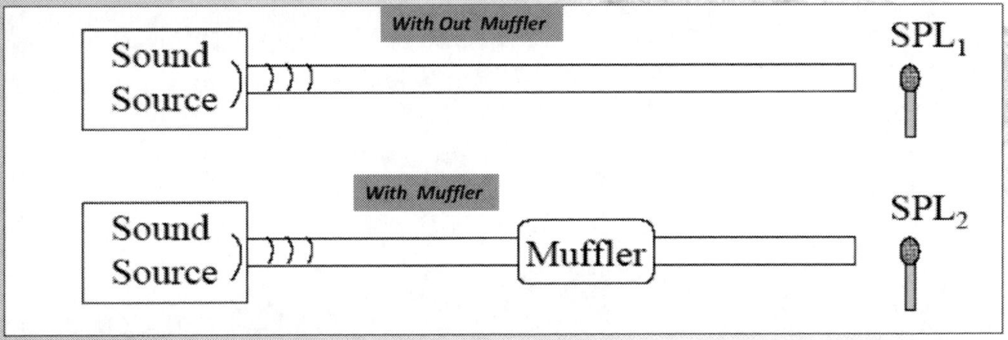

Noise Reduction (NR)

2013 Automotive Simulation World Congress
October 29 - 30 • Frankfurt am Main, Germany

- Noise Reduction (NR) is difference in sound pressure level between one point upstream of the muffler and one point downstream

$$NR = 10 \cdot \log_{10}\left(\frac{p^2_{RMS,before}}{p^2_{RMS,after}}\right)$$

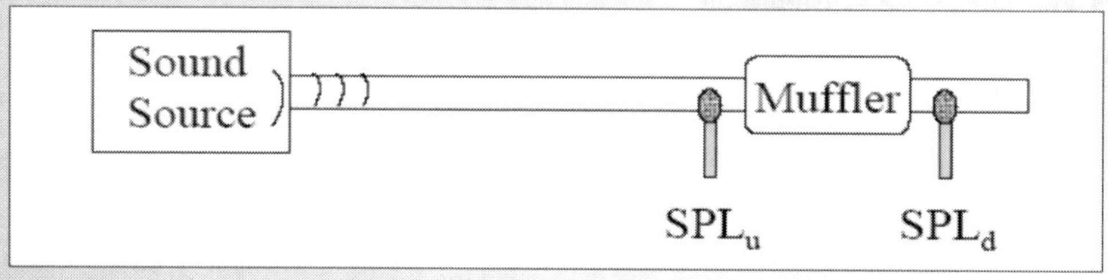

2013 Automotive Simulation World Congress

Simulation Challenges and ANSYS Advantages

Simulation Challenges

2013 Automotive Simulation World Congress
October 29 - 30 • Frankfurt am Main, Germany

1. Creating fluid volume from the complex CAD geometry for 3D analysis
2. Mesh to capture the details of Fluid cavity as per max frequency range requirement
3. Ease of use in setting up the model
4. Capability to solve Sound propagation and Vibro Acoustic
5. Accuracy of solver against experimental tests and with theory
6. Fast solver to quick solution

ANSYS Advantage

2013 Automotive Simulation World Congress
October 29 - 30 • Frankfurt am Main, Germany

1. DesignModeler has bi-directional CAD connectivity, Fluid volume extraction, geometry repair and simplification tools to handle dirty CAD geometries
2. If not Hex, Quadratic tetrahedral meshing is enough, hence quick meshing
3. Customized Acoustic ACT Extension for easy setup and post-processing
4. Distributed Parallel (HPC) solver to reduce the solve time.
5. Capability to do Parametric design study to optimize muffler design

2013 Automotive Simulation World Congress

Validation Using ANSYS Acoustics ACT

Validation Study

2013 Automotive Simulation World Congress
October 29 - 30 • Frankfurt am Main, Germany

- **Validation study of Muffler Transmission loss analysis is done for below configurations;**
  1. **Simple expansion muffler**
  2. **Three pass perforated reflective muffler**
  3. **Single pass absorptive muffler**

- **ANSYS Mechanical R14.5 with Acoustic ACT extension is used for these studies**

Acoustics ACT Extension

2013 Automotive Simulation World Congress
October 29 - 30 • Frankfurt am Main, Germany

- Expose 3D acoustic features in Mechanical without the need for APDL
  - Define acoustics properties
  - Apply acoustic boundary conditions & loads
  - Postprocess acoustic results

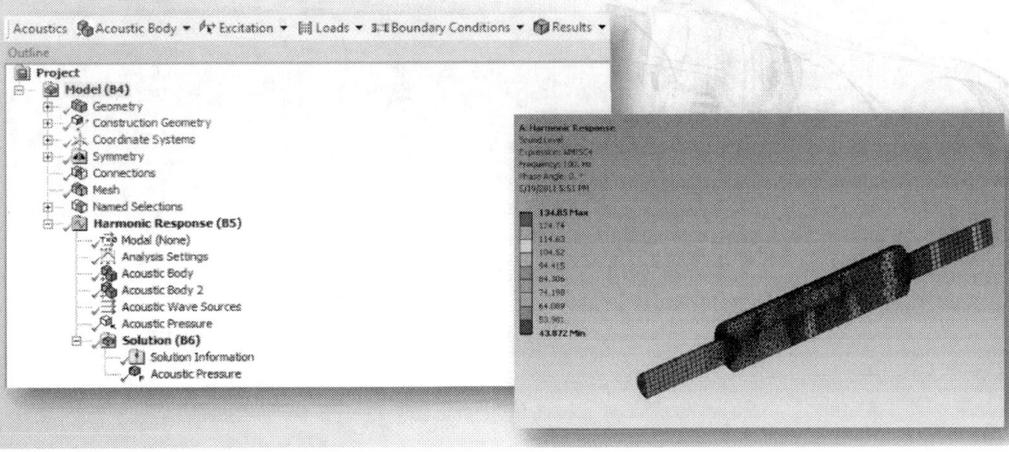

Features in Acoustics ACT Extension

2013 Automotive Simulation World Congress
October 29 - 30 • Frankfurt am Main, Germany

- Acoustics ACT extension can be used to create acoustic boundary conditions and defining fluid bodies:
  - Pressure
  - Mass Flow
  - Impedance
  - FSI Interface
  - ...many more
  - Fluid body (FLUID30, 220 or 221 elements according to the mesh)

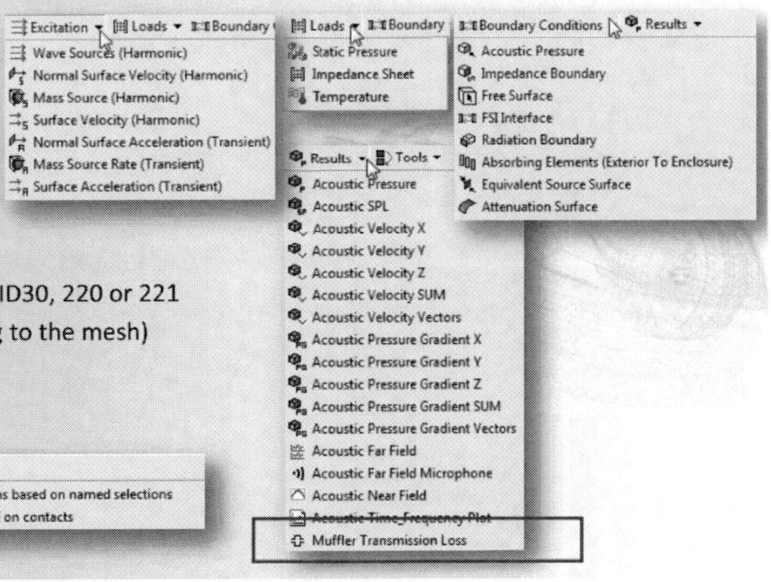

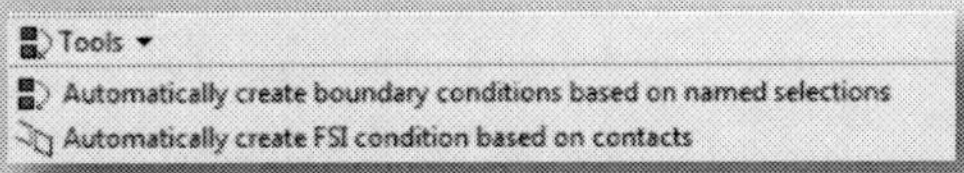

#1. Simple Expansion Muffler

•**Objective:**

Muffler transmission loss using FEM will be validated with Analytical and experimental results.

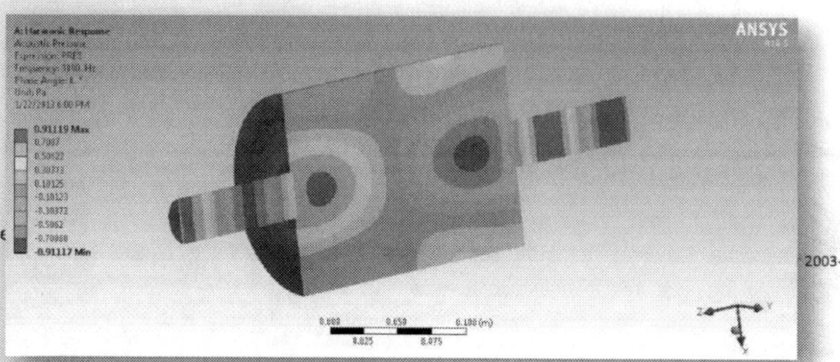

- Re

2003-01-

Geometry & Meshing

2013 Automotive Simulation World Congress

- Muffler geometry is created using cross section mentioned in Figure 2 of reference paper-1
- Muffler will be meshed with uniform mesh size.
- Mesh size is calculated from smallest wave length. Is advisable to have at least 6 elements in one wave length
- Mesh size calculation:
  - Max frequency , f= 3000Hz
  - Sonic speed in air, C = 343m/s
  - Wavelength $\lambda = C/f = 0.1143m$
  - Element size = $\lambda/6 = 0.019$ m
  - On safer size we have used 0.01m

In Inch

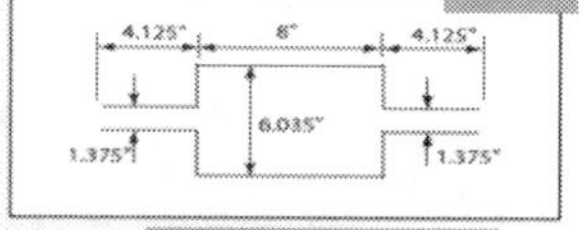

Muffler cross section

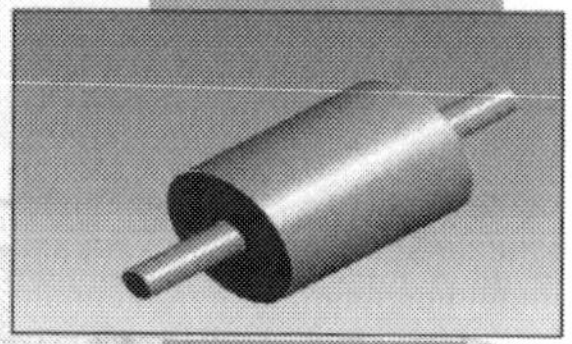

Muffler geometry

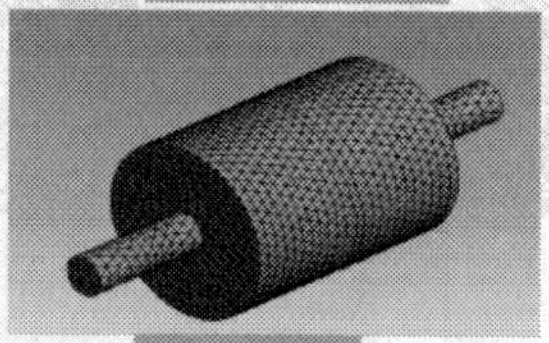

Muffler mesh

Cross-section of Muffler mesh

Validation: Comparison of TL with Experiment and Theory

2013 Automotive Simulation World Congress
October 29 - 30 • Frankfurt am Main, Germany

- Comparison shows higher degree of solver accuracy with Analytical and experimental TL

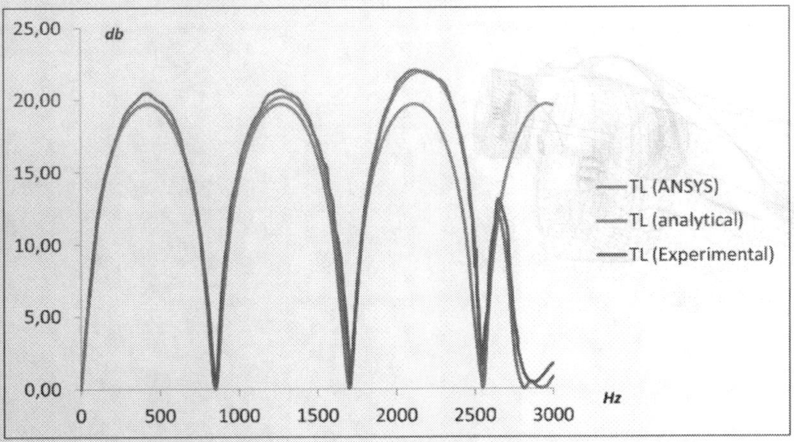

#2. Three Pass Perforated Reflective Muffler

2013 Automotive Simulation World Congress
October 29 - 30 • Frankfurt am Main, Germany

•Objective:

Muffler transmission loss using FEM will be validated with experimental and theory results. Parametric study is done to evaluate performance of muffler for different design.

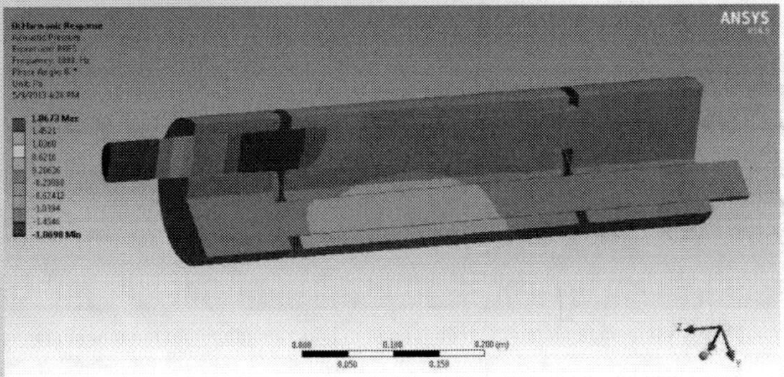

•References:

1. A. Selamet and V. Easwaran, Three-pass mufflers with uniform perforations, © 1999 Acoustical Society of America

Geometry & Meshing

2013 Automotive Simulation World Congress
October 29 - 30 • Frankfurt am Main, Germany

- Muffler geometry is created using cross section shown in Figure 2 and 5 of reference paper-1
- Muffler will be meshed with uniform mesh size.
- Mesh size is calculated from smallest wave length. It is advisable to have at least 6 elements in one wave length
- Mesh size calculation:
    - Max frequency , f= 1200Hz
    - Sonic speed in air, C = 344.7m/s
    - Wavelength λ = C/f = 0.28725m
    - Element size = λ/6 = 0.047875 m
    - On safer size we have used 0.01m

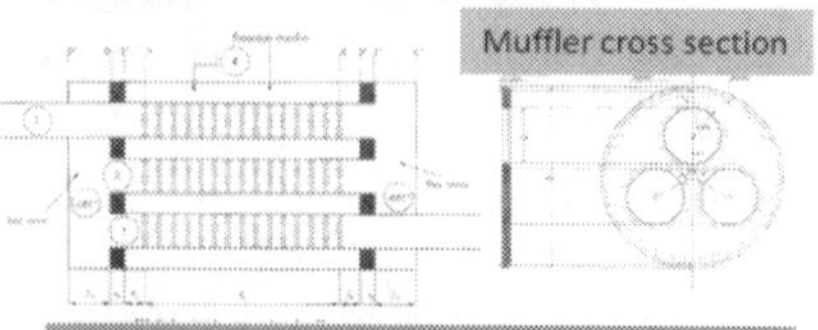

Muffler cross section

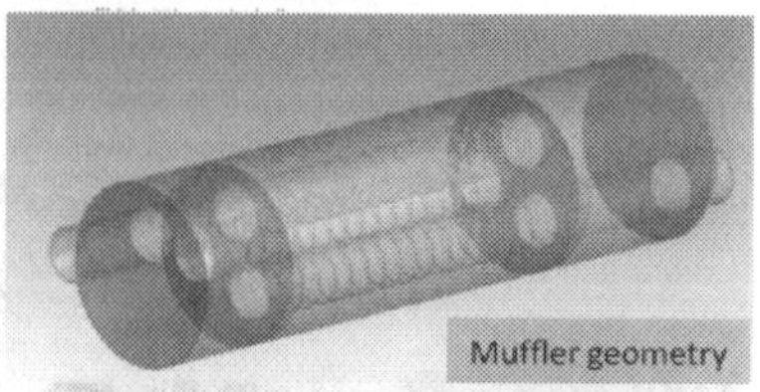

Muffler geometry

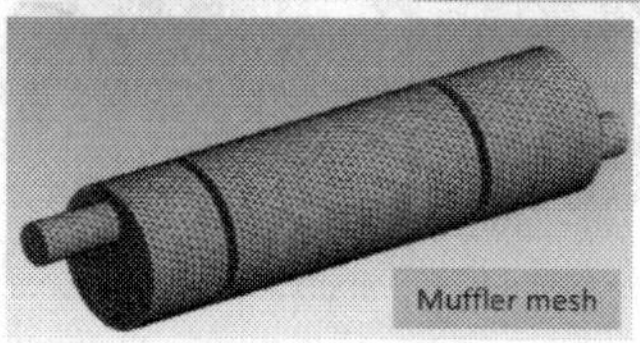

Muffler mesh

Cross-section of Muffler mesh

Validation: TL Comparison with Experiment and Theory

2013 Automotive Simulation World Congress
October 29 - 30 • Frankfurt am Main, Germany

- Comparison shows higher degree of solver accuracy with Analytical and experimental TL
- TL calculation using Theory matches very well with Experiment and ANSYS upto 800Hz, which is the frequency range of design.

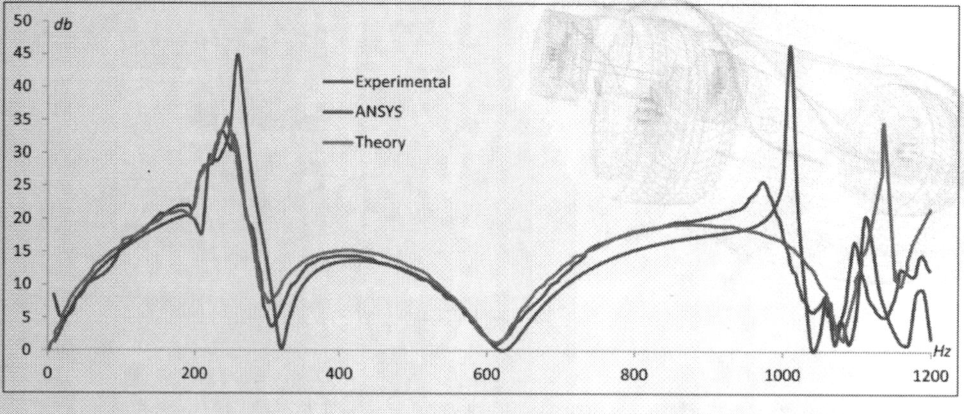

Parametric Design Study

2013 Automotive Simulation World Congress
October 29 - 30 • Frankfurt am Main, Germany

Parameter Definition

- Parametric study is done for design changes and compared with theory
  - Lc = Outlet chamber length = 0.15m
  - Φ = Porosity/open-area ratio in the central region of pipe = 0.045
- Lc (Outlet chamber length) is made equal to the inlet Chamber length Ld of 0.102 m
- Porosity changes with change in diameter of the holes, so porosity of 0.02 and 0.06 is modeled with changing diameter from 0.00234 m to 0.0017 m and 0.0027m respectively.

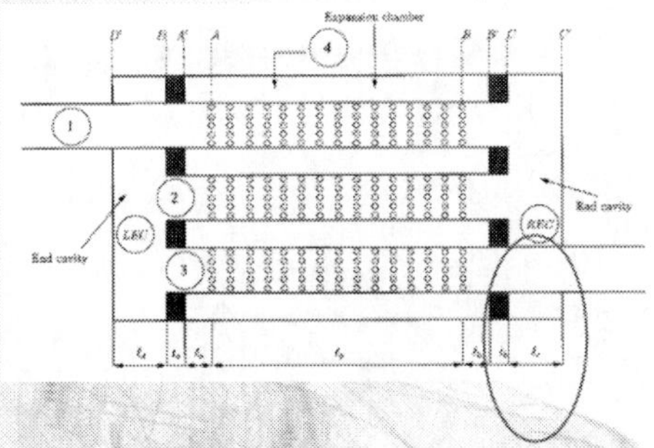

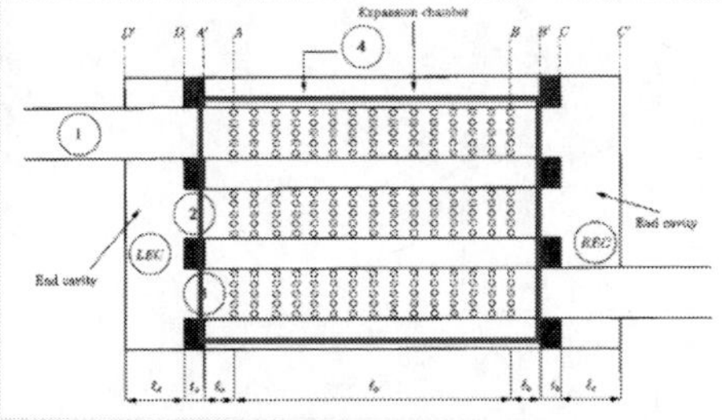

Validation: TL Comparison with Theory for Lc (outlet chamber length)

2013 Automotive Simulation World Congress
October 29 - 30 • Frankfurt am Main, Germany

- Transmission loss using ANSYS matches very well with Theory upto 800Hz, which is an area of interest in design on Muffler.

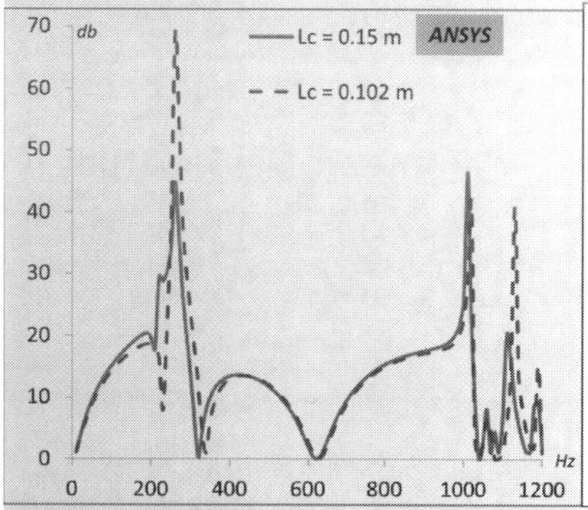

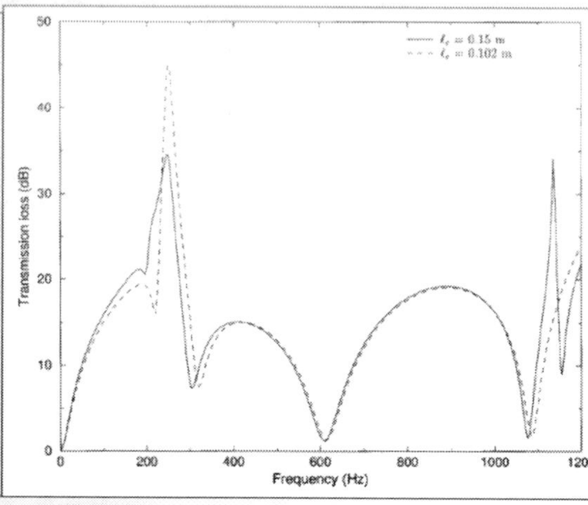

Validation: TL Comparison with Theory for Φ (porosity) parameter

2013 Automotive Simulation World Congress
October 29 - 30 • Frankfurt am Main, Germany

- Transmission loss using ANSYS matches very well with Theory upto 800Hz, which is an area of interest in design on Muffler.

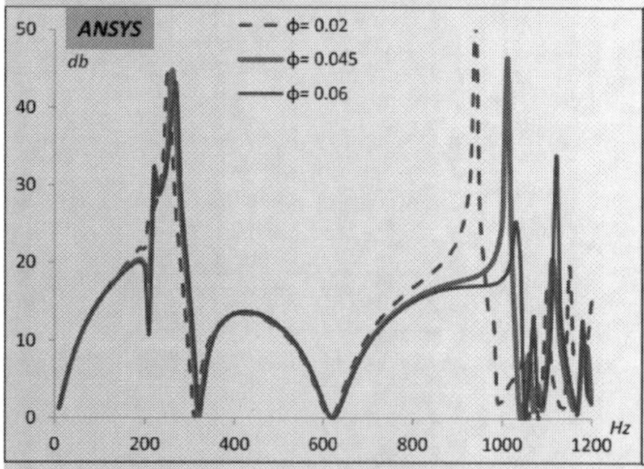

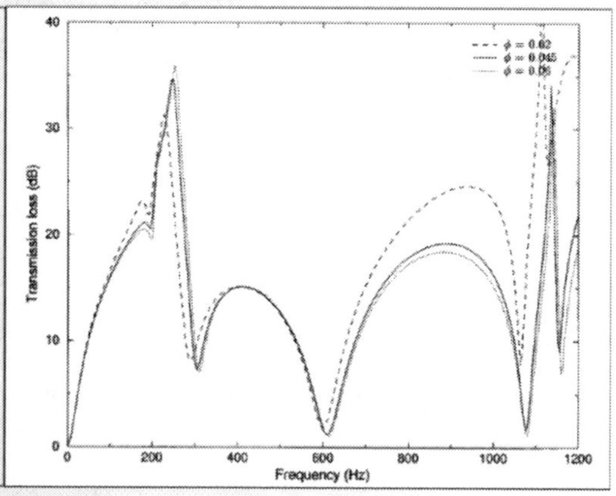

#3. Single Pass Absorptive Muffler

2013 Automotive Simulation World Congress
October 29 - 30 • Frankfurt am Main, Germany

Objective:

Muffler transmission loss using FEM will be validated with experimental and theory results. Parametric study is done to evaluate performance of muffler for different design.

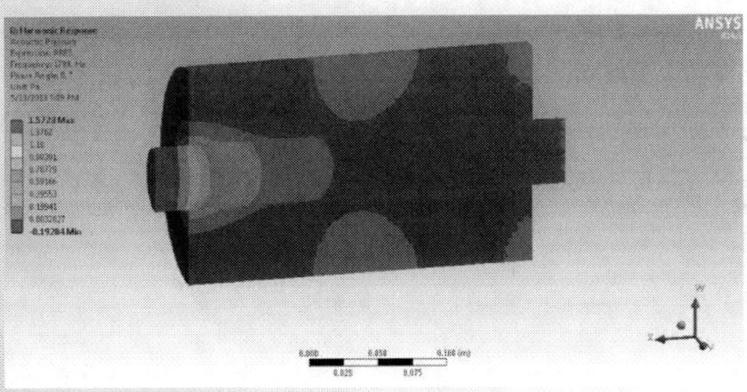

- **References:**
  1. Iljae Lee, M.S. , Acoustic characteristics of perforated dissipative and hybrid silencers, 2005

Geometry & Meshing

2013 Automotive Simulation World Congress
October 29 - 30 • Frankfurt am Main, Germany

- Muffler geometry is created using cross section shown in Figure 7.15 of reference paper-1
- Fluid volume is split at the holes to create absorptive fluid volume.
- Muffler is meshed with uniform mesh size.
- Mesh size is calculated from smallest wave length. Is advisable to have at least 6 elements in one wave length
- Mesh size calculation:
  - Max frequency , f= 3000Hz
  - Sonic speed in air, C = 343m/s
  - Wavelength λ = C/f = 0.1143m
  - Element size = λ/6 = 0.019 m
  - On safer size we have used 0.01m

Muffler cross section

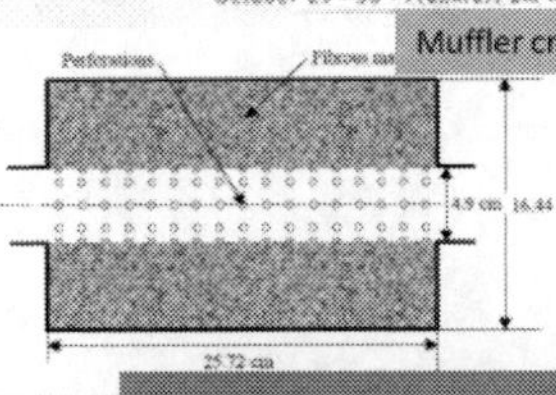

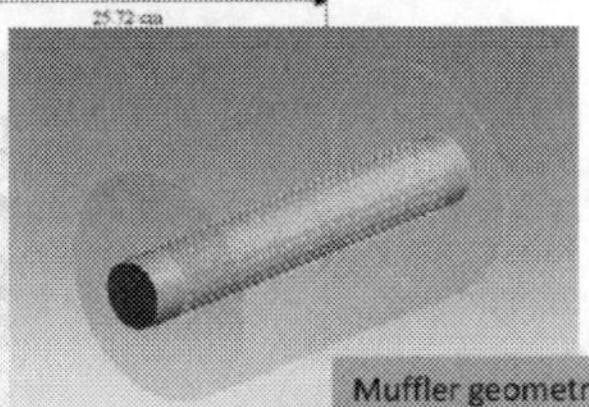

Muffler geometry

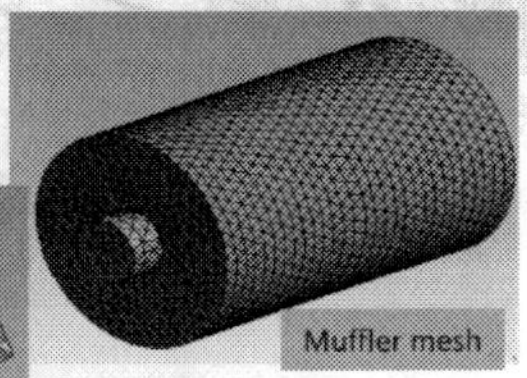

Muffler mesh

Cross-section of Muffler mesh

Validation: TL Comparison with Experiment

2013 Automotive Simulation World Congress
October 29 - 30 • Frankfurt am Main, Germany

- Comparison shows higher degree of solver accuracy with experimental TL

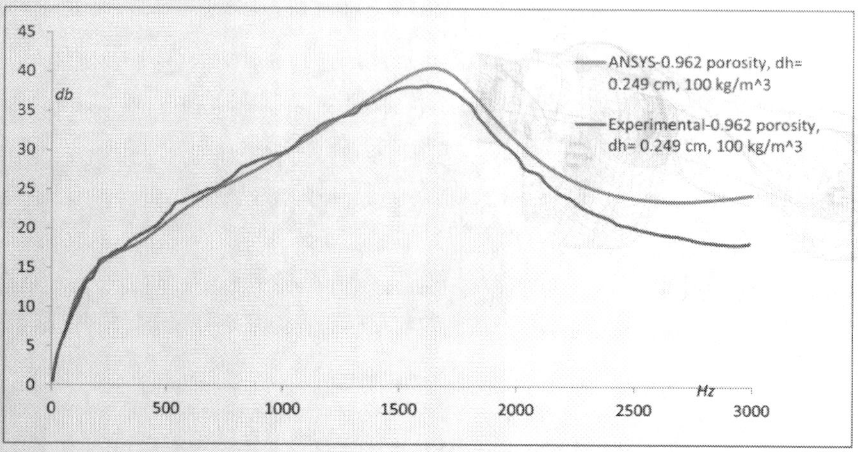

2013 Automotive Simulation World Congress
October 29 - 30 • Frankfurt am Main, Germany

Parametric Design Study

Introduction

2013 Automotive Simulation World Congress
October 29 - 30 • Frankfurt am Main, Germany

- **Parametric study is done for design changes and compared with experiment**
  - **dh = Hole diameter of the perforated plate**
- **Hole diameter of the perforated plate is changed from 0.249cm to 0.498 cm.**

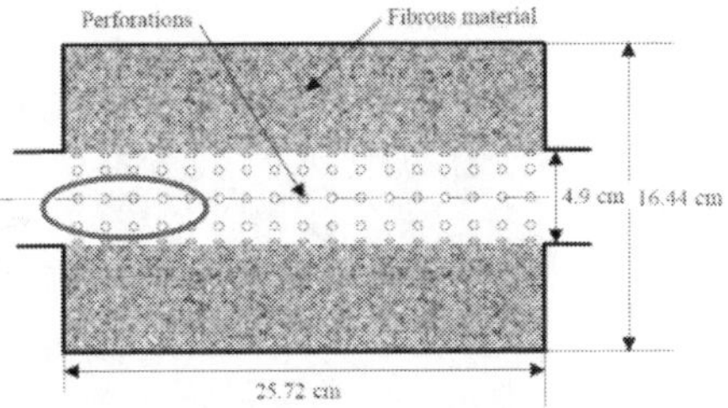

Validation: TL Comparison with Experiment for dh (Perforation hole Dia)

2013 Automotive Simulation World Congress
October 29 - 30 • Frankfurt am Main, Germany

- Transmission loss using ANSYS matches very close with Experiment at peak TL

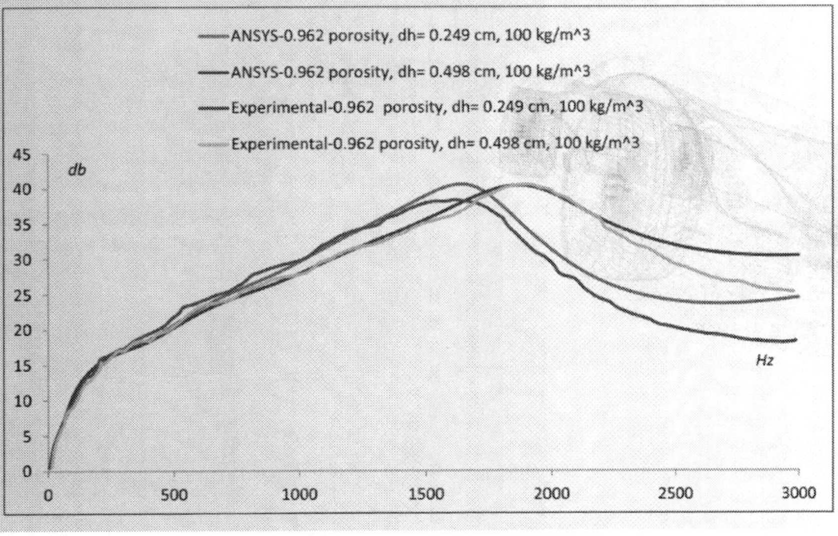

2013 Automotive Simulation World Congress

Other Similar Applications

2013 Automotive Simulation World Congress

October 29 - 3[illegible]

Industrial products where sound propagation can be studies

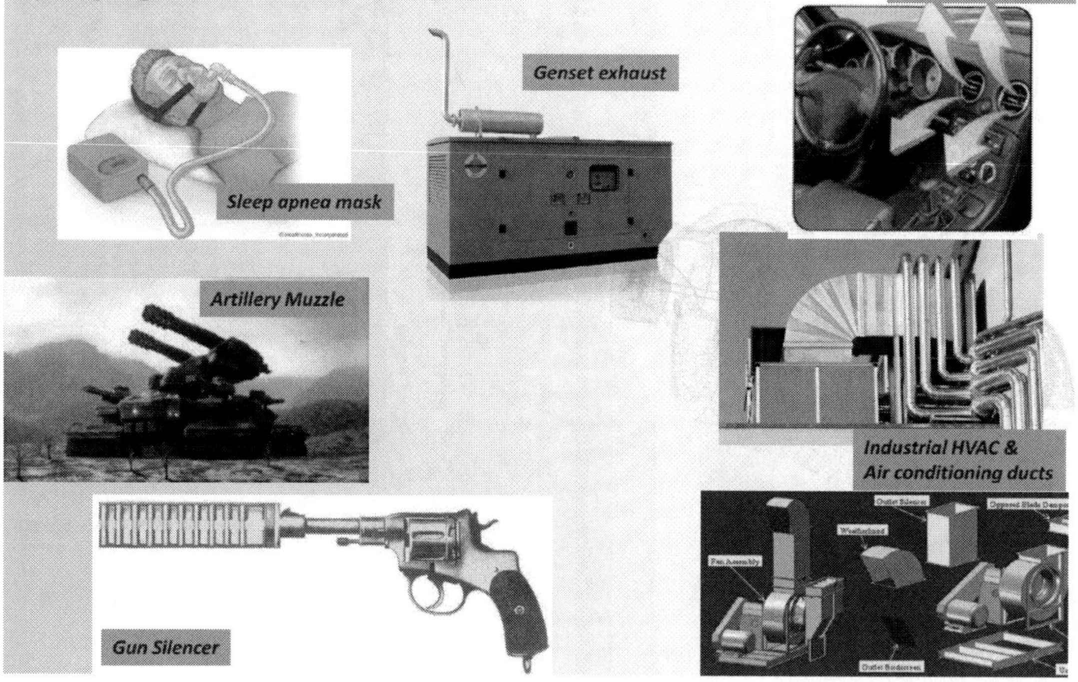

Summary

2013 Automotive Simulation World Congress
October 29 - 30 • Frankfurt am Main, Germany

- ANSYS FEM acoustic solver helps in accurately predicting transmission loss
- Parametric design study helps in design optimization
- Ansys can also study other Muffler design criteria , such as
  - Insertion loss (IL) prediction
  - Pass by noise prediction
  - Vibro acoustics of Muffler
- ANSYS CFD solver can help in study of,
  - Aero acoustics
  - Back pressure prediction

2013 Automotive Simulation World Congress
October 29 - 30 • Frankfurt am Main, Germany

Thank You !

Q&A

2013 Automotive Simulation World Congress

CDF-based evaluation of automotive exhaust system

Ralf Löffler, ANSYS Germany GmbH

Agenda

- Motivation
- Model definition
  - Geometry
  - Boundary Conditions & Assumptions
- Workflow for Simulation
- Results
- Conclusion & Discussion

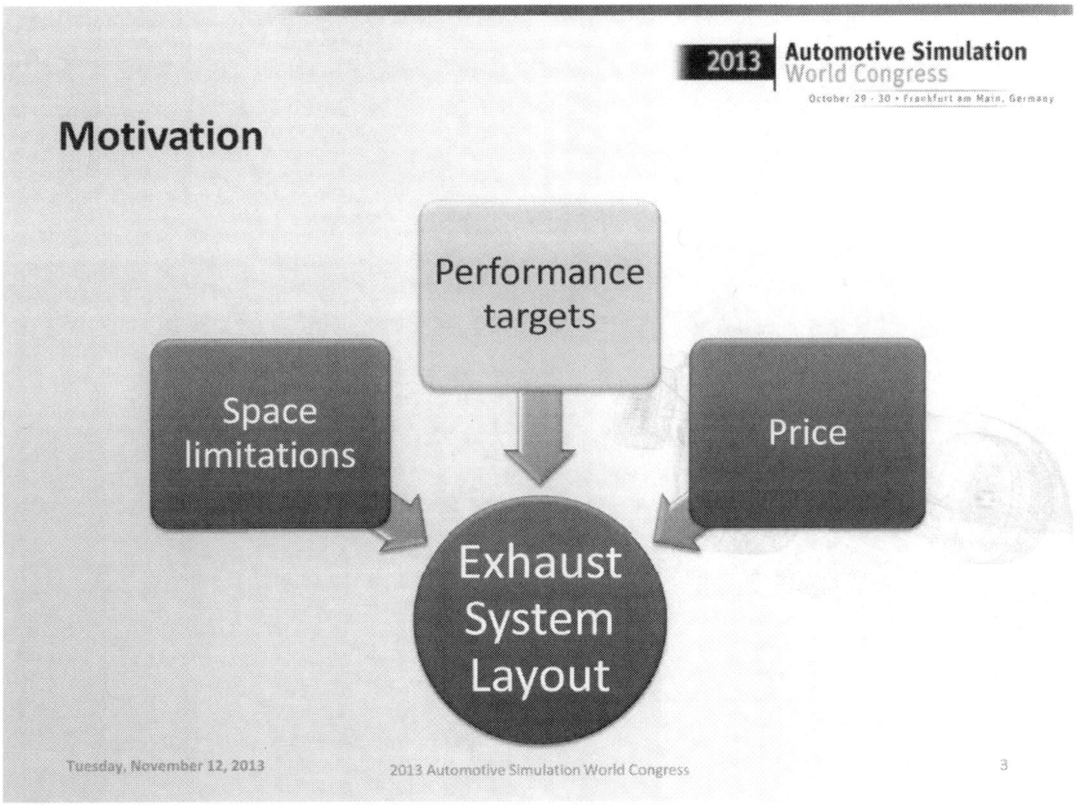
2013
Automotive Simulation
World Congress
October 29 - 30 • Frankfurt am Main, Germany
Motivation
Performance targets
Space limitations
Price
Exhaust System Layout
Tuesday, November 12, 2013
2013 Automotive Simulation World Congress
3

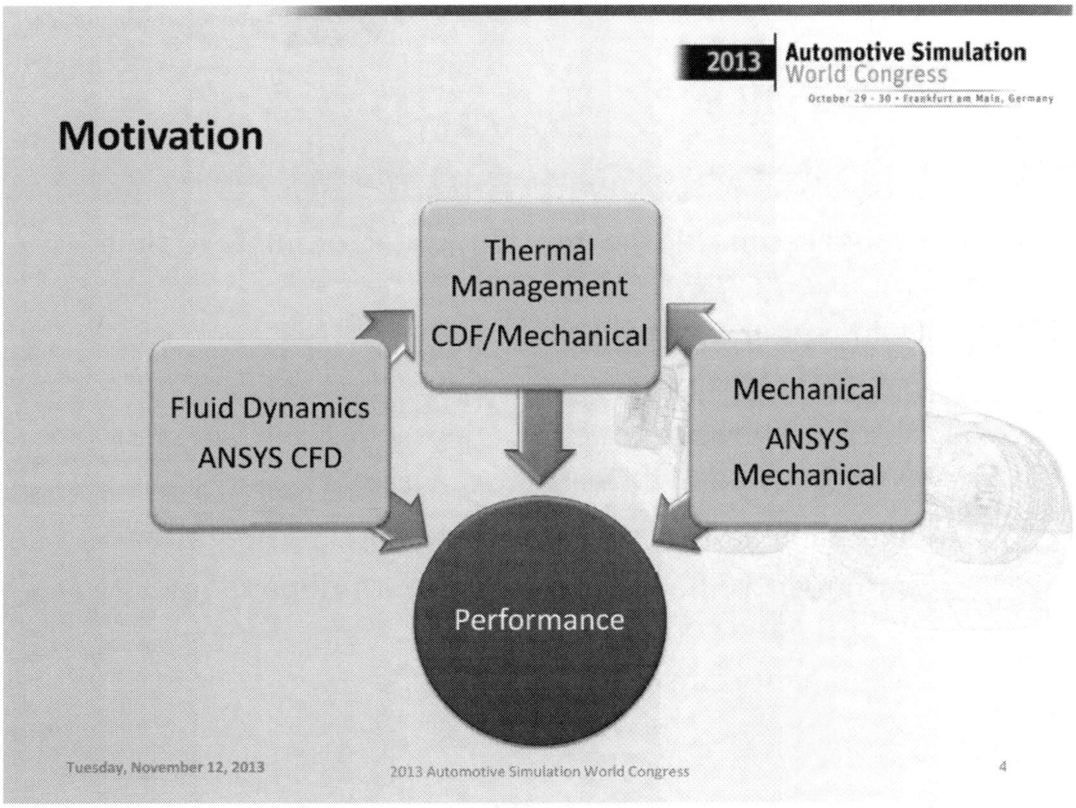
2013
Automotive Simulation
World Congress
October 29 - 30 • Frankfurt am Main, Germany
Motivation
Thermal Management
CDF/Mechanical
Fluid Dynamics
ANSYS CFD
Mechanical
ANSYS Mechanical
Performance
Tuesday, November 12, 2013
2013 Automotive Simulation World Congress
4

2013 Automotive Simulation World Congress
October 29 - 30 • Frankfurt am Main, Germany

Motivation

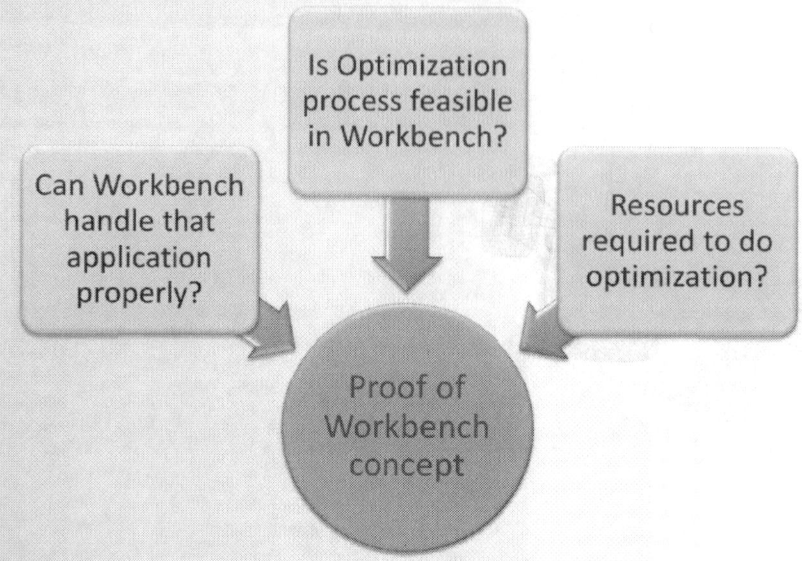

2013 Automotive Simulation World Congress
October 29 - 30 • Frankfurt am Main, Germany

Model Definition

- Four-stroke engine exhaust system
  - Model geometrical extension: collector inlets to fluid zone after system exit
  - Fluid region without solids
  - Catalytic Converter: porous zone
  - Absorber: porous zone
- Domain & Bondary Conditions
  - Inlet BC: steady-state massflow + total temperature
  - Compressible flow with temperature-dependent viscosity and conductivity
  - Mixed heat transfer bcs to ambient

2013 Automotive Simulation World Congress
October 29 - 30 • Frankfurt am Main, Germany

Model (about 2000 Solid Parts)

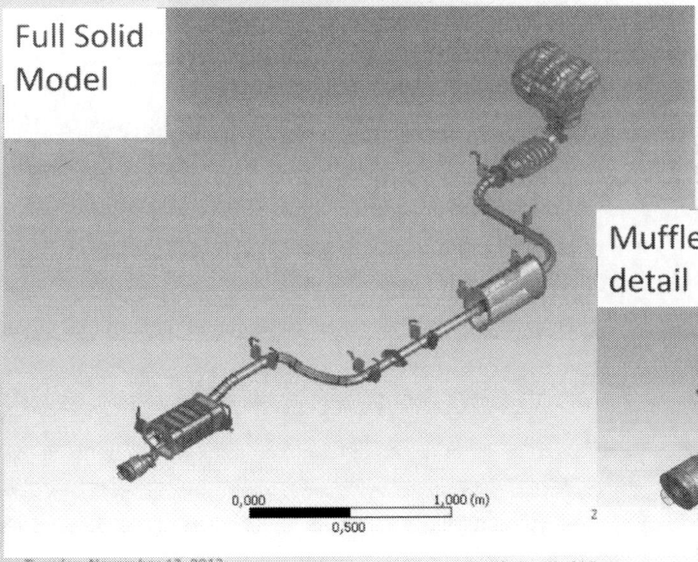

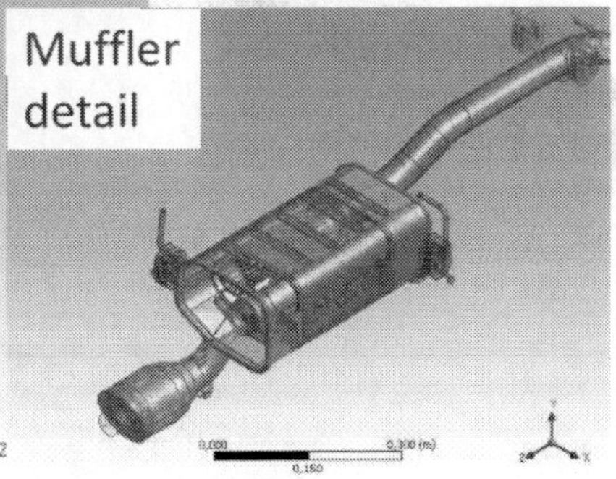

2013 Automotive Simulation World Congress
October 29 - 30 • Frankfurt am Main, Germany

Initial Workbench Workflow for Simulation

Design modeller → **ANSYS Meshing** → **ANSYS Fluent**

Design modeller
- Geometry import
- flow volume Extraction
- Adding outlet region

ANSYS Meshing
- general setup
- local refinements
- Zone definitions

ANSYS Fluent
- Physical & Numerical Setup
- Calculation

Variational Workbench Workflow

Design modeller
- Introduce geometry variations for optimization
- Can be parameterized

ANSYS Meshing
- Remeshing
- Pushbutton for minor geometry changes

ANSYS Fluent
- Re-Calculation
- Pushbutton for identical zonal definitions

Mesh Details

≈30mio cells in meshing
≈10mio Polyhedral cells in Fluent
Hybrid mesh including boundary layers

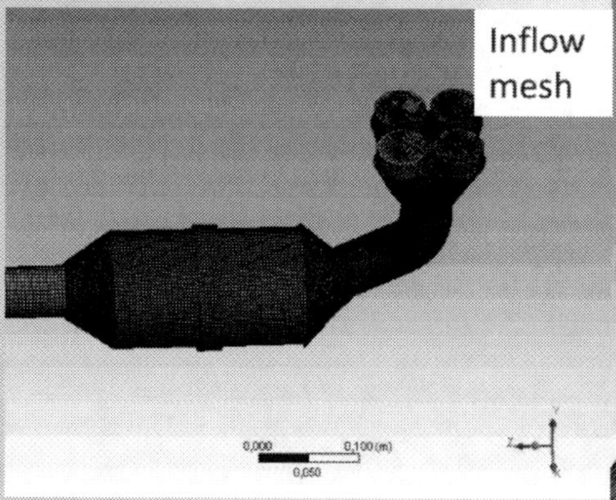

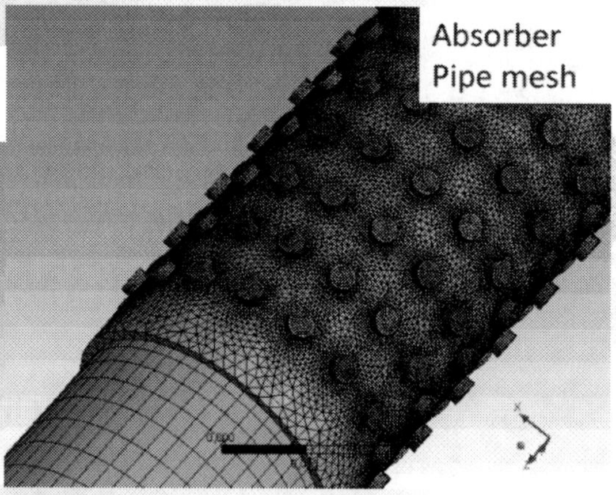

Results

Optimization Criteria used:

- Pressure drop minimization
- Maximization of CAT volume utilization
  - i.e. improvement of uniformity of CAT flow
- Keep modified geometry within original geometry
  - Changes extending the volume covered by the exhaust system may be unaceptable due to packaging constraints
- Mechanical constraints currently not considered
  - Only fluid volume is subject to considerations

2013 Automotive Simulation World Congress
October 29 - 30 • Frankfurt am Main, Germany

Results: Collector & cat

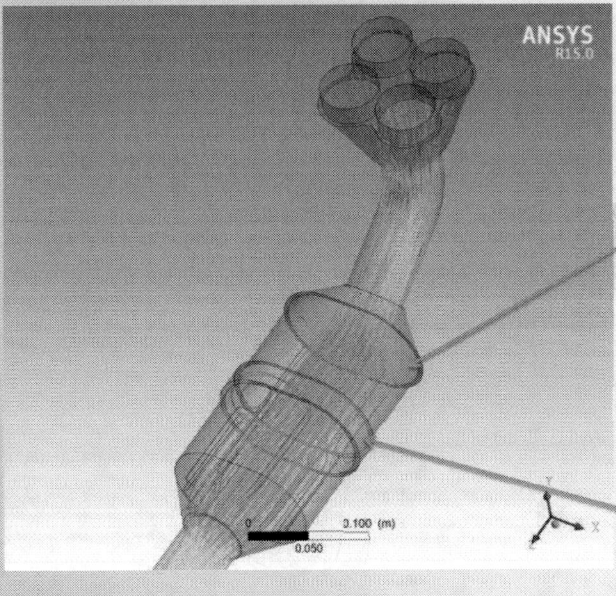

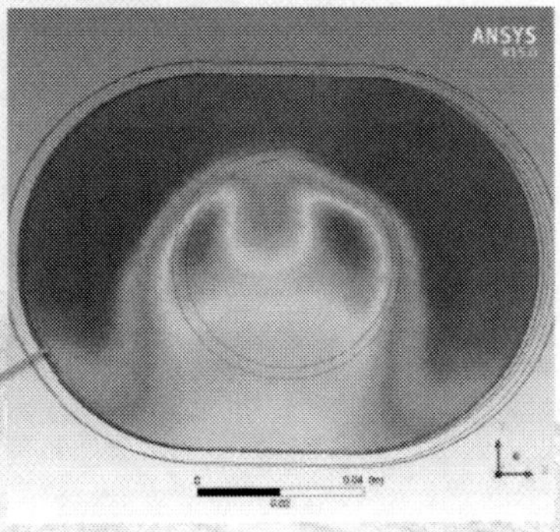

CAT: at cat region inlet, uneven loading of cat

Redistribution gap works, tough

2013 Automotive Simulation World Congress
October 29 - 30 • Frankfurt am Main, Germany

Results: absorber

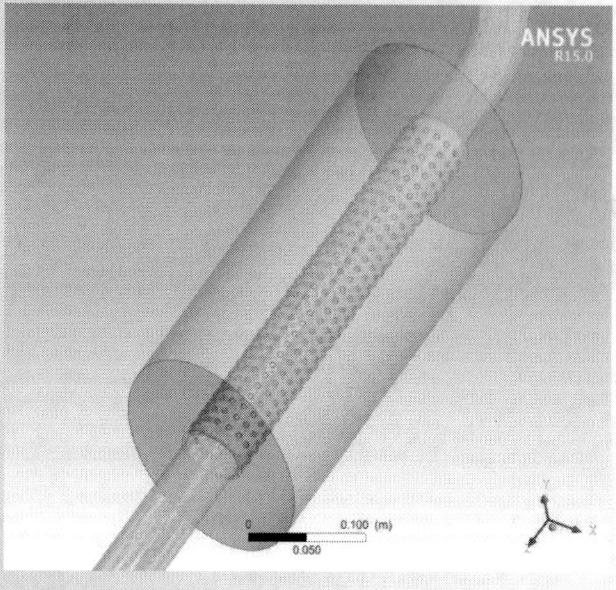

- will be interesting for transient (Aeroacoustics)

2013 Automotive Simulation World Congress
October 29 - 30 • Frankfurt am Main, Germany

Results: muffler

- Flow pattern reasonable
- Later: aeroacoustices

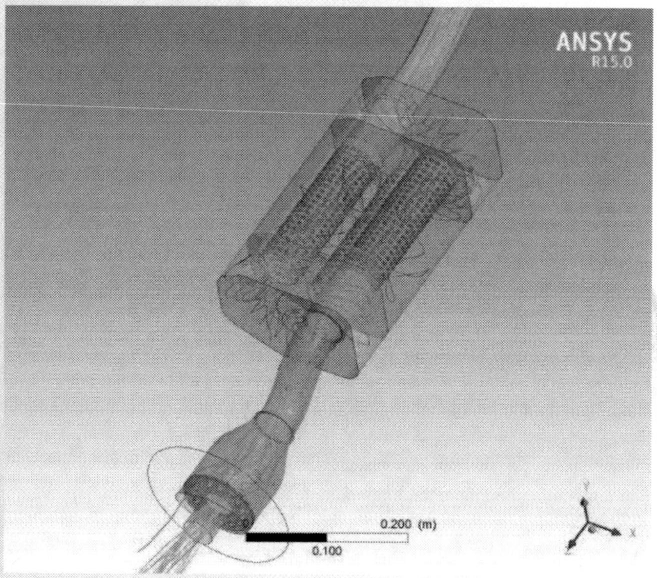

Results: variations

Bump from bottom

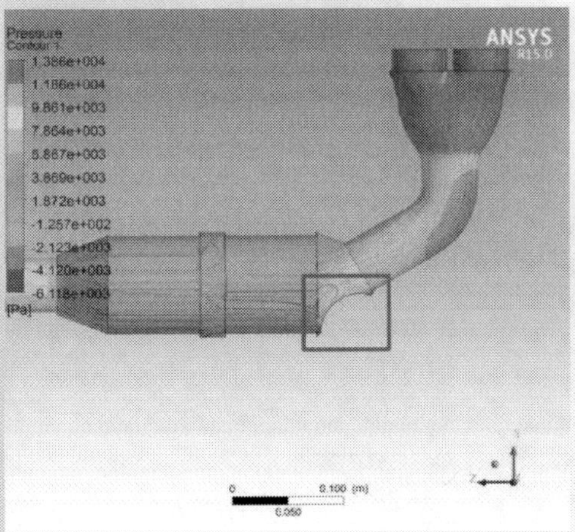

Bump from top

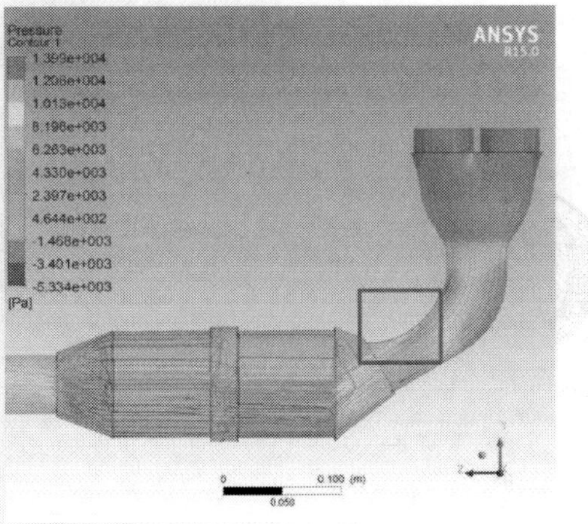

2013 Automotive Simulation World Congress
October 29 - 30 • Frankfurt am Main, Germany

Optimization parameters

Variations of:

- Massflow
  - To check robustness of solution
- Geometry
  - Modify Cat inlet to homogenize Cat flow, if possible
- Cat Resistivity
  - Since resistivity dominates pressure drop and eventually impacts flow distribution

Target Variables

- Total pressure drop
- Homegenity of Cat flow/utilization

Long-term

- Acoustic behavior
- Thermal effects
- Mechanical stresses and fatigue

2013 Automotive Simulation World Congress
October 29 - 30 • Frankfurt am Main, Germany

Optimization effort

Potential variations

- 5 massflows
- 10 geometric modifications
- 3 resistivities

- Transient heat up for 5min

Effort

- 5X10 + 5X3 steady-state runs totals to 65 runs

- Four-stroke engine with 2000rpm
  - 1 round resolved by 40 timesteps
  - 2000 x 40 x 5 = 400,000 timesteps

2013 Automotive Simulation World Congress
October 29 - 30 • Frankfurt am Main, Germany

Runtimes

500 iterations staedy run

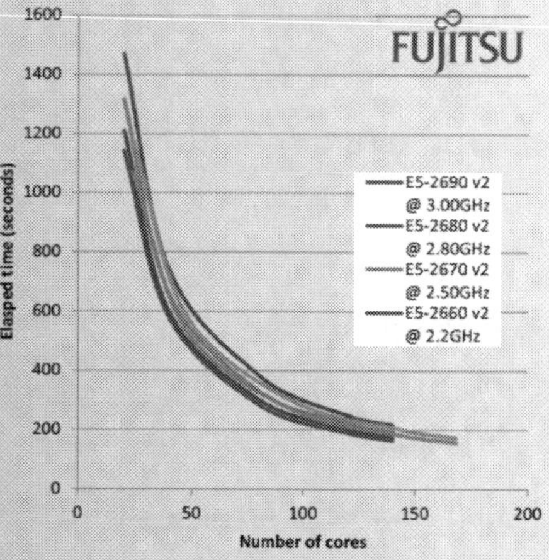

Speedup on Infiniband cluster

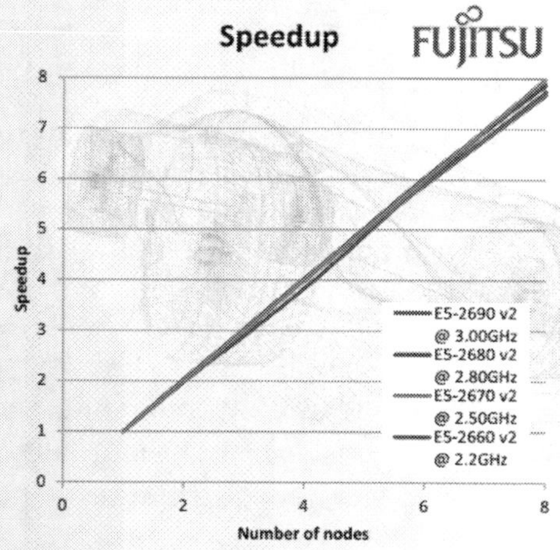

Runtimes

Cases	Runtimes
Steady	
– 65 variations	– Solver runs serially done on 160 cores in 145s * 65 ≈ 3h wall-clock
Transient (assumed)	**transient**
– 20 iterations per timestep	– Trasient on 1 node: 240 days
– Overhead of 10% for data output in transient	– Transient on 8 nodes: 30 days
– i.e. 500 iterations steady is equivalent about 22 timesteps	– Transient on 12 nodes: 20 days
	– Transient on 48 nodes: 5 days

Outlook

Future work

- Automatized optimization by Design Exporation
- Automatized optimization by Adjoint Solver
- Transient cases
- Mechanical side + FSI
- Aeroacoustics

2013 Automotive Simulation World Congress
October 29 - 30 • Frankfurt am Main, Germany

Conclusion

- Proof of Concept was successful
  - Workbench Workflow works with such models
    - Meshing produces reasonably good results
    - Desgin modeller allows for parametric modifications
- Optimization is feasible with moderate haradware
  - FLUENT scaleup is very good for this application
- Transient evaluations requires reasonable hardware for acceptable runtimes

- Special Thanks to Ian Godfrey from Fujitsu for providing HW benchmarks!

ASWC, October 29-30, 2013

NUMERICAL SIMULATION OF JET NOISE OF AUTOMOTIVE EXHAUST SYSTEMS USING SCALE-RESOLVING TURBULENCE MODELS AND AN ACOUSTIC ANALOGY.

Jan Hillenbrand, BMW Group, München
Thomas Sailer, Dr. Wolfram Kirchweger, Dr. Martin Wetzel, Dr. Oliver Hausner, BMW Group, München
Prof. Dr. Stefan Becker, Friedrich-Alexander Universität Erlangen-Nürnberg

BMW
GROUP

CONTENTS.

CONTENTS.

ACOUSTIC SOURCES IN SUBSONIC TURBULENT JETS.

Jet noise of automotive exhaust systems:

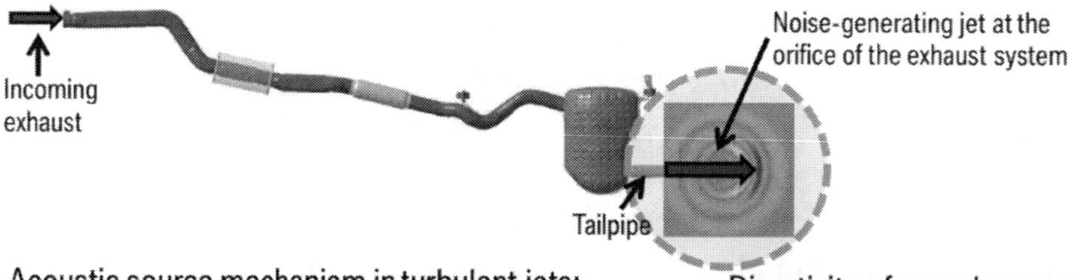

Acoustic source mechanism in turbulent jets:

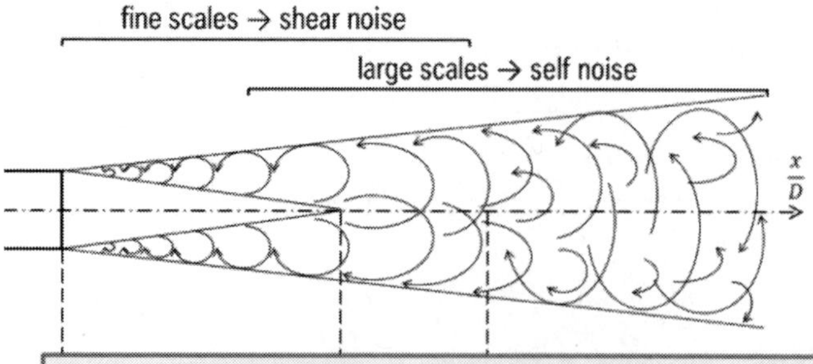

Directivity of sound pressure field:

(Jet as point source)

Andersson, N., 2005, Dissertation

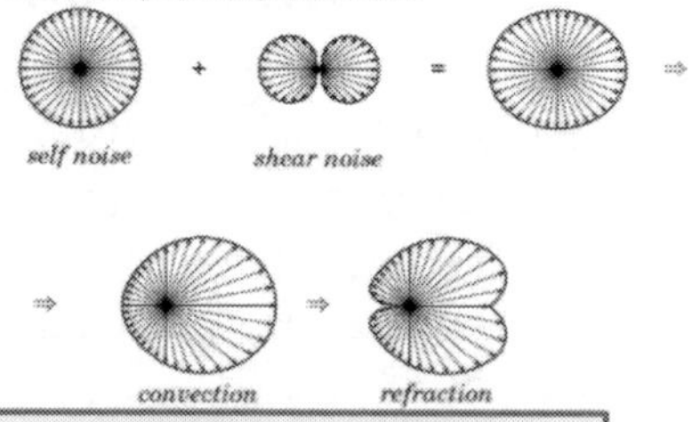

⇒ Use of scale-resolving turbulence models essential for the simulation of jet noise!

CONTENTS.

I. Acoustic sources in subsonic turbulent jets

II. Simulation workflow

III. Pre-processing of scale-resolving jet simulation

IV. A method to avoid numerical reflections

V. Scale-resolving simulation results

VI. Outlook

Automotive Simulation World Congress, October 30, 2013

Page 5

SIMULATION WORKFLOW.

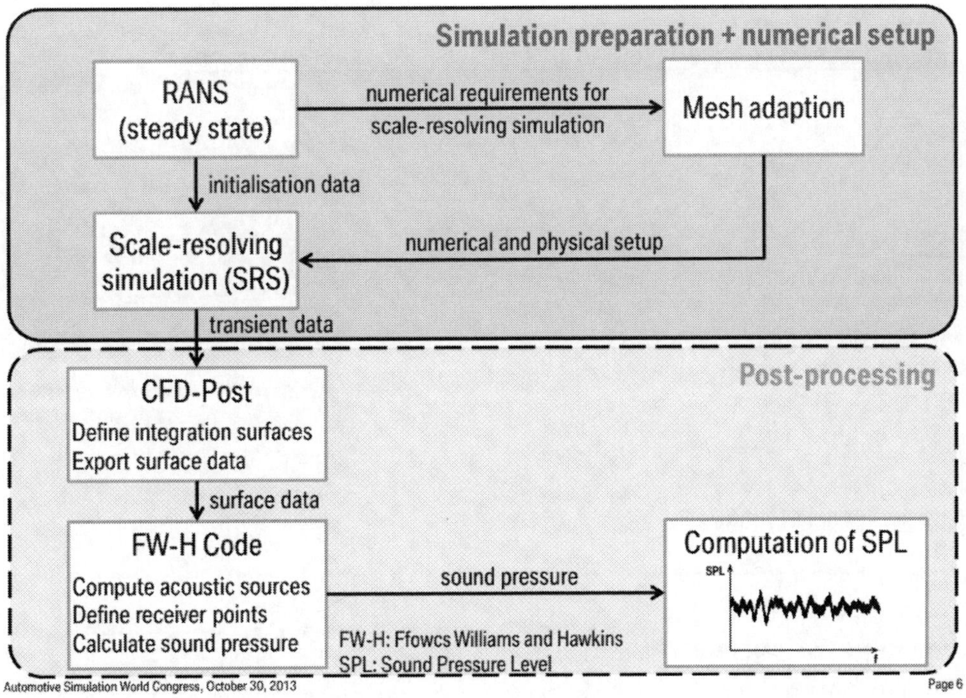

Automotive Simulation World Congress, October 30, 2013

Page 6

CONTENTS.

PRE-PROCESSING OF SCALE-RESOLVING JET SIMULATION.

Pre-processing of scale-resolving simulations needs special care on

- Model selection
- Grid generation
- Numerical settings

PRE-PROCESSING OF SCALE-RESOLVING JET SIMULATION.

Pre-processing of scale-resolving simulations needs special care on

- Model selection ➔ Categorization of the flow type
- Grid generation
- Numerical settings

FLOW TYPE CATEGORIZATION.

Globally unstable flows

Locally unstable flows

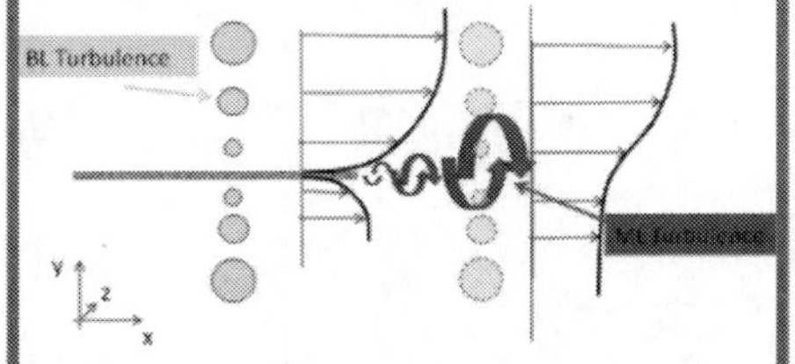

$\Rightarrow$ Jet flow is a locally unstable flow

Stable flows

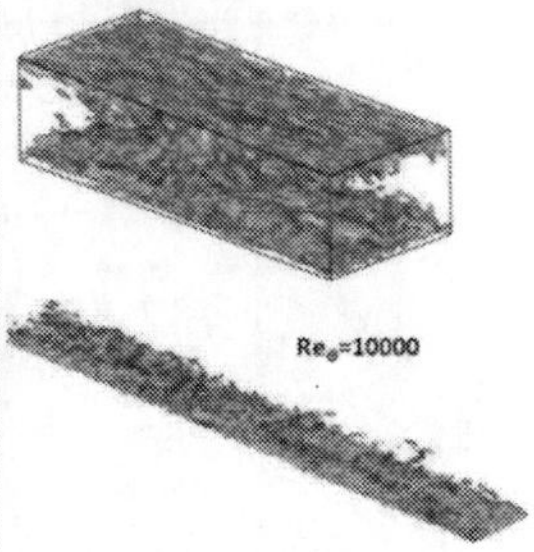

Suitable models for locally unstable flows:

- Embedded LES (ELES)
- DES-SST
- SAS-SST
  - Limited suitability: Instabilities too weak → produced eddy-viscosity too large.
  - Possible remedy: Application of synthetic turbulence generator.

Preferences of SAS-SST:

- Same underlying turbulence model in the whole simulation domain (no model-switch like in DES or ELES).
- Always RANS solution as a fall-back result for insufficient resolution in space or time (no „grey zones“).

$\Rightarrow$ First investigations were carried out with SAS-SST model.

 Menter, F.R., 2012, „Best Practice: Scale-Resolving Simulations in ANSYS CFD“

PRE-PROCESSING OF SCALE-RESOLVING JET SIMULATION.

Pre-processing of scale-resolving simulations needs special care on

- Model selection ➔ Categorization of the flow type
- Grid generation ➔
- Numerical settings

Automotive Simulation World Congress, October 30, 2013

Page 11

PRE-PROCESSING OF SCALE-RESOLVING JET SIMULATION.

Pre-processing of scale-resolving simulations needs special care on

- Model selection ➔ Categorization of the flow type
- Grid generation ➔

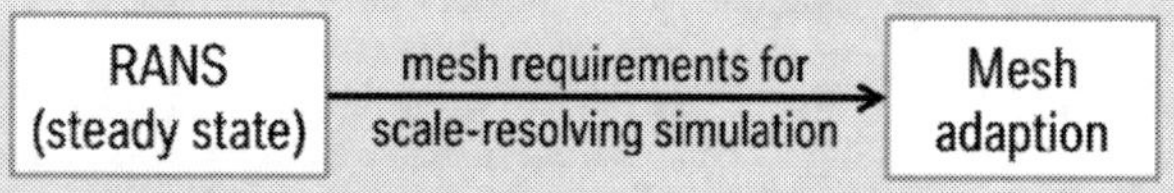

- Numerical settings

Automotive Simulation World Congress, October 30, 2013

Page 12

PRE-PROCESSING OF SCALE-RESOLVING JET SIMULATION.

Pre-processing of scale-resolving simulations needs special care on

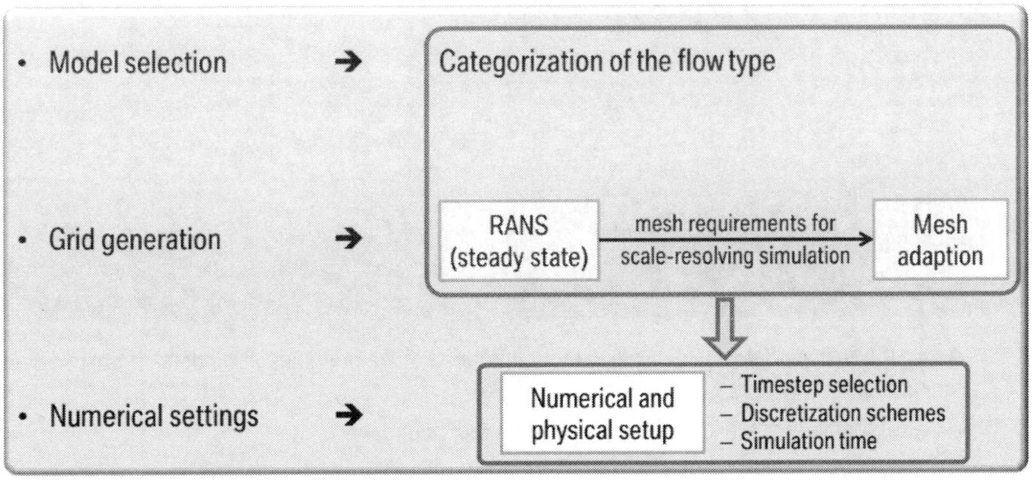

CONTENTS.

ANALYSIS OF BOUNDARY CONDITION INFLUENCES.

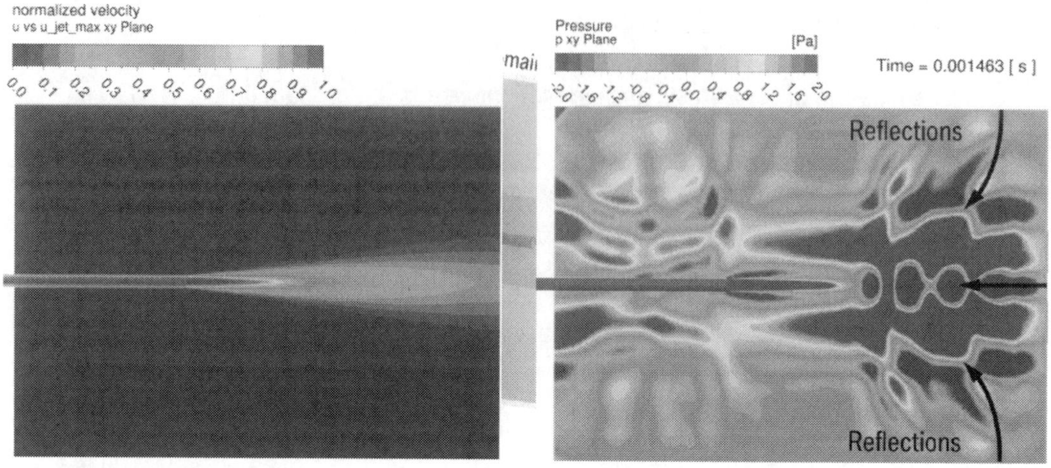

⇒ While the velocity field can develop without influence of the boundary conditions, there is a deep impact on the pressure propagation:

- Pressure field of the jet collapses at the domain boundary.
- Boundary condition yields to strong unphysical reflections of propagated pressure waves.

METHODS TO AVOID NUMERICAL REFLECTIONS.

Properties of investigated Methods:

Method	Efficiency	Robustness
Grid stretching	– Very low – Damps only weak fluctuations	High
Characteristic Non-Reflective Boundary Conditions (NRBC) (beta feature in ANSYS CFX)	– Medium – Reduce reflections – Only 1-D → inefficient for skew incident waves	Very instable
Absorbing Boundary Conditions (ABC) via implicit damping	High	High

ABSORBING BOUNDARY CONDITIONS.

Principle of Absorbing Boundary Conditions:

– Extension of the normal inner domain with an encasing absorbing domain (sponge layer).

– Goal: Damping of fluctuations before they can reach the domain boundary without influence on the flow in the inner domain.

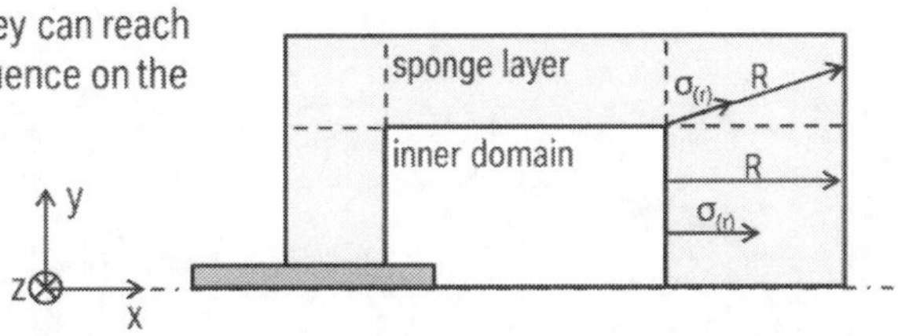

Implementation of Absorbing Boundary Conditions:

– Physical effect:

$$p \cdot V = m \cdot R \cdot T \Longrightarrow \Delta p \cdot V = \Delta m \cdot R \cdot T$$

(Ideal gas equation)

$\Rightarrow$ Damping of pressure fluctuations by locally adding or subtracting fluid mass.

– Implemented by location-depending sink or source term in the continuity equation.

ABSORBING BOUNDARY CONDITIONS.
SINE OSCILLATION 120DB 3000HZ.

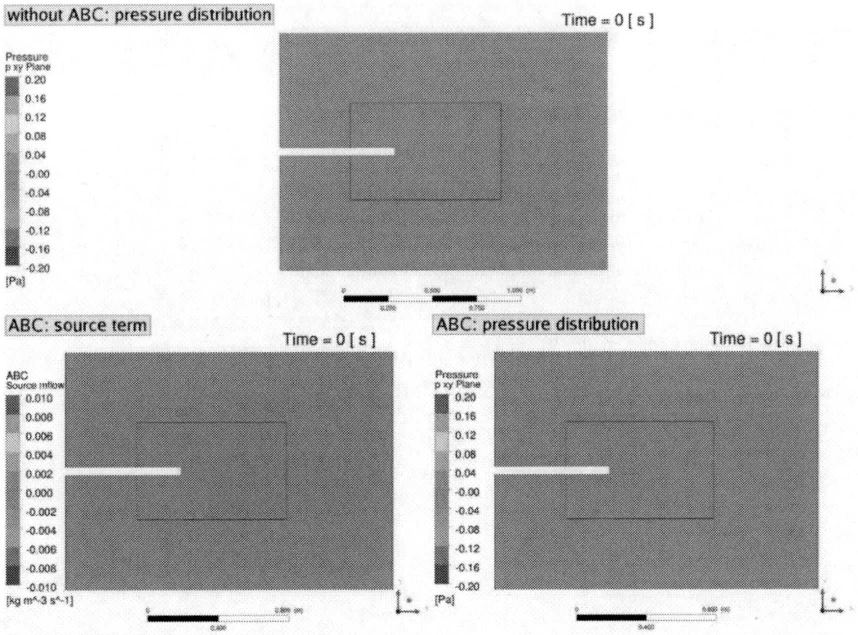

ABSORBING BOUNDARY CONDITIONS. SINE OSCILLATION 120DB 3000HZ.

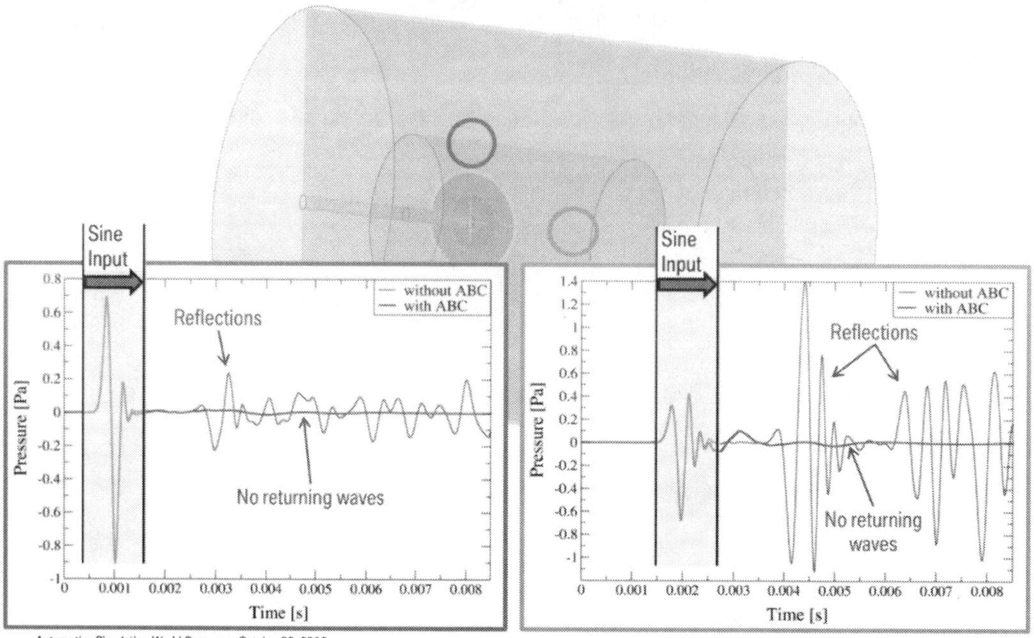

CONTENTS.

SCALE-RESOLVING JET SIMULATION.

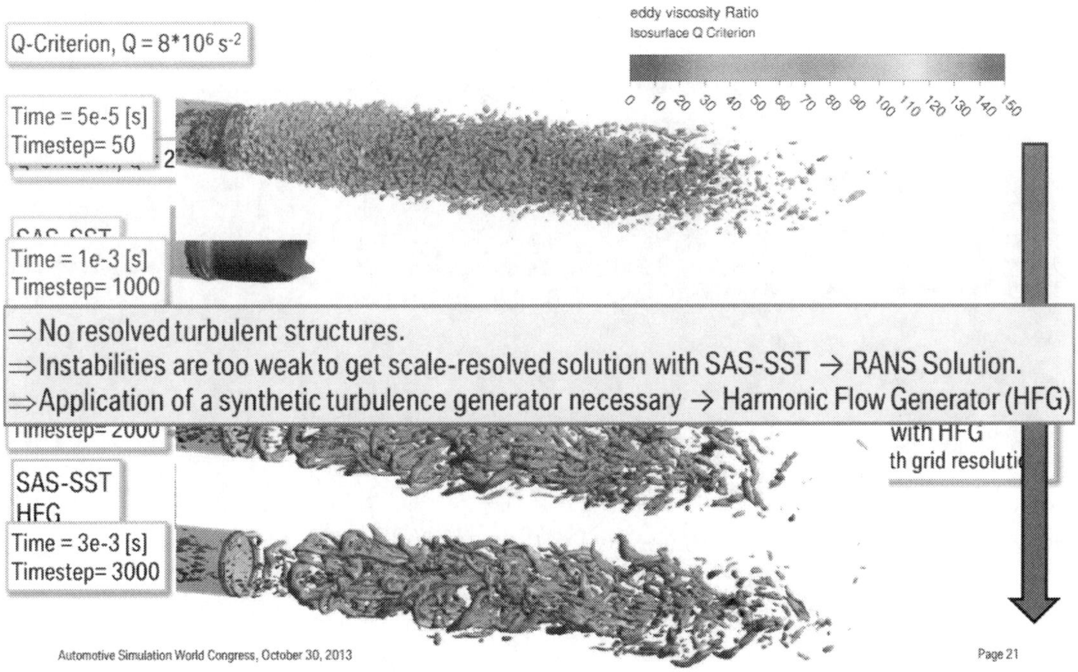

SCALE-RESOLVING JET SIMULATION. URANS VS. SAS-SST (HFG).

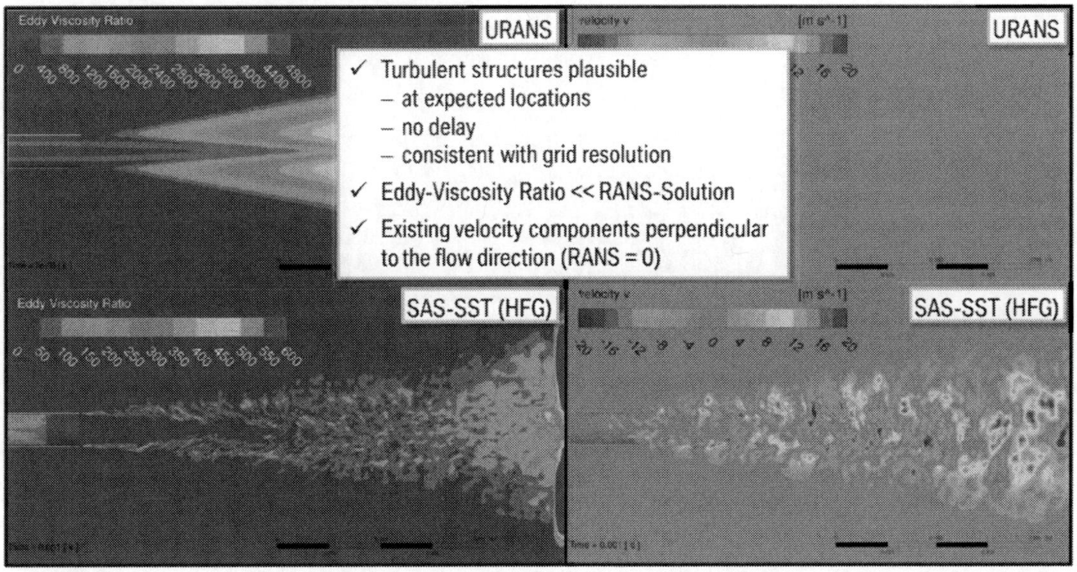

SCALE-RESOLVING JET SIMULATION. URANS VS. SAS-SST (HFG).

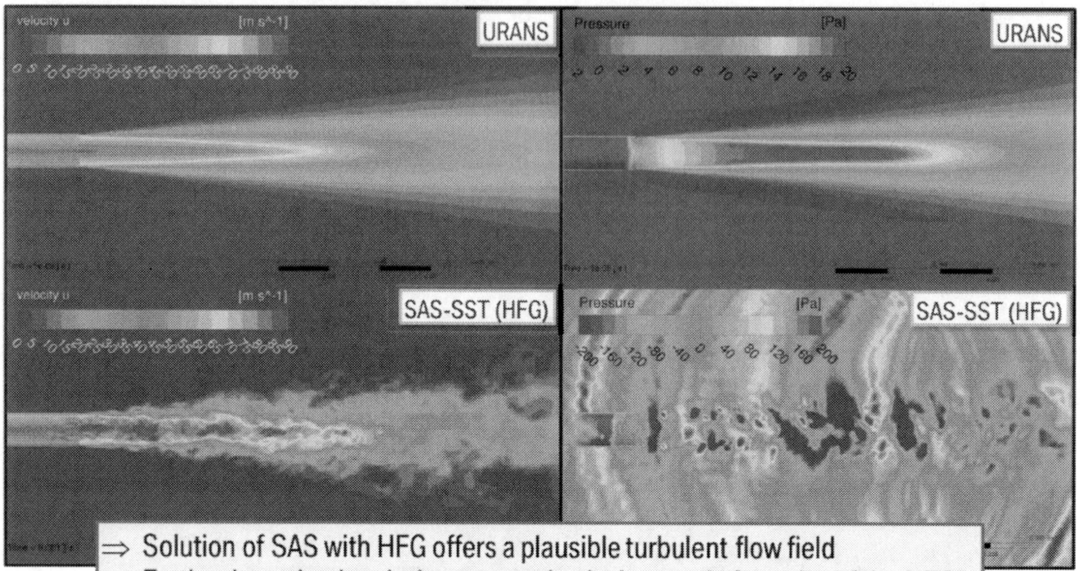

⇒ Solution of SAS with HFG offers a plausible turbulent flow field
⇒ Further investigation: Is there an unphysical acoustic footprint of the HFG?

CONTENTS.

OUTLOOK.

- Coupling of the SAS-SST results with the FW-H Code in order to compute the acoustic sources and the directivity of the jet noise.
- Execution of DES-SST jet simulation in order to compare the DES results with the SAS results and acoustic results.
- Investigation of different coupling methods between the flow field solutions and the FW-H Code.
- Validation of the calculated acoustic results by experimental data.

THANK YOU FOR YOUR ATTENTION.

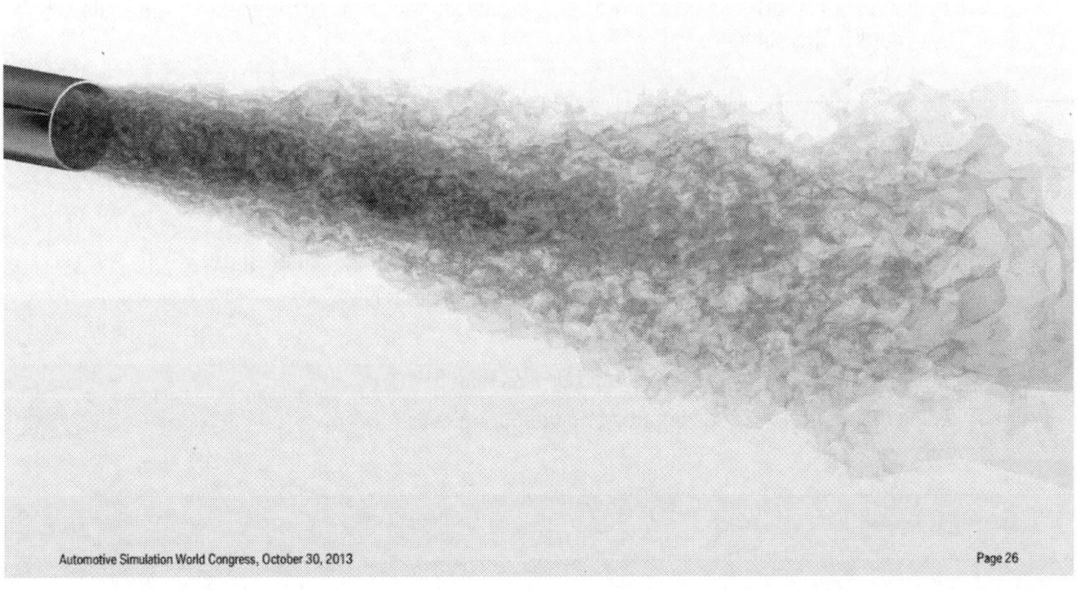

BACKUP.

BACKUP.
GRID GENERATION.

Mesh resolution:

– Quantification of the required mesh resolution in the jet region
– Calculation of the maximum edge length Δ_{max} by investigation of the turbulent length scale L_t^{RANS} of a previous RANS solution:

$$R_L = \frac{\Delta_{max}}{L_t^{RANS}}; \quad L_t^{RANS} = \left(\frac{k^{3/2}}{\varepsilon}\right)^{RANS} = \left(\frac{k^{1/2}}{C_\mu \omega}\right)^{RANS} \qquad R_L \le 0.2 - 0.1$$

(0.2: extrem lower limit; 0.1: desirable lower limit)

– Received maximum edge length should be provided in all 3 space dimensions

Additional aspects:

– Aspect ratio in the relevant region should be less than 5
– Avoiding of jumps in the edge length at neighbouring cells

BACKUP.
NUMERICAL AND PHYSICAL SETUP.

Timestep selection:

Requirement for the Courant Number in the LES zone (jet region): CFL < 1 → $\Delta t = \frac{CFL \cdot \Delta x}{u}$ (Δx ← mesh; u ← RANS solution)

Discretization schemes:

– Advection: Central Difference (can be unstable or produce unphysical wiggles)
alternative: Bounded Central Difference

– Turbulence: First order upwind

– Time: Second order backward Euler

Simulation time:

1. Time period: Reaching the point of statistical steady state
2. Time period: Acquisition of enough information for acoustic computation

(1. and 2. together: Overall simulation time)

BACKUP.
IMPLEMENTATION OF ABC.

Realization of the damping function by using a source term in the continuity equation:

$$S_C = -\sigma_1 \underbrace{(p - p_{spec})}_{p_{rel} = \Delta p}$$

S_C: mass source
σ: scaling factor (damping strength)
p: local pressure
p_{spec}: specified pressure (ambient pressure)

Physical effect:

– Local flow state defined by ideal gas equation

$$p \cdot V = m \cdot R \cdot T \quad \Longrightarrow \quad \Delta p \cdot V = \Delta m \cdot R \cdot T$$

(V, R, T: const.)

⇒ Damping of pressure fluctuations by locally adding or subtracting fluid mass

– Requirement: no impact on the other state variables and ensure a free flow development

⇒ The added mass has to be defined with the local fluid state

⇒ Specification of the following variables with the particular local value

- Temperature
- Turbulence Eddy Dissipation
- Velocity u,v,w
- Turbulence Kinetic Energy

BACKUP.
IMPLEMENTATION OF ABC.

Implementation of a blending function for the damping strength σ:

$$S_C = -\sigma_1(p - p_{spec})$$

– Smooth transition from the inner domain into the sponge layer

n interface

oundary

ngth with an exponential increase:

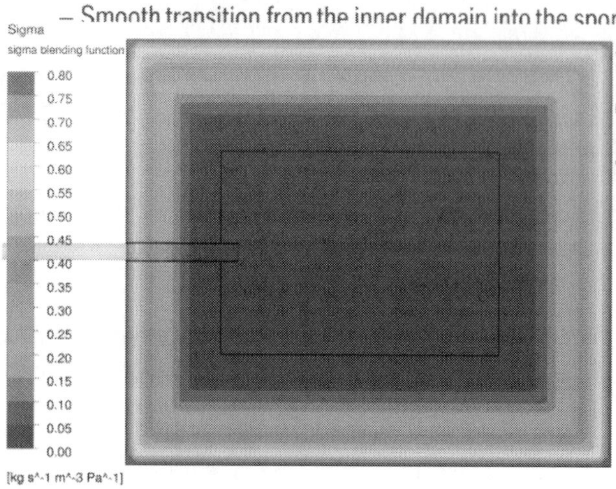

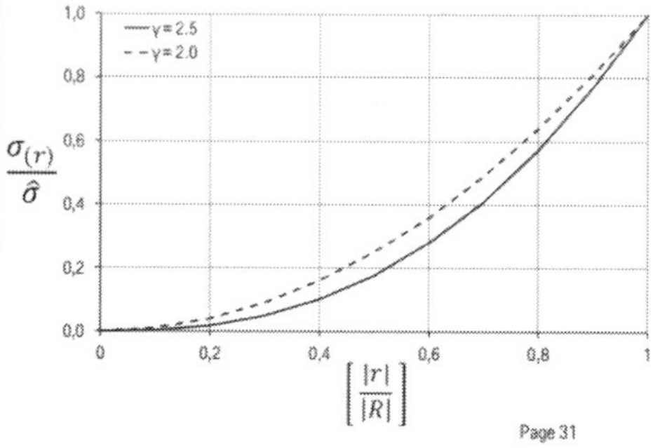

ABSORBING BOUNDARY CONDITIONS.
JET.

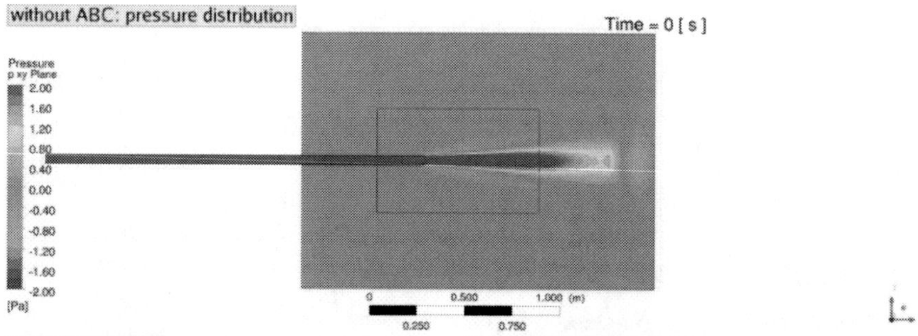

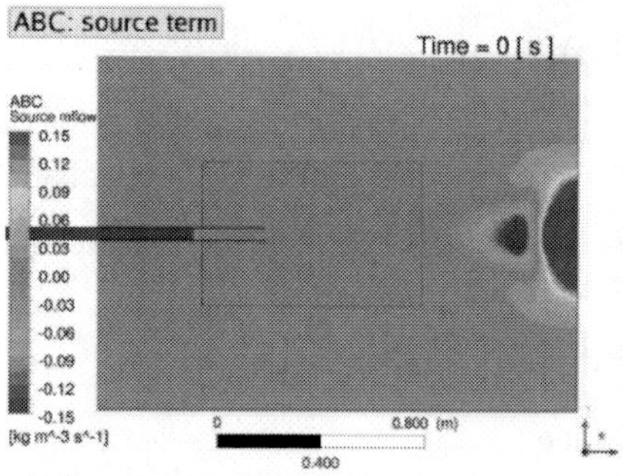

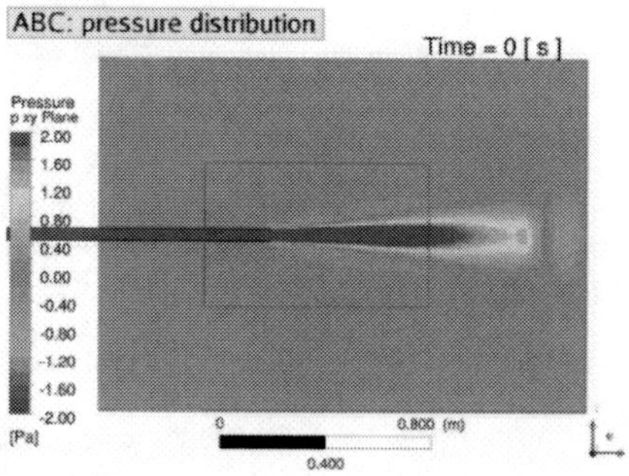

ABSORBING BOUNDARY CONDITIONS.
JET.

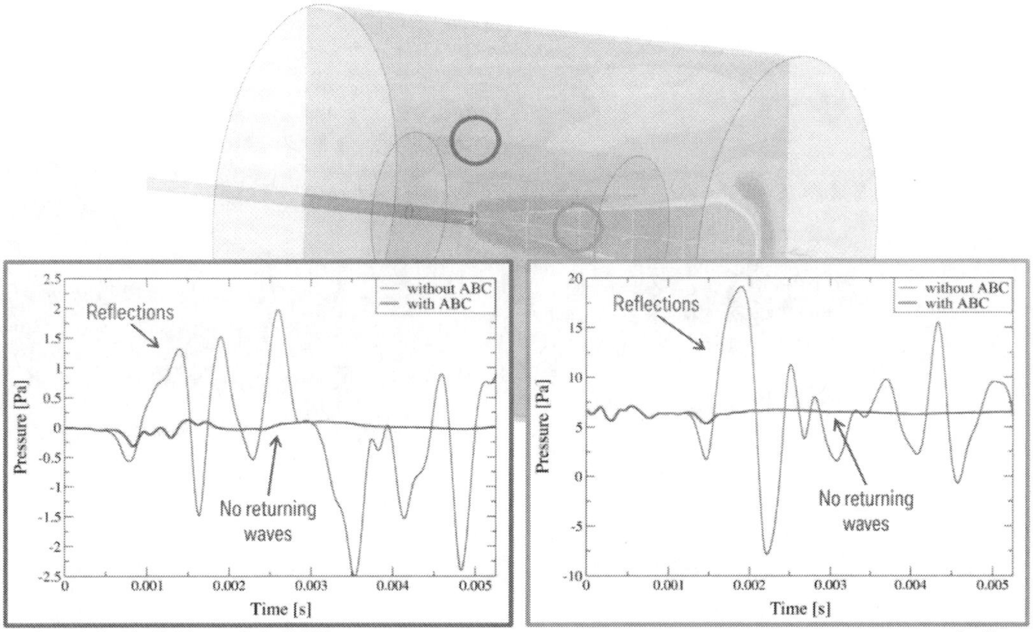

BACKUP.
SCALE-RESOLVING JET SIMULATION – SETUP.

Physical setup:

- D = 47 mm
- mflow = 517 kg/h
- v_{mean} ≈ 70 m/s
- Re ≈ 220000
- Ma ≈ 0.25

Numerical setup:

– Advection:	central difference
– Turbulence:	first order upwind
– Time:	second order backward euler
– Turbulence Model:	SAS-SST
– Timestep:	$1*10^{-6}$ s

Mesh:

- Type: hexahedral
- Nodes: ca. 35 mio.

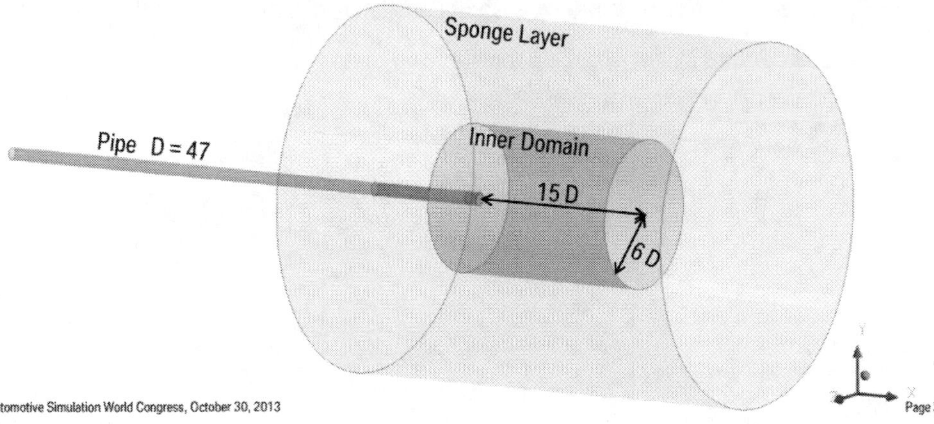

SCALE-RESOLVING JET SIMULATION. SST VS. SAS-SST.

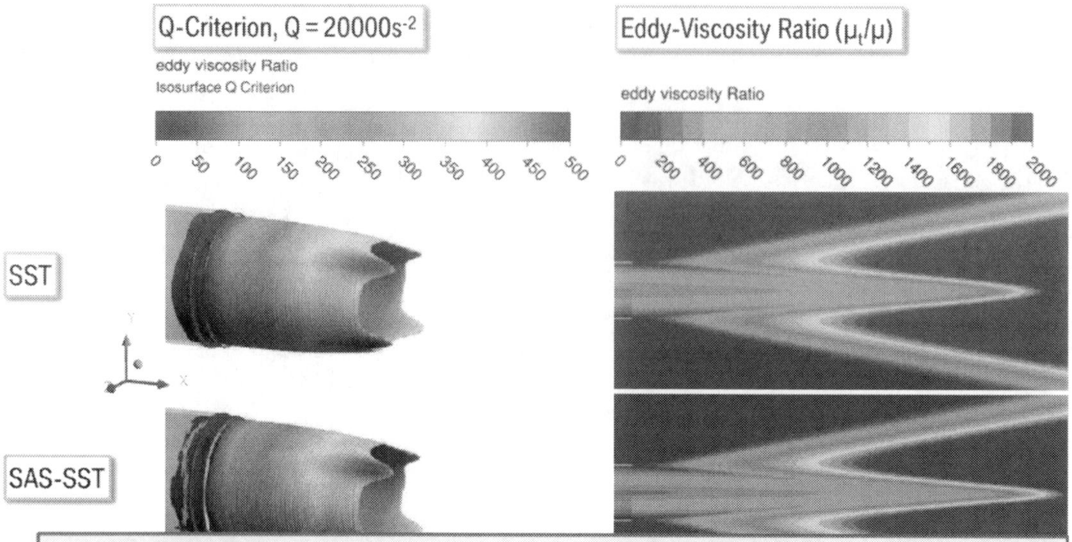

⇒ Instabilities are too weak to get scale-resolved solution with SAS-SST → RANS Solution
⇒ Application of a synthetic turbulence generator is necessary to get resolved fluctuations

BACKUP. SCALE-RESOLVING JET SIMULATION – HFG.

Enable LES behaviour by synthetic turbulence forcing:
(Menter et al., 2009; beta feature in ANSYS CFX)

- Forcing of unsteady structures by introducing synthetic turbulence
- Explicit transfer of modelled turbulent kinetic energy into resolved fluctuations
  - ⇒ Introduction of velocity fluctuations by stochastic source terms in the momentum equations (based on the length and time scale of the RANS turbulence model)
  - ⇒ Corresponding sink term in the k-transport equation (energy preservation)
- Isotopic source terms in user-specified zone (forcing zone):

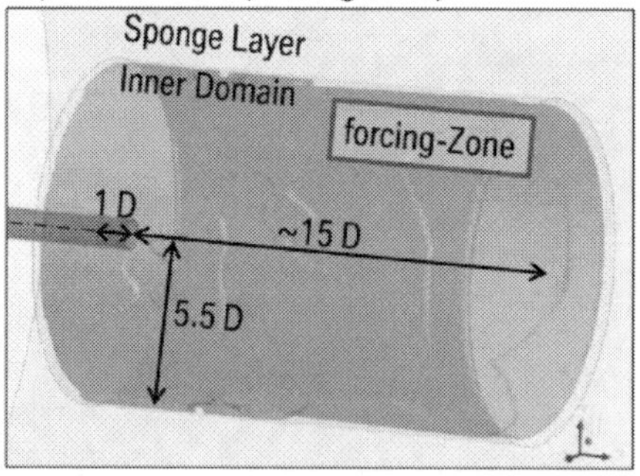

An Automated Routine for Postprocessing and Report Generation in CFD-Post

Friedemann J. Hahn
Martin J. Lehmann

Welcome to MANN+HUMMEL. From Ludwigsburg around the world.

▶▶ Global presence means being there for our customers, wherever and whenever they need us. At more than 50 locations in all five continents.

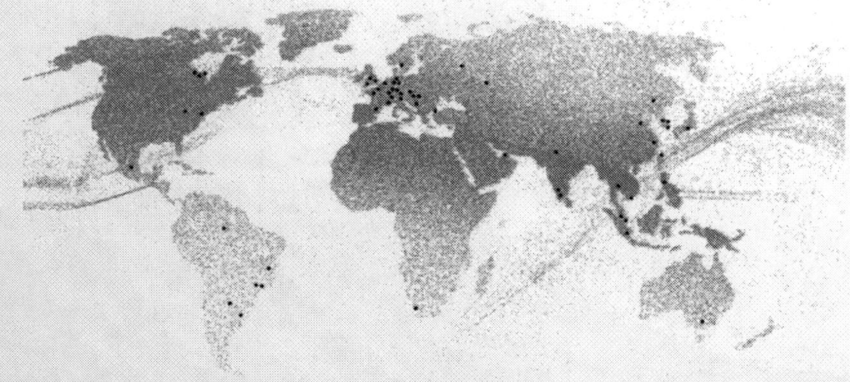

Welcome to Our Product World

▶▶ MANN+HUMMEL divisions:

Automotive Original Equipment	Automotive Aftermarket	Industrial Filtration	Water Filtration

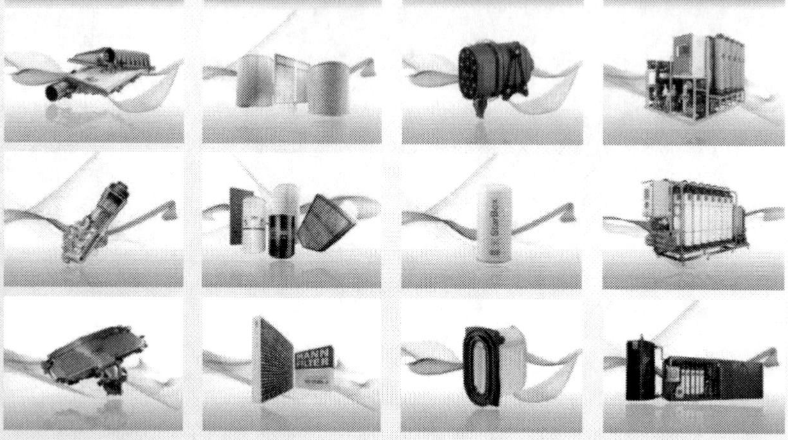

4

An Automated Routine for Postprocessing and Report Generation in CFD-Post

▶▶ A CFD simulation report – more than just pictures?

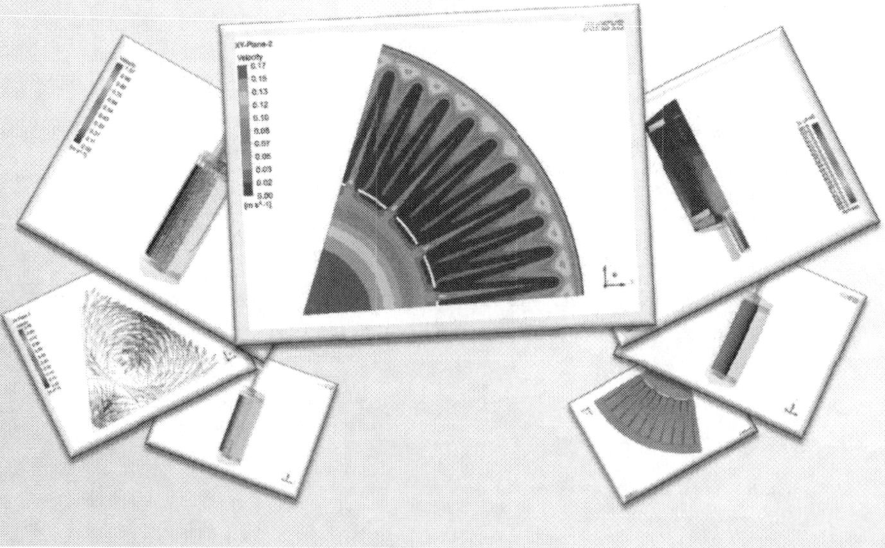

An Automated Routine for Postprocessing and Report Generation in CFD-Post

MANN+HUMMEL

▶▶ Motivation

Each CFD simulation is performed following a typical workflow:

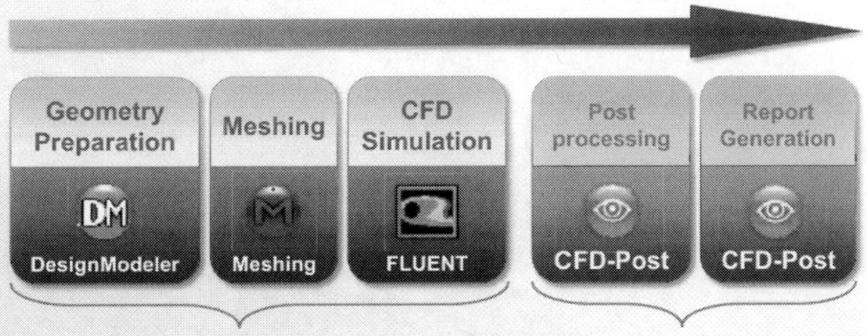

Has to be done manually for each individual case or variant of a simulation

Always the same actions
➔ Time consuming and tedious
➔ „Why don't we automate this?"

6

An Automated Routine for Postprocessing and Report Generation in CFD-Post

MANN+HUMMEL

▶▶ Motivation

We would like to create an automated routine which is able to:

- Generate all plots and views necessary to visualize the CFD results
  ➔ standardized content
- Allow user to add custom information: text, pictures, etc.
  ➔ user-customized content
- Merge all this in a single report document looking „presentable"

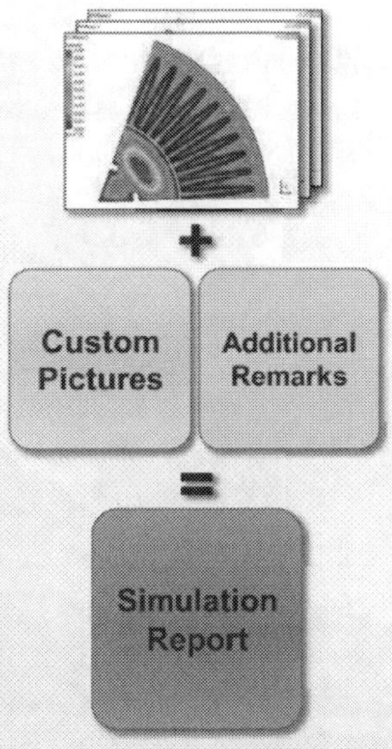

7

An Automated Routine for Postprocessing and Report Generation in CFD-Post

▶▶ Creating a simulation report in CFD-Post

CFD-Post provides reporting features already:

- Standard and custom report templates
- Report templates are script files that can perform automated task when called

➔ how can we achieve our goal using such a report template?

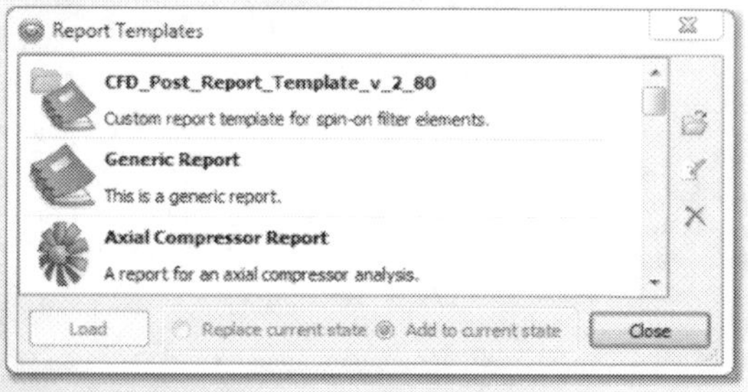

8

An Automated Routine for Postprocessing and Report Generation in CFD-Post

▶▶ Machinery available in CFD-Post

CFD-Post report templates are written in a script language called **CCL**, offering all functionality we need:

CFX Command Language (CCL)

- **Object Creation**
  - Define and create objects: Surface Groups, Volumes, Plots, etc.
- **Perform Actions**
  - E.g. shift camera angle, create figures objects, save report to file etc.
- **Power Syntax**
  - Enables PERL programming in combination with the upper two features: variables, if-else expressions, loops, subroutines, etc.

9

An Automated Routine for Postprocessing and Report Generation in CFD-Post

MANN+HUMMEL

▶▶ Machinery available in CFD-Post

CFD-Post report templates are written in a script language called **CCL**, offering all functionality we need:

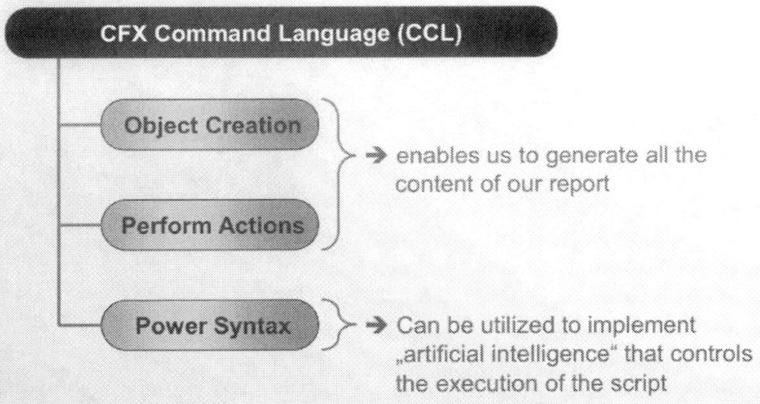

10

An Automated Routine for Postprocessing and Report Generation in CFD-Post

MANN+HUMMEL

▶▶ Examples

MANN+HUMMEL

Title

Exemplary Filter Element Simulation for ASWC 2013

Author

Friedemann J. Hahn

Date

2013/10/30 10:15 am

Contents

11

An Automated Routine for Postprocessing and Report Generation in CFD-Post

MANN+HUMMEL

▶▶ Example 1: Automatic Figure Generation

What we want:

- Velocity contours in three section planes, distributed over the height of the geometry (z direction)

What can be automated here:

1. Calculate positions of planes
2. Create section planes
3. Create velocity contour plots
4. Create figure objects from plots

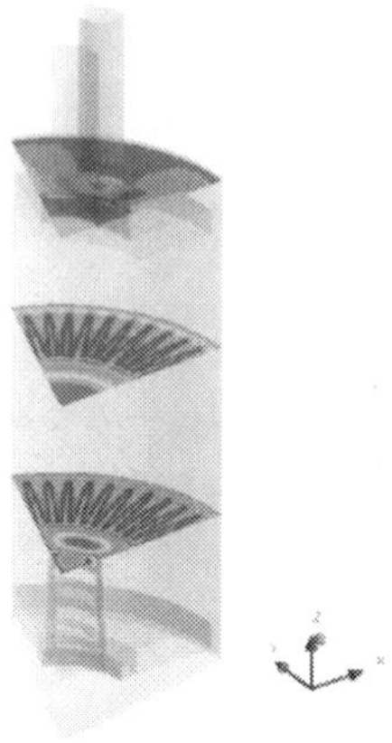

Velocity contours in z-normal section planes over the heigth of the geometry

12

An Automated Routine for Postprocessing and Report Generation in CFD-Post

MANN+HUMMEL

▶▶ Example 1: Automatic Figure Generation

1. Calculate positions of planes
   - Create a geometry object we can measure, such as a volume

```
VOLUME: VolumeObject
  Domain List = /DOMAIN GROUP:All Domains
  Element Type List =
  Tet,Pyramid,Wedge,Hex,Polyhedron
END
```

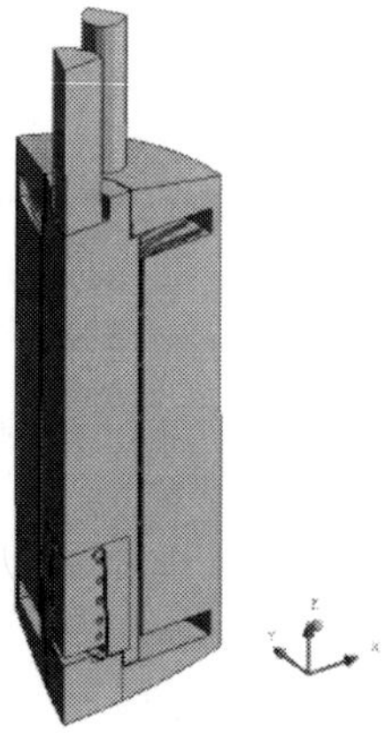

Volume object created from the simulation geometry

13

An Automated Routine for Postprocessing and Report Generation in CFD-Post

MANN+HUMMEL

▶▶ Example 1: Automatic Figure Generation

1. Calculate positions of planes
   - Create a geometry object we can measure, such as a volume
   - Determine min/max dimensions

```
! $z_min    = minVal("Z", "VolumeObject");
! $z_max    = maxVal("Z", "VolumeObject");
! $height   = abs($z_max-$z_min);
```

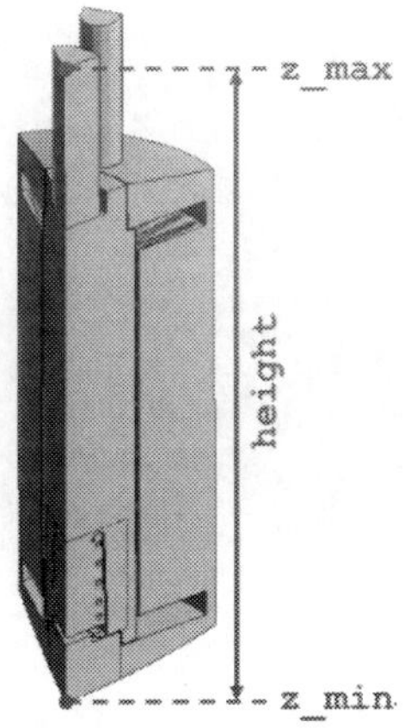

Volume object created from the simulation geometry

14

An Automated Routine for Postprocessing and Report Generation in CFD-Post

MANN+HUMMEL

▶▶ Example 1: Automatic Figure Generation

1. Calculate positions of planes
   - Create a geometry object we can measure, such as a volume
   - Determine min/max dimensions
   - Calculate three different positions over height of geometry

```
! $z_min    = minVal("Z", "VolumeObject");
! $z_max    = maxVal("Z", "VolumeObject");
! $height   = abs($z_max-$z_min);

! $z_pos_1 = $z_min + 0.25*$height;
! $z_pos_2 = $z_min + 0.50*$height;
! $z_pos_3 = $z_min + 0.75*$height;
```

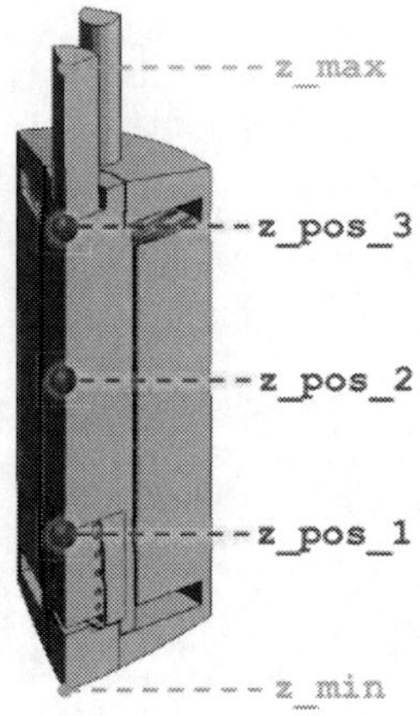

Volume object created from the simulation geometry

15

An Automated Routine for Postprocessing and Report Generation in CFD-Post

MANN+HUMMEL

▶▶ Example 1: Automatic Figure Generation

2. Create section planes
   - Create three individual XY planes in previously calculated positions

```
PLANE:Plane XY1
  Colour = 1, 0.333333, 0
  Domain List = /DOMAIN GROUP:All Domains
  Option = XY Plane
  Z = $z_pos_1 [m]
END
```

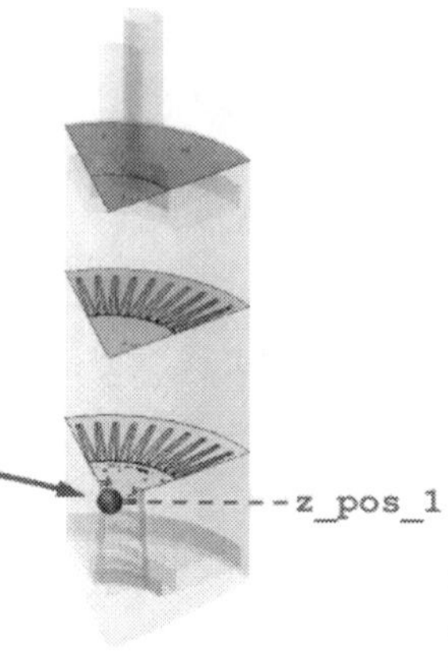

Three z-normal section planes over the height of the geometry

16

An Automated Routine for Postprocessing and Report Generation in CFD-Post

MANN+HUMMEL

▶▶ Example 1: Automatic Figure Generation

3. Create velocity contour plots
   - Create a contour plot of velocity in each of the generated section planes

```
CONTOUR: Velocity Contour XY1
  Clip Contour = Off
  Colour Variable = Velocity
  Contour Range = Local
  Domain List = /DOMAIN GROUP:All Domains
  Location List = /PLANE:Plane XY1
  Number of Contours = 11
END
```

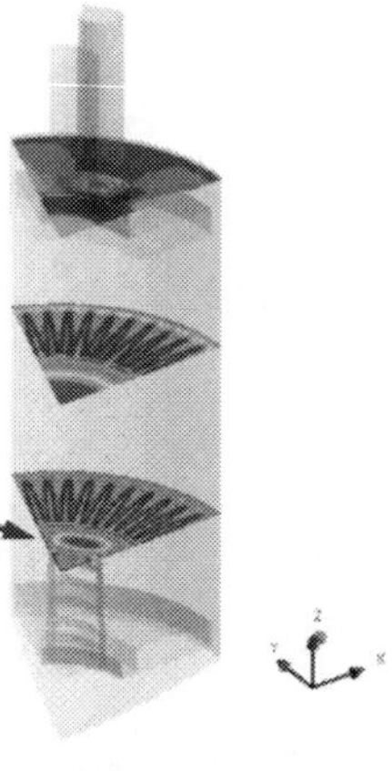

Velocity contours in z-normal section planes over the height of the geometry

17

An Automated Routine for Postprocessing and Report Generation in CFD-Post

MANN+HUMMEL

▶▶ Example 1: Automatic Figure Generation

4. Create figure objects from plots
   - Shift the viewer orientation to Z direction

```
>setcamera viewport=1, camera=+Z
```

Empty viewer, oriented in Z direction

18

An Automated Routine for Postprocessing and Report Generation in CFD-Post

MANN+HUMMEL

▶▶ Example 1: Automatic Figure Generation

4. Create figure objects from plots
   - Rotate the viewer parallel to Z direction
   - Display the desired contour plot

```
>setcamera viewport=1, camera=+Z
>show /CONTOUR:Velocity Contour XY1,
      view=/VIEW:View 1
```

Velocity contour in XY1 plane

19

An Automated Routine for Postprocessing and Report Generation in CFD-Post

MANN+HUMMEL

▶▶ Example 1: Automatic Figure Generation

4. Create figure objects from plots
   - Rotate the viewer parallel to Z direction
   - Display the desired contour plot
   - Fit the plot to the viewer frame

```
>setcamera viewport=1, camera=+Z
>show /CONTOUR:Velocity Contour XY1,
     view=/VIEW:View 1
>centerview 1
```

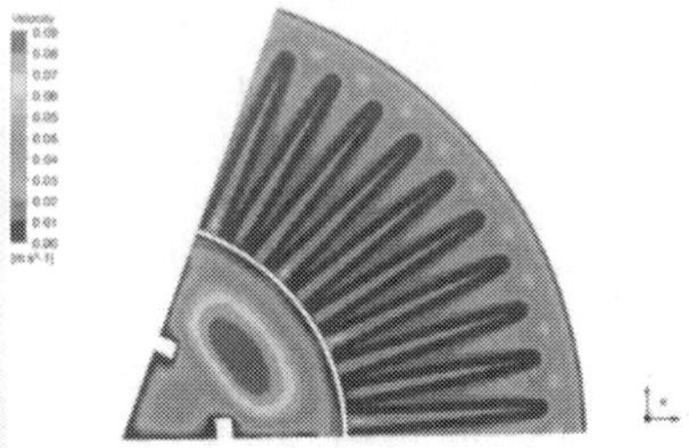

Velocity contour in XY1 plane, now fitted to viewer frame

20

An Automated Routine for Postprocessing and Report Generation in CFD-Post

MANN+HUMMEL

▶▶ Example 1: Automatic Figure Generation

4. Create figure objects from plots
   - Rotate the viewer parallel to Z direction
   - Display the desired contour plot
   - Fit the plot to the viewer frame
   - Create the figure object

```
VIEW:Figure Velocity Contour XY1
  OBJECT REPORT OPTIONS:
    Report Caption = Velocity
    Contour on XY-Plane 1
  END
END
```

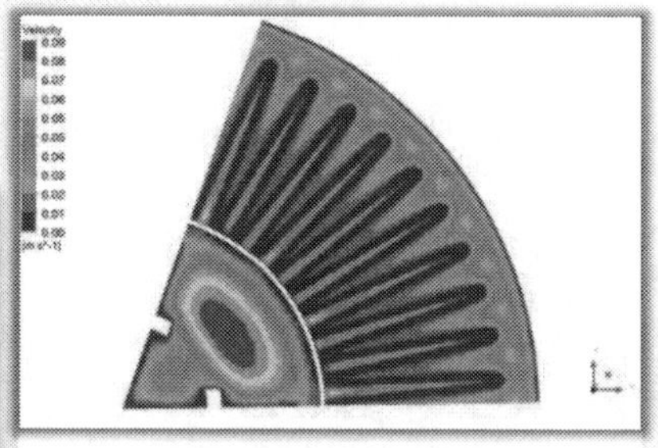

Finalized figure object

21

An Automated Routine for Postprocessing and Report Generation in CFD-Post

MANN+HUMMEL

▶▶ Example 2: Create comments allowing for user customization

What we want:

- Comment objects with standardized title and text
- User just needs to fill in case specific content

What can be automated here:

- Create comment objects
- Initial setup of comment objects

5. Convergence Report:

Solver Type: Pressure-based

Solver Time: Steady

Solution Scheme: SIMPLE

Under-Relaxation Factors:
- Pressure: 0.3
- Density: 1.0
- Momentum: 0.7

Example of a comment:
Convergence Report

22

An Automated Routine for Postprocessing and Report Generation in CFD-Post

MANN+HUMMEL

▶▶ Example 2: Create comments allowing for user customization

Create a comment object:

- Create comment object and initialize it with standard content

```
COMMENT: Convergence Text
  Comment Exists = True
  Comment Heading = Convergence Report
  Heading Level = 1
  Comment Text = Add information on
  convergence behavior here
END
```

5. Convergence Report:

Add information on convergence behavior here

Convergence Report comment
with standard content

23

An Automated Routine for Postprocessing and Report Generation in CFD-Post

MANN+HUMMEL

▶▶ Example 2: Create comments allowing for user customization

Create a comment object:

- Create comment object and initialize it with standard content
- User can now open and edit comment in CFD-Post

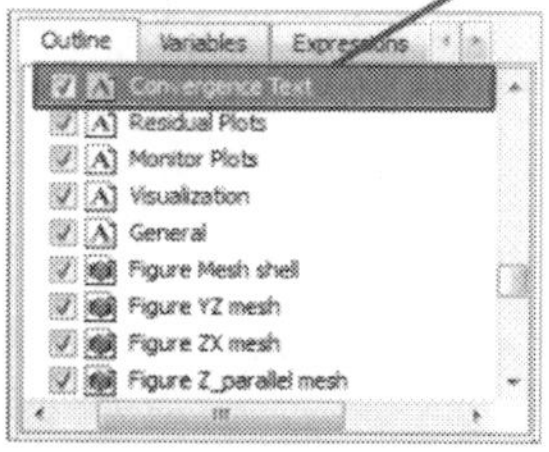

Created comment in the CFD-Post object browser

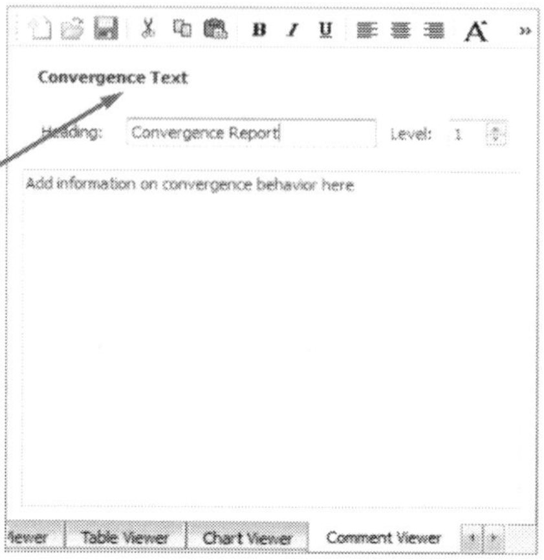

Comment viewer window in CFD-Post

24

An Automated Routine for Postprocessing and Report Generation in CFD-Post

MANN+HUMMEL

▶▶ Example 3: Including additional pictures from files

What we want:

- Additional pictures to be included in the report automatically

What can be automated here:

1. Detect additional pictures of project automatically
2. Include figures in comment objects via HTML code strings

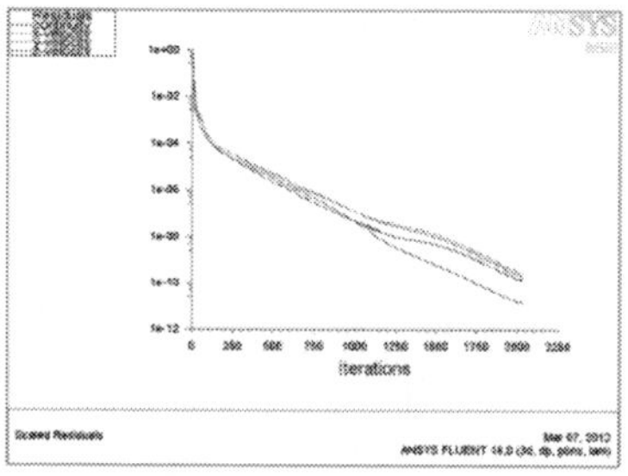

Example for an additional picture: residual monitor

25

An Automated Routine for Postprocessing and Report Generation in CFD-Post

MANN+HUMMEL

▶▶ Example 3: Including additional pictures from files

1. Detection of additional figures:
   - Define PERL packages to be used

```
! use File::Find;
! use File::Basename;
! use Cwd;
```

26

An Automated Routine for Postprocessing and Report Generation in CFD-Post

MANN+HUMMEL

▶▶ Example 3: Including additional pictures from files

1. Detection of additional figures:
   - Define PERL packages to be used
   - Specify a directory to be searched for picture files

```
! use File::Find;
! use File::Basename;
! use Cwd;

! $InputDir = *****;
```

27

An Automated Routine for Postprocessing and Report Generation in CFD-Post

MANN+HUMMEL

▶▶ Example 3: Including additional pictures from files

1. Detection of additional figures:
   - Define PERL packages to be used
   - Specify a directory to be searched for picture files
   - Subroutine writing all .png files to the array „@images"

   ➔ Now we have the pictures – how to get them to show up in CFD-Post?

```
! use File::Find;
! use File::Basename;
! use Cwd;

! $InputDir = *****;

! sub filter {
  ! if($_ =~ /\.png$/i) {
    ! push(@images, $_);
  ! }
! }

! my @images;
! find(\&filter, $InputDir);
```

28

An Automated Routine for Postprocessing and Report Generation in CFD-Post

MANN+HUMMEL

▶▶ Example 3: Including additional pictures from files

2. Including pictures via comment objects
   - Define a variable to hold an HTML string

```
! my $FigString;
```

29

An Automated Routine for Postprocessing and Report Generation in CFD-Post

MANN+HUMMEL

▶▶ Example 3: Including additional pictures from files

2. Including pictures via comment objects
   - Define a variable to hold an HTML string
   - Write HTML code with path to each picture in string

```
! my $FigString;

! foreach $file (@images) {
  ! $FigString = $FigString
      . "<img src=\""
      . $InputDir."/".$file
      . "\" /></p> ";
! }
```

30

An Automated Routine for Postprocessing and Report Generation in CFD-Post

MANN+HUMMEL

▶▶ Example 3: Including additional pictures from files

2. Including pictures via comment objects
   - Define a variable to hold an HTML string
   - Write HTML code with path to each picture in string
   - Create a comment object and put the HTML string as text

   → Now the pictures are also visible to the user in CFD-Post!

```
! my $FigString;

! foreach $file (@images) {
  ! $FigString = $FigString
      . "<img src=\""
      . $InputDir."/".$file
      . "\" /></p> ";
! }

COMMENT: Residual Plots
  Comment Exists = True
  Comment Heading = Plots of
  Residuals
  Comment Text = $FigString
  Heading Level = 2
END
```

31

An Automated Routine for Postprocessing and Report Generation in CFD-Post

▶▶ Example 3: Including additional pictures from files

Created comment object displayed inside CFD-Post:

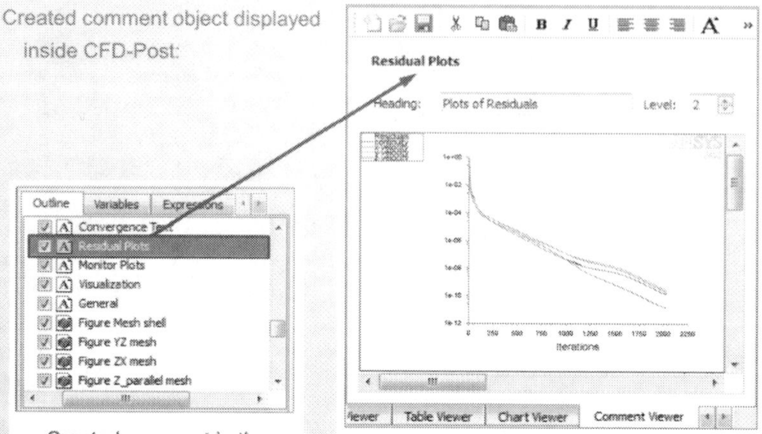

Created comment in the CFD-Post object browser

Comment viewer window in CFD-Post

32

An Automated Routine for Postprocessing and Report Generation in CFD-Post

▶▶ Example 4: Managing the complexity of different cases

What we want:

- Some plots are not required for all simulation cases, e.g. plots of y_pos variable – not needed in laminar cases!

What can be automated here:

- Check for existence of the variable Yplus using Power Syntax!

➔ YES: generate plot and figure

➔ No: skip this and do nothing!

```
! my $yplus_exists =
  objectExists("Yplus");

! if( $yplus_exists )
! {

  #Generate y_plus
  #plot and figure here

! }
```

33

An Automated Routine for Postprocessing and Report Generation in CFD-Post

MANN+HUMMEL

▶▶ Conclusions

Making use of these features we created a generic report script for CFD-Post that can be applied to all simulation geometries relevant for us

Characteristics of our current report script:

- 5000+ lines of CCL code
- About 40 standard figures, showing geometry, mesh and plots
- Multiple custom sections for user defined comments and custom figures

The benefits are striking:

- Script execution time: < 3 minutes!
- High quality of produced report: increased productivity
- Standardized report format
- Reports contain both project information and results: suitable as a document to be communicated with customers

34

An Automated Routine for Postprocessing and Report Generation in CFD-Post

MANN+HUMMEL

▶▶ Acknowledgements

We would like to thank all our colleagues for:

- Supportive feedback
- Improvement suggestions
- Bug reports

35

An Automated Routine for Postprocessing and Report Generation in CFD-Post

MANN+HUMMEL

▶▶ A CFD simulation report – more than just pictures!

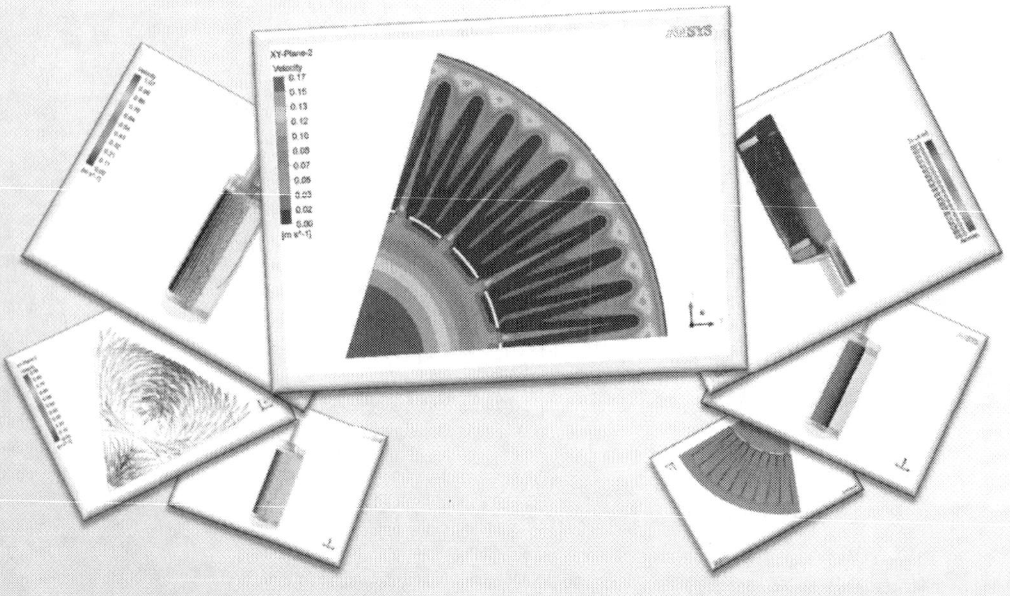

2013 Automotive Simulation World Congress

Simulation Driven Design of EGR Coolers

Padmesh Mandloi
ANSYS Inc.

EGR Cooler Design Process

- Optimizing Heat Exchange
  - Optimizing Inlet Diffuser
    - Shape Optimization
    - Topological Optimization
  - Tube Bundle Optimization
    - Parametric Optimization of Tube Shapes
    - Optimum Tube Arrangement
- Designing Against Failure
  - Thermal Loading
    - High Cycle Thermal Loading
    - Low Cycle Loading (Boiling)
  - Other Mechanical Loading
    - Random Vibrations
    - Mode Shapes
- Improving Performance
  - Condensation
  - Fouling

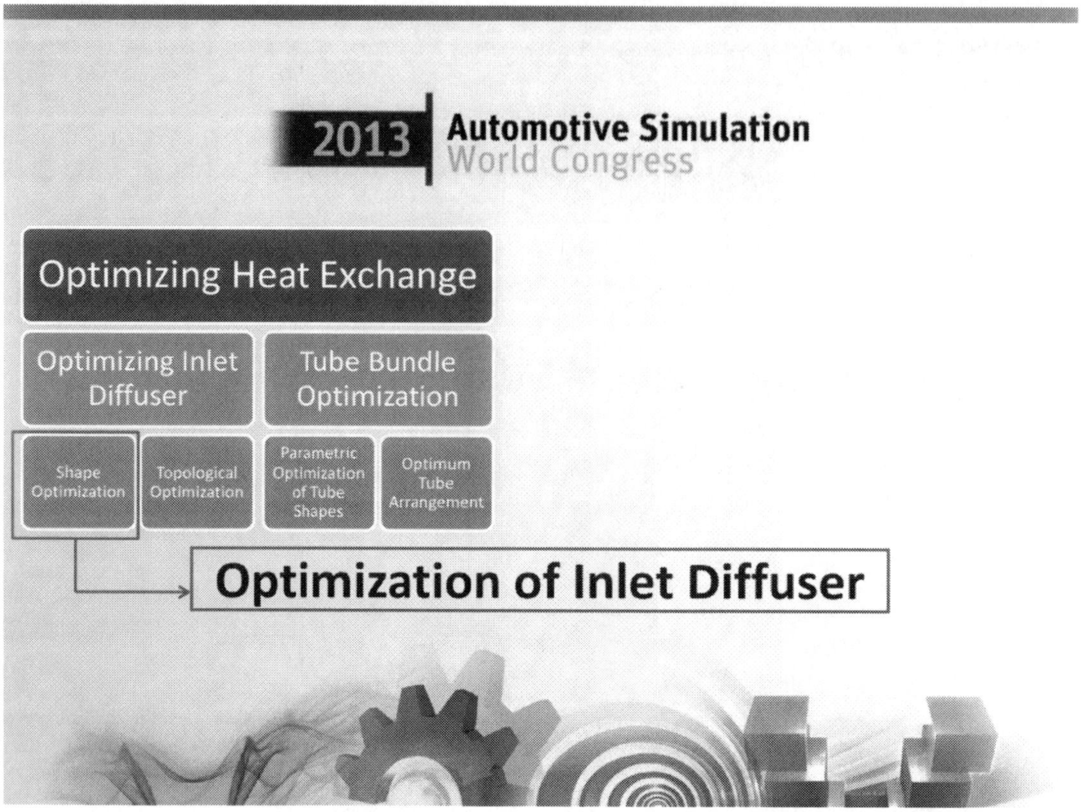

2013 Automotive Simulation World Congress
October 29 - 30 • Frankfurt am Main, Germany

Optimization of Inlet Diffuser

- Uniform distribution of exhaust gas into each tube is desired
- This would reduce non-uniform cooling of the tubes
- ..leading to more uniform stress distribution on end plates

Shape Optimization of Inlet Diffuser

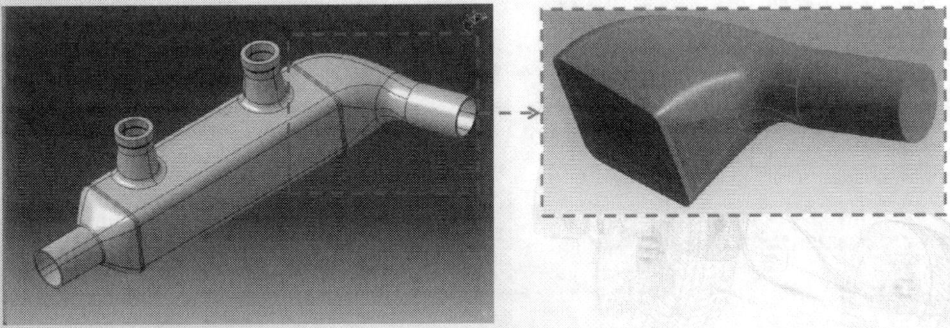

Goals:
- Maximize flow uniformity at outlet of manifold
- Minimize pressure drop

2013 Automotive Simulation World Congress
October 29 - 30 • Frankfurt am Main, Germany

Shape Parameters

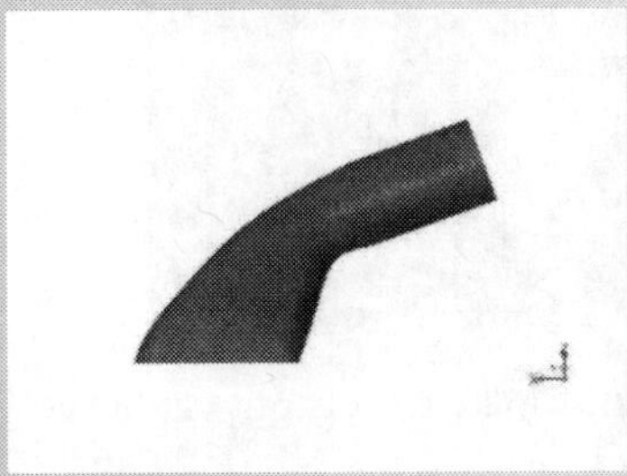

Parameter-1

Parameter-2

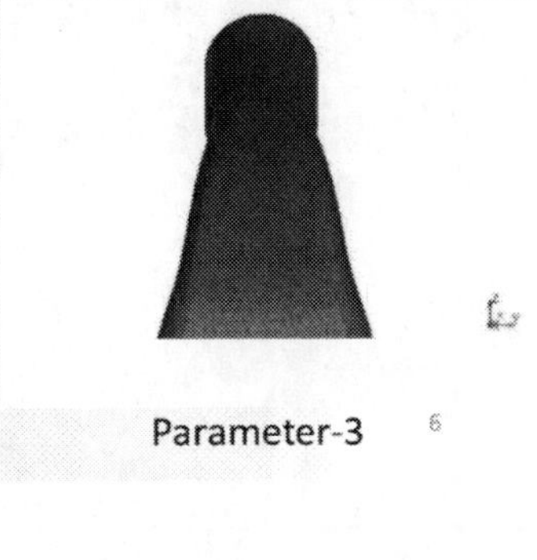

Parameter-3

2013 Automotive Simulation World Congress
October 29 - 30 • Frankfurt am Main, Germany

DOE and Response Surfaces

Table of Schematic C2: Design of Experiments (Custom)

	A	B	C	D	E	F
1	Name	P1 - shape_change_01	P2 - shape_change_02	P3 - shape_change_03	P4 - uniformity_outlet	P5 - pressure_inlet (Pa)
2	1 (DP 9)	0	-1.5	0	0.33681	-4991.5
3	2	2.4391	-0.28045	-2.4391	0.40866	-6636.5
4	3	-2.4391	-2.7196	2.4391	0.32701	-213.08
5	4	-2.4391	-0.28045	2.4391	0.30205	3175.1
6	5	2.4391	-0.28045	2.4391	0.31401	-951.22
7	6 (DP 2)	-3	-1.5	0	0.33863	-2585.2
8	7 (DP 15)	3	-1.5	0	0.36907	-5795.5
9	8 (DP 7)	0	-3	0	0.34385	-5125
10	9 (DP 0)	0	0	0	0.32621	-3399.3
11	10 (DP 8)	0	-1.5	-3	0.39434	-6491.7
12	11 (DP 10)	0	-1.5	3	0.30177	248.07
13	12	-2.4391	-2.7196	-2.4391	0.36387	-4992.7
14	13	2.4391	-2.7196	-2.4391	0.42458	-7219.8
*	New Design Point					

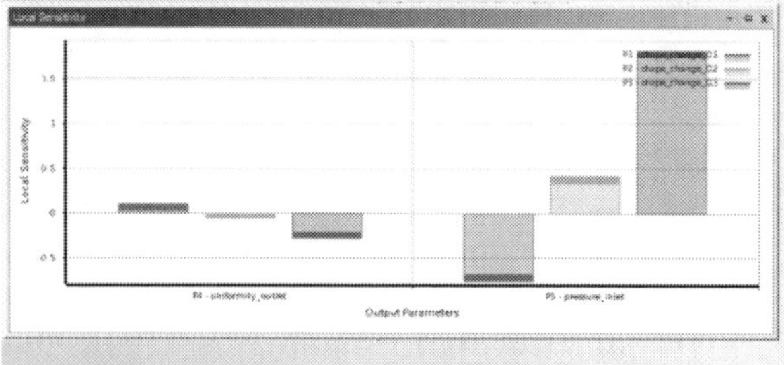

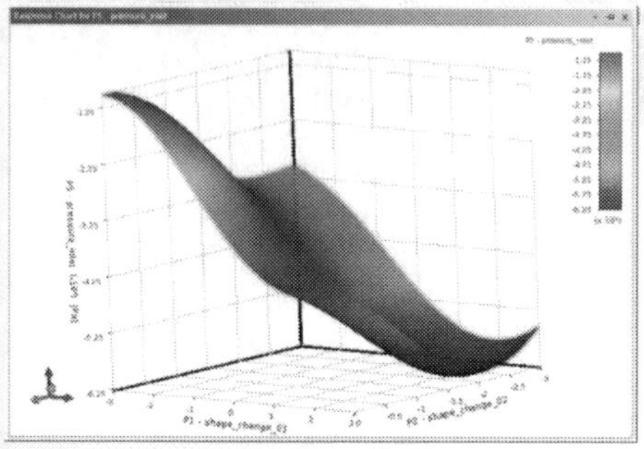

2013 Automotive Simulation World Congress
October 29 - 30 • Frankfurt am Main, Germany

Goal Driven Optimization

Outline of Schematic C4: Optimization

	A	B	C
1		Enabled	Monitoring
2	✓ Optimization		
3	Objectives and Constraints		
4	Maximize P4		
5	Minimize P5		
6	Domain		
7	Fluent (A1)		
8	P1 - shape_change_01	☑	
9	P2 - shape_change_02	☑	
10	P3 - shape_change_03	☑	

Table of Schematic C4: Optimization

	A	B	C	D
1	Optimization Study			
2	Maximize P4	Goal, Maximize P4 (Higher importance)		
3	Minimize P5	Goal, Minimize P5 (Default importance)		
4	Optimization Method			
5	MOGA	The MOGA method (Multi-Objective Genetic Algorithm) is a variant of the popular NSGA-II (Non-dominated Sorted Genetic Algorithm-II) based on controlled elitism concepts. It supports multiple objectives and constraints and aims at finding the global optimum.		
6	Configuration	Generate 100 samples initially, 100 samples per iteration and find 3 candidates in a maximum of 20 iterations.		
7	Status	Converged after 604 evaluations.		
8	Candidate Points			
9		Candidate Point 1	Candidate Point 2	Candidate Point 3
10	P1 - shape_change_01	2.063	2.0873	2.9906
11	P2 - shape_change_02	-2.2405	-2.4103	-2.3769
12	P3 - shape_change_03	-2.4669	-2.4483	-2.3528
13	P4 - uniformity_outlet	★★★ 0.42501	★★★ 0.42538	★★★ 0.42578
14	P5 - pressure_inlet (Pa)	★★★ -7610.5	★★★ -7555	★★★ -7432.2

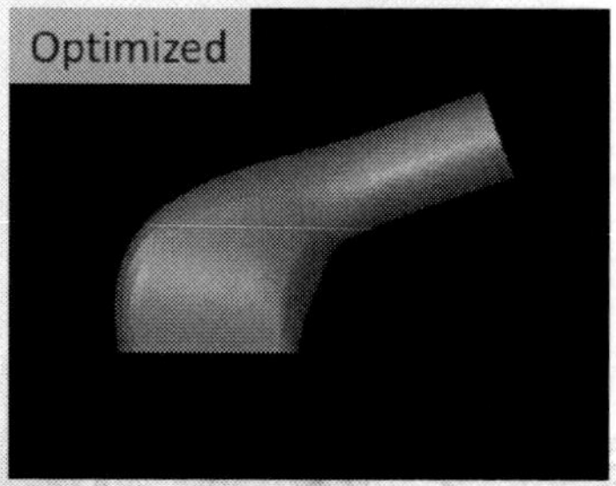

0.6 % and 6% error in actual output values Vs values provided by MOGA in uniformity and pressure drop respectively

~ 25% improvement in uniformity and marginal reduction in pressure drop

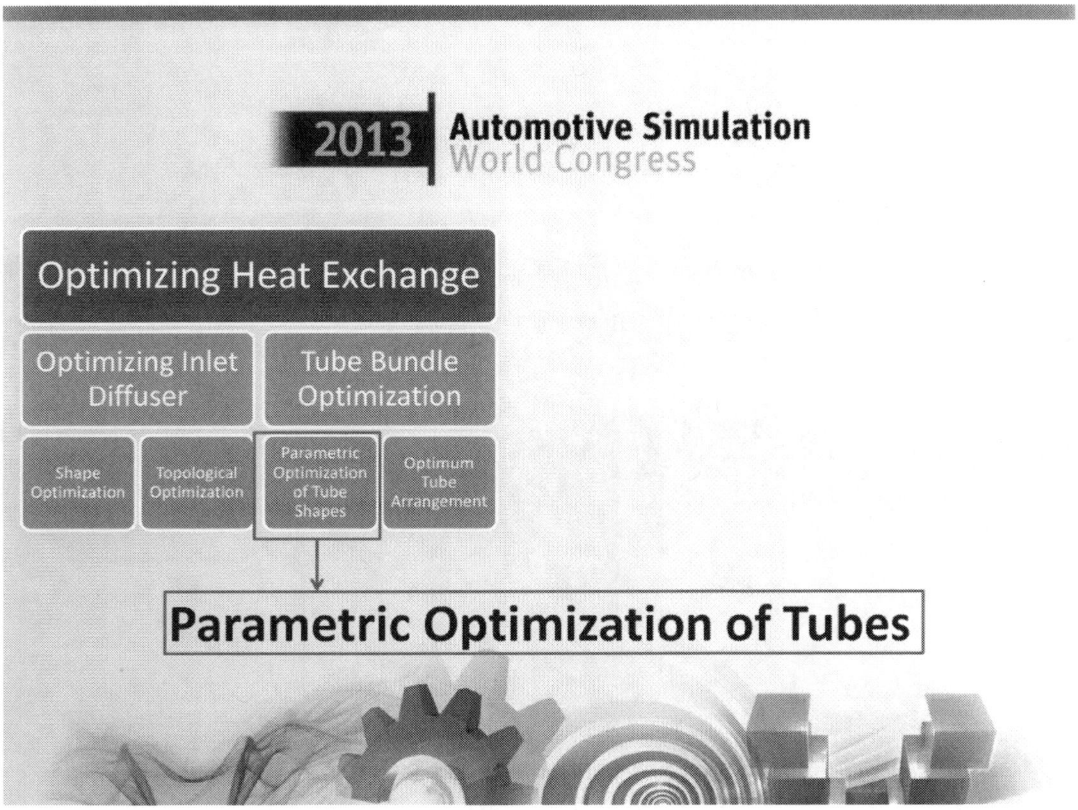
2013
Automotive Simulation
World Congress
Optimizing Heat Exchange
Optimizing Inlet Diffuser
Tube Bundle Optimization
Shape Optimization
Topological Optimization
Parametric Optimization of Tube Shapes
Optimum Tube Arrangement
Parametric Optimization of Tubes

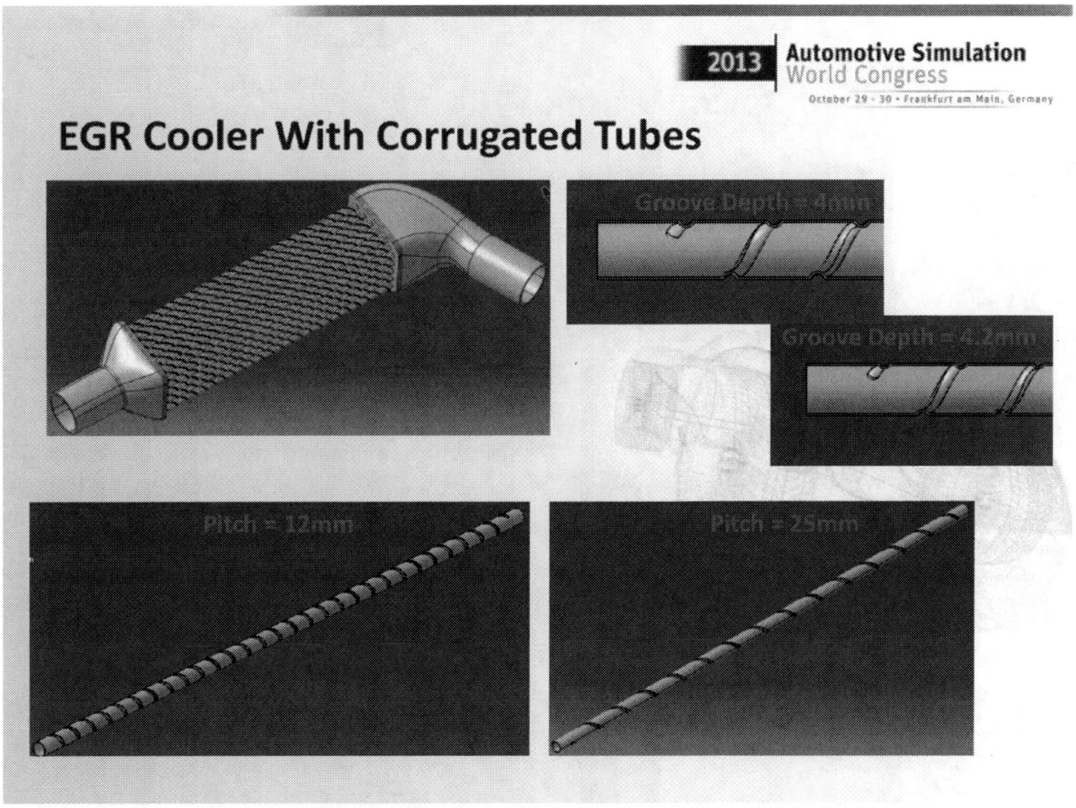
2013
Automotive Simulation
World Congress
October 29 - 30 • Frankfurt am Main, Germany
EGR Cooler With Corrugated Tubes
Groove Depth = 4mm
Groove Depth = 4.2mm
Pitch = 12mm
Pitch = 25mm

Meshing Workflow

- Meshing Corrugated Tubes
  - Thin-Volume-Mesh generation using Fluent Meshing R15
  - Two elements in the thickness
  - Mesh Quality
  - Total cell count in tube solids ~162 K

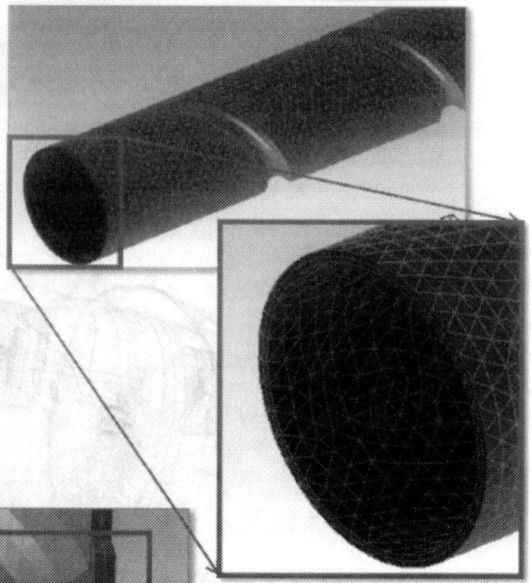

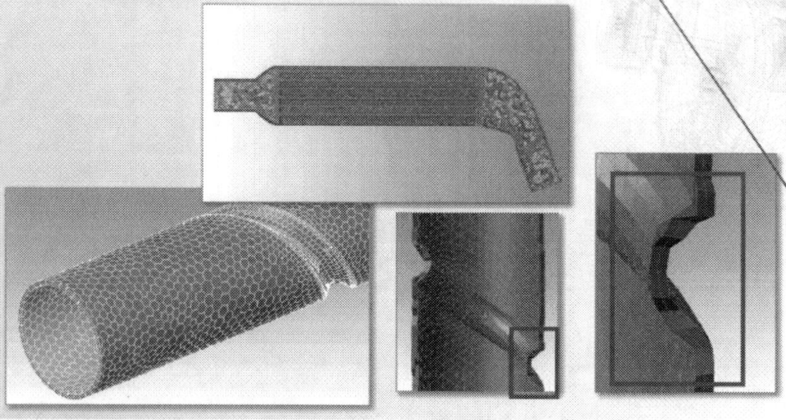

2013 Automotive Simulation World Congress
October 29 - 30 • Frankfurt am Main, Germany

Parametric Optimization: WIP

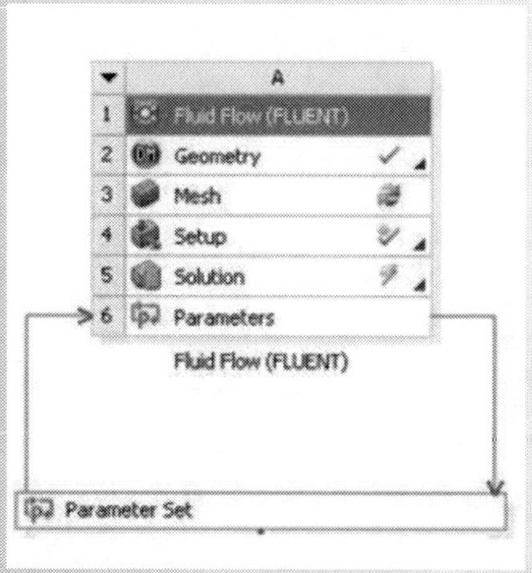

- Maximize Heat Transfer
- Minimize Pressure Drop

2013 Automotive Simulation World Congress

Designing Against Failure

- Thermal Loading
  - High Cycle Thermal Loading
  - Low Cycle Loading (Boiling)
- Other Mechanical Loading
  - Random Vibrations
  - Mode Shapes

High Cycle Thermal Loading

Thermal Fatigue in EGR Coolers

- Thermal loading in EGR coolers are of two types
  - Thermal Fatigue (high cycle)
    - Cyclic thermal loading due to load cycle variations
  - Thermal shock (low cycle)
    - Uneven, abrupt change due to part failure

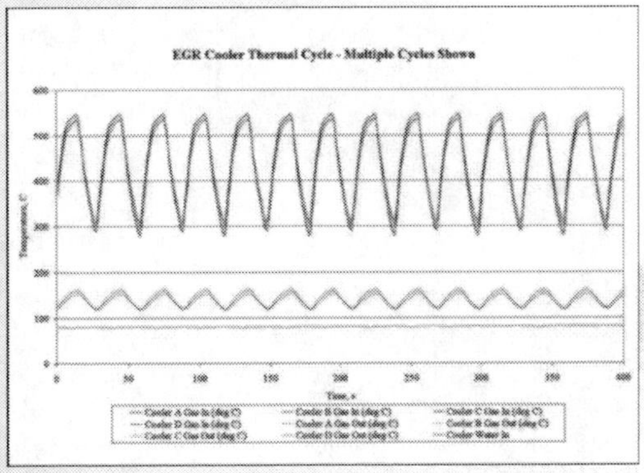

http://www.swri.org/3pubs/brochure/d03/EGR/EGRFlyer.pdf

Workflow

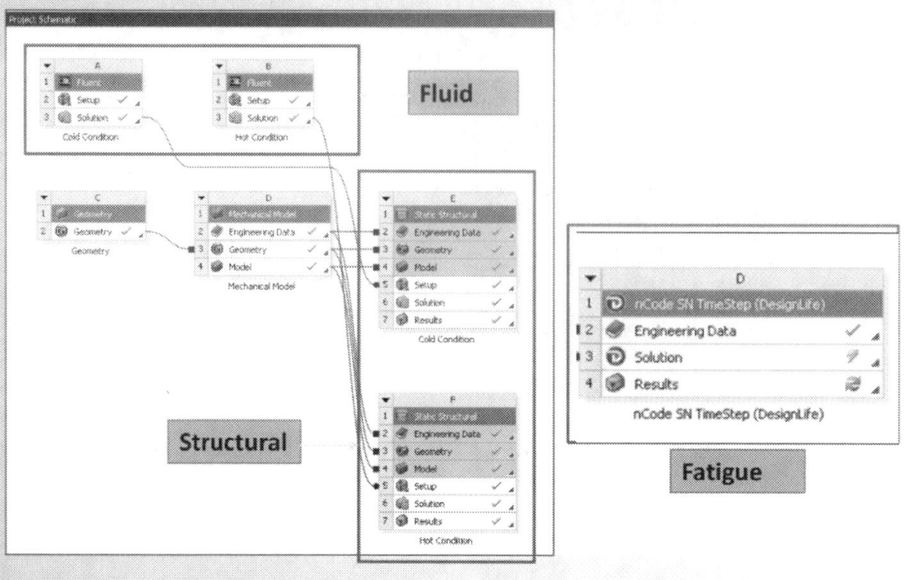

2013 Automotive Simulation World Congress
October 29 - 30 • Frankfurt am Main, Germany

Workbench Meshing Process

Non-conformal Meshing Process

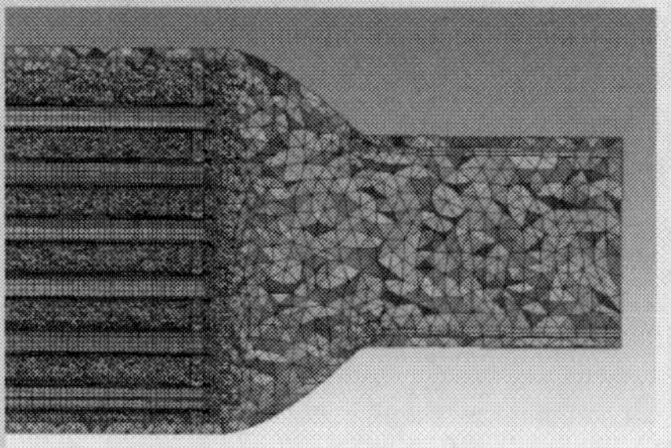

Process:
- Cleanup, flow-volume extraction, decomposition to allow sweeping
- Sweep mesh in tube channels
- Map mesh for tube solids
- 3 Inflation layers from all surface

Statistics:
- Geometry operations: 1 hour
- Meshing operations: 1 hour
- Cell count: 6 million

CHT Simulation

Hot condition

EGR inlet	flow rate (kg/min)	Temp K
	15	948.15

coolant inlet	flow rate (LPS)	Temp K
	4.3	363.15

Cold condition

EGR inlet	flow rate (kg/min)	Temp K
	2.5	533.15

coolant inlet	flow rate (LPS)	Temp K
	1.9	363.15

EGR Out pressure (Pa)	120000
Coolant Out pressure (Pa)	112000

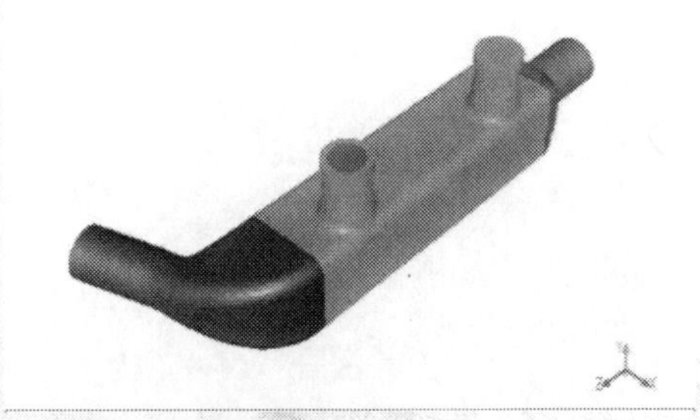

- Solver Settings
  - Pressure Based Coupled Solver
  - Green Gauss Node Based Gradient Scheme
  - Second Order for Momentum, and Energy
  - High Order Term Relaxation

Results: CHT Simulation

Cold Cycle

Hot Cycle

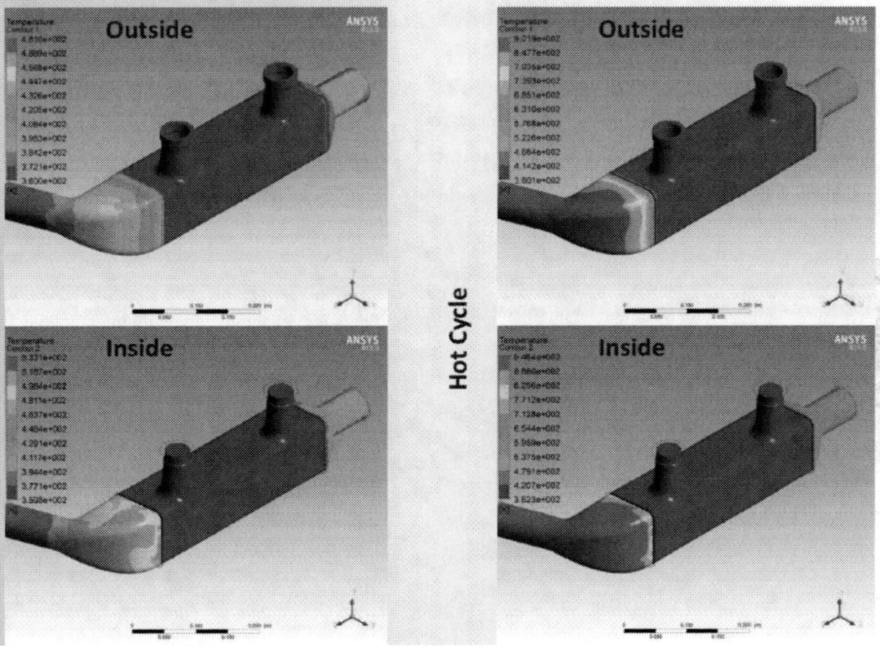

2013 Automotive Simulation World Congress
October 29 - 30 • Frankfurt am Main, Germany

Results: CHT Simulation

Cold Cycle

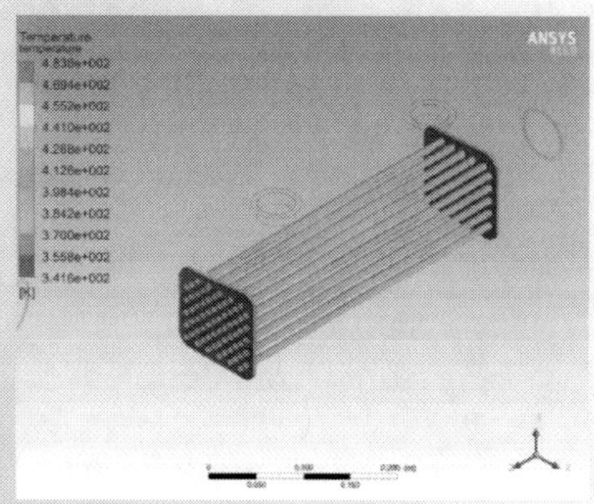

Hot Cycle

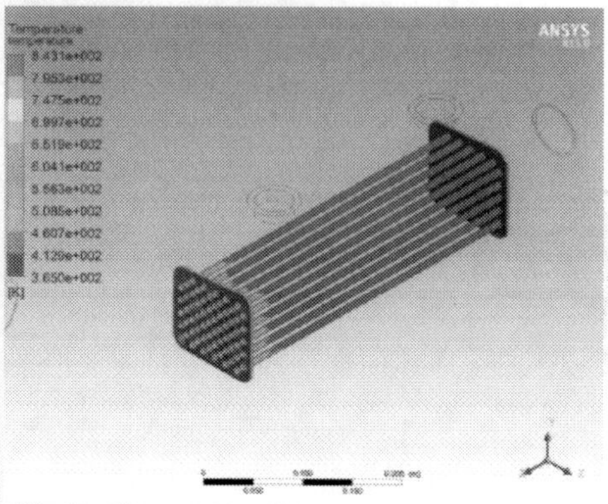

2013 Automotive Simulation World Congress
October 29 - 30 • Frankfurt am Main, Germany

Temperature – Iso Surface

Cold Cycle

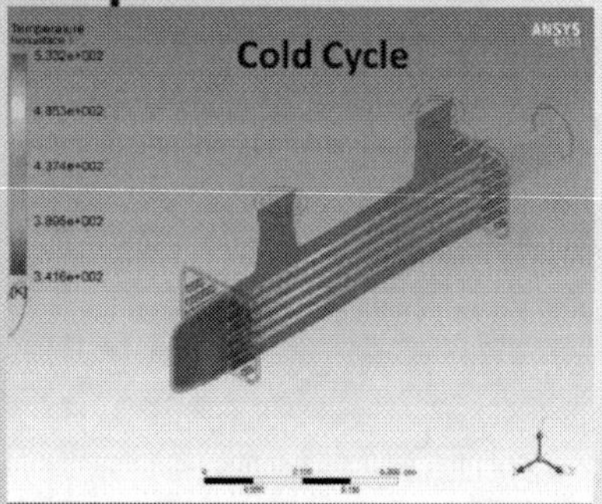

Hot Cycle

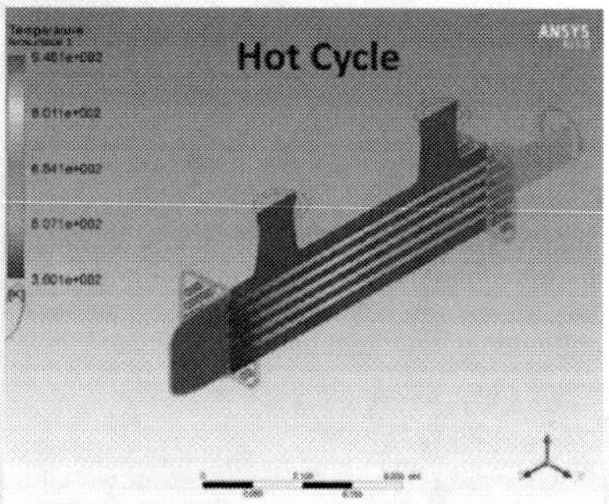

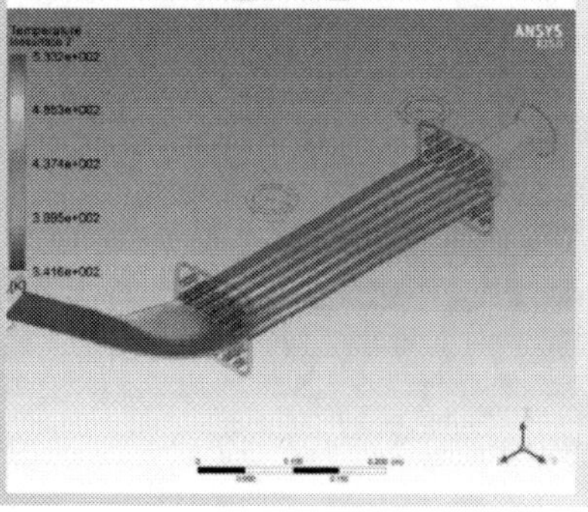

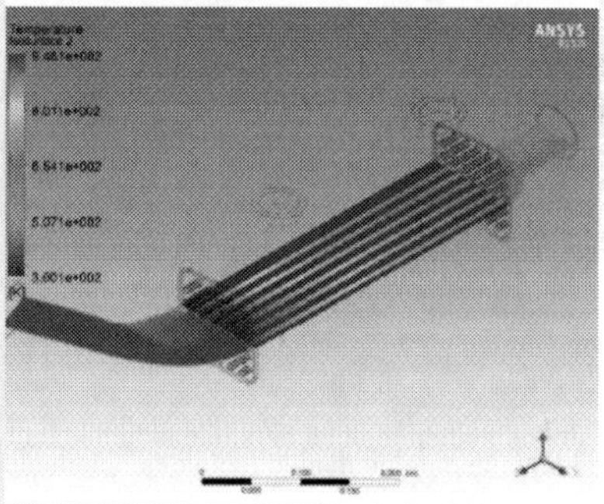

2013 Automotive Simulation World Congress
October 29 - 30 • Frankfurt am Main, Germany

Thermal Structural Setup

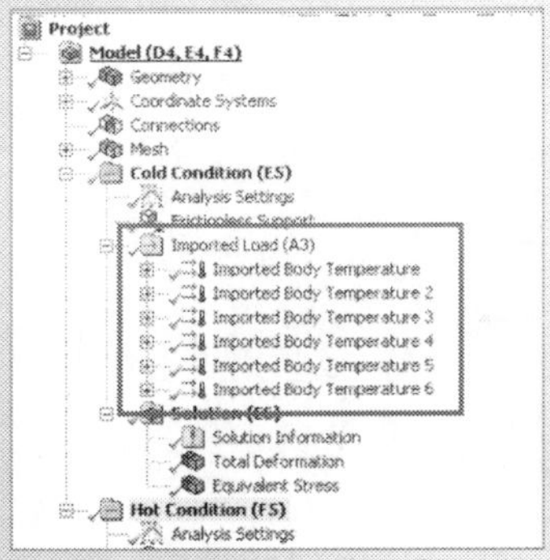

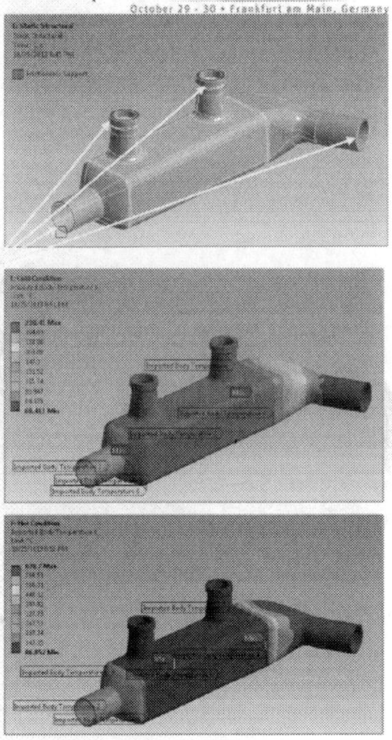

Frictionless Support

Imported Temperature

2013 Automotive Simulation World Congress
October 29 - 30 • Frankfurt am Main, Germany

Results – Von-Mises Stress

- Equivalent Von-Mises Stress is obtained for both the cold and hot condition.
- Due to consideration of linear properties, the stress is higher in hot condition

Von-Mises Stress (Cold Condition)

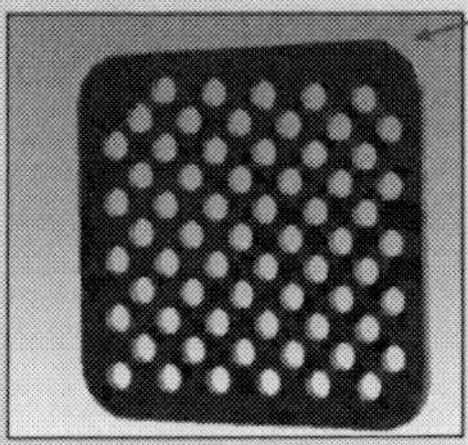

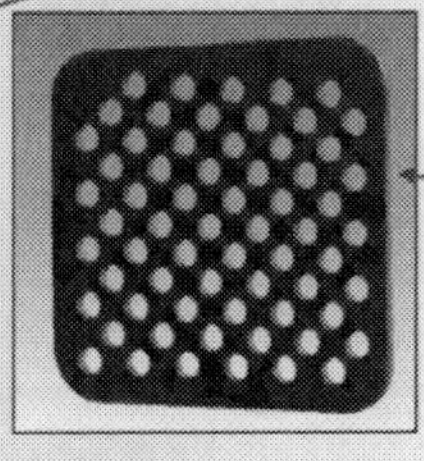

Von-Mises Stress (Hot Condition)

Fatigue Analysis

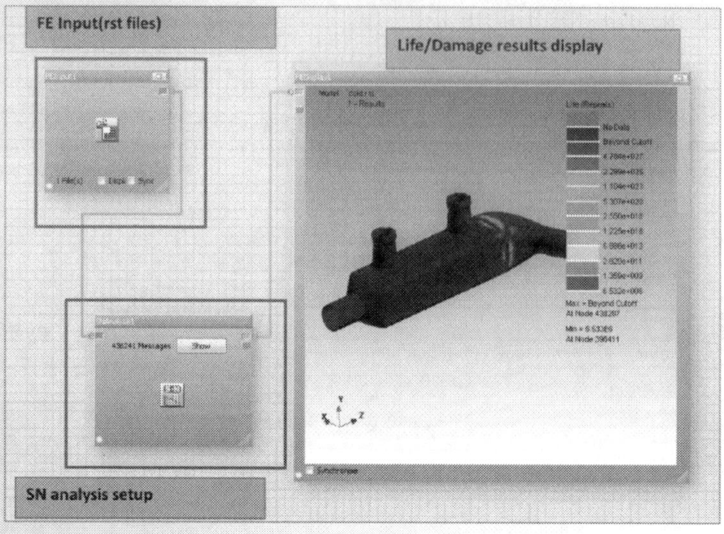

nCode DesignLife Project Schematic

2013 Automotive Simulation World Congress
October 29 - 30 • Frankfurt am Main, Germany

Setup

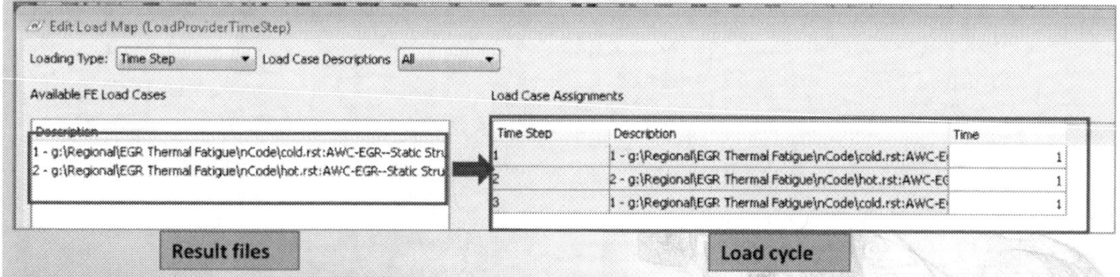

- The stress cycle for fatigue analysis is created in DesignLife itself using the stress states at the cold and the hot condition

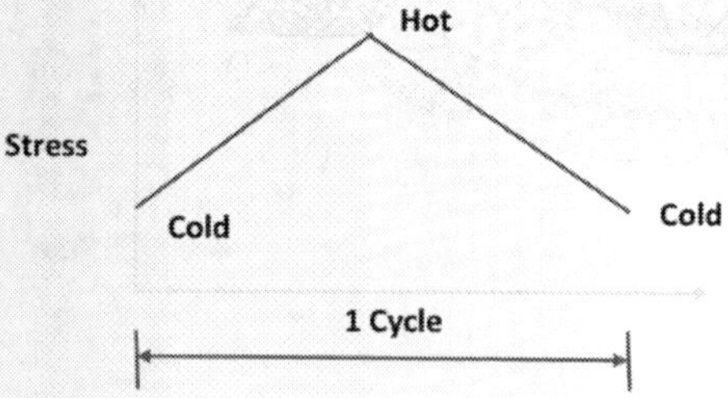

2013 Automotive Simulation World Congress
October 29 - 30 • Frankfurt am Main, Germany

Setup

SN Engine

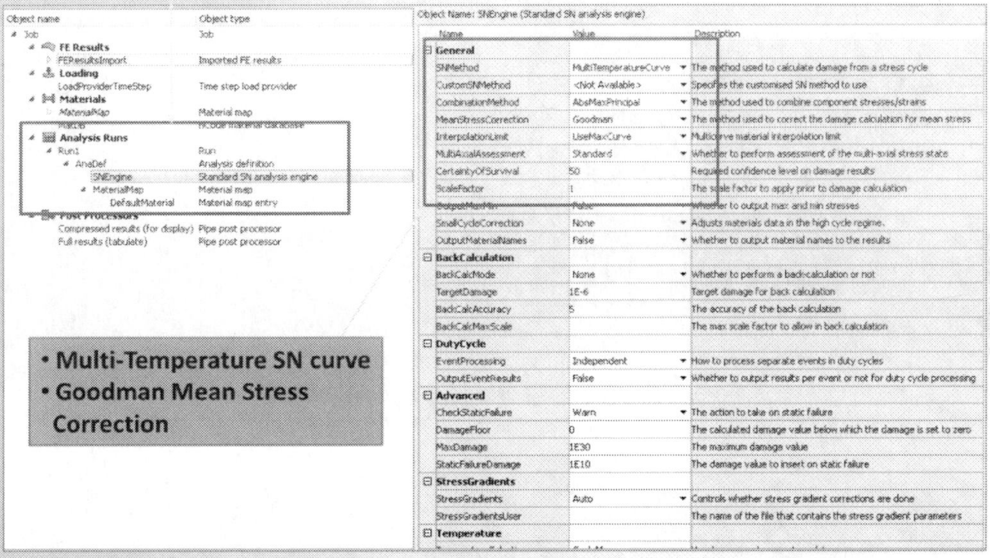

- Multi-Temperature SN curve
- Goodman Mean Stress Correction

2013 Automotive Simulation World Congress
October 29 - 30 • Frankfurt am Main, Germany

Results and Summary

- The minimum life in cycles 6.5 e6 is observed for the inlet plate
- Accuracy of the analysis can be increased by considering plasticity and EN (Strain) method for fatigue analysis for the areas where plasticity is significant.

Life (Repeats)

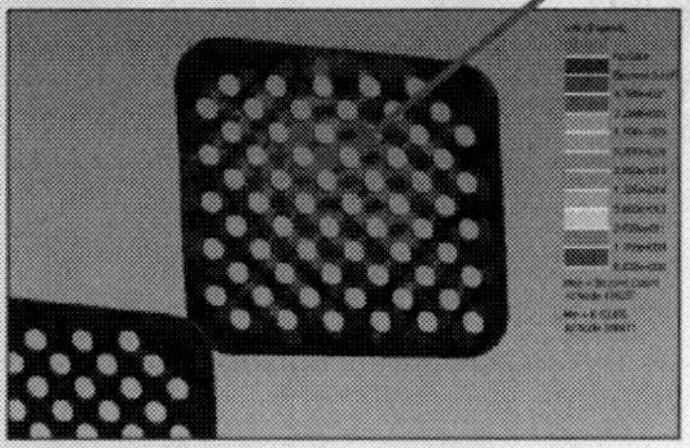

2013 Automotive Simulation World Congress

Designing Against Failure

- Thermal Loading
  - High Cycle Thermal Loading
  - Low Cycle Loading (Boiling)
- Other Mechanical Loading
  - Random Vibrations
  - Mode Shapes

Mode Shapes → **Pre-stressed Modal Analysis**

2013 Automotive Simulation World Congress
October 29 - 30 • Frankfurt am Main, Germany

Pre-stressed Modal Analysis

Frictionless Support

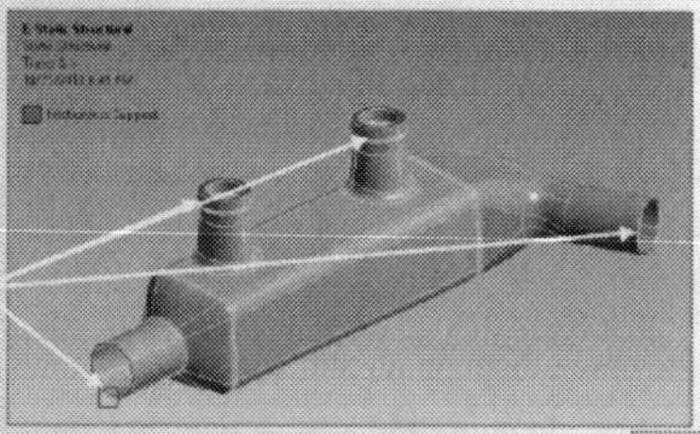

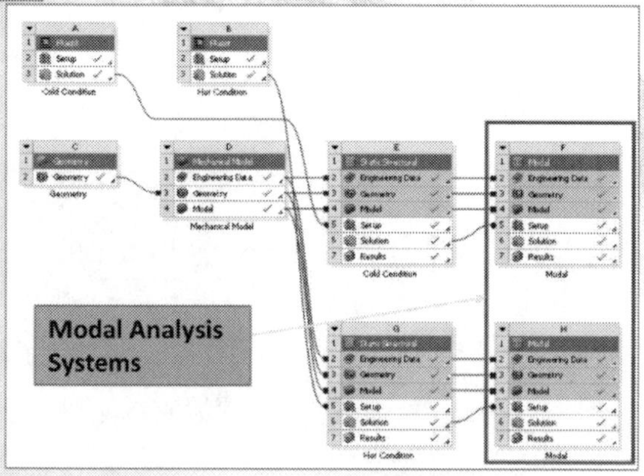

Results and Summary

- First 6 modes are extracted. The 1st mode is a global mode. 2nd to 6th are local inner pipe modes as seen from the mode shapes.
- The global mode frequency and mode shape is same for both the cold and hot condition.
- The local pipe modes and frequencies slightly differ depending upon the local stress state.

Frequencies (Cold)

Mode	Frequency [Hz]
1.	138.52
2.	337.7
3.	338.11
4.	338.71
5.	339.13
6.	339.82

Frequencies (Hot)

Mode	Frequency [Hz]
1.	138.51
2.	316.95
3.	318.43
4.	324.43
5.	326.48
6.	326.64

Cold Condition modes

Mode 1

Mode 2

Mode 3

Mode 4

2013 Automotive Simulation World Congress

Designing Against Failure

Thermal Loading

Other Mechanical Loading

High Cycle Thermal Loading

Low Cycle Loading (Boiling)

Random Vibrations

Mode Shapes

Boiling in EGR Coolers

Boiling in EGR Coolers

- Puking of "coolant" out of the degas bottle is an indicator of boiling in EGR coolers
- Boiling in EGR coolers can happen both due to poor design or due to loss of coolant flow
- This can lead of overheating of tubes which can result into mechanical failure

2013 Automotive Simulation World Congress
October 29 - 30 • Frankfurt am Main, Germany

Geometry and Boundary Condition

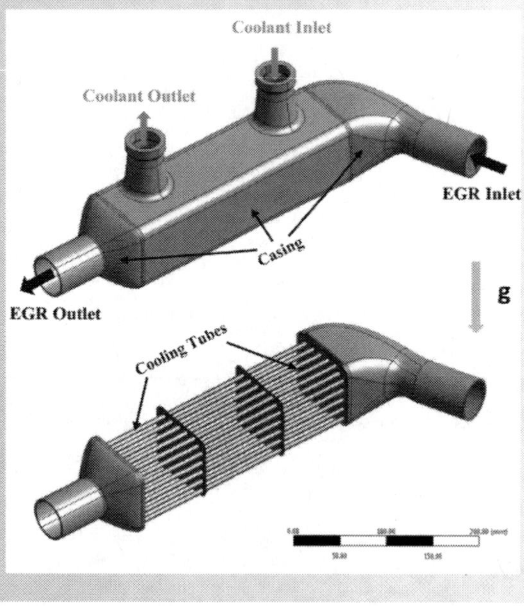

Coolant:
50-50 vol. ethylene glycol
Operating Pressure : 2 bar
Inlet mass flow rate = 4.3 Kg/s
Inlet Temperature = 363.15 K

Exhaust Gas:
Inlet mass flow rate = 0.25 kg/s
Inlet Temperature = 948.15

Outer casing:
Exposed to ambient
h = 25 W/m-K
T_inf = 313.15 K

2013 Automotive Simulation World Congress
October 29 - 30 • Frankfurt am Main, Germany

Semi-Mechanistic Boiling Model

- One equation nucleate boiing model based on Chen's correlation
- Fast and accurate
- Works with mixture-multiphase model
- Valid at lower pressure levels and for both horizontal and vertical channels

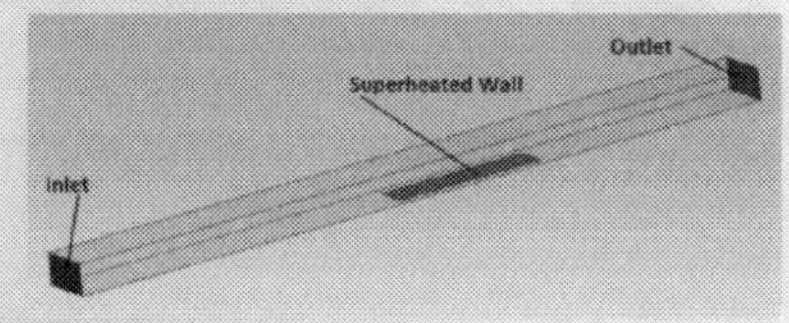

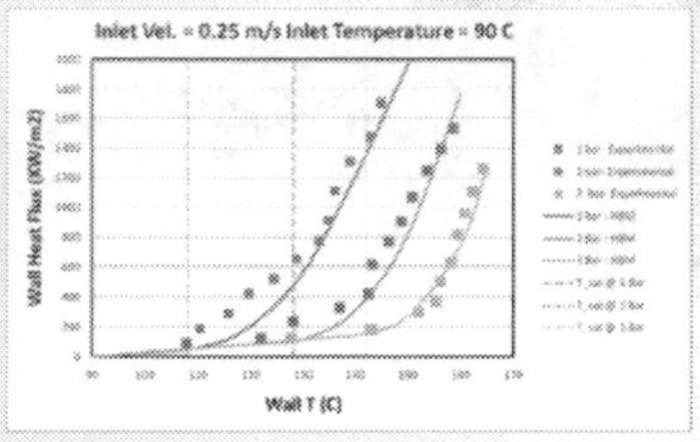

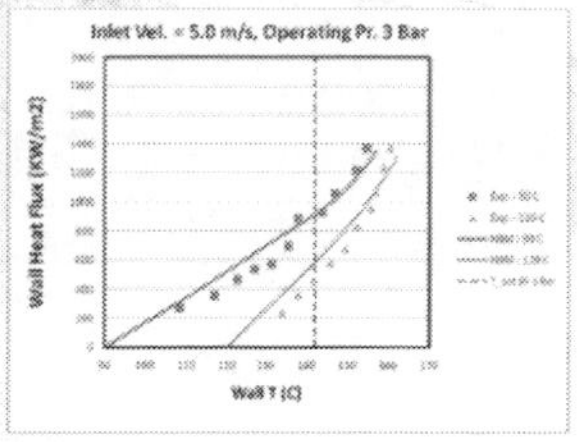

Robinson, K., Hawley, J. G. and Campbell, N. A. F., 2003. Experimental and modeling aspects of flow boiling heat transfer for application to internal combustion engines. Proceedings of the Institution of Mechanical Engineers, Part D: Journal of Automobile Engineering, 217 (10), pp. 877-889.

2013 Automotive Simulation World Congress
October 29 - 30 • Frankfurt am Main, Germany

Results : Temperature

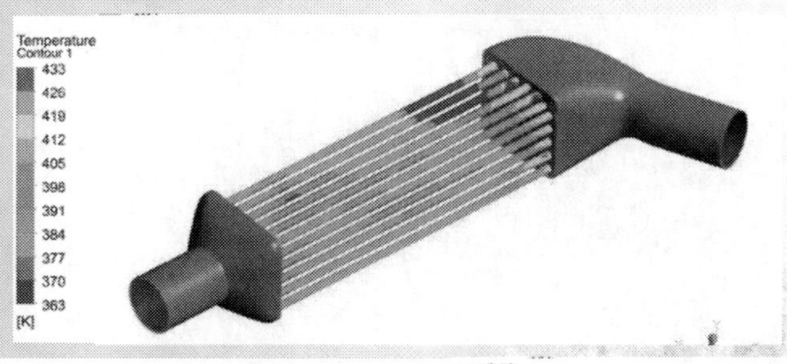

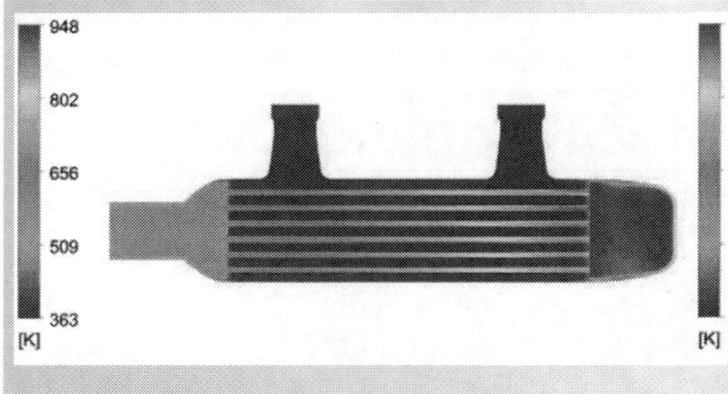

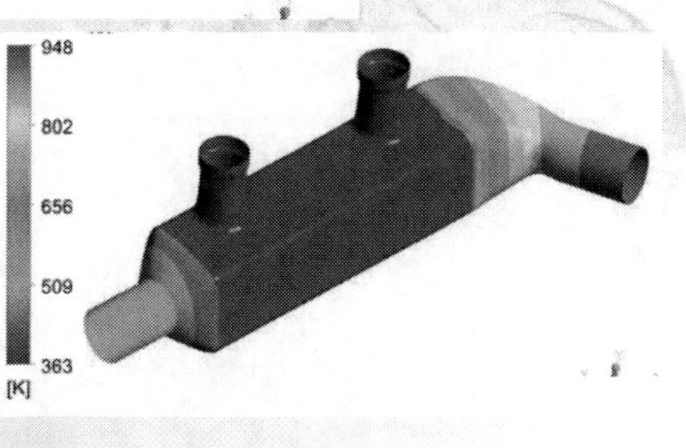

2013 Automotive Simulation World Congress
October 29 - 30 • Frankfurt am Main, Germany

Results : Vapor Fraction

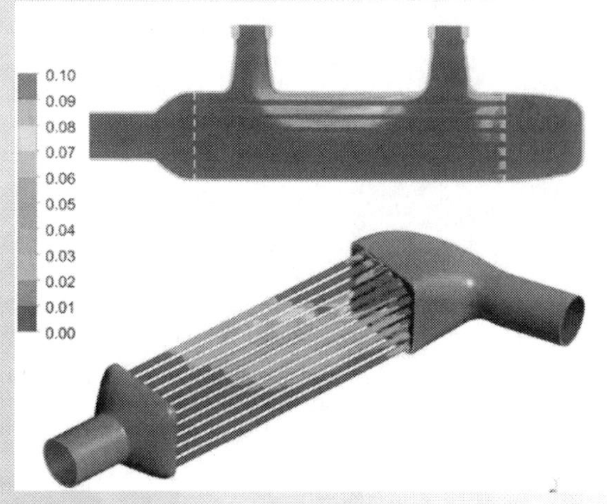

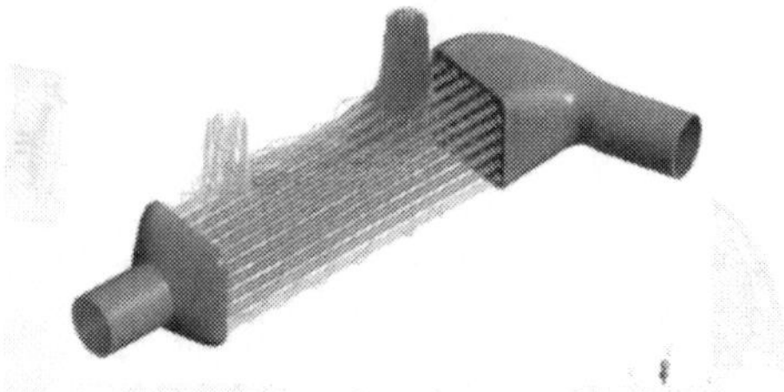

Stream Lines of coolant flow

2013 Automotive Simulation World Congress

Improving Performance

Condensation | Fouling

Condensation in EGR Coolers/CAC

Condensation in Air Coolers

- At lower operating temperature conditions (EGR coolers and CACs) result in condensation of water vapor in the exhaust gas/air
- Condensation in EGR coolers/CACs can lead to two issues
  - In CACs in particular, condensate film can flow towards the engine and get inside it
  - In EGR coolers, condensate reacts with exhaust gases to form sulphuric acid (fouling) which causes corrosion

Geometry and Boundary Condition

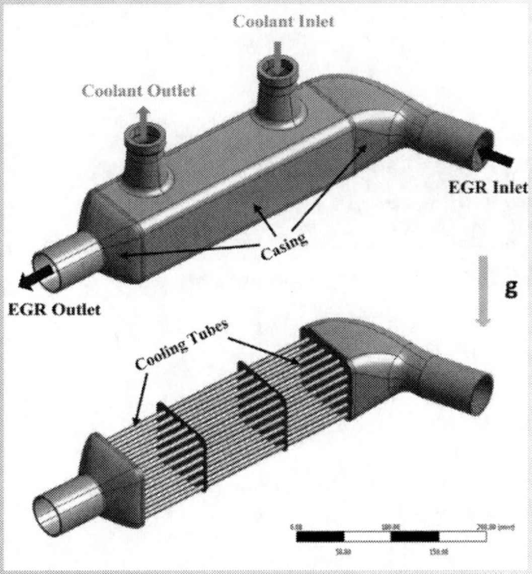

Coolant:
50-50 vol. ethylene glycol
Operating Pressure : 2.2 bar
Inlet mass flow rate = 4.3 Kg/s
Inlet Temperature = 313.15 K

Exhaust Gas:
Inlet mass flow rate = 0.042 kg/s
Inlet Temperature = 533.15

Outer casing:
Exposed to ambient
h = 5 W/m-K
T_inf = 300 K

2013 Automotive Simulation World Congress
October 29 - 30 • Frankfurt am Main, Germany

Eulerian Wall Film Model

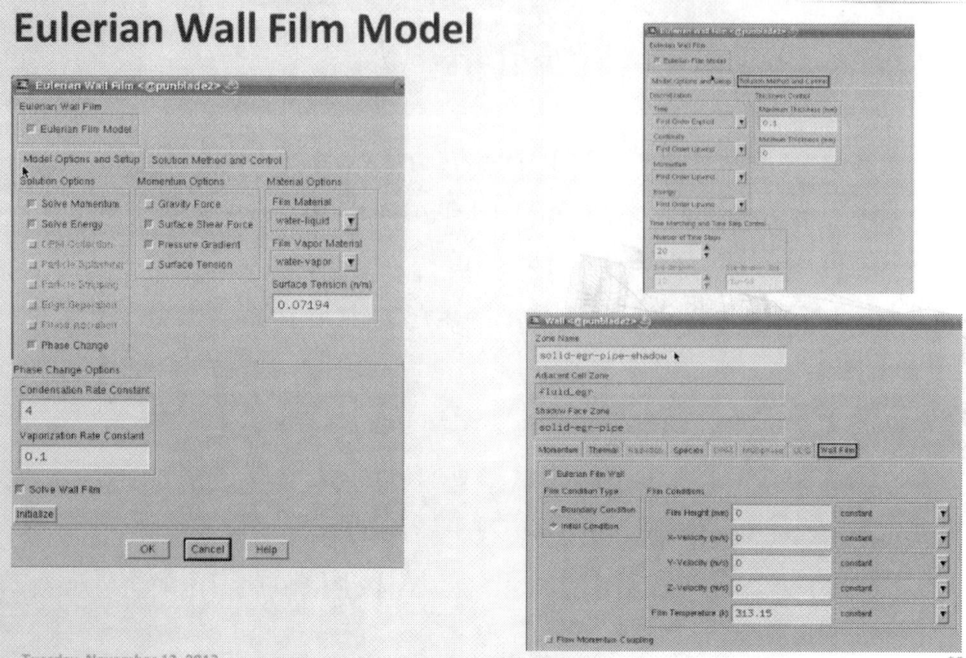

2013 Automotive Simulation World Congress
October 29 - 30 • Frankfurt am Main, Germany

Condensation Pattern

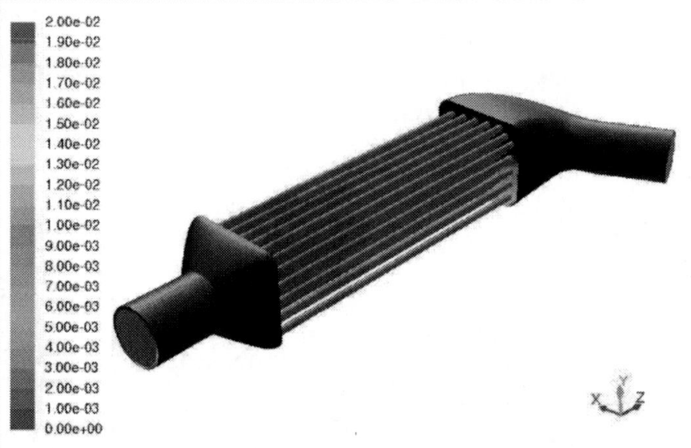

Contours of Film Thickness (mm) (Time=3.0000e-02) Apr 25, 2013
ANSYS Fluent 14.5 (3d, dp, pbns, spe, ske, transient)

2013 Automotive Simulation World Congress

Improving Performance

Condensation | Fouling

Fouling in EGR Coolers

Fouling in EGR Coolers

Courtesy: John.Hoard.UM.COE.PS21.2012.Presentation.pdf

Summary

- **Optimizing Heat Exchange**
  - Optimizing Inlet Diffuser
    - Shape Optimization
    - Topological Optimization
  - Tube Bundle Optimization
    - Parametric Optimization of Tube Shapes
    - Optimum Tube Arrangement
- **Designing Against Failure**
  - Thermal Loading
    - High Cycle Thermal Loading
    - Low Cycle Loading (Boiling)
  - Other Mechanical Loading
    - Random Vibrations
    - Mode Shapes
- **Improving Performance**
  - Condensation
  - Fouling

Team

- CAD model creation and meshing – **Sudesh Palase**
- Shape optimization using RBF morph – **Ashok Khondge**
- Thin sweep meshing process in Fluent Meshing – **Aditya Kshatriya**
- Condensation in CAC/EGR – **Saurish Das, Seyed Hosseini**
- Boiling in EGR Cooler – **Saurish Das, Hemant Punekar**
- CHT analysis of EGR Cooler – **Sourabh Shrivastava**
- Thermal Stress and Fatigue Analysis – **Chetan Patil, Ashutosh Shrivastava, Santosh Kottalgi**
- Parametric optimization of EGR cooler with corrugated tubes – **Sudesh Palase, Jayesh Mutyal**
- Discussions – **Laz Foley**

2013 Automotive Simulation World Congress
October 29 - 30 • Frankfurt am Main, Germany

References

- Cooled EGR - A Key Technology for Future Efficient HD Diesels, SAE Technical Paper Series, 980190
- *Effects of the Internal Shape of EGR Cooler on Heat Exchanger Efficiencies,* SAE Technical Paper 2007-01-1252, 2007
- *CFD Optimization of an EGR Cooler for Heavy-Duty Diesel Engines*, SAE Technical Paper 2001-01-1755, 2001
- *Transient CFD Simulation of Exhaust Gas Recirculation Coolers for Further Structural Analyses*, SAE Technical Paper 2009-01-1228, 2009
- CFD Numerical Simulation of HP-EGR Cooler Performance Under Pulsating Engine Conditions, SAE International 2011-26-0073, 2011
- http://www1.eere.energy.gov/vehiclesandfuels/pdfs/deer_2010/wednesday/presentations/deer10_styles.pdf
- http://www.senior-flexonics.com/EGRCoolers/EGRCoolers.html
- http://www.swri.org/3pubs/brochure/d03/EGR/EGRFlyer.pdf
- http://www.senior-flexonics.com/EGRCoolers/EGRCoolers.html

Leveraging SGI's UV™ and VizServer® Platform to Enable Multiple CAE Users to Run ANSYS® Software with Remote Visualization in a Data Center

Tony DeVarco
Director, Manufacturing Solutions
tdevarco@sgi.com

Customers Challenges and Needs

CHALLENGES:

- Maintaining multiple geographically distributed CAE engineering groups.
- Engineering visualization of large models requires high cost workstations.
- Remote workstations can't interact with the compute resources and data in the data center.
- Remote collaboration requires transfer of large data sets.

NEEDS:

- Consolidate hardware, applications, and data into centralized data centers.
- Replace workstations with low-cost remote clients.
- Enable remote collaboration.

Typical HPC Scenario

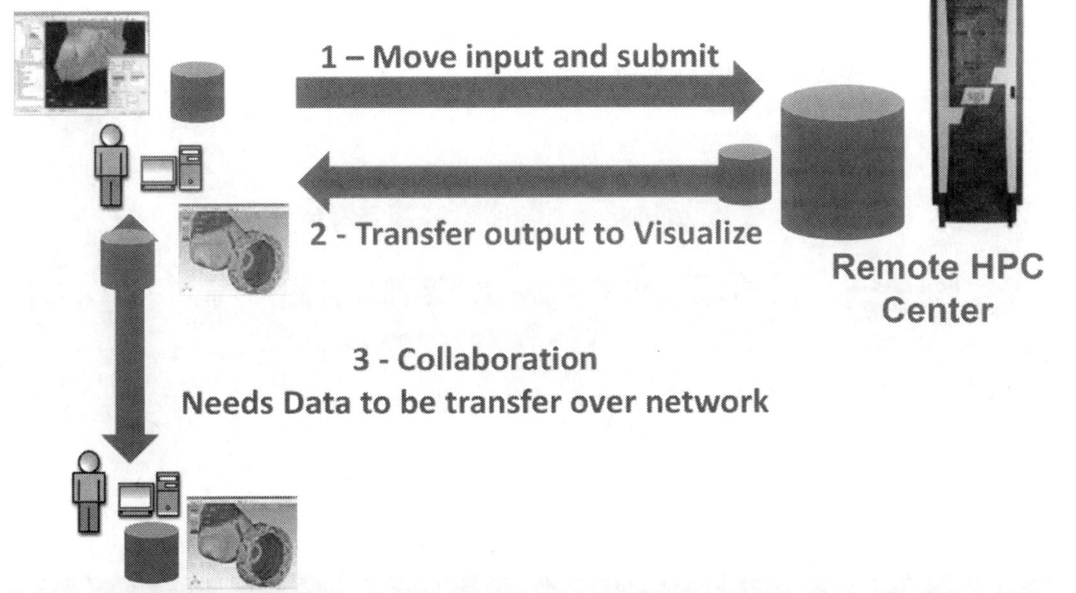

Problems With This Model

NETWORK
Network overload leading to poor performance and response times all round

COST
Expensive, dedicated workstations (GPU, memory, ...) with short lifecycle

IT MANAGEMENT
Support, update and replace tens to tens of thousands of workstations

WORKSTATION SIZING
Workstations have to be sized for the largest expected models

SECURITY
Moving sensitive data around (in/out organization) is always risky

WORKFORCE WORKING METHODS
Current models do not support a diverse mobile workforce

SGI® VizServer® with NICE Desktop Cloud Visualization™ (DCV) and EnginFrame™ Views Software

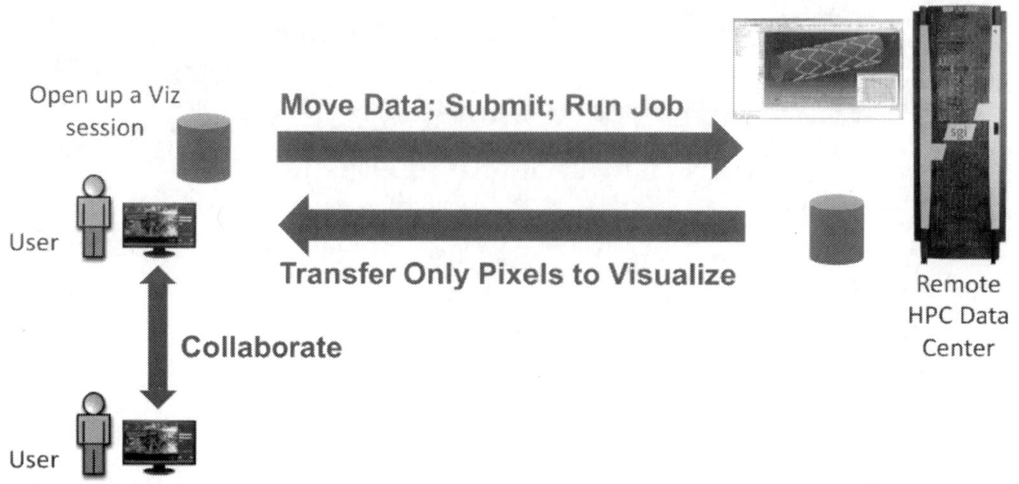

What the Solution Solves

NETWORK
Network is no longer a bottleneck and data loads faster

COST
Centralized & Shared servers are less expensive to buy & manage

IT MANAGEMENT
Support, update and replacement are more efficient & do not affect users

WORKSTATION SIZING
Resources are dynamically sized based on users needs

SECURITY
Sensitive data remain within protected data center with full access control

WORKFORCE
Users can virtually connect & collaborate from anywhere with any client

CAE Enterprise Solution Example

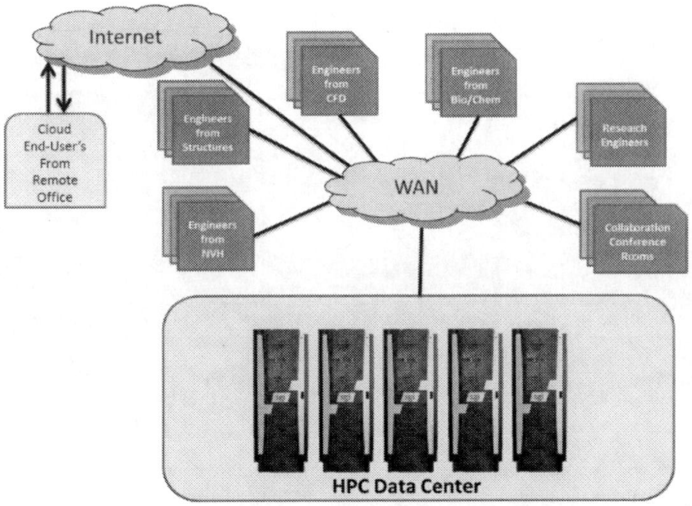

SGI® VizServer® with DCV and EnginFrame Views Software Stack

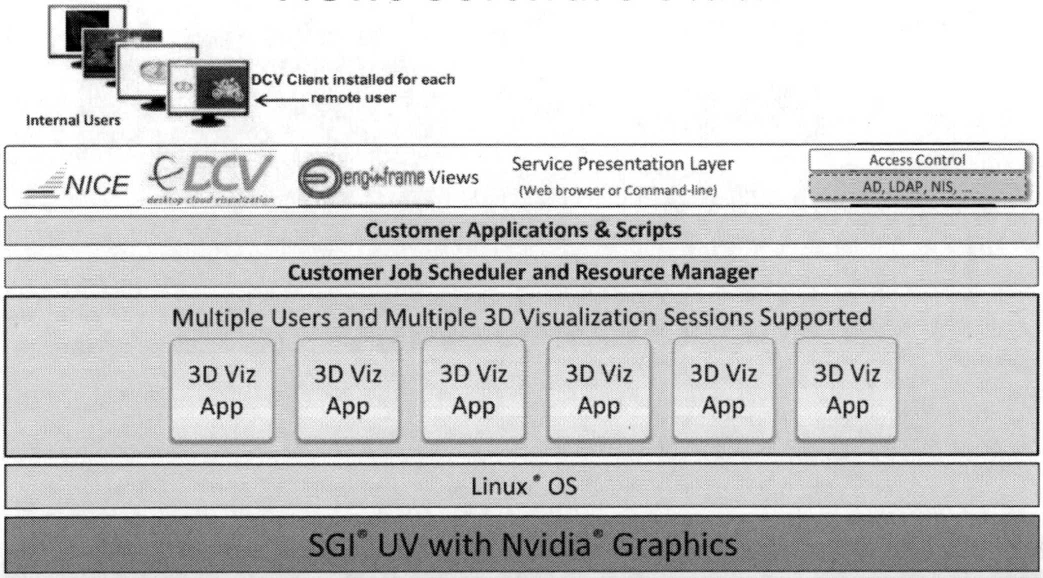

Characterization of ANSYS Applications

Typical HPC Behavior of a Single Job for Various Applications

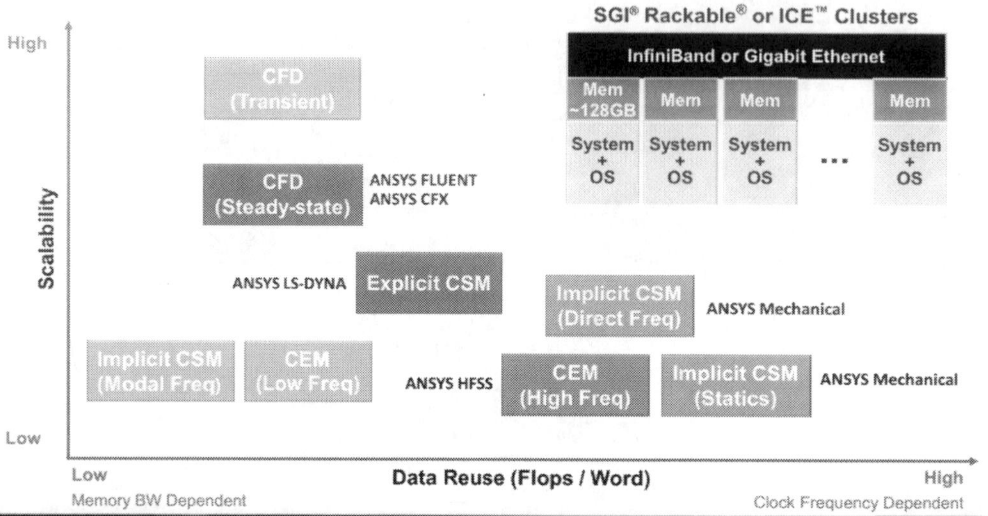

Characterization of ANSYS Applications

Typical HPC Behavior of a Single Job for Various Applications

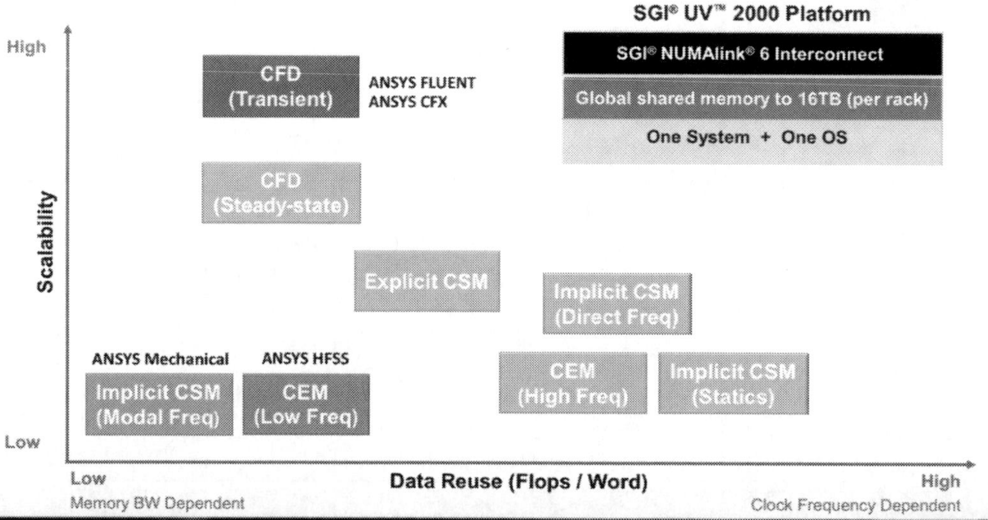

Characterization of ANSYS Applications

SGI UV can provide the environment for both

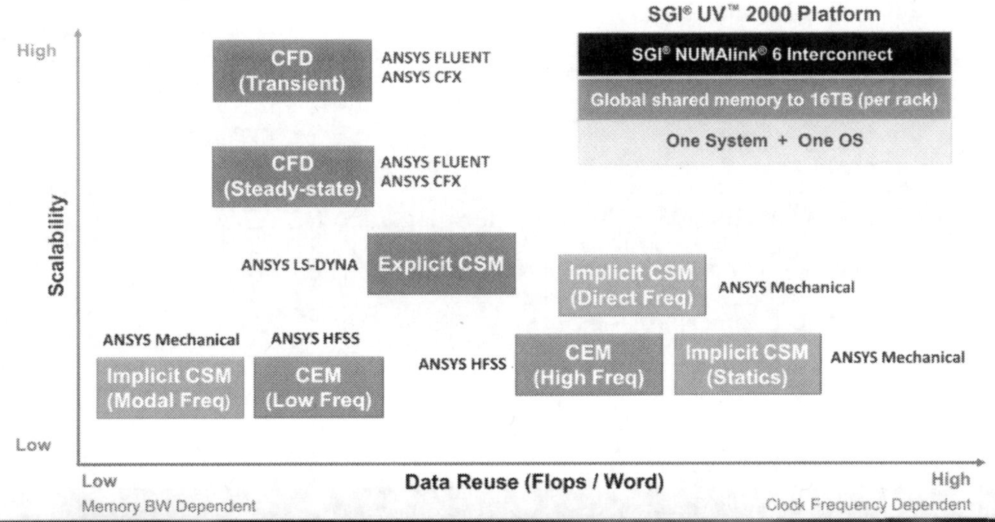

CAE Workflow Using VizServer and ANSYS Workbench

- CAD model creation
- Mesh-generation
- Model decomposition
- Running solvers
- Viewing results
- Adjust and repeat

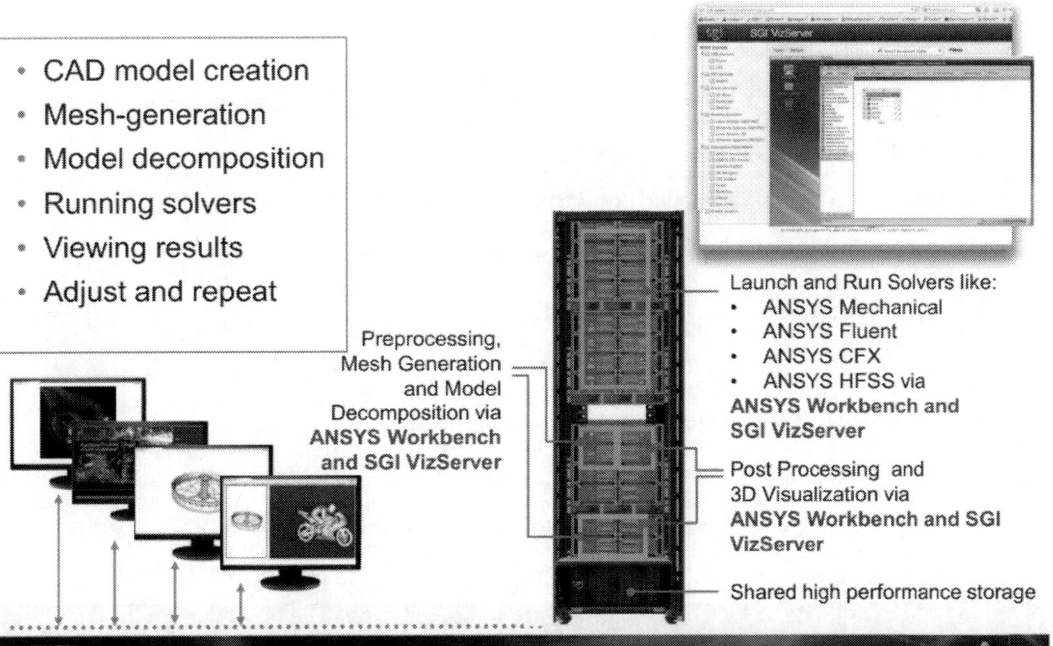

Multiple Users and Multiple Jobs

EnginFrame Views at a Glance

- User friendly Web based session access management
- Flexible service offering for end users
  - Full 2D / 3D desktop
  - Individual application
- Multiple remote display protocols support
  - TurboVNC
  - NICE Desktop Cloud Visualization (including RealVNC VE)
- Comprehensive scheduler-based load balancing policies
- Built-in collaboration capabilities
- Resource usage monitoring
- Comprehensive authentication options

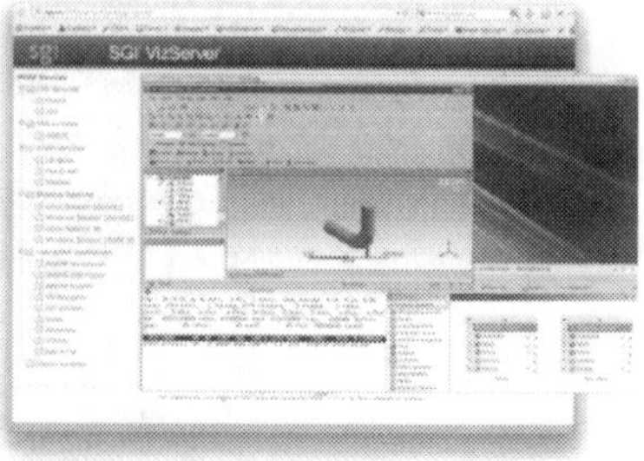

NICE

intel inside XEON sgi

NICE Desktop Cloud Visualization (DCV)

- Remote 3D visualization technology for OpenGL applications
- Broad OS and device support
  - Server:SUSE® Linux® Enterprise Server 11; Red Hat® Enterprise Linux 6
  - Client: Windows, Linux, Mac OS X, iOS, Android, applets, ...
- Support for **Virtualization**, with or without GPU pass-through
- **Bandwidth optimized** balancing of quality Vs. frame rate
- Encrypted communication, proxy option
- **Collaborative** working and visualization
- High-end GPU sharing across **multiple users, multiple OS**
- Validated and optimized for Technical Computing applications
- Bundled with NICE EnginFrame Views for session management
- SGI PS Services- Factory installation and remote or onsite testing

intel inside XEON sgi

Remote Visualization Session Management Tools

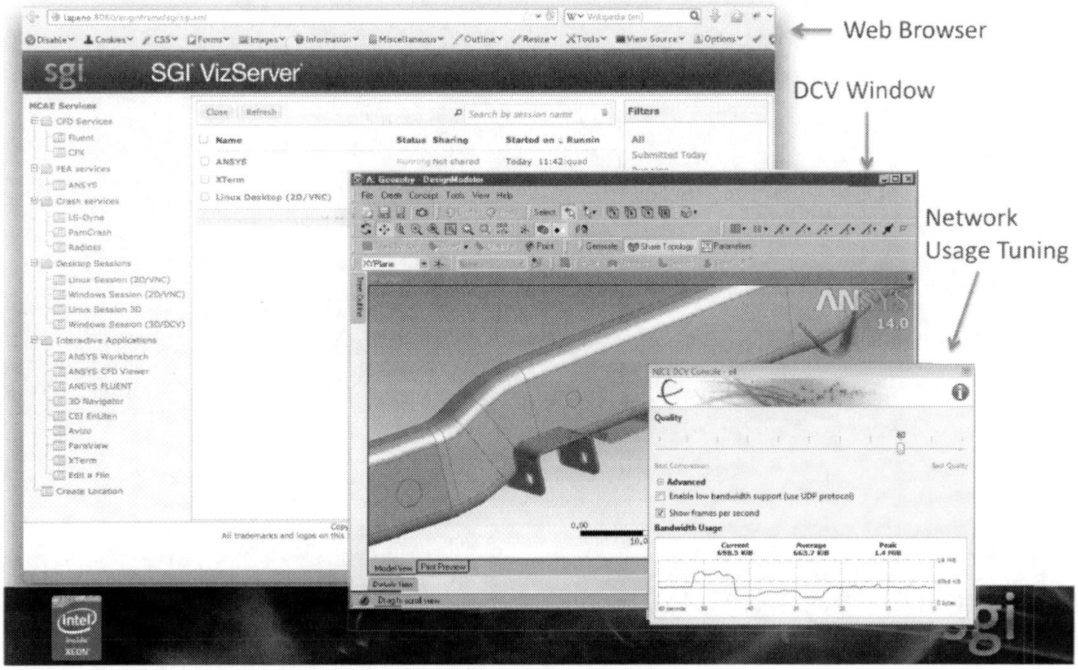

Web Browser

DCV Window

Network Usage Tuning

Built-in collaboration

CAE / HPC Workflow Integration with EnginFrame

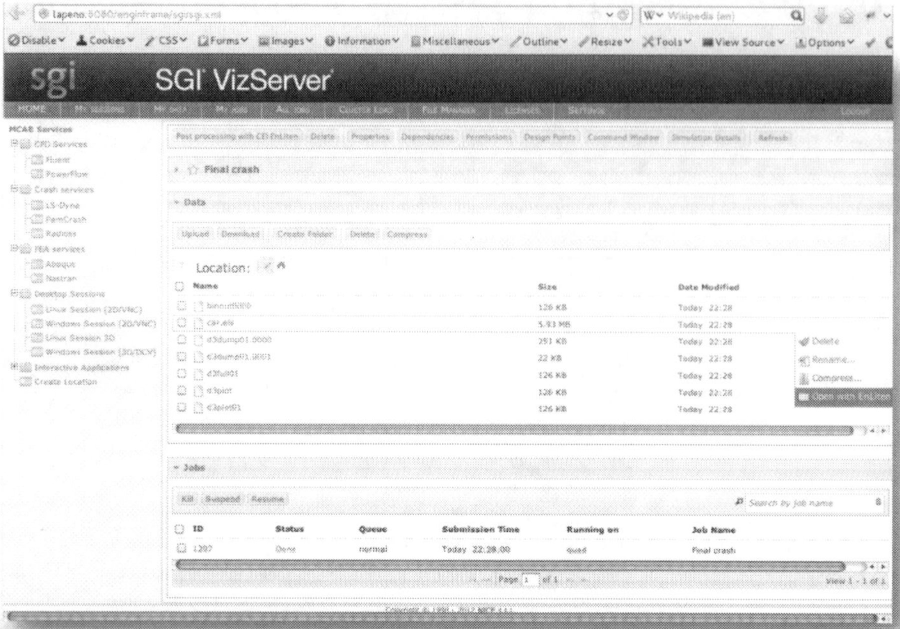

Linux Application Consolidation

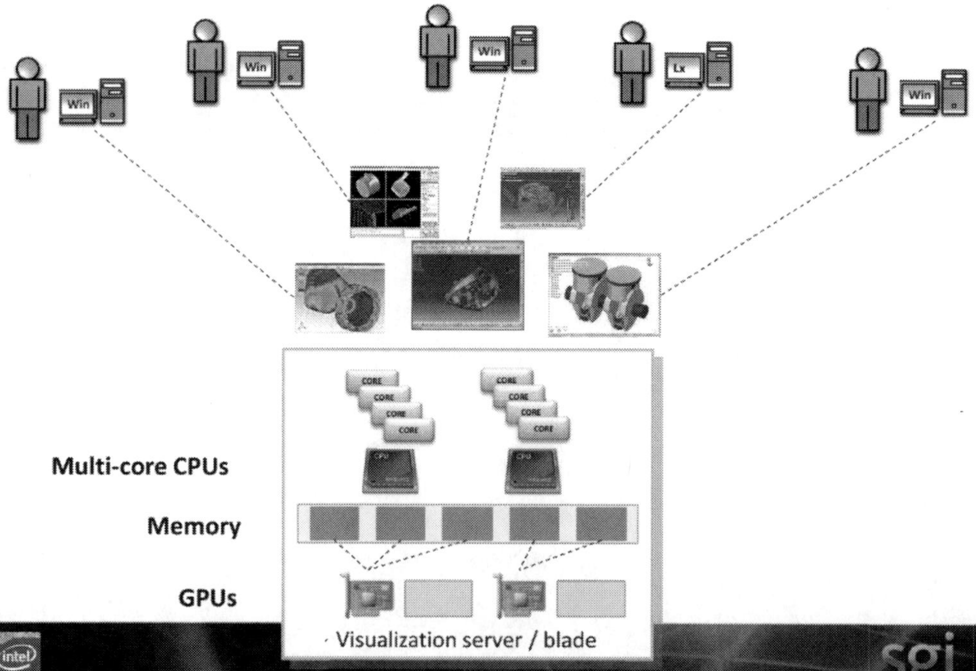

Windows Application Consolidation

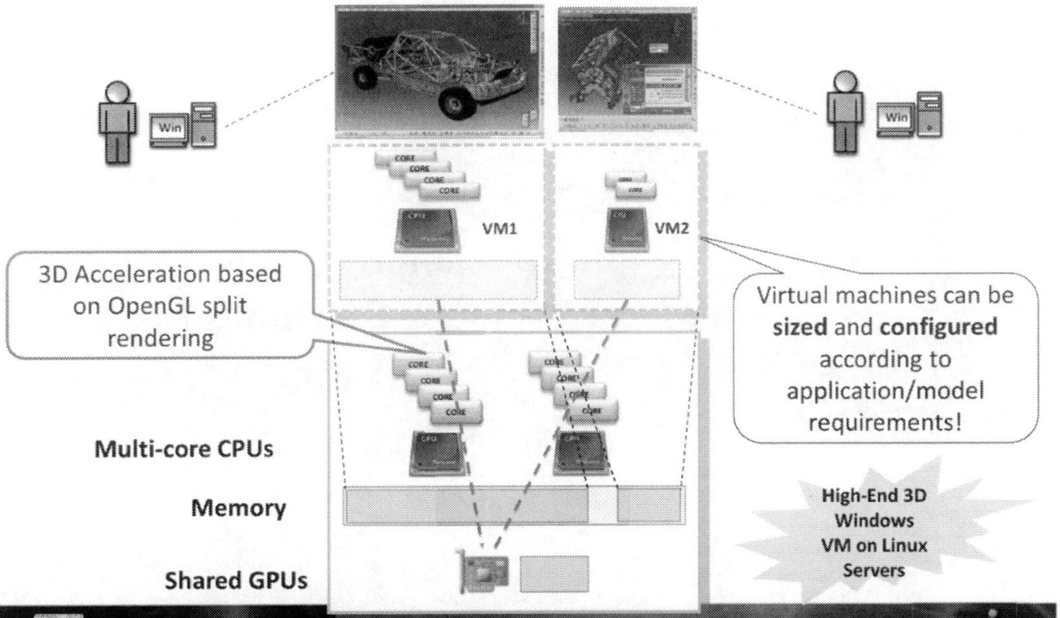

Questions

PIERBURG PUMP TECHNOLOGY

A Complete Loop of Numerical Analyses about a Variable Oil Pump

2013 Automotive Simulation World Congress

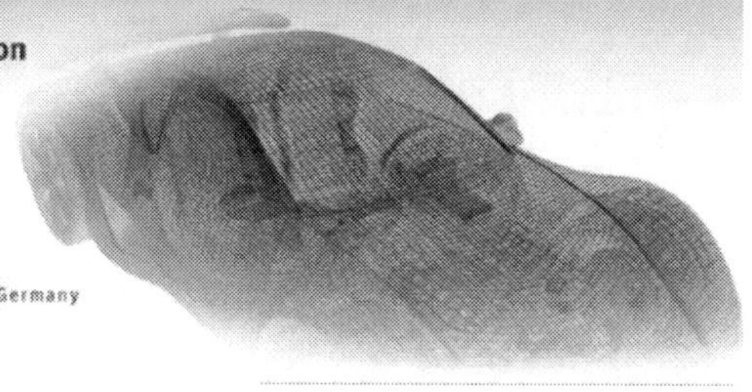

October 29 - 30 • Frankfurt am Main, Germany

KSPG Automotive

The Pierburg Pump Technology Company

- PPT is a division of the Kolbenschmidt Pierburg Group (KSPG Automotive)
- Worldwide leader in the design and in the production of oil pumps, water pumps and vacuum pumps for the automotive (light-duty & heavy-duty applications) market.

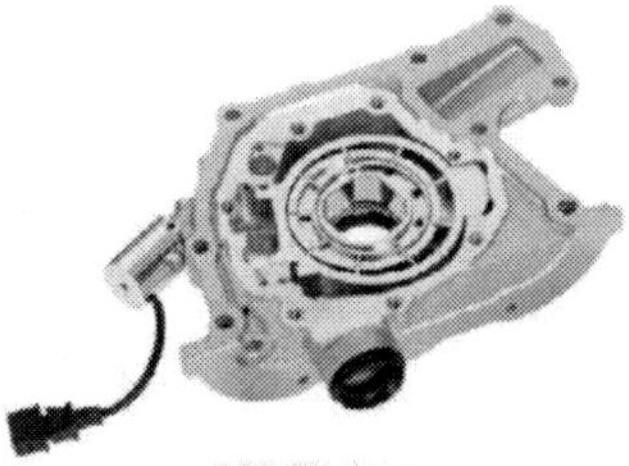

Oil Pump

Water Pump

Vacuum Pump

Pierburg Pump Technology – Engineering Centers

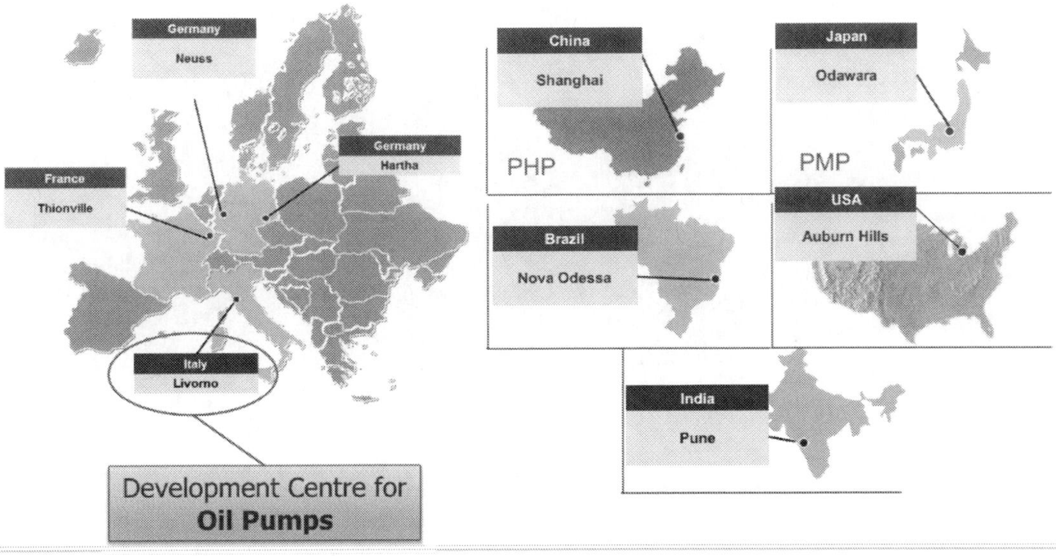

Pierburg Pump Technology – Some Customers

TATA FIAT OPEL GM

RENAULT NISSAN

Ford Ferrari

A driver for technological progress

CO_2 regulations worldwide

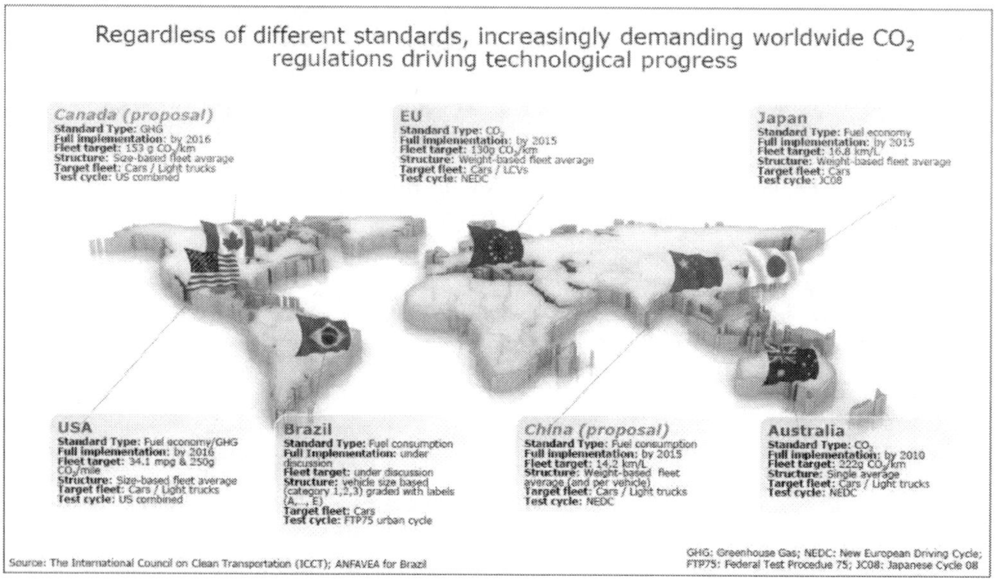

Variable Oil Pump – Advantage

- ✓ **Reduction** of the energy consumption: this is also valid for engines' accessories!
- ✓ **VOP**: an innovative concept of oil pump

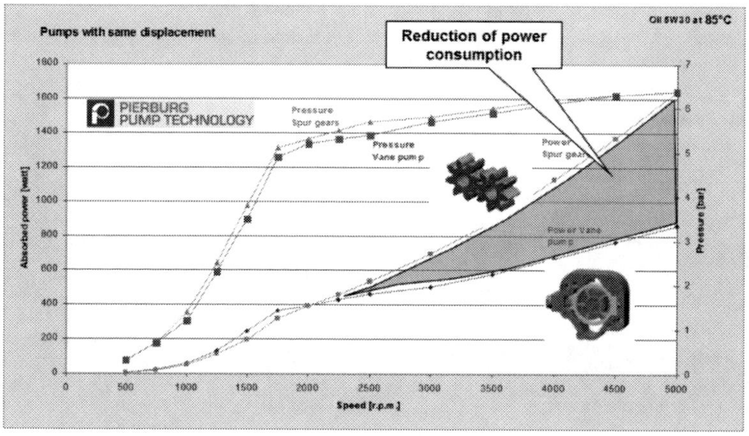

Variable Oil Pump – Advantage

- ✓ **Reduction** of the energy consumption: this is also valid for engines' accessories!
- ✓ **VOP**: an innovative concept of oil pump

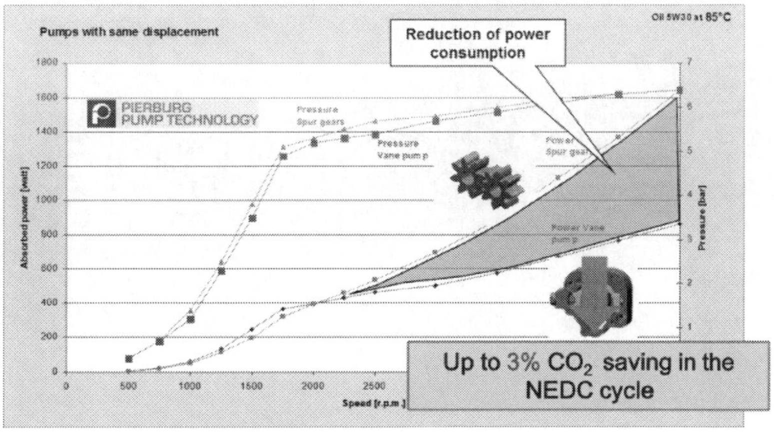

Variable Oil Pump – Conventional Type

Features:

- Vane pump.
- Displacement controlled by the linear (or pivoting) movement of the control ring driven by the pressure signal.
- Continuous control of the volume of the working chamber.
- Simple design, few components.
- Pressure working directly on the volume control system.

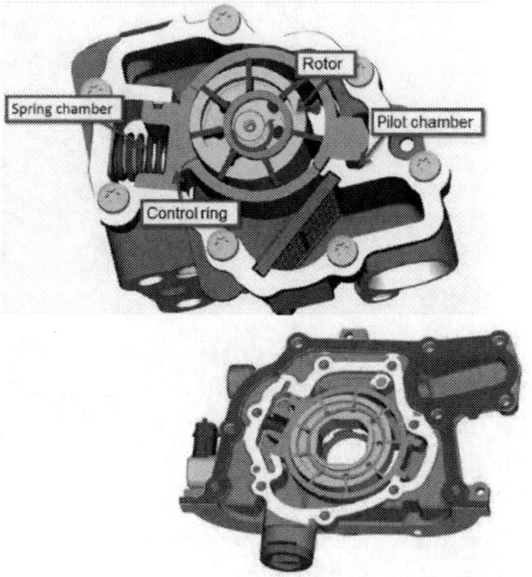

Variable Oil Pump – Functionality

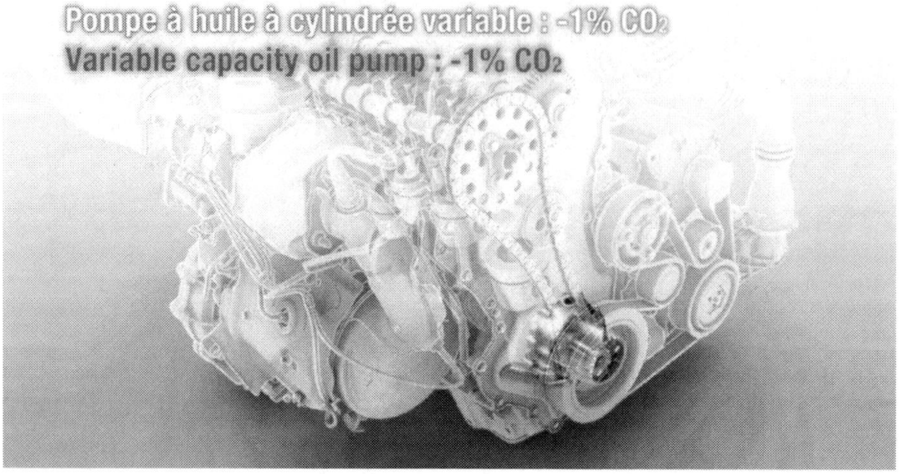

Variable Oil Pump – Genesis of a Product

Customer SOR

- Oil Flow Rate Requirement
- Oil Pressure Requirement
- The **Minimum** Absorbed Energy

Variable Oil Pump – Genesis of a Product

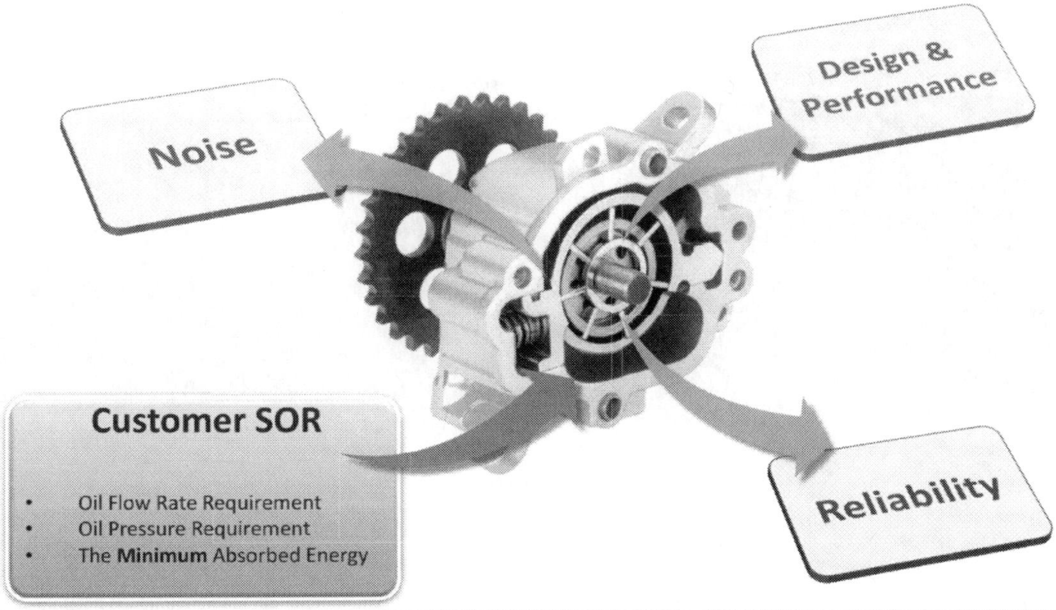

A Complete Loop of Numerical Analyses about a Variable Oil Pump PIERBURG PUMP TECHNOLOGY

Variable Oil Pump – Pre-Design

Customer Inputs

- **Oil Flow Rate Requirement**
- **Engine Idle Speed**
- **Max Engine Speed**
- **Transmission Ratio**

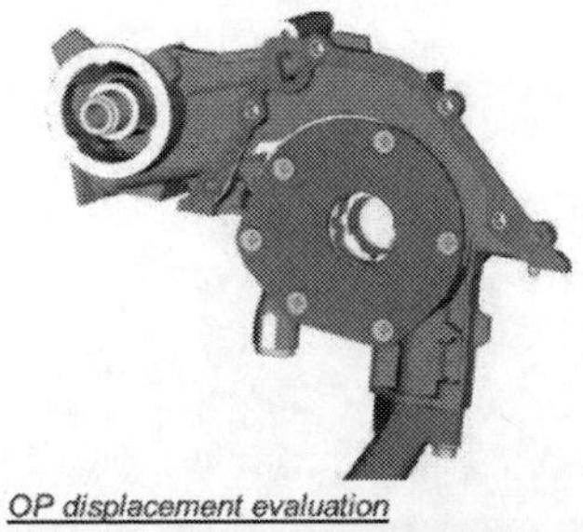

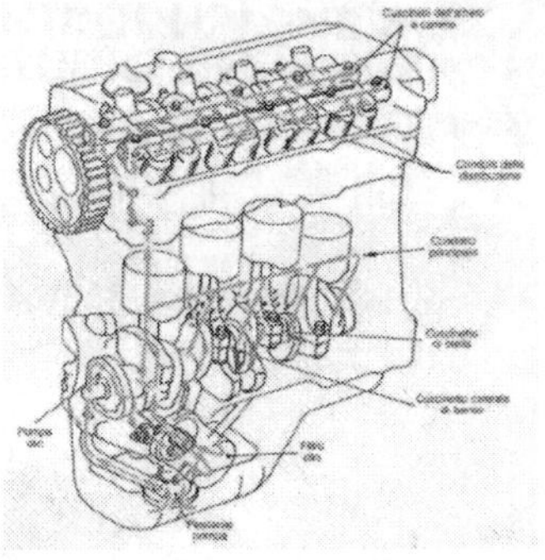

OP displacement evaluation

[illegible]equested delivery flow rate	Q	[L/min]	2.50	12.00
Engine speed	n_{engine}	[illegible]	600	1500
Engine crankshaft sprocket teeth	Z_{engine}		33	19
Pump shaft sprocket teeth	Z_{pump}		[illegible]	21
[illegible]x engine speed	[illegible]	[illegible]	7000	10000
T[illegible]nsmission ratio	$\tau = n_{pump} / n$[illegible]		1.650	0.905
Pu[illegible]p speed	n_{pump}	[rpm]	990	[illegible]357
Volu[illegible]tric efficienc[illegible] (or pump)			99%	98%
Max d[illegible]cay of performance [illegible]	η_p		1%	1%
Total pu[illegible]p efficiency	$\eta = \eta_v/(1+\eta_p)$		[illegible]	[illegible]7.0%
Max pum[illegible] speed	Max n[illegible]	[illegible]	11550	9048
Required t[illegible]oretical displacement	C	[cm³/rev]	2.53	8.84
Real needed required displacement	$C_{theoretical}$	[cm³/rev]	**2.58**	**9.11**

Calculation of the Oil Pump Displacement

Variable Oil Pump – Design Loop

OP displacement evaluation

Requested delivery flow rate	Q	[L/min]	2.50	12.00
Engine speed	n_{engine}	[rpm]	600	1500
Engine crankshaft sprocket teeth	Z_{engine}		33	19
Pump shaft sprocket teeth	Z_{pump}		[illegible]	21
Max engine speed	[illegible]	[illegible]	7000	10000
Transmission ratio	[illegible]		1.650	0.905
Pump speed	n_{pump}	[rpm]	990	1357
Volumetric efficiency [illegible] pump	[illegible]		[illegible]	[illegible]
Max decay of performance [illegible]	%		1%	1%
Total pump efficiency	$\eta = \eta_v(1-m_d)$		[illegible]	[illegible]
Max pump speed	[illegible]	[illegible]	11550	9048
Required theoretical displacement	C	[cm³/rev]	2.53	8.84
Real needed required displacement	$C_{theoretical}$	[cm³/rev]	2.58	9.11

Calculation of the Oil Pump Displacement

Variable Oil Pump – Design Loop

OP displacement evaluation

Requested delivery flow rate	Q	[L/min]	2.50	12.00
Engine speed	n_{engine}	[rpm]	600	1500
Engine crankshaft sprocket teeth	Z_{engine}		33	19
Pump shaft sprocket teeth	Z_{pump}		[illegible]	21
Max engine speed	[illegible]	[illegible]	7000	10000
Transmission ratio	[illegible]		1.650	0.905
Pump speed	n_{pump}	[rpm]	990	1357
Volumetric efficiency [illegible] pump	[illegible]		[illegible]	[illegible]
Max decay of performance [illegible]	%		1%	1%
Total pump efficiency	$\eta = \eta_v(1-m_d)$		[illegible]	[illegible]
Max pump speed	[illegible]	[illegible]	11550	9048
Required theoretical displacement	C	[cm³/rev]	2.53	8.84
Real needed required displacement	$C_{theoretical}$	[cm³/rev]	2.58	9.11

Calculation of the Oil Pump Displacement

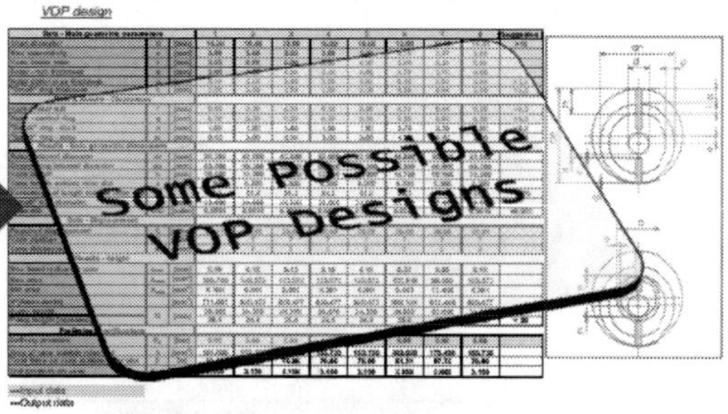

Variable Oil Pump – Design Loop

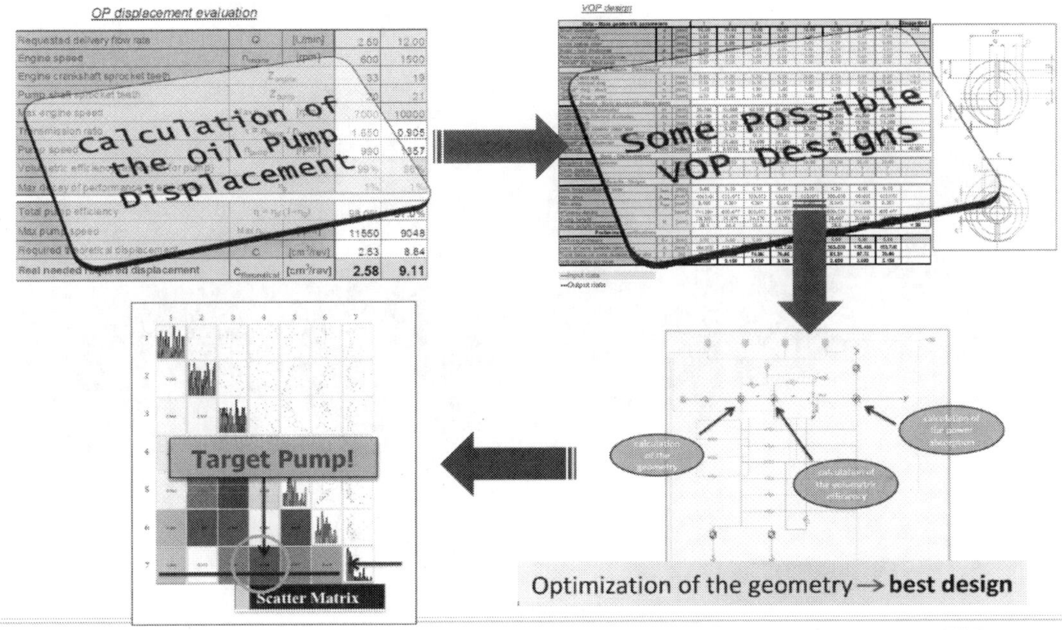

Optimization of the geometry → **best design**

Variable Oil Pump – Performance

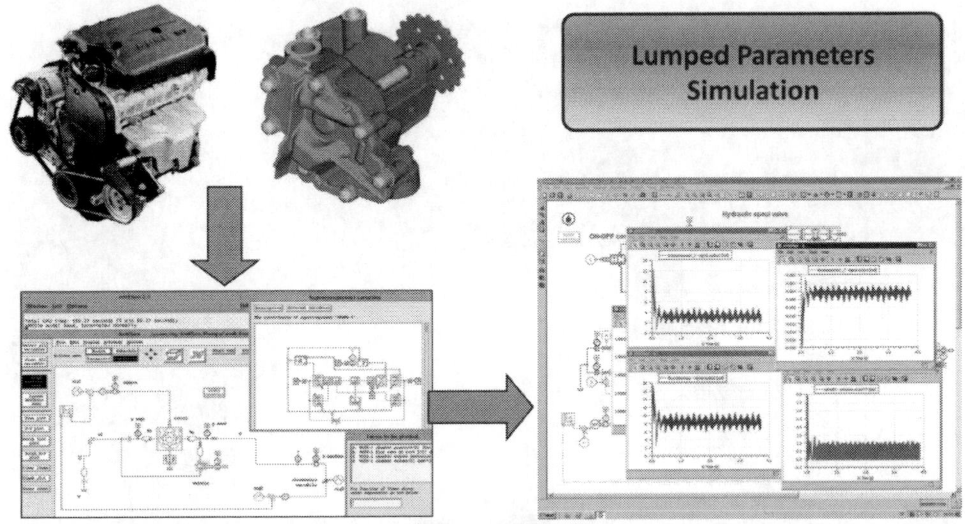

Equivalent hydraulic circuit build wi[redacted] **custom** & standard sub-models.

[redacted]**utput** results: instantaneous pressure - flow - torque values in different pump areas

Variable Oil Pump – Performance

From 1D to 3D Simulation → **CFD Analyses**

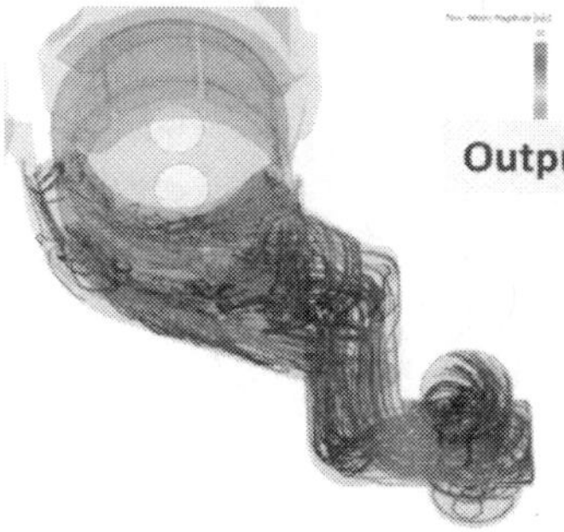

Output result: flow speed streamlines

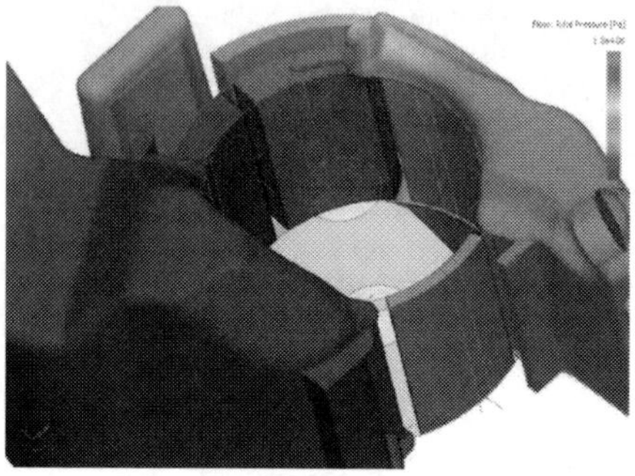

Output result: pressure distribution

A Complete Loop of Numerical Analyses about a Variable Oil Pump PIERBURG PUMP TECHNOLOGY

Variable Oil Pump – Performance

CFD Multi-Phase Approach

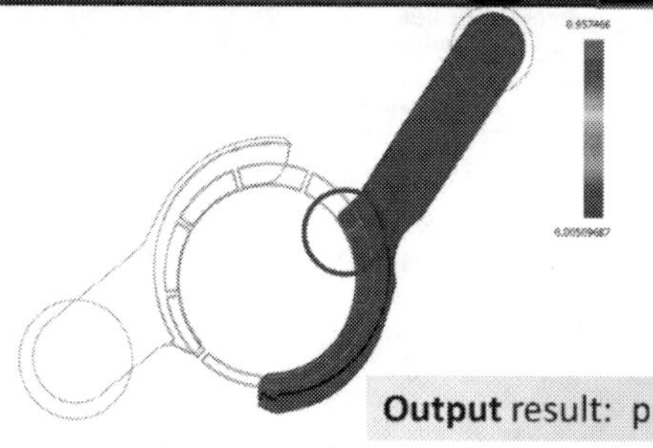

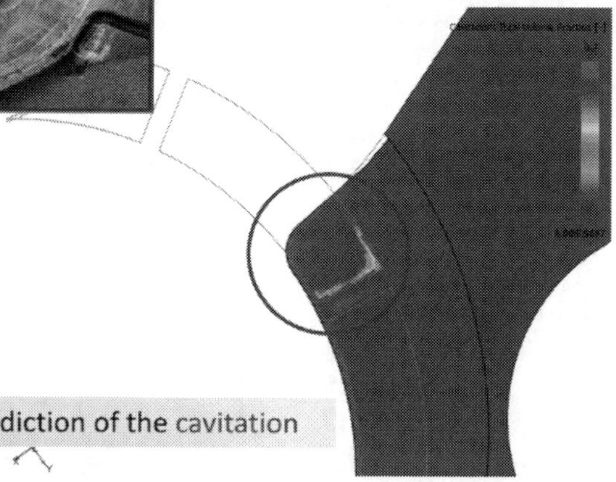

Output result: prediction of the cavitation

A Complete Loop of Numerical Analyses about a Variable Oil Pump

PIERBURG PUMP TECHNOLOGY

Variable Oil Pump – Reliability

Multi-Body Analyses

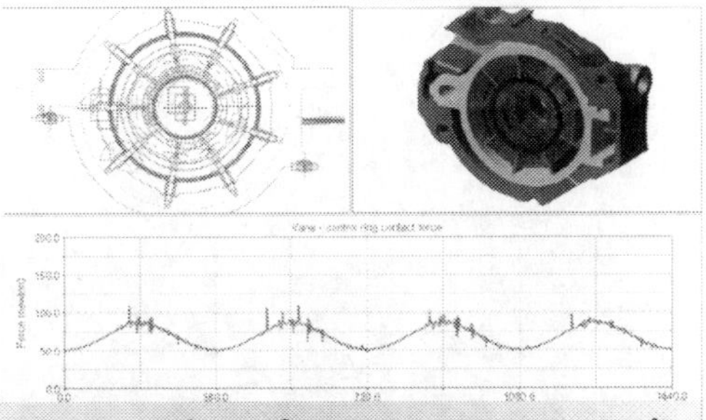

Output results: **forces, torques, speeds, accelerations.**

FEA Structural Analyses

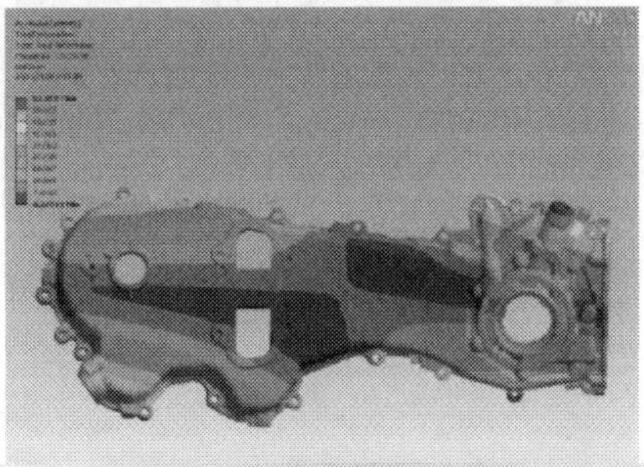

Output results: **stresses, deformations, hertzian pressures, fatigue lifetime.**

A Complete Loop of Numerical Analyses about a Variable Oil Pump

PIERBURG PUMP TECHNOLOGY

Variable Oil Pump – Reliability

An Example of Study: Shape Tolerance

The internal diameter of the control Ring

Deformed shape1 **Process-A**

Deformed shape2 **Process-B**

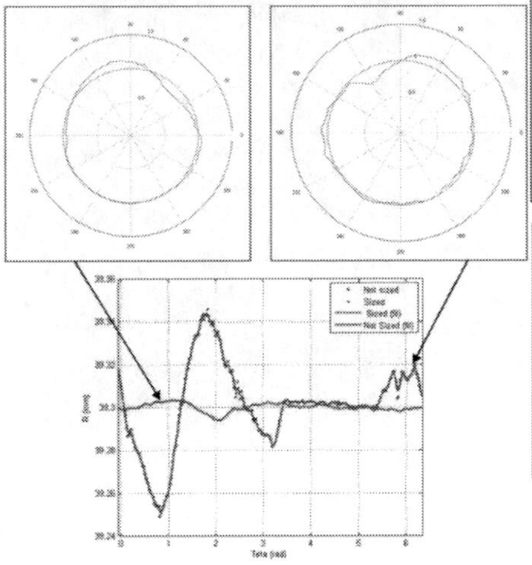

Multi-Body Analyses

Results

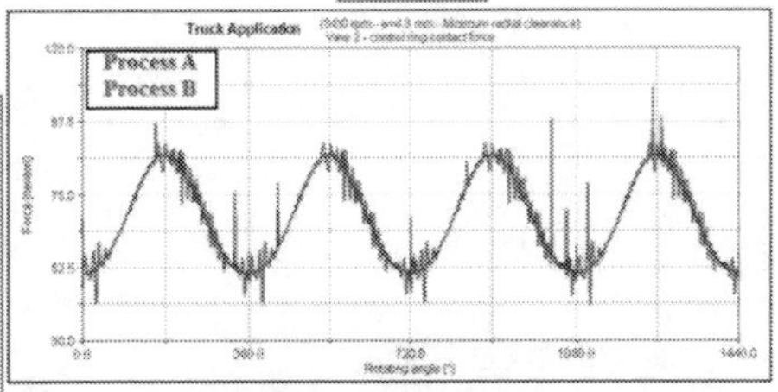

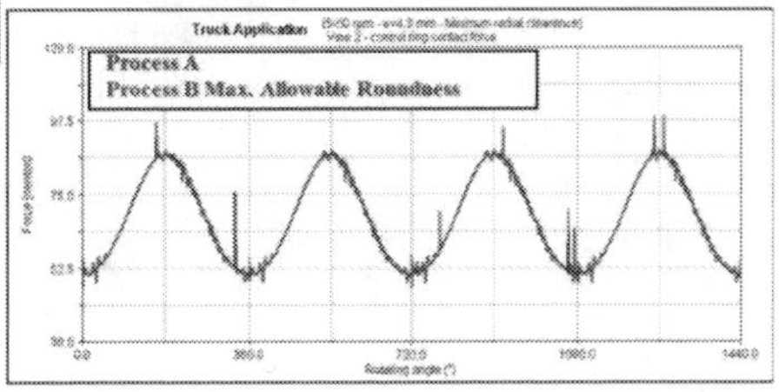

Output:

Maximum allowable Roundness

0.0XX

A Complete Loop of Numerical Analyses about a Variable Oil Pump PIERBURG PUMP TECHNOLOGY

Variable Oil Pump – Reliability

FEA Structural Analyses

ANSYS

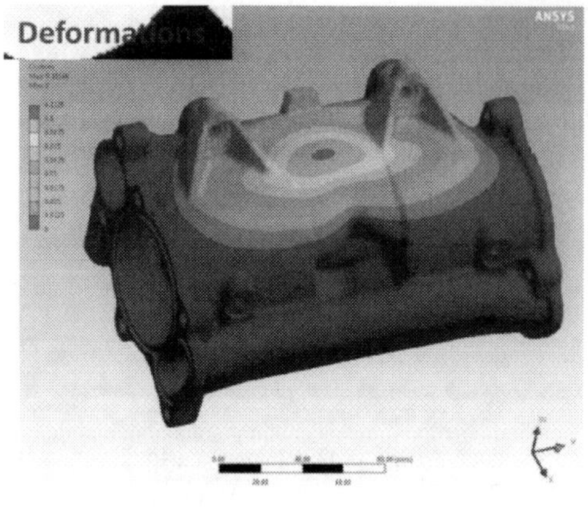

Simulation of **component strength.**

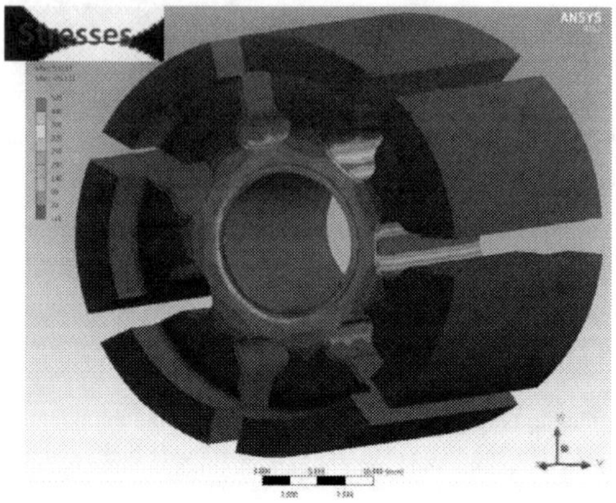

www.kspg.com © Pierburg Pump Technology 2012 KSPG Automotive 21

A Complete Loop of Numerical Analyses about a Variable Oil Pump PIERBURG PUMP TECHNOLOGY

Variable Oil Pump – Reliability

FEA Structural Analyses

ANSYS

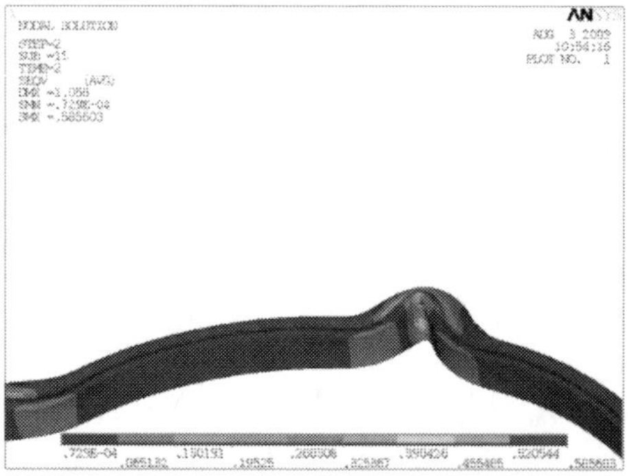

Simulation of **gaskets/sealing rings.**

www.kspg.com © Pierburg Pump Technology 2012 KSPG Automotive 22

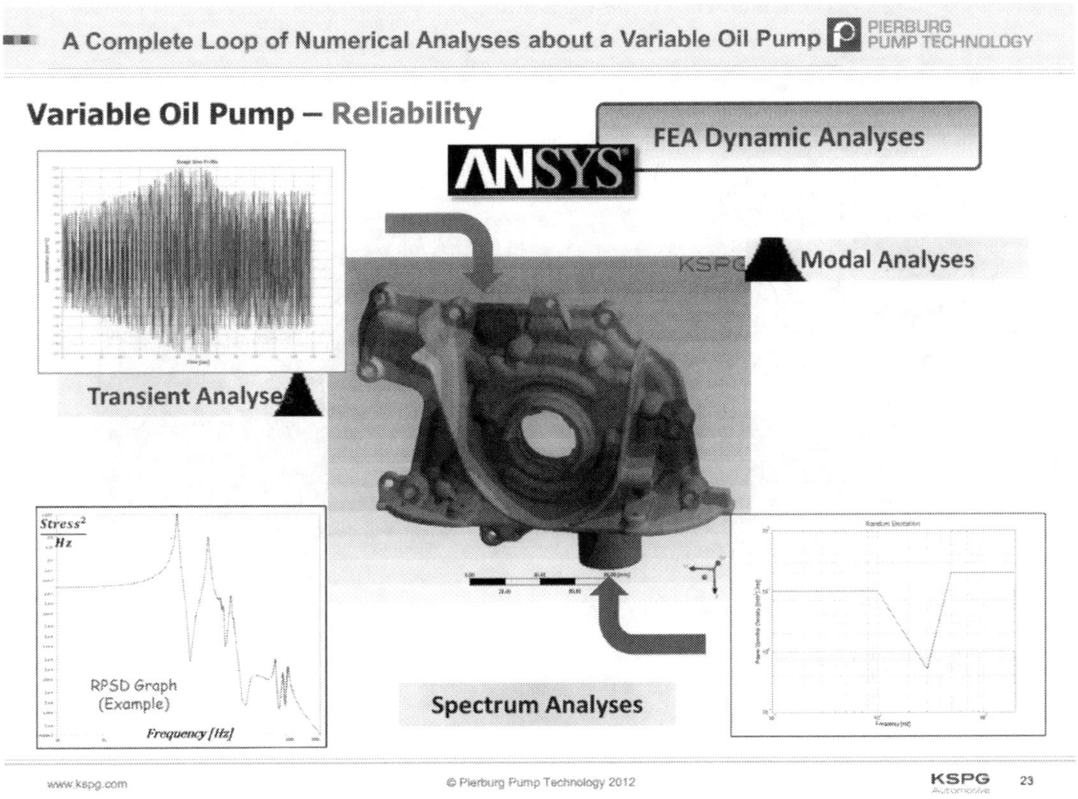
A Complete Loop of Numerical Analyses about a Variable Oil Pump
PIERBURG PUMP TECHNOLOGY
Variable Oil Pump – Reliability
ANSYS
FEA Dynamic Analyses
Modal Analyses
Transient Analyses
Stress²
Hz
RPSD Graph
(Example)
Frequency [Hz]
Spectrum Analyses
www.kspg.com
© Pierburg Pump Technology 2012
KSPG
Automotive
23

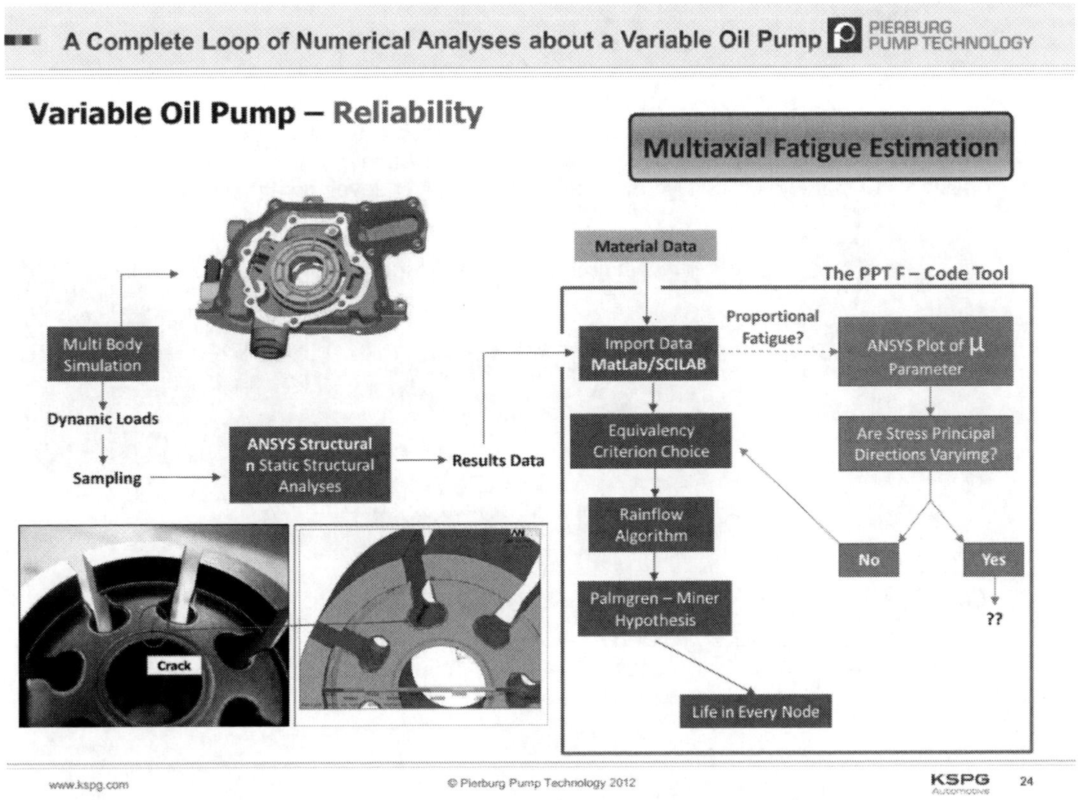
A Complete Loop of Numerical Analyses about a Variable Oil Pump
PIERBURG PUMP TECHNOLOGY
Variable Oil Pump – Reliability
Multiaxial Fatigue Estimation
Material Data
The PPT F – Code Tool
Multi Body Simulation
Dynamic Loads
Sampling
ANSYS Structural n Static Structural Analyses
Results Data
Import Data MatLab/SCILAB
Proportional Fatigue?
ANSYS Plot of μ Parameter
Equivalency Criterion Choice
Are Stress Principal Directions Varyimg?
Rainflow Algorithm
No
Yes
??
Palmgren – Miner Hypothesis
Life in Every Node
Crack
www.kspg.com
© Pierburg Pump Technology 2012
KSPG
Automotive
24

Variable Oil Pump – Noise

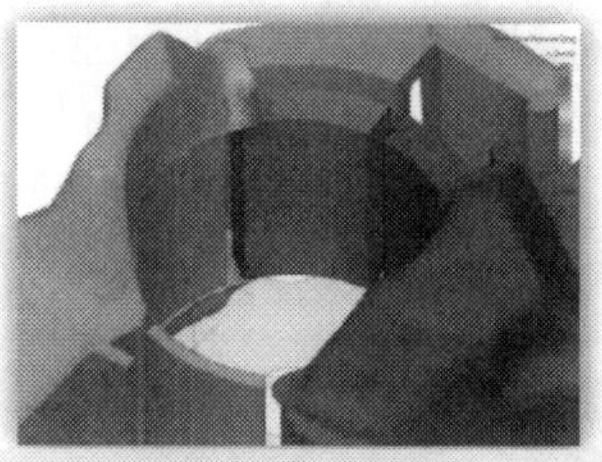

CFD Results → INPUT SIGNAL

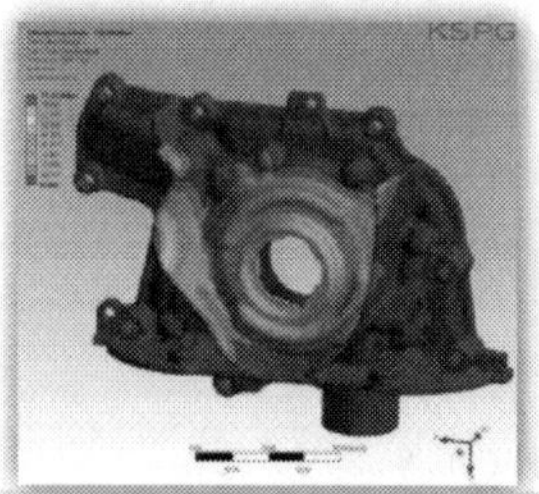

FEM Modal Analysis → Vibrational Modes

Acoustic Simulation

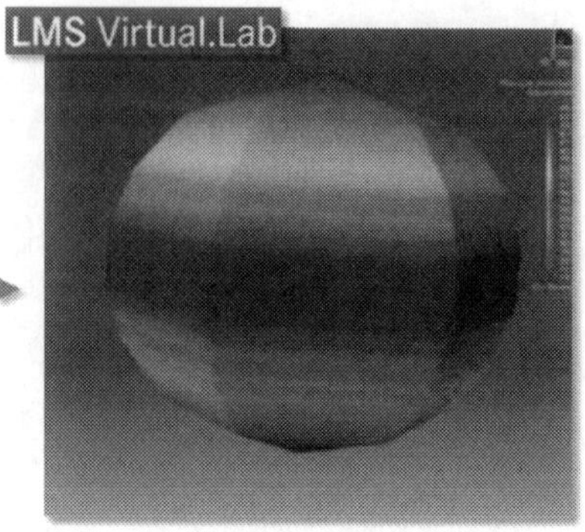

Output results

- **dB Sound Power**
- **Emission Type**
- **Noise Radiation**

A Complete Loop of Numerical Analyses about a Variable Oil Pump PIERBURG PUMP TECHNOLOGY

Variable Oil Pump – Example of Combined Simulation

The increased "know how" gained in different simulation areas like the FEM, CFD and MBA has allowed to run complex combined simulations. These skills gives the possibility to manage difficult situation during the product development.

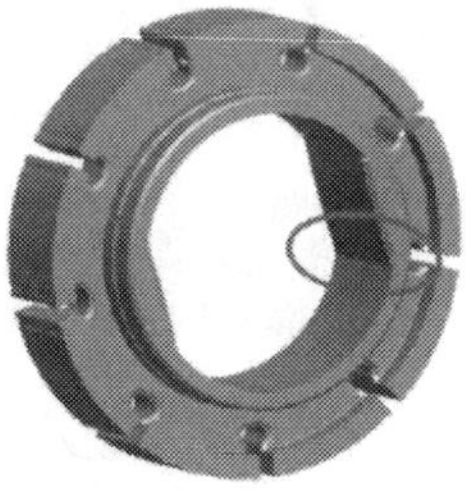

PROBLEM: crack on a VOP rotor.

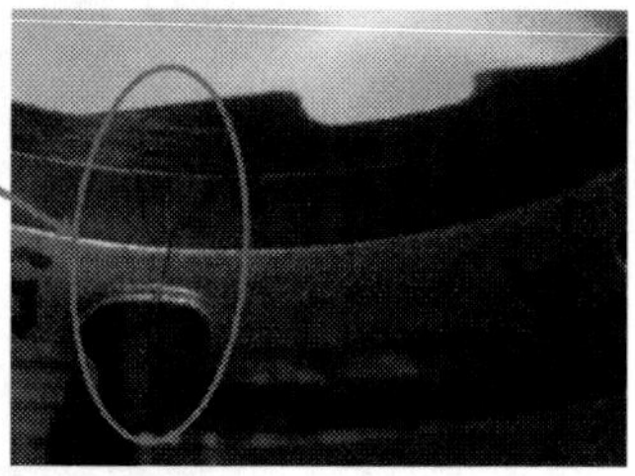

Cause of the failure ?

weak design or engine conditions

Calculation areas:

- fluid dynamics analysis (**CFD**) → internal pressure peaks
- dynamical analysis (**MBA**) → contact forces crankshaft-rotor
- structural analysis (**FEM**) → fatigue life

Variable Oil Pump – Example of Combined Simulation

CFD Analysis

CFD analysis has been conducted in order to evaluate **local pressure peaks** into the chambers and therefore the loads on vanes.

A Complete Loop of Numerical Analyses about a Variable Oil Pump PIERBURG PUMP TECHNOLOGY

Variable Oil Pump – Example of Combined Simulation

Advanced Engine Tests

Torsional vibrations of the crankshaft have been measured by an **encoder** in the whole engine speed range.

ENCODER

Acquired data has b implemented in th **multi body** model of the variable oil pump.

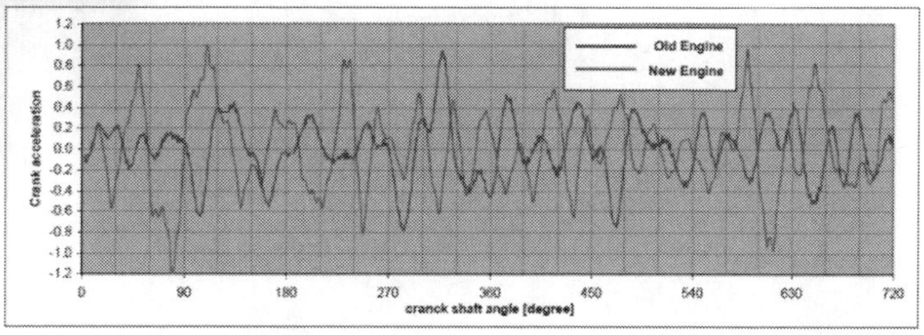

Variable Oil Pump – Example of Combined Simulation

Dynamical Analysis

Dynamic analysis was run in order to evaluate **contact forces** between crankshaft and rotor under crankshaft torsional vibration. Clear effect of high unexpected vibration of the new engine against an existing application were highlighted.

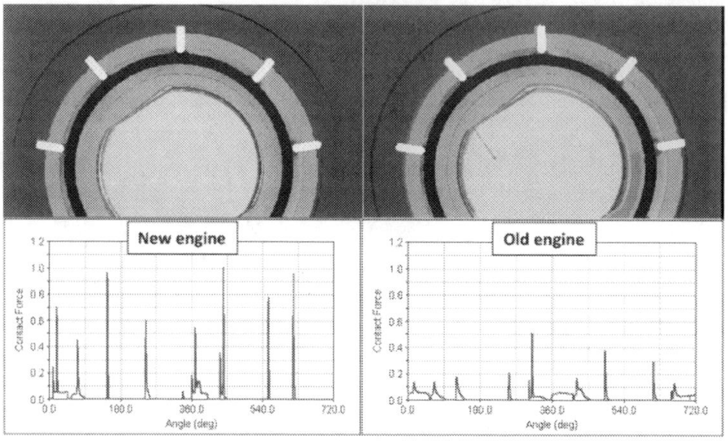

Crankshaft mounted camera

A Complete Loop of Numerical Analyses about a Variable Oil Pump PIERBURG PUMP TECHNOLOGY

Variable Oil Pump – Example of Combined Simulation

Structural Analysis

Calculated loads on shaft has been used to conduct **fatigue life analysis**.

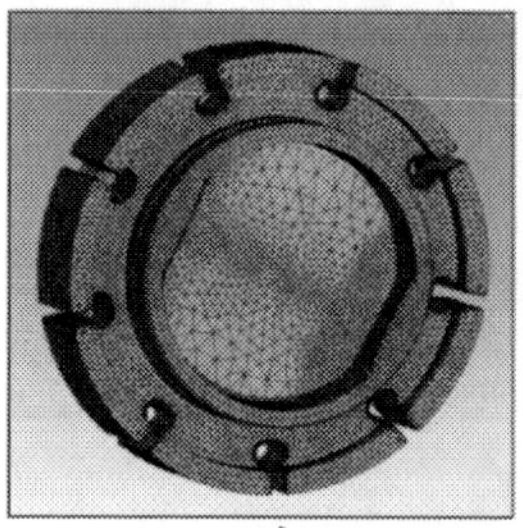

mesh

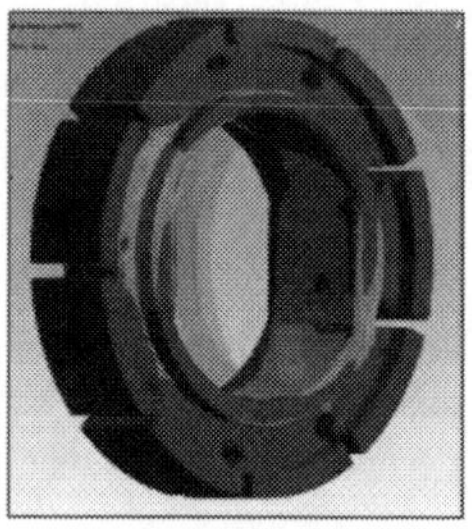

internal stresses

submodeling

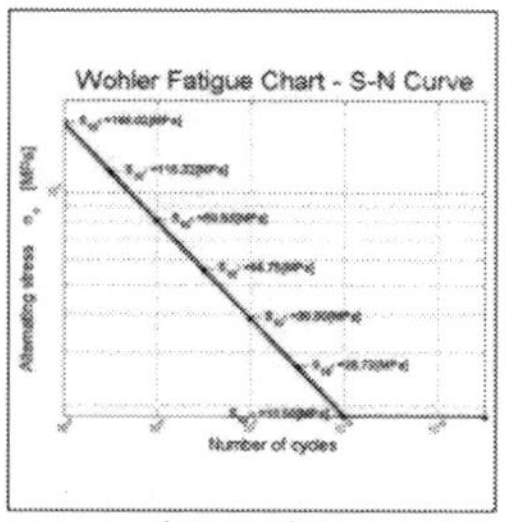

fatigue life

At the end of this activity it was shown that the problem was engine related (excessive crankshaft torsional vibrations).

Variable Oil Pump – Example of Combined Simulation

Structural Analysis

Calculated loads on shaft has been used to conduct **fatigue life analysis**.

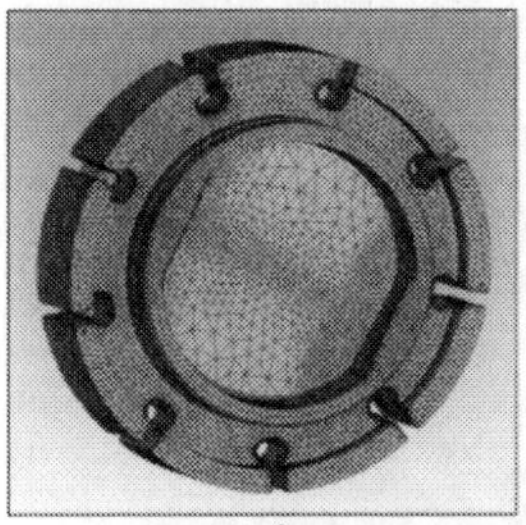

mesh

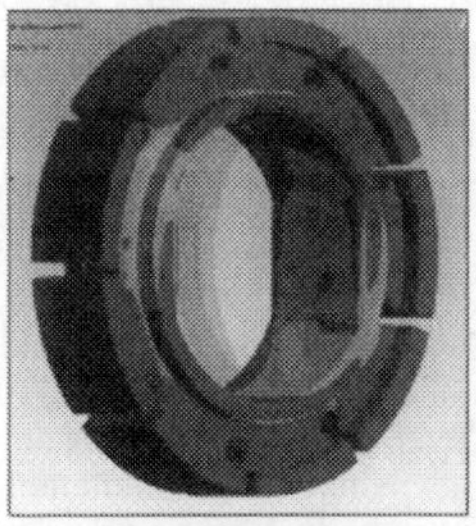

internal stresses

submodeling

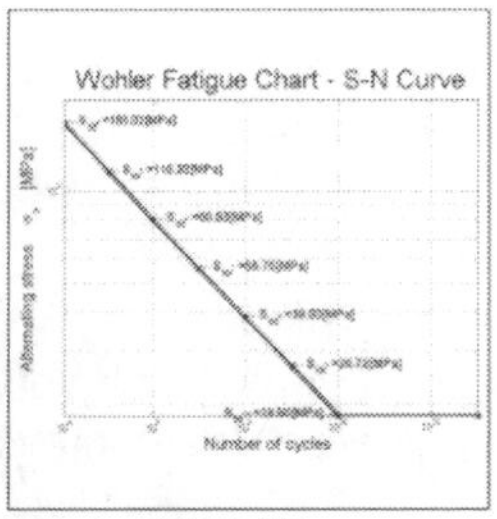

fatigue life

At the end of this activity it was shown that the problem was engine related (excessive crankshaft torsional vibrations).

Customer worked to reduce the amplitude of torsional vibrations by tuning the engine (crankshaft modifications, new damper, etc).

A Complete Loop of Numerical Analyses about a Variable Oil Pump PIERBURG PUMP TECHNOLOGY

Variable Oil Pump – Example of Combined Simulation

Structural Analysis

Calculated loads on shaft has been used to conduct **fatigue life analysis**.

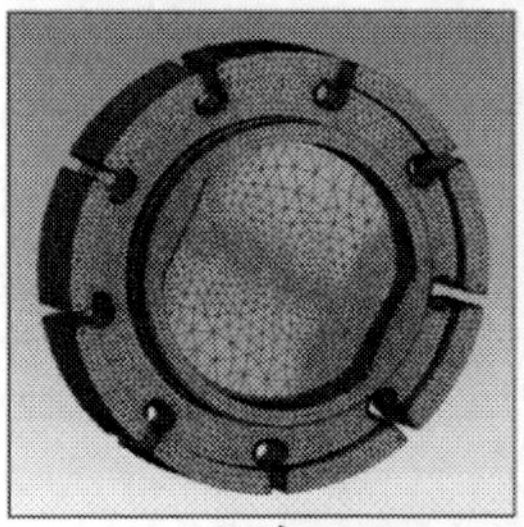

mesh

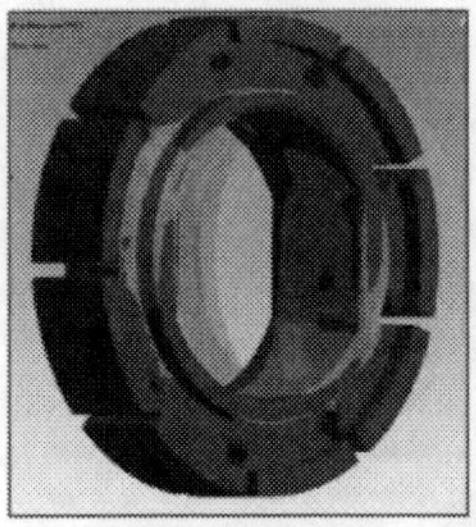

internal stresses

submodeling

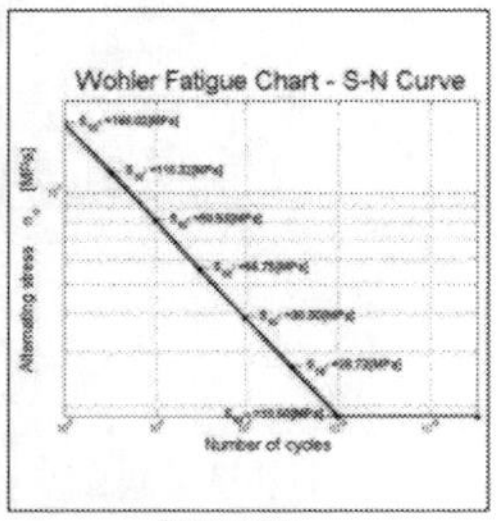

fatigue life

At the end of this activity it was shown that the problem was **engine** related (excessive crankshaft torsional vibrations).

Customer worked to reduce the amplitude of torsional vibrations by tuning the engine (crankshaft modifications, new damper, etc).

No re-design of the pump was necessary.
No costs for PPT.

www.kspg.com
© Pierburg Pump Technology 2012
KSPG Automotive

Conclusions

- The present work has shown some possible numerical analyses which can be performed to design, optimize and verify a generic variable oil pump, in order to have a successful product with a reasonable cost.
- Thanks to CAE software, all of the numerical evaluations are executed without the building of any prototype, with a great economy in terms of materials and money.
- Thanks to the virtual prototyping it's possible to explore "unusual" working loads, not reachable with experimental tests, in order to verify the pump also outside from the nominal conditions.
- Finally, it is worth noting the great flexibility of the current CAE software, like **ANSYS**, which permit a complete multidisciplinary approach in designing and verifying whatever mechanical component, providing reliable results in short time.

PIERBURG PUMP TECHNOLOGY

Thank you for your attention!

Riccardo Maccherini
R&D Department - Livorno Plant
☎ +39 0586 255740
riccardo.maccherini@it.kspg.com

KSPG Automotive

Disclaimer

This presentation contains “forward-looking statements”. Forward-looking statements are sometimes, but not always, identified by their use of a date in the future or such words as “will”, “anticipates”, “aims”, “could”, “may”, “should”, “expects”, “believes”, “intends”, “plans” or “targets”. By their nature, forward-looking statements are inherently predictive, speculative and involve risk and uncertainty because they relate to events and depend on circumstances that will occur in the future. There are a number of factors that could cause actual results and developments to differ materially from those expressed or implied by these forward-looking statements. In particular, such factors may have a material adverse effect on the costs and revenue development of the KSPG group.

All written or oral forward-looking statements attributable to KSPG AG or any group company of KSPG AG or any persons acting on their behalf contained in or made in connection with this presentation are expressly qualified in their entirety by factors of the kind referred to above. No assurances can be given that the forward-looking statements in this presentation will be realized. Except as otherwise stated herein and as may be required to comply with applicable law and regulations, KSPG AG does not intend to update these forward-looking statements and does not undertake any obligation to do so.

Initialization Techniques in Parametric CFD Design Studies

Giles Endicott

A Typical Engineering Map

- Building maps of aerodynamic performance is common in engineering design and analsyis.
- Few components operate a single conditon, rather a range must be studied.
- In this presentation, we study how to build these maps quicker, cheaper and more efficiently.

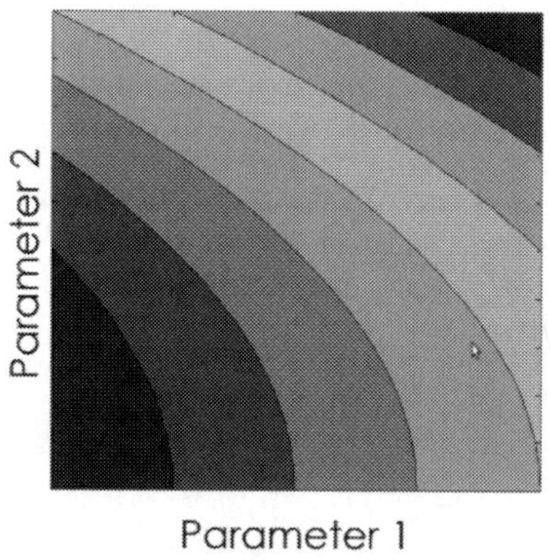

A Design Problem

- Find the design parameters that give the best:
  - Drag, lift, pressure loss, efficiency ....etc.
- Parameter counts vary from 1 – 100s
- Optimization / exploration methods for the problem:
  - Surrogate assisted requires ~10x variable count simulations
  - Direct optimization requires ~50x variable count simulations

Design Evaluation with CFD

- The performance of each design is evaluated using a CFD solver e.g. ANSYS Fluent.
- Huge meshes must be built to well resolve the flow-fields.
- This can be very expensive, proc-hours can be in tens of thousands of hours in some cases, e.g F1.
- Thus parameteric design studies become incredibly expensive processes.

The CFD Solution Process

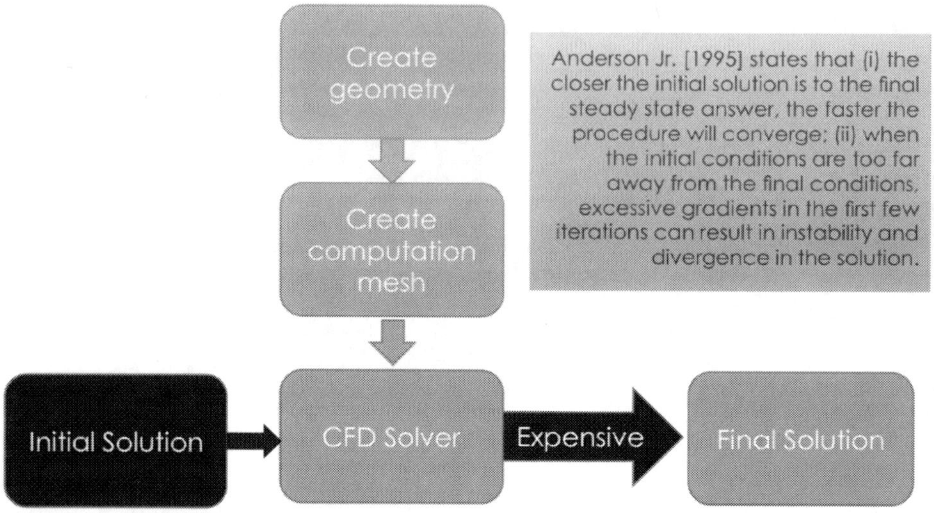

Anderson Jr. [1995] states that (i) the closer the initial solution is to the final steady state answer, the faster the procedure will converge; (ii) when the initial conditions are too far away from the final conditions, excessive gradients in the first few iterations can result in instability and divergence in the solution.

1D Design Study

A one variable design study was developed to allow simple analysis of the effects of initial solutions on CFD simulations.

- 1m airfoil in 56m/s air at 3 degs AOA
- Modelled as turbulent incompressible gas.
- ~60,000 CFD cells
- 5 primitive flow-field variables.

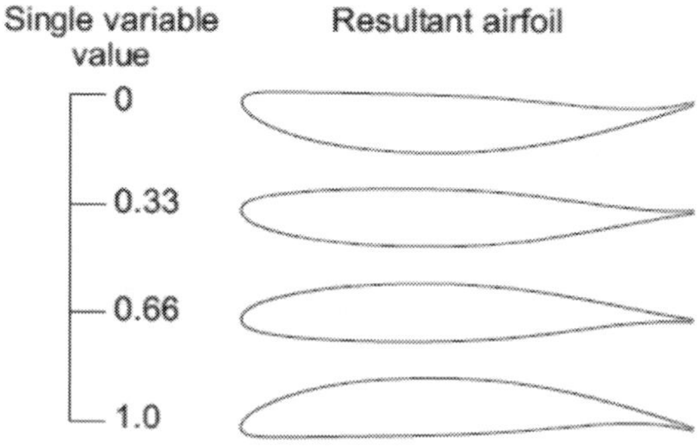

10D Design Study

Transferring the same simulation in to a 10 variable design study provides a more realistic enviroment for development.

(Geometry and mesh modified by Free Form Deformation)

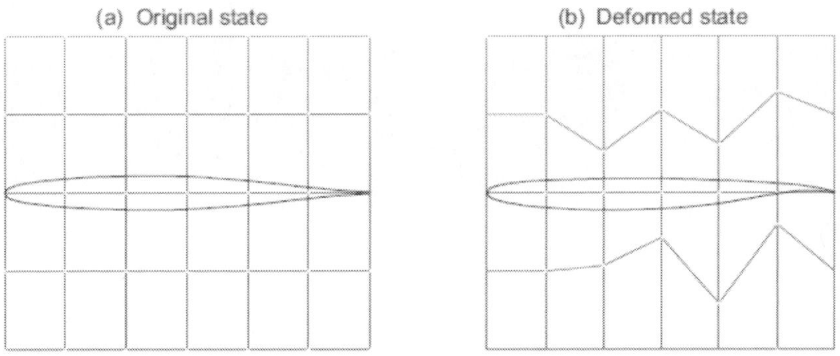

Typical Initialialization Techniques

Baseline Method: Use the flowfield from the first simulated design as the initial solution to all further simulations.

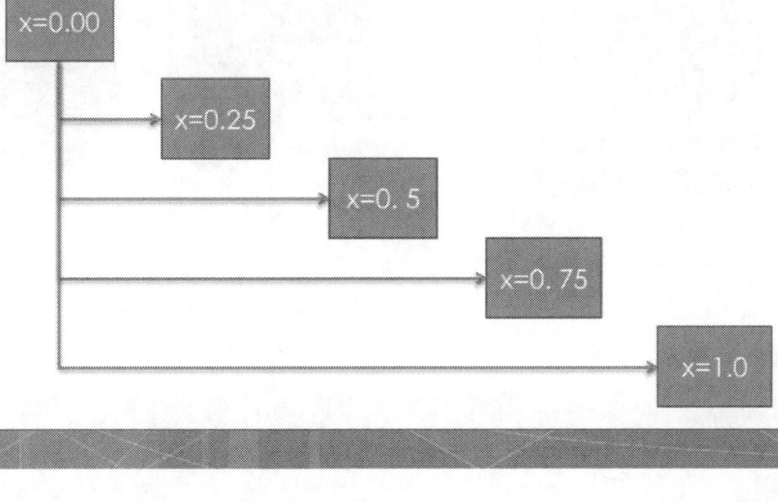

Typical Initialialization Techniques

Nearest Neighbour Method: Use the flow-field from closest completed simulation in the parameter space to initilize further simulations. Optimize the order of simulations accordingly (TSP)

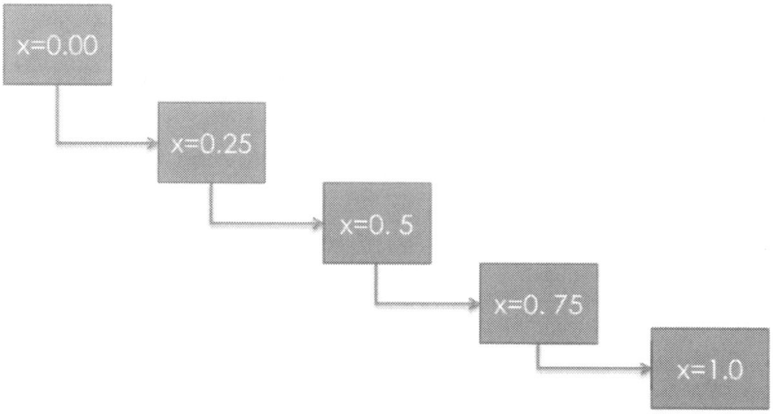

Testing Initilization Methods in a Ten Variable Design Space.

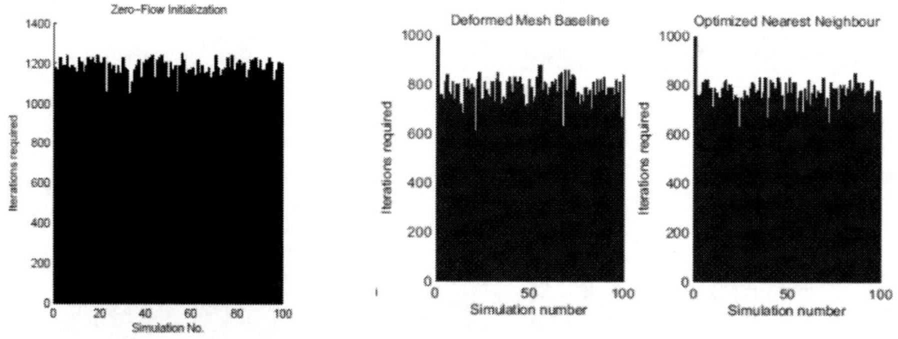

Zero-flow initial solutions are those with zero flow velocity and zero pressure throuhout the flow domain

Zero-flow ~ 1200 iterations
Baseline method ~800
Nearest Neighbour w/ TSP ~800

Why is NN so similar to baseline?

- Iterations required seems to scale with the log(Δx) from initial solution to final solution, tested in 1D space.
- Mean Δx between baseline and nearest neighbour methods are similar in sparse hyperspaces with small sample sizes.
- To improve significanlty, initial solutions must be orders of magnitude more accurate

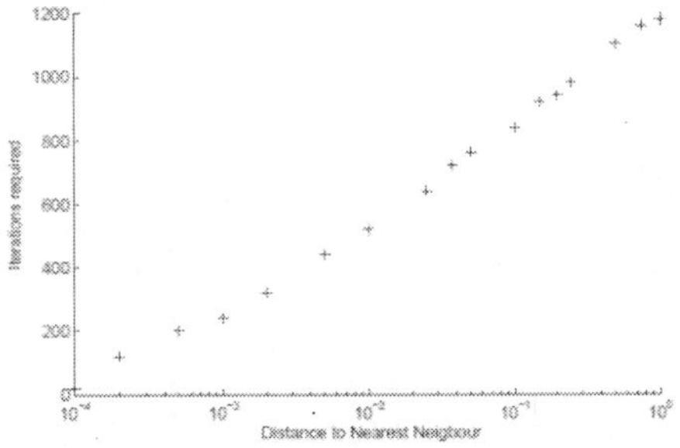

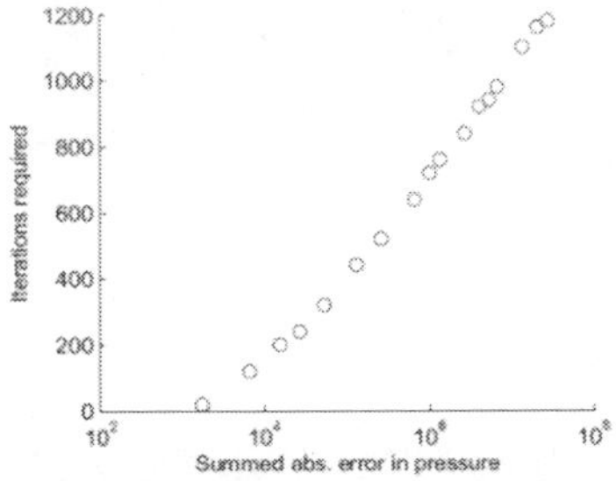

A Different Approach

Suppose three shapes simulated separately:

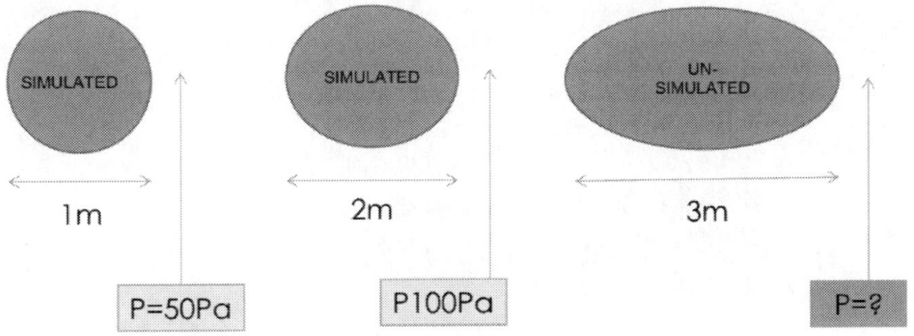

In a single parmater space, nearest neighbour approaches seem inferior to simple interpolation or extrapolation

Re-thinking the Initialization

- With any two simulations (in 1D) a simple model could be built of the primitive variables in each cell.
  - Huge datasets
  - Data must be physical
- Using such a model could form more accurate initial solutions.
- If the model is accurate, the convergence will be improved.

A New Initialialization Technique

Flowfield Prediction Method: Build a predicted flow-field for each further simulation, interpolating from previous completed simulations.

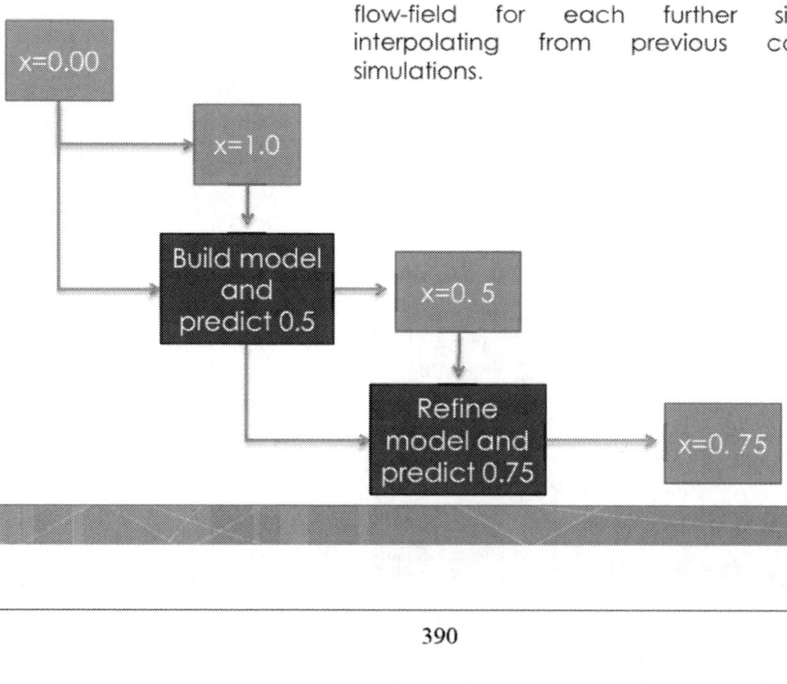

Performance in 1D

- Target simulation @ x=1.0
- Zero-flow initialization requires ~1200 iterations.
- Even at larger step sizes, up to 75% reduction in cost
- Quadratic and cubic methods perform best as could be expected.

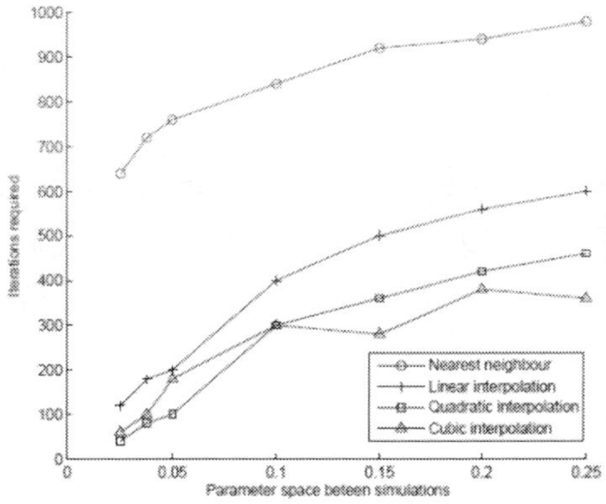

The Limit of Polynomial Modelling

- Polynomial modelling worked well with one variable but it does not scale well with variable count.
  - Linear: N+1
  - Quadratic: N^2 +N+1
  - Cubic: N^3+N^2+N+1
- Sample points required for ten variable model:
  - Linear: 11points → Feasible
  - Quadratic: 111 points → Unfeasible!
  - Cubic: 1111 points → Unfeasible!
- Limited to linear polynomial models for realistic situations.

The RBF based Interpolator

- RBF models use an arbitrary number of data samples to create a non-linear response model.
- Just one tuning parameter affects the shape.
- Model creates a set of weights to multiply the existing sample by.
  - Very efficient scaling for large CFD meshes.

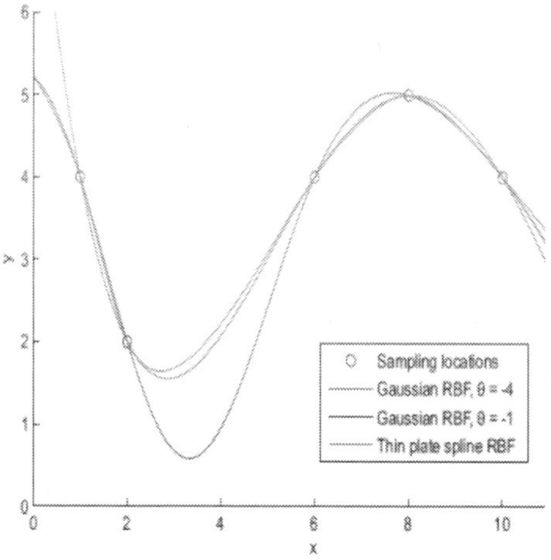

RBF Model Tuning

- Traditional tuning of the shape parameter is not feasible on huge datasets.
- Instead a single value must be chosen for all cell/flow-variable combinations.
- This can be chosen by minimizing the error in creating a now-known previous solution.

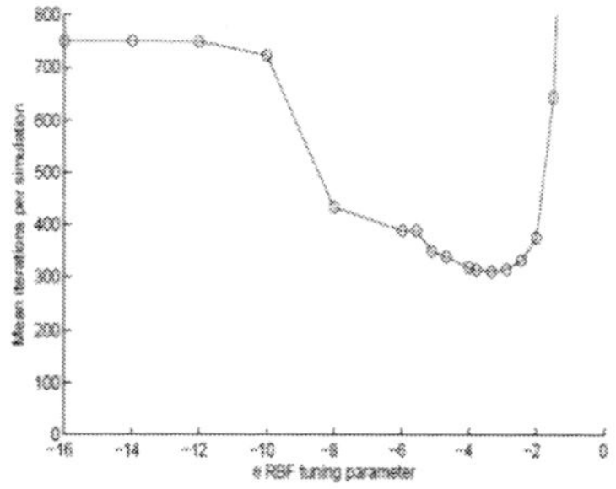

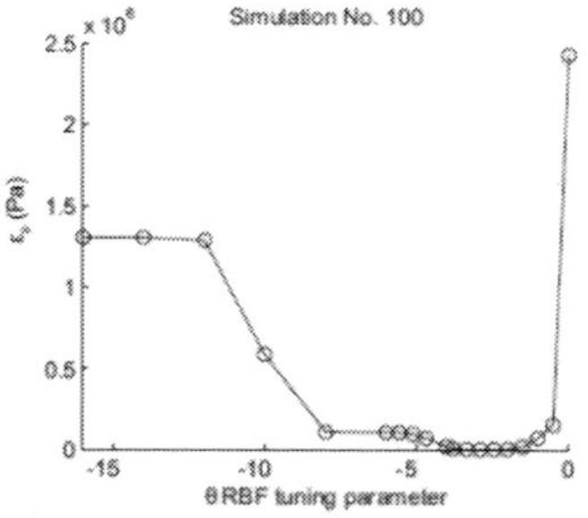

Performance in 10D: RBF

1st simulation: zero flow init
2-100 simulations: RBF prediction init.

- Excellent performance of RBF prediction.
- Large improvement after ~11 samples.
- Simulations required
  - ~30% of zero-flow init
  - ~50% of baseline init
- Overall design study cost
  - 30% of zero flow init
  - 44% of baseline init

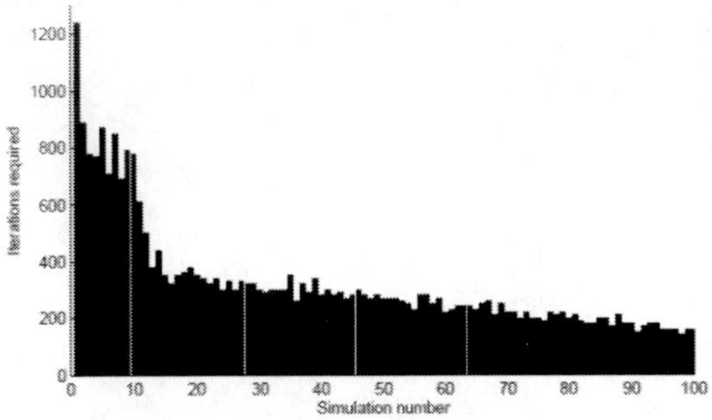

Case Studies with More Complexity: F1 Nose

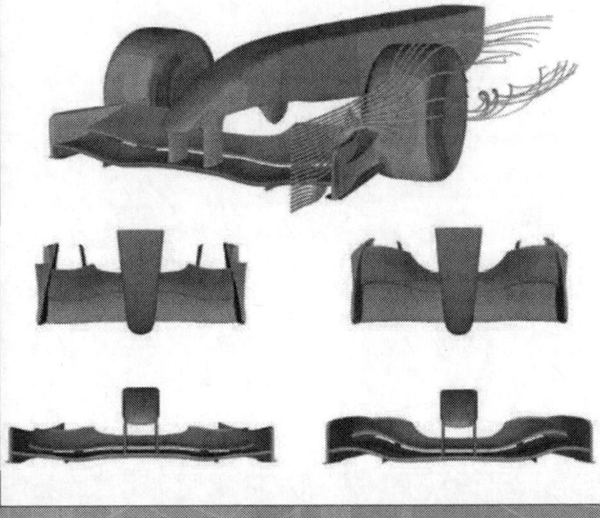

- Stress testing of the RBF predictor system.
- Complex Geometry
- Complex Flowfield
  - Strong wakes
  - Strong vortices
- Computationally expensive to simulate
- DF variation 800N-940N
- 10M cell mesh
- 8 design variables

F1 Nose: Initialization Test.

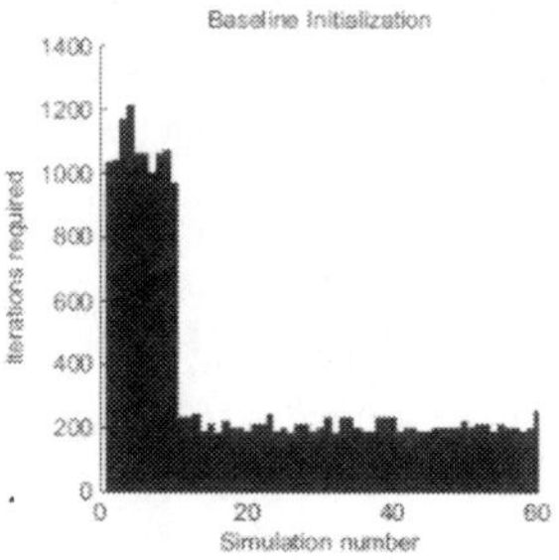

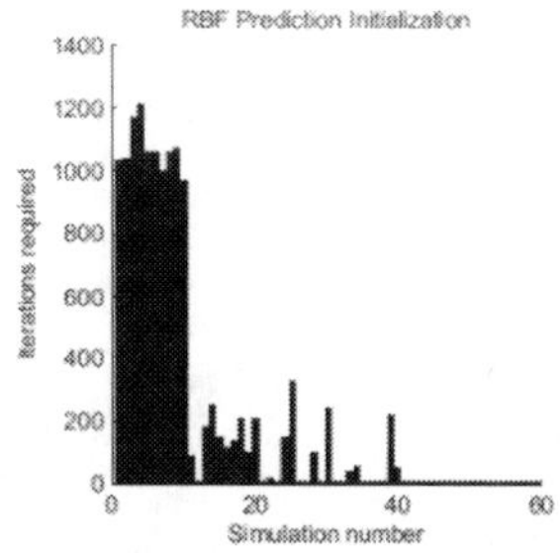

- 10x parallel running of simulations
- Good performance demonstrated again
- Convergence criteria reached very fast for many simulations.
- Demonstrated here:
  - Zeroflow 61200
  - Baseline 20500
  - RBF prediction 13180
- Optimally, using simulations 2-9 with baseline method:
  - RBF prediction ~5800

Computational Cost Analysis

- Cost of creating the RBF flow-field prediction is low.
- Scales linearly with mesh size and sample number.
- Easily paralellized for big meshes.
- For 99 data-files, cost ~4 iterations of CFD solver using a single CPU.
- Benefit far outweights cost.

Conclusions

- The computational cost of a design study is dramatically affected by the choice of initial solution.
- Using prediction techniques over an entire CFD mesh can create an accurate initial solution.
- This enables savings of up to 90% per simulation.
- Applicable to boundary conditions as well as geometry deformation demonstrated here.

Many Thanks for Listening

- Additional thanks to:
  - Andy J. Keane and Neil W. Bressloff at the University of Southampton.
- To be presented at 2013 ANSYS World Automotive Simulation congress.

ISUZU

2013 Automotive Simulation World Congress

OPTIMIZATION OF THE DEFROSTER DUCTS OF A COACH BUS

Barış TAVUKCU , Mechanical Engineer (MSc)
Kaan ALTAY , Mechanical Engineer (MSc)
ANADOLU ISUZU R&D DEPARTMENT

Altan ONGUN , Mechanical Engineer (Bsc.)
Desica Ltd. (Engineering Consultant Company)

ANADOLU ISUZU

Anadolu Isuzu Otomotiv Sanayi ve Ticaret A.Ş. (AIOS) is a joint stock company open to public in the partnership of ISUZU Motors Limited and Itochu. Its main functioning areas are production and marketing of commercial vehicles such as light trucks, trucks, buses and pick-up.

2

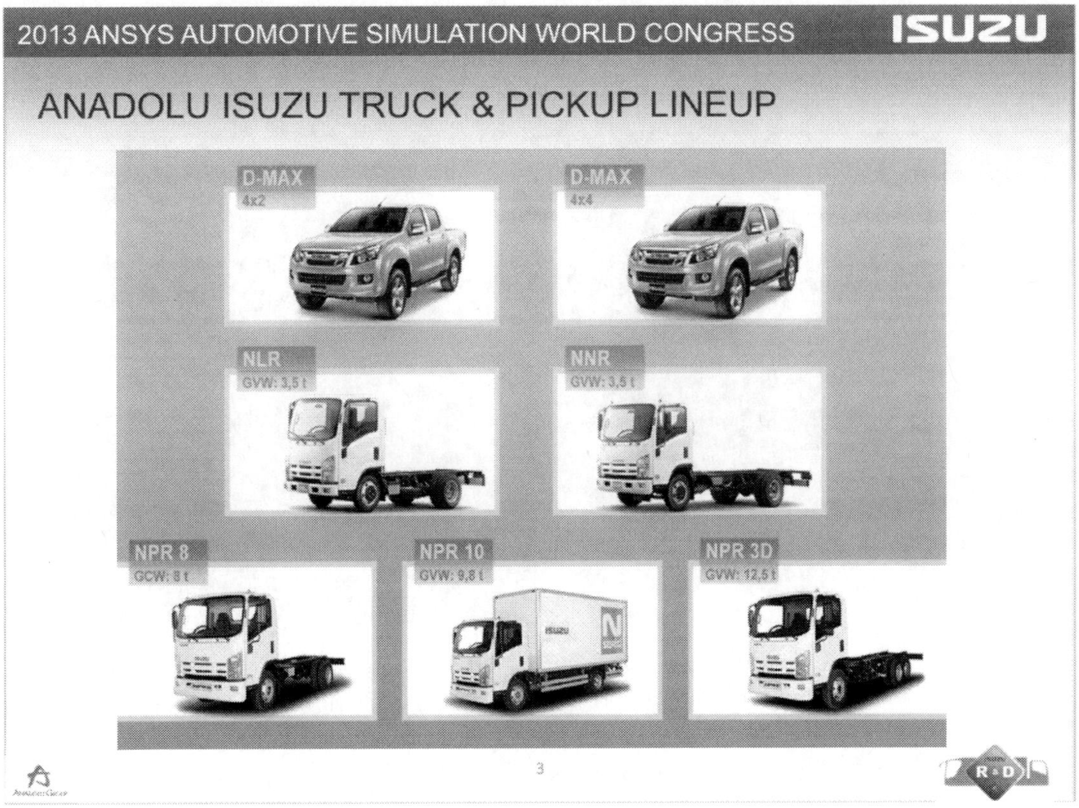
2013 ANSYS AUTOMOTIVE SIMULATION WORLD CONGRESS
ISUZU
ANADOLU ISUZU TRUCK & PICKUP LINEUP
D-MAX
4x2
D-MAX
4x4
NLR
GVW: 3,5 t
NNR
GVW: 3,5 t
NPR 8
GCW: 8 t
NPR 10
GVW: 9,8 t
NPR 3D
GVW: 12,5 t
3

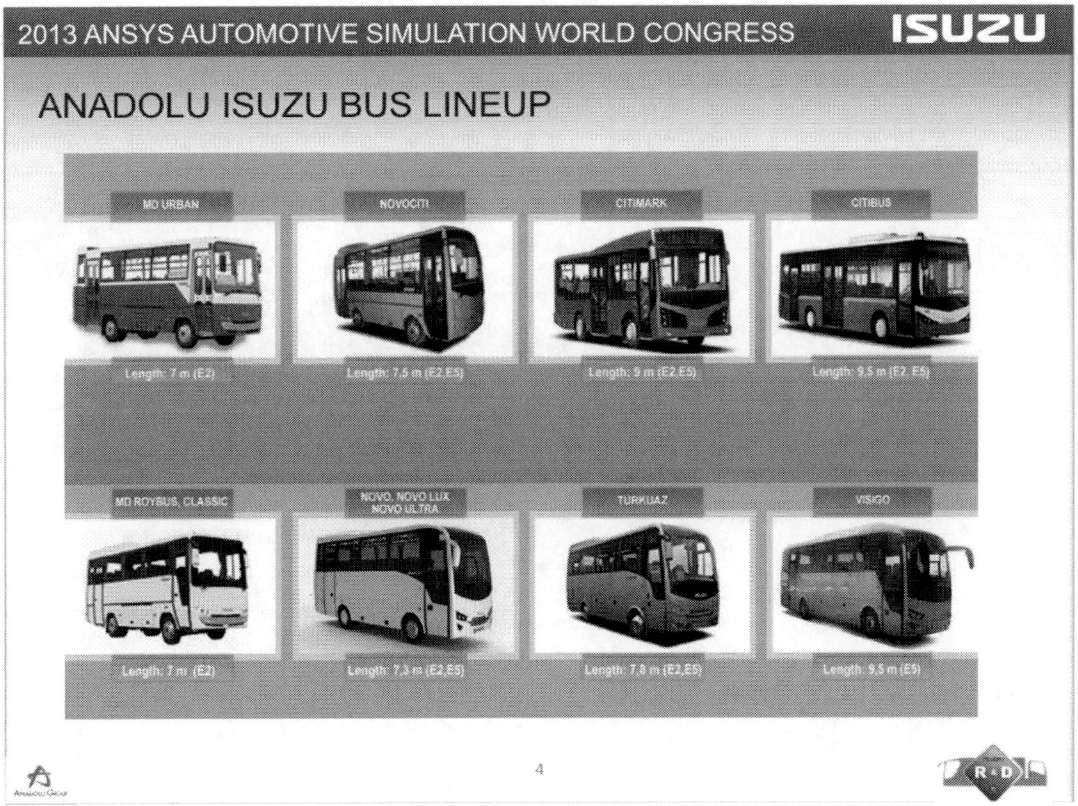
2013 ANSYS AUTOMOTIVE SIMULATION WORLD CONGRESS
ISUZU
ANADOLU ISUZU BUS LINEUP
MD URBAN
NOVOCITI
CITIMARK
CITIBUS
Length: 7 m (E2)
Length: 7,5 m (E2,E5)
Length: 9 m (E2,E5)
Length: 9,5 m (E2, E5)
MD ROYBUS, CLASSIC
NOVO, NOVO LUX
NOVO ULTRA
TURKUAZ
VISIGO
Length: 7 m (E2)
Length: 7,3 m (E2,E5)
Length: 7,8 m (E2,E5)
Length: 9,5 m (E5)
4

ANADOLU ISUZU / VISIGO

✓ This study is performed on ' VISIGO ' coach bus developed by Anadolu ISUZU

CONTENTS

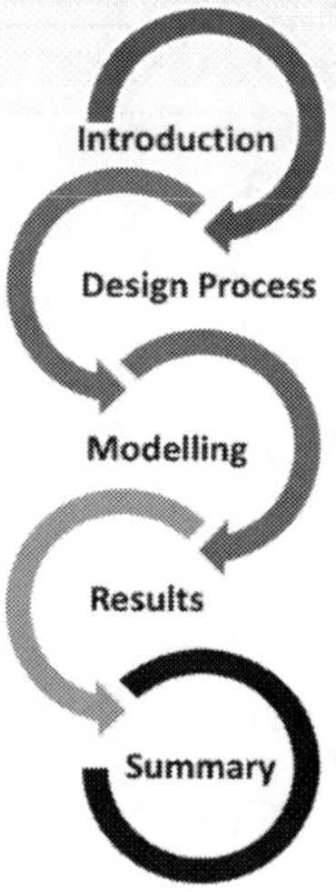

Introduction → Overview, Motivations / Goals

Design Process → Problem Description/ Standards, Methodology, Air Channel Geometry

Modelling → Meshing, Boundary Conditions, Solving The Model

Results → Simulation Results, Test Results, Comparison of SR and TR

Summary → Final Remarks, Further Developments

2013 ANSYS AUTOMOTIVE SIMULATION WORLD CONGRESS ISUZU

OVERVIEW

- ✓ Visibility in the automotive industry is a major source of concern for the bus manufacturers and design engineers
- ✓ Water film (fog) that forms on the windshield during winter times would reduce and disturb the driver's visibility
- ✓ Thus, defrost system is a critical element of the vehicle design

7

2013 ANSYS AUTOMOTIVE SIMULATION WORLD CONGRESS ISUZU

MOTIVATIONS

- ✓ The aim is to remove the ice/mist from the windscreen and maintain a clear screen under driving conditions
- ✓ The short timetable of the bus development programs makes not feasible the work being done exclusively on an experimental tests basis
- ✓ Thus, numerical simulation plays a very important role through an extensive number of virtual testing and a better understanding of the flow

GOALS

- ✓ Development defrost/defogging air ducts in order to achieve a uniform flow field in windscreen with high velocities

8

PROBLEM DESCRIPTION

- ✓ Vapor layer that generally forms at the zero degree temperature on the front and side windows of the busses, is one of the main problem to be solved
- ✓ A fast and uniform defog/defrost of bus windscreen is very important for driving safety and occupant's comfort

STANDARDS

- ✓ The clearance of the windscreen has to meet legislative requirements
- ✓ The defrost/defogging tests are normalized in United States by SAE J381 and in Europe by 78/317/EEC

STANDARDS

- ✓ Benefiting from the relevant standards, for the windshield of the bus, required regions that should be primarily defrosted during condensation are specified
- ✓ 1st region is the area that should be defrosted at first, and 2nd Region is the area that should be defrosted secondly

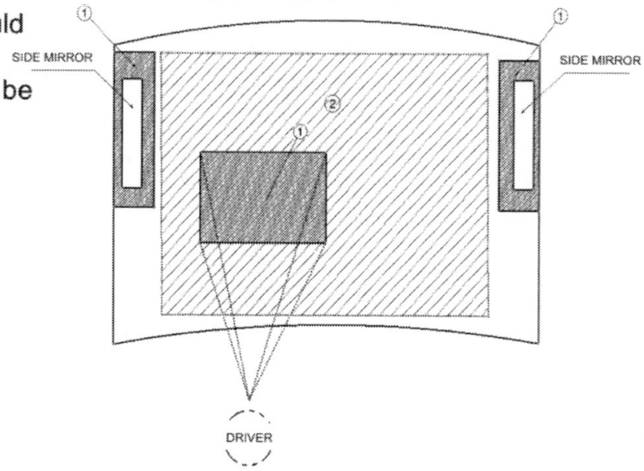

2013 ANSYS AUTOMOTIVE SIMULATION WORLD CONGRESS ISUZU

METHODOLOGY

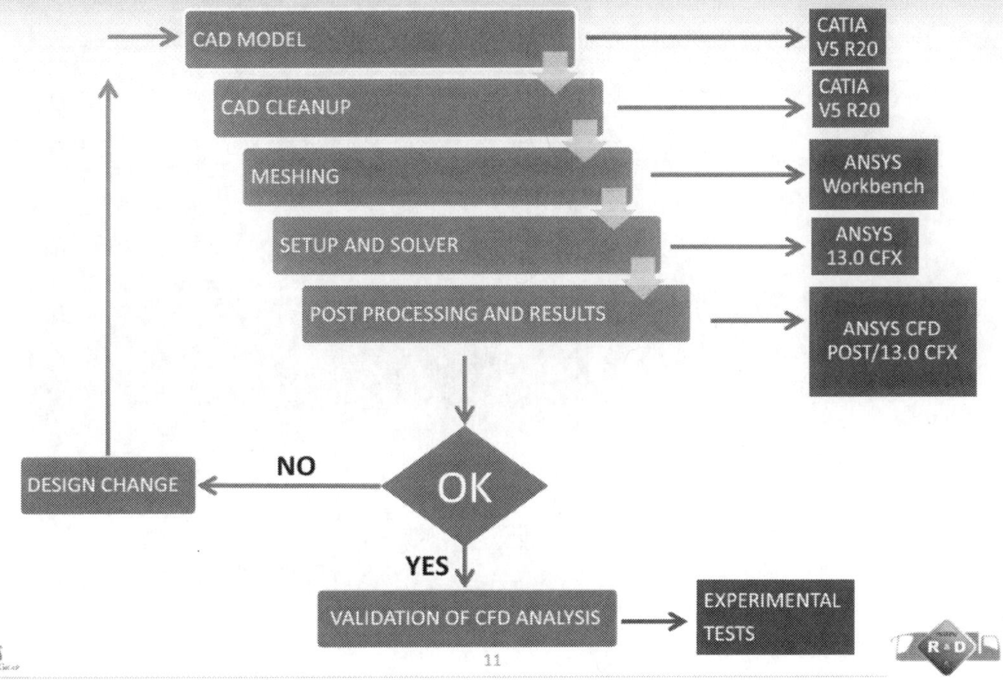

11

2013 ANSYS AUTOMOTIVE SIMULATION WORLD CONGRESS ISUZU

AIR CHANNEL GEOMETRY

- ✓ The defrost system design begins when the industrial design is finished
- ✓ The initial concept ducting is then modelled and refined to achieve a design that is feasible and acceptable to design
- ✓ to achieve the correct air flow distribution across the windscreen, CFD calculations are initially run

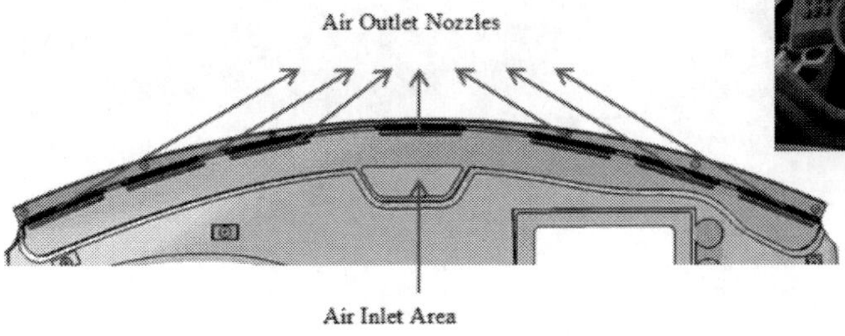

12

2013 ANSYS AUTOMOTIVE SIMULATION WORLD CONGRESS ISUZU

AIR CHANNEL GEOMETRY

- ✓ The design of the defrost system is driven through the use of ANSYS CFX
- ✓ The example illustrates the design evolution
- ✓ (a) Improved Design
  (b) Initial Design

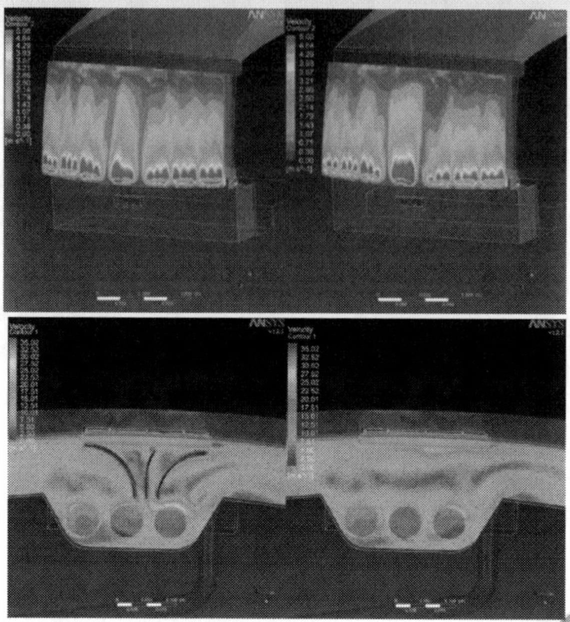

13 a b

2013 ANSYS AUTOMOTIVE SIMULATION WORLD CONGRESS ISUZU

MESHING

- ✓ The defrost channel numerical domain with 14,624,098 element

14

2013 ANSYS AUTOMOTIVE SIMULATION WORLD CONGRESS ISUZU

MESHING

- ✓ The whole cabin is modelled (driver region)
- ✓ Mesh Type; tetrahedron, wedge and prismatic surface elements
- ✓ Includes detailed geometry, including A posts

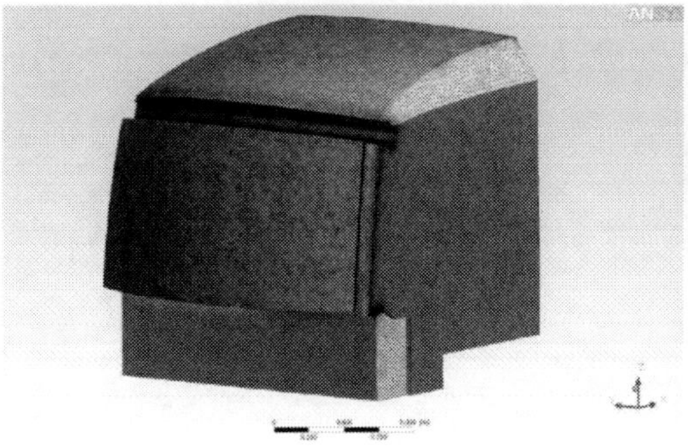

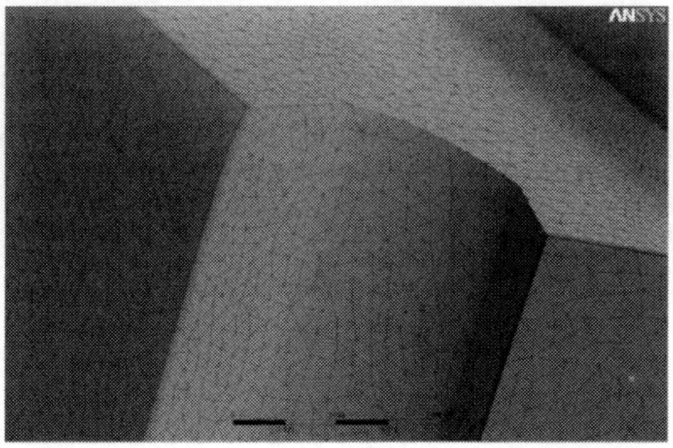

15

2013 ANSYS AUTOMOTIVE SIMULATION WORLD CONGRESS ISUZU

BOUNDARY CONDITIONS

Ambient Temperature at out of windscreen	0 °C
Material Selection of solid calculation volume glass	Glass plate
Cabin Initial Temperature	12 °C
Turbulance Model	Shear Stress Transport
Cabin Rear Side	Opening Boundary Condition, 0 Pa @12 °C
Inlet Temperature of Defroster Duct	50 °C
Fluid	Air
Windscreen / cabin interface	Conjugate heat transfer surface
Other Surfaces	Adiabatic
Fan Flow	A second order function was obtained from fan curve of the inlet pressure to the inlet speed.
Total Mesh Number	14,624,098

16

SOLVING THE MODEL

- ✓ The solution is run at steady state
- ✓ In the analysis windscreen is simulated as a glass plate
- ✓ For the numeric solution for flow and windscreen conjugated heat transfer solution, defroster duct is modelled, and meshed and obtaining steady state phase,
- ✓ And this results is used as a boundary condition for the 2 dimensional solution

HYPOTHESIS FOR THE FLOW

- ✓ Steady State
- ✓ Incompressible
- ✓ Turbulance Model;Shear Stress Transport (SST)
- ✓ The initial thickness of the mist layer=0.12 mm

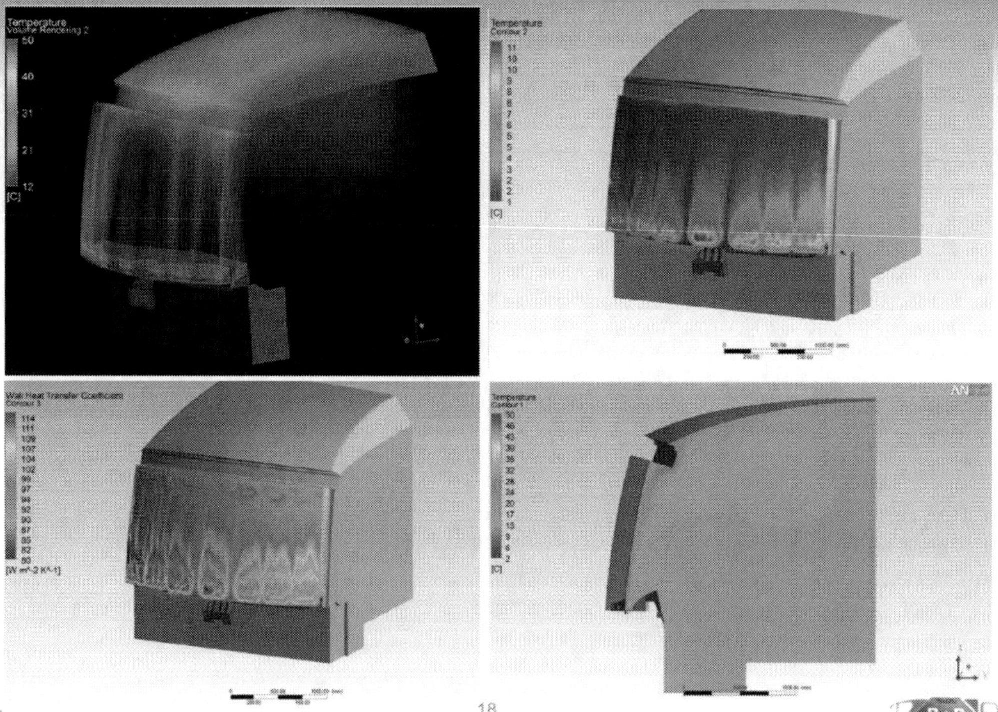

2013 ANSYS AUTOMOTIVE SIMULATION WORLD CONGRESS ISUZU

SOLVING THE MODEL

Mist Formation on Glass Surface (a) Visual Represantation – (b) Macro shooting

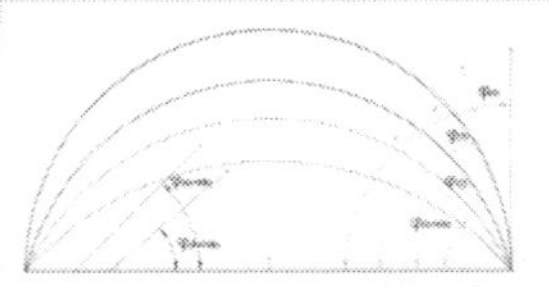

Contact Angle Change to the Time domain

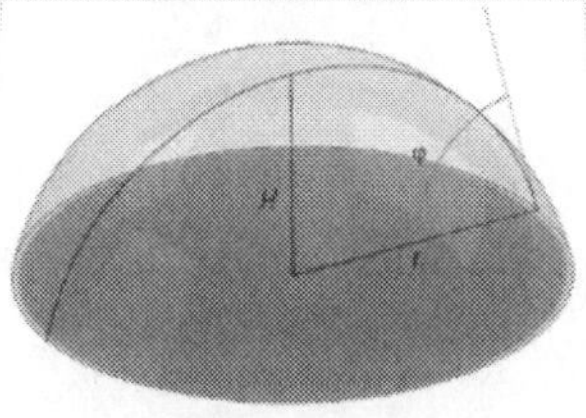

Representative of the particle – φ, Contact Angle- **r**, The radius of the base– **H**, Particle Height

- ✓ It is very critical to describe the physical structure of the mist particle
- ✓ There is two models,
  - ✓ Film Model
  - ✓ Drop Model

19

2013 ANSYS AUTOMOTIVE SIMULATION WORLD CONGRESS ISUZU

COLD CHAMBER AND MIST SIMULATION

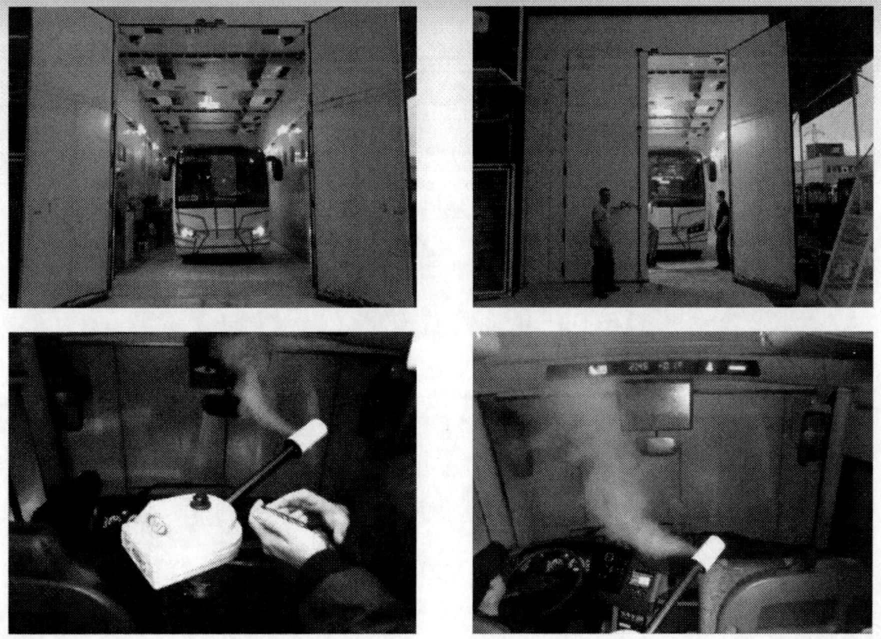

20

SIMULATION RESULTS vs. TEST RESULTS

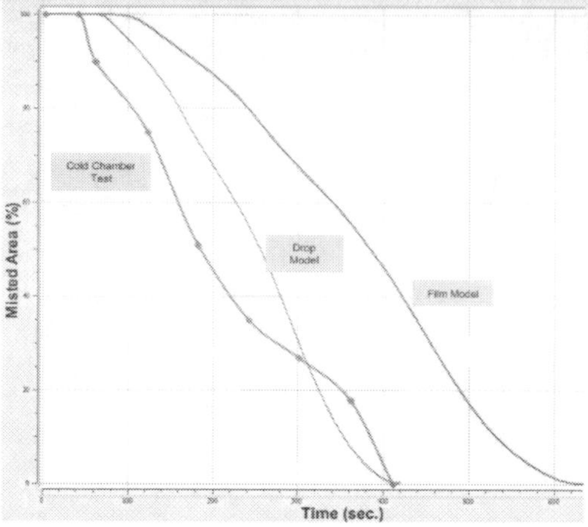

Misted Area Change according to time parameter

✓ Drop model graph is closer to the test result graph.

2013 ANSYS AUTOMOTIVE SIMULATION WORLD CONGRESS ISUZU

SIMULATION RESULTS vs. TEST RESULTS

Cold Chamber Test	Drop Model CFD Result	Film Model CFD Result
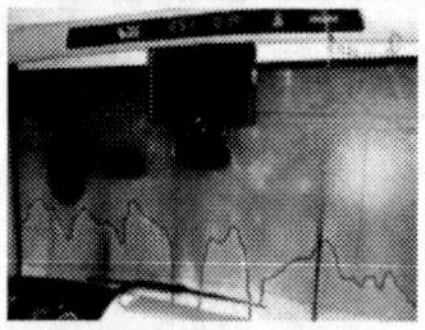	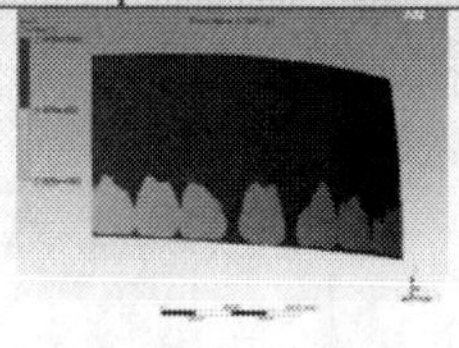	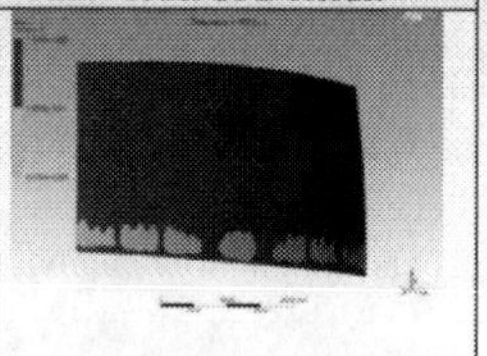
Opened Area in the 120-180 seconds	Opened Area in the 167.second	Opened Area in the 167.second
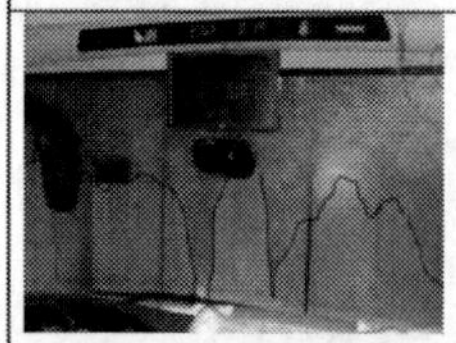	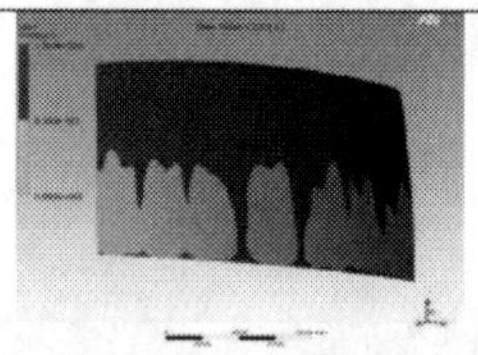	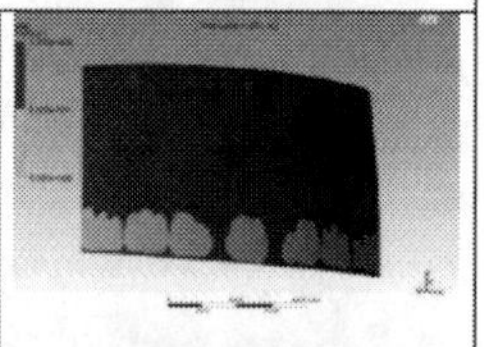
Opened Area in the 180-240 seconds	Opened Area in the 220.second	Opened Area in the 220. second
	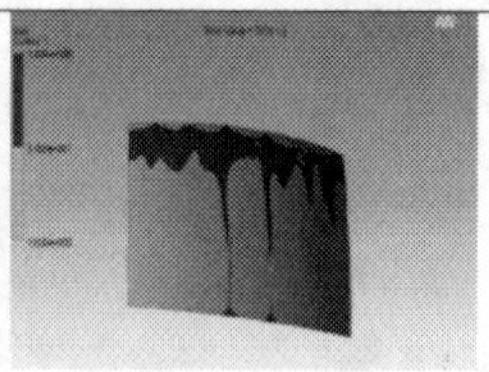	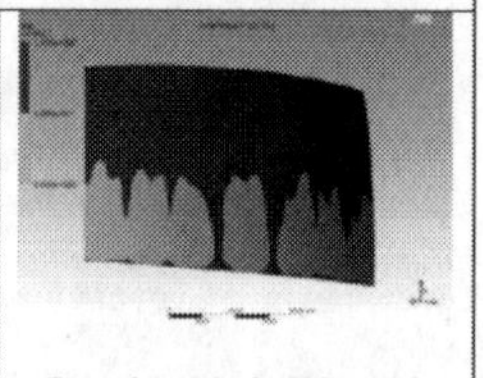
Opened Area in the 300-360 seconds	Opened Area in the 320.second	Opened Area in the 320.second

2013 ANSYS AUTOMOTIVE SIMULATION WORLD CONGRESS ISUZU

FINAL REMARKS

- ✓ CFD OPTIMIZED DEFROST DUCT DEMISTING WINDSCREEN AFTER 6 MINUTES.
  - ✓ DROP MODEL RESULTS APPROACH THE TEST RESULTS MORE, COMPARED WITH THE FILM MODEL. SO IT CAN BE USED EASILY BEFORE THE DESIGN.
    - ✓ CFX EXPRESSION LANGUAGE (C.E.L) AND ADDITIONAL VARIABLES ARE USED TO EMBEDDING EMPIRICAL CORRELATION TO SOLVER.

      (THANKS TO FLEXIBILITY OF CFX)

FURTHER DEVELOPMENTS

- ✓ SAME SOLUTION METHODOLOGY CAN BE USED OPPOSITE DIRECTION, MAKING ESTIMATION MIST CONDENSATION FORMING AND USED IN THE NUMERICAL WINDSCREEN ICE DEFROSTING.

23

2013 ANSYS AUTOMOTIVE SIMULATION WORLD CONGRESS ISUZU

Use of the Discrete Ordinate Model for Temperature Calculation in Light Systems at Volkswagen

Martin Kleimeyer, Volkswagen AG

Agenda

1 Introduction
2 Properties of the Discrete Ordinate Model
3 Results for Volkswagen Golf Tail Lamp
4 Summary

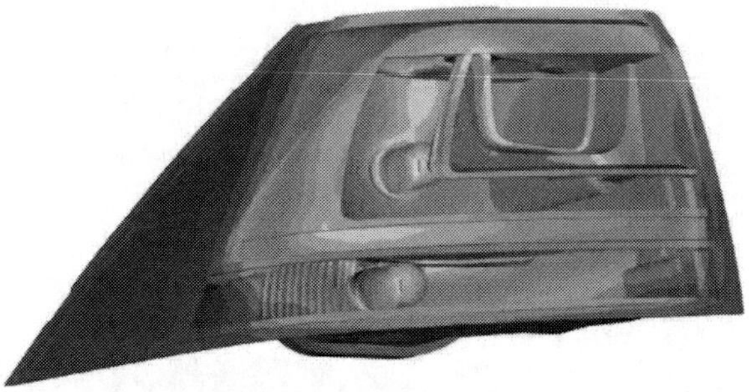

Introduction
Why Thermal Simulations?

- thermal simulation in concept development
- test drives late in development process for changes

goal: reducing the number of test drives

damages due to insufficient choice of material

Introduction
Why Thermal Simulations?

- analysis of reasons for high temperatures
- properties of plastics necessary for counter measures
- choice of material for extrem ambient conditions
- quantification of geometry changes impossible to estimate without simulation

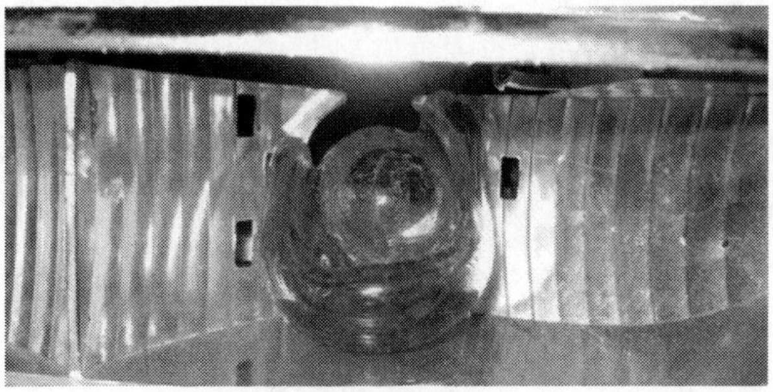

damage due to incorrect choice of material in test part

Introduction
CFD-Simulations at Volkswagen Lighting Department

- temperature simulations
  - *head lamps*
  - fog lamps
  - tail lamps
- defogging simulations
  - head and fog lamps
  - tail lamps

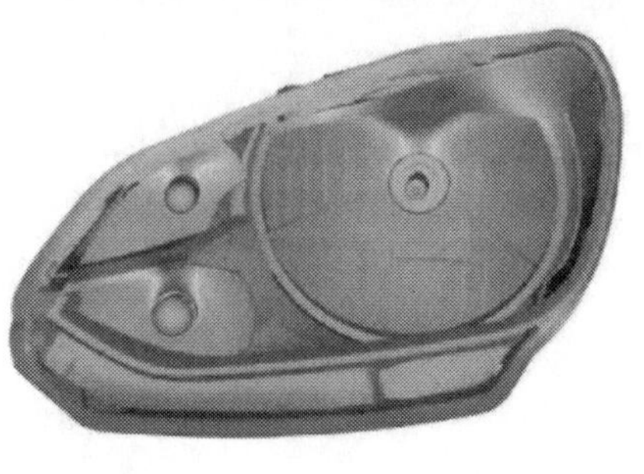

Introduction
CFD-Simulations at Volkswagen Lighting Department

- temperature simulations
  - head lamps
  - *fog lamps*
  - tail lamps
- defogging simulations
  - head and fog lamps
  - tail lamps

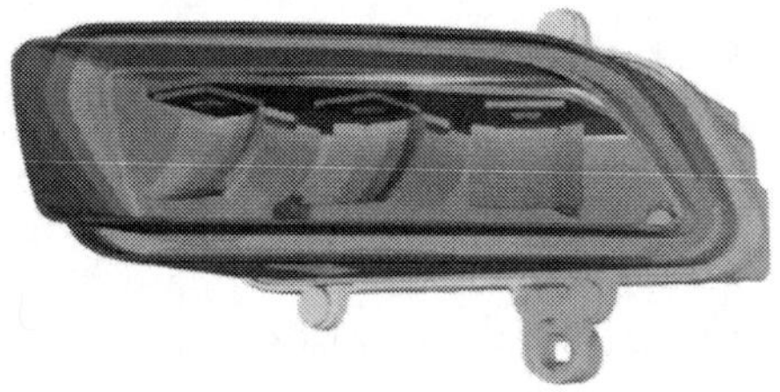

Introduction
CFD-Simulations at Volkswagen Lighting Department

- temperature simulations
  - head lamps
  - fog lamps
  - *tail lamps*
- defogging simulations
  - head and fog lamps
  - tail lamps

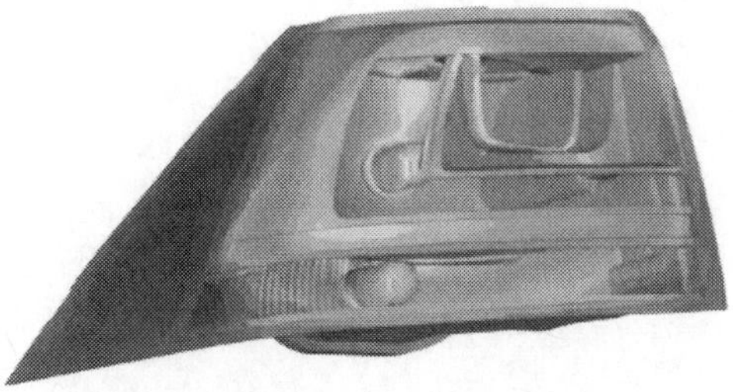

Page 7 29. October 2013 Martin Kleimeyer, Volkswagen AG

Introduction
CFD-Simulations at Volkswagen Lighting Department

- temperature simulations
  - head lamps
  - fog lamps
  - tail lamps
- defogging simulations
  - *head and fog lamps*
  - tail lamps

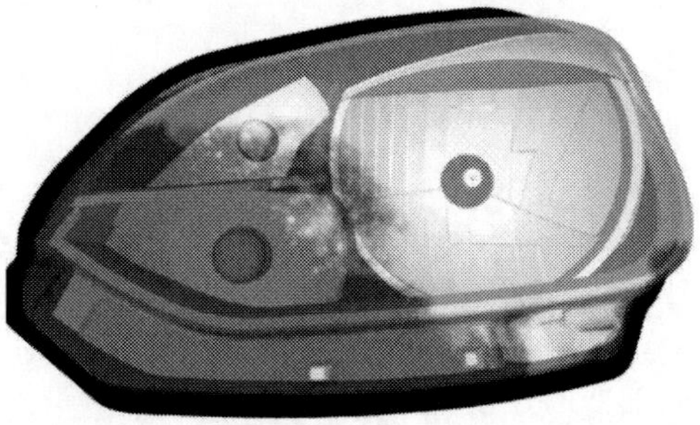

Page 8 29. October 2013 Martin Kleimeyer, Volkswagen AG

Introduction
CFD-Simulations at Volkswagen Lighting Department

- **temperature simulations**
  - head lamps
  - fog lamps
  - tail lamps
- **defogging simulations**
  - head and fog lamps
  - *tail lamps*

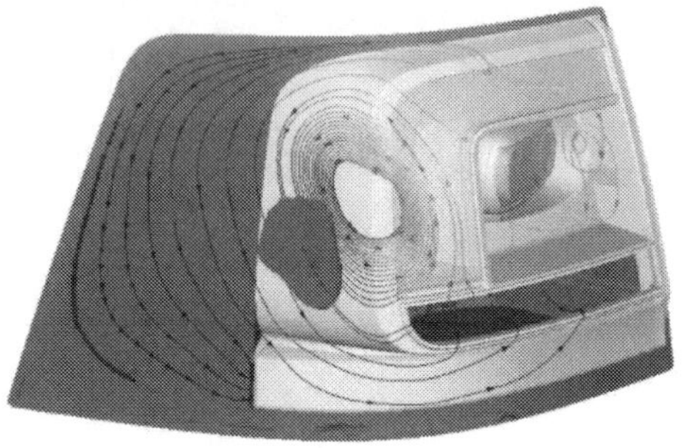

Introduction
Development Process

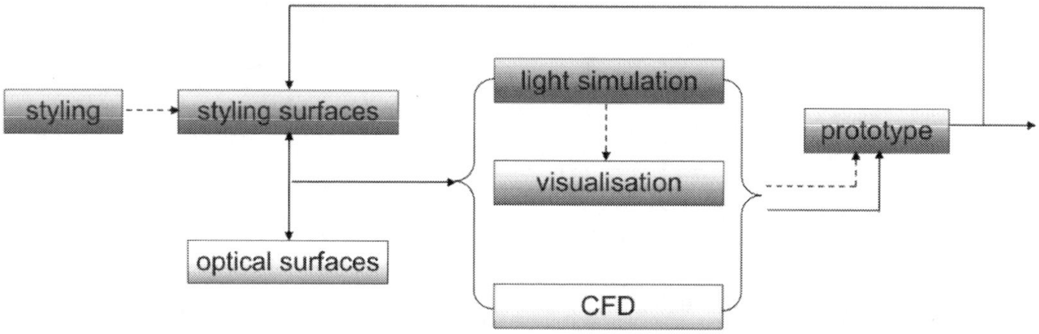

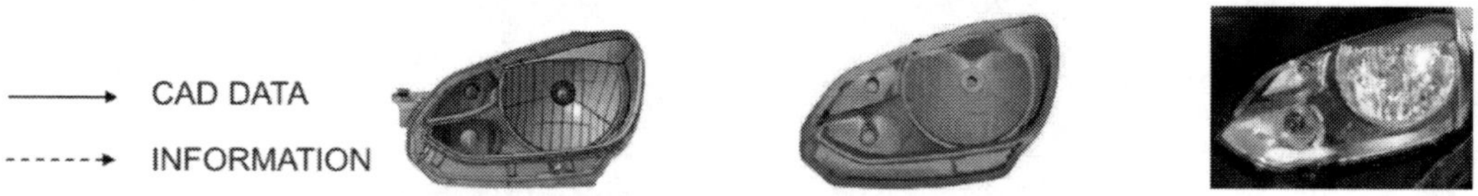

Introduction
Heat Sources and Heat Transfer Mechanisms

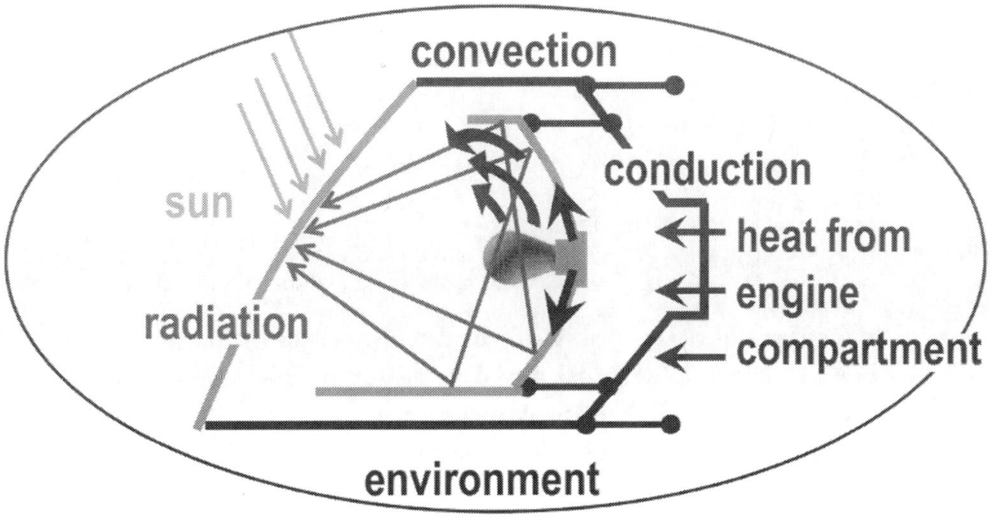

Discrete Ordinate Model
Necessary Properties of Radiation Model

- transmission
- absorption
- specular and diffuse reflection

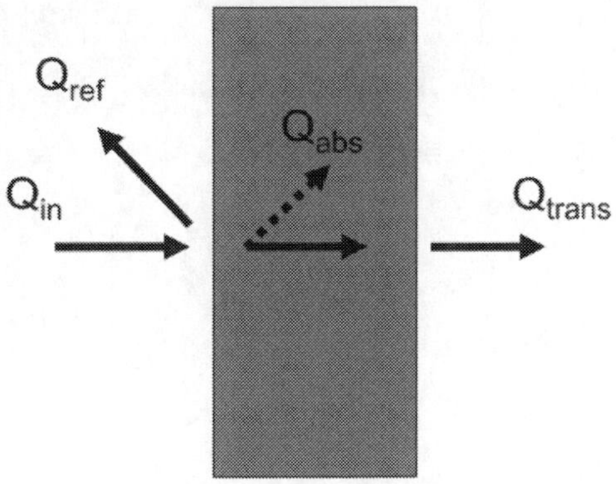

Discrete Ordinate Model
Properties of the DO-Model

equation of transport for radiation intensity:

$$\frac{dI(s,\omega)}{ds} = -(\alpha+\sigma_s)I(s,\omega) + \alpha\frac{\sigma T^4}{\pi} + \frac{\sigma_s}{4\pi}\int_0^{4\pi} I^{in}(s,\omega)\Phi d\omega$$

change along the path = - (absorption+scattering) + emission (increase due to radiation of material) + scattering (increase due to neighbouring cells)

- 1 transport equation is solved for each discrete direction
- space is split up in oktants: e.g. 6x6 directions per oktant yields 6x6x8 = 288 additional transport equations

⇨ huge increase in computational memory and calculation time

Discrete Ordinate Model
Test Model

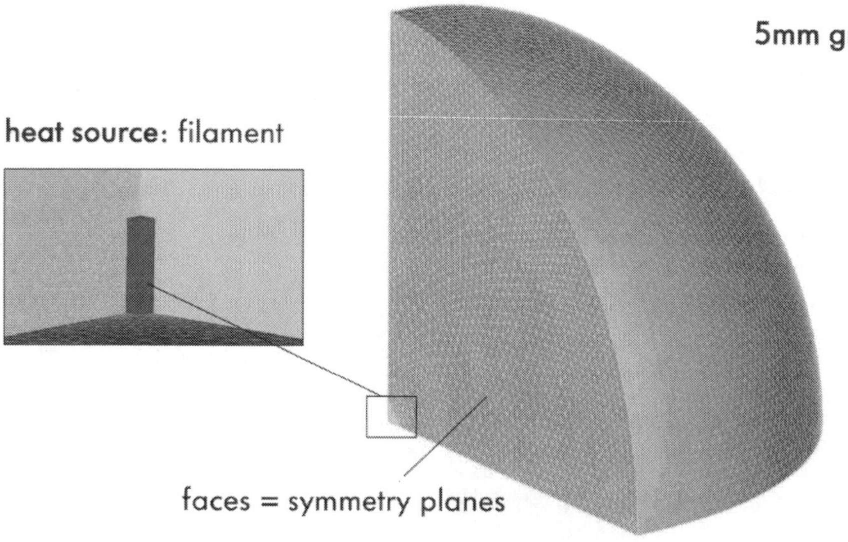

Discrete Ordinate Model
Grid

surface resolution	5 mm	3 mm	1 mm

no. of cells	236361	1008042	2908869

Discrete Ordinate Model
Spatial Discretisation

grid

5 mm

3 mm

1 mm

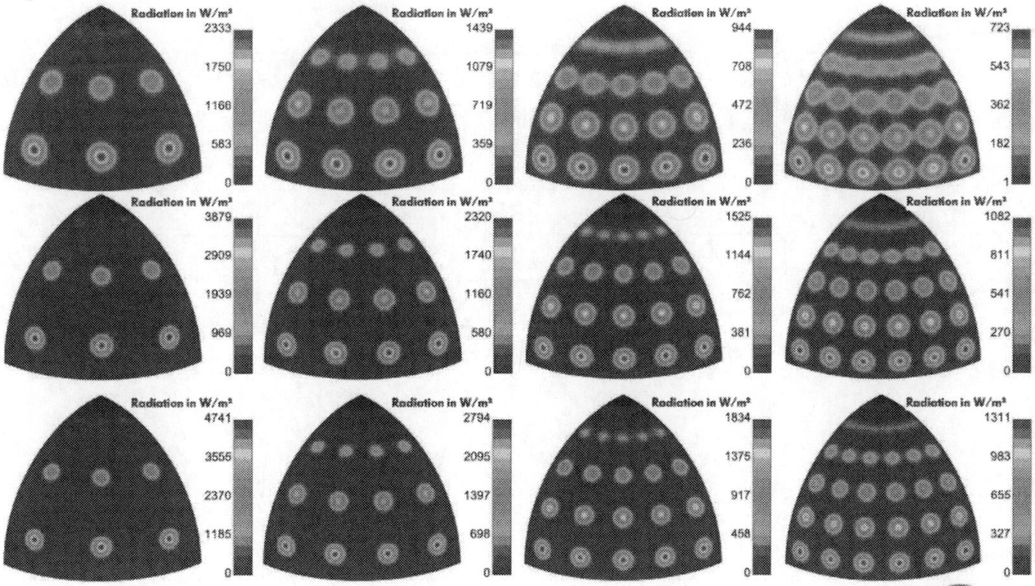

Discrete Ordinate Model
Spatial Discretisation

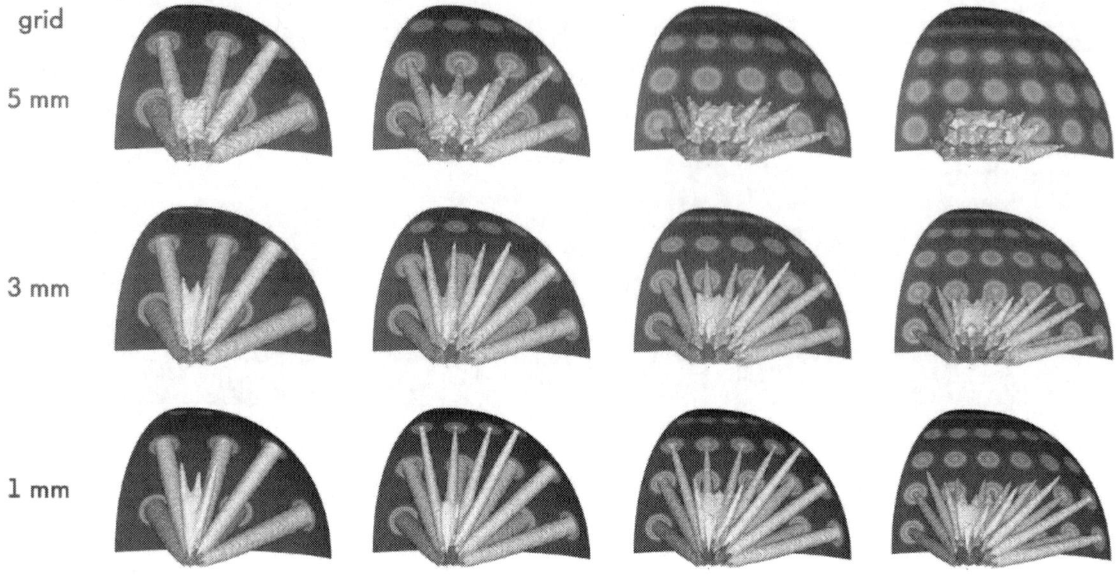

isosurfaces of radiation intensity

Discrete Ordinate Model
Results

surface resolution	5 mm	3 mm	1 mm
no. of cells	236361	1008042	2908869

q_{max} 3x3	2333	3879	4747
q_{max} 4x4	1439	2320	2794
q_{max} 5x5	944	1525	1834
q_{max} 6x6	723	1082	1311

Discrete Ordinate Model
Effects of Grid Rotation

grid 5 mm

3x3

3x3

15° rotation
on z-axis

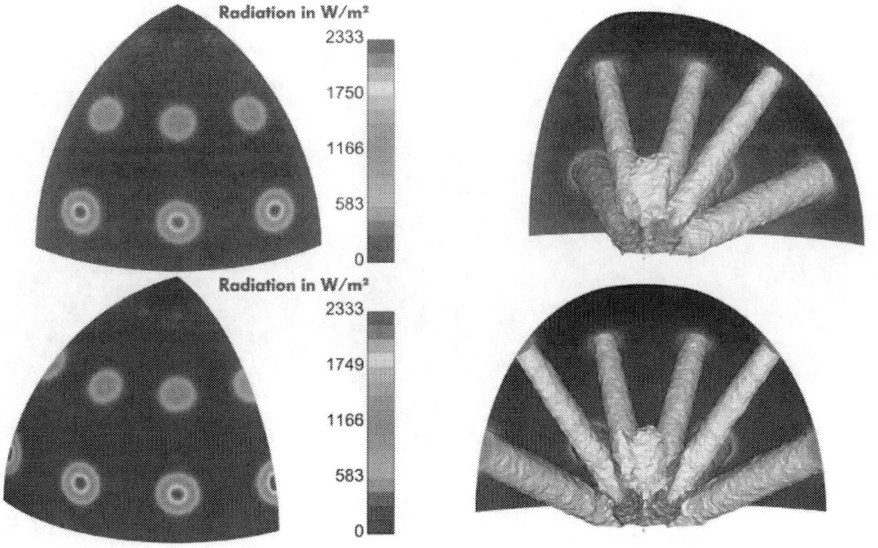

Discrete Ordinate Model
Grid Rotation – Results for *Volkswagen up*!

3x3

3x3 – 45° rotation

radiation
distribution

temperature
distribution

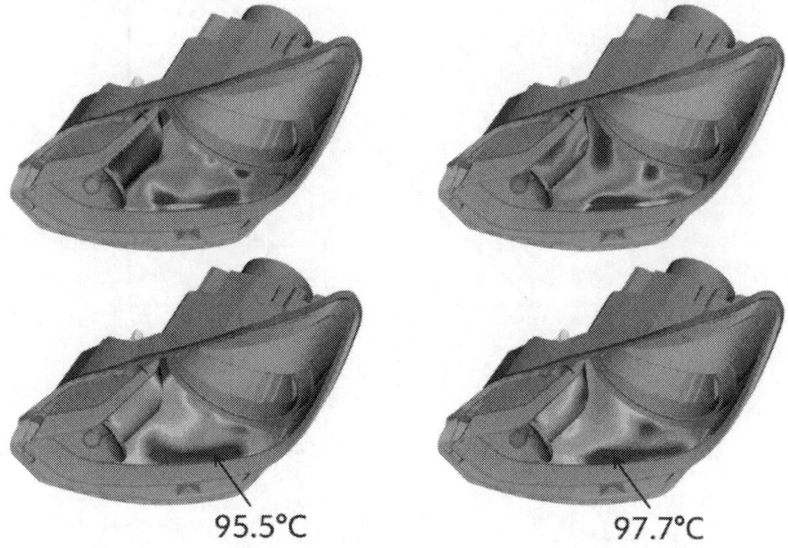

Discrete Ordinate Model
Grid Rotation – Results for *Volkswagen up!*

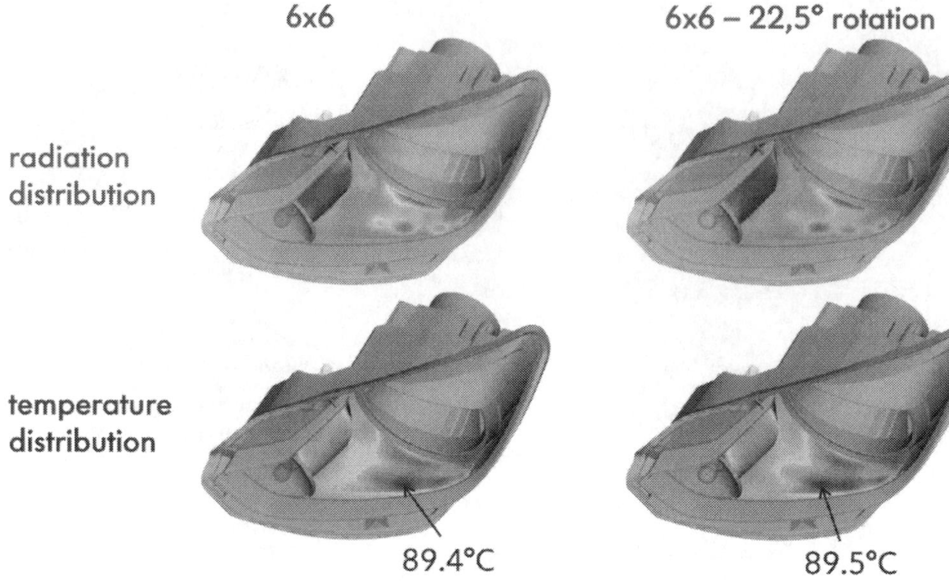

Discrete Ordinate Model
Specular Reflection

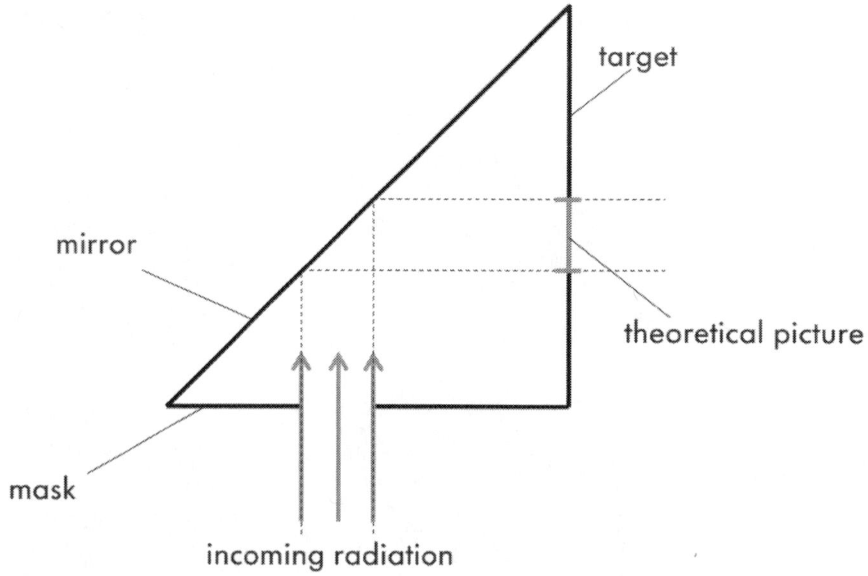

Discrete Ordinate Model Specular Reflection

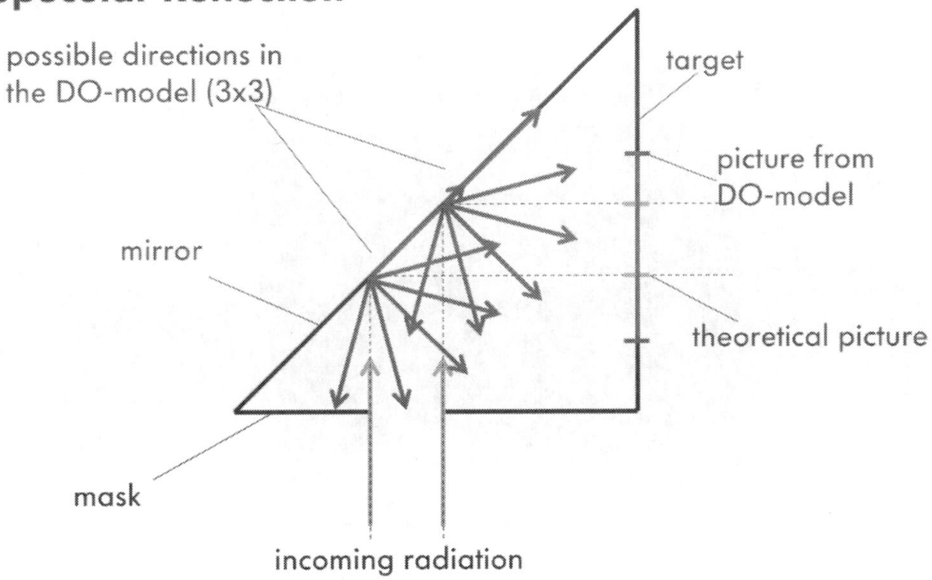

Page 23 29. October 2013 Martin Kleimeyer, Volkswagen AG

Discrete Ordinate Model Specular Reflection

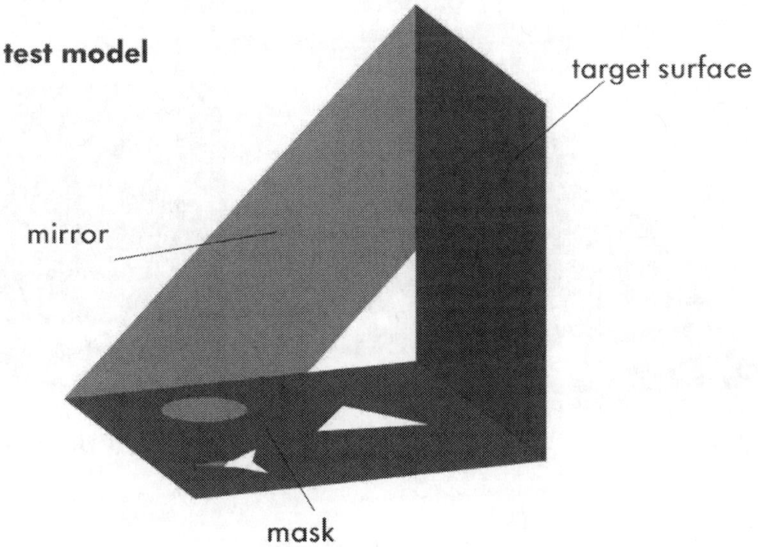

Page 24 29. October 2013 Martin Kleimeyer, Volkswagen AG

Discrete Ordinate Model
Specular Reflection

radiation distribution on the mirror

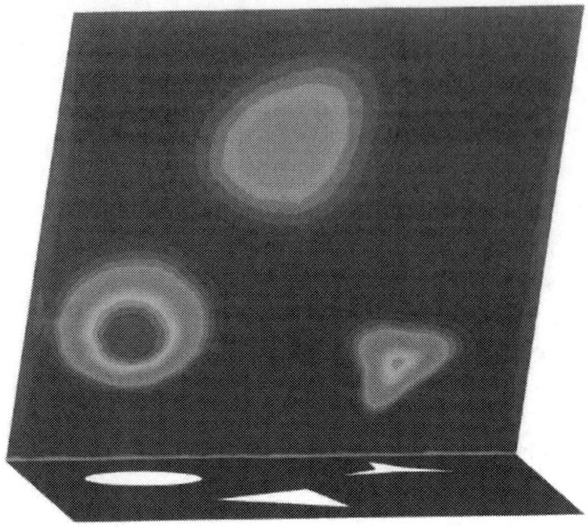

Discrete Ordinate Model
Specular Reflection

radiation distribution on the target surface

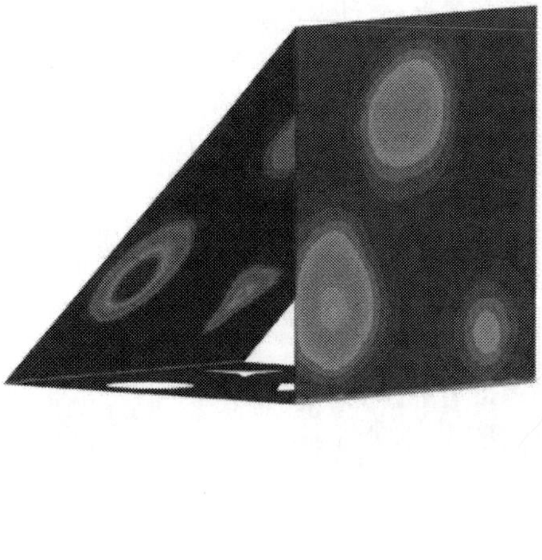

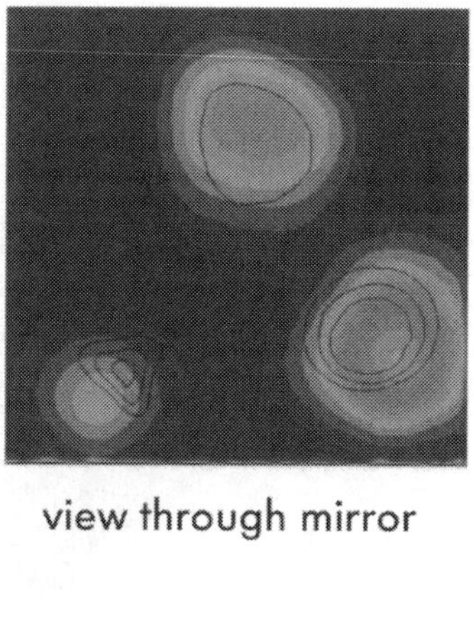

view through mirror

Discrete Ordinate Model
Summary

spatial resolution:

- the finer the grid, the finer the rays (ray effect)
- the finer the spatial discretisation, the lower the energy maximum (energy conservation)

discrete ordinates:

- results dependent on the spatial discretization (i.e. number of ordinates)

specular reflection

- increase in the size of the pictures due to limited spatial discretisation
- change in the positions of the pictures
- 3 seperate spots instead of one large blur!

Comparison between Simulation and Measurement
Example Golf Tail Lamp

3 functions:

- tail/stop light
- turn indicator
- approx. 35W

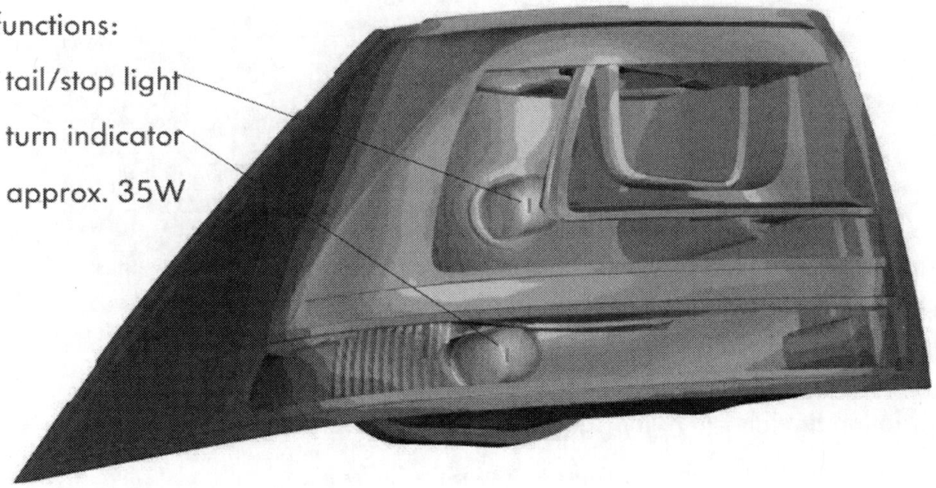

Comparison between simulation and measurement Example Golf Tail Lamp

temperatures on bezel

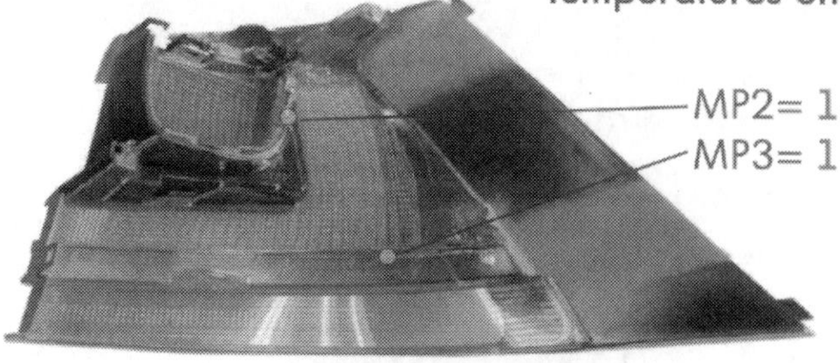

stop/tail bulb in section

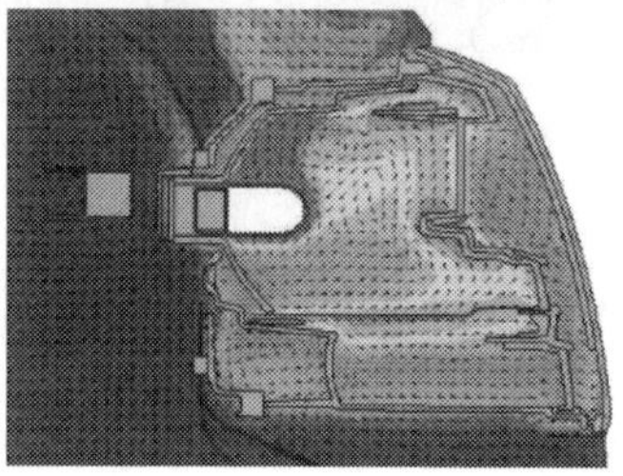

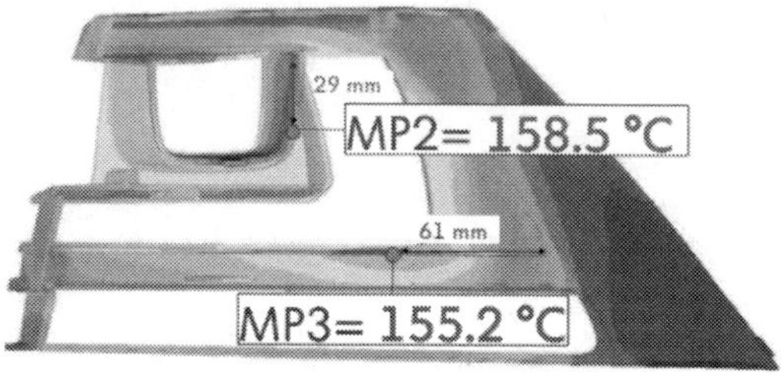

Comparison between simulation and measurement Example Golf Tail Lamp

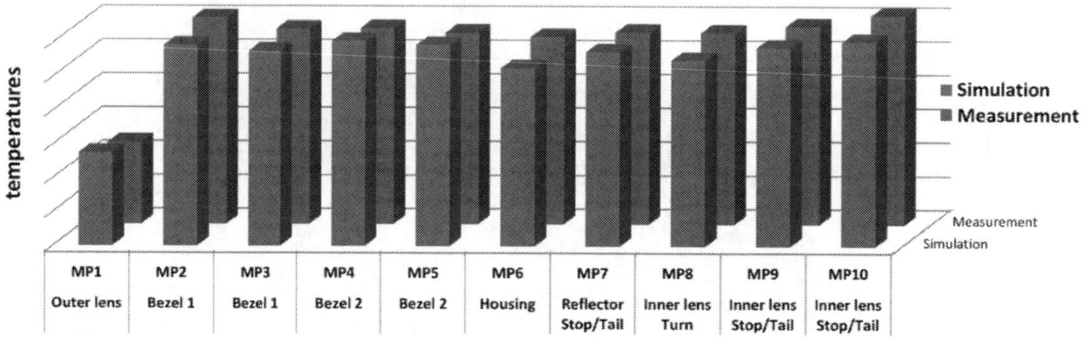

- simulation deviates in higher and lower temperatures
- simulation is close to measurements - within 5K-range

Summary

DO-model

- great increase in computational resources
- limitations due to spatial discretisation for specular reflection

temperature simulations

- due to specular reflexion limits in calculating hot spots
- very good agreement between simulations and measurements possible

Thank you for your attention!

DÜRR

LEADING IN PRODUCTION EFFICIENCY

PAINT AND FINAL ASSEMBLY SYSTEMS

Heat-Up Simulation of Car Bodies in Paint Shop Ovens

Christof Knüsel, R&D

Frankfurt, 29.10.2013

www.durr.com

DÜRR GROUP STRUCTURE WITH 4 GLOBAL DIVISIONS

DÜRR

Overview 2012 (2011)

DÜRR-Group			
Paint and Assembly Systems Sales: € 1,125.2 m *(€ 878.7 m)*	**Application Technology** Sales: € 531.2 m *(€ 406.8 m)*	**Measuring and Process Systems** Sales: € 647.9 m *(€ 550.4 m)*	**Clean Technology Systems** Sales: € 95.5 m *(€ 86.1 m)*

Paintshops and Final Assembly	Paint Application Technology	Balancing and Assembly Products	Air Purification Systems
	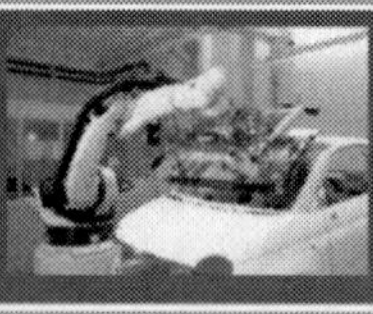		
Aircraft Production Technology	Gluing and Sealing Technology	Cleaning and Filtration Systems	Energy Efficiency Systems

OVERVIEW: CFD AT DÜRR

DÜRR

- Use of CFD: Since 1998
- Software: Fluent
- Capacity: 4 people

OPEL AIRBUS RENAULT HONDA KARMANN JAGUAR valmet automotive DAIMLER HYUNDAI PORSCHE Audi

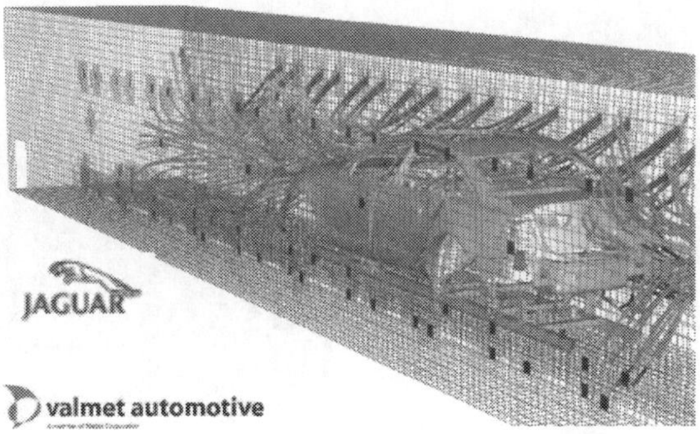

PAINT SHOP OVEN

DÜRR

Overview

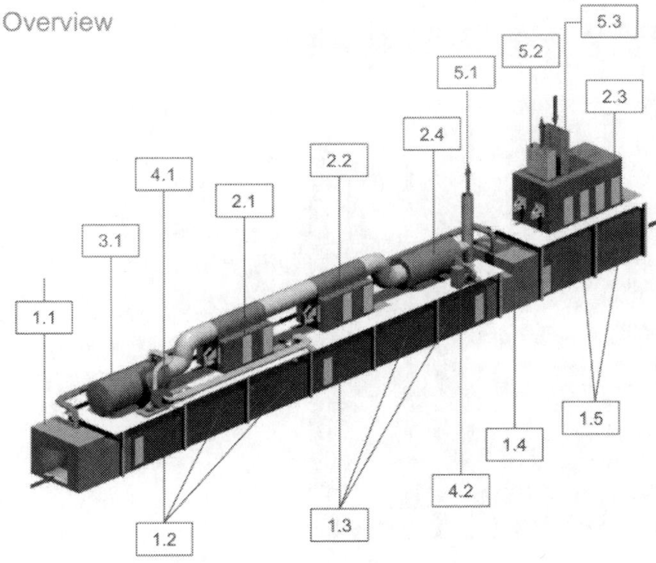

1. Tunnel
   - 1.1 Entrance air seal
   - 1.2 Heat-up zone
   - 1.3 Hold zone
   - 1.4 Exit air seal
   - 1.5 Cooling zone
2. Aggregates
   - 2.1 Hot box (heat-up zone)
   - 2.2 Hot box (hold zone)
   - 2.3 Cooling unit
   - 2.4 Fresh air unit
3. Abatement
   - 3.1 TAR
4. Fans (stand alone)
   - 4.1 Exhaust air fan
   - 4.2 Fresh air fan
5. Stacks
   - 5.1 Exaust air (oven)
   - 5.2 Extract air (cooling)
   - 5.3 Supply air (cooling)

HEAT-UP SIMULATION

Goal:

- Flow and turbulence
- Heat transfer (convection, conduction, radiation)
- Heat-up curves
- Stress analysis

- No calibration

© PFS,R&D, Automotive Simulation World Congress 2013

HEAT-UP SIMULATION

Benefits:

- Improve performance at critical measuring points
- Optimize nozzle design
- Optimize nozzle arrangement
- Optimize process parameters

- Air balance of complete oven
- Optimize air seals

References:

- Audi
- BMW
- Daimler
- FAW
- Ford
- Honda
- Opel

© PFS,R&D, Automotive Simulation World Congress 2013

HEATUP SIMULATION

Decoupled simulation

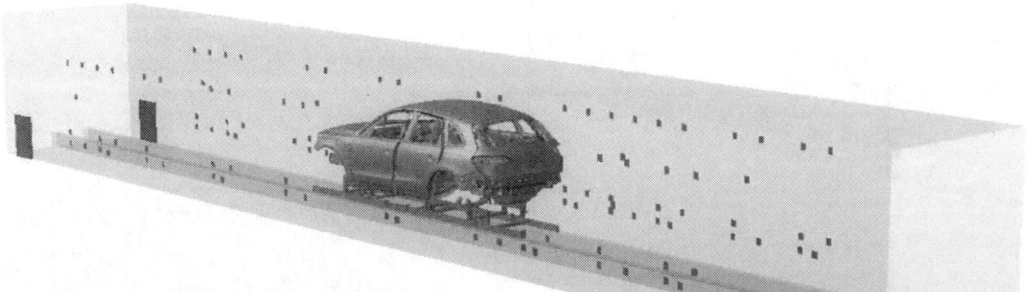

1. Flow simulation (sliding mesh)
2. Calculating average flow field (velocity, turbulence, heat transfer coefficients) for each oven zone
3. Heat-up simulation (convection, heat conduction, radiation) based on average flow field

MODELING OF THE CAR BODY

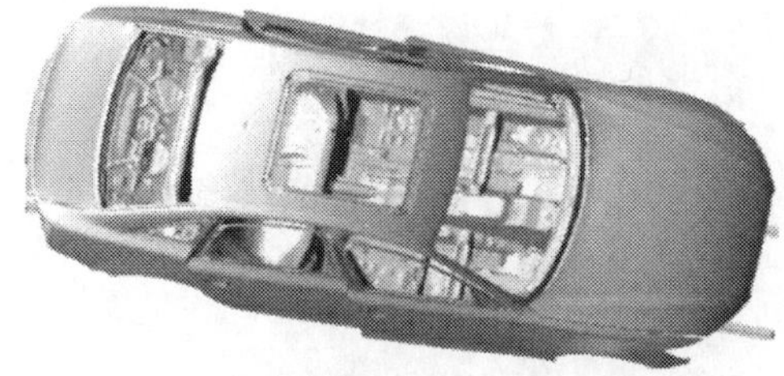

- Not only outer shell, but any interior plate has to be resolved
- 0-thickness walls (shell conduction)
- Gaps < 3 mm are closed

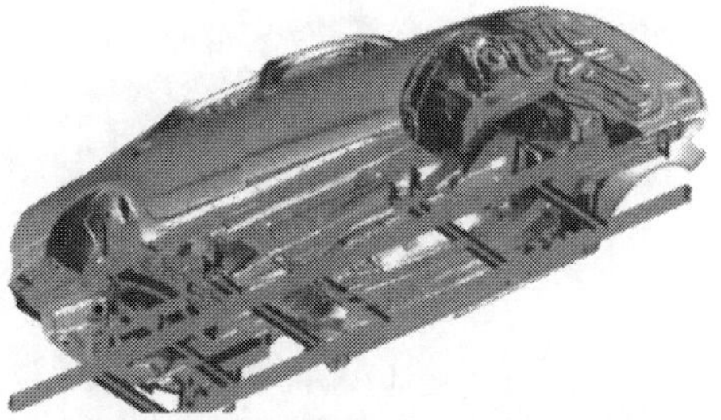

MODELING OF THE CAR BODY

Boundary zones

- Individual boundary zone for any plate thickness and any material (or any mixed material, if plates are sewn together)
- 300 boundary zones are possible

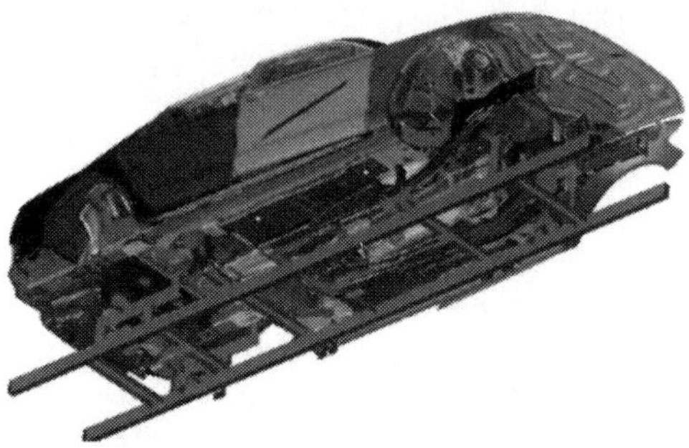

FLOW SIMULATION

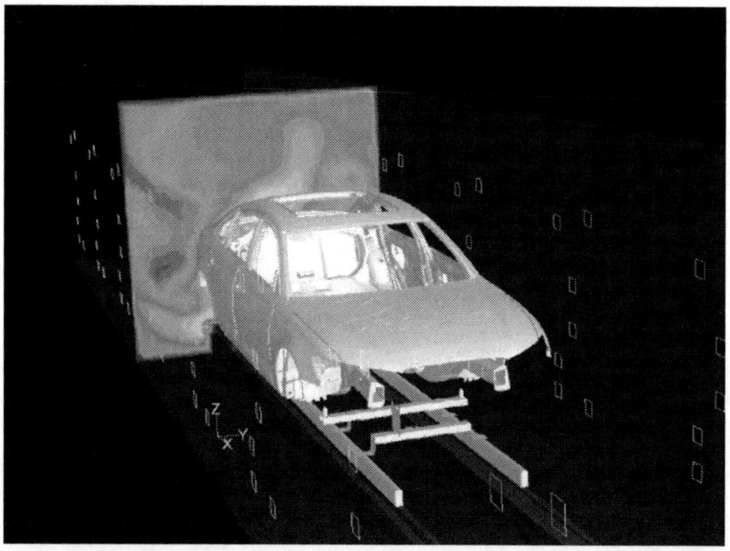

DÜRR

FLOW SIMULATION

Velocity [m/s] at time x

DÜRR

FLOW SIMULATION

Averaged velocity [m/s] (user-defined function)

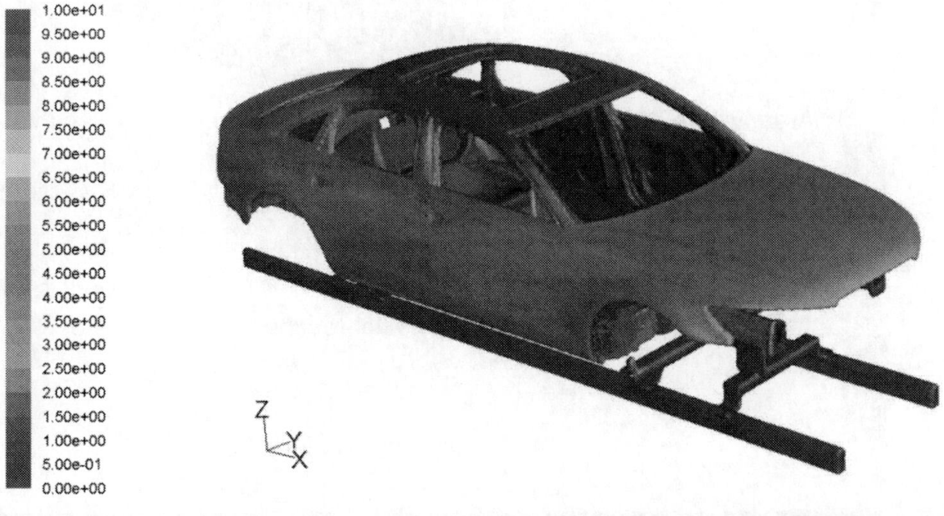

HEAT-UP SIMULATION

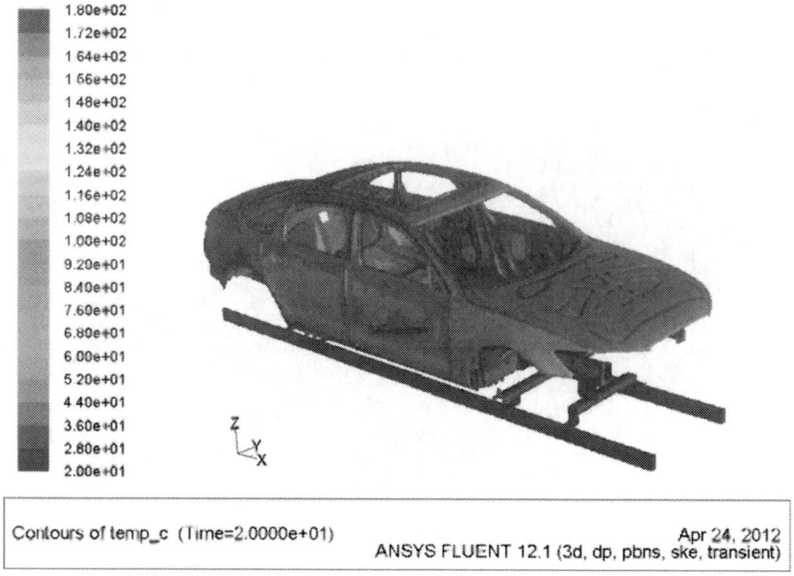

HEAT-UP SIMULATION

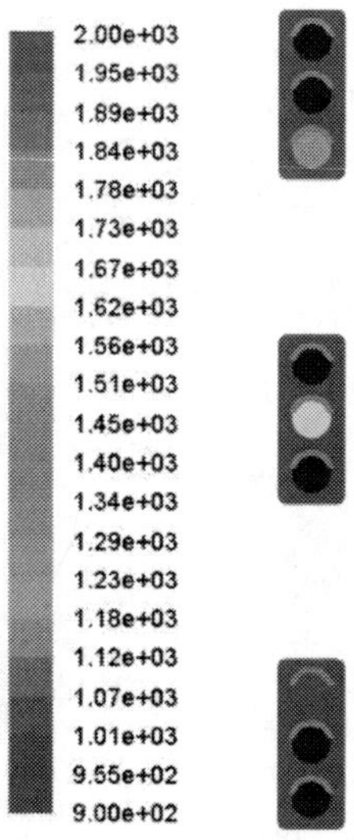

Area, where the heat-up criterion is reached.
Scale: Time above T_{min} [s]

Area, where the heat-up criterion is hardly reached.

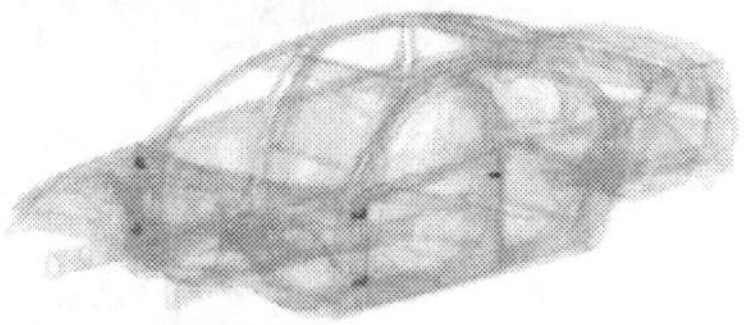

Area, where the heat-up criterion is not reached.

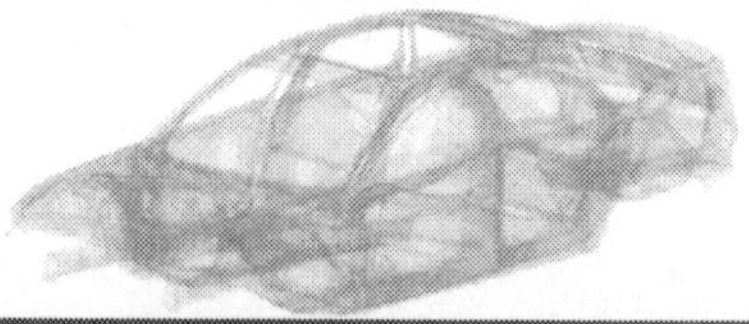

STRESS ANALYSIS

- Based on the temperature distribution within the car body at each time step, a stress and deformation analysis may follow.
- This will be done by an external supplier and with a different software (Abaqus).

SUMMARY

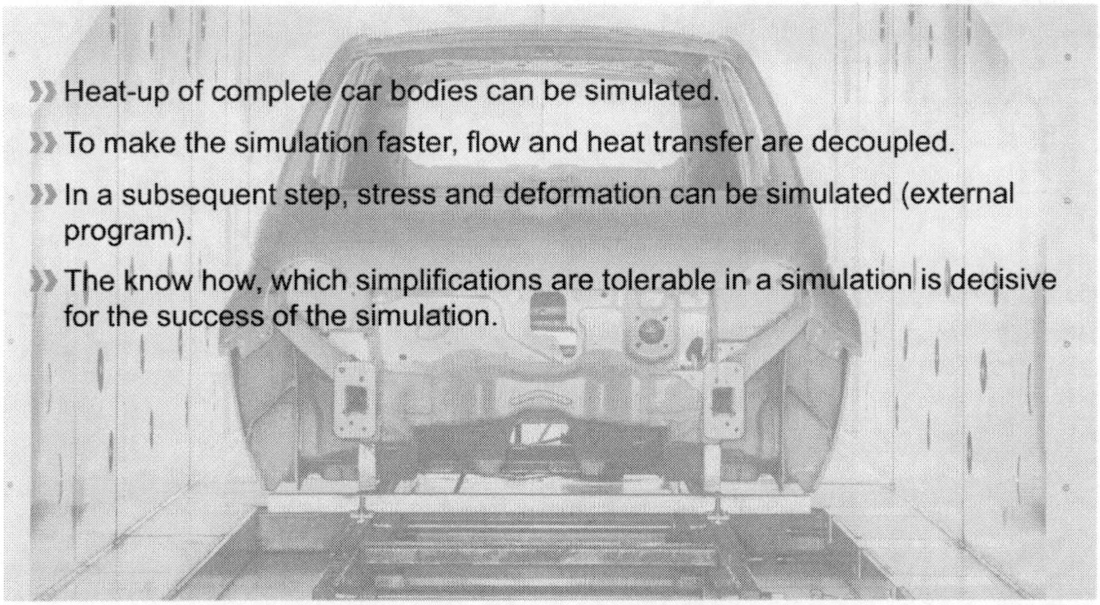

- Heat-up of complete car bodies can be simulated.
- To make the simulation faster, flow and heat transfer are decoupled.
- In a subsequent step, stress and deformation can be simulated (external program).
- The know how, which simplifications are tolerable in a simulation is decisive for the success of the simulation.

DÜRR

LEADING IN PRODUCTION EFFICIENCY

PAINT AND FINAL ASSEMBLY SYSTEMS

Heat-Up Simulation of Car Bodies in Paint Shop Ovens

Christof Knüsel, R&D

Frankfurt, 29.10.2013

www.durr.com

2013 Automotive Simulation World Congress

Quality meets Efficiency

Developing Best Practice Guidelines for Cooling applications in Automotive Manufacturing Tool Design

Marco Lanfrit, ANSYS Germany GmbH

Content

- The three pillars of Quality & Efficiency
- How to develop a best practice methodology
- Application Example: Hot Forming Tool
- Summary

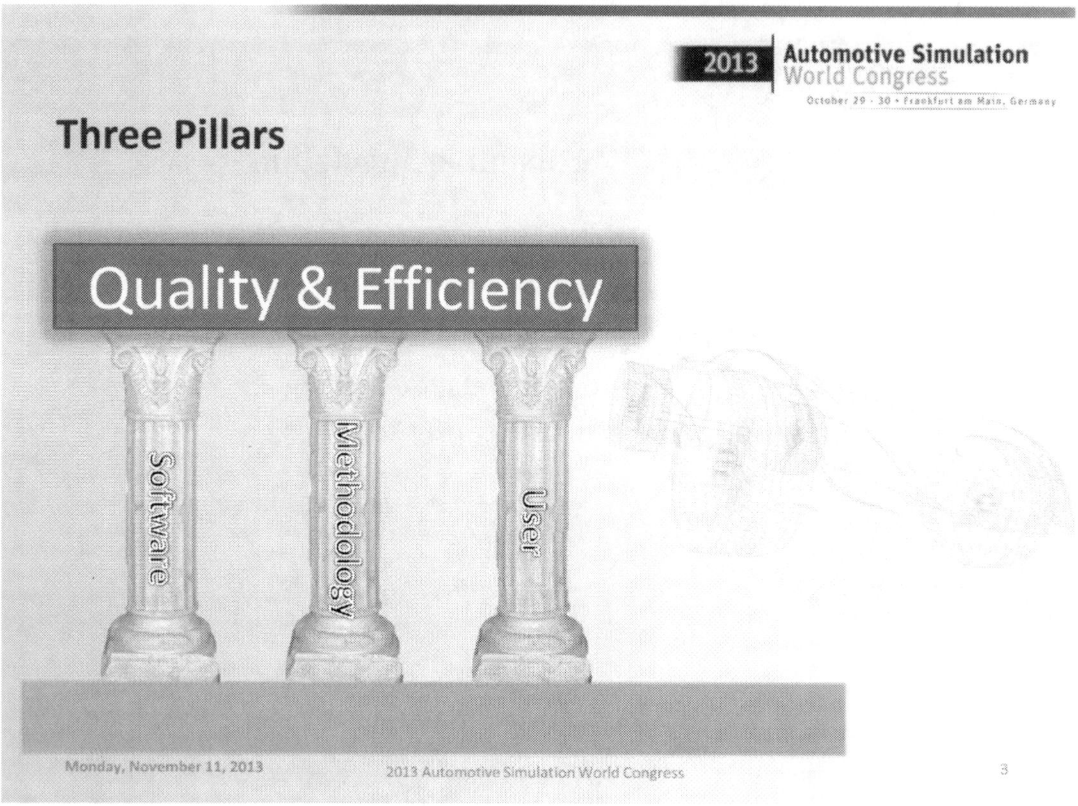
2013 Automotive Simulation World Congress
October 29 - 30 • Frankfurt am Main, Germany
Three Pillars
Quality & Efficiency
Software
Methodology
User
Monday, November 11, 2013
2013 Automotive Simulation World Congress
3

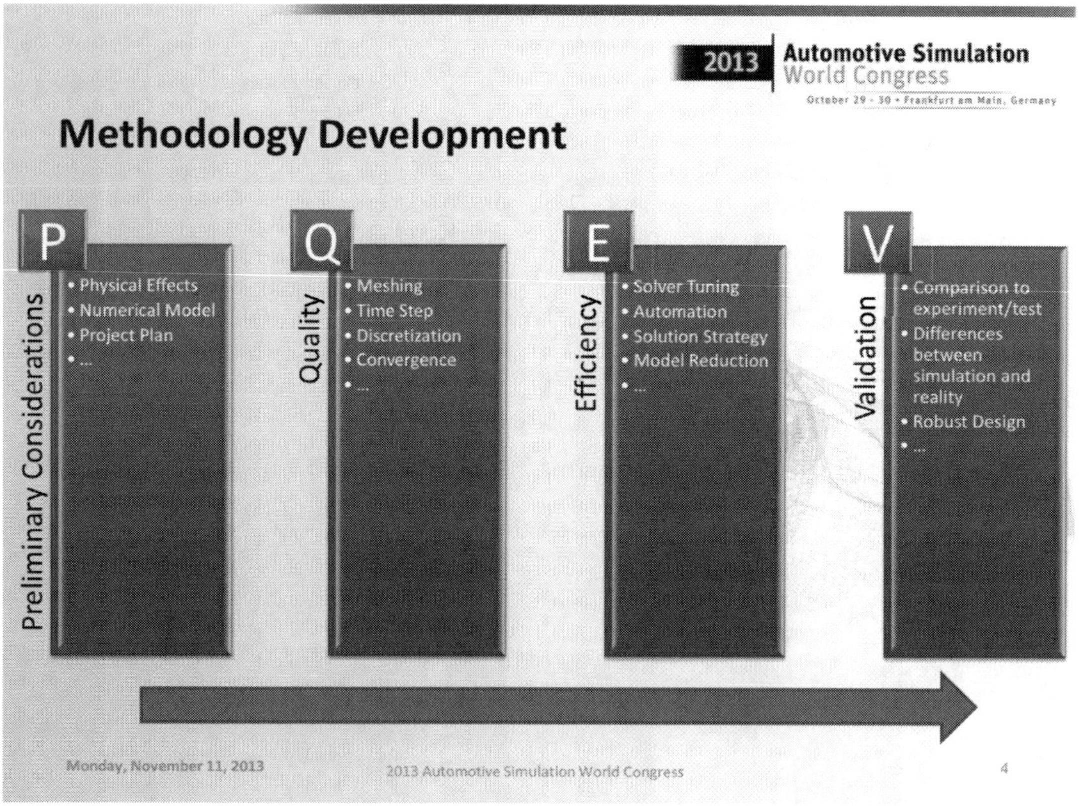
2013 Automotive Simulation World Congress
October 29 - 30 • Frankfurt am Main, Germany
Methodology Development
P
Preliminary Considerations
• Physical Effects
• Numerical Model
• Project Plan
• …
Q
Quality
• Meshing
• Time Step
• Discretization
• Convergence
• …
E
Efficiency
• Solver Tuning
• Automation
• Solution Strategy
• Model Reduction
• …
V
Validation
• Comparison to experiment/test
• Differences between simulation and reality
• Robust Design
• …
Monday, November 11, 2013
2013 Automotive Simulation World Congress
4

Methodology Development

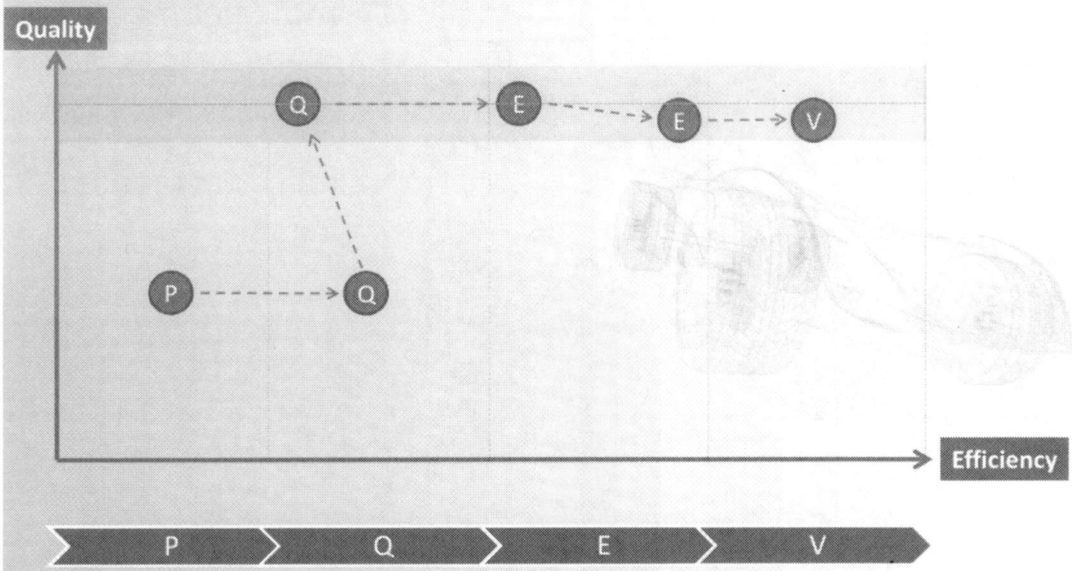

2013 Automotive Simulation World Congress
October 29 - 30 • Frankfurt am Main, Germany

Application: Hot Forming Tool Design

- Process:
  - 1-Step
  - 2-Step
- Advantages:
  - Improved Safety
  - Weight Reduction
- Challenges:
  - Short cycle times
  - Long tool life
  - Maintenance costs

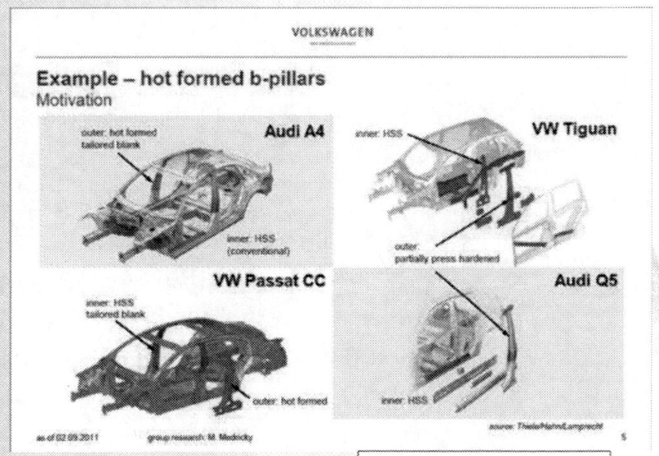

courtesy of Volkswagen AG

Hot Forming Test Facility

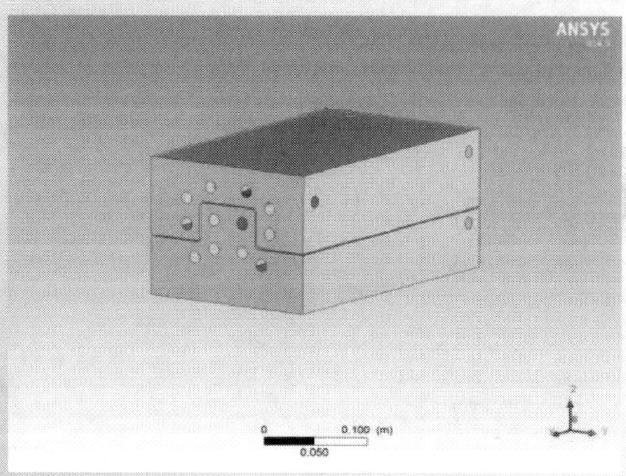

courtesy of Volkswagen AG

- Simple geometry
- Realistic cooling scenario
- Experimental data available

2013 Automotive Simulation World Congress
October 29 - 30 • Frankfurt am Main, Germany

Potential Weaknesses and Best Practice

- Error types influencing quality:
  - Numerical errors
  - Model errors
  - Systematic errors
  - User errors

- Best Practice means:
  - Reduction of numerical errors by deterministic and rational procedures
  - Reduction of model and systematic errors by validation
  - Well documented methodology and high degree of automation to avoid user errors

2013 Automotive Simulation World Congress
October 29 - 30 • Frankfurt am Main, Germany

Preliminary Considerations

- Physical Effects to be considered:
  - turbulent flow
  - conjugate heat transfer
  - transient fluid behaviour

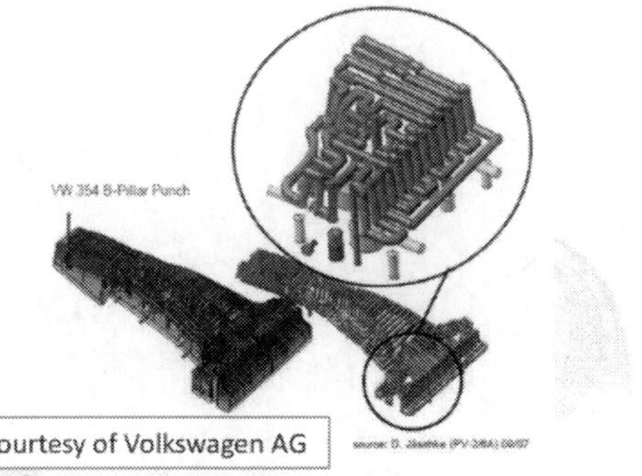

- Challenges:
  - complex geometry
  - minimum user interaction expected
  - tool heating ➔ multi cycle simulation > 20 minutes

2013 Automotive Simulation World Congress
October 29 - 30 • Frankfurt am Main, Germany

Computational Mesh

Goal: Find the best compromise between accuracy, efficiency and easiness to generate

- Resolution
  - Near Wall: modeling or resolving (y-plus)
  - Core Region: degree of refinement
- Quality
- Meshing Method: easiness to generate

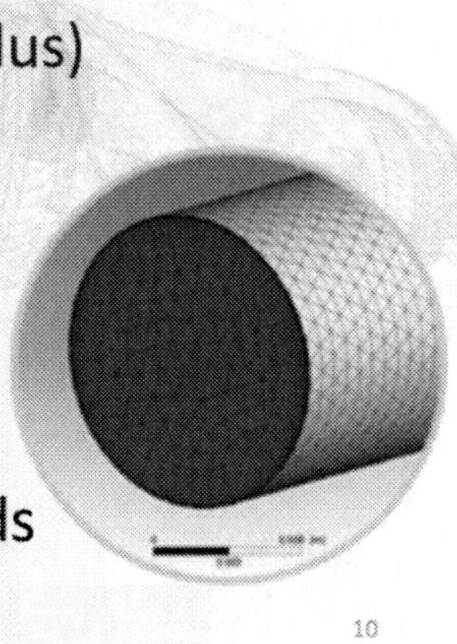

➔ Comparison of results on different grids

Computational Mesh – Near Wall Resolution

- Simple formula to estimate y-plus for duct flows:

$$\frac{y}{D} \cong 7\mathrm{Re}_D^{-7/8}\, y^+ \qquad \approx \frac{10}{\mathrm{Re}_D}$$

- Or use an excel based turbulence calculator (example available at ANSYS Customer Support)

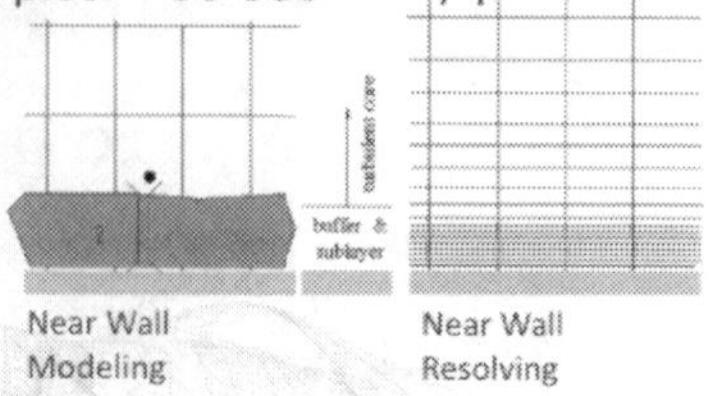

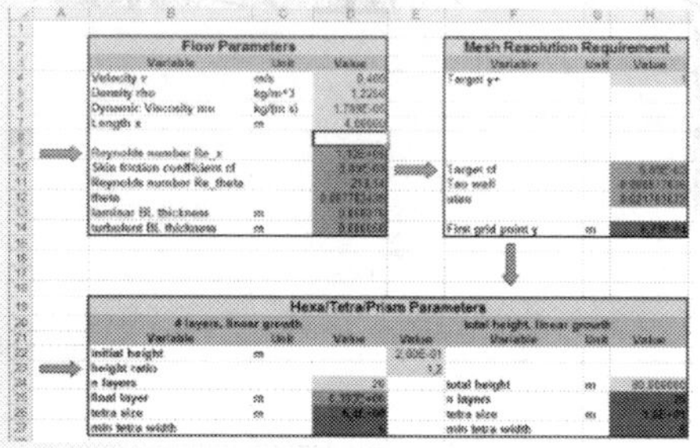

2013 Automotive Simulation World Congress
October 29 - 30 • Frankfurt am Main, Germany

Computational Mesh – Near Wall Resolution

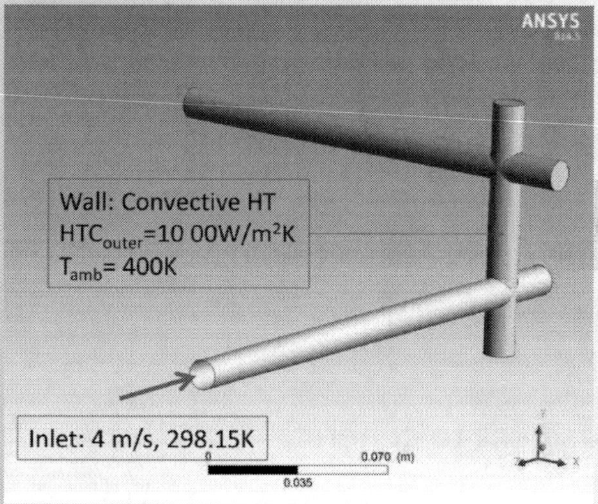

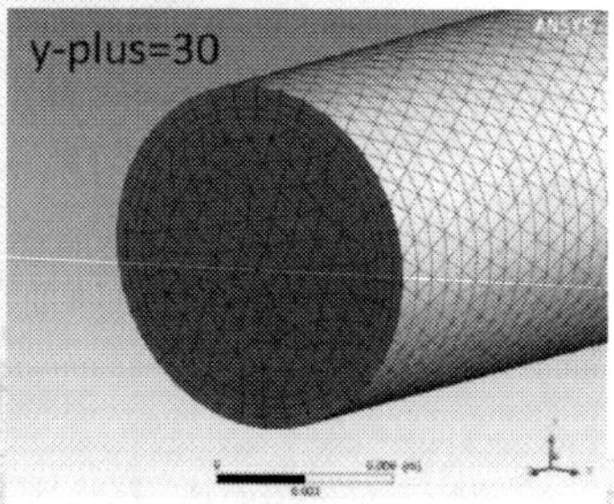

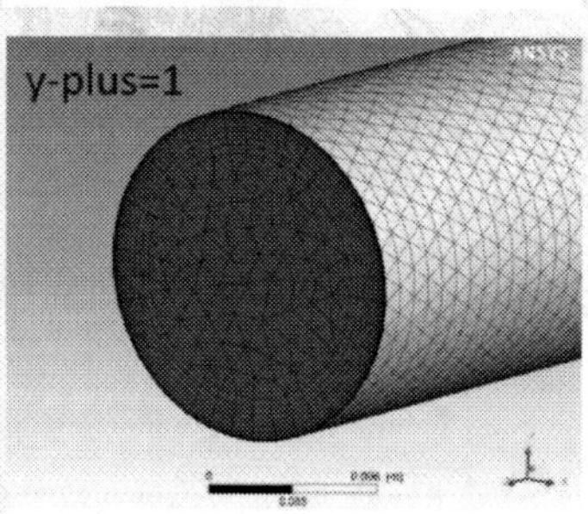

Computational Mesh – Near Wall Resolution

- 970.000 cells
- $\Delta p_{90°}$=6.982 Pa
- THF=1.653 W

y^{+}=1

- 475.000 cells
- $\Delta p_{90°}$=6.411 Pa
- THF=1.667 W

y^{+}=30

50%

>8%

<1%

2013 Automotive Simulation World Congress
October 29 - 30 • Frankfurt am Main, Germany

Computational Mesh – Near Wall Resolution

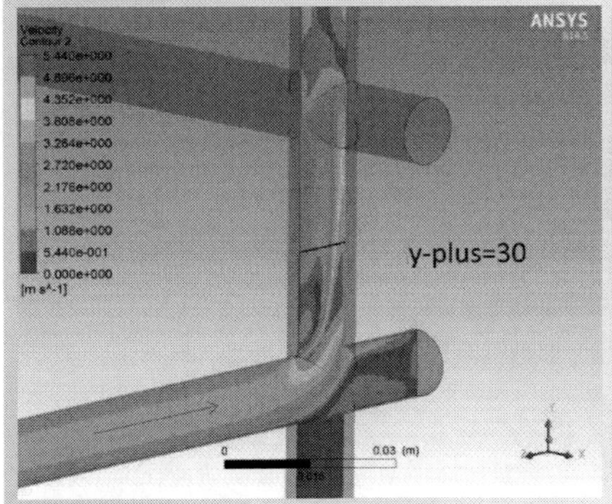

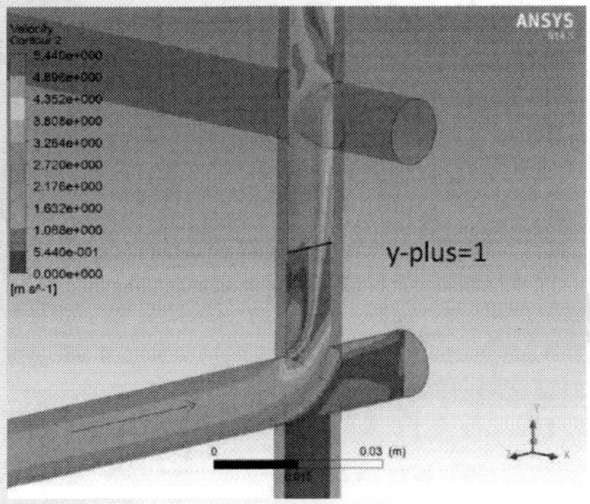

Computational Mesh – Near Wall Resolution

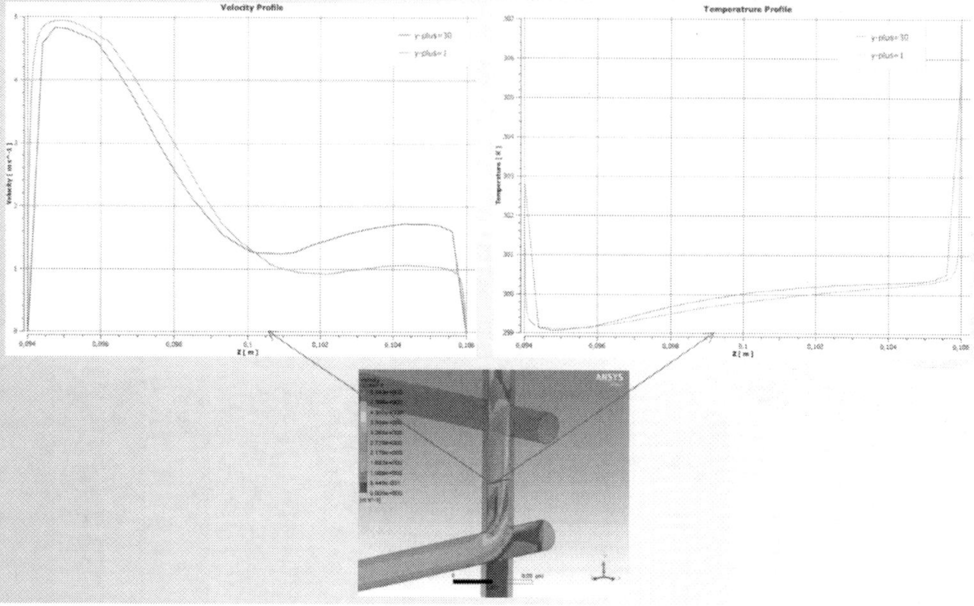

2013 Automotive Simulation World Congress
October 29 - 30 • Frankfurt am Main, Germany

Best Practice Decisions

Mesh

- Near wall resolving mesh (y-plus=1)

Solution Strategy

Transient Solution

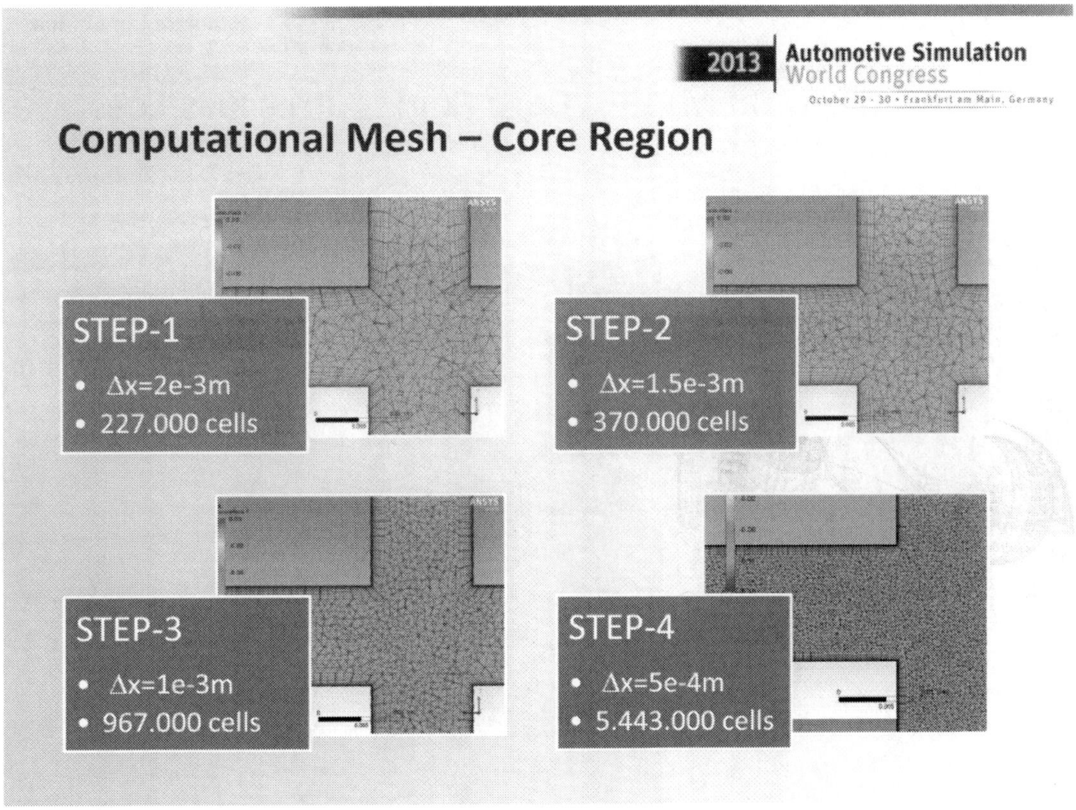
2013 Automotive Simulation World Congress
October 29 - 30 • Frankfurt am Main, Germany
Computational Mesh – Core Region
STEP-1
• Δx=2e-3m
• 227.000 cells
STEP-2
• Δx=1.5e-3m
• 370.000 cells
STEP-3
• Δx=1e-3m
• 967.000 cells
STEP-4
• Δx=5e-4m
• 5.443.000 cells

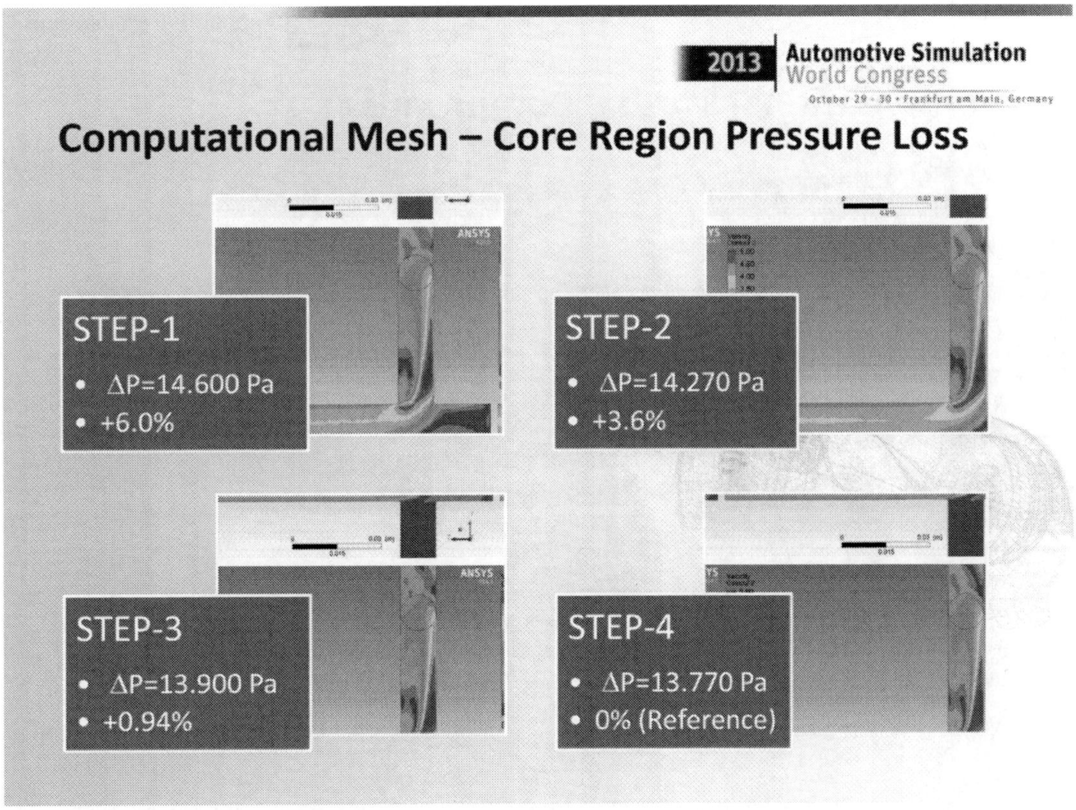
2013 Automotive Simulation World Congress
October 29 - 30 • Frankfurt am Main, Germany
Computational Mesh – Core Region Pressure Loss
STEP-1
• ΔP=14.600 Pa
• +6.0%
STEP-2
• ΔP=14.270 Pa
• +3.6%
STEP-3
• ΔP=13.900 Pa
• +0.94%
STEP-4
• ΔP=13.770 Pa
• 0% (Reference)

2013 Automotive Simulation World Congress
October 29 - 30 • Frankfurt am Main, Germany

Computational Mesh – Core Region Pressure Loss

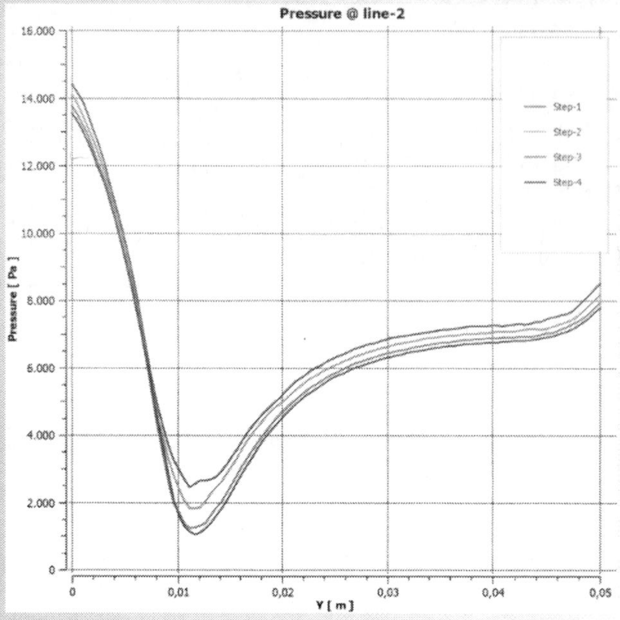

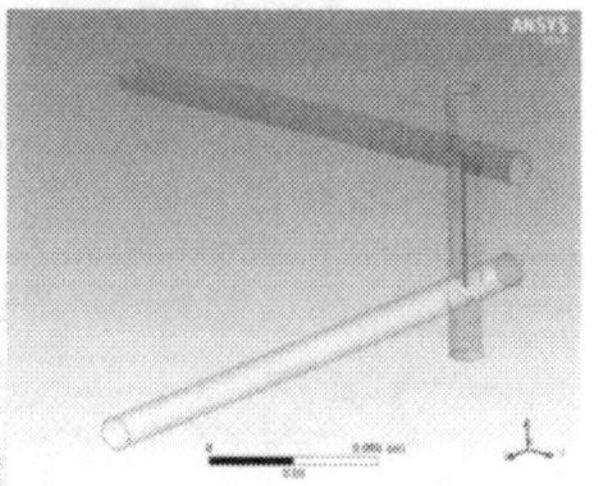

2013 Automotive Simulation World Congress
October 29 - 30 • Frankfurt am Main, Germany

Computational Mesh – Compromise

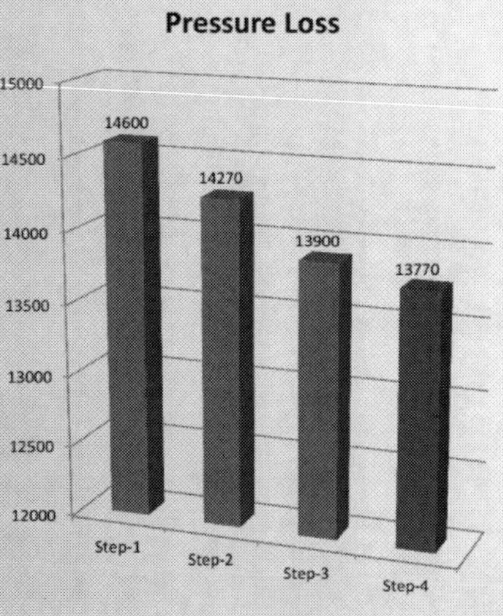

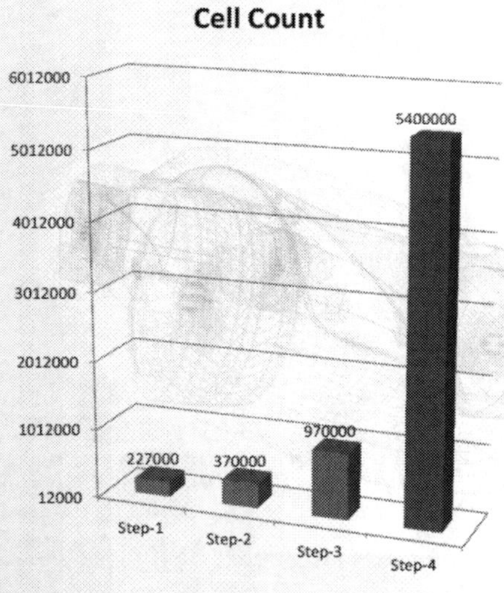

Monday, November 11, 2013 2013 Automotive Simulation World Congress 20

2013 Automotive Simulation World Congress
October 29 - 30 • Frankfurt am Main, Germany

Computational Mesh – Efficiency

Automation

Low Cell Count

Adequate Accuracy

Cut Cell Method

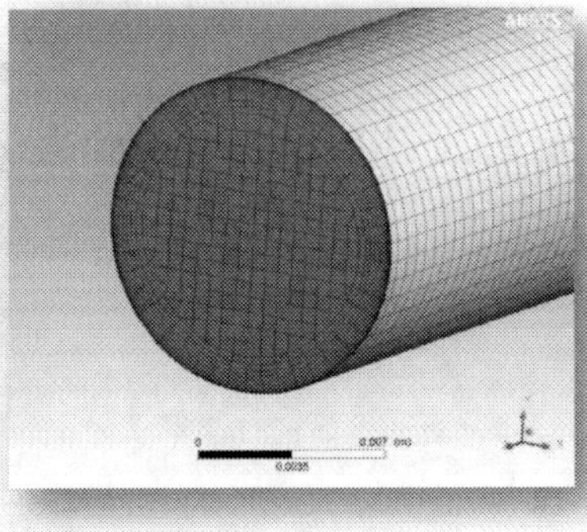

2013 Automotive Simulation World Congress
October 29 - 30 • Frankfurt am Main, Germany

Computational Mesh – Efficiency

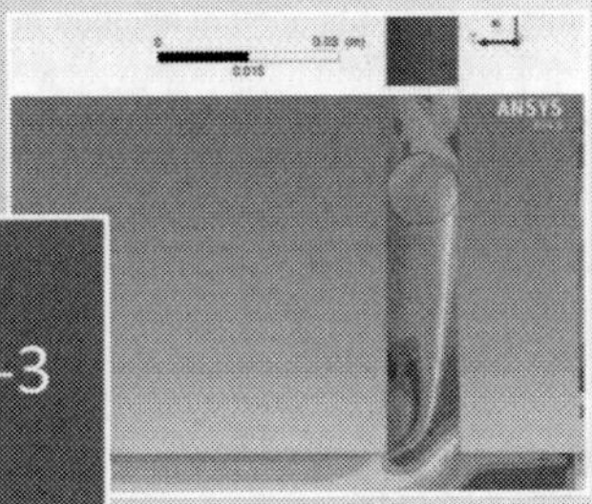

STEP-3

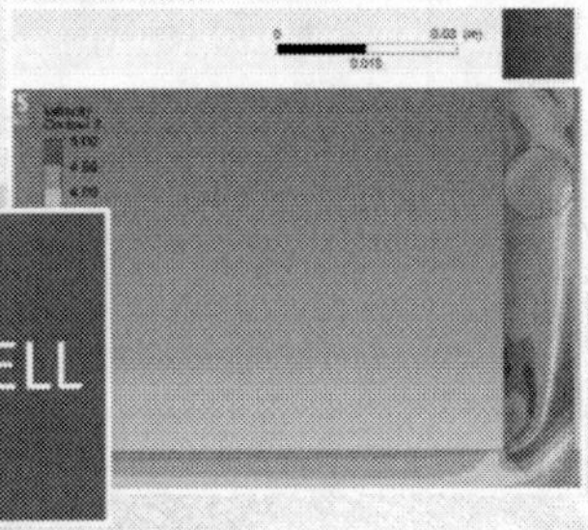

CUTCELL

	CELL COUNT	BODY SIZE [m]
TETPRISM (STEP-3)	967.000	1e-3 (0.08*D)
CUTCELL	581.000	7.5e-4 (0.0625*D)

Computational Mesh – Near Wall Resolution

- 970.000 cells
- $\Delta p_{90°}$=6.982 Pa
- THF=1.653 W

TETPRISM

- 581.000 cells
- $\Delta p_{90°}$=7.010 Pa
- THF=1.663 W

CUTCELL

40%

<1%

<1%

2013 Automotive Simulation World Congress
October 29 - 30 • Frankfurt am Main, Germany

Best Practice Decisions

Mesh

- Near wall resolving mesh (y-plus=1)
- Core Resolution (Step-3, 0.08*D)
- CutCell is an option

Solution Strategy

Transient Solution

Solution Strategy

1. Meshing applied to complete model (6.7 mio cells)

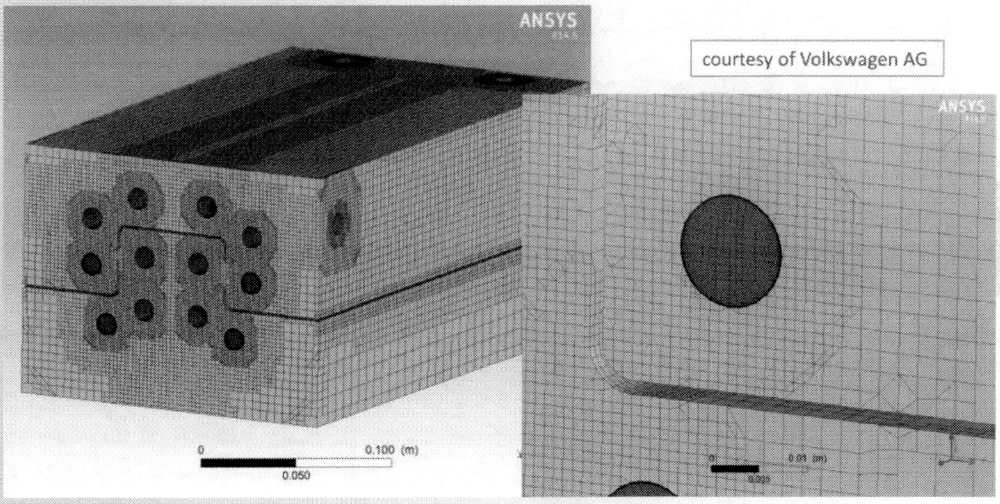

2013 Automotive Simulation World Congress
October 29 - 30 • Frankfurt am Main, Germany

Solution Strategy

Challenge: Multi cycle simulation with physical time of >20 minutes to be simulated

Simultaneously solve FLOW & ENERGY	Steady State FLOW solution Freeze Flow Field Transient ENERGY solution In Fluid and Solid
Solve all equations transient FLOW determines time-step size	Solve only energy equation transient Allows bigger time-steps Assumes constant flow-field

Solution Strategy

Transient Flow Field

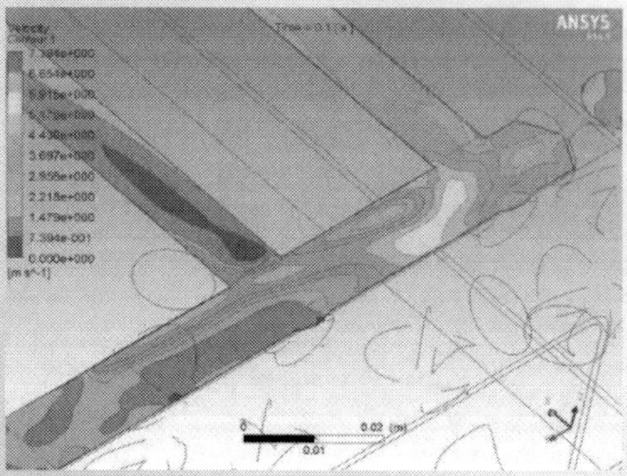

Transient Temperature Field

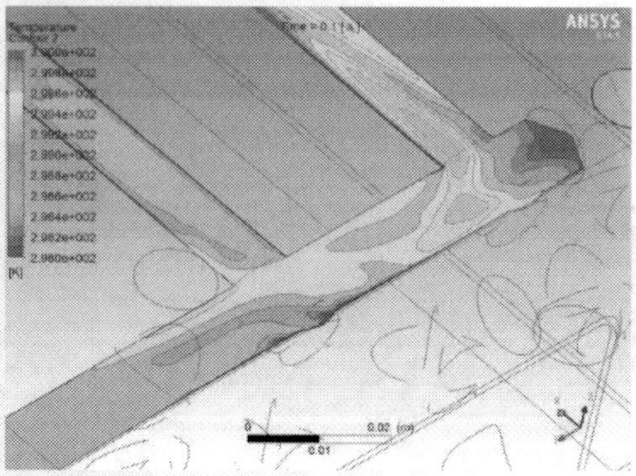

courtesy of Volkswagen AG

Solution Strategy

- Process:
  - @ t=0s the blank-solid is „patched“ to 600°C
  - Cool down (blank) is calculated for 15s (cycle-time)

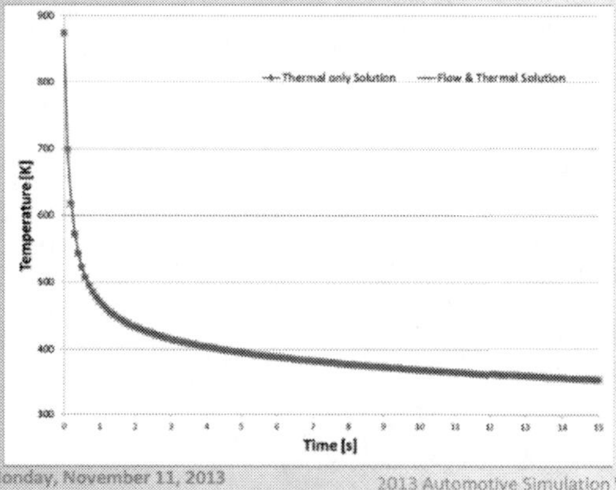

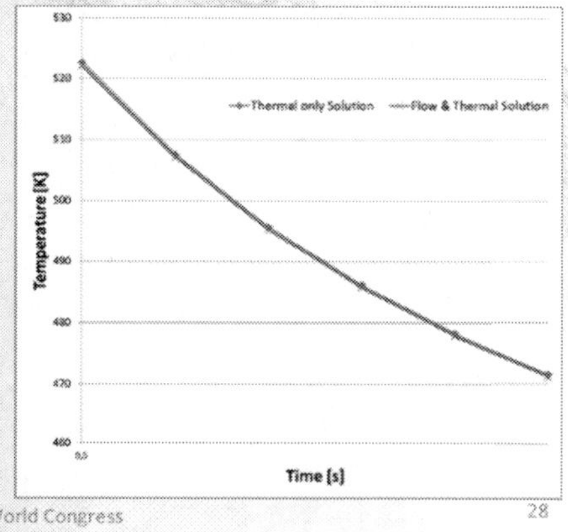

Solution Strategy

- Computational Time for 1 cycle (15s) @ 64 cores : **150 minutes**
- Comp. Time for 50 cycles (15s) @ 64 cores : **125 hours**

FLOW+THERMAL

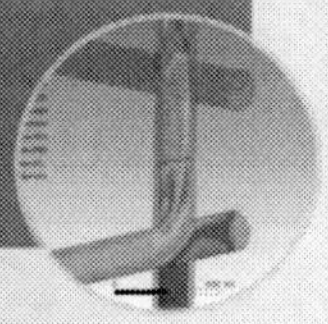

- Computational Time for 1 cycle (15s)@ 64 cores : **8 minutes**
- Comp. Time for 50 cycles (15s)@ 64 cores : **7 hours**

THERMAL ONLY

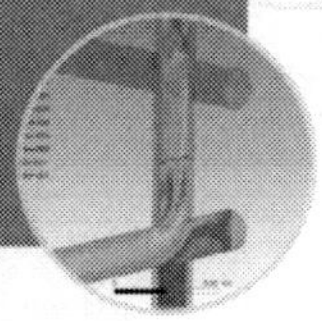

Best Practice Decisions

Mesh

- Near wall resolving mesh (y-plus=1)
- Core Resolution (Step-3, 0.08*D)
- CutCell is an option

Solution Strategy

- Solve steady state Flow
- Freeze Flow field
- Calculate Thermal Solution

Transient Solution

Transient Simulation – Time Step

Time Step Size	# of Time Steps per cycle	Time per cycle @ 64 CPUs	Time for 50 cycles @ 64 CPUs
1s	15	6 minutes	5 hours
0.1s	150	38 minutes	31 hours
0.01s	1500	349 minutes	291 hours

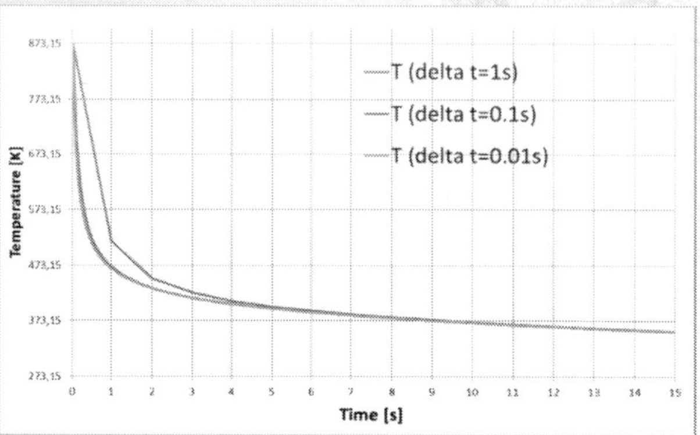

2013 Automotive Simulation World Congress
October 29 - 30 • Frankfurt am Main, Germany

Transient Simulation – Solution Tuning

- Maintain Quality
- Speed-Up Solution

➔ Tune solver settings
  - Multigrid Settings
  - Convergence Criterion

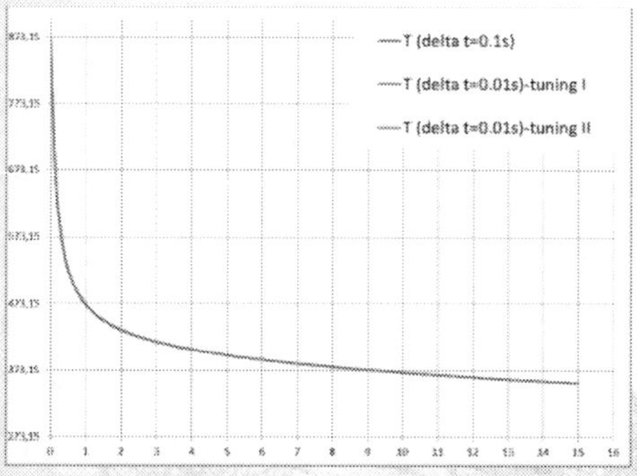

	Time per cycle @ 64 CPUs	Time for 50 cycles @ 64 CPUs	Speed-Up
Basic	38 minutes	31 hours	
Tuned	8 minutes	7 hours	> 400%

Best Practice Decisions

Mesh
- Near wall resolving mesh (y-plus=1)
- Core Resolution (Step-3, 0.08*D)
- CutCell is an option

Solution Strategy
- Solve steady state Flow
- Freeze Flow field
- Calculate Thermal Solution

Transient Solution
- Use Time-Step of 0.1s
- Check optimal solver settings

Validation

- Compare numerical model and physical prototype and analyze differences in results
- Review both numerical model and physical prototype:
  - Initial conditions (especially thermal)
  - Boundary conditions
  - Contact between solids (e.g. depending on contact pressure)
  - Material properties
  - Additional physical effects to be considered (e.g. radiation)

ANSYS supports you ...

Customer Portal
- NEW: support.ansys.com
- Best Practice Documents, Training Material, ...

Customer Support
- Industry-, Application- & Software- Know-How
- Support & Training

Consulting
- Methodology Development, Pilot Projects
- Jump-Start Trainings

Summary & Recommandations

- Reccuring simulation tasks ➔ Best Practice Guideline
- Invest in a good methodology is valued by:
  - High quality results ➔ confidence
  - massive process speed-up ➔ performance
- ANSYS supports you:
  - Best Practice Material
  - Customer Support
  - Training
  - Consulting

Mechatronic Systems

Shock Absorber: A Study on the Dynamic Performance of a Solenoid Hydraulic Valve for Automotive Semi-Active Suspensions Coupling ANSYS Fluent with Lumped Parameter Modeling

Matteo Pelosi
Öhlins Racing AB

ANSYS AWSC 2013

October 29th, Frankfurt, Germany

Contents

3

Öhlins Racing AB

ADVANCED SUSPENSION TECHNOLOGY

4

Mechatronic Systems

Öhlins Racing Systems: Conventional, EC & CES

Like many other systems on vehicles, today **electronically controlled suspensions** are entering the market

Conventional passive are systems offering multiple levels of damping by fixed steps adjustements based on ***clicks***. *Clicks* through externally accessible screws and knobs.

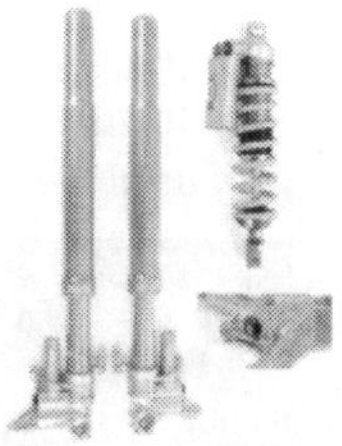

electronically controlled

EC is a system where the damping can be adjusted discretely or in real-time by **stepper motors**. The hydraulic metering is electronically controlled in fixed steps.

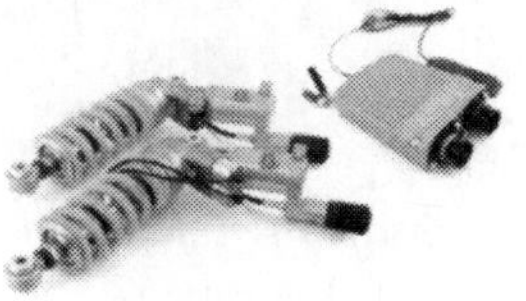

CES BY ÖHLINS RACING

CES is a **semi-active suspension** system where the damping is continuously adjusted based on state information measured with sensors such as stroke sensors and accelerometers.

Versatility

Mechatronics

ÖHLINS 5

Mechatronic Systems

CES by Öhlins Racing

- Proprietary technology developed since 1984
- **Semi-Active Damping**
- **Feedback:** accelerometers, height sensors, steering sensors ...

ÖHLINS

CES TECHNOLOGIES

Video

ÖHLINS 6

Öhlins Modeling & Simulation Toolchain

To improve design & understanding of its products, Öhlins is moving towards CAE integration within the standard trial & error work

3D modeling (CFD, FSI) of components and sub-components:

1D modeling of components (shock absorbers, valves):

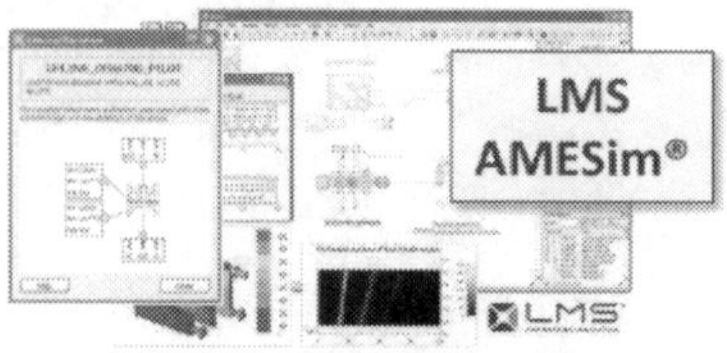

Controls, signal processing:

Vehicle dynamics simulations:

7

Öhlins CES Valve Modeling

1D Model - CES8700

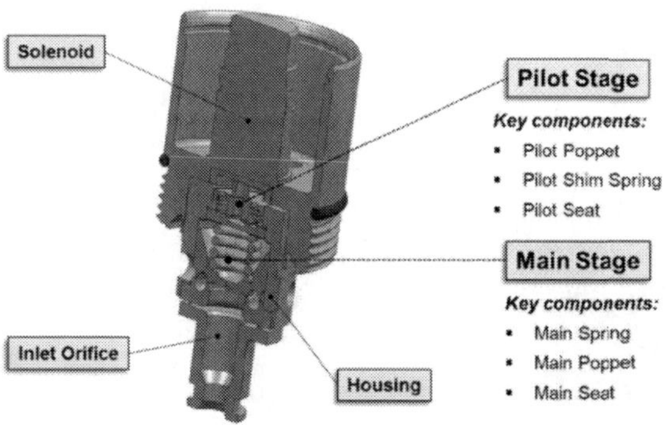

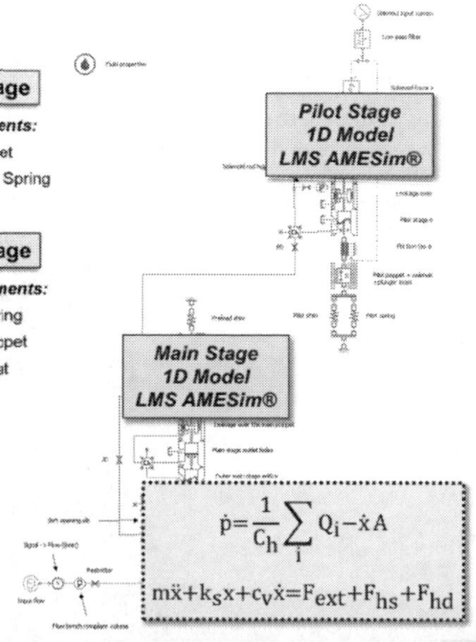

$$\dot{p} = \frac{1}{C_h} \sum_i Q_i - \dot{x} A$$

$$m\ddot{x} + k_s x + c_v \dot{x} = F_{ext} + F_{hs} + F_{hd}$$

- 1D model implements **detailed** CES mechanical & hydraulic features
- 1D model solves **transient second order dynamics** considering external, spring, friction & pressure forces

8

Öhlins CES Valve Modeling

ANSYS Fluent Off-line CFD Analysis - Overview

- **ANSYS Fluent off-line CFD analysis** to obtain more accurate results from the start & improve physical understanding provided by 1D model
- Derive **critical parameters** to improve 1D models prediction:

Flow Coefficients: $Q = C_q A \sqrt{\frac{2|\Delta p|}{\rho}} \operatorname{sign}(\Delta p)$

Flow Forces: $F_f = 2C_q A \Delta p \cos\theta$

Solver	ANSYS Fluent 14.5
Fluid model	Newtonian, incompressible
Turbulence model	RANS, realizable k-ε
Near wall model	Enhanced Wall Treatment
Schemes Accuracy	Second order
Mesh type	Hybrid (Teth + Prism + Hex)
Mesh size	~5E6 cells

- **Main stage** & **pilot stage** analyzed separately
- **Steady-state 3D CFD** at different combinations of opening position & pressure drop to cover known operating range

Öhlins CES Valve Modeling

Off-line CFD Analysis - Validation

- 3D CFD model setup **validated** before off-line simulation campaign
- p-Q measurement comparison with **main stage set at fixed positions**

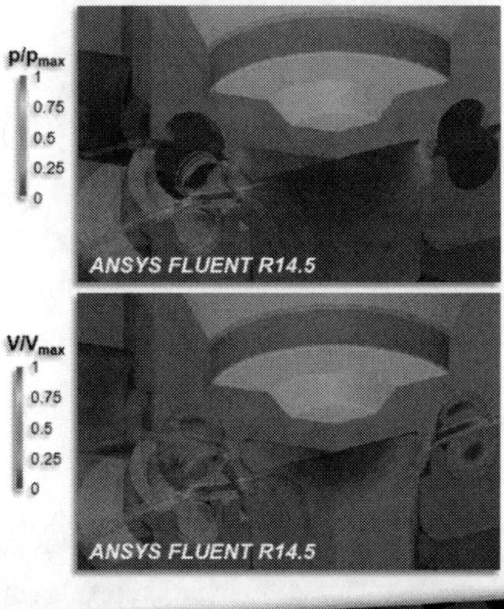

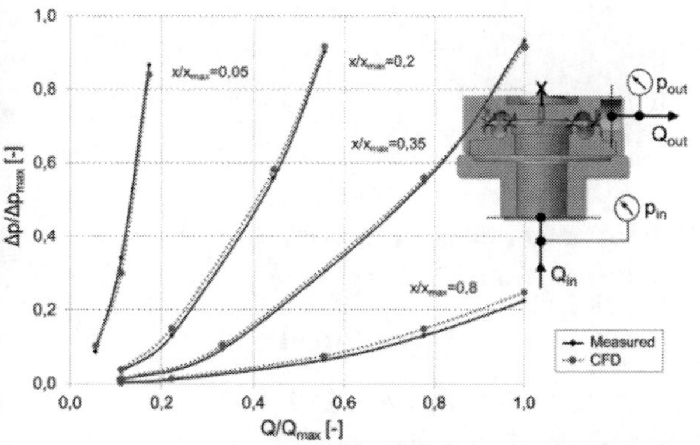

Öhlins CES Valve Modeling

Pilot Stage - Overview

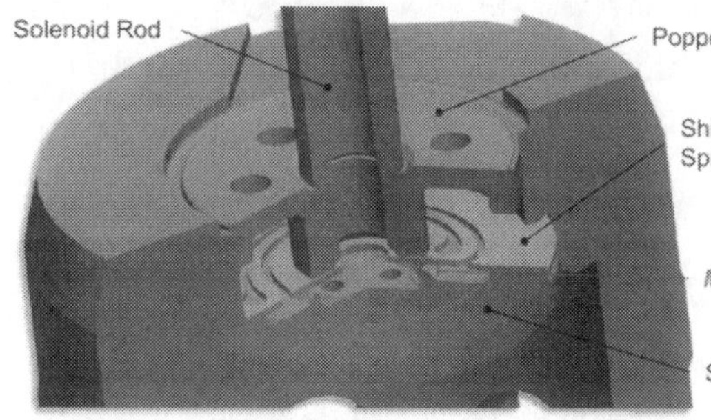

Shim spring **simultaneously** fulfills:

- Hydraulic metering
- Elastic function

Mesh

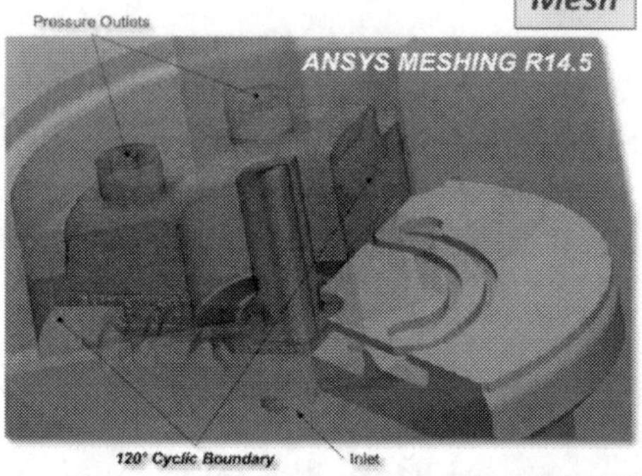

Static pressure distribution on compliant moving surfaces affecting pilot stage dynamics

40 CFD simulations:

- 8 positions
  - 5 pressure drops

Öhlins CES Valve Modeling

Pilot Stage - Flow Coefficients

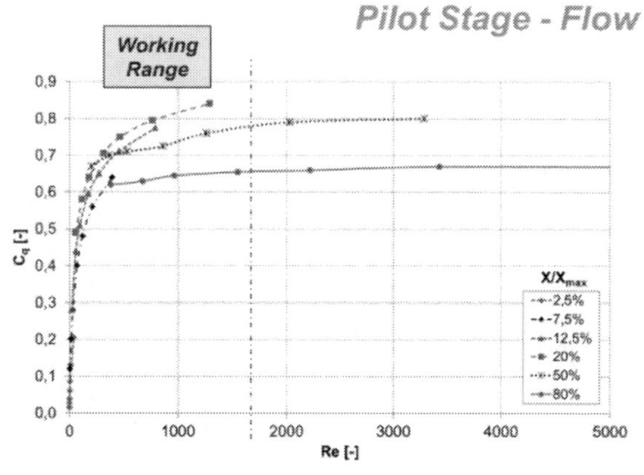

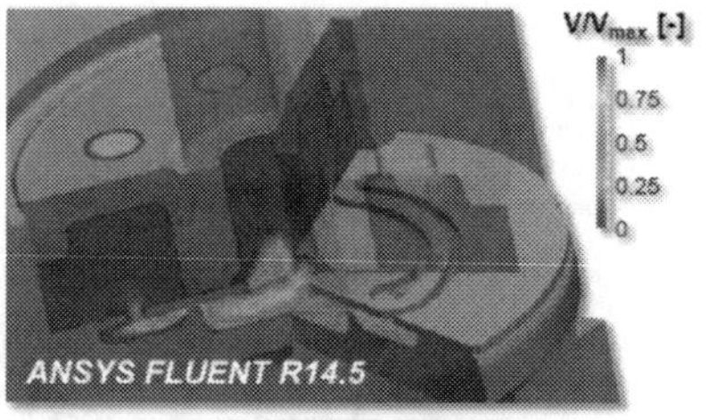

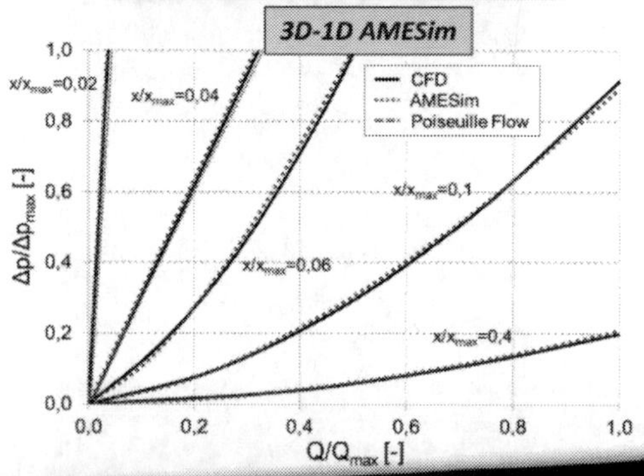

Pilot metering mainly works in **laminar/transition** region, where:

$$C_q = C_q(Re, x)$$

Method to implement **CFD derived C_q** in AMESim 1D model

Öhlins CES Valve Modeling

Pilot Stage - Flow Forces

Static **pressure stagnation phenomena** on shim spring **compliant** moving surface

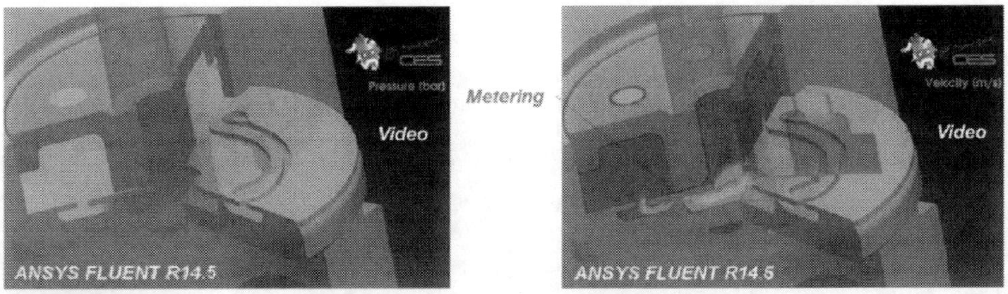

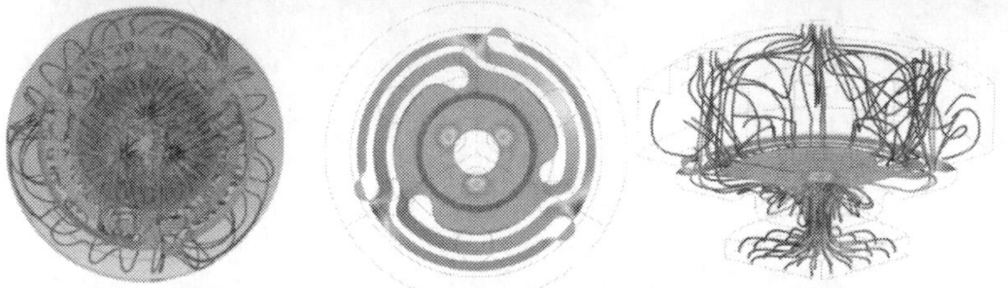

A_{eq}
Equivalent pressurized area extracted from 3D CFD results

Clavier, A., Alirand, M., Vernant, F. and Sagot, B. 2003. *Local Approach to Improve the Global Approach of Hydraulic Forces in Ball Poppet Valves.*

Öhlins CES Valve Modeling

Pilot Stage - Flow Forces

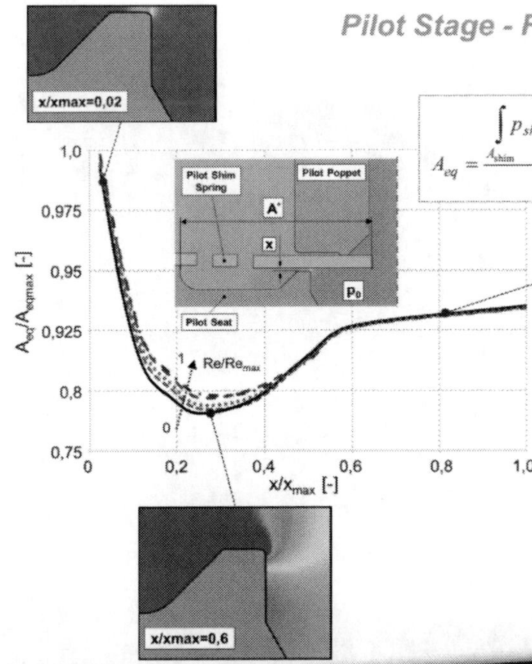

$$A_{eq} = \frac{\int_{A_{shim}} p_{shim\,CFD} \cdot \hat{i}\, dA}{p_0}$$

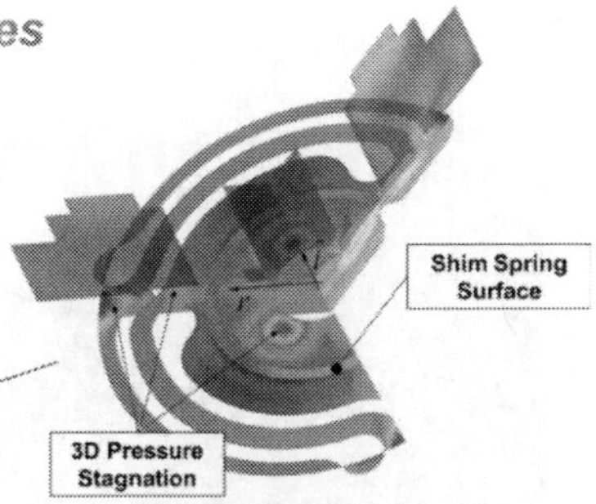

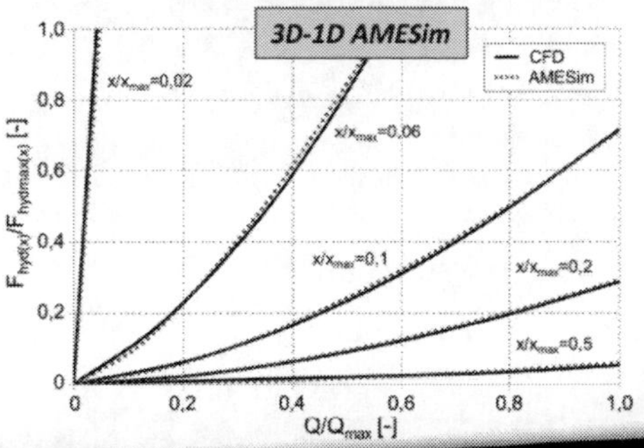

Öhlins CES Valve Modeling

3D-1D Weak Coupling

- A **special AMESim submodel** for the pilot stage orifice is developed

Shim Orifice Submodel

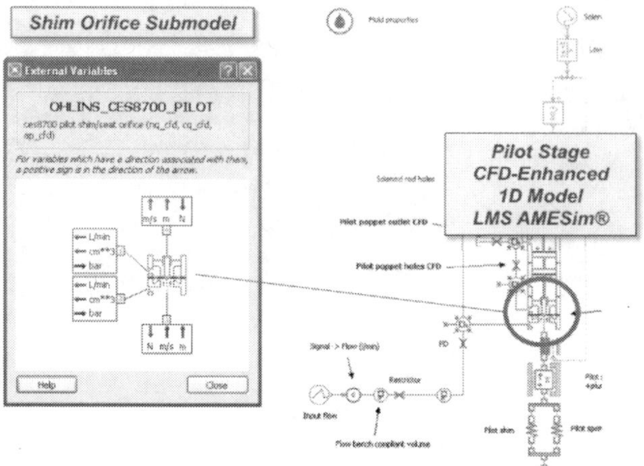

The C++ model expects as input ASCII tables of C_q and A_{eq} function of the poppet position and of the Reynolds number

- The C++ model calculates flow rate and poppet instantaneous force balance

Öhlins CES Valve Modeling

Main Stage - Overview

- Similar analysis performed on CES8700 **main stage**...

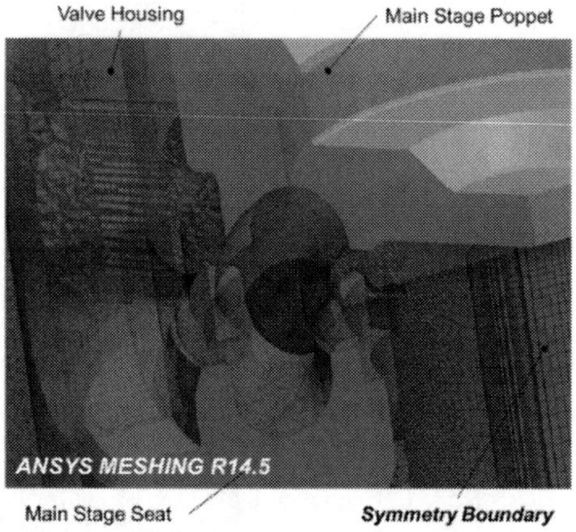

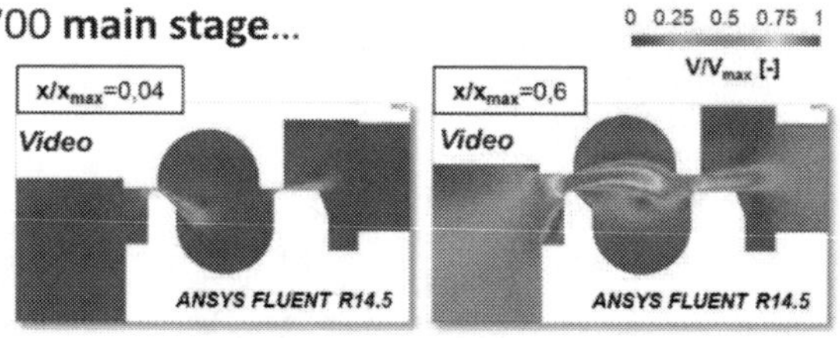

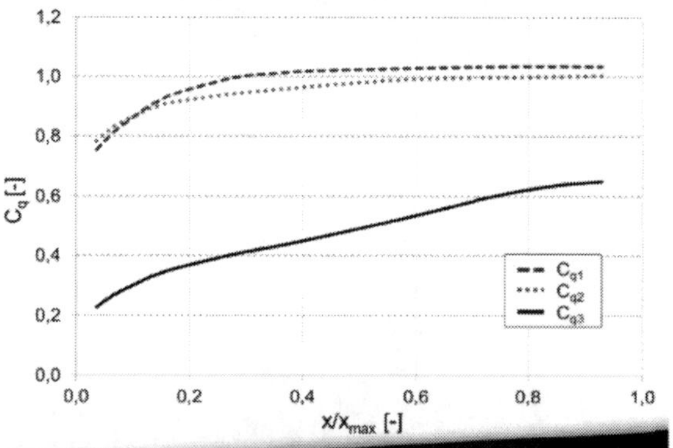

40 CFD simulations:

- 8 positions
  - 5 pressure drops

Öhlins CES Valve Modeling

Numerical Results - Inputs

- A **p-Q test cycle** is simulated with the 3D/1D model
- Pilot stage fed by successive ramps of flow rate at different levels of current

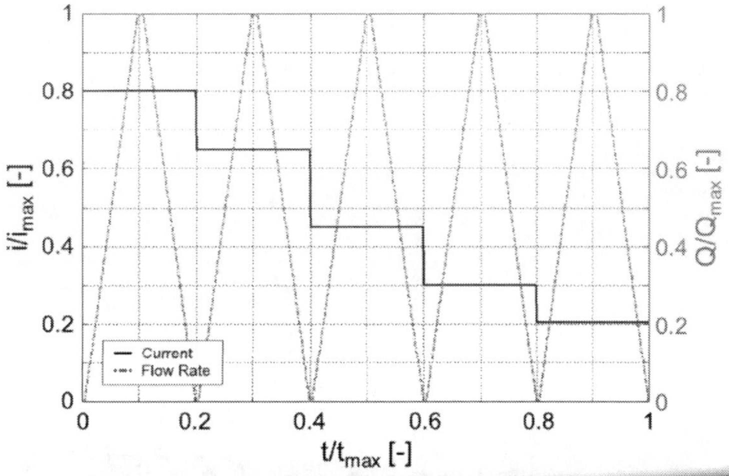

17

Öhlins CES Valve Modeling

Numerical Results - Pilot Stage

- 3D-1D model compares well with measurements
- Improved prediction over a standard AMESim model

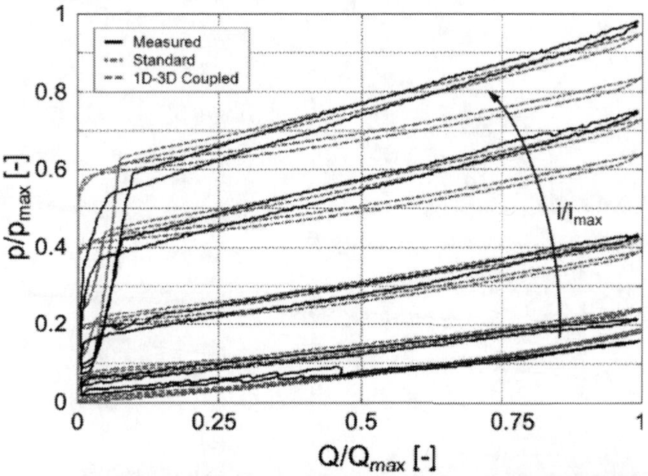

18

Öhlins CES Valve Modeling

Numerical Results - Pilot Stage

- 3D-1D modeling gives insight on pilot stage **entire systems dynamics**

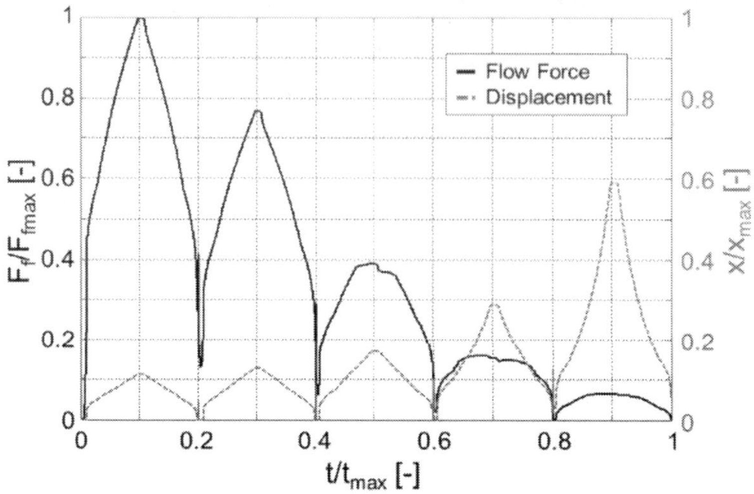

Öhlins CES Valve Modeling

Numerical Results - Full CES Valve

- Low speeds & high speed regions compare well at all current levels

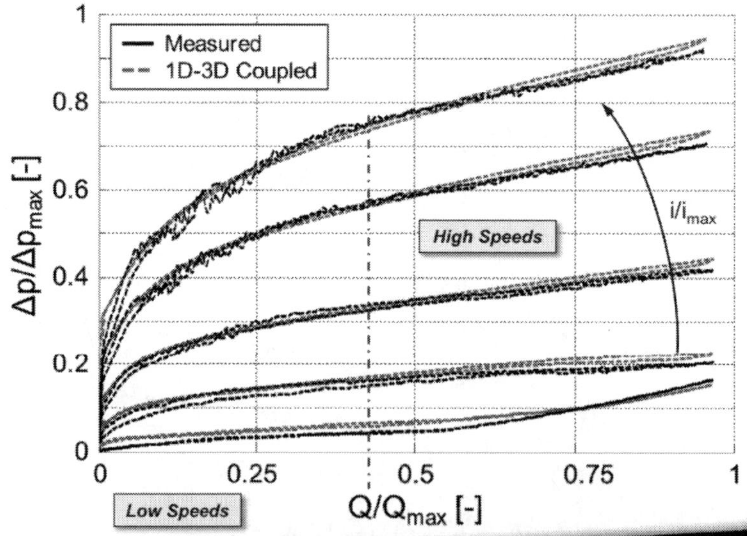

Öhlins CES Valve Modeling

Numerical Results - Full Valve

- 3D-1D modeling gives insight on CES full valve **entire systems dynamics**

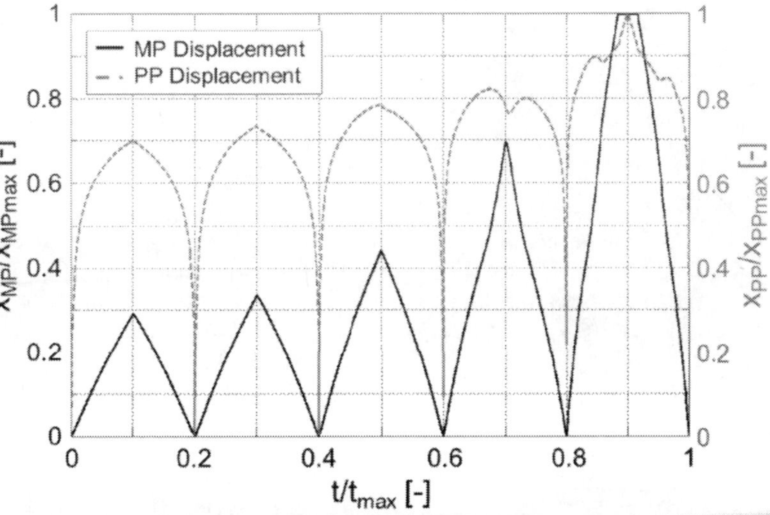

21

Öhlins CES Valve Modeling

ANSYS Fluent 3D Flow Driven Rigid Body Motion

3D-1D models cannot easily capture **intimate fluid flow/structure behavior**

3D RBM simulation: **Fluent UDF** poppet dynamics library, **transient** CFD & **Moving Deforming Mesh**

Bringing 2nd order system simulation into Fluent

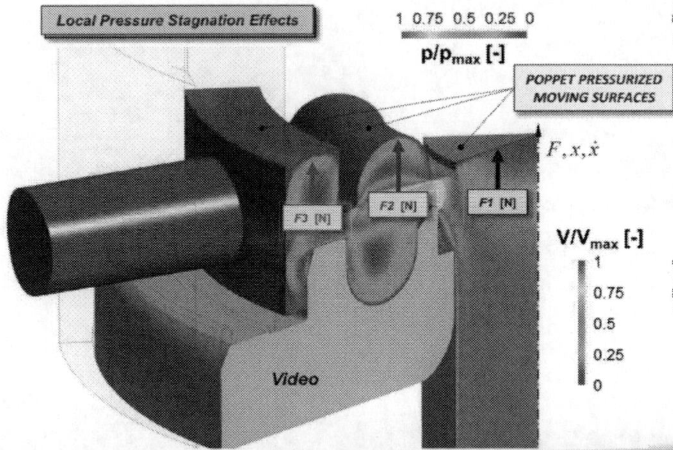

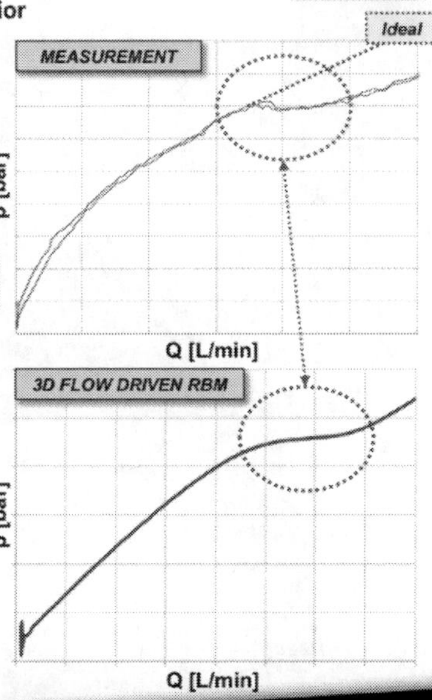

22

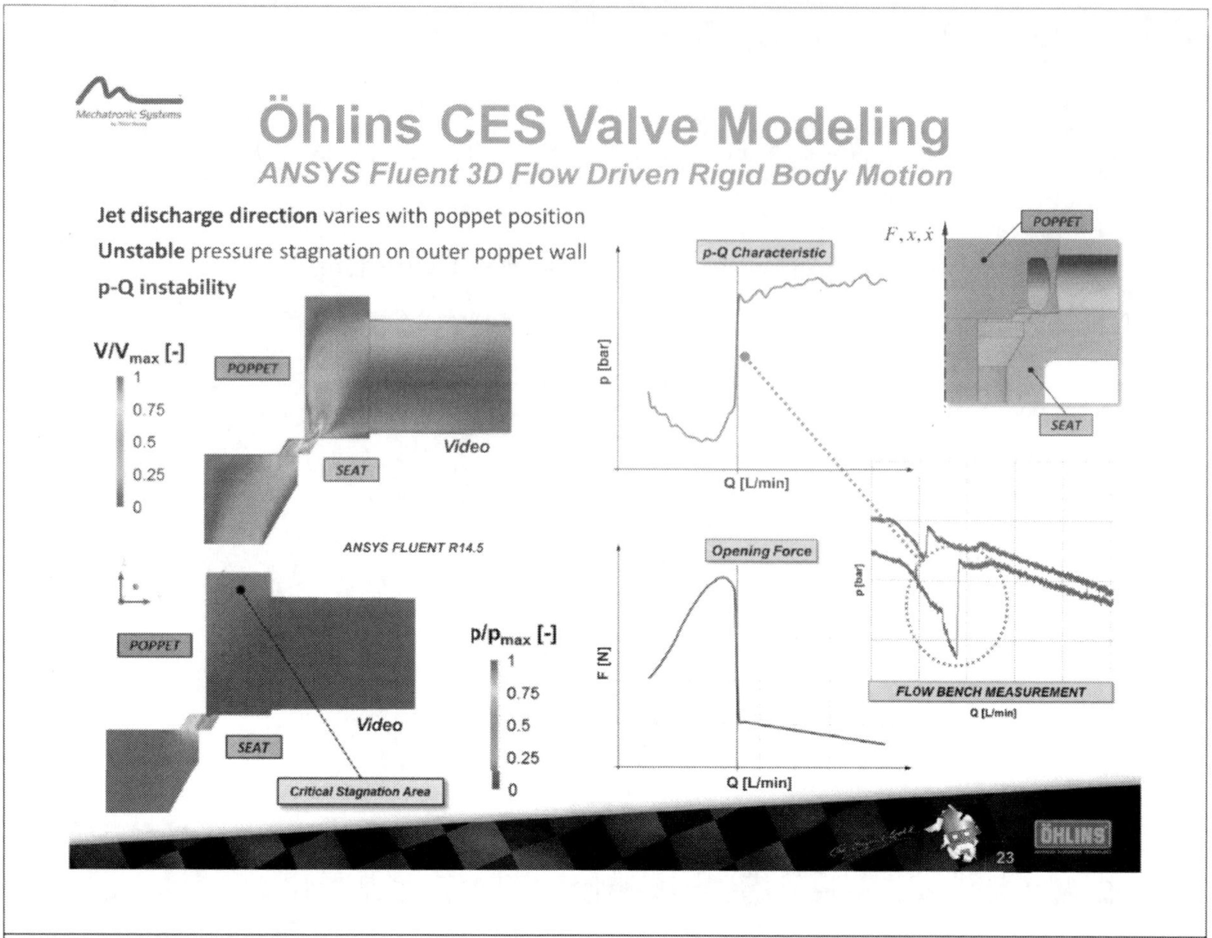
Mechatronic Systems
Öhlins CES Valve Modeling
ANSYS Fluent 3D Flow Driven Rigid Body Motion
Jet discharge direction varies with poppet position
Unstable pressure stagnation on outer poppet wall
p-Q instability
V/Vmax [-]
1
0.75
0.5
0.25
0
POPPET
SEAT
Video
ANSYS FLUENT R14.5
POPPET
SEAT
Video
Critical Stagnation Area
p/pmax [-]
1
0.75
0.5
0.25
0
p-Q Characteristic
p [bar]
Q [L/min]
F, x, ẋ
POPPET
SEAT
Opening Force
F [N]
Q [L/min]
p [bar]
FLOW BENCH MEASUREMENT
Q [L/min]
23
ÖHLINS

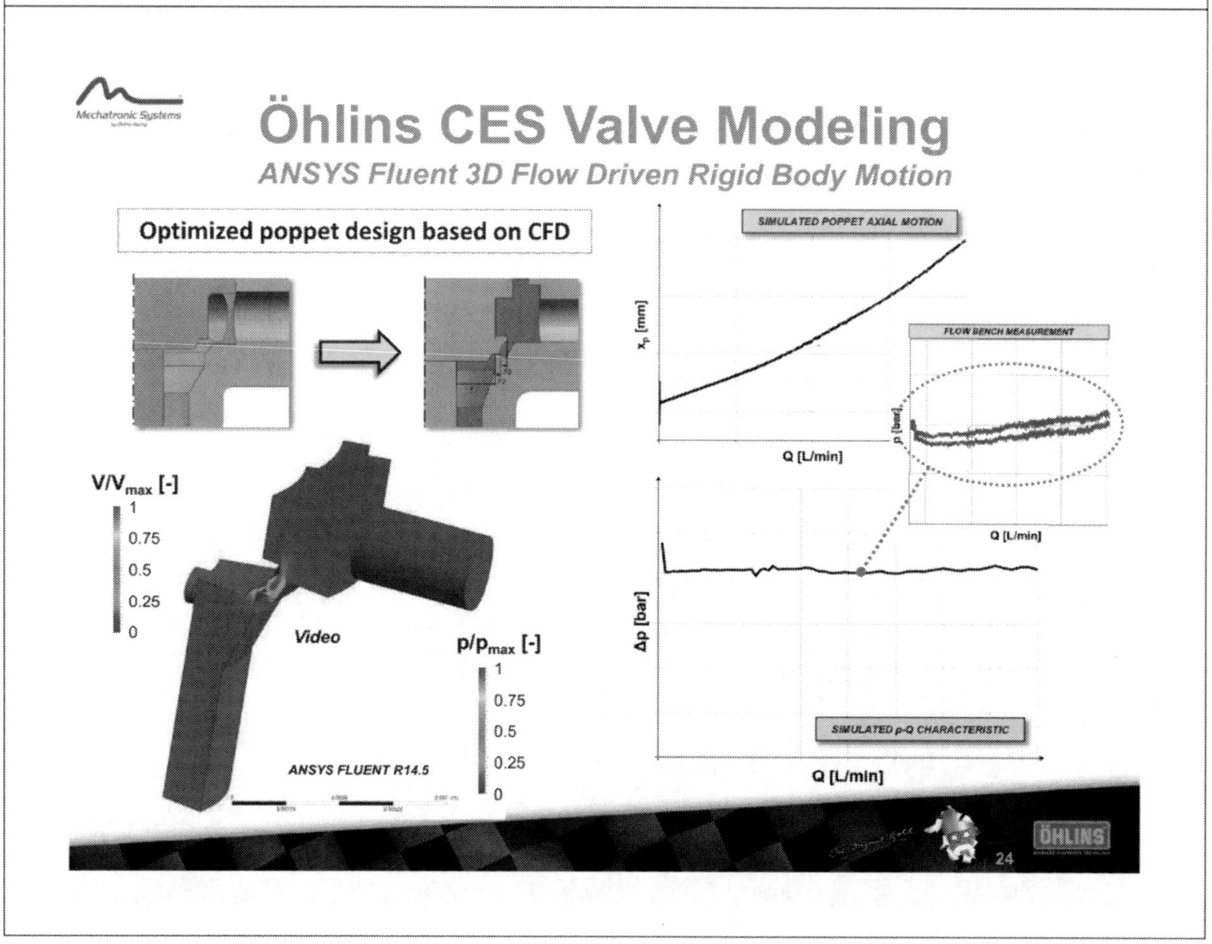
Mechatronic Systems
Öhlins CES Valve Modeling
ANSYS Fluent 3D Flow Driven Rigid Body Motion
Optimized poppet design based on CFD
V/Vmax [-]
1
0.75
0.5
0.25
0
Video
p/pmax [-]
1
0.75
0.5
0.25
0
ANSYS FLUENT R14.5
SIMULATED POPPET AXIAL MOTION
xp [mm]
Q [L/min]
FLOW BENCH MEASUREMENT
p [bar]
Q [L/min]
Δp [bar]
SIMULATED p-Q CHARACTERISTIC
Q [L/min]
24
ÖHLINS

Mechatronic Systems

Future Challenges & Vision

Öhlins Racing Numerical Modeling Spectrum

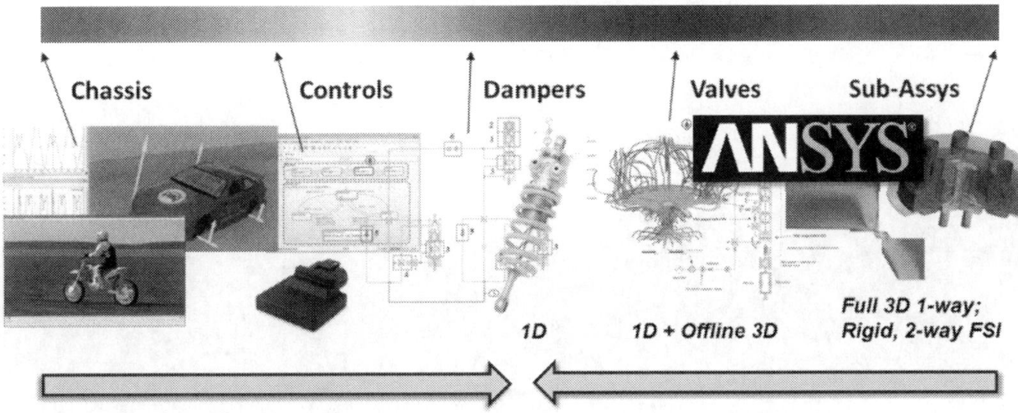

Co-simulations to predict Öhlins Racing systems physical behavior

Opportunity to perform accurate **design optimization** studies

ÖHLINS 25

Mechatronic Systems

Conclusions

Öhlins' strategy to make **Computer Aided Engineering** and **Simulation Driven Product Development** integrated parts of its standard development processes has been presented

Design, size and dynamics complexity of Öhlins' products, make accurate **3D simulations** paramount for physical understanding at a **components level**

1D modeling is a fundamental modeling component as well to deepen the understanding at a **systems level**

A key success factor is to strive for convergence of the modeling extremes through FSI, flow driven RBM and **co-simulation**

ÖHLINS 26

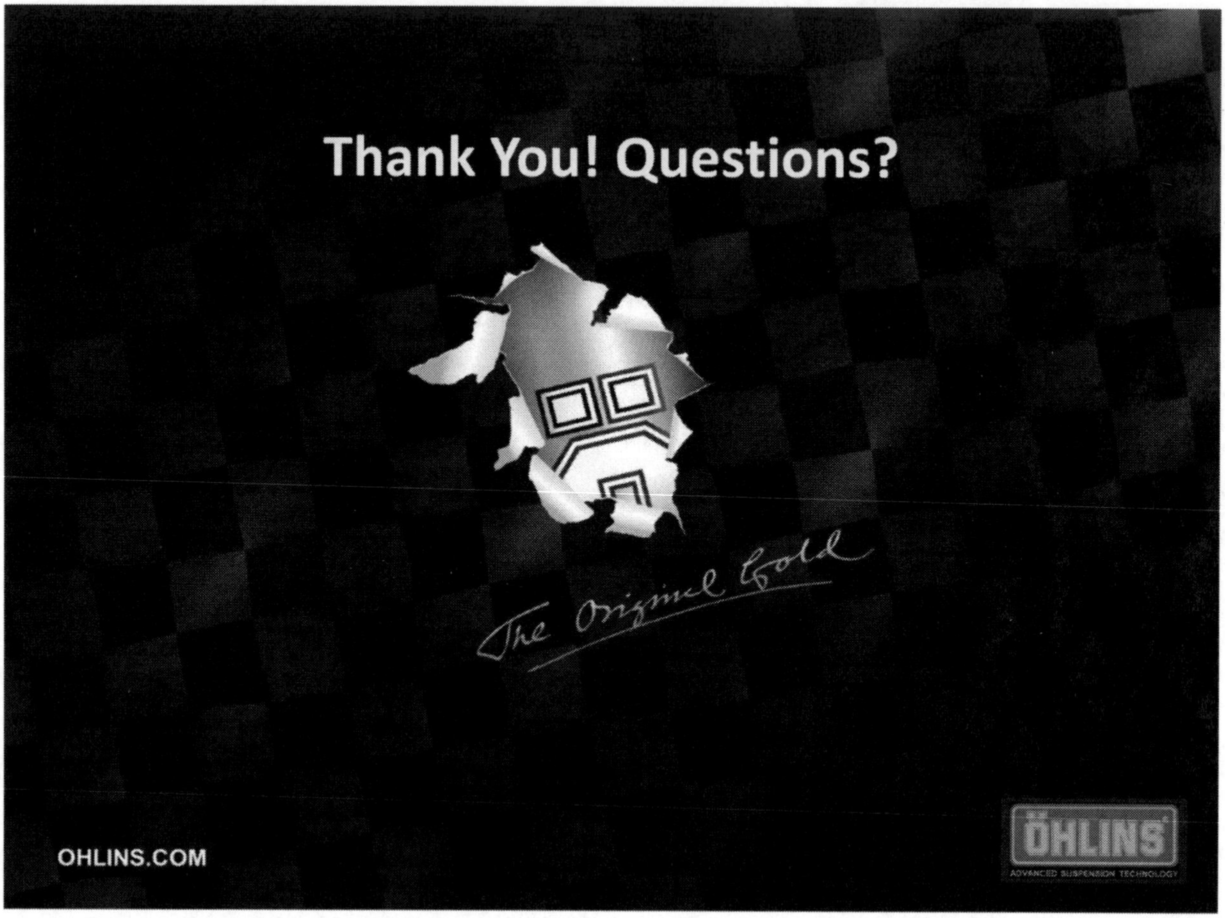
Thank You! Questions?
The Original Gold
OHLINS.COM
ÖHLINS
ADVANCED SUSPENSION TECHNOLOGY

2013 Automotive Simulation World Congress

Significance of Numerical Simulations in the Journey of Tire from Rubber to Road

Chetan Patil, Santosh Kottalgi, Peiran Ding, Ashok Khondge, Ankit Adhiya, Lubeena R. & Raman Babu
ANSYS

Outline

- Introduction
- Tire Manufacturing & Design
- Simulating Tire Manufacturing Processes
- Simulating Tire Design & Performance
- Summary

Introduction

- Tires are surprisingly Complex – Geometry, Materials & Service conditions
- New tire development is a highly complex process consisting of many engineering activities such as
  - Tire pattern design,
  - Material selection,
  - Tire mold manufacture, and mass production
  - Tribological analysis,
  - Prototyping,

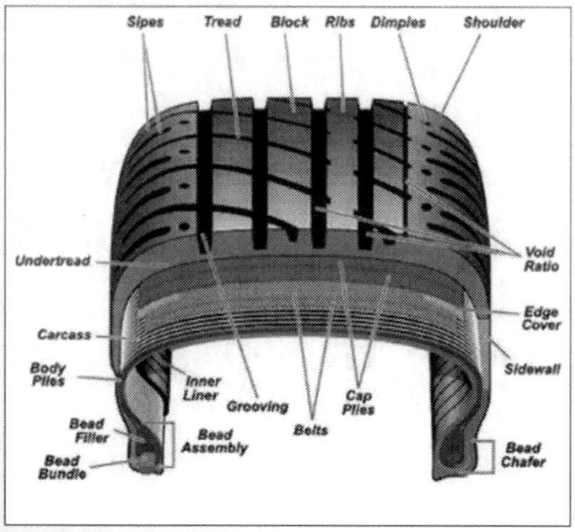

2013 Automotive Simulation World Congress
October 29 - 30 • Frankfurt am Main, Germany

Typical Tire Cross-Section [Line Diagram]

There are total 9 parts in the model

1. Tread
2. Side Wall
3. Inner Liner
4. Body Ply
5. Belt1
6. Belt2
7. Bead Steel
8. Apex
9. Rim Strip

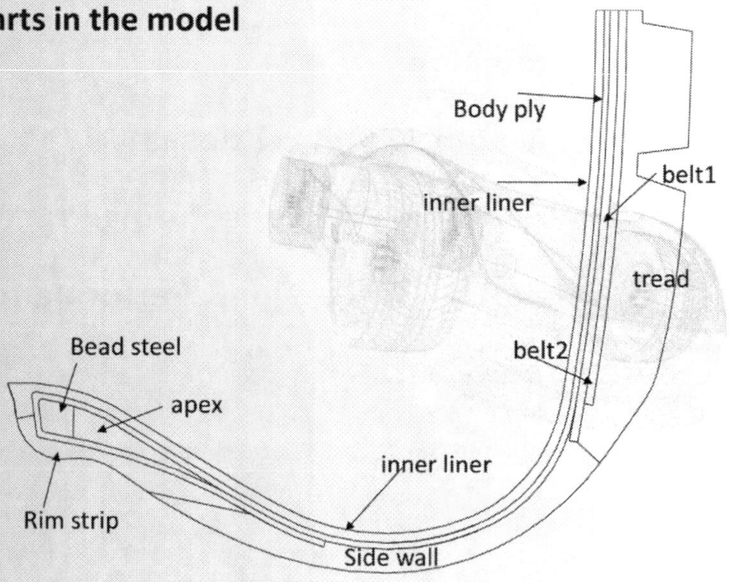

Tire Manufacturing & Design

Raw Materials
1. Rubber Bales
2. Textile
3. Steel
4. Chemicals
5. Etc ...

1. Rubber Compound Mixing
2. Tread & Side Wall Extrusion
3. Steel Belt / Fabric Cord Calendaring
4. Assembly Tire components
5. Tire Curing

1. Performance Verification
2. Inflation, footprint, rolling analysis
3. Aquaplaning, water splashing
4. NVH Performance
5. Etc ...

Rubber → Manufacturing → Road

TIRE Manufacturers

Automotive / Off Highway OEMs

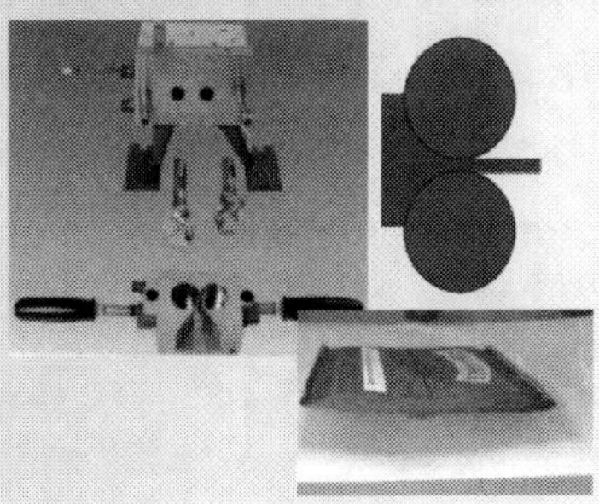

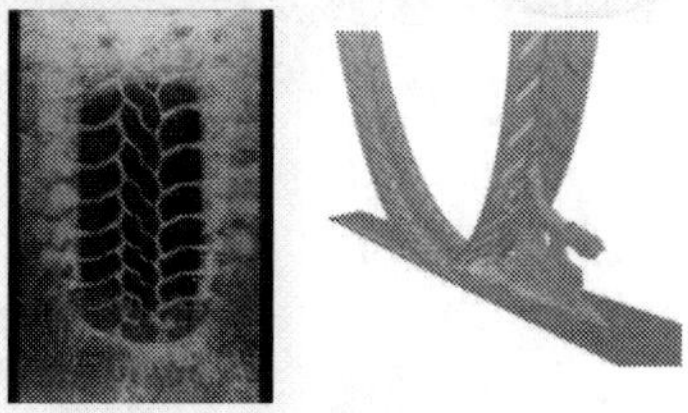

2013 Automotive Simulation World Congress
October 29 - 30 • Frankfurt am Main, Germany

Simulation of Tire Manufacturing Processes

1. Mixing of Rubber Compound
2. Tire Tread Shaping & Extrusion
3. Tire Curing

1. Mixing of Rubber Compound

- Efficient Mixing is very important for processing rubber compounds
- Quality of Mixing is ensured by
  - Uniformity of mixture
  - Homogeneous properties of the resulting compound
- Most mixing devices generate a complex, truly 3D and transient flow pattern
  - Difficult to study with classical numerical technique
  - Various degrees of simplification are usually necessary
- ANSYS Polyflow Solution: Mesh Superposition Technique
  - Complemented by statistical analysis of the computation results.
  - Very powerful technique to simulate flow with moving and intermeshing parts, such as batch mixers and Twin Screw Extruders

ANSYS Polyflow : Batch Mixer Simulation

- Case study: Batch mixing of rubber compounds
- Analysis of the effect of the rotor shape on the distributive mixing of an internal mixer by comparing configurations with 2, 3 and 4 wings.
  - Carrying fluid : white compound, SBR solution 60 *phr* silica/10 *phr* TiO^2
  - Tracer (SBR solution 80 *phr* carbon black), negligible mass

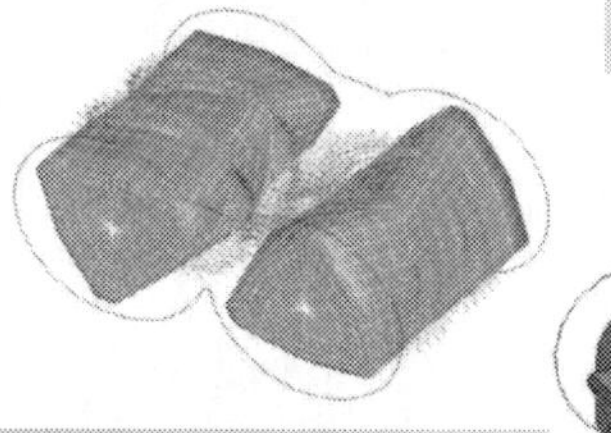

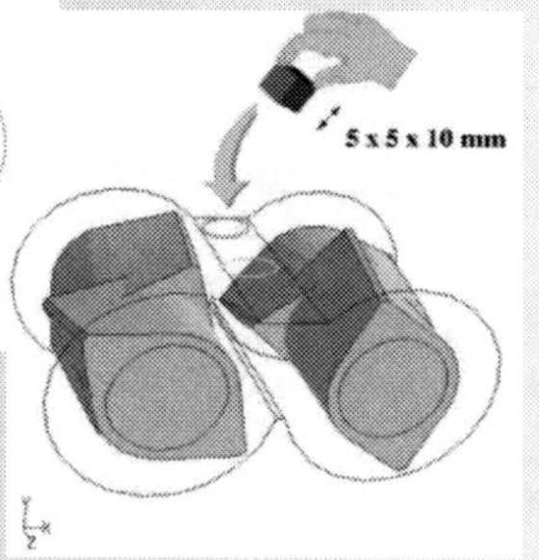

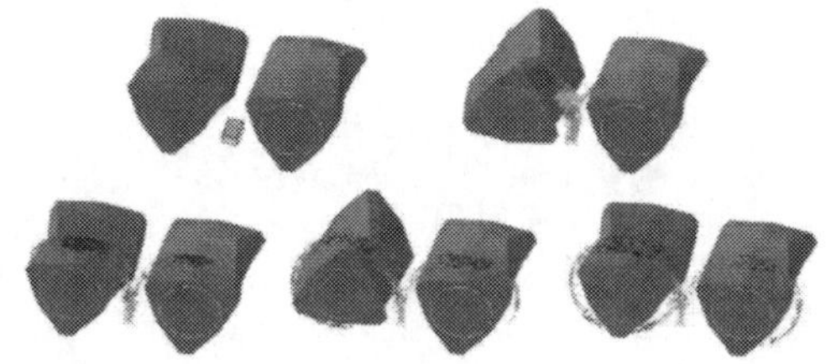

2013 Automotive Simulation World Congress
October 29 - 30 • Frankfurt am Main, Germany

2. Tire Tread Shape & Extrusion

- Tire treads are produced by continuous extrusion
  - It is very important to control this process in order to obtain the prescribed shapes and properties at reasonable expense
- Numerical Simulations Can help
  - Improve geometrical die data
  - Improve processing conditions
- ANSYS POLYFLOW Solution
  - Automatic die design functionality.
  - Reduce number of experiments on production line to obtain the right product

2013 Automotive Simulation World Congress
October 29 - 30 • Frankfurt am Main, Germany

ANSYS Polyflow : Die Design for Tire Tread Extrusion

Simulation Condition

- Temperatures measured about 95~105 °C, The first approach : isothermal model
- Take-away conveyor speed Vout=8.7 m/min
- Flow rate Q = 1.817x10-4 m3/sec
- Slip along walls evaluated and considered

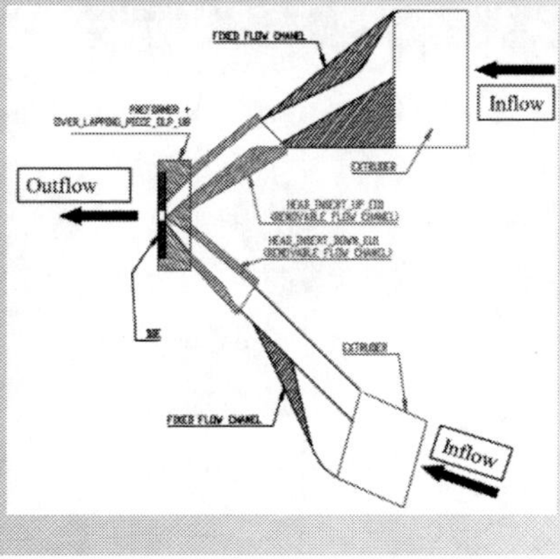

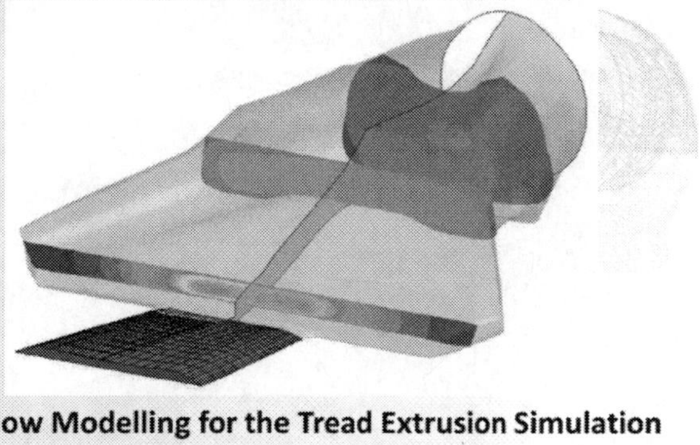

Flow Modelling for the Tread Extrusion Simulation with several cutting planes

3. Tire Curing Process

- After the tire is formed to the desired pattern in a press, the compound is converted to a strong elastic material to meet tire performance needs by vulcanization
- The Curing Process is Energy-consuming and has a Strong Effect on Product Quality.
- The Cure Cycle should be Optimized, Minimizing Capital and Energy Expenditures.
- Numerical Modelling can be used to give a Valuable Insights into the Process Prior to Manufacturing.

Pressing the pattern sculpture before curing
(Photo: TTN Management Group)

ANSYS Polyflow : Tire Curing Process Modeling

Model set-up

Schematic cross-section of the model tire and cure press

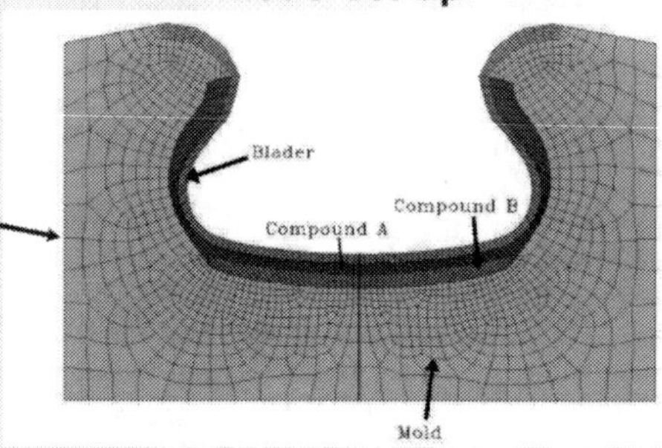

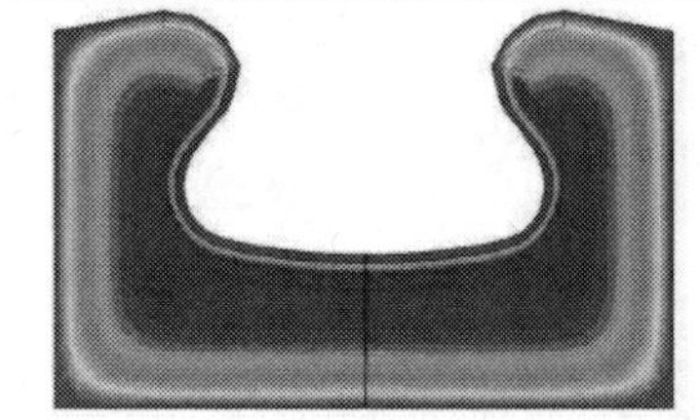

Initial temperature

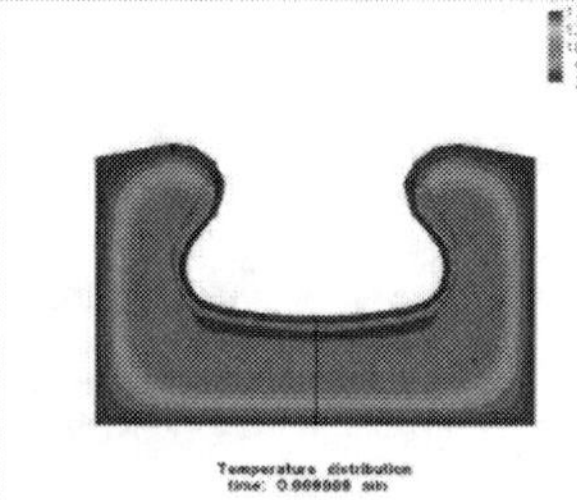

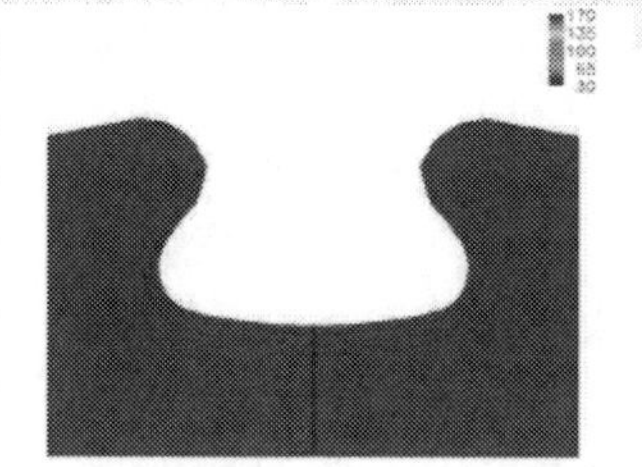

Temperature after 6 min

2013 Automotive Simulation World Congress
October 29 - 30 • Frankfurt am Main, Germany

ANSYS Polyflow : Tire Curing Process Modeling

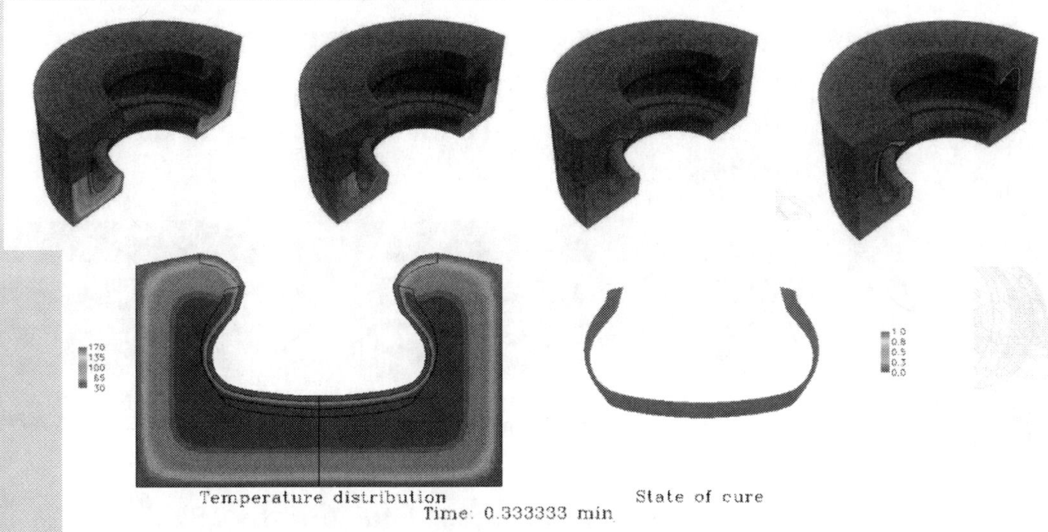

Comparison between the evolution of the temperature and the state of cure

2013 Automotive Simulation World Congress
October 29 - 30 • Frankfurt am Main, Germany

Simulation of Tire Design Performance

1. **Tire Inflation, Foot-print & Rolling**
2. **Water Splash By Tire**
3. **Tire Subjected to Blast Wave**

2013 Automotive Simulation World Congress
October 29 - 30 • Frankfurt am Main, Germany

Tire Inflation, Foot-Print & Rolling Analysis using ANSYS Help understand Tire performance

- These are important tests to understand the tire Tread Design performance
- These can affect the vehicle fuel economy, tire wear, durability & vehicle handling etc.

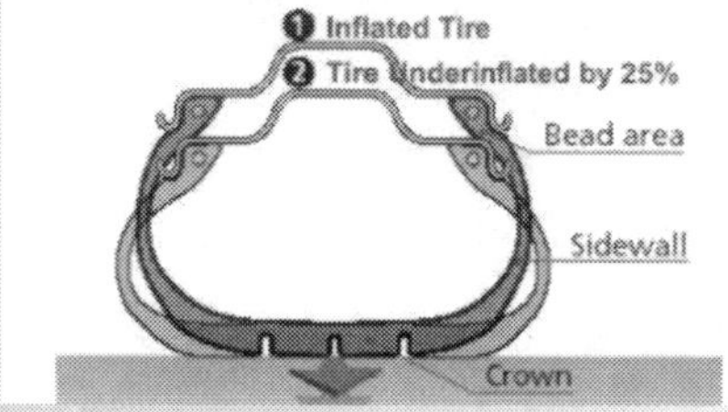

32 psi, Area = 36.5 in^2

24 psi, Area = 43.7 in^2

25% Pressure Loss Increases Footprint by 20% [1]

[1] http://www.arb.ca.gov/cc/tire-pressure/meetings/060408/060408exxonpresentation.pdf

2013 Automotive Simulation World Congress
October 29 - 30 • Frankfurt am Main, Germany

Inflation, Foot-Print & Rolling Simulation Flow chart

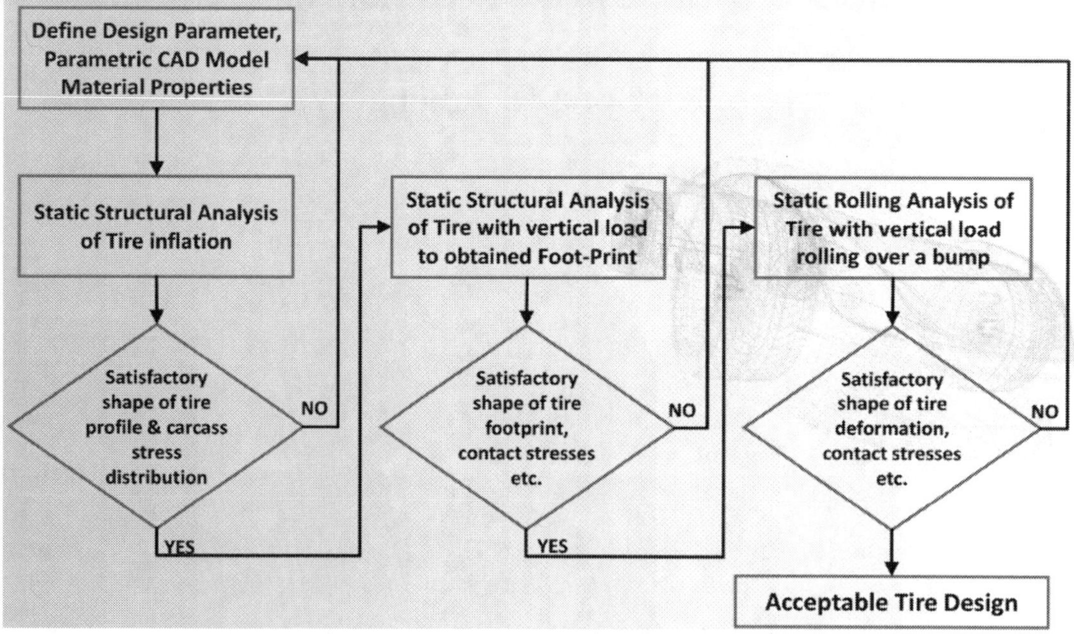

Geometry/Meshing & Material Setup

Mesh Count :
Nodes = 157035 , Elements = 219921

Rubber Components : Mooney Rivlin			
	C10	C01	D1
Tread Cap	0.406962	0.03468	0.0145
Side wall	0.02372	0.29765	0.0804
Belt	1.01025	0.07532	0.009712
Rimstrip	2.637214	0.0500535	0.0014804
Apex	2.637214	0.0500535	0.0014804
Body-ply	0.2652	0.1397	0.03349
Inner Liner	0.14458	0.14281	0.04861

Material Properties for "Bead Steel"
E = 200 GPa, Poisson's Ratio = 0.3

Material Properties for Rebar Elements		
Body Ply Rebar	Young's Modulus	60 GPa
	Possion's Ratio	0.3
Belts Rebar	Young's Modulus	200 GPa
	Possion's Ratio	0.3

	Cross- Section Area	Spacing
Body Ply Rebar	0.25	1.0
Belts Rebar	0.25	1.0

2013 Automotive Simulation World Congress
October 29 - 30 • Frankfurt am Main, Germany

Reinforcement & Hydrostatic Element Setup

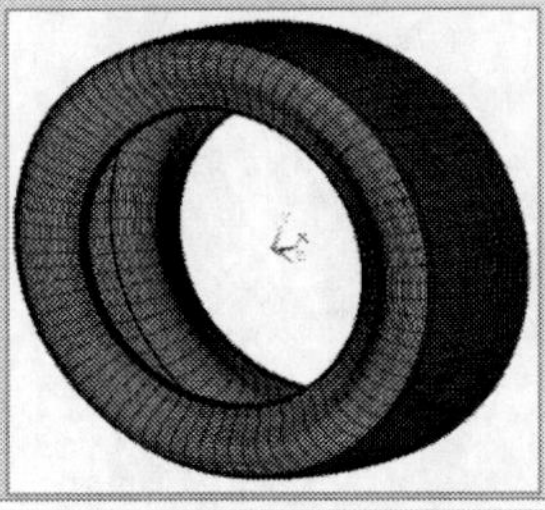

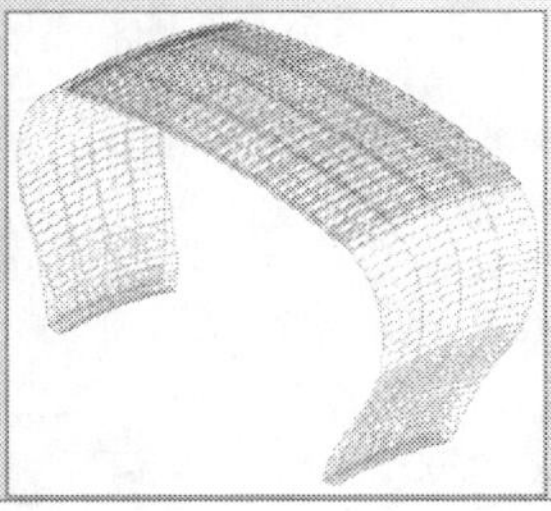

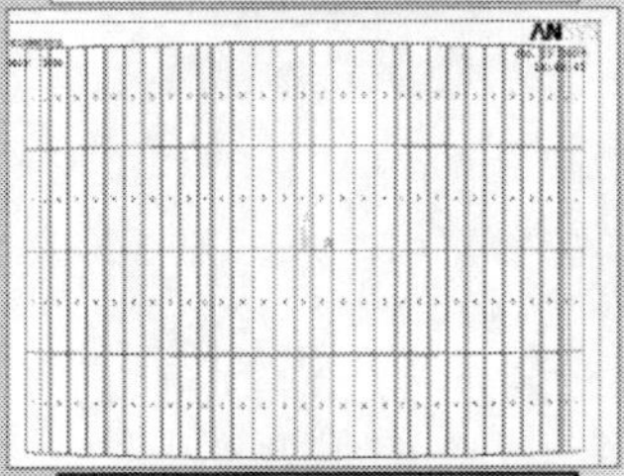

+/- 70 degree rebars in belts

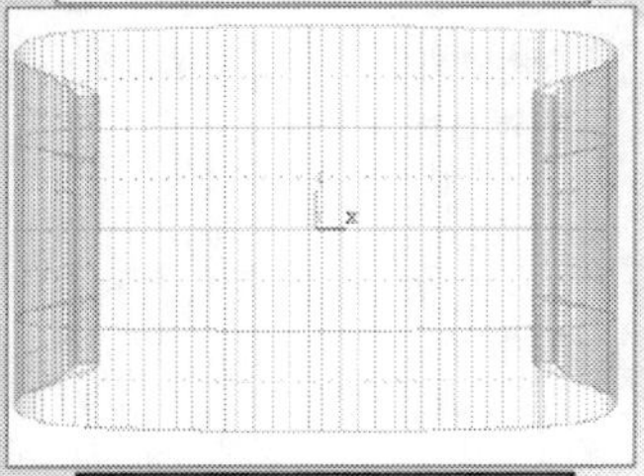

0 degree rebars in body-ply

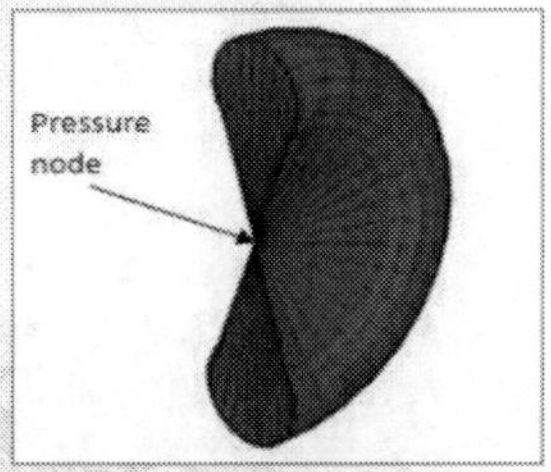

Hydrostatic Fluid Element with Positive Volume

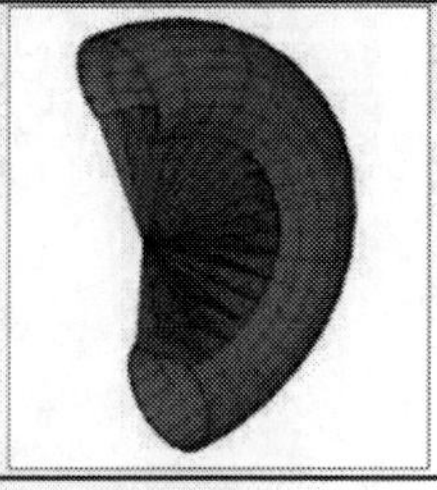

Hydrostatic Fluid Element with Negative Volume

Inflation Analysis Loading & Results

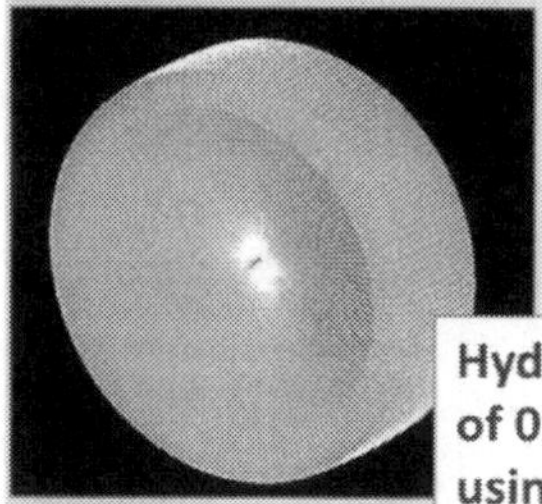

Hydrostatic Pressure of 0.24 MPa is applied using Hydrostatic Elements (HSFLD242)

Fixed Support applied to the Rim hub

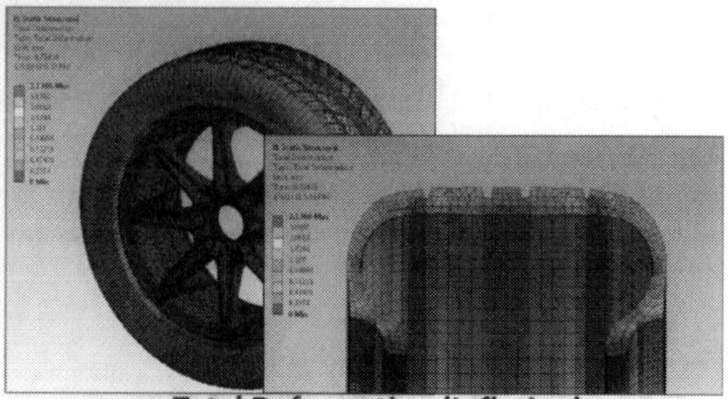

Total Deformation (Inflation)

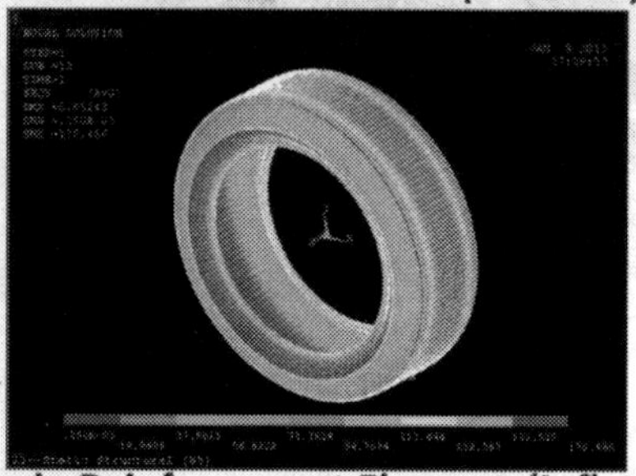

Stress in Reinforcement Elements(Inflation)

Foot-print Analysis Setup & Results

1. The tire is inflated to internal pressure of 0.24 MPa (34.8 psi)
2. Then displacement is applied to the hub of the tire to close the gap between the road & the tire and To create a force of 3000 N in the Spring
3. Assuming 3000 N = ¼ th Weight of the vehicle

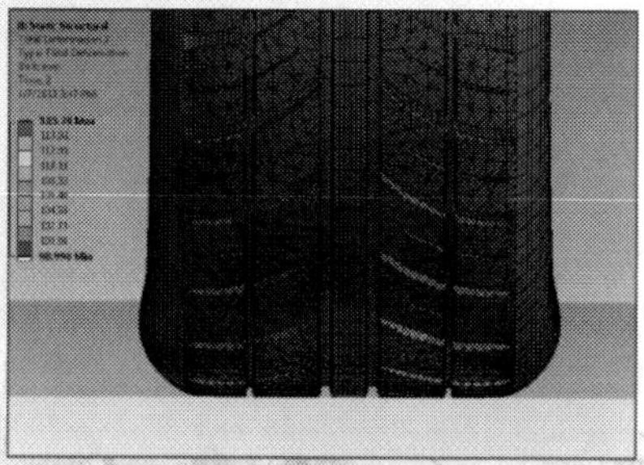

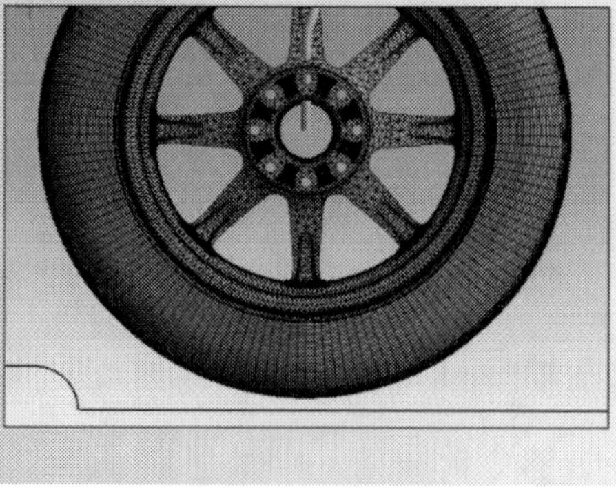

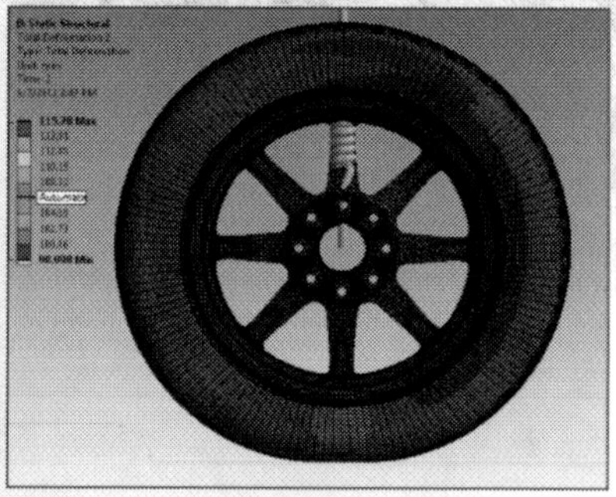

2013 Automotive Simulation World Congress
October 29 - 30 • Frankfurt am Main, Germany

Footprint analysis – Contact Pressure

Tire internal pressure = 24 psi

Tire internal pressure = 34.8 psi

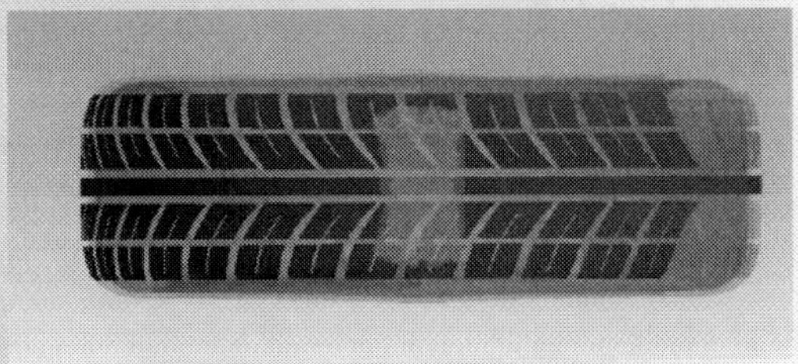

Tire internal pressure = 37 psi

Tire internal pressure = 40 psi

2013 Automotive Simulation World Congress
October 29 - 30 • Frankfurt am Main, Germany

Rolling Analysis Setup

- Steady State Rolling Simulation
- The tire is inflated in the first step to an internal pressure of 0.24 MPa
- Displacement is applied to the hub of the tire to close the gap between the road and the tire that creates a force of 3000 N in the Spring
- 3000 N is ¼ th weight of the vehicle & the road surface is moved in the positive X direction so that the tire rolls over the bump

Rolling Analysis Results

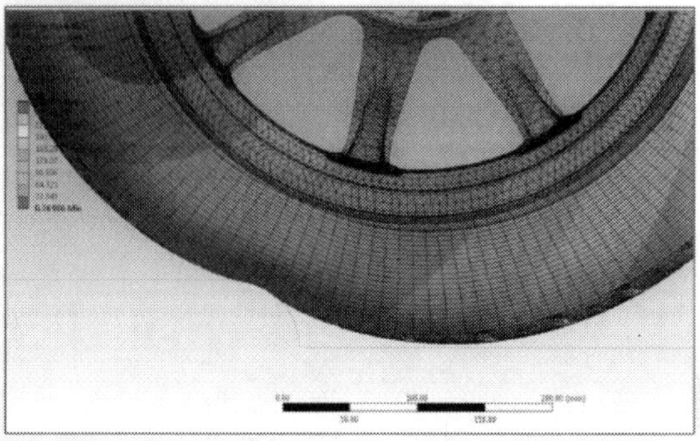

Total Deformation

2013 Automotive Simulation World Congress
October 29 - 30 • Frankfurt am Main, Germany

2. Water Splash by Tire

- **Hydroplaning and Water splash are important parameters to be considered during the tire design**
- **Other than road resurfacing or vehicle body design, proper design of tire profile can reduce the water splash**
- **The tire has a unique shoulder/sidewall profile (chine) which acts as a splash deflector**
- **It is important to minimize the water splash by tire**
  - **To minimize the contamination in the car's engine bay and other critical components**
  - **To protect other vehicles and pedestrians from water/mud**

Mesh Details : Water Splash by Tire Simulation

- **Tire Radius : 0.31 m**
- Tunnel Size : 2.6 m x 1.0 m x 1.0 m
- **Mesh Type : Tetrahedral**
- **Mesh Count : 34 Million**
- **Mesh Size : 5mm near the tire, 20mm in the domain**

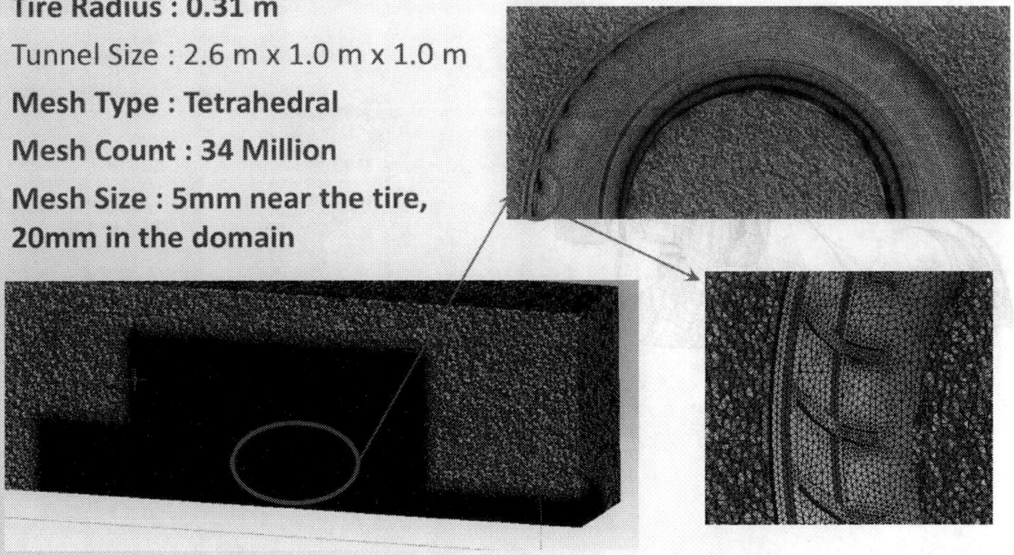

2013 Automotive Simulation World Congress
October 29 - 30 • Frankfurt am Main, Germany

Problem Setup : Water Splash by Tire Simulation

- **Multiphase Volume of Fluid (VOF) model to track the air-water interface**
- **SST K-Omega Turbulence Model**
- **Air-Water surface tension and wall adhesion are included in the model**
- **Tire rotation is modeled by moving wall where rpm is computed from the car velocity and the tire radius**
- **Computational Time taken : 20 hours for running the complete simulation (0.15s) in 96 CPUs**

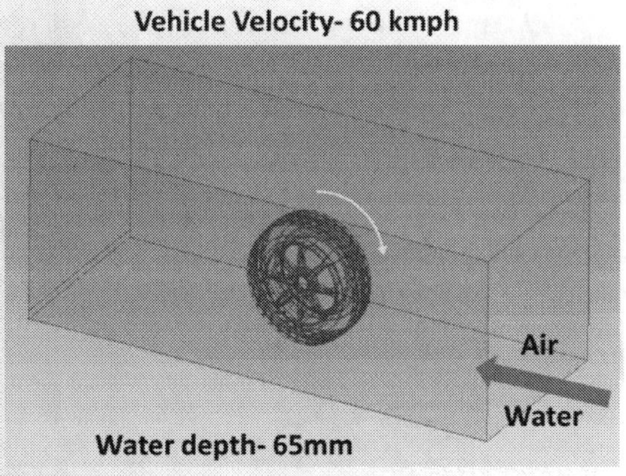

2013 Automotive Simulation World Congress
October 29 - 30 • Frankfurt am Main, Germany

Results : Water Splash by Tire Simulation

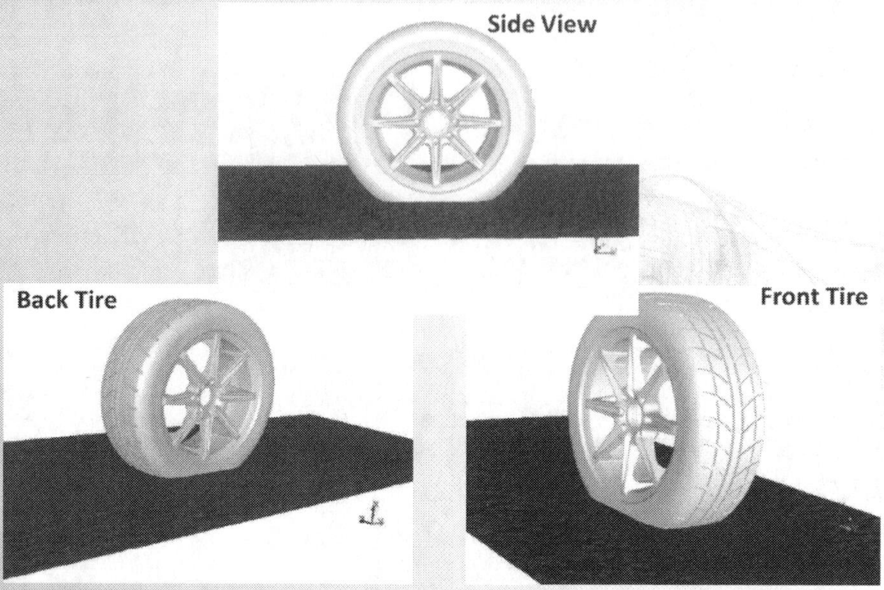

2013 Automotive Simulation World Congress
October 29 - 30 • Frankfurt am Main, Germany

Results : Water Splash by Tire Simulation

1. **This Methodology Captures the Water Splash at the Front and the Back of the tire**
2. It provides very useful insights required for tire profile design to minimize water splash
3. **All the vehicle parts close to the tire and a large portion behind the back tire are wetted**
4. **Side splashing is not captured as it would need a transient moving mesh methodology which is** computationally expensive

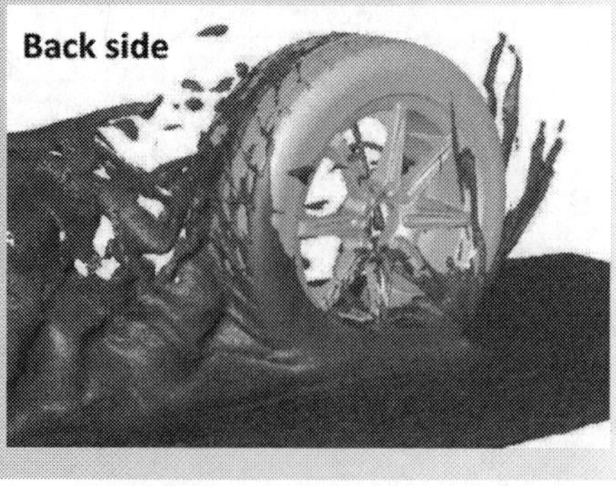

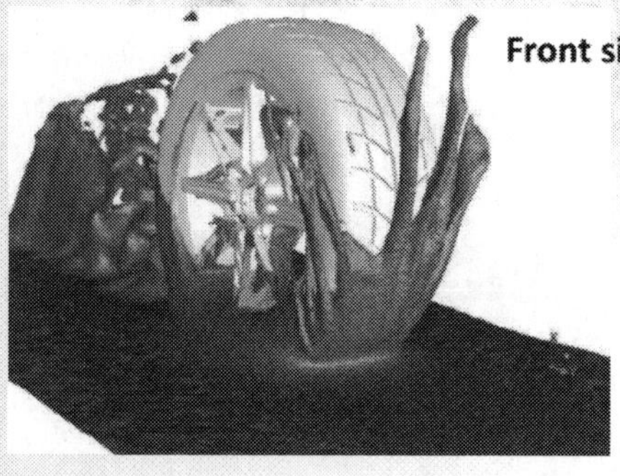

2013 Automotive Simulation World Congress
October 29 - 30 • Frankfurt am Main, Germany

3. Tire subjected to Blast Wave

- Simulating Tire damage when it is subjected to blast wave using explicit analysis in ANSYS AUTODYN
- Complex composite sub-layers of tire made of nonlinear rubber and steel reinforcements
- Complex definition, with Euler domain of tire air, Lagrange domain of tire parts, TNT explosives and surrounding Euler domain of air
- High fidelity blast solver to solve interaction of compressed tire air with Blast wave of TNT and Tire bodies
- Accurate erosion model to predict the tire damage

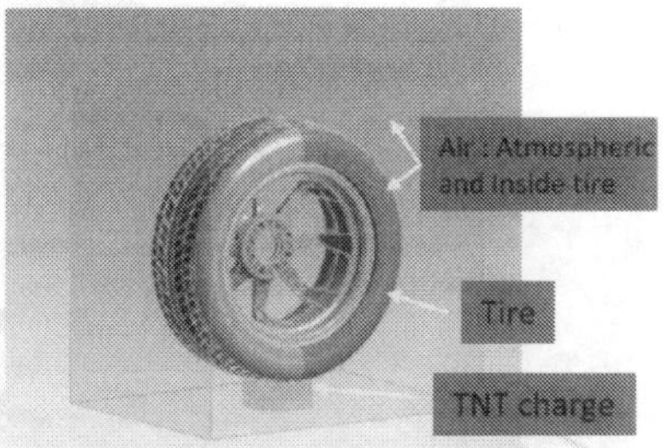

2013 Automotive Simulation World Congress
October 29 - 30 • Frankfurt am Main, Germany

3. Tire subjected to Blast Wave

- TNT explosion damage Tire considerably by eroding the rubber at the road interface.
- Tire damage leads to interaction of blast wave with Air inside Tire, which try to compress it as well as push it out to engage with Atmospheric air.

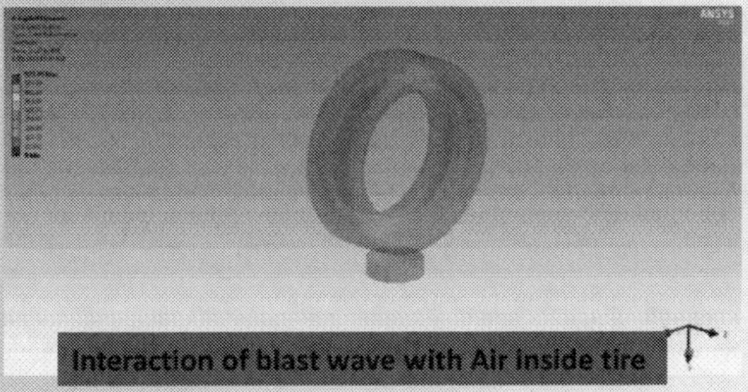

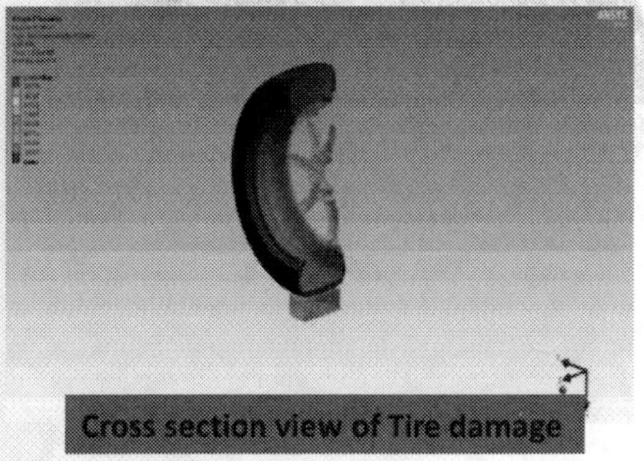

2013 Automotive Simulation World Congress
October 29 - 30 • Frankfurt am Main, Germany

Summary

- Simulations can play a vital role in Journey of Tire from Rubber to Road
- At every stage simulation adds value by providing valuable insights
- ANSYS has breadth of solutions to gain insights of Tire manufacturing and design process
  - ANSYS Polyflow [Tire manufacturing process]
  - ANSYS Mechanical [Tire structural performance]
  - ANSYS Explicit Structural [Blast]
  - ANSYS Fluent [Water splash]

High frequency structural vibrations on ANSYS meshes - the DEA tool for modelling vibrations and acoustics in automotive engineering structures

G. Tanner[1], D J Chappell[2], D Löchel[3], N Søndergaard[3]

[1] *University of Nottingham, United Kingdom, email: gregor.tanner@nottingham.ac.uk*

[2]*Nottingham Trent University, United Kingdom, email: david.chappell@ntu.ac.uk*

[3]*inuTech GmbH, Fürther Str. 212, 90429 Nürnberg, Germany,*

Abstract

We present a simulation tool for modelling noise and vibrations in complex mechanical structures in the mid-to-high frequency regime. The method - Dynamical Energy Analysis (DEA) - extends standard tool such as Statistical Energy Analysis (SEA) towards non-diffusive wave fields. DEA takes into account the full directionality of the wave field and makes sub-structuring obsolete. It can thus be implemented on standard ANSYS FEM grids [1] and gives detailed information about the vibro-acoustic response with high resolution. Interfaces between different components are described in terms of coupling matrices obtained by solving the wave equation locally with the help of full wave solution methods. The method is applied to benchmark model problems the automotive industries in collaboration with Land Rover Ltd. We demonstrate that DEA can handle multi-mode wave propagation effectively taking into account mode conversion between shear, pressure and bending waves at interfaces and on curved surfaces. DEA is thus the first high-frequency tool which can directly couple into ANSYS workbench FEM solver.

1 Introduction

A large range of numerical methods has been developed for solving noise and vibrations problems for mechanical structures. Popular tools include finite or boundary element methods. There are, however, basic limitations when approximating the solutions of wave equations directly: the size of the associated linear system increases with decreasing wavelength and numerical schemes become inefficient when the local wavelengths are orders of magnitude smaller than typical dimensions of the physical system. One therefore moves to high frequency methods, such as ray tracing or statistical approaches. Both these methods have disadvantages, however. Tracking rays including multiple reflections on boundaries can become cumbersome, in particular on curved surfaces and including mode conversion at interfaces and has not yet found widespread use in mechanical engineering. Statistical methods normally start by dividing the structure into a set of substructures and assuming diffuse wave fields and quasi-equilibrium conditions in each of the subsystems. These ideas forms the basis of Statistical Energy Analysis (SEA) [2], which has found widespread applications in the automotive and aviation industry as well as in building acoustics. The underlying assumptions are, however, often hard to verify a-priori or are only justified when an additional averaging over 'equivalent' subsystems is considered. These shortcomings have been addressed by Langley [3] and more recently by Le Bot [4,5].

A computational tool based on a linear operator approach for propagating ray densities – *Dynamical Energy Analysis* (DEA) – has been proposed in [6,7]. DEA systematically interpolates between SEA and full ray tracing. The name points at the similarities with SEA but stresses at the same time the importance of non-diffusive transport along the ray dynamics. In particular, in DEA we have much more freedom in sub-structuring the total system and variations of the energy density across sub-structures can be resolved. An efficient implementation of DEA on meshes has been presented in [1] called *Discrete Flow Mapping* (DFM) making it possible to deal with high-frequency vibro-acoustics on complex mechanical structures including multi-mode wave propagation and curved surfaces. Coupling at material interfaces is described in terms of reflection/transmission matrices. These matrices are obtained by solving the wave equation locally in the coupling region.

The method is applied to benchmark problems provided by Germanischer Lloyd (part of a double-hull structure of a large ship) [8] and Jaguar Land Rover (cast aluminum shock tower of a Range Rover).

2 Dynamical Energy Analysis – a brief overview

2.1 From ray tracing to flow equations – an operator formulation

We give here a brief overview over the set-up of the method; for more details, see [9,10]. We consider a linear wave problem driven by a distribution of sources at a fixed angular frequency ω; a generalisation to frequency band excitation is straightforward as discussed below. We split the total system into sub-domains $j = 1,\ldots,N$, (such as the elements of an FEM grid), whereby the material parameters and thus the local wave speeds c_j are assumed to be constant in each sub-domain, but may vary between domains. Damping is incorporated through a complex wavenumber with an imaginary part proportional to a damping coefficient μ_j, which may depend on ω. In general, one wants to determine the solution u of a wave equation, (where u is, for example, the displacement within a solid or pressure variations within a fluid), that is,

$$(H + \omega^2)u(r) = -F(r). \tag{1}$$

Here, H corresponds to a linear operator containing, for example, mass and stiffness matrices describing the vibro-acoustic dynamics including damping and $F(r)$ represents a force distribution driving the system. Appropriate boundary conditions apply at the outer boundaries and the interfaces between sub-domains. The wave energy density ε at a point r is then proportional to the modulus-square of the wave amplitude u, that is,

$$\epsilon(r,\omega) \propto |u(r,\omega)|^2. \tag{2}$$

The linear wave equation (1) can in a natural way be associated with a ray dynamics via the Eikonal approximation expressing the wave function in (2) in terms of ray contributions [5]. The mean wave energy density is proportional to the density of rays $\rho(r,p)$ going through a point r, that is,

$$\epsilon(r,\omega) \propto \int \rho(r,p;\omega)\, dp, \tag{3}$$

where p is the direction (or momentum) vector which is related to the wave number. In the stationary case, rays emerge uniformly from source points and undergo reflections at boundaries as well as absorption processes. This makes it possible to relate wave energy densities to classical flow equations and thus thermodynamical concepts, which are at the heart of SEA or DEA treatments. Note that different mode types such as shear, pressure or bending modes in plates are treated as rays with different local wave speed. Mode coupling at boundaries or interfaces leads to 'mode-conversion' of rays, that is, the classical flow of rays will undergo 'ray-splitting'. The conversion rates between rays corresponding to different modes are related to the modulus square of the entries of an interface scattering matrix. Likewise, the transmission/reflection probabilities between sub-domains are given by the ratios of the outgoing normal power fluxes to the incoming normal power flux at the interface, which can again be obtained from the scattering matrix.

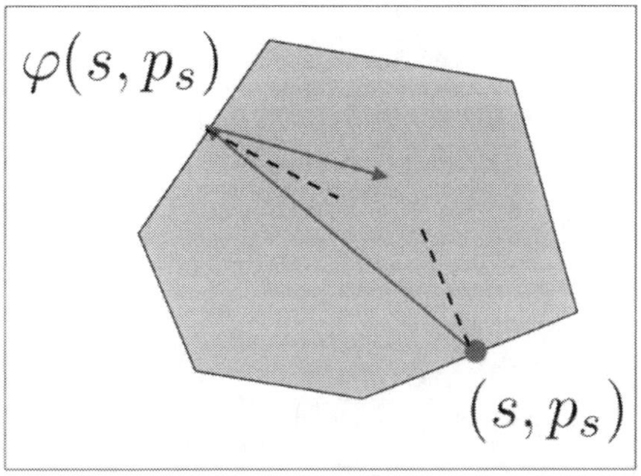

Figure 1: Boundary map in a single subsystem.

DEA is based on the observation that these flow equations for ray densities can again be written in terms of linear partial differential equations, the so-called Liouville equation (LE). In order to solve the stationary flow problem, we rewrite the LE in boundary integral form; the boundary can be the physical boundary of the system or the boundary of sub-domain. We first consider a problem with a single domain with boundary Γ, see Fig. 1. The stationary ray density in the interior can be mapped onto a boundary layer density ρ^0, ie the boundary density producing the same ray field in the interior as the original field before the first reflection. Ray densities ρ□emanating from the boundary are transported to the next intersection with the boundary by an operator $\boldsymbol{B}$,

$$\boldsymbol{B}[\rho](X^s) = w(X^s)\,\rho(\varphi^{-1}(X^s)), \tag{4}$$

see Fig. 1. Here $X^s = (s; p_s)$ represents phase-space coordinates on the boundary, that is s parameterises the boundary Γ□and p_s denotes direction or momentum components tangential to Γ□at s. Also φ is the invertible boundary map; it takes a ray from a boundary point s with tangential direction p_s along a straight ray to the next intersection with the boundary. The weight function $w(X^s)$ contains absorption factors. The stationary density on the boundary induced by the initial boundary distribution $\rho^0\,(X^s)$ is then obtained using

$$\rho = \sum_{n=1}^{\infty} \boldsymbol{B}^n[\rho^0] = (\boldsymbol{I} - \boldsymbol{B})^{-1}[\rho^0], \tag{5}$$

where $\boldsymbol{B}^n$ contains trajectories undergoing n reflections at the boundary. The density distribution in the interior region can then be obtained from the boundary density $\rho(X^s)$. One obtains the density (3) after projecting down onto coordinate space. Note that this treatment is formally equivalent to ray tracing - we therefore call $\boldsymbol{B}$ the *ray tracing boundary operator*.

In a next step, we consider several subsystems Ω_i, $i= 1,...,M$. Coupling between sub-elements are treated as losses in one subsystem and source terms in another. Typical subsystem interfaces are surfaces of reflection/transmission due to sudden changes in material parameters or local boundary conditions. However, as DEA takes into account the full ray dynamics, the final results will not depend critically on the choice of the sub-systems; in fact, we can always split a given sub-system into smaller sub-domains if necessary – a big advantage compared to SEA where the choice of sub-domains is often critical. We describe the full ray dynamics in terms of subsystem boundary operators $\boldsymbol{B}_{ij}$, flow between Ω_i□and Ω_j is possible only if subsystem i and j have a common boundary, see Fig. 2. We introduce a weight function w_{ij} in Eq. (4), which contains (apart from the usual damping term) reflection and transmission coefficients characterising the coupling between subsystems i and j at the interface.

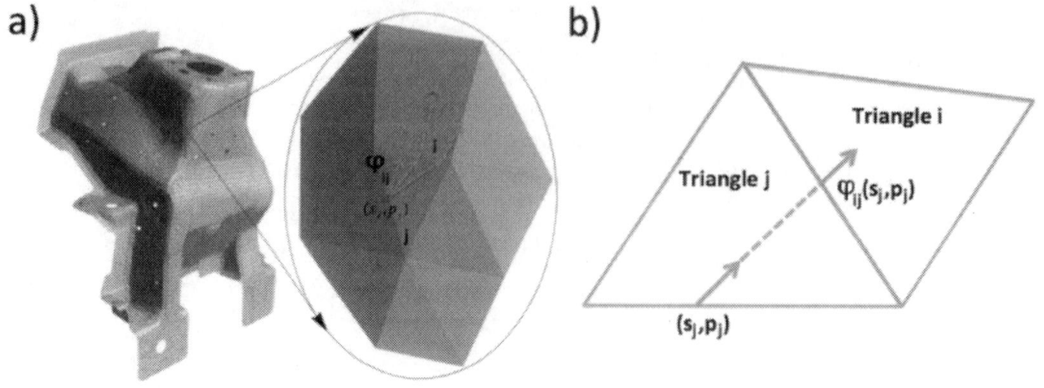

Figure 2: Discrete flow mapping defined on FEM grid (a); the boundary map for adjacent triangles (b)

2.2 Discrete flow mapping

In order to evaluate Eq. (5), a finite dimensional approximation of the operator $\boldsymbol{B}$ is constructed. We are facing two fundamental problems at this stage; firstly, one needs to choose a subdivision of the full structure into sub-domains together with the associate flow-map for each sub-domain; and secondly, one needs to construct the operator $\boldsymbol{B}$ in an appropriate approximation. The first point needs to address the complexity of the structure, the second needs to deal with the fact, that we doubled the dimension of the problem by going from position space to phase space. Using the discrete flow mapping technique, we can overcome both problems: to address the complexity issue, we choose the mesh-elements of a grid as the sub-domains. This allows us to deal with very complex structures but keeps the local dynamics simple with flow maps defined on triangles or rectangles. The dimensionality problem is tackled by constructing B on the boundary of the mesh elements in terms of basis functions [1,9,10]; for the spatial variable s, piecewise constant boundary element functions on the (discretised) boundary are used. For the approximation in the momentum argument we choose a Legendre polynomial basis; note that in 2D, the variable $p_s \in (-|p|;|p|)$ with $|p|$ on the 'energy surface' defined through $H(p,r) = \omega^2$, where the Hamiltonian H is related to the wave operator $\boldsymbol{H}$ in (1). We thus expand the density ρ in the form

$$\rho(X^s) \cong \sum_{\alpha=1}^{n} \sum_{\beta=1}^{N} \rho_{(\alpha,\beta)} b_\alpha(s) P_\beta(p_s). \tag{6}$$

Here, N is the order of the basis expansion, n is the number of elements of the boundary mesh, P_β is the Legendre polynomial of order β and b_α denotes a boundary element basis function being constant in the boundary segment $\alpha \in \Gamma$ and zero elsewhere. The coefficient vector $\rho_{(\alpha,\beta)}$ in (6) is labeled in terms of the multi-index (α, β). In the case of multi-domain problems, we need to sum in Eq. (6) in addition over the domain index j. The matrix approximation B of the ray tracing operator $\boldsymbol{B}$ is obtained in the same way by representing the integral in (4) in the basis approximation Eq. (6), see [1,9]. The simplicity of the flow map makes it possible to solve some of the integrals (necessary for computing B) analytically, which effectively halves the dimensionality of the problem, see [1] for details.

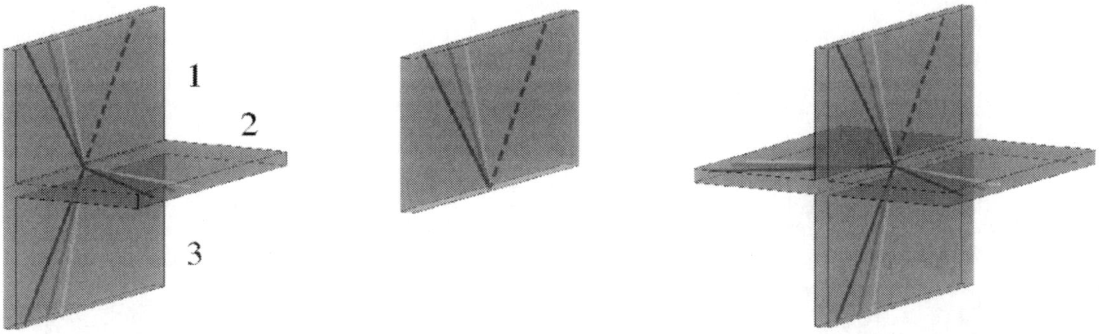

(a) T-junction (b) open end (c) X-junction

Figure 3: Dashed line: incoming ray, solid lines: reflected/transmitted rays of the three mode types.

Once the matrix B has been computed, the values of $\rho_{(\alpha,\beta)}$ in Eq. (6) are evaluated using Eqn. (5) by solving

$$(1 - B)\rho = \rho^0. \tag{7}$$

This yields an approximation for the density distribution ρ on the boundary given by (6). This distribution is then mapped back into the interior region and after projecting onto position space we obtain an approximation for the energy density in (3), see [1] for details. Frequency band calculations are obtained by sampling the results over the frequency band considered.

Note that transmission or reflection at boundaries depends in general on the angle of incidence of the incoming/outgoing ray. DEA incorporates directed transmission through interfaces, in contrast to SEA, which assumes diffusive wave fields in each subsystem – i.e. wave fields consisting of a uniform superposition of waves (and thus rays) from all directions. We will discuss the details about the implementation of interface scattering matrices in the next section.

2.3 Reflection, transmission and curvature

Reflection/transmission coefficients are obtained by calculating interface scattering matrices. We will focus here on interfaces forming junctions between plates of varying thickness, a setting, which is relevant for our benchmark problems discussed in Sec. 3. In DEA, we find local wave solutions at the interface taking into account the angle dependence of the corresponding wave scattering. To find the transmission/reflection coefficients of a set of plates being coupled at a common interface (such as depicted in Fig. 3), we follow [11,12]. In particular, we consider the connection between plates as line junctions, that is, the interior properties of the junction are not modeled and the mass and moment of inertia are neglected.

Let us consider a line junction, which couples n different plates assuming semi-infinite plates for simplicity. The boundary conditions at the line junction correspond here to dynamic conditions involving stresses and moments and kinematic conditions for displacement and rotations for all plates. To construct the transmission coefficients, we calculate the response of the system with respect to excitation due to an incoming plane wave. The incoming wave has a fixed wave number and a characteristic mode, that is, it is of b, p or s (bending, pressure and shear) type. The outgoing waves typically have components in all plates and are a mixture of all mode types. An evanescent bending mode is included to complete the description. Thus, for each plate we have 4 unknown amplitudes for the 4 different wave types. Possible material differences between the plates can lead to different wave numbers in different plates. For a given forcing with a particular incoming mode in a particular plate, we can solve for the unknown modal coefficients in all plates. In practice, we find directly the transmission probabilities by calculating the ratio of outgoing to incoming normal power fluxes.

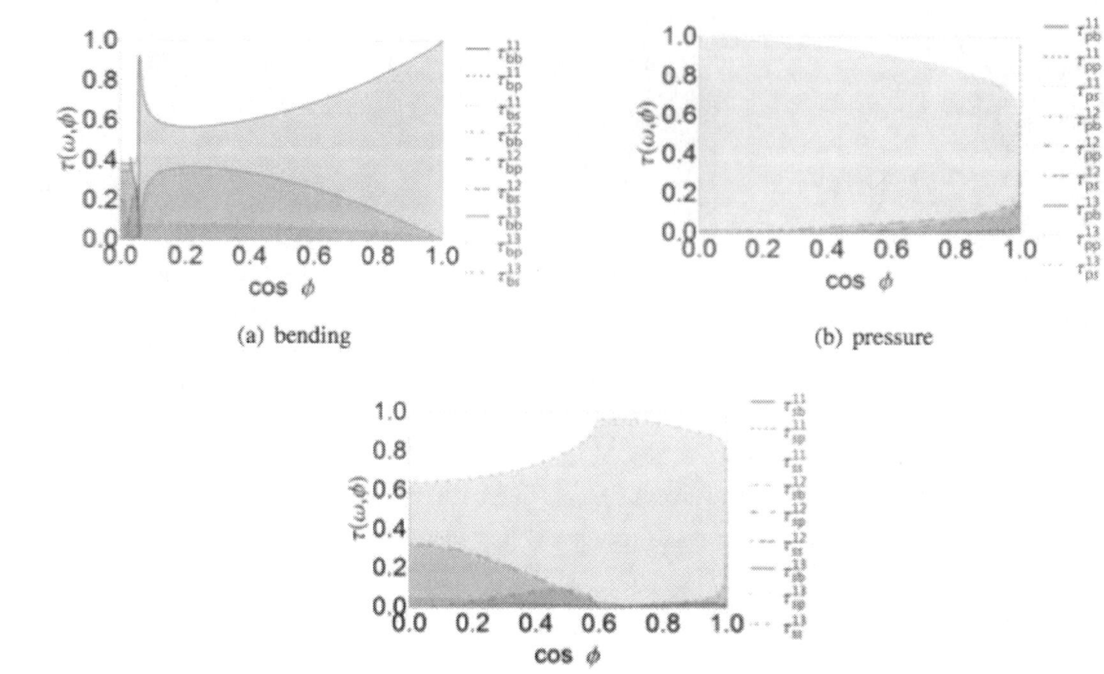

(a) bending

(b) pressure

(c) shear

Figure 4: Reflection/transmission coefficients for a T-joint with material parameters given in Sec. 3 at 500 Hz. Shown is the normal flux τ_{xy}^{ij} of the outgoing mode $y = b, p, s$ in plate $j = 1, 2, 3$ for an in-coming wave of specified type $x = b, p, s$ in plate 1. The wave incoming at an angle θ is of bending (a), pressure (b) and shear (c) type, respectively.

Fig. 4 shows transmission coefficients calculated with the method developed in [11], here for the T-joint in Fig.3 (a) with material properties given [1]. In particular, we plot the individual transmission coefficients for bending, pressure or shear forcing on the vertical plate in Fig. 2 (a). The coefficients are labeled as τ_{xy}^{ij} with $x, y = b, p, s$ and i,j = plate 1, 2, and 3. The plot shows the dependence of the coefficients with respect to the cosine of the angle of the incoming ray to the interface. The results, as displayed in Fig. 4, are now incorporated as part of the weight function w into the boundary kernel, Eq. (4), or its finite dimensional approximation.

For modelling the vibration of thin shells we need to consider ray tracing on curved surfaces. Applying thin shell theory in the high frequency limit as in [13], curved rays follow the geodesics of the surface. This approximation is valid for wavelengths shorter than the radii of curvature of the shell, but larger than the thickness. Mimicking geodesic paths on the corresponding triangulated surface, we employ the theory outlined in [14] and [15], making use of the fact that for ray paths on triangulated surfaces, the notions of shortest and straightest (discrete) geodesic are equivalent, see [15]. Hence the straightest geodesic choice of $\theta_i = \theta_j$ will approximate the direction of the geodesic flow on homogeneous regions of the triangulated surface, and abrupt changes of surface thickness or material will result in a Snell's Law effect, see [14].

Corrections need to be considered if the radius of curvature is of order of the wavelength. It is possible to construct modified ray paths from the dispersion curves given by thin shell theory, see [13], but this requires a detailed knowledge of the local curvature. In the interests of keeping the model as simple as possible we will follow a different approach here. We treat the meshed structure as a set of plate-like elements and estimate reflection/transmission properties due to the finite angle between mesh elements using the plate-junction theory sketched above and detailed in [11,12]. This enables us to mimic wave barriers due to regions of high curvature as demonstrated in the next section.

3 Numerical results

3.1 Double-hull structure

We apply the method to a stiffened double bottom test structure provided by Germanischer Lloyd [8], material parameters are specified in [1]. For our DFM treatment, we divide the structure along plate intersections; this leads to N = 212 subsystems for the test structure in Fig. 5. We further subdivide the boundary of each subsystem; the boundary mesh is chosen such that the mesh size is 0.2m or less. For the momentum basis we use up to 8th order Legendre polynomials. The transmission/reflection coefficients are obtained as discussed in Sec. 2.3 and computed directly when evaluating the integrals in Eq. (4) in the basis expansion. Note that here, due to the regularity of the test structure, we only need to consider three different types of line joints, namely those displayed in Fig. 3. The structure is excited by 'rain on the roof', that is, random excitation in the bending degrees of freedom at 200Hz on the upper part of the structure. The computation was done on a quad-core processor machine (Intel core i5-2540M cpu 2.60GHz) with a total computation time of 6.5mins. The computation of the matrix coefficients in the 194,562×194,562 sparse matrix with 309,434,904 nonzero matrix entries took 3.5mins and solving the linear system with Gauss-Seidel iteration took 2.5mins.

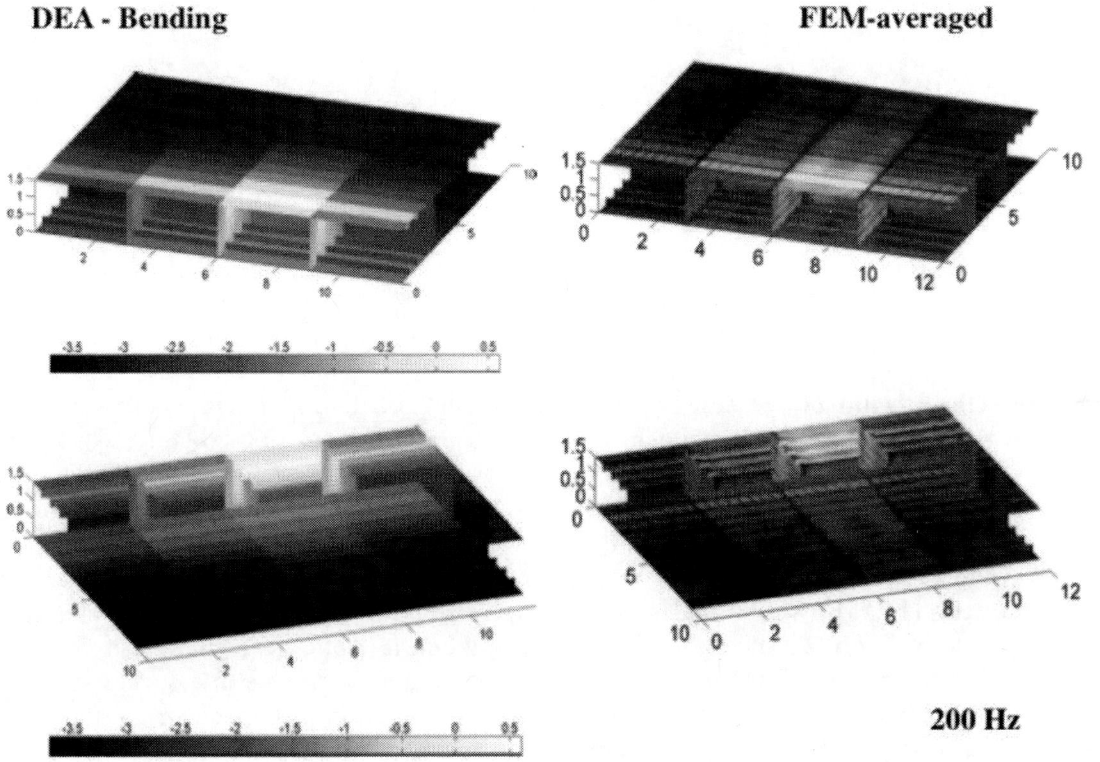

Figure 5: Germanischer Lloyd benchmark example with 'rain on the roof' excitation. Computed energy density (logarithmic scale) of the bending mode with DEA (left hand side), compared with an FEM calculation averaged over 1/3 octave band (right hand side). The axis scaling is in meters.

Fig. 5 shows the structure and the wave energy density of the bending modes on a logarithmic colour scale. The first column displays the energy stored in the bending mode including the excitation region. This is compared with an FEM calculation averaged over a frequency band (1/3 octave band). As expected, the energy density decreases with the distance to the excitation points due to damping; the overall agreement with the FEM calculation is good.

3.2 Range Rover shock tower

Next, we consider the transport of high frequency wave energy in thin shells. High frequency vibro-acoustic models based on an SEA treatment will be unsuitable in these circumstances since complex geometrical features are difficult to include in SEA and subdivision of the model into sub-systems is far less clear cut for large molded castings. DEA using the DFM method can overcome these problems, since it can be easily applied in the framework of existing grids for finite element models, requires no choice of sub-system division and incorporates the full geometry and directionality of the energy flow.

The response of a thin molded aluminum car component (shock tower of a Range Rover), see Fig. 6, to a unit point force applied normal to the surface is modeled and compared with a finite element simulation for the full wave model performed using Nastran. In order to maintain a tractable model size for the finite element simulation and to study frequency ranges of industrial interest, the computation is performed at frequencies between 8kHz and 10kHz. This approximately corresponds to a third of an octave band centered at 9kHz.

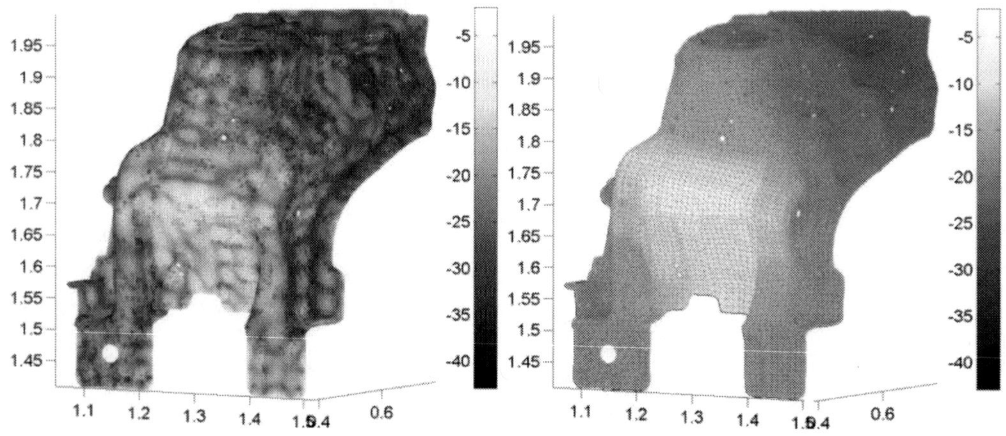

Figure 6: Energy density on a thin aluminum shell (Range Rover shock tower) estimated using an averaged full wave finite element model (left) and our DEA model (right) for 3% hysteretic damping.

For the DFM treatment, we now consider the elements of the FEM grid as our sub-domains. Here, the power of the DEA-method becomes obvious – we can immediately work with the meshes provided by the manufacturer, here Jaguar Land Rover, to perform the calculations; any preprocessing in terms of finding adequate subsystems becomes obsolete. Assuming thin shell theory [13], rays travel along geodesics on the curved surface (as long as the radius of curvature is large compared to the wavelength). Ray tracing along geodesics can be implemented on the mesh following [14,15] by choosing incoming angles equal to outgoing angles at each interface as discussed in Sec. 2.3. To incorporate mode mixing and reflection effects in regions of strong curvature, we interpret the FEM model as a set of plane plate segments intersecting at the mesh boundaries. The angles of intersection enter into the computation of the scattering matrix giving rise to reflection, transmission and mode conversion as described in Sec. 2.3.

Fig. 6 shows the ray model result compared with a full FEM calculation of the wave equation using shell elements in Nastran; here, the kinetic energy averaged over 41 evenly spaced frequencies spanning the prescribed range and with a typical hysteretic damping level of 3% are plotted. The ray computation is performed using a triangulated surface consisting of 11623 triangles and with a 10th order Legendre polynomial basis in direction space. The Nastran grid contains 40670 elements comprising a mix of piecewise linear triangles and quadrilaterals. As would be expected one sees more oscillation in the full wave model.

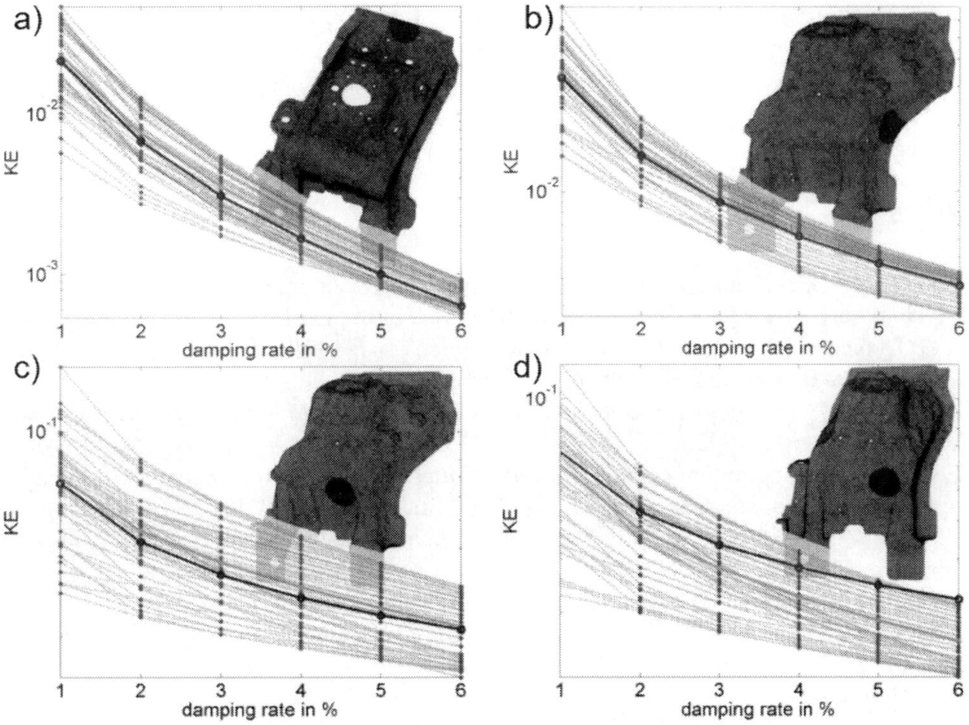

Figure 7: Detailed comparison of DFM and FEM results; kinetic energy integrated over the blue circular region at different postions. Red line: energy density at different frequencies as a function of the damping parameter; Green line: FEM averaged over frequency band; Blue line: DFM result.

The prediction of the overall energy flow is good both in the regions of high and low curvature as well as close and far away from the source, see Fig. 7. In particular, it can be noted in both models that high curvature regions act as barriers for the energy flow. Such geometric features would be entirely absent from SEA-type models and represent a major advance in the simulation of large-scale high frequency vibro-acoustics. The computations in this section were performed with the same quad-core processor machine as above in 15mins.

4 Conclusions

We present an approach for modeling the transport of wave energies through complex structures. The proposed method – DEA – interpolates between SEA and ray tracing, the latter being based on the ray dynamics within the structure including reflection/transmission and mode conversion at interfaces. For more complicated junctions between subsystems, we calculate coupling coefficients based on a scattering approach, which are then incorporated into the DFM framework. We explain how to determine the scattering coefficients for junctions connecting several plates and how to incorporate the parameters into a DFM model both for flat and curved surfaces. We apply DEA with DFM methodology to a model for a ship structure provided by Germanischer Lloyd and for a car-body part from a Range Rover. Results for a system with several thousand sub-domains and including all three wave modes in plates are obtained on a competitive computing time scale of a few minutes. They clearly show large-scale effects due to non-diffusive wave transport.

Our numerical results demonstrate the advantages of a DEA treatment compared to SEA; damping and directionality in the wave transport between subsystems as well as curvature effects are incorporated into DEA using DFM. The spatial resolution in each wave mode allows for a detailed analysis of the response of the structure both near and far from the sources.

Acknowledgement
The authors wish to thank Christian Cabos and Mark Wilken from Germanischer Lloyd and Steve Fisher from Jaguar Land Rover for providing the benchmark problems and for helpful discussions. Support from the EU (FP7IAPP grant MIDEA) is gratefully acknowledged.

References

[1] D.J. Chappell, G. Tanner, Dominik Löchel and Niels Søndergaard, *Discrete flow mapping: transport of phase space densities on triangulated surfaces, Proc. R. Soc. A* 2013 **469**, 20130153.
[2] R.H. Lyon and R.G. DeJong, *Theory and Application of statistical energy analysis* (2nd edn.). Butterworth-Heinemann, Boston MA, 1995.
[3] R.S. Langley, *A wave intensity technique for the analysis of high frequency vibrations*, J. Sound. Vib.**159**, 483–502, 1992.
[4] Le Bot and A. Bocquillet, *Comparison of an integral equation on energy and the ray-tracing technique for room acoustics*, J. Acoust. Soc. Am. **108**, 1732-1740, 2000.
[5] Le Bot, *Energy transfer for high frequencies in built-up structures*, J. Sound. Vib. **250**, 247–275, 2002.
[6] G. Tanner, *Dynamical energy analysis – Determining wave energy distributions in vibro-acoustical structures in the high-frequency regime*. J. Sound Vib. **320**, 1023-1038, 2009.
[7] D.J. Chappell, S. Giani and G. Tanner, *Dynamical energy analysis for built-up acoustic systems at high frequencies*, J. Acoust. Soc. Am., **130**, 1420-1429, 2011.
[8] C. Cabos, H.G. Matthies, *The Energy Finite Element Method Noise FEM*. in Proc. IUTAM Symp. Vib. Analysis of Structures with Uncertainties, St Petersburg, 2009, Eds.: A.K. Belyaev and R.S. Langley, IUTAM series **27** (Springer, Heidelberg, 2011).
[9] D.J. Chappell and G. Tanner, *Solving the stationary Liouville Equation via a Boundary Element Method*, submitted to J. Comp. Phys., 2012.
[10] D.J. Chappell, S. Giani and G. Tanner, *Boundary element dynamical energy analysis: a versatile method for solving two or three dimensional wave problems in the high frequency limit*, J. Comp. Phys., 6181 – 6191, 2012.
[11] R.S. Langley and K.H. Heron, *Elastic wave transmission through plate/beam junctions*, J. Sound Vib. **143**, 241, 1990.
[12] R.J.M. Craik, I. Bosmans, C. Cabos, K.H. Heron, E. Sarradj, J.A. Steel and G. Vermeir, *Structural transmission at line junctions: a benchmarking exercise*, J. Sound Vib. **272**, 1086, 2004.
[13] A.N. Norris and D.A. Rebinsky, *Membrane and Flexural Waves on Thin Shells*, ASME J. Vib. Acoust. **116**, 457-467, 1994.
[14] R. Kimmel and J.A. Sethian, *Computing geodesic paths on manifolds*, Proceedings of the National Academy of Sciences of the USA **95**, 8431-8435, 1998.
[15] D. Martinez, D.L. Vehlo, and P.C. Carvalho, Computing geodesics on triangular meshes, Computers & Graphics **29**, 667-675, 2005.

Fracture Mechanical Approach for Fatigue of Adhesive Joints

Dr.-Ing. Stephan Vervoort
Hottinger Baldwin Messtechnik GmbH

Agenda

1. Background/Motivation
2. Analysis Process
3. FEM Guidelines
4. Adhesive Joints Calculation Method
5. Example
6. Correlation
7. Acknowledgements

Background/Motivation

nCode ...an HBM brand

- Increased use of adhesive joints in automotive engineering:
  - 15 kg adhesives in average car
- Adhesive joining technique build base for material substitution (sheet metals -> composites)
  - Lightweight design
- Examples of automotive engineering:
  - Audi A6: Bond seam approx. 90 m
  - Daimler A-Class: Bond seam approx. 120 m
  - BMW 7-Series: Bond seam approx. 150 m

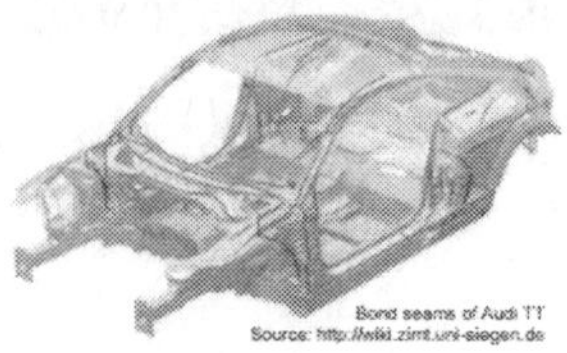

Bond seams of Audi TT
Source: http://wiki.zimt.uni-siegen.de

Background/Motivation

nCode ...an HBM brand

- Example: Adhesive joints in automotive engineering

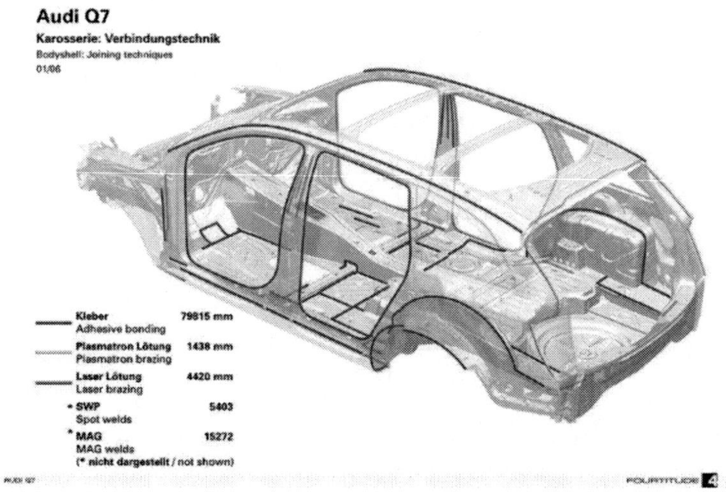

Source: www.volkswagen-forum.com

Automation Simulation World Congress, October 29 – 30, 2013

Background/Motivation

nCode ...an HBM brand

- Advantages
  - Nondestructive joining technology
  - Constant stress distribution across adherent
  - Elastic characteristic (cold-heat equalization)
  - Damping characteristic during crash

- Disadvantages
  - Long-term set-up times (e.g. for cleaned surfaces)
  - Long cure times (could be improved by use of weld/adhesive- or rivet/adhesive-combinations)

- Need clarification in respect of fatigue
  - Aging
  - Repairing of bonded car

Analysis Process

nCode ...an HBM brand

Global model

Simple joint modelling

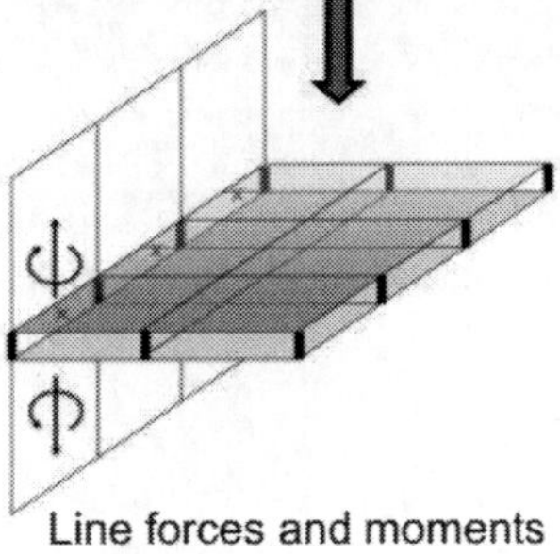

Line forces and moments on flange edge

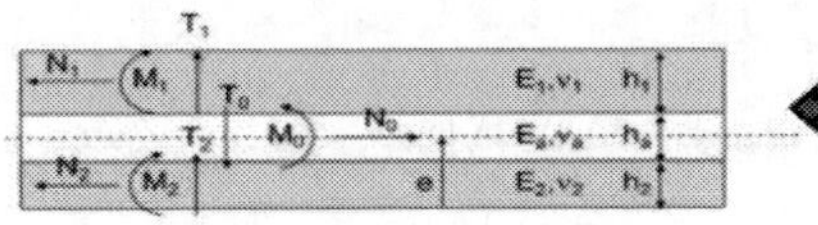

Transfer line forces to analytical sandwich model

FEM Guidelines

nCode ...an HBM brand

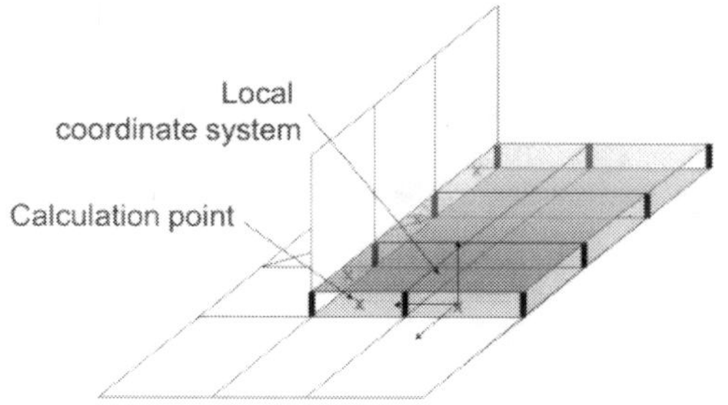

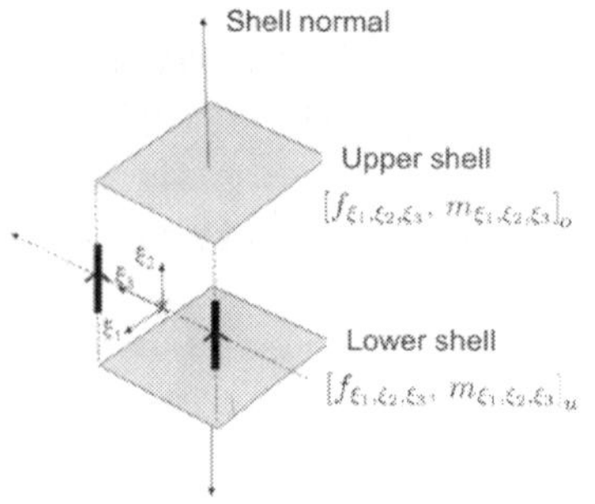

- FE Guidelines:
  - Congruent mesh
  - Linear shell elements
  - 4-node shell elements are preferred (approx. 10 mm)
  - Adhesive joint is modeled by beam elements (radius approx. 5 mm)
- FE results: Grid Point Forces

Adhesive Joint Calculation Process

nCode ...an HBM brand

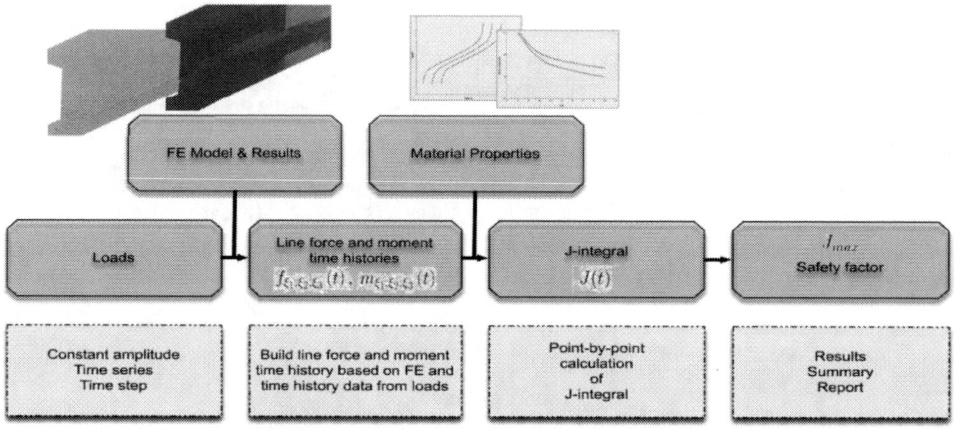

Adhesive Joint Calculation Process

nCode ...an HBM brand

Sandwich model (schematic)

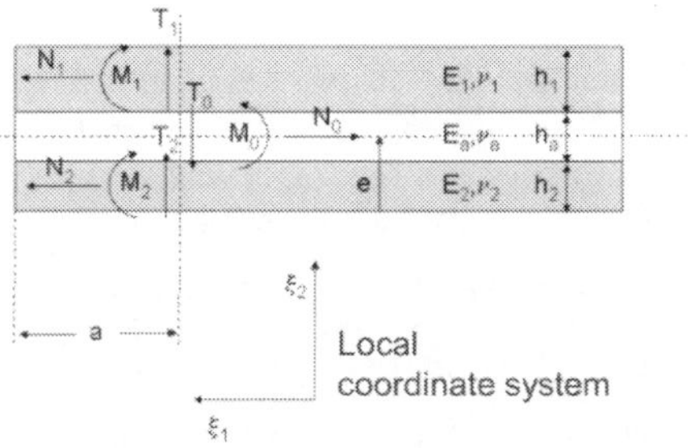

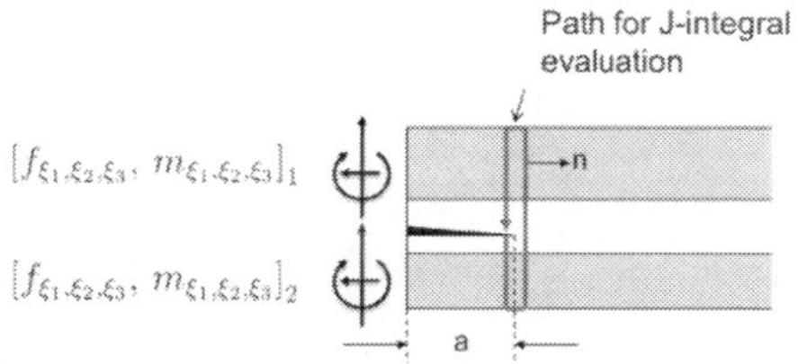

$h_1,\, h_2,\, h_a$: Thickness

$E_1,\, E_2,\, E_a$: Moduli

$\nu_1,\, \nu_2,\, \nu_a$: Poisson Ratios

$$
\begin{aligned}
N_1 &= f^o_{\xi_1} & N_2 &= f^u_{\xi_1}\\
T_1 &= f^o_{\xi_2} & T_2 &= f^u_{\xi_2}\\
F_{z1} &= f^o_{\xi_3} & F_{z2} &= f^u_{\xi_3}\\
M_1 &= m^o_{\xi_3} + a \cdot f^o_{\xi_2} & M_2 &= m^u_{\xi_3} + a \cdot f^u_{\xi_2}
\end{aligned}
$$

Initial Condition: Initial flaw size a_0 has to be specified

Adhesive Joint Calculation Process

nCode ...an HBM brand

- Shell elements

$$
\left.
\begin{aligned}
\overline{G_i} &= \overline{G_i}\,(E_i,\, \nu_i)\\
Q_i &= Q_i\,(E_i,\, \nu_i)\\
H_i &= H_i\,(h_i)\\
\phi_i &= \phi_i\,(h_i,\, h_a,\, E_i,\, E_a,\, M_i,\, T_i)
\end{aligned}
\right\} \quad i = 1, 2
$$

- Sandwich model

$$
\begin{aligned}
E_0 &= E_1\\
h_0 &= h_0\,(h_1,\, h_2,\, h_a,\, E_1,\, E_2,\, E_a)\\
\overline{G_0} &= \overline{G_1}\\
Q_0 &= Q_1\\
H_i &= H_i\,(h_1,\, h_2,\, h_a,\, Q_1,\, Q_2,\, Q_a)\\
\phi_0 &= 0
\end{aligned}
$$

$\overline{G}, Q$ ⋙ Shear moduli

H ⋙ Second moment of area

ϕ ⋙ Rotation of the cross beam section

Adhesive Joint Calculation Process

nCode ...an HBM brand

- J-integral equation:
  - for each point of loading time series

$$J_i = \frac{N_i^2}{2\,E_i\,h_i} + \frac{1{,}2\,T_i^2}{2\,G_i\,h_i} + \frac{F_{zi}^2}{2\,G_i\,h_i} + \frac{M_i^2}{2\,Q_i\,H_i} + T_i\,\phi_i$$

$i = 0, 1, 2$

$$J_T = J_1 + J_2 - J_0$$

$$J_{max} = max\,(J_T)$$

- Calculation of safety factor:

$$S_f = \sqrt{\frac{G_{th}}{J_{max}}}$$

G_{th} = Threshold strain energy release rate

Automation Simulation World Congress, October 29 – 30, 2013

Example

nCode ...an HBM brand

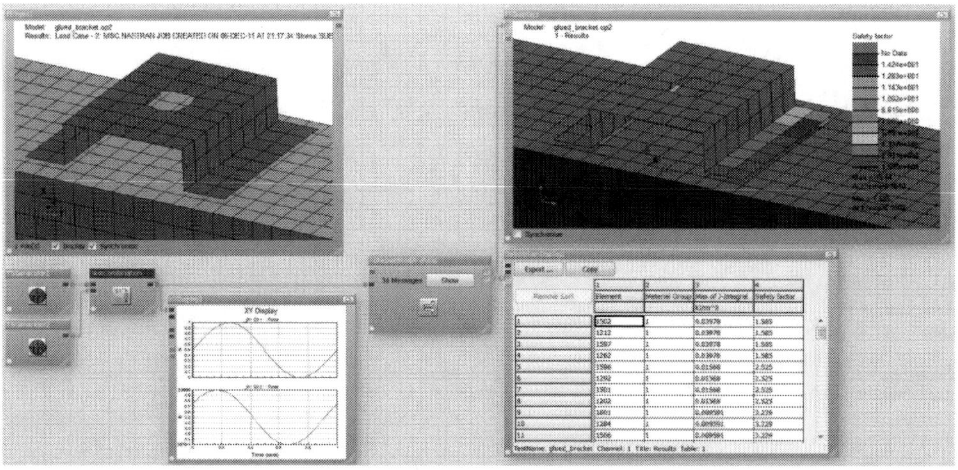

Automation Simulation World Congress, October 29 – 30, 2013

Correlation

nCode ...an HBM brand

Test rig with Jaguar XJ body-in-white

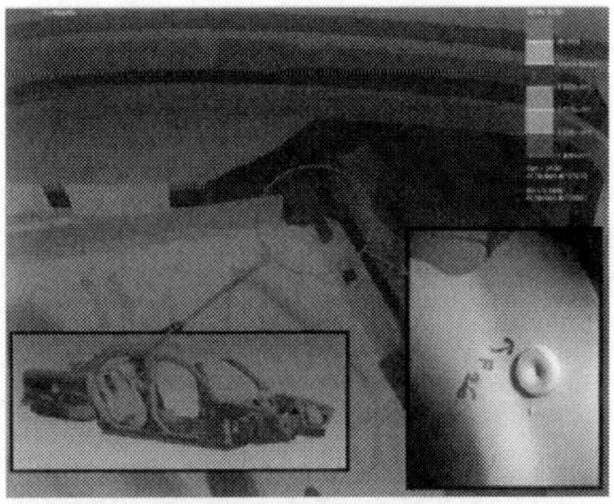

Fatigue test simulation results and corresponding cracking on test

SAE Paper 2012-01-731

A Fracture Mechanics Approach to Durability Calculations for Adhesive Joints

Peter Heyes (HBM UK, Ltd.),
Gunnar Björkman (Volvo Technology),
Andrew Blows and Tim Mumford (Jaguar/Land Rover Cars),
Paul Briskham (Coventry University)

Acknowledgements

nCode ...an HBM brand

- The testing and software implementation work described in this paper was carried out as part of the "Bonded Car" collaborative research project, the partners being Jaguar Land Rover, HBM UK Ltd (nCode), Coventry University, Warwick University, Innoval Technology, Henrob and Stoke Golding Applied Research.
- The project was co-funded by the Technology Strategy Board's Collaborative Research and Development program, following an open competition. The Technology Strategy Board is an executive body established by the UK Government to drive innovation. It promotes and invests in research, development and the exploitation of science, technology and new ideas for the benefit of business - increasing sustainable economic growth in the UK and improving quality of life.
- The theoretical basis of the work presented here was developed as part of an earlier project conducted by the Volvo Group, whose contribution is gratefully acknowledged.

Thanks for Your Attention

Dr.-Ing. Stephan Vervoort
HBM nCode Produkte

Tel: +49 (0)89 9605372 18
Fax: +49 (0)89 9605372 21
Email: stephan.vervoort@hbmncode.com

Suitable simulation tools for the development of a CFRP sports car

Günther Lindbichler, Katharina Fischer, Martin Perterer, KTM Technologies
Matthias Alberts, CADFEM US
Madhukar Chatiri, Phelippe Pereira, Matthias Hörmann, CADFEM GmbH

ANSYS ANSYS Competence Center FEM CADFEM

KTM X-Bow

- Chassis:
  - Torsional rigidity 35,000 Nm/°
- Engine:
  - Audi 2.0 TFSI Turbo-charged, 4-cylinder
  - Max power: 285 hp (210 kW)
  - Max torque: 420 Nm
- Weight
  - Dry weight: 847 kg
- Performance:
  - 0 to 100 km/h in 4.10 sec

KTM TECHNOLOGIES

CADFEM

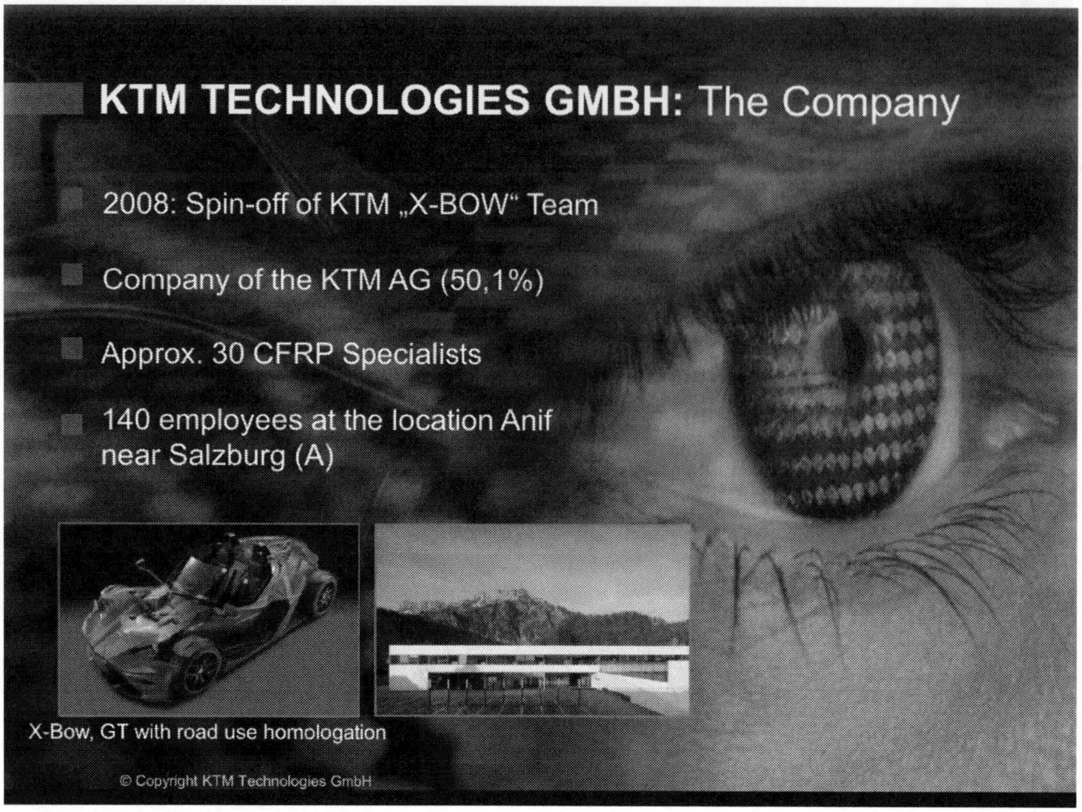

KTM TECHNOLOGIES GMBH: The Company
2008: Spin-off of KTM „X-BOW“ Team
Company of the KTM AG (50,1%)
Approx. 30 CFRP Specialists
140 employees at the location Anif near Salzburg (A)
X-Bow, GT with road use homologation
© Copyright KTM Technologies GmbH

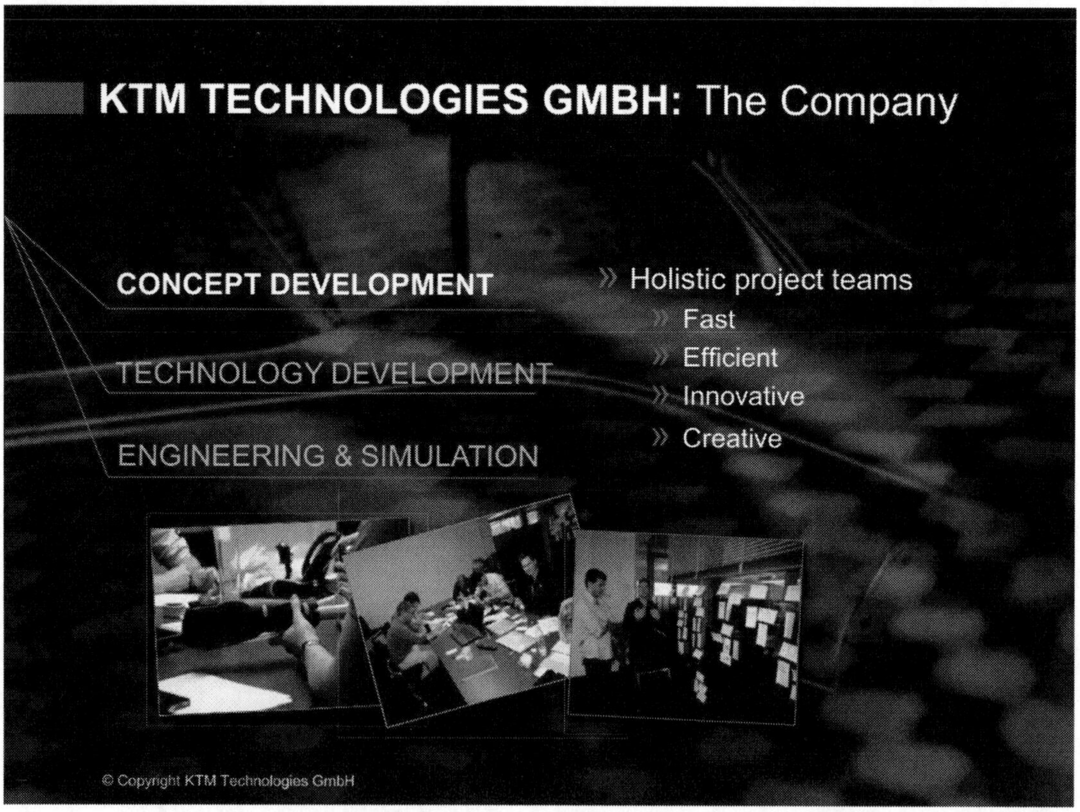

KTM TECHNOLOGIES GMBH: The Company
CONCEPT DEVELOPMENT
TECHNOLOGY DEVELOPMENT
ENGINEERING & SIMULATION
» Holistic project teams
» Fast
» Efficient
» Innovative
» Creative
© Copyright KTM Technologies GmbH

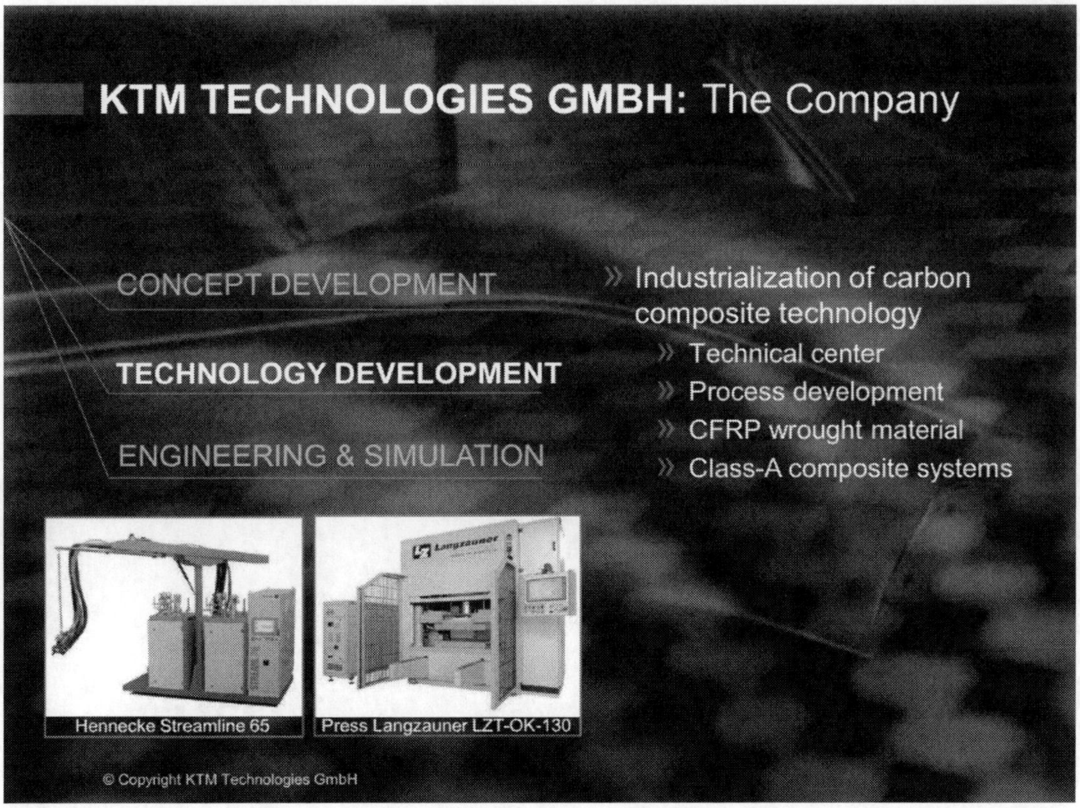
KTM TECHNOLOGIES GMBH: The Company
CONCEPT DEVELOPMENT
TECHNOLOGY DEVELOPMENT
ENGINEERING & SIMULATION
» Industrialization of carbon composite technology
» Technical center
» Process development
» CFRP wrought material
» Class-A composite systems
Hennecke Streamline 65
Press Langzauner LZT-OK-130
© Copyright KTM Technologies GmbH

KTM TECHNOLOGIES GMBH: The Company
CONCEPT DEVELOPMENT
TECHNOLOGY DEVELOPMENT
ENGINEERING & SIMULATION
» Concept, product and car development
» Structure, system, complete car
» Carbon composite Know-how
» Manufacturing process for lightweight structures
© Copyright KTM Technologies GmbH

CFD-Simulation for Comfort in Passenger Compartment

Günther Lindbichler, KTM Technologies

ANSYS ANSYS Competence Center FEM CADFEM

Boundary Conditions X-Bow & Investigated Designs

- Motivation
  - Ensure passenger comfort
  - Avoid highly turbulent & high velocity flow in cockpit
- BC of Simulation:
  - Driving velocity: 140 km/h
  - Ground: free-slip-wall

With Windblocker

Without Windblocker

KTM_TECHNOLOGIES

CADFEM

Results of Fluid Simulation

Velocity at y0-plane: with windblocker reduced velocity

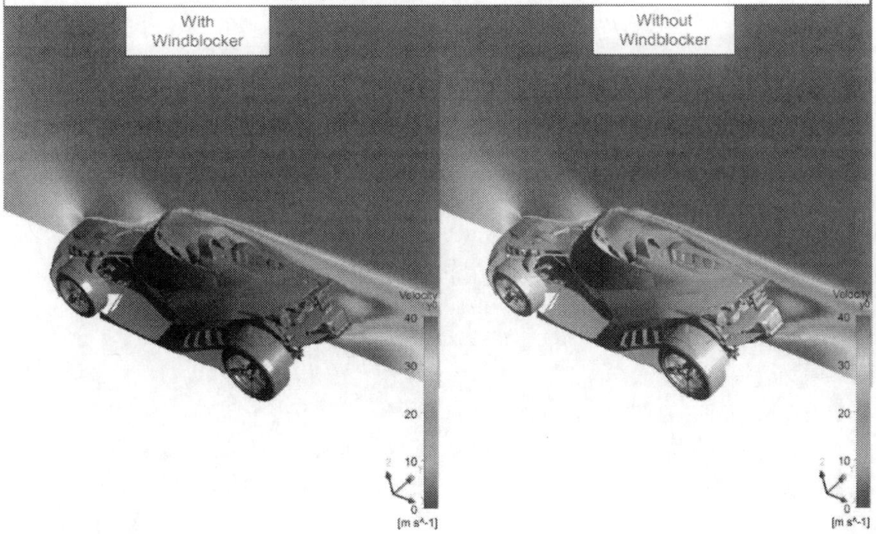

KTM TECHNOLOGIES

CADFEM

Results of Fluid Simulation

Detail Cockpit – Velocity at y0-plane: with windblocker reduced velocity in passenger compartment

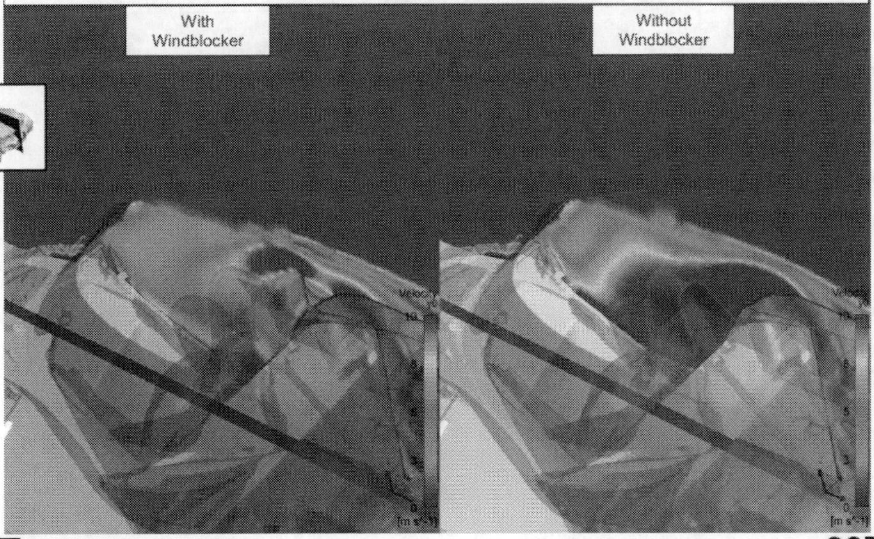

KTM TECHNOLOGIES

CADFEM

Results of Fluid Simulation

Detail Cockpit – Velocity-vector at y0-plane: with windblocker reduced velocity in passenger compartment

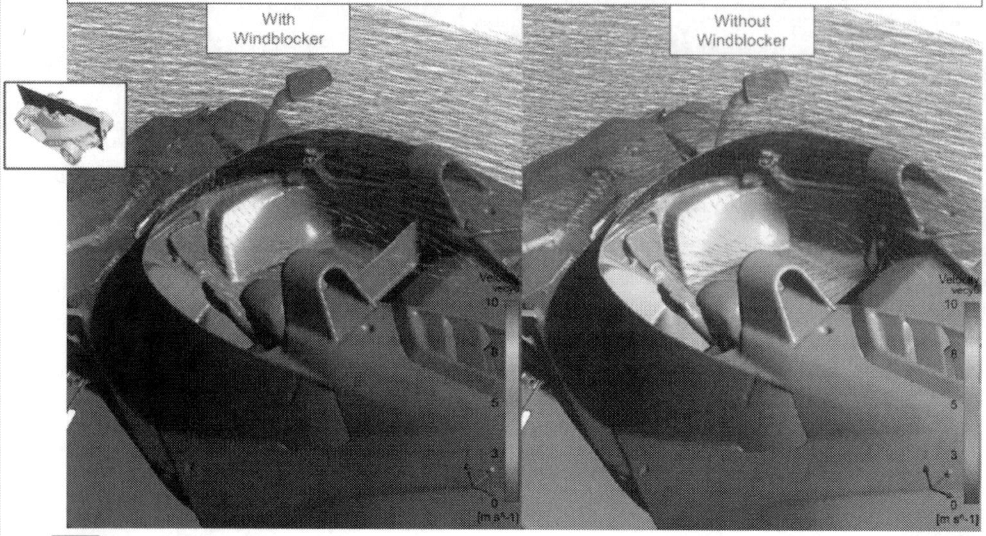

KTM_TECHNOLOGIES

CADFEM

Results of Fluid Simulation

Detail Cockpit – Velocity at z0.2-plane: with windblocker reduced velocity in passenger compartment

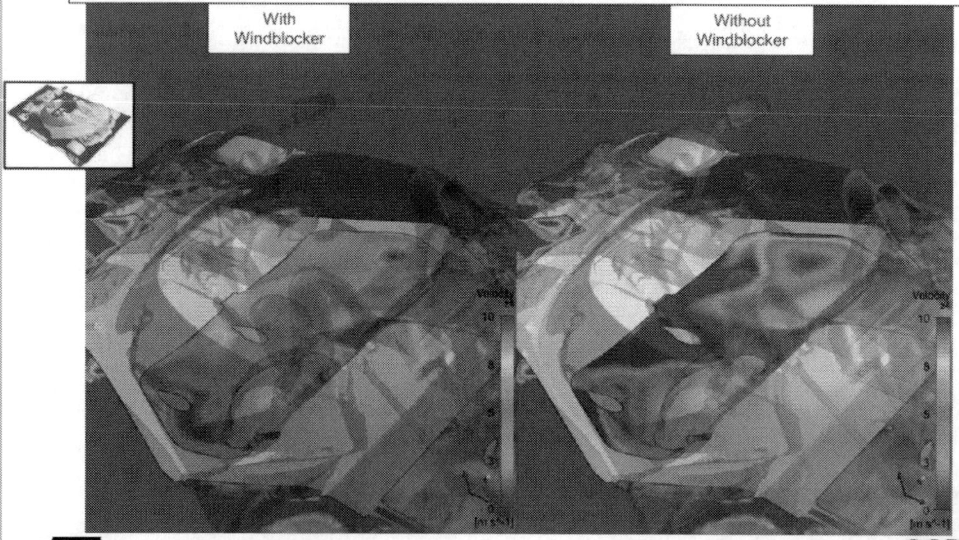

KTM_TECHNOLOGIES

CADFEM

Results of Fluid Simulation

Detail Cockpit – Velocity at z0.4-plane: with windblocker reduced velocity in passenger compartment

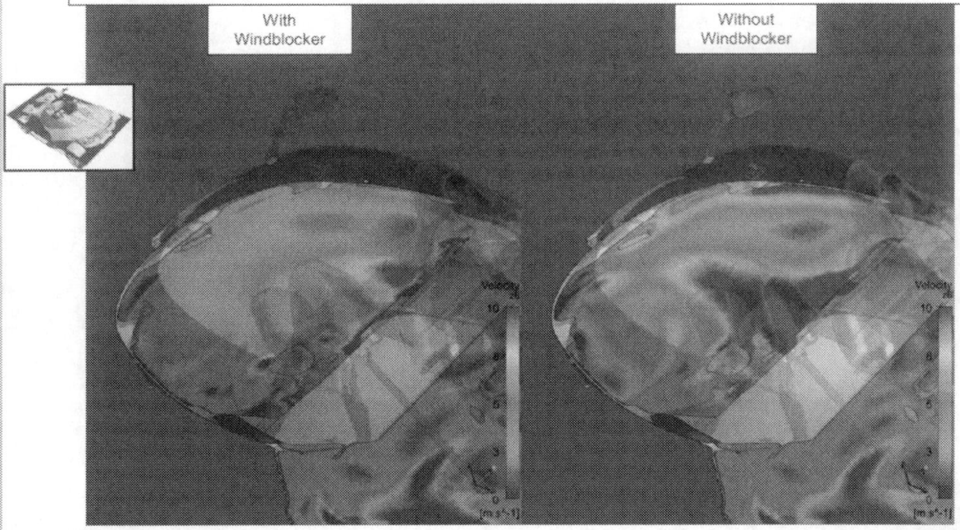

KTM TECHNOLOGIES

CADFEM

CFD – Conclusion

- Efficient use of CFD-Simulation to investigate flow field at the car
- Identification of backflow at antiroll bar
- Enhanced passenger comfort due to effective windblocker

→ Developed in virtual world – tested on the road

KTM TECHNOLOGIES

CADFEM

Stiffness and Strength Simulation of Monocoque

Matthias Alberts, CADFEM US

ANSYS

ANSYS Competence Center FEM

CADFEM

Monocoque

- Challenge
  - Develop an exciting sports car using a monocoque made from carbon composites
  - High level of functional integration → external skin provides vehicle's structural support
  - Analyze different layups, fiber orientations and composites materials.
  - Investigate failure behavior of composites design under different load scenarios.

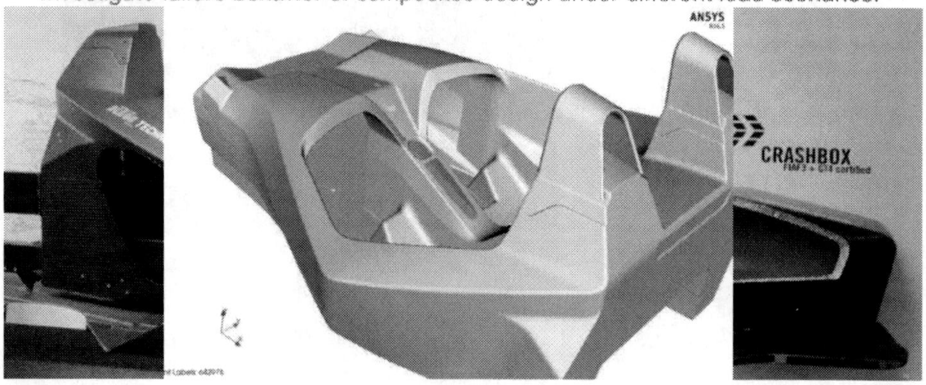

KTM TECHNOLOGIES

CADFEM

Monocoque – Load Case "Torsion Stiffness Performance"

- Boundary Condition
  - Fixed support on all rear frame attachment points
- Loads
  - Application of load at fixation points of crashbox
  - Fz = 4 kN at y = 1,000 mm
  - Evaluation of z-deformation at load application point
- Criteria
  - Torsional stiffness > 25,000 Nm/°
  - No initial cracks, i.e. First-ply failure

KTM TECHNOLOGIES

CADFEM

ANSYS Composite Solution & Workflow

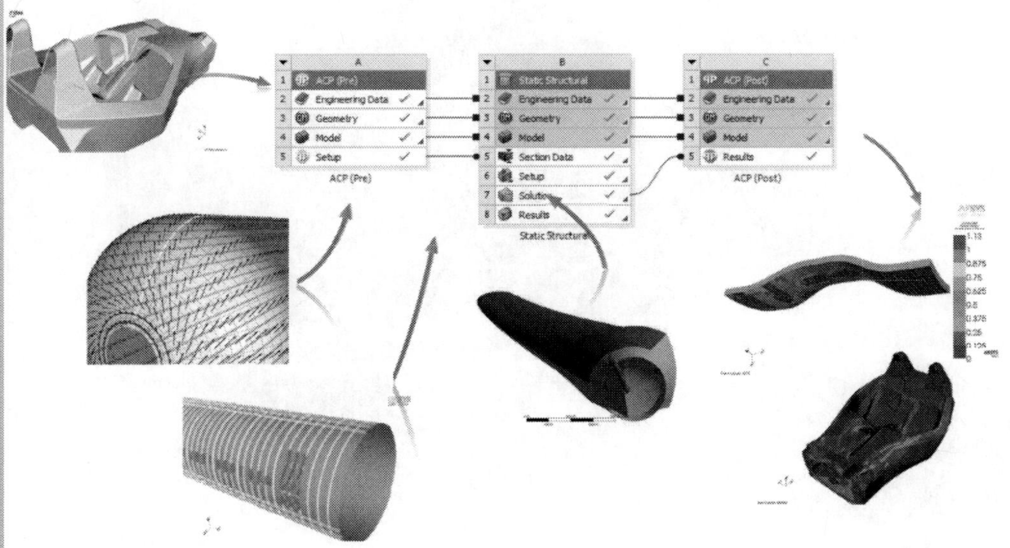

KTM TECHNOLOGIES

© 2013 CADFEM GmbH

CADFEM

Three Steps to a Composite Layup

Material and Fabrics | **Oriented Element Sets** | **Composite Layup**

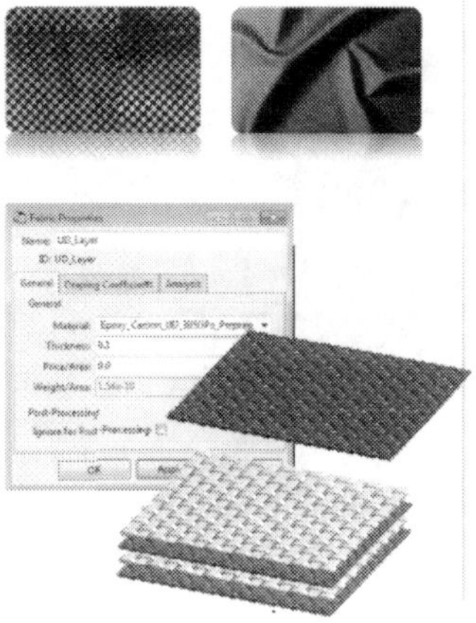

Layup Surface

Layup Direction

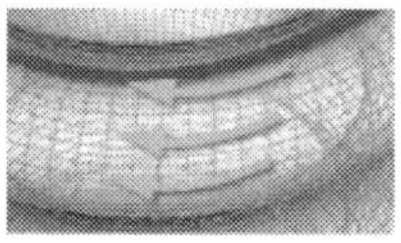

Fiber Orientation

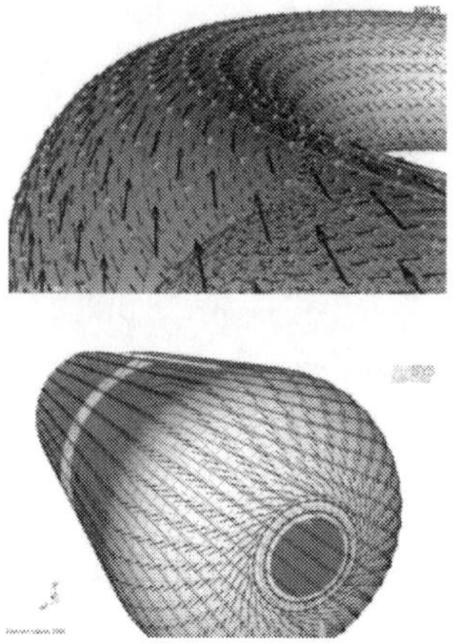

KTM TECHNOLOGIES

CADFEM

Monocoque – Layup

- Real composite design
- Made from 300 precut composite patches
- Various shapes and sizes
- Lots of taperings for material savings

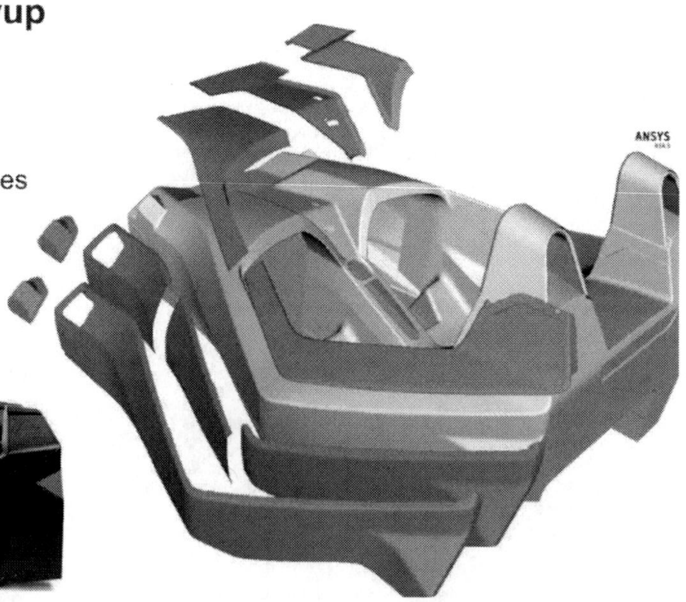

KTM TECHNOLOGIES

CADFEM

Monocoque – Layup

- Ply layup and fiber orientations for the complex structure are optimized to reduce weight

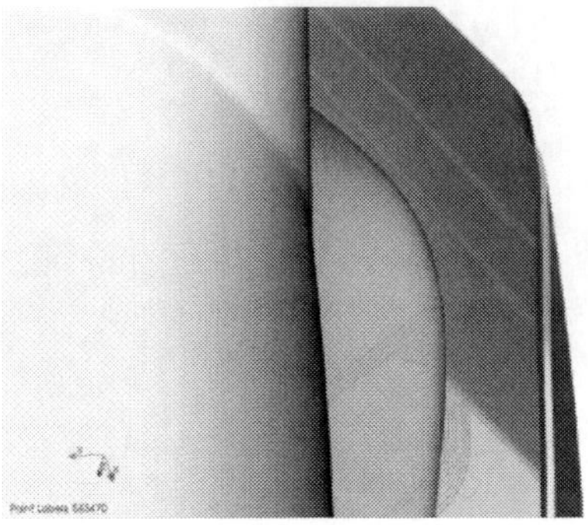

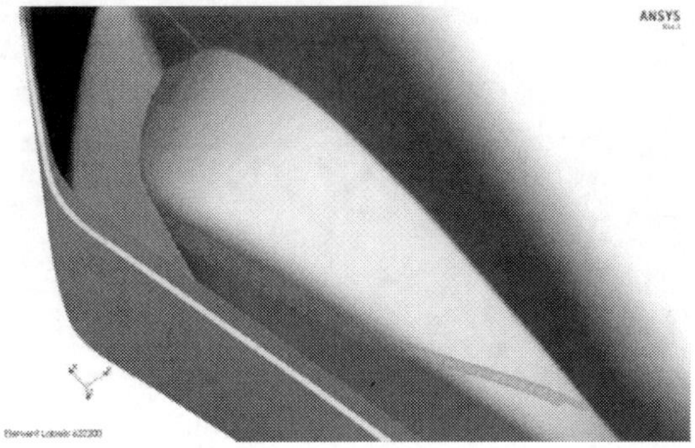

KTM TECHNOLOGIES

CADFEM

Monocoque – Result Evaluation

- Failure evaluated through all layers and load cases in one view
- And also ply wise
- Overall no failure indicated through First-Ply-Failure criteria

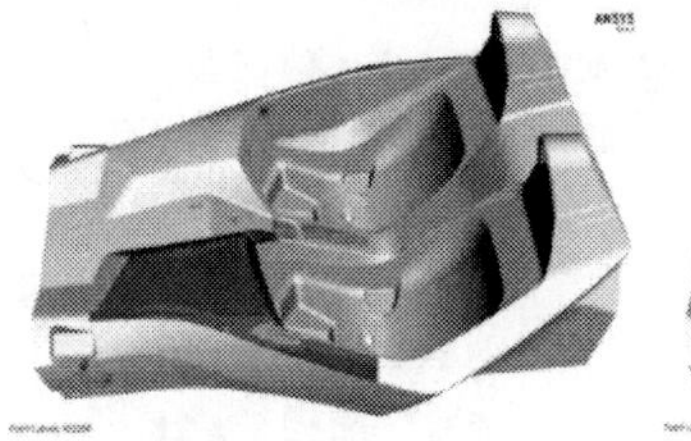

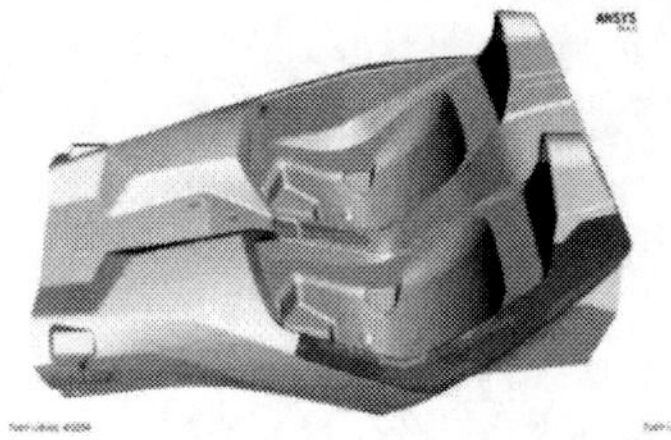

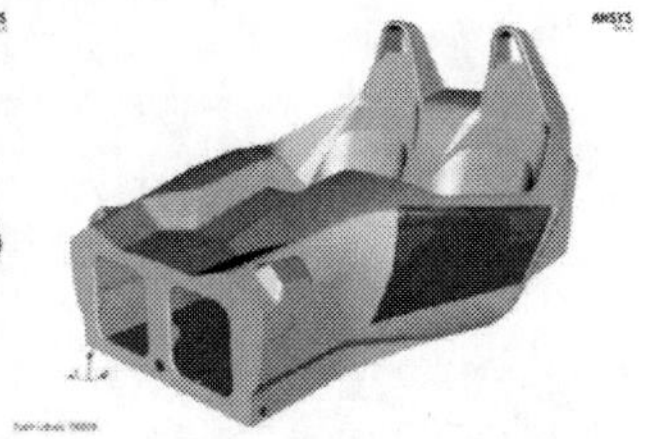

KTM TECHNOLOGIES

CADFEM

Monocoque – Summary

- Customer's Voice

"The first monocoque designs of the KTM X-Bow were engineered without the use of ANSYS Composite PrepPost. When we did employ ANSYS software for composites, we were able to reduce the monocoque's weight - a very important aspect of sports car design - by 20 percent."

Peter Martin
CEO
KTM Technologies

KTM TECHNOLOGIES

CADFEM

Stiffness and Strengh Simulation of A-Pillar

Katharina Fischer, KTM Technologies

ANSYS

ANSYS Competence Center FEM

CADFEM

A-Pillar – Functions

- Fixture for frontal wind screen
- Fixture for side window (foldable)
- Used from passenger as handle bar to get in and out
- Misuse load case "Lean on Door"

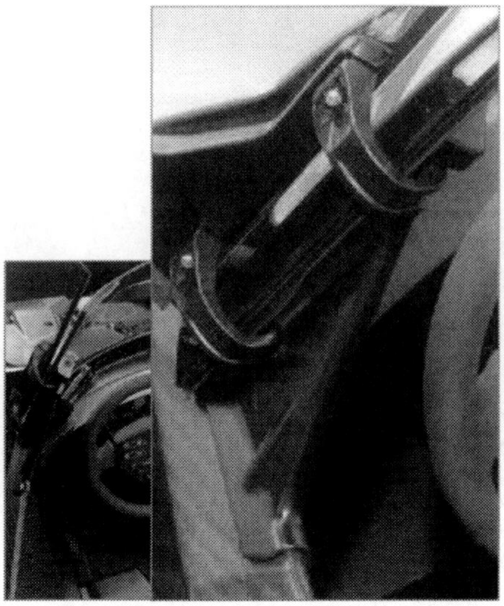

KTM TECHNOLOGIES

CADFEM

A-Pillar – Load Case "Lean on Door"

- Load case & Boundary Condition
  - A-Pillar bonded to monocoque
  - Monocoque frame with fixed support on boundaries
  - Load applied on the far end point of slightly opened door F_z = -1000 N
  - Load transfer from door to A-Pillar by hinge
- Criteria:
  - A-Pillar with no damage → Design for First Ply Failure

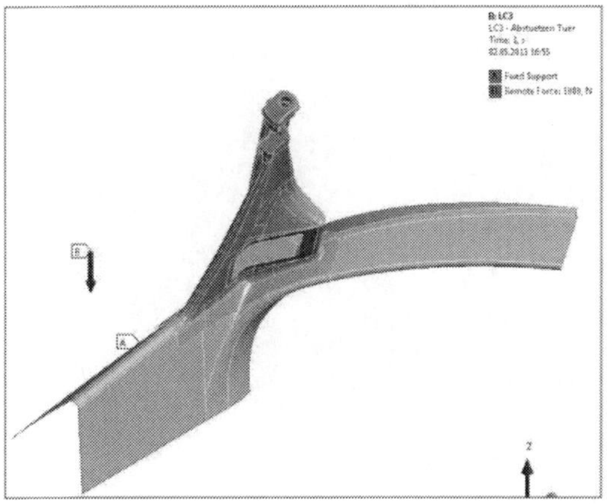

KTM TECHNOLOGIES

CADFEM

A-Pillar – Composite Modeling

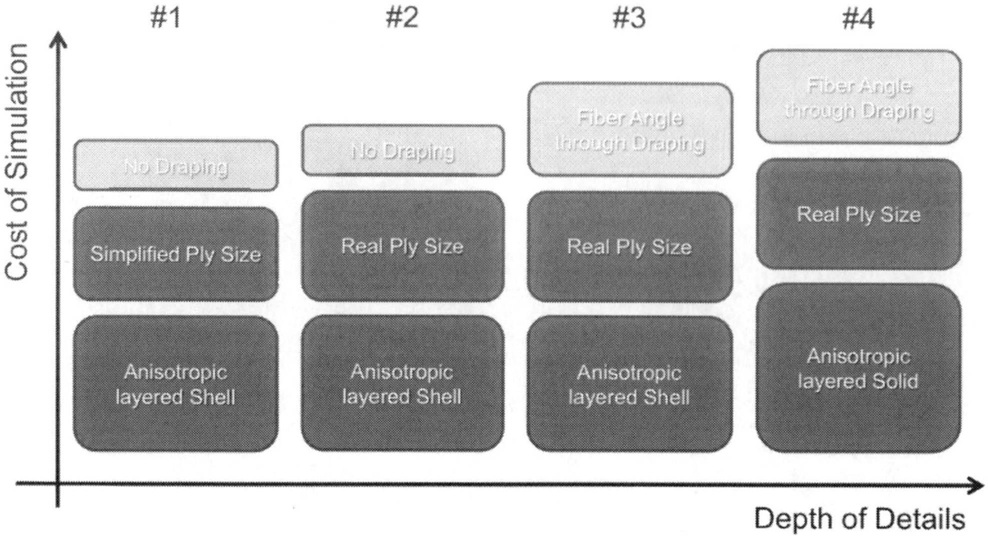

KTM TECHNOLOGIES

CADFEM

A-Pillar – Modeling Variant #1

- Simplified ply size
- Simplified layup
- Fiber angles from plybook
- One constant thickness
  → no overlaps

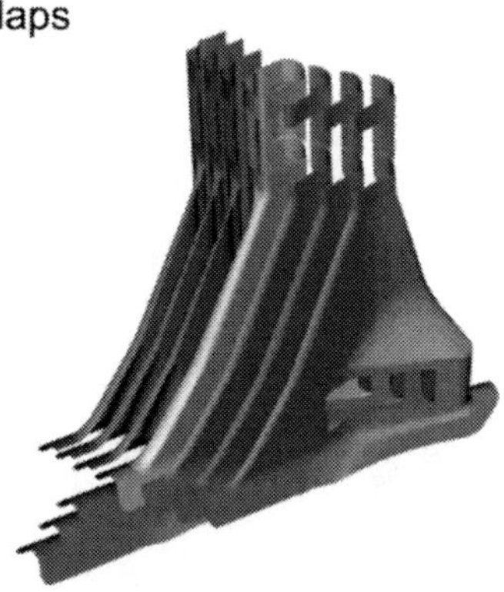

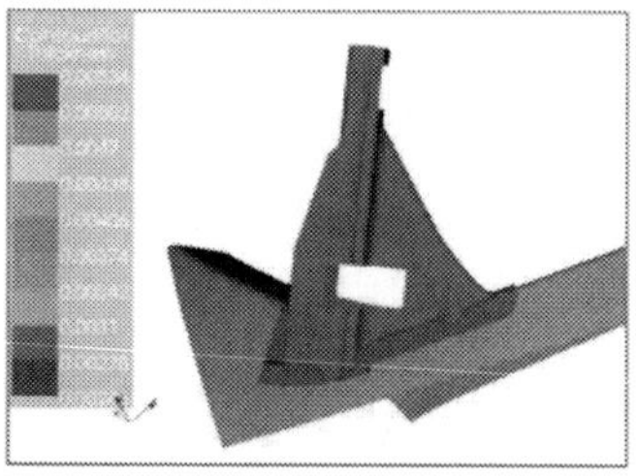

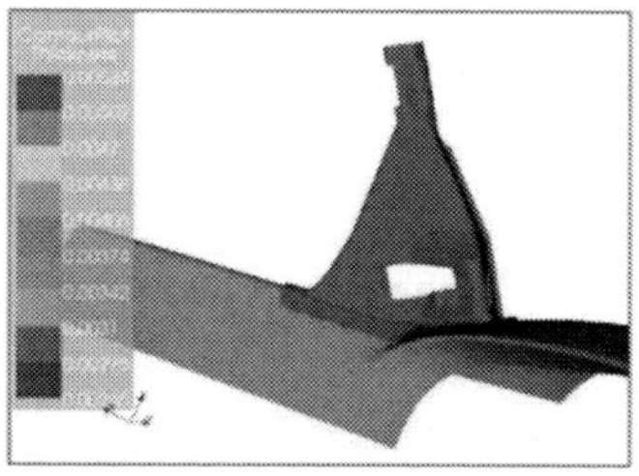

KTM TECHNOLOGIES

CADFEM

A-Pillar – Modeling Variant #2

- Real ply size
- Real layup
- Fiber angles from plybook
- Definition of overlaps with rules

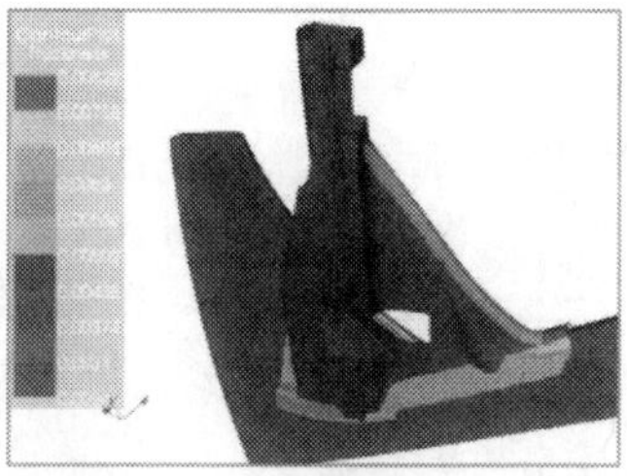

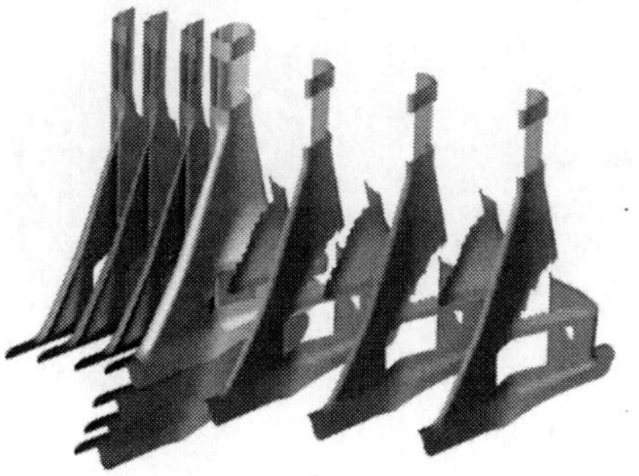

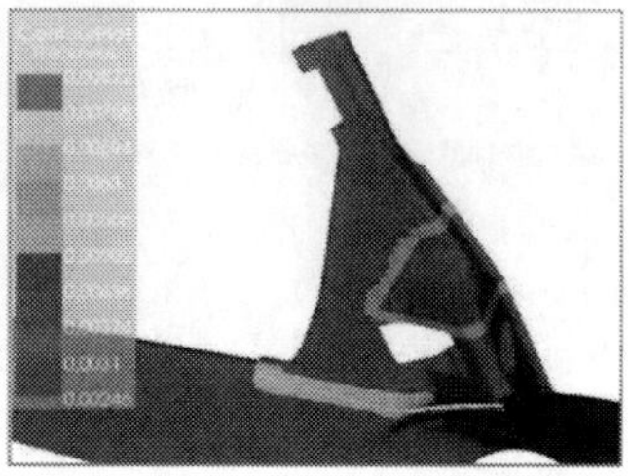

KTM TECHNOLOGIES

CADFEM

A-Pillar – Modeling Variant #3

- Real ply size & real layup
- Fiber angles from draping
- Draping simulation
  - Realistic prediction of fiber angles
  - Careful definition of Seed Point important → big influence on fiber angle
- Challenge:
  - Draping algorithm needs computation time and may also stop without finding a solution

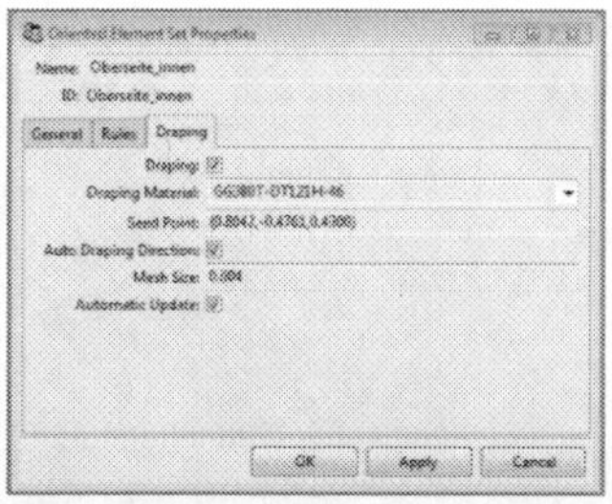

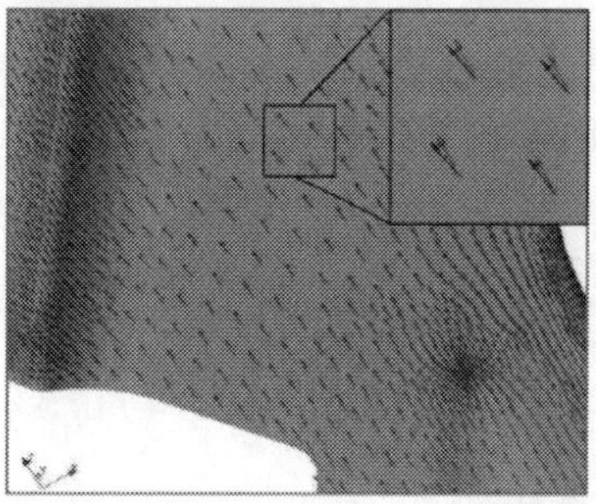

KTM TECHNOLOGIES

CADFEM

A-Pillar – Modeling Variant #4 – Solid Model

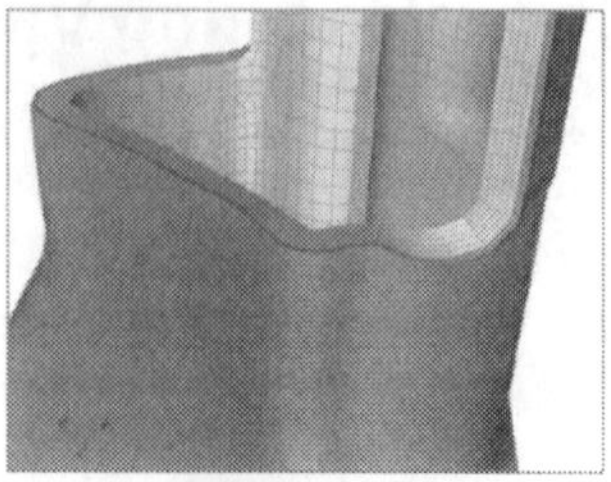

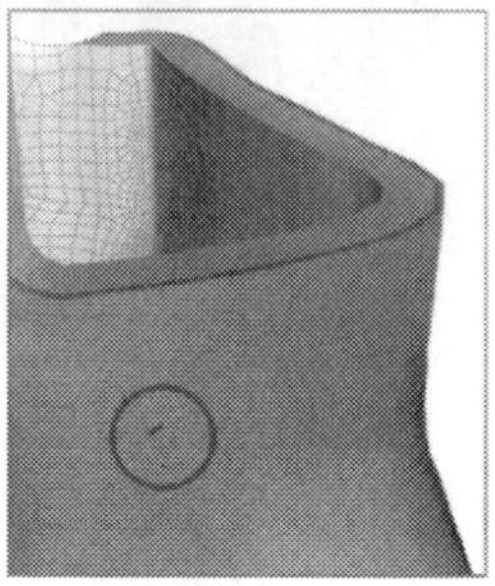

- Function solid extrusion in ANSYS with different choices
  - Monolithic
  - Analysis Ply Wise
  - Modeling Ply Wise
  - Material Wise
- Challenge:
  - Solid-Extrusion problematic in areas with changing outside-inside radii
    → ACP deletes degenerated solid elements
- Potential solutions:
  - Coarsen mesh in radius region and defeaturing
    → Helps but is not perfect

KTM TECHNOLOGIES

CADFEM

A-Pillar – Comparison of Shell Results

	#1: Simplified	#2: w/o Draping	#3: with Draping
	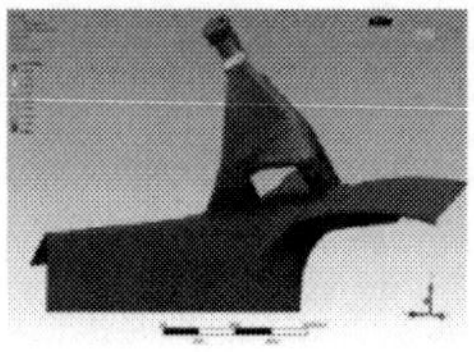	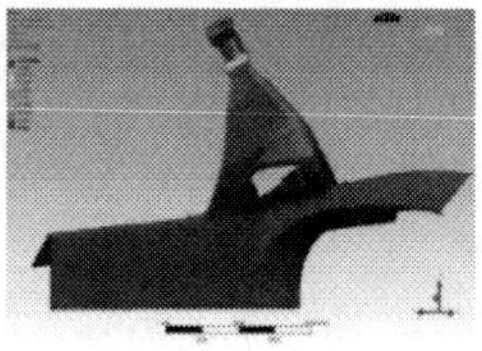	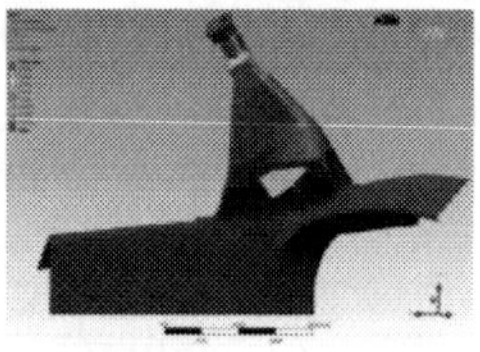
U_max	2,79 mm	2,65 mm (-5%)	2,66 mm (-4,7%)
CPU time	100 %	103,6%	106,4%
Material	1,18 m^2	1,22 m^2 (+3,3%)	1,22 m^2 (+3,3%)

KTM TECHNOLOGIES

CADFEM

A-Pillar – First Ply Failure

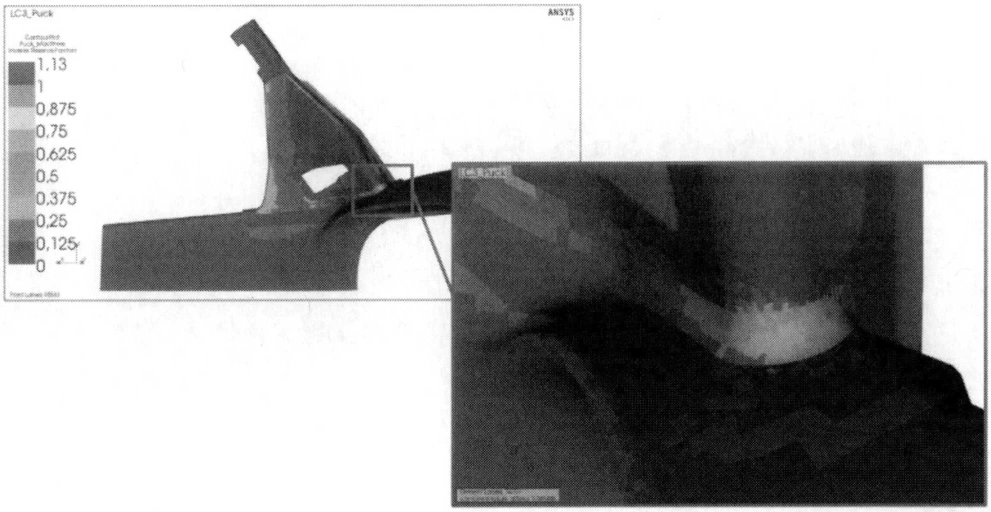

KTM TECHNOLOGIES

CADFEM

A-Pillar – Conclusion

- Flexibility to adjust depth of details is crucial for composite structures within the design process
- Needed depth of details dependent on application case and expected result
- First results of solid model indicate that shell modeling is sufficient in most cases

Next Steps

- Finalization of solid model and detailed comparison with shell variants

KTM TECHNOLOGIES

CADFEM

Crashworthiness Simulation of KTM X-Bow Crashbox

Madhukar Chatiri, Phelippe Pereira, Matthias Hörmann, CADFEM GmbH

ANSYS   ANSYS Competence Center FEM   CADFEM

Crashbox

- Ensure driver's safety and limit the decelerations acting on the human body
- Crash behavior of the composite crash box unit

Objective of Simulation

- 0° impact test against a rigid wall of KTM "X-Bow" Front Impact Structure
- Non linear dynamic impact test simulation
- General deformation of crash box and deceleration curves

KTM TECHNOLOGIES   CADFEM

Crashbox – Test Setup

"Crashbox – Test Setup": Crash scenario according to FIA Art. 259 Technical regulations for production sports cars

KTM TECHNOLOGIES

CADFEM

Crashbox – Test Results

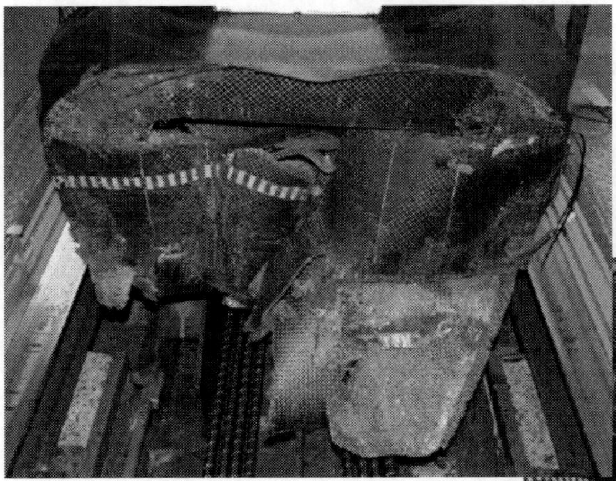

KTM TECHNOLOGIES

Crashbox – Modeling using Shell-Solid-Shell

- Core is modeled with solid elements and skins are modeled with shell elements, connected by a contact formulation
- Advantages to use this formulation:
  - Full 3D stress state in core
  - Would allow for cover skin/core debonding
  - Representation for core shear deformations and core indentation in thickness direction

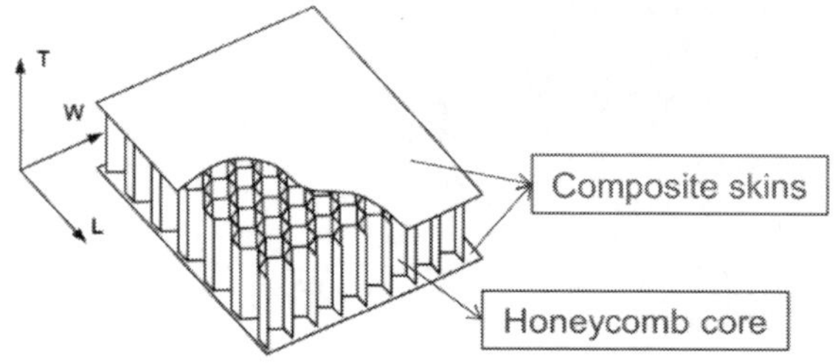

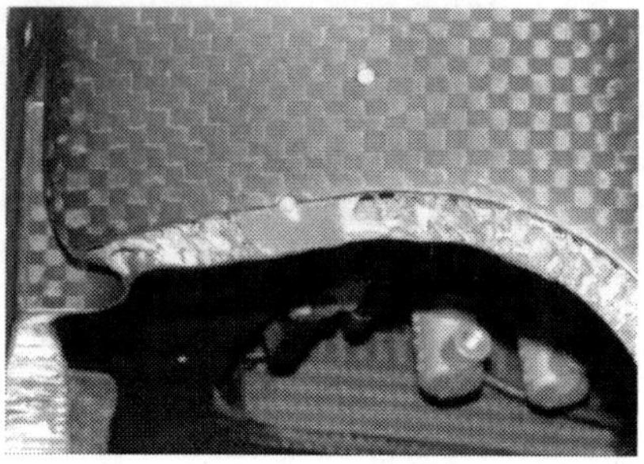

Crashbox – Workflow

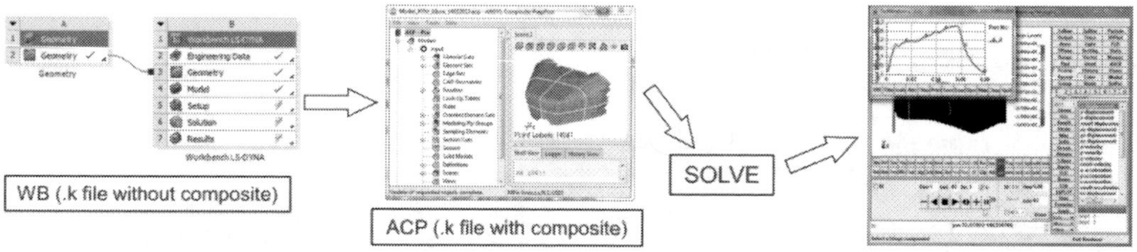

KTM TECHNOLOGIES
© 2013 CADFEM GmbH
CADFEM

Crashbox – CAD-data in detail

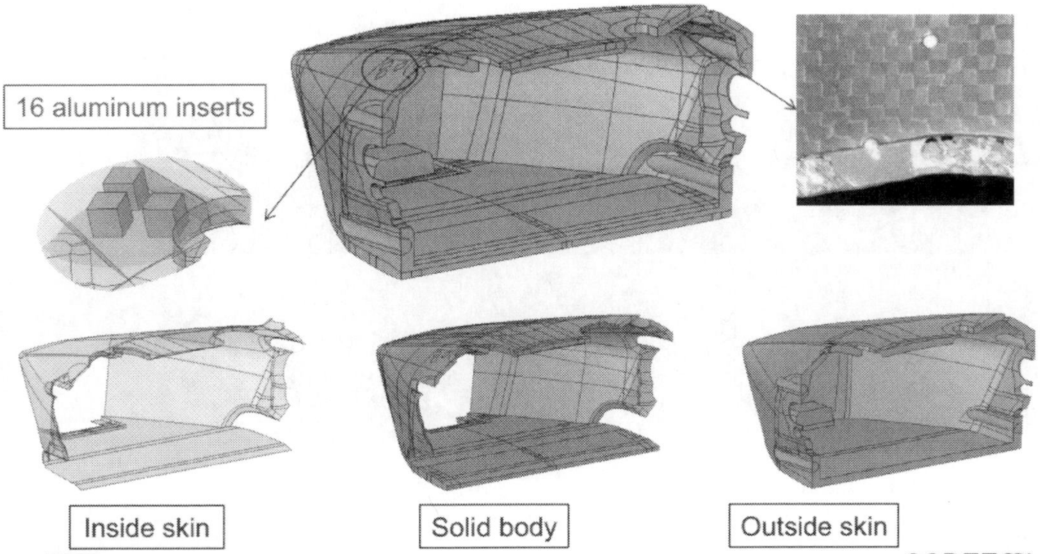

Crashbox – Intrusion over time

- Comparison between experimental and simulation results

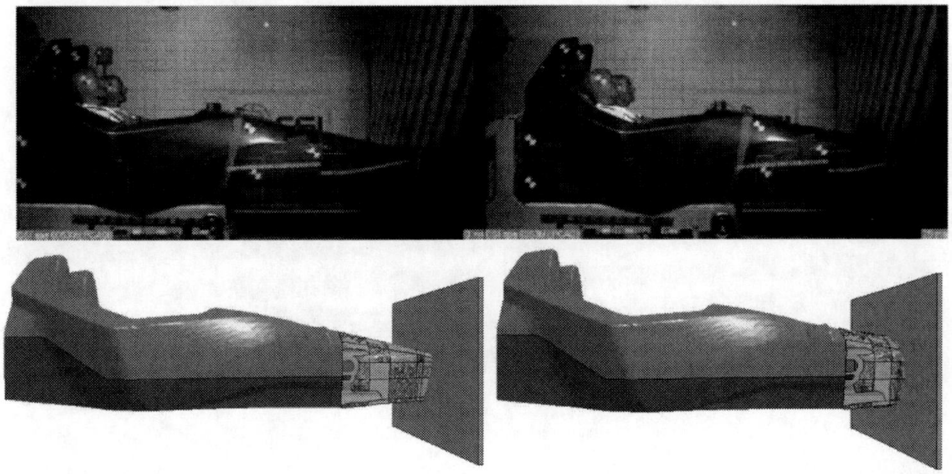

Crashbox – Intrusion over time

- Comparison between experimental and simulation results

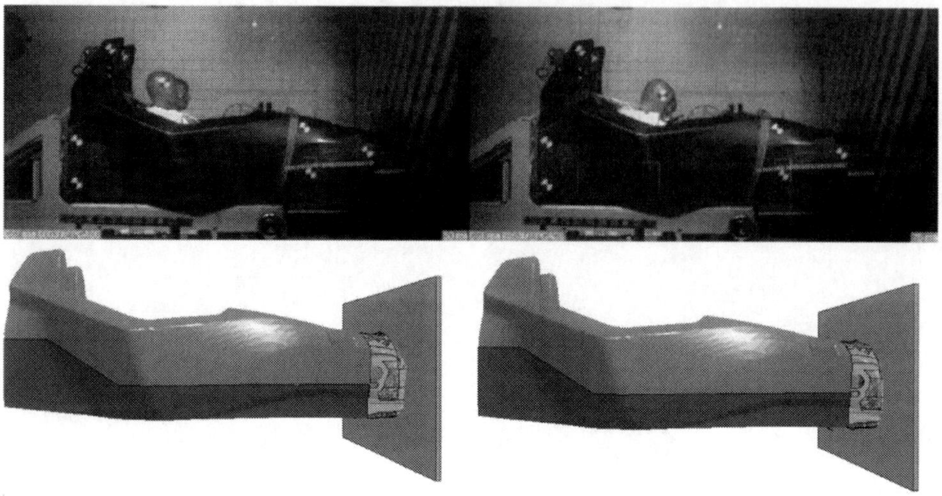

KTM TECHNOLOGIES

CADFEM

Crashbox – Intrusion over time

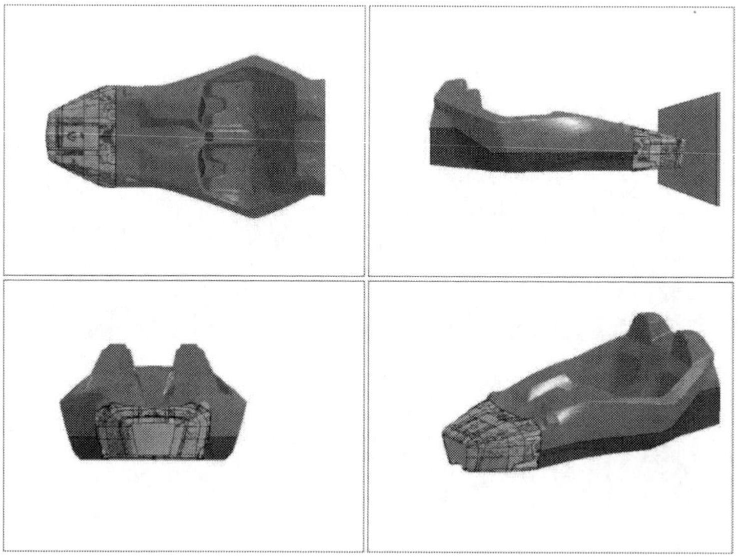

KTM TECHNOLOGIES

CADFEM

Crashbox – Failure

- Total tensile matrix failure

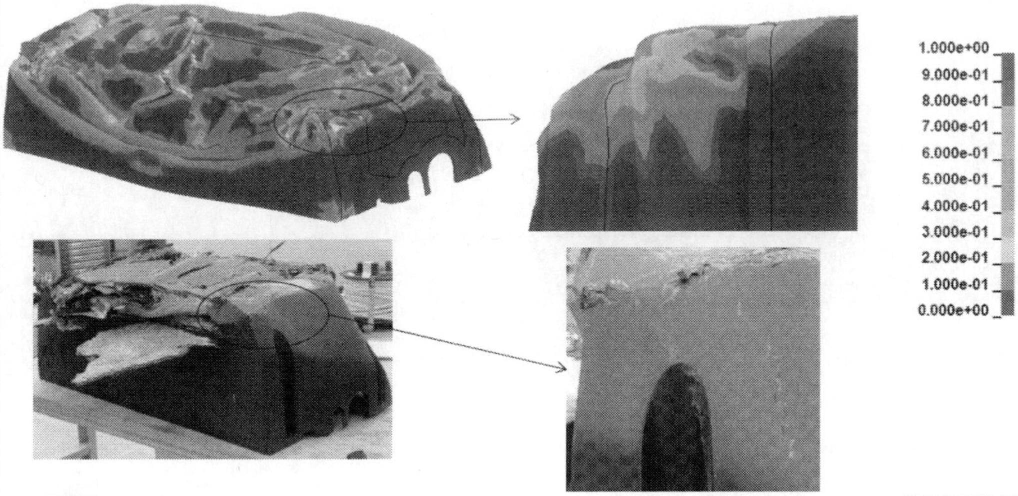

KTM TECHNOLOGIES

CADFEM

Crashbox – Deceleration over time

- Comparison of deceleration over time

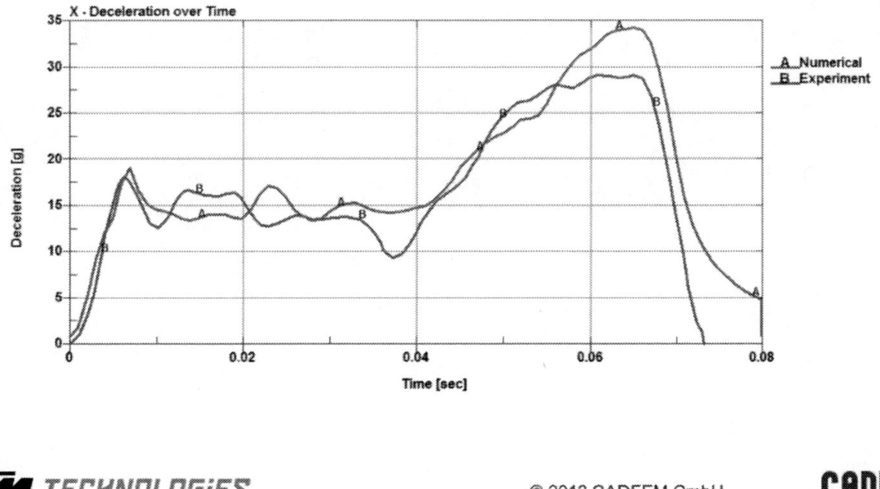

KTM TECHNOLOGIES

CADFEM

Crashbox – Conclusion

- Crashbox behavior and intrusion were captured well
- Failure in specific regions was achieved like in the experiment
- The deceleration, velocity and displacement response presented close results compared to the experiment

Next Steps

- Other material models for composite skin should be investigated

KTM TECHNOLOGIES

CADFEM

CAE Process- and Crash Simulation of Fiber-Reinforced Composite Vessels

2013 Automotive Simulation World Congress

October 29th-30th, Frankfurt am Main, Germany

Madhukar Chatiri
CADFEM GmbH, Grafing/Munich, Germany

Wir leben Autos.

CADFEM

Company portfolio

CADFEM

Products

- Full scale CAE solution based on ANSYS, complementary software and IT infrastructure

ANSYS Competence Center FEM

  - ANSYS Software
  - Complementary Software
  - Hardware
  - IT infrastructure

Service

- Create confidence and reliability in optimized product development by simulation driven engineering
  - Customer support
  - Training & deployment
  - Simulation-to-order
  - Customization & Automation

Knowledge

- Comprehensive and innovative background knowledge for users, supervisors and officers
  - Qualification
  - Exchange of experiences
  - Networking
  - CAE media

CADFEM

Outline

- Introduction and motivation
- CAE process for filament-wound high pressure composite vessels
- Mechanism-based modeling of failure and damage in thick multi-layered composite structures
- Verification and validation
- Summary

Motivation

Numerical Simulation

- Complex 3D stress state with varying thickness stresses
- Complex failure of composite material and damage process
- Accurate and CPU efficient solution needed

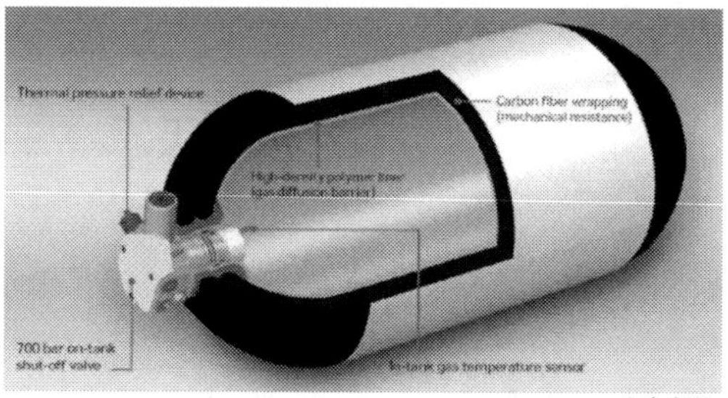

Ref: OPEL

CAE-Process Integration

- Virtual vessel generation using OPELs MatWind solution
- Interaction with ANSYS and LS-DYNA for automatic solid model generation including layering information

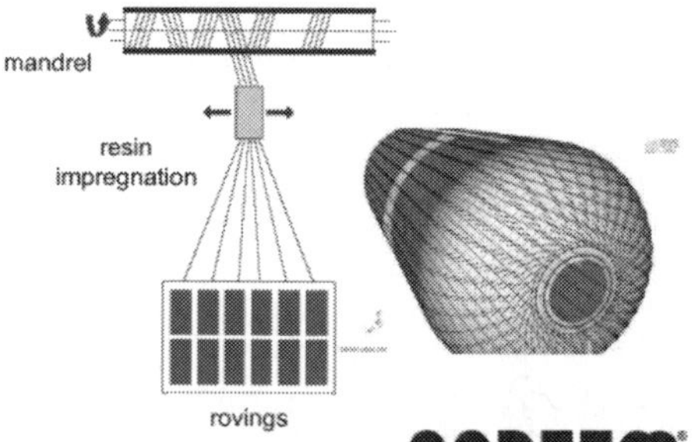

Motivation – Continued

2 Ingrediens – FEM Technology & Material Model

FEM-Technology

- Complex 3D nature with 3D stress state
- Varying normal thickness stress caused by inner pressure
- Accurate FE technology essential

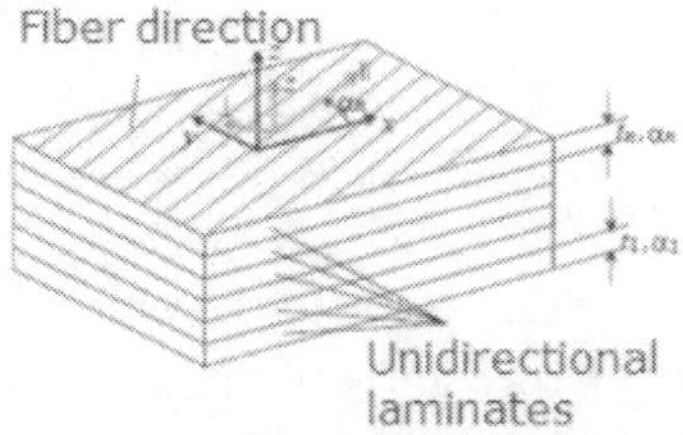

Material Model

- Account for non-linear stress-strain relations
- State-of the art Puck failure criteria
- Post-failure behavior based on anisotropic continuum damage mechanics

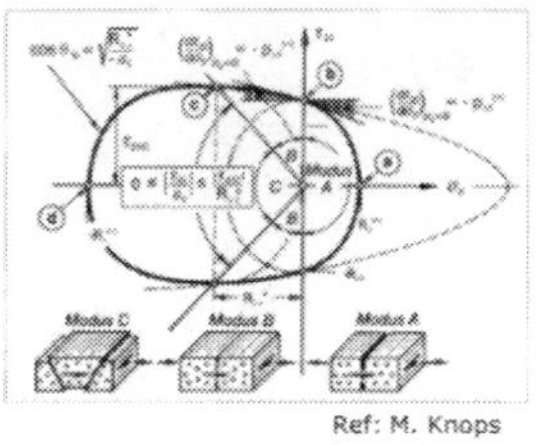

Ref: M. Knops

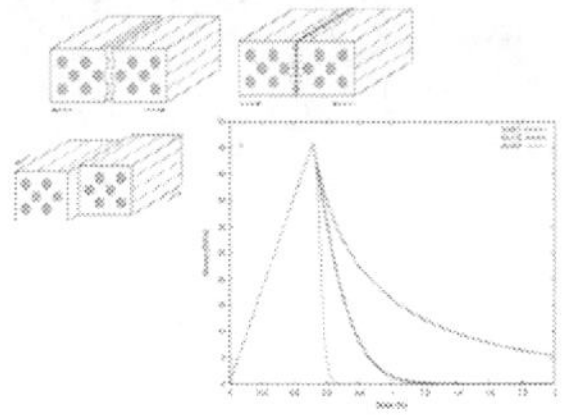

CADFEM

OPEL HydroGen4

Specifications

- Power → 73 kW
- Acceleration → (0-100 km/h in 12s
- Top speed → 160 km/h
- Range →~ 320 km
- Fuel → 4.2 kg CGH_2 (700 bar)
- Hydrogen4 storage system→ Three 700 bar CFRP tanks

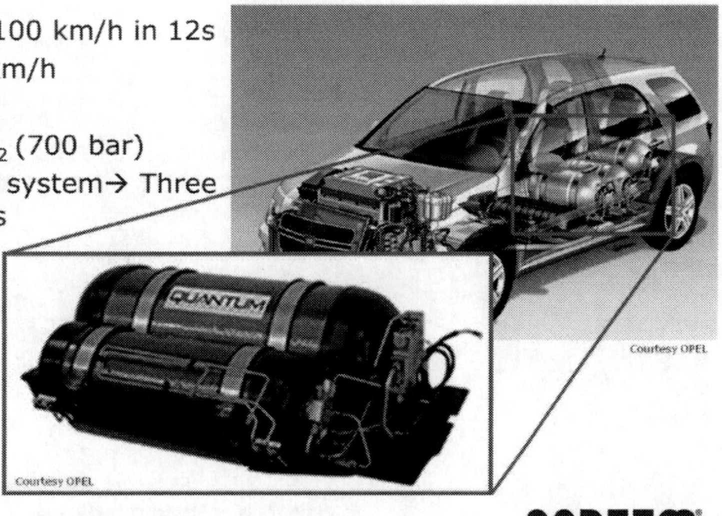

Courtesy OPEL

Courtesy OPEL

CADFEM

Filament winding

Characteristics

- Promising and most widely used manufacturing approach for CFRP tanks
- Highly reproducible and automated process
- Continuous fiber over the entire part
- Range → ~ 320 km
- Ability to orient fibers in the load direction

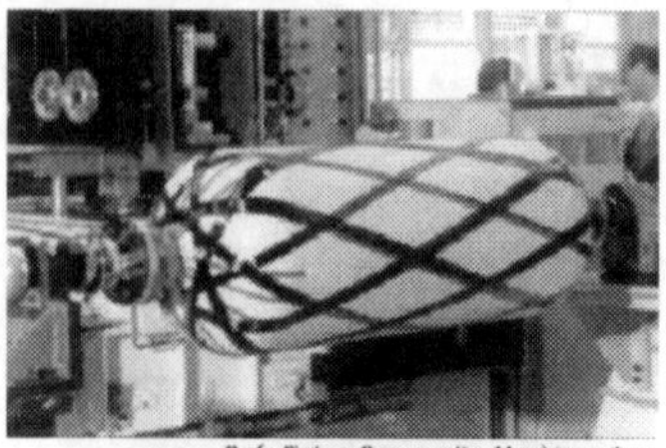

Ref: Entec Composite Machines Inc.

CADFEM

CAE- process Chain

Different steps involved

Netting Theory + CLT	Mat Wind	ANSYS Composite PrepPost	LS-DYNA
Design of the cylindrical part of the composite shell of the vessel	Virtual manufacturing of the entire composite shell of the vessel	Generation of an entire FE vessel model in ANSYS/ LS-DYNA (Liner, composite shell, valve block)	FEM Model of the HSS Module in order to investigate Static & Crash Loads
1	2	3	4

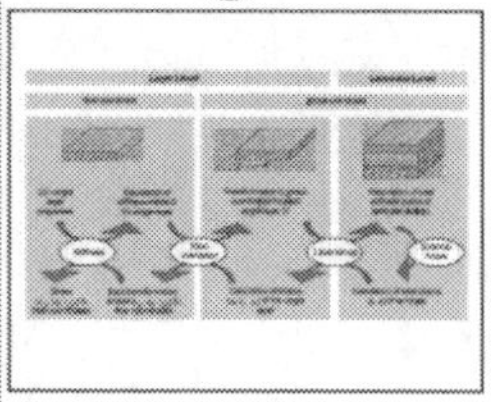

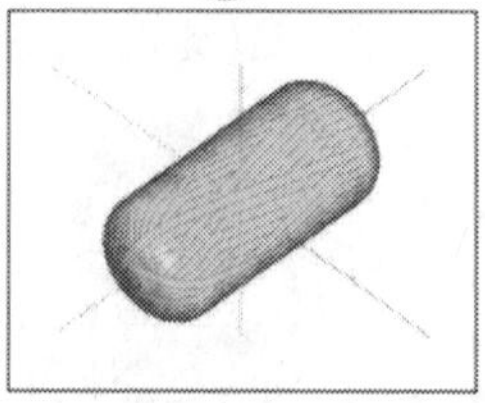

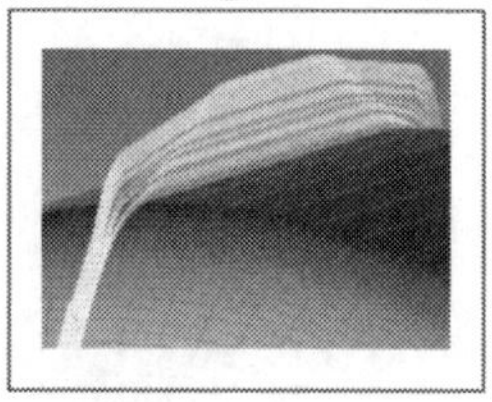

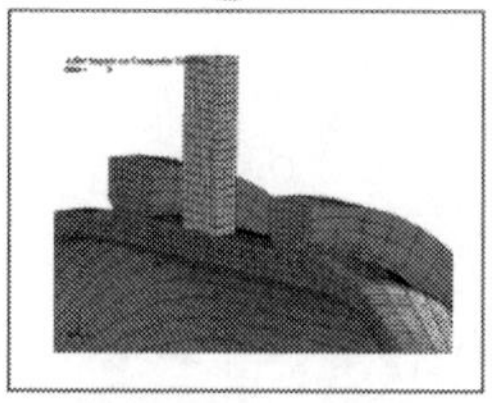

CADFEM

CAE- process Chain

Efficient Composite Pre-Processing

- Varying ply thickness → thicker in dome regions
- Changing fiber orientation in dome regions
- Thicknesses & fiber directions directly from virtual filament winding process through ANSYS Composite PrepPost (ACP)
- Automatic generation of LS-DYNA Multi-Layered Solid FE model

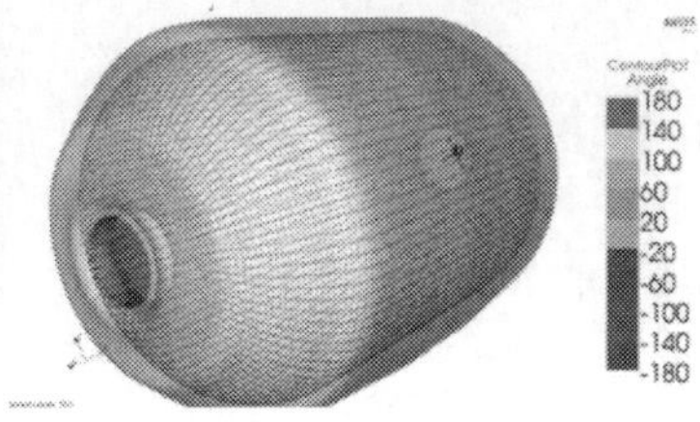

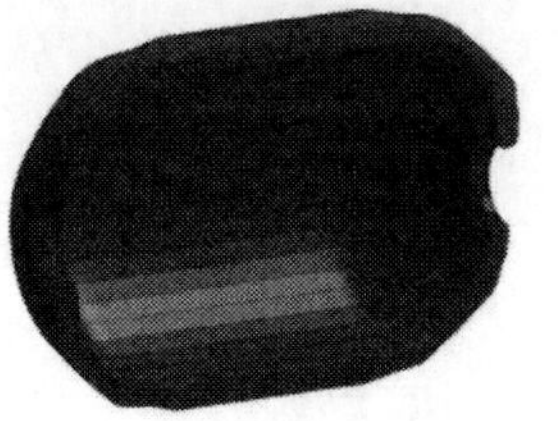

Modeling possibilities – FE Technology

Intralaminar

- Element → Layered thin/thick shells one solid per ply
- Material → failure and damage model

Agarwal/Broutman (1990)

Interlaminar

- Cohesive zone modeling
  - Cohesive interface elements

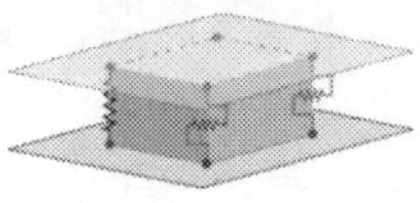

  - Tiebreak

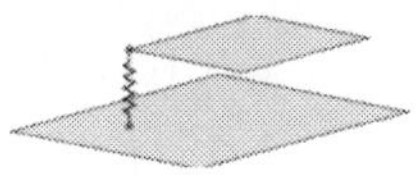

Modeling possibilities – FE Technology

Multi-Layered Solid Element

- Several layers modeled inside one solid element
- Applicable for thin and thick composite structures featuring normal thickness stresses σ_{33}
- Improved resolution of normal thickness and transverse shear stresses
- Remedy for transverse shear and volumetric locking
- Verification and validation for several examples based on OPEL's bottom up pyramid

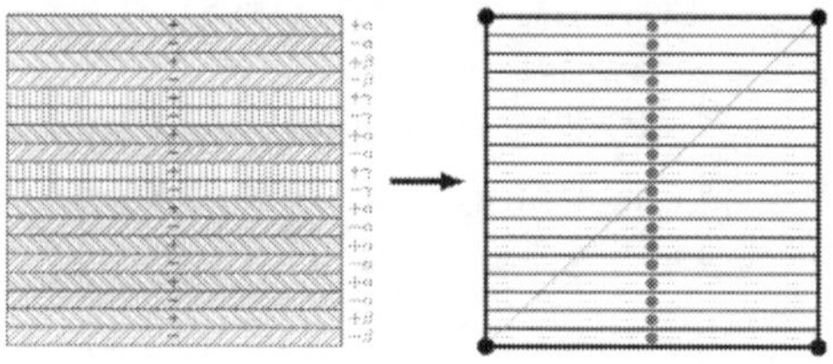

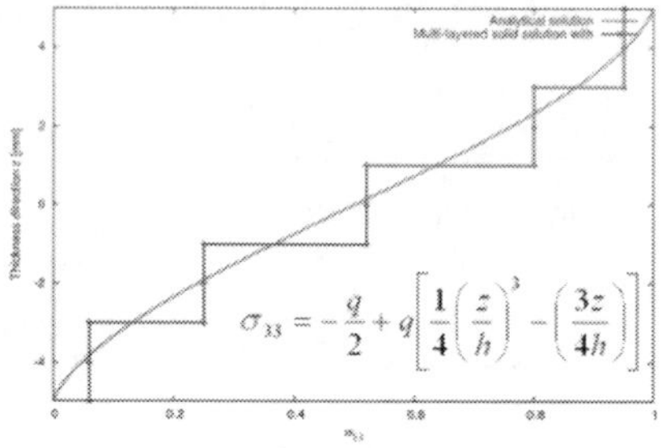

→ **Sufficient accurate and still efficient**

Modeling possibilities – FE Technology

Interlaminar Delamination

Cohesive/Interface elements with material model

- Model the bonding material with cohesive elements
- Volume oriented solid elements
- Finite / Zero thickness, time step computation without thickness
- Connection between solid or shell elements
- The traction-separation law describes the cohesion zone behavior and the energy release

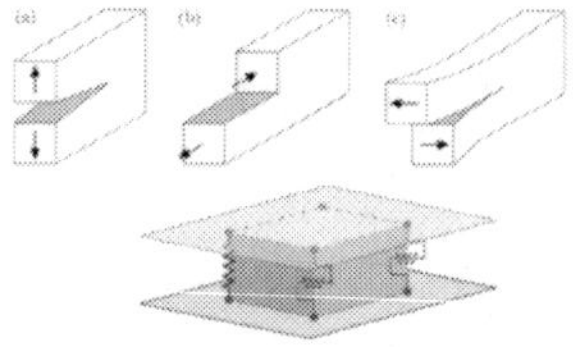

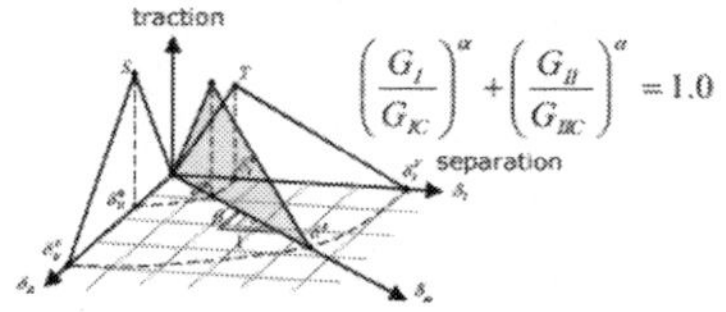

Tied contact with integrated material model

- Use the contact-tiebreak definition between the plies
- Kinematics → 3 relative displacements between slavenodes and master segments
- Material → stress computations similar to cohesive elements

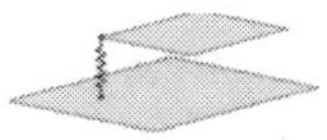

User-defined material model

Identification and characterization of material properties

T/C-T test

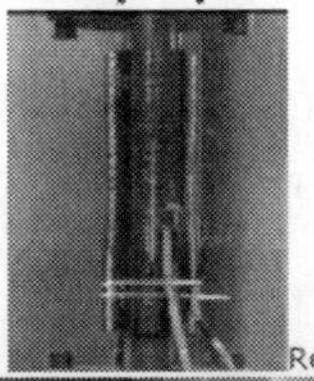

Ref: OPEL

Elip test

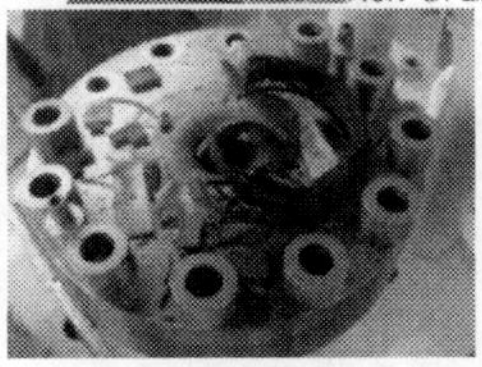

Model necessities
- Non-linear elasticity?
- Orthotropic?
- 3D Puck failure?
- Damage ?
- Delamination?

Verification

Non-linear elasticity

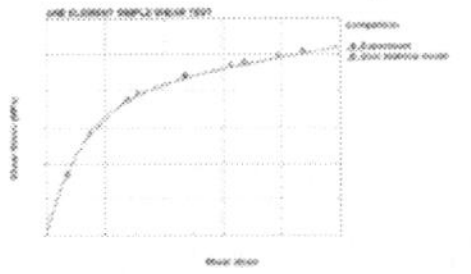

3D Puck Failure

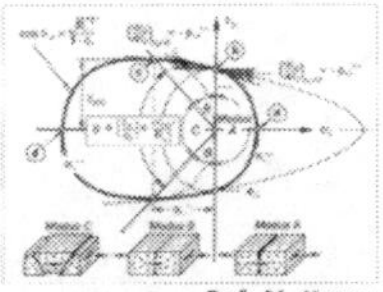

Ref: M. Knops

Anisotropic Continuum Damage Model

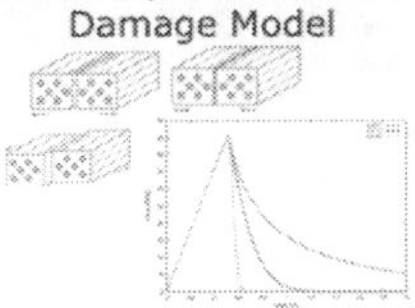

Validation

Test

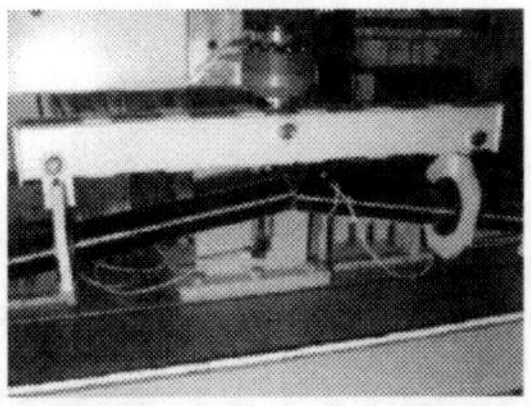

Ref: OPEL

Simulation

User-defined material model

- Constitutive relation

$$\epsilon = \mathbf{H}_0\hat{\boldsymbol{\sigma}},\ \mathbf{H}_0 = \begin{bmatrix} \frac{1}{E_{11}} & -\frac{\nu_{21}}{E_{22}} & -\frac{\nu_{31}}{E_{33}} & 0 & 0 & 0 \\ -\frac{\nu_{12}}{E_{11}} & \frac{1}{E_{22}} & -\frac{\nu_{32}}{E_{33}} & 0 & 0 & 0 \\ -\frac{\nu_{13}}{E_{11}} & -\frac{\nu_{23}}{E_{22}} & \frac{1}{E_{33}} & 0 & 0 & 0 \\ 0 & 0 & 0 & \frac{1}{G_{12}(\gamma_{22})} & 0 & 0 \\ 0 & 0 & 0 & 0 & \frac{1}{G_{23}} & 0 \\ 0 & 0 & 0 & 0 & 0 & \frac{1}{G_{13}} \end{bmatrix},\ \epsilon = \begin{bmatrix} \epsilon_{11} \\ \epsilon_{22} \\ \epsilon_{33} \\ 2\epsilon_{12} \\ 2\epsilon_{23} \\ 2\epsilon_{13} \end{bmatrix}$$

$$\epsilon = \mathbf{H}_0\hat{\boldsymbol{\sigma}} = \mathbf{H}_0\mathbf{M}\boldsymbol{\sigma}$$

$$\mathbf{H}(\boldsymbol{\omega}) = \begin{bmatrix} \frac{1}{(1-\omega_{11})E_{11}} & -\frac{\nu_{21}}{E_{22}} & -\frac{\nu_{31}}{E_{33}} & 0 & 0 & 0 \\ -\frac{\nu_{12}}{E_{11}} & \frac{1}{(1-\omega_{22})E_{22}} & -\frac{\nu_{32}}{E_{33}} & 0 & 0 & 0 \\ -\frac{\nu_{13}}{E_{11}} & -\frac{\nu_{23}}{E_{22}} & \frac{1}{(1-\omega_{33})E_{33}} & 0 & 0 & 0 \\ 0 & 0 & 0 & \frac{1}{(1-\omega_{12})G_{12}(\gamma_{22})} & 0 & 0 \\ 0 & 0 & 0 & 0 & \frac{1}{(1-\omega_{23})G_{23}} & 0 \\ 0 & 0 & 0 & 0 & 0 & \frac{1}{(1-\omega_{13})G_{13}} \end{bmatrix}$$

Verification: non-linear shear stress

- One-element simple shear test
- UD layer configuration

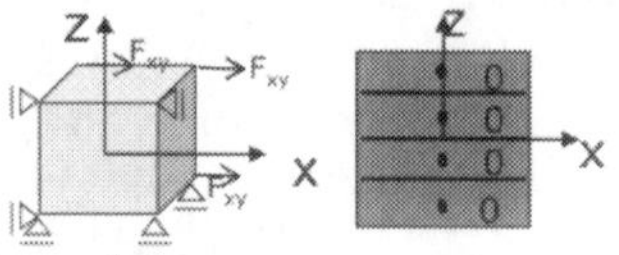

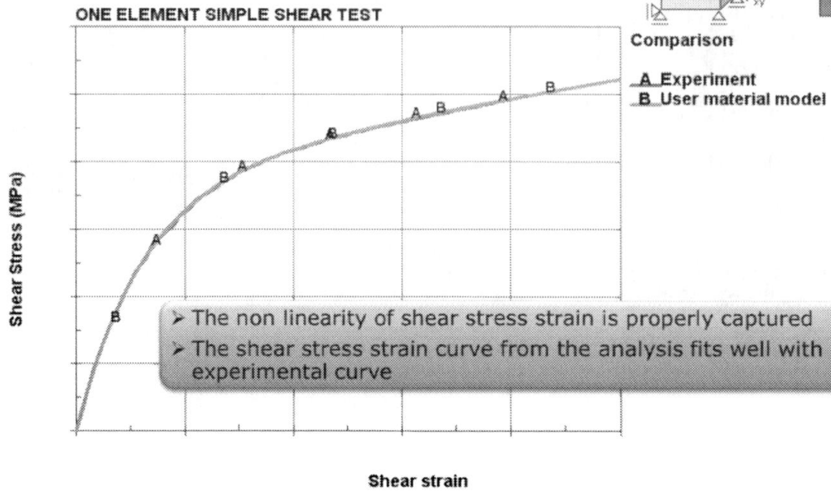

- ➢ The non linearity of shear stress strain is properly captured
- ➢ The shear stress strain curve from the analysis fits well with experimental curve

Verification: non-linear shear stress

- Component test (T/C-T) → Applied Torsion
- UD layer specimen

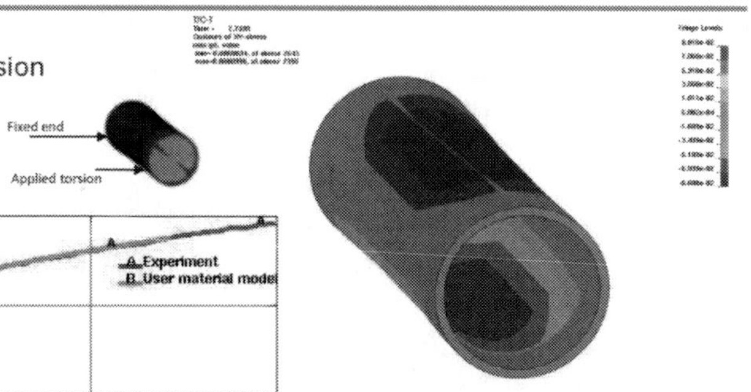

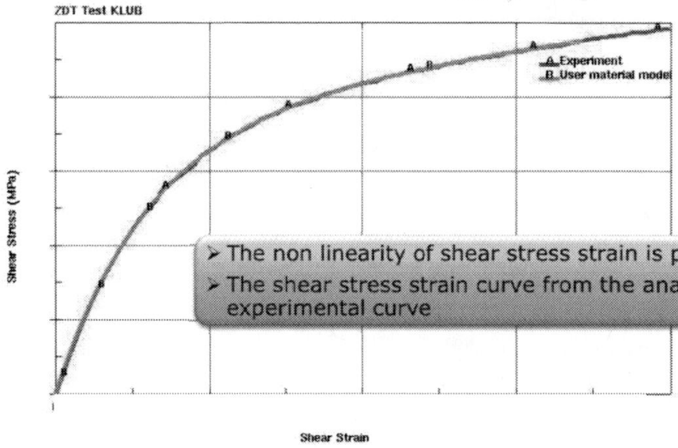

- ➢ The non linearity of shear stress strain is properly captured
- ➢ The shear stress strain curve from the analysis fits well with experimental curve

User defined material model: 3D Puck failure

- Different failure modes

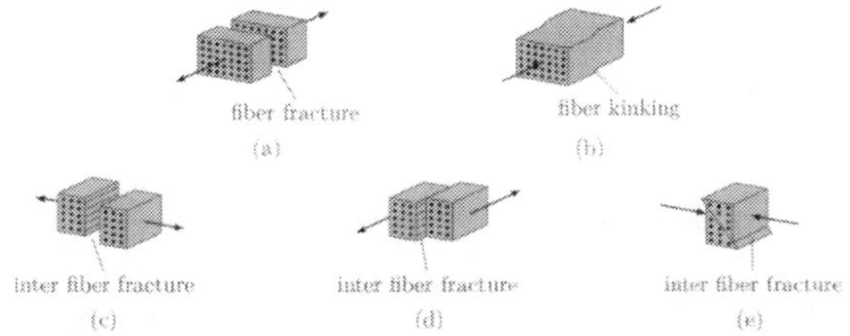

- State of the art 3D PUCK failure
  - Fiber failure & Inter fiber failure
  - Action plane, fracture angle, mode of failure
    - ➢ Mode A, B: non-critical
    - ➢ Mode C: critical failure

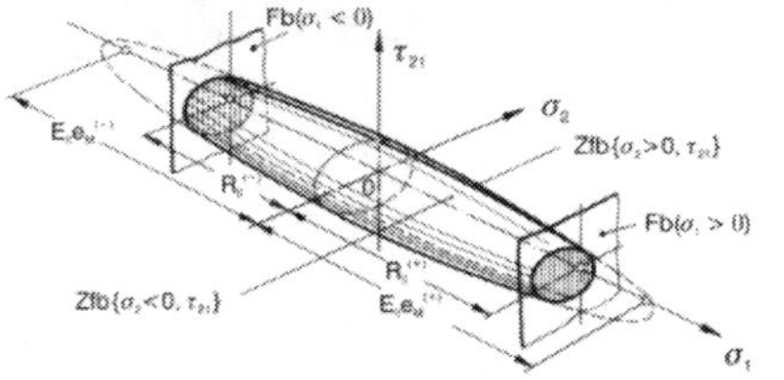

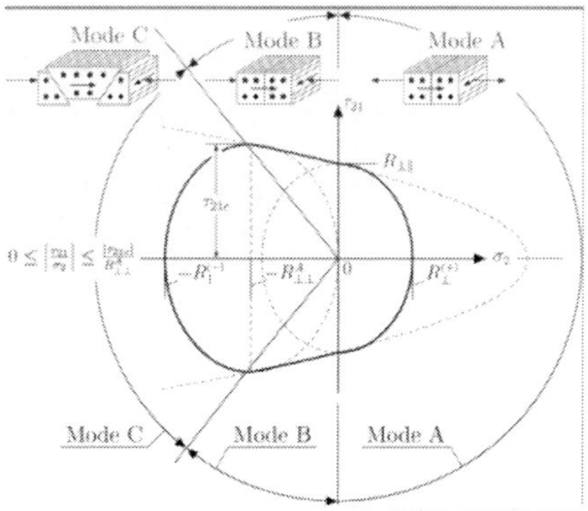

ANSYS CONFERENCE & 31. CADFEM USERS' MEETING
June, 19-21, 2013 - Mannheim, Germany

CADFEM

User defined material model: 3D Puck failure

- 3D PUCK failure: Inter-fiber failure
- Action plane fracture criteria
- If σ_n is tensile stress → assists shear stresses τ_{nt}, τ_{n1} in bringing IFF
- If σ_n is compressive stress → impedes the IFF brought about by the shear stresses τ_{nt}, τ_{n1}
- Action plane stresses

$$\begin{bmatrix} \cos^2\theta & \sin^2\theta & 2\sin\theta\cos\theta & 0 & 0 \\ \sin^2\theta & \cos^2\theta & -2\sin\theta\cos\theta & 0 & 0 \\ -\sin\theta\cos\theta & \sin\theta\cos\theta & \cos^2\theta-\sin^2\theta & 0 & 0 \\ 0 & 0 & 0 & \cos\theta & -\sin\theta \\ 0 & 0 & 0 & \sin\theta & \cos\theta \end{bmatrix} \cdot \begin{Bmatrix} \sigma_2 \\ \sigma_3 \\ \tau_{23} \\ \tau_{31} \\ \tau_{21} \end{Bmatrix} = \begin{Bmatrix} \sigma_n \\ \sigma_t \\ \tau_{nt} \\ \tau_{t1} \\ \tau_{n1} \end{Bmatrix} \begin{matrix} \leftarrow IFF \\ \\ \leftarrow IFF \\ \\ \leftarrow IFF \end{matrix}$$

- Inverse reserve factors, failure effort

$$f_{E,IFF}(\theta) = \sqrt{\left(\frac{1}{R_\perp^{(+)}} - \frac{p_{\perp\psi}^{(+)}}{R_{\perp\psi}^A}\right)^2 \sigma_n^2 + \left(\frac{\tau_{nt}}{R_{\perp||}}\right)^2 + \left(\frac{\tau_{n1}}{R_{\perp||}}\right)^2} + \frac{p_{\perp\psi}^{(+)}}{R_{\perp\psi}^A}\sigma_n \text{ for } \sigma_n \geq 0$$

$$f_{E,IFF}(\theta) = \sqrt{\left(\frac{\tau_{nt}}{R_{\perp||}}\right)^2 + \left(\frac{\tau_{n1}}{R_{\perp||}}\right)^2 + \left(\frac{p_{\perp\psi}^{(-)}}{R_{\perp\psi}^A}\sigma_n\right)^2} + \frac{p_{\perp\psi}^{(-)}}{R_{\perp\psi}^A}\sigma_n \text{ for } \sigma_n < 0$$

with

$$\frac{p_{\perp\psi}^{(\pm)}}{R_{\perp\psi}^A} = \begin{cases} \frac{1}{\tau_{nt}^2+\tau_{n1}^2}\left(\frac{p_{\perp\perp}^{(\pm)}}{R_{\perp\perp}^A}\tau_{nt}^2 + \frac{p_{\perp||}^{(\pm)}}{R_{\perp||}}\tau_{n1}^2\right) & \text{for } \tau_{nt}^2+\tau_{n1}^2 > 0 \\ 0 & \text{for } \tau_{nt}^2+\tau_{n1}^2 = 0 \end{cases}$$

2013 Automotive Simulation World Congress
October, 29-30, Frankfurt am Main, Germany

CADFEM

Verification: fracture curve based on 3D Puck

- One-element→ σ_y/τ_{xy} stress combination
- Verified with test values and analytical solution (AlfaLam)

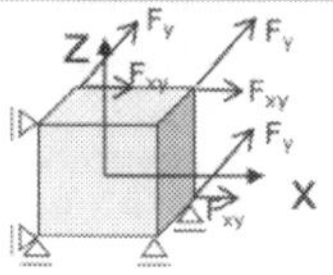

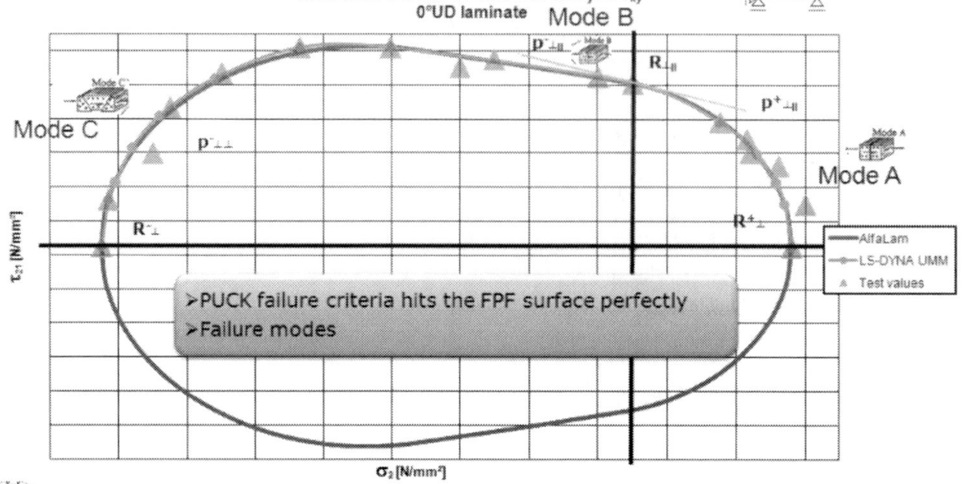

User-defined material model: progressive failure analysis

- Damage modes

1-Fiber mode

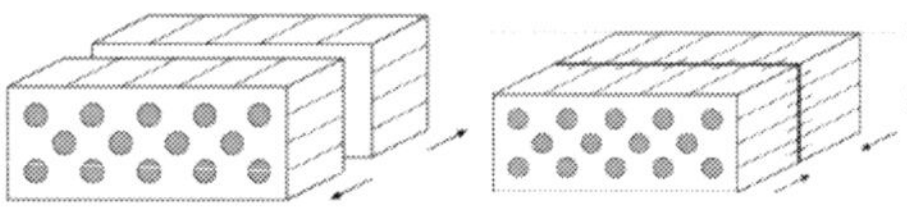

For $\sigma_1 \geq 0$: $\quad f_1^+ = \frac{\sigma_{11}^2}{(1-\omega_{11}^+)^2 R_\parallel^{+2}} - r_1^+ = 0$

and

for $\sigma_1 < 0$: $\quad f_1^- = \frac{\sigma_{11}^2}{(1-\omega_{11}^-)^2 R_\parallel^{-2}} - r_1^- = 0,$

2-Perpendicular Matrix mode

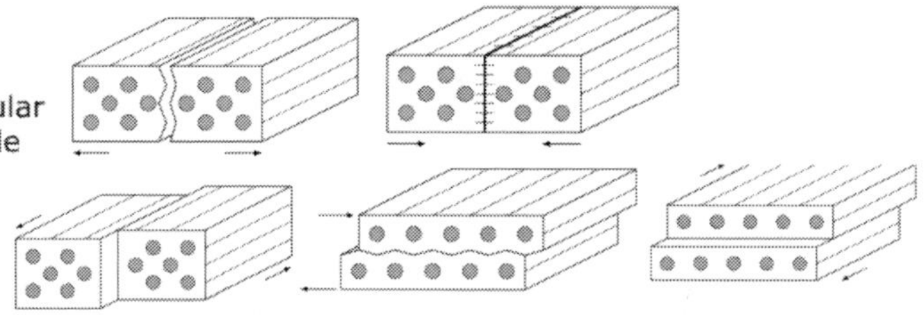

For $\sigma_n \geq 0$: $$f_2^+ = \sqrt{\left[\frac{1}{R_\perp^{(+)}} - \frac{p_{\perp\psi}^{(+)}}{R_{\perp\psi}^{(A)}}\right]^2 \sigma_n^2 + \left[\frac{\tau_{nt}}{R_{\perp\perp}^{(A)}}\right]^2 + \left[\frac{\tau_{n1}}{R_{\perp\parallel}}\right]^2} + \frac{p_{\perp\psi}^{(+)}}{R_{\perp\psi}^{(A)}}\sigma_n - r_2^+ = 0$$

and

for $\sigma_n < 0$: $$f_2^- = \sqrt{\left[\frac{\tau_{nt}}{R_{\perp\perp}^{(A)}}\right]^2 + \left[\frac{\tau_{n1}}{R_{\perp\parallel}}\right]^2 + \left[\frac{p_{\perp\psi}^{(-)}}{R_{\perp\psi}^{(A)}}\sigma_n\right]^2} + \frac{p_{\perp\psi}^{(-)}}{R_{\perp\psi}^{(A)}}\sigma_n - r_2^- = 0.$$

User-defined material model: progressive failure analysis

- Damage vector $\boldsymbol{\varpi} = \omega \mathbf{n}$

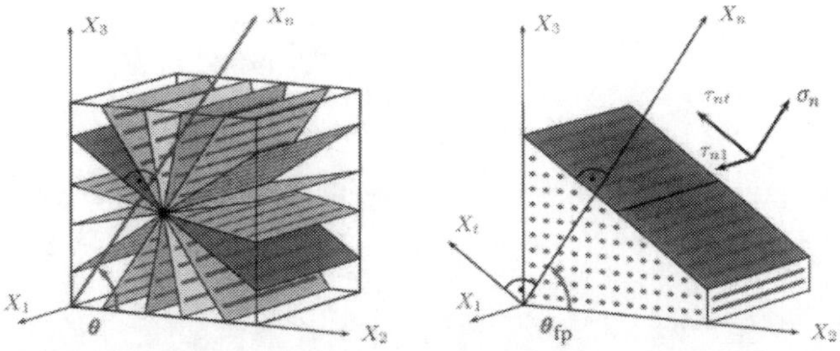

- Fracture angle is frozen in damage modeling

$$\mathbf{n}_{\mathrm{IFF}} = \begin{bmatrix} 0 \\ 1 \\ 0 \end{bmatrix} \qquad \mathbf{n}_{\perp} = \mathbf{T}\mathbf{n}_{\mathrm{IFF}} = \begin{bmatrix} 0 \\ \cos\theta_{\mathrm{fp}} \\ \sin\theta_{\mathrm{fp}} \end{bmatrix}$$

$$\boldsymbol{\varpi}_{\perp} = \omega_{\mathrm{IFF}} \begin{bmatrix} 0 \\ \cos\theta_{\mathrm{fp}} \\ \sin\theta_{\mathrm{fp}} \end{bmatrix}$$

User-defined material model: progressive failure analysis

- Growth rate of damage variables ω_i → damage rule of the form [7] =

$$\dot{\boldsymbol{\omega}} = \sum_i \varphi_i \mathbf{q}_i = \varphi_1 \mathbf{q}_1 + \varphi_2 \mathbf{q}_2$$

φ_i → controls the amount of growth

q_i → coupling between the individual damage modes

- The damage criteria $f_i - r_i = 0$ and $g_i - r_i = 0$ provide the damage functions in stress and strain space

- Damage growth $\dot{\boldsymbol{\omega}} > 0$ → will occur when strain path crosses the updated damage surface $g_i - r_i = 0$ and strain increment has a non-zero component in the direction of the normal to the damage surface

$$\sum_i \frac{\partial g_i}{\partial \epsilon} \dot{\epsilon} > 0$$

User-defined material model: progressive failure analysis

- Damage variable ω and damage coupling vector $\mathbf{q}_i$

$$\dot{\boldsymbol{\omega}} = \sum_i \varphi_i \mathbf{q}_i = \varphi_1 \mathbf{q}_1 + \varphi_2 \mathbf{q}_2$$

Fiber mode — Perpendicular Matrix mode

$$\dot{\boldsymbol{\omega}} = \begin{bmatrix} \dot{\omega}_{11} \\ \dot{\omega}_{22} \\ \dot{\omega}_{33} \\ \dot{\omega}_{12} \\ \dot{\omega}_{23} \\ \dot{\omega}_{13} \end{bmatrix} = \varphi_1 \begin{bmatrix} 1 \\ l_{\parallel\sigma_{22}} \\ l_{\parallel\sigma_{33}} \\ l_{\parallel\tau_{12}} \\ l_{\parallel\tau_{23}} \\ l_{\parallel\tau_{13}} \end{bmatrix} + \varphi_2 \begin{bmatrix} 0 \\ l_{\perp\sigma_{22}} \\ l_{\perp\sigma_{33}} \\ l_{\perp\tau_{12}} \\ l_{\perp\tau_{23}} \\ l_{\perp\tau_{13}} \end{bmatrix}$$

- Utilizing the damage coupling function and growth functions, a damage variable can be obtained for an individual failure mode as $\varphi_i = 1 - e^{\frac{1}{m_i}(1-[r_i]^{m_i})}$, $r_i \geq 1$

- Compliance tensor of the damage variable

$$\mathbf{H}(\boldsymbol{\omega}) = \begin{bmatrix} \frac{1}{(1-\omega_{11})E_{11}} & -\frac{\nu_{21}}{E_{22}} & -\frac{\nu_{31}}{E_{33}} & 0 & 0 & 0 \\ -\frac{\nu_{12}}{E_{11}} & \frac{1}{(1-\omega_{22})E_{22}} & -\frac{\nu_{32}}{E_{33}} & 0 & 0 & 0 \\ -\frac{\nu_{13}}{E_{11}} & -\frac{\nu_{23}}{E_{22}} & \frac{1}{(1-\omega_{33})E_{33}} & 0 & 0 & 0 \\ 0 & 0 & 0 & \frac{1}{(1-\omega_{12})G_{12}(\gamma_{22})} & 0 & 0 \\ 0 & 0 & 0 & 0 & \frac{1}{(1-\omega_{23})G_{23}} & 0 \\ 0 & 0 & 0 & 0 & 0 & \frac{1}{(1-\omega_{13})G_{13}} \end{bmatrix}$$

Verification test: UD laminate cube under tension

- Degradation-tension transverse to fiber direction
- Damage evolution law $\omega_i = 1 - e^{\frac{1}{m}\left(1-\left[\frac{\varepsilon}{\varepsilon_y}\right]^m\right)}$

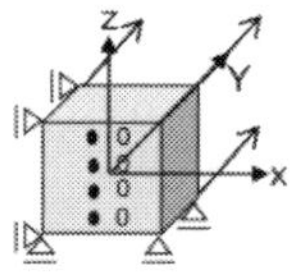

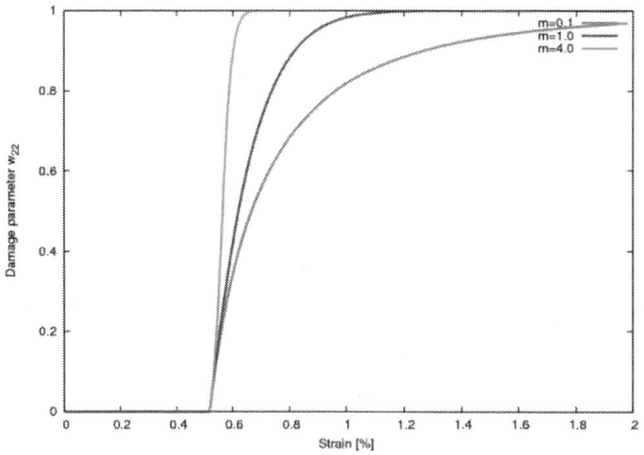

Verification test: UD laminate cube under tension

- Degradation-tension transverse to fiber direction
- Comparison with LS-DYNA material models

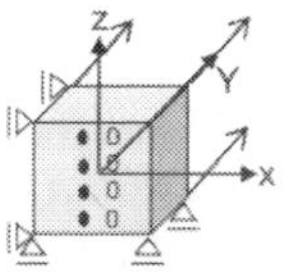

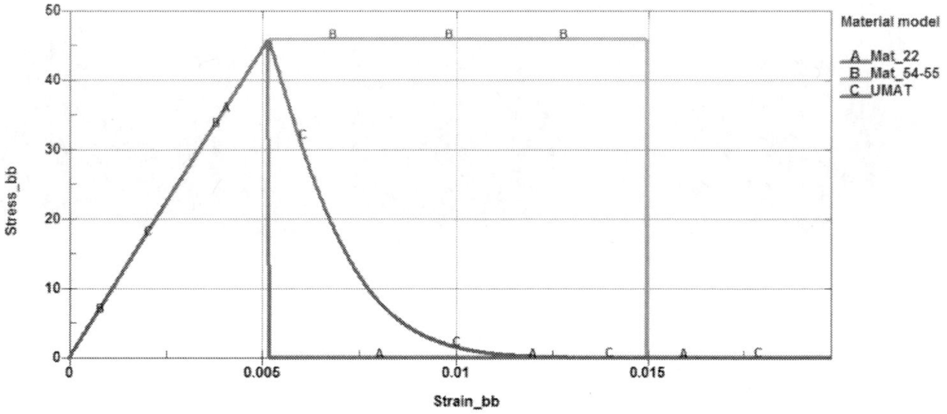

Verification test: [0/90/90/0] laminate under tension

- Failure inside [0-90-90-0] laminate under tensile load
- Failure process

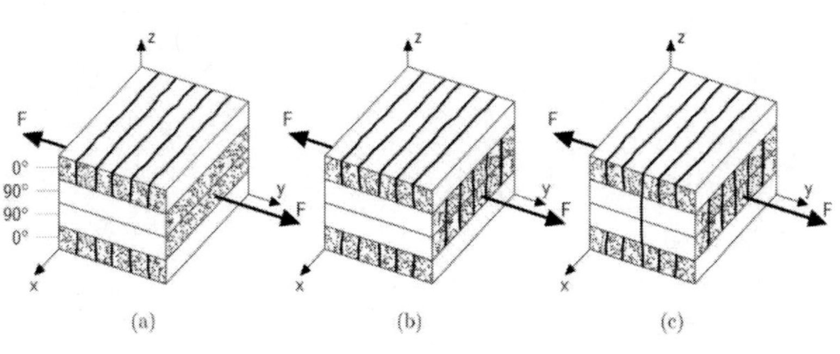

Section of a laminate $[0^\circ/90^\circ/90^\circ/0^\circ]$ under increasing load: (a) first ply failure IFFs in the 0°-laminae, (b) second ply failure IFFs in the 90°-laminae and (c) third ply failure FFs in the 90°-laminae

2013 Automotive Simulation World Congress
October, 29-30, Frankfurt am Main, Germany

CADFEM

Verification test: [0/90/90/0] laminate under tension

- Failure inside [0-90-90-0] laminate under tensile load
- Failure process

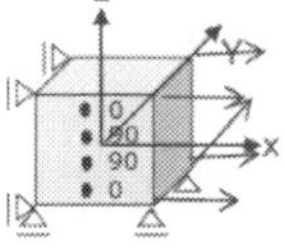

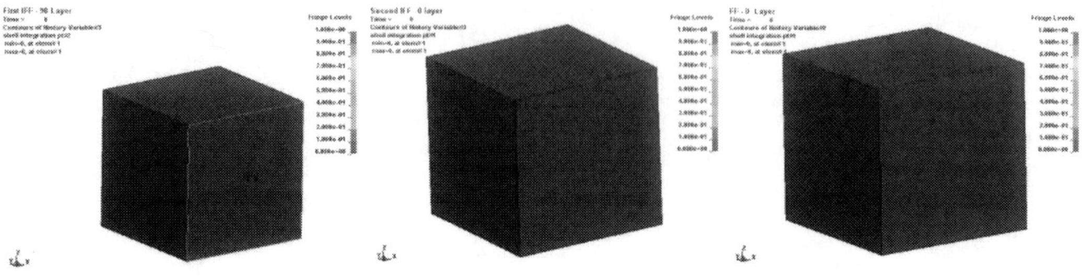

CADFEM

2013 Automotive Simulation World Congress
October, 29-30, Frankfurt am Main, Germany

Verification test: [0/90/90/0] laminate under tension

- Degradation of stresses

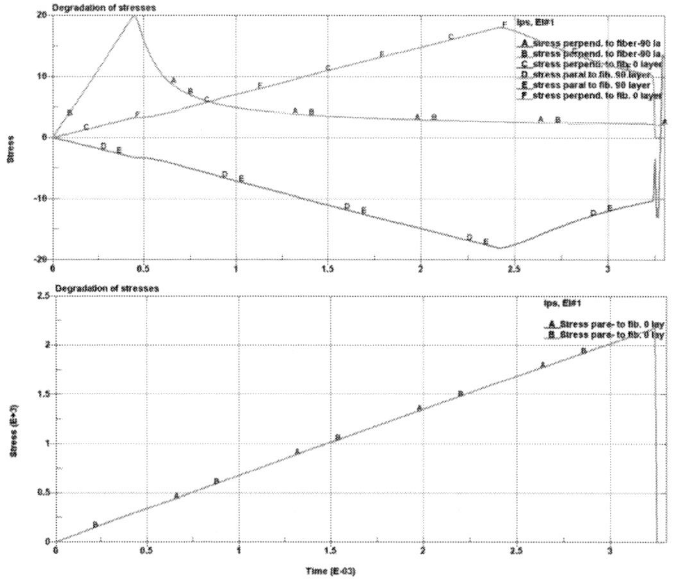

CADFEM

2013 Automotive Simulation World Congress
October, 29-30, Frankfurt am Main, Germany

Validation: 4-point bending test

- 4-point bending wi
- Multi-directional la

4-point bending

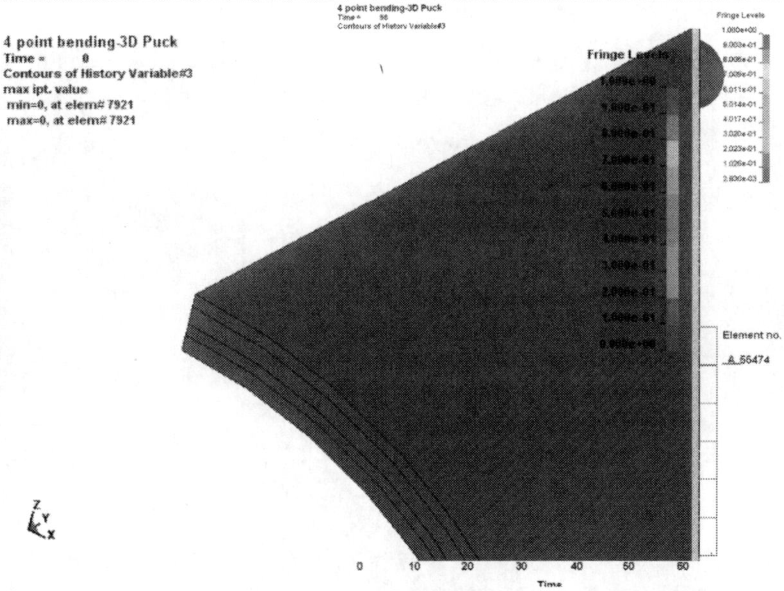

Failure index inter fiber failure ($f_{e,iff}$)1=failed

ANSYS CONFERENCE & 31. CADFEM USERS' MEETING
June, 19-21, 2013 – Mannheim, Germany

CADFEM

Validation: 4-point bending test

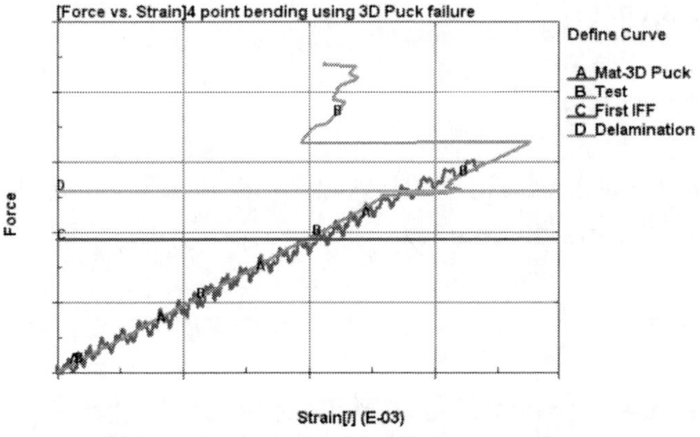

- Importance of fiber volume content in thick composites
- Force vs. Strain simulation curve (slope) fits well with the experimental curve
- First inter-fiber failure (IFF) in test and simulation
- Needs further investigation in delamination modeling

ANSYS CONFERENCE & 31. CADFEM USERS' MEETING
June, 19-21, 2013 – Mannheim, Germany

CADFEM

CAE- process Chain

- Fall Tower Impact Testing and Simulation Results

Test Boundary Conditions

Fall Tower:

- Impactor Geometry: 50 x 50mm
- Sledge Weight: 188 kg
- Drop Height: 2442 mm
- Velocity: 6,88 m/s
- Impact Energy: 4.5 kJ
- Pressure: ambient

Vessel:

l = 717 mm

d = 497 mm

t_{comp} = 31 mm

Impact on cylindrical part o

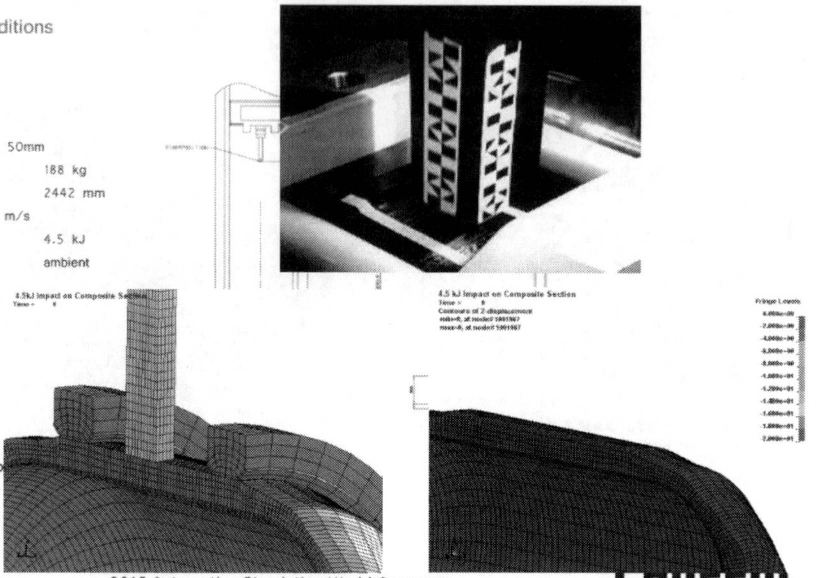

Summary

- CAE process chain
- Non-linear elasticity
- Failure initiation based on 3D Puck and progressive damage based on continuum damage mechanics
- Delamination
- Verification one element level & Validation at component level

CADFEM

Acknowledgements

- Funding by GM APCE/Opel, Germany (Dr.-Ing. Stephan Fell and Dipl.-Ing. Thorsten Schütz)
- CADFEM support (Dr.-Ing. Uli Stelzmann & Dr.-Ing. Matthias Hörmann)
- Prof. Matzenmiller (University of Kassel)

Thank you for your attention

2013 Automotive Simulation World Congress
October, 29-30, Frankfurt am Main, Germany

References

[1] **Kachanov, L. M.** (1958): Time of the rupture process under creep conditions. *Isv. Akad. Nauk S.S.S.R., Otd. Tekh. Nauk* 8, pp. 26–31.

[2] **Chaboche, J. -L.** (1995): A continuum damage theory with anisotropic and unilateral damage. *Recher. Aerospat.* 2, pp. 139–147.

[3] **Simo, J. C.; Ju, J. W.** (1987a): Strain and stress-based continuum damage models-I. Formulat. *Int. J. Solids Struct.* vol. 23, pp. 821–840.

[4] **Simo, J. C.; Ju, J. W.** (1987b): Strain and stress-based continuum damage models-II. Comput. Aspects. *Int. J. Solids Struct.* vol 23, pp. 841–869.

[5] **Cauvin, A.; Testa, R. B.** (1999): Damage mechanics: basic variables in continuum theories. *Int. J. Solids Struct.* vol 36, pp. 747–761.

[6] **Kurnatowski, B.** (2010): Zweiskalensimulation von mikroheterogenen Strukturen aus spröden Faserverbund-werkstoffen, Institut für Mechanik, Fachbereich Maschinenbau, Universität Kassel, Ph. D. Thesis.

[7] **Matzenmiller, A.; Lubliner, J.; Taylor, R. L.** (1995): A constitutive model for anisotropic damage in fibre composites, *Mechanics of Materials*, vol. 20, pp. 125-152.

[8] **Williams, K.V.; Vaziri, R.; Poursartip, A.** (2003): A physically based continuum damage mechanics model for thin laminated composite structures. *Int. J. Solids Struct.* vol. 40, pp. 2267–2300.

[9] **Hashin, Z.** (1980): Failure criteria for unidirectional fibre composites, *J. Appl. Mech.* vol. 47, pp. 329-334.

[10] **Rosen, B.W.; Dow, N. E.** (1972): Mechanics of failure of fibrous composites, in: H. Liebowitz, ed., *Fracture - An Advanced Treatise*, Academic Press, p. 661.

[11] **Knops, M.** (2008): *Analysis of Failure in Fibre Polymer Laminates: The Theory of Alfred Puck.* Springer-Verlag: Berlin, Heidelberg.

[12] **Puck, A.; Schürmann, H.** (2004): Failure analysis of FRP laminates by means of physically based phenomenological models, in *Failure Criteria in Fibre Reinforced Polymer Composites: The World-Wide Failure Exercise*, P. D. Hinton, A. S. Kaddour, and M. L. Soden, Eds. Elsevier: Oxford, pp. 832-876.

[13] **Deuschle, H. M.; Kröplin, B. H.** (2012): Finite element implementation of Puck's failure theory for fibre-reinforced composites under three-dimensional stress, *J. Composite Materials*, vol. 46, pp. 2485-2513.

2013 Automotive Simulation World Congress
October, 29-30, Frankfurt am Main, Germany

References

[14] **Ramberg, W.; Osgood, W. E.** (1943): Description of stress–strain curves by three parameters. *National Advisory Committee for Aeronautics, Technical Note No., 902.*

[15] **Ortiz, M.** (1985): A constitutive theory for the inelastic behavior of concrete, *Mech. Mater.* vol. 4, pp. 67-93.

[16] **Ju, J. W.** (1989): On energy-based coupled elastoplastic damage theories: constitutive modeling and computational aspects, *Int. J. Solids Struct.* vol. 25, No. 7, pp. 803-833.

[17] **Fiolka, M.; Matzenmiller, A.** (2007): On the resolution of transverse stresses in solid-shells with a multi-layer formulation, *Commun. Numer. Meth. Engg.*, vol. 23, pp. 313-326.

[18] **Hallquist, J. O.** (2010): *LS-DYNA Theoretical Manual*, Livermore Software Technology Corporation, Livermore, California.

[19] **Soden, P. D.; Hinton, M. J.; Kaddour, A. S.** (2004): Lamina properties, lay-up configurations and loading conditions for a range of fibre reinforced composite laminates, in *Failure Criteria in Fibre Reinforced Polymer Composites: The World-Wide Failure Exercise*, Soden, P. D.; Hinton, M. J.; Kaddour, A. S., Eds. Elsevier: Oxford, pp. 30-51.

[20] **Voigt, W.** (1889): Über die Beziehung zwischen den beiden Elastizitätskonstanten isotroper Körper, *Wied. Ann.*, vol. 38, pp. 573-587.

[21] **Reuss, A.** (1929): "Berechnung der Fließgrenze von Mischkristallen auf Grund der Plastizitätsbedingung für Einkristalle," *Z. Angew. Math. Mech.*, vol. 9, pp. 49-58.

[22] **Gotsis, P. K.; Chamis, C. C.; Minnetyan, L.** (2004): Application of progressive fracture analysis for predicting failure envelopes and stress-strain behaviors of composite laminates: a comparison with experimental results, in *Failure Criteria in Fibre Reinforced Polymer Composites: The World-Wide Failure Exercise*, Soden, P. D.; Hinton, M. J.; Kaddour, A. S., Eds. Elsevier: Oxford, pp. 703-723.

[23] **Cuntze, R. G.** (2006): Efficient 3D and 2D failure conditions for UD laminae and their application within the verification of the laminate design, *Composites Science and Technology*, vol. 66, pp. 1081-1096.

[24] **Chatiri, M.; Schuetz, T,; Matzenmiller, A.** (2010): An assessment of the new LS-DYNA® multi-layered solid element: basics, patch simulation and its potential for thick composite structural analysis, *Proc. 11th Intl. LS-DYNA® Users Conf.* Detroit, USA.

Wir leben Autos.

2013 Automotive Simulation World Congress
October, 29-30, Frankfurt am Main, Germany

CADFEM®

Numerical Computation of the Flow around a Racing Car

Dipl.-Ing. Bastian Schoeneberg
CFX Berlin Software GmbH
bastian.schoeneberg@cfx-berlin.de

Patrick Dekeyser
HAWKS Racing e. V. / HAW Hamburg
p.dekeyser@hawksracing.de

Automotive Simulation World Congress, October 29-30, 2013 - Frankfurt

Outline

Part 1: Introduction HAWKS Racing

- Formula Student
- Vehicles and success
- Structure of the team
- Technical data

Part 2: Flow calculation H08

- Geometry
- Mesh
- Boundary conditions
- Calculation
- Results
- Outlook

Outline

CFX BERLIN — HAWKS RACING

Part 1: Introduction HAWKS Racing
- Formula Student
- Vehicles and success
- Structure of the team
- Technical data

Part 2: Flow calculation H08
- Geometry
- Mesh
- Boundary conditions
- Calculation
- Results
- Outlook

Formula Student

CFX BERLIN — HAWKS RACING

Formula Student
- International design contest
- Approx. 550 teams worldwide
  80 teams in Germany
- 10 Events around the world
- 8 disciplines – 1000 scores
- Best overall concept wins

Goals within one year
- Finding of the concept
- Design
- Manufacture
- Testing
- Events

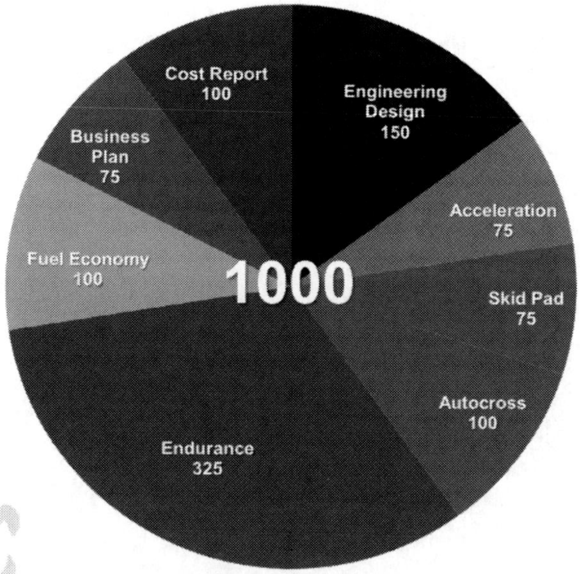

The garage

Success

2011

- 6th place Over-All Germany
- 2nd place Autocross Hungary
- 1st place Endurance Hungary
- 3rd place Design Report Hungary
- 3rd place Over-All Hungary
- 1st place Statics Italy
- 3rd place Over-All Italy
- Winner „Engineering Excellence Award“ Italy

2012

- 9th place Over-All Germany
- 3rd place Autocross Germany
- 5th place Endurance Germany
- 1st place Skid Pad Italy
- 1st place Autocross Italy
- 1st place Overall Dynamics Italy

Structure of the team

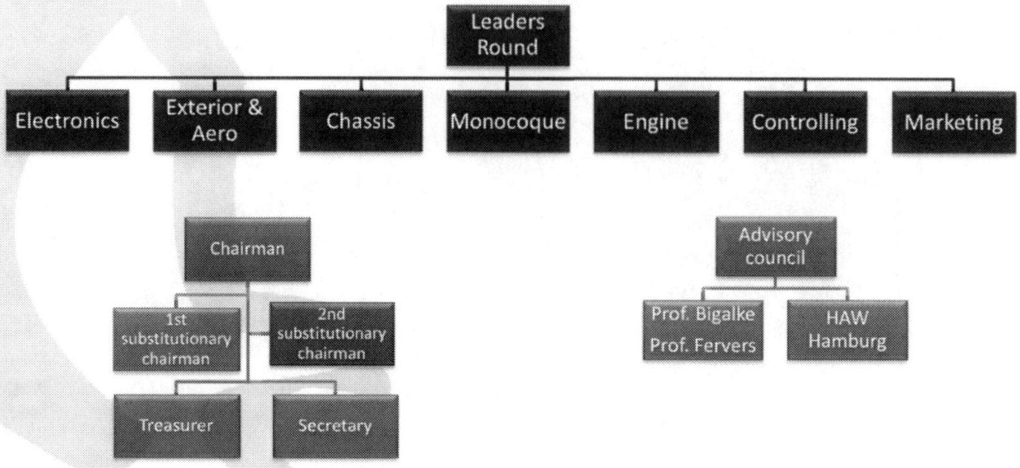

Many and varied tasks

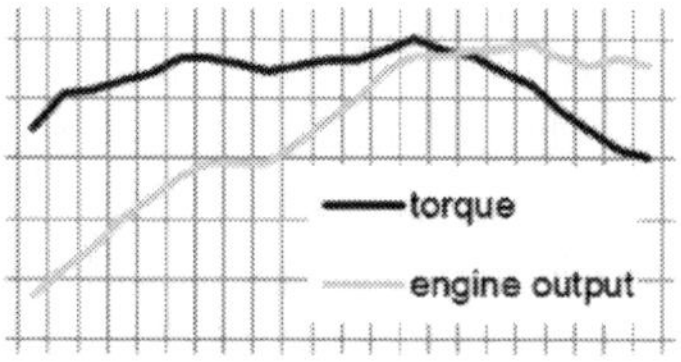

Exterior & Aero

- Marketing
- Controlling
- Electronics
- Chassis
- Engine
- Monocoque
- Exterior & Aero

- Unique Design
- CFD-validated aerodynamics package
- Optimised sidepod for better cooling and higher downforce
- 400 N downforce at 50 km/h

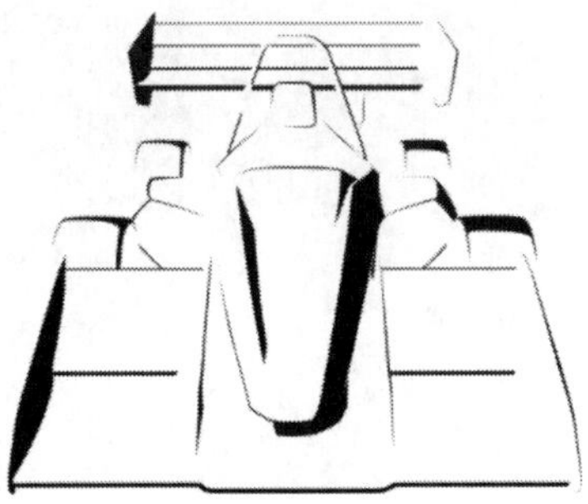

Season 2012/13 – The H09

- **Mass**: 205 kg
- **Acceleration**: 0 – 100 km/h in 3.5 s
- c_A = 2.4 up to 2.6
  - Theoretically possible to drive on ceiling at approx. 140 km/h

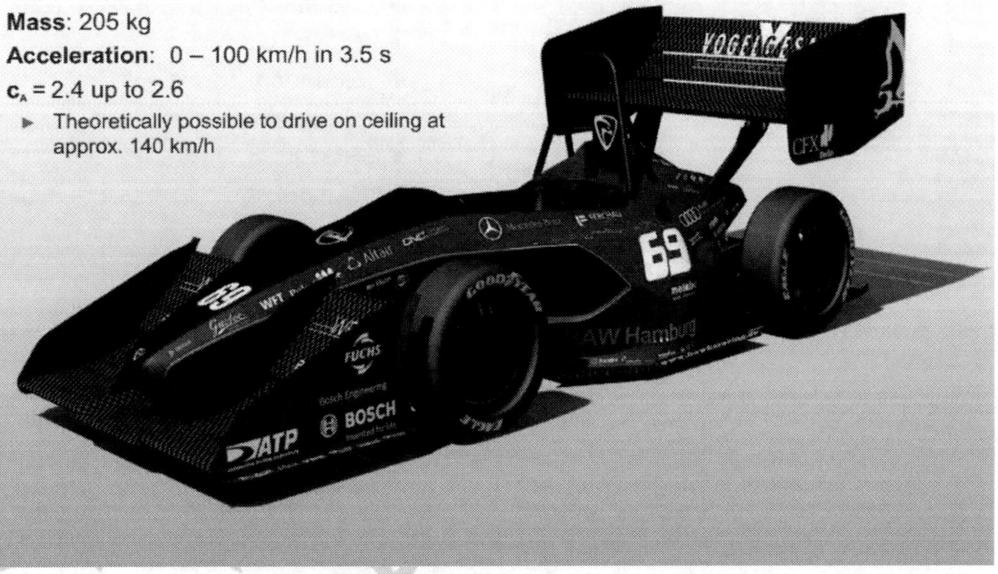

CFX BERLIN HAWKS RACING

Outline

CFX BERLIN HAWKS RACING

Part 1: Introduction HAWKS Racing

- Formula Student
- Vehicles and Success
- Structure of the team
- Technical data

Part 2: Flow calculation H08

- Geometry
- Mesh
- Boundary conditions
- Calculation
- Results
- Outlook

Overview

CFX BERLIN HAWKS RACING

Object of study:

- Racing car H08 of the team HAWKS Racing

Objectives:

- Examine flow through the radiator in the sidepod
- Effect of different cooling unit designs on aerodynamics
- Determination of resistance and downforce

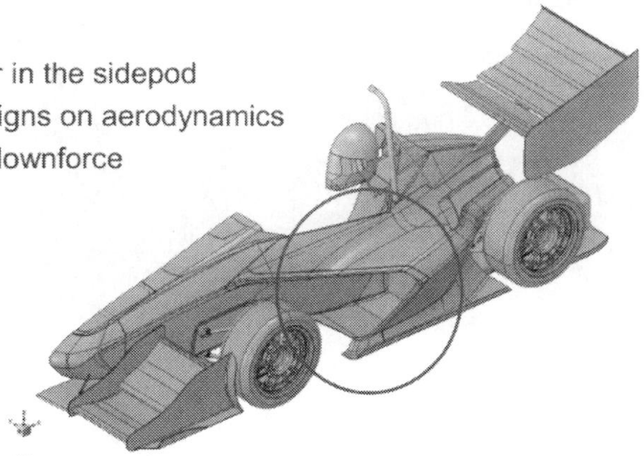

Geometry

CFX BERLIN HAWKS RACING

Geometry clean-up:

- Removed irrelevant details (screws, small rounds)
- Closed wholes and gaps, added simple driver body

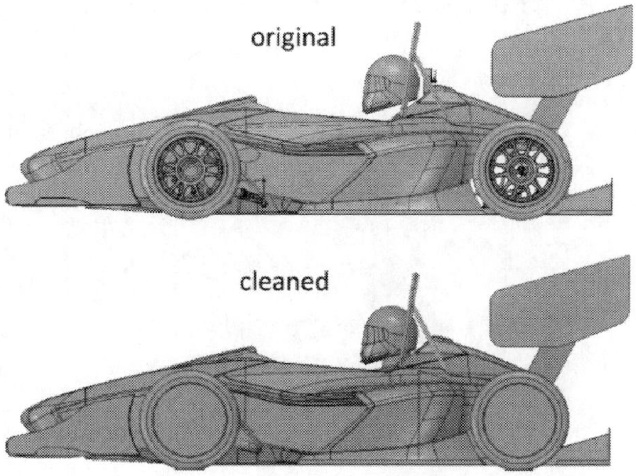

Geometry

Simplification of complex details
- Wheel suspension, rims
- Powertrain, engine
- Contact of wheel to ground

Tools:
- ANSYS SpaceClaim Direct Modeler
- ANSYS ICEM CFD

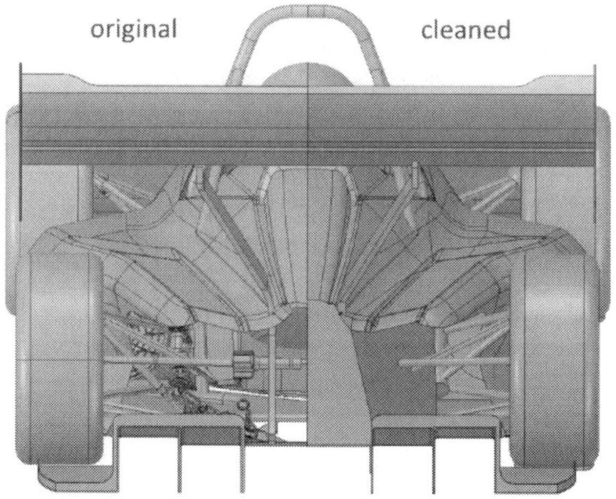

Computational mesh

Automatic meshing with ANSYS ICEM CFD Tetra:
- Wedge elements for resolution of boundary layers
- 30 million tetrahedral and 5 million wedge elements (7.6 million nodes)

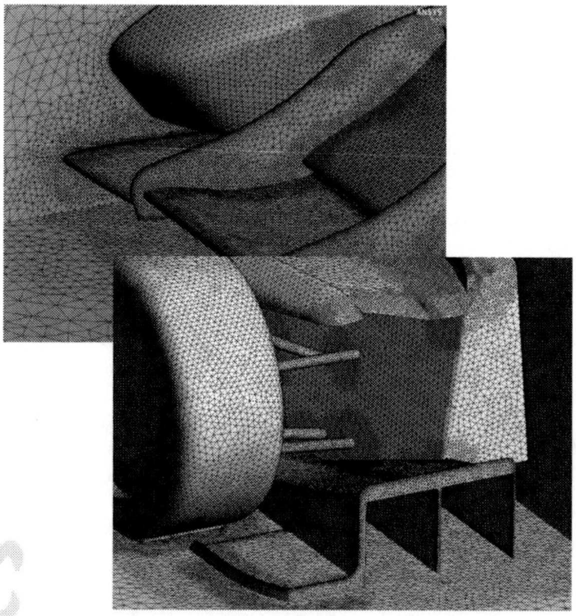

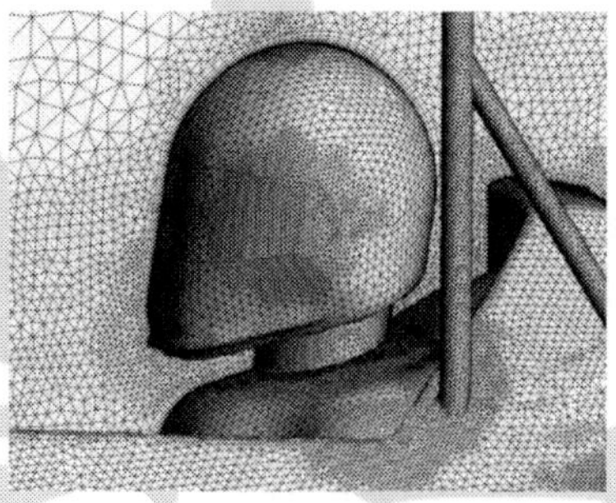

Boundary Conditions

All velocity statements relative to vehicle

Symmetry plane

- Through vehicle mid plane

Inlet

- Velocity 10 m/s

Surrounding

- Ambient pressure 1 atm

Ground

- Velocity 10 m/s

Rotating wheels

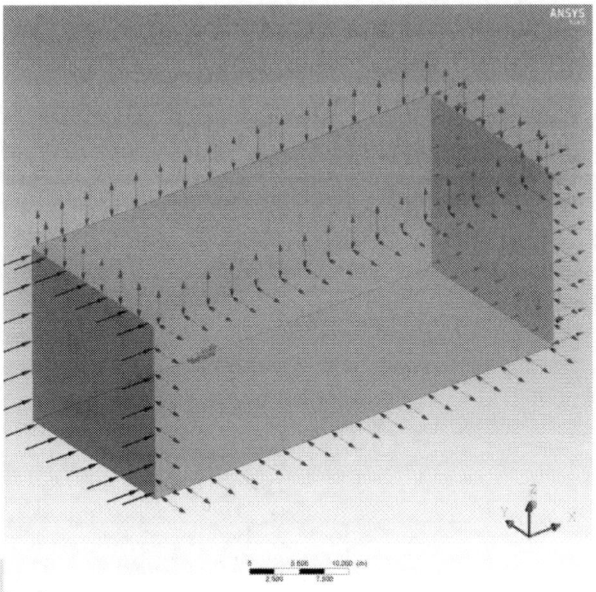

Boundary Conditions

High complexity of cooling ribs and fan

- Not taken into account in mesh
  → Radiator modeled as porous medium with pressure loss
- Loss depends on local flow velocity

$$\Delta p_{tot} = \zeta \frac{\rho}{2} v^2$$

- Pressure loss coefficient ζ of the cooling unit was varied

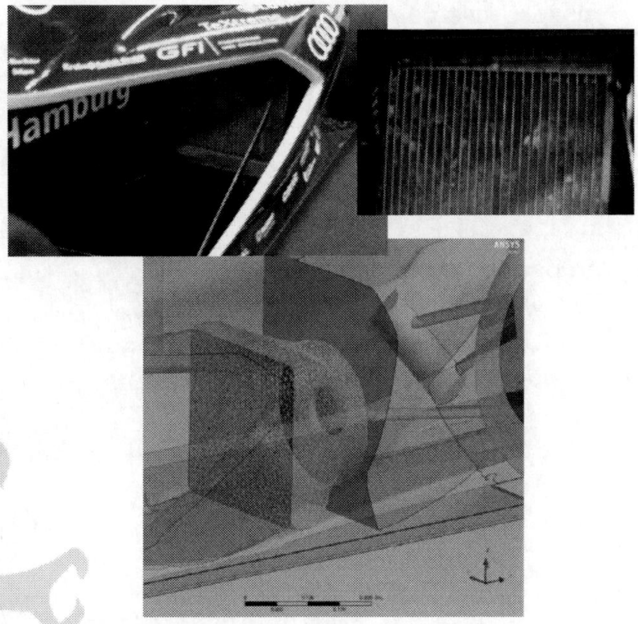

Solution

Solver:
- ANSYS CFX 14.5

Turbulence model:
- Shear Stress Transport (two-equation-model)

Stationary computation:
- 68 GB memory usage
- 2x8 cores
- 200 iterations
- 20 h solution time

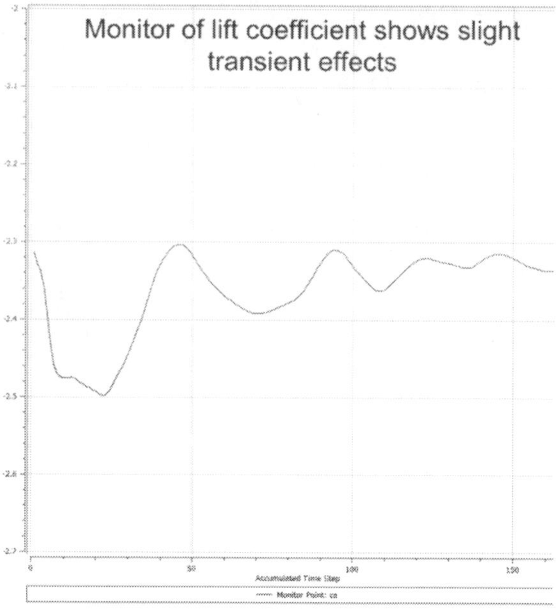

Results

Pressure loss coefficient ζ of the cooling unit was varied
- ζ = 0 means free flow through sidepod
- ζ = 5
- ζ = 50

Comparison of results:
- Flow distribution in radiator?
- Impact on vehicle aerodynamics?

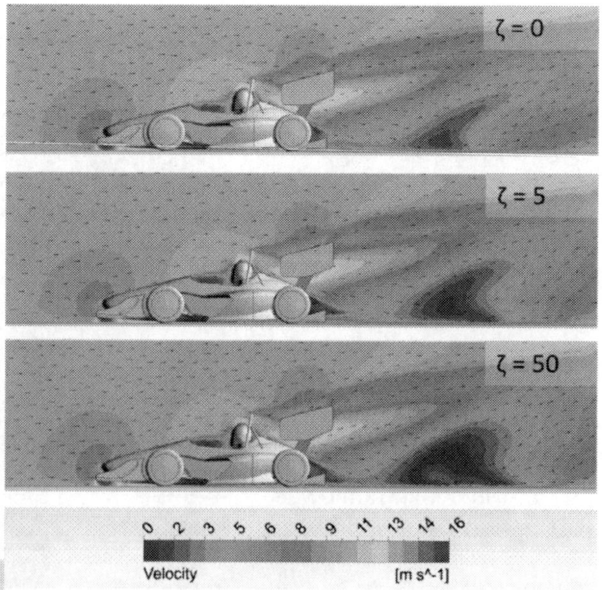

Results

Velocity in rear direction at radiator inlet

- Inhomogenous flow into the cooling unit
- Flow rate decreases with larger pressure loss, but becomes more even

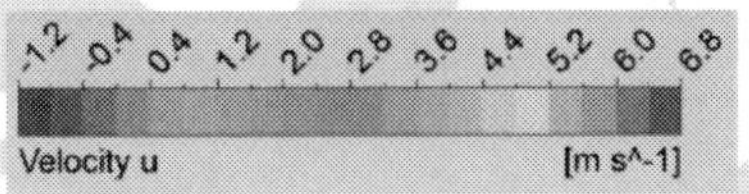

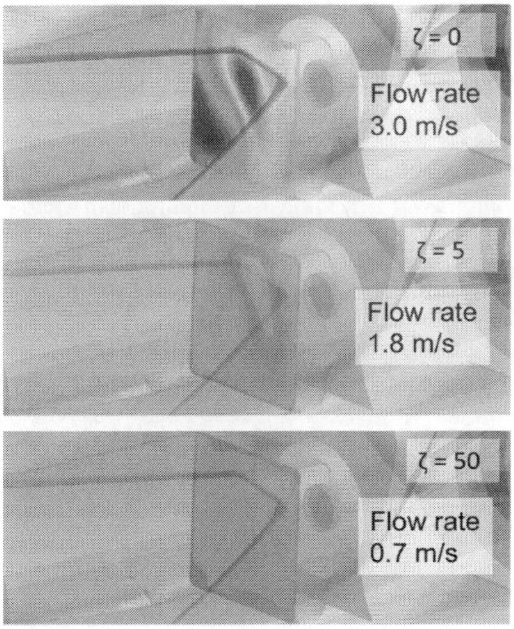

Results

Definition of aerodynamic forces and coefficients:

- Stagnation force $F_{stag} = A_x \frac{\rho}{2} v^2$
- Drag coefficient $c_d = {F_x}/{F_{stag}}$
- Lift coefficient $c_l = {F_z}/{F_{stag}}$
- Force on front axle $F_F = {M_{y,R}}/{L}$
- Force on rear axle $F_R = {M_{y,F}}/{L}$

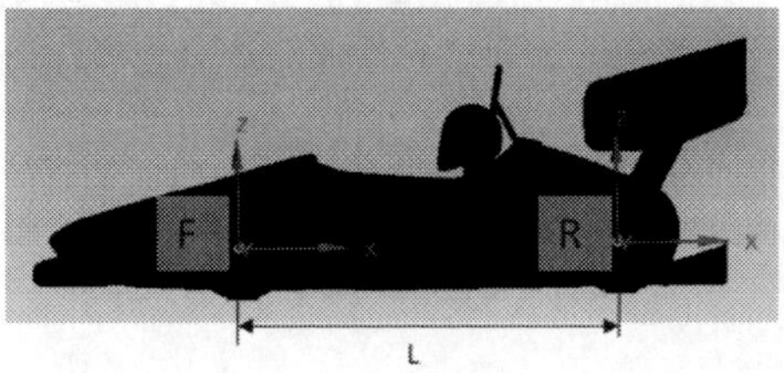

Results

CFX BERLIN HAWKS RACING

Pressure loss coefficient ζ of cooling unit	0	5	50
Mean flow rate [m/s]	3.0	1.8	0.7
Drag coefficient	1.13	1.12	1.11
Lift coefficient	-2.32	-2.21	-2.19
Aerodynamic force on front axle [N]	72.48	72.12	71.24
Aerodynamic force on rear axle [N]	83.02	75.38	74.60

Results

CFX BERLIN HAWKS RACING

Reason for decreased downforce:

- Lower flow rate of cooling unit also leads to less velocity in underbody region
- Decreased downforce output of diffusor

ζ = 0 ζ = 50

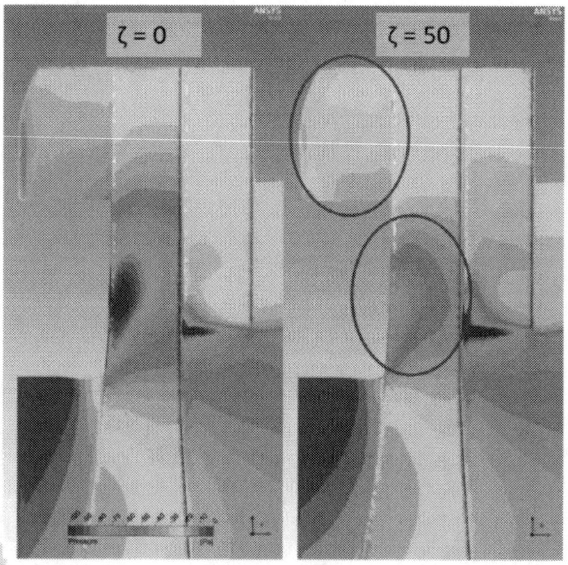

Results

CFX BERLIN | HAWKS RACING

The pressure loss coefficient of the cooling unit interacts with the aerodynamics of the car.

- Effect on resistance is negligible
- In case of ζ = 50 …
  - the downforce drops by 5.6 % compared to free flow through sidepod.
  - the force on the rear axle decreases disproportionately high by 10%.

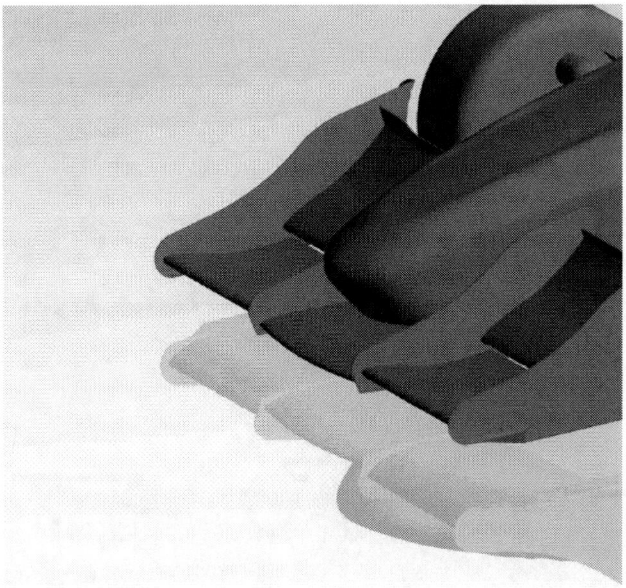

2013 Automotive Simulation World Congress

ANSYS Solutions in Competition: Formula Student

Tammo Grundmann, ANSYS Germany

Outline

- What is Formula Student?
- Formula Student in Europe
- ANSYS Sponsoring Formula Student Teams
- Increasing Speed

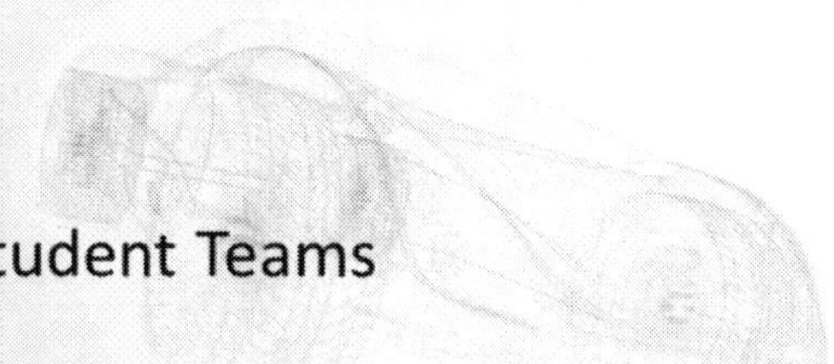

2013 Automotive Simulation World Congress
October 29 - 30 • Frankfurt am Main, Germany

What is Formula Student?

- Student Design Competition
- Building a Formula Race Car
- Fictional manufacturing company

2013 Automotive Simulation World Congress
October 29 - 30 • Frankfurt am Main, Germany

Judgement rules

Static Events	325
Engineering Design	150
Cost Analysis	100
Business Presentation	75
Dynamic Events	675
Skidpad	75
Acceleration	75
Autocross	100
Endurance	325
Fuel Economy	100
Total	1000

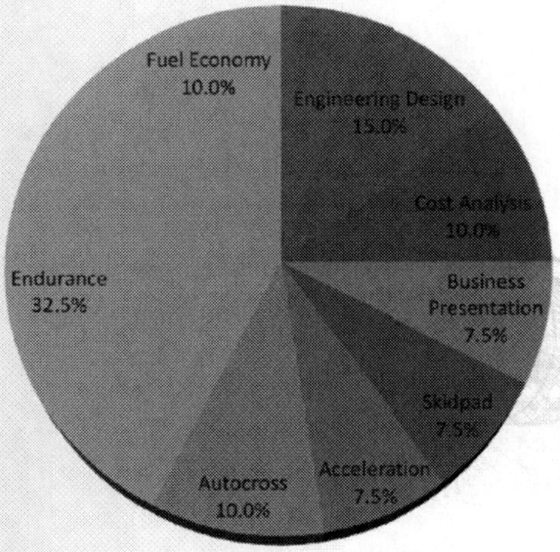

Formula Student in Europe

Events in 2013

- Formula Student UK
- Formula Student Germany
- Formula Student Italy
- Formula Student Austria
- Formula Student Hungary
- Formula Student Spain
- Formula Student Czech

Teams participating at Formula Student Germany

2013 Automotive Simulation World Congress
October 29 - 30 • Frankfurt am Main, Germany

Teams participating at Formula Student Germany

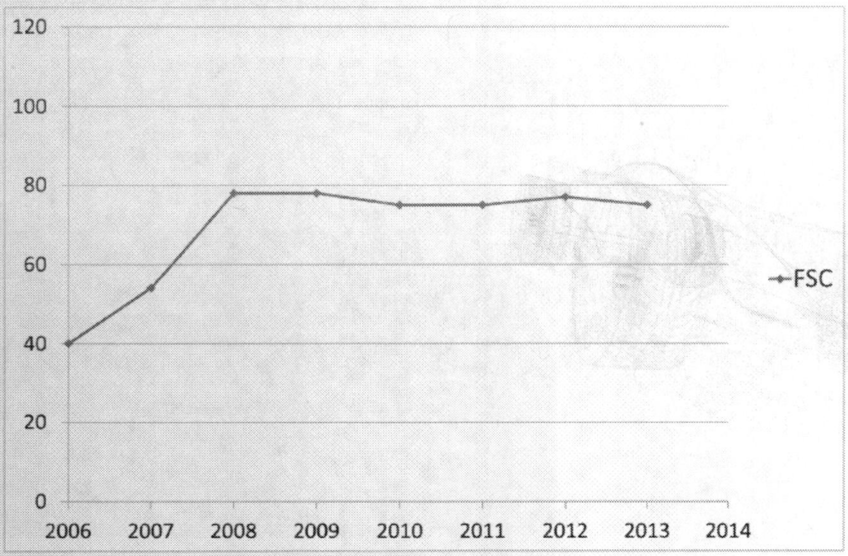

2013 Automotive Simulation World Congress
October 29 - 30 • Frankfurt am Main, Germany

Teams participating at Formula Student Germany

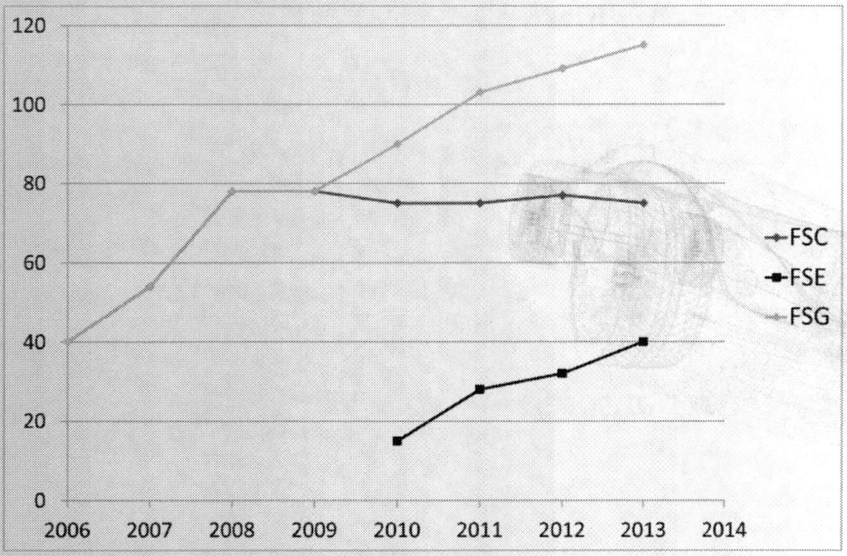

Formula Student Germany 2013

2013 Automotive Simulation World Congress
October 29 - 30 • Frankfurt am Main, Germany

ANSYS Sponsoring Formula Student Teams

- Sponsoring Teams with
  - Licenses
  - Kick-Off-Training
    - CFD
    - FEM
    - EM
  - Support
- 24 Teams sponsored

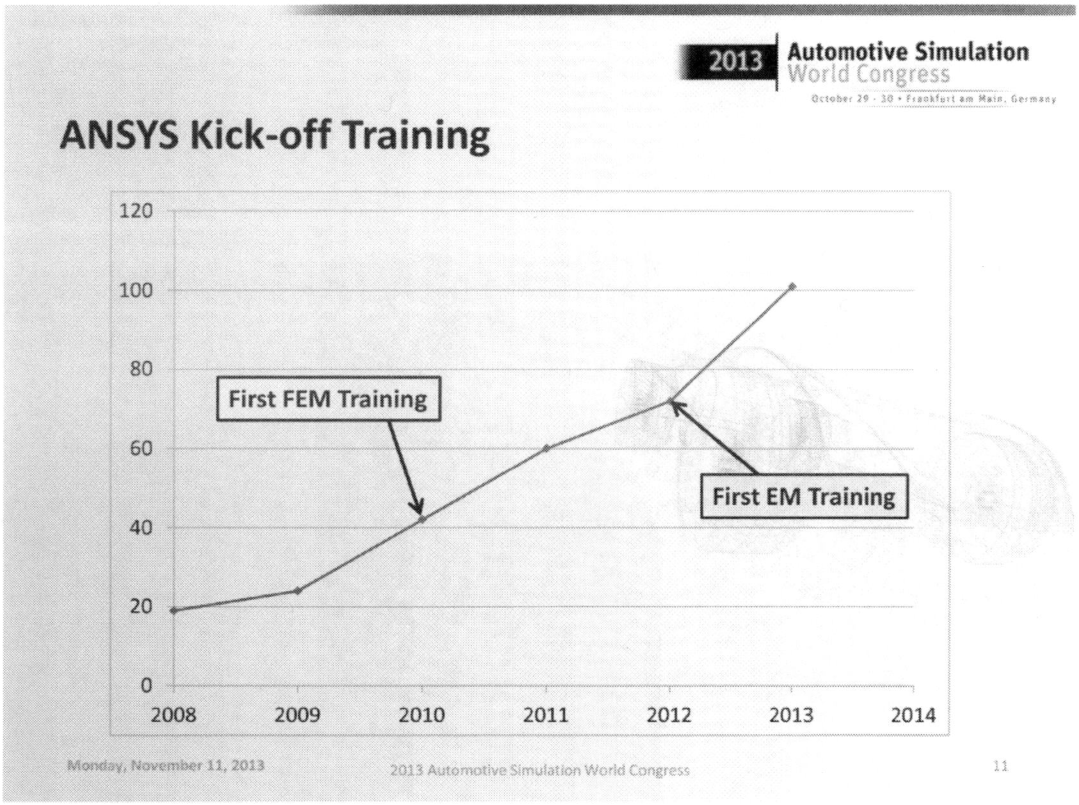
2013 Automotive Simulation World Congress
October 29 - 30 • Frankfurt am Main, Germany
ANSYS Kick-off Training
120
100
80
60
40
20
0
2008 2009 2010 2011 2012 2013 2014
First FEM Training
First EM Training
Monday, November 11, 2013
2013 Automotive Simulation World Congress
11

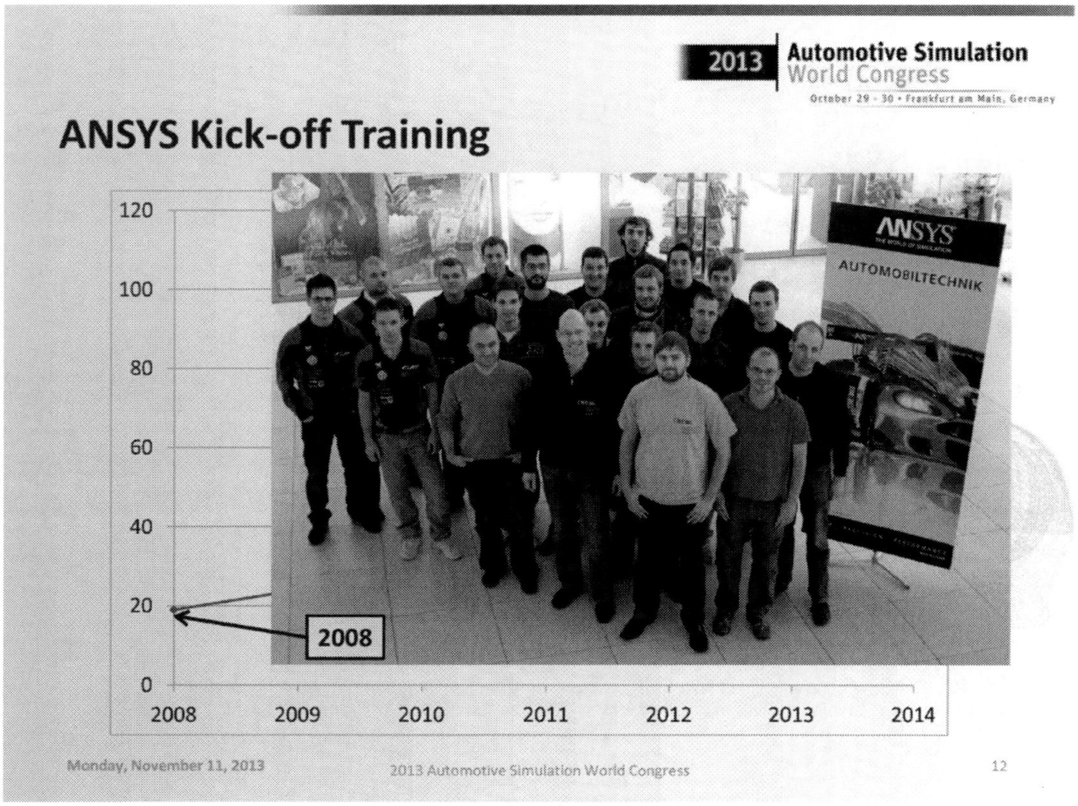
2013 Automotive Simulation World Congress
October 29 - 30 • Frankfurt am Main, Germany
ANSYS Kick-off Training
120
100
80
60
40
20
0
2008 2009 2010 2011 2012 2013 2014
2008
ANSYS
AUTOMOBILTECHNIK
Monday, November 11, 2013
2013 Automotive Simulation World Congress
12

2013 Automotive Simulation World Congress
October 29 - 30 • Frankfurt am Main, Germany

ANSYS Kick-off Training

Monday, November 11, 2013 2013 Automotive Simulation World Congress 13

2013 Automotive Simulation World Congress
October 29 - 30 • Frankfurt am Main, Germany

ANSYS Kick-off Training

Monday, November 11, 2013 2013 Automotive Simulation World Congress 14

2013 **Automotive Simulation** World Congress
October 29 - 30 • Frankfurt am Main, Germany

ANSYS Kick-off Training

Monday, November 11, 2013 2013 Automotive Simulation World Congress 15

2013 **Automotive Simulation** World Congress
October 29 - 30 • Frankfurt am Main, Germany

ANSYS Kick-off Training

Monday, November 11, 2013 2013 Automotive Simulation World Congress 16

2013 Automotive Simulation World Congress
October 29 - 30 • Frankfurt am Main, Germany

ANSYS Kick-off Training

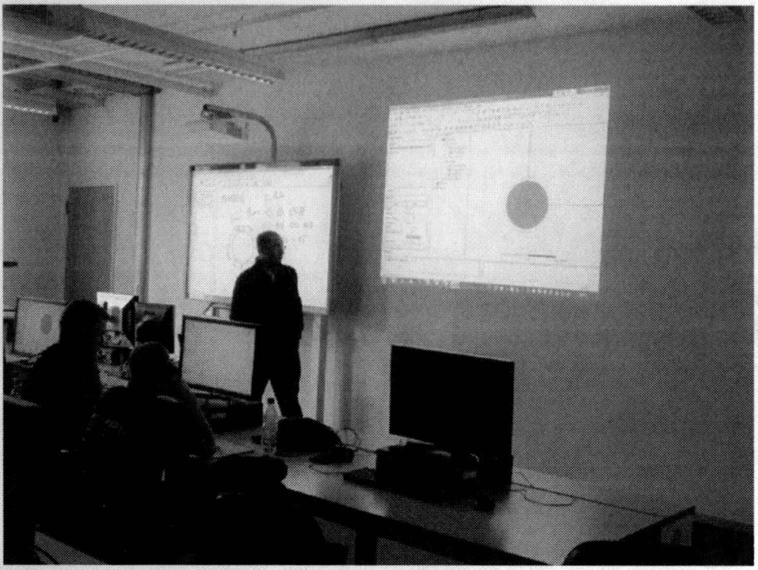

2013 Automotive Simulation World Congress
October 29 - 30 • Frankfurt am Main, Germany

A Season in Formula Student

September
October
November
December
- Form the team
- Design and construction period

→ Frontloading with heavy use of simulation tools

January
February
March
April
- Building the car
- Roll-out

May
June
- Testing
- Formula SAE,

July
August
- Formula Student events in Europe

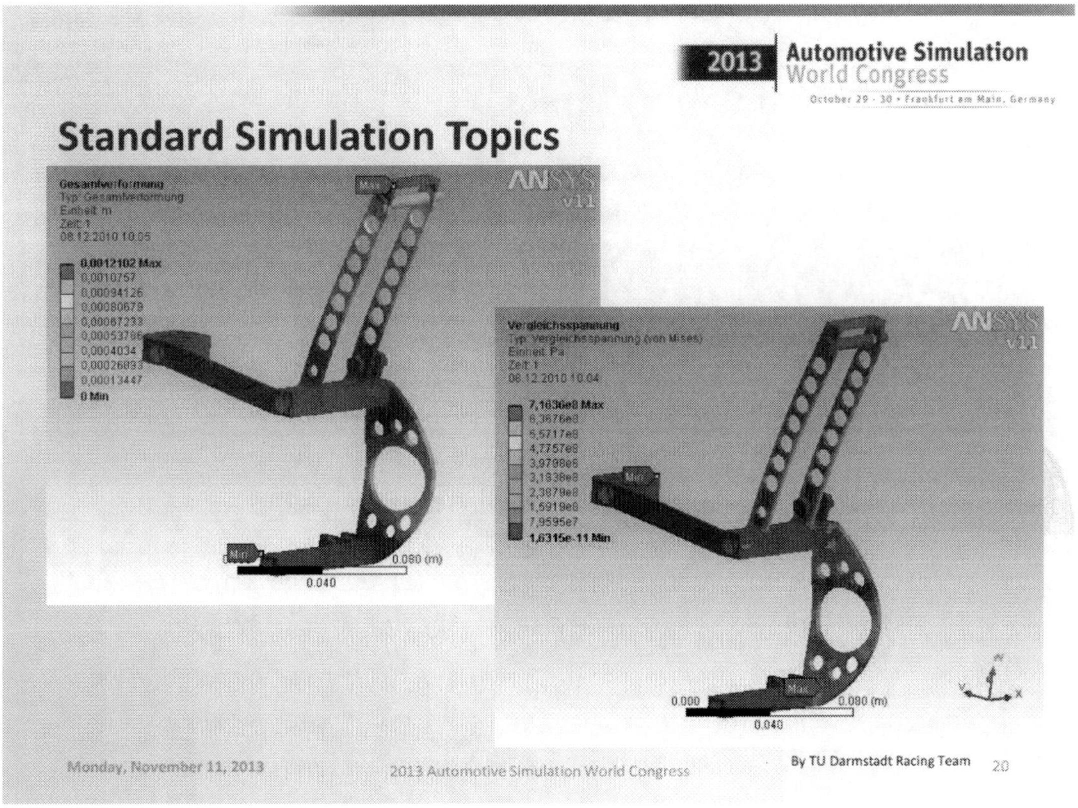
2013
Automotive Simulation
World Congress
October 29 - 30 • Frankfurt am Main, Germany
Standard Simulation Topics
Gesamtverformung
Typ: Gesamtverformung
Einheit: m
Zeit: 1
08.12.2010 10:05
0,0012102 Max
0,0010757
0,00094126
0,00080679
0,00067233
0,00053786
0,0004034
0,00026893
0,00013447
0 Min
Max
Min
0.080 (m)
0.040
Vergleichsspannung
Typ: Vergleichsspannung (von Mises)
Einheit: Pa
Zeit: 1
08.12.2010 10:04
7,1636e8 Max
6,3676e8
5,5717e8
4,7757e8
3,9798e8
3,1838e8
2,3879e8
1,5919e8
7,9595e7
1,6315e-11 Min
0.000
0.080 (m)
0.040
Monday, November 11, 2013
2013 Automotive Simulation World Congress
By TU Darmstadt Racing Team
20

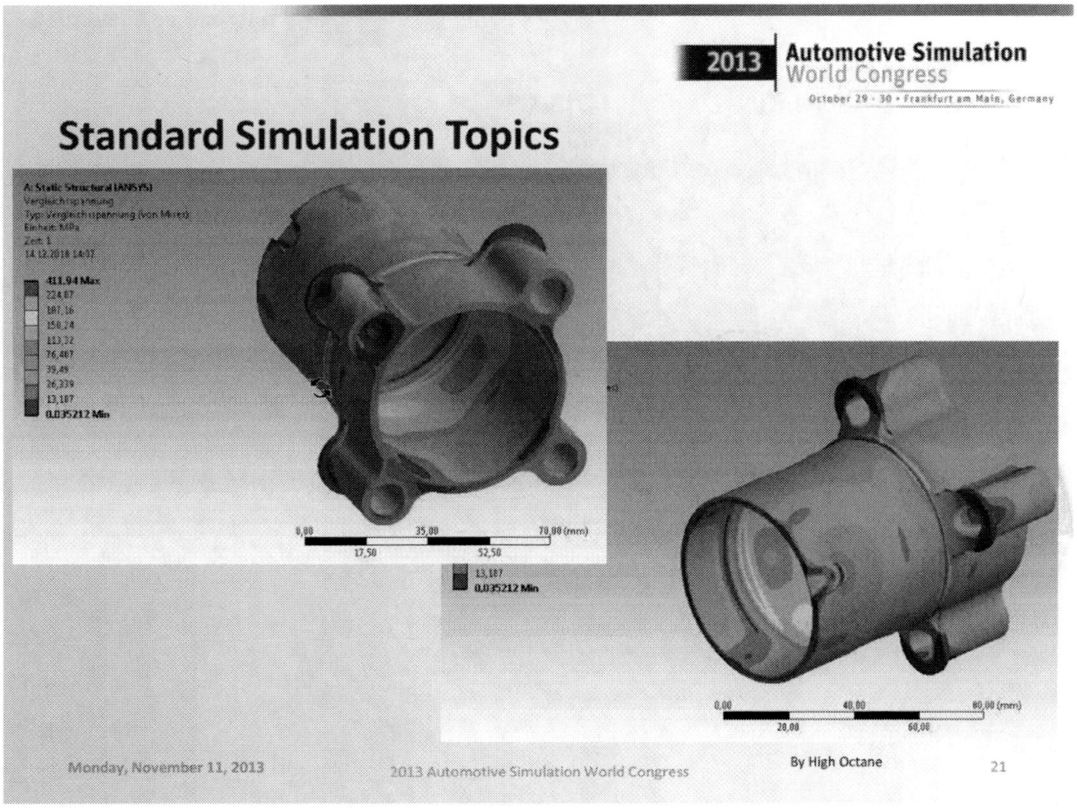
2013
Automotive Simulation
World Congress
October 29 - 30 • Frankfurt am Main, Germany
Standard Simulation Topics
A: Static Structural (ANSYS)
Vergleichsspannung
Typ: Vergleichsspannung (von Mises)
Einheit: MPa
Zeit: 1
411,94 Max
224,07
187,16
150,24
113,32
76,407
39,49
26,339
13,187
0,035212 Min
0,00
35,00
70,00 (mm)
17,50
52,50
0,00
40,00
80,00 (mm)
20,00
60,00
Monday, November 11, 2013
2013 Automotive Simulation World Congress
By High Octane
21

2013 Automotive Simulation World Congress
October 29 - 30 • Frankfurt am Main, Germany

Standard Simulation Topics

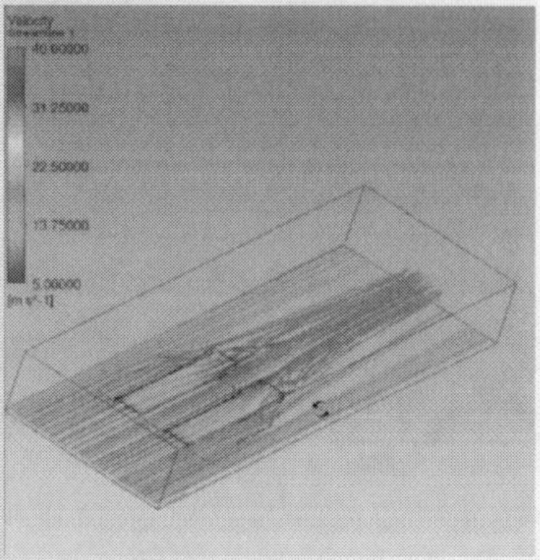

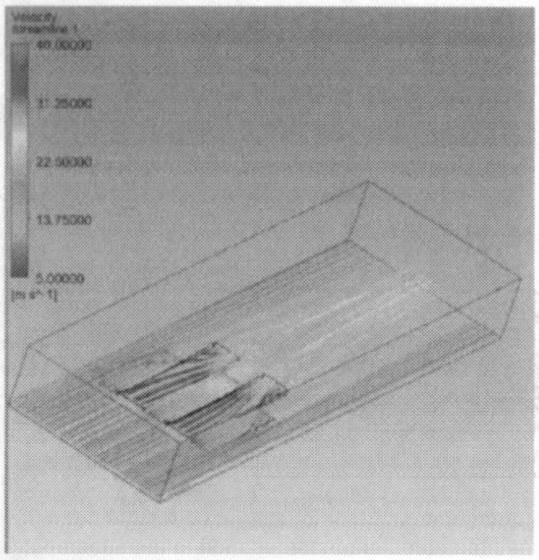

By Running Snail

2013 Automotive Simulation World Congress
October 29 - 30 • Frankfurt am Main, Germany

Standard Simulation Topics

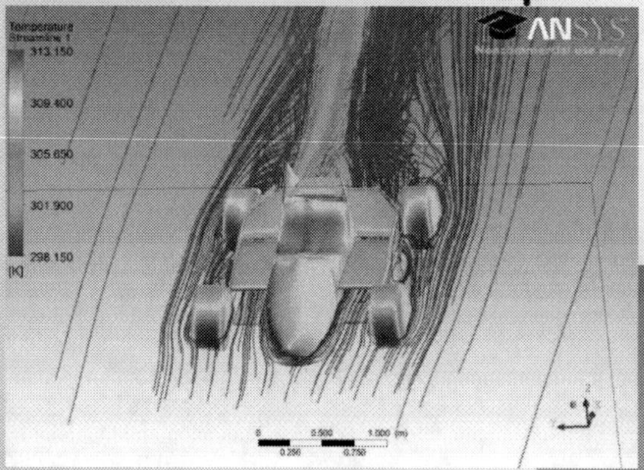

By FastDA Racing

Standard Simulation Topics

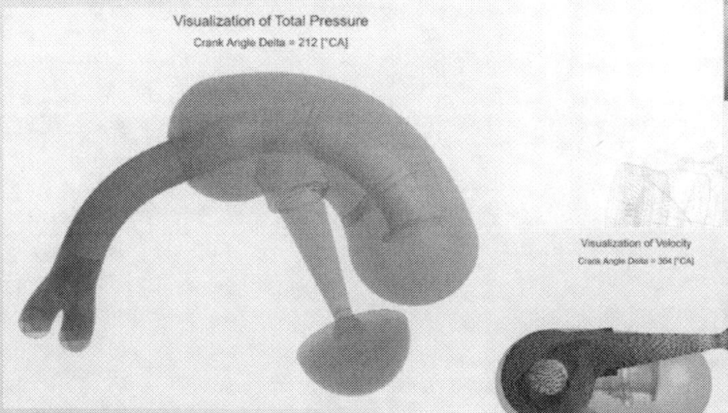

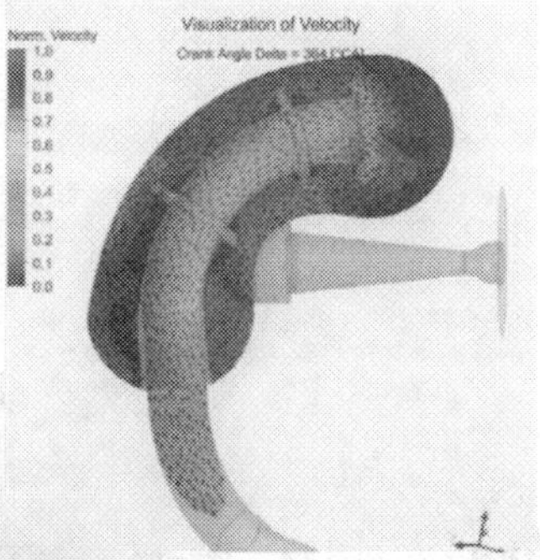

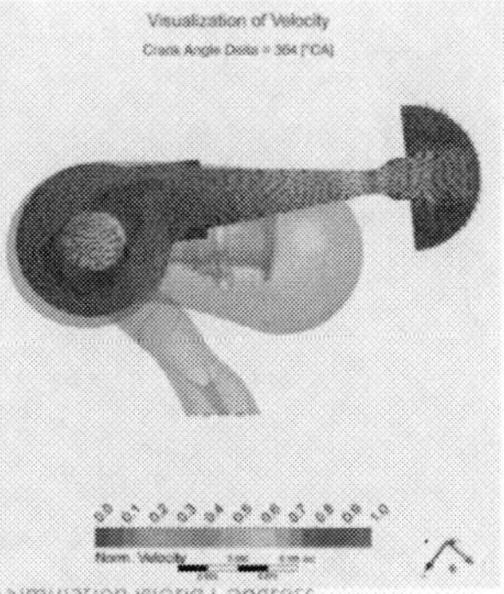

Standard Simulation Topics

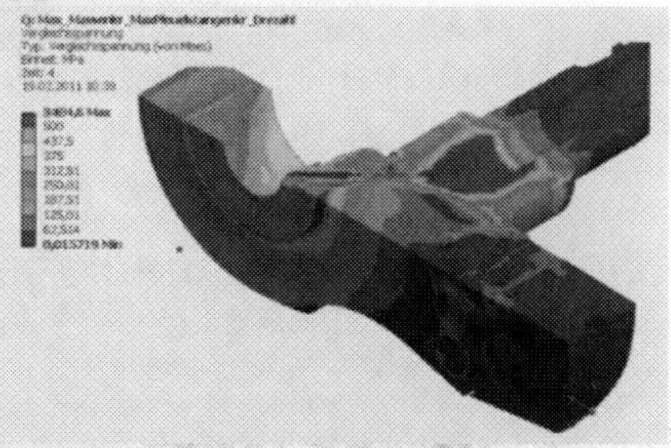

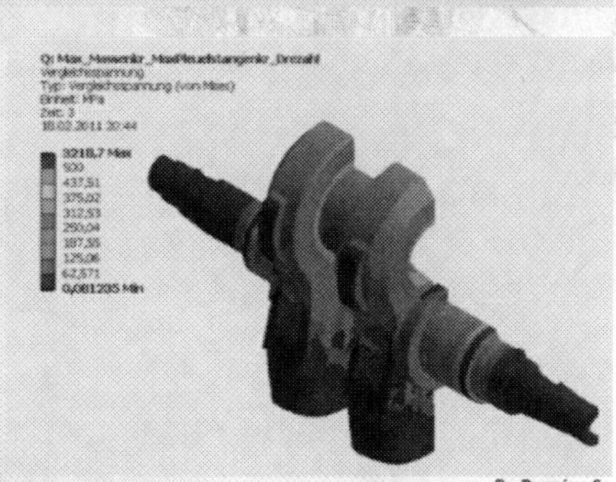

Advance Simulation Topics

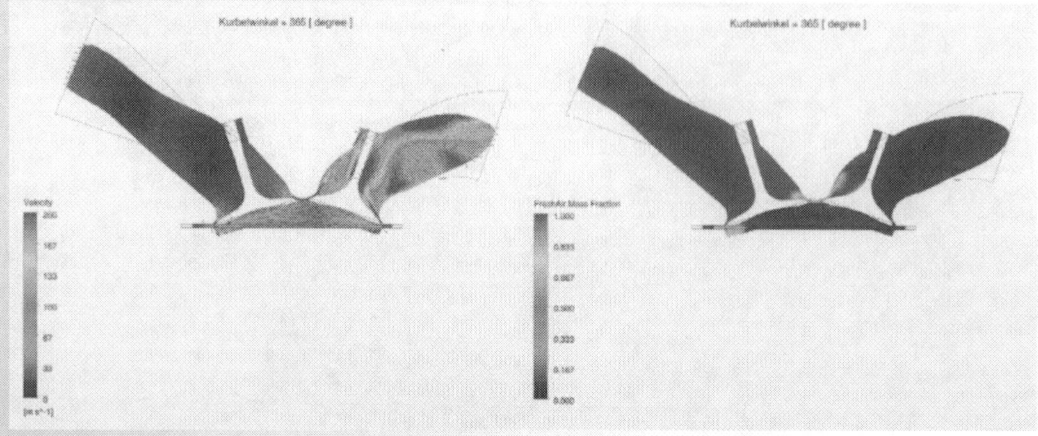

2013 Automotive Simulation World Congress
October 29 - 30 • Frankfurt am Main, Germany

Summary

- Highly motivated students
- Collecting experience in simulation
- Facing practical relevance of their simulations
- Focus on simulation technologies

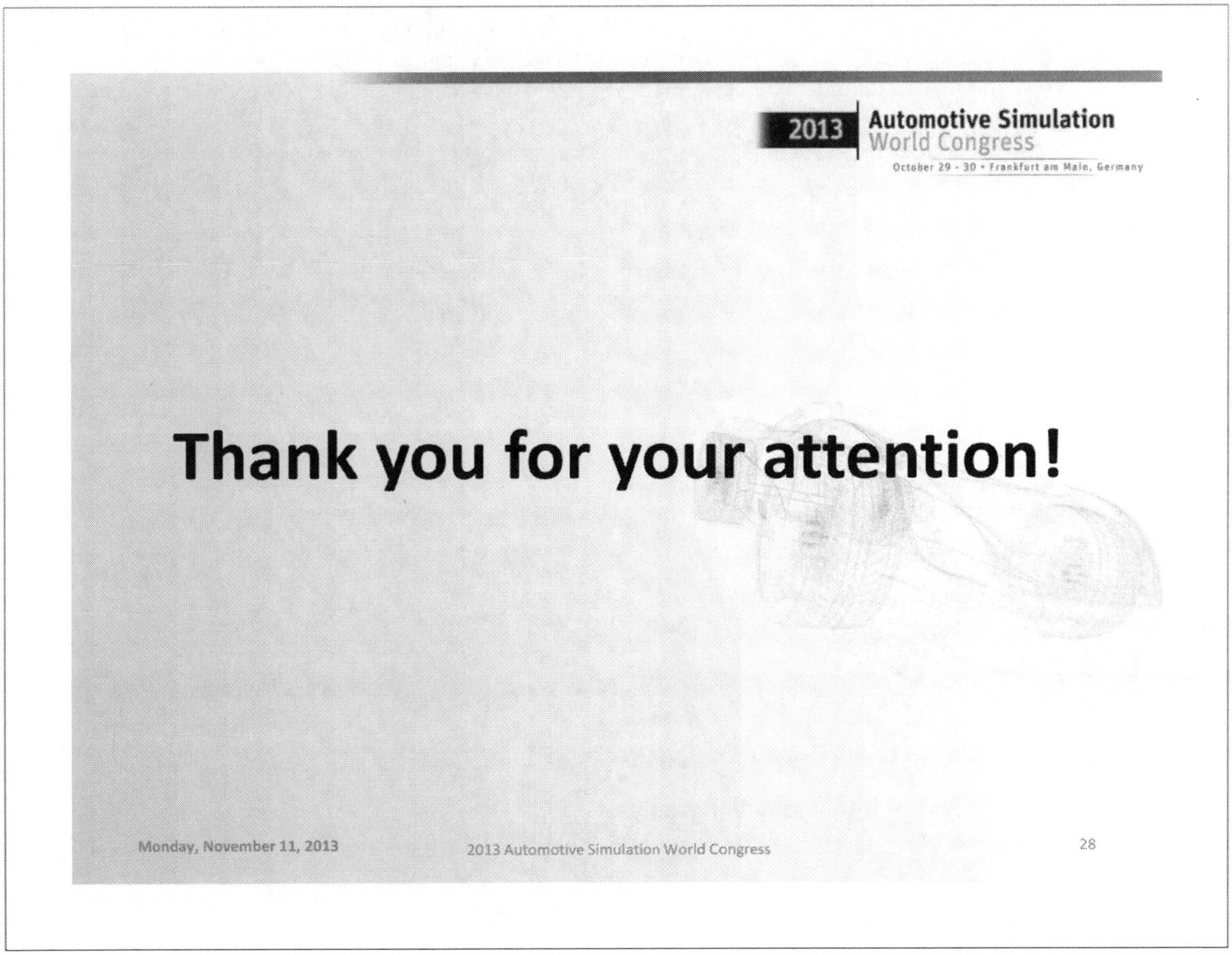
2013 Automotive Simulation World Congress
October 29 - 30 • Frankfurt am Main, Germany
Thank you for your attention!
Monday, November 11, 2013
2013 Automotive Simulation World Congress
28

UNIVERSITY OF CAMBRIDGE

Designing a revolutionary solar vehicle

How CFD enabled CUER to design a revolutionary solar vehicle in under a year

Richard Stephens

Contents

1. Introduction – CUER & The World Solar Challenge
2. Change of strategy & design process
3. Project Methodology and Analysis
4. Conclusion
5. Race Performance
6. Q&A

UNIVERSITY OF CAMBRIDGE

1. CUER and the World Solar Challenge

- CUER - Cambridge University Eco Racer
- World Solar Challenge
- Bring ecological technologies into prominence

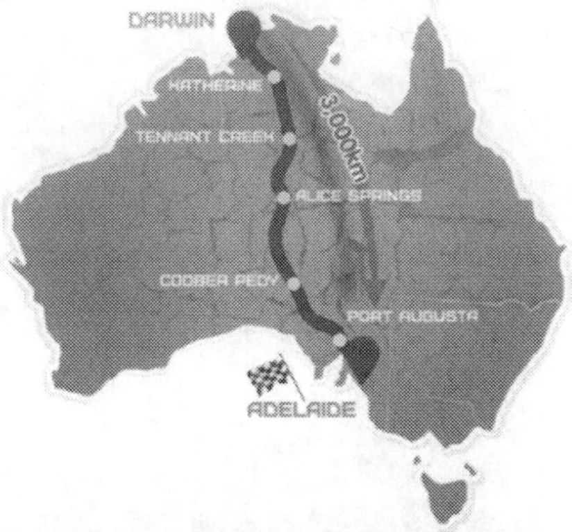

UNIVERSITY OF CAMBRIDGE

1. World Solar Challenge

Tokai University "Tokai Challenger"

Nuon "Nuna 6"

University of Michigan "Infinium"

UNIVERSITY OF CAMBRIDGE

1. World Solar Challenge

UNIVERSITY OF CAMBRIDGE

1. World Solar Challenge

- $6m^2$ silicon photovoltaic cells

OR

- $3m^2$ gallium arsenide photovoltaic cells

UNIVERSITY OF CAMBRIDGE

2. Change of Strategy

First optimise the aerodynamics of the vehicle before considering the solar performance

UNIVERSITY OF CAMBRIDGE

2. Design Process

- Starting completely from scratch
- Race rules released July 2012 - work began.

UNIVERSITY OF CAMBRIDGE

2. Design Process

UNIVERSITY OF CAMBRIDGE

2. Design Process

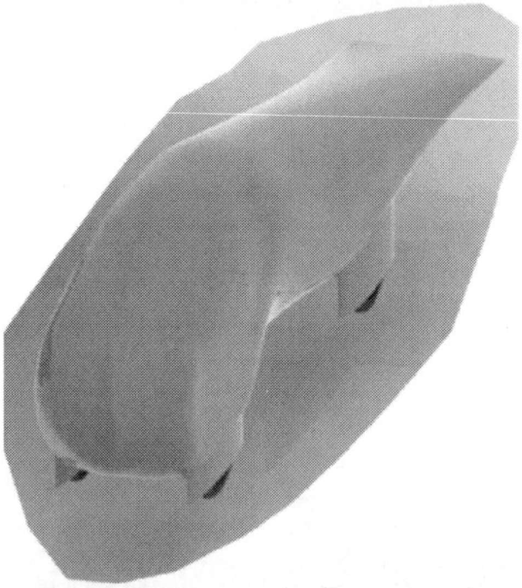

UNIVERSITY OF CAMBRIDGE

2. 2013 Concept Vehicle

UNIVERSITY OF CAMBRIDGE

2. Aerodynamic Optimisation

- Obtain value for the C_DA
- Seek ways of improving the C_DA to meet a target of $0.08m^2$
- 3 Month Deadline

UNIVERSITY OF CAMBRIDGE

3. Project methodology

Wind tunnel testing

- Key measurements: C_DA, C_LA.
- Flow visualisation: oil flow, tufts.

CFD Analysis

- ANSA used for meshing
- FLUENT used for solving flow

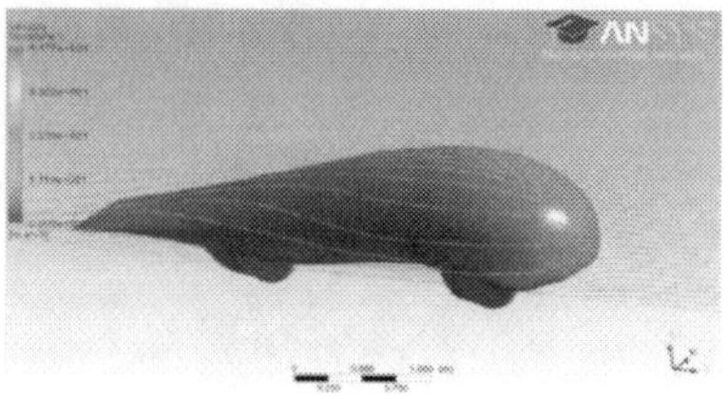

Aim: use combination of CFD and wind tunnel tests to accelerate design process

3. Mesh

- Combination of structured and unstructured mesh
- Mesh resolution study

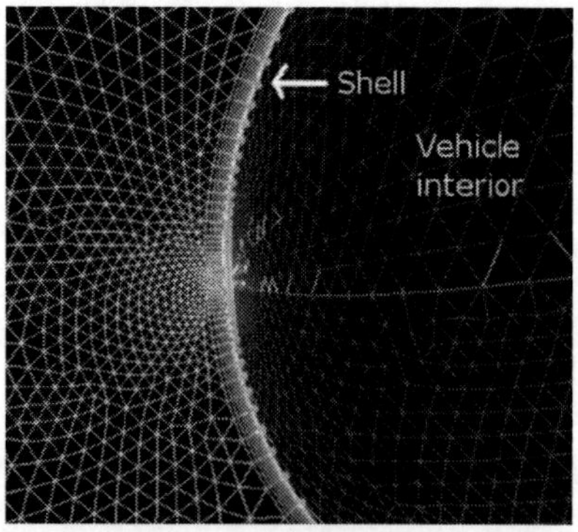

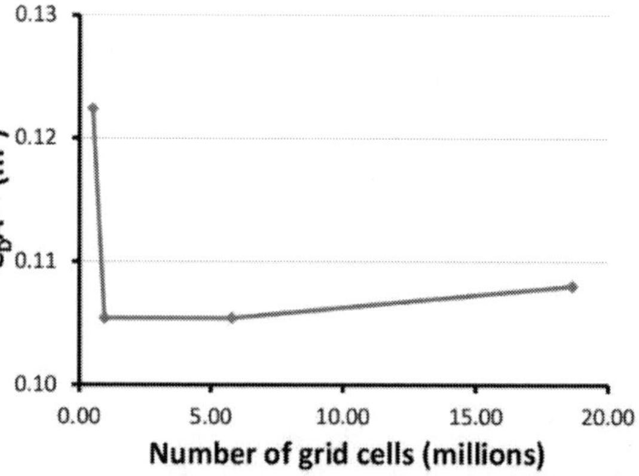

3. Mesh Resolution Study

- k-ε model used
- Rolling ground plane, stationary wheels
- Surface roughness modeled
- Solutions run until C_D, C_L converge

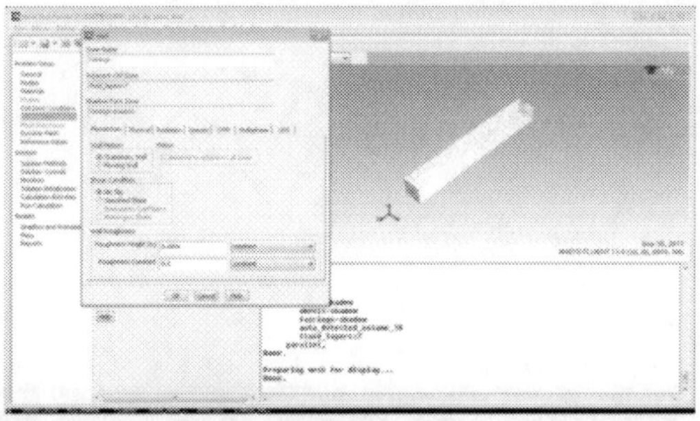

UNIVERSITY OF CAMBRIDGE

15

3. Project methodology

1. Analysis of main body design

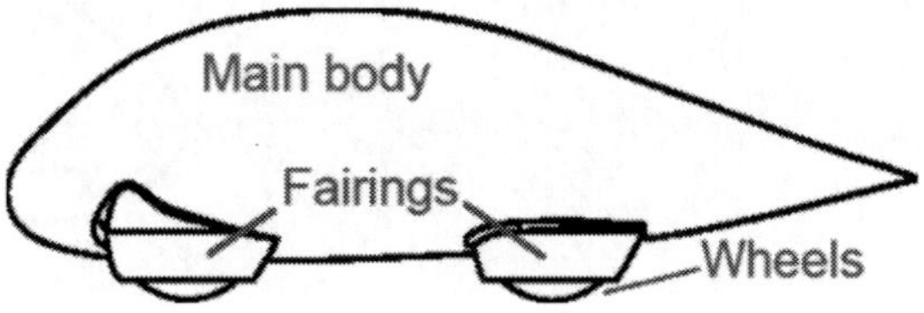

UNIVERSITY OF CAMBRIDGE

3. Project methodology

1. Analysis of main body design
2. Fairing design

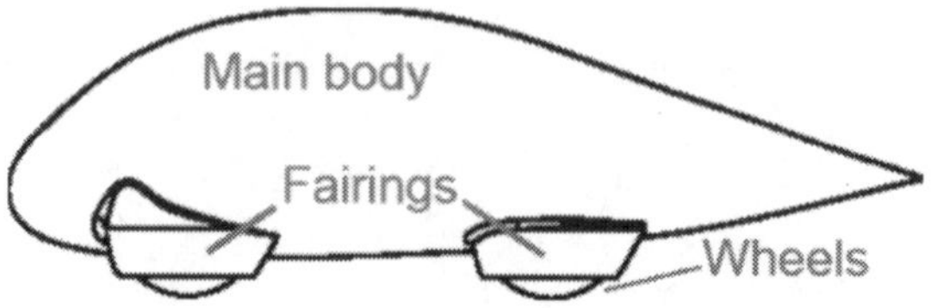

UNIVERSITY OF CAMBRIDGE

3. Project methodology

1. Analysis of main body design
2. Fairing design
3. Cross-wind analysis

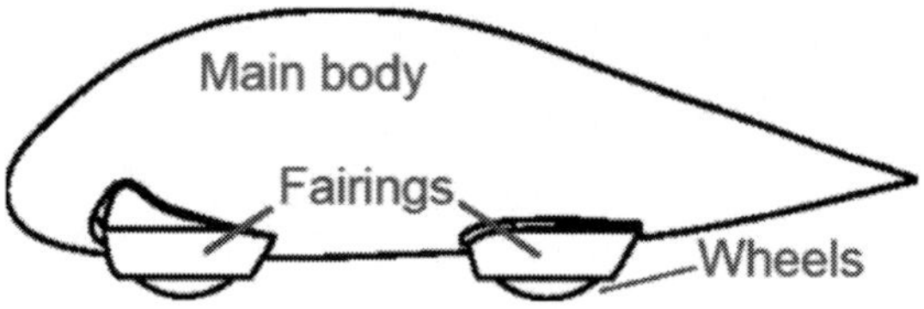

UNIVERSITY OF CAMBRIDGE

3. Main Body Analysis

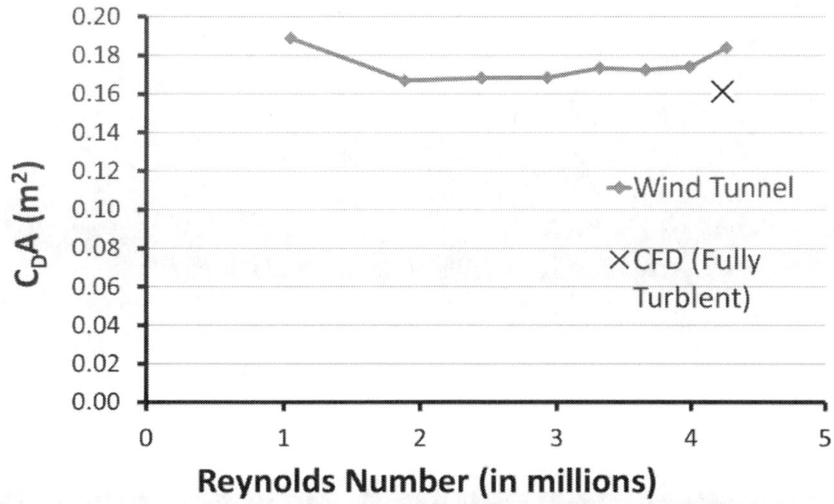

UNIVERSITY OF CAMBRIDGE

19

3. Main Body Analysis

- Fine adjustment with careful consideration of rules
- 6 CFD iterations per week
- Total of about 30 CFD iterations

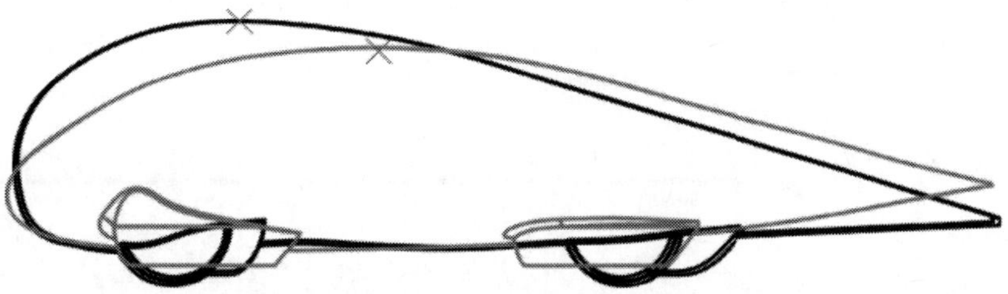

UNIVERSITY OF CAMBRIDGE

20

3. Main Body Analysis

Final design of main body, **without** fairings

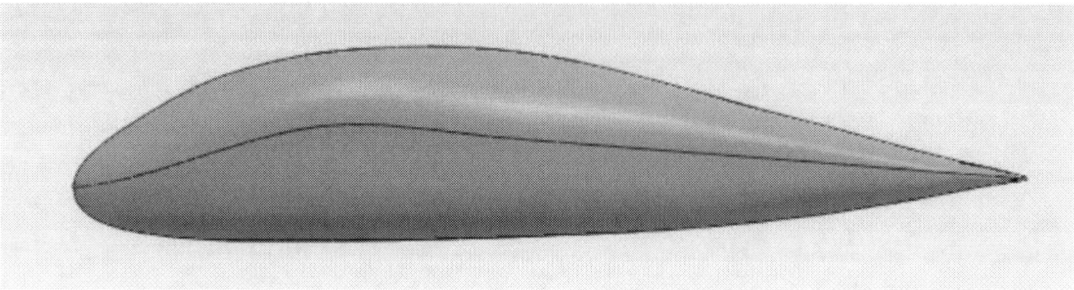

3. Main Body Analysis

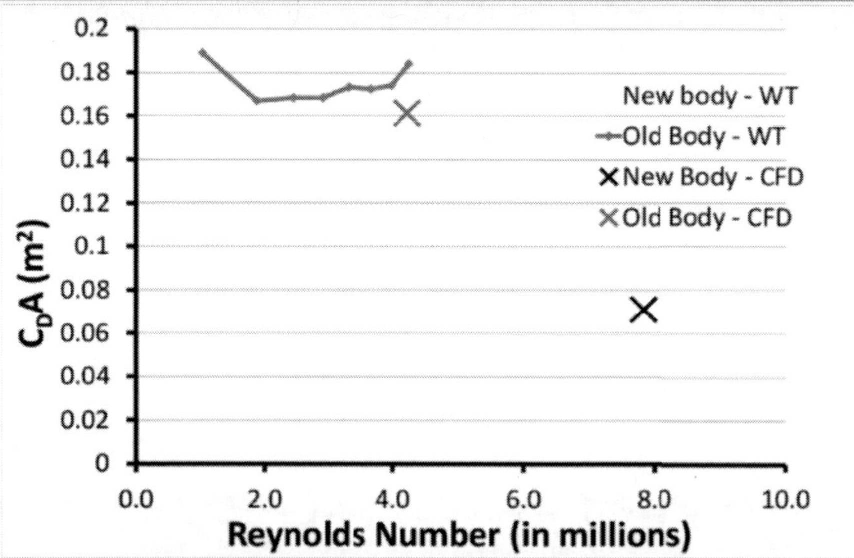

3. Main Body Analysis

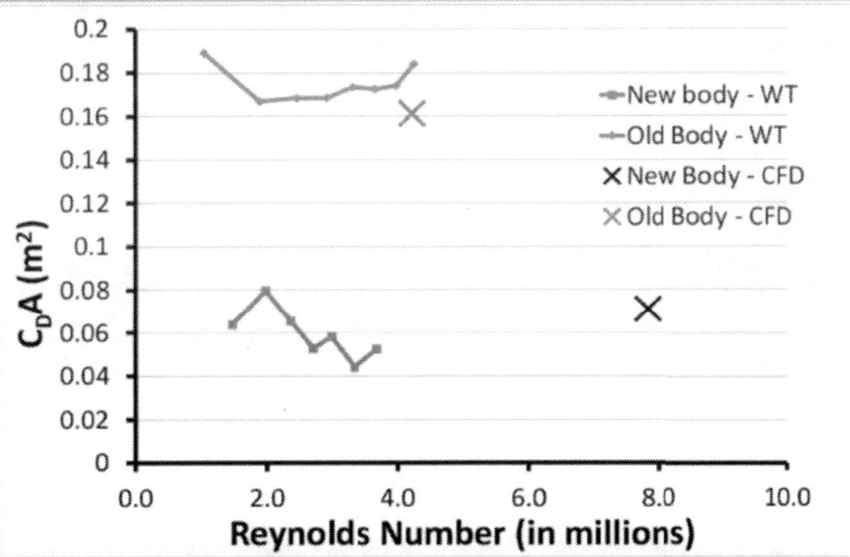

UNIVERSITY OF CAMBRIDGE

22

3. Fairing Design

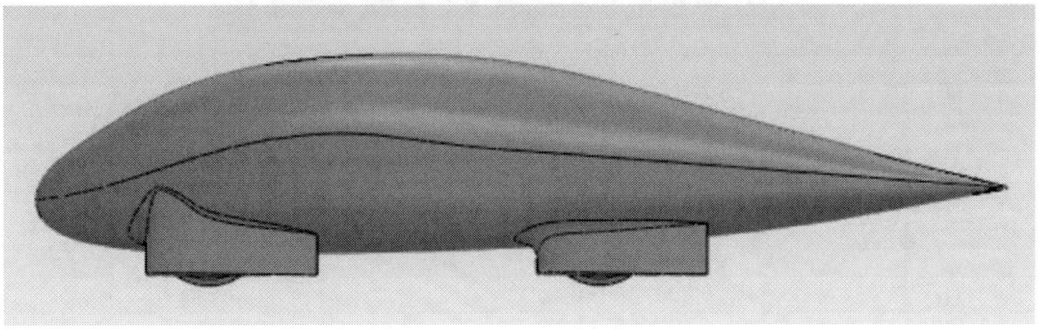

UNIVERSITY OF CAMBRIDGE

3. Fairing Design

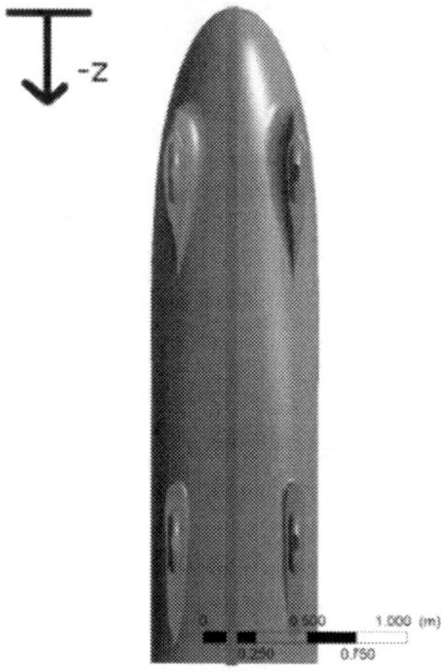

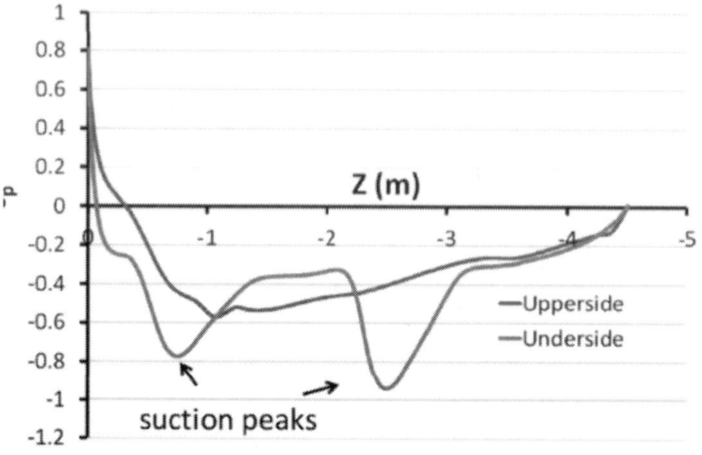

UNIVERSITY OF CAMBRIDGE

3. Fairing Design

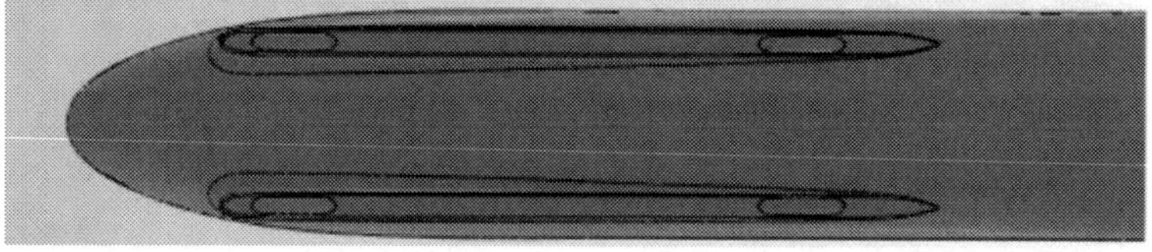

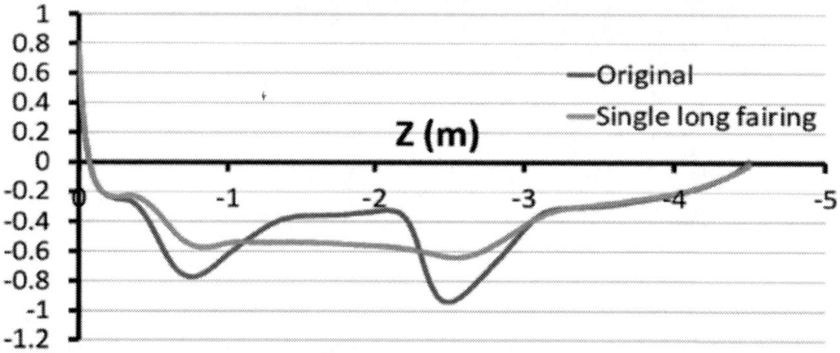

UNIVERSITY OF CAMBRIDGE

3. Fairing Design

Drag line

3. Fairing Design – DESIGN A

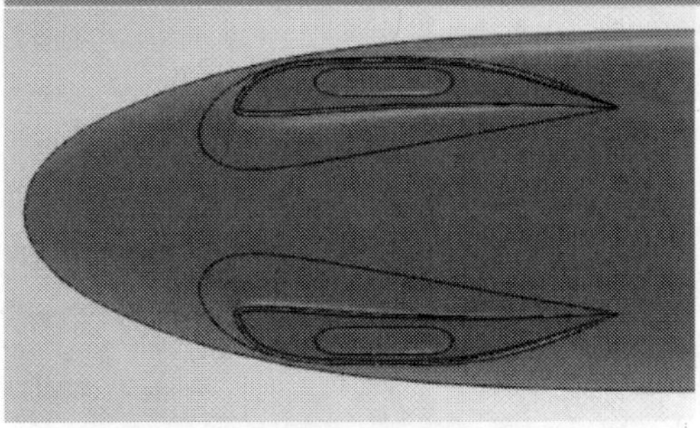

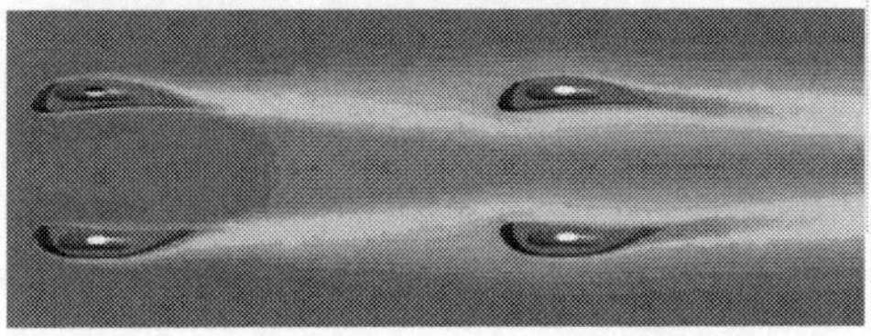

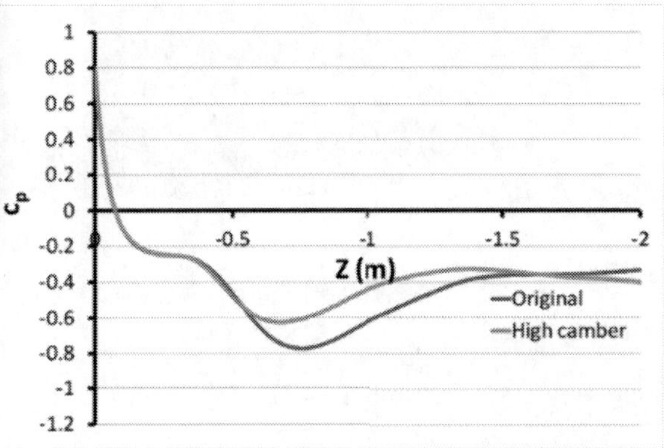

3. Fairing Design – DESIGN A

Drag line

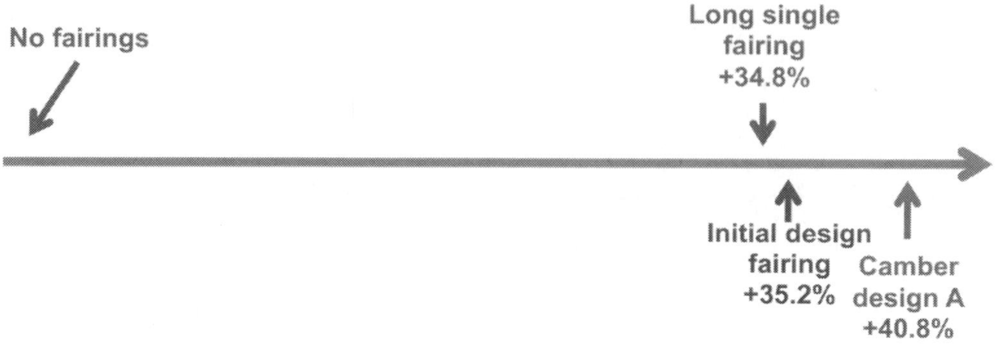

UNIVERSITY OF CAMBRIDGE

3. Fairing Design – DESIGN B

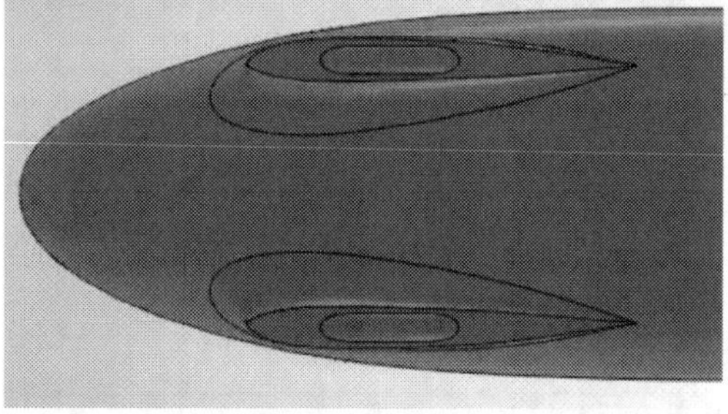

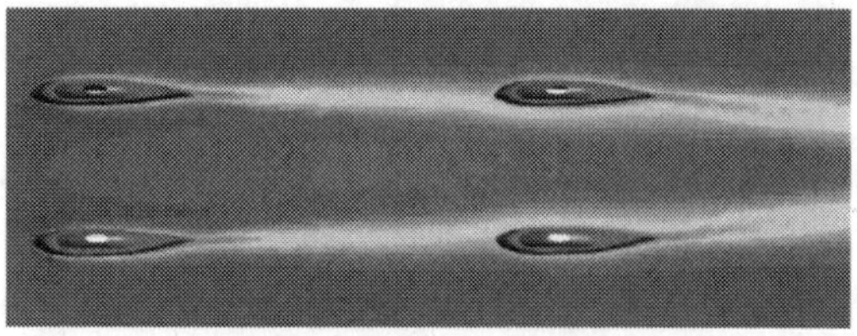

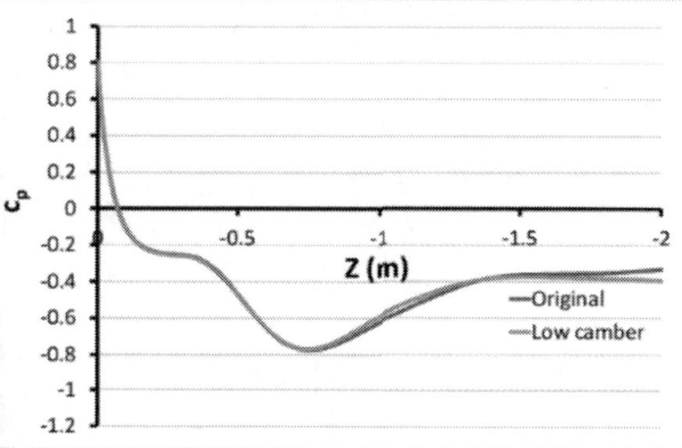

UNIVERSITY OF CAMBRIDGE

3. Fairing Design – DESIGN B

Drag line

No fairings

Long single fairing +34.8%

Camber design B +33.5%

Initial design fairing +35.2%

Camber design A +40.8%

UNIVERSITY OF CAMBRIDGE

3. Fairing Design

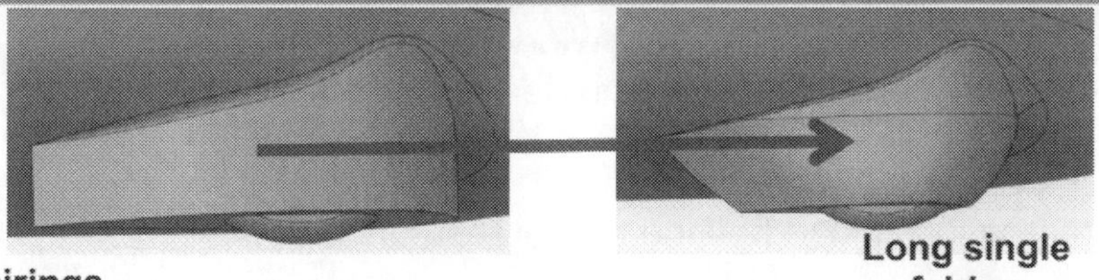

No fairings

Long single fairing +34.8%

Camber design B *with sweep* +28.1%

Camber design B +33.5%

Initial design fairing +35.2%

Camber design A +40.8%

UNIVERSITY OF CAMBRIDGE

3. Fairing Design

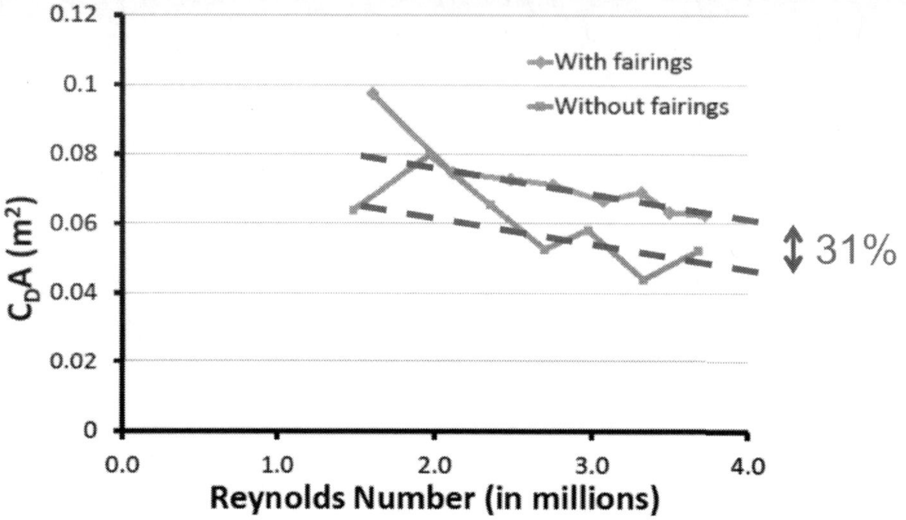

UNIVERSITY OF CAMBRIDGE

4. Conclusion

1. Using a combination of CFD and wind-tunnel tests allowed CUER to make rapid progress in the vehicle aerodynamics
2. CFD gave results close to wind tunnel results
3. Vehicle had capability of winning race…

UNIVERSITY OF CAMBRIDGE

5: Race Performance

- At the time of submission, the race had not yet taken place... Details will be added here for presentation at the conference.

UNIVERSITY OF CAMBRIDGE

Thank you for listening

UNIVERSITY OF CAMBRIDGE

(rbf-morph)™
Welcome to the World of Fast Morphing!

2013 Automotive Simulation World Congress

Advanced Mesh Morphing for Automotive Applications using RBF Morph

Marco Evangelos Biancolini
University of Rome Tor Vergata
info@rbf-morph.com
biancolini@ing.uniroma2.it

Tor Vergata

(rbf-morph)™

29-30 October 2013
Automotive Simulation World Congress

ANSYS
www.rbf-morph.com
RBF Morph, an ANSYS Inc. Partner

(rbf-morph)™
Welcome to the World of Fast Morphing!

2013 Automotive Simulation World Congress

Outline

- RBF Morph tool presentation
- Industrial Applications
- Advanced features

29-30 October 2013
Automotive Simulation World Congress

ANSYS
www.rbf-morph.com
RBF Morph, an ANSYS Inc. Partner

(rbf-morph)™
Welcome to the World of Fast Morphing!

2013 Automotive Simulation World Congress

RBF Morph tool presentation

29-30 October 2013
Automotive Simulation World Congress

www.rbf-morph.com
RBF Morph, an ANSYS Inc. Partner

ANSYS

(rbf-morph)™
Welcome to the World of Fast Morphing!

2013 Automotive Simulation World Congress

Morphing & Smoothing

- A mesh morpher is a tool capable to perform **mesh modifications**, in order to achieve arbitrary shape changes and related volume smoothing, without changing the mesh topology.
- In general a morphing operation can introduce a reduction of the **mesh quality**
- A **good** morpher has to minimize this effect, and maximize the possible shape modifications.
- If mesh quality is well preserved, then using the same mesh structure it's a **clear benefit** (remeshing introduces **noise**!).

29-30 October 2013
Automotive Simulation World Congress

www.rbf-morph.com
RBF Morph, an ANSYS Inc. Partner

ANSYS

RBF Morph Features

- **Add on** fully integrated within **Fluent** (GUI, TUI & solving stage) and **Workbench**
- **Mesh-independent** RBF fit used for surface mesh morphing and volume mesh smoothing
- **Parallel** calculation allows to morph **large size** models (many millions of cells) in a short time
- Management of **every kind of mesh** element type (tetrahedral, hexahedral, polyhedral, etc.)
- Support of the **CAD re-design** of the morphed surfaces
- **Multi fit** makes the Fluent case truly parametric (only 1 mesh is stored)
- **Precision**: exact nodal movement and exact feature preservation (**RBF** are better than **FFD**).

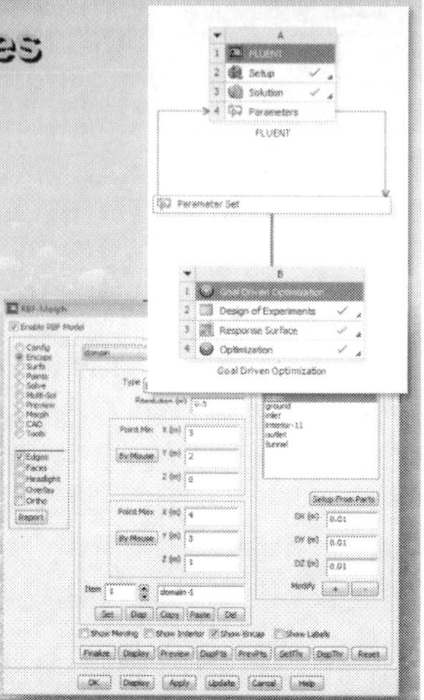

(rbf-morph)™
Welcome to the World of Fast Morphing!
2013 Automotive Simulation World Congress

Mesh Morphing with Radial Basis Functions

- A system of **radial functions** is used to **fit** a **solution** for the mesh movement/morphing, from a list of **source points** and their **displacements**.
- The RBF problem definition does not depend on the mesh
- Radial Basis Function interpolation is used to derive the displacement in **any location** in the space, each component of the displacement is interpolated:

$$\begin{cases} v_x = s_x(\mathbf{x}) = \sum_{i=1}^{N} \gamma_i^x \phi\left(\left\|\mathbf{x} - \mathbf{x}_{k_i}\right\|\right) + \beta_1^x + \beta_2^x x + \beta_3^x y + \beta_4^x z \\ v_y = s_y(\mathbf{x}) = \sum_{i=1}^{N} \gamma_i^y \phi\left(\left\|\mathbf{x} - \mathbf{x}_{k_i}\right\|\right) + \beta_1^y + \beta_2^y x + \beta_3^y y + \beta_4^y z \\ v_z = s_z(\mathbf{x}) = \sum_{i=1}^{N} \gamma_i^z \phi\left(\left\|\mathbf{x} - \mathbf{x}_{k_i}\right\|\right) + \beta_1^z + \beta_2^z x + \beta_3^z y + \beta_4^z z \end{cases}$$

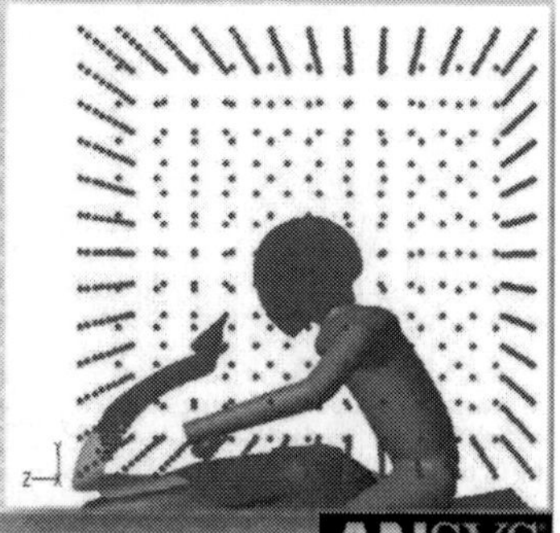

29-30 October 2013
Automotive Simulation World Congress
www.rbf-morph.com
ANSYS
RBF Morph, an ANSYS Inc. Partner

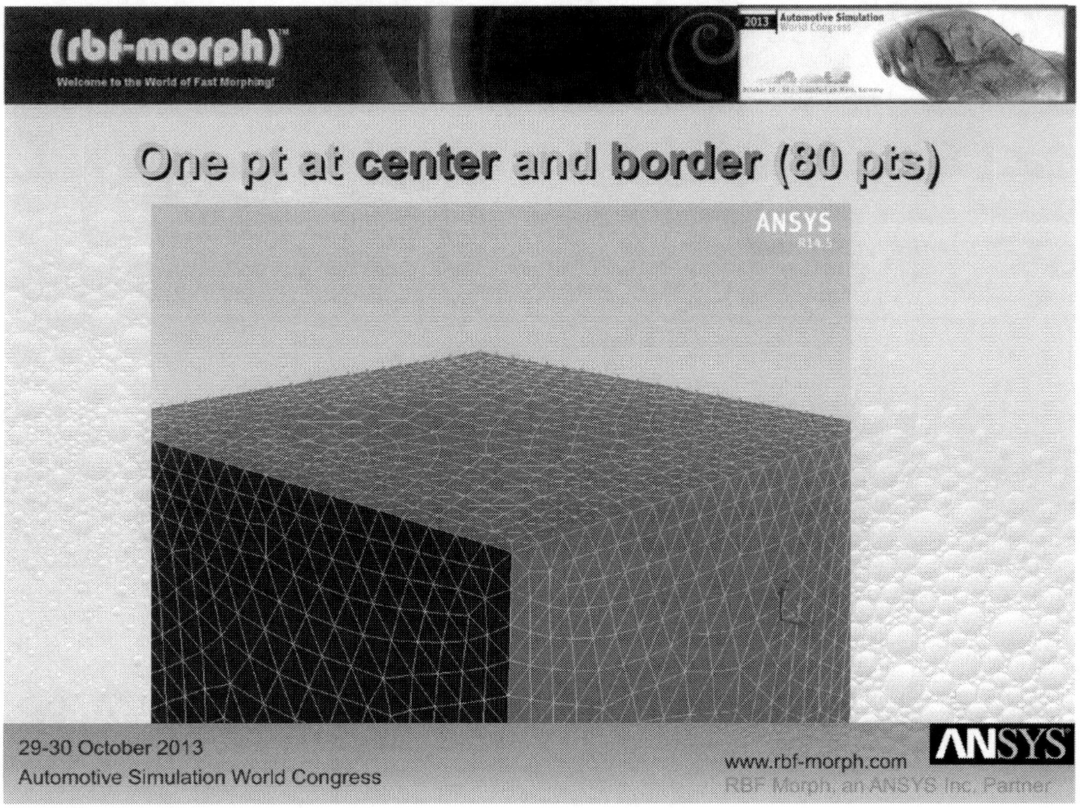
(rbf-morph)™
Welcome to the World of Fast Morphing!
2013 Automotive Simulation World Congress
One pt at center and border (80 pts)
ANSYS
R14.5
29-30 October 2013
Automotive Simulation World Congress
www.rbf-morph.com
RBF Morph, an ANSYS Inc. Partner
ANSYS

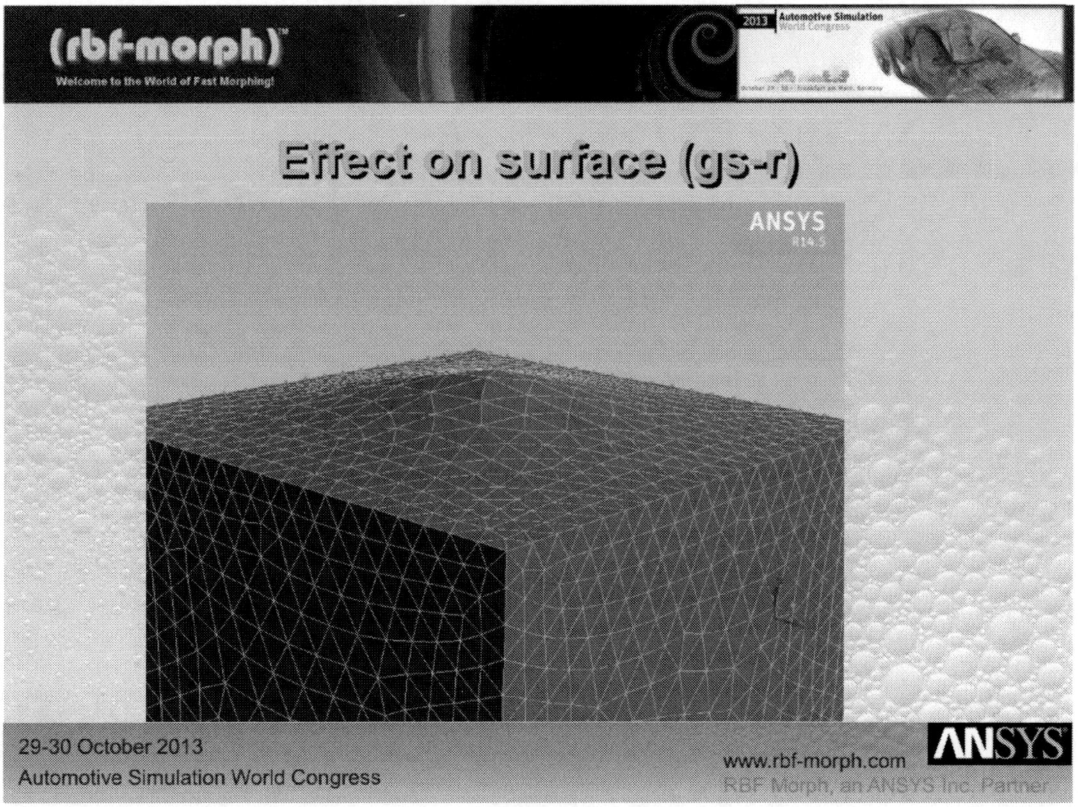
(rbf-morph)™
Welcome to the World of Fast Morphing!
2013 Automotive Simulation World Congress
Effect on surface (gs-r)
ANSYS
R14.5
29-30 October 2013
Automotive Simulation World Congress
www.rbf-morph.com
RBF Morph, an ANSYS Inc. Partner
ANSYS

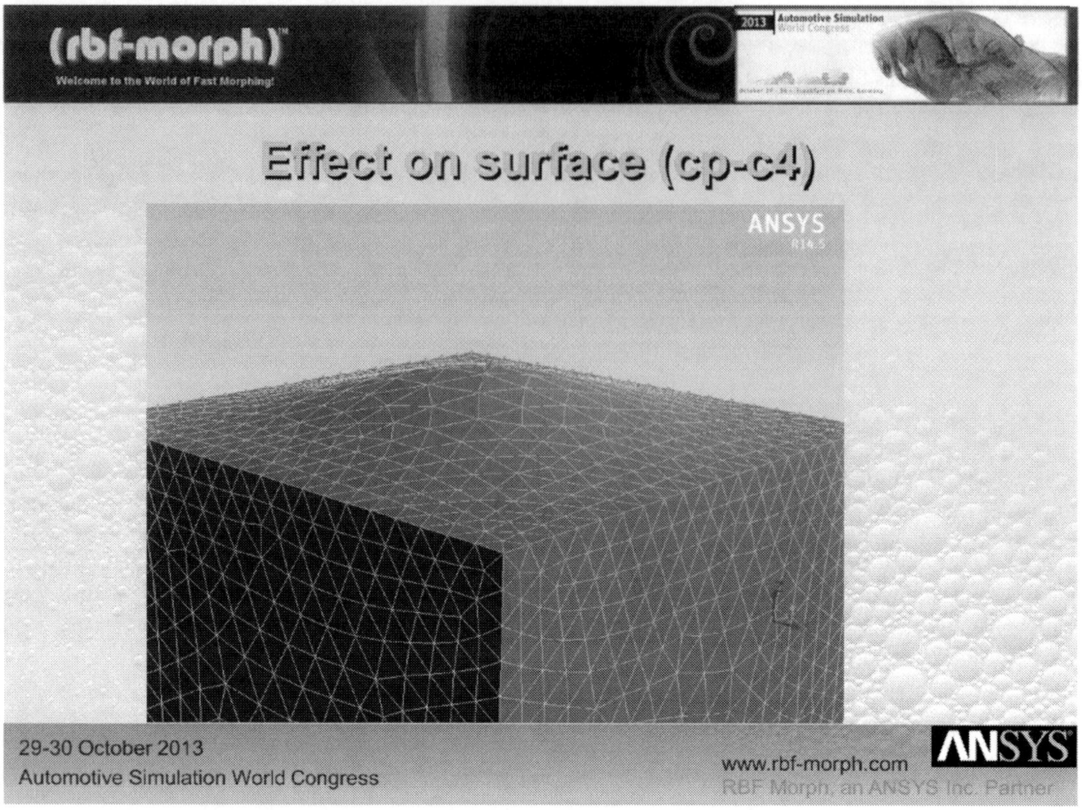
(rbf-morph)™
Welcome to the World of Fast Morphing!
2013 Automotive Simulation World Congress
Effect on surface (cp-c4)
ANSYS
R14.5
29-30 October 2013
Automotive Simulation World Congress
www.rbf-morph.com
RBF Morph, an ANSYS Inc. Partner
ANSYS

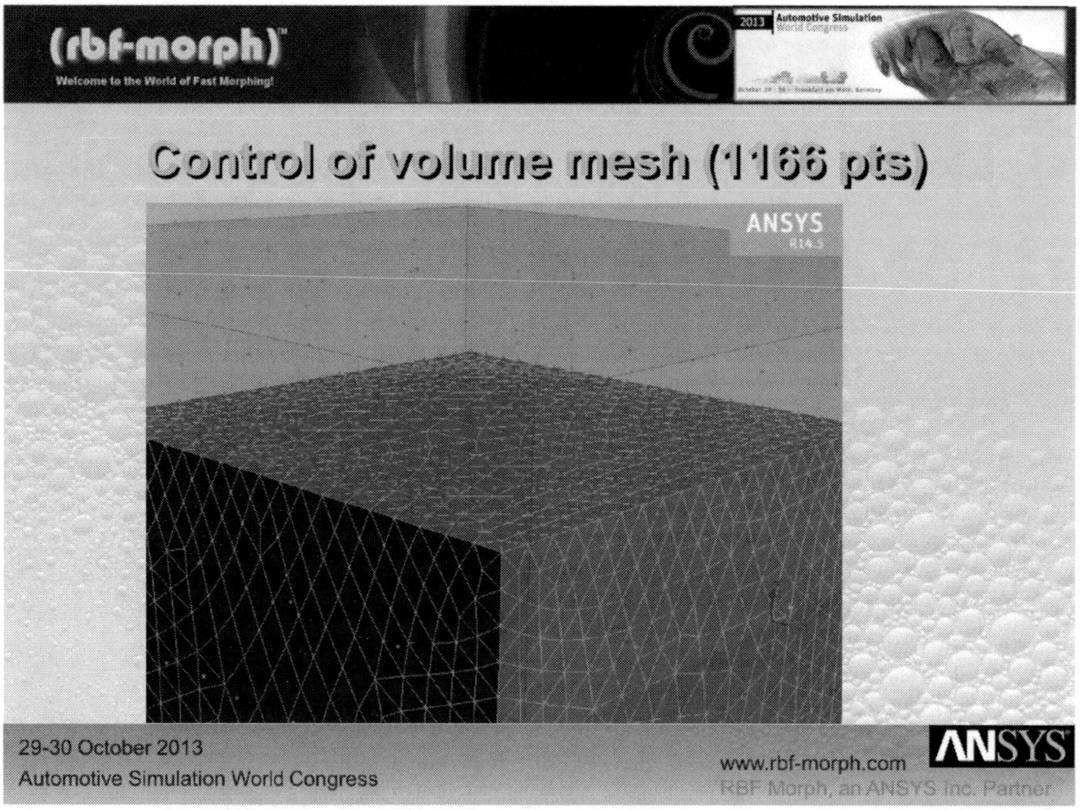
(rbf-morph)™
Welcome to the World of Fast Morphing!
2013 Automotive Simulation World Congress
Control of volume mesh (1166 pts)
ANSYS
R14.5
29-30 October 2013
Automotive Simulation World Congress
www.rbf-morph.com
RBF Morph, an ANSYS Inc. Partner
ANSYS

(rbf-morph)™
Welcome to the World of Fast Morphing!

2013 Automotive Simulation World Congress

Morphing the volume mesh

29-30 October 2013
Automotive Simulation World Congress

www.rbf-morph.com
ANSYS
RBF Morph, an ANSYS Inc. Partner

(rbf-morph)™
Welcome to the World of Fast Morphing!

2013 Automotive Simulation World Congress

How it Works: the problem setup

- The problem must describe correctly the **desired changes** and must **preserve exactly** the fixed part of the mesh.
- The prescription of the **source points** and their displacements fully defines the *RBF Morph* problem.
- Each problem and its fit define a mesh **modifier** or a **shape parameter**.

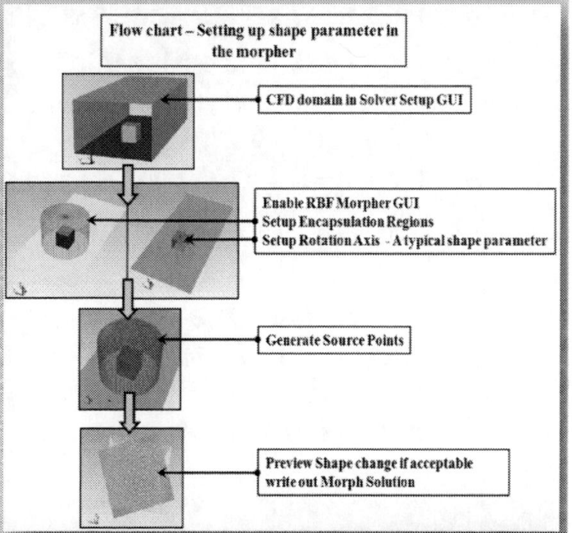

29-30 October 2013
Automotive Simulation World Congress

www.rbf-morph.com
ANSYS
RBF Morph, an ANSYS Inc. Partner

Background: accelerating the solver

- The evaluation of RBF at a point has a cost of order N
- The fit has a cost of order **N^3** for a direct fit (full populated matrix); this limit to ~10.000 the number of source points that can be used in a practical problem
- Using an iterative solver (with a good pre-conditioner) the fit has a cost of order **N^2**; the number of points can be increased up to ~70.000
- Using also space partitioning to accelerate fit and evaluation the number of points can be increased up to **~300.000**
- The method can be further accelerated using fast pre-conditioner building and FMM RBF evaluation…

Background: solver performances escalation

- 10.000 RBF centers FIT
  - 120 minutes Jan 2008
  - 5 seconds Jan 2010
- Largest fit **2.600.000** 133 minutes
- Largest model morphed **300.000.000** cells
- Fit and Morph a **100.000.000** cells model using **500.000** RBF centers within **15 minutes**

#points	2010 (Minutes)	2008 (Minutes)
3.000	0 (1s)	15
10.000	0 (5s)	120
40.000	1 (44s)	Not registered
160.000	4	Not registered
650.000	22	Not registered
2.600.000	133	Not registered

(rbf-morph)™
Welcome to the World of Fast Morphing!

2013 Automotive Simulation World Congress

Coming soon: GPU acceleration!

NVIDIA.

- Single RBF complete evaluation
- Unit random cube
- **GPU**: Kepler 20 2496 CUDA Cores GPU Clock 0.71 GHz
- **CPU**: quad core Intel(R) Xeon(R) CPU E5-2609 0 @ 2.40GHz

#points	CPU	GPU	speed up
5000	0,098402	0,004637	21,2
10000	0,319329	0,011746	27,2
15000	0,667639	0,024982	26,7
20000	1,135127	0,038352	29,6
25000	1,721781	0,054019	31,9
30000	2,451661	0,079459	30,9
35000	3,306897	0,108568	30,5
40000	4,286706	0,134978	31,8
45000	5,390029	0,181181	29,7
50000	6,707721	0,2135	31,4
100000	26,13633	0,745482	35,1
150000	58,96981	1,735367	34,0
200000	115,3628	2,861737	40,3

29-30 October 2013
Automotive Simulation World Congress

www.rbf-morph.com
RBF Morph, an ANSYS Inc. Partner
ANSYS

(rbf-morph)™
Welcome to the World of Fast Morphing!

2013 Automotive Simulation World Congress

Scaling plot

- Complexity is expected to grow as N^2
- GPU observed as $N^{1.87}$
- CPU observed as $N^{2.174}$
- Estimation at one million points:

  GPU: 59 s

  CPU: 2783 s

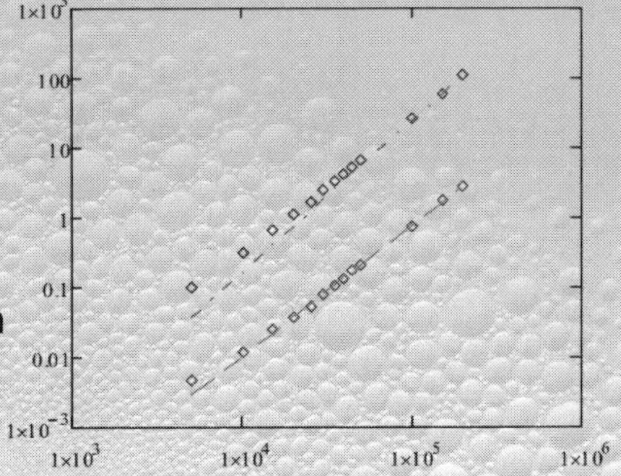

29-30 October 2013
Automotive Simulation World Congress

www.rbf-morph.com
RBF Morph, an ANSYS Inc. Partner
ANSYS

(rbf-morph)™
Welcome to the World of Fast Morphing!

2013 Automotive Simulation World Congress

Industrial Applications

29-30 October 2013
Automotive Simulation World Congress

www.rbf-morph.com
RBF Morph, an ANSYS Inc. Partner

ANSYS

(rbf-morph)™
Welcome to the World of Fast Morphing!

2013 Automotive Simulation World Congress

Formula 1 Front Wing

(rbf-morph)™
Welcome to the World of Fast Morphing!

www.rbf-morph.com

29-30 October 2013
Automotive Simulation World Congress

www.rbf-morph.com
RBF Morph, an ANSYS Inc. Partner

ANSYS

(rbf-morph)™
Welcome to the World of Fast Morphing!
2013 Automotive Simulation World Congress
Motorbike Windshield
(Bricomoto, MRA)
(rbf-morph)™
Welcome to the World of Fast Morphing!
BRICO MOTO
MOTORCYCLE WINDSHIELDS
www.rbf-morph.com
29-30 October 2013
Automotive Simulation World Congress
www.rbf-morph.com
ANSYS
RBF Morph, an ANSYS Inc. Partner

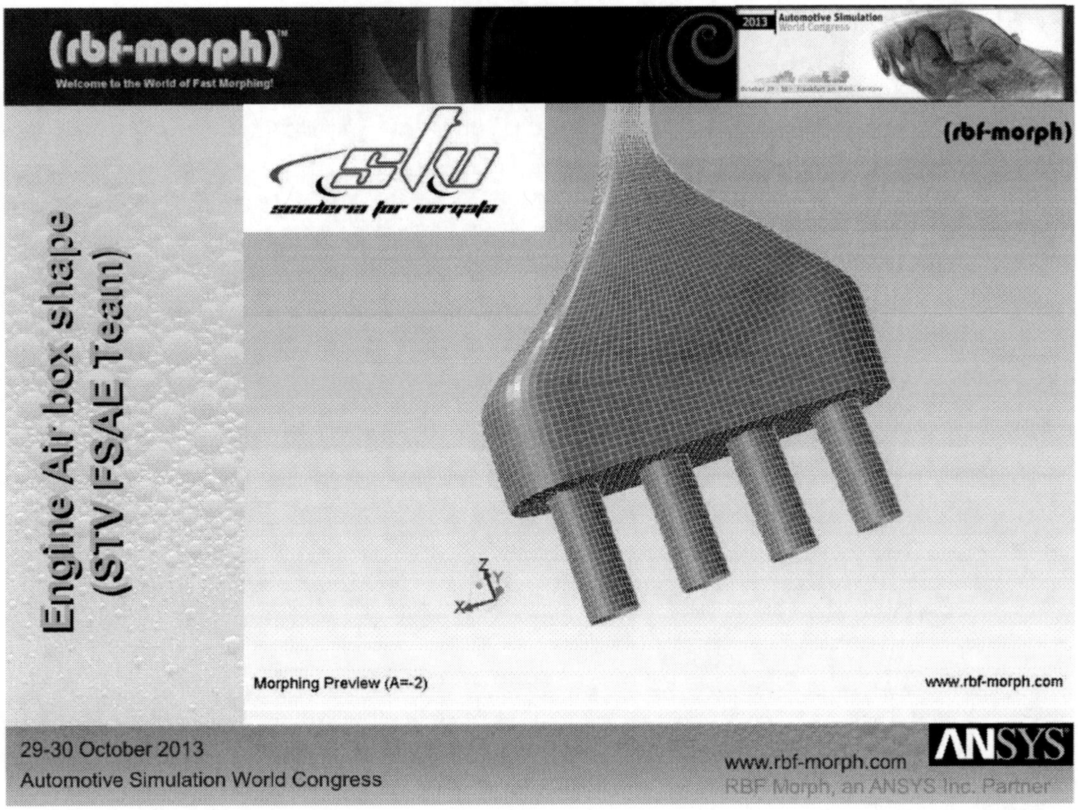

(rbf-morph)™
Welcome to the World of Fast Morphing!
2013 Automotive Simulation World Congress
Engine Air box shape
(STV FSAE Team)
scuderia tor vergata
(rbf-morph)
Morphing Preview (A=-2)
www.rbf-morph.com
29-30 October 2013
Automotive Simulation World Congress
www.rbf-morph.com
ANSYS
RBF Morph, an ANSYS Inc. Partner

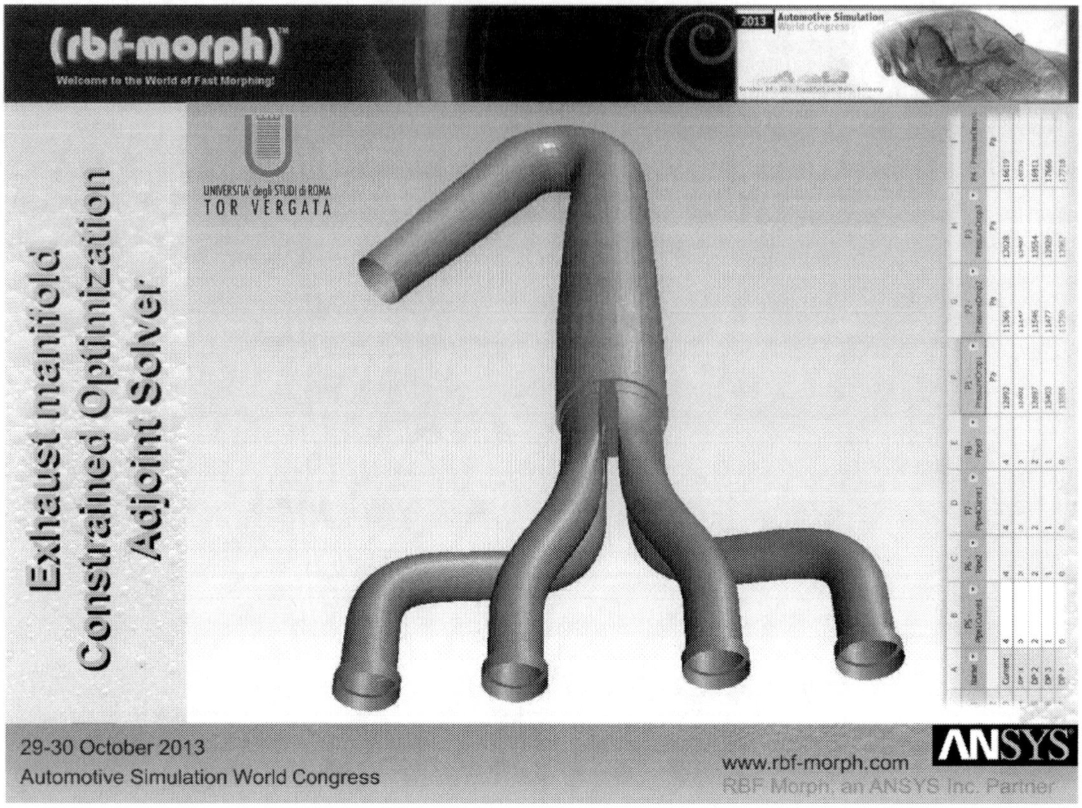
(rbf-morph)™
Welcome to the World of Fast Morphing!
2013 Automotive Simulation World Congress
UNIVERSITA' degli STUDI di ROMA
TOR VERGATA
Exhaust manifold
Constrained Optimization
Adjoint Solver
29-30 October 2013
Automotive Simulation World Congress
www.rbf-morph.com
RBF Morph, an ANSYS Inc. Partner
ANSYS

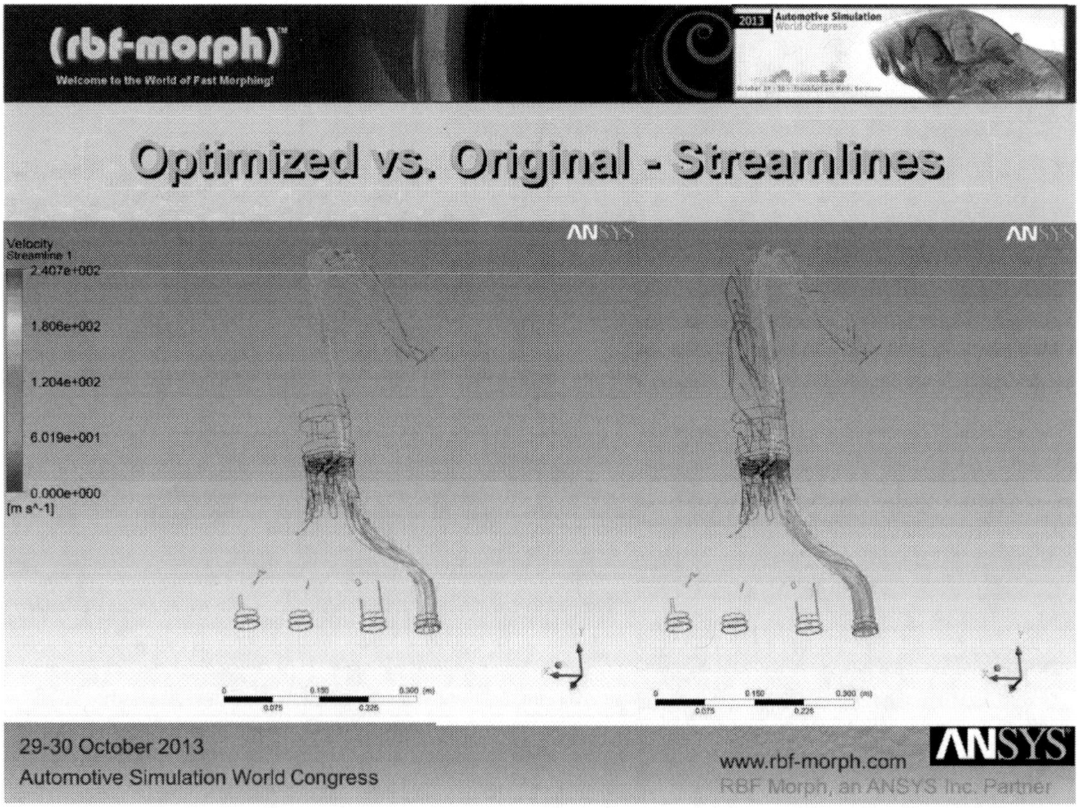
(rbf-morph)™
Welcome to the World of Fast Morphing!
2013 Automotive Simulation World Congress
Optimized vs. Original - Streamlines
Velocity
Streamline 1
2.407e+002
1.806e+002
1.204e+002
6.019e+001
0.000e+000
[m s^-1]
ANSYS
ANSYS
29-30 October 2013
Automotive Simulation World Congress
www.rbf-morph.com
RBF Morph, an ANSYS Inc. Partner
ANSYS

(rbf-morph)™
Welcome to the World of Fast Morphing!
2013 Automotive Simulation World Congress
MIRA
MIRA Reference car
(MIRA ltd)
MIRA Reference Car
Shape Optimisation using RBF-Morph
Smarter Thinking.
29-30 October 2013
Automotive Simulation World Congress
www.rbf-morph.com
RBF Morph, an ANSYS Inc. Partner
ANSYS

(rbf-morph)™
Welcome to the World of Fast Morphing!
2013 Automotive Simulation World Congress
50:50:50 Project Volvo XC60
(Ansys, Intel, Volvo)
VOLVO
intel
50 Million Cells
25 Million Cells
5 Million Cells
ACCURACY
EXTENT
SPEED
5 Design Pts
25 Design Pts
50 Design Pts
Prior aerodynamics optimization processes have either achieved speed at the expense of accuracy and extent or vice versa
The goal of the current work is to achieve speed without compromising accuracy or extent
29-30 October 2
Automotive Simu

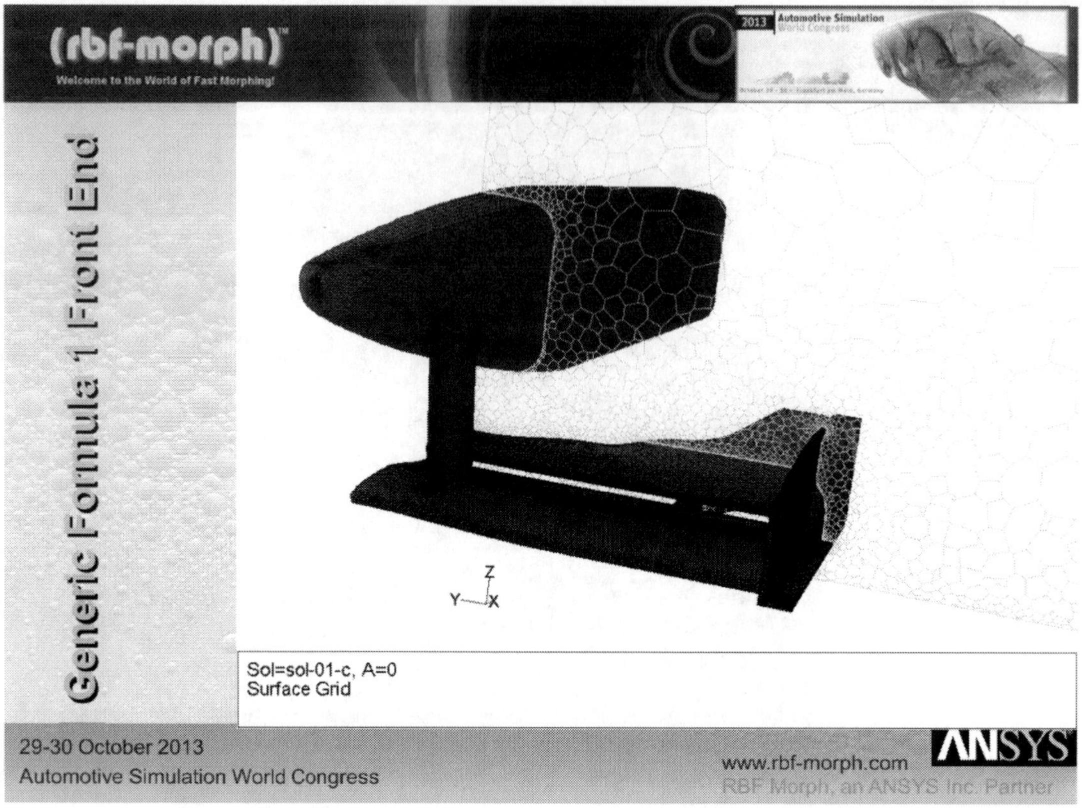
(rbf-morph)
Welcome to the World of Fast Morphing!
2013 Automotive Simulation World Congress
Generic Formula 1 Front End
Z
Y
X
Sol=sol-01-c, A=0
Surface Grid
29-30 October 2013
Automotive Simulation World Congress
www.rbf-morph.com
RBF Morph, an ANSYS Inc. Partner

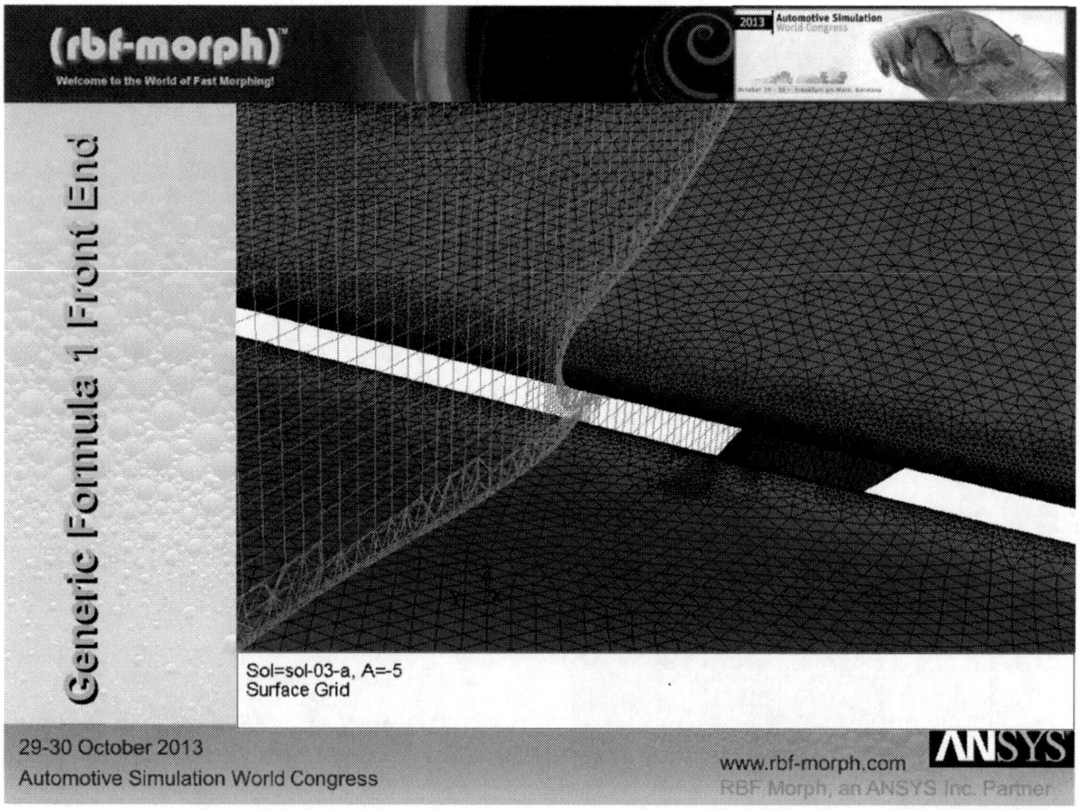
(rbf-morph)
Welcome to the World of Fast Morphing!
2013 Automotive Simulation World Congress
Generic Formula 1 Front End
Sol=sol-03-a, A=-5
Surface Grid
29-30 October 2013
Automotive Simulation World Congress
www.rbf-morph.com
RBF Morph, an ANSYS Inc. Partner

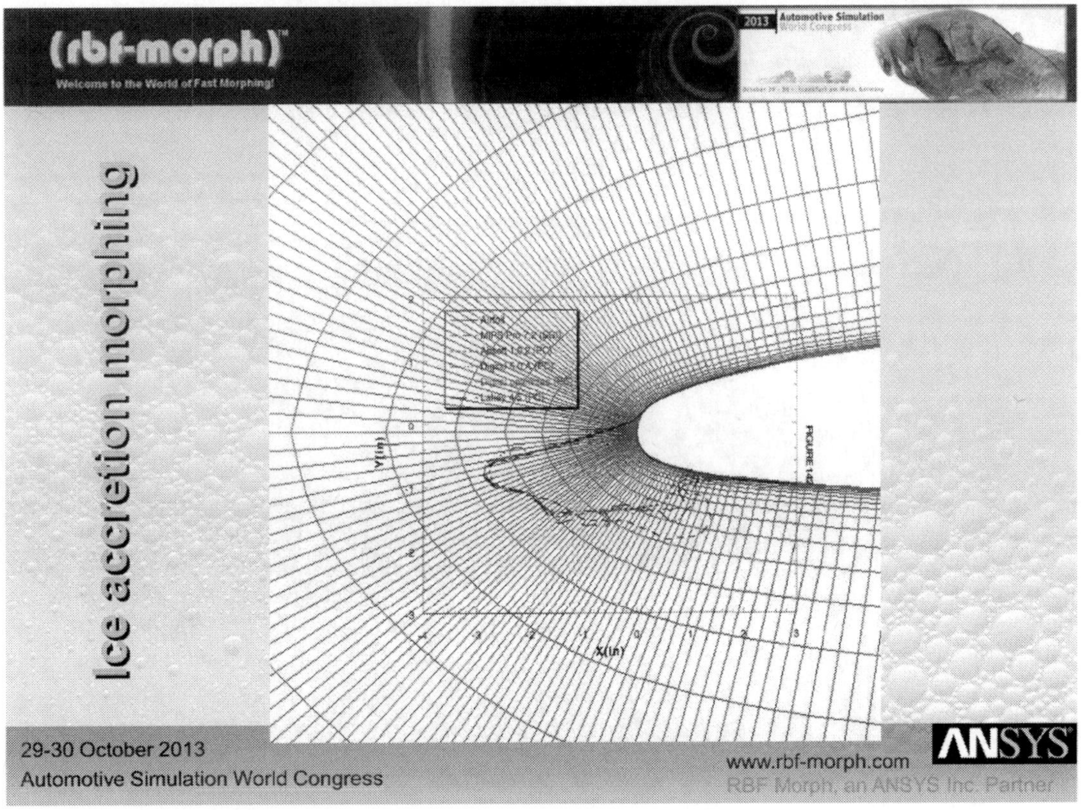
(rbf-morph)™
Welcome to the World of Fast Morphing!
2013 Automotive Simulation World Congress
Ice accretion morphing
X(in)
Y(in)
29-30 October 2013
Automotive Simulation World Congress
www.rbf-morph.com
RBF Morph, an ANSYS Inc. Partner
ANSYS

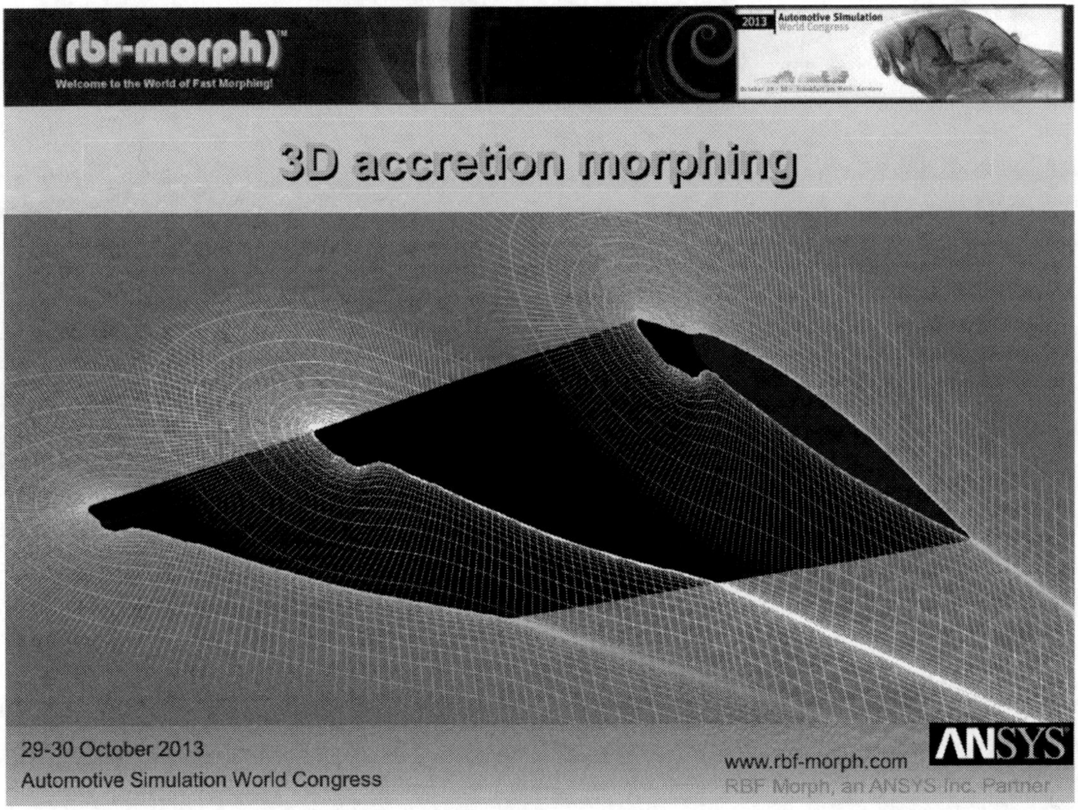
(rbf-morph)™
Welcome to the World of Fast Morphing!
2013 Automotive Simulation World Congress
3D accretion morphing
29-30 October 2013
Automotive Simulation World Congress
www.rbf-morph.com
RBF Morph, an ANSYS Inc. Partner
ANSYS

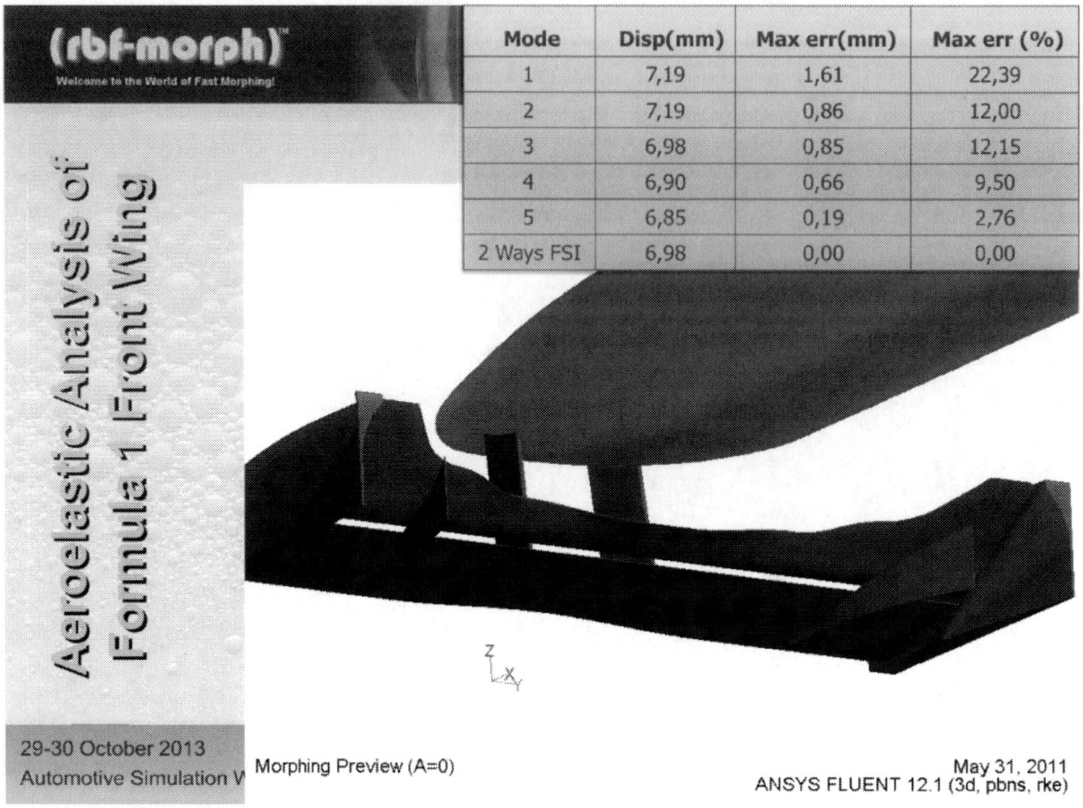

Mode	Disp(mm)	Max err(mm)	Max err (%)
1	7,19	1,61	22,39
2	7,19	0,86	12,00
3	6,98	0,85	12,15
4	6,90	0,66	9,50
5	6,85	0,19	2,76
2 Ways FSI	6,98	0,00	0,00

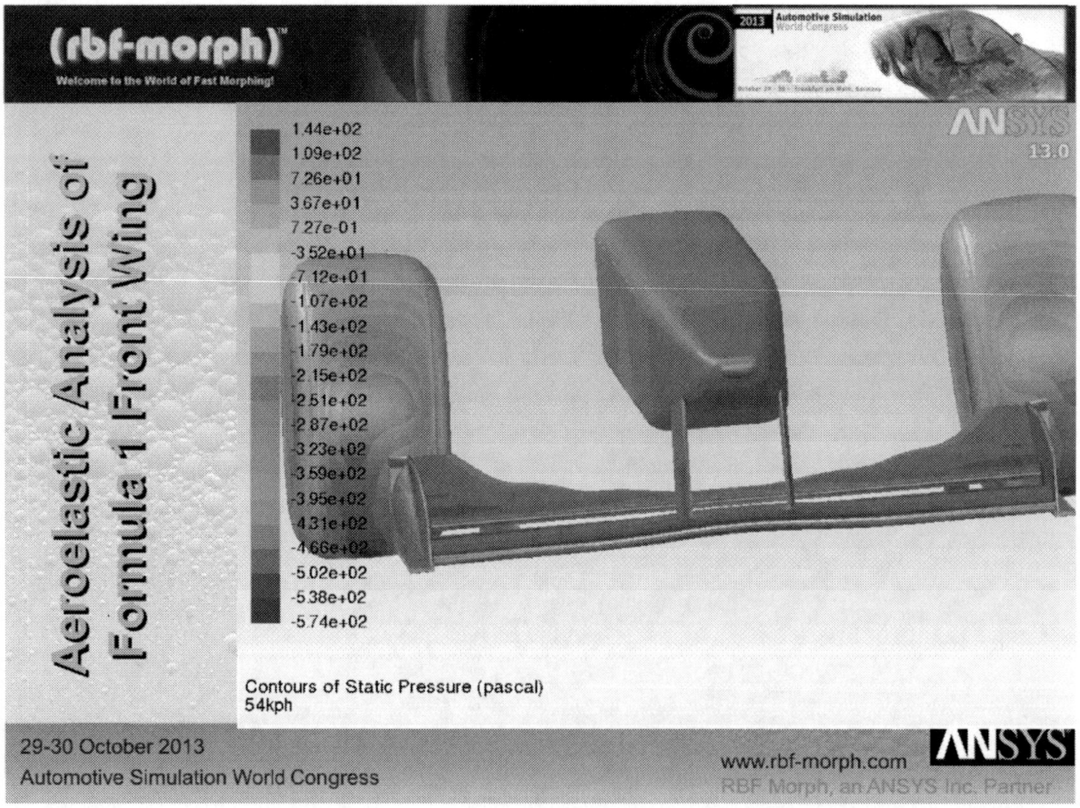

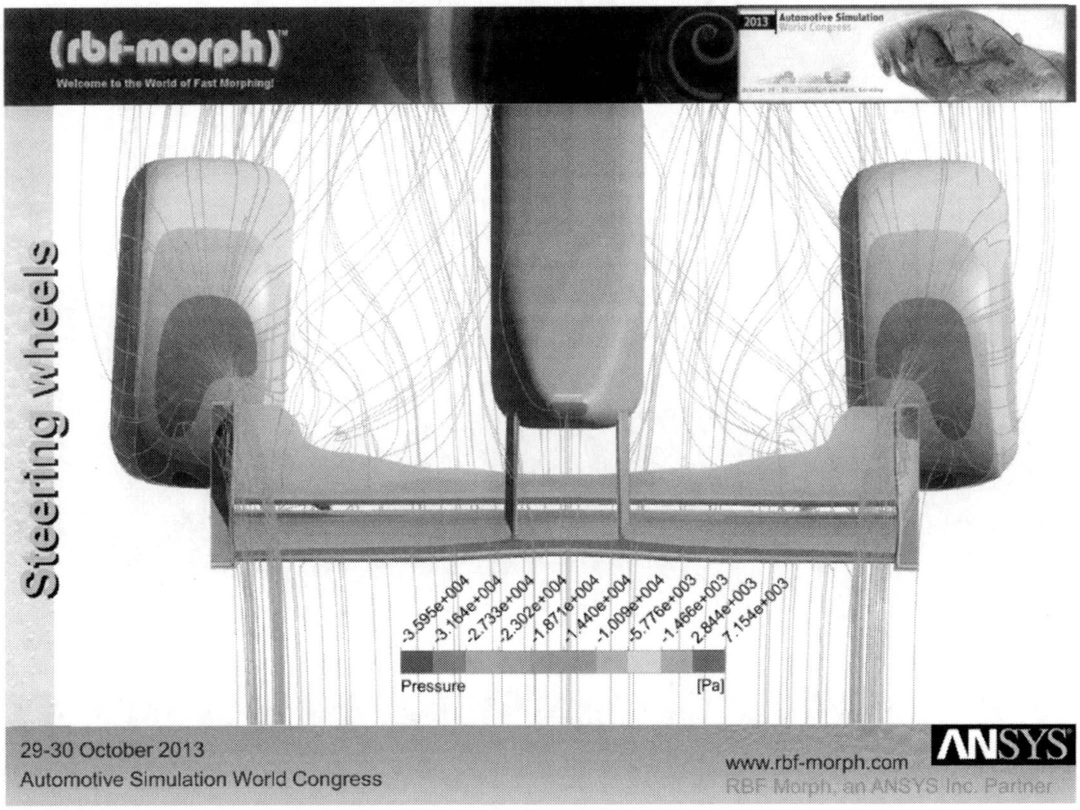
(rbf-morph)™
Welcome to the World of Fast Morphing!
2013 Automotive Simulation World Congress
Steering wheels
-3.595e+004
-3.164e+004
-2.733e+004
-2.302e+004
-1.871e+004
-1.440e+004
-1.009e+004
-5.776e+003
-1.466e+003
2.844e+003
7.154e+003
Pressure
[Pa]
29-30 October 2013
Automotive Simulation World Congress
www.rbf-morph.com
ANSYS
RBF Morph, an ANSYS Inc. Partner

(rbf-morph)™
Welcome to the World of Fast Morphing!
2013 Automotive Simulation World Congress
MORPH lab
dallara
ANSYS FLUENT
hp
NVIDIA
PIAGGIO AERO
(rbf-morph)™
CINECA
SCAI
SuperComputing Applications and Innovation
D'APPOLONIA
29-30 October 2013
Automotive Simulation World Congress
www.rbf-morph.com
ANSYS
RBF Morph, an ANSYS Inc. Partner

(rbf-morph)™
Welcome to the World of Fast Morphing!

2013 Automotive Simulation World Congress

EU Project RBF4AERO – FP7Transport

- *"Innovative Benchmark Technology for Aircraft Engineering Design and Efficient Design Phase Optimisation"* – **GA no. ACP3-GA-2013-605396**

RBF4AERO

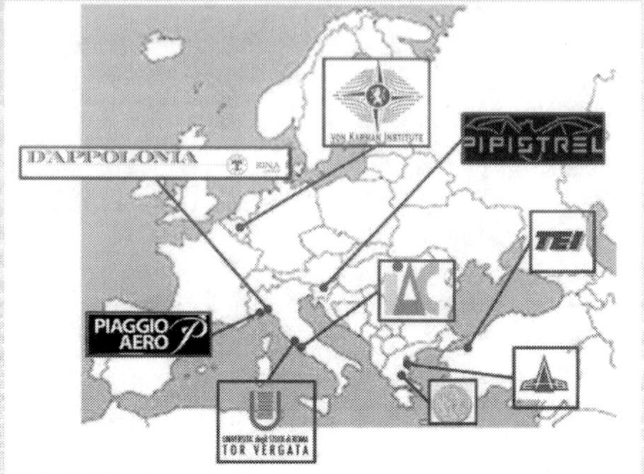

29-30 October 2013
Automotive Simulation World Congress

www.rbf-morph.com
RBF Morph, an ANSYS Inc. Partner
ANSYS

(rbf-morph)™
Welcome to the World of Fast Morphing!

2013 Automotive Simulation World Congress

Advanced features

29-30 October 2013
Automotive Simulation World Congress

www.rbf-morph.com
RBF Morph, an ANSYS Inc. Partner
ANSYS

Two steps method

- **STEP-1** Specific set-up can be defined for accurate **surface control** (high order RBF)
- **STEP-2** Resulting shape can be used in the final set-up to control the **volume mesh** (preservation of other surfaces, fast and smooth bi-harmonic function)

(rbf-morph)™
Welcome to the World of Fast Morphing!
2013 Automotive Simulation World Congress

Adjoint sculpting

- Shape information coming from the **Fluent Adjoint** morpher can be used to control a surface
- A single adjoint (baseline shape) is used to define shape modifications in **multiple locations**
- New shape modifications can be **combined** as usual

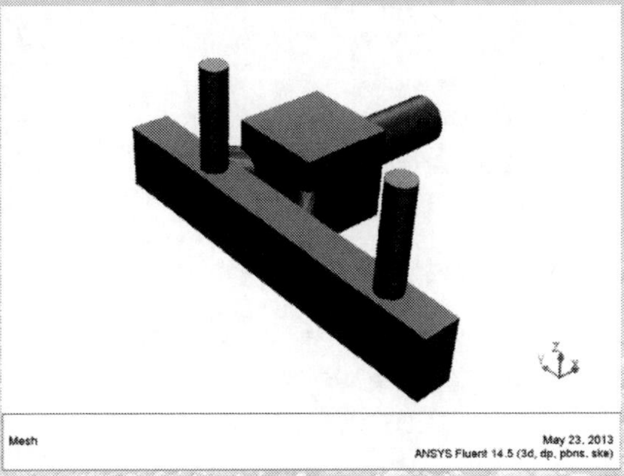

29-30 October 2013
Automotive Simulation World Congress
www.rbf-morph.com
ANSYS
RBF Morph, an ANSYS Inc. Partner

(rbf-morph)™ Welcome to the World of Fast Morphing!

2013 Automotive Simulation World Congress

CAD controlled surfaces

- A new shape known in advance can be inserted using an **STL target**
- In the example a **fillet** with radius in the range 20-30 mm is applied to one edge of the 1000 mm side cube
- Shape **blending** allows a continuous variation

29-30 October 2013
Automotive Simulation World Congress

www.rbf-morph.com
RBF Morph, an ANSYS Inc. Partner
ANSYS

(rbf-morph)™ Welcome to the World of Fast Morphing!

2013 Automotive Simulation World Congress

Box sculpting

- **Arbitrary points distribution** can be used to control surfaces
- Effect typical of **FFD boxes** can be obtained using special shapes to generate control points
- In the example a **tube** is **reshaped** using 20 points computes on a box (SP → Points command)

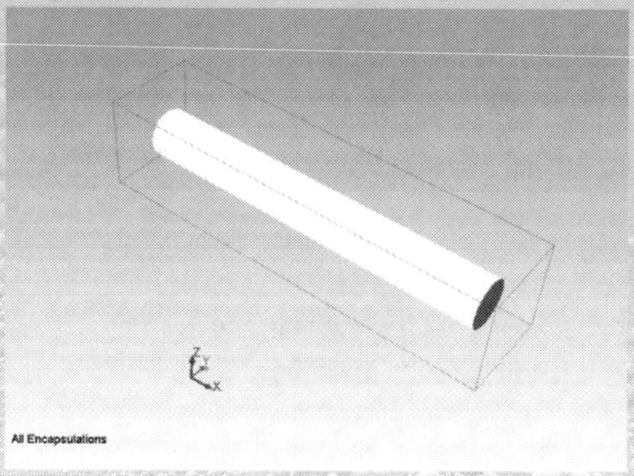

29-30 October 2013
Automotive Simulation World Congress

www.rbf-morph.com
RBF Morph, an ANSYS Inc. Partner
ANSYS

(rbf-morph)™
Welcome to the World of Fast Morphing!

2013 Automotive Simulation World Congress

Constrained optimization

- When several shape parameters are combined it's difficult to control the **final position** of morphed shape
- RBF Morph allows to defined **list of planes** and to control the minimum distance of **monitored surfaces** with respect to all the plane in the list (**value exposed to WB**)
- Constraints are shown projecting the points of monitored threads

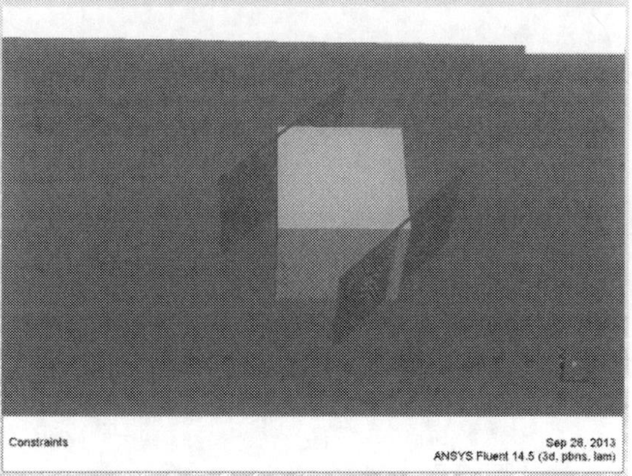

29-30 October 2013
Automotive Simulation World Congress
www.rbf-morph.com
RBF Morph, an ANSYS Inc. Partner
ANSYS

(rbf-morph)™
Welcome to the World of Fast Morphing!

2013 Automotive Simulation World Congress

Surfaces intersection

- Some times to accommodate morphing some undesired **deformations** are introduced
- Re-projection onto original surfaces (using **STL target** or self-projection) is then required
- In the example a multi step RBF sequence is used to **preserve** the shape of the flat surface

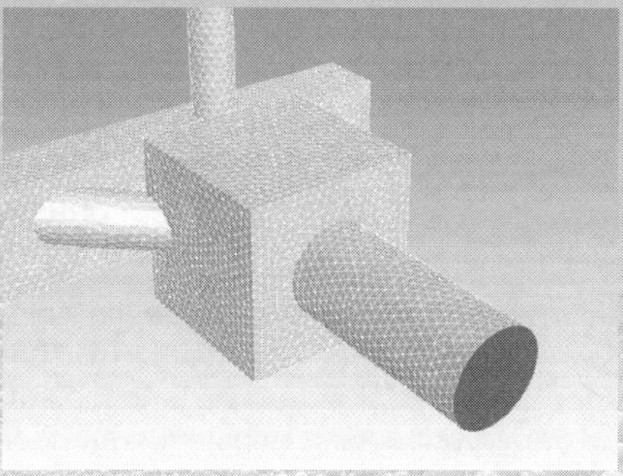

29-30 October 2013
Automotive Simulation World Congress
www.rbf-morph.com
RBF Morph, an ANSYS Inc. Partner
ANSYS

(rbf-morph)™
Welcome to the World of Fast Morphing!

Importing in the CAD the new design

- Solution 07-b with ampli = 1 has to be reversed (**nose rotation** 1 deg)
- STEP file of original shape is loaded (points overlap within **Fluent GUI**)
- **Morphed STEP** file is generated

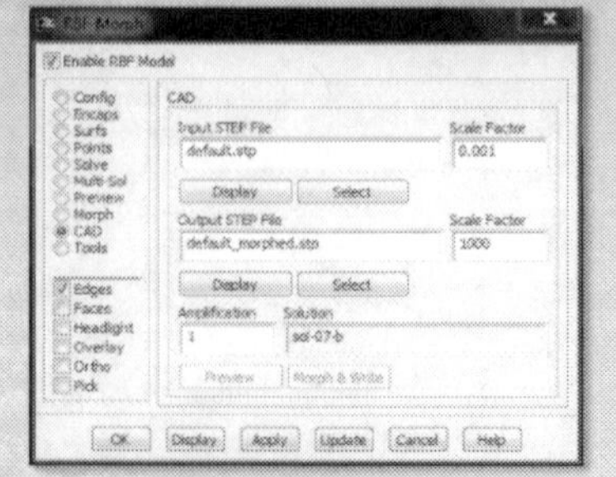

29-30 October 2013
Automotive Simulation World Congress

www.rbf-morph.com
RBF Morph, an ANSYS Inc. Partner

(rbf-morph)™
Welcome to the World of Fast Morphing!

Conclusions

- A **shape parametric** CFD model can be defined using ANSYS Fluent and *RBF Morph*.
- Such **parametric CFD model** can be easily coupled with preferred optimization tools to steer the solution to an **optimal design** that can be imported in the preferred **CAD** platform (using **STEP**)
- Proposed approach **dramatically** reduces the man time required for set-up widening the CFD calculation capability (**50:50:50**).
- ***M.E. Biancolini***, *Mesh morphing and smoothing by means of Radial Basis Functions (RBF): a practical example using Fluent and RBF Morph* in Handbook of Research on Computational Science and Engineering: Theory and Practice (http://www.cse-book.com/).

29-30 October 2013
Automotive Simulation World Congress

www.rbf-morph.com
RBF Morph, an ANSYS Inc. Partner

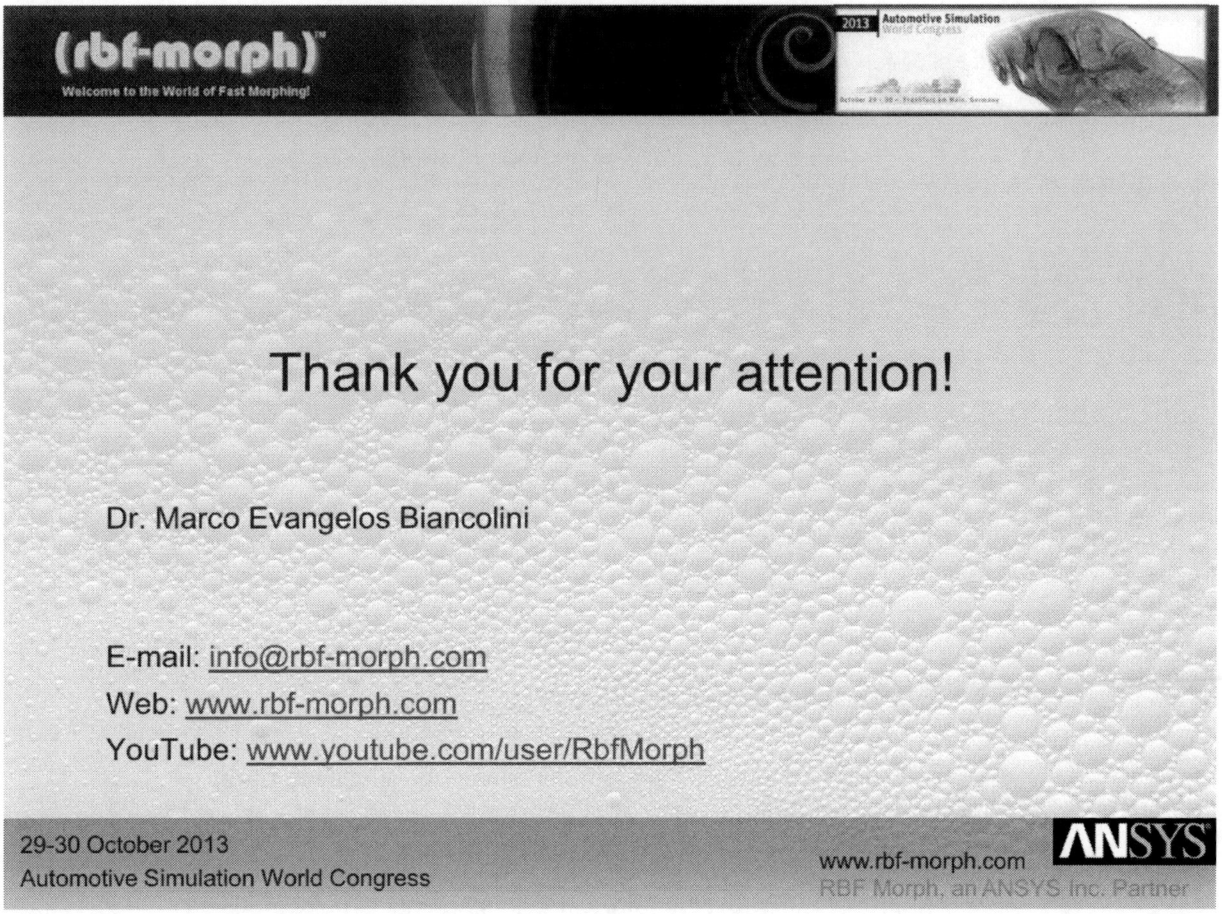
(rbf-morph)™
Welcome to the World of Fast Morphing!
2013
Automotive Simulation
World Congress
Thank you for your attention!
Dr. Marco Evangelos Biancolini
E-mail: info@rbf-morph.com
Web: www.rbf-morph.com
YouTube: www.youtube.com/user/RbfMorph
29-30 October 2013
Automotive Simulation World Congress
www.rbf-morph.com
ANSYS
RBF Morph, an ANSYS Inc. Partner

"Numerical study of aerodynamic characteristics of sport car with movable and deformable aerodynamic elements"

J.Piechna[1], T.Janson[1], P.Sadowski[1], S.Tudruj[1], A.Piechna[2], L.Rudniak[3]

[1]Faculty of Power & Aeronautical Engineering,

[2]Faculty of Mechatronics

[3]Department of Chemical & Process Engineering

Warsaw University of Technology, Poland

1

Agenda

1. Overview of sport car movable aerodynamic elements
2. Generic car with movable aerodynamic elements simulation using ANSYS Fluent software
3. Discussion and suggestions of optimal flaps configurations
4. Real sport car with movable aerodynamic elements simulation using ANSYS Fluent software
5. Deflecting flaps and airbrake efficiency discussion
6. Implementation of home made FSI code to simulate deformable car elements and airbags protecting pedestrians

Agenda 2

Fast cars are moving at varying traffic conditions.

During acceleration high downforce with low aerodynamic resistance is desirable.

Braking requires a lot of drag and high downforce, mainly a the rear axle, also cornering requires a significant downforce.

To meet all these different requirements moving bodyworks are needed to change the aerodynamic characteristics of the car.

Sports car shapes, for obvious reasons, must also fullfilled designer requirements. This limits the range of applicable design solutions. Elastic deformable elements of bodywork and rigid flaps have been considered. Moving flaps and elastic deformable surfaces can be hidden in the contour of the body not deforming the shape generated by the stylists.

SymKom Warsaw University of Technology POLAND

Flaps, after opening at high speed, ensure changes of car aerodynamic characteristics.

The numerical study was performed using ANSYS-Fluent software. Impact on the flow around the car of various flaps, located in different body areas was analyzed.

Characteristic differences of flow structure around flaps, with front and back position of rotation axis, was found.

Significant influence of flaps position on the car body was recognized.

UDF integration routine, tracking deformation of the flexible shell, is used for dynamic prediction of shape of car flexible elements.

An example of the application of movable flaps on the proposed supercar was presented.

Typical aerodynamic elements of the sports car body. Some of them can be movable or deformable

1. Deflector
2. Venturi channels
3. Side skirts
4. Front splitter

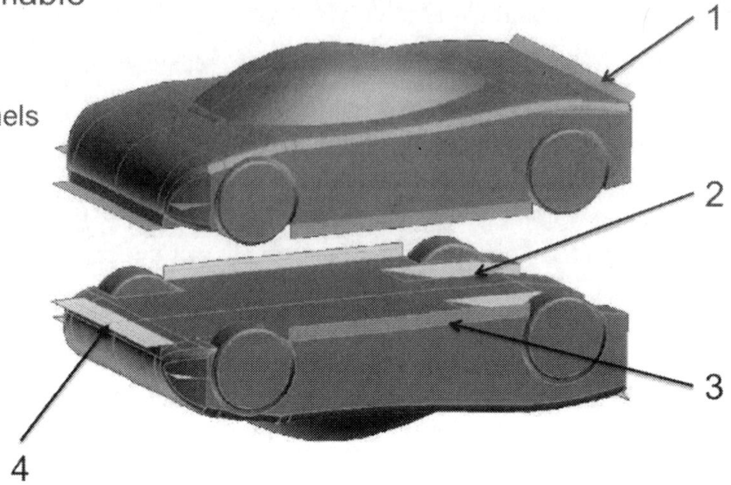

Use of additional aerodynamic elements can have significant influence on the car aerodynamic characteristics

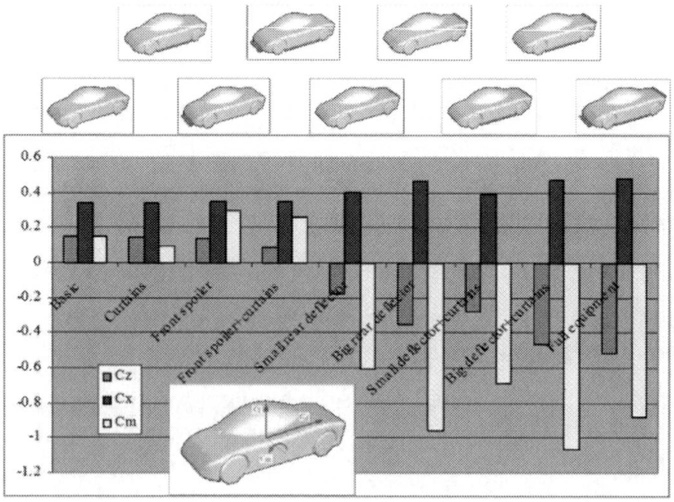

Movable flaps - known examples

Mitsubishi SHR II experimental car had a large flaps located at the rear part of the body rotating around the front flap's edges.

1955 Mercedes 300 SLR had a single large flap having the axis of rotation located at the rear and was also used as airbreak.

In the supercar Pagani Huayra movable flaps have been used in the front and rear part of the vehicle.

Generic car geometry

Considered movable elements

Small movable flaps on the rear part of the body

Big movable flaps on the rear part of the body

Flaps over the body

Movable flaps on the front part of the body

Movable flaps on the rear edge of the body

Computational domain

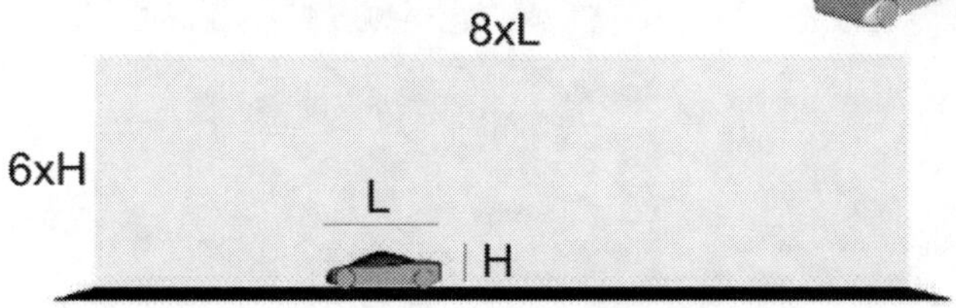

Surface pressure changes caused by motion of rear flaps.

Animation :

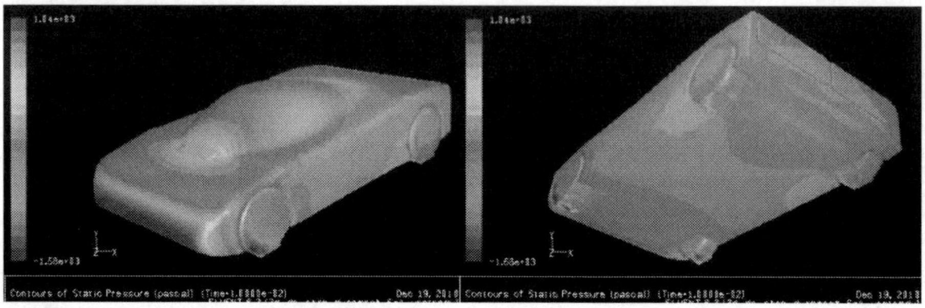

Warsaw University of Technology POLAND

Changes of flow structure and pressure distribution by extended rear flaps

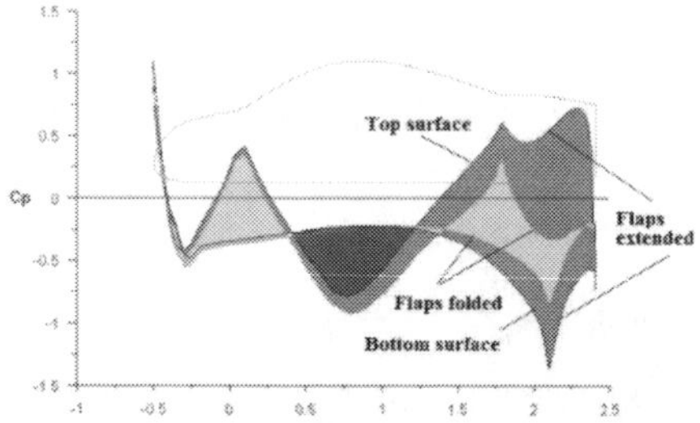

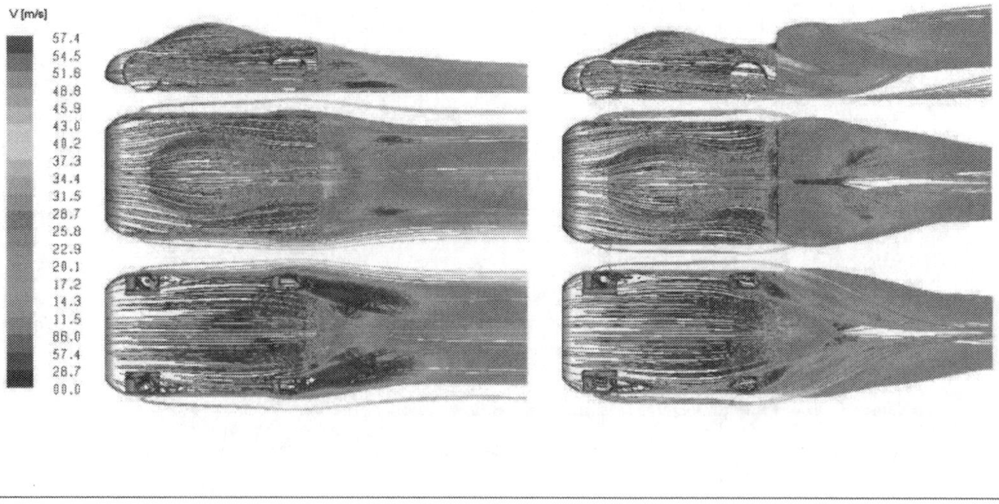

Front position of flap rotation axis

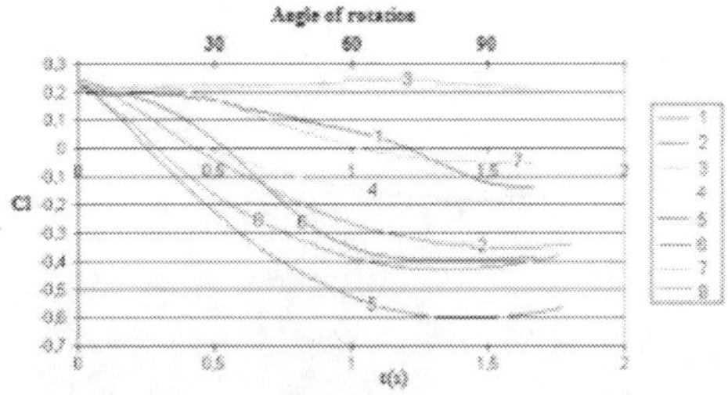

Back position of flap rotation axis

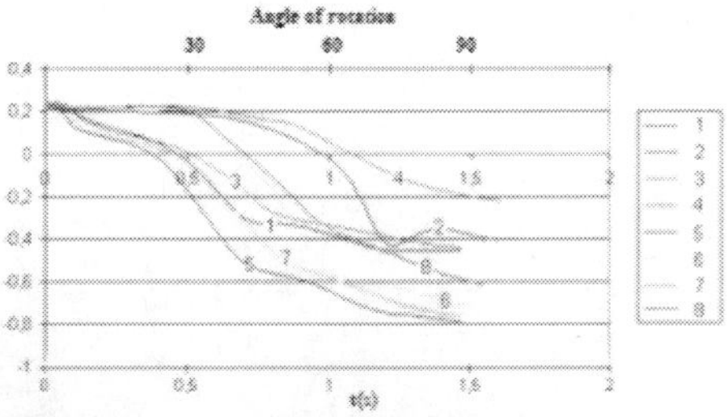

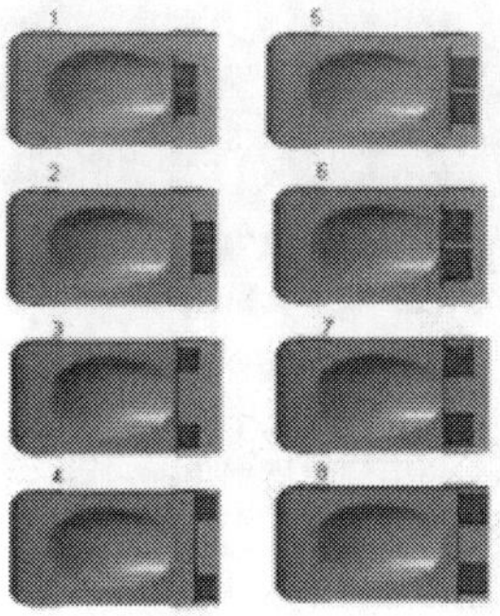

Moving flaps placed on the fast car body can greatly affect its aerodynamic characteristics.

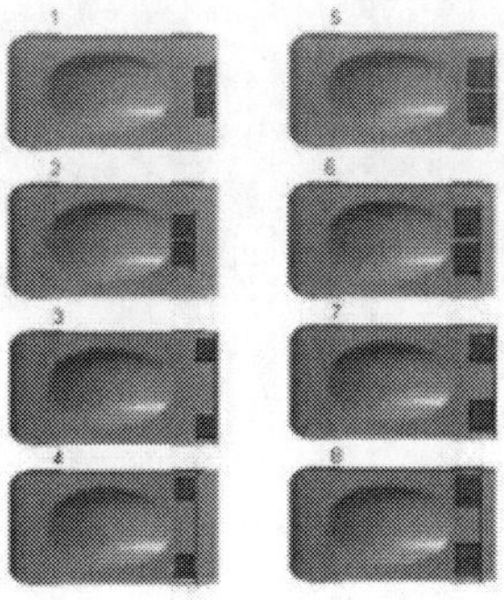

The way the flaps generate the aerodynamic forces strongly depends on the position of the flap on the car body and the position of its axis of rotation.

Depending on whether the rotation axis is located, at the front or rear edge of the flap, there are two types of characteristics. When the rotation axis is arranged at the front edge of the flap, lift force changes are smooth, and proportional to the opening angle of the flap. When the rotation axis is disposed at the rear flap edge, there is a delay in changing the lift force.

It have been recognized that the flow reattachment on the flap top surface is the reason of the delayed generation of downforce.

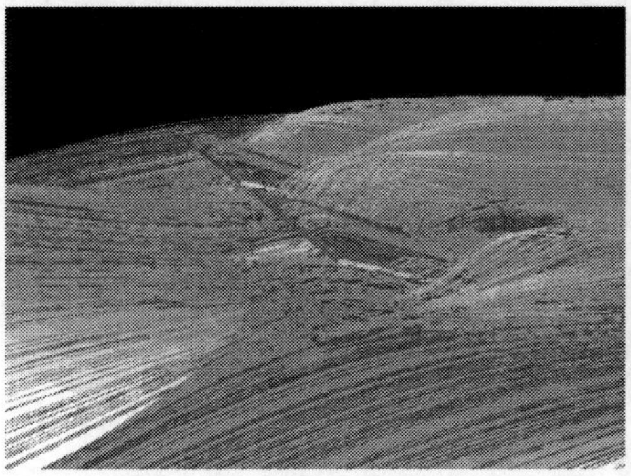

Flap rotation = 23 [deg] attached flow on the top flap surface

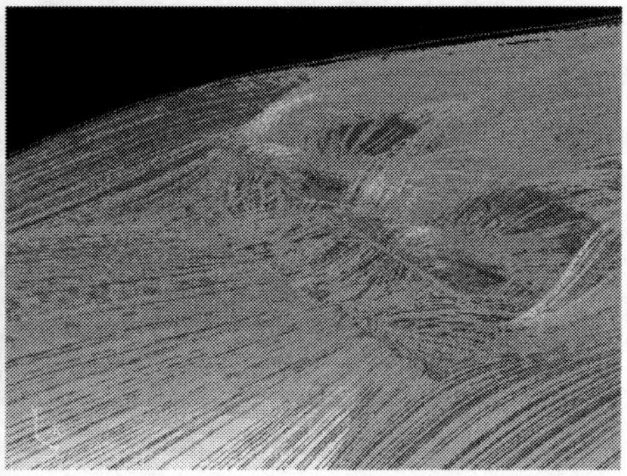

Flap rotation = 29 [deg] separated flow on the top flap surface

Flaps located on a front car area seem not to be efficient as downforce generators, they work much better as air brakes.

Flaps - front position of rotation

Flaps - back position of rotation

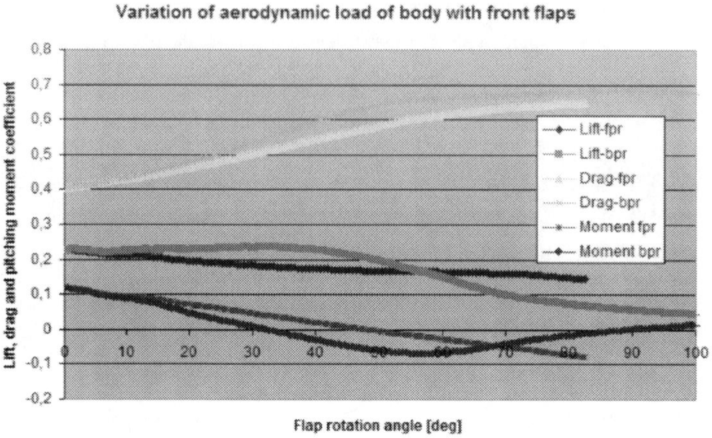

Real sport car simulation

Computational domain and exemplary mesh around the model of sport car

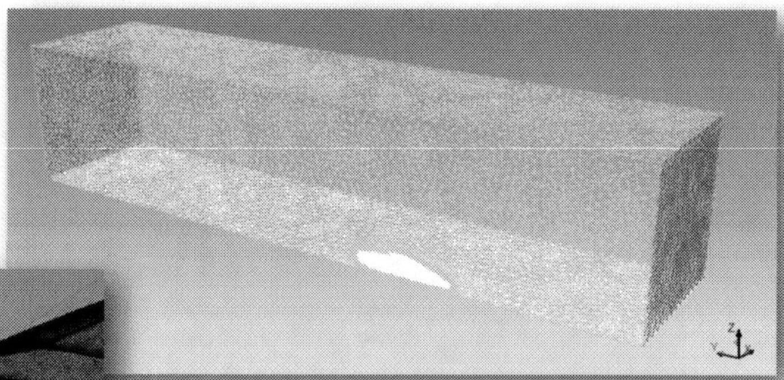

Most efficient rear flaps geometry (nothing new)

Porsche 917
(year 1969)

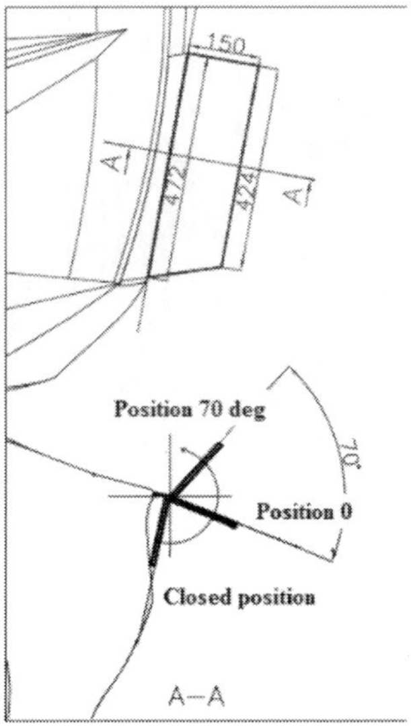

Changes of pressure distribution caused by deflecting flaps

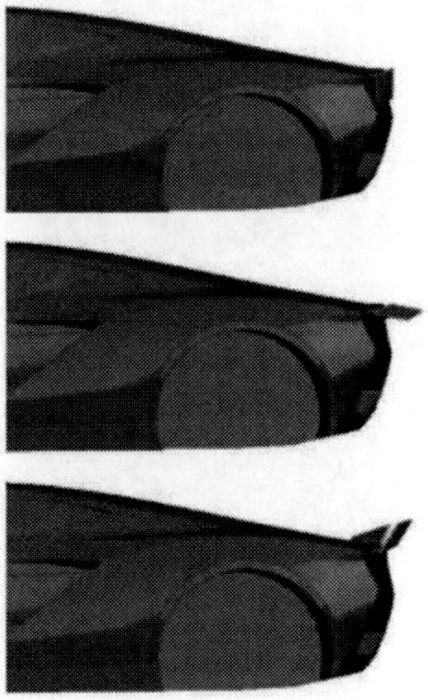

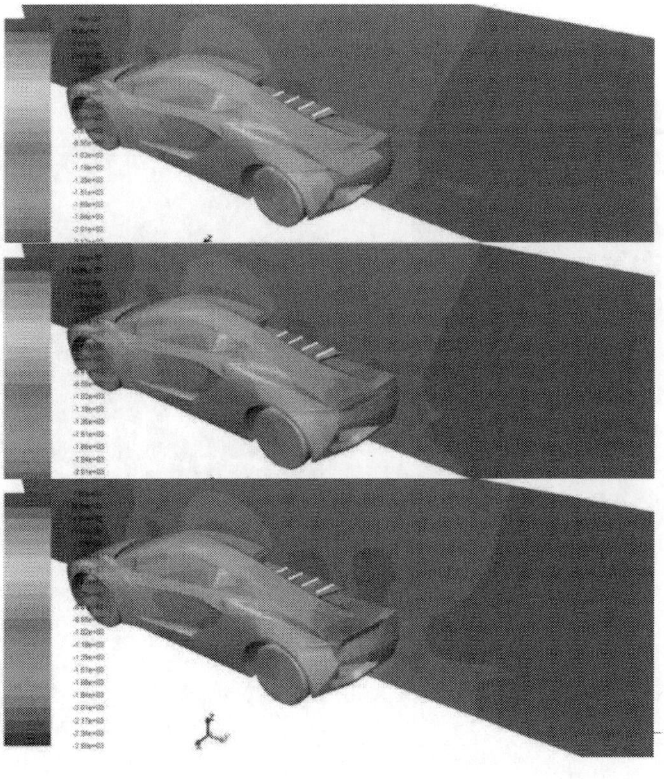

Unsteady simulation of flaps operation

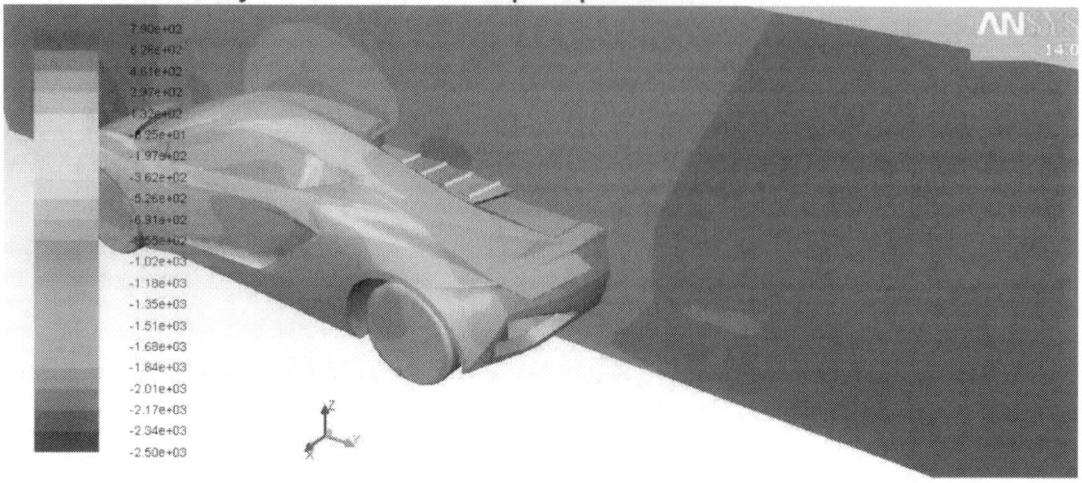

Contours of Static Pressure (pascal) (Time=7.1428e-03)

Feb 22, 2013
ANSYS FLUENT 14.0 (3d, pbns, dynamesh, S-A, transient)

Animation

SymKom
Warsaw University of Technology POLAND

Unsteady simulation of flaps operation

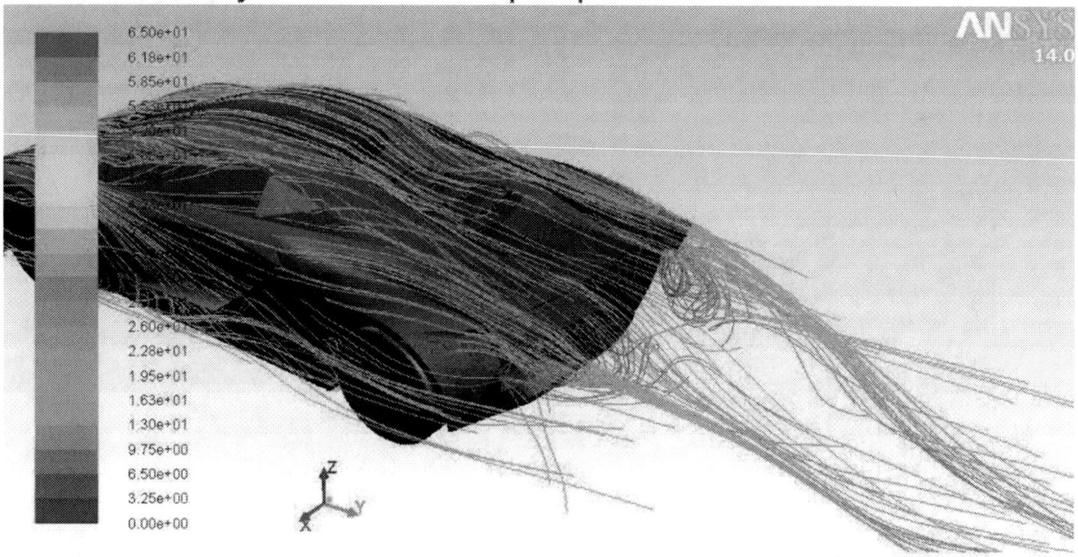

Pathlines Colored by Velocity Magnitude (m/s) (Time=7.1428e-03)

Feb 22, 2013
ANSYS FLUENT 14.0 (3d, pbns, dynamesh, S-A, transient)

Animation

SymKom Warsaw University of Technology POLAND

Unsteady simulation of flaps operation

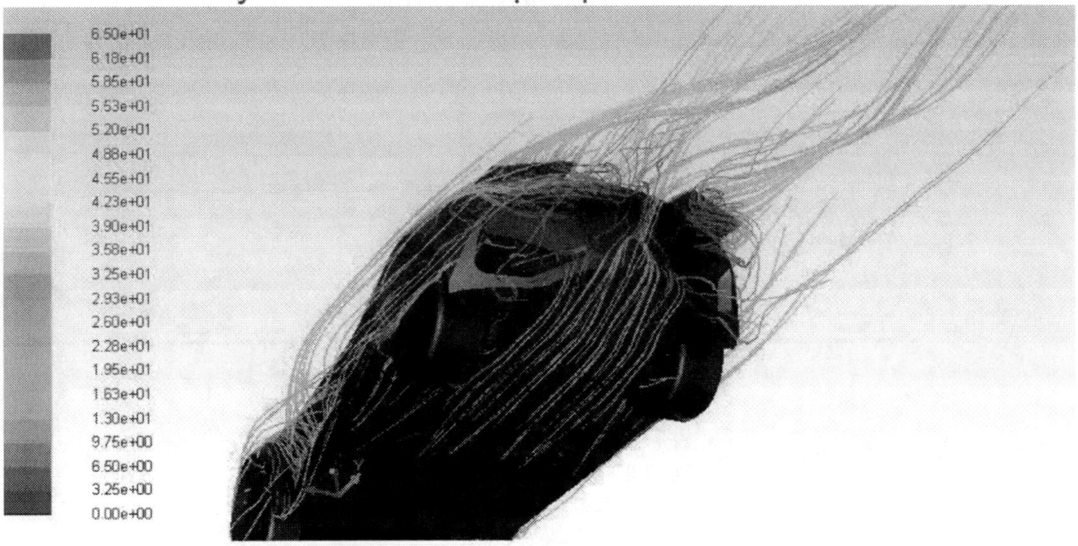

Colored by Velocity Magnitude (m/s) (Time=7.1428e-03)

ANSYS FLUENT 14.0 (3d,

Animation Sport car simulation 19

SymKom Warsaw University of Technology POLAND

Flaps can be used as airbrakes.

Mercedes 300 SLR (year 1955)

Airbrake geometry considered

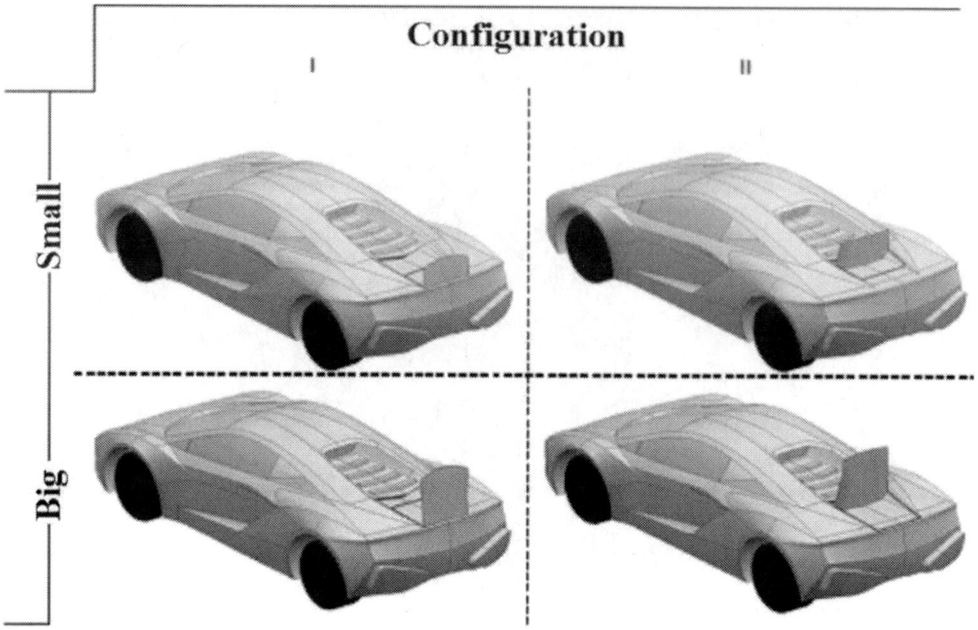

SymKom
Warsaw University of Technology POLAND

Comparison of pressure distribution (closed and opened airbrakes)

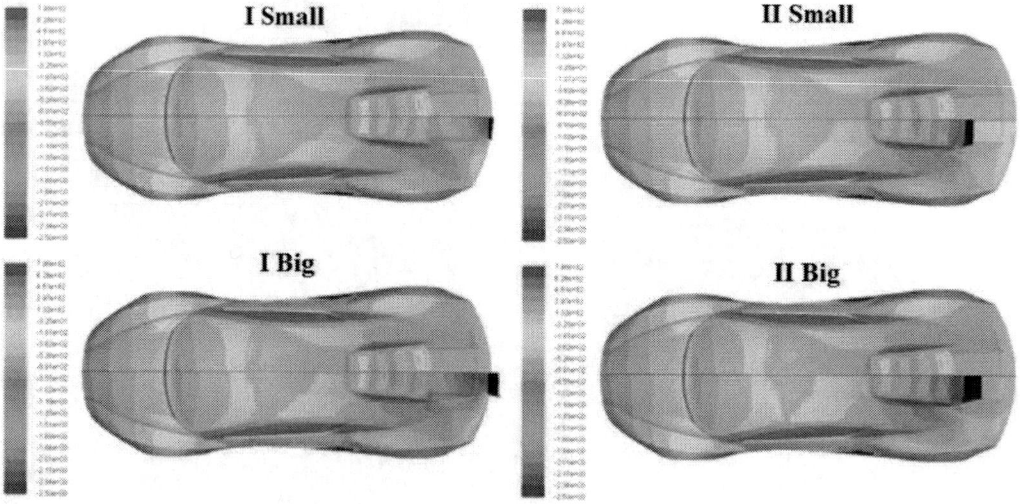

Aerodynamic effects of opening airbrake flaps of different shape and position

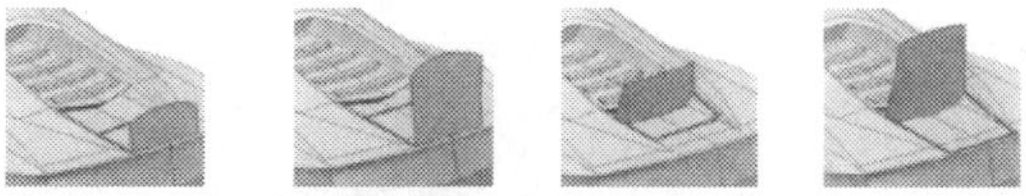

Configuration

	I Small	I Big	II Small	II Big
C_x	1,39	1,85	1,31	1,82
C_z	-0,68	-1,81	-0,51	-2,12
C_{m_y}	0,26	-0,55	-0,63	-2,5

Flaps can be used as an airbrakes. In this case, flaps can be used with the axis of rotation on the rear edge. In that way more braking power and downforce can be obtained in the back. The delay in the occurrence of downforce is not so important.

Deformable elements

Special UDF was developed for integration in time, equations of motion of deformable aerodynamic elements described by lumped parameters model, using fast semi-analytical scheme.

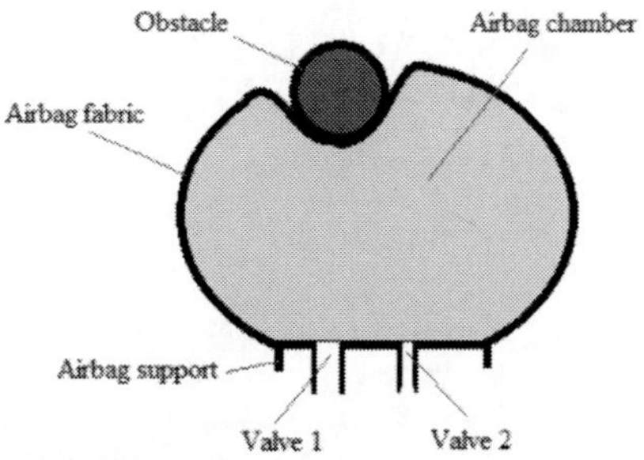

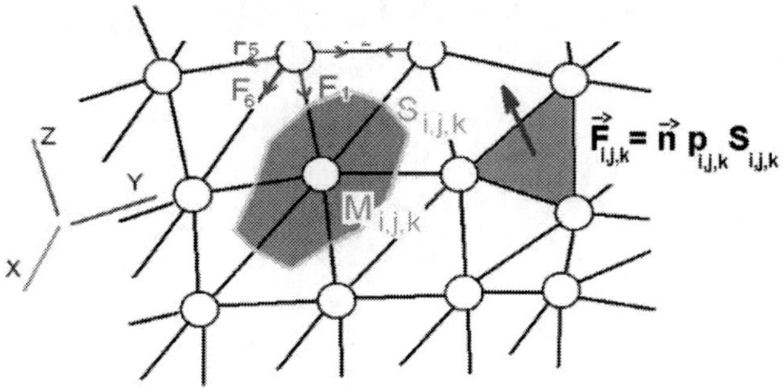

Warsaw University of Technology POLAND

Aerodynamic effects of deploy of flexible side skirts located under the sport car simulated in Fluent

Animation :

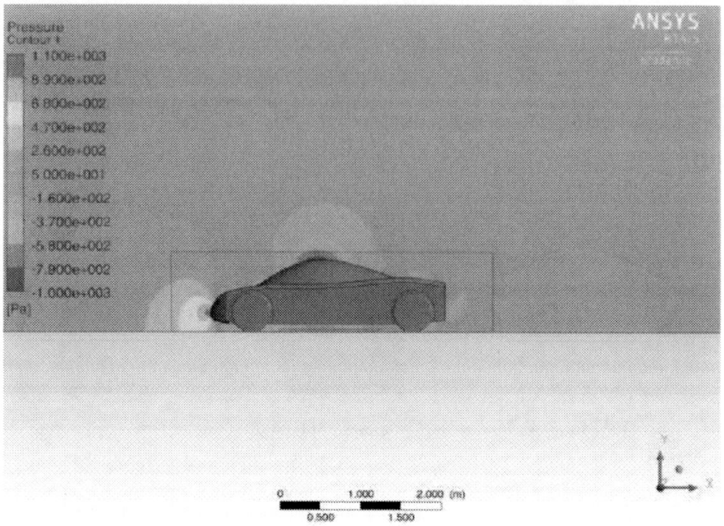

Warsaw University of Technology POLAND

Car front protection by external airbag system

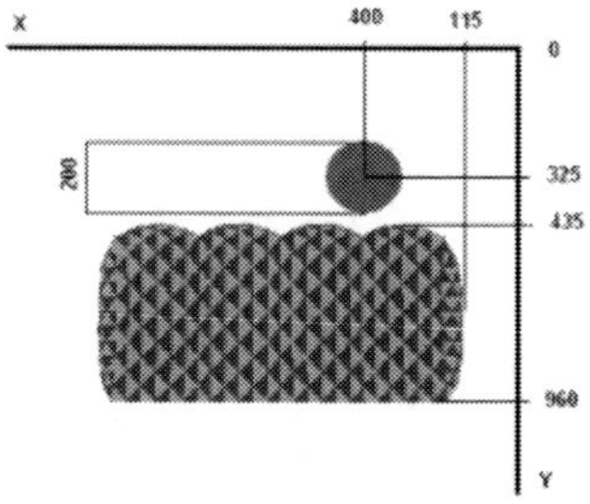

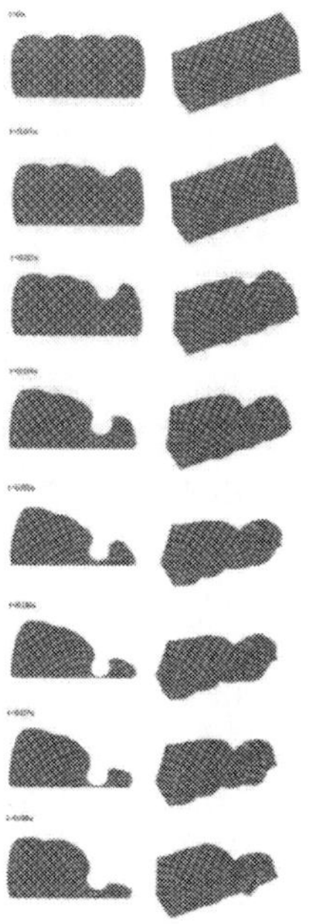

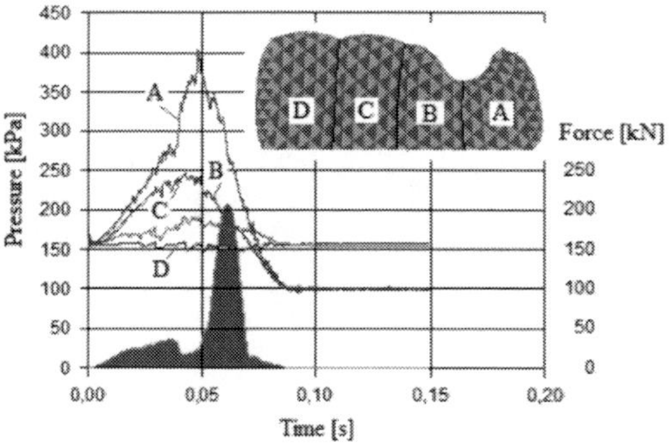

Warsaw University of Technology POLAND

Car front airbag dynamics during collision with cylindrical obstacle

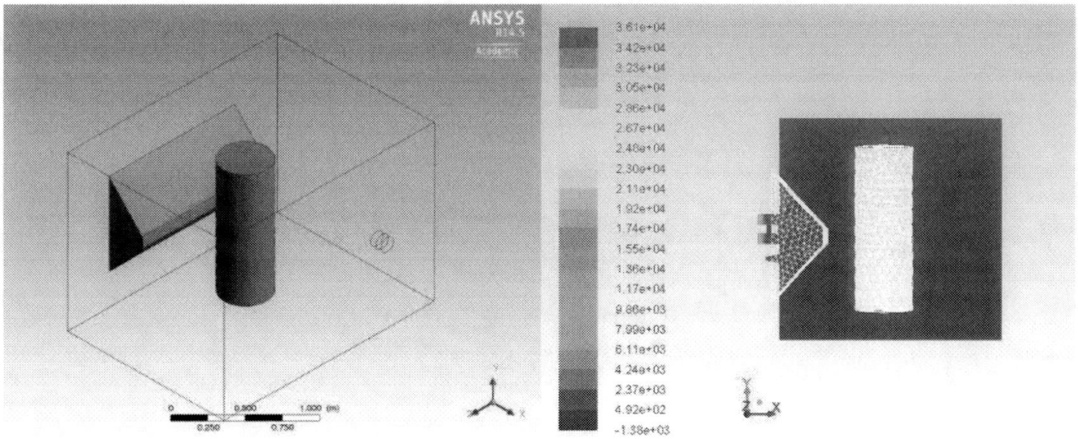

Contours of Static Pressure (pascal) (Time=4.0000e-05)
ANSYS FLUENT 14.0 (3d, dbns imp, c

Animation

Sport car simulation 27

Warsaw University of Technology POLAND

References

Barnard, R.H., Road Vehicle Aerodynamic Design. MechAero Publishing, 2001

Katz J., Race Car Aerodynamics, Robert Bentley, Inc. 1995

Piechna J., Podstawy aerodynamiki pojazdów, (Introduction to Vehicle Aerodynamics) WKŁ, Warszawa 2000.

Piechna J., Rudniak L., Piechna A., 2009, „CFD Analysis of the Central Engine Generic Sports Car Aerodynamics", Proceedings Book -Simulation for Innovative Design, 4th European Automotive Simulation Conference Munich, pp. 95-106

Tudruj S., Piechna J. " New model of the simulation of the airbag operation in the case of the out-of-position occupant - the comparison with existing models" *The Archive of Mechanical Engineering*, Vol. L, 2003, Nr. 2 pp. 201-223

Tudruj S., Piechna J., Dziewoński T., 2004, Influence of the outside air inertia effect on airbag deployment, IRCOBI , Graz.

Tudruj S., Piechna J., Numerical analysis of the possibility of using an external air bag to protect a small urban vehicle during a collision, *The Archive of Mechanical Engineering*, Vol. LIX 2012 Number 3 pp. 257-281

References 28

SymKom

Warsaw University of Technology POLAND

Thank you for attention

The End

29

The Design of Formula SAE Car Body Aerodynamics Using ANSYS Fluent Optimisation Tools

Mateusz Gugala

Warsaw University of Technology,
Supervisor: Prof. Janusz Piechna

Queen Mary University of London,
Supervisor: Dr Jens-Dominik Müller

October 26, 2013

Contents

MSc work overview

NX
- Wireframe model ⟷
- 2D models
  - Front wing profiles ⟷
  - Rear wing profiles
  - Diffuser profile ⟶
- 3D model

MSC Adams * Done by other team members
- Multibody dynamics

ANSYS
- Flow calculation
- The design of experiments
- Sensitivity fields
- Gradient based optimisation

MSc Overview Design of Experiments Adjoint Based Optimisation Formula SAE Analysis Other Sensitivity Fields Examples

NX
- Wireframe model ⟷
- 2D models
  - Front wing profiles ⟷
  - Diffuser profile ⟶
- 3D model

MSC Adams *
- Multibody dynamics

ANSYS
- Flow calculation
- The design of experiments
- Sensitivity fields
- Gradient based optimisation

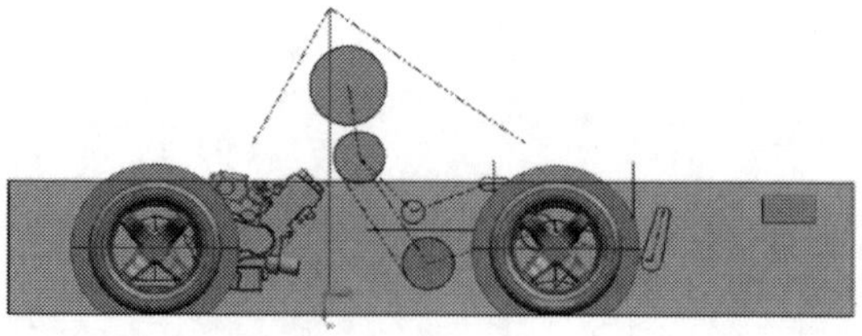

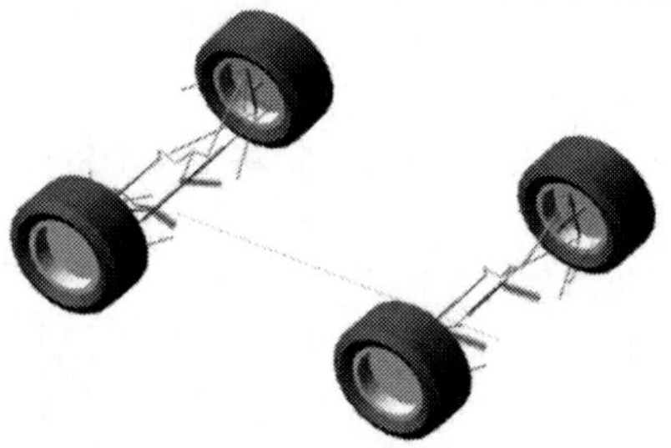

NX

- Wireframe model ⟷ Multibody dynamics
- 2D models
  - Front wing profiles
  - Rear wing profiles
  - Diffuser profile
- 3D model

MSC Adams *

- Multibody dynamics

ANSYS

- Flow calculation
- The design of experiments
- Sensitivity fields
- Gradient based optimisation

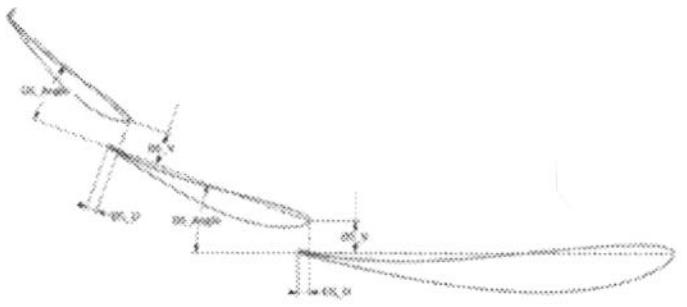

NX

- Wireframe model ⟷ Multibody dynamics
- 2D models ⟷ ANSYS
  - Front wing profiles
  - Rear wing profiles
  - Diffuser profile
- 3D model

MSC Adams *

- Multibody dynamics

ANSYS

- Flow calculation
- The design of experiments
- Sensitivity fields
- Gradient based optimisation

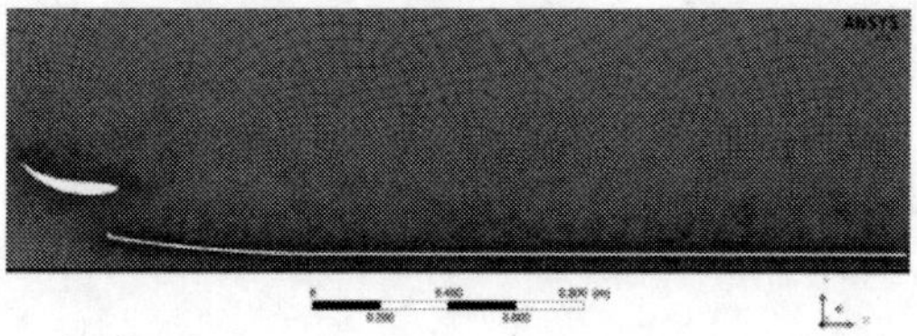

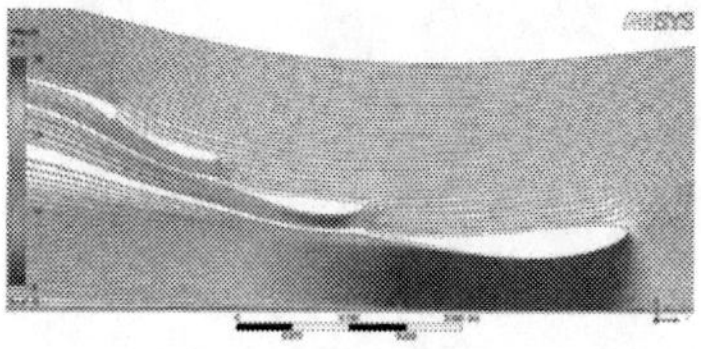

NX

- Wireframe model
- 3D model

MSC Adams *

- Multibody dynamics

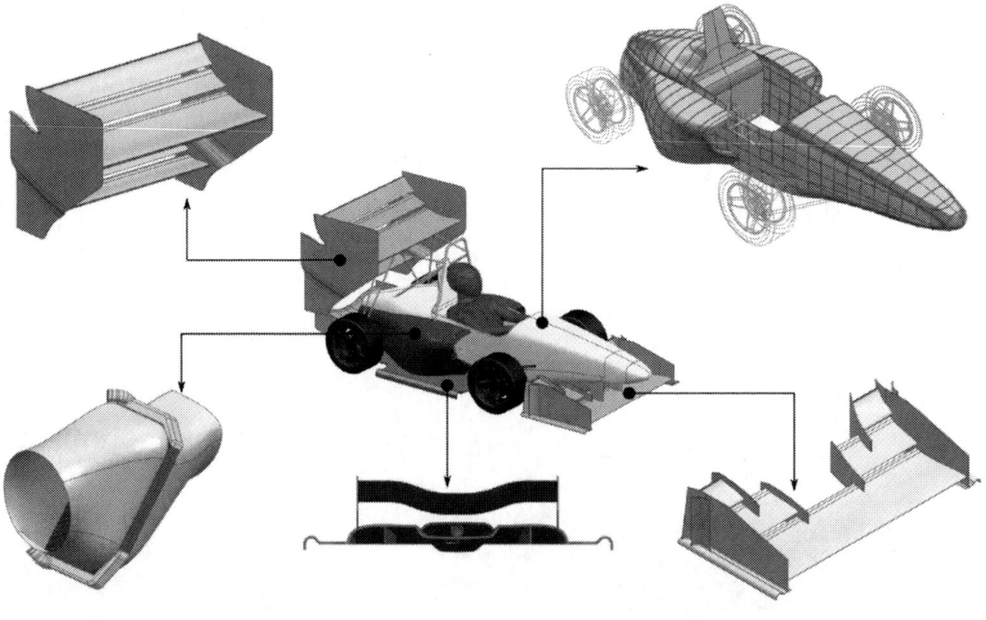

NX
- Wireframe model ⟷ Multibody dynamics
- 2D model
  - Front wing profiles
  - Rear wing profiles
  - Diffuser profile
- 3D model →

MSC Adams *
- Multibody dynamics

ANSYS
- Flow calculation
- The design of experiments
- Sensitivity fields
- Gradient based optimisation

Design of Experiments

Running CFD case:

Typical attitude:

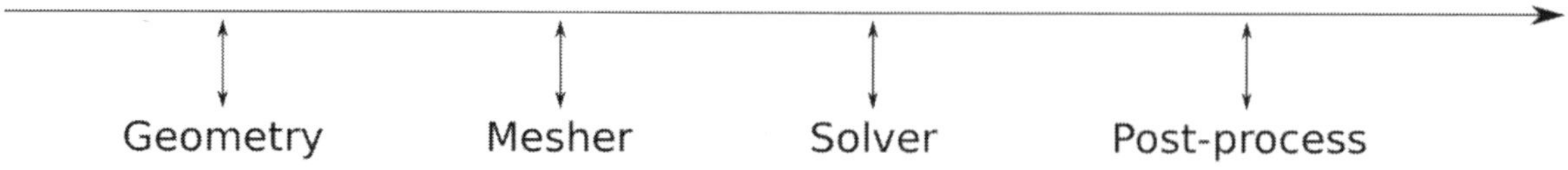

The Design of Experiment attitude:

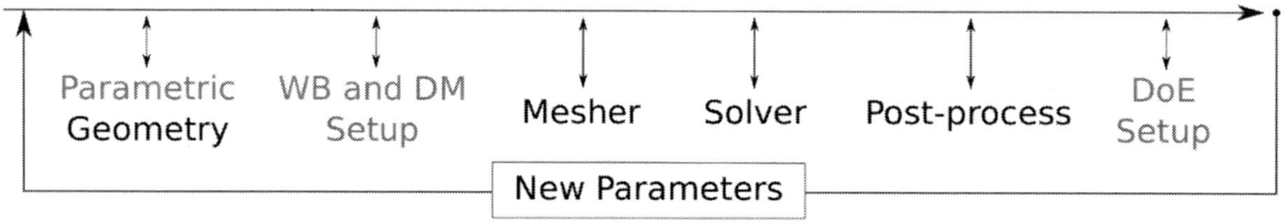

The loop can be automated with ANSYS WB!

DoE 2D example - case description

Three independent design parameters:

- *DS_U* - distance between leading edge of current aerofoil and trailing edge of preceding aerofoil (|| to its chord)
- *DS_V* - distance between leading edge of current aerofoil and trailing edge of preceding aerofoil (⊥ to its chord)
- *DS_Angle* - angle between aerofoil chords ($10° - 40°$, *increment* $10°$)

Goal: Maximize downforce

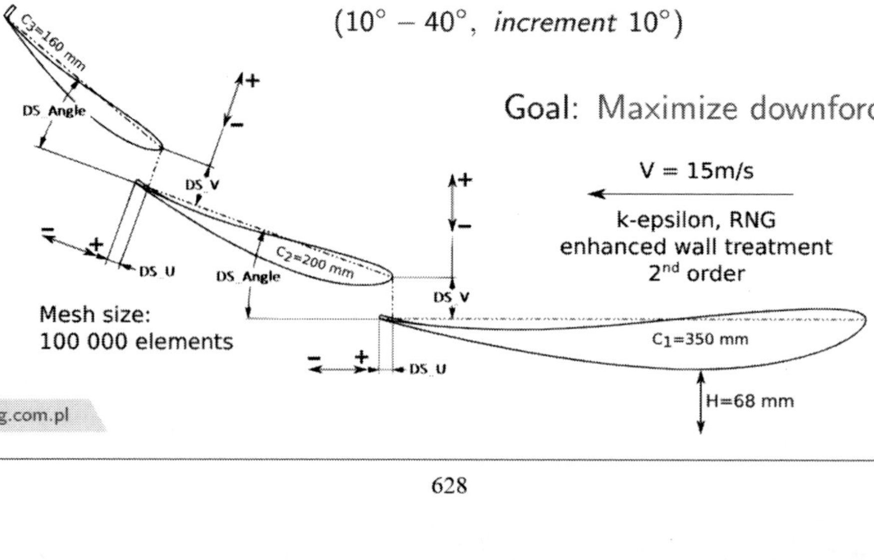

DoE 2D example - results

260 cases within 5 days !

The table gather all Quantitative results for three varying parameters: U, V, Angle.

Red color shows best result

The best result for parameter set:
$U = -20mm, V = 25mm, Angle = 20°$

10°	C_{LIFT} - lift coefficient									
U \ V	10	15	20	25	30	40	50	60	70	90
0	-3.9	-4.6	-5.0	-5.1	-5.2	-5.0	-5.0	-4.8	-4.6	-4.1
10	-2.8	-3.9	-4.6	-4.9	-5.0	-5.1	-5.0	-4.8	-4.6	-4.1
20	-2.5	-3.2	-4.2	-4.6	-4.8	-4.9	-4.8	-4.7	-4.5	-4.1
30	-2.5	-2.9	-4.0	-4.4	-4.6	-4.7	-4.7	-4.6	-4.4	-4.0
40	-2.2	-2.8	-3.7	-4.2	-4.3	-4.5	-4.6	-4.4	-4.3	-3.9
20°										
U \ V	10	15	20	25	30	40	50	60	70	90
-30	-4.3	-4.5	-4.4	-4.2	-4.6	-4.2	-4.0	-3.9	-3.7	-3.3
-20	-6.3	-6.5	-6.6	-6.6	-6.5	-5.3	-4.8	-4.4	-4.2	-3.5
-10	-5.5	-5.9	-6.0	-6.2	-6.3	-6.3	-5.7	-4.9	-4.4	-3.7
0	-4.4	-5.1	-5.4	-5.6	-5.7	-6.0	-6.0	-5.4	-4.7	-3.9
10	-3.1	-4.5	-4.9	-5.1	-5.2	-5.6	-5.8	-5.6	-4.9	-4.0
20	-2.9	-4.2	-4.5	-4.7	-4.9	-5.2	-5.4	-5.5	-5.0	-4.1
30	-3.0	-4.0	-4.3	-4.4	-4.6	-4.9	-5.2	-5.3	-5.0	-4.1
40	-3.3	-3.8	-4.0	-4.2	-4.4	-4.7	-4.9	-5.1	-4.8	-4.1
30°										
U \ V	10	15	20	25	30	40	50	60	70	90
-30	-2.9	-3.2	-3.3	-3.4	-3.3	-2.9	-2.8	-3.0	-3.0	-2.5
-20	-3.5	-3.7	-3.9	-3.7	-4.2	-3.5	-3.4	-3.2	-3.2	-2.8
-10	-5.4	-5.5	-5.6	-5.7	-5.9	-4.1	-3.9	-3.4	-3.3	-2.7
0	-4.4	-4.7	-4.9	-5.1	-5.3	-4.4	-3.9	-3.6	-3.2	-3.0
10	-3.8	-4.2	-4.4	-4.6	-4.8	-5.0	-4.0	-3.6	-3.1	-3.2
20	-3.7	-3.9	-4.1	-4.3	-4.4	-4.7	-3.7	-3.6	-3.5	-3.4
30	-3.5	-3.7	-3.9	-4.0	-4.2	-4.4	-3.6	-3.2	-3.6	-2.9
40	-3.4	-3.6	-3.7	-3.8	-4.0	-4.2	-3.3	-3.3	-3.1	-3.2
40°										
U \ V	10	15	20	25	30	40	50	60	70	90
0	-3.9	-4.2	-4.4	-4.6	-4.6	-3.9	-3.2	-2.8	-3.1	-3.2
10	-3.5	-3.7	-3.9	-4.1	-4.1	-3.2	-4.1	-3.6	-3.1	-3.1
20	-3.3	-3.5	-3.6	-3.7	-3.8	-2.8	-3.8	-3.8	-2.8	-3.6
30	-3.2	-3.3	-3.4	-3.5	-3.7	-3.4	-4.2	-3.8	-2.6	-3.4
40	-3.1	-3.1	-3.3	-3.4	-3.5	-2.7	-4.3	-3.6	-3.8	-3.6

DoE 2D example - results

$$U = -20mm, V = 25mm, Angle = 20°, C_{LIFT} = -6.62, \frac{C_{LIFT}}{C_{DRAG}} = 15.44$$

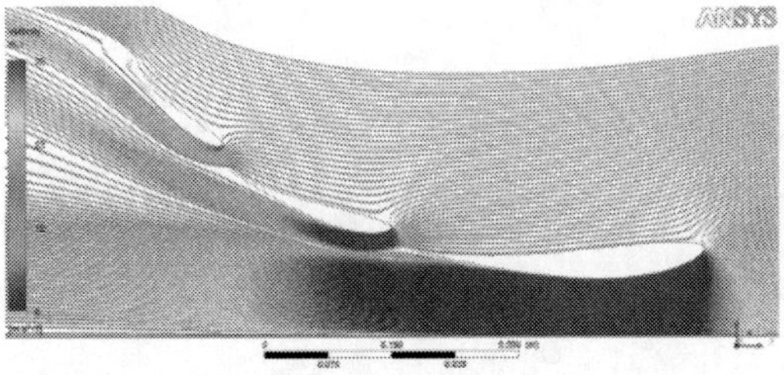

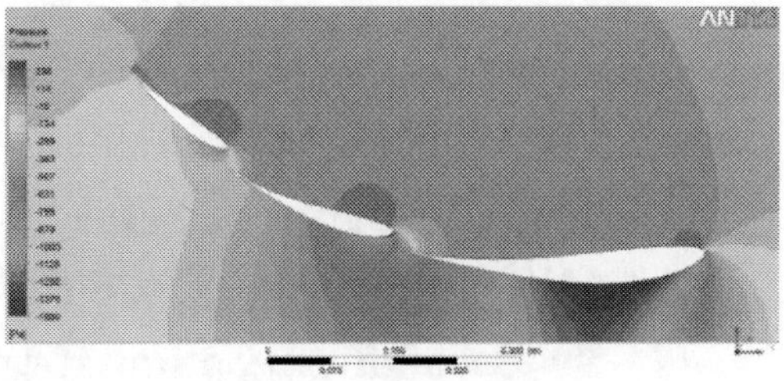

$$U = -10mm, V = 30mm, Angle = 30°, C_{LIFT} = -5.86, \frac{C_{LIFT}}{C_{DRAG}} = 9.32$$

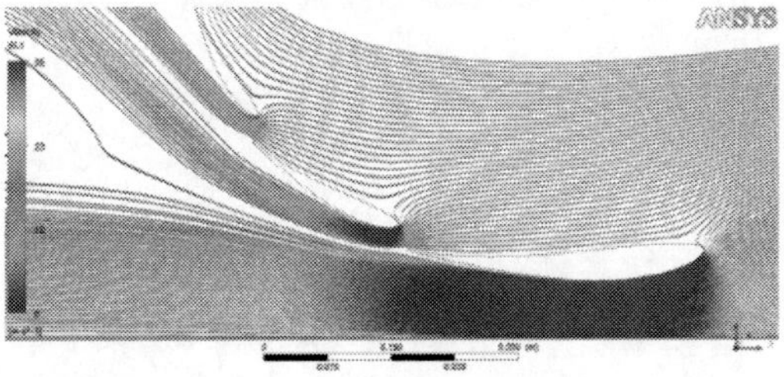

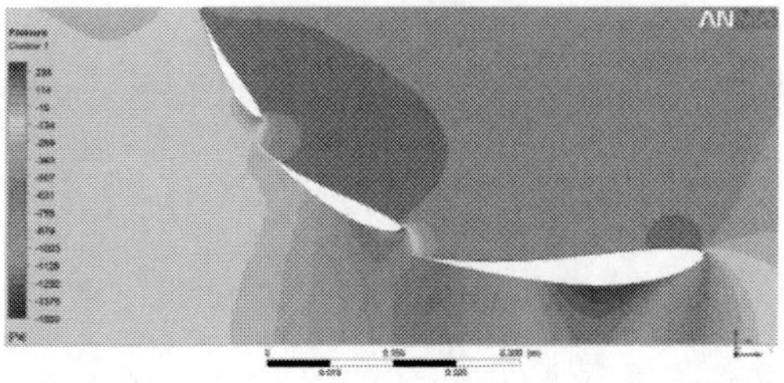

DoE 2D example - results

$$U = 0mm, V = 25mm, Angle = 40^\circ, C_{LIFT} = -4.55, \frac{C_{LIFT}}{C_{DRAG}} = 5.52$$

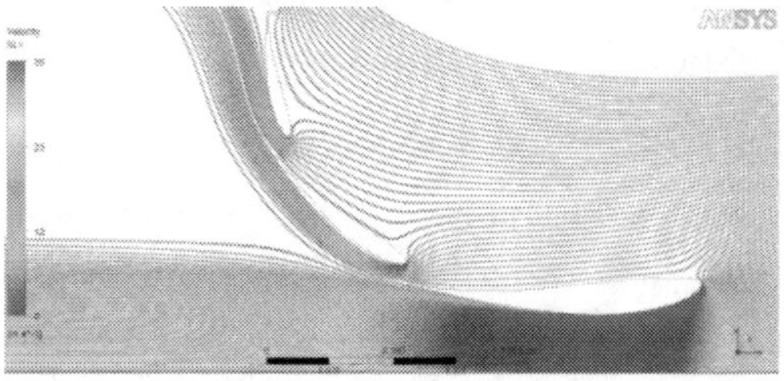

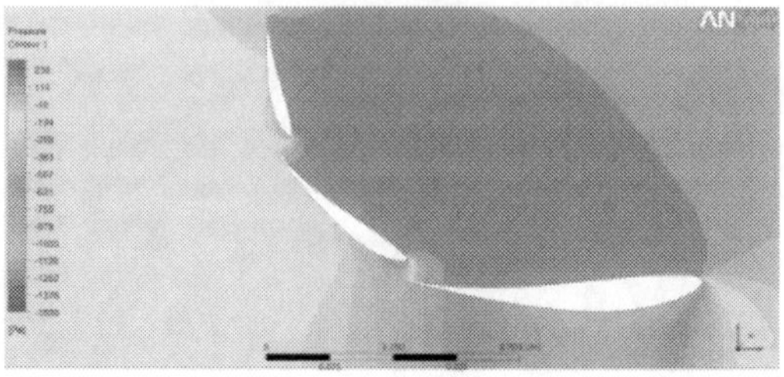

$$U = 20mm, V = 60mm, Angle = 40^\circ, C_{LIFT} = -3.81, \frac{C_{LIFT}}{C_{DRAG}} = 4.11$$ NOT RELIABLE

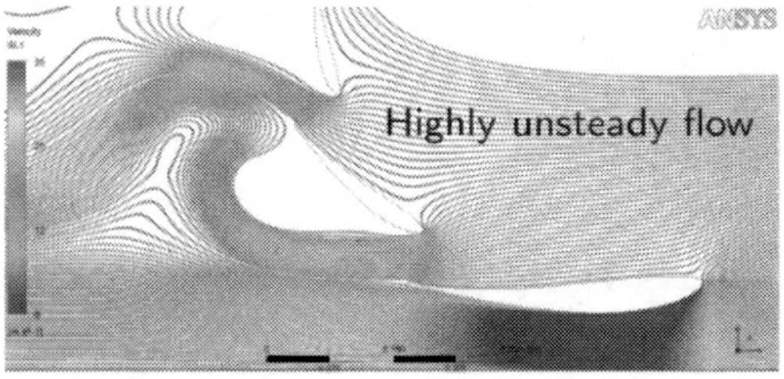

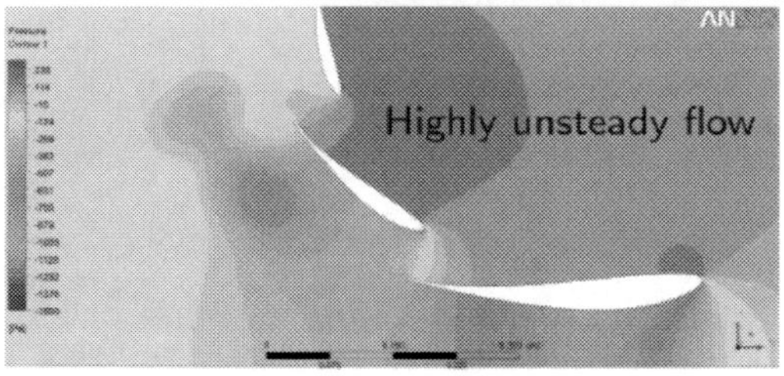

Summary

1. ANSYS WB allows to automate design processes through recording and replying scripts
2. DoE applicable for limited number of design parameters
3. DoE can be very useful in initial design stage
4. ANSYS WB offers additional features like:
   - Goal driven optimisation
   - Parameters correlation
   - Response surface
   - other...

   which can make DoE more efficient.

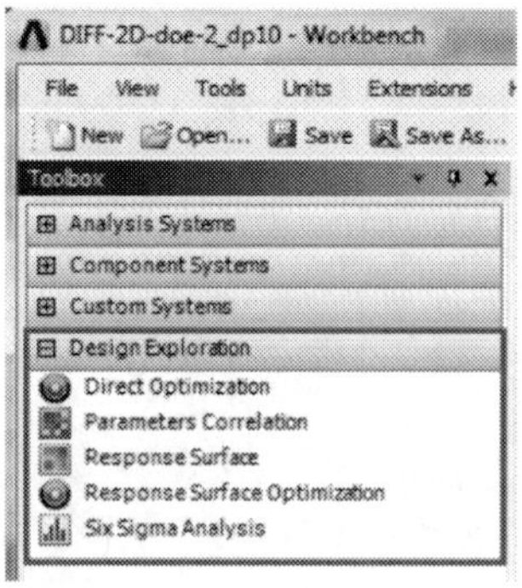

Adjoint Based Optimisation

MSc Overview

Design of Experiments

Adjoint Based Optimisation

Formula SAE 2014 Analysis

Other Sensitivity Fields Examples

Introduction and nomenclature

Naming convention:

- R(U,α), the system of flow equations (N-S equations)
- U, vector of flow variables (e.g. ρ, u, v, w, p, e)
- α, design variable (e.g. mesh node coordinates)
- J, cost function (e.g. drag, lift)
- $\frac{dJ}{d\alpha}$, gradient (shape sensitivity)

Primal (flow) solution:

$$R(U, \alpha) = 0$$

How to **efficiently** calculate the gradient?

Gradient: $$\frac{dJ(U,\alpha_i)}{d\alpha_i} = \frac{\partial J}{\partial \alpha_i} + \boxed{\frac{\partial J}{\partial U}\frac{\partial U}{\partial \alpha_i}}$$

Evaluation cost:
N flow evaluations for
N design variables

Adjoint equation: $$\left(\frac{\partial R}{\partial U}\right)^T \frac{\partial J}{\partial R} = \left(\frac{\partial J}{\partial U}\right)^T \longrightarrow \; A^T v_a = g$$

$$\frac{\partial J}{\partial U}\frac{\partial U}{\partial \alpha_i} = g^T u_p = (\underbrace{A^T v_a}_{A^T v_a = g})^T u_p = (v_a)^T \underbrace{A u_p}_{A u_p = f} = (v_a)^T f$$

$$\frac{dJ(U,\alpha_i)}{d\alpha_i} = \frac{\partial J}{\partial \alpha_i} + v_a^T f \quad \longrightarrow$$

$\sim$ **1.5-2** *flow evaluations*
for **N** design variables

ANSYS Fluent adjoint solver

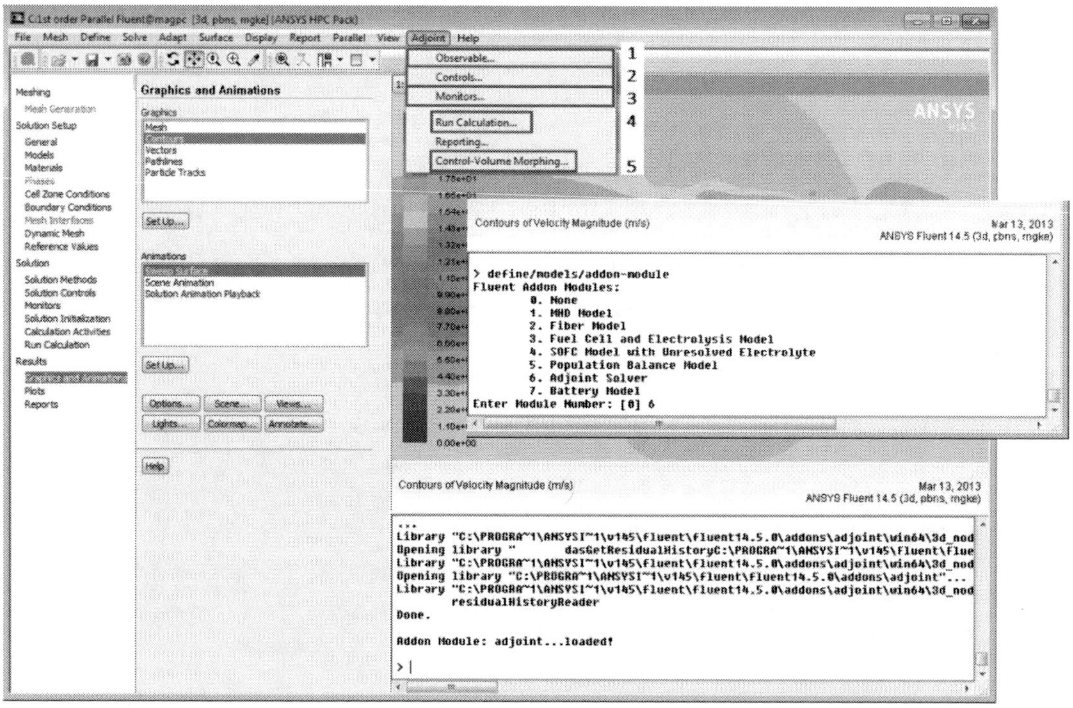

Adjoint solver setup

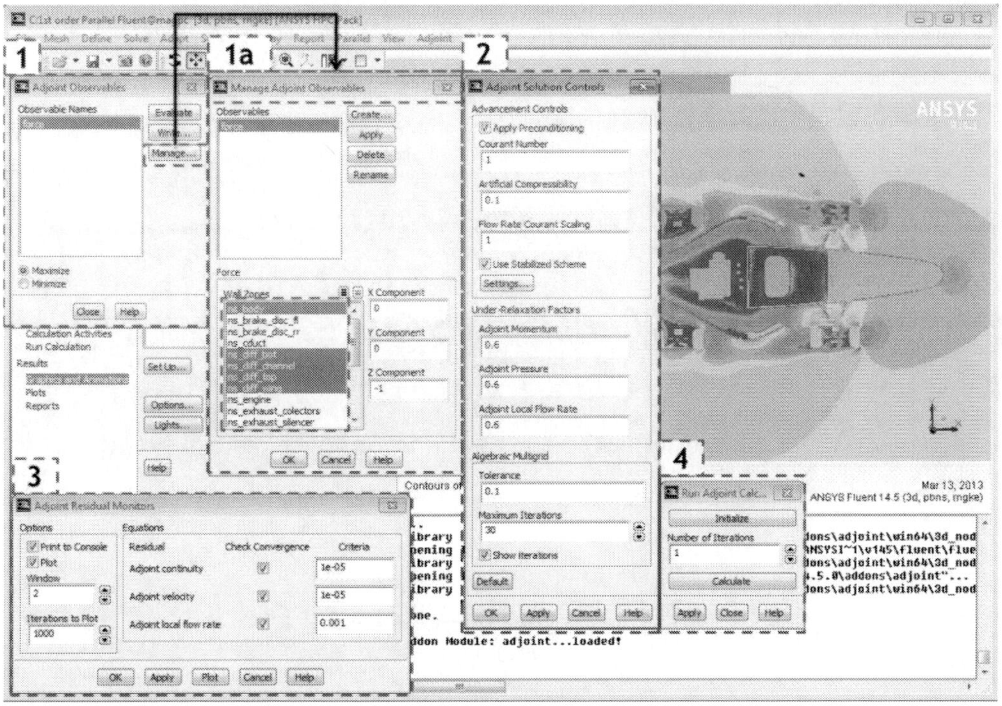

2D optimisation example - case description

Design variables: each diffuser and aerofoil mesh node ($\sim$ 1800 degrees of freedom)

Goal: Minimize lift (Maximize downforce)

Constraints:

- Diffuser total length projected on X-axis
- Aerofoil chord projected on X-axis

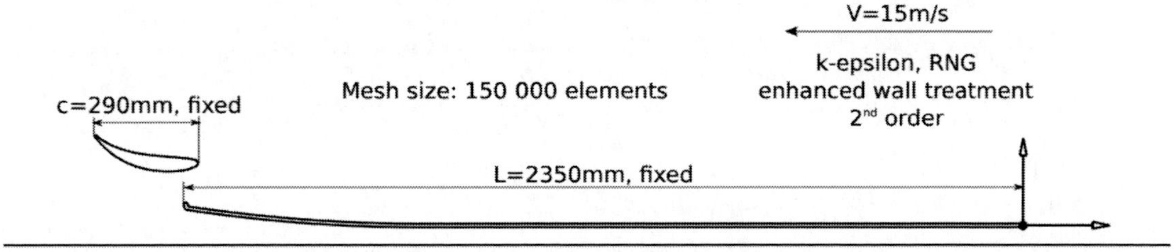

2D optimisation procedure

PRIMAL ⟶ *ADJOINT* ⟶ *ASSEMBLE GRADIENT*

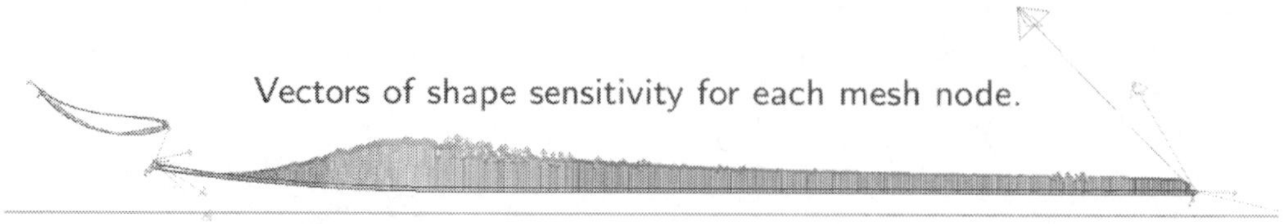

Vectors of shape sensitivity for each mesh node.

INIT MESH MORPHER ⟶ *STEEPEST DESCENT*

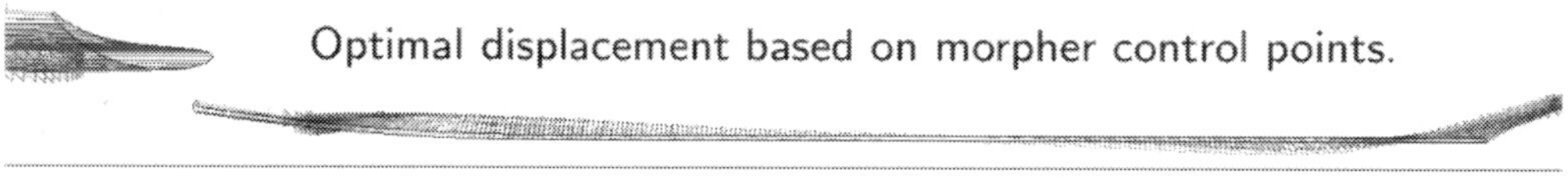

Optimal displacement based on morpher control points.

NEXT OPTIMISATION STEP

Mesh morpher

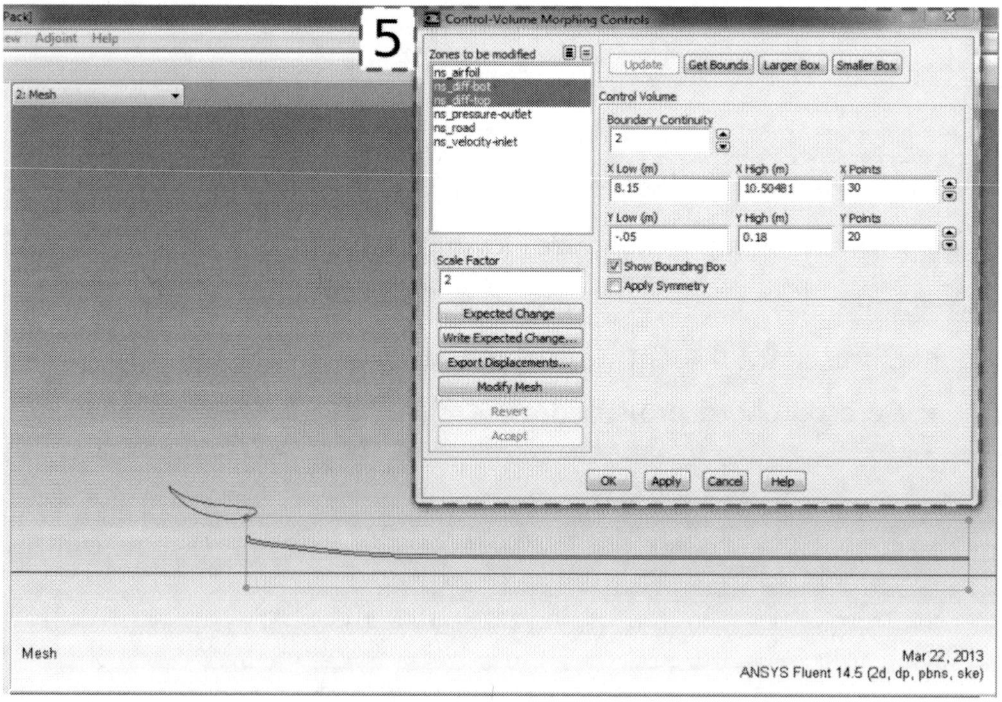

2D optimisation - results

DOWNFORCE:

Step 0: -1081.8 N

Step 6: -1281.3 N

Increased by: 18.4 %

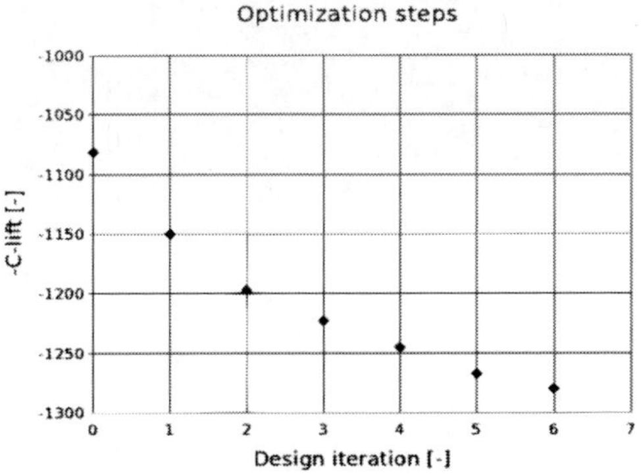

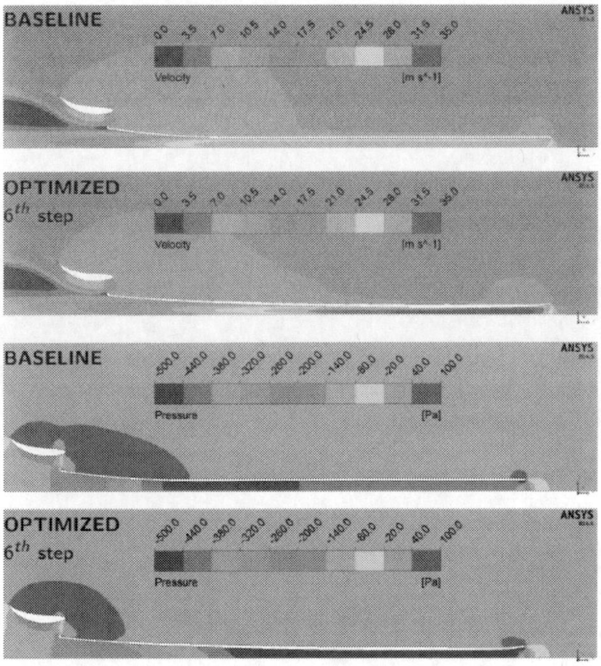

Summary

Advantages of using gradient based optimisation in industry:

- Sensitivity and adjoint solution fields gives interesting insight in flow physics
- Help engineers improve their design
- Could be useful at technological and manufacturing level. Sensitivity fields can provide an information on where high manufacturing accuracy is required and where the tolerances can be lower (reduce manufacturing costs).

Disadvantages:

- Significant effort required for code development

Cost of gradient calculation:

- Only 1.5-2 times the cost of typical flow calculation for adjoint mode

Formula SAE Analysis

MSc Overview

Design of Experiments

Adjoint Based Optimisation

Formula SAE 2014 Analysis

Other Sensitivity Fields Examples

3D analysis - case description

Main boundary conditions:
Inlet: V = 15 m/s, T = 27°C
Outlet: P = 0 Pa (OTOCZENIE: 1 atm.)
Ground: V = 15 m/s, T = 45°C
Wheels: ω = 59.055 rad/s
Exchust outlet: V = 30.3 m/s
Engine intake: m_f = 0.072kg/s

Temperatures:
Wheel tires: T_t = 45 °C
Brakes: T_b = 150 °C
Ground: T_g = 45 °C
Exchaust gases: T_{exg} = 480 °C
Engine block: T_e = 90 °C
Exchaust pipes: T_{ex-1}=720°C, T_{ex-2}=680°C, T_{ex-3}= 250 °C

Other settings:
Turbulence: k-epsilon, RNG
Enhenced wall treatment
Material: Air, ideal gas
Energy eqn.: On

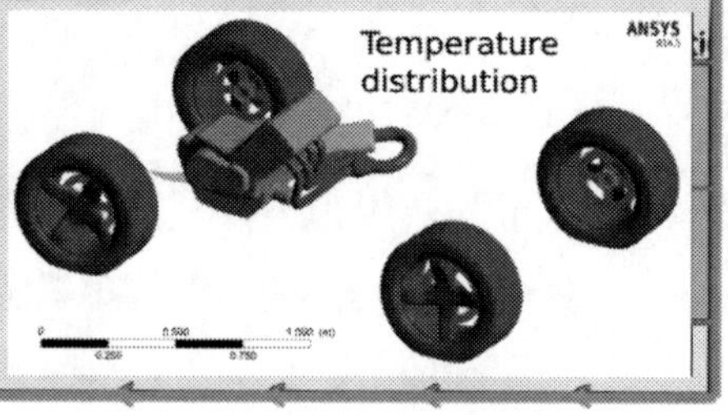

Mesh

Computational domain: 20x10x7m
Mesh type: Tetrahedron + Prisms
Number of boundary layers: 10
First layer thickness: 6e-5
Total size: ~ 20 mln cells

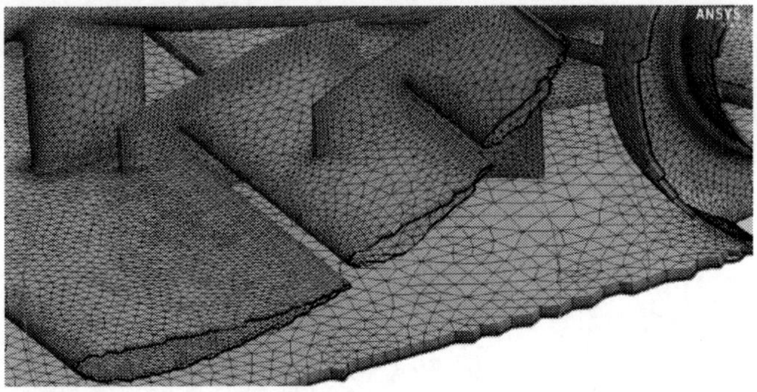

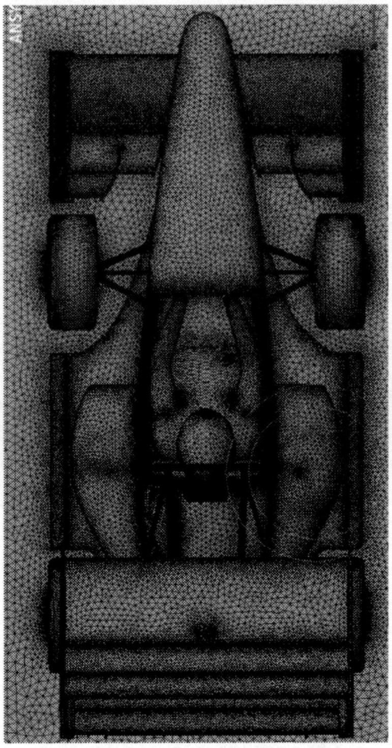

Quantitative results

Constants	
A_{ref} [m^2]	1.2
ρ_{AIR} [kg/m^3]	1.2
V [m/s]	15.0

Component	C_X [-]	F_X [N]	C_{DF} [-]	F_{DF} [N]	C_{DF} / C_X
Front wing	0.15	24.79	0.78	125.85	5.08
Main rear wing	0.56	91.39	2.07	335.23	3.67
Diffuser	0.05	7.35	0.63	102.81	13.98
Supproting wing	0.01	1.79	0.04	6.21	3.47
Body and sidepods	0.20	31.94	-0.61	-99.03	-3.10
Suspension-front	0.11	17.30	-0.07	-11.48	-0.66
Suspension-rear	0.11	18.09	-0.05	-8.88	-0.49
Engine and exhaust	0.00	0.71	0.001	0.12	0.17
Mountings and other	0.05	7.59	-0.01	-1.35	-0.18
Human	0.02	3.84	-0.03	-5.57	-1.45
Total	1.26	204.78	2.74	443.90	2.17
	Pressure Force	194.13	Pressure Force	444.72	
	Viscous Force	10.66	Viscous Force	-0.82	

The table presents force generated by each main car component .

Results

THREE MOVIES

Formula SAE car body - sensitivity fields

Cost function: Lift $\longrightarrow$ Minimize

RED: pull out, BLUE: push in

Results not reliable due to highly unsteady flow nature

Other Sensitivity Fields Examples

MSc Overview

Design of Experiments

Adjoint Based Optimisation

Formula SAE 2014 Analysis

Other Sensitivity Fields Examples

Fuel efficient vehicle 2012 - Drag minimization

Primal solution (pressure)

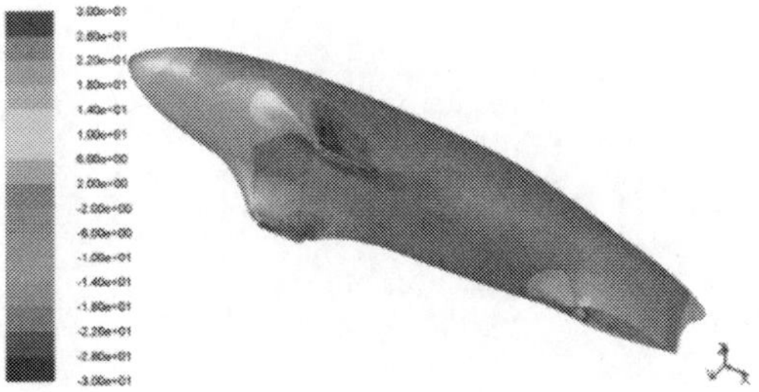

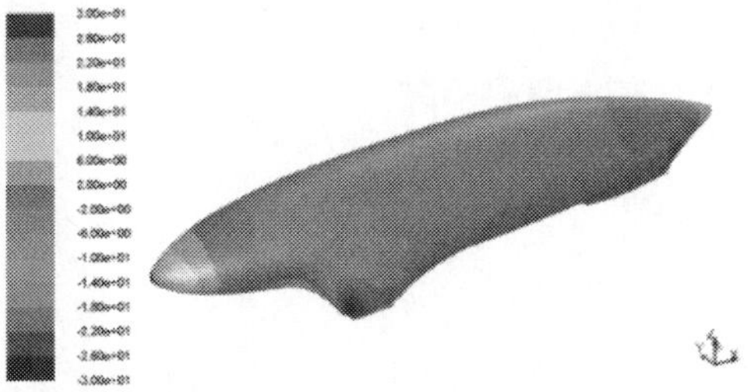

Sensitivity fields

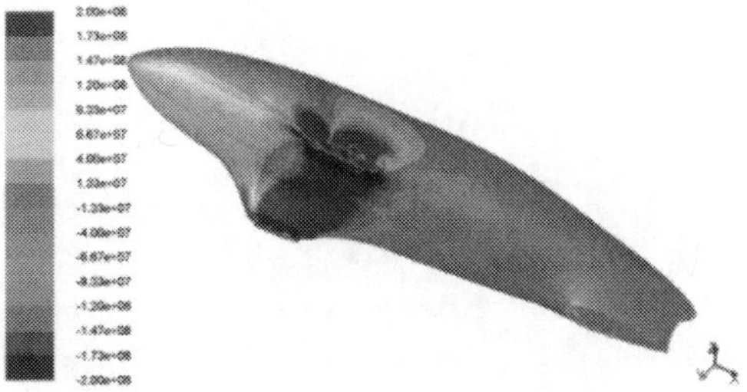

RED: pull out, BLUE: push in

Fuel efficient vehicle 2013 - Drag minimization

Primal solution (pressure)

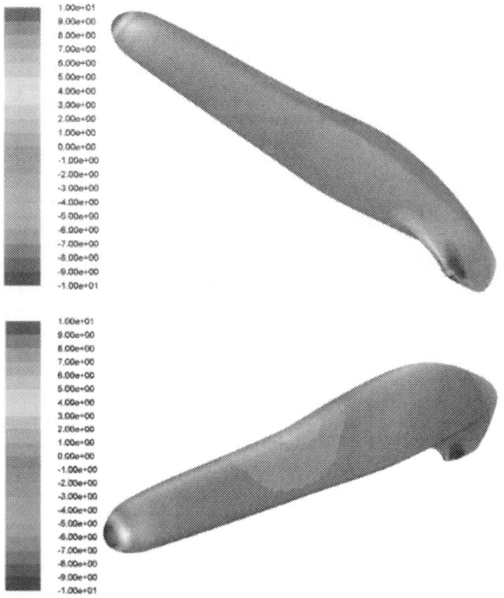

Sensitivity fields

RED: pull out, BLUE: push in

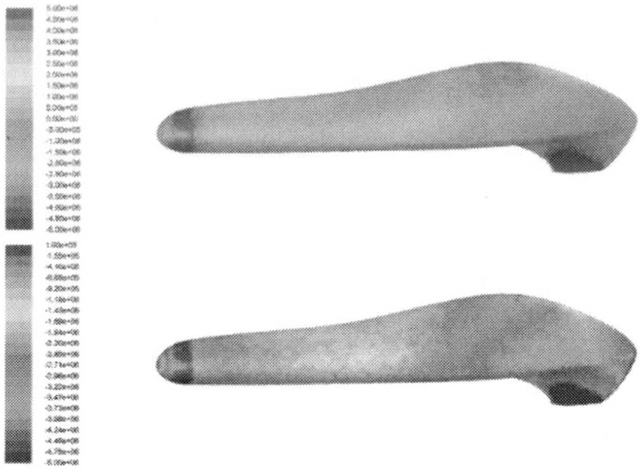

Thank you for your attention.

Ostbayerische Technische Hochschule
Amberg-Weiden

Continuous Improvements on the Wheel Hub of a Formula Student Race Car

Prof. Dr.-Ing. Horst Rönnebeck

Table of Contents

Ostbayerische Technische Hochschule
Amberg-Weiden

Major Criteria at Race Car Development

- Rules of formula student competition
  - Maximum safety
  - Maximum space for different developments
  - Every year a new car
- Performance
  - Low mass
  - Low friction
  - Good drivability
  - High stiffness
  - High power
- Reliability
- Costs
- Deadlines
  - Competition
  - Own

Enter the Competition 2005

Major goals:

- Meet all formula student rules
- Meet all deadlines
- See checkered flag

Basic conditions (or limiting factors):

- Starting from scratch
- Rough and only theoretical load assumptions

Engineering tools:

- 3D-CAD-program Pro/Engineer Wildfire
- Simple FEA´s integrated within CAD

Enter the Competition 2005

Design decision for wheel hubs:
- (Re)use wheel hubs from Golf II

Results:
- Quick "development"
- Cheap
- Heavy

- and...

Best European Newcomer Award!

Wheel Hub Design 2006/2007

Goals:
- Improve knowledge about loads using strain gauges on suspension members
- Self designed wheel hubs
- Reduce weight drastically

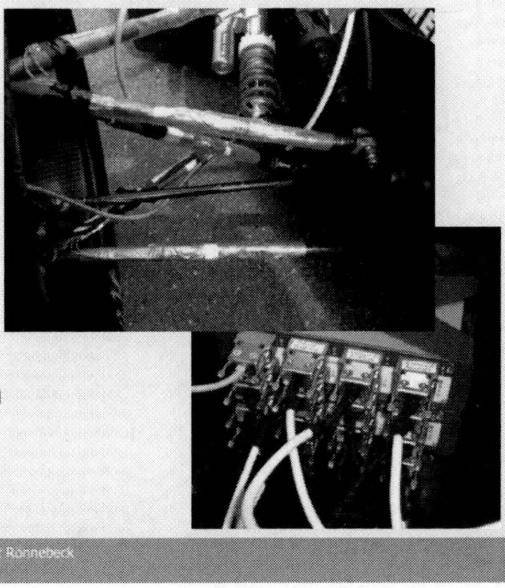

Engineering Tools:
- 3D-CAD-program Pro/Engineer Wildfire
- FEA´s integrated within CAD
- First steps on using ANSYS Workbench
- Suspension kinematic simulation program

Wheel Hub Design 2006/2007

Design decision for wheel hubs:
- Simple design
- Light
- Tapper roller bearings (low priced)
- Radial shaft sealing rings

Axle Leg and Wheel Hub Assembly RS06

Results:
- Very light design
- Weaknesses in design (notch effect)
- "Alignments" during FEA´s lead to underestimation of loads
- Wheel hubs crack during testing
- Use of fall-back-solution (steel instead of aluminum)

Wheel Hub Design 2008/2009

Goals:
- Increase strength and stiffness
- Reduce friction

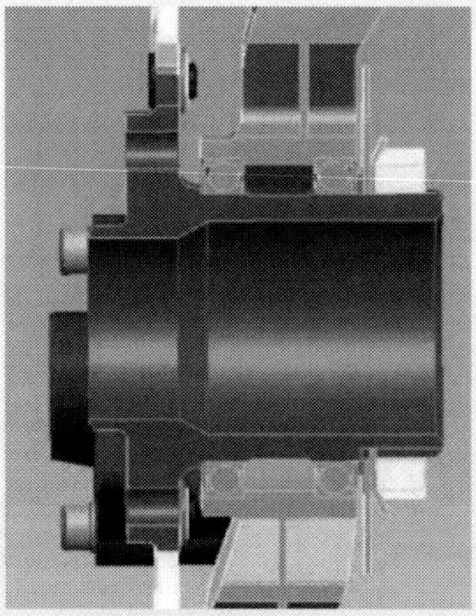

Design decision for wheel hubs:
- Hollow shaft with high diameter
- Angular contact ball bearings
- Hybrid bearings (Si_3N_4 balls)
- Contactless seals

Results:
- No failure of wheel hubs during tests and competitions (till today)
- First season with successful ending all endurance events

Wheel Hub Design 2010/2011

Goals:
- Reduce weight
- Improve stiffness

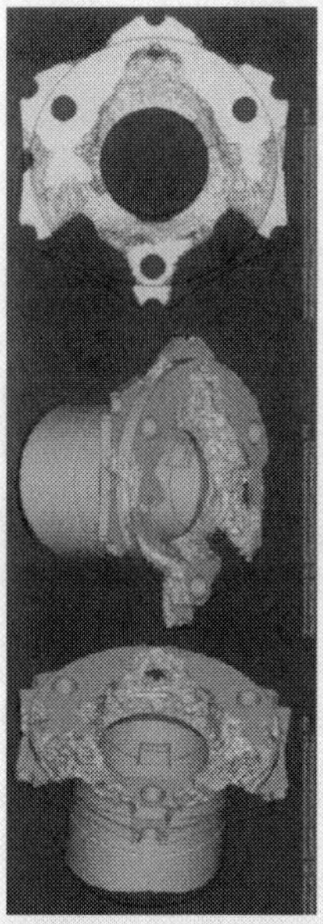

Engineering Tools:
- 3D-CAD-Program Pro/Engineer Wildfire
- FEA´s integrated within CAD
- ANSYS Workbench
- Suspension kinematic simulation program
- Topology optimization program

Wheel Hub Design 2010/2011

Design decision for wheel hubs:
- Similar design to last seasons
- Angular contact ball bearings
- Hybrid bearings (Si_3N_4 balls)
- Contactless seals
- Three wheel bolts

2007 – 359.4 g 2011 – 334.0 g

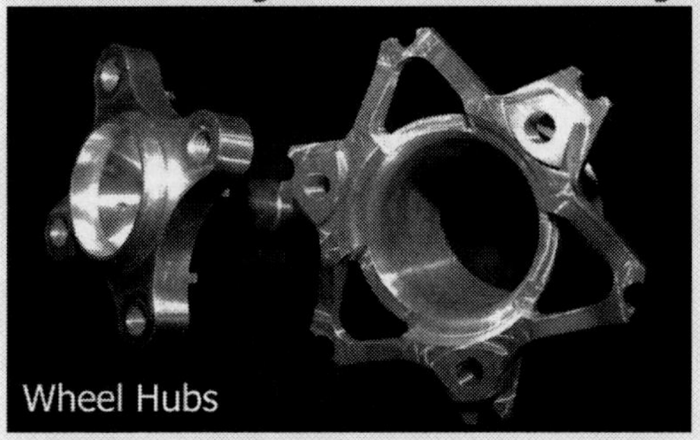

Results:
- Lighter and stiffer design
- Formula student Hungary, 2010: 1. Place overall
- Airbus light weight award in Silverstone, 2010
- Engineering award in Michigan, 2011

Conclusion

The Formula Student Competition is an excellent possibility for students to
- Improve their engineering knowledge
- Learn from experiences

and especially to train engineering and managing skills like

- Teambuilding
- Project management
- Meet deadlines
- Customer relations
- Networking
- Communication skills
- Competition thinking

Conclusion

RS05 – 350 kg
RS06 – 250 kg
RS07 – 235 kg
RS08 – 232 kg
RS09 – 190 kg
RS10 – 168 kg
RS11 – 163 kg
RS12 – 160 kg
RS13 – 200 kg
1. E-Car of Running Snail

Conclusion

Ostbayerische Technische Hochschule Amberg-Weiden

What about the wheel hubs of seasons 2012 and 2013?

That is another story...

Thank you for your attention!

Continuous Improvements on the Wheel Hub of a Formula Student Race Car

Prof. Dr.-Ing. Horst Rönnebeck 30.10.2013 / 13

The Benefit of SpaceClaim for Ansys Users

Markus Huebner, Sales Director Germany and Austria

"Our work is not to draw plans for machining; we are here to imagine and conceive new systems…. SpaceClaim lets me bring designs to fruition four to five times faster than previous methods."

- Paul Gateau, Founder of SYNGAS

SPACECLAIM CORPORATION

Ansys SpaceClaim Direct Modeler Value for CAE experts who…

- **are not a skilled CAD users**
- **have very complex models that need to be defeatured**
- **do not always receive native CAD files**
- **do not have native CAD system installed and licensed**
- **want to build conceptual models**
  - Wants to be liberated from the CAD group!

SPACECLAIM CORPORATION

CAE Model Preparation

➢ **Benefits**

– Improve time-to-result by 10x
– Empower CAE analysts to work in 3D
– Simulation proceeds CAD
– Strong ROI, very fast pay back

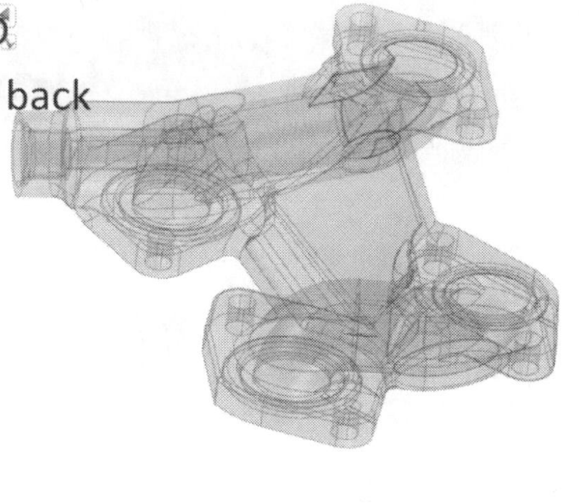

SPACECLAIM CORPORATION

SpaceClaim Corporation

- **Founded in late 2005 by industry leaders**
  - Founders at leading CAD companies PTC and SolidWorks
- **Initial product released in 2008, followed by 5 years of triple- or double-digit revenue growth**
- **Head-quartered in the US, with regional offices in UK, Germany, France, Denmark, and Japan**
  - Sales in 30 countries through global reseller network
  - SpaceClaim software is translated into 10 languages

SPACECLAIM CORPORATION

Embraced by Top Companies Around the World

SpaceClaim – the right tool for CAE users

Join our booth at the conference

SPACECLAIM

Live Demo
Markus.huebner@spaceclaim.com
+49 176 32246656
With SpaceClaim, I have no limitations for design, our communications with customers are improved, and we've saved 60% to 70% of our time, enabling us to take on more business.
- Ed Swets, General Manager at Batelaan Plastics
SPACECLAIM
CORPORATION
© Copyright SpaceClaim 2013 – Confidential Material
© Copyright 2013

2013 **Automotive Simulation**
World Congress

Optimization with Adjoint Solver

Dipl.-Ing. Oliver Zühlke

Outline

- Introduction to Optimization Methods
- Introduction to the Fluent Adjoint Solver
- Fundamentals
- Typical Workflow
- Capabilities
- Mesh to CAD
- Outlook R15+
- Summary

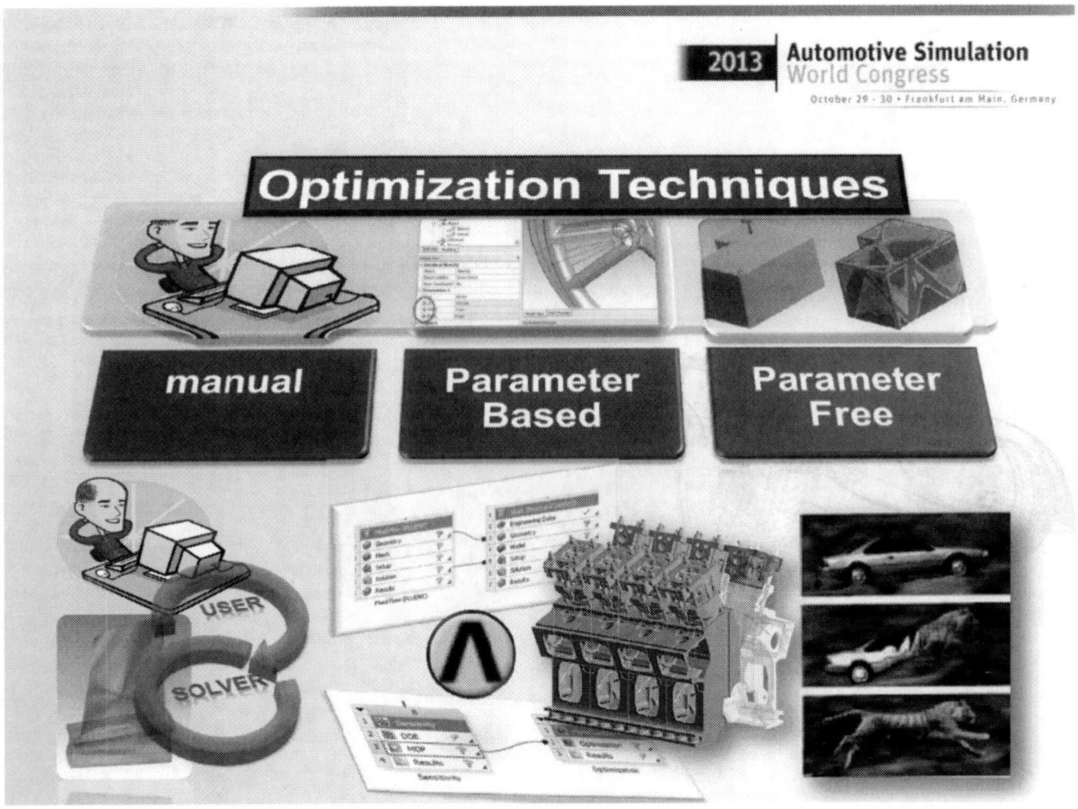
2013
Automotive Simulation
World Congress
October 29 - 30 • Frankfurt am Main, Germany
Optimization Techniques
manual
Parameter Based
Parameter Free
USER
SOLVER

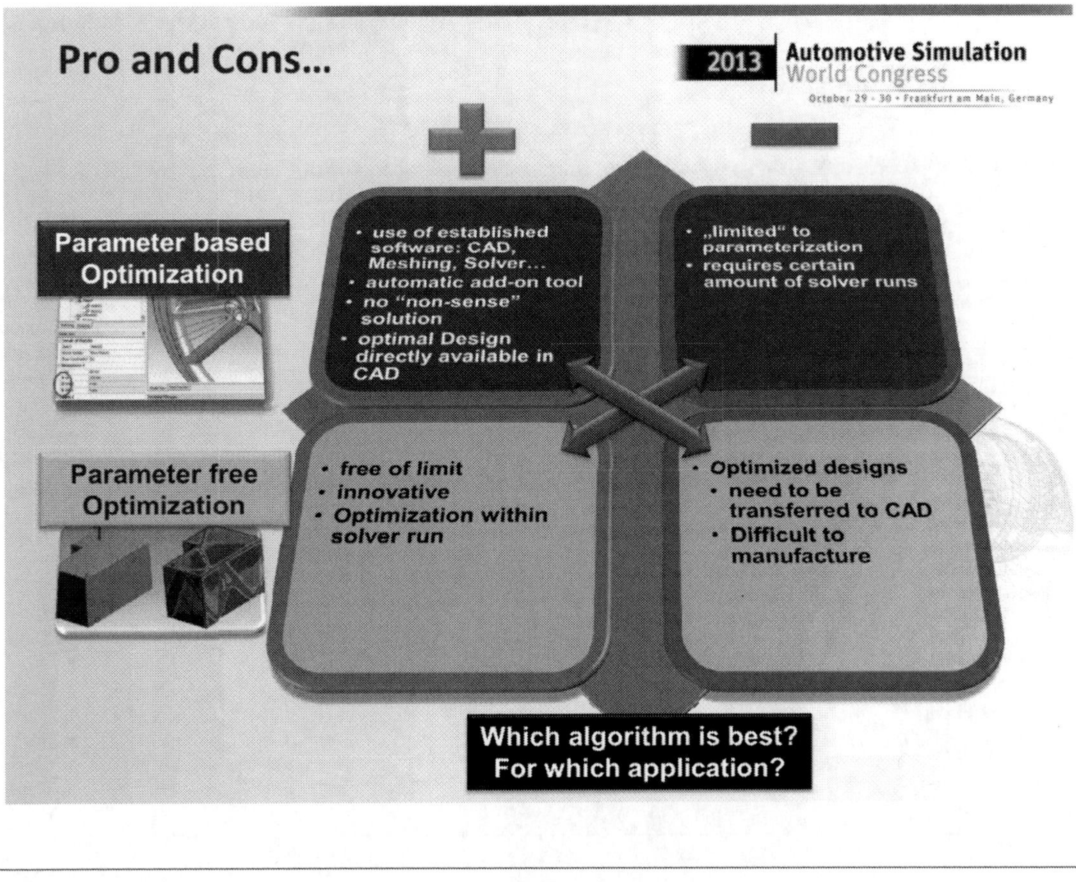
Pro and Cons...
2013
Automotive Simulation
World Congress
October 29 - 30 • Frankfurt am Main, Germany
Parameter based Optimization
Parameter free Optimization
• use of established software: CAD, Meshing, Solver...
• automatic add-on tool
• no "non-sense" solution
• optimal Design directly available in CAD
• „limited" to parameterization
• requires certain amount of solver runs
• free of limit
• innovative
• Optimization within solver run
• Optimized designs
• need to be transferred to CAD
• Difficult to manufacture
Which algorithm is best?
For which application?

2013 **Automotive Simulation** World Congress
October 29 - 30 • Frankfurt am Main, Germany

Introduction to the Fluent Adjoint Solver

What is the Adjoint Solver?

2013 **Automotive Simulation** World Congress
October 29 - 30 • Frankfurt am Main, Germany

An Adjoint Solver (AS) is a specialized tool that extends the scope of the analysis provided by a conventional flow solver by providing detailed sensitivity data for the performance of a fluid system.

An Adjoint Solver can be used to compute the derivative of an engineering quantity with respect to <u>all</u> of the inputs for the system.

Once the adjoint solution is computed it can be used to guide intelligent design modifications to a system by a simple gradient algorithm for design optimization.

Motivation

2013 Automotive Simulation World Congress
October 29 - 30 • Frankfurt am Main, Germany

Why is derivative data important?

Insight	Identification of the most important factors affecting performance.
Design Exploration	How a prescribed change will alter the performance.
Optimization	Systematic improvement of performance using gradient information.
Robust Design	Comprehensive identification of the most influential design parameters.
Robust Simulation	Analysis of the sensitivity of the result to the mesh and numerical schemes.

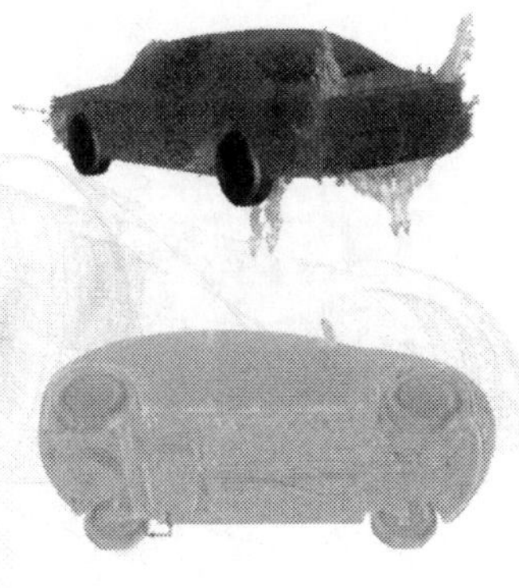

$\log_{10}$(magnitude of shape sensitivity)

2013 Automotive Simulation World Congress
October 29 - 30 • Frankfurt am Main, Germany

Fundamentals of the Adjoint Solver

"I THINK YOU SHOULD BE MORE EXPLICIT HERE IN STEP TWO."

Key Ideas - Fundamentals

2013 Automotive Simulation World Congress
October 29 - 30 • Frankfurt am Main, Germany

Discrete or continuous adjoint?

- Continuous
  - Mathematically formal.
  - Adjoint is constructed at PDE level.
  - Easier initial implementation.
  - Wall functions, boundary conditions and expansion to richer physics can all be problematic.
- Discrete — **CHOSEN METHOD**
  - Numerically formal.
  - Adjoint is constructed at the level of the discretized equations.
  - Mechanical process to construct the adjoint – somewhat challenging.
  - Easier to test.

Key Ideas - Fundamentals

2013 Automotive Simulation World Congress
October 29 - 30 • Frankfurt am Main, Germany

High-level "system" view of a conventional flow solver

Inputs

- Boundary mesh
- Interior mesh
- Material properties
- Boundary condition 1
  - Flow angle
  - Inlet velocity
  - ...
- ...

FLOW SOLVER

Outputs

- Field data
  - Contour plots
  - Vector plots
- xy-plots
- Scalar values
  - Lift
  - Drag
  - Total pressure drop

Key Ideas - Fundamentals

2013 Automotive Simulation World Congress
October 29 - 30 • Frankfurt am Main, Germany

How are changes to key outputs dependent on changes to the inputs?

Inputs

- Boundary mesh
- Interior mesh
- Material properties
- Boundary condition 1
  - Flow angle
  - Inlet velocity
  - ...
- ...

?

ADJOINT SOLVER

Outputs

- Field data
  - Contour plots
  - Vector plots
- xy-plots
- Scalar values
  - Lift
  - Drag
  - Total pressure drop

2013 Automotive Simulation World Congress
October 29 - 30 • Frankfurt am Main, Germany

Typical Workflow

Closed Optimization Workflow

2013 Automotive Simulation World Congress
October 29 - 30 • Frankfurt am Main, Germany

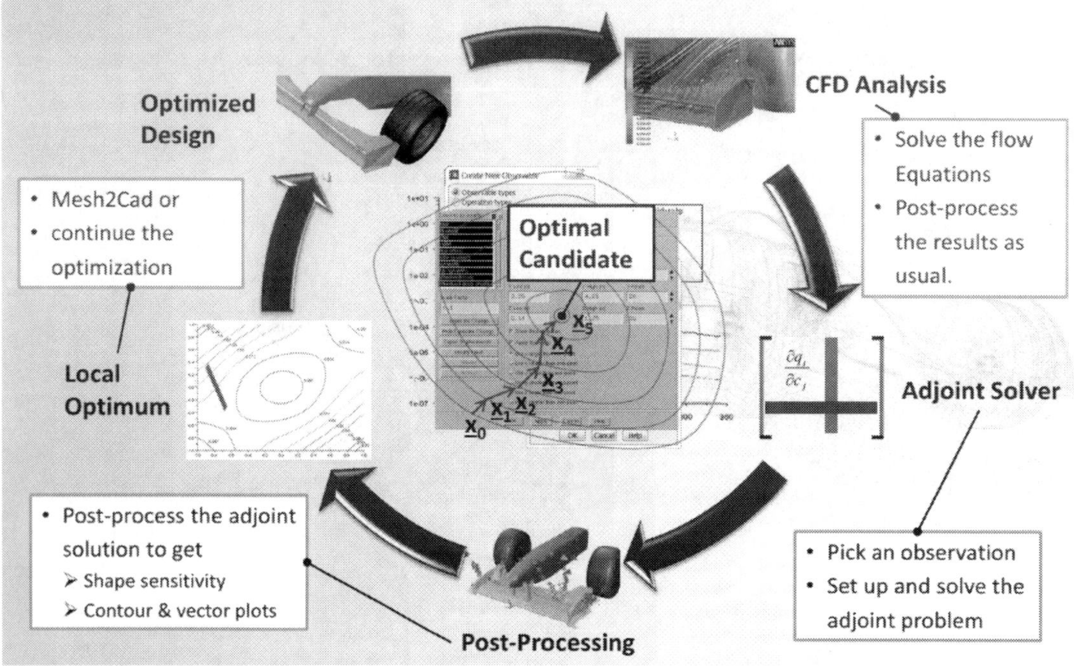

Morphing – Optimization loop

2013 Automotive Simulation World Congress
October 29 - 30 • Frankfurt am Main, Germany

- To be close to the global optimum candidate, the complete optimization can run by a script automatically
- A *Journal file* or *Scheme routine* can be used

```
(define start 1)
(define stop 30)
(do ((i start (+ i 1)))
   ((> i stop) i)
   (ti-menu-load-string "/adjoint/observable/write y observable y\n")
   (ti-menu-load-string "/adjoint/run/initialize\n")
   (ti-menu-load-string "/adjoint/run/iterate 2000\n")
   (ti-menu-load-string "/adjoint/morphing/control-point-update\n")
   (ti-menu-load-string "/adjoint/morphing/modify-mesh\n")
   (ti-menu-load-string "/solve/initialize/hyb-initialization yes\n")
   (ti-menu-load-string "/solve/iterate 3000\n")
   (ti-menu-load-string (format #f "/file/write-case-data ANSYS_is_great.cas.gz\n" i))
```

2013 Automotive Simulation World Congress
October 29 - 30 • Frankfurt am Main, Germany

Applications / Capabilities

Applications / Capabilities

2013 Automotive Simulation World Congress
October 29 - 30 • Frankfurt am Main, Germany

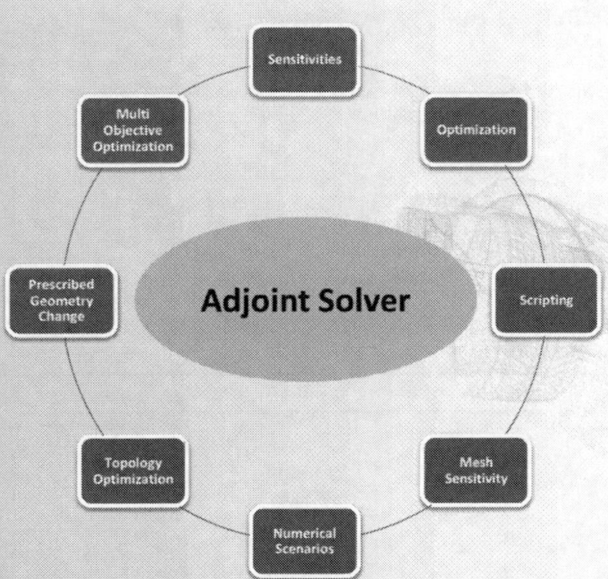

Sensitivities - Sedan

2013 Automotive Simulation World Congress
October 29 - 30 • Frankfurt am Main, Germany

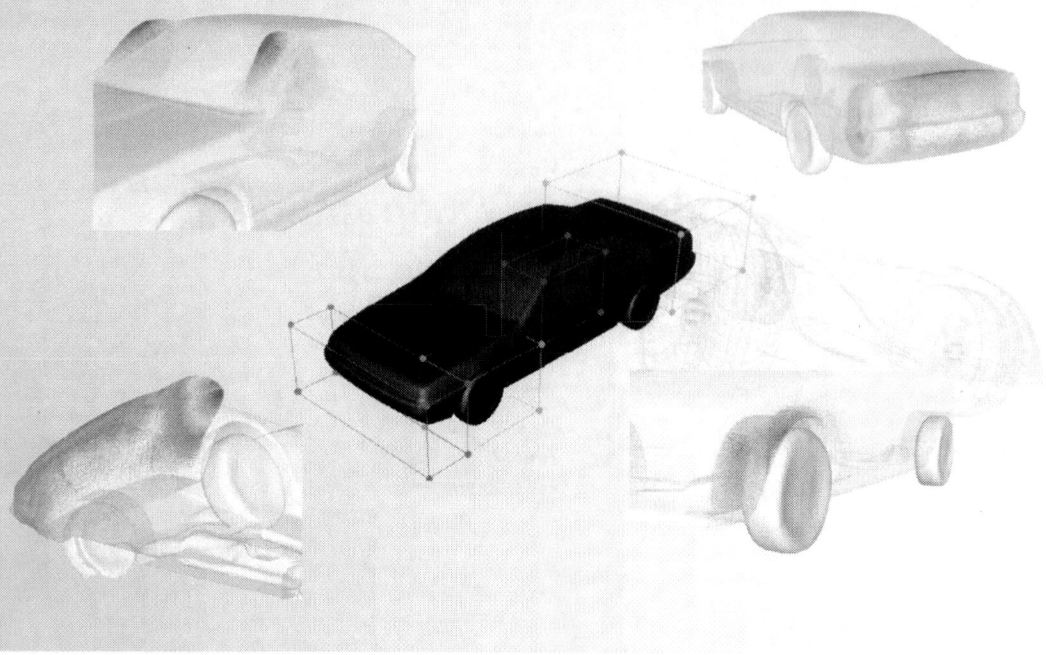

Sensitivities - Sedan

2013 Automotive Simulation World Congress
October 29 - 30 • Frankfurt am Main, Germany

- Sensitivity contour plot of ALL geometry parts
- get an overview of ... ties

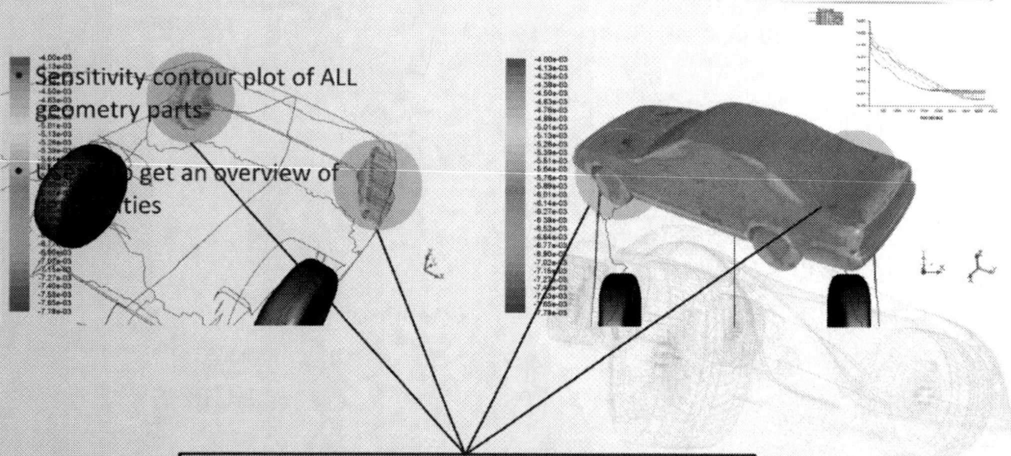

Identification of the most influential design parameters

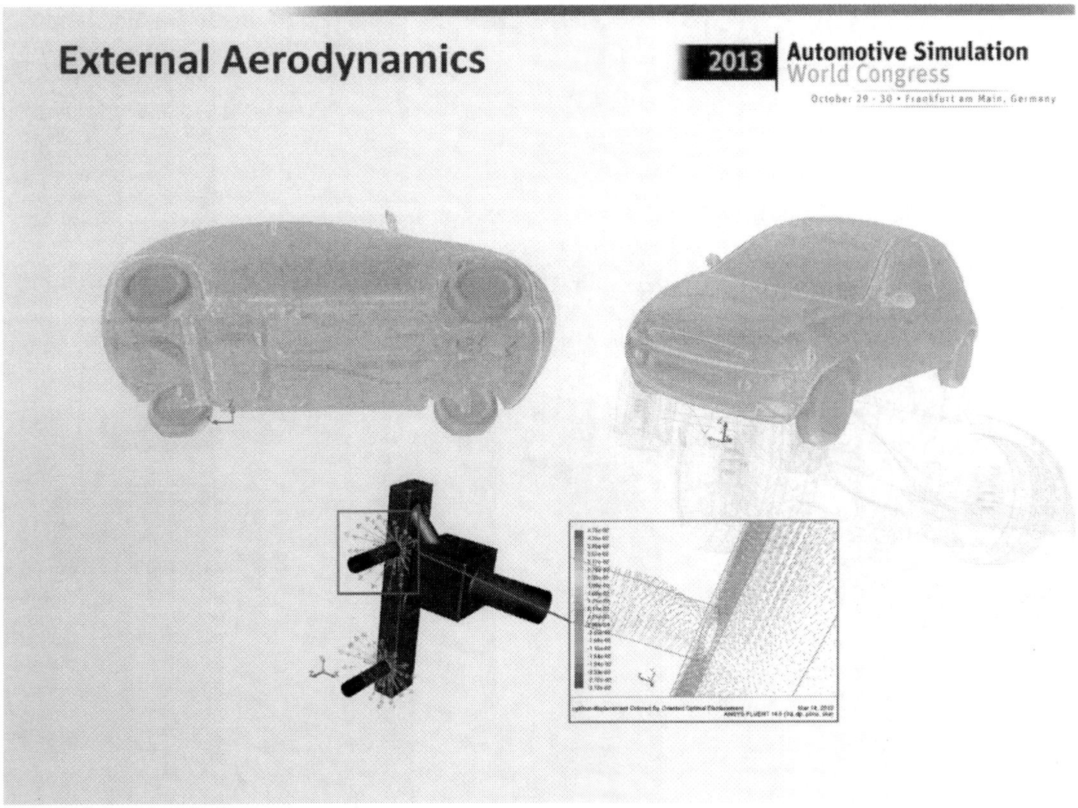
External Aerodynamics
2013 Automotive Simulation World Congress
October 29 - 30 • Frankfurt am Main, Germany

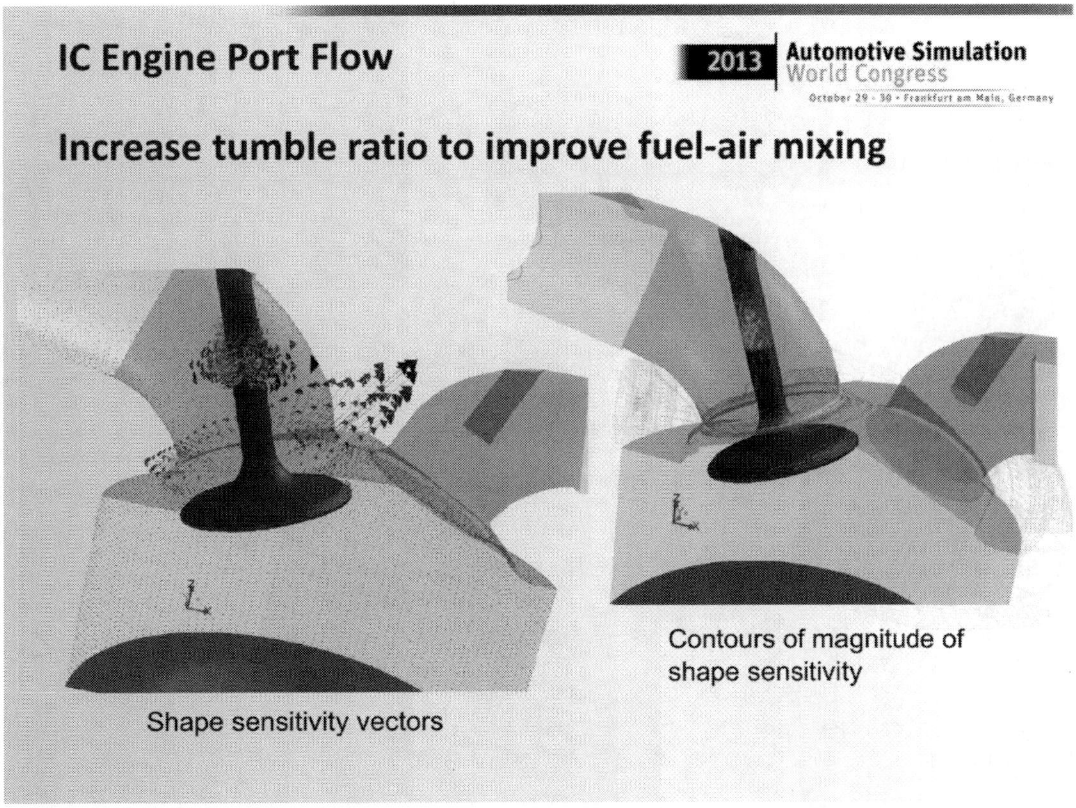
IC Engine Port Flow
2013 Automotive Simulation World Congress
October 29 - 30 • Frankfurt am Main, Germany
Increase tumble ratio to improve fuel-air mixing
Contours of magnitude of shape sensitivity
Shape sensitivity vectors

What does it mean for you?

2013 Automotive Simulation World Congress
October 29 - 30 • Frankfurt am Main, Germany

Better insight into your application

Identification of sensitive areas

Gives you new ideas for improved designs

Applications / Capabilities

2013 Automotive Simulation World Congress
October 29 - 30 • Frankfurt am Main, Germany

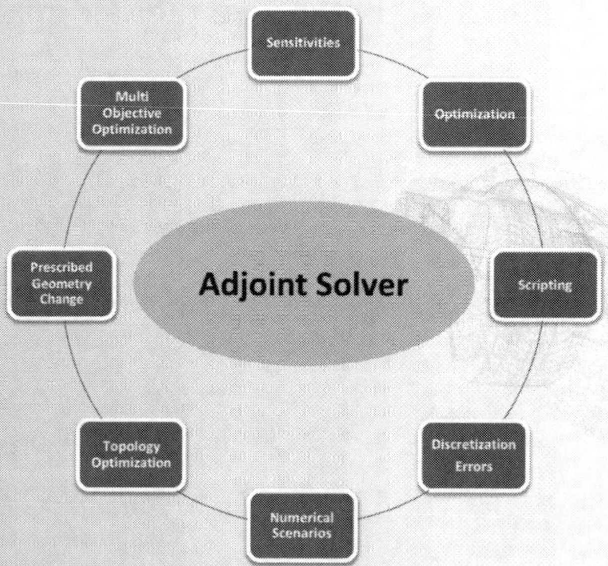

F1 Front Wing – Max Downforce

2013 Automotive Simulation World Congress
October 29 - 30 • Frankfurt am Main, Germany

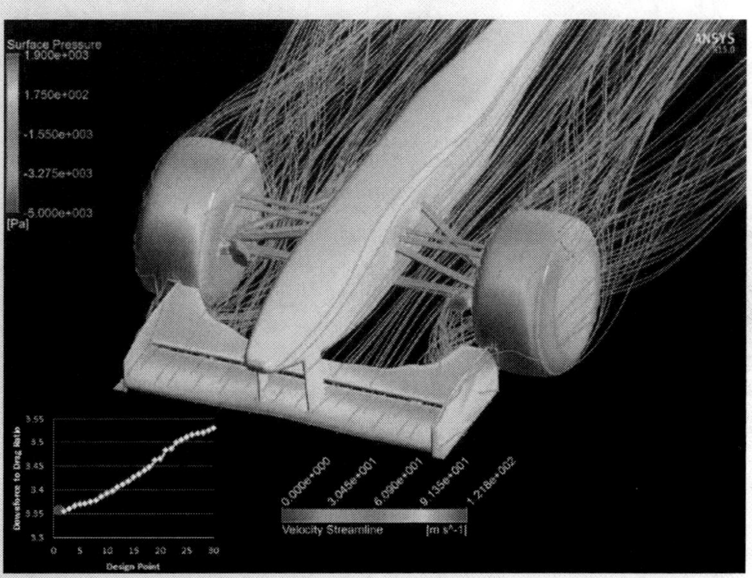

F1 Front & Rear Wing
Maximize Downforce

2013 Automotive Simulation World Congress
October 29 - 30 • Frankfurt am Main, Germany

Geometry	Predicted [N]	Result [N]
Original	-	**425.7**
Modified	447.4	**451.1**

Reduction in Lift

6% 14%

Total = 20%

Geometry	Predicted [N]	Result [N]
Original	-	**425.7**
Modified	481.3	**492.5**

Sedan – Minimize Drag

2013 Automotive Simulation World Congress
October 29 - 30 • Frankfurt am Main, Germany

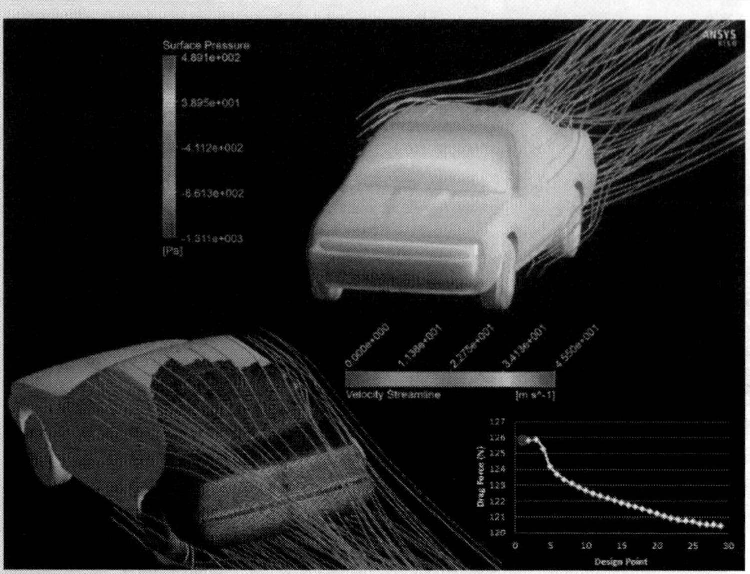

S809 HAWT Blade - Max Lift/Drag

2013 Automotive Simulation World Congress
October 29 - 30 • Frankfurt am Main, Germany

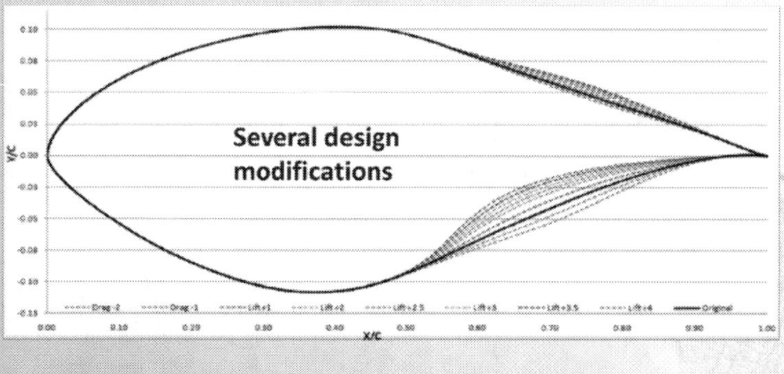

	Original [N]	Optimized Shape [N]
Drag	7.02	7.57
Lift	71.63	98.53
Ratio	10.20	13.02

30% increased lift

Rear Cabin Automobile HVAC Duct
Minimize Total Pressure Drop

2013 **Automotive Simulation** World Congress
October 29 - 30 • Frankfurt am Main, Germany

- Rear cabin automobile HVAC duct morphed to decrease pressure drop
- The pressure drop was reduced significantly

33% reduced pressure drop

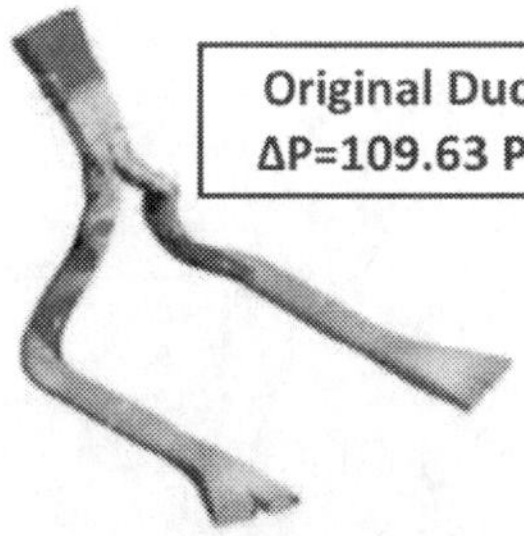

Original Duct
ΔP=109.63 Pa

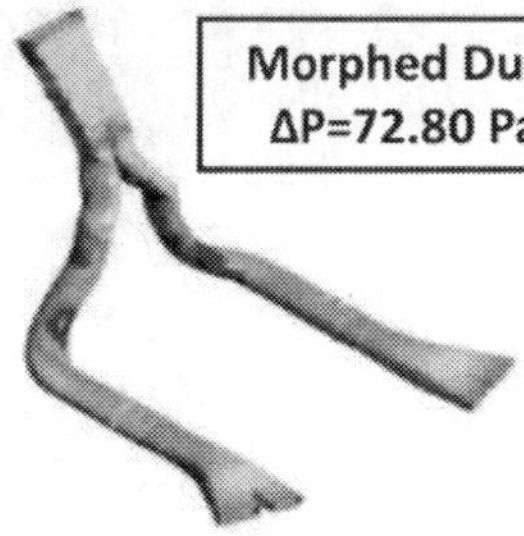

Morphed Duct
ΔP=72.80 Pa

Duct System
Reduce Mass Flux Variance at Outlets

2013 **Automotive Simulation** World Congress
October 29 - 30 • Frankfurt am Main, Germany

Goal

Get an insight into the flow field – identify causing regions	Reduce variance of mass flux at outlets

Post-Processing

Geometry far upstream is dictating the flow split.	Manufacturing variances at the inlet region can cause flow variance

Optimization

Strong variations in sensitivity imply that non-uniformity in the inflow can adversely affect the quantity of interest.

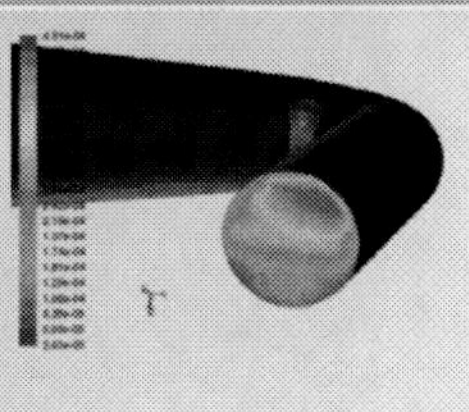

Total Pressure [Pa]

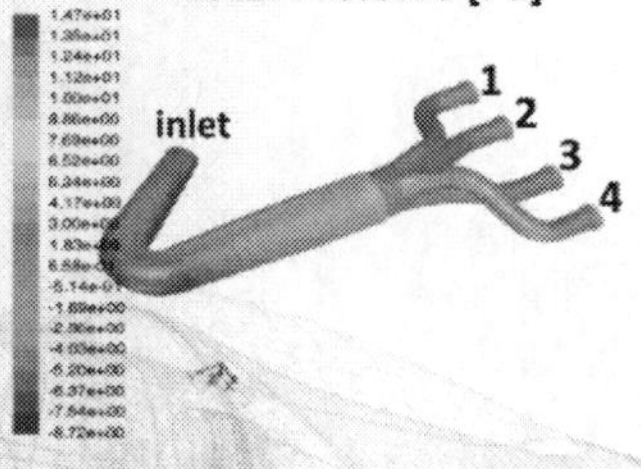

Surface normal displacements that Induce O(1) change in variance

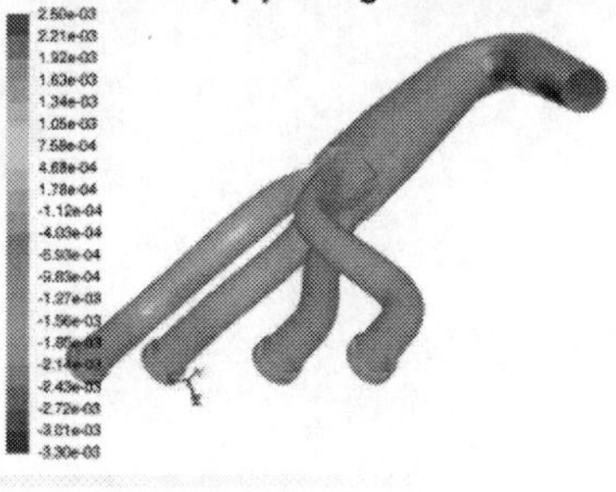

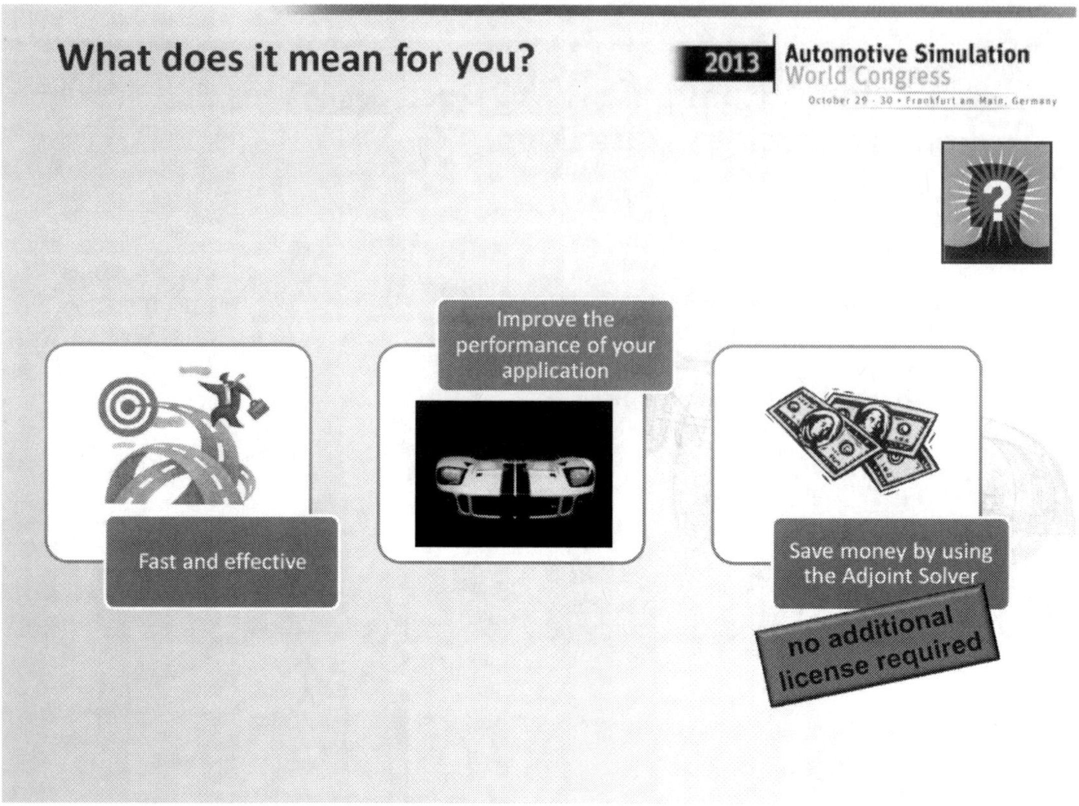
What does it mean for you?
2013 Automotive Simulation World Congress
October 29 - 30 • Frankfurt am Main, Germany
Improve the performance of your application
Fast and effective
Save money by using the Adjoint Solver
no additional license required

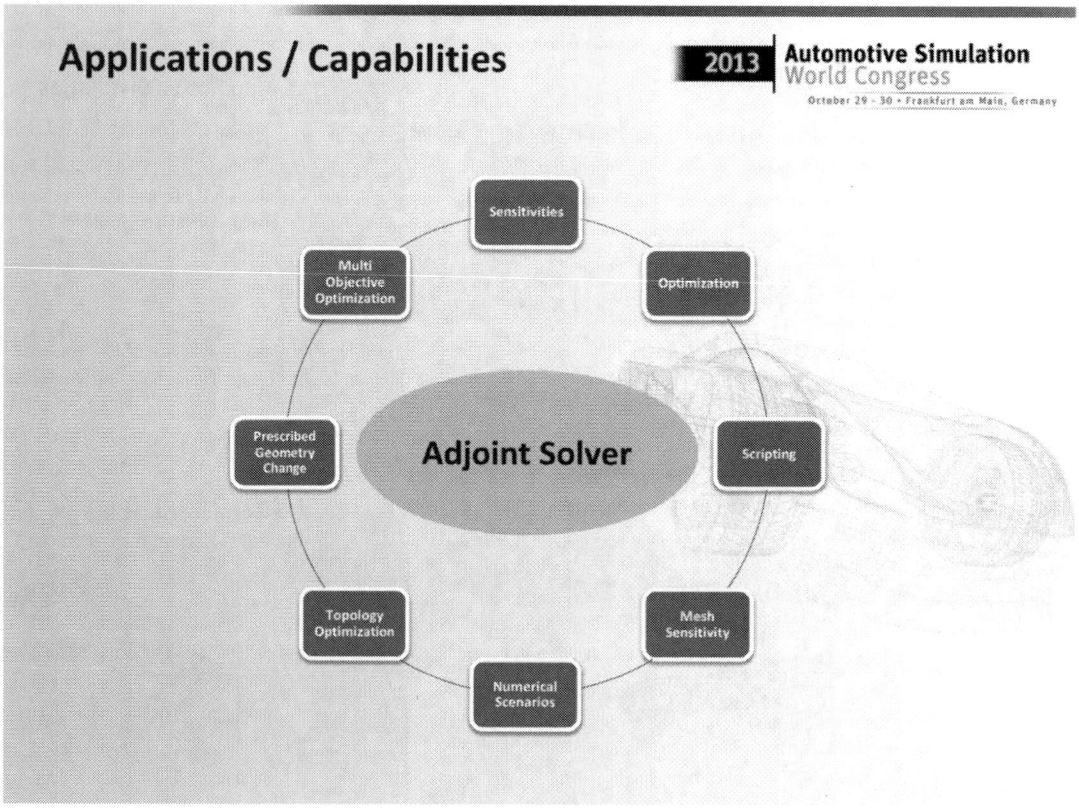
Applications / Capabilities
2013 Automotive Simulation World Congress
October 29 - 30 • Frankfurt am Main, Germany
Adjoint Solver
Sensitivities
Optimization
Scripting
Mesh Sensitivity
Numerical Scenarios
Topology Optimization
Prescribed Geometry Change
Multi Objective Optimization

Function enhancement by scripting

2013 Automotive Simulation World Congress
October 29 - 30 • Frankfurt am Main, Germany

Content

- Script: Adjoint Solver Loop
  - Improved Procedure
    - Set Scaling Factor within each iteration
    - Check mesh quality and use for improvement
    - Terminate Fluent if solution does not converge

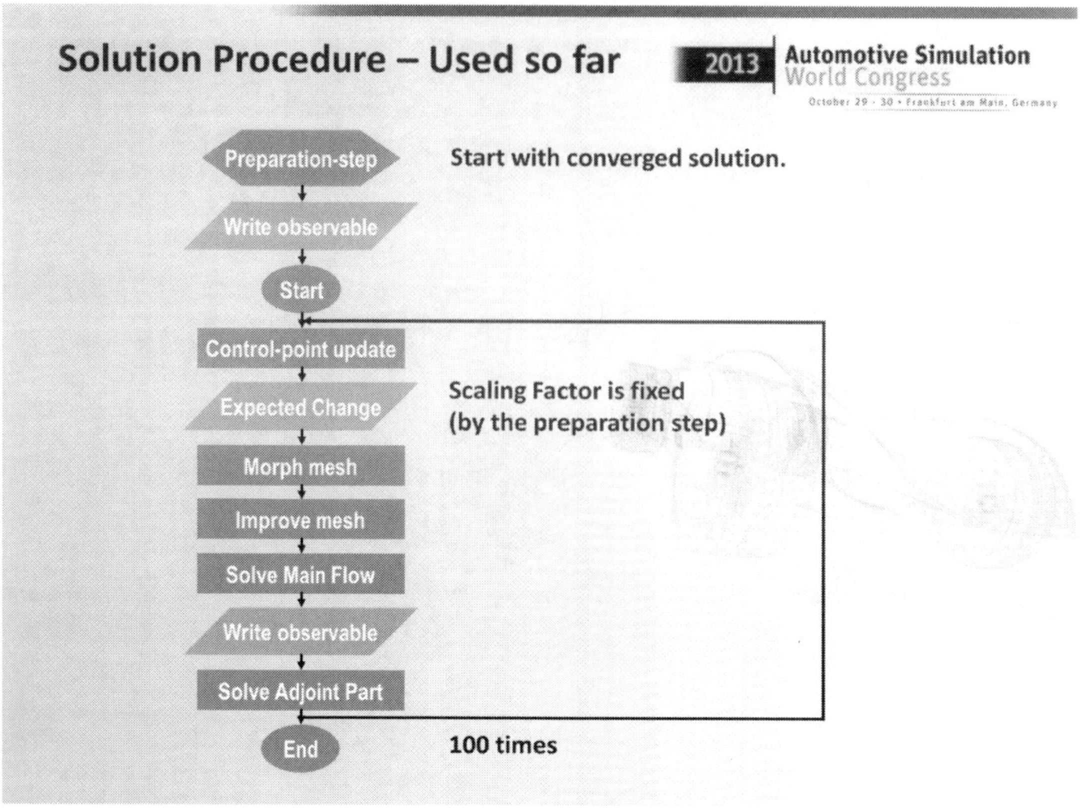
Solution Procedure – Used so far
2013
Automotive Simulation
World Congress
October 29 - 30 • Frankfurt am Main, Germany
Preparation-step
Start with converged solution.
Write observable
Start
Control-point update
Expected Change
Scaling Factor is fixed
(by the preparation step)
Morph mesh
Improve mesh
Solve Main Flow
Write observable
Solve Adjoint Part
End
100 times

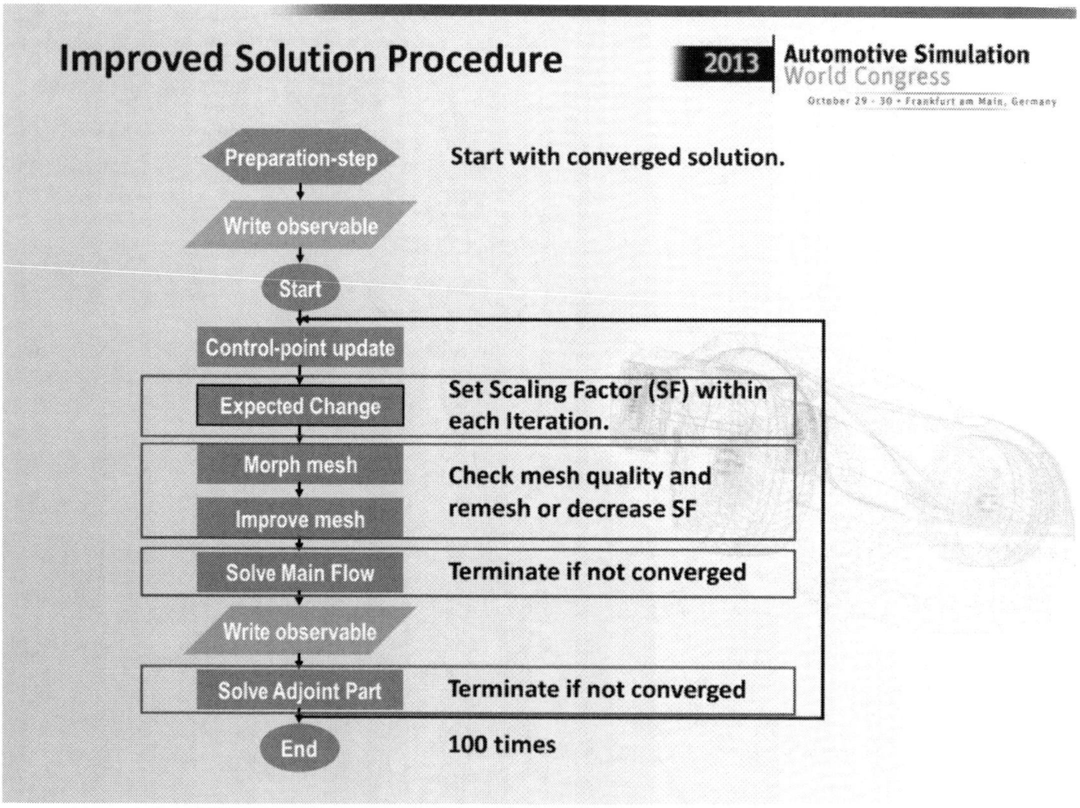
Improved Solution Procedure
2013
Automotive Simulation
World Congress
October 29 - 30 • Frankfurt am Main, Germany
Preparation-step
Start with converged solution.
Write observable
Start
Control-point update
Expected Change
Set Scaling Factor (SF) within
each Iteration.
Morph mesh
Improve mesh
Check mesh quality and
remesh or decrease SF
Solve Main Flow
Terminate if not converged
Write observable
Solve Adjoint Part
Terminate if not converged
End
100 times

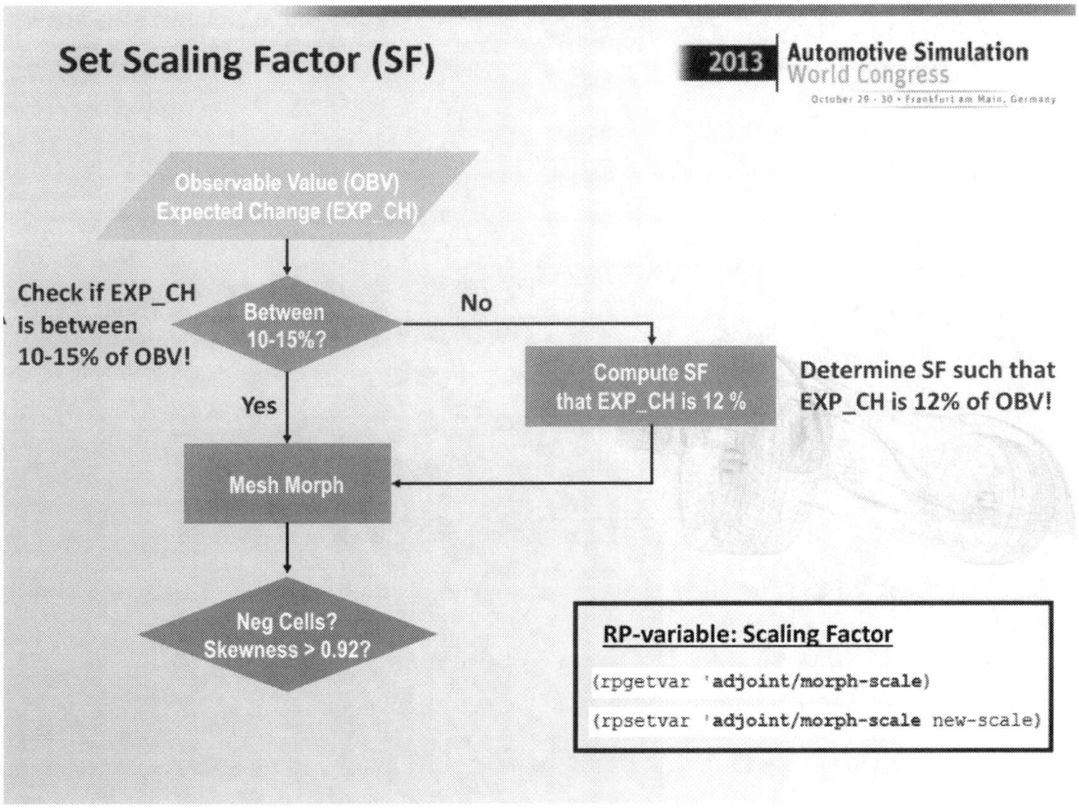
Set Scaling Factor (SF)
2013 Automotive Simulation World Congress
October 29 - 30 • Frankfurt am Main, Germany
Observable Value (OBV)
Expected Change (EXP_CH)
Check if EXP_CH is between 10-15% of OBV!
Between 10-15%?
No
Compute SF that EXP_CH is 12 %
Determine SF such that EXP_CH is 12% of OBV!
Yes
Mesh Morph
Neg Cells? Skewness > 0.92?
RP-variable: Scaling Factor
(rpgetvar 'adjoint/morph-scale)
(rpsetvar 'adjoint/morph-scale new-scale)

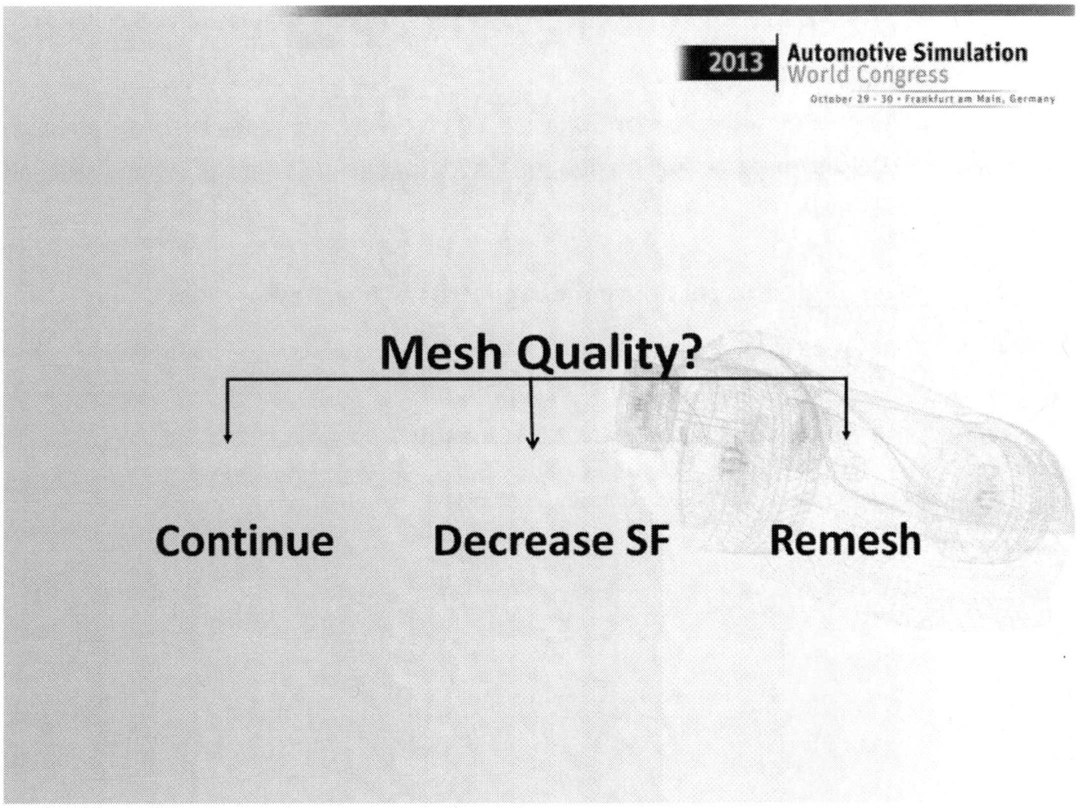
2013 Automotive Simulation World Congress
October 29 - 30 • Frankfurt am Main, Germany
Mesh Quality?
Continue
Decrease SF
Remesh

Mesh Quality – Decrease SF

2013 Automotive Simulation World Congress
October 29 - 30 • Frankfurt am Main, Germany

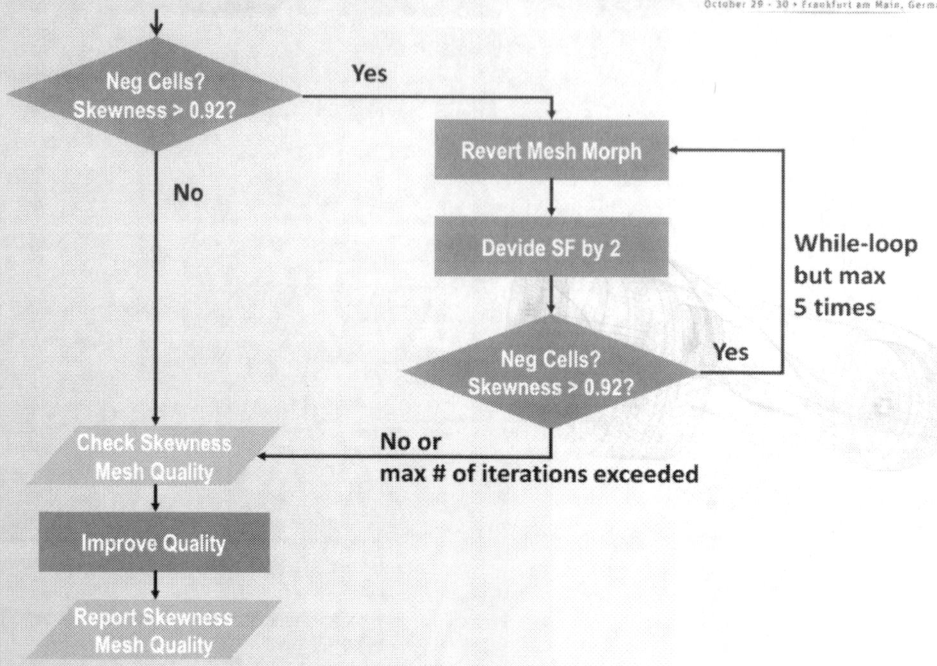

Mesh Quality – Remeshing, Part 1

2013 Automotive Simulation World Congress
October 29 - 30 • Frankfurt am Main, Germany

1) **Enable dynamic mesh option**

/define/dynamic-mesh/dynamic-mesh?

and remesh

<u>Reminder</u>: Not possible if mesh consists of hexahedral cells!

Check cell-element-type

```
> /report/volume-integrals/max (fluid) cell-element-type no

                              Max
               Cell Element Type
-------------------------------- --------------------
                            fluid                    4
                ---------------- --------------------
                              Net                    4
```

```
;  triangle       1
;  tetrahedron    2
;  quadrilateral  3
;  hexahedron     4
;  pyramid        5
;  wedge          6
```

Scheme

```
(define volu (map thread-name (append (get-threads-of-type 'fluid) (get-threads-of-type 'solid))))
(define cmd (format #f "/report/volume-integrals/max ~a cell-element-type no" volu))
(define max-element-type (pick-a-real cmd))
```

Mesh Quality – Remeshing, Part 2

2) **Start external program to remesh, e.g. ICEM**

External program can be started with "!"

```
!"C:\Program Files\ANSYS
Inc\v145\icemcfd\win64_amd\bin\icemcfd.bat"
```


Terminate Fluent

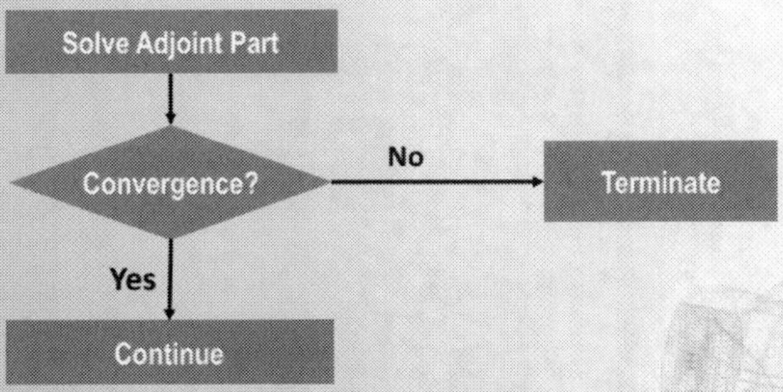

Scheme

```
(if (not(rpgetvar 'solution/converged?))
    (begin ;
        (display "Flow Solver NOT CONVERGED \n STOP Fluent STOP")
        terminate-fluent ;; alias-command
    )
    ;; if false-value can be ommitted
)
```

```
(if ( = (rpgetvar 'adjoint/converged) 0)
    ( begin ;; if true-value
        (display "Adjoint Solver NOT CONVERGED \n STOP Fluent STOP")
        terminate-fluent ;; alias-command
    )
    ;; if false-value can be ommitted
)
```

What does it mean for you?

2013 Automotive Simulation World Congress
October 29 - 30 • Frankfurt am Main, Germany

Adapt the process to your needs

Speed-Up your workflow

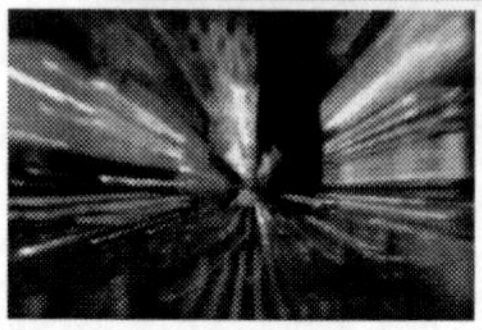

Be more flexible

Applications / Capabilities

2013 Automotive Simulation World Congress
October 29 - 30 • Frankfurt am Main, Germany

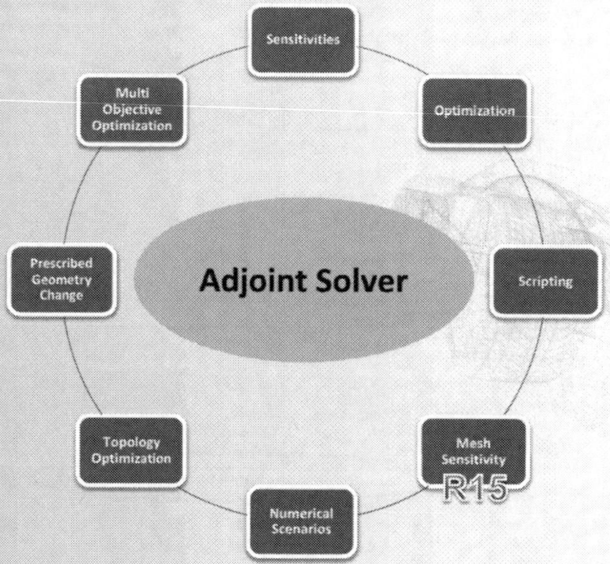

Robust Simulation

2013 Automotive Simulation World Congress
October 29 - 30 • Frankfurt am Main, Germany

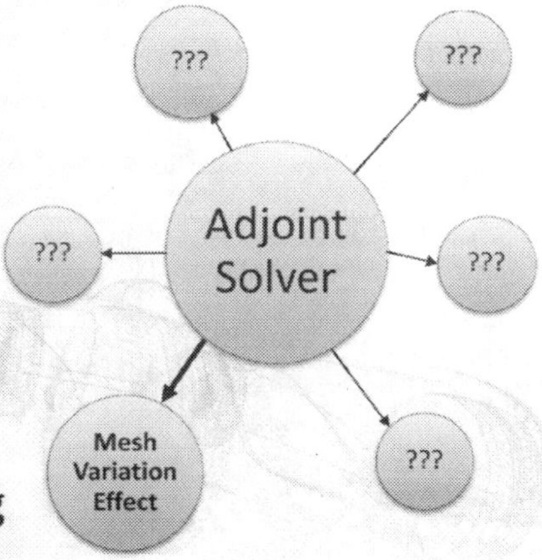

- The Adjoint Solver offers new ways to analyze mesh numerics.
- One strategy is to use the new feature "Mesh Variation Effect"
- This can be used to report the effect on the observable of moving mesh nodes locally.
- Progressive refinement of the mesh on the basis of this quantity can improve results.

Mesh Variation Effect - Example

2013 Automotive Simulation World Congress
October 29 - 30 • Frankfurt am Main, Germany

- 2D domain with cylinder, incoming flow on the left side

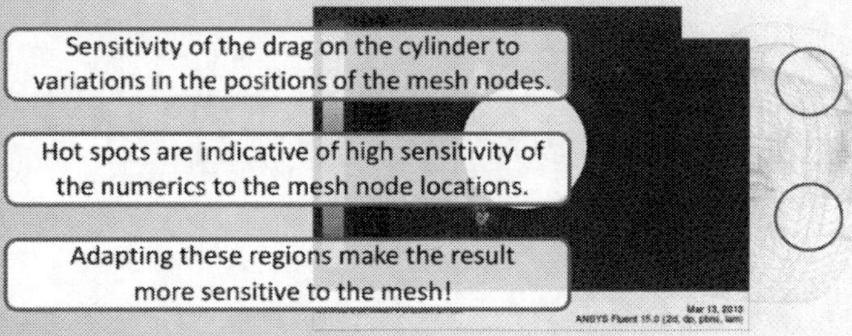

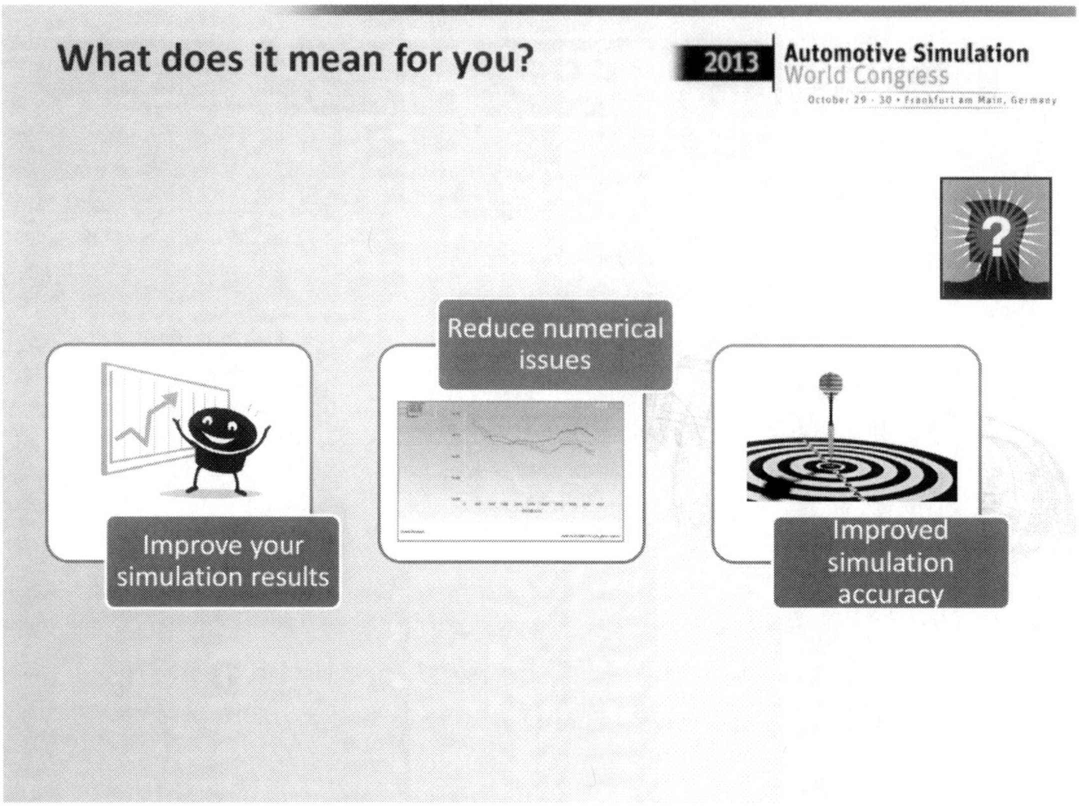
What does it mean for you?
2013 Automotive Simulation World Congress
October 29 - 30 • Frankfurt am Main, Germany
Reduce numerical issues
Improve your simulation results
Improved simulation accuracy

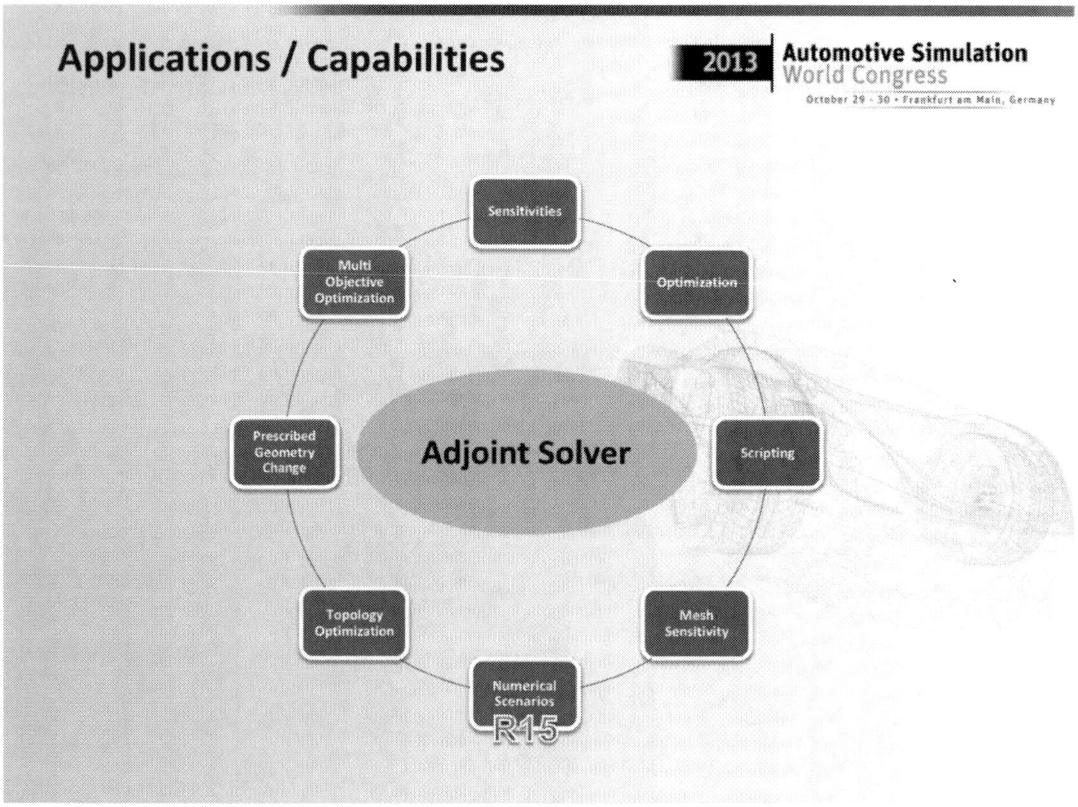
Applications / Capabilities
2013 Automotive Simulation World Congress
October 29 - 30 • Frankfurt am Main, Germany
Adjoint Solver
Sensitivities
Optimization
Scripting
Mesh Sensitivity
Numerical Scenarios
R15
Topology Optimization
Prescribed Geometry Change
Multi Objective Optimization

Numerical Scenarios

2013 Automotive Simulation World Congress
October 29 - 30 • Frankfurt am Main, Germany

- Additional discretization schemes have been added to the adjoint solver as part of R15 development.
- The numerical scenario analysis estimates fully-converged values for observables of interest, such as drag.
- The effect of swapping between different discretization schemes is explored as a fast post-processing step.
- The effect of eliminating artificial dissipation is estimated.

Numerical Scenarios - Example

2013 Automotive
World Co

Flow past a half vehicle (3.6M Cells)
- Flow solved second order
- Adjoint solved first order
- Flow solver reports drag as 126.01N

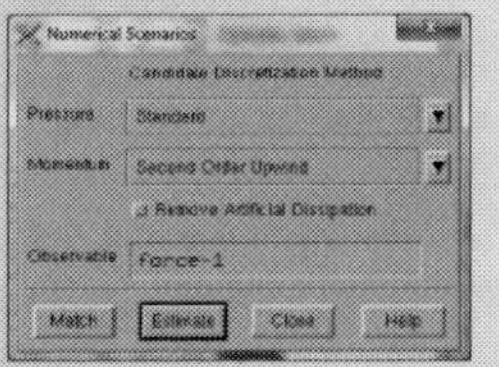

- Estimating for discretizations -
Pressure Scheme : Standard
Momentum Scheme : Second Order Upwind
Artificial Dissipation : Included

Observable estimated value at full convergence: 125.60 N

Shows proximity to drag convergence.

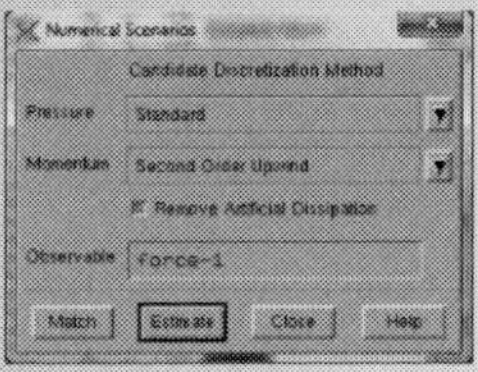

- Estimating for discretizations -
Pressure Scheme : Standard
Momentum Scheme : Second Order Upwind
Artificial Dissipation : **Removed**

Observable estimated value at full convergence: 126.39 N

Effect of artificial dissipation eliminated from the result!

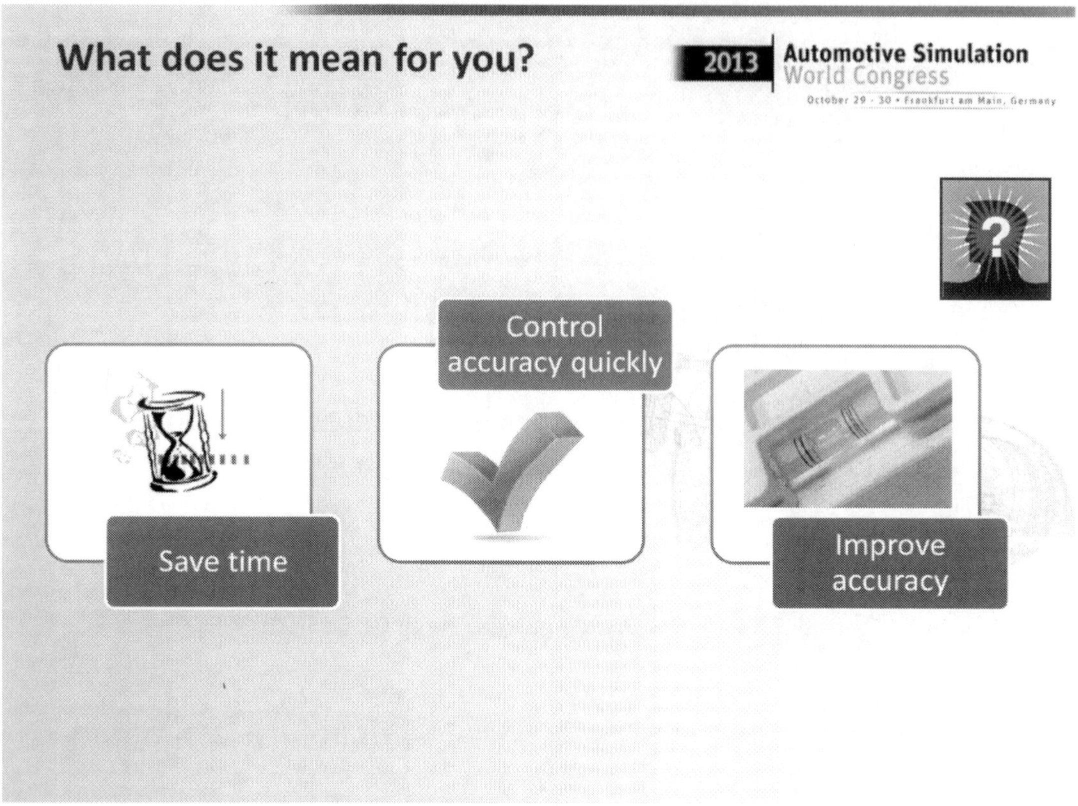
What does it mean for you?
2013
Automotive Simulation
World Congress
October 29 - 30 • Frankfurt am Main, Germany
Control accuracy quickly
Save time
Improve accuracy

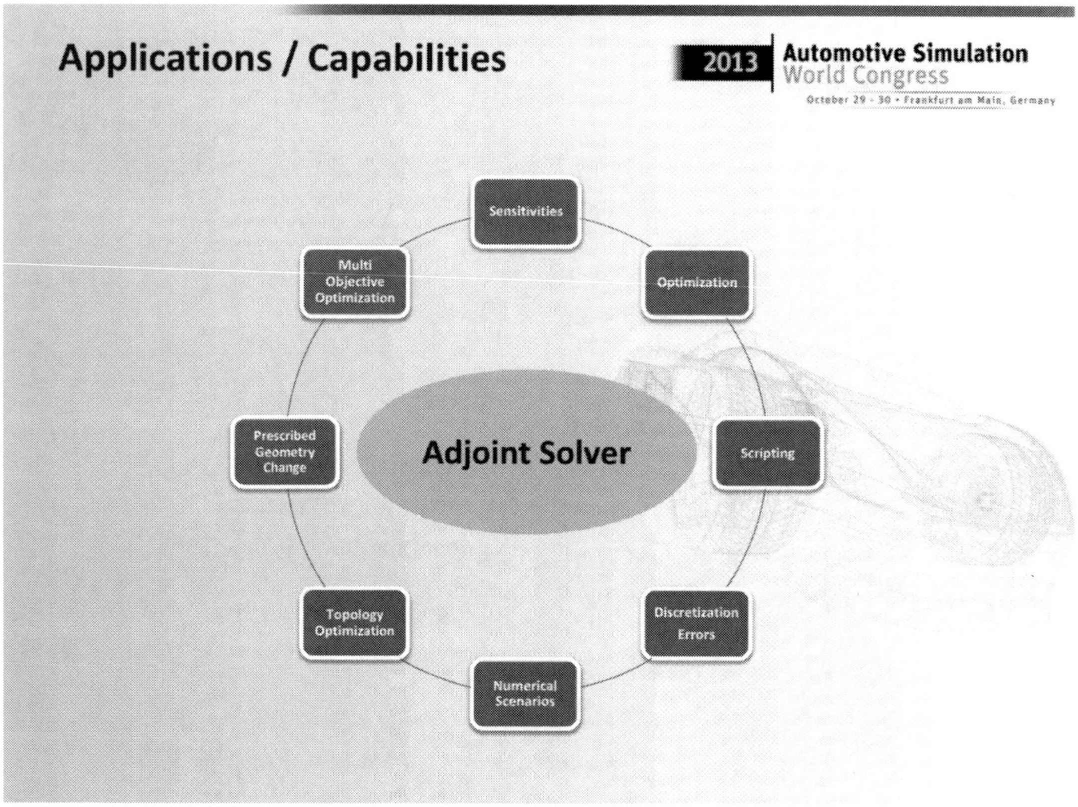
Applications / Capabilities
2013
Automotive Simulation
World Congress
October 29 - 30 • Frankfurt am Main, Germany
Adjoint Solver
Sensitivities
Optimization
Scripting
Discretization Errors
Numerical Scenarios
Topology Optimization
Prescribed Geometry Change
Multi Objective Optimization

Topology Optimization

Background - Flow Blockage

2013 Automotive Simulation World Congress
October 29 - 30 • Frankfurt am Main, Germany

- This field is provided as a convenient tool for identifying portions of the flow domain where the introduction of blockages or obstructions in the flow can affect the observation of interest.
- Consider a blockage in the flow that generates a reaction force on the flow that is proportional to the local flow speed, and acting in the opposite direction to the local flow. The local contribution of this force on the observation of interest is determined by the vector product of this force with the adjoint velocity field.
- The flow blockage field that is plotted is the negative of the vector product of the flow velocity and the adjoint velocity (Cell Value).

Theoretical foundation

2013 Automotive Simulation World Congress
October 29 - 30 • Frankfurt am Main, Germany

- Adjoint solution indicates how the action of body-forces upon the flow modifies the observation.

- Consider the body force as being the result of changing the porosity of the medium.

$$\underline{F} = -\alpha \underline{u} V$$

Cell volume (V); Porosity (α); Flow velocity ($\underline{u}$)

- The sensitivity of the result to changes in porosity is then

Adjoint velocity ($\underline{\tilde{u}}$)

$$J = \underline{\tilde{u}} \cdot \underline{F} = -\alpha \underline{u} \cdot \underline{\tilde{u}} V \qquad \partial J / \partial \alpha = -\underline{u} \cdot \underline{\tilde{u}} V$$

- This is the post-processed quantity Flow Blockage Effect

Topology Optimization - Workflow

2013 Automotive Simulation World Congress
October 29 - 30 • Frankfurt am Main, Germany

- An indication for an optimal internal flow is the avoidance of flow recirculation.

- For many technical flows a consequence of this change is the reduction of pressure drop.

- The flow blockage variable can be used to analyze the flow field.

- These regions can deleted by Fluent

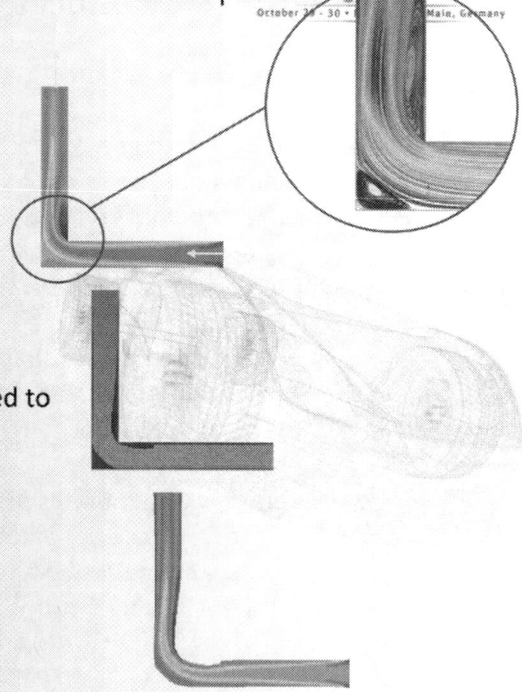

Topology Optimization - Meander

2013 Automotive World Congress

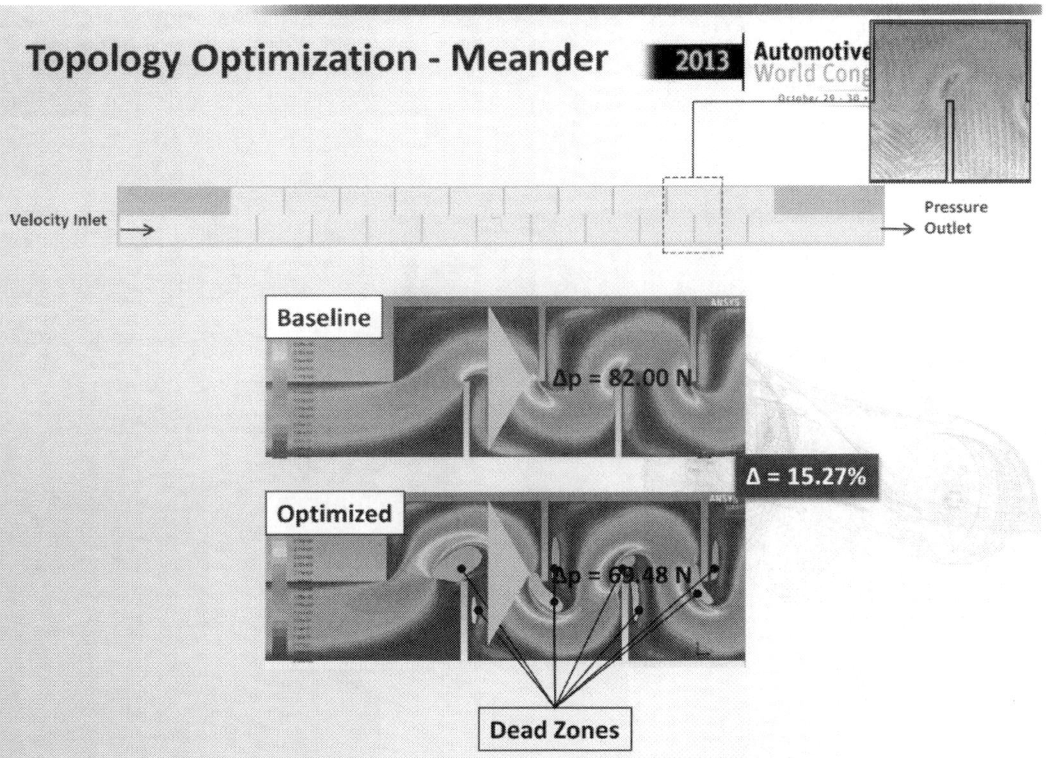

Example 2 – 3D case w/ two outlets

2013 Automotive Simulation World Congress
October 29 - 30 • Frankfurt am Main, Germany

Video Presentation in
Original e-media

What does it mean for you?

2013 Automotive Si World Congres October 29 - 30 • Frankf

ANSYS FLUENT

No need of additional tools

Easy post-process step to see areas with flow blockage

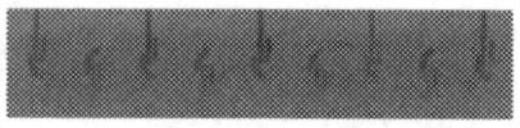

Improved designs with higher product performance

Applications / Capabilities

2013 Automotive Simulation World Congress October 29 - 30 • Frankfurt am Main, Germany

Adjoint Solver

- Sensitivities
- Optimization
- Scripting
- Discretization Errors
- Numerical Scenarios
- Topology Optimization
- Prescribed Geometry Change R15
- Multi Objective Optimization

Prescribed Geometry Changes

- The adjoint allows an extrapolation of the performance measure for the system for changes to the geometry and boundary conditions.
- The general mesh morpher provides an "optimal design change" recommendation in which changes in a high-dimension design space are defined.
- In some cases only a few design parameters may be of importance. The adjoint can be used to predicted the effect of changes to these parameters.

Examples of applications

- The effect of changing the radius of some part of a system needs to be understood.

How does the drag on the cylinder depend upon its radius?

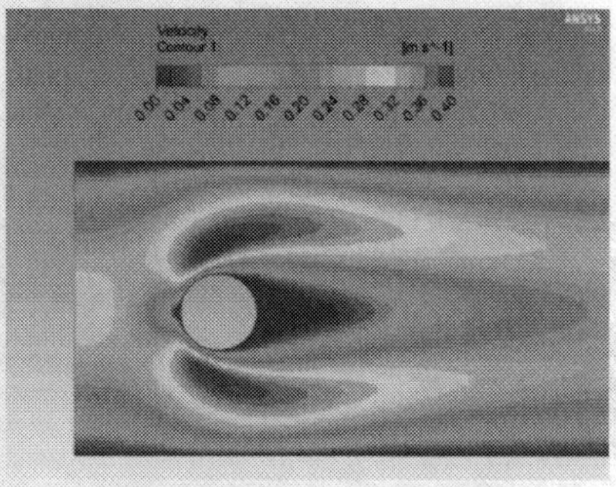

cylinder in crossflow

- The effect of changing the height, and width of a heat-exchanger fin upon its heat-transfer rate is to be computed.

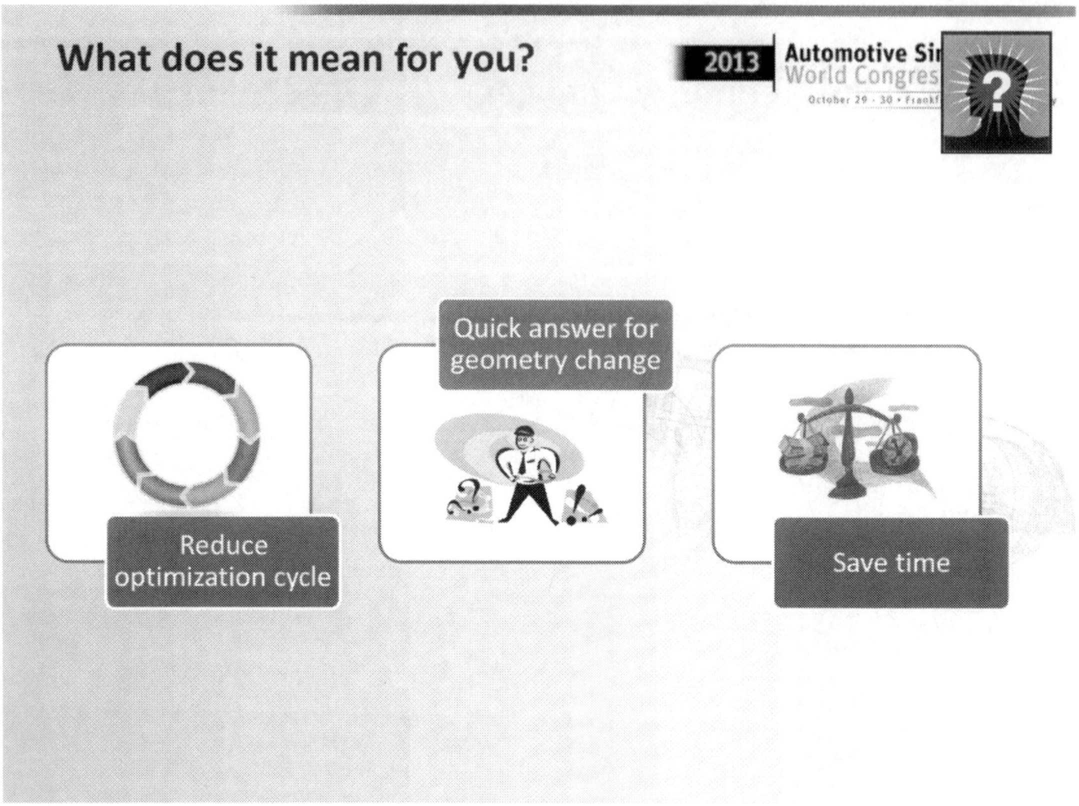
What does it mean for you?
2013 Automotive Simulation World Congress
Quick answer for geometry change
Reduce optimization cycle
Save time

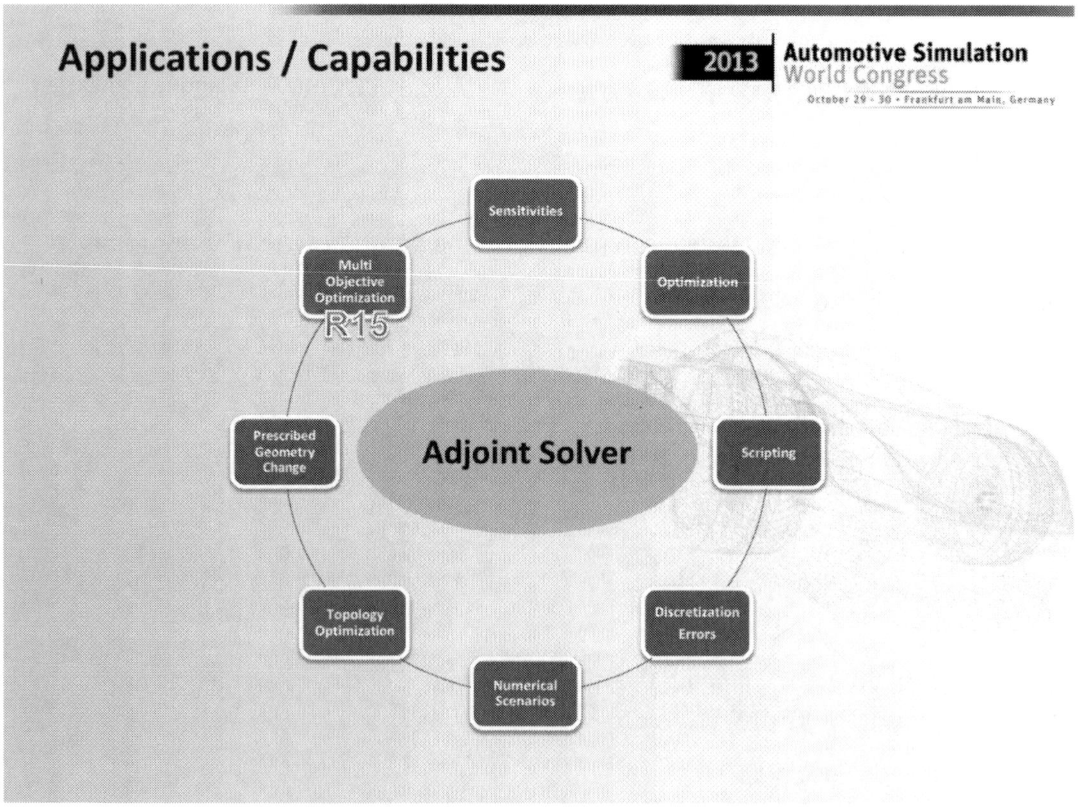
Applications / Capabilities
2013 Automotive Simulation World Congress
October 29 - 30 • Frankfurt am Main, Germany
Sensitivities
Optimization
Scripting
Discretization Errors
Numerical Scenarios
Topology Optimization
Prescribed Geometry Change
Multi Objective Optimization
R15
Adjoint Solver

Multi-Objective Design

2013 Automotive Simulation World Congress
October 29 - 30 • Frankfurt am Main, Germany

- Consider a problem with multiple objectives

$$\underline{J} = \begin{bmatrix} J_1 & J_2 & \ldots & J_N \end{bmatrix}$$

- Goal is to optimize performance while allowing tradeoffs between the different objectives.
- There may also be constraints

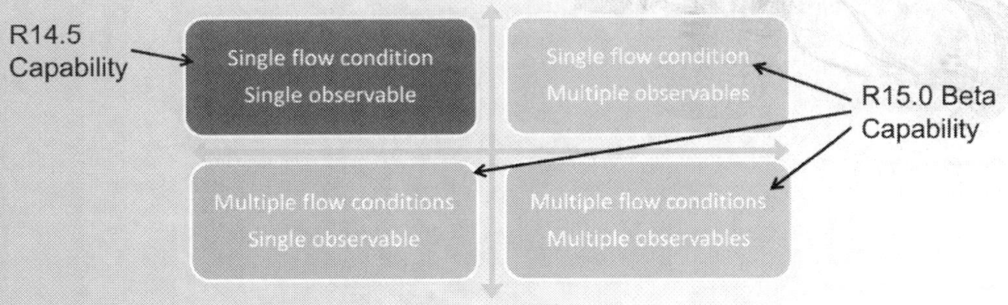

Multi-Objective Design

2013 Automotive Simulation World Congress
October 29 - 30 • Frankfurt am Main, Germany

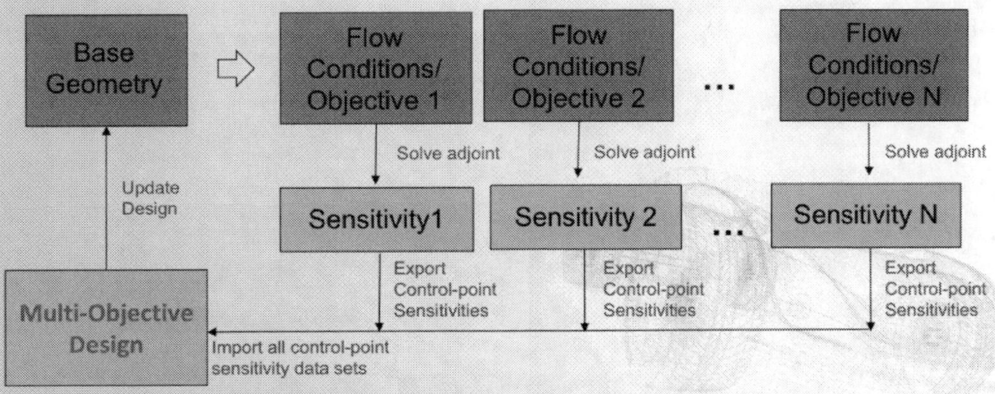

- Weighted linear combination of objectives
- Some objectives may be constrained to be constant
- Explore the design space – estimates of how each objective changes
- Decide on a design change and apply to all cases

Multi-Objective Design

- Optimal design change will be computed based on optimizing a linear combination of the observables

Maximize the change to $\sum_{i=0}^{N-1} w_i J_i$ **where** $\Delta J_\nu = 0$ **for some ν**

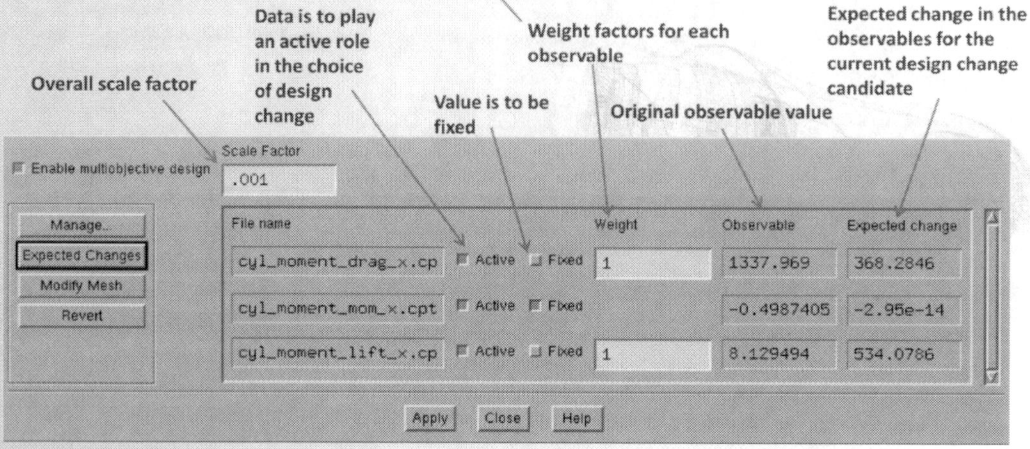

Multi-Objective Design - Examples

2013 Automotive Simulation World Congress
October 29 - 30 • Frankfurt am Main, Germany

Reduce the drag on a wing at a variety of different flight speeds

Reduce the drag on a wing while holding the lift fixed.

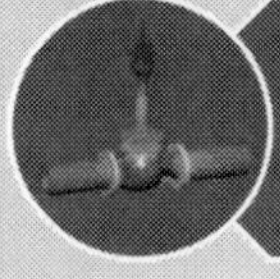

Trade off the minimization of pressure drop and mass-flow distribution for an internal flow system.

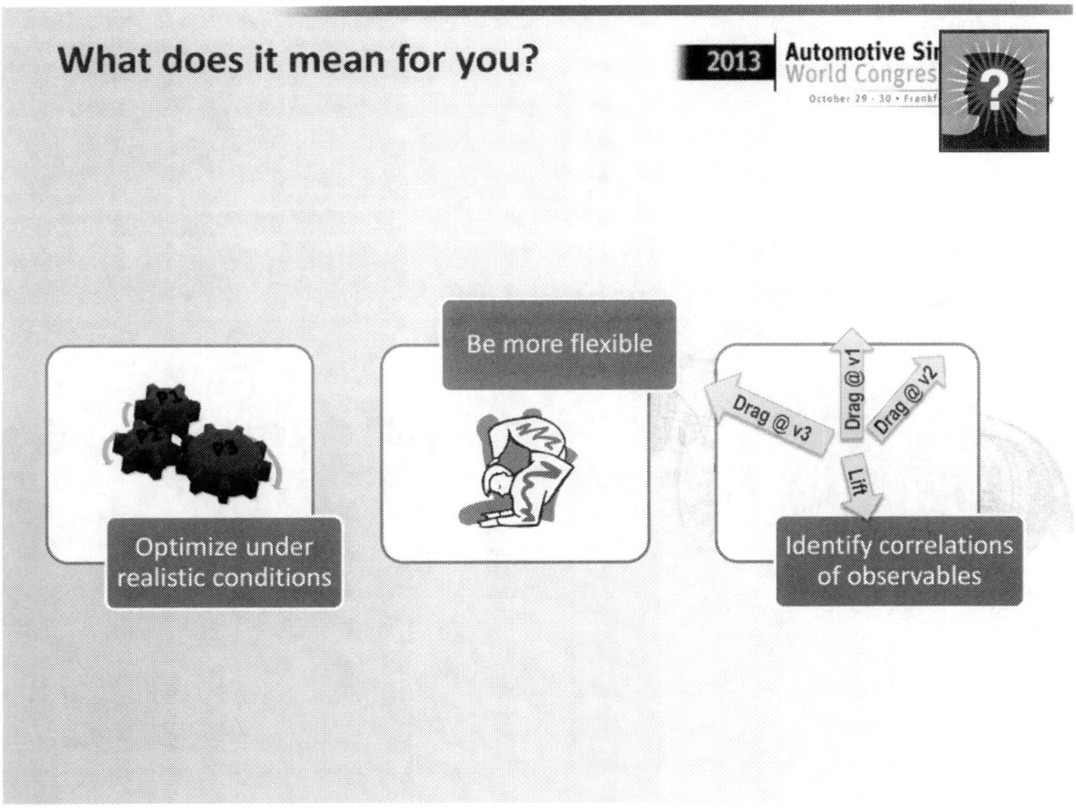
What does it mean for you?
2013
Automotive Si
World Congres
Be more flexible
Drag @ v1
Drag @ v2
Drag @ v3
Lift
Optimize under realistic conditions
Identify correlations of observables

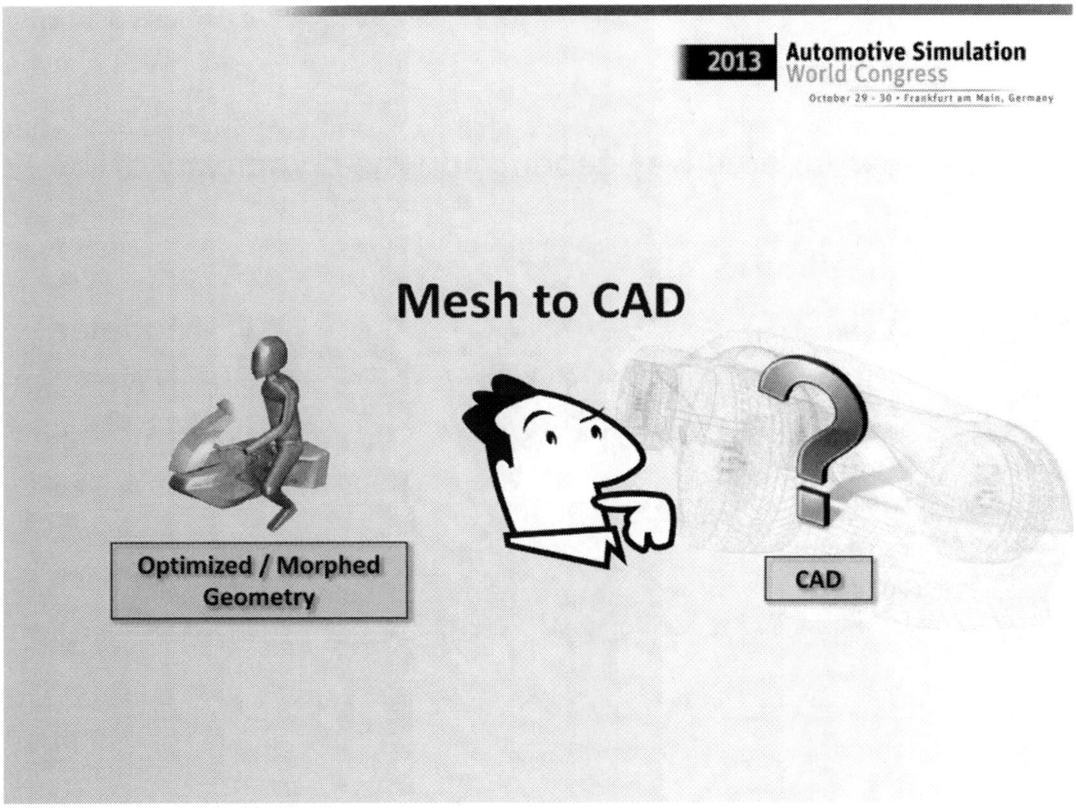
2013
Automotive Simulation
World Congress
October 29 - 30 • Frankfurt am Main, Germany
Mesh to CAD
Optimized / Morphed Geometry
CAD

Mesh to CAD

2013 Automotive Simulation World Congress
October 29 - 30 • Frankfurt am Main, Germany

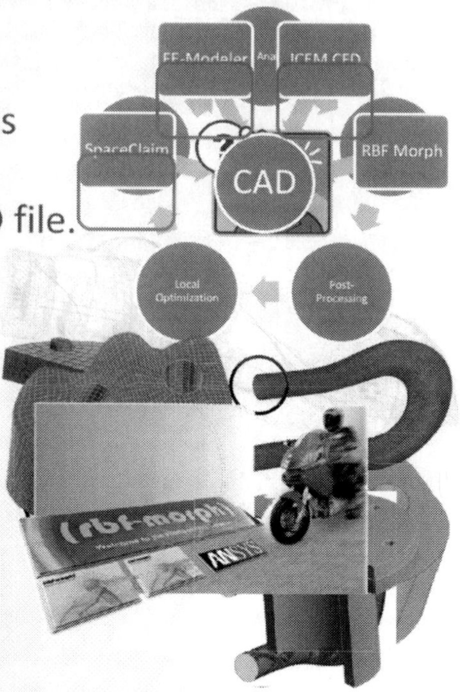

A closed optimization cycle needs an effective way to translate the optimized mesh in a generic CAD file.

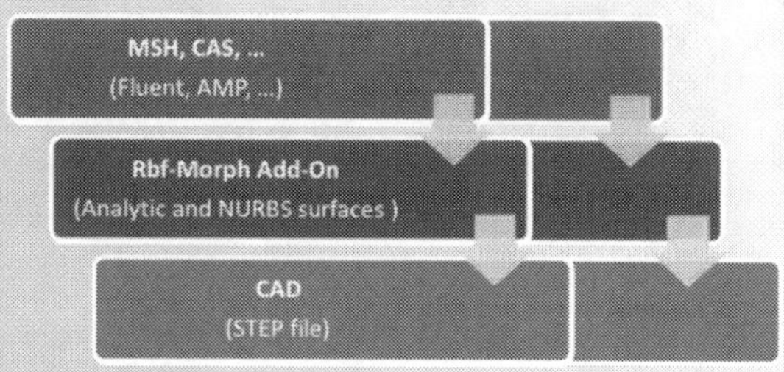

Mesh to CAD

2013 Automotive Simulation World Congress
October 29 - 30 • Frankfurt am Main, Germany

The appropriated way depends on the complexity of your geometry!

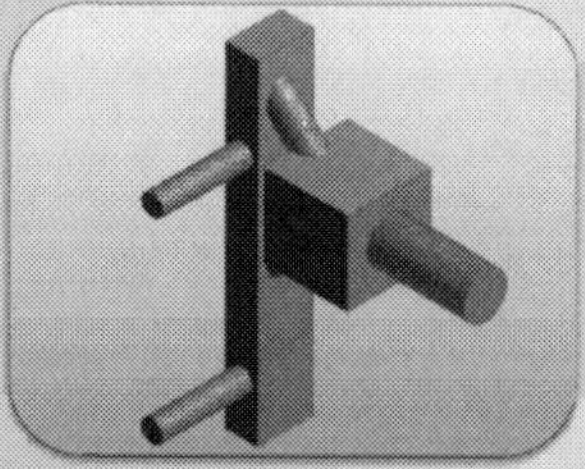

- Simple geometries / primitives
  - **FE-Modeler**
  - **RBF Morph**
  - **ICEM CFD**

- Organic shapes
  - **SpaceClaim**

Outlook

Outlook R15+

2013 Automotive Simulation World Congress
October 29 - 30 • Frankfurt am Main, Germany

R15	R14.5 / 15 Beta Feature	R15+
Stabilization for Large Cases	Multi Objective Optimization	More models (MRF, transient, ...)
Energy Equation	Prescribed Geometry Change	More options
Robust Simulation	Topology Optimization	More Optimization Variables

Summary

2013 Automotive Simulation World Congress
October 29 - 30 • Frankfurt am Main, Germany

- The Adjoint solution is carried out as an addition to the primary flow solution.
- The output from the Adjoint is a field of the sensitivity of the observable of interest to the inputs such as the geometry and boundary conditions.
- Provide extremely valuable engineering insight, optimize system performance and detecting areas in the flow where discretization errors can potentially have a strong effect.
- Capability to use scheme / journal scripts to automize the global optimization and extend functionalities of the workflow.
- Can be coupled with external tool RBF Morph for advanced morphing

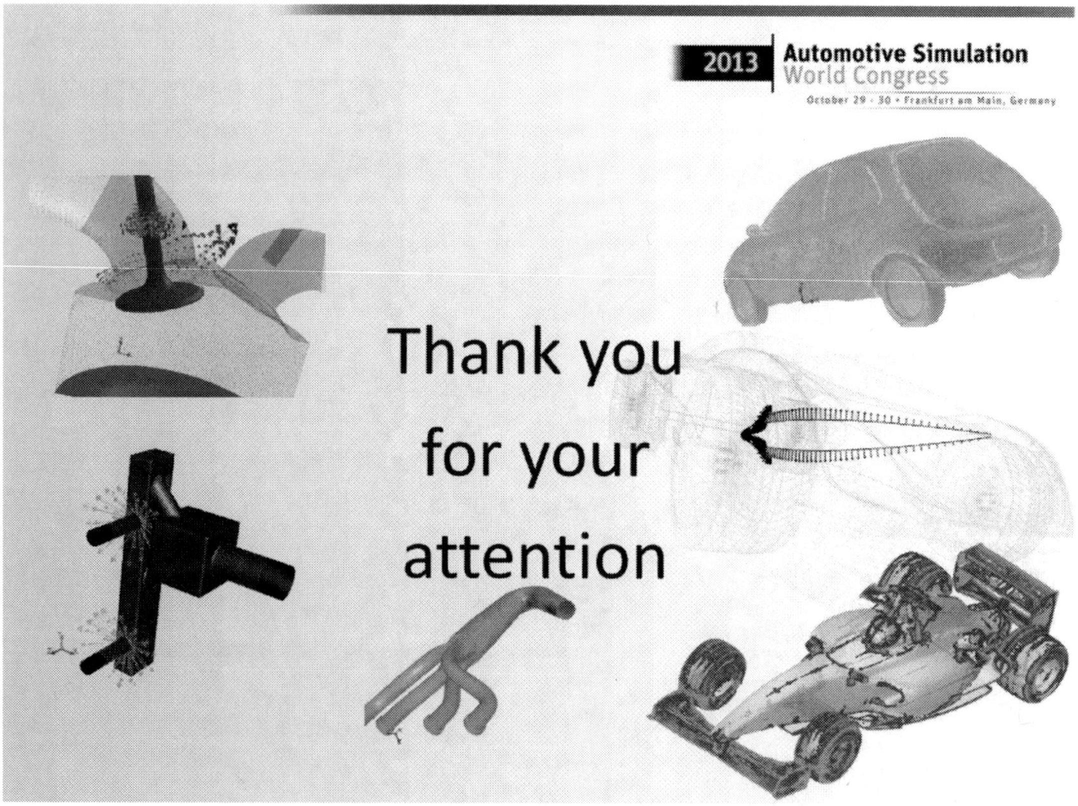

ANSYS

Realize Your Product Promise™

DrivAer - Aerodynamic Investigations for a New Realistic Generic Car Model using ANSYS CFD

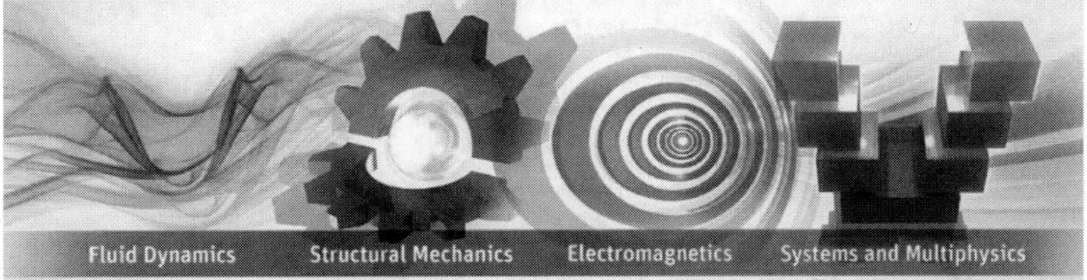

Thomas Frank(*), Benedikt Gerlicher(*), Juan Abanto(**)

(*) ANSYS Germany, Otterfing, Germany

(**)ANSYS Inc., Lebanon, NH, USA

Thomas.Frank@ansys.com

ANSYS

Contents

- The DrivAer Benchmark by TUM, Inst. Aerodynamics
  - Investigated DrivAer car model variants
- The meshing process
- CFD investigations for the DrivAer fastback car:
  - F_S_woM_wW
  - F_D_wM_wW
- Comparison to TU Munich wind tunnel data
- Cross-comparison of ANSYS CFX and ANSYS Fluent
- Summary & Outlook

  September 19, 2013

ANSYS Objectives

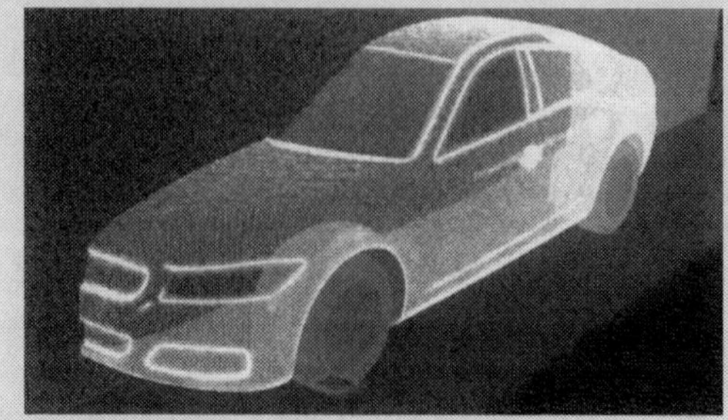

- Automotive Aerodynamics Validation of ANSYS CFD
- Generic reference model with modern car geometry
- Investigation of meshing process and technologies for contemporary and complex car body geometry
  - Including wheels
  - Including mirrors
  - Including detailed floor with exhaust system
- Validation of ANSYS CFX & ANSYS Fluent
- Comparison to TU Munich wind tunnel data
- Turbulence model validation and data comparison
  → steady/transient SST and SAS-SST

ANSYS DrivAer Geometry

Development of the DrivAer model by TU Munich

BMW 3 Series
Limousine

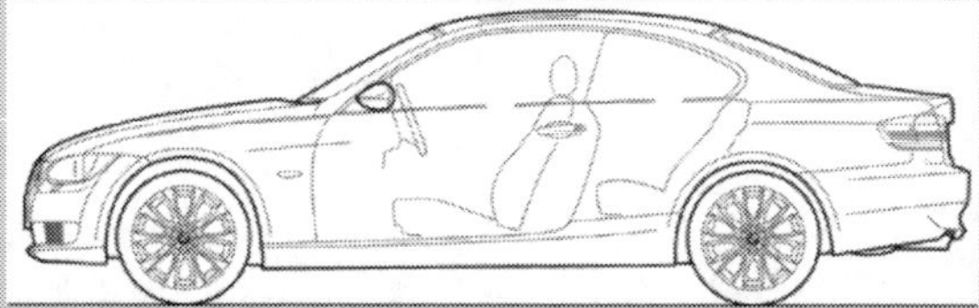

Audi A4
Limousine

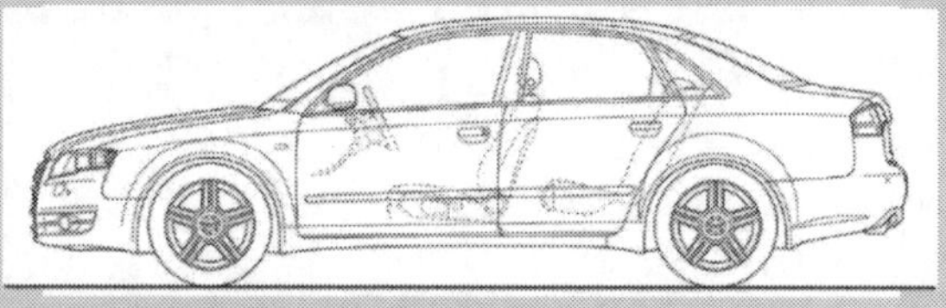

Courtesy by TU Munich, Inst. of Aerodynamics

ANSYS DrivAer Geometry

Development of the DrivAer model by TU Munich

BMW 3 Series Limousine

DriveAer Car Body

Audi A4 Limousine

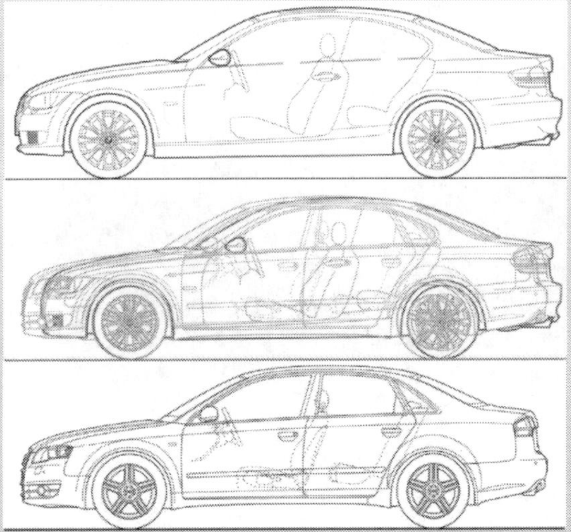

Courtesy by TU Munich, Inst. of Aerodynamics

5 © 2011 ANSYS, Inc. September 19, 2013

ANSYS DrivAer Geometry

Development of the DrivAer model by TU Munich

Total length	4613mm
Total width	1820mm
Total height	1418mm
Wheelbase	2786mm
Track width front	1520mm
Track width back	1527mm

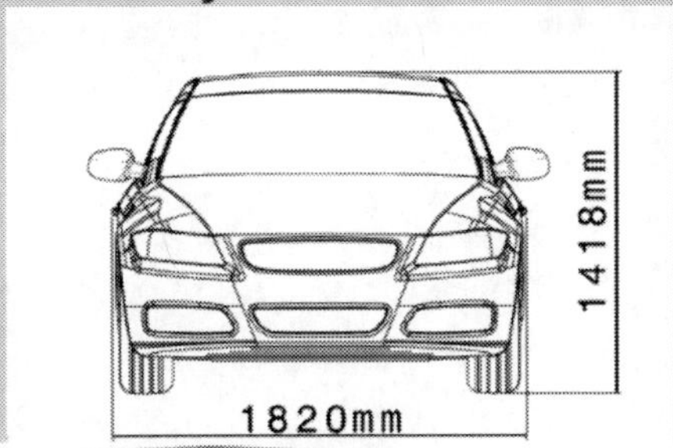

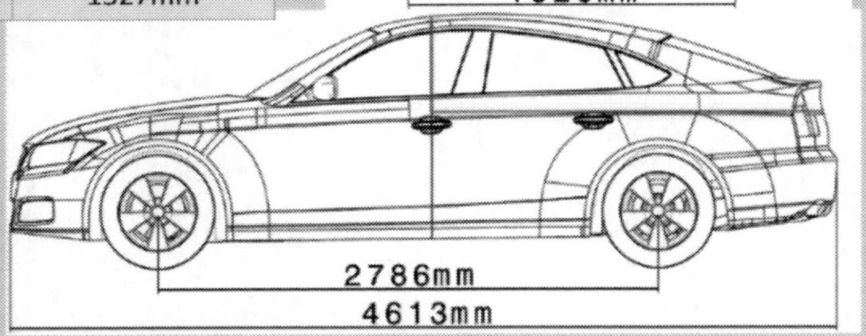

Courtesy by TU Munich, Inst. of Aerodynamics

6 © 2011 ANSYS, Inc. September 19, 2013

ANSYS Experimental Facility and Data

- The experimental data is provided by the Institute of Aerodynamics and Fluid Mechanics, TU Munich
- Experiments are performed in a wind tunnel including a moving belt @ 1:2.5 model scale

Courtesy by TU Munich, Inst. of Aerodynamics

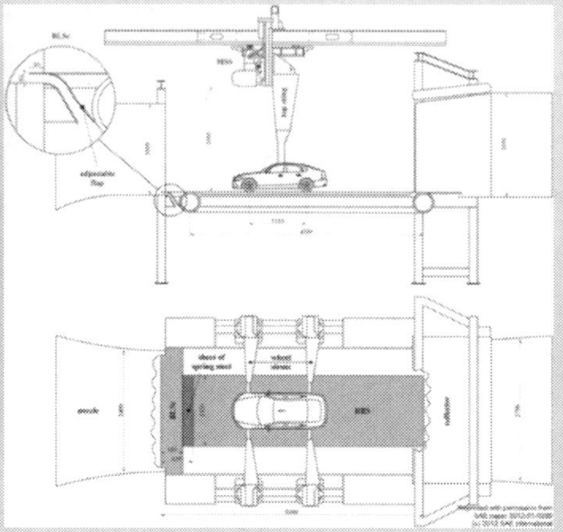

ANSYS Test Case Conditions

Model Scale	1:2.5
Inlet velocity	40 m/s
Air Temperature	20°C
Air Pressure	1.013 bar
Air Density	1.2047 kg/m^3
Reference Length (Length of car model)	1.84 m
Resulting Reynolds number	4.87*10^6
Ground velocity	**40 m/s**

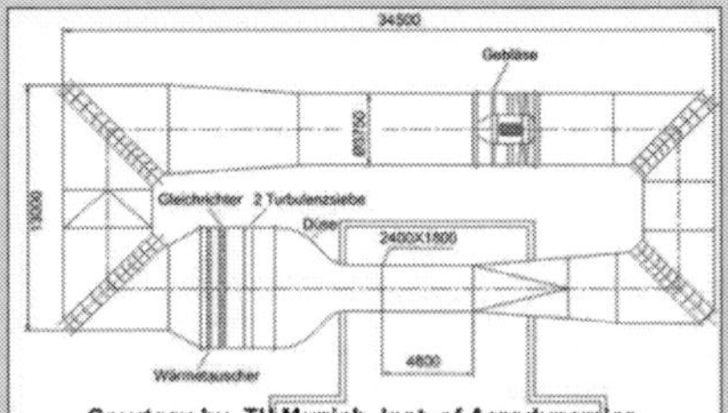

Courtesy by TU Munich, Inst. of Aerodynamics

ANSYS Investigated DrivAer Car Models

E_S_woM_woW
Estate_
Smooth underbody_
without Mirrors_
without Wheels

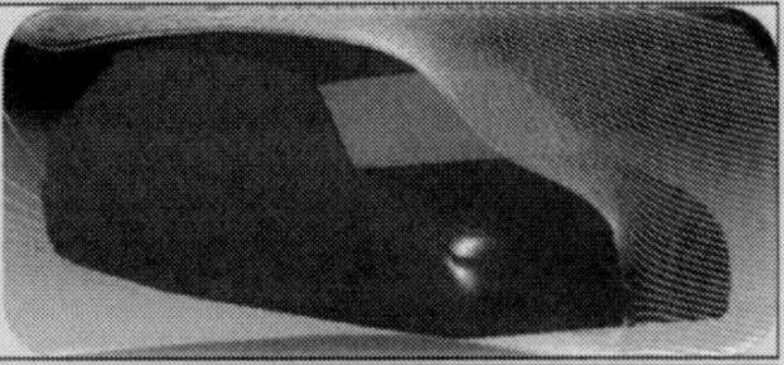

F_S_woM_wW
Fastback_
Smooth underbody_
without Mirrors_
with Wheels

F_D_wM_wW
Fastback_
Detailed underbody_
with Mirrors_
with Wheels

9 © 2011 ANSYS, Inc. September 19, 2013

ANSYS Investigated DrivAer Car Models

E_S_woM_woW
Estate_
Smooth underbody_
without Mirrors_
without Wheels

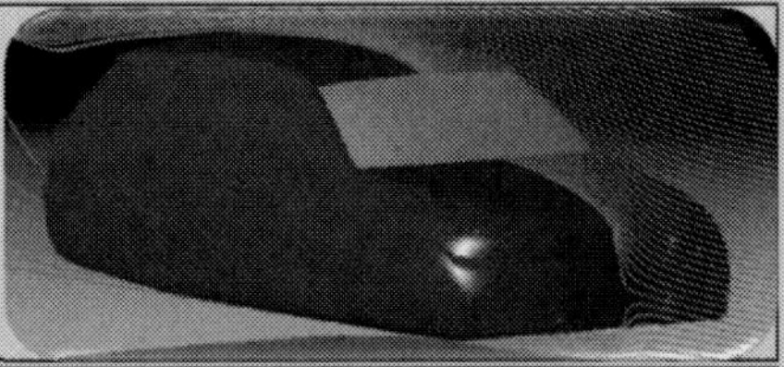

F_S_woM_wW
Fastback_
Smooth underbody_
without Mirrors_
with Wheels

F_D_wM_wW
Fastback_
Detailed underbody_
with Mirrors_
with Wheels

10 © 2011 ANSYS, Inc. September 19, 2013

ANSYS Geometry & Computational Domain

Model Scale to Car Size	1:1
Inlet velocity	16 m/s
Air Temperature	20°C
Air Pressure	1.013 bar
Air Density	1.2047 kg/m^3
Reference Length (Car Length)	4.6
Resulting Reynolds number	4.87*10^6
Ground velocity	16 m/s

Dimensions of the Bounding Box		
Model scale	1 : 1	1 : 1
Total length	10L	46.13 m
Total width	11B	20.02 m
Total height	8H	11.34 m

13 September 19, 2013

ANSYS Meshing Process

- **Meshing process using:**
  - **ANSYS DesignModeler 14.5**
  - **ANSYS TGrid in Fluent 14.5**

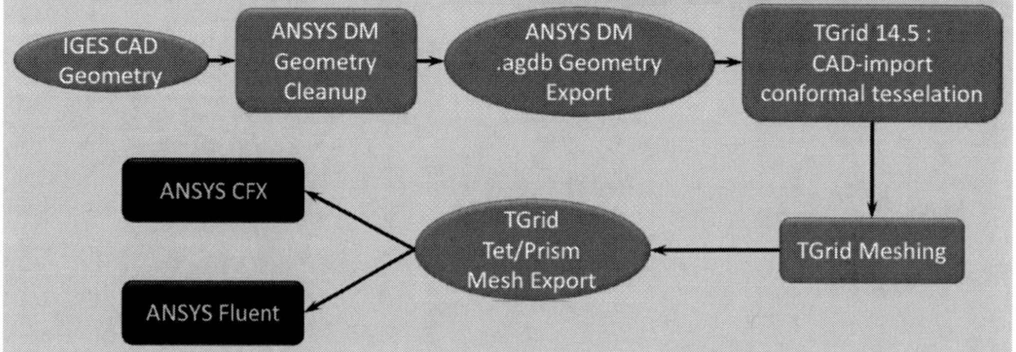

14 September 19, 2013

ANSYS

Geometry Clean-up

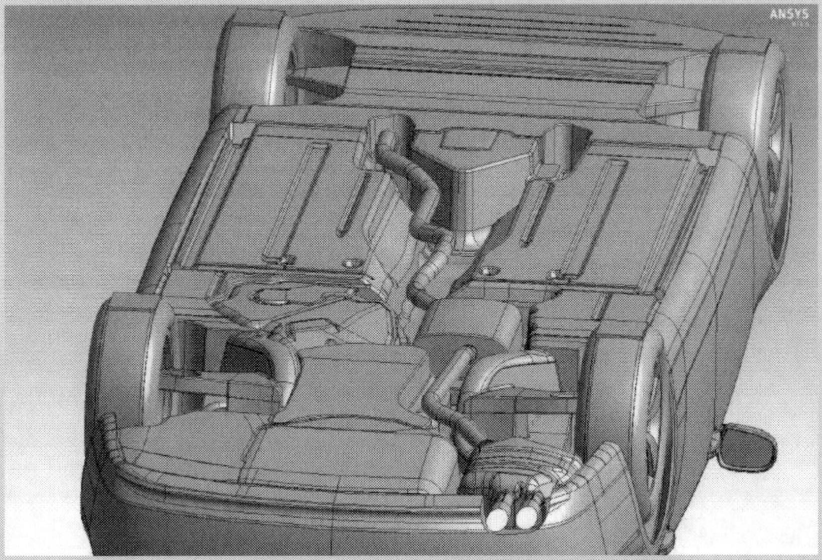

ANSYS

F_D_wM_wW: Computational Mesh2

- Full 3d model → SAS-SST
- ~110 Mill. Cells
- Four refinement zones
- 20 Inflations on the car
- 15 Inflations on the road
- y^+<1 on the car body
- MRF-Zones for the rims (MRF=Moving Reference Frame)

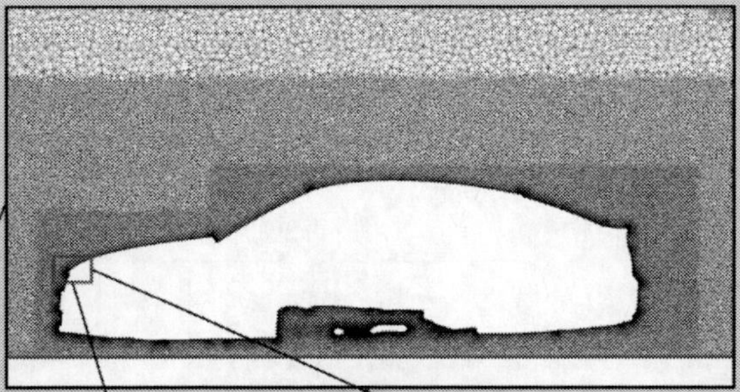

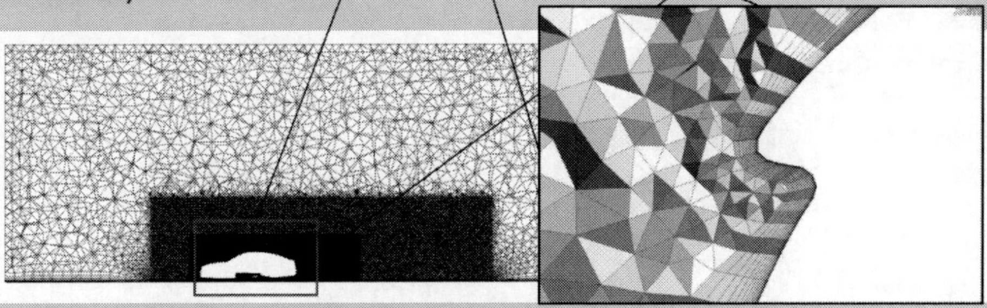

ANSYS TGrid Meshing Details

Simulation Matrix

Timestep		Mesh 1	Mesh 2	Mesh 2 Full Domain
Steady SST	Δt = 0.1ms	X		
Steady SST	Δt = 1 ms	X	X X	X X
Steady SST	Δt = 10 ms	X		
Transient SST	Δt = 1 ms	X		X X
Transient SAS-SST	Δt = 1 ms			X X
Transient SAS-SST	Δt = 0.2 ms			X X

X – ANSYS CFX investigation
X – ANSYS Fluent investigation

ANSYS Investigation Results F_S_woM_wW

F_S_woM_wW - Fastback_Smooth underbody_without Mirrors_with Wheels

ANSYS ANSYS CFD, SAS-SST – C_D Histories

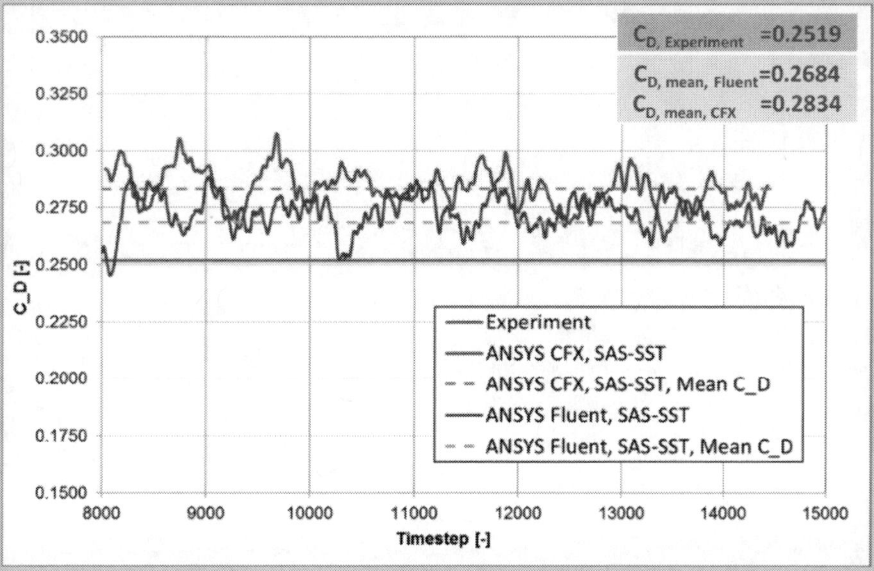

Experimental data by courtesy of TU Munich, Inst. of Aerodynamics

ANSYS

ANSYS CFD, SAS-SST – C_p at Symmetry Plane y=0mm (top)

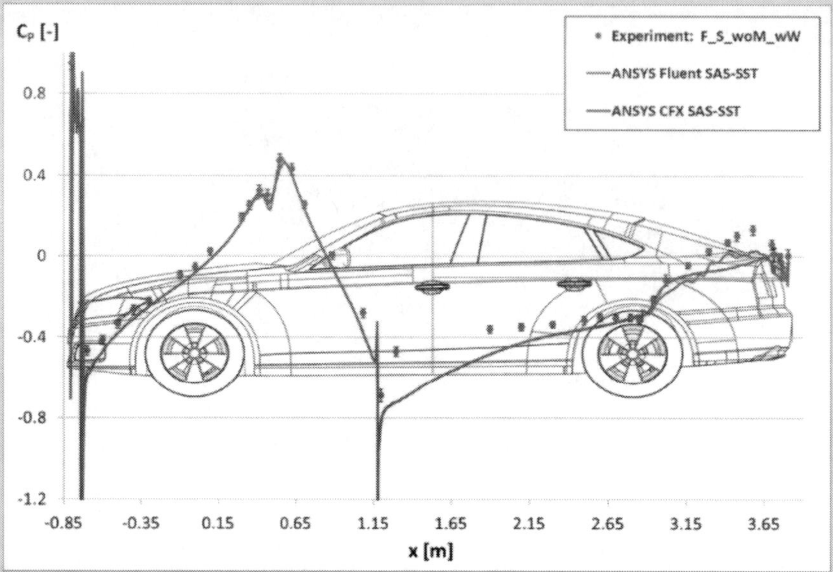

21 © 2011 ANSYS, Inc. September 19, 2013

Experimental data by courtesy of TU Munich, Inst. of Aerodynamics

ANSYS

ANSYS CFD, SAS-SST – C_p at Symmetry Plane y=0mm (bottom)

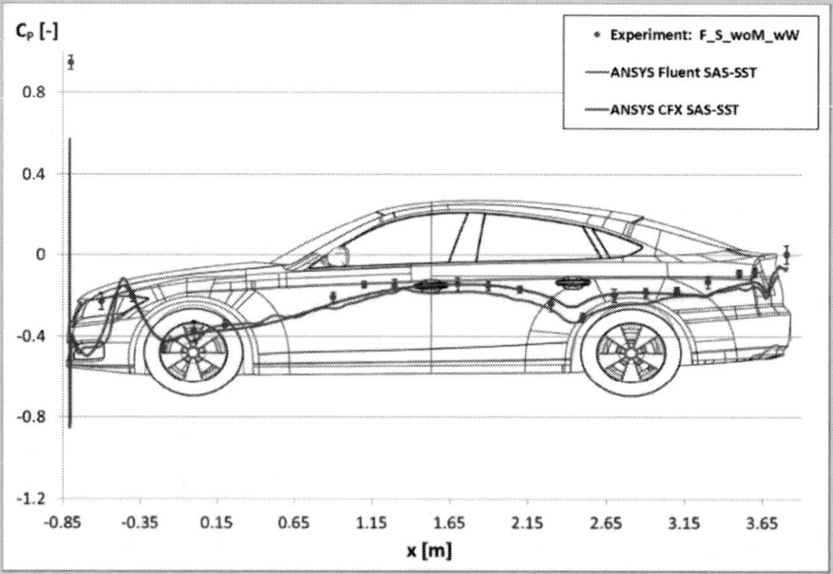

22 © 2011 ANSYS, Inc. September 19, 2013

Experimental data by courtesy of TU Munich, Inst. of Aerodynamics

ANSYS CFD, SAS-SST – C_p at z=0.15m

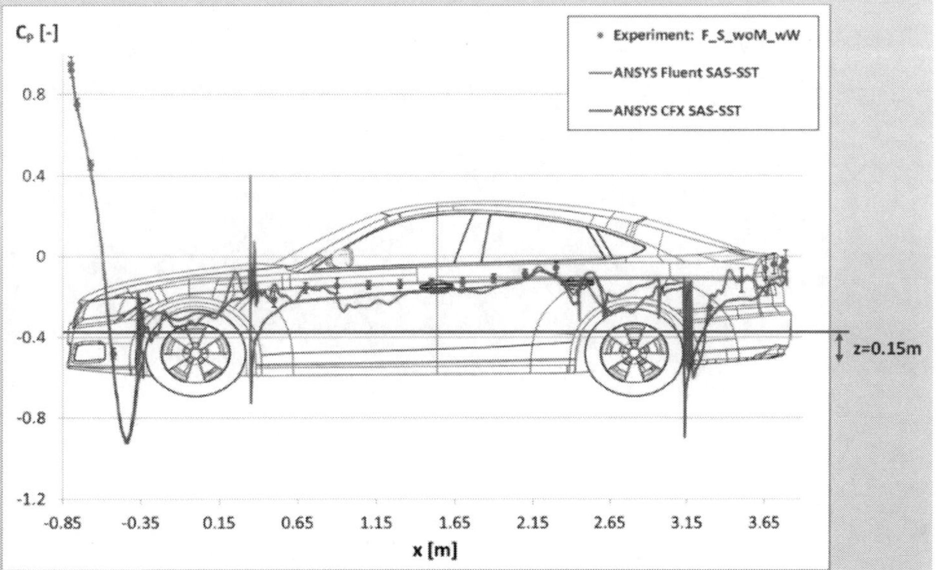

 Experimental data by courtesy of TU Munich, Inst. of Aerodynamics

Comparison of Q-Criterion

Fluent, SAS-SST
Δt=0.2 ms
15000 time steps
3.0 s total time

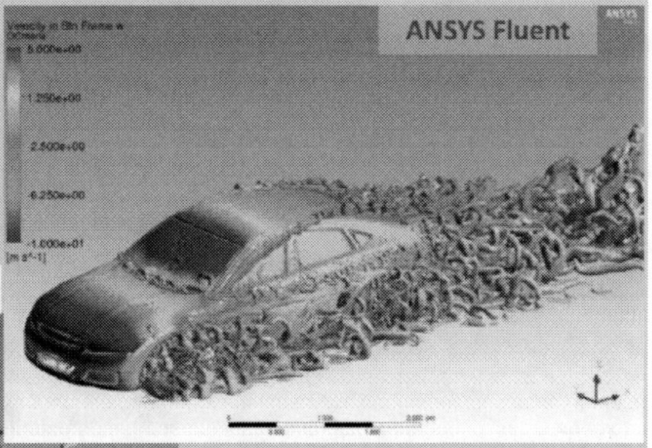

Q criterion level = 0.0005

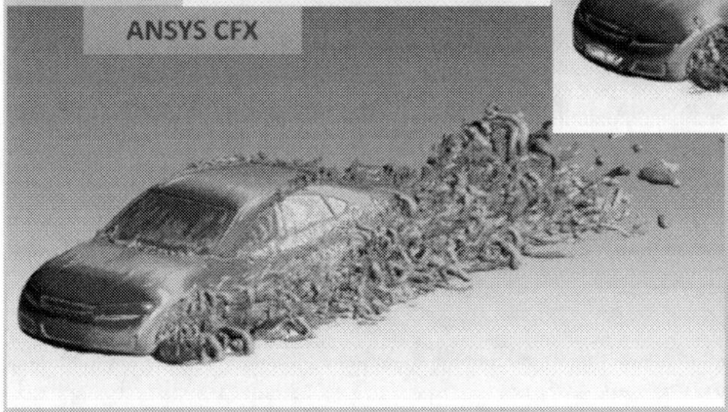

CFX, SAS-SST
Δt=0.2 ms
14419 time steps
2.884s total time

ANSYS

Vortex Structure from Transient Simulation, SAS-SST, Δt=0.2ms

25 © 2011 ANSYS, Inc. September 19, 2013

ANSYS

Vortex Structure from Transient Simulation, SAS-SST, Δt=0.2ms

26 © 2011 ANSYS, Inc. September 19, 2013

Investigation Results F_D_wM_wW

F_D_wM_wW - Fastback_Detailed underbody_with Mirrors_with Wheels

27 © 2011 ANSYS, Inc. September 19, 2013

ANSYS CFD, URANS SST & SAS-SST – C_D Histories –

$C_{D, Experiment}$ =0.2927

$C_{D, mean, Fluent}$=0.3114

$C_{D, mean, CFX}$ =0.3158

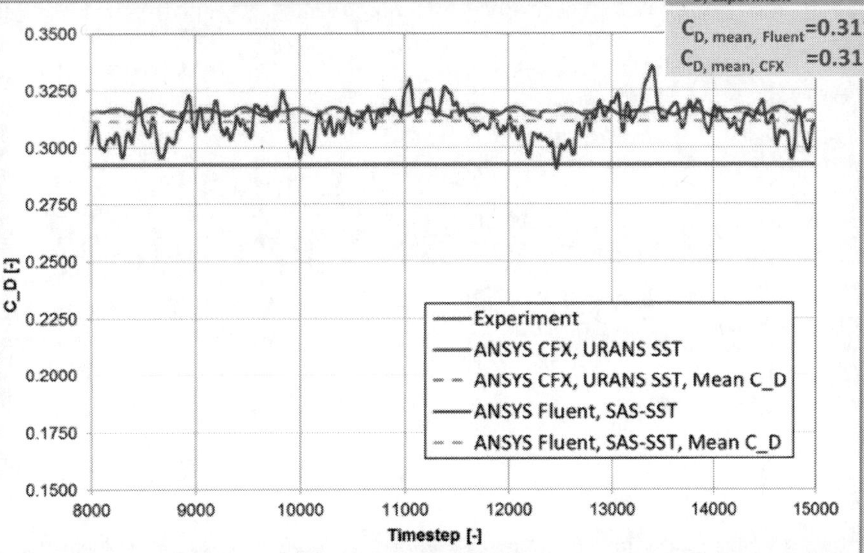

28 © 2011 ANSYS, Inc. September 19, 2013 **Experimental data by courtesy of TU Munich, Inst. of Aerodynamics**

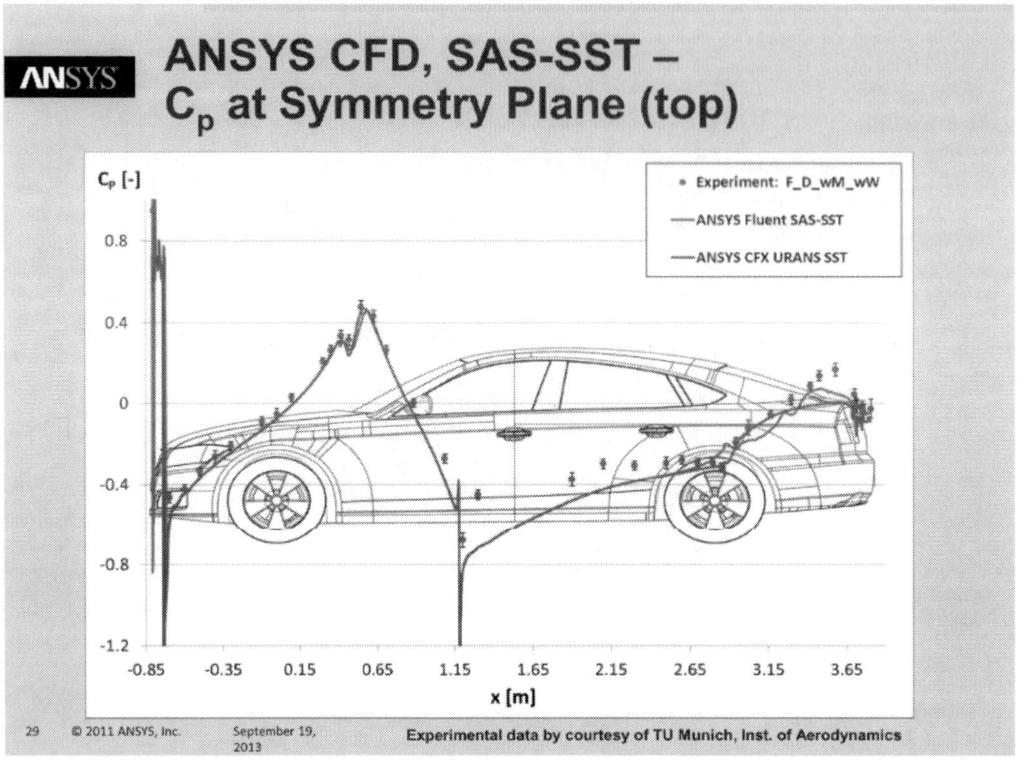
ANSYS
ANSYS CFD, SAS-SST –
C_p at Symmetry Plane (top)
Cp [-]
0.8
0.4
0
-0.4
-0.8
-1.2
-0.85
-0.35
0.15
0.65
1.15
1.65
2.15
2.65
3.15
3.65
x [m]
Experiment: F_D_wM_wW
ANSYS Fluent SAS-SST
ANSYS CFX URANS SST
29
© 2011 ANSYS, Inc.
September 19, 2013
Experimental data by courtesy of TU Munich, Inst. of Aerodynamics

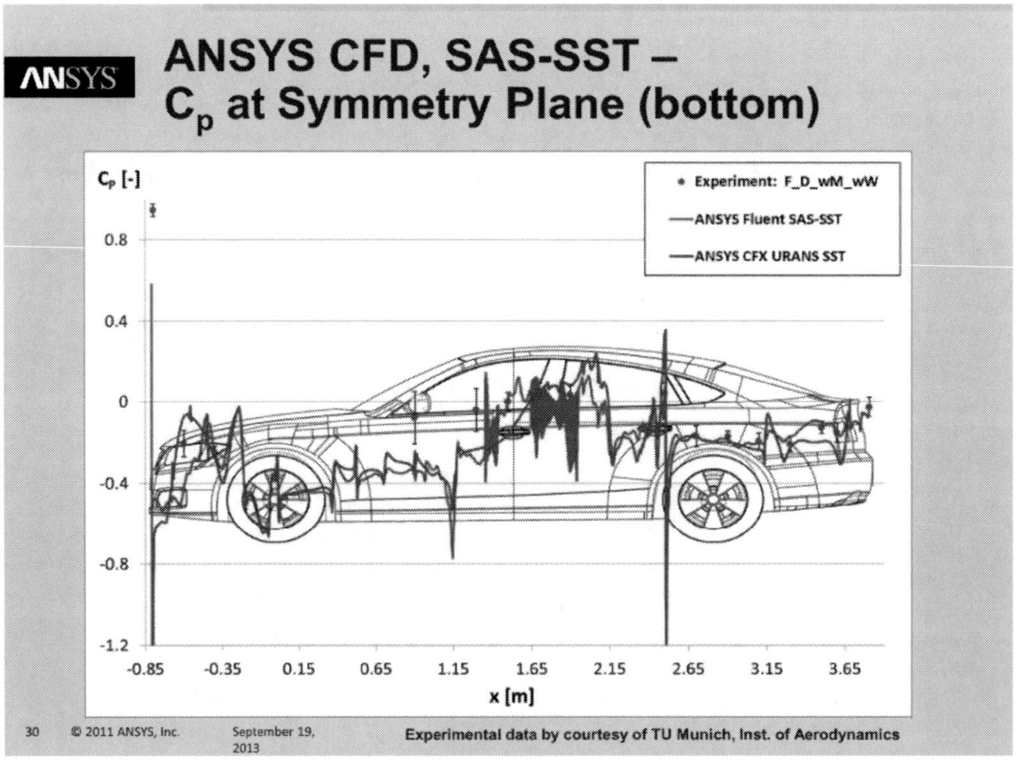
ANSYS
ANSYS CFD, SAS-SST –
C_p at Symmetry Plane (bottom)
Cp [-]
0.8
0.4
0
-0.4
-0.8
-1.2
-0.85
-0.35
0.15
0.65
1.15
1.65
2.15
2.65
3.15
3.65
x [m]
Experiment: F_D_wM_wW
ANSYS Fluent SAS-SST
ANSYS CFX URANS SST
30
© 2011 ANSYS, Inc.
September 19, 2013
Experimental data by courtesy of TU Munich, Inst. of Aerodynamics

ANSYS CFD, SAS-SST – C_p at z=0.15m (left)

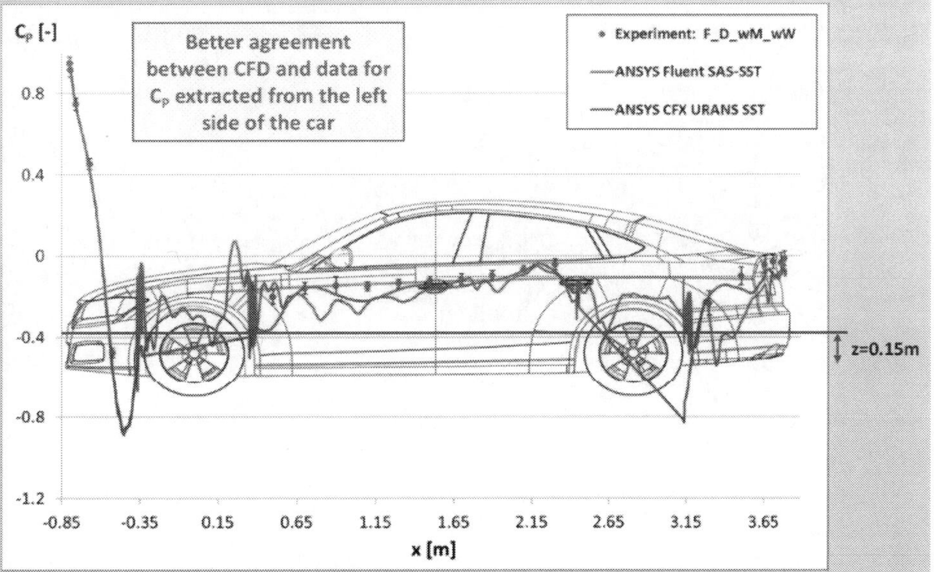

31 © 2011 ANSYS, Inc. September 19, 2013

Experimental data by courtesy of TU Munich, Inst. of Aerodynamics

ANSYS CFD, SAS-SST – C_p at z=0.15m (right)

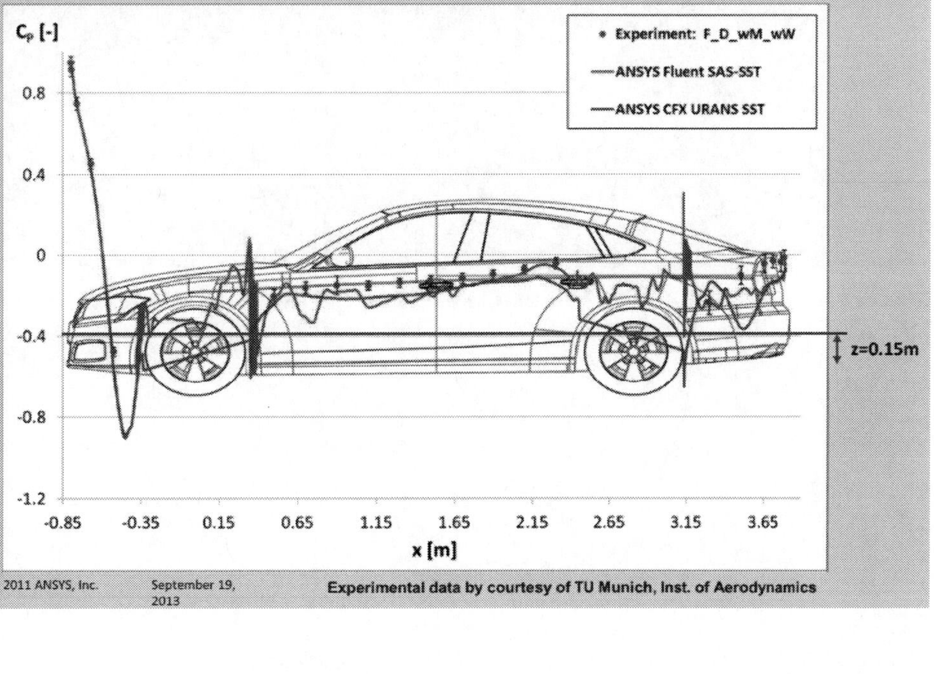

32 © 2011 ANSYS, Inc. September 19, 2013

Experimental data by courtesy of TU Munich, Inst. of Aerodynamics

ANSYS

ANSYS CFX, URANS SST, Δt=0.2ms – Asymmetric wake –

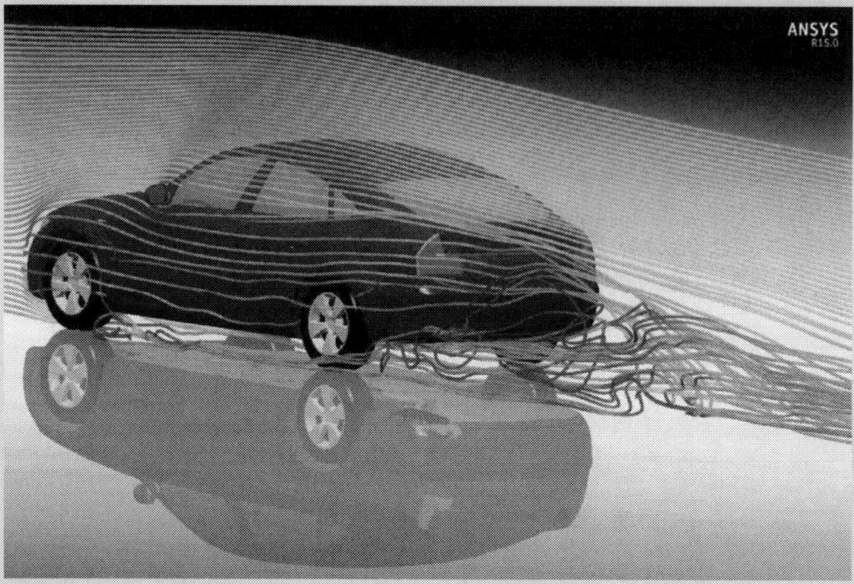

ANSYS

ANSYS CFX, URANS SST, Δt=0.2ms – Asymmetric wake –

ANSYS CFX, URANS SST, Δt=0.2ms – Asymmetric wake –

ANSYS

ANSYS Fluent SAS-SST, Δt=0.2ms – Q-Criterion Isosurfaces –

ANSYS

Q criterion level = 0.0005

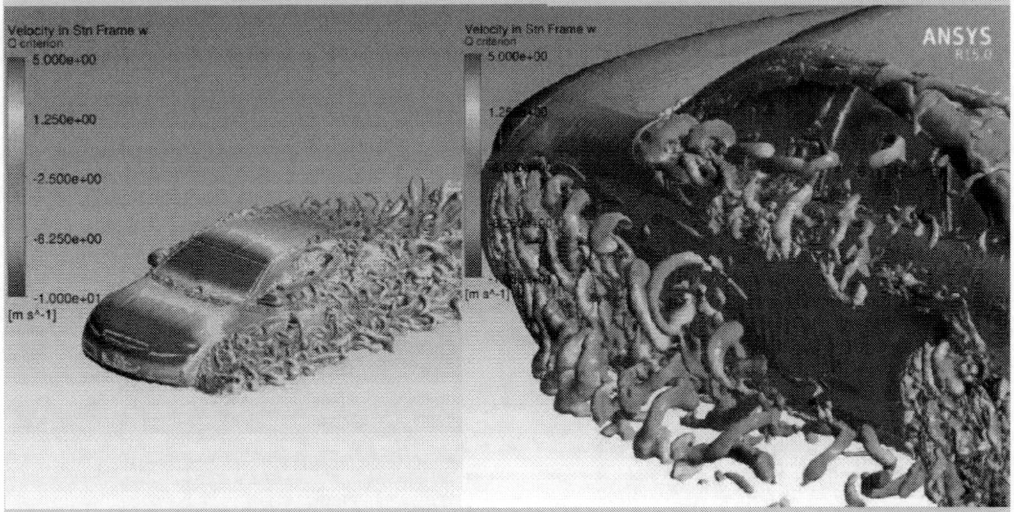

ANSYS Fluent SAS-SST, Δt=0.2ms

Lambda-2 criterion level = 0.01
Bottom view of F_D_wM_wW

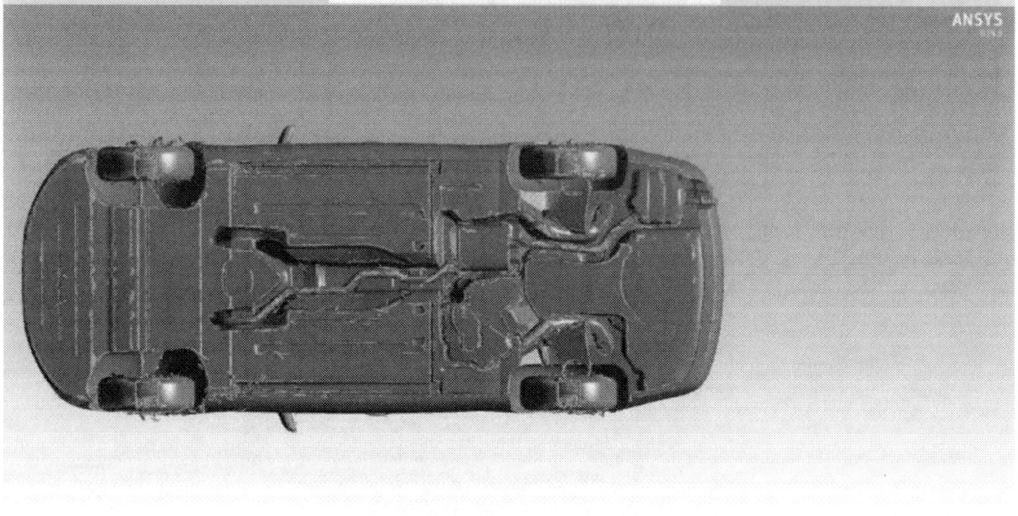

ANSYS Fluent SAS-SST, Δt=0.2ms

Lambda-2 criterion level = 0.001
Bottom view of F_D_wM_wW

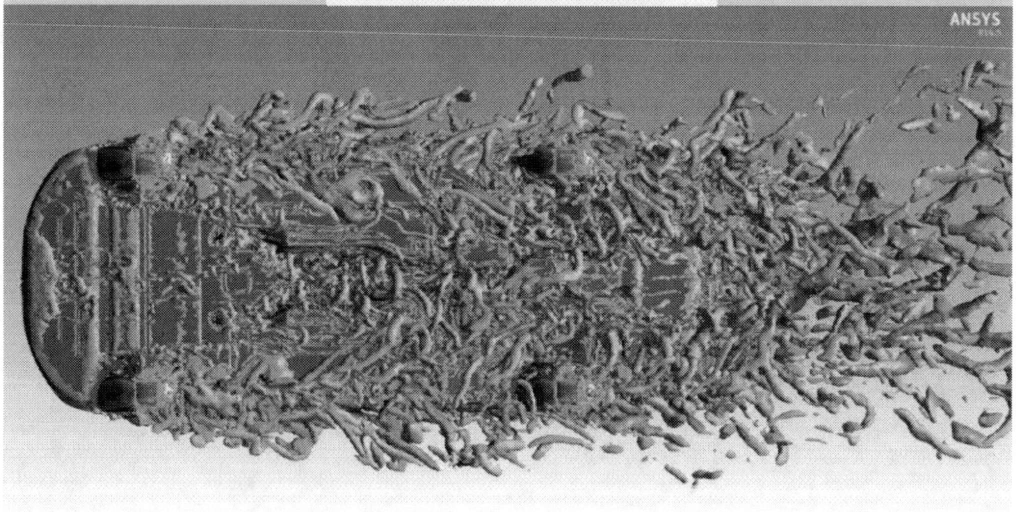

ANSYS

Realize Your Product Promise™

Summary & Conclusions

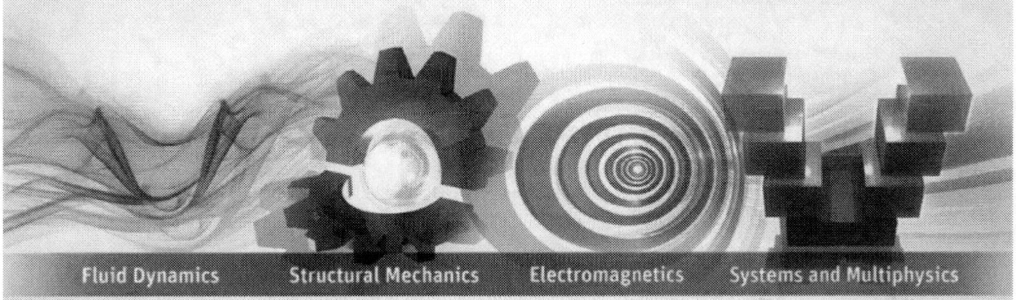

ANSYS

Summary & Outlook

- Simulating the DrivAer car is first of all a meshing challenge!
- Established a meshing process, where ANSYS TGrid in Fluent 14.5 and direct CAD model tessellation was applied
- Three different DrivAer cars meshed and simulated
  → (U)RANS SST and SAS-SST comparison
- Applied feasible amount of CFD best practice related investigations:
  - mesh and timestep dependence
  - iteration error → convergence
  - steady vs. transient
  - (U)RANS vs. scale-resolving turbulence modeling

ANSYS Summary & Outlook (cont.)

- Reasonable good agreement for C_P value comparison to data
- Influence from the model support system (MSS) on C_P on the top of the car roof observable
- Differences at point of vortex impingement in the rear of the car
- CFD predicted slightly higher C_D values in comparison to data
  - Influence from wind tunnel geometry
  - Quite high blocking ratio for this large model in TUM wind tunnel
  - Influence from road simulator vs. entirely moving road (CFD)

  → **Desirable to have PIV data for flow field comparison**
- Good and very consistent comparison between ANSYS CFX and ANSYS Fluent for investigated DrivAer car models
- **Further streamlining and refinement of the ANSYS TGrid in Fluent based meshing process possible**

  → e.g. longer extension of refined zones behind the car

ANSYS

Questions?

References

1. http://www.aer.mw.tum.de/abteilungen/automobilaerodynamik/drivaer/
2. http://www.cfd-online.com/Wiki/DrivAer_Model
3. A. Heft, T. Indinger, N. Adams: *"Experimental and Numerical Investigation of the DrivAer Model"*, ASME 2012, July 8-12, 2012, Puerto Rico, USA, FEDSM2012-72272
4. A. Heft, T. Indinger, N. Adams: *"Introduction of a New Realistic Generic Car Model for Aerodynamic Investigations"*, SAE 2012 World Congress, April 23-26, 2012, Detroit, Michigan, USA, Paper 2012-01-0168
5. P. Nathen: *"Investigation of the Complex Turbulent Flow around a Generic Vehicle"*, MSc Thesis, TU Munich, Inst. Aerodynamics and Fluid Dynamics, April 2012

43 September 19, 2013

BETA CAE Systems SA
www.beta-cae.gr

The influence of mesh characteristics on CFD simulations for automotive applications

Vangelis Skaperdas, Aristotelis Iordanidis
BETA CAE Systems S.A.

BETA CAE Systems SA
www.beta-cae.gr

ANSA and META integration with CFD

ANSA has been used in pre-processing for Fluent since the mid 90s for geometry clean up and surface mesh generation. Currently, ANSA and META provide complete support for the most robust and high quality use of CFD in industrial scale and complexity simulations

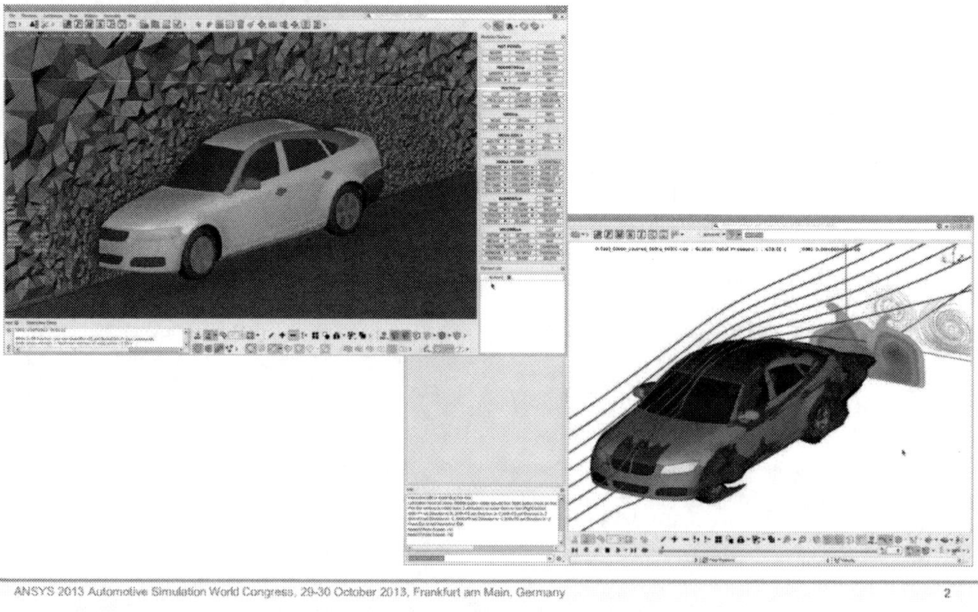

The DrivAer model of the Technical University of Munich

Experimental setup:
1:2.5 scale wind tunnel model
Re = 4.87×10^6
L = 1.84 m
U inlet = 40 m/sec
free stream turbulence = 0.4%

Acknowledgments to:
Mrs. Angelina Heft from the
Institute of Fluid Mechanics and Aerodynamics of the
Technical University of Munich
for providing the model geometries
in IGES and STEP formats

Sources:
1) "Introduction of a New Realistic Generic Car Model for Aerodynamic Investigations", A. Heft, T. Indinger and N. Adams, SAE 2012-01-0168
2) "Experimental and Numerical Investigation of the DrivAer Model", A. Heft, T. Indinger and N. Adams, ASME 2012 Fluids Engineering Summer Meeting, July 8-12, 2012, Rio Grande, Puerto Rico

Software and hardware used

- ANSA v14.2.0 for pre-processing
- ANSYS Fluent v14.5 for solving on 16 nodes
- μETA v14.2.0 for post-processing

Fedora Linux 64
24 core Intel Xeon X5680 at 3.33GHz
96 Gb RAM

βBETA CAE Systems SA www.beta-cae.gr

Geometry preparation: STEP file input and property assignment on certain areas

Geometries that included detailed underbody and mirrors were selected

βBETA CAE Systems SA www.beta-cae.gr

Model management in ANSA

Managing CFD model properties while tracking Part and Property from PDM system

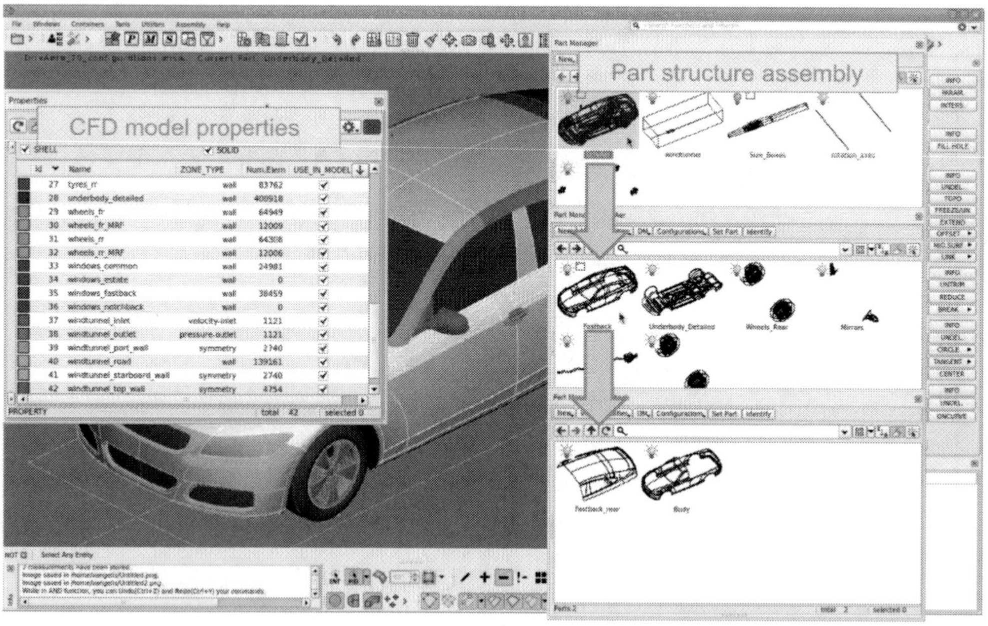

β BETA CAE Systems SA
www.beta-cae.gr

Configurations management in ANSA handling three variants in one file

By activating/de-activating the use of certain variant parts the user can automatically obtain a different model

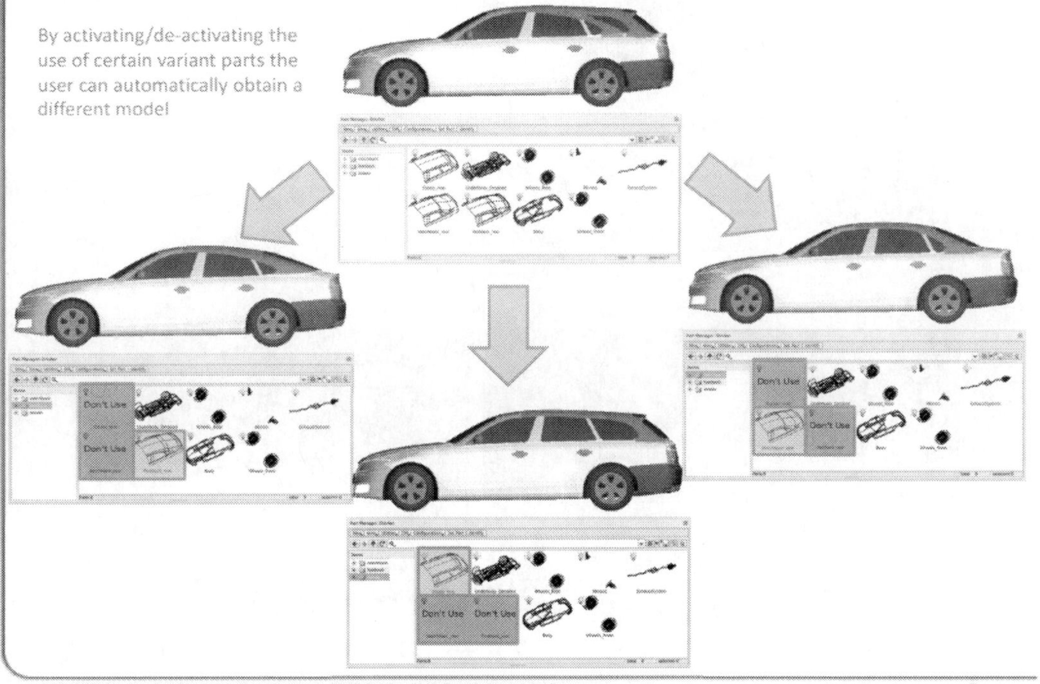

β BETA CAE Systems SA
www.beta-cae.gr

Flexible Size Boxes controlling mesh refinement aligned to the flow

Batch Meshing setup: automation and consistency in meshing

βBETA CAE Systems SA www.beta-cae.gr

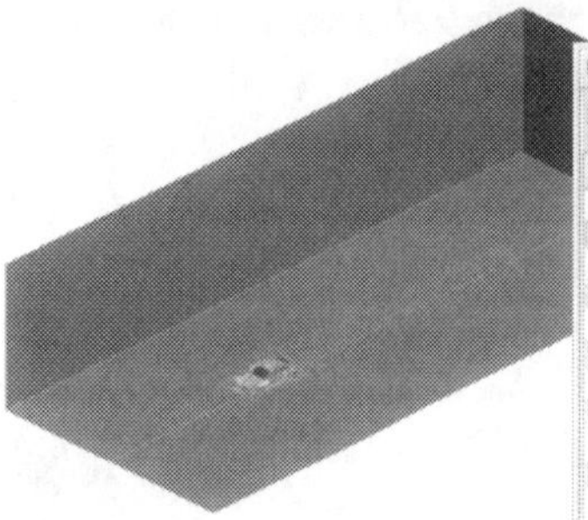

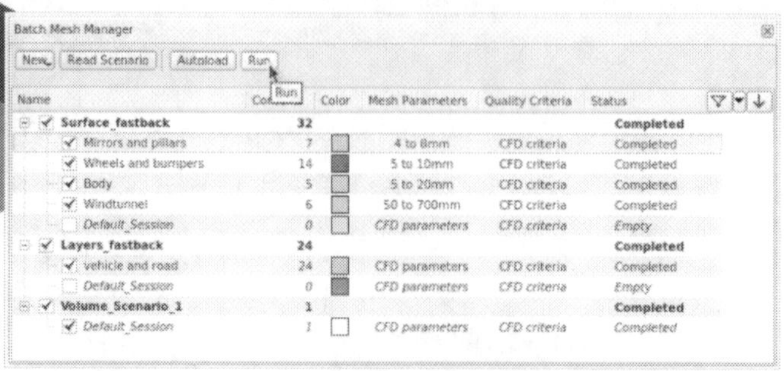

Batch Mesh Manager

New | Read Scenario | Autoload | Run

Name	Contents	Color	Mesh Parameters	Quality Criteria	Status
Surface_fastback	32				Completed
Mirrors and pillars	7		4 to 8mm	CFD criteria	Completed
Wheels and bumpers	14		5 to 10mm	CFD criteria	Completed
Body	5		5 to 20mm	CFD criteria	Completed
Windtunnel	6		50 to 700mm	CFD criteria	Completed
Default_Session	0		CFD parameters	CFD criteria	Empty
Layers_fastback	24				Completed
vehicle and road	24		CFD parameters	CFD criteria	Completed
Default_Session	0		CFD parameters	CFD criteria	Empty
Volume_Scenario_1	1				Completed
Default_Session	1		CFD parameters	CFD criteria	Completed

Model Properties (Fluent zones)

Surface batch mesh sessions

Batch mesh generated surface mesh

βBETA CAE Systems SA www.beta-cae.gr

Automatic curvature and sharp edge refinement, in combination with the use of Size Boxes ensure the efficient and accurate capturing of all details of the model.
Quality according to Fluent skewness < 0.5

Coarse
650 k trias

Medium
1.4 million trias

Fine
2.5 million trias

Re-using the same mesh for other variants

Batch mesh generated (medium size)
1.4 million tria elements in 14 minutes

Using configurations management the model is updated to the new variant
Size boxes are also updated respectively
Original mesh of common body is retained unaffected

Batch mesh tool is then used to mesh only the updated unmeshed area at a fraction of the time

Boundary layer generation

Hi Reynolds layers
first height 2 mm
growth rate = 1.2
4 layers
+3 layers in aspect mode
last aspect ratio 40% of length

total layer height ≈ 30mm

Low Reynolds layers
first height 0.05 mm
growth rate = 1.2
10 layers in absolute
+15 layers in aspect mode
last aspect ratio 40% of length
(25 layers in total)

total layer height ≈ 30 mm

Boundary layer generation: squeezing and collapsing of layers at proximities

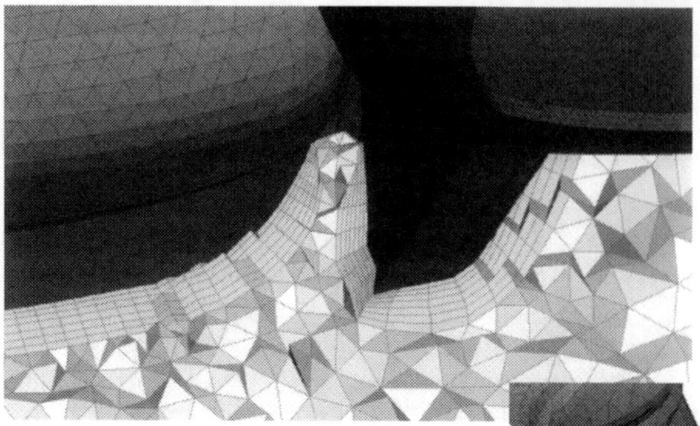

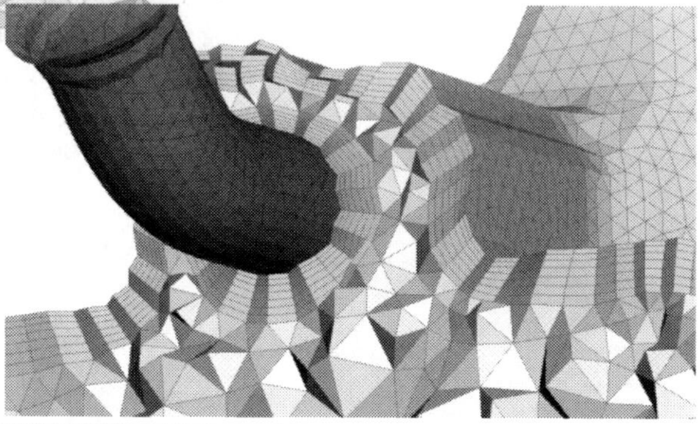

Batch mesh generated volume mesh

Automatic generation of layers and volume mesh for all variants and mesh densities (15 combinations)
Image below of medium size mesh with Hi Re layers (26.5 million cells) generated in under 20 minutes

βBETA CAE Systems SA www.beta-cae.gr

Indicative mesh quality statistics – Coarse Tetra Hi Re layers

Minimum orthogonality = 0.05
Maximum cell squish = 0.93

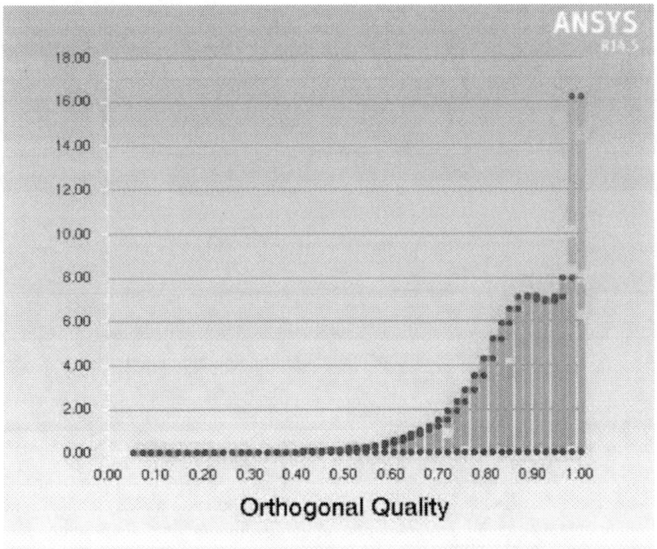

βBETA CAE Systems SA www.beta-cae.gr

Mesh refinement study for Hi Re layer tetra case

βBETA CAE Systems SA www.beta-cae.gr

Mesh type variations – medium cases Hi Re layers

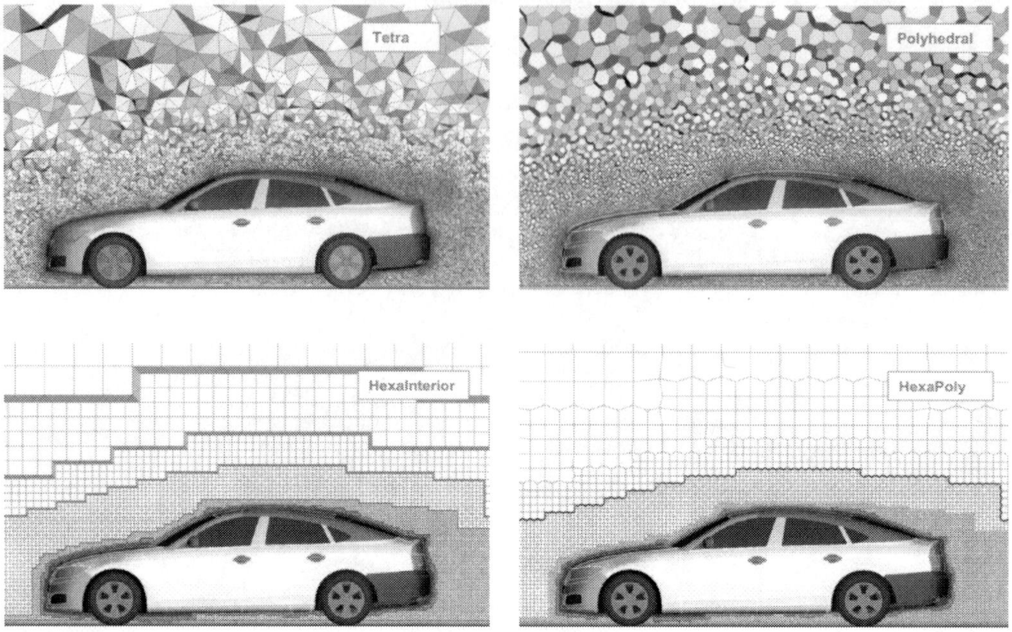

βBETA CAE Systems SA www.beta-cae.gr

Summary of mesh models for different variants

		Coarse	Medium	Fine
Fastback	without layers	Tetra (8.5 million)	Tetra (18.3 million)	-
	Hi Re layers	Tetra (12.4 million)	Tetra (26.5 million)	Tetra (54 million)
		Hexa Interior (11.7 million)	Hexa Interior (23.6 million)	-
		Hexa Poly (9.3 million)	Hexa Poly (19.2 million)	-
		Polyhedral (6.8 million)	Polyhedral (14.8 million)	-
	Low Re layers	Tetra (24.4 million)	Tetra (51.3 million)	-
Notchback	Hi Re layers	-	Tetra (26.4 million)	-
Estate	Hi Re layers	-	Tetra (26.8 million)	-

Fluent simulations

Model was scaled up to full size L = 4.612 m
Domain size 50 x 20 x 11.5 m
blockage ratio= 1%
domain sides set to symmetry

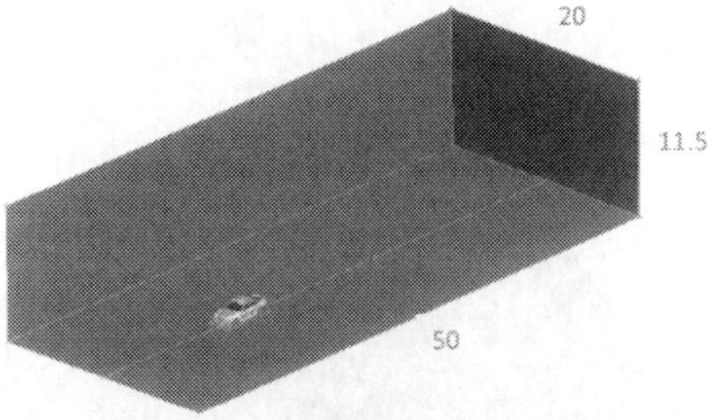

Steady State simulations
U free stream = 15.96 m/sec
Re = 4.87×10^6
Turbulence model: k-omega SST
2nd order discretization scheme for all variables

Cases with and without moving ground simulation with MRF modeling of rotating wheels

Forces averaged for last 5000 iterations

Mapping of results from one mesh to another for 18 different runs

Transient simulations
time step 0.00125 sec
10 iterations per time step
run for 1.5 sec real time
starting from steady state solution

Post-processing in μETA: y+ results for Hi and Low Re meshes

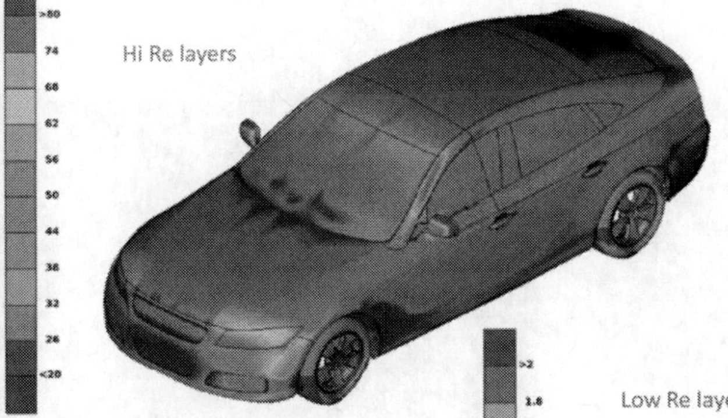

Post-processing was performed manually for one CFD run and then META run in batch mode for the other 18 simulations producing automatically the same images

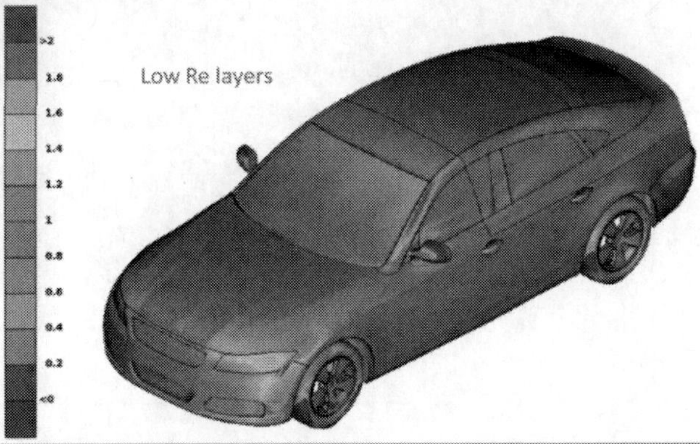

The models for Hi Re and Low Re layers give y+ values within the recommended guidelines

Effect of layers on skin friction coefficient and wall flow (tetra coarse meshes)

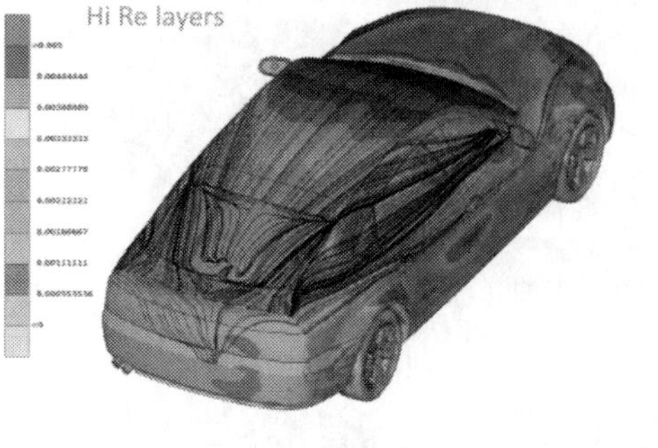

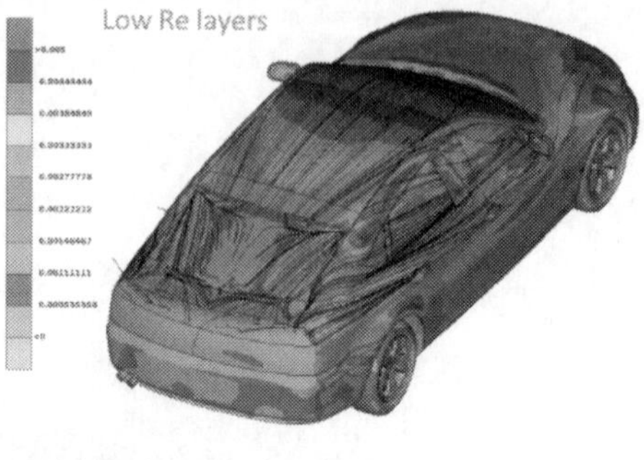

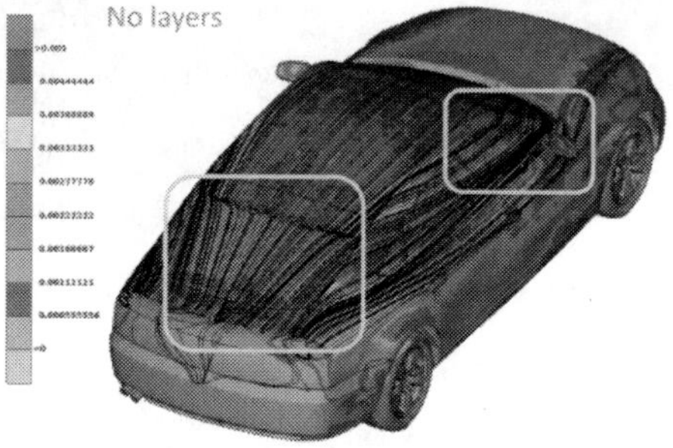

Both Hi and Low Re layers depict similar skin friction and separation patterns, while the case without layers does not predict any vortex fomation and recirculation around the A-pillar and the rear window

C_p contours and streamlines on symmetry plane (tetra medium meshes)

Low RE layer mesh seems to predict some flow separation and recirculation areas that the Hi Re layer models seem to miss

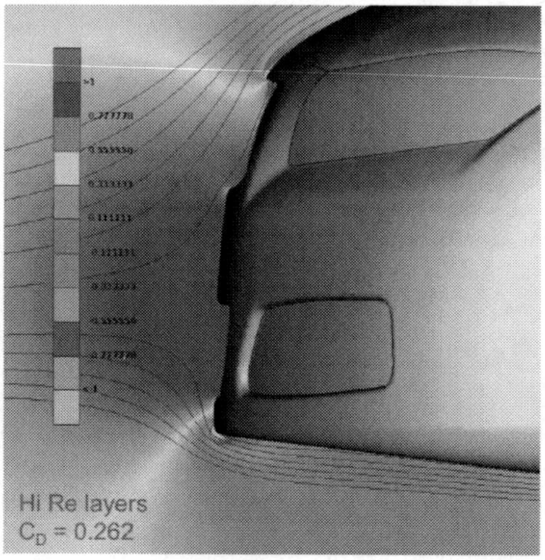

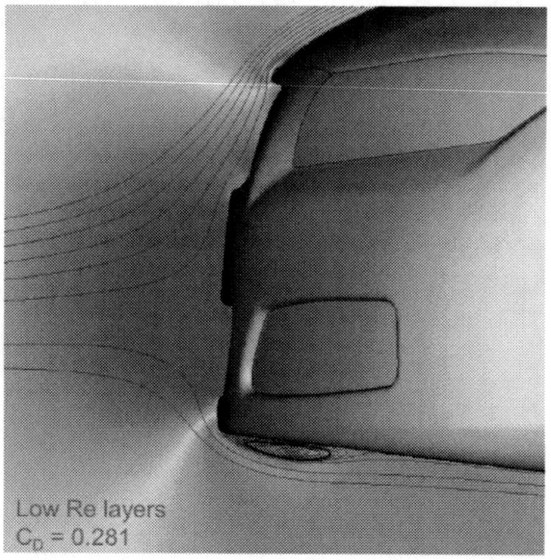

Steady state vs Transient drag monitoring for fastback tetra coarse Hi Re layers

Averaged over 1000 iterations / time steps
steady state C_d = 0.271
transient C_d = 0.269

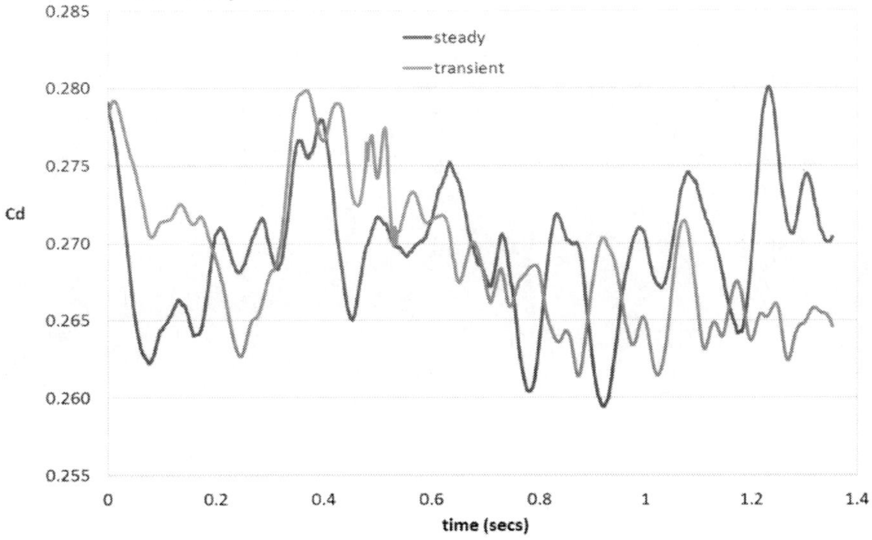

Comparison of transient vs steady state runs for fastback with moving ground

Tetra coarse with Hi RE layers

Steady state

Transient

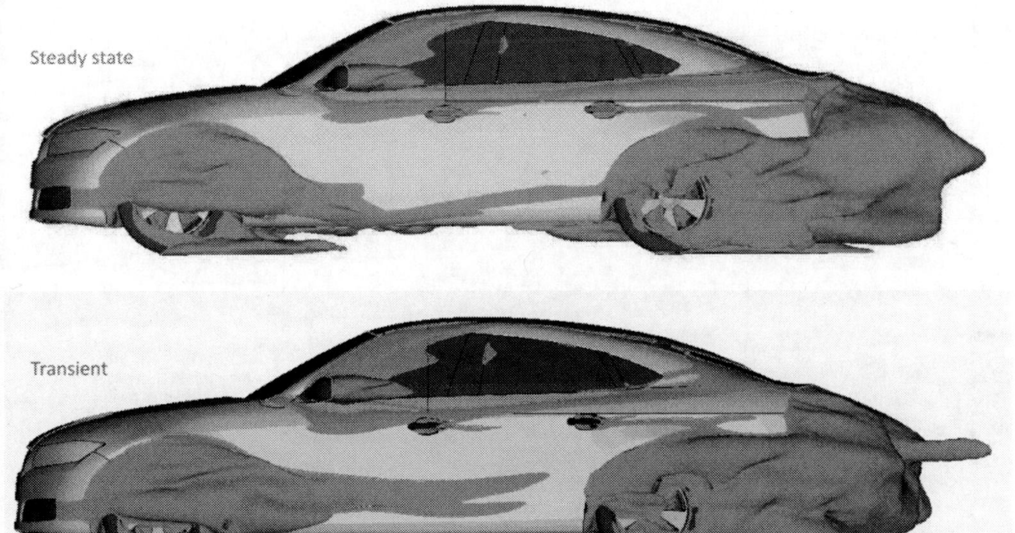

BETA CAE Systems SA
www.beta-cae.gr

Convergence plots: notchback tetra medium Hi Re without moving ground

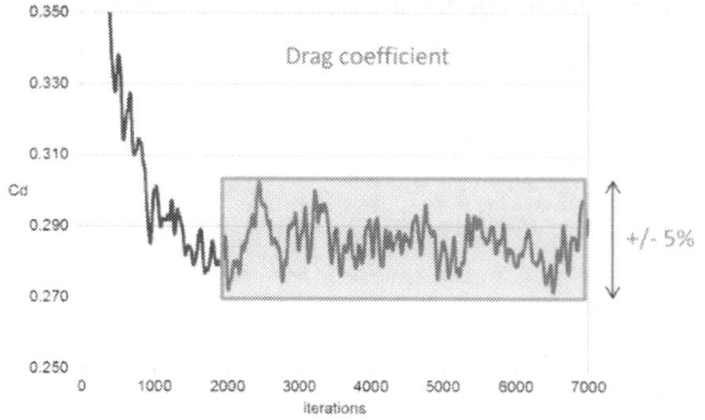

Averaged over 5000 last iterations
C_D = 0.286
C_L = 0.0389

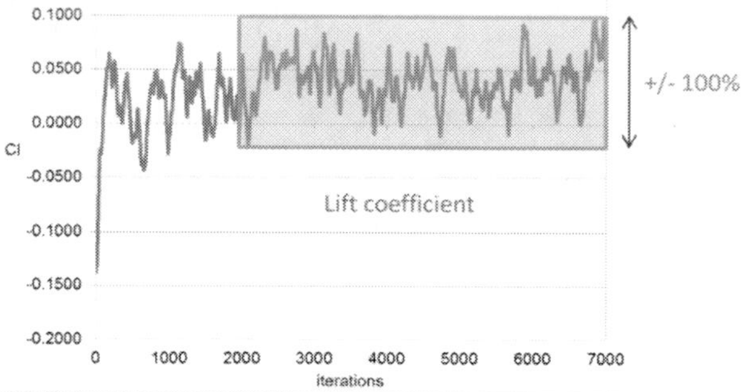

BETA CAE Systems SA
www.beta-cae.gr

Drag and Lift: Mesh refinement results for tetra Hi Re layer meshes

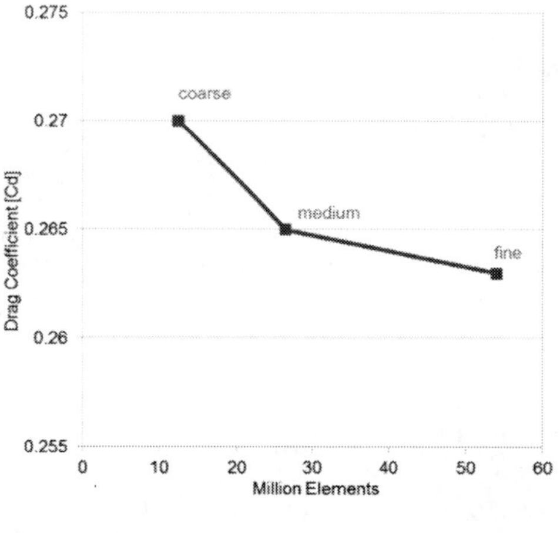

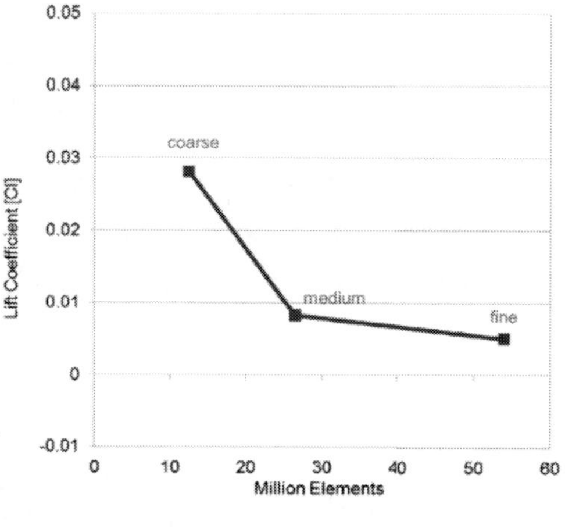

Coefficients calculated based on fastback projected frontal area = 2.169 m2

βBETA CAE Systems SA www.beta-cae.gr

Effect of moving ground on velocity magnitude field – medium tetra Hi Re layers

Stationary ground

Moving ground: higher velocities under the vehicle, less drag and lift produced

βBETA CAE Systems SA www.beta-cae.gr

Velocity magnitude for stationary ground runs (tetra medium Hi Re meshes)

C_D = 0.288
C_L = 0.0583

C_D = 0.286
C_L = 0.0389

C_D = 0.306
C_L = -0.0493

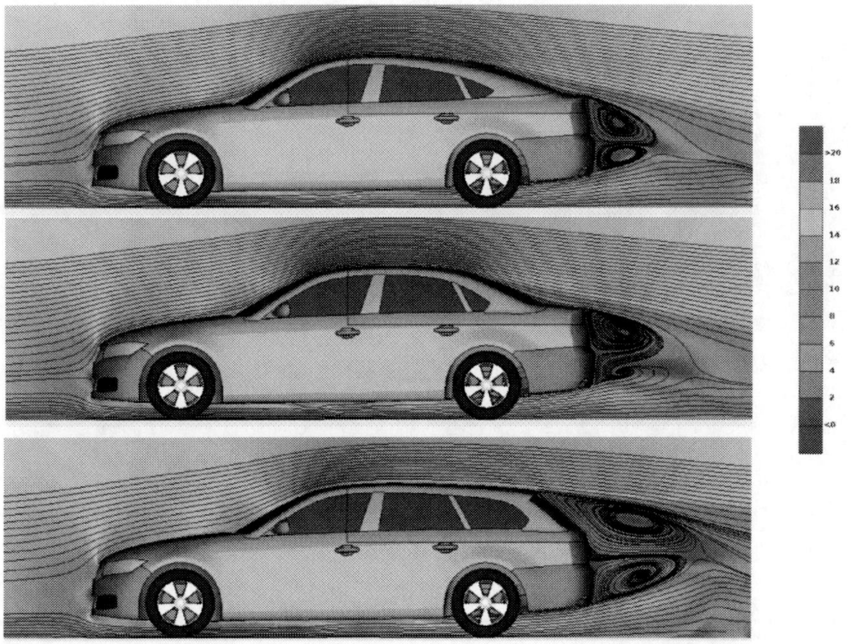

Effect of Hi/Low Re layers on force prediction for fastback with moving ground

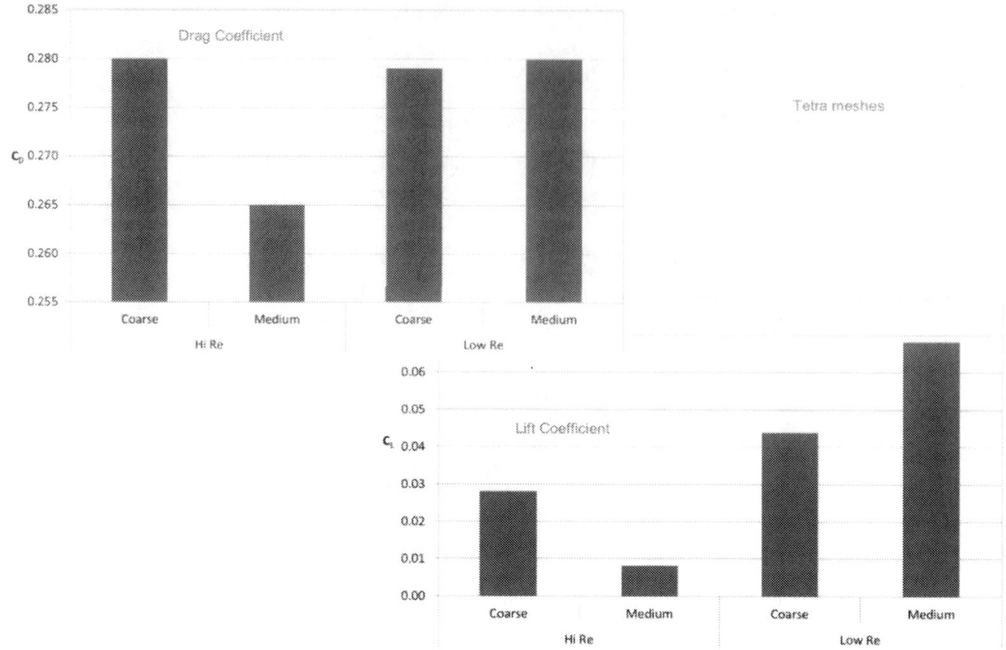

Tetra meshes

Effect of mesh type on force prediction for fastback with moving ground

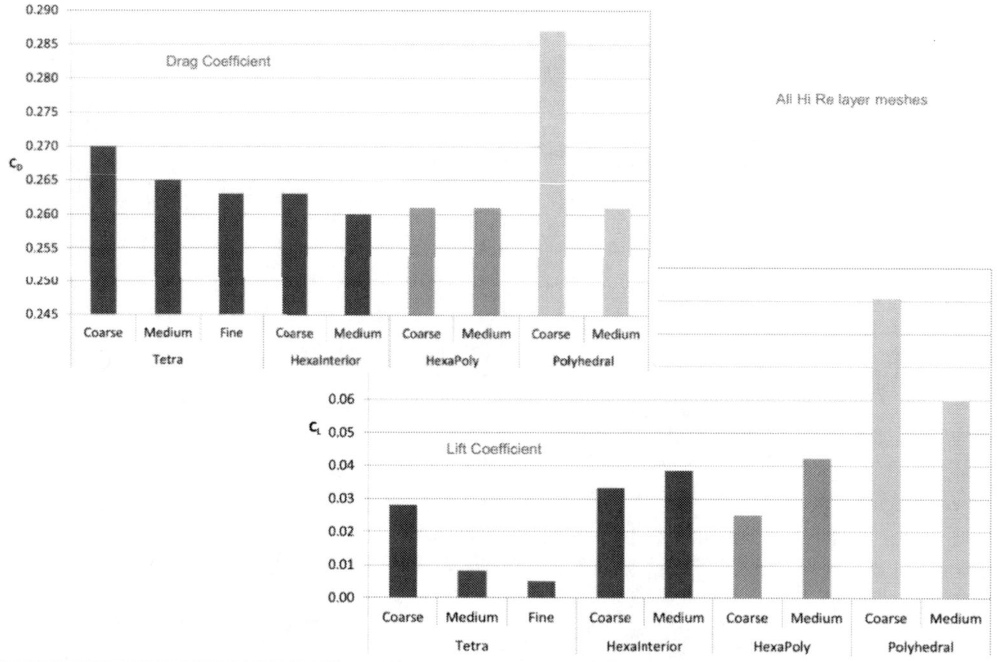

All Hi Re layer meshes

BETA CAE Systems SA
www.beta-cae.gr

Comparison with experiment: C_p field on side window

fastback medium tetra Hi Re with moving ground simulation

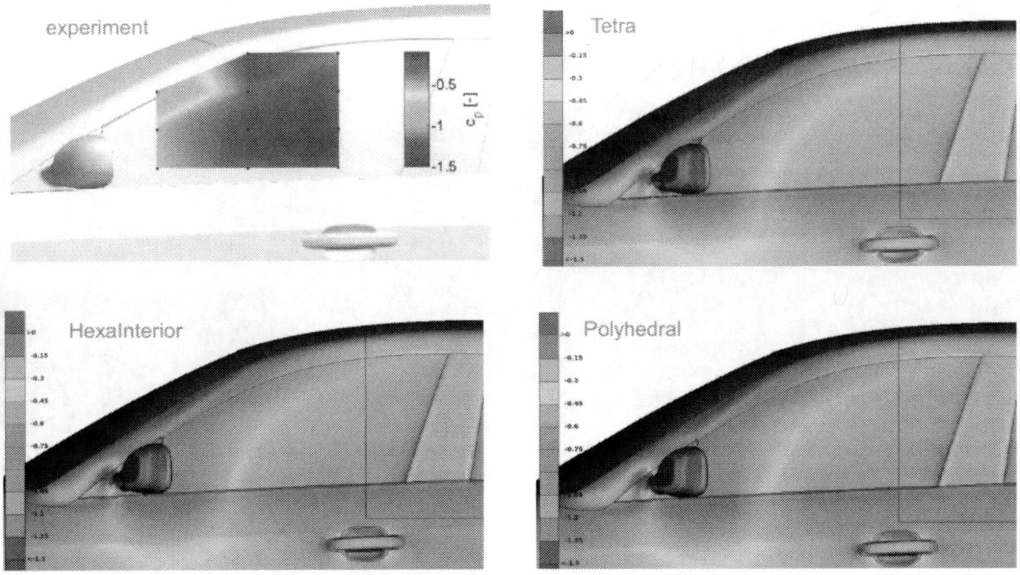

BETA CAE Systems SA
www.beta-cae.gr

Comparison with experiment: C_p on rear fastback window

fastback medium tetra Hi Re with moving ground simulation

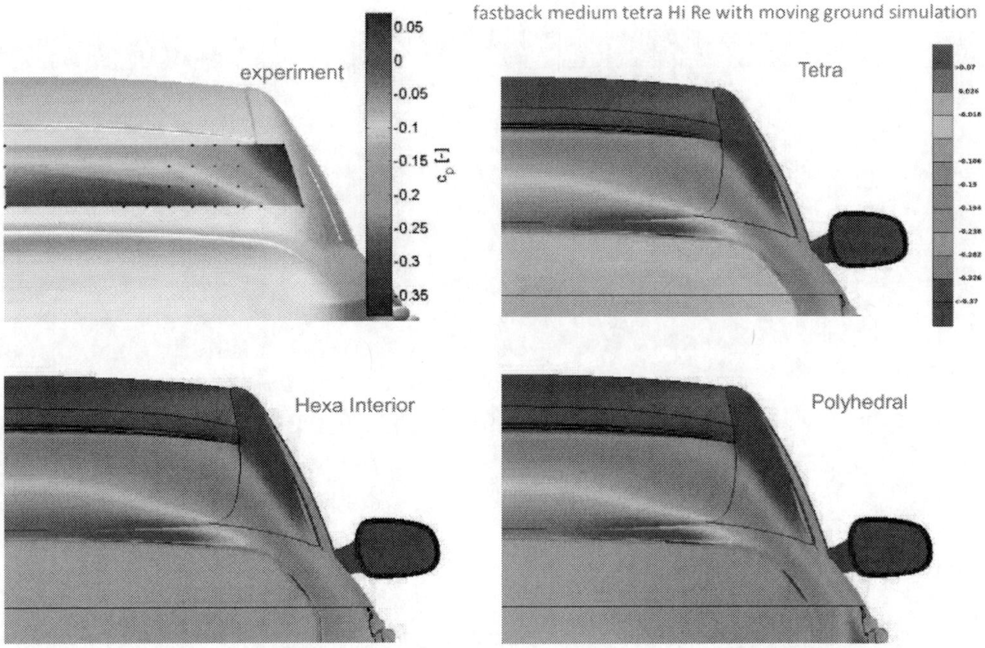

βBETA CAE Systems SA www.beta-cae.gr

Comparison with experiment: C_p along upper symmetry line

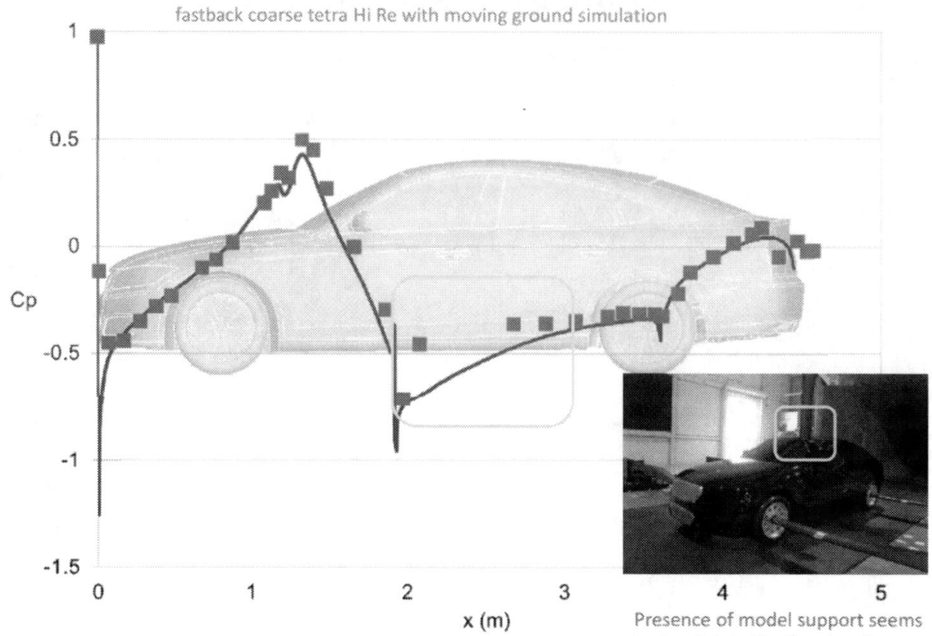

Presence of model support seems to decelerate the flow locally

βBETA CAE Systems SA www.beta-cae.gr

Comparison with experiment: C_D values for fastback with moving ground

Percentage deviation from experimental value of C_D = 0.275

	Coarse	Medium
No layers	Tetra 0.291 (+6%)	Tetra 0.292 (+6%)
Hi Re layers	Tetra 0.270 (-2%)	Tetra 0.262 (-5%)
	Hexa Interior 0.263 (-4%)	Hexa Interior 0.260 (-5%)
	Hexa Poly 0.261 (-5%)	Hexa Poly 0.261 (-5%)
	Polyhedral 0.287 (+4%)	Polyhedral 0.261 (-5%)
Low Re layers	Tetra 0.279 (+1%)	Tetra 0.281 (+2%)

Summary of C_D values for 3 variants for tetra medium Hi Re meshes

βBETA CAE Systems SA www.beta-cae.gr

	experiment	CFD Coarse	CFD Medium
fastback (with moving ground)	0.275	0.270 (-2%)	0.265 (-4%)
fastback (no moving ground)	0.284	0.302 (+6%)	0.288 (+1%)
notchback (no moving ground)	0.286	-	0.286 (0%)
estate (no moving ground)	0.318	-	0.306 (-4%)

Concluding remarks

βBETA CAE Systems SA www.beta-cae.gr

- ANSA pre-processing for ANSYS Fluent was demonstrated with key points like:
  - High quality complete surface and volume meshing, fully automated through batch mesh tool, ensuring mesh consistency and minimizing user manual operations
  - Effective management of multiple variants of a vehicle geometry in one file, taking advantage of already meshed areas for quick model updates
- μETA provides automated post-processing for CFD through the use of recorded and played back session files in batch mode
- The k-omega turbulence model provides highly transient flow patterns even in steady state runs
- The presence of layers is necessary to capture flow details. Low Re layers were also more accurate in this field than Hi Re layers. Low Re predicted higher drag and also lift than Hi Re ones, however due to limited resources in this study we mainly used the Hi Re approach
- Mesh refinement study showed that acceptable mesh independence can be reached at medium size. All medium refinement size meshes gave similar C_D values within 2%. However lift prediction shows very low values and significant scatter that cannot be clarified due to lack of experimental measurements of lift force
- Small deviation of drag calculations from experimental measurements could be attributed to the steady state approach. In addition, the CFD analysis does not include the model supports and other wind tunnel details that may affect the flow field
- The study will be extended in order examine the effect of the supports on the model in fully transient runs with Low Re meshes

βBETA
CAE Systems SA
www.beta-cae.gr
stay connected:
www.beta-cae.gr
You
Tube
GET IT ON
Google play
Available on the
App Store
© Copyright 2013 BETA CAE Systems S.A. All rights reserved
ANSYS 2013 Automotive Simulation World Congress, 29-30 October 2013, Frankfurt am Main, Germany
37

2013 Automotive Simulation World Congress

A Total Li-Ion Battery Simulation Solution (Part 1)*

Scott Stanton, Xiao Hu

ANSYS, Inc.

*Part 2: Structural Integrity Assessment of a Li-Ion Battery Pack Using Simulation

---

A Total Li-Ion Battery Simulation Solution

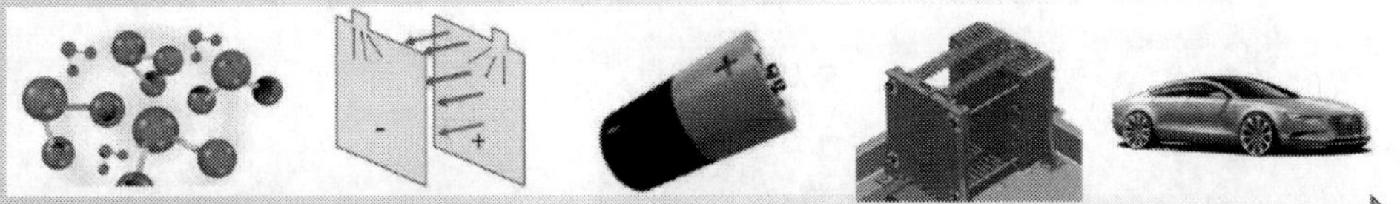

Small Scale → Large Scale

Molecular	Electrode	Cell	Module/Pack	Powertrain
Materials	Layout Process Life SEI	Charging Discharging Heating Safety	Thermal Mgt Durability NVH EMI/EMC	System Integration
	Electro-chemistry	ECM CFD FEA	ECM CFD ROM FEA	ECM ROM

A Total Li-Ion Battery Simulation Solution

2013 Automotive Simulation World Congress
October 29 - 30 • Frankfurt am Main, Germany

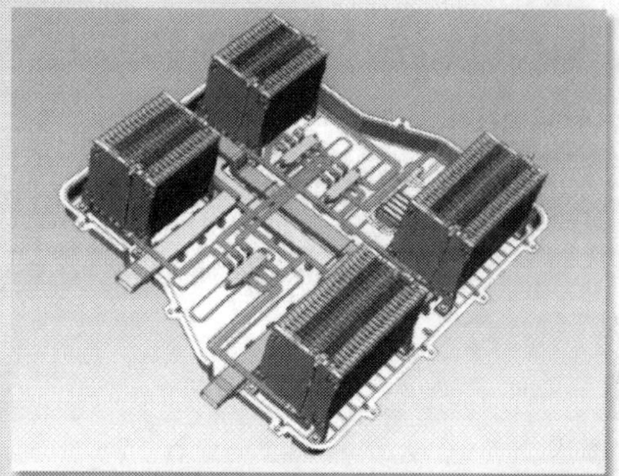
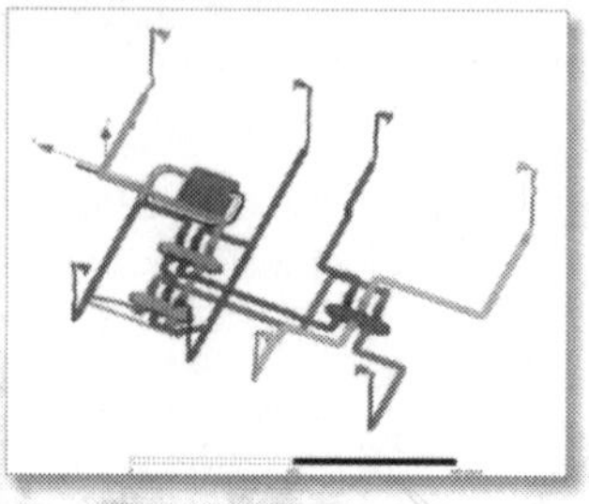
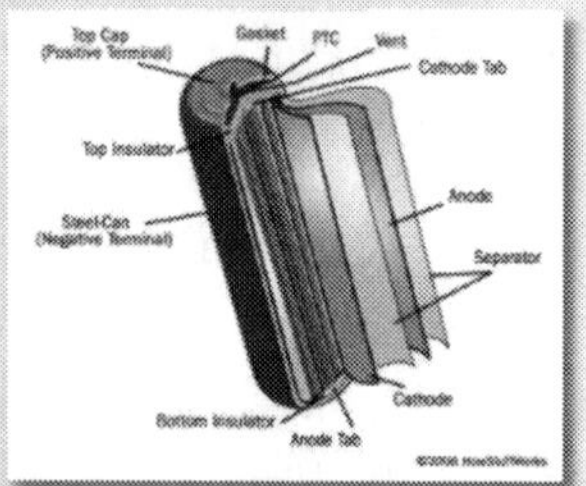

Electrode Level - Electrochemistry

2013 Automotive Simulation World Congress
October 29 - 30 • Frankfurt am Main, Germany

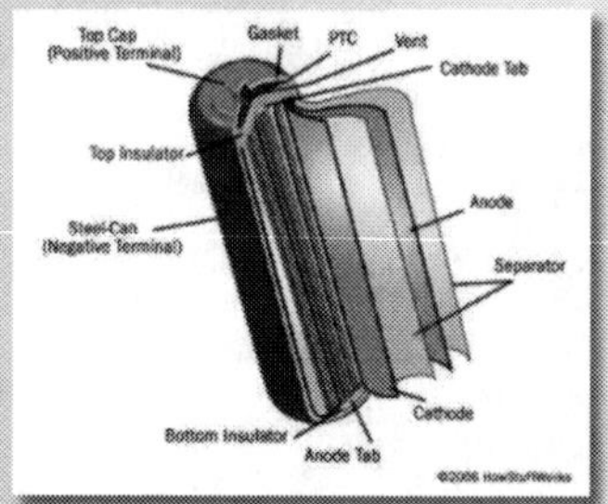

- Electrochemical Kinetics
- Solid-State Li Transport
- Electrolytic Li Transport
- Charge Conservation/Transport
- (Thermal) Energy Conservation

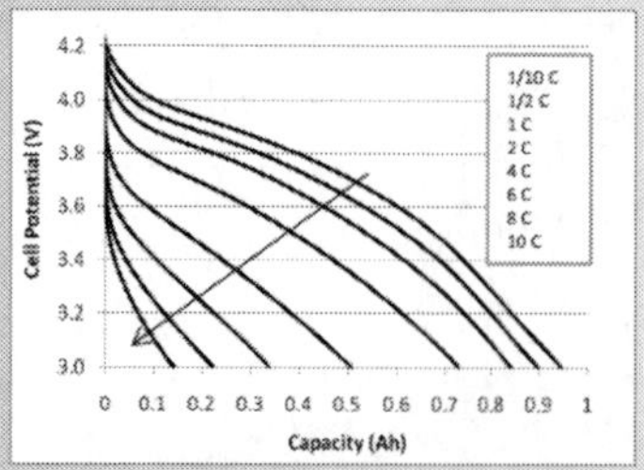

Impact of Discharge Rate on Capacity

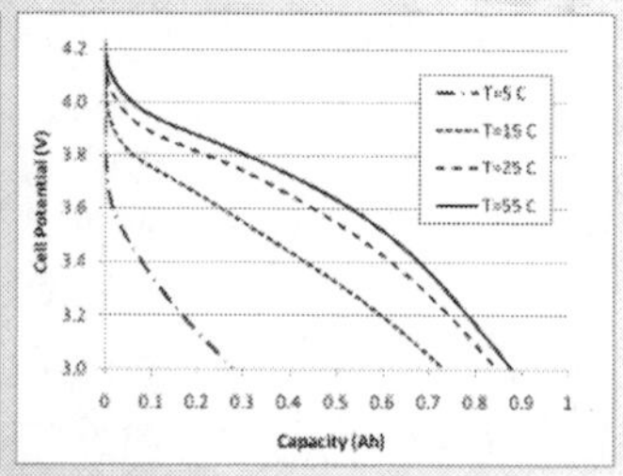

Temperature Impact on Capacity

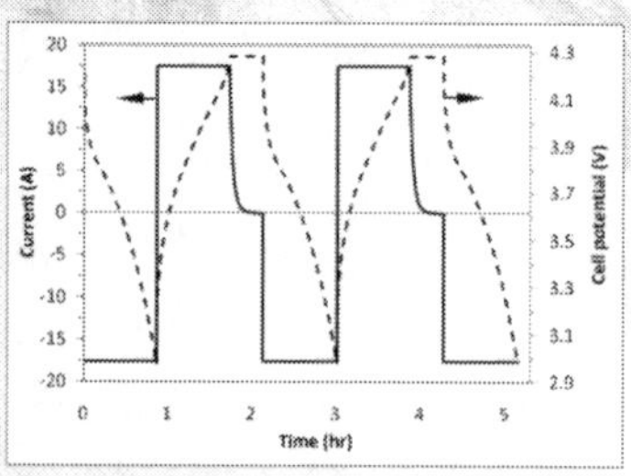

Charge Discharge Cycles

X. Hu, S. Lin, S. Stanton, SAE paper 2012-01-0665

Electrode Level - Electrochemistry

2013 Automotive Simulation World Congress
October 29 - 30 • Frankfurt am Main, Germany

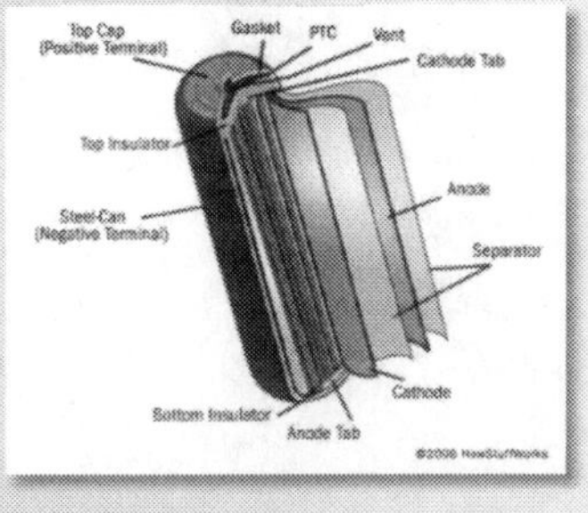

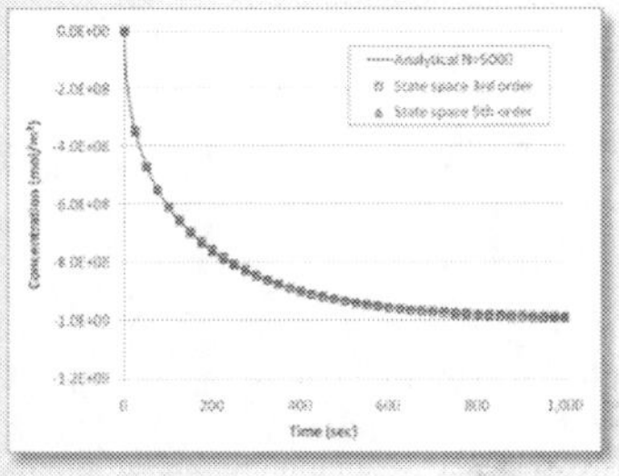

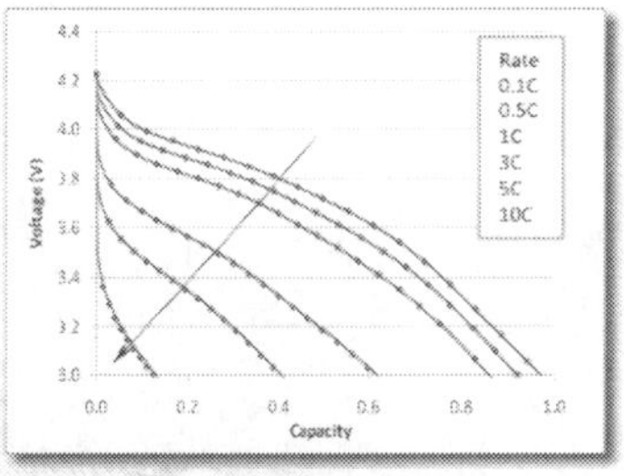

Validation of Reduced Order Electrochemistry

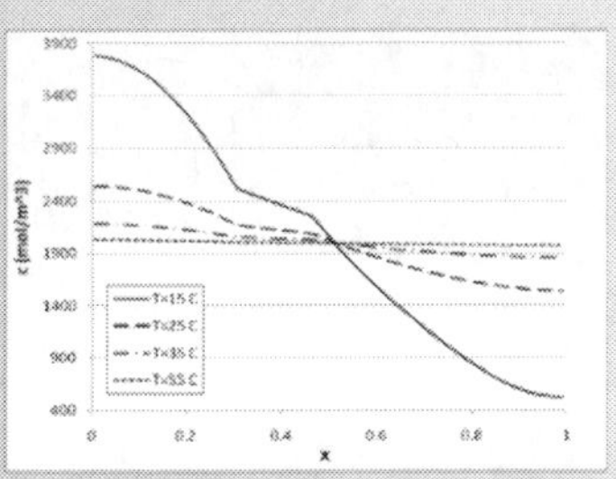

Impact of Temperature on Concentration Distribution

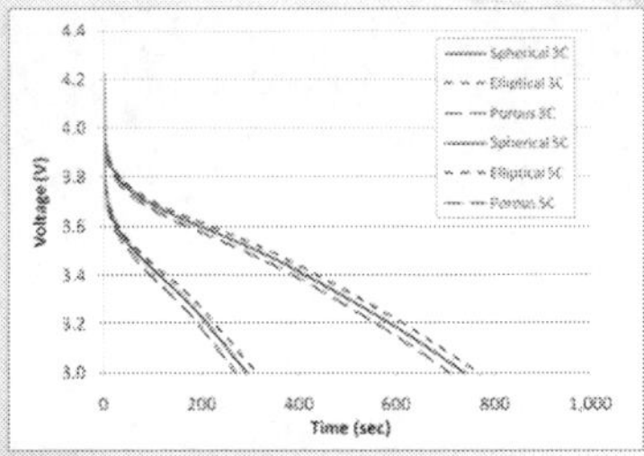

Impact of Particle Shape on Capacity

[1] X. Hu, S. Stanton, L. Cai, R.E. White, J. Power Sources 214, 40-50, 2012.
[2] X. Hu, S. Stanton, L. Cai, R.E. White, J. Power Sources 218, 212-220, 2012.

Newman P2D Electrochemistry Model in Fluent (CAEBAT project)

2013 Automotive Simulation World Congress
October 29 - 30 • Frankfurt am Main, Germany

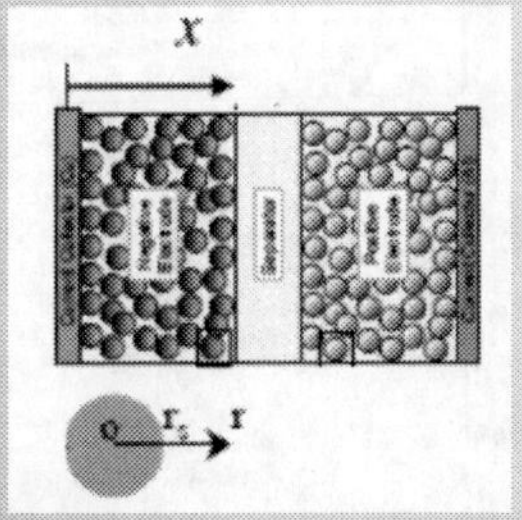

Domains

- negative electrode
- separator
- positive electrode
- spherical particles

$$\nabla\cdot(\sigma\nabla\phi_s)-j^{Li}=0 \qquad \nabla\cdot(k\nabla\phi_e)+\nabla\cdot(k_D\nabla\ln c_e)+j^{Li}=0 \qquad j^{Li}=\xi_a i_0\left\{\exp\left(\frac{\alpha_a F}{RT}\eta\right)-\exp\left(-\frac{\alpha_c F}{RT}\eta\right)\right\}$$

$$\frac{\partial(\varepsilon_e c_e)}{\partial t}=\nabla\cdot(D_e\nabla c_e)+\frac{1-t^+}{F}j^{Li} \qquad \frac{\partial c_s}{\partial t}=\frac{D_s}{r^2}\frac{\partial}{\partial r}\left(r^2\frac{\partial c_s}{\partial r}\right)$$

A system of DAEs are solved for every CFD cells:
- Total number of equations: (n_Cs+2) *n_NE + 2n_SP+(N_Cs+2) n_PE
- Memory requirements: N_cell*{(n_Cs+1) *n_NE+ n_SP +(N_Cs+1) n_PE)}

L. Cai and R.E. White, "Reduction of Model Order Based on Proper Orthogonal Decomposition for Lithium-Ion Battery Simulations" J. of Electrochemical. Soc. 156(3) A154-A161 (2009).

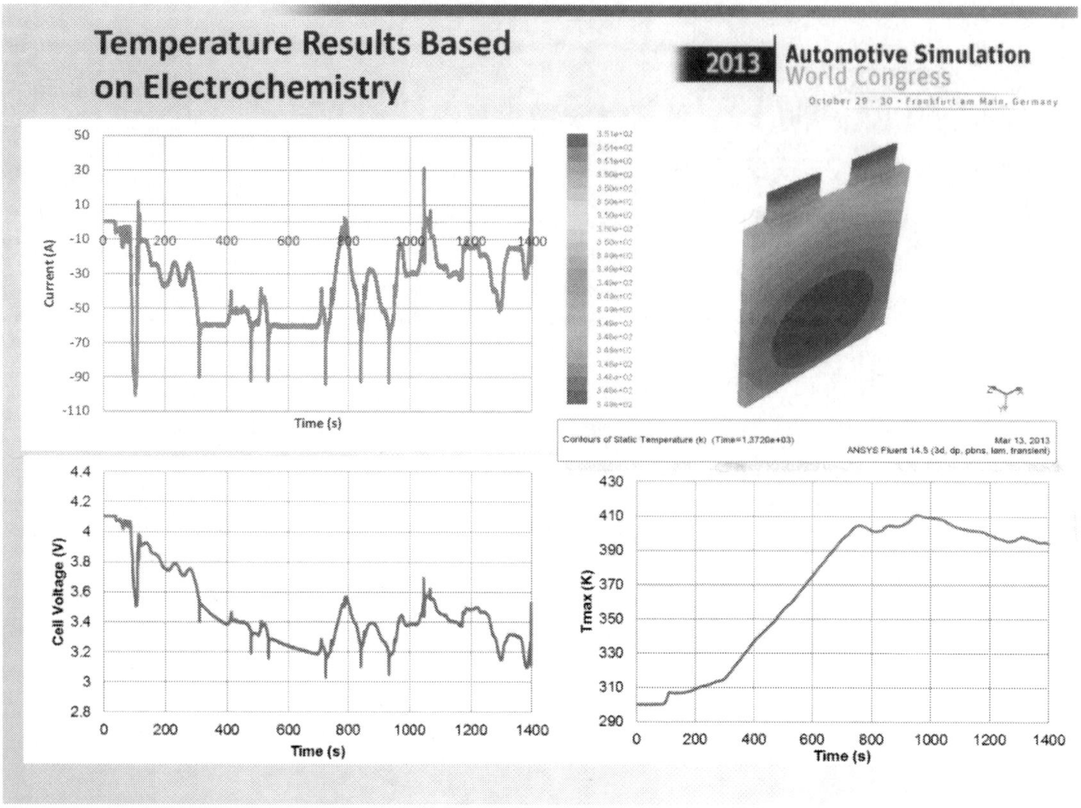
Temperature Results Based on Electrochemistry
2013 Automotive Simulation World Congress
October 29 - 30 • Frankfurt am Main, Germany
Current (A)
Time (s)
Contours of Static Temperature (k) (Time=1.3720e+03)
Mar 13, 2013
ANSYS Fluent 14.5 (3d, dp, pbns, lam, transient)
Cell Voltage (V)
Time (s)
Tmax (K)
Time (s)

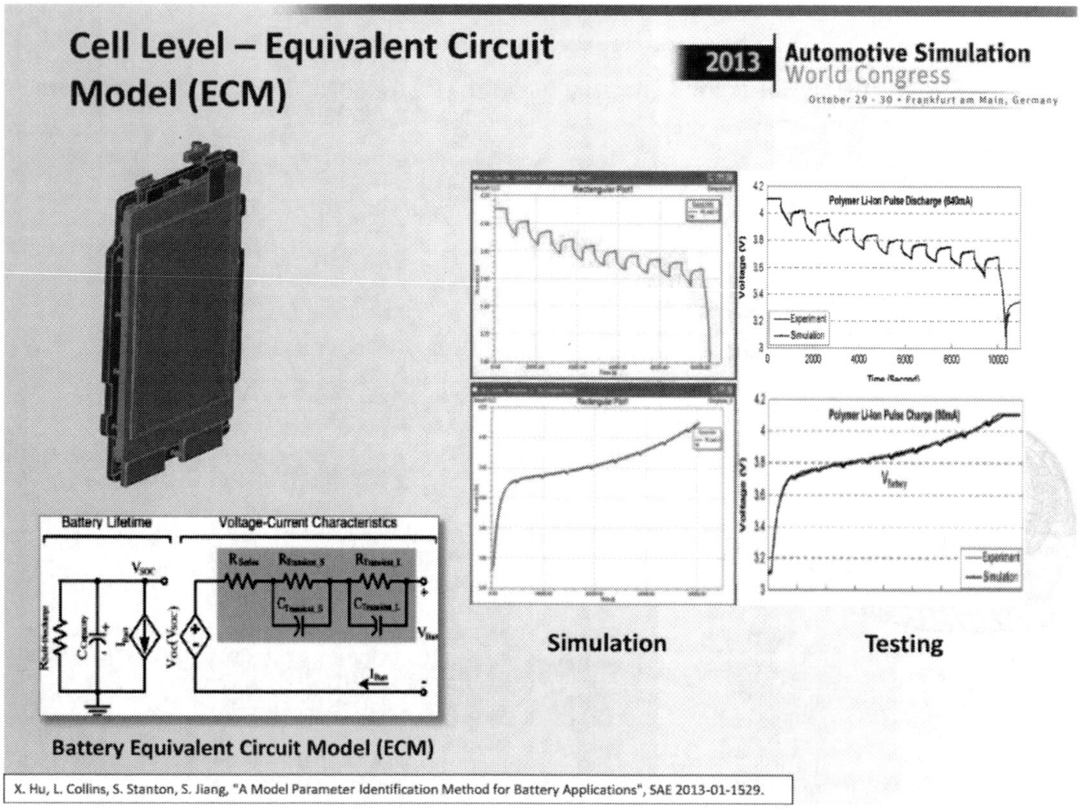
Cell Level – Equivalent Circuit Model (ECM)
2013 Automotive Simulation World Congress
October 29 - 30 • Frankfurt am Main, Germany
Polymer Li-Ion Pulse Discharge (640mA)
Voltage (V)
Experiment
Simulation
Polymer Li-Ion Pulse Charge (80mA)
Battery Lifetime
Voltage-Current Characteristics
Simulation
Testing
Battery Equivalent Circuit Model (ECM)
X. Hu, L. Collins, S. Stanton, S. Jiang, "A Model Parameter Identification Method for Battery Applications", SAE 2013-01-1529.

Module/Pack Level – ECM

- 28 Cells are connected in a module
- 4 modules are connected to the final configuration

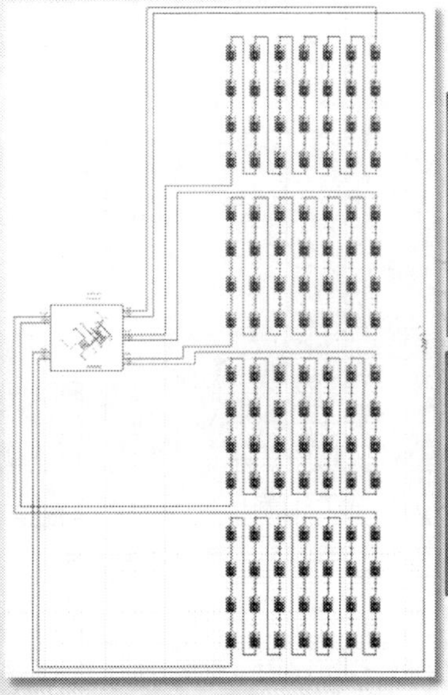

Battery Pack ECM

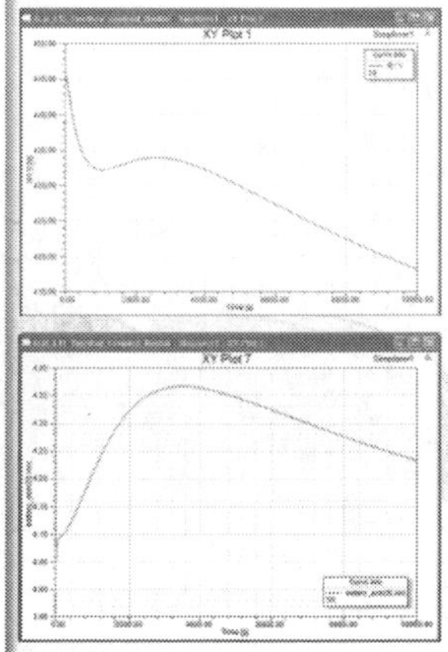

Simulation Results

X. Hu, L. Collins, S. Stanton, S. Jiang, "A Model Parameter Identification Method for Battery Applications", SAE 2013-01-1529.

Cell Level – ECM Extraction Tool

2013 Automotive Simulation World Congress
October 29 - 30 • Frankfurt am Main, Germany

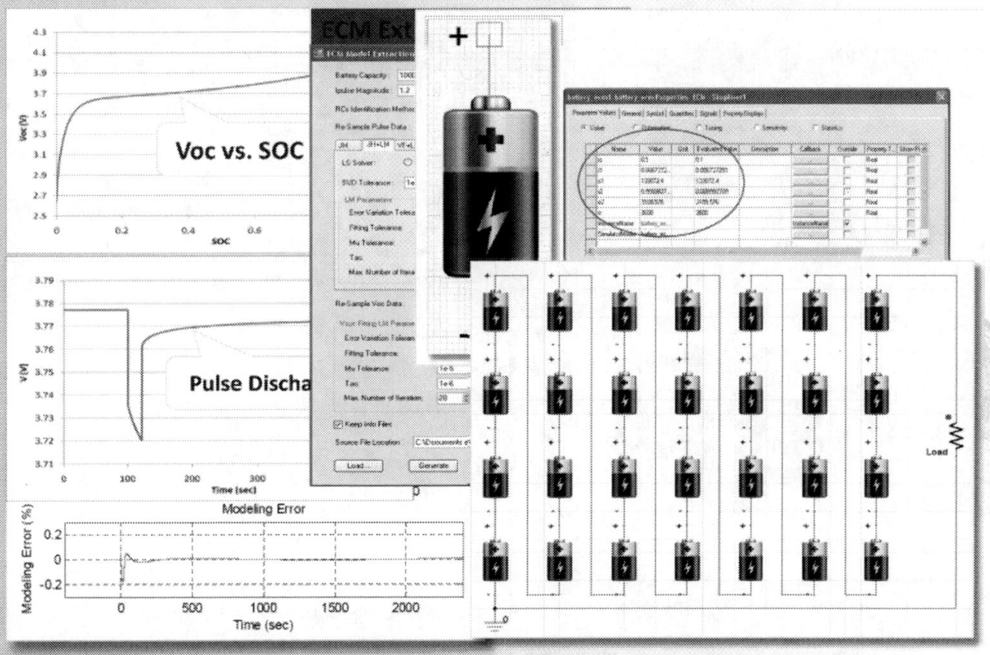

Cell Level – ECM Extraction Tool

2013 Automotive Simulation World Congress
October 29 - 30 • Frankfurt am Main, Germany

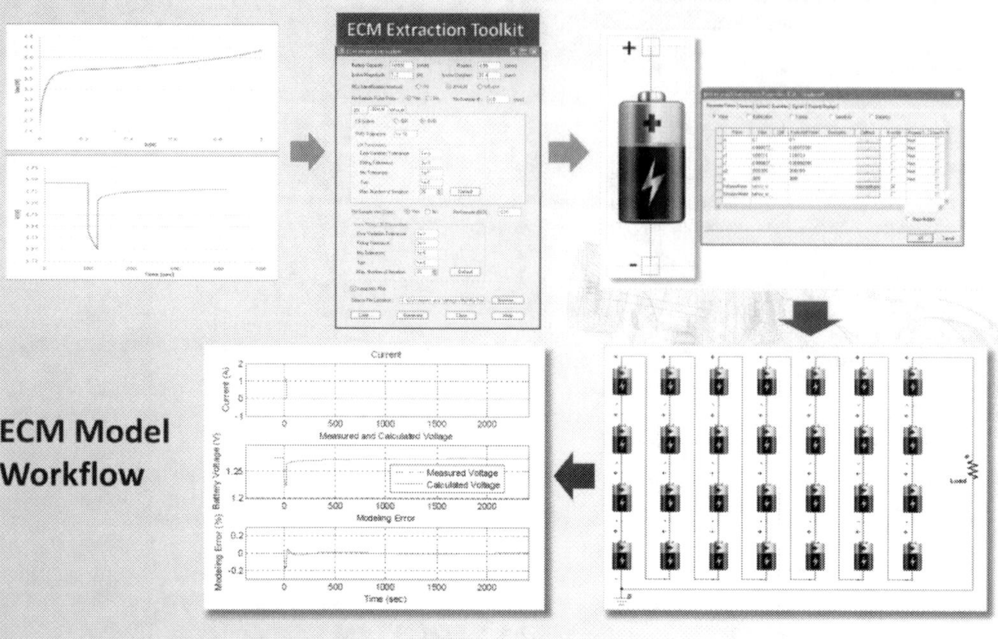

ECM Model Workflow

X. Hu, L. Collins, S. Stanton, S. Jiang, "A Model Parameter Identification Method for Battery Applications", SAE 2013-01-1529.

Module Level – CFD Thermal

2013 Automotive Simulation World Congress
October 29 - 30 • Frankfurt am Main, Germany

Temperature Distribution

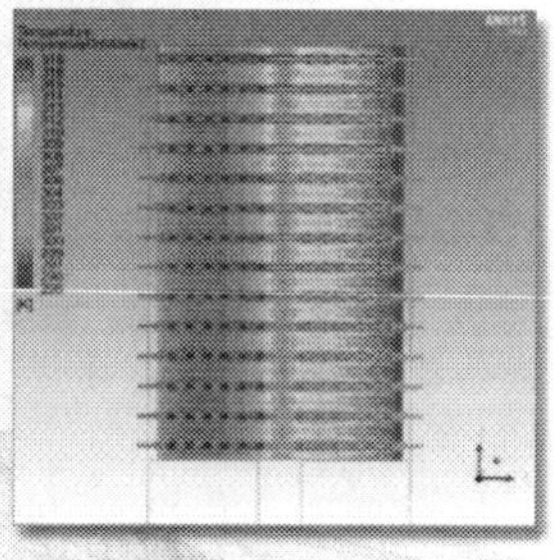

Temperature Distribution

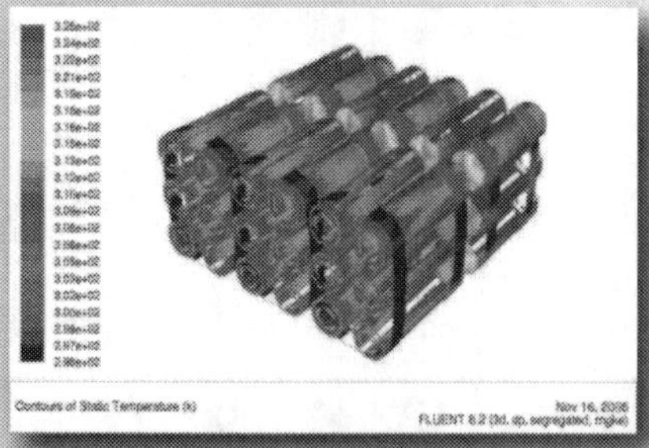

Temperature Distribution

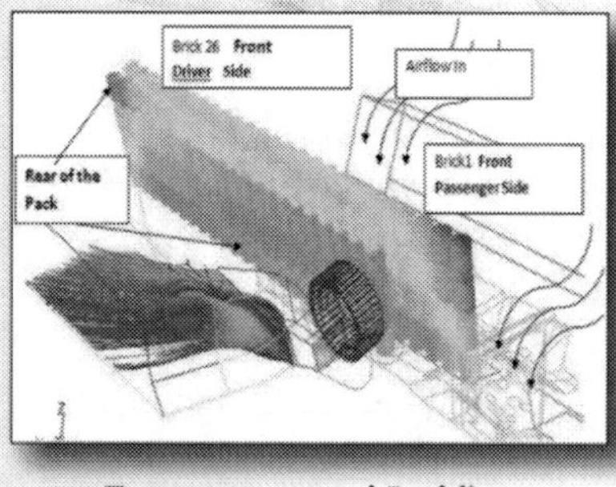

Temperature and Pathlines

Motivation of Reduced Order Model(ROM)

2013 Automotive Simulation World Congress

Germany

CFD as a general thermal analysis tool is accurate.

- Can be computationally expensive for system level analysis

ROM can significantly reduce the model size and simulation time.

ROM is an import tool for system level simulation.

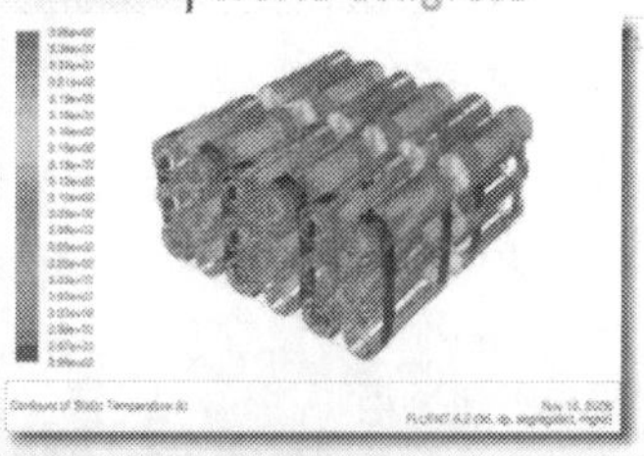

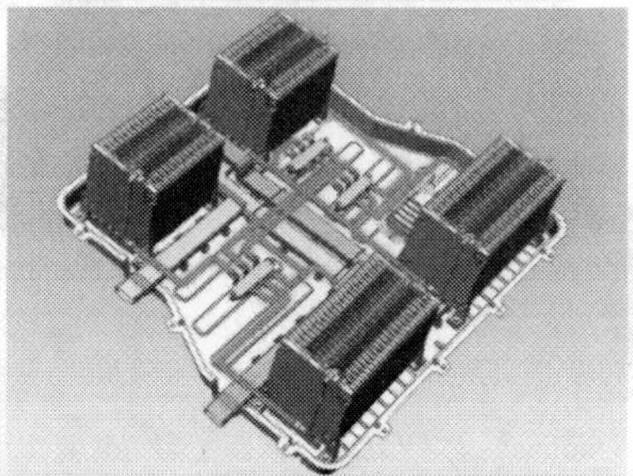

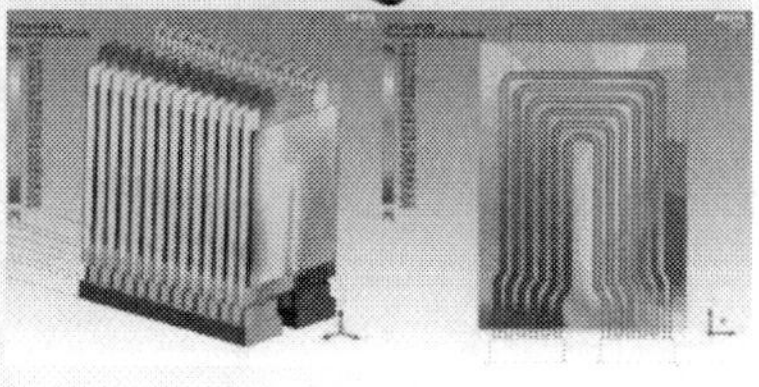

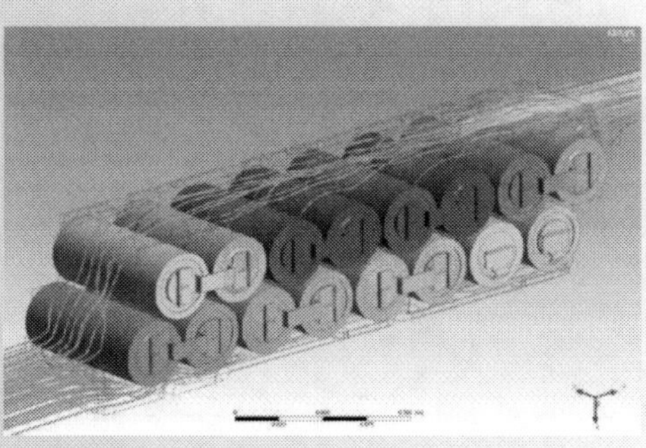

Battery Thermal Behavior via ROM

2013 Automotive Simulation World Congress

October 29 - 30 • Frankfurt am Main, Germany

Type 1: Thermal Network

- Careful calculation and calibration needed
- Accuracy compromises

Type 2: LTI – state space

- Can be as accurate as CFD
- No calibration

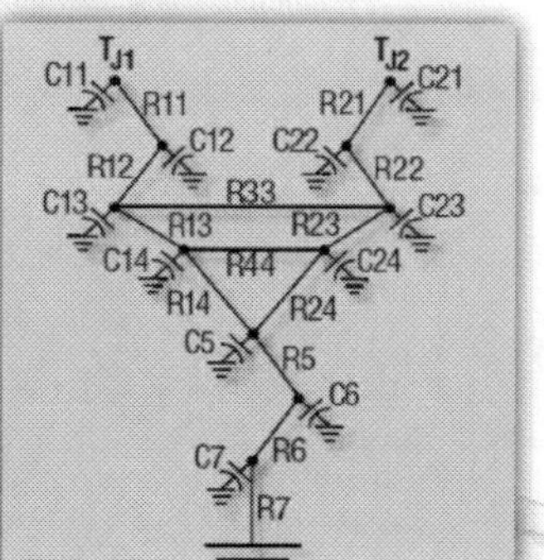

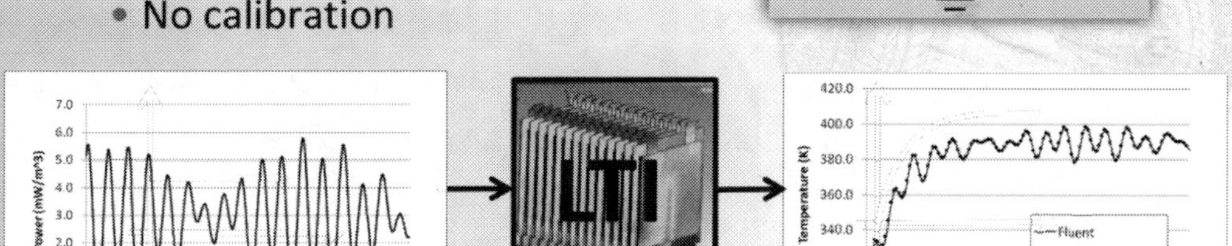

LTI ROM for Battery Cooling

1. Create the CFD model
2. **Generate step responses**
   - Icepak has specialized tools
   - Other CFD codes are fine
3. **Extract LTI ROM**
   - Use Simplorer
4. **Simulate inside Simplorer**

1

2

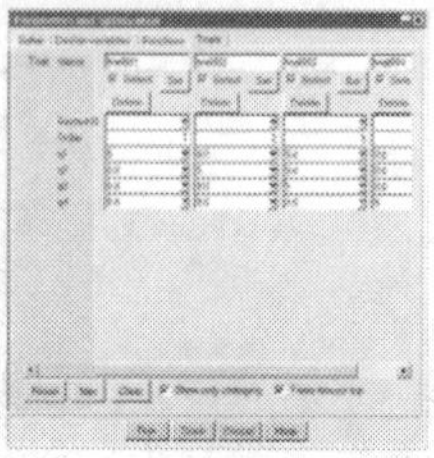

4

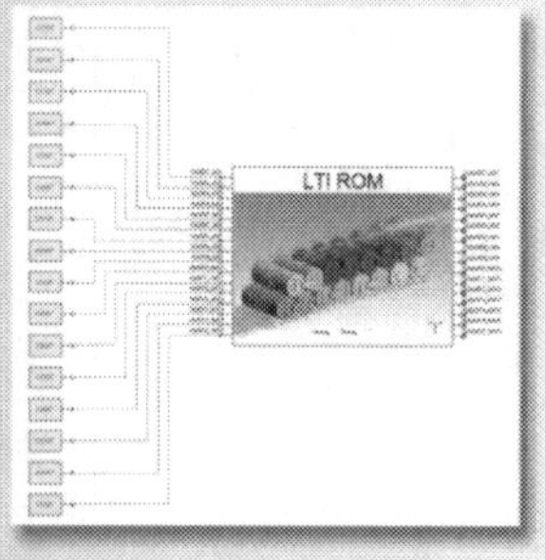

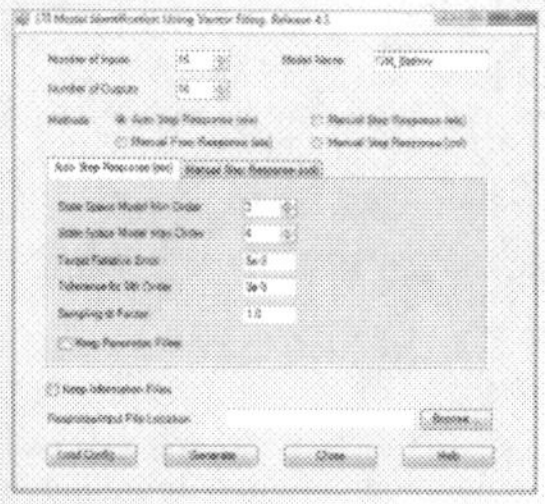

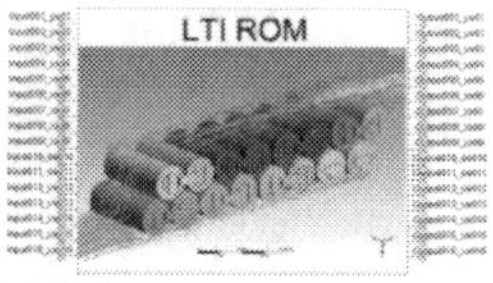

3

2013 Automotive Simulation World Congress
October 29 - 30 • Frankfurt am Main, Germany

LTI ROM for GM Battery Module

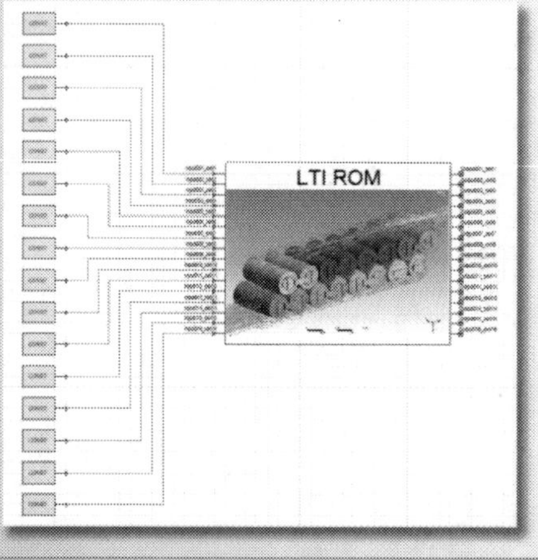

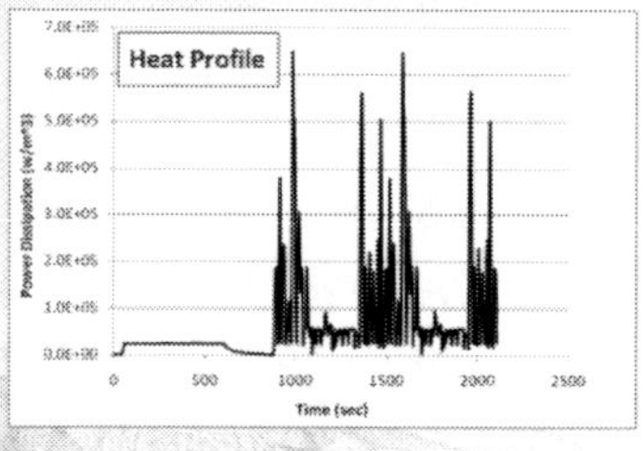

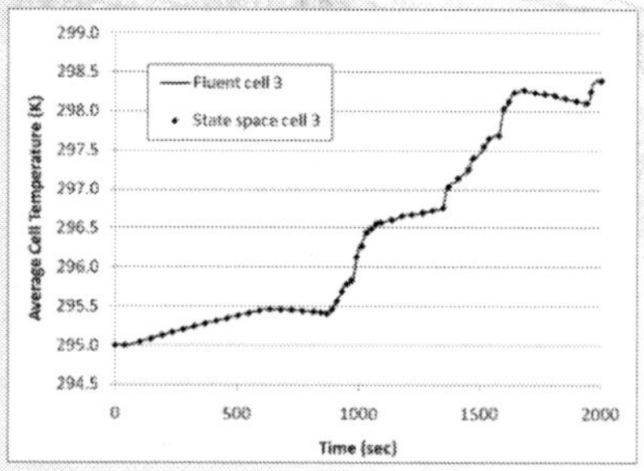

LTI ROM gives the same results as CFD. LTI ROM runs in less than 2 seconds while the CFD runs 2 hours on one single CPU.

X. Hu, S. Lin, S. Stanton, W. Lian, "A Foster Network Thermal Model for HEV/EV Battery Modeling," IEEE TRANSACTIONS ON INDUSTRY APPLICATIONS, VOL. 47, NO. 4, JULY/AUGUST 2011

X. Hu, S. Lin, S. Stanton, W. Lian, "A State Space Thermal Model for HEV/EV Battery Modeling", SAE 2011-01-1364

Module Level – ROM for Thermal

2013 Automotive Simulation World Congress
October 29 - 30 • Frankfurt am Main, Germany

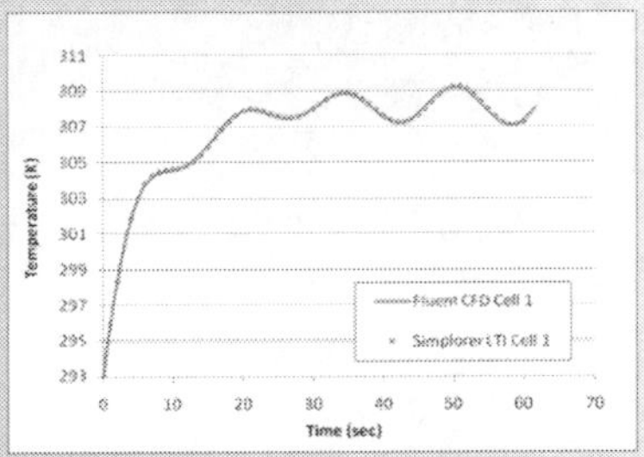

ROM vs CFD

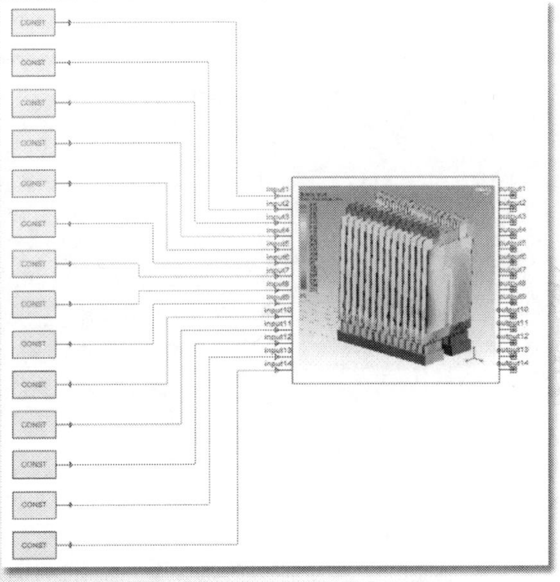

ROM for the Battery Module

SVD+LTI ROM : GM 16 Cell Test Case

2013 Automotive Simulation World Congress
October 29 - 30 • Frankfurt am Main, Germany

SVD+LTI technology allows for quick temperature distribution calculation in addition to average temperature calculation.

Using a heat source from GM, SVD+LTI ROM is applied to the GM 16 cell case.

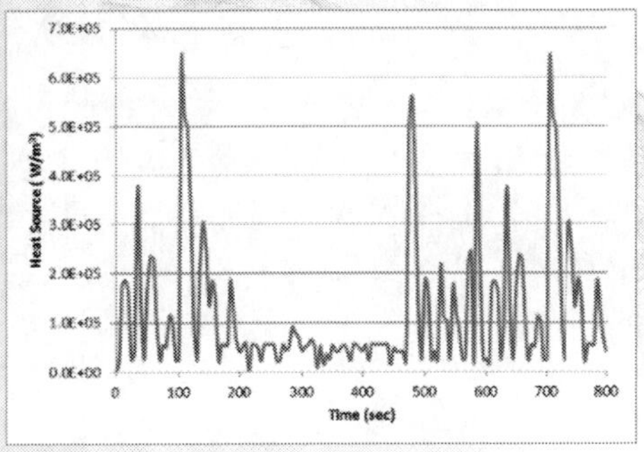

Heat source used

SVD+LTI ROM for Battery Cooling

2013 **Automotive Simulation** World Congress
October 29 - 30 • Frankfurt am Main, Germany

1. Create the Fluent CFD model
2. Generate step responses
3. Extract SVD+LTI ROM
   - Use Simplorer
4. Simulate inside Simplorer

1

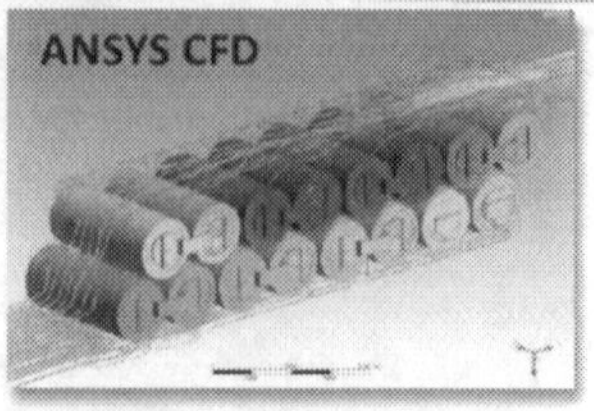

2

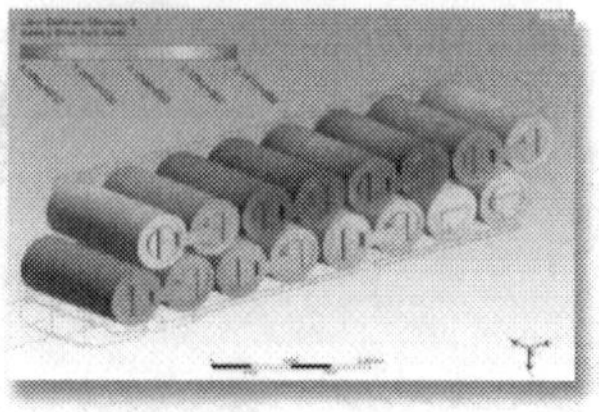

3

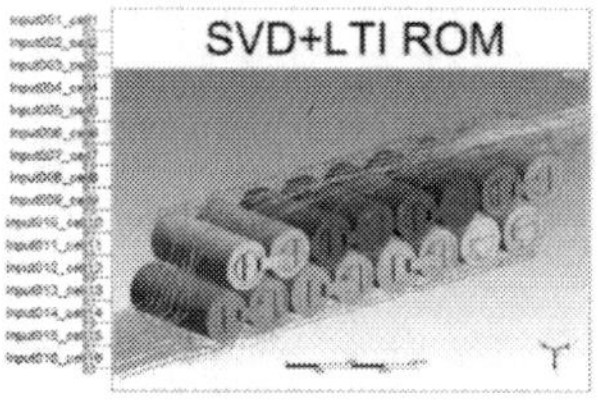

4

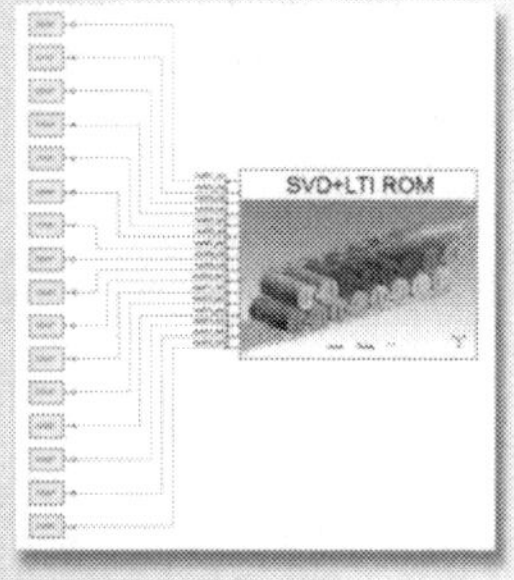

SVD+LTI ROM Validation: GM 16 Cell Test Case

2013 **Automotive Simulation** World Congress
October 29 - 30 • Frankfurt am Main, Germany

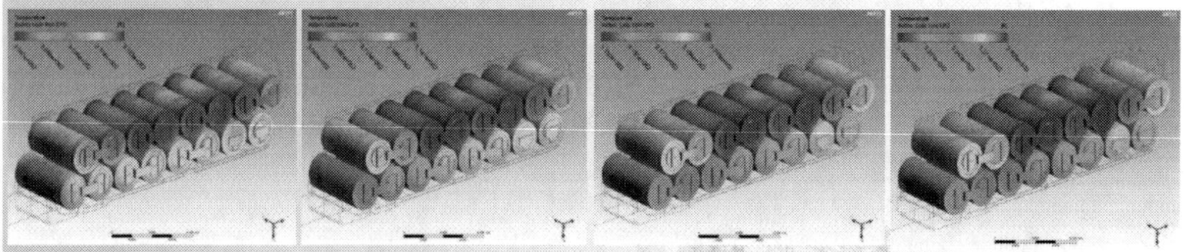

CFD (200 sec) **CFD (400 sec)** **CFD (600 sec)** **CFD (800 sec)**

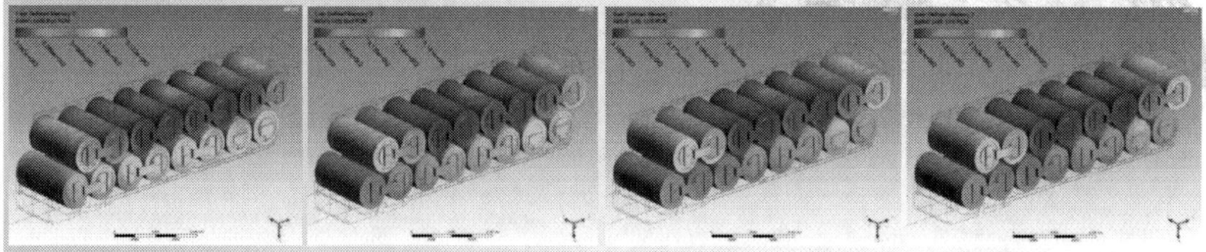

SVD+LTI ROM (200 sec) **SVD+LTI ROM (400 sec)** **SVD+LTI ROM (600 sec)** **SVD+LTI ROM (800 sec)**

SVD+LTI ROM Statistics for GM 16 Cell Test Case

2013 Automotive Simulation World Congress
October 29 - 30 • Frankfurt am Main, Germany

CFD mesh size	**3,025,067**
Number of steps in step response run	180
Size of matrix ***A***	3,025,067×180
Number of snap shots used for SVD calculation	180
SVD calculation time	4 min 50 sec
SVD calculation memory usage	8.1 G
SVD+LTI ROM extraction time	5 min
CFD validation simulation time on single CPU	**5 hr**
SVD+LTI ROM simulation time on the same single CPU	**a few seconds**

SVD+LTI ROM Validation

2013 Automotive Simulation World Congress
October 29 - 30 • Frankfurt am Main, Germany

CFD (10 sec)

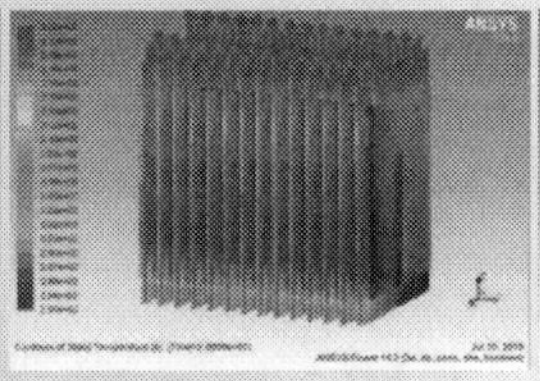

CFD (20 sec)

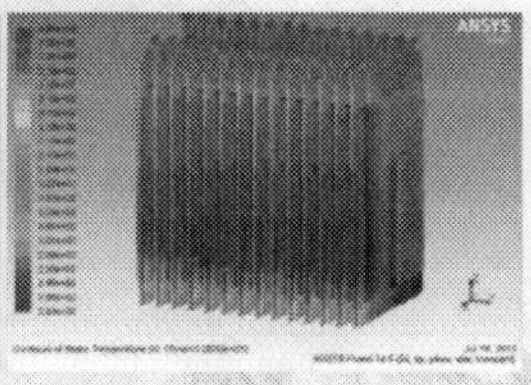

CFD (30 sec)

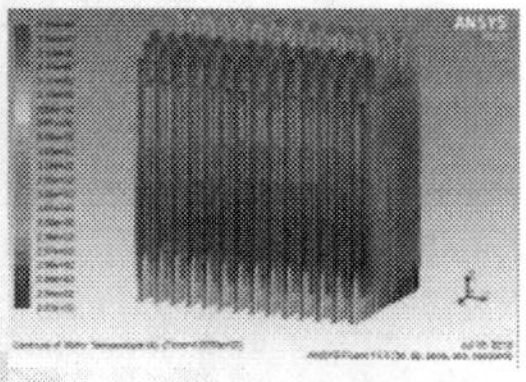

CFD (40 sec)

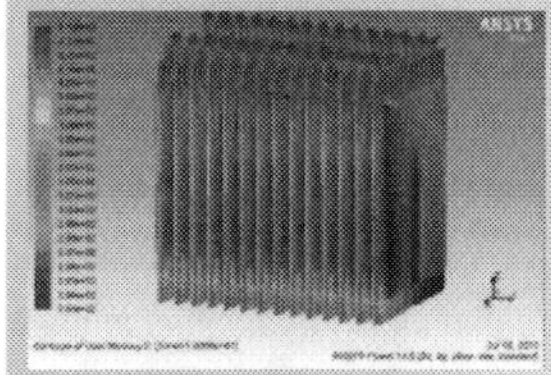

SVD+LTI ROM (10 sec)

SVD+LTI ROM (20 sec)

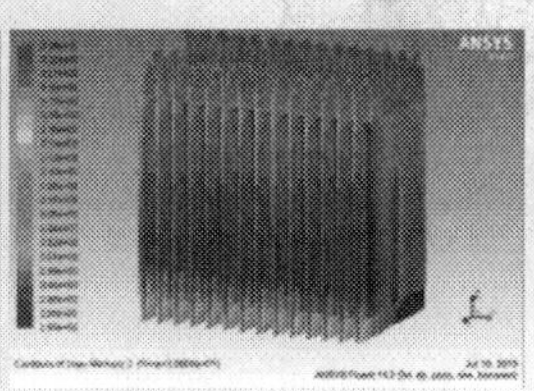

SVD+LTI ROM (30 sec)

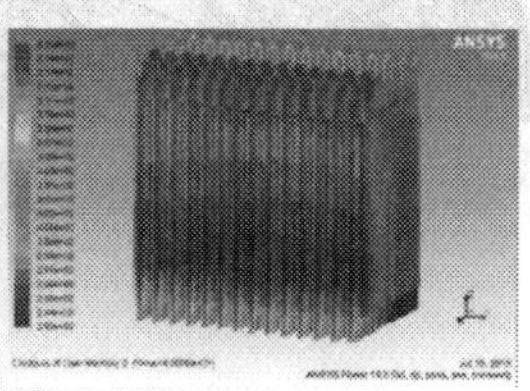

SVD+LTI ROM (40 sec)

SVD+LTI ROM Statistics for the Test Case

CFD mesh size	**12,209,486**
Number of steps in step response run	120
Size of matrix ***A***	12,209,486×120
Number of snap shots used for SVD calculation	60
SVD calculation time	3 min
SVD calculation memory usage	18 G
SVD+LTI ROM extraction time	5 min
CFD validation simulation time on six CPUs	**20 hr**
SVD+LTI ROM simulation time single CPU	**a few seconds**

GM Battery Module – ECM Coupled with ROMs

2013 Automotive Simulation World Congress
October 29 - 30 • Frankfurt am Main, Germany

ECM calculates heat source and sends it to the two ROMs.
LTI ROM calculates average temperature and sends it to ECM.
SVD+LTI ROM *calculates* temperature distribution.

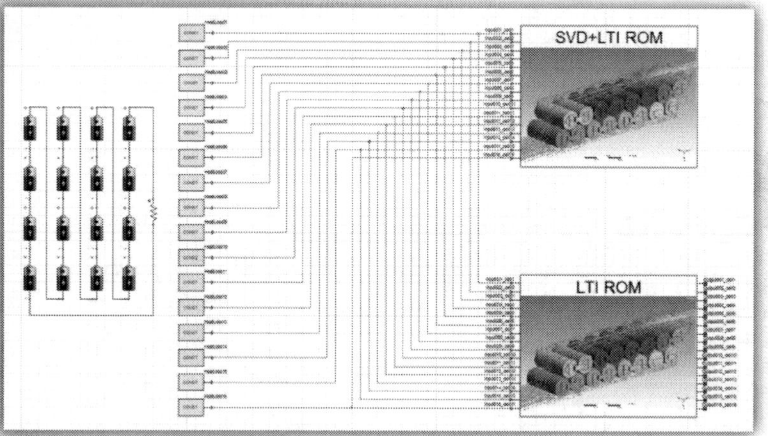

ROM Results

2013 Automotive Simulation World Congress
October 29 - 30 • Frankfurt am Main, Germany

- LTI ROM calculates average temperature.
- SVD+LTI ROM *calculates* temperature field.

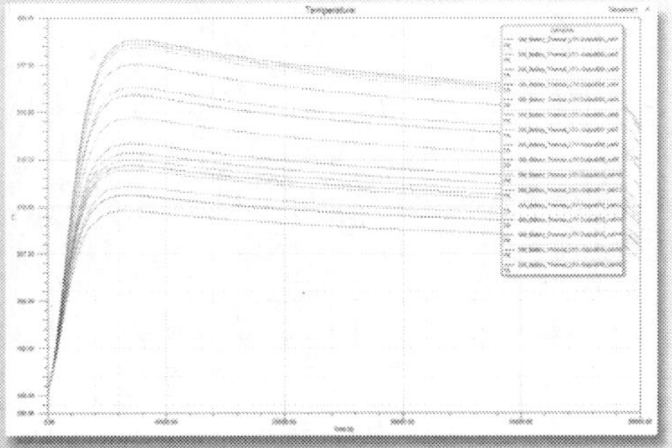

Average Temperature

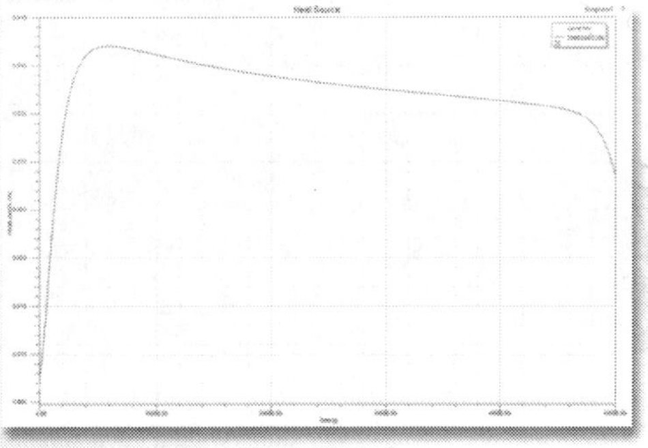

Heat Source for One Cell

Temperature Distribution - Animation

2013 Automotive Simulation World Congress
October 29 - 30 • Frankfurt am Main, Germany

Temperature calculated from CFD.
7 hr simulation time with 6 CPUs.

Temperature calculated from SVD+LTI ROM.
0.5 hr simulation time with 1 CPU.

Module/Pack Level – Bus Bar Model

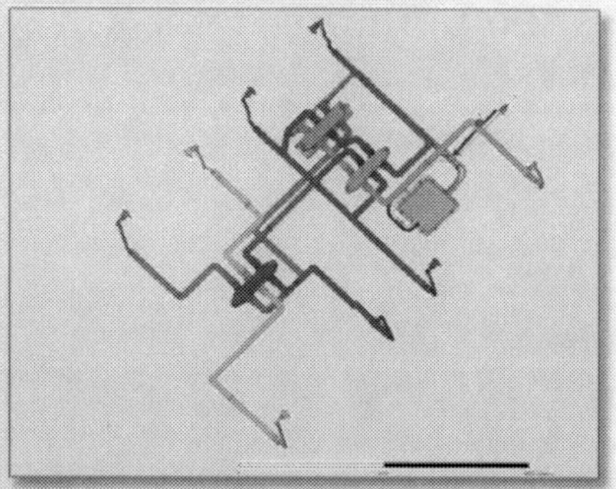

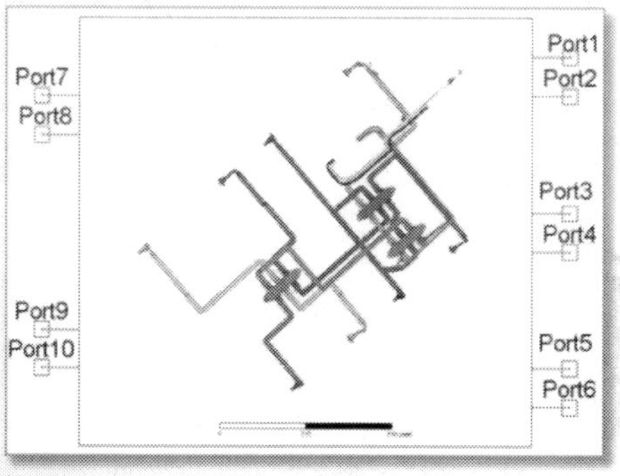

Electromagnetic FEA Analysis for Busbar RLC Network Extraction

2013 Automotive Simulation World Congress
October 29 - 30 • Frankfurt am Main, Germany

Module/Pack Level – Bus Bar Model

Full Battery Simulation

2013 **Automotive Simulation** World Congress
October 29 - 30 • Frankfurt am Main, Germany

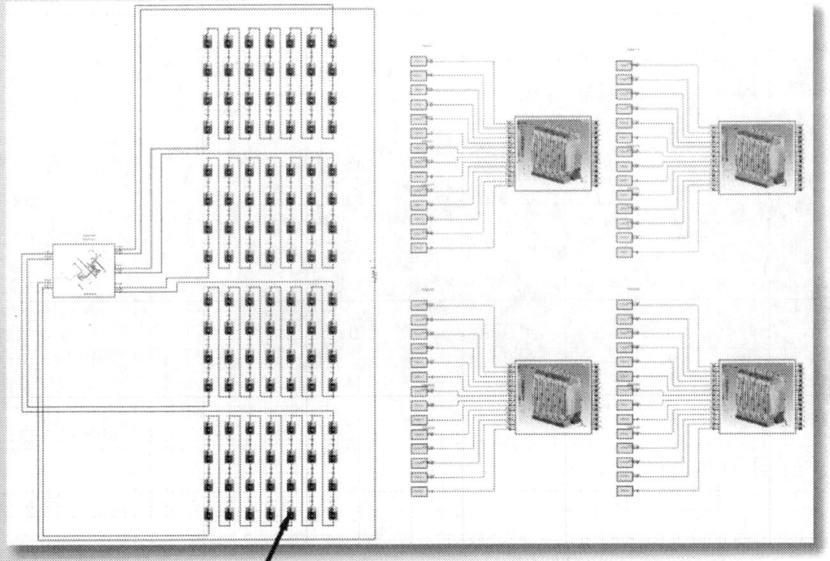

Voc = -1.031*exp(-35*(abs(IBatt.V/Vinit))) + 3.685 + 0.2156*(abs(IBatt.V/Vinit)) - 0.1178*(abs(IBatt.V/Vinit))^2 + 0.3201*(abs(IBatt.V/Vinit))^3 + 0.3/30.0*(U1.Temp_block_1-273)

Full Battery Simulation

2013 **Automotive Simulation** World Congress
October 29 - 30 • Frankfurt am Main, Germany

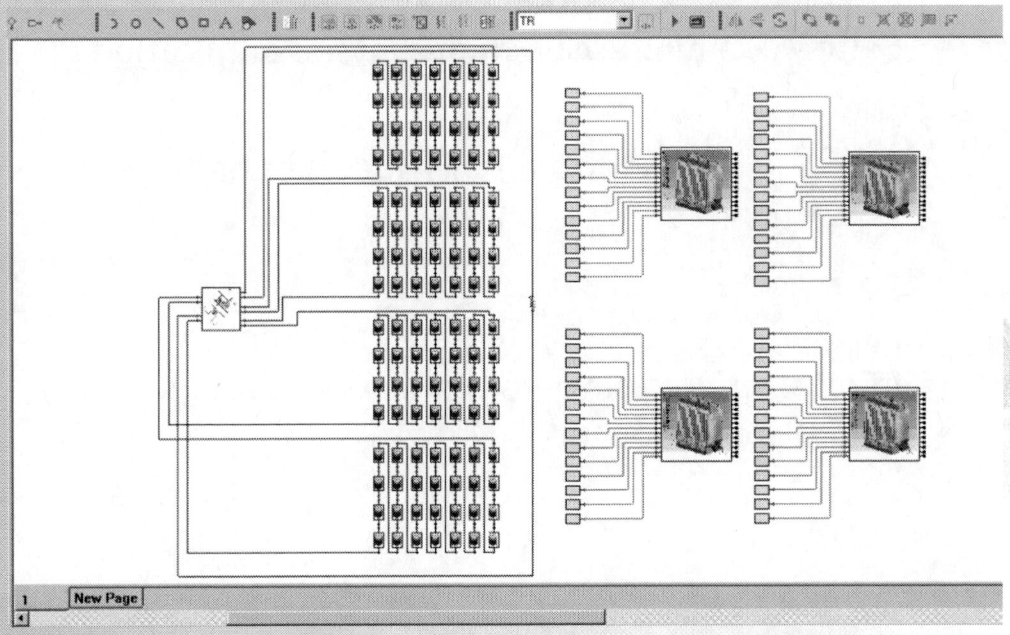

Part 2: Mechanical Analysis

15:00 – Structural Integrity Assessment of a Li-Ion Battery Pack using Simulations. By S. Kottargi, ANSYS India

Conclusions

2013 Automotive Simulation World Congress
October 29 - 30 • Frankfurt am Main, Germany

- Battery is a multi-scale multi-physics application.
- ANSYS provides tools for all aspects of battery simulation.
- Furthermore, ANSYS integrates different models into battery system level simulation through model order reduction.

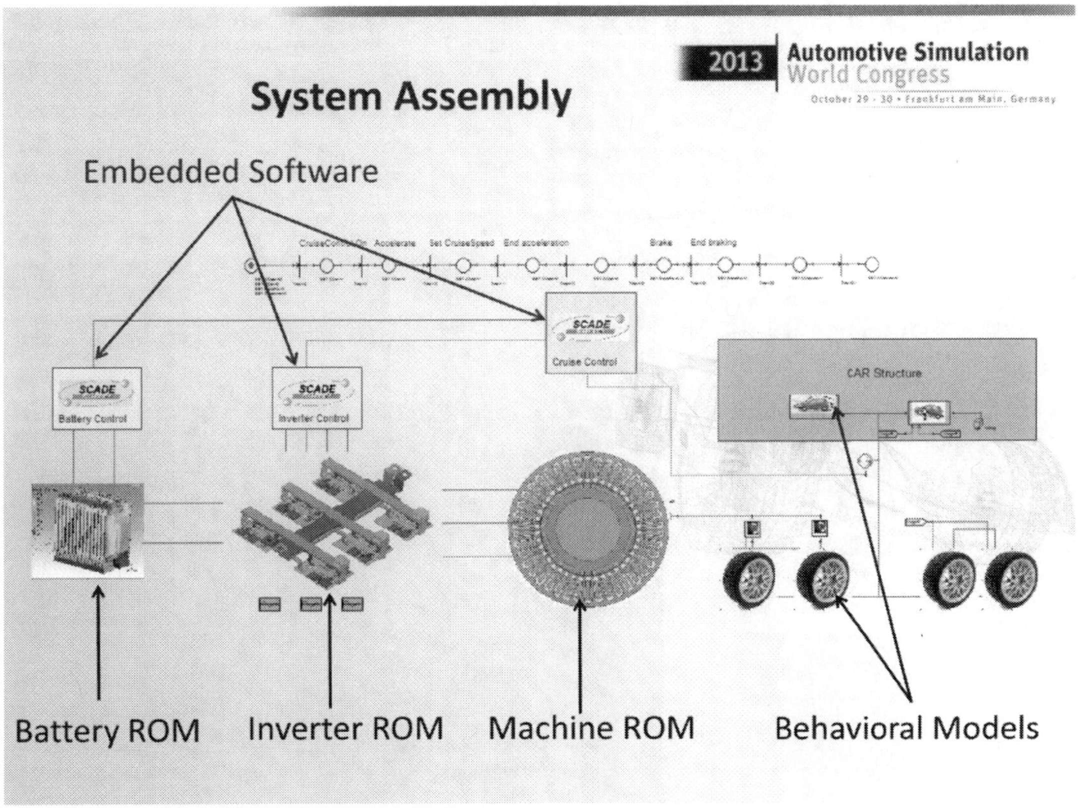
2013 Automotive Simulation World Congress
October 29 - 30 • Frankfurt am Main, Germany
System Assembly
Embedded Software
SCADE
Cruise Control
SCADE
Battery Control
SCADE
Inverter Control
CAR Structure
Battery ROM
Inverter ROM
Machine ROM
Behavioral Models

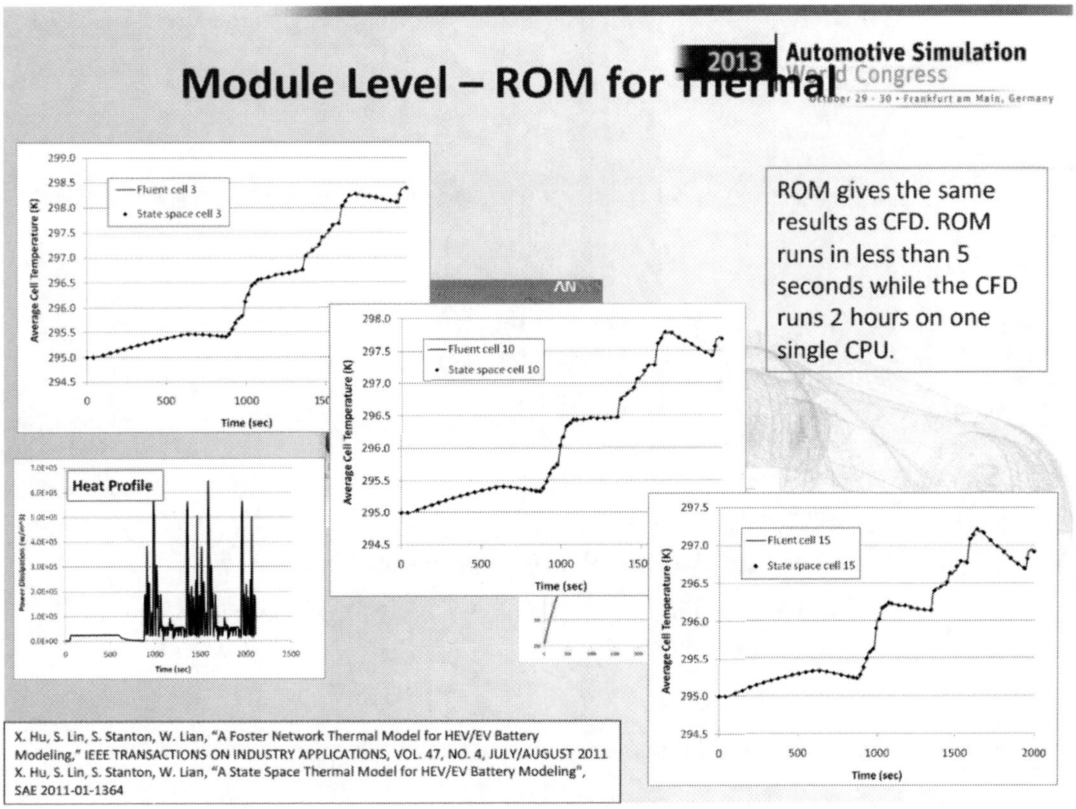
2013 Automotive Simulation World Congress
October 29 - 30 • Frankfurt am Main, Germany
Module Level – ROM for Thermal
Fluent cell 3
State space cell 3
Average Cell Temperature (K)
Time (sec)
Fluent cell 10
State space cell 10
Fluent cell 15
State space cell 15
Heat Profile
ROM gives the same results as CFD. ROM runs in less than 5 seconds while the CFD runs 2 hours on one single CPU.
X. Hu, S. Lin, S. Stanton, W. Lian, "A Foster Network Thermal Model for HEV/EV Battery Modeling," IEEE TRANSACTIONS ON INDUSTRY APPLICATIONS, VOL. 47, NO. 4, JULY/AUGUST 2011
X. Hu, S. Lin, S. Stanton, W. Lian, "A State Space Thermal Model for HEV/EV Battery Modeling", SAE 2011-01-1364

Electrochemistry for Thermal Runaway

2013 Automotive Simulation World Congress
October 29 - 30 • Frankfurt am Main, Germany

- **Energy Equation + Heat Source**
  - **Energy Equation (Default)**

$$\frac{\partial(\rho c_p T)}{\partial t} = \nabla \cdot \lambda \nabla T + q$$

  - **Heat Source (reaction heat + Joule heating)**

$$q = a_{sj}\bar{i}_{nj}\left(\phi_s - \phi_e - U_j + T\frac{\partial U_j}{\partial T}\right)$$

$$+\sigma^{eff}\nabla\phi_s \cdot \nabla\phi_s + \left(\kappa^{eff}\nabla\phi_e \cdot \nabla\phi_e + \kappa_D^{eff}\nabla \ln c_e \cdot \nabla\phi_e\right)$$

W. B. Gu, C. Y. Wang, Journal of Electrochem. Soc., 147 (8), 2910-2922 (2000)

Electrochemistry for Thermal Runaway

2013 Automotive Simulation World Congress
October 29 - 30 • Frankfurt am Main, Germany

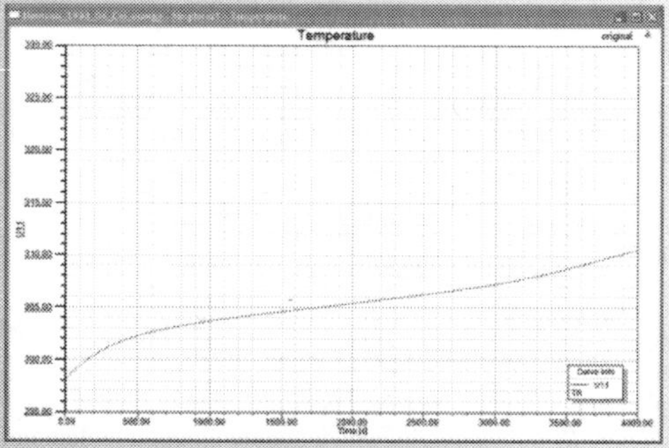

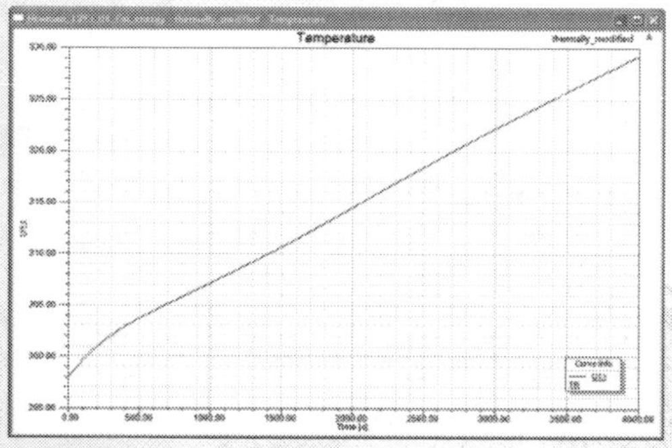

- **Shows impact of material properties on thermal behavior.**
- **Can also investigate impact of design on thermal behavior, say thickness of the separator.**

W. B. Gu, C. Y. Wang, Journal of Electrochem. Soc., 147 (8), 2910-2922 (2000)

Numerical Method

- Two user defined scalars are used to solve for ϕ_+ and ϕ_- on CFD mesh
- ϕ_+ and ϕ_- are loosely coupled through source term, special treatments are used to enhance the convergence for different current, voltage or power boundary condition
- Sundials IDA solver is used to solve E-chemistry.
  - Newman model: 300~2000 equations.
- The key is try to reduce the CPU cost!

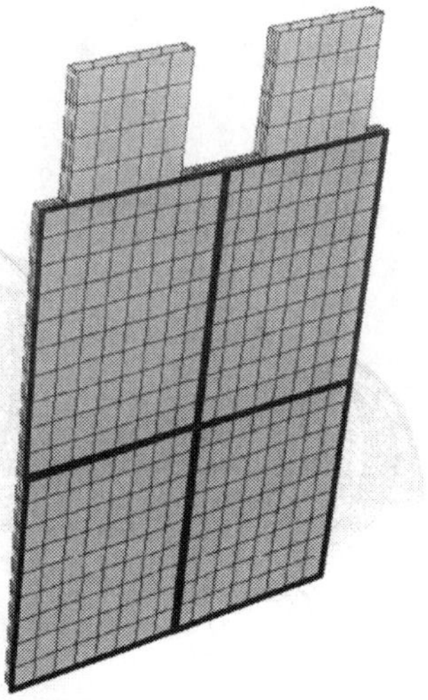

2013 Automotive Simulation World Congress
October 29 - 30 • Frankfurt am Main, Germany

Cell Level – CFD Thermal

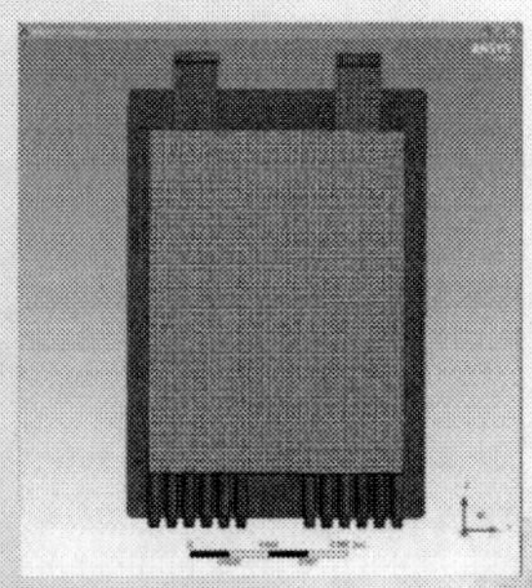

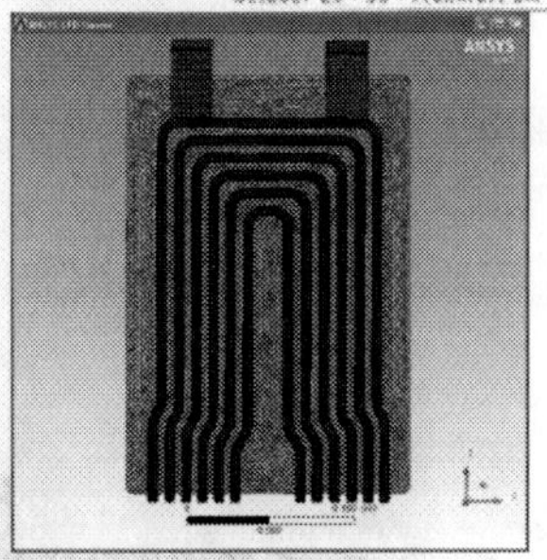

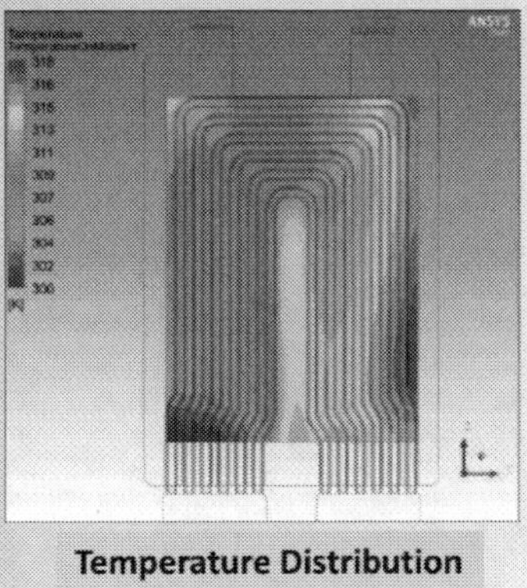

Temperature Distribution

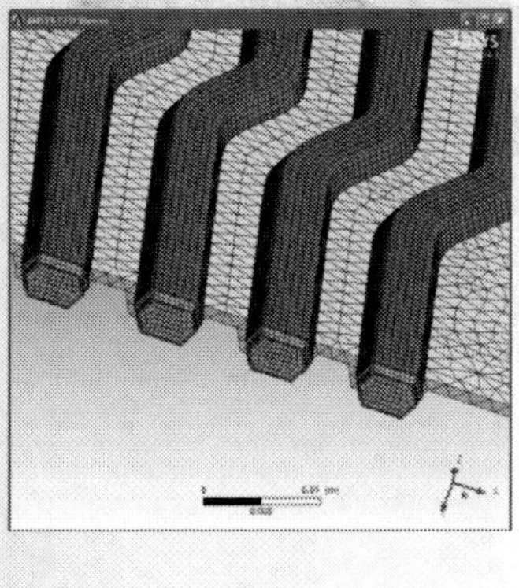

Cell Level –Electro-Thermal

2013 **Automotive Simulation** World Congress
October 29 - 30 • Frankfurt am Main, Germany

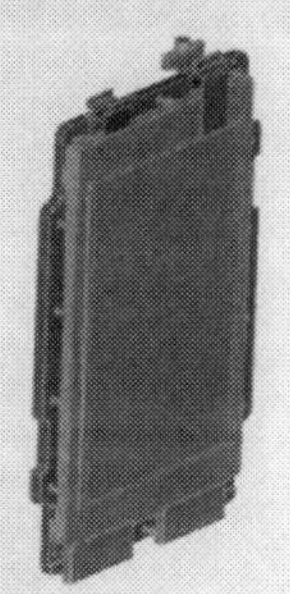

$$\nabla\cdot(\sigma\nabla\phi)=J$$

$$J=Y(\phi_c-\phi_a-U)$$

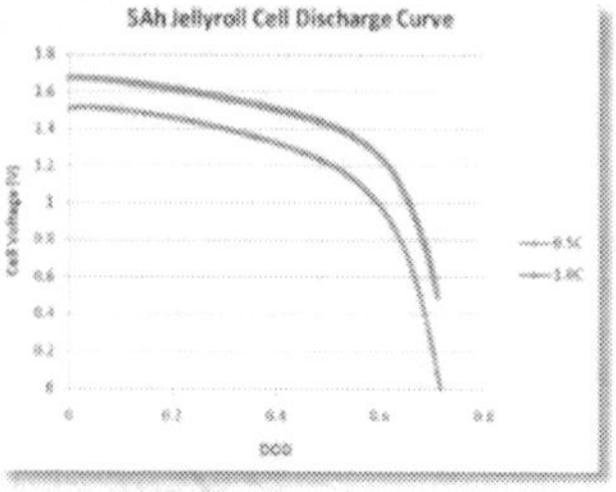

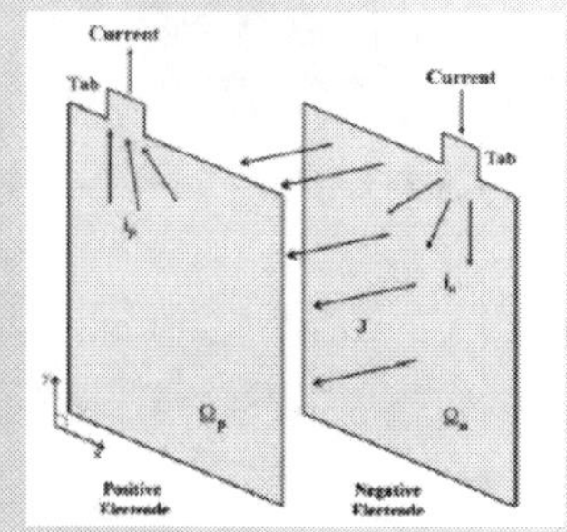

Cathode Anode

Current Density

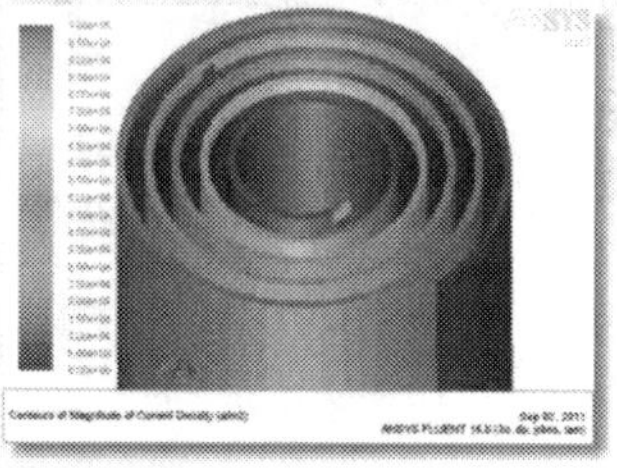

Current Density

Temperature Distribution - Animation

2013 **Automotive Simulation** World Congress
October 29 - 30 • Frankfurt am Main, Germany

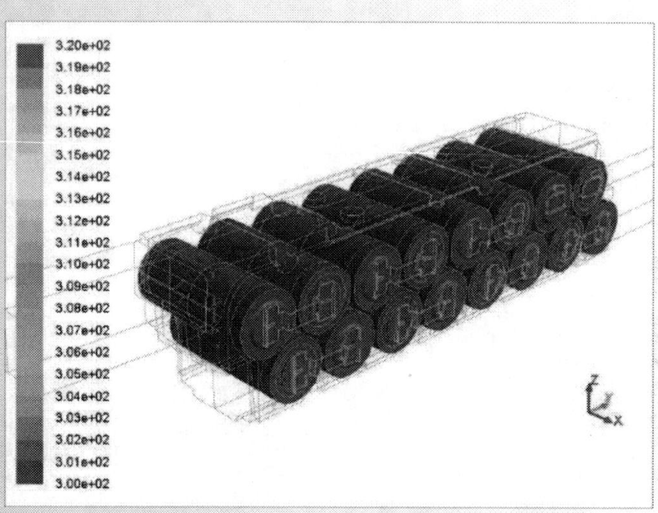

Temperature calculated from SVD+LTI ROM.
0.5 hr simulation time with 1 CPU.

PhilonNet
Engineering Solutions

ANSYS
Certified Reseller

Active Force Feedback Pedal
The Smart Way to Save Fuel

Johannes Heydenreich
PhilonNet Engineering Solutions

nest
nest
72

10:47 AM
hue
PHILIPS
PHILIPS
hue
App Store

AFFP
Continental

Disassembling and assembling accelerator pedal and cable

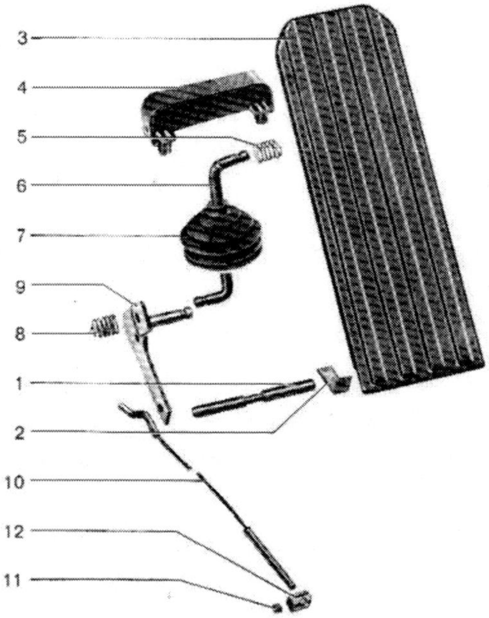

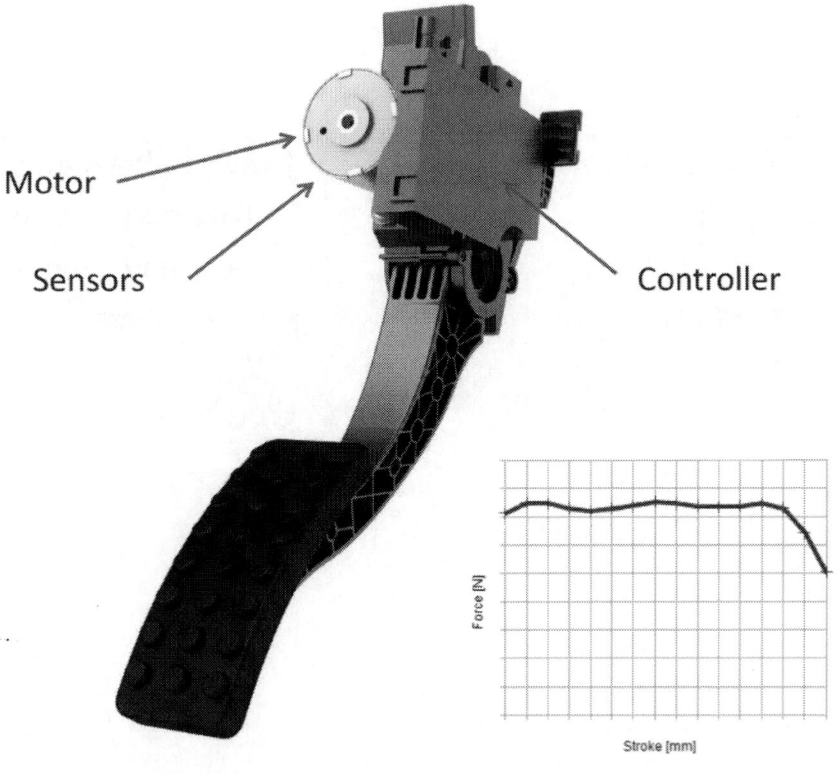

Motor
Sensors
Controller
Force [N]
Stroke [mm]

A Simple Gas Peddle

Where you used to have

- Joints
- Springs
- Wires

A "Simple" Gas Peddle

Where you used to have

- Joints
- Springs
- Wires

Now you have on top

- Rare earth magnets
- Coils and windings
- Controllers
- Buses
- Sensors

More Physics

Where you used to have

- Mechanical

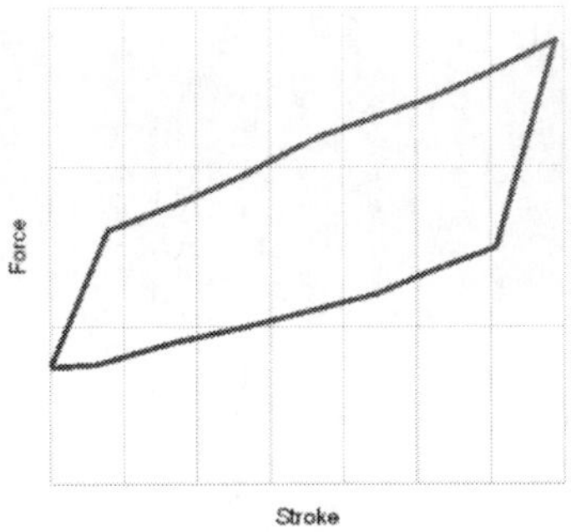

More Physics

Where you used to have

- Mechanical

Now you have on top

- Permanent Magnets
- Electromagnetics
- Electronics

Different Time Scales

Where you used to have

- Mechanical

Now you have on top

- Permanent Magnets [msec]
- Electromagnetics [msec]
- Electronics [µsec]

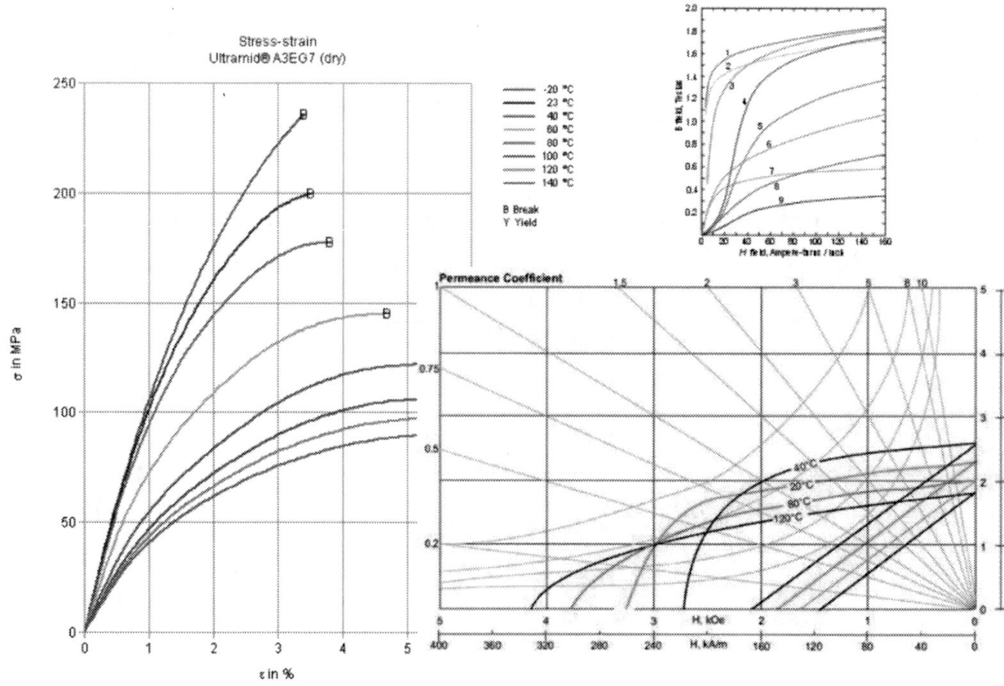

Small

Cheap

→ Hot 

→ EMI 

→ Constraints

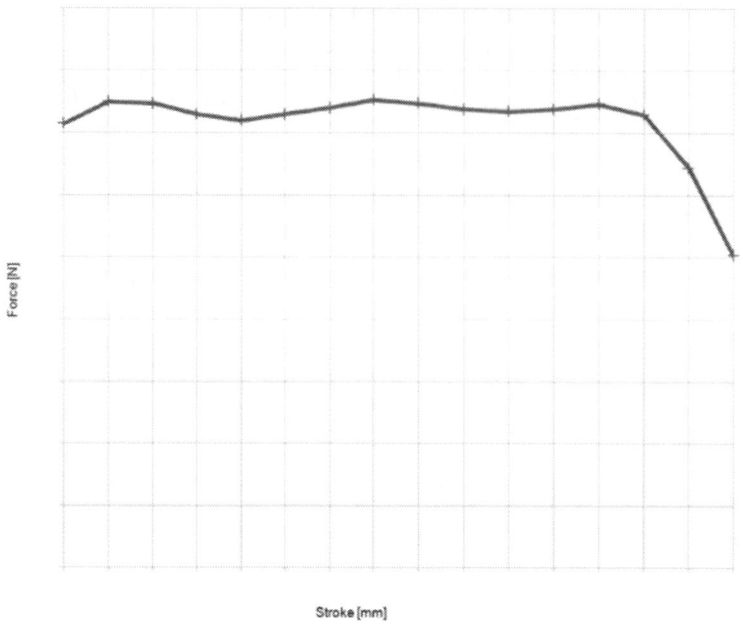
Force [N]
Stroke [mm]

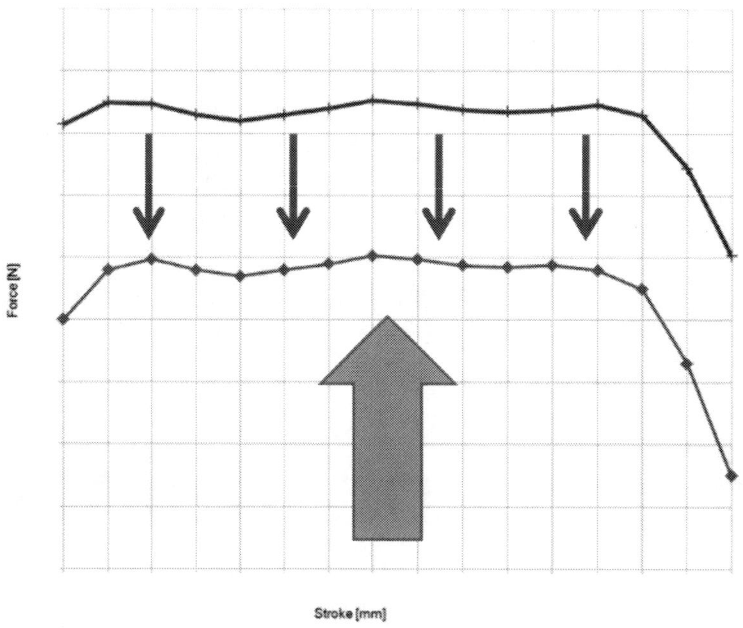
Force [N]
Stroke [mm]

Optimized Design

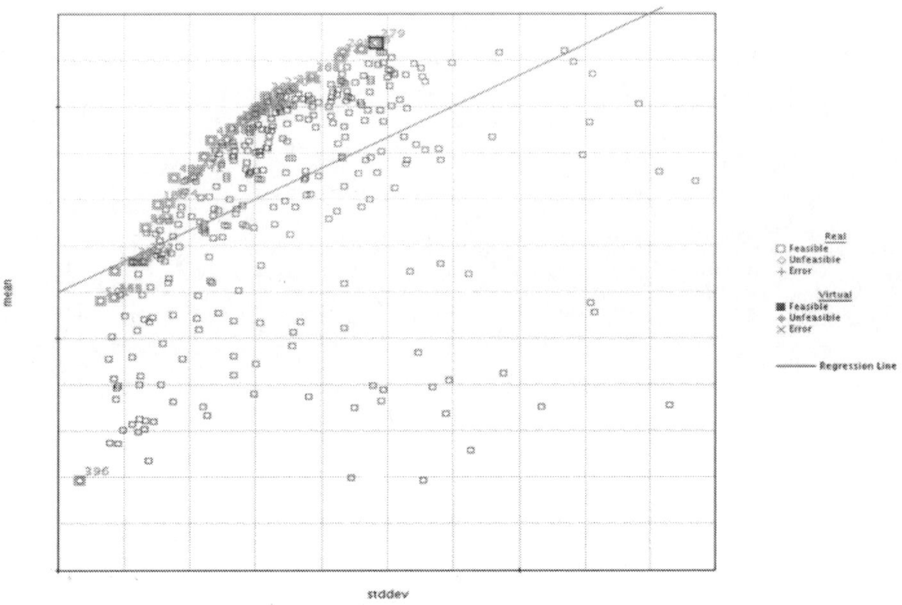

Why ANSYS ?

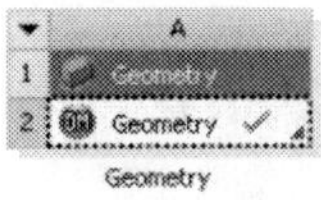

Why ANSYS ?

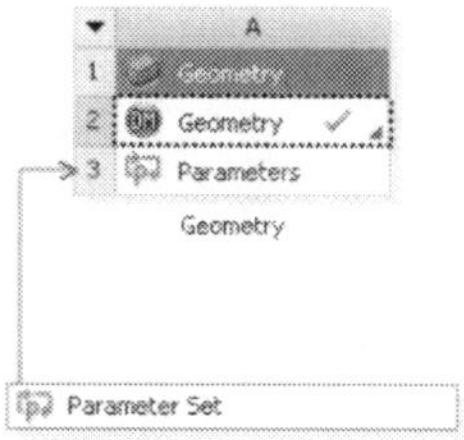

Why ANSYS ?

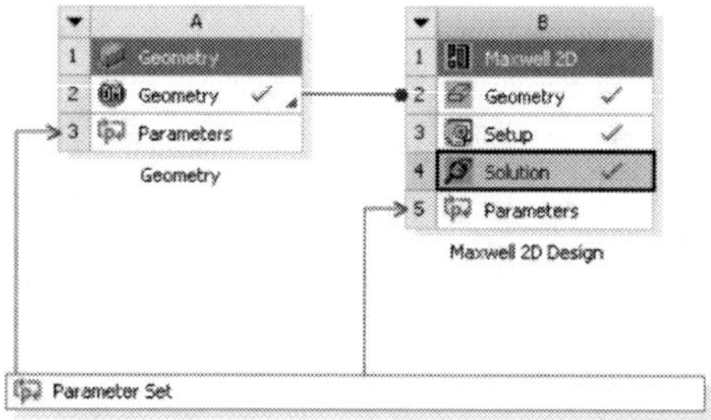

Why ANSYS ?

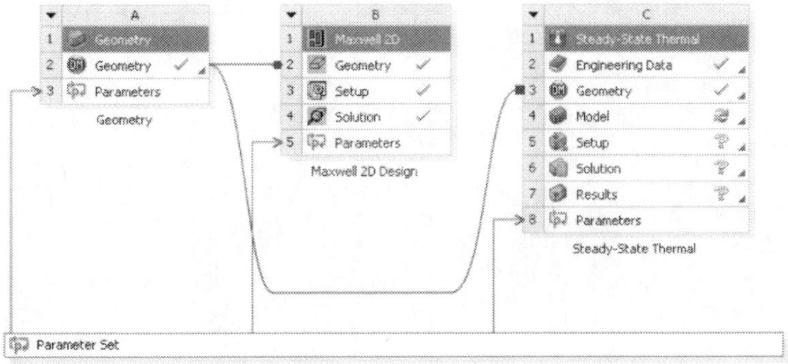

Why ANSYS ?

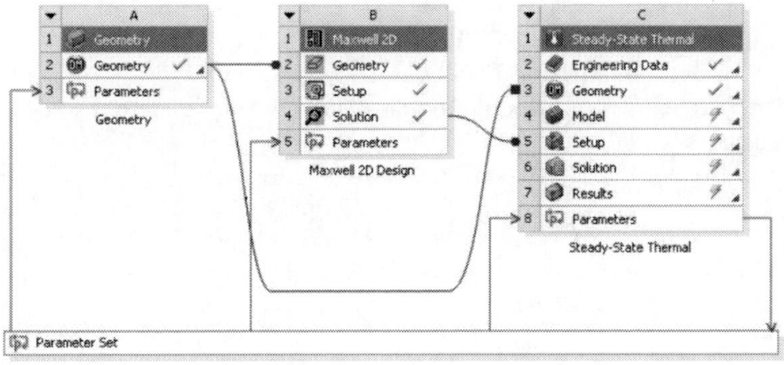

Why ANSYS ?

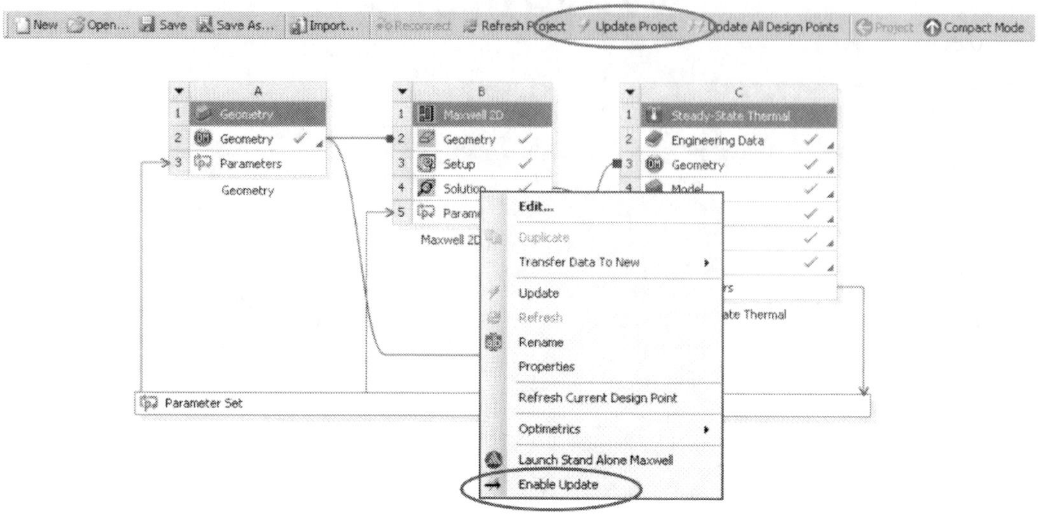

Journal Result

```
# encoding: utf-8
SetScriptVersion(Version="14.5")
system1 = GetSystem(Name="Maxwell")
Ansoft.ForceSolutionIntoUpdateRequiredState(System=system1)
Update()
```

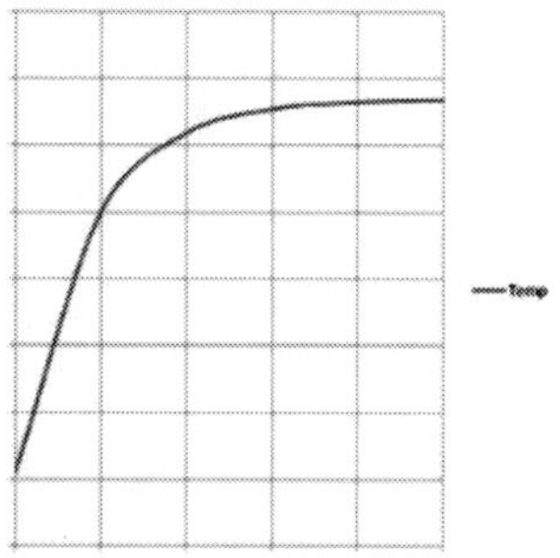

```
# encoding: utf-8
SetScriptVersion(Version="14.5")
system1 = GetSystem(Name="Maxwell")

tempNow = 22.0
for nowPass in range(1,10):
        Ansoft.ForceSolutionIntoUpdateRequiredState(System=system1)
        Update()
        tempBefore = tempNow
        tempNow = float(Parameters.GetParameter(Name='P3').Value.Value)
        if abs(tempBefore - tempNow) < 1.0:
                break
```


```
# encoding: utf-8
SetScriptVersion(Version="14.5")
system1 = GetSystem(Name="Maxwell")

tempNow = 22.0
for nowPass in range(1,10):
        Ansoft.ForceSolutionIntoUpdateRequiredState(System=system1)
        Update()
        tempBefore = tempNow
        tempNow = float(Parameters.GetParameter(Name='P3').Value.Value)
        if abs(tempBefore - tempNow) < 1.0:
                break
```

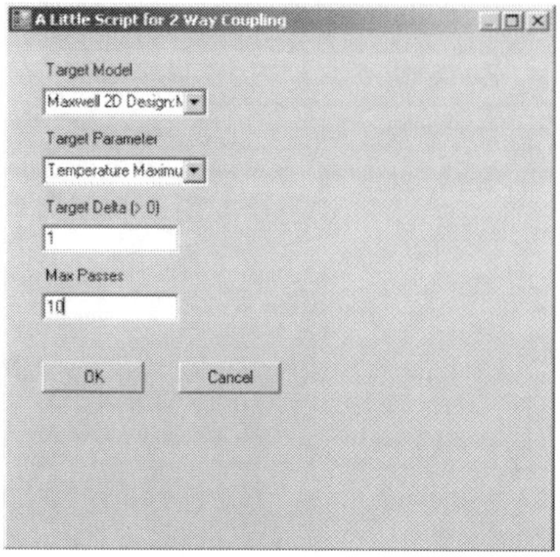

A Little Script for 2 Way Coupling
Target Model
Maxwell 2D Design:M
Target Parameter
Temperature Maximu
Target Delta (> 0)
1
Max Passes
10
OK
Cancel

Why ANSYS ?

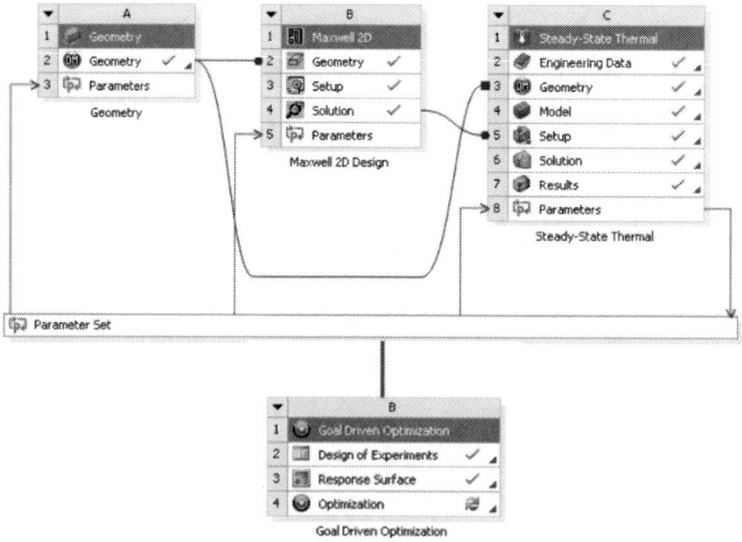

A
Geometry
Geometry
Parameters
Geometry
B
Maxwell 2D
Geometry
Setup
Solution
Parameters
Maxwell 2D Design
C
Steady-State Thermal
Engineering Data
Geometry
Model
Setup
Solution
Results
Parameters
Steady-State Thermal
Parameter Set
B
Goal Driven Optimization
Design of Experiments
Response Surface
Optimization
Goal Driven Optimization

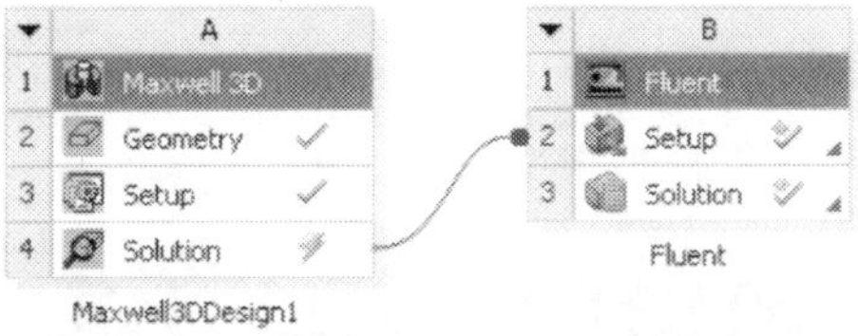

System Level Design

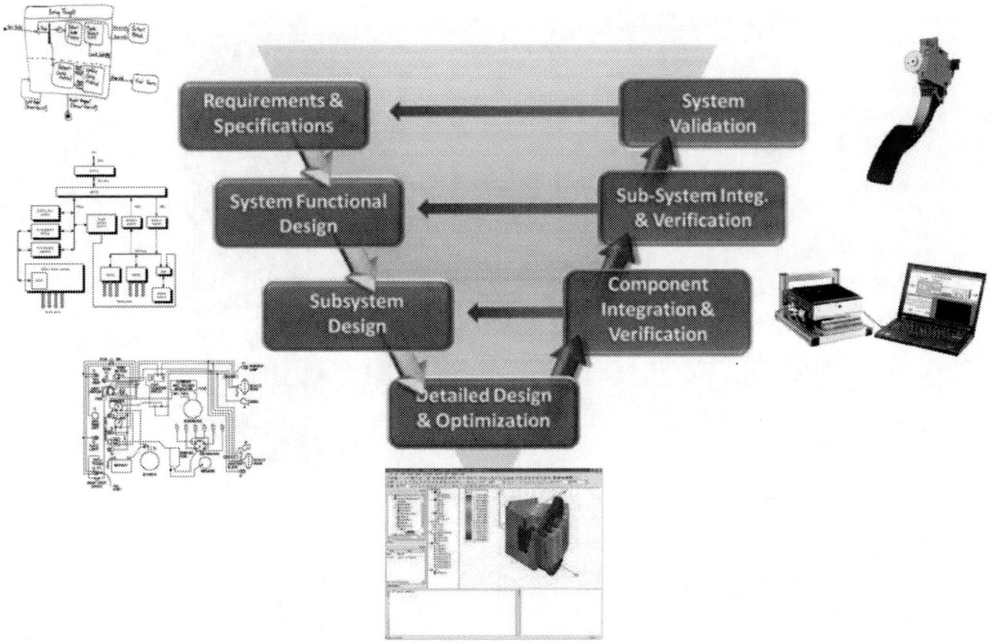

- Smart Products
- Smart People
- Smart Tools

PhilonNet
Engineering Solutions

ANSYS
Certified Reseller

Thank You

2013 Automotive Simulation World Congress

Structural Integrity Assessment Of a Li-Ion Battery Pack Using Simulation

Santosh Kottalgi, Ashok Khondge, Aditya Kshatriya,
Peng Yuan and Rajiv Rath
ANSYS Inc.

Outline

- Introduction to Structural integrity study in Battery Design
- Challenges and ANSYS advantages in structural integrity study
- Smart solution to structural integrity study
- Summary

Why study Li-Ion Battery?

- Li-ion Battery delivers high Power and Energy, making it ideal to address increasing expectation from Electric vehicles

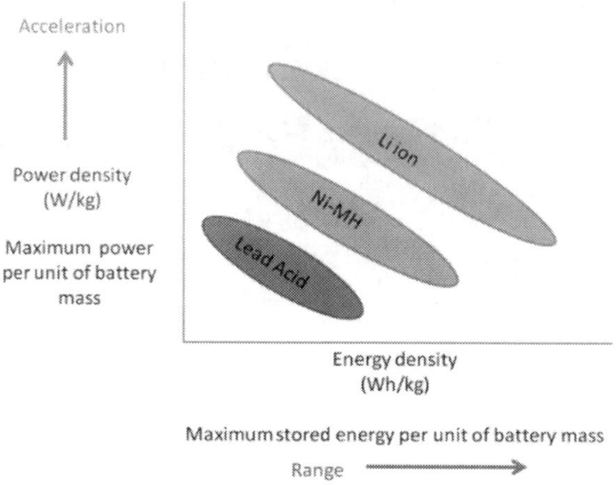

Source: CGGC based on (Abuelsamid, 2007)

Battery Design Challenge

High Energy Performance
(light weight and compact)

Cost	Long Life	Safety and Durability
• Battery adds ~$10,000 to car cost • Battery cost is big barrier to adoption of hybrid/electric vehicles	• Battery capacity = amount of energy stored in a battery • Determines how far an H/EV can go on one charge.	• Batteries store a large amount of energy and can potentially explode • Li-ion batteries a particularly vulnerable to safety issues

ANSYS

Scope of Battery Design

2013 Automotive Simulation World Congress
October 29 - 30 • Frankfurt am Main, Germany

Small Scale → Large Scale

	Molecular	Electrode	Cell	Module/Pack	Powertrain
Physics	Material	Electro-Chemistry	Electrical Thermal Mechanical	Electrical Thermal Mechanical	System Integration
Design Factor	Cost	Long life, Cost	Safety/Reliability	Safety/Reliability	Safety/Reliability
	R&D Centers, Universities	Suppliers	Suppliers	Suppliers, OEM	OEM

Standard Tests in Battery Design

2013 Automotive Simulation World Congress
October 29 - 30 • Frankfurt am Main, Germany

1. External Short Circuit Protection
2. Overcharge Protection
3. Over Discharge Protection
4. Fire Resistance – Short/Long Duration
5. Thermal Shock
6. Over/Under Temperature Protection
7. Water Intrusion Test
8. Vibration/NVH
9. Crush/Impact Test
10. Drop Test
11. Durability analysis

- Electrical and Electronics Integrity (1–4)
- Thermal and Fluid Flow Integrity (4–7)
- Structural Integrity (8–11)

Major Global Standards: SAE , UN 38.3, JIS, FMVSS, ECS

Outline

2013 Automotive Simulation World Congress
October 29 - 30 • Frankfurt am Main, Germany

- Introduction to Structural integrity study in Battery Design
- **Challenges and ANSYS advantages in Structural Integrity Study**
- Smart solution to structural integrity study

Monday, November 11, 2013 2013 Automotive Simulation World Congress 7

Where are challenges in Structural Integrity study?

2013 Automotive Simulation World Congress
October 29 - 30 • Frankfurt am Main, Germany

- Studying entire breadth of design and testing, in reasonable time and cost , with least amount of manufacturing and field testing
  - Virtual Simulation has emerged as reliable alternative to reduce these efforts
- Two major challenges in adaptation of Virtual Simulation
  1. Geometry, meshing and multiple design parameter study
  2. Multiple physics and their interdependence

Monday, November 11, 2013 2013 Automotive Simulation World Congress 8

Details of Challenges in Design and Testing

2013 Automotive Simulation World Congress
October 29 - 30 • Frankfurt am Main, Germany

- Significant Design and Testing efforts at Cell, Module and Assembly level, spread across teams and organizations
- Performing Design and Testing experimentally is very time consuming, costly and vulnerable to safety hazards
- Design and Testing require study of multiple physics individually or together
- Multiple Design and Testing iterations increase lead time from paper to on road usage

Monday, November 11, 2013 2013 Automotive Simulation World Congress 9

Virtual Simulation Advantages using ANSYS

2013 Automotive Simulation World Congress
October 29 - 30 • Frankfurt am Main, Germany

- Deploy ANSYS at teams and organizations for detailed individual level study and connect them together with smart processes
- Simulation can follow Design Standards for both destructive and non destructive study, reducing experimental study significantly
- ANSYS Multiphysics capabilities helps in studying individual as well as coupled physics design and testing requirements
- Simulation helps in studying multiple design and tests simultaneously reducing lead time significantly

Monday, November 11, 2013 2013 Automotive Simulation World Congress 10

Challenges in Geometry, Meshing & Multiple parameter study

2013 Automotive Simulation World Congress
October 29 - 30 • Frankfurt am Main, Germany

- Retaining geometry quality as good as CAD software
- Difficulty in maneuvering large assemblies for geometry simplification
- Efficiently cleaning up of CAD file for meshing
- Defining common mesh parameters to address various simulation goals
- Meshing complete assembly in one go
- Connecting components efficiently
- Total effort required for meshing
- Possibility of Human errors

ANSYS Advantages

2013 Automotive Simulation World Congress
October 29 - 30 • Frankfurt am Main, Germany

- Bi-directional CAD connectivity reduces data loss and increases import speed
- Efficient (B-rep) representation of CAD reduces graphics requirement
- DesignModeler has features to clean up, repair and simplify geometries
- Advance meshing and range of element types , helps in creating common mesh for different simulation
- ACT based Assembly Manager Tool helps to tackle geometry at module and cell level
- Auto contacts/joints generation reduces time for interface definition
- Smart meshing process reduces CAD to Meshing time significantly.
- Usage of assembly manager tool, auto contact definition and worksheet based meshing reduces human error

Details of Challenges in Multiphysics Solvers

2013 Automotive Simulation World Congress
October 29 - 30 • Frankfurt am Main, Germany

- Ability to study body weight, Thermal, Vibration and destructive design and tests
- Ability to Study multiple physics and effect of one physics on another, by avoiding multiple Simulation Software
- Reduction of interdependency of engineers for knowledge transfer
- Ability to study entire breadth with least amount of computer resource and time.

Virtual Simulation Advantages using ANSYS

2013 Automotive Simulation World Congress
October 29 - 30 • Frankfurt am Main, Germany

- ANSYS Static, Transient, Thermal , Dynamics/NVH and Explicit Dynamics solves different Design and Tests requirements
- Workbench Platform helps to solve multiphysics problem in single project, reducing use of multiple software
- Intuitive project page helps to understand interdependency of workflow, bringing engineers on board quickly
- Interconnected workflow and HPC parametric pack, helps in solving simultaneous multiple parametric designs and tests

Outline

2013 Automotive Simulation World Congress
October 29 - 30 • Frankfurt am Main, Germany

- Introduction to Structural integrity study in Battery Design
- Challenges and ANSYS advantages in Structural Integrity Study
- **Smart solution to structural integrity study**

Scope of Structural Integrity Study

2013 Automotive Simulation World Congress
October 29 - 30 • Frankfurt am Main, Germany

	Standard process	Present Work	Standard
1	Effect of body weight on Support	✓	SAE J2380
2	Pre-stressed natural frequency analysis	✓	SAE J2380
3	Pre-stressed Random Vibration analysis in Vertical, longitudinal and lateral direction	✓	SAE J2380
4	Random Vibration fatigue analysis	×	
5	Drop Test from 2 meter at an angle	✓	SAE J2464
6	Thermal Stress analysis	✓ (Module)	SAE J2464
7	Impact/Crush analysis	×	SAE J2464
8	Shock Spectrum analysis in Vertical, longitudinal and lateral direction	×	SAE J2464
9	Nail Penetration analysis	×	SAE J2464

How Smart Process Works?

2013 Automotive Simulation World Congress
October 29 - 30 • Frankfurt am Main, Germany

- Link all the Simulation in workbench for better understanding of work flow
- Have common geometry, meshing and contacts for all type of Simulations
- Have common Master Material library for all type of physics
- User Smart geometry cleanup and meshing process using Assembly Manager ACT

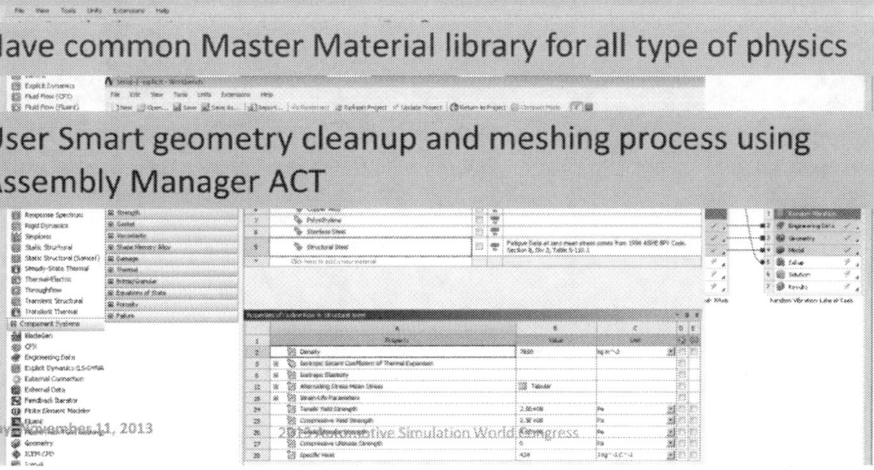

Smart Geometry Cleanup and Meshing Process

2013 Automotive Simulation World Congress
October 29 - 30 • Frankfurt am Main, Germany

Video Presentation in Original e-media

How Assembly Manager ACT tool works?

2013 Automotive Simulation World Congress
October 29 - 30 • Frankfurt am Main, Germany

Assembly Manager capability is exposed in ANSYS Workbench Mechanical only via an ACT extension

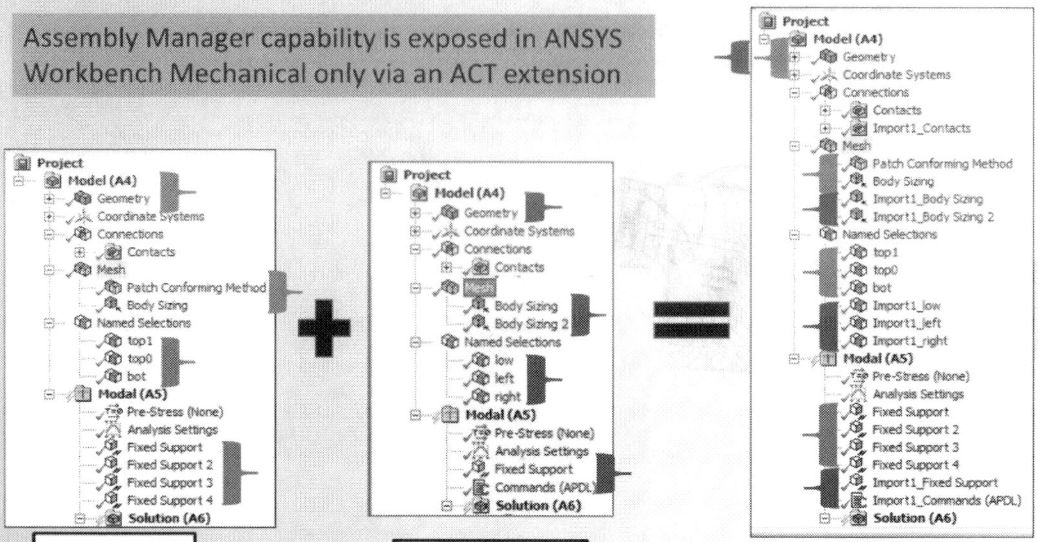

Base Model | Imported Part | Assembly

Self Weight Analysis

2013 Automotive Simulation World Congress
October 29 - 30 • Frankfurt am Main, Germany

- Objective:
  - Study of effect of battery self weight on support structure
- Loading:
  - Standard earth gravity
- Boundary condition:
  - Fix all the mounting location of battery bottom
- Results:
  - Deformation is as expected high on top of the heavy module
  - Stress is within limit and is max at opposite corner support surface to max deformation location

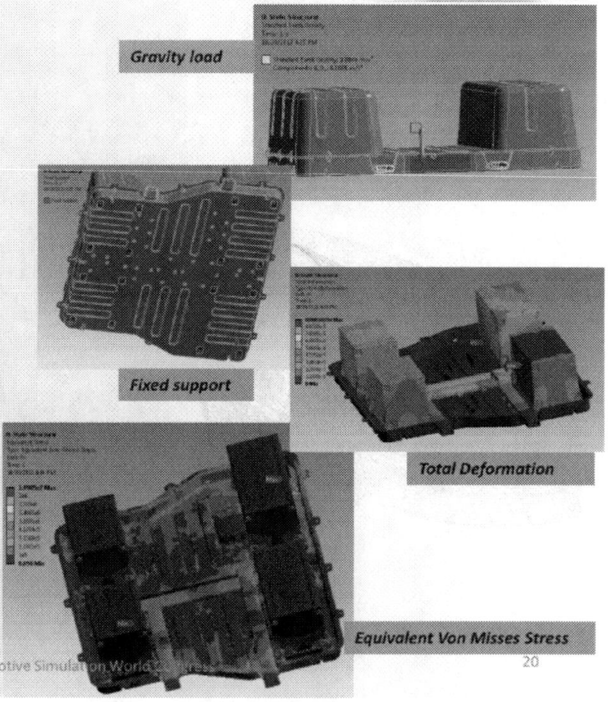

Pre-stressed Natural Frequency Analysis

- Objective:
  - Study of pre-stressed natural frequency of battery assembly
- Analysis setting:
  - Since random vibration load is upto 190 Hz, we have to solve upto 1.5 times frequency band, so solve upto 250 Hz.
- Boundary condition:
  - Since its pre-stressed analysis, boundary condition will obtained from static structural
- Results:
  - 32 modes extracted
  - Mass upto 80 % of total in all three directions, which is enough to fair accuracy in Random vibration

Analysis setting

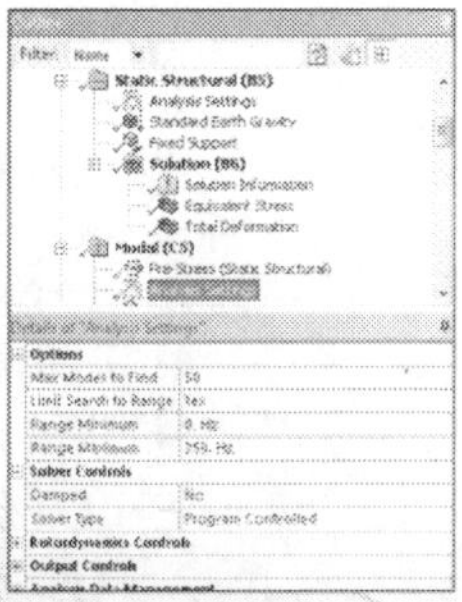

Mode shape 1st mode = 36Hz

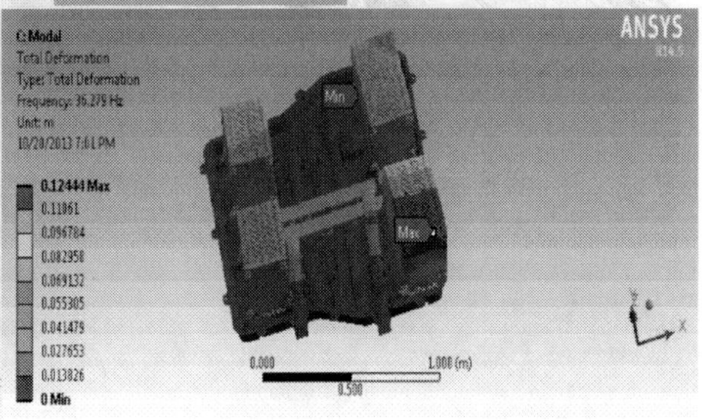

Random Vibration Analysis

- Objective:
  - Perform random vibration analysis of prestressed modal in vertical, longitudinal and lateral direction
- Loading:
  - Follow loading as per SAE J2380 standard
  - Use same base excitation for longitudinal and lateral direction
- Boundary condition:
  - Fixed support definition will be used from static analysis

Vertical		Longitudinal and Lateral	
Frequency (Hz)	1.9 grms Amplitude (G^2/Hz)	Frequency (Hz)	1.9 grms Amplitude (G^2/Hz)
10	0.113	10	0.064
15	0.113	13	0.064
18	0.083	22	0.032
25	0.037	45	0.016
35	0.037	80	0.01
45	0.021	120	0.0057
80	0.021	190	0.0057
120	0.0092		
170	0.0052		
190	0.0052		

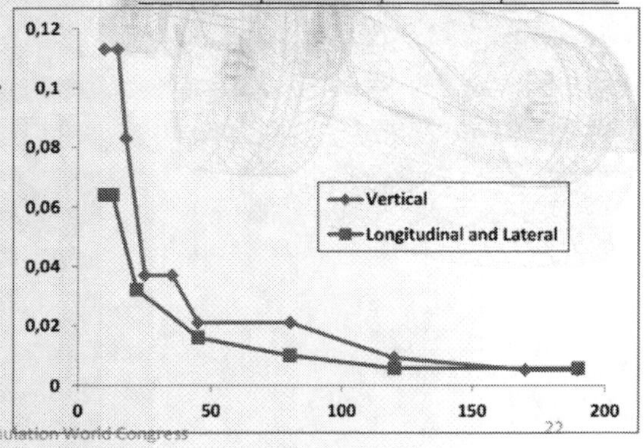

Random Vibration Analysis contd..

2013 Automotive Simulation World Congress

- Results:
  - 1 sigma stress in all three direction is safe and is max in vertical direction
  - Deformations are also small and is maximum in longitudinal direction
  - Though excitation is same in both longitudinal and lateral direction, results changes significantly

Direction	Deformation (m)	Equivalent Stress (Pa)
Vertical	6.93e-4	1.78e8
Longitudinal	2.4e-3	1.24e8
Lateral	1.19e-3	8.58e7

Directional deformation *1 sigma equivalent stress*

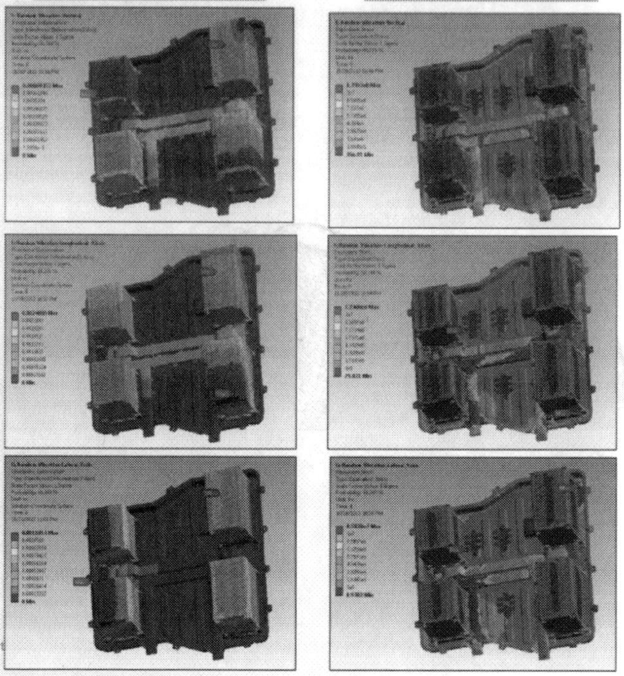

Thermal Stress Analysis

2013 Automotive Simulation World Congress
October 29 - 30 • Frankfurt am Main, Germany

- Objective:
  - Perform thermal stress analysis for a battery module,
- Loading
  - Thermal loads are calculated from Electromagnetic Analysis using MAXWELL
  - Module is fixed at the bottom
- Results:
  - Thermal Stress and deformation are studies for failure of component due to high stresses

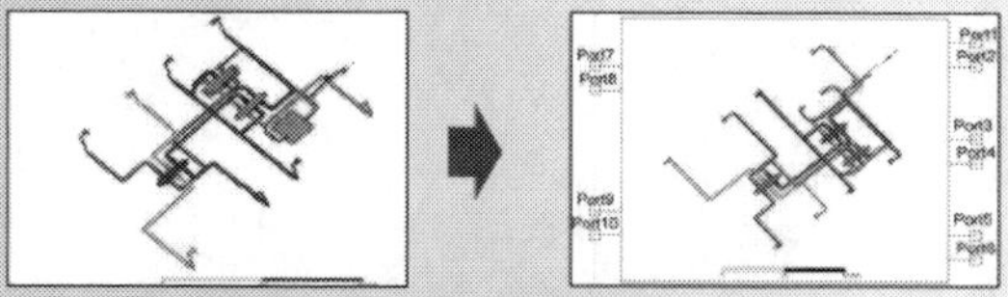

Electromagnetic FEA Analysis for Busbar RLC Network Extraction

Structural FEA: Total Deformation

Structural FEA: Equivalent Stress

Drop Test Analysis

2013 Automotive Simulation World Congress
October 29 - 30 • Frankfurt am Main, Germany

- Objective:
  - Perform drop test analysis of Battery assembly at most vulnerable angle as per SAE J2464 standard
- Product:
  - Explicit dynamics
- Meshing & contacts:
  - While meshing mid side nodes are dropped and pyramid elements are avoided to help have a common mesh for all type of simulation
  - Breakable bonded contacts are defined to separate parts after 150MPa load
- Loading & boundary condition:
  - Drop the battery from 2 meter height at 45 degree on heavy side
  - To save time of free fall from 2 meter, velocity just before touching to the ground is applied as initial load
- Result:
  - The top and bottom part of casing are separating

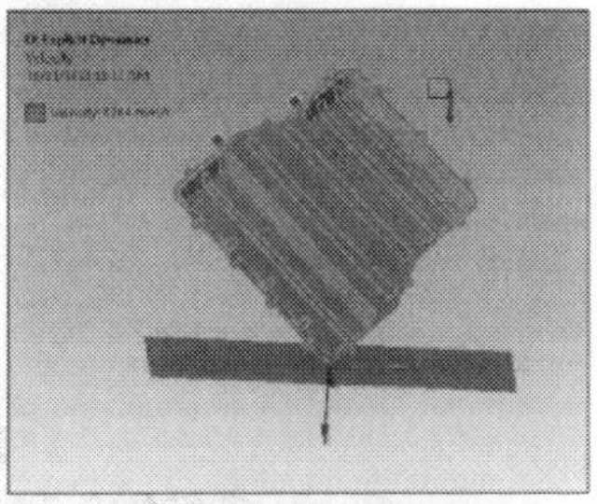

Initial velocity

Total deformation after 5% solve time

Monday, November 11, 2013 — 2013 Automotive Simulation World Congress — 25

Summary

2013 Automotive Simulation World Congress
October 29 - 30 • Frankfurt am Main, Germany

- ANSYS Multiphysics capability helps to solve different simulation types in structural integrity analysis
- Smart process helps in saving time and efforts, while increasing efficiency
- Indirectly it also helps in reducing R&D and Testing cost

Monday, November 11, 2013 — 2013 Automotive Simulation World Congress — 26

2013 Automotive Simulation World Congress

Oct. 29, 2013

Collaborative Co-Simulation Cloud for Functional Mockup Design Assessments

Mikio Nagasawa

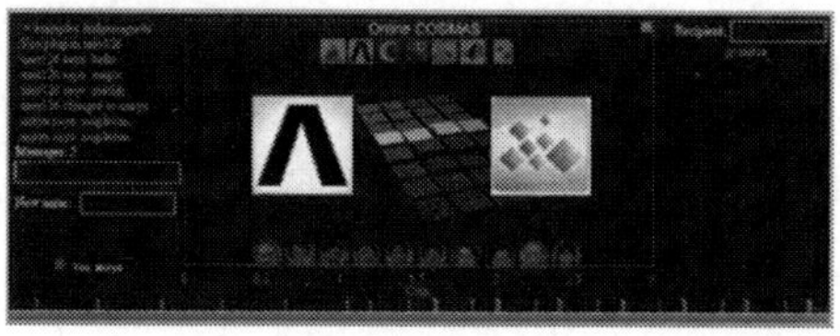

CYBERNET SYSTEMS CO., LTD.
Energy for your Innovation

About Our Company

- History in Cybernetic and Supercomputing
  - 1985 CYBERNET SYSTEMS Co. Ltd. takes over CYBERNET business from Control Data Japan
- CYBERNET "Legacy Cloud"
  - Remote TSS Service (1980)
- Multi-Domain CAE Software
  - ANSYS, MapleSim, and more

CYBERNET CLOUD

CYBERNET SYSTEMS CO., LTD.
Energy for your Innovation

Our Multi-Domain Expertise

- Know-Who
  - intranet search engine for Engineers collaboration
- Know-What
  - multi-physics solvers available for Co-simulation

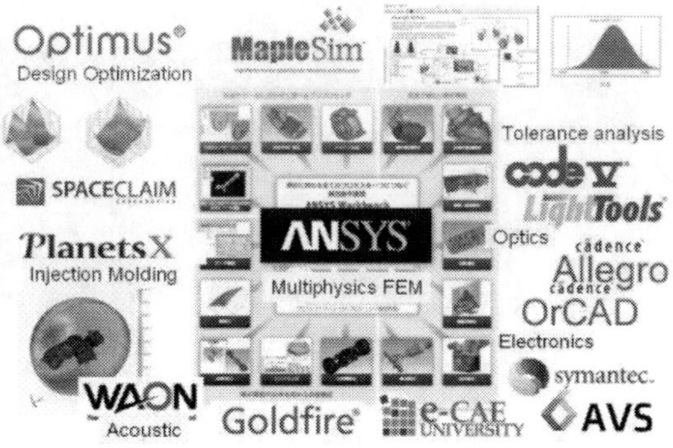

CYBERNET SYSTEMS CO., LTD.
Energy for your Innovation

Problems for Automotive Simulation

- Q1. Software License
  - Multi-Product purchase
  - Multi-OS installation
    - ✓ A1. Web online simulation

- Q2. Program Interface
  - Multi-Program connection : N(N-1)/2
  - Multi-Compiler knowledge
    - ✓ A2. FMI standard co-simulation

- Q3. Result Validation
  - Multi-Domain perspective
  - Multi-Expertise communication
    - ✓ A3. UPC multi-player collaboration

CYBERNET SYSTEMS CO., LTD.
Energy for your Innovation

Co-Simulation or Coupled Simulation

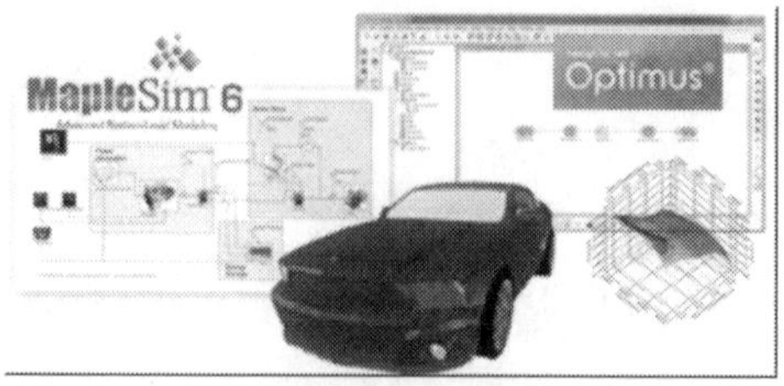

Problem	Dimension	Product	Interface	Process Thread
D-o-E Optimization	0D (Objective)	Optimus	File Batch	A B
Control Co-Simulation	1D x 3D (Time)(Space)	MapleSim Simulink	Communication Protocol	A B
Coupled Analysis	3D x 3D $(Space)^2$	MpCCI	Shared (Big) Memory	A B

Co-Simulation Tools and Users

- ♦ Good Master and Good Players wanted

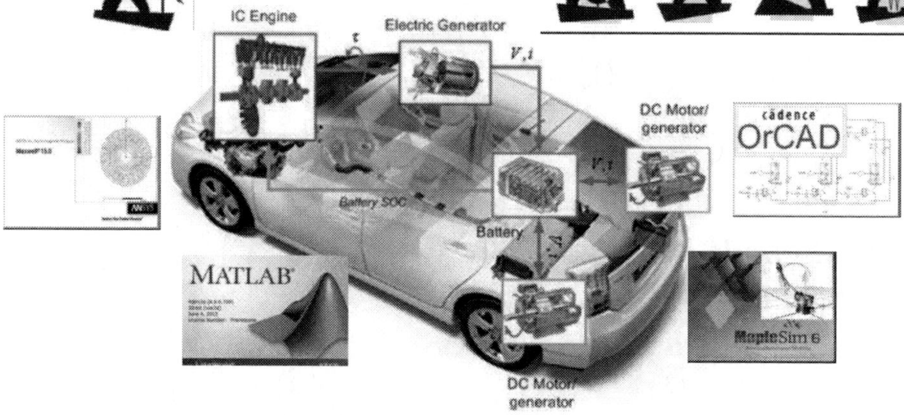

Nominate Online Game Players

- First, Enjoy Real-time Simulations
- Next, Analyze

CYBERNET SYSTEMS CO., LTD.
Energy for your Innovation

Standards and Communications

- Union Platform Protocol
  - —join the online game stage
    - http://unionplatform.com/specs/upc/
- Functional Mockup Interface
  - —concert the simulations
    - http://functional-mockup-interface.org

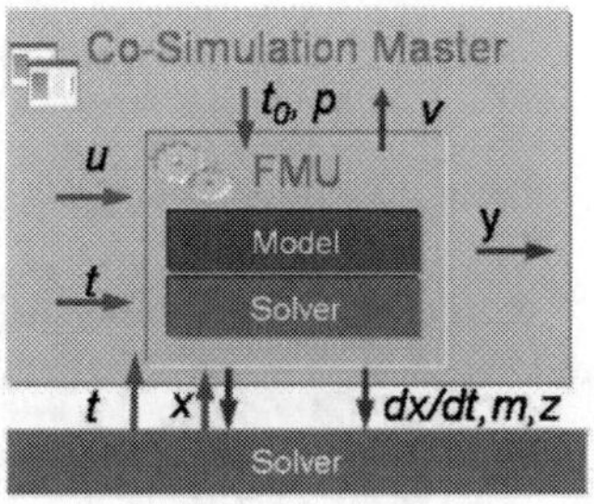

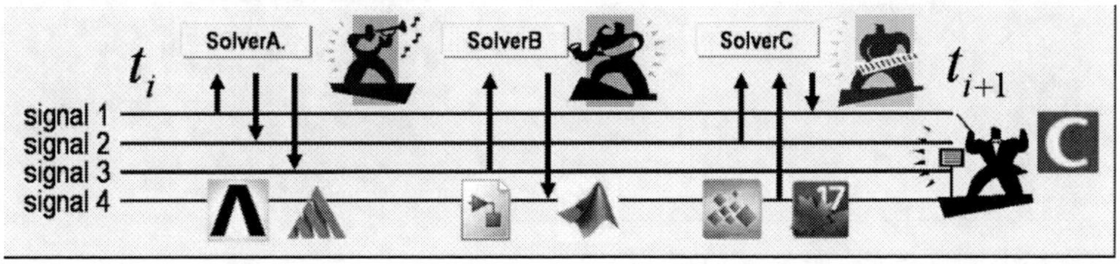

CYBERNET SYSTEMS CO., LTD.
Energy for your Innovation

Collaborative Co-Simulation Architecture

- Socket Connections among Simulation Tools
- UPC: Multiplayer Online Game Union Protocol
- FMI: Standard Functional Mockup Interface

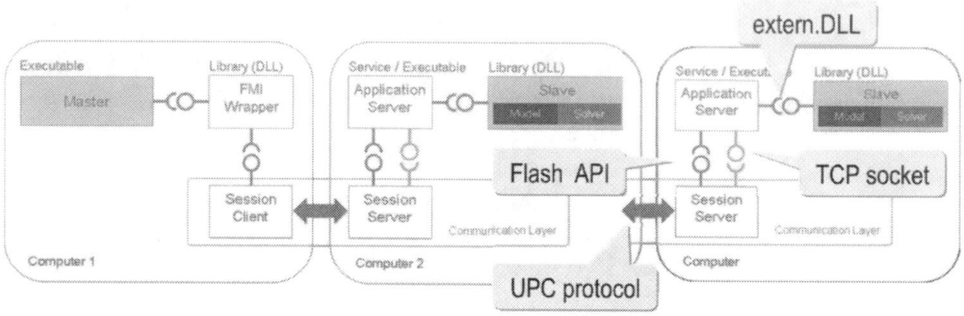

Collaborative Co-Simulation Demo

- Delayed and Synchronized bouncing balls
- Visual Interaction

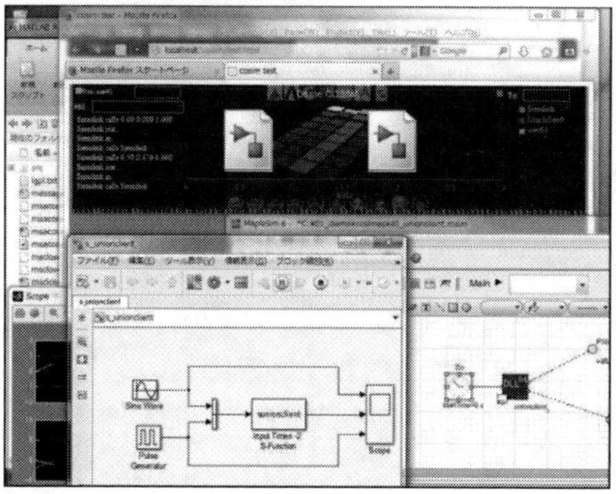

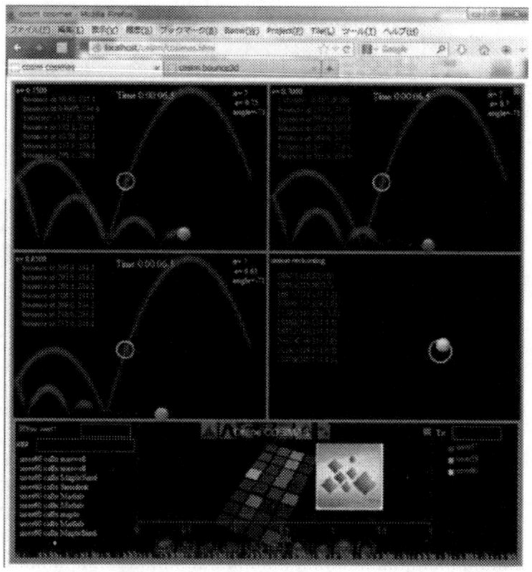

Case study: Solenoid Plunger Control

- ANSYS/Maxwell Plant model
- MapleSim Controller model

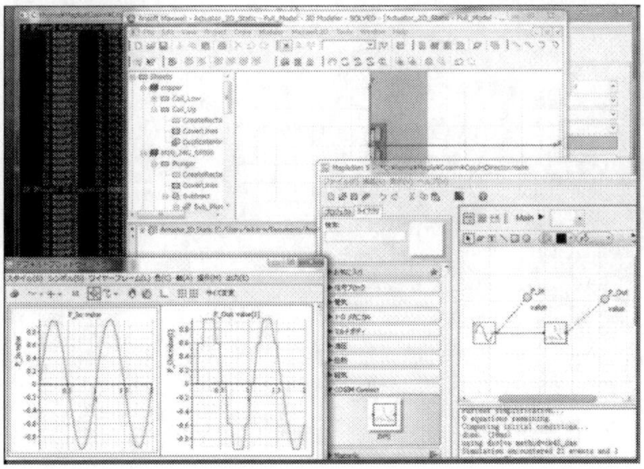

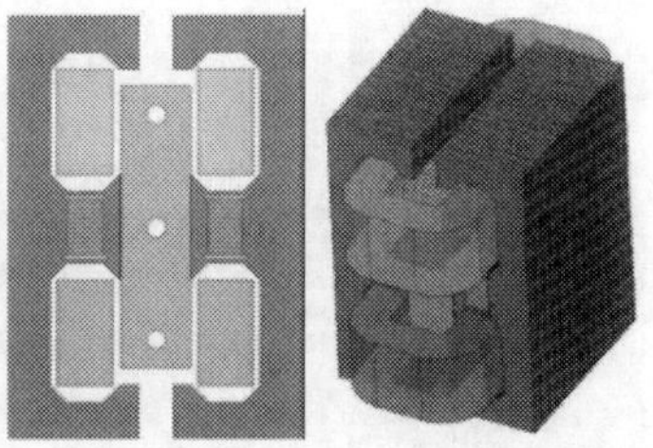

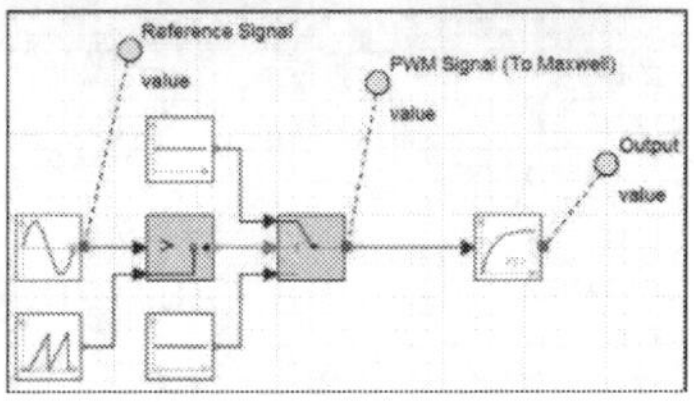

CYBERNET SYSTEMS CO., LTD.
Energy for your Innovation

Case Study: Multi-Method Integration

- Power Window Control
- Rigid body and Elastic body co-simulation
  - —1st order Symplectic Euler
  - —4th order Runge-Kutta

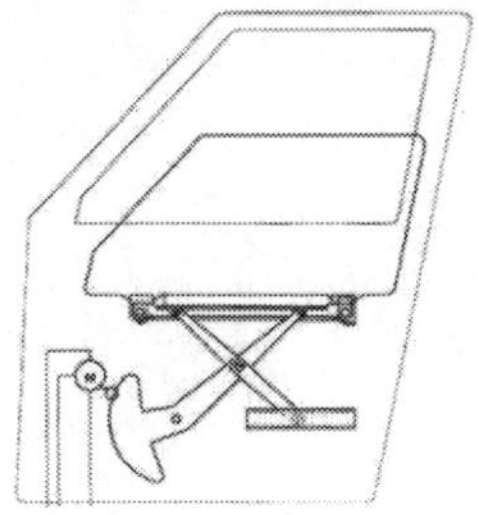

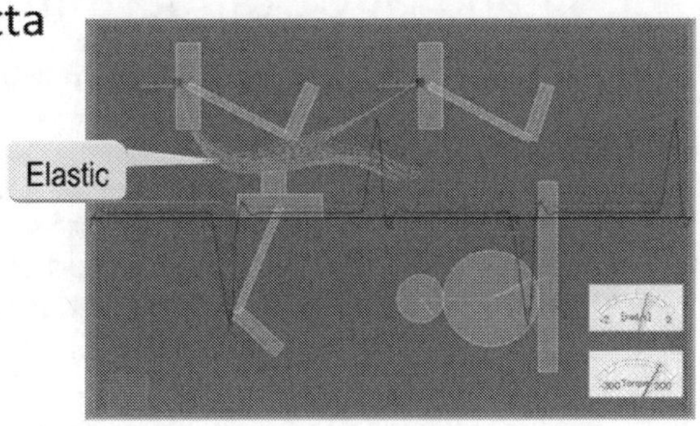

CYBERNET SYSTEMS CO., LTD.
Energy for your Innovation

Multi-Method-In-the-Loop

- Switch for Model, Software, and Hardware
- Visual and Interactive Assessment

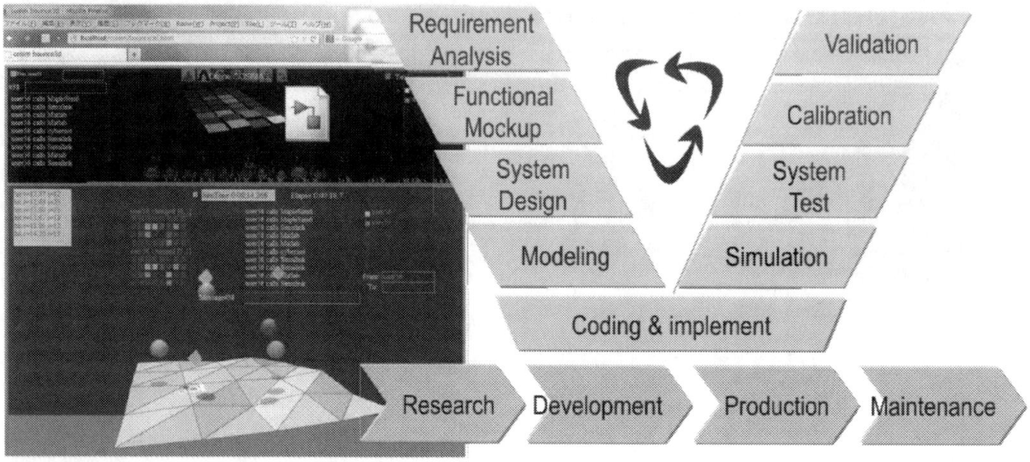

CYBERNET SYSTEMS CO., LTD.
Energy for your Innovation

Summary

- Cybernet = Control + Communication
- Collaboration among CAE and Non-CAE Users
- Online Co-Simulation Platform
- FMI and UPC Standard Integration
    - Multi-X Cloud will be realized

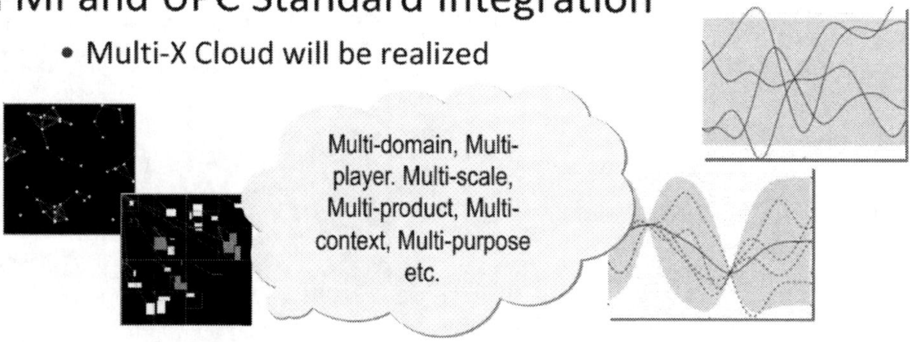

CYBERNET SYSTEMS CO., LTD.
Energy for your Innovation

Parametric Multiphysics Simulation

Martin Hanke, CADFEM GmbH

ANSYS   ANSYS Competence Center FEM   CADFEM

Technical Challenge: Power Connectors

- Guarantee current carrying capacity
- Maintain acceptable temperature range
  - Insulation strength
  - Handling
- Reliability of connection
  - Mechanical stability
  - Push/pull forces
  - Endurance
- Optimization in design space
- Material usage / coating

CADFEM

Physical Effects

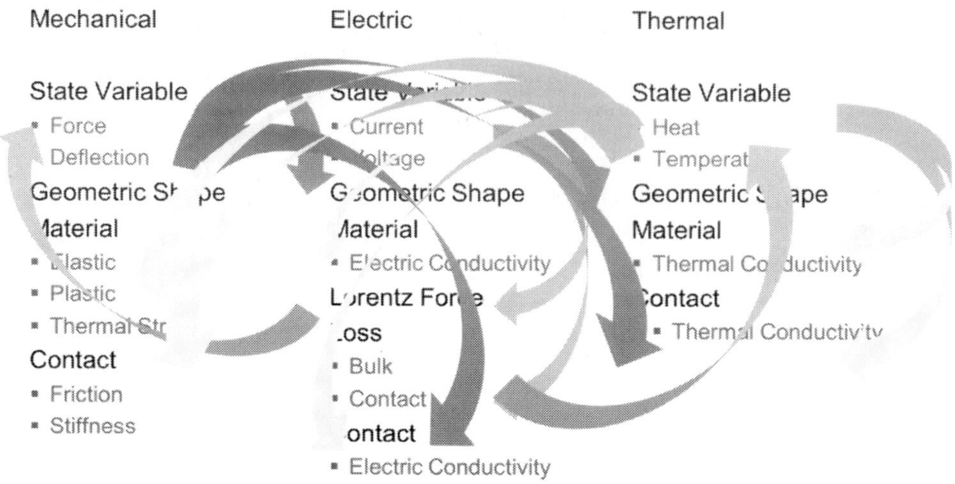

CADFEM

Physical Effects

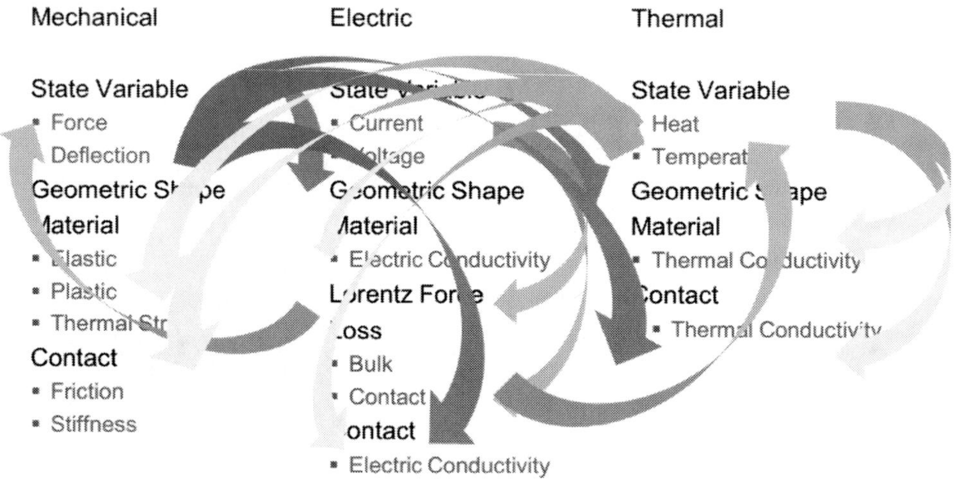

CADFEM

Simulation Model

Mechanical

State Variable
- Force
- Deflection

Geometric Shape

Material
- Elastic
- Plastic
- Thermal Strain

Contact
- Friction
- Stiffness

Electric

State Variable
- Current
- Voltage

Geometric Shape

Material
- Electric Conductivity

Loss
- Bulk
- Contact

Contact
- Electric Conductivity

Thermal

State Variable
- Heat
- Temperature

Geometric Shape

Material
- Thermal Conductivity

Contact
- Thermal Conductivity

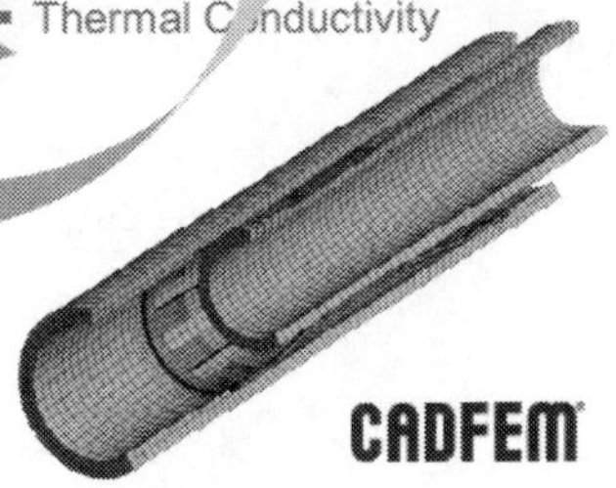

CADFEM

Simulation Model Coupling Details

- Pressure dependent **E**lectric (ECC) and **T**hermal **C**ontact **C**onductivity (TCC) approximated with square root law (constriction resistance according to R. Holm)

$$R_E = \frac{\rho}{2}\sqrt{\frac{\pi H_K}{F_K}}$$

- Temperature dependent bilinear material model to represent plasticity

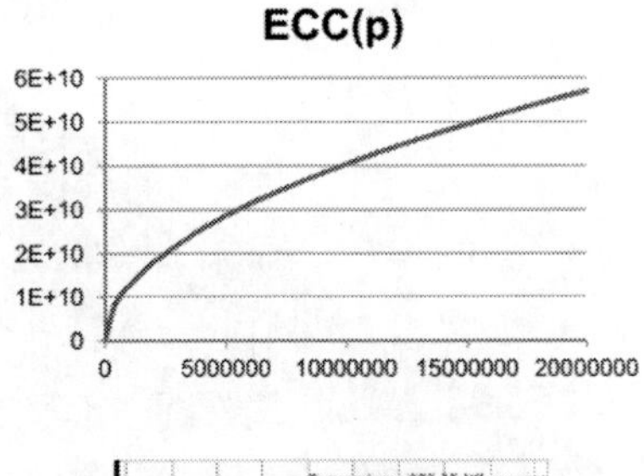

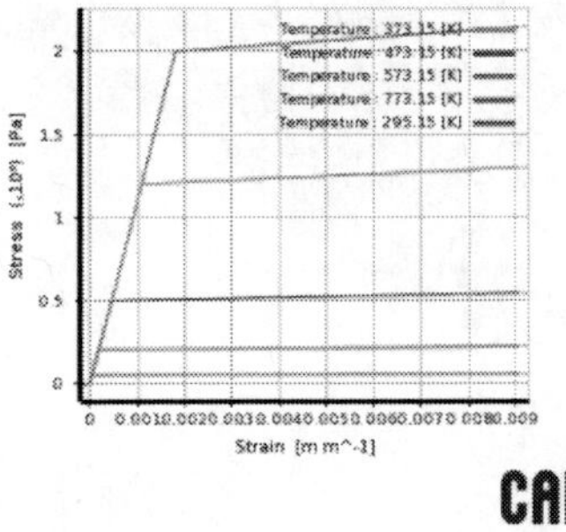

CADFEM

Simulation Model

- 3D solid model with coupled field elements SOLID22x (multiple DOF)
  - Common mesh
- 3D contact for coupled field CONTA174 (multiple DOF)
  - Mechanical contact stiffness including friction
  - Thermal conductivity as a function of pressure
  - Electric conductivity as a function of pressure
- Analysis workflow:

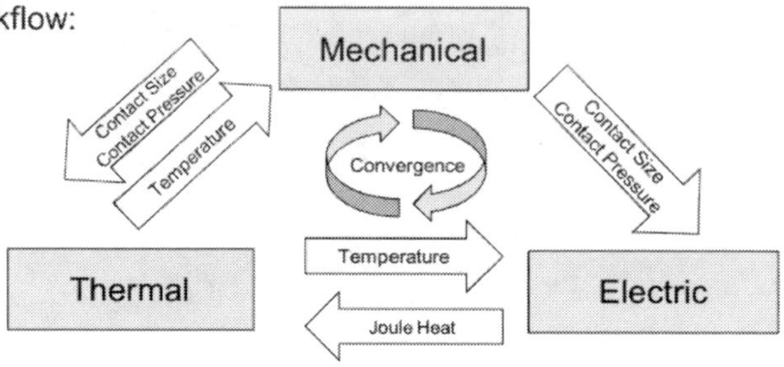

- 6 -

CADFEM

FEM-Model of Power Connector

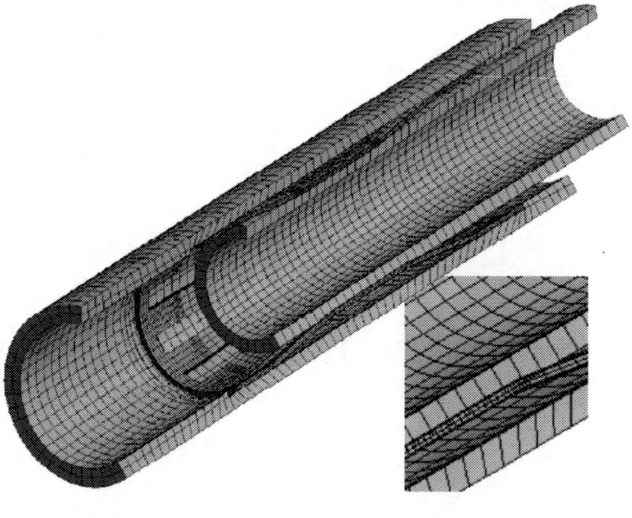

Variable overlap

Current load:
Ground and current input at terminal ends

Thermal boundary:
Cable acts as thermal resistance (convection on end face)

Mesh density:
Correct stiffness
Contact zone
Plasticity

- 7 -

CADFEM

Validation of Simulation Method

- Voltage drop along current conduction path
- Voltage drop across contact: ΔU=ECC*j

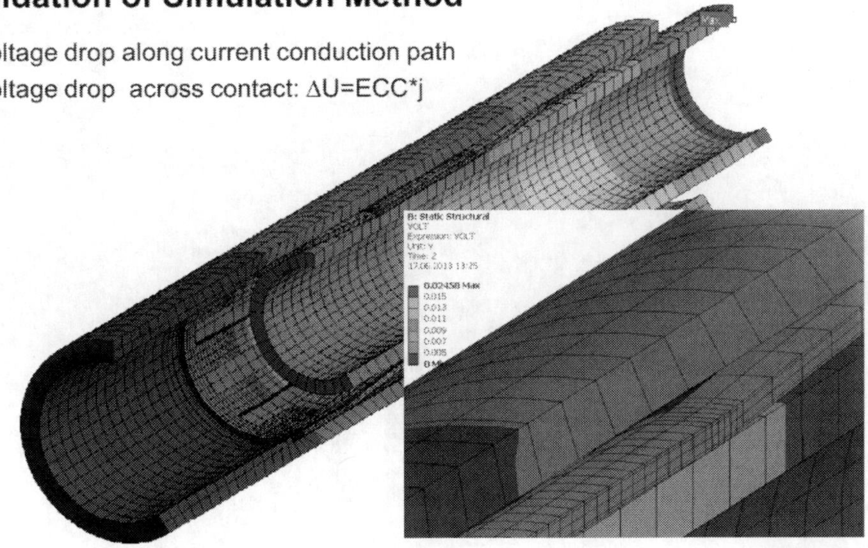

CADFEM

Validation of Simulation Method

- Heat generation rate inside conductor = $\rho*j^2$
- Contact as surface heat source = ΔU*j

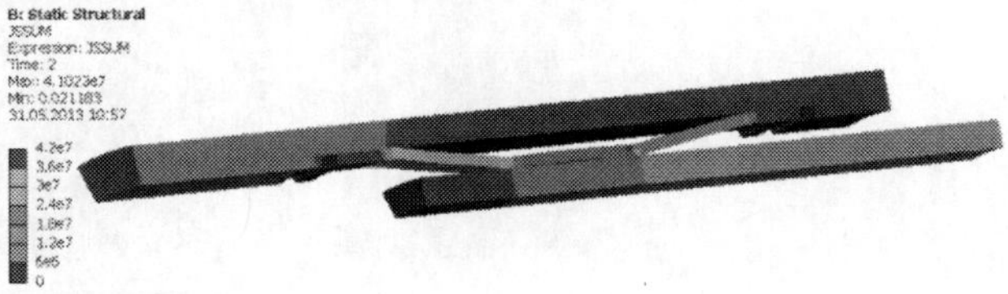

- Maximum temperature

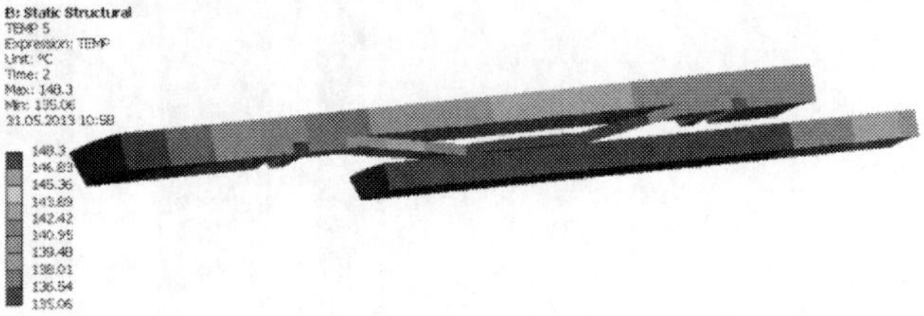

CADFEM

Validation of Simulation Method (Contact Pressure)

Local contact

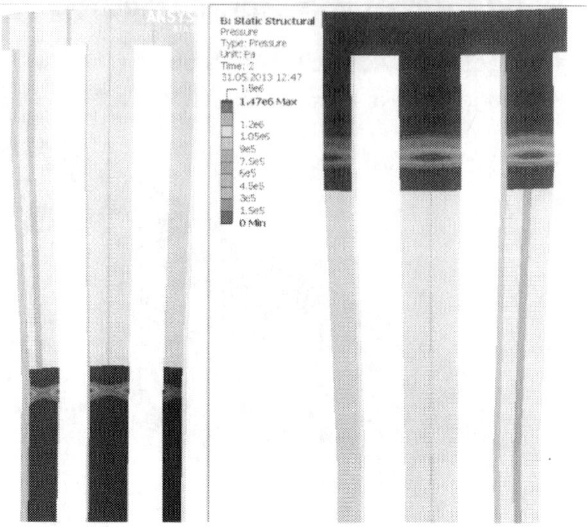

- 10 -

CADFEM

Validation of Simulation Method (Deformation)

Bending stress

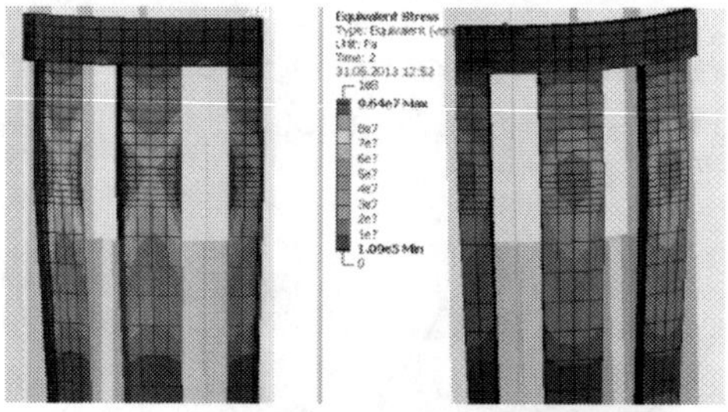

Local plasticity

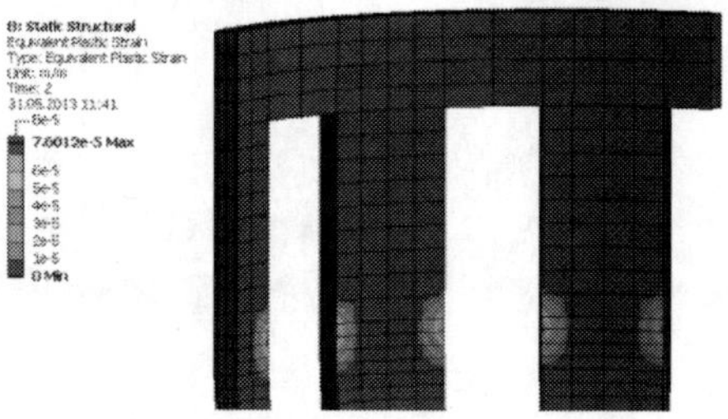

- 11 -

CADFEM

Validation of Simulation Method

- Variation of physical parameter
  - Temperature dependent plasticity
  - Temperature dependent resistivity
- Variation of overlap
- 100 Design samples
  - 92 successful
  - 8 failed due to model setup or convergence

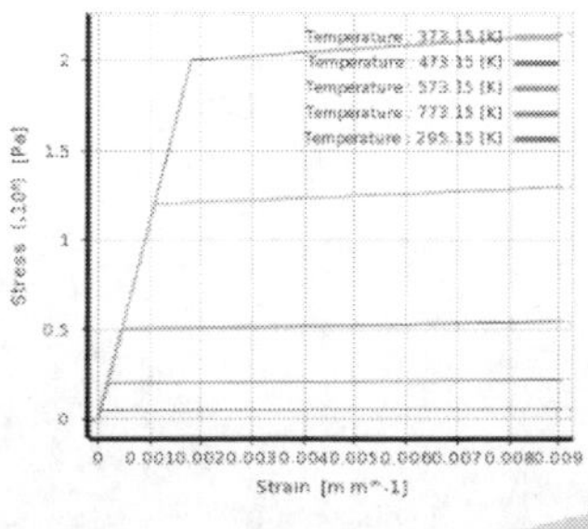

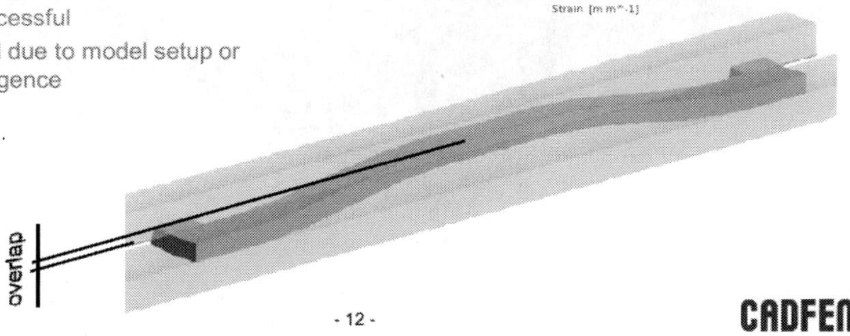

- 12 -

CADFEM

Validation of Simulation Method

- Increasing overlap → increasing plasticity

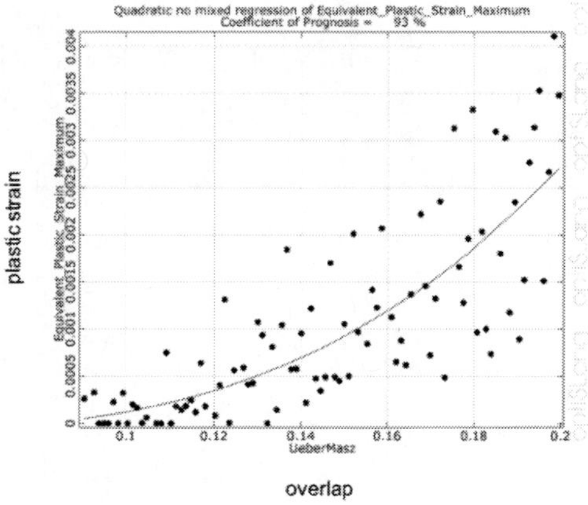

- 13 -

CADFEM

From Method to Application

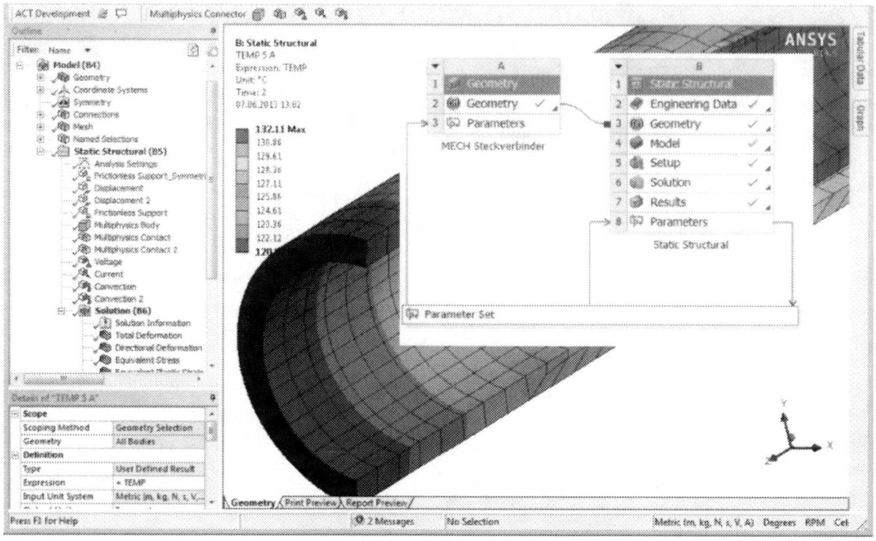

CADFEM

Workflow

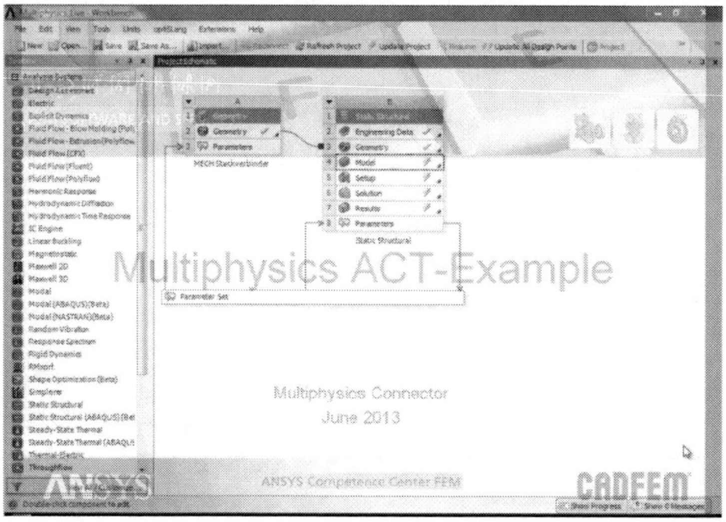

CADFEM

Power Connector Design

- Requirements
  - Low voltage drop
  - Low force
  - Max temperature < 110 °C
- Ranges
  - Length: 6-10 mm
  - Angle: 8-12 °
  - Thickness: 0.2-0.35 mm
  - Overlap: 0.09-0.2 mm
- Optimization based on sensitivity results
  - Mathematical representation of behavior
  - Verified fit for optimal prognosis

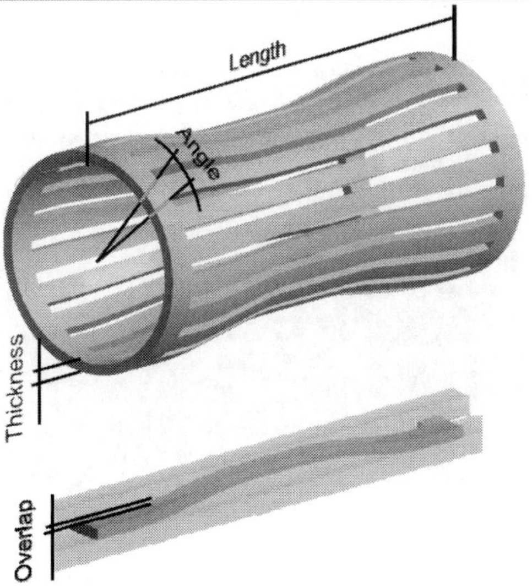

- 19 -

CADFEM

Power Connector Design

- Setup

Objectives

Name	Criterion	Expression	Value
Volt	MIN	VOLT_Maximum	1.69828e-05
Force	MIN	Force_max_Cur...	2.50881
new			

Constraints

Name	Left side expression	Criterion	side expre	Value
Temp	TEMP_5_Maximum	≤	115	113.375 ≤ 115
new				

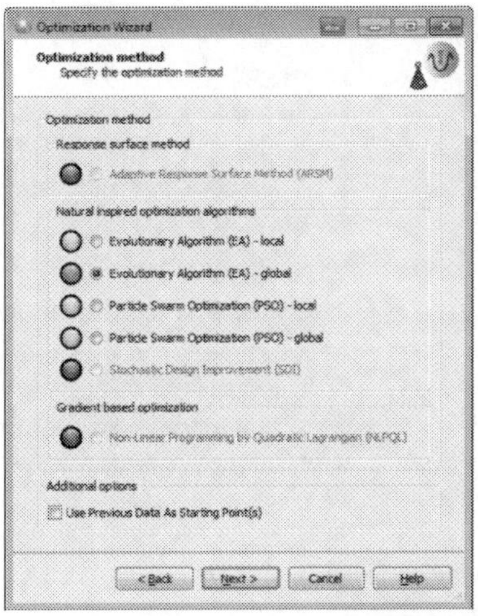

- 20 -

CADFEM

Power Connector Design

- 210 designs and 90 seconds later…

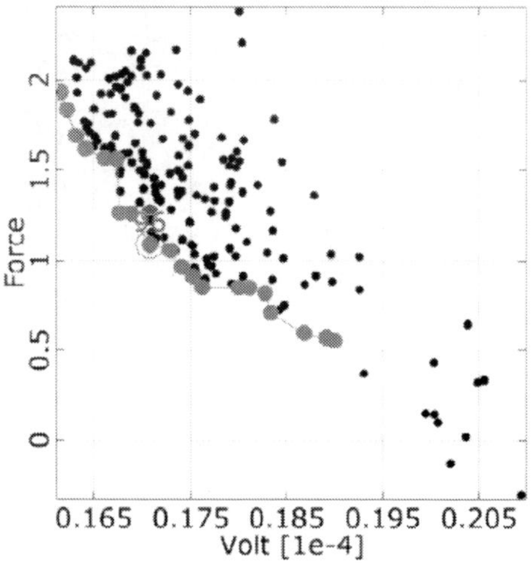

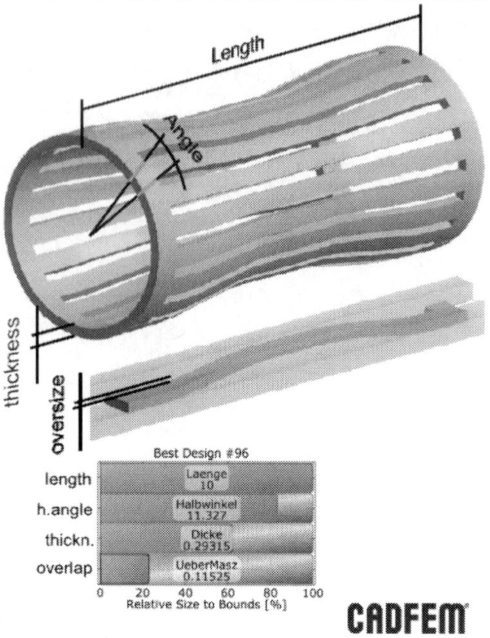

- 21 -

CADFEM

Summary of Multiphysics Power Connector Simulation

- Advanced numerical models
  - Dependency of temperature, stiffness resistivity, contact pressure, thermal and electric conduction
  - Nonlinearities in material, load and geometry
  - Scalable coupling on solver and element level
- Customization & automation
  - Adopted multiphysics simulation in Workbench look & feel
  - Individual functionality encapsulated in user friendly Workbench functions
- Improved product design
  - Parametric variation supports engineering
  - Sensitivity studies for
    - model verification
    - calibration of physical models
    - understanding of driving parameters
  - Goal driven optimization
  - Robustness analysis to consider scatter or unknown data

Many Thanks to:

Daniel Bachinski Pinhal
Christof Gebhardt
Jens Otto
Christian Römelsberger

- 22 -

CADFEM

Design of a FSPM Bridged Motor Rotor Using ANSYS Simulation Tools

Xuefeng Wu, Gary McDonald TriVista Engineering Limited
Adam Pride, IMRA Europe SAS.
September 2013

Abstract

The use of permanent magnet electrical machines has increased dramatically in recent years especially in the automotive industry because of the high efficiency of PM motors. Conventional PM motors have magnets on the rotor but to retain the magnets on the rotor at high rotational speed is difficult. A flux switching permanent magnet (FSPM) motor has the magnets placed in the stator and therefore has a very simple and robust rotor which is capable of running at high speed. The electrical design of FSPM rotor will have equal spaced poles which allow magnetic flux going through in a controlled manner.

The drawback of the rotor with poles is that significant windage loss will be generated at high rotational speed. Therefore, efforts have been made to reduce the windage loss at the motor rotor, which results in bridges being designed between the rotor poles. ANSYS simulations were extensively used in this design process. The windage loss of the rotor was evaluated using ANSYS CFD and its was compared with the windage loss of un-bridged rotor

Different rotor bridges were designed, and analysed using ANSYS Workbench. The final design was tested and proved to be a viable rotor bridge design, which is capable of running at high speed and with significantly reduced windage loss.

Background

The use of permanent magnet electrical machines has increased dramatically in recent years especially in the automotive industry because of the high efficiency of PM motors. Conventional PM motors have magnets on the rotor but the task of retaining the magnets on the rotor at high rotational speed can be difficult.

A flux switching permanent magnet (FSPM) motor has the magnets placed in the stator and therefore has a very simple and robust rotor which is capable of running at high speed. The electrical design of FSPM rotor will have equally spaced poles which allow magnetic flux going through in a controlled manner.

TriVista Engineering Limited
www.trivista.co.uk
TriVista

Windage loss evaluation of non-bridged rotor and bridged rotor

The drawback of the rotor with poles is significant windage loss will be generated at high rotational speed. In this design, the FSPM motor rotor is required to run up to 30,000 rpm. The poles on the rotor cause significant windage loss.

Images below show the velocity plot and windage loss in the air gap between the FSPM rotor and stator. At a rotational speed of 30,000 rpm, a windage loss of over 7kW was predicted for the FSPM rotor comparing with a windage loss of around 1kW for a rotor with cylindrical external rotor surface.

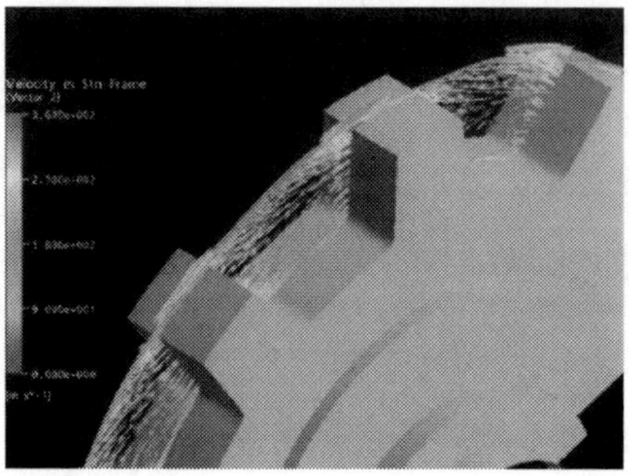

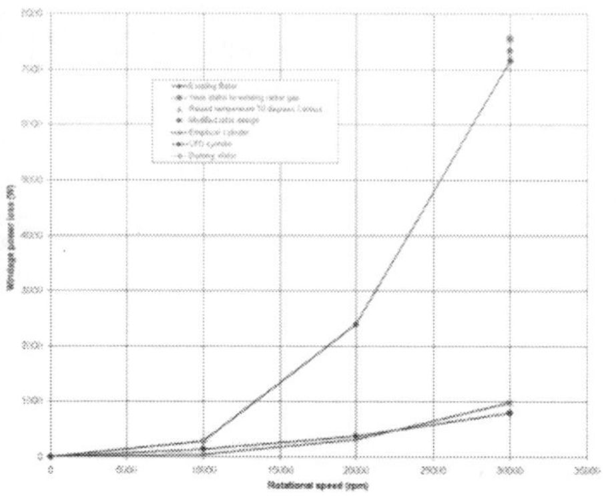

Figure 1: Velocity vector plot and windage loss at the air gap between the FSPM rotor and stator.

Geometry

Bridge pieces (rotor bridges) were designed and fitted between the rotor poles for the purpose of forming a smooth cylindrical surface at the outer rotor surface therefore reducing windage loss.

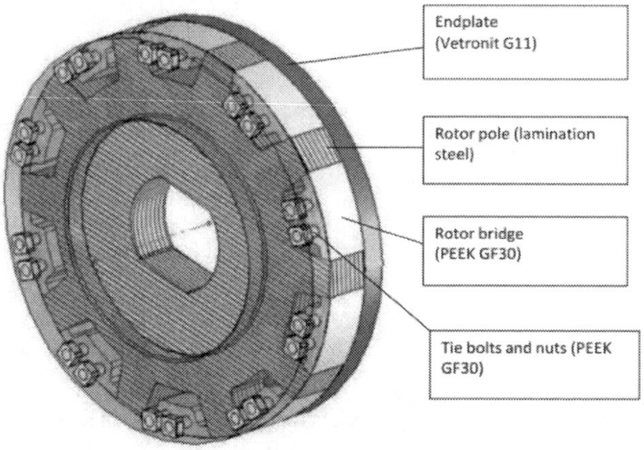

Figure 2: FSPM rotor with bridge pieces between the rotor poles.

Materials

Properties	Units	PEEK GF 30	Vetronit G11	Lamination steel	PEEK 450CA30
Young's Modulus	GPa	9.7	24	210	22.3
Density	kg/m^3	1490	1900	7650	1400
Maximum Service Temperature (long term)	°C	260	Not available		Not available
Maximum Service Temperature (short term)	°C	300	Not available		Not available
Melting point	°C	343	Not available		343
Yield strength	MPa			450	
Tensile strength	MPa	157	375	565	220
Thermal expansion coefficient	1/K	2.20E-05	1.50E-05		1.5E-05

Figure 3: Various materials were evaluated for the bridge piece. The material properties being evaluated are shown in this table.

PEEK GF30 was considered to be a preferred choice for rotor bridge and tie bolts for the purpose of avoiding eddy current loss as it is non-metal (non-conductive) . PEEK GF30 is a strong engineering plastic which is made of PEEK filled with 30% glass fibre. This material has lower density and relatively high tensile strength.

FE model

A FE model of a quarter geometry of rotor assembly was used because of the nature of symmetry in geometry and loading. The bridge pieces were held at these holes by tie bolts against the centrifugal force.

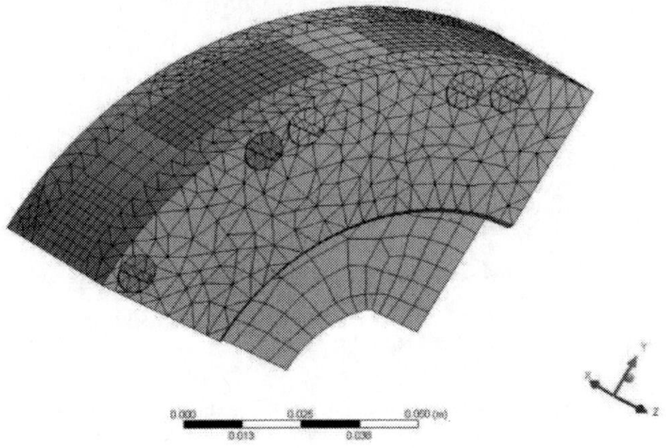

Figure 4: A FE model of a quarter rotor assembly geometry.

TriVista Engineering Limited
www.trivista.co.uk

Results

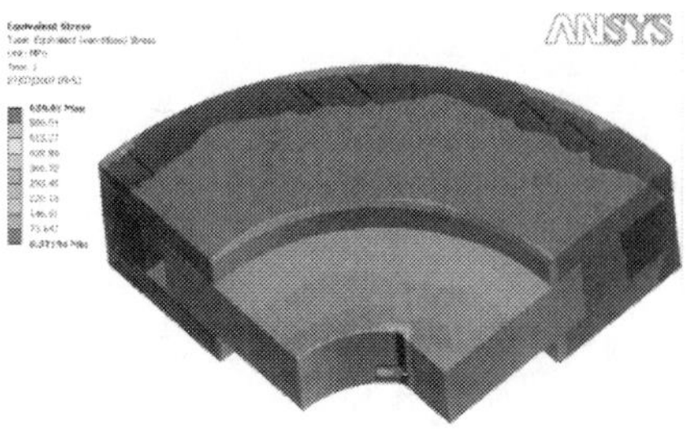

Figure 5: Equivalent stresses of the rotor sub assembly at 30,000 rpm.

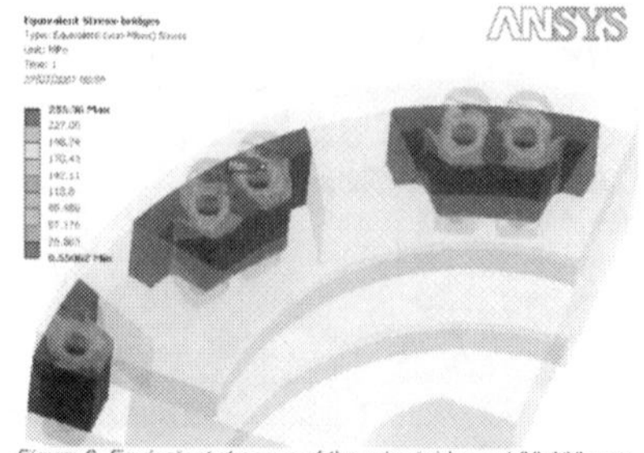

Figure 6: Equivalent stresses of the rotor bridges at 30,000 rpm.

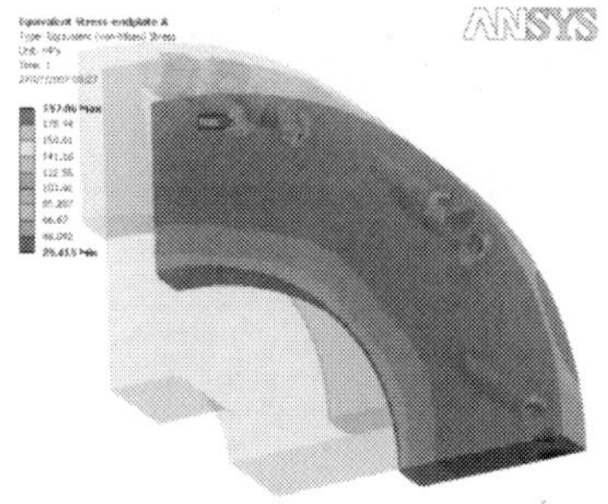

Figure 7: Equivalent stresses of the endplate at 30,000 rpm.

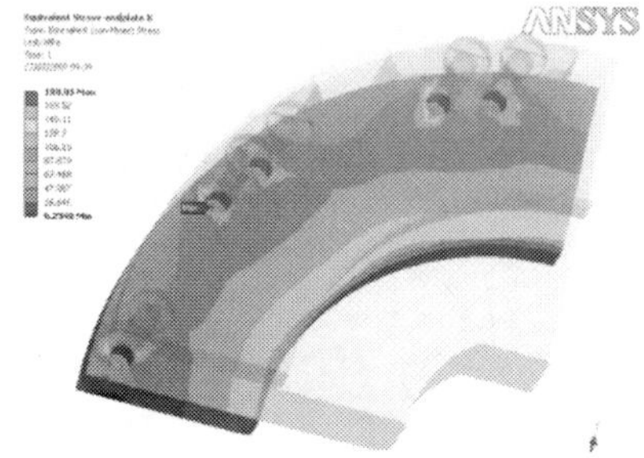

Figure 8: Equivalent stresses of the endplate B at 30,000 rpm.

Results

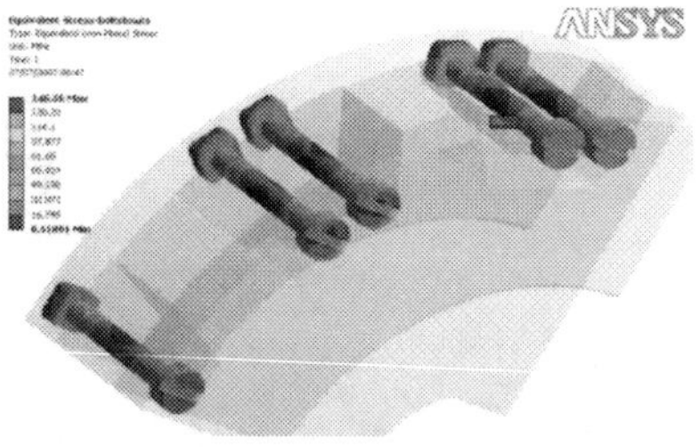

Figure 9: Equivalent stresses of the tie bolts and nuts at 30,000 rpm.

Observations

1. The maximum equivalent stress of 197 MPa at the endplates, which were made from Vetronit G11, was low comparing with the tensile strength 375 MPa of the material.

2. The maximum equivalent stress of 659.8 MPa at the corner of the central bore of the laminations, which were made in lamination steel, exceeded the tensile strength 565 MPa for the material.

3. The maximum equivalent stress of 255 MPa at the bridges, which were made of PEEK GF30, exceeded the tensile strength 157 MPa for the material.

4. The maximum equivalent stress of 146 MPa at the bolts, which were made of PEEK GF30, was under the material strength of 157 MPa for the material. However, IMRA reported the failure of the thread of the PEEK GF30 bolts. It has been established now that it is not recommended to cut threads on PEEK GF30 material because of its brittleness.

Observations Cont.

This design showed the following weaknesses:

(1) Tie bolts made of PEEK material do not have sufficient strength to hold back the rotor bridges.

(2) The masses from the rotor bridges and the tie bolts led to higher stresses at the central bore of the rotor lamination.

Effort was made to design lighter rotor bridges and locate them into the endplates therefore removing the need for tie bolts.

Various options for the bridge structures were investigated and the following guides were followed in the bridge designs:

- **Minimise the weight of the bridges to minimise the centrifugal loading.**
- **Use appropriate enforcements to avoid excessive deformation of the bridge.**
- **Avoid stress concentrations**

The bridge design shown in the images below are an optimised design selection.

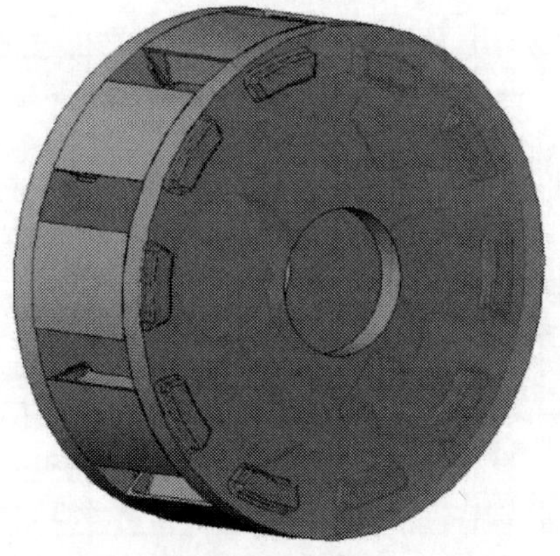

Figure 10: Rotor sub assembly where the rotor bridges were held by the end plates (rotor lamination hidden).

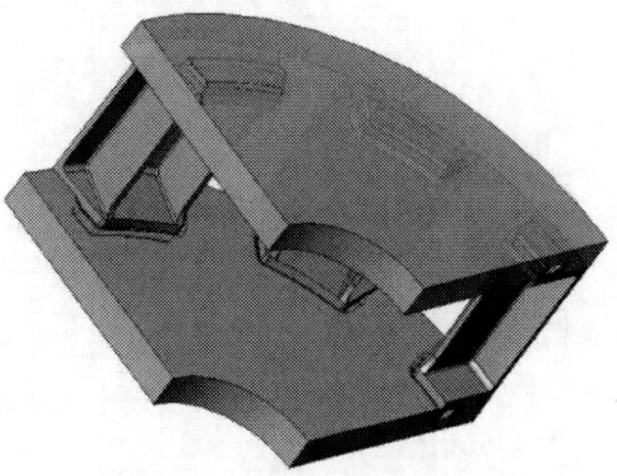

Figure 11: A quarter model of the rotor sub assembly (rotor lamination hidden), which was used in the FE stress analysis.

TriVista Engineering Limited
www.trivista.co.uk
TriVista

Figure 12: Contour plot of the equivalent stress of the PEEK GF30 bridges.

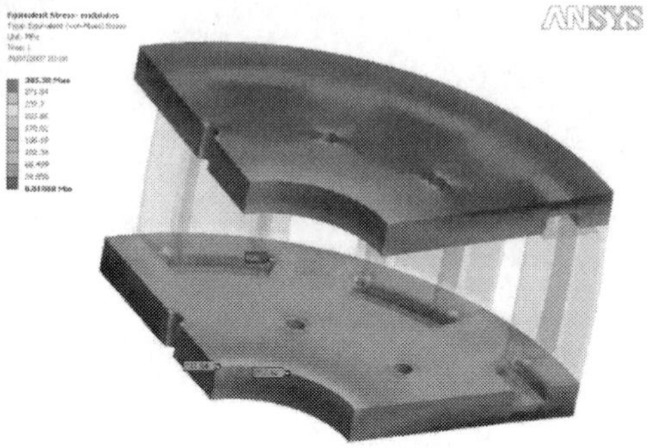

Figure 13: Contour plot of the equivalent stress of the endplates.

TriVista Engineering Limited
www.trivista.co.uk

TriVista

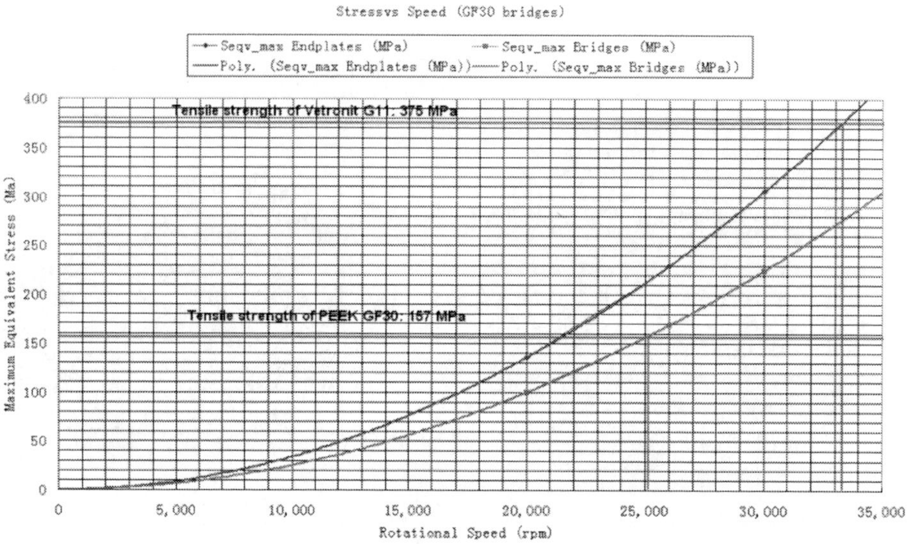

Figure 14: Graph of the maximum equivalent stress versus rotational speed (PEEK GF30 bridges).

Due to the limit of the material strength of PEEK GF30, the rotor will not be able to run beyond 25,000 rpm.

TriVista Engineering Limited
www.trivista.co.uk

TriVista

The design option of using PEEK 450CA30 for rotor bridge was investigated. PEEK 450CA30 is a PEEK material filled with 30% of carbon fibre, and is a stronger material.

Figure 15: The contour plot of the equivalent stress of the PEEK 450CA30 bridges.

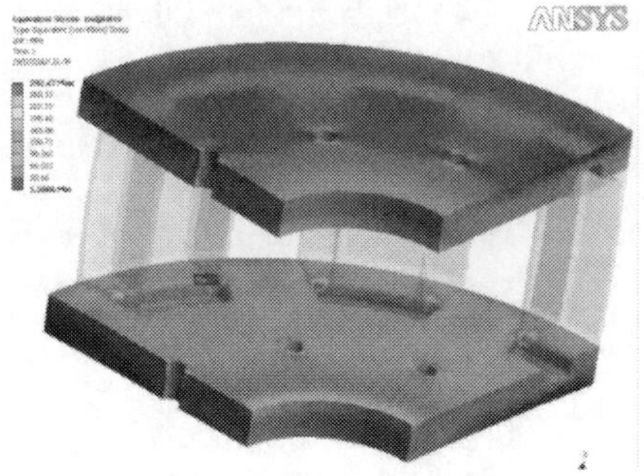

Figure 16: Graph of maximum equivalent stress versus rotational speed (PEEK GF30 bridges).

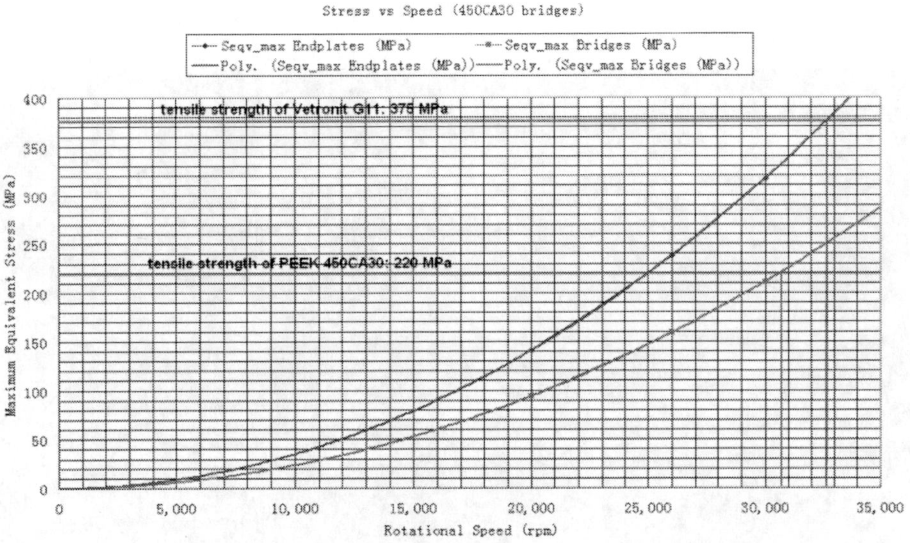

Figure 17: Graph of the maximum equivalent stress versus rotational speed (PEEK 450CA30 bridges).

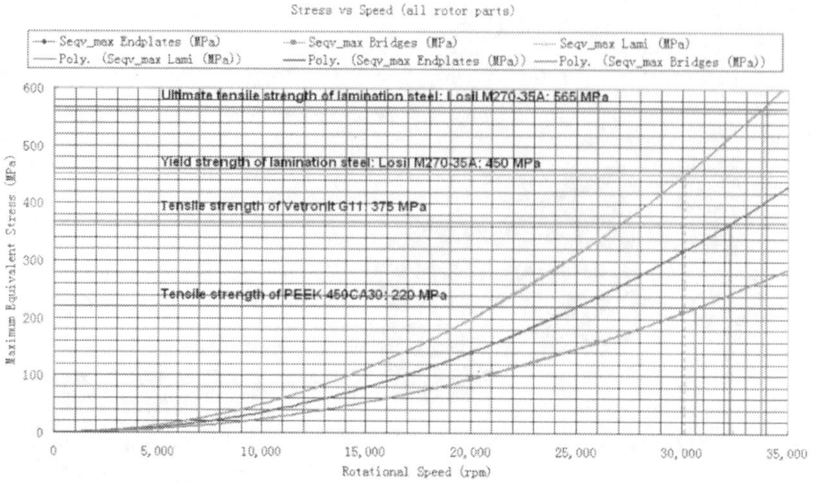

Figure 18: Graph of the stress versus speed including all the rotor components.

The rotor pole bridges were designed to be thin shell structures with appropriate support to avoid excessive deformation. Bolts and nuts were not used and therefore this avoided the bolt cracking (PEEK GF30) problems experienced in the early prototype and this also made the system simpler to manufacture in mass production and assembly. The glass fibre filled PEEK GF30 used in the early prototype did not have sufficient strength to achieve the required speed of 30,000 rpm. It was expected the bridges made of carbon fibre filled PEEK 450CA30 should achieve the 30,000 rpm speed. However, one should give a certain margin to allow factors such as inconsistency of material data and material quality, fatigue failure etc. Therefore, it was recommended to rate the rotor to run up to 25,000 rpm.

Bench Test Results.

The prototype motor with rotor bridges made of PEEK 450CA30 was made and tested. The rotor run up to 18,400 rpm and failed. The photos of the test rig below show the failure of the PEEK rotor bridge.

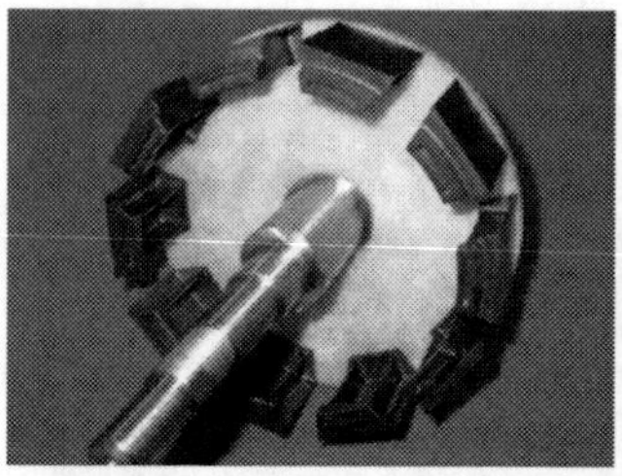

PEEK Bridge pieces trial assembly

Rotor as manufactured (Vetronit end-plate not yet fitted)

PEEK debris after shutdown

PEEK debris after removal from test bench

Rotor with bridge pieces completely destroyed (NB Red colour is copper smeared over laminations) The rotor seems to be relatively undamaged.

Stator-wires pulled out of slot and PEEK debris

The rotor bridges, which were made of PEEK 450CA30, was predicted to run up to near 30,000 rpm but only achieved less than a two thirds of the predicted speed. Three factors might contribute to this early failure:

1. The material strength data may be obtained under a simple loading condition. When the loadings were complicated (combined bending and shearing), the material strength may be substantially lower.

2. The PEEK 450CA30 is a brittle material. The prototype rotor bridge pieces were made via machining. The machining process left some machining marks which might also contribute to the early failure.

3. The UTS of PEEK is very near the proof stress so there is very little margin for differences between calculation and actual service conditions.

Figure 19: Bridge piece showing machining marks

Metal Rotor Bridge Design and Analysis

After the test failure of the non-metal rotor bridges, it was followed by the design and evaluation of the metal rotor bridge. The drawback of metal bridge is it is associated with some eddy current loss. Effort should be made to minimise this loss.

Table 1 lists the stress results of "T shape" rotor bridge. The rotor bridge was made of stainless steel. The effects of gap or interference at the bridge locking positions were investigated.

Table 2 lists the stress results of "bridge shape" rotor bridge. The effects of different materials were investigated.

Table 3 lists the stress results of "bridge shape" rotor bridge. The rotor bridge was made of stainless steel. The effects of different rotor pole heights were investigated.

Table 4 lists the stress results of "bridge shape" rotor bridge with different locking option and weight reduction designs. The rotor bridge was made of stainless steel. The effects of different rotor pole heights were investigated.

Table 1 - "T shape" rotor bridge

Iteration: Base

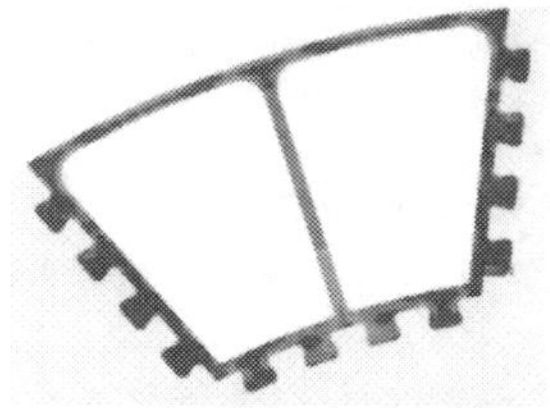

Max Equiv. Stress = 254.51 MPa
Max Deformation = 0.16 mm

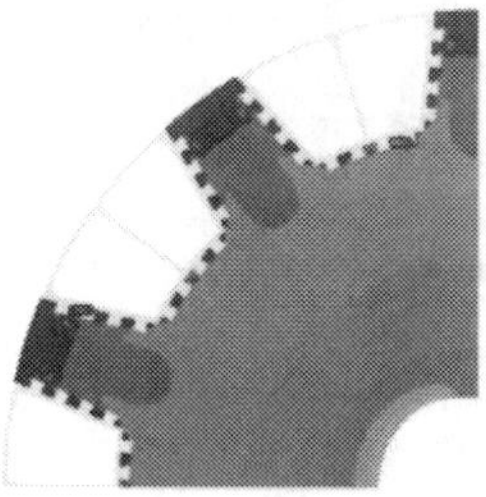

Max Equiv. Stress = 492.84 MPa
Bore Equiv. Stress = 462.64 MPa

Iteration: Thermal

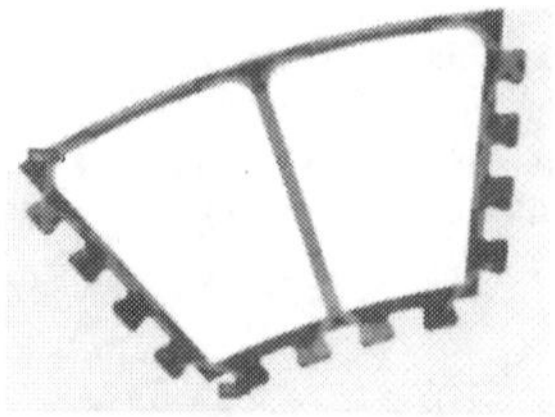

Max Equiv. Stress = 252.06 MPa
Max Deformation = 0.301 mm

Max Equiv. Stress = 489.81 MPa
Bore Equiv. Stress = 463.3 MPa

Table 1 - "T shape" rotor bridge

Iteration: 10micron gap

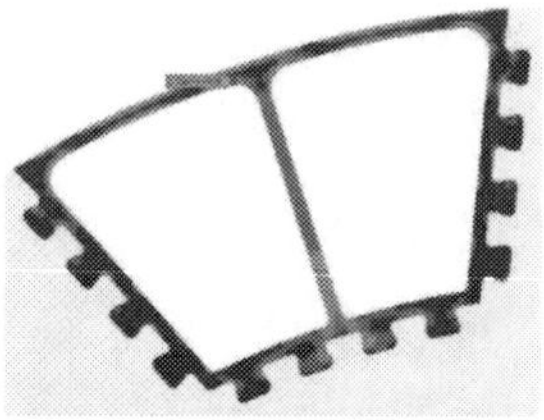

Max Equiv. Stress = 263.95 MPa
Max Deformation = 0.168 mm

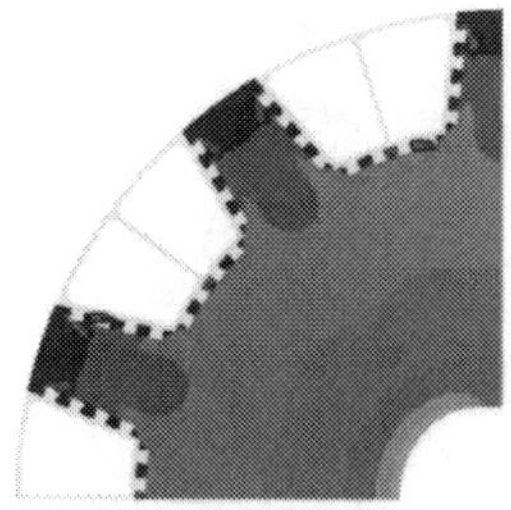

Max Equiv. Stress = 500.56 MPa
Bore Equiv. Stress = 462.9 MPa

Iteration: 20micron gap

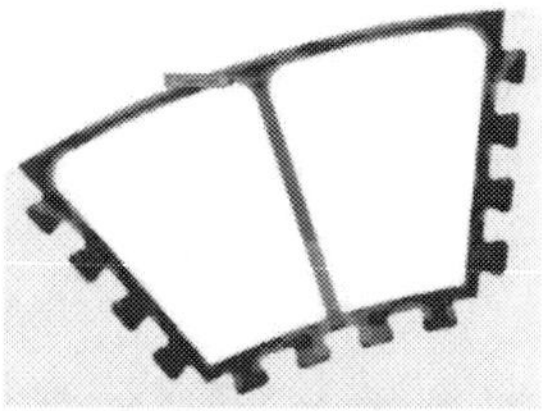

Max Equiv. Stress = 279.02 MPa
Max Deformation = 0.179 mm

Max Equiv. Stress = 505.8 MPa
Bore Equiv. Stress = 463.09 MPa

Iteration: 50micron gap

Max Equiv. Stress = 578.09 MPa
Max Deformation = 0.21 mm

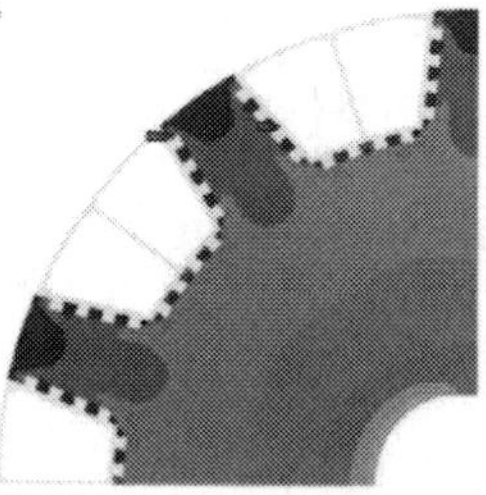

Max Equiv. Stress = 594.95 MPa
Bore Equiv. Stress = 463.64 MPa

TriVista Engineering Limited
www.trivista.co.uk

Table 1 - "T shape" rotor bridge

Iteration: 10micron interference

Max Equiv. Stress = 257.08 MPa
Max Deformation = 0.152 mm

Max Equiv. Stress = 500.56 MPa
Bore Equiv. Stress= 462.44 MPa

Iteration: 20micron interference

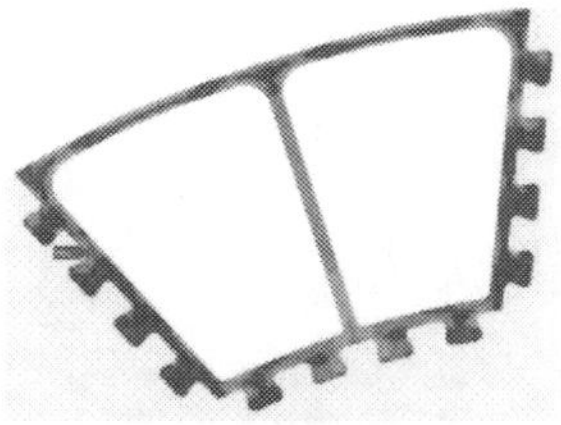

Max Equiv. Stress = 279.02 MPa
Max Deformation = 0.179 mm

Max Equiv. Stress = 505.8 MPa
Bore Equiv. Stress = 463.09 MPa

TriVista Engineering Limited
www.trivista.co.uk

Table 2 - "Bridge Shape" rotor bridge (different materials)

Iteration 1 Initial Geometry from interation16 of previous investigations. Bridge Material: Stainless Steel

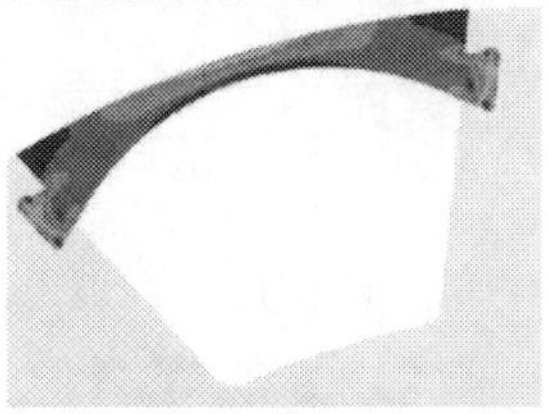

Max Equiv. Stress = 1797.8 MPa
Max Deformation = 0.283 mm

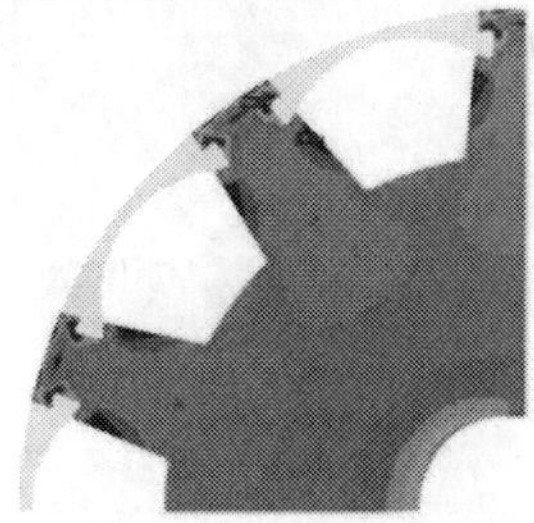

Max Equiv. Stress = 2675.9 MPa
Bore Equiv. Stress = MPa

Iteration 3A Changes from previous iteration: Rotor tip angle changed from 10.5° to 12.75°. Bridge material: Stainless Steel

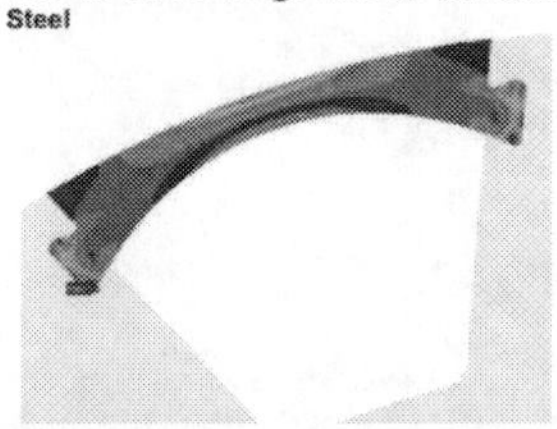

Max Equiv. Stress = 1887.3 MPa
Max Deformation = 0.276 mm

Max Equiv. Stress = 2483.7 MPa
Bore Equiv. Stress = 517.85 MPa

Iteration 3B Changes from previous iteration: Rotor tip angle changed to 15° Bridge Material: Stainless Steel

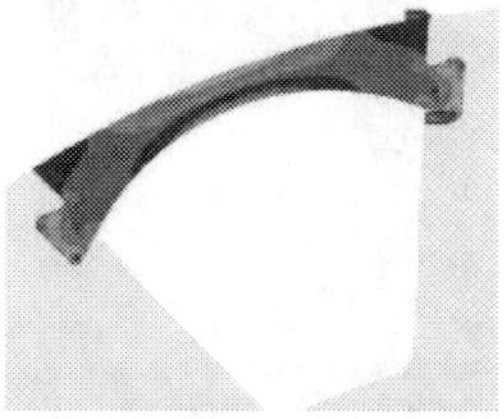

Max Equiv. Stress = 1577.8 MPa
Max Deformation = 0.229 mm

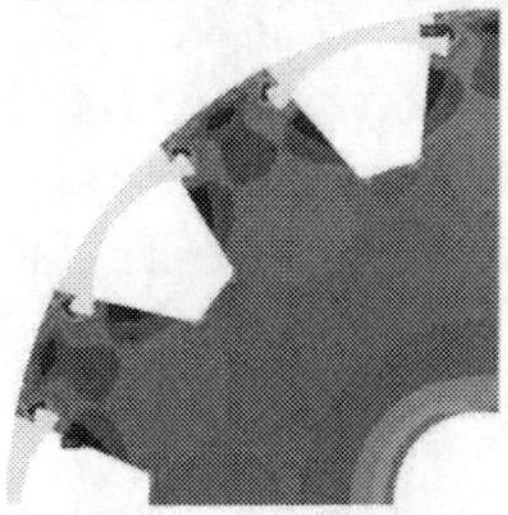

Max Equiv. Stress = 2094.6 MPa
Bore Equiv. Stress = 539.57 MPa

TriVista Engineering Limited
www.trivista.co.uk

Table 2 - "Bridge Shape" rotor bridge (different materials)

Iteration 3C Changes from previous iteration: Rotor tip angle changed to 16.5° Bridge Material: Stainless Steel

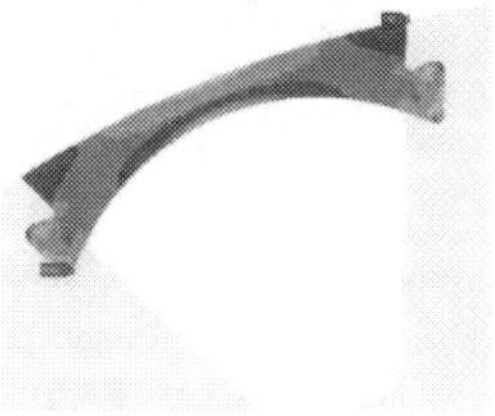

Max Equiv. Stress = 1406.4 MPa
Max Deformation = 0.202 mm

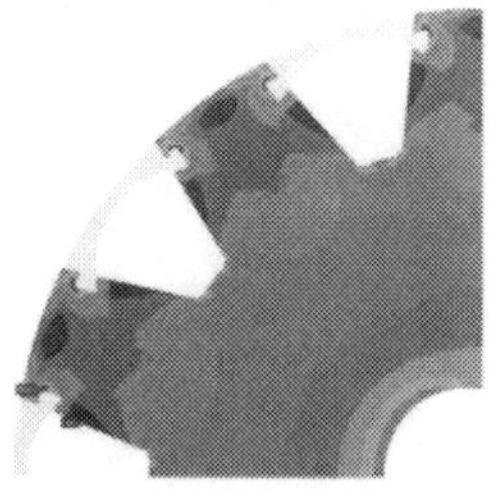

Max Equiv. Stress = 1883.8 MPa
Bore Equiv. Stress = 554.8 MPa

Iteration 3D Changes from previous iteration: Rotor tip angle changed to 18° Bridge Material: Stainless Steel

Max Equiv. Stress = 1188.8 MPa
Max Deformation = 0.177 mm

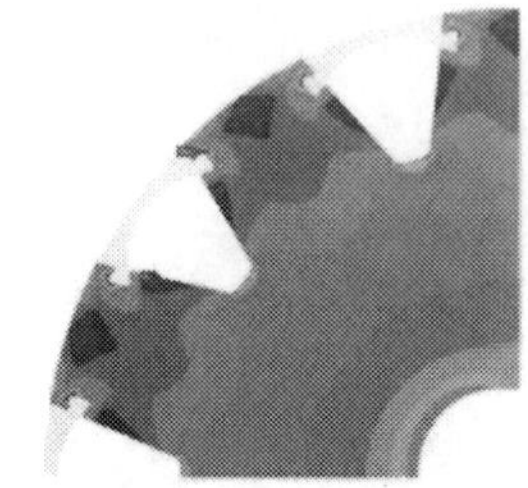

Max Equiv. Stress = 1668.7 MPa
Bore Equiv. Stress = 569.68 MPa

Iteration 2A Changes from previous iteration: Return to 10.5° and apply material change. Bridge Material: Titanium

Max Equiv. Stress = 864.86 MPa
Max Deformation = 0.236 mm

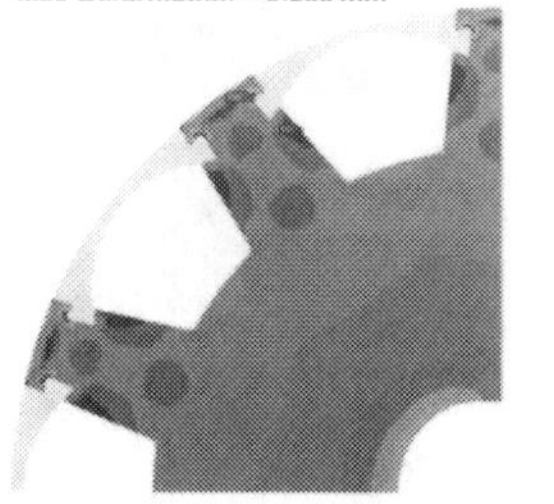

Max Equiv. Stress = 1653.9 MPa
Bore Equiv. Stress = 467.13 MPa

Table 2 - "Bridge Shape" rotor bridge (different materials)

Iteration 4B Changes from previous iteration: Rotor tip angle changed from 10.5° to 15°. Bridge Material: Titanium

Max Equiv. Stress = 754.48 MPa
Max Deformation = 0.192 mm

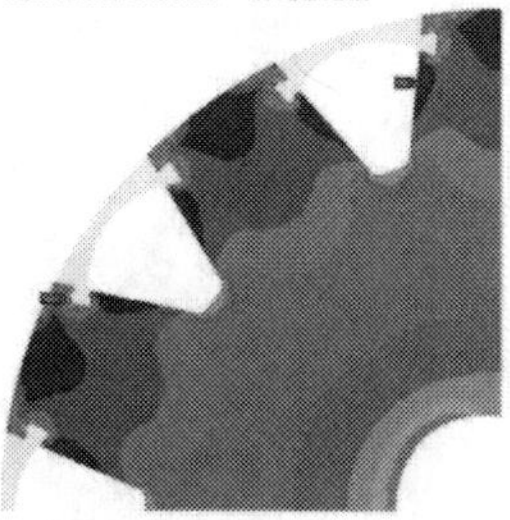

Max Equiv. Stress = 1336.5 MPa
Bore Equiv. Stress = 520.7 MPa

Iteration 4D Changes from previous iteration: Rotor tip angle changed from 15° to 18°. Bridge Material: Titanium

Max Equiv. Stress = 566.61 MPa
Max Deformation = 0.149 mm

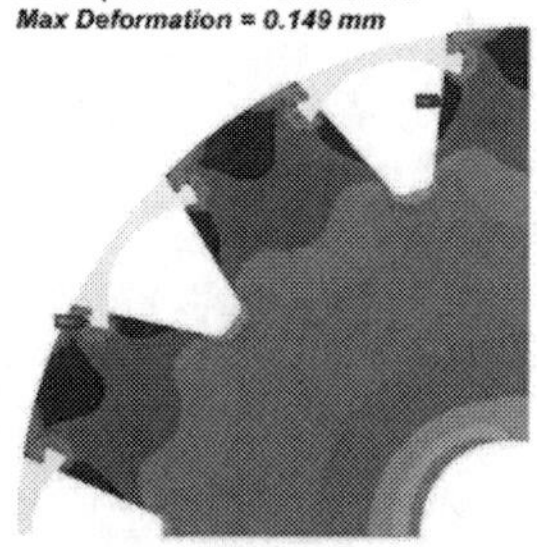

Max Equiv. Stress = 1078.2 MPa
Bore Equiv. Stress = 554.08 MPa

Iteration 2B Changes from previous iteration: Return to 10.5° and apply material change. Bridge Material: Aluminium

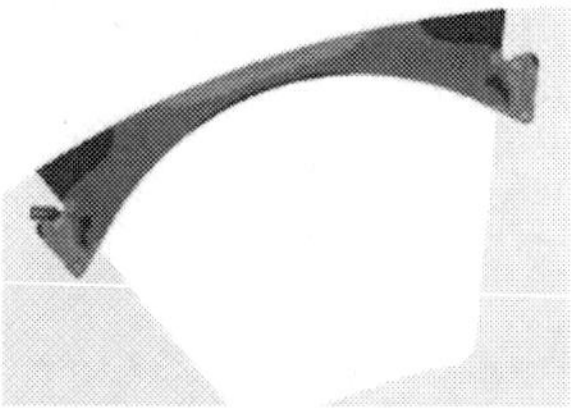

Max Equiv. Stress = 514.58 MPa
Max Deformation = 0.176 mm

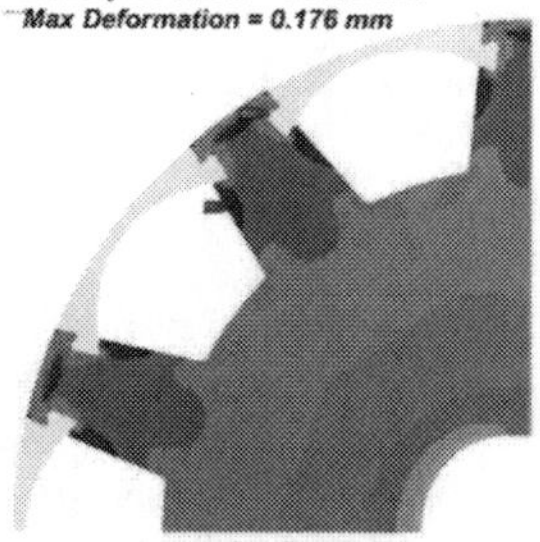

Max Equiv. Stress = 1051.1 MPa
Bore Equiv. Stress = 455.36 MPa

TriVista Engineering Limited
www.trivista.co.uk

Table 2 - "Bridge Shape" rotor bridge (different materials)

Iteration 5B Changes from previous iteration: Rotor tip angle changed from 10.5° to 15° Bridge Material: Aluminium

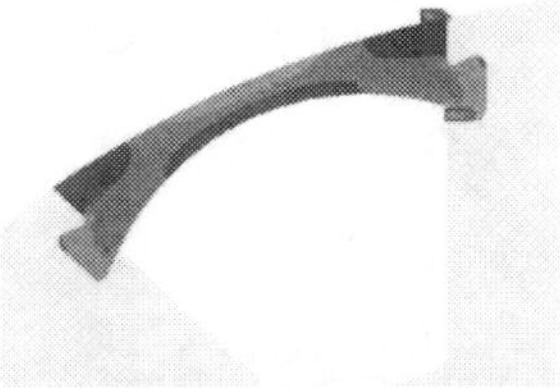

Max Equiv. Stress = 419.83 MPa
Max Deformation = 0.145 mm

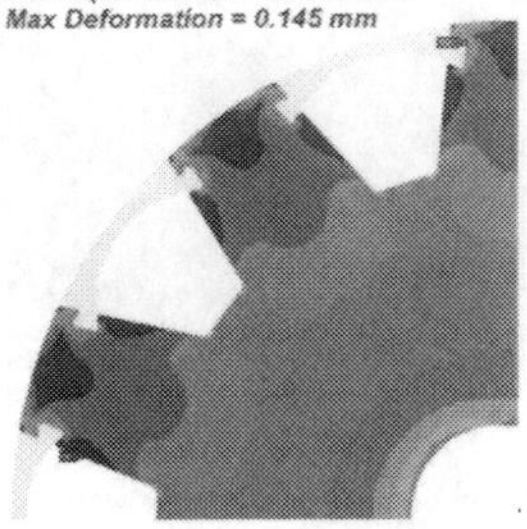

Max Equiv. Stress = 869.05 MPa
Bore Equiv. Stress = 510.8 MPa

Iteration 5D Changes from previous iteration: Rotor tip angle changed from 15° to 18°. Bridge Material: Aluminium

Max Equiv. Stress = 315.12 MPa
Max Deformation = 0.115 mm

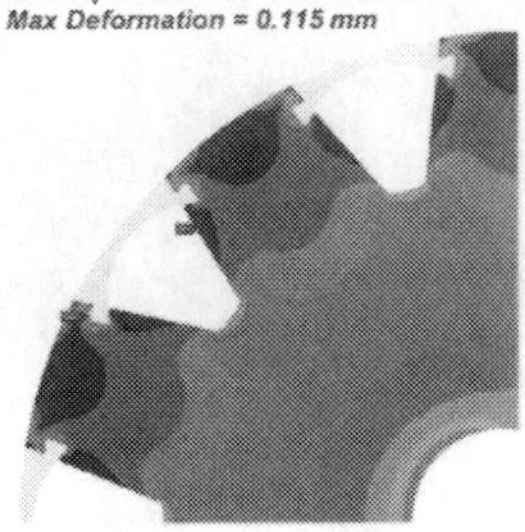

Max Equiv. Stress = 713.13 MPa
Bore Equiv. Stress = 545.89 MPa

TriVista Engineering Limited
www.trivista.co.uk

TriVista

Table 3 - "Bridge Shape" rotor bridge (Stainless Steel)

Iteration 2 Changes from previous iteration: Using iteration 1 from Investigation 1 as base, rotor pole height decreased from 26mm to 20mm. Bridge Material: Stainless Steel

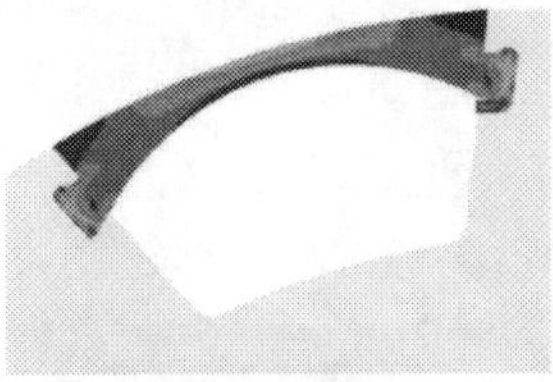

Max Equiv. Stress = 1788.4 MPa
Max Deformation = 0.284 mm

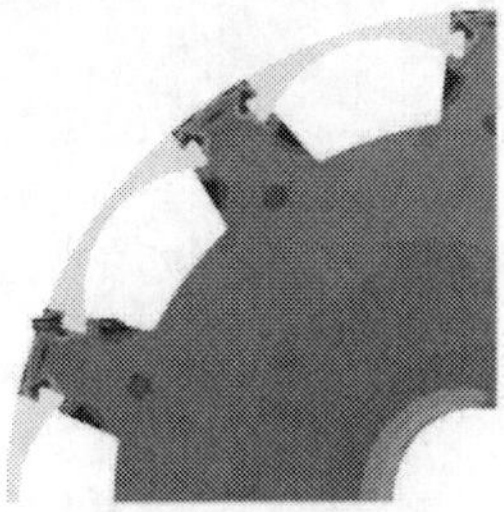

Max Equiv. Stress = 2600.4 MPa
Bore Equiv. Stress = 490.57 MPa

Iteration 3 Changes from previous iteration: Rotor pole height decreased from 20mm to 14mm. Bridge Material: Stainless Steel

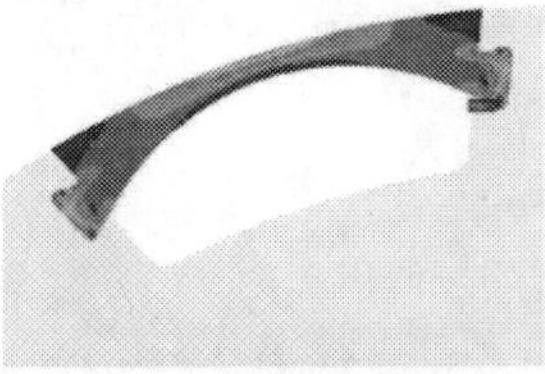

Max Equiv. Stress = 1817.6 MPa
Max Deformation = 0.281 mm

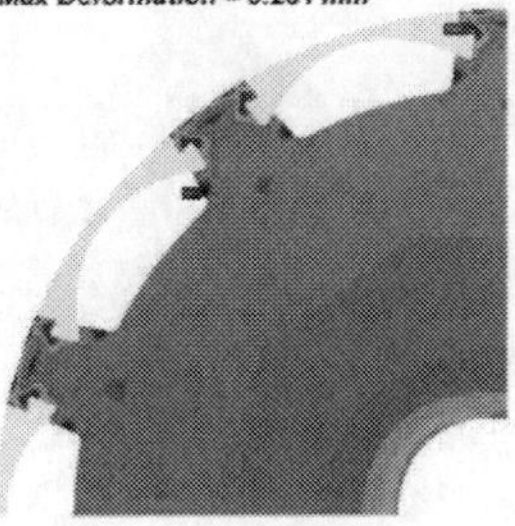

Max Equiv. Stress = 2740 MPa
Bore Equiv. Stress = 505.96 MPa

Iteration 4 Changes from previous iteration: Rotor pole height decreased from 14mm to 10mm. Bridge Material: Stainless Steel

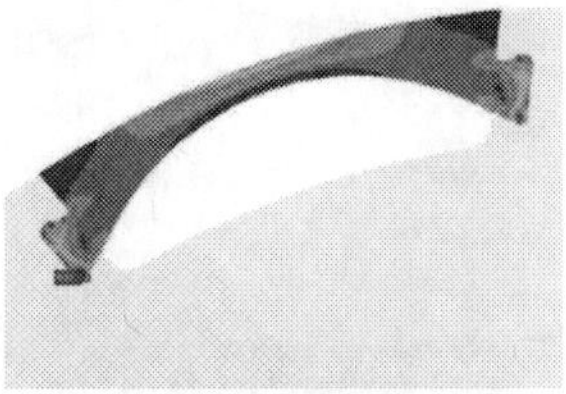

Max Equiv. Stress = 1842 MPa
Max Deformation = 0.285 mm

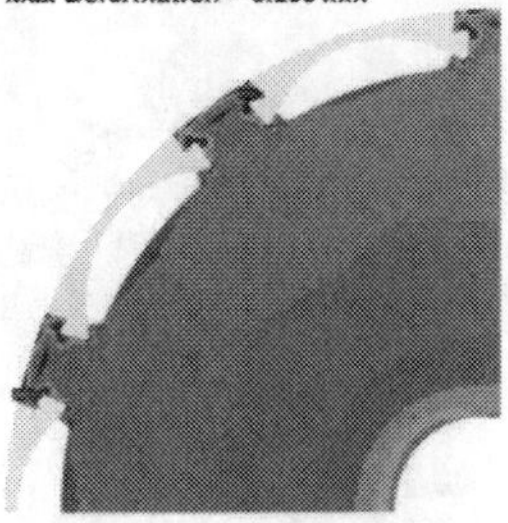

Max Equiv. Stress = 2515 MPa
Bore Equiv. Stress = 524.94 MPa

TriVista Engineering Limited
www.trivista.co.uk

TriVista

Table 4 - "Bridge Shape" rotor bridge (Locking Option & Weight Reduction)

Iteration 1 Changes from previous iteration: Bridge Material: Aluminium Alloy

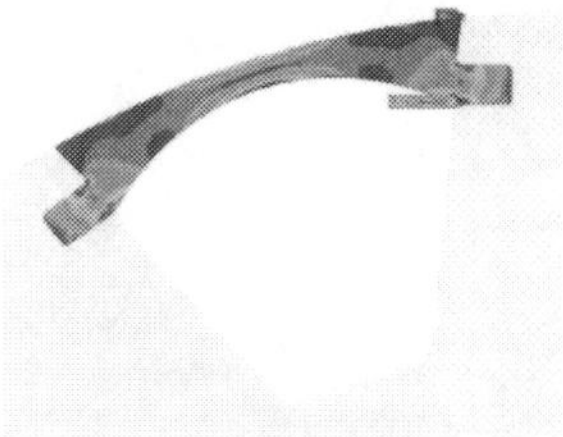

Max Equiv. Stress = 841.8 MPa
Max Deformation = 0.342 mm

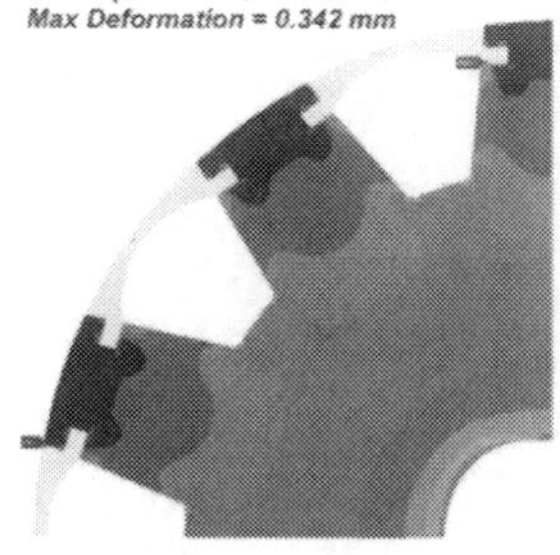

Max Equiv. Stress = 680.01 MPa
Bore Equiv. Stress = 516.6 MPa

Iteration 3 Changes from previous iteration: Depth of side tooth increased from 4.5mm to 5.5mm. Bridge Material: Aluminium Alloy

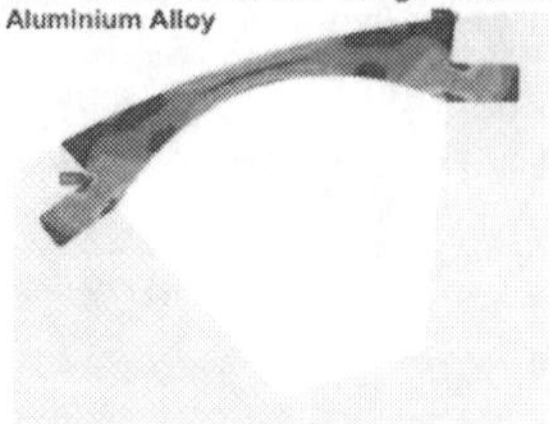

Max Equiv. Stress = 831.38 MPa
Max Deformation = 0.343 mm

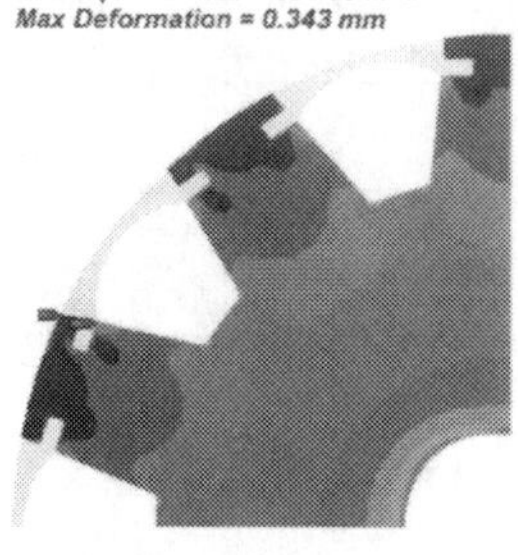

Max Equiv. Stress = 676.64 MPa
Bore Equiv. Stress = 515.23 MPa

Iteration 3A Changes from previous iteration: Change of contact type to more realistic constraints. No geometry change. Bridge Material: Aluminium Alloy

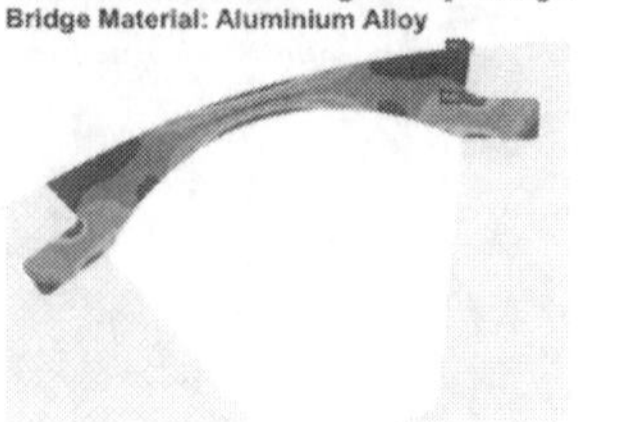

Max Equiv. Stress = 780.83 MPa
Max Deformation = 0.507 mm

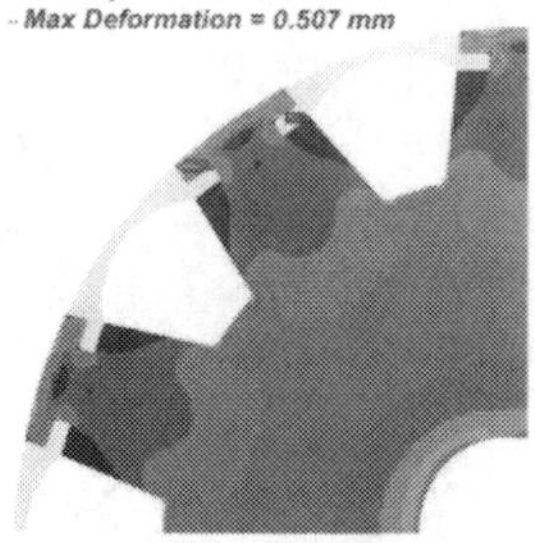

Max Equiv. Stress = 1155.9 MPa
Bore Equiv. Stress = 514.53 MPa

TriVista Engineering Limited
www.trivista.co.uk

Table 4 - "Bridge Shape" rotor bridge (Locking Option & Weight Reduction)

Iteration 5A Changes from previous iteration: New geometry based on investigations. Rotor bridge angle is 15°. Bridge Material: Aluminium Alloy

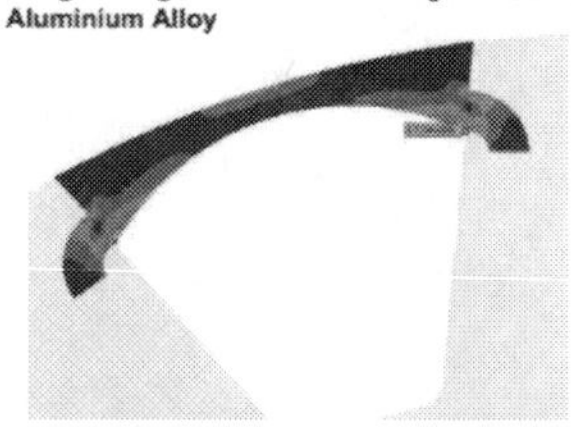

Max Equiv. Stress = 567.3 MPa
Max Deformation = 0.100 mm

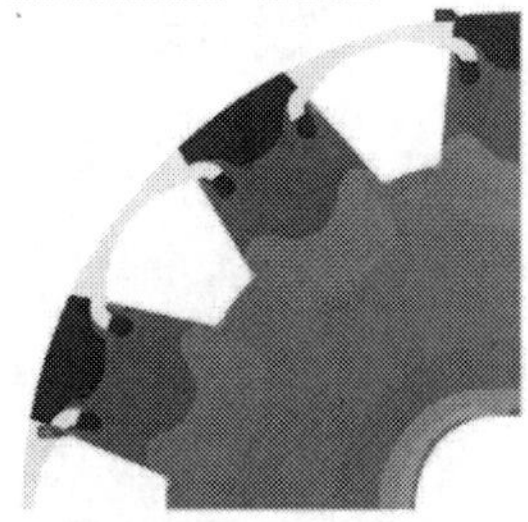

Max Equiv. Stress = 1259.4 MPa
Bore Equiv. Stress = 512.65 MPa

Iteration 5B Changes from previous iteration: Radius added onto lamination at bridge support position. Bridge base curve made smoother. Bridge Material: Aluminium Alloy

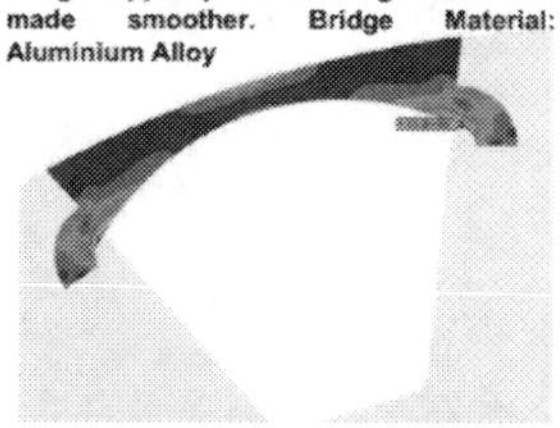

Max Equiv. Stress = 485.46 MPa
Max Deformation = 0.13 mm

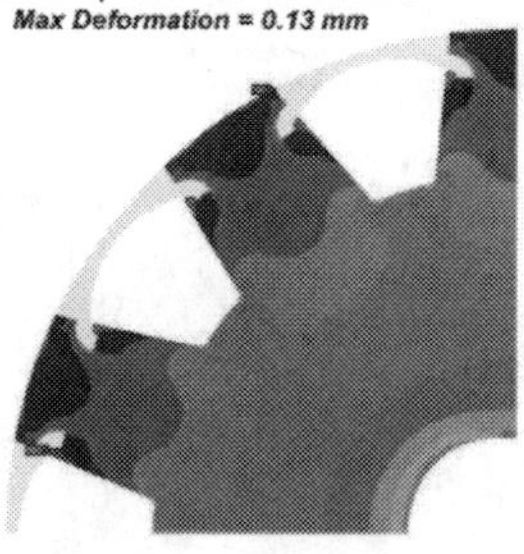

Max Equiv. Stress = 634.14 MPa
Bore Equiv. Stress = 511.89 MPa

Iteration 5C Changes from previous iteration: Holes added to bridge in attempt to reduce mass causing stress. Bridge Material: Aluminium Alloy

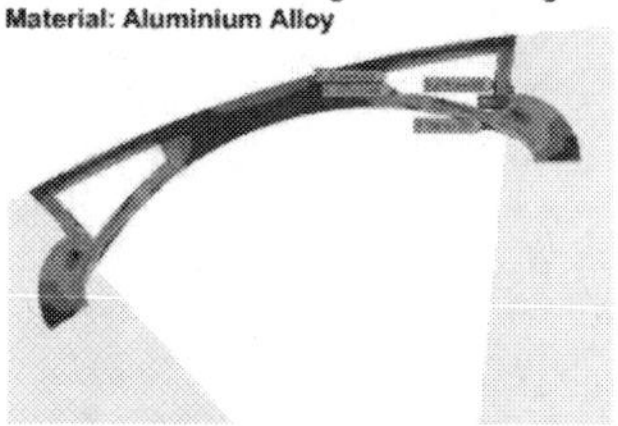

Max Equiv. Stress = 541.4 MPa
Max Deformation = 0.177 mm

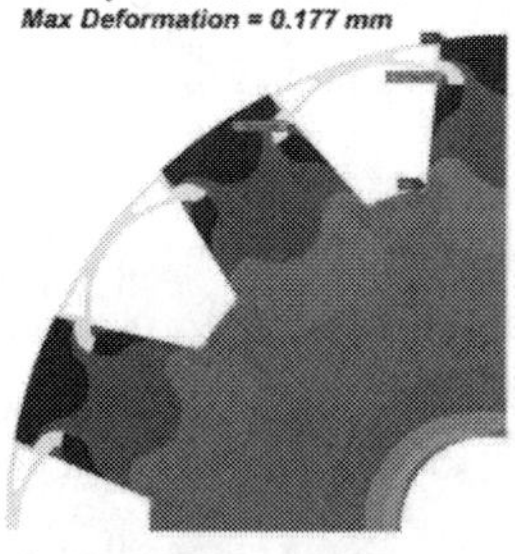

Max Equiv. Stress = 564.7 MPa
Bore Equiv. Stress = 507.64 MPa

TriVista Engineering Limited
www.trivista.co.uk

Table 4 - "Bridge Shape" rotor bridge (Locking Option & Weight Reduction)

Iteration 5D Changes from previous iteration: Larger fillet added to base of rotor pole at high stress location. Bridge Material: Aluminium Alloy

Max Equiv. Stress = 541.36 MPa
Max Deformation = 0.177 mm

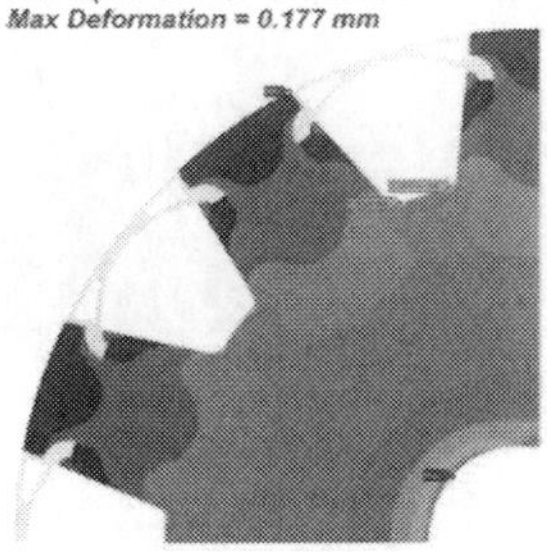

Max Equiv. Stress = 507.14 MPa
Bore Equiv. Stress = 507.14 MPa

Iteration 5D2 Changes from previous iteration: 2nd radius added to lamination at bridge support location. Bridge Material: Aluminium Alloy

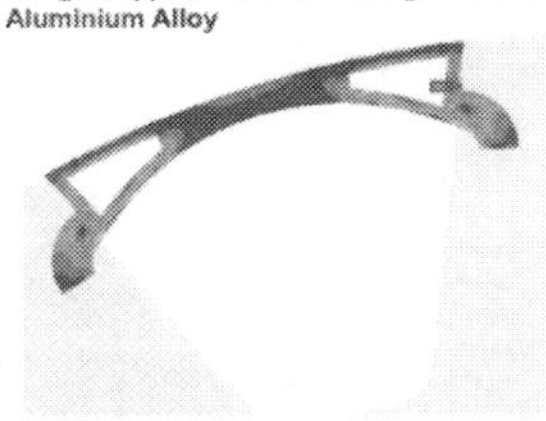

Max Equiv. Stress = 550.42 MPa
Max Deformation = 0.178 mm

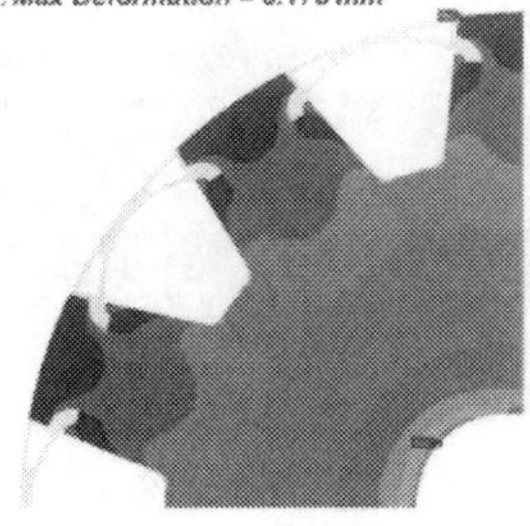

Max Equiv. Stress = 507.1 MPa
Bore Equiv. Stress = 507.1 MPa

Iteration 6A Changes from previous iteration: Curved cutouts added to bridge at high stress locations. Bridge Material: Aluminium Alloy

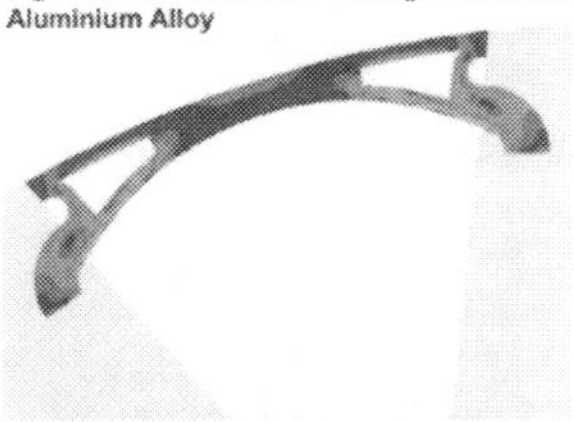

Max Equiv. Stress = 565.71 MPa
Max Deformation = 0.158 mm

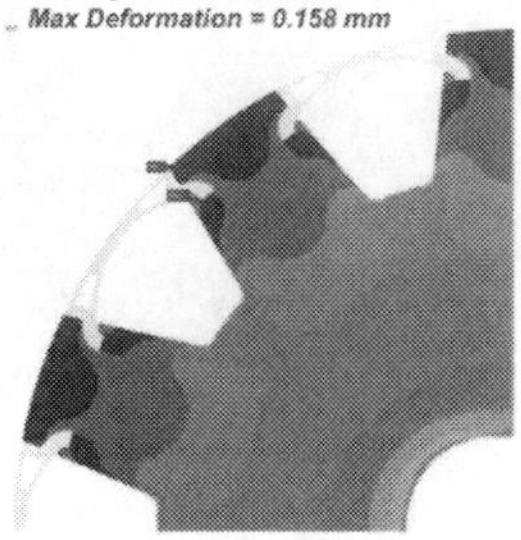

Max Equiv. Stress = 527.14 MPa
Bore Equiv. Stress = 507.61 MPa

TriVista Engineering Limited
www.trivista.co.uk

TriVista

Table 4 - "Bridge Shape" rotor bridge (Locking Option & Weight Reduction)

Iteration 6B Changes from previous iteration: No geometry change. Contact on bridge support edited to give realistic freedom of movement. Bridge Material: Aluminium Alloy

Max Equiv. Stress = 565.3 MPa
Max Deformation = 0.163 mm

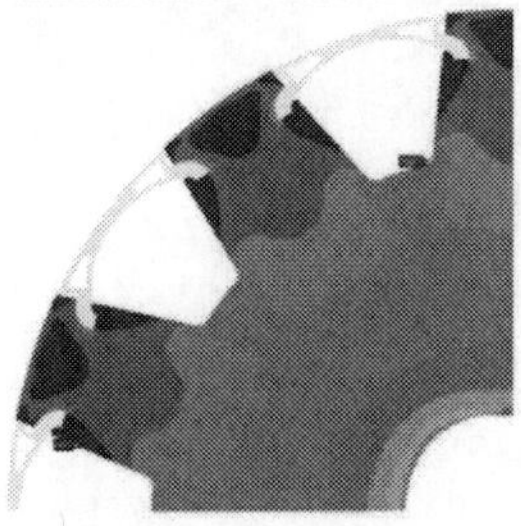

Max Equiv. Stress = 517.6 MPa
Bore Equiv. Stress = 507.36 MPa

Iteration 6C Changes from previous iteration: Increase lamination fillet at rotor pole base. Bridge Material: Aluminium Alloy

Max Equiv. Stress = 565.43 MPa
Max Deformation = 0.163 mm

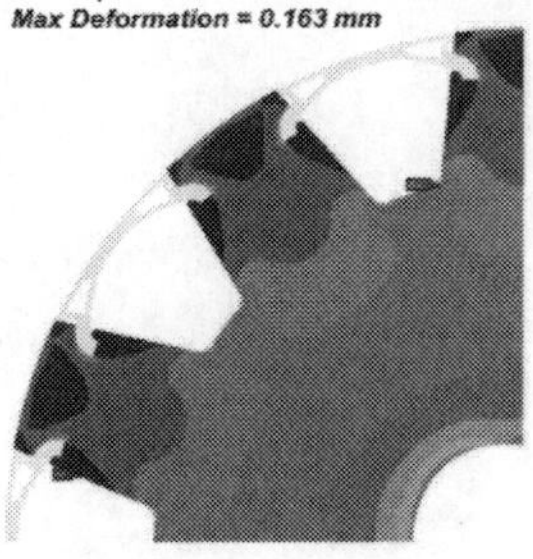

Max Equiv. Stress = 506.55 MPa
Bore Equiv. Stress = 506.55 MPa

Iteration 6D Changes from previous iteration: Add fillet to bridge hole at high stress location.
Bridge Material: Aluminium Alloy

Max Equiv. Stress = 535.64 MPa
Max Deformation = 0.161 mm

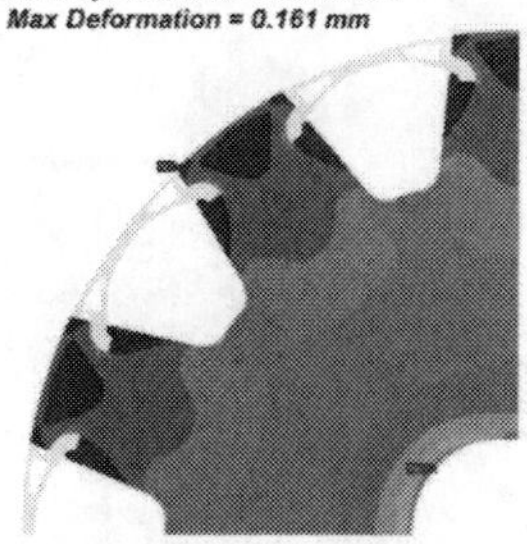

Max Equiv. Stress = 506.56 MPa
Bore Equiv. Stress = 506.56 MPa

TriVista Engineering Limited
www.trivista.co.uk

TriVista

Table 4 - "Bridge Shape" rotor bridge (Locking Option & Weight Reduction)

Iteration 6D5 Changes from previous iteration: Further fillets added to bridge holes. Bridge Material: Aluminium Alloy

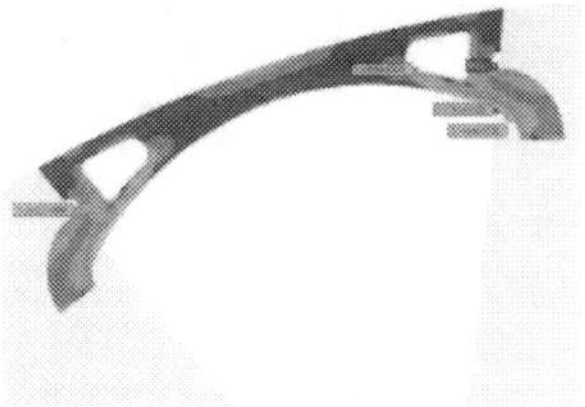

Max Equiv. Stress = 349.01 MPa
Max Deformation = 0.141 mm

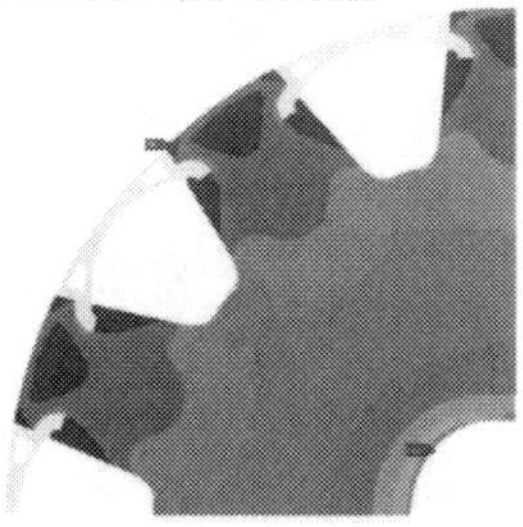

Max Equiv. Stress = 507.67 MPa
Bore Equiv. Stress = 507.67 MPa

Iteration 7A Changes from previous iteration: Increase rotor pole tip angle to 16.5° (based on investigation 1 results). Bridge Material: Aluminium Alloy

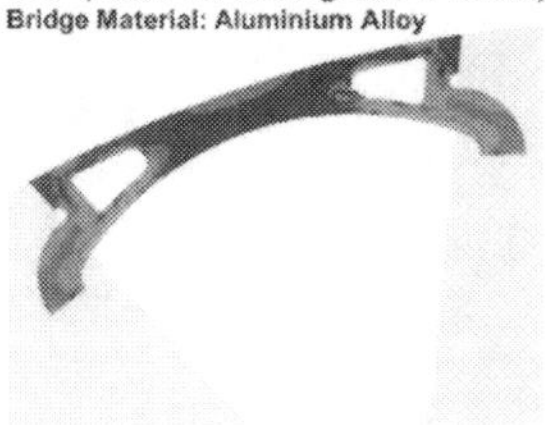

Max Equiv. Stress = 424.05 MPa
Max Deformation = 0.137 mm

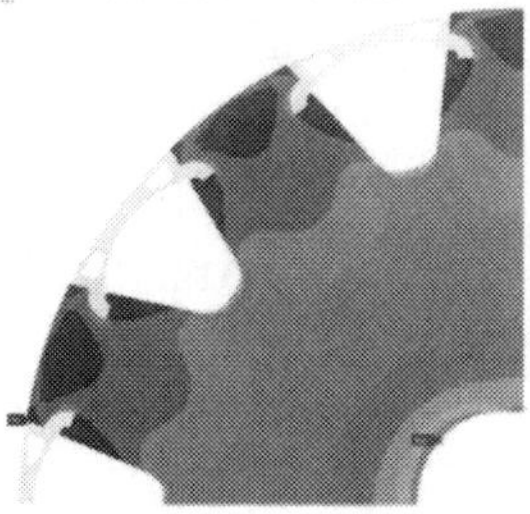

Max Equiv. Stress = 525.51 MPa
Bore Equiv. Stress = 525.51 MPa

Iteration 7B Changes from previous iteration: Decrease rotor pole tip angle to 12°. Bridge Material: Aluminium Alloy

Max Equiv. Stress = 379.85 MPa
Max Deformation = 0.169 mm

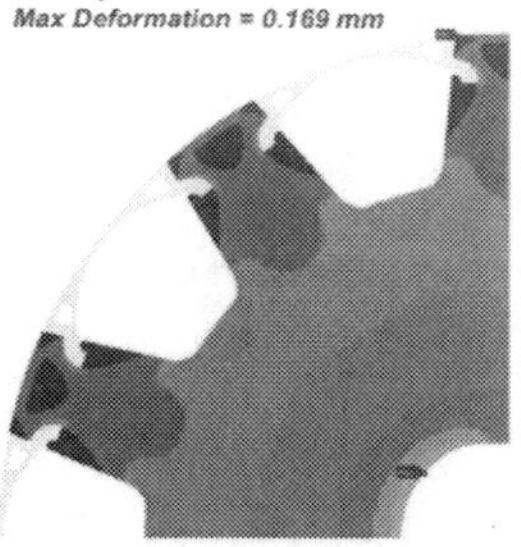

Max Equiv. Stress = 470.16 MPa
Bore Equiv. Stress = 470.16 MPa

Table 4 - "Bridge Shape" rotor bridge (Locking Option & Weight Reduction)

Iteration 7C Changes from previous iteration: Increase cut out at bridge support location. Bridge Material: Aluminium Alloy

Max Equiv. Stress = 333.09 MPa
Max Deformation = 0.175 mm

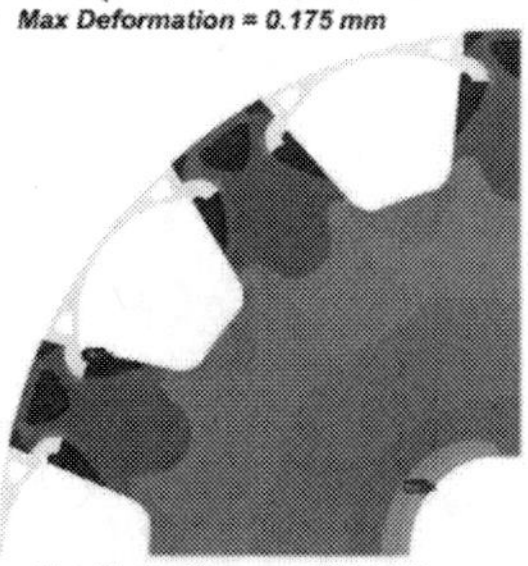

Max Equiv. Stress = 470.02 MPa
Bore Equiv. Stress = 470.02 MPa

Iteration 7C2 Changes from previous iteration: No geometry change. Testing change in material. Bridge Material: Duralunium

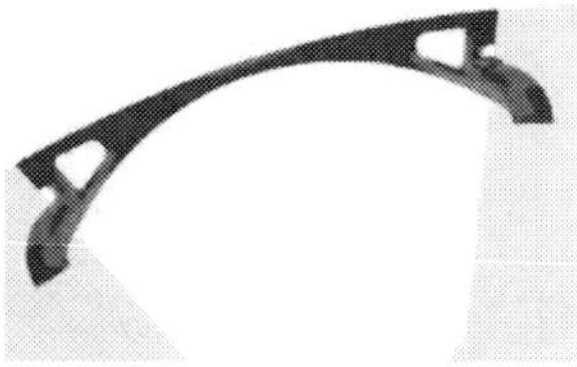

Max Equiv. Stress = 344.17 MPa
Max Deformation = 0.177 mm

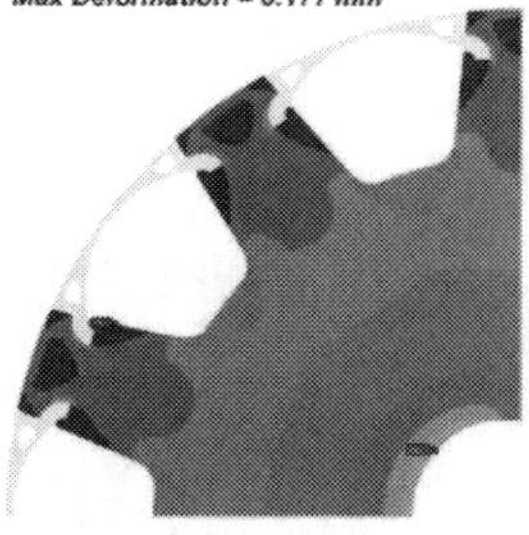

Max Equiv. Stress = 470.16 MPa
Bore Equiv. Stress = 470.16 MPa

TriVista Engineering Limited
www.trivista.co.uk

Final Design

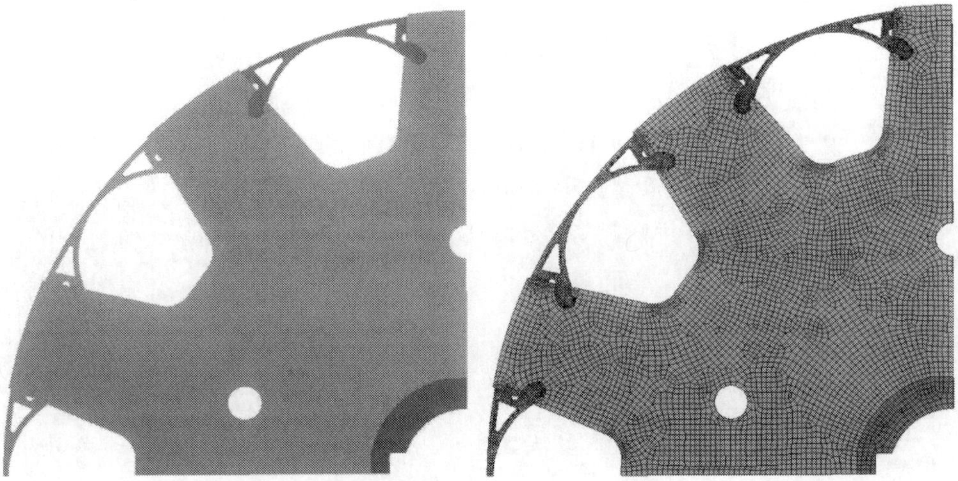

Figure 20: Shows the geometry and meshes of the motor rotor with light weight rotor bridges locked into the rotor poles by legs of a special shape.

Final Design

Material	Young's Modulus (GPa)	Poisson's Ratio	Density (kg/m3)	Thermal expansion coefficients (1/C)	Thermal conductivity (W/(m*C))	Yield Strength (MPa)	UTS (MPa)
Lamination Steel	200	0.27	7650	1.10E-005	18	450	565
7075 T6 (bridge)	71.7	0.33	2810	2.30E-005	200	503	572
EN8 (shaft sleeve)	210	0.3	7860	1.10E-005	60.5	420	610

Figure 21: Material properties

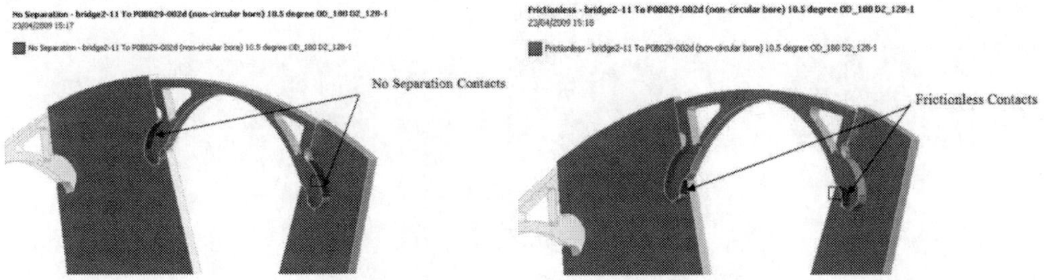

Figure 22: Contacts at the locking legs of the rotor bridge.

Results

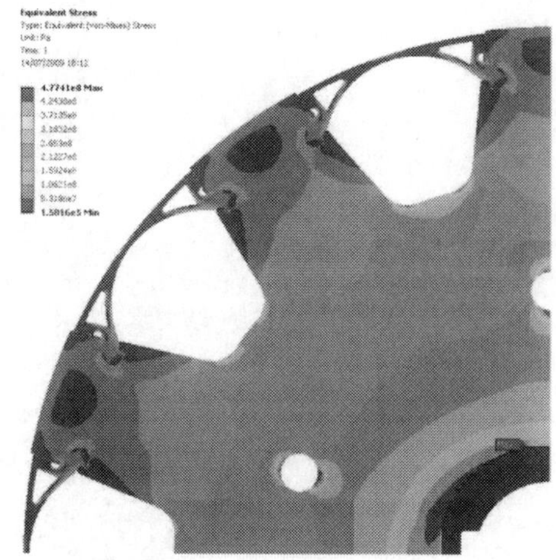

Figure 23: Image shows the contour plot of the equivalent stress of the rotor assembly at 30,000 rpm.

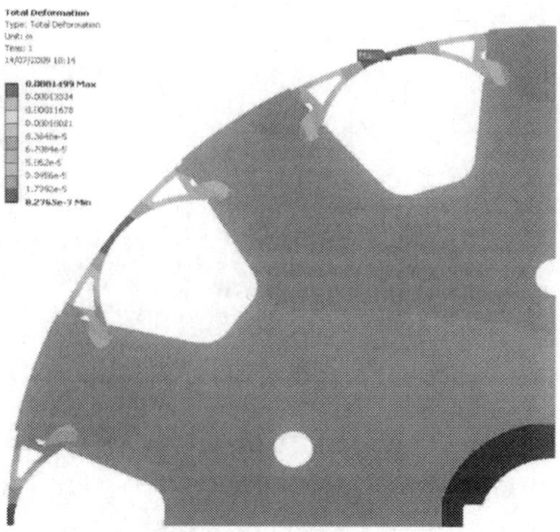

Figure 24: Image shows the contour plot of the total deformation of the rotor assembly at 30,000 rpm. Maximum deformation predicted to be 0.1mm in radial direction.

The stress in the rotor bridges does not exceed 212 MPa, which is less than the yield strength 503 MPa for aluminium alloy 7075 T6. The stress in lamination at the locking location does not exceed 318 MPa, which is significantly lower than the yield strength 450 MPa for lamination steel.

A prototype FSPM motor with this rotor bridge was built and test successfully up to 30,000 rpm.

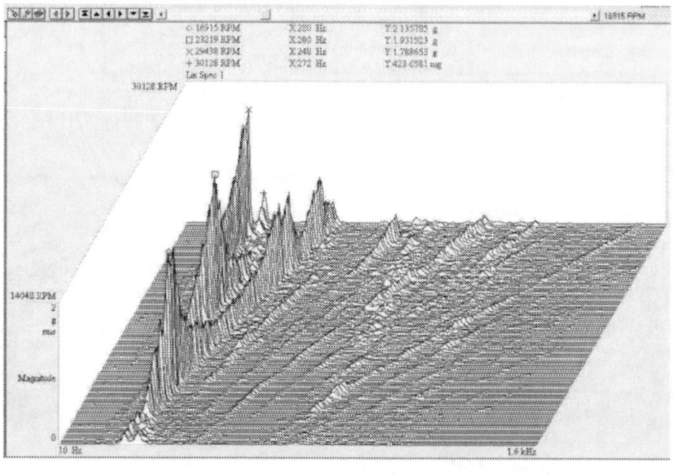

Figure 25: This image shows the recorded waterfall plot of measured motor vibration up to 30,128 rpm.

Manufactured Bridge and Rotor

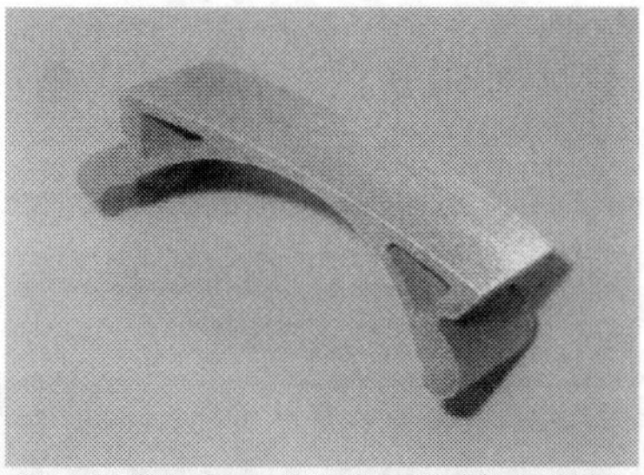

Figure 26: Bridge Pieces (All hard-anodized for insulation)

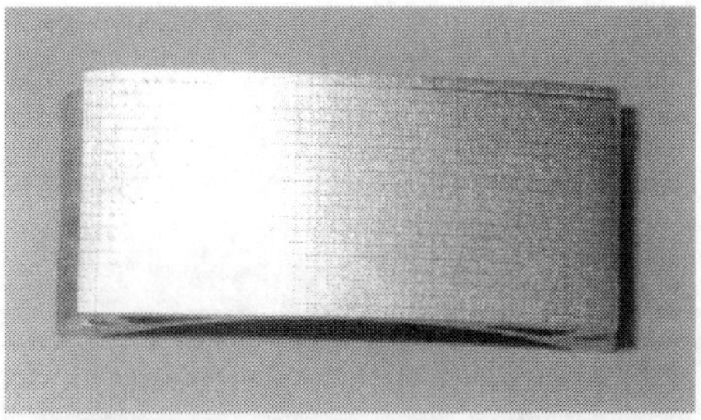

Figure 27: Stack of Laminated Bridge Pieces

Figure 28: Fully Assembled Rotor as tested to 30,000 rpm

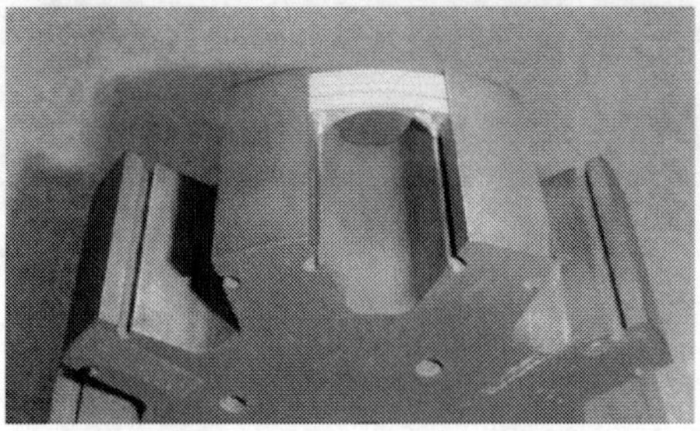

Figure 29: Bridge Pieces in Rotor

Comments and Conclusions

The FSPM motor rotor has a robust structure for high speed application but it is inherently associated with high windage loss. A rotor with rotor bridges provides the correct electromagnetic requirements for rotor poles and also a smooth cylindrical external surface to minimise windage loss. A bridge design for this high speed application was challenging. The bridge design was obtained after extensive investigations to be mechanically strong and light weight. The bridges were made in laminated pieces to minimise eddy current loss.

ANSYS

Realize Your Product Promise™

From Radar Traps to Navigation Systems - How Simulations Can Help Reduce Cost and Improve Efficiency

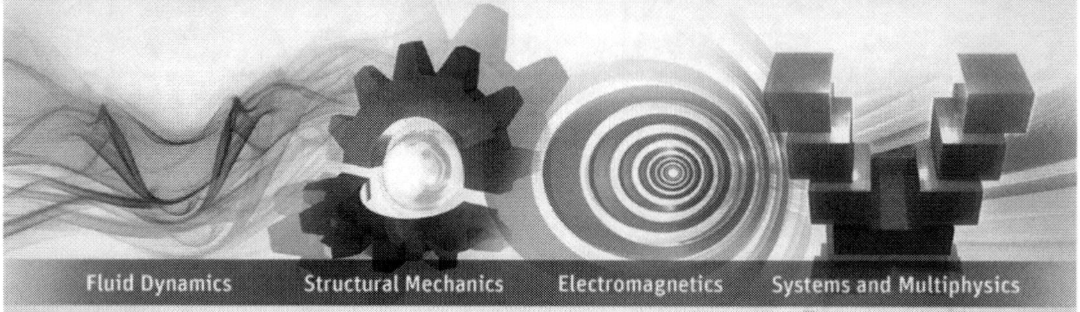

Simon Muff
ANSYS Germany GmbH

ANSYS

Overview

ANSYS

Trends in Automotive Industry

- Added value of electronics
- High frequency applications in automotive applications

Examples

simulation challenges: Ansys solutions

- FEBI
- Satellite systems
- Radar Traps

Summary

  November 11, 2013

ANSYS Two Trends in Automotive Industry

1) **Added Value of Electronic components in Automotive production (M. Wissmann , VDA Pres ; IAA 2011)[1] :**

1990 : 16 %

2011 : 30...40%

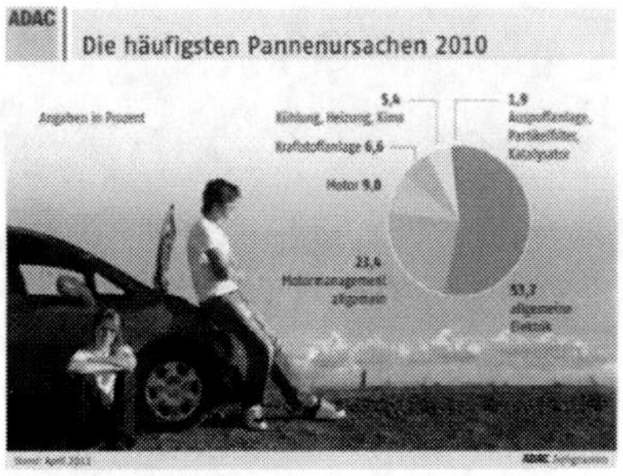

1) **Statistic of Automotive Breakdowns shows major contributions of electric and electronic parts in automotive [2]**

1) http://www.presseportal.de/pm/32847/2117580/das-mitdenkende-auto-ist-keine-zukunftsmusik-mehr
2) http://www.handelsblatt.com/downloads/4102524/1/ADAC Pannenstatistik 2010: Die häufigsten Ursachen

ANSYS High Frequency applications for Automotive

Car Entertainment
- FM /Radio
- SDARS
- DVBT

Positioning
- GPS / Navigation
- Stolen vehicle tracking
- Asset tracking / fleet management
- Car sharing systems

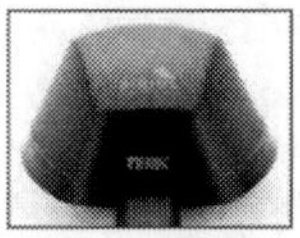

Mobile Communication
- GSM
- UMTS
- Bluetooth

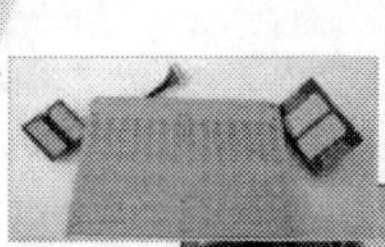

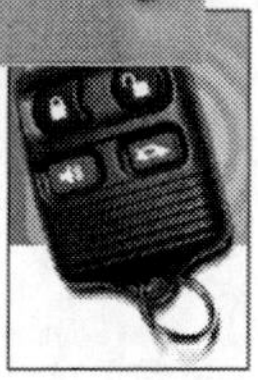

Telematics
- ETC (Electronic Toll collection)
- C2C (Car to Car Communication)
- C2I (Car to infrastructure / Traffic guidance systems)
- Ecall (obligatory f. 2015)

Car Safety
- TPM Tire pressure monitoring
- Keyless Entry
- Proximity / collision avoidance radar

GPS Reception in-Vehicle

ANSYS

5 © 2011 ANSYS, Inc. November 11, 2013

Animations perfomed by ESSS

GPS Reception in-Vehicle

ANSYS

A GPS telematics ECU receives the signal broadcasted by satellites at L1 frequency (1.575 GHz), calculates the actual position of the vehicle and send this information through a data network (usually an EDGE or GPRS network).

How can we evaluate the GPS reception to chose the proper placement of the ECU?

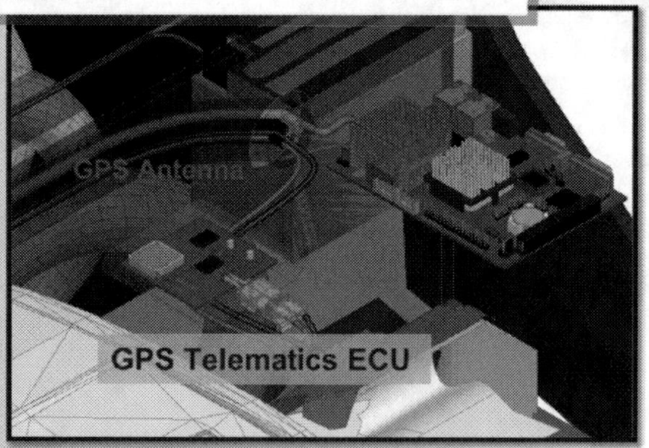

6 © 2011 ANSYS, Inc. November 11, 2013

Animations perfomed by ESSS

GPS Reception in-Vehicle

ANSYS

Video Presentation in Original e-media

Intuitively one can use a transient solver, having a incident plane wave coming from above, RHCP polarized at 1.575GHz, flowing normally towards the vehicle surface to simulate a GPS signal.

The electric field can be visualized anywhere in time, showing the reflections due to vehicle´s structures which will cause multipath and also the attenuation and phase shift.

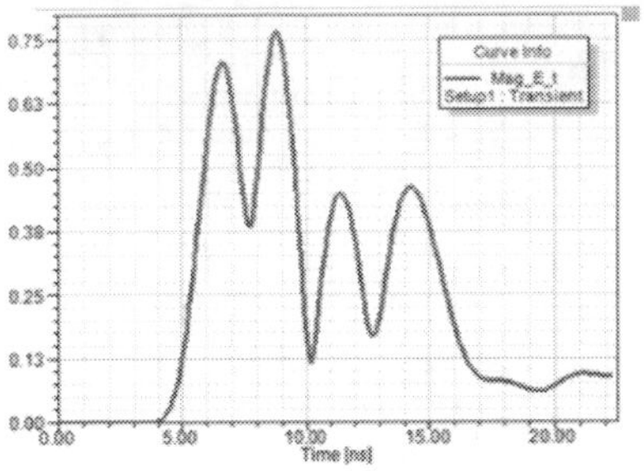

Voltage on GPS Antenna

Animations perfomed by ESSS

GPS Reception in-Vehicle

ANSYS

A GPS signal is received by numerous incident angles. Transient analysis for numerous waves becomes very time consuming. One alternative is to use Radiation Efficiency by having the GPS antenna of the ECU transmitting a L1 signal instead of receiving.

Radiation Efficiency is the ratio of the radiated power to the accepter power. Radiating Power is the amount of time-averaged power (in watts) exiting a radiating antenna structure through a radiation boundary (the lateral walls of the airbox).

Radiation Efficiency

The radiation efficiency is the ratio of the radiated power to the accepted power given by

$$e = \frac{P_{rad}}{P_{acc}}$$

where

- P_{rad} is the radiated power in watts.
- P_{acc} is the accepted power in watts.

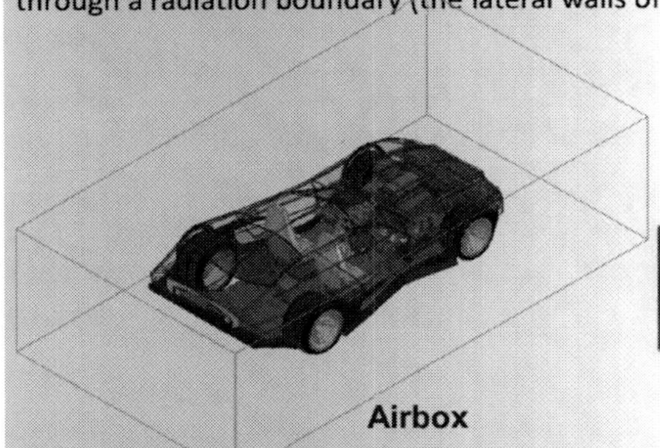

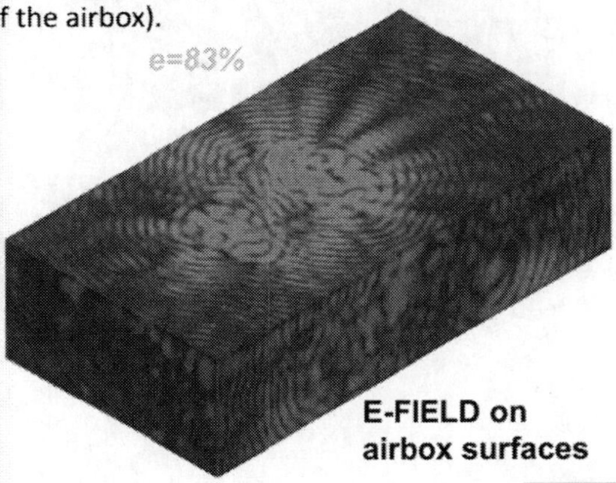

Animations perfomed by ESSS

ANSYS

GPS Reception in-Vehicle

Radiation Efficiency can give us fast results in frequency domain indicating the best candidates for GPS antenna placement inside the vehicle. The example below shows a comparison between the ECU installed in the current position (position 1) and a new position (position 2) under the dashboard near the throttle.

Near field plots are shown (3D polar plot and radiation pattern) as well as the radiation efficiency.

Video Presentation in Original e-media

Animations perfomed by ESSS

ANSYS

ANSYS

Automotive EMC Standards – ISO 11451-2

Picture taken at INPE. Courtesy of Volvo Brasil.

Animations perfomed by ESSS

Automotive EMC Standards – ISO 11451-2

ANSYS

Work Developed by FIAT Brazil (FIASA)

ISO 11451-2

The international standard ISO 11451-2 is applied to road vehicles and describes a vehicle test methods for electrical disturbances from narrowband radiated electromagnetic energy. It determines the immunity of passenger cars and commercial vehicles to electrical disturbances from off-vehicle radiation sources, regardless of the vehicle propulsion system. It can also be readily applied to other types of vehicles including hybrid electric vehicles (HEV). The test should be performed in an absorber-lined shielded enclosure, trying to create an indoor electromagnetic compatibility testing facility that simulates open field testing.

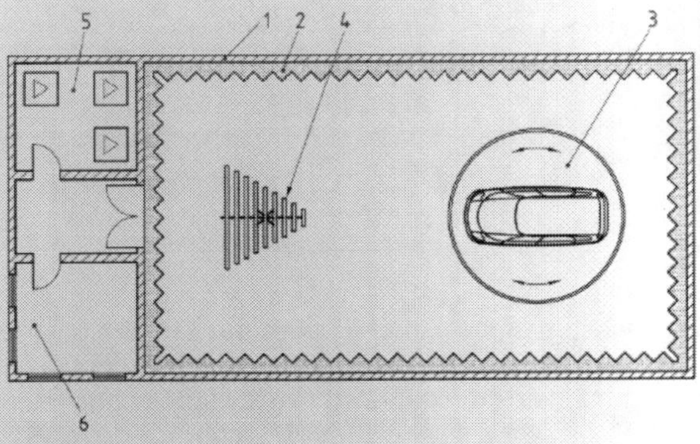

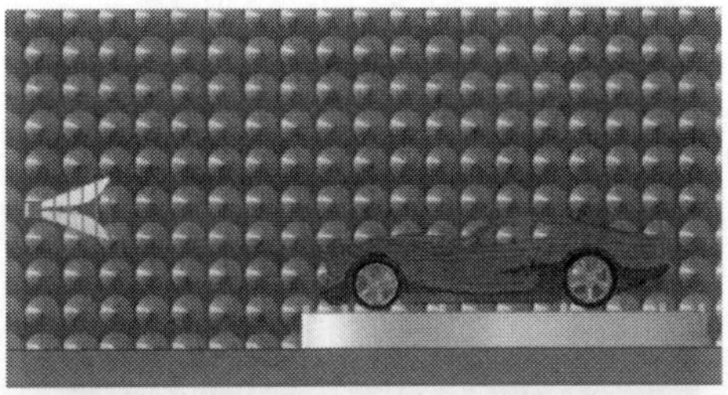

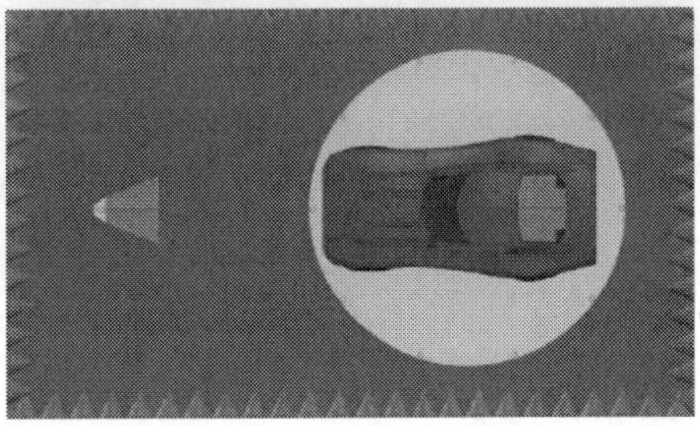

Animations perfomed by ESSS

Automotive EMC Standards – ISO 11451-2

ANSYS

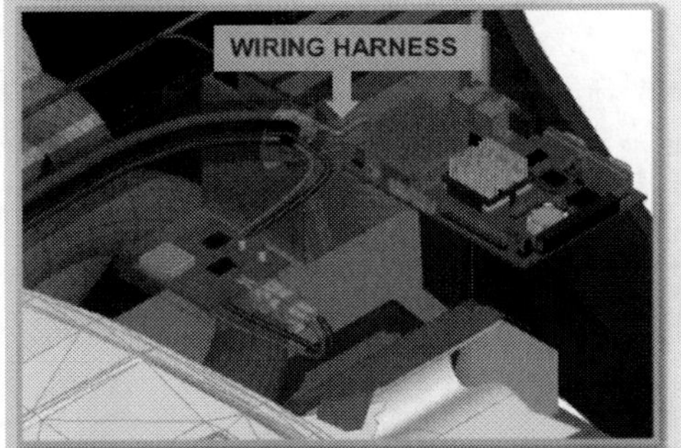

The antenna illuminates the vehicle and two simulations are performed:

1- The ECU is placed inside the vehicle without wiring harness and a clock signal is applied to the connector and goes to the uprocessor through a PCB trace.
2-The same ECU is now connected to the wiring harness and the same clock signal is now applied to the end of the harness near the motor that is connected to the ECU through the connector.

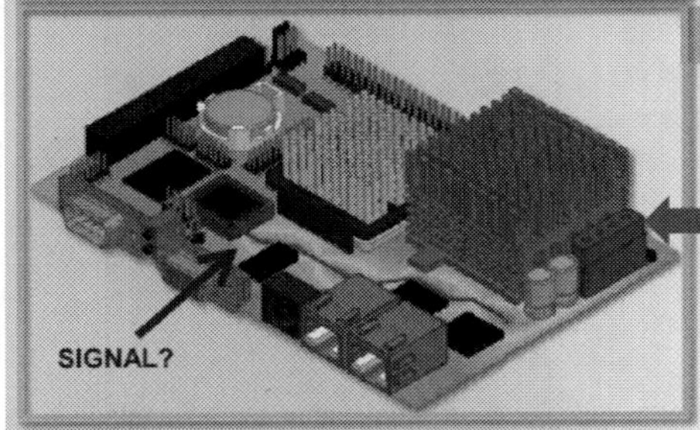

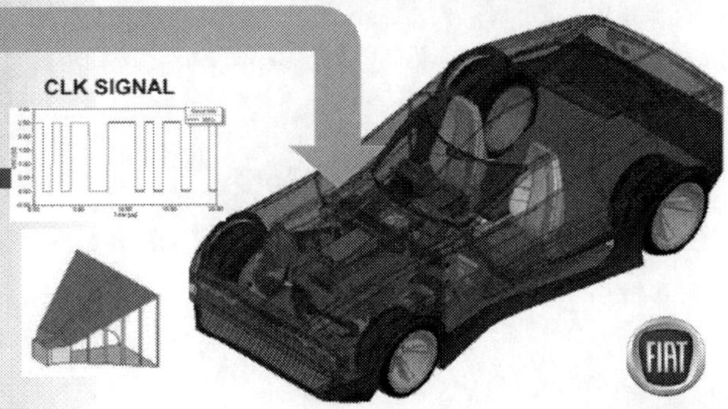

Animations perfomed by ESSS

Automotive EMC Standards – ISO 11451-2

ANSYS

Video Presentation in Original e-media

This simulation shows that when the ECU is connected to the wiring harness the EMI is higher for a given bandwidth . Bit Error Rate (BER) for 165MHz is 1E-3 when the ECU is connected to the cable harness and it is 1E-17 when there is no cables and the clock is applied directly to the connector

Animations perfomed by ESSS

Full Vehicle EMC tests

ANSYS

PAPER 2011-36-0085

Domain Decomposition Method

HPC – High Performance Computing

DOMAIN DECOMPOSITION TECHNIQUE ENABLES THE SIMULATION OF VERY LARGE FIELD PROBLEMS BY SHARING THE ORIGINAL MESH INTO SUB DOMAINS USING PARALLEL PROCESSING COMPUTING

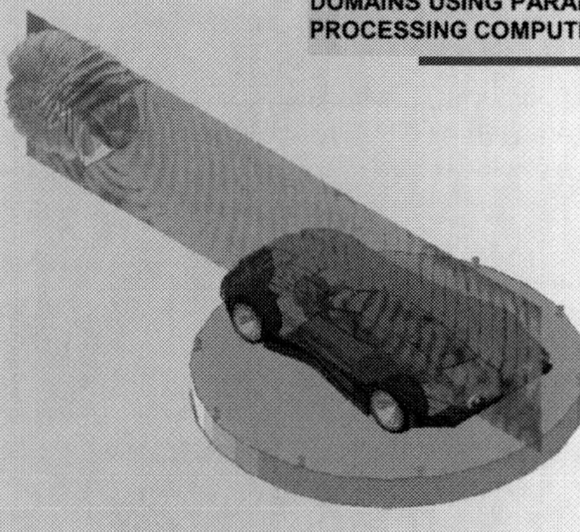

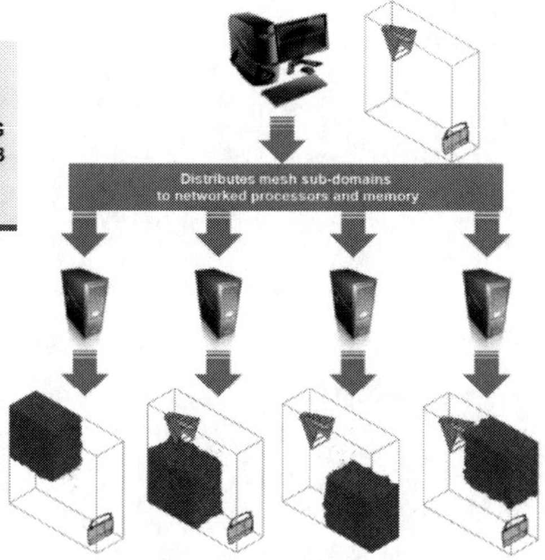

Animations perfomed by ESSS

ANSYS

Full Vehicle EMC tests

Domain Decomposition Method

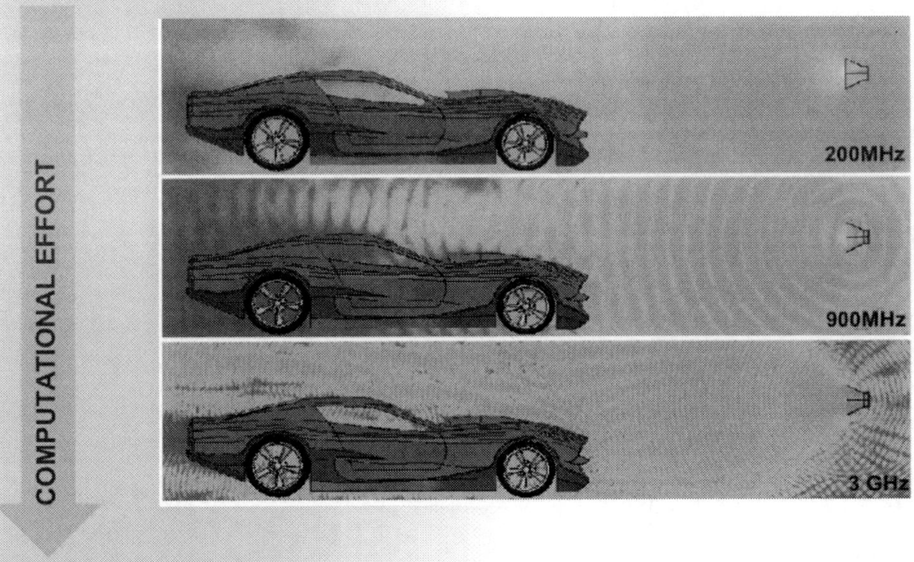

15 © 2011 ANSYS, Inc. November 11, 2013

Animations perfomed by ESSS

ANSYS

Full Vehicle EMC tests

FEBI – Finite Element Boundary Integral

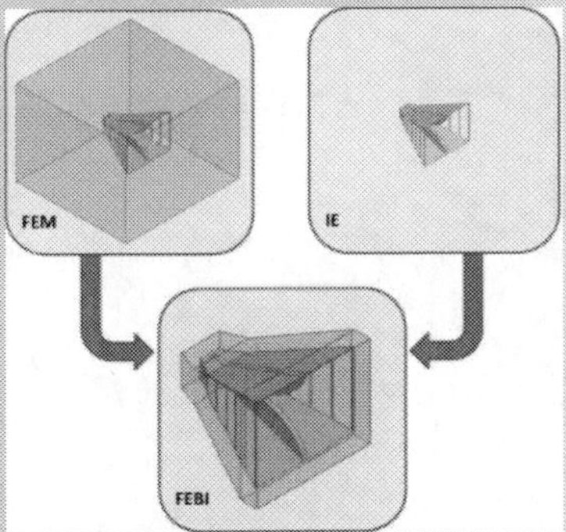

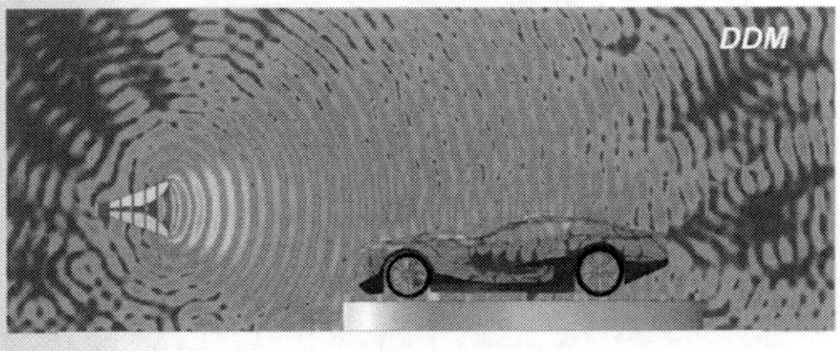

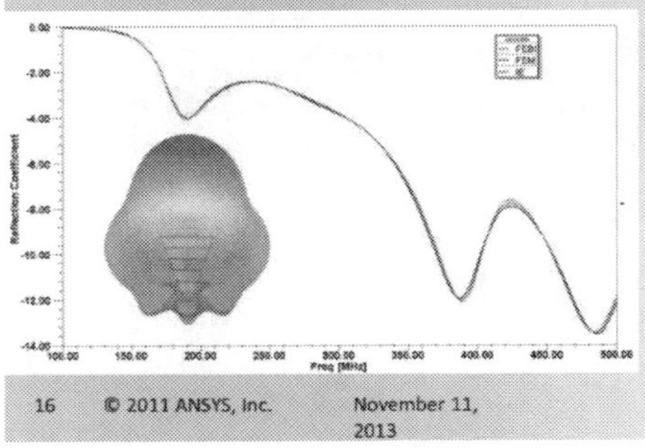

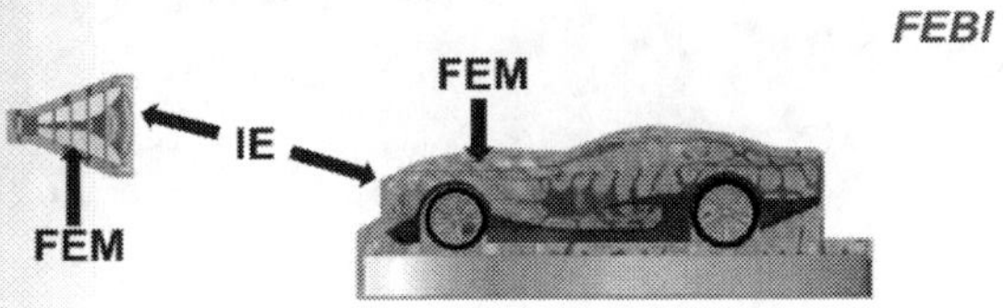

16 © 2011 ANSYS, Inc. November 11, 2013

Animations perfomed by ESSS

Hybrid IE-Regions

- Metal objects can be solved directly with an IE solution applied to surface
  - Removes the need for air box to surround metal objects
- Homogeneous dielectric objects can be replaced with IE-Region
  - Dielectric is solved using IE to surface

Assignment:

- Select Objects
  - HFSS → IE Regions → Assign As IE Region

IE-Regions shows ~70% RAM reduction vs. ABC

FEM Solution (20G RAM)

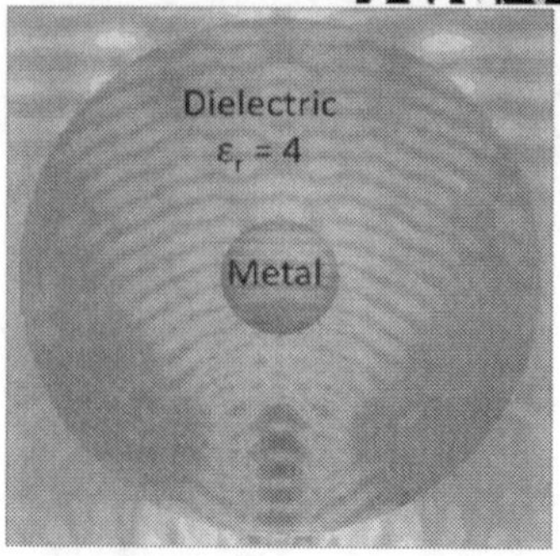

Hybrid FEM-IE Solution (6.2G RAM)

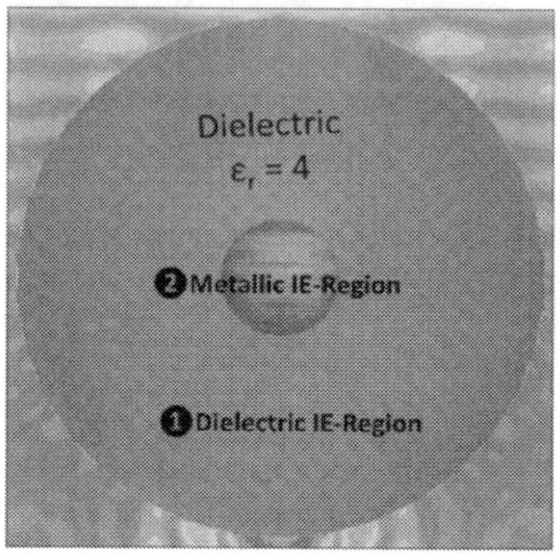

Full Vehicle EMC tests

FEBI – Finite Element Boundary Integral

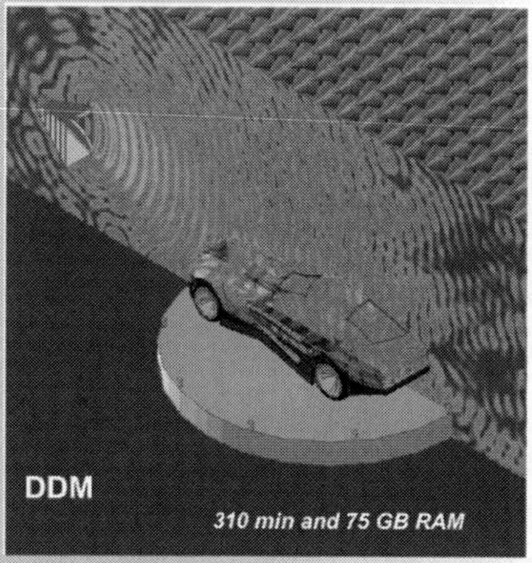

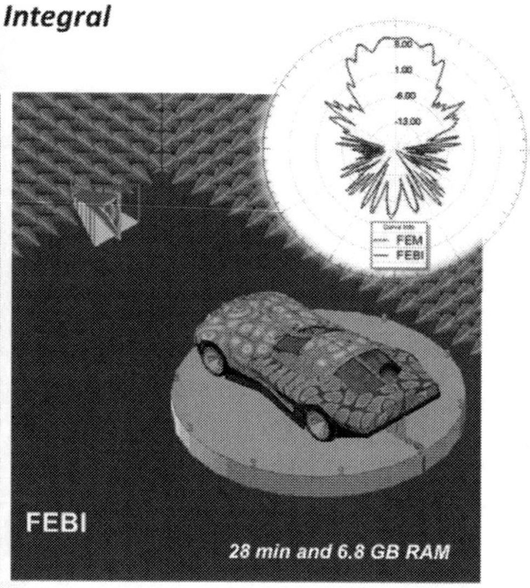

Animations perfomed by ESSS

Case study in HFSS: car body with remote keyless entry (RKE) system

•Overall Project geometry

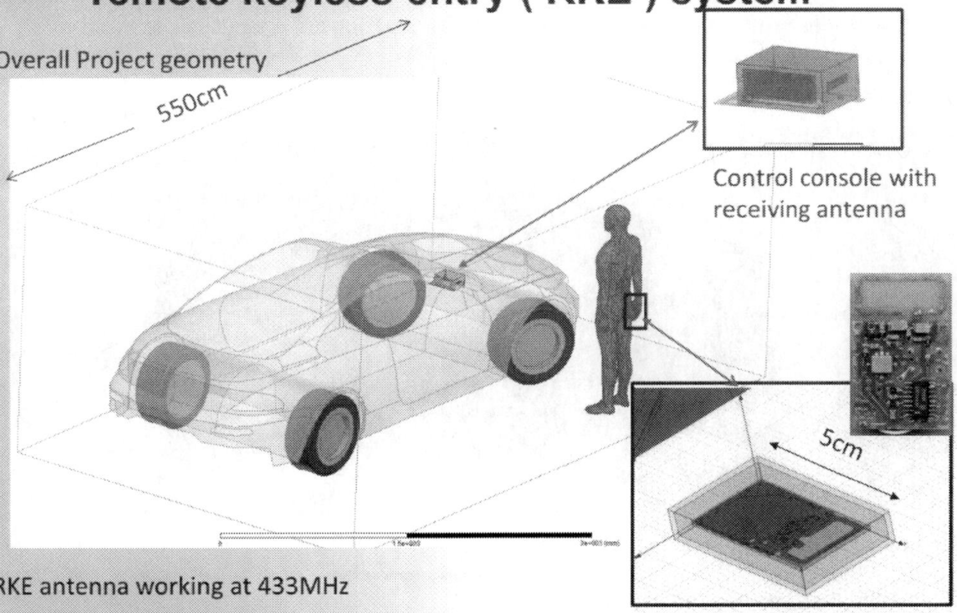

RKE antenna working at 433MHz

Key transmitter

DDM : car body with RKE

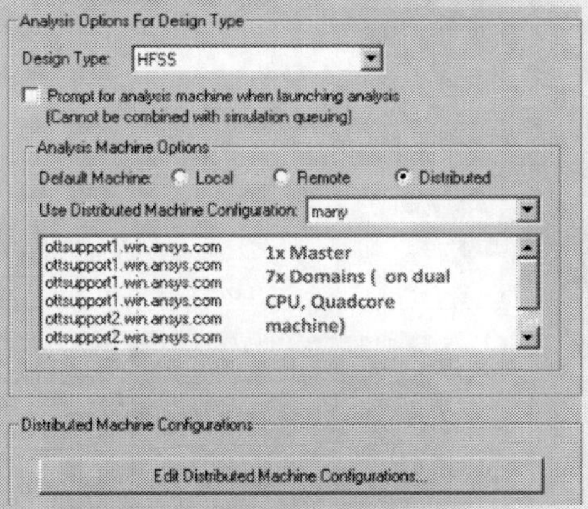

Dom7 on ottsupport2...				
Matrix Assembly/Solv...	00:20:24	00:14:58	1.46 G	Disk = 2514 KBytes, 44347 tetrahedra ,
Distributed Solve for: ...	00:21:10	01:01:38	0 K	Maximum domain memory: 1.859 GB
				Average domain memory: 1.417 GB
				Total memory for all domains: 9.916 GB

Automatic separation to domains

Parallel simulation of seven domains

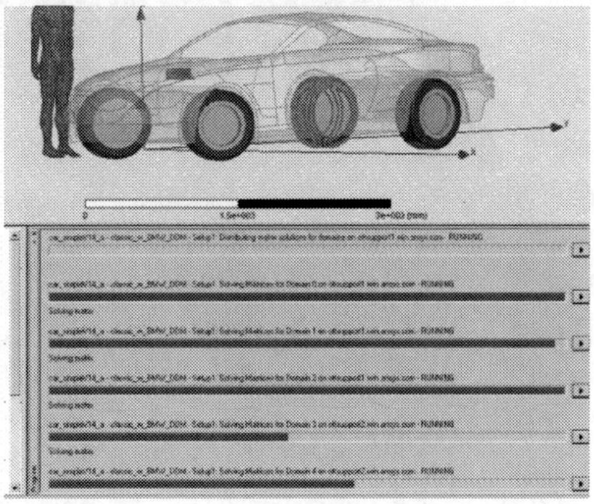

ANSYS Case study : car body with RKE

•Field solution (433 MHz) : E-Field from transmitting key-circuit in the hand of a person near a car

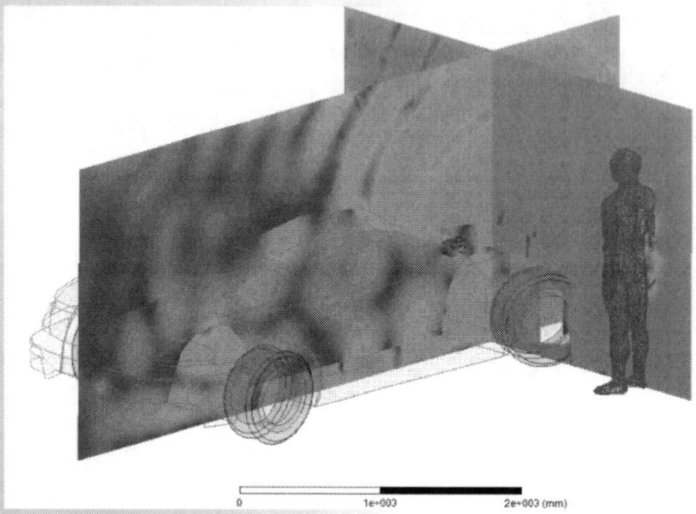

ANSYS Tire Pressure Monitoring System (TPMS)

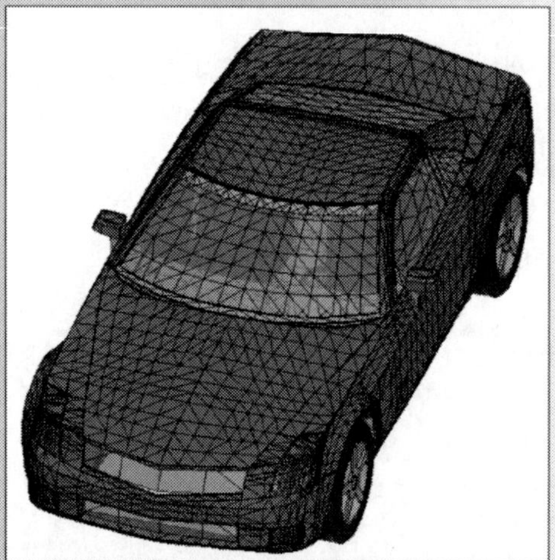

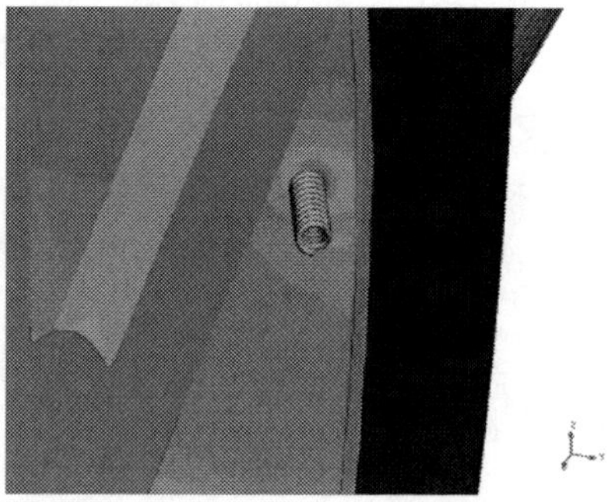

TPMS - Mesh Generation

- **Geometrically conformal mesh automatically created and adapted for each iteration**
  - Engineer not required to create each mesh
  - E-field at IP vs. rotation angle

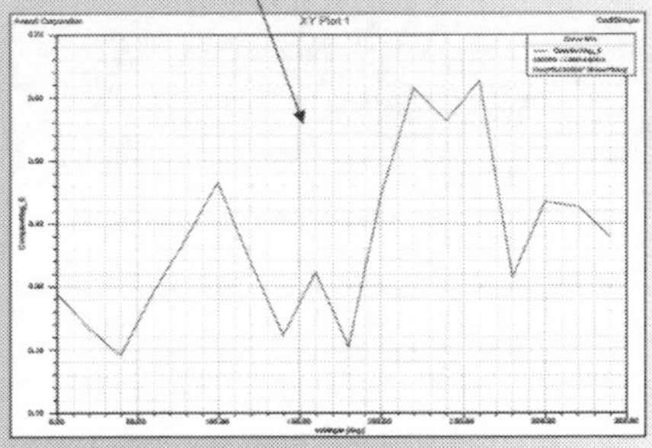

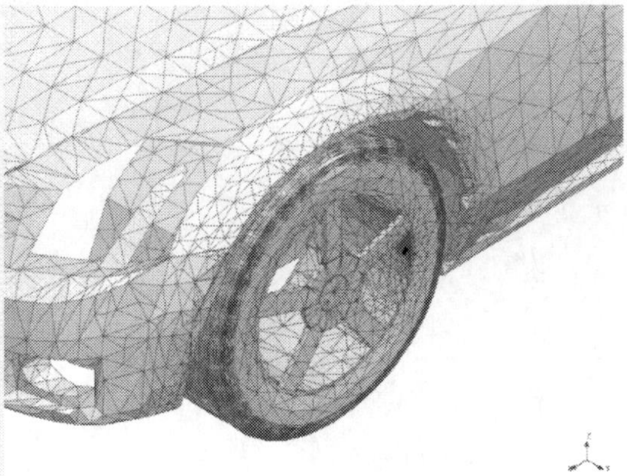

Roof Mounted GPS

- GPS antenna mounted on roof
  - Circular patch
- Volume = 2500 λ^3

ANSYS EMI/EMC: Shielding Enclosure Standard

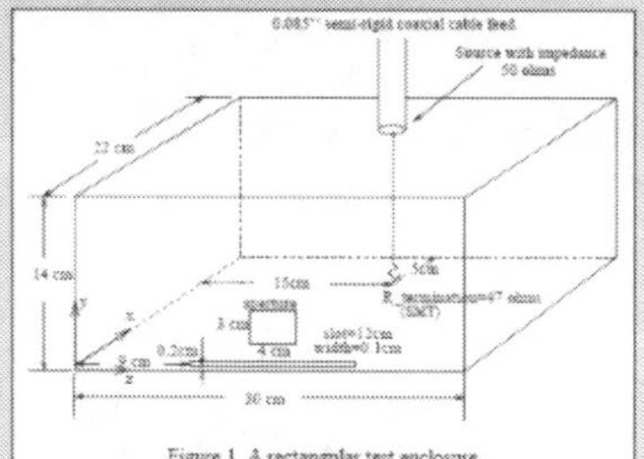

Figure 1. A rectangular test enclosure.

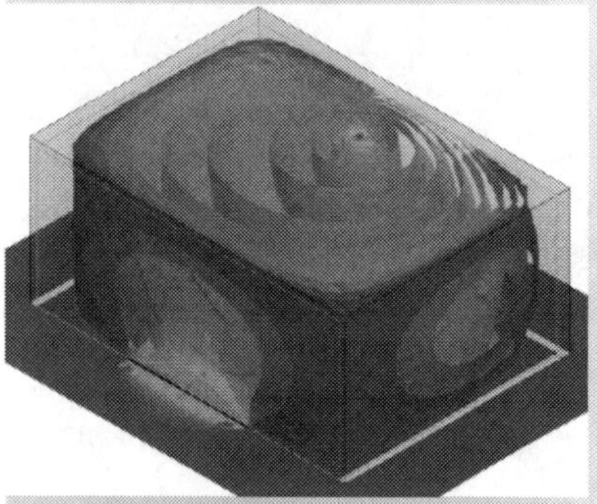

EMC standard from ACES (Applied Computational Electromagnetic Society)

- Shielded enclosure
- Good test for accuracy

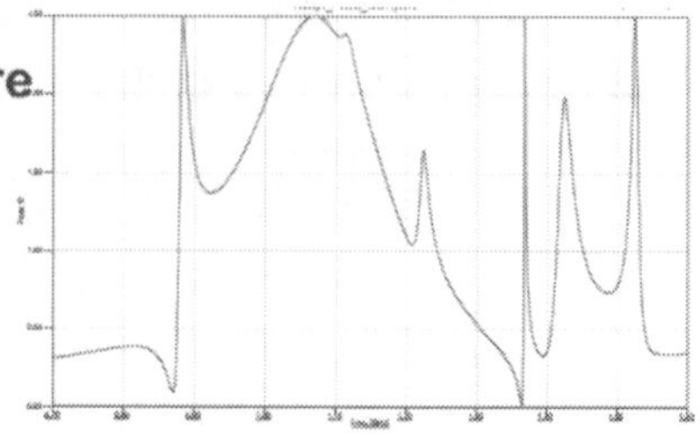

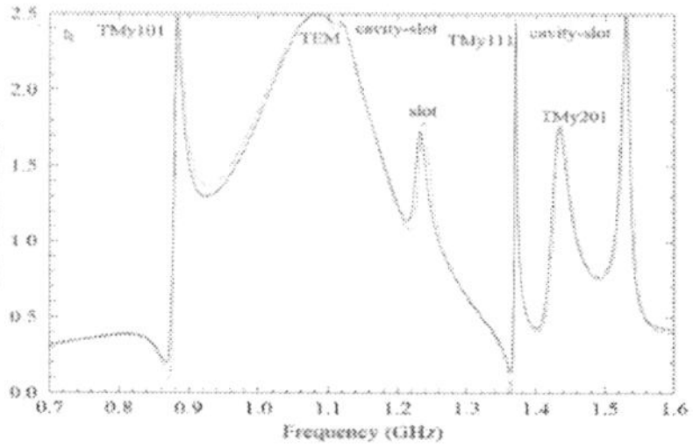

25 © 2011 ANSYS, Inc. November 11, 2013

ANSYS Wire Harness

- common mode signal on wire harness at 50 MHz
- "Bundle of wires" simulated as one "wire" with single ended common mode excitation ports.

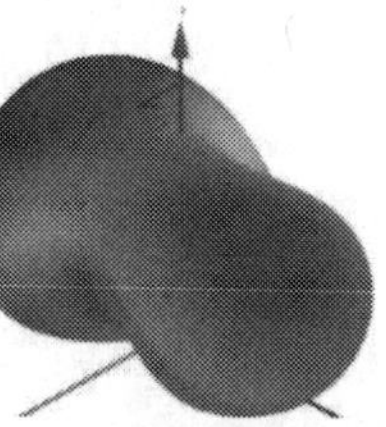

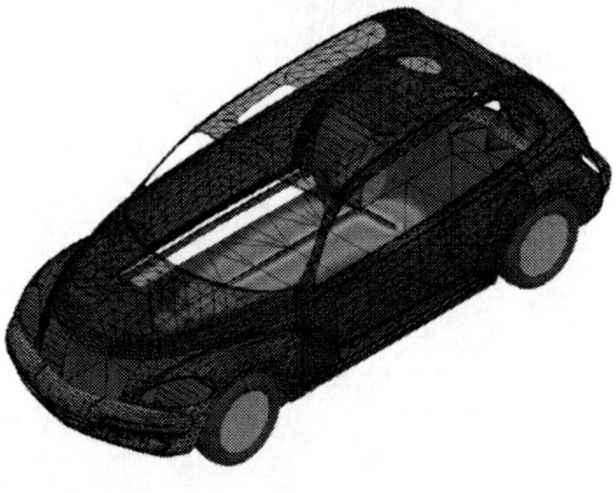

26 © 2011 ANSYS, Inc. November 11, 2013

Mobile Phone in Auto: an antenna in a Farady cage

ANSYS

- What if radiator not connected to enclosure
  - E.g. mobile phone in auto
- Solve mobile device alone and impress fields from this solution into simulation car 900 MHz
- HFSS to HFSS link.

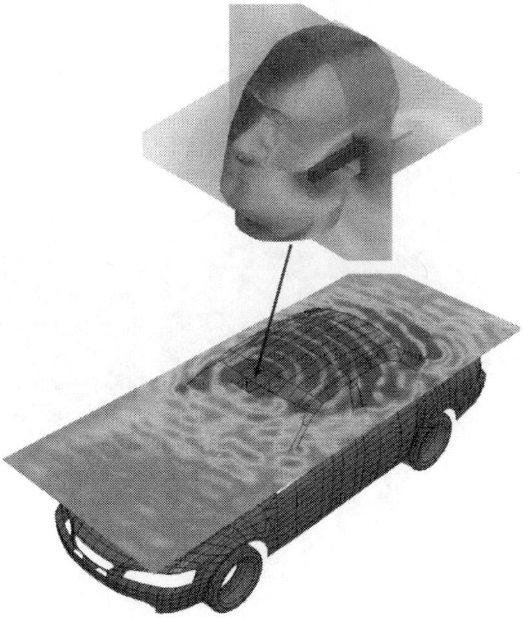

27 © 2011 ANSYS, Inc. November 11, 2013

Hybrid Solution for Antenna Placement Analysis Using IE-Regions : Results

ANSYS

No need to solve space in between antenna and car body

Solution Type	Total RAM (GB)	Elapsed Time (hours)
FEM w/ DDM	160G	8
Hybrid Solution	**11**	**2.7**
	15X Less	3X Faster

28 © 2011 ANSYS, Inc. November 11, 2013

ANSYS Fields from DVD Player

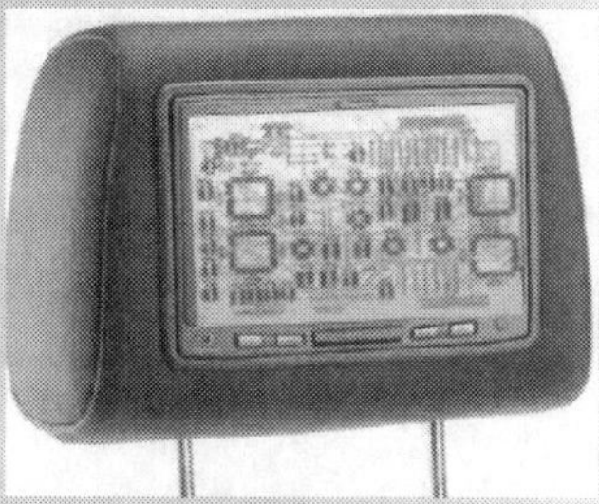

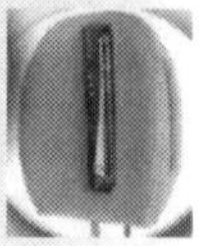

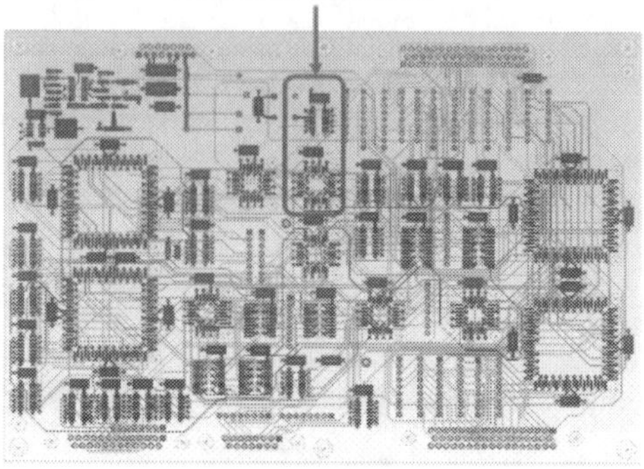

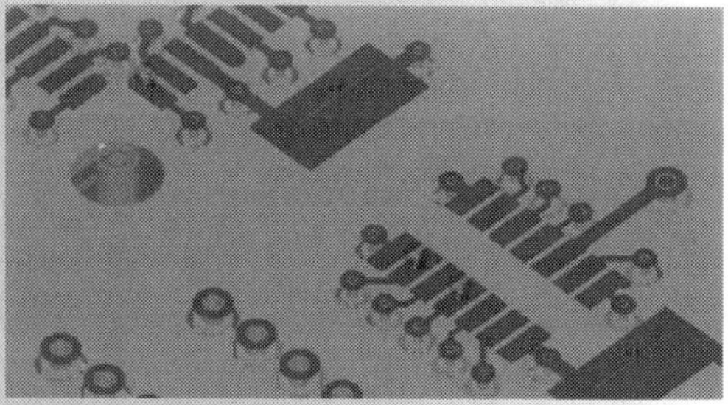

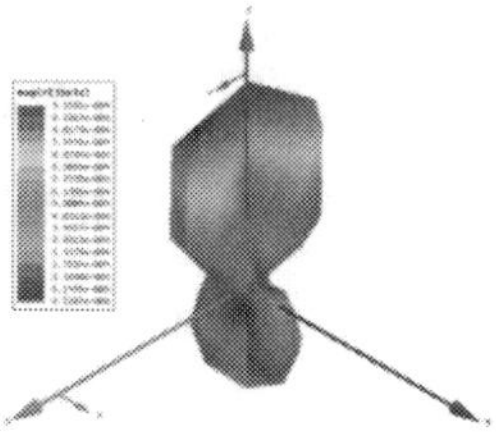

ANSYS SIwave continued

- **electromagnetic analysis tool for Complex PCBs**
- **How does it work?**
  - application of full-wave solvers breaks a complex problem into a collection of simpler problems.
  - Radiated fields included.
- **Radiated fields from PCB can be impressed into an HFSS simulation → the SIwave - HFSS link**

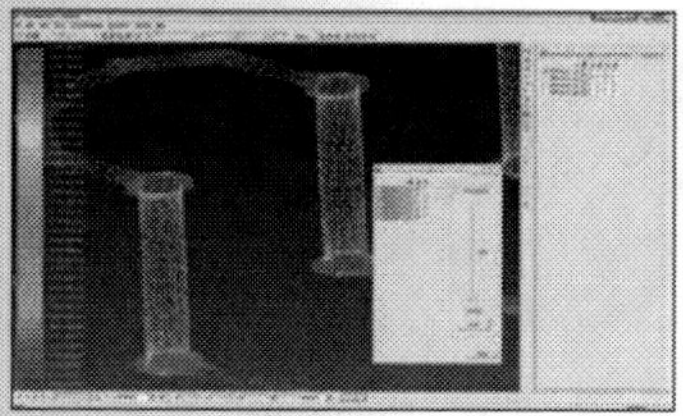

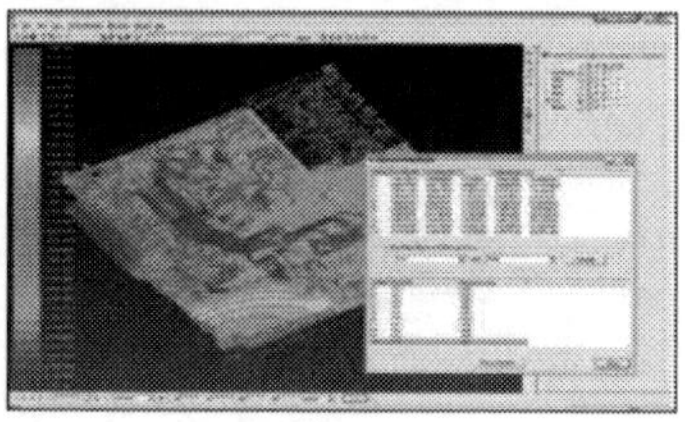

ANSYS Power Integrity with SIwave

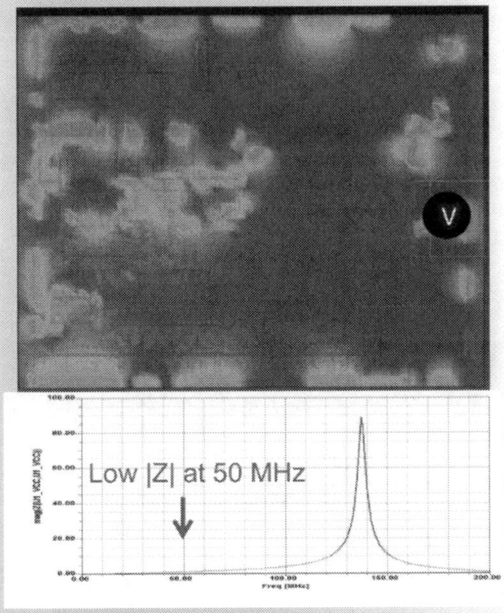

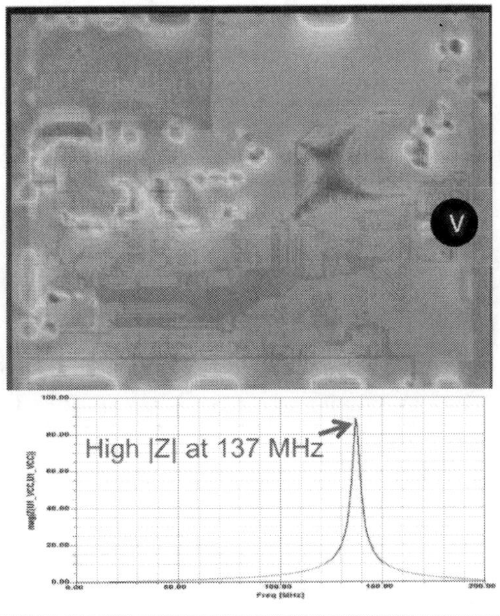

ANSYS More on SIwave

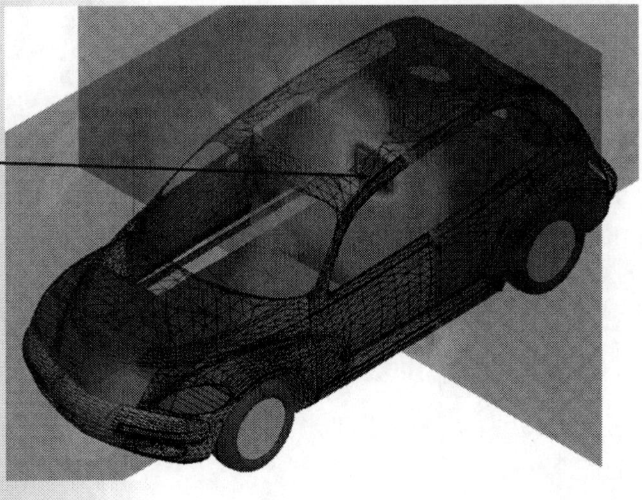

SIwave linked data:
Radiated source fields from DVD at 125 MH

ANSYS Back to problem size: Reflector Antennas

- Design of large structures
- Efficient in memory, simulation time

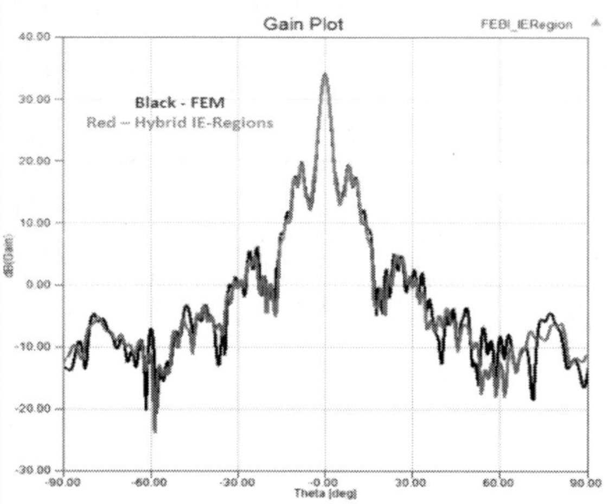

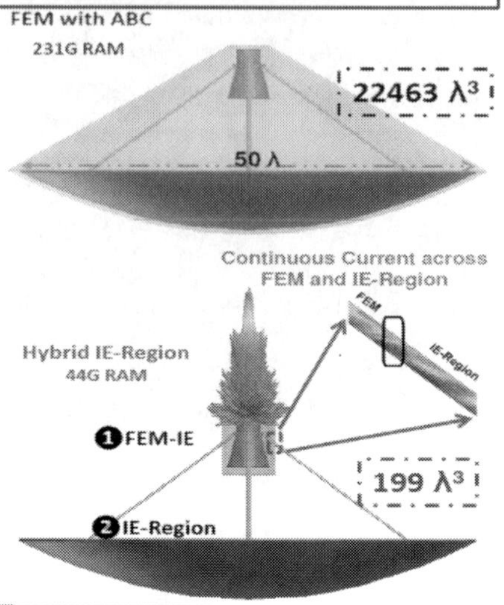

81% Less RAM: Enhanced IE-Region vs. FEM

ANSYS Satellite Systems

- Simulation of satellite systems
- Efficient in memory, simulation time

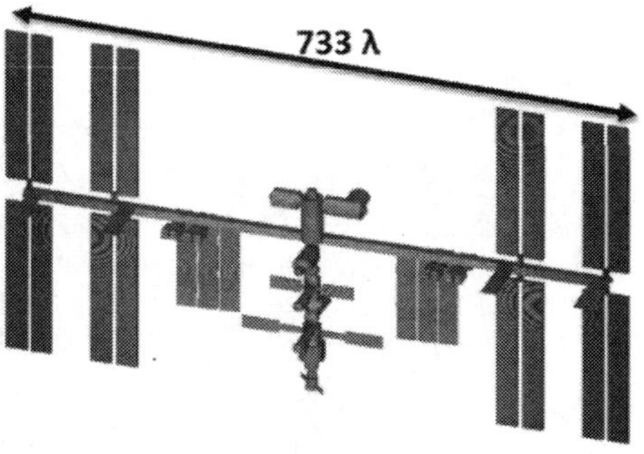

47 GB RAM
Simulation time: 57 mins

ca. 4.4 Mio λ³

ΛNSYS Radar trap problem

~ 7000 x 5000 x 2000 mm

40.5 Mio λ^3

fradar ~ 25 GHZ
→ λ ~ 12 mm

ΛNSYS Leveraging High Performance Computing Hardware

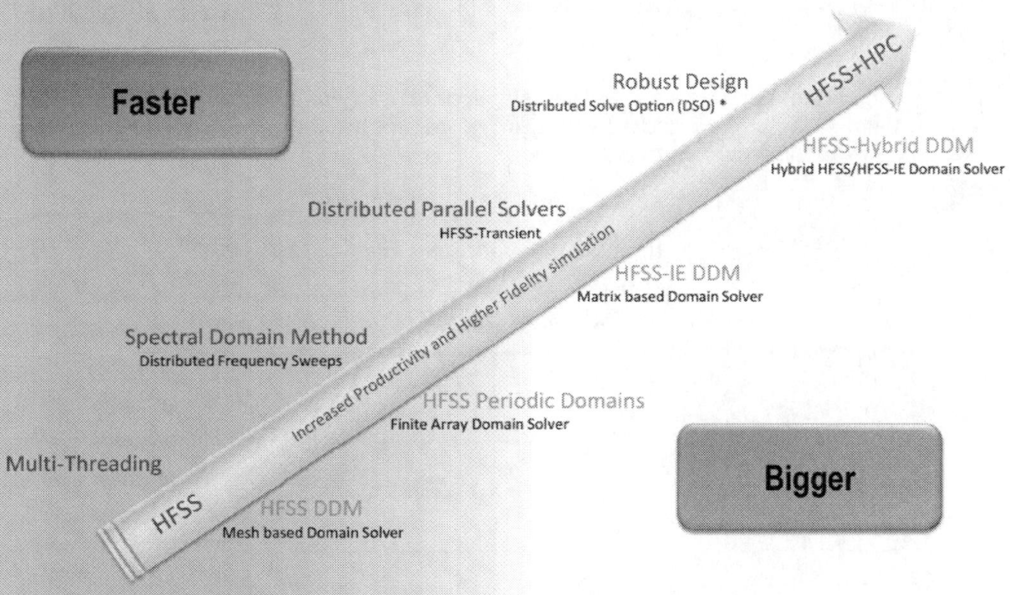

*Distributed Solve Option is a separate license and is not included with HPC

ANSYS

ANSYS

Back to Radar Traps:
What are Metamaterials?

- **A Metamaterial is essentially a material whose bulk electromagnetic properties arise from the interaction of structures, or structural voids, that are placed in arrays that are spaced on the order of a wavelength**
  - **Therefore, the design of the individual structures/periodicity can be used to modify the properties**
  - **Gives possibility for many opportunities as the limitation of electrical properties of natural materials is well known**

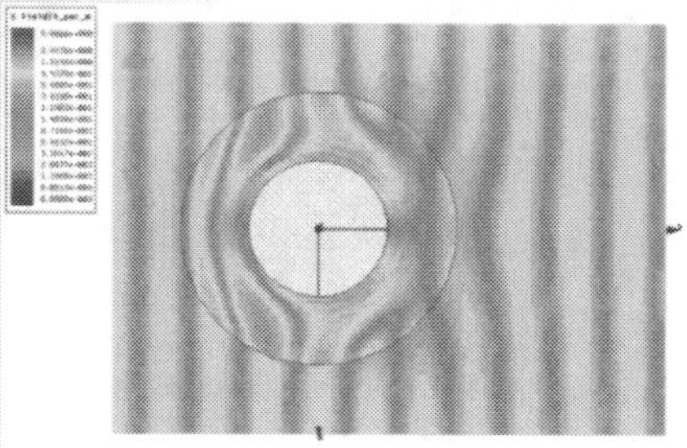

ANSYS

ANSYS

What are Metamaterials?

- **On left, see how natural materials behave in a bulk manner when an external field is incident. In actuality, the effect is due to the superposition of many very small induced *atomic or molecular* dipoles**
- **On right, a metamaterial uses a periodically spaced array of structures to induce a similar bulk effect, but via structure design. Note that the periodicity of the structures is smaller than the wavelength.**

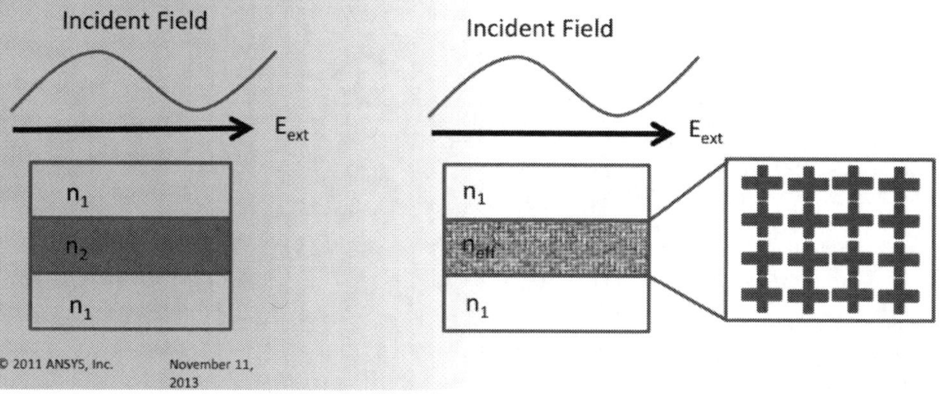

What are Metamaterials?

- It is of note that reflection is related to the Relative Permittivity and the Relative Permeability

$$\tilde{n} = \sqrt{\tilde{\varepsilon}_r \tilde{\mu}_r}$$

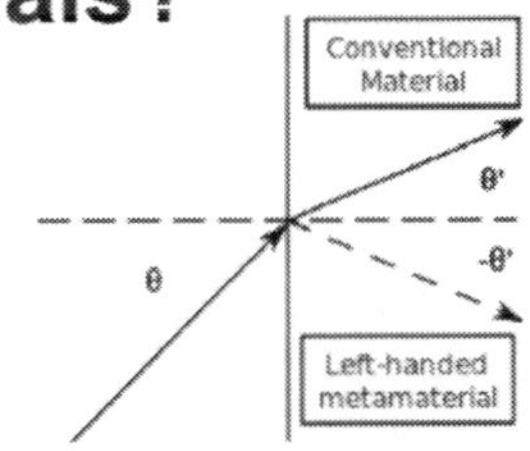

- What do the signs of the real parts of and mean?
  - If *both* real parts are greater than zero, or if *both* real parts are less than zero, a propagating wave is allowed
  - If either of the real parts are less than zero with the remaining being greater than zero, only evanescent fields are allowed

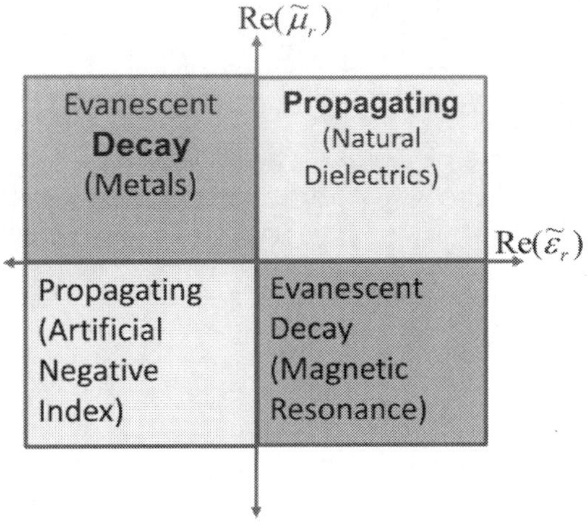

Implementation of MIN into Macro-simulation

- Note that the PEC (Red) scatters strongly in many directions
- The Cloaked PEC (Blue) has nearly no backscatter

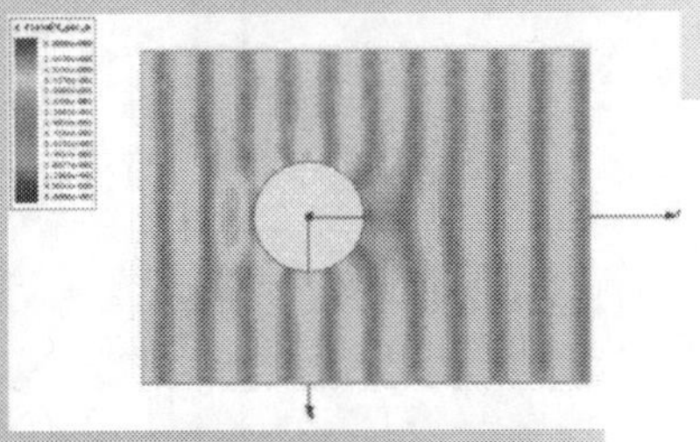

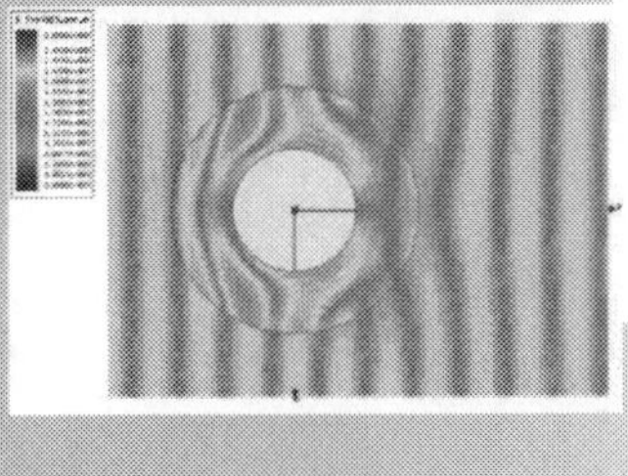

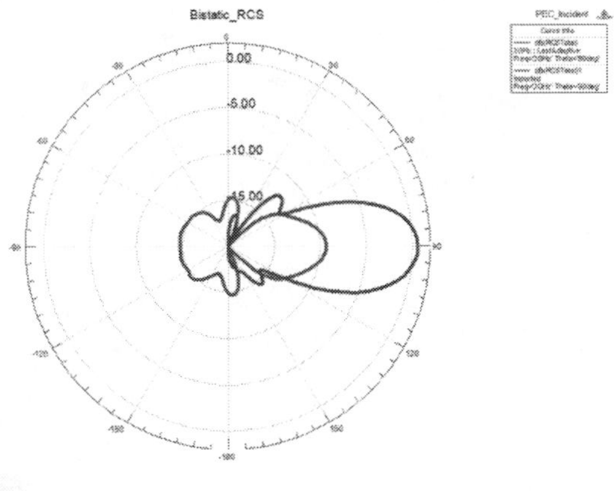

ANSYS

ANSYS HFSS results

PEC = ideal | NIM-material

Einfallende Welle

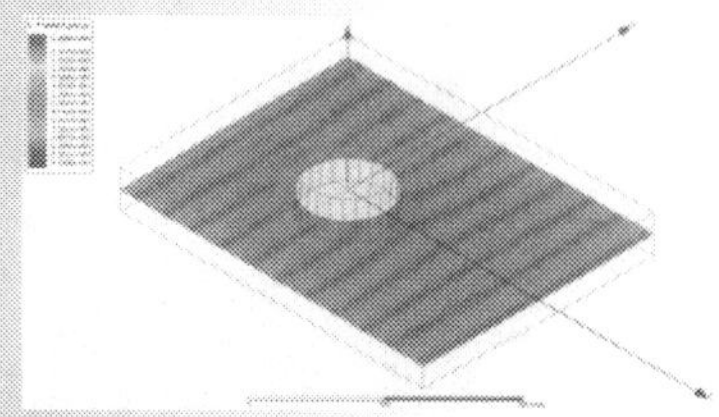

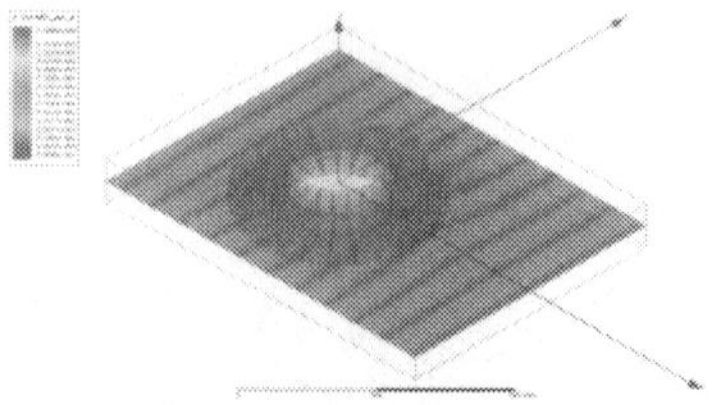

Gestreute Welle

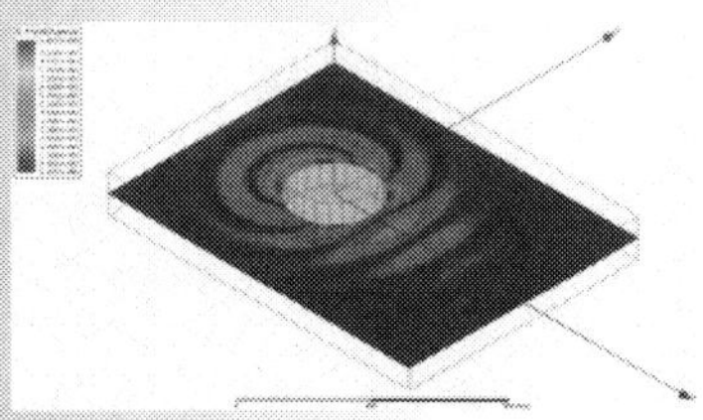

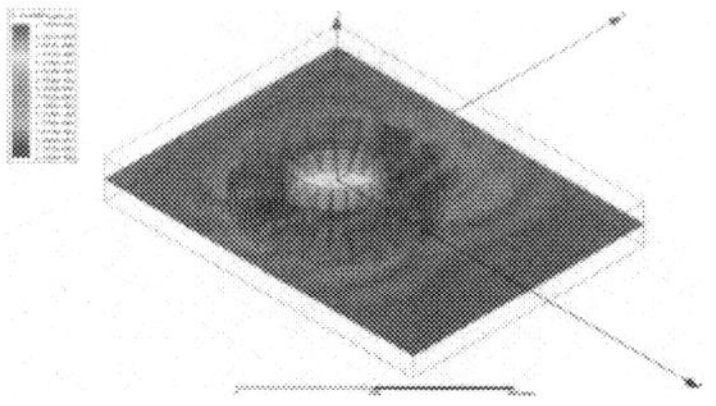

ANSYS

ANSYS „ Solutions for 40 Mio λ^3 „

ΛNSYS

Summary

HPC scaling

ANSYS HFSS → quick & efficient simulation solution

Large structures

Whole systems

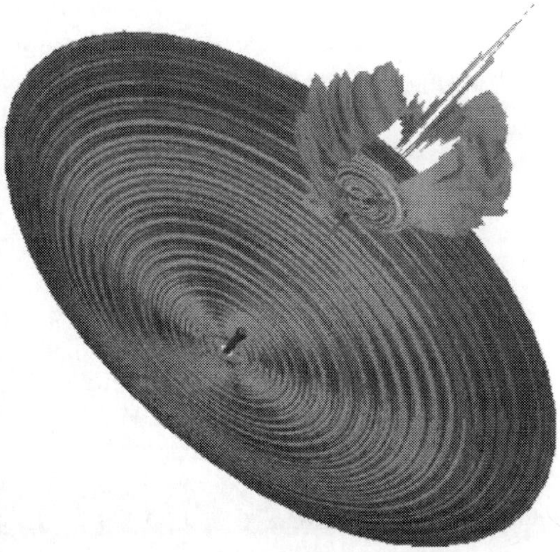

43 © 2011 ANSYS, Inc. November 11, 2013

Integrated Carbon Fiber Antennas for Vehicular Communication

ASWC 2013, Frankfurt

Christoph Kienmayer, PIDSO
Christian Römelsberger, CADFEM GmbH

ANSYS

PIDSO
> Propagation Ideas & Solutions

CADFEM
ANSYS Competence Center FEM

Brief company introduction

PIDSO
> Propagation Ideas & Solutions

Company

- 2006: founded in Vienna, Austria.
- Focus on high frequency antennas for WLAN, UAV, and other applications
- Founder: Christoph Kienmayer
  applying years of experience from working at Infineon Technologies in Munich (research, product engineering and marketing)
- 2010: Organized as GmbH company
- 2010: U.S. operation established in California, USA.
- Focus for the future: carbon based light-weight antennas and custom products
- Mercur 2010 Award Winner Innovation
- Nomination for Austrian National Prize

CADFEM

Brief company overview

PIDSO
> Propagation Ideas & Solutions

Company

- Measurement equipment up to 26 GHz and microwave absorber lab
- Strong relation with Vienna University of Technology (access to additional measurement equipment, antenna measurement chambers,…)
- Cooperation with Seibersdorf-Laboratories
- Member of AUVSI*
- Patented technologies for antennas available (RFID, carbon-fiber,…)
- Offering technical support in Munich (Germany) → currently supporting BMW research and development.

* Association for Unmanned Vehicle Systems International

CADFEM

Current pillars of our business

PIDSO
> Propagation Ideas & Solutions

PIDSO
> Propagation Ideas & Solutions

Standard and customized antennas

Products

- Customized antenna manufacturing
- Standard antennas

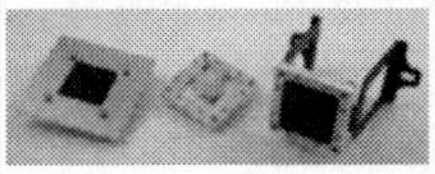

Developments

- High-frequency developments of antennas and circuit designs

BOEING

PORSCHE DESIGN

BMW Group

Services

- Field and lab measurements
- Consulting

Distribution

- Strengthen relationship with customers

HyperLink Wireless

indi<ot

LUXUL

Samso HITE

CADFEM

CADFEM

Products

- Full scale CAE solution based on ANSYS, complementary software and IT infrastructure
  - Requirement analysis
  - Configuration
  - Integration
  - Increase the breadth and depth of simulation usage

ANSYS Competence Center FEM

Service

- Create confidence and reliability in optimized product development by simulation driven engineering
  - Qualified training
  - Reliable user support
  - Transfer know-how to break the hurdles of adoption

Knowledge

- Comprehensive and innovative background knowledge for users, supervisors, and managers
  - Information exchange
  - Quality assurance
  - Life long learning

Motivation

- Future vehicles tend to increase their capabilities to connect wirelessly to a large variety of services
- Next-generation communications techniques and networks for automotive use-cases are currently being researched intensively
- Ad-hoc based communications like ETSI ITS G5 are of prime interest to the automotive industry
- Inter-vehicle communication systems will be based on the ETSI ITS G5 technology centered around the 5.9 GHz communication band
- Antenna system as an interface between propagation channel and vehicular on-board unit exceptionally defines the performance of the related ITS applications
- Carbon fiber materials are already used to replace current metal vehicle parts

- This leads to a paradigm change: carbon fiber vehicles and high frequency applications → new combinations of antenna and placing have to be investigated

5.9GHz Antenna in the 3rd Break Light

- 5.9GHz antenna for car to car communication is to be integrated into the 3rd break light of a passenger vehicle

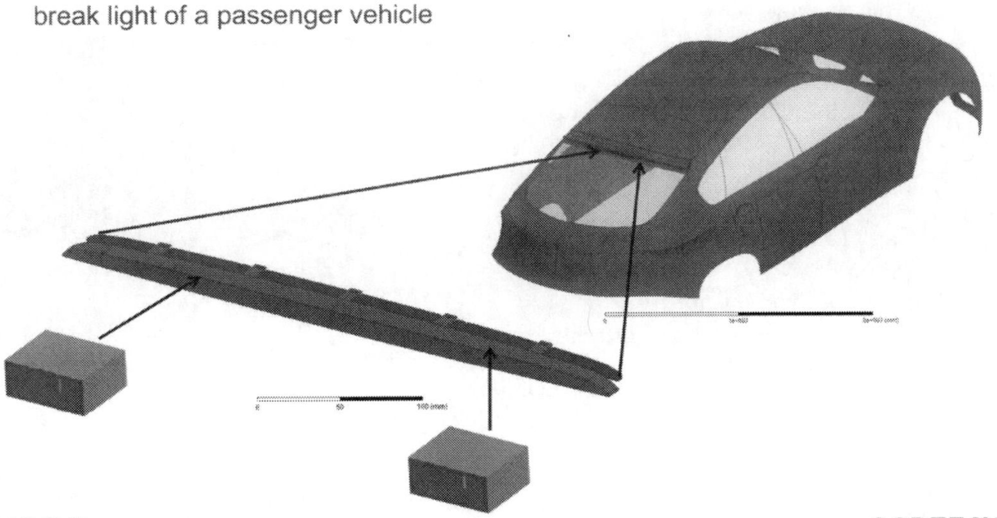

PIDSO

CADFEM

Basic Design

- The basic antenna design is a monopole antenna with an aperture

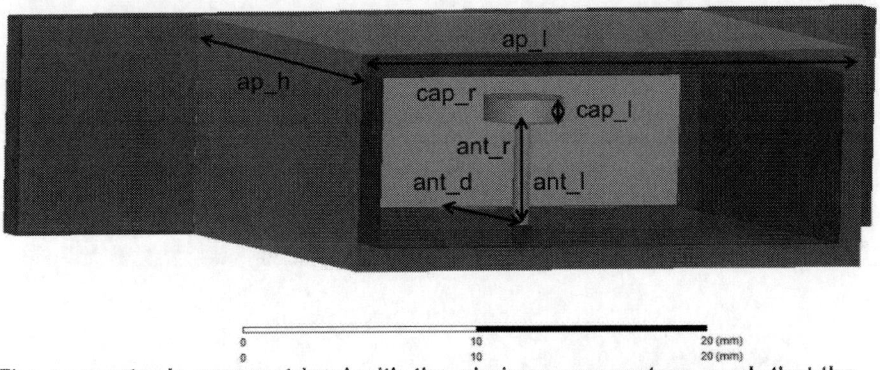

- The geometry is parametrized with the obvious parameters, such that the antenna conforms with the constraints of the break light geometry
- The antenna design is optimized for minimal return loss at 5.9GHz

PIDSO

CADFEM

optiSLang: Robust Design Optimization (RDO) in Virtual Product Development

- Sensitivity - Understand a Design
- Optimization - Use the full potential of a design
- Robustness - Ensure a safe design in real world conditions and don't get surprised by uncertainties
- Easy to Use - Safe to Use – Inside Workbench
- Efficient Processing of
  - Systems with nonlinearities and multiple physic domains
  - Design spaces including design failures
  - Low and high number of parameters

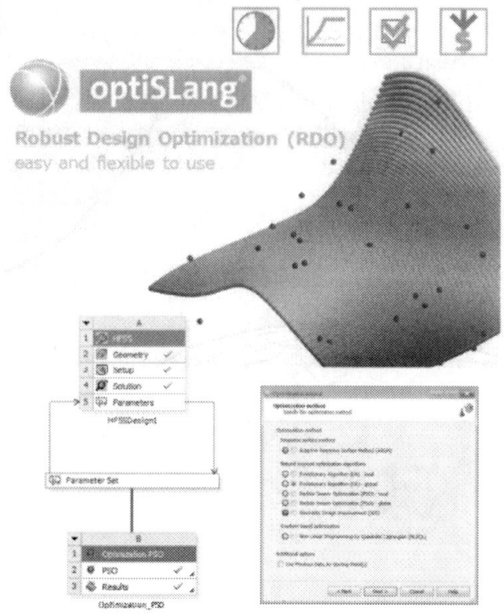

Basic Design

- Using a global Evolutionary algorithm an optimal antenna design was found after 22 generations

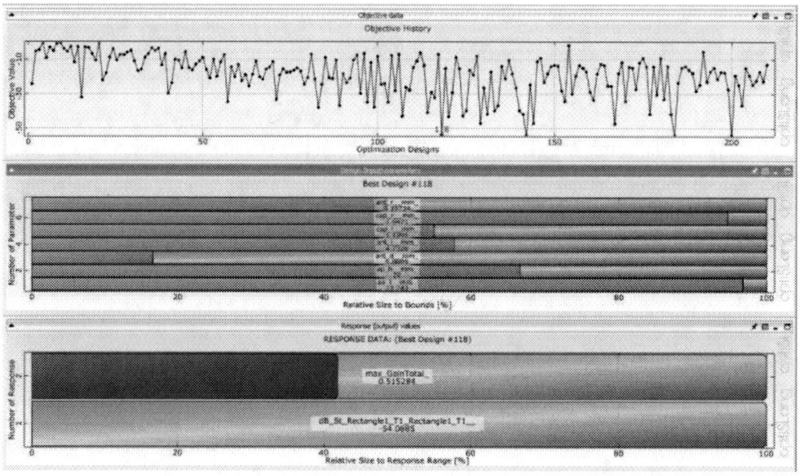

Basic Design

- The plot of the return loss of the best design shows that the antenna is actually tuned for 5.9GHz!

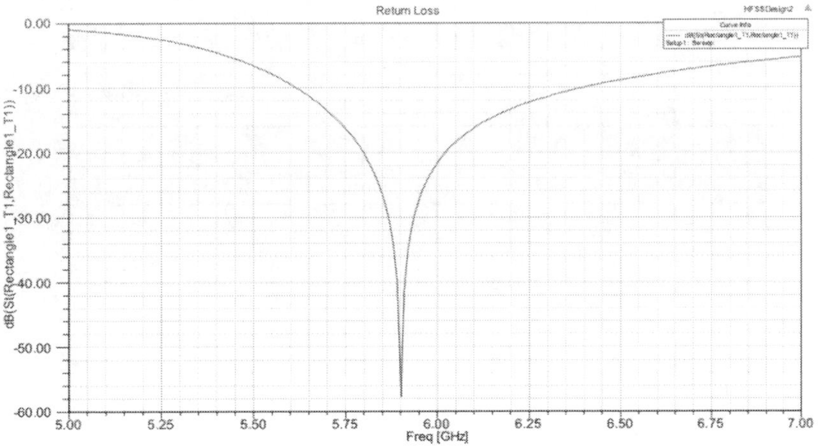

Basic Design

- The radiation pattern is typical for an aperture antenna

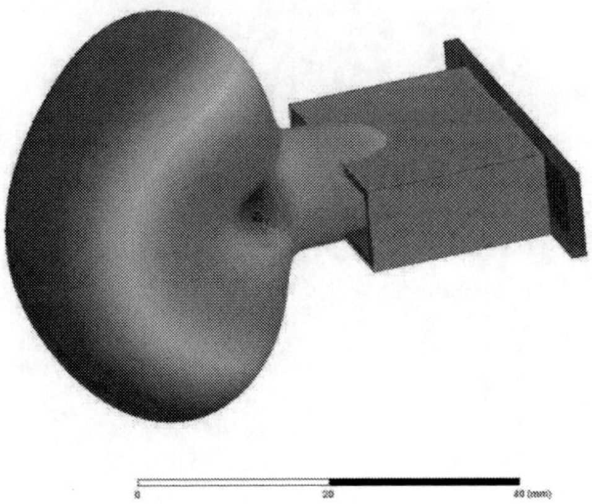

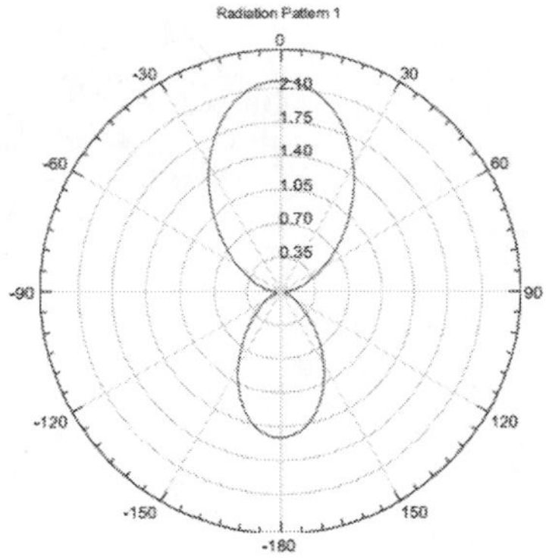

Basic Design

- The resonance can be nicely seen in the electric field

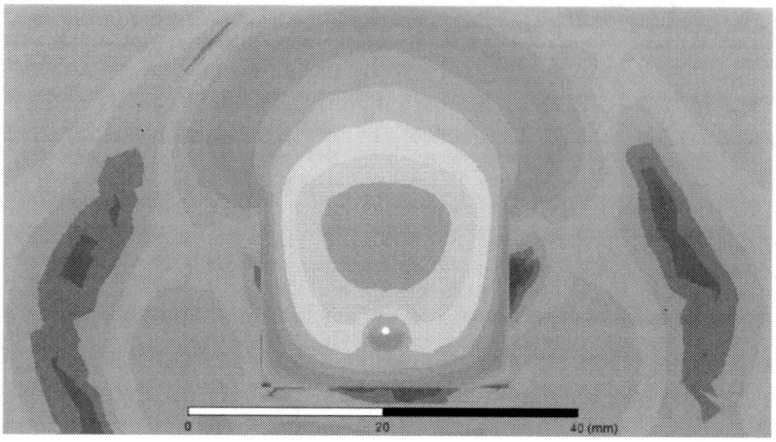

PIDSO
> Propagation Ideas & Solutions

- 12 -

CADFEM

Design with cover

- The plastic cover of the break light reflects waves back into the antenna and destroys the resonance

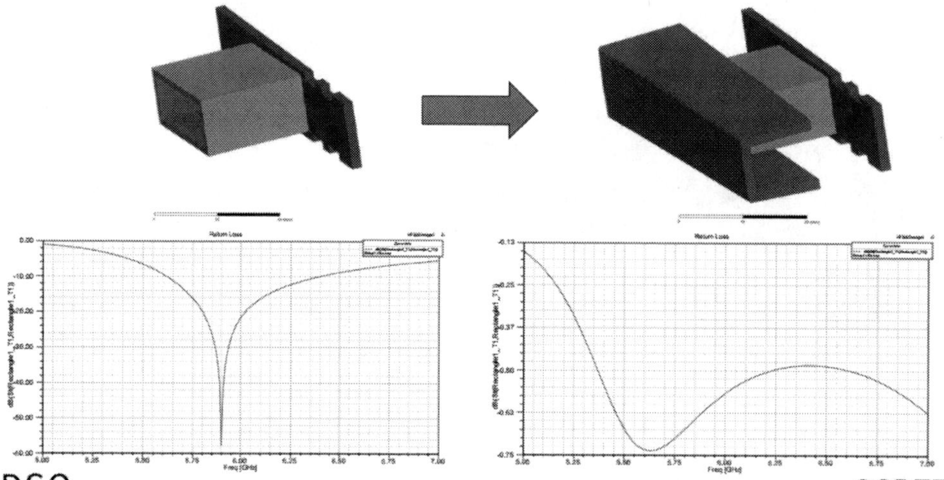

PIDSO
> Propagation Ideas & Solutions

- 13 -

CADFEM

Basic Design with Cover

- The antenna can be tuned with the cover to get a good return loss

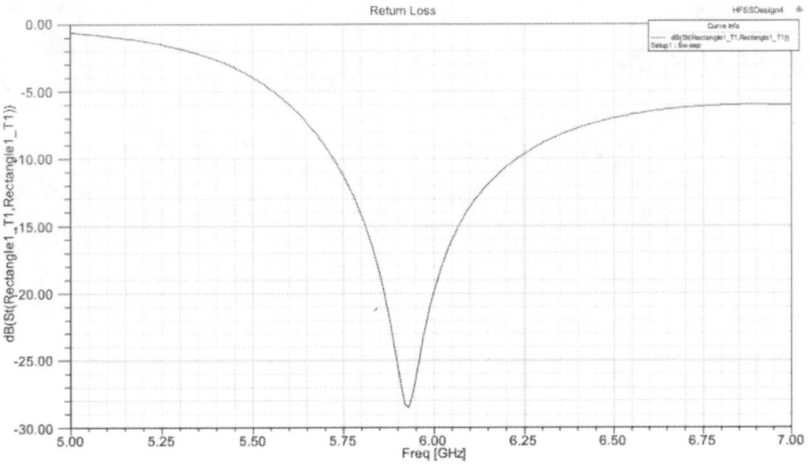

CADFEM

Comparison of different car body materials

Steel
- Conductivity : 1.1e6Si/m
- Return Loss @5.9GHZ: -20.7dB

Carbon
- Conductivity: 1000Si/m
- Return Loss @5.9GHZ: -20.9dB

Plexi Glass
- Relative Permittivity: 3.4
- Return Loss @5.9GHZ: -31.6dB

Wave is reflected

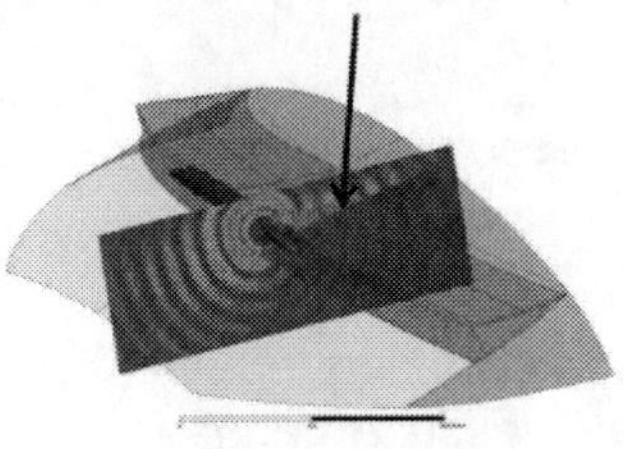

Wave is reflected

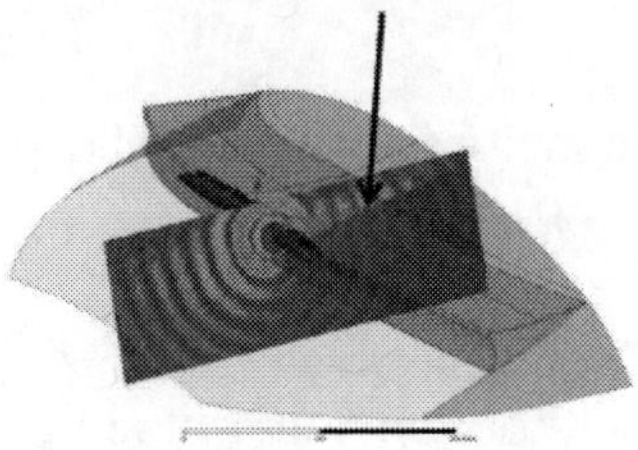

Wave goes through

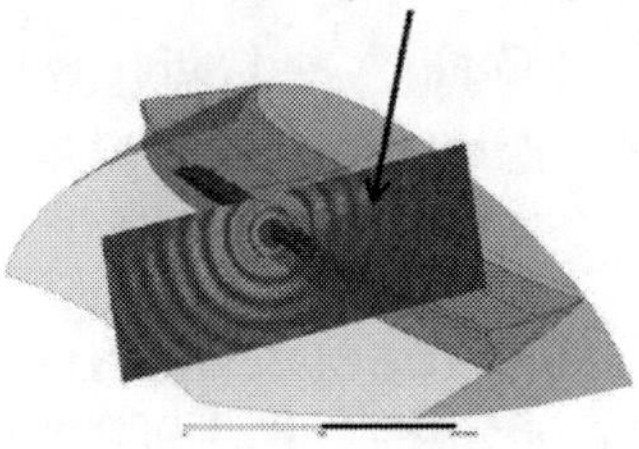

CADFEM

Comparison of different car body materials

Steel | Carbon | Plexi Glass

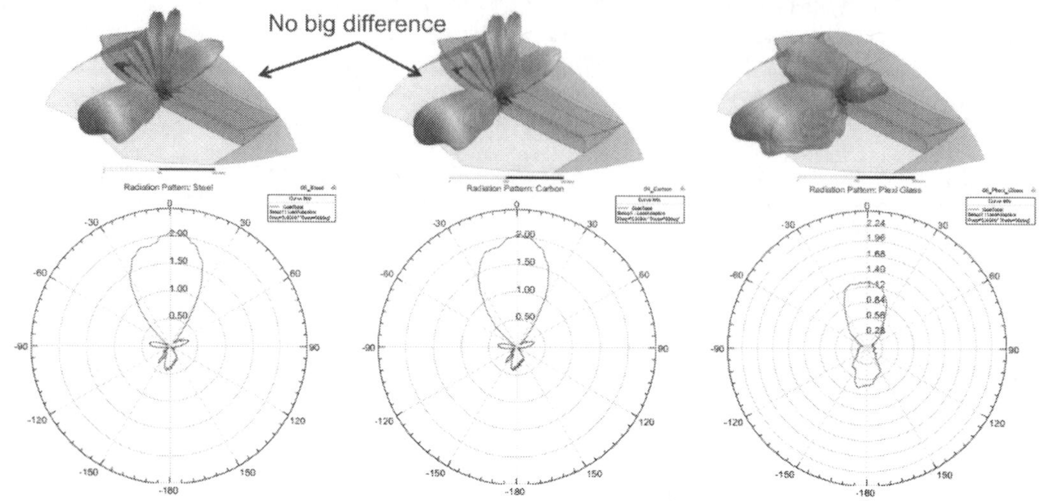

Conclusions

- Investigation of 3rd brake light a new construction space was successful for ITS G5 operation
- Diversity antenna system could be integrated but with restricted radiation performance
- Carbon fiber roof tops haven't shown big influence on current simulations results

Outlook and future work:

- Further modeling of carbon fiber and material capabilities is planned to improve simulation accuracy
- PIDSO, BMW and the Technical University of Vienna are working together in a Christian-Doppler Laboratory on "Integrated Vehicle Chassis Antennas in Carbon Fiber"

Signal, Power integrity and EMI analysis of a high voltage lithium battery charger for full electric vehicle

MetaSystem

History

1973

META SYSTEM IS BORN.

Our first entrepreneurial activities involved the development and realization of electronic devices used in automated control.

1974 - 1984

Production of radio frequency broadcasting equipment, in particular, high frequency transmitters and power amplifiers, begins. In the same period, we launch into the automotive industry with the production of electronic antennae and car alarm systems. We develop the first custom integrated circuit of codified signals for radio command. The Energy branch is created, producing the first line of Meta System branded uninterruptible power supplies (UPS) products.

1985 - 1995

Expansion of our car alarm division, creating specific accessory line products for prominent car manufacturers. Launch of our first series of modular and redundant UPS products.

1998 - 2009

Collaboration with the major European based car companies for the supply of various specialised original equipment manufacturing (OEM) devices. We enter the automotive Telematics world with the innovative platform, Clear Box®. We are focusing our innovative efforts on those areas which have historically been our strong points: automotive, telematics, and power supply systems.

2010-2012

Design and production using solar energy.

All Meta System's plants are equipped with PV Systems for the production and use of clean energy.

Total yearly energy production: 629.695 kWh CO_2 saved per year: 315 ton.

META SYSTEM

Philosophy

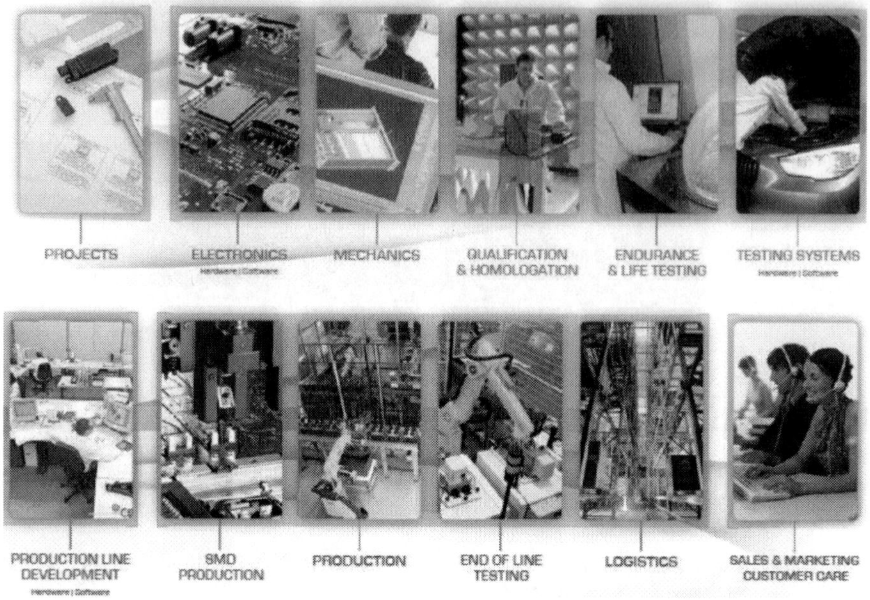

META SYSTEM

Product range

Product range

Battery Charger

DUT description:

- 3.5kW battery charger
- From 250 to 400 VDC output
- Input 110/230 VAC
- Efficiency up to 95%
- PFC for AC input
- LLC Type Converter for DC Output
- Water cooling

EMC Requirements for CE mark

EMC standards for Battery Charger:

- Dedicated standards from Automotive Manufacturer
- EN 55022 Conducted and radiated emissions
- CISPR 25 Limits and methods of measurement for the protection of on board receivers
- ISO 11452-2 Radiated immunity
- EN 61000-4-4 Fast Transient
- EN 61000-4-5 Surge

Among the various tests to be performed to obtain the CE mark, the conducted emissions, in accordance with EN 55022, are a critical test for the type of product.

Proper EMI analysis and design of EMI filters are mandatory

MetaSystem

Pag. 7

Battery charger electronic boards

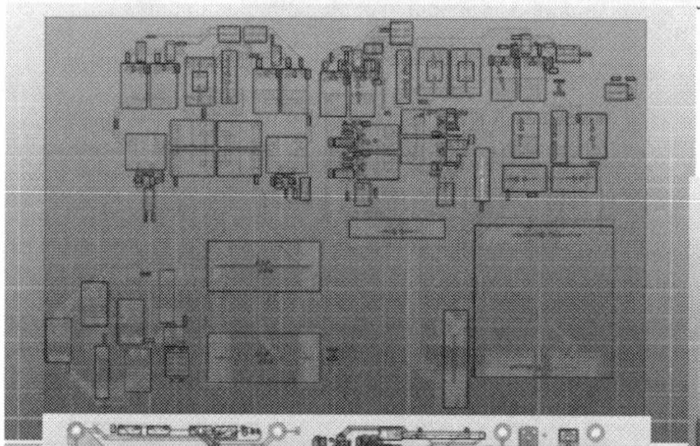

IMS

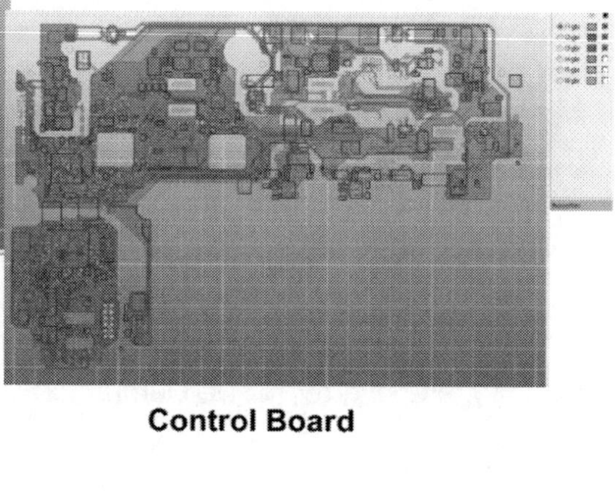

Control Board

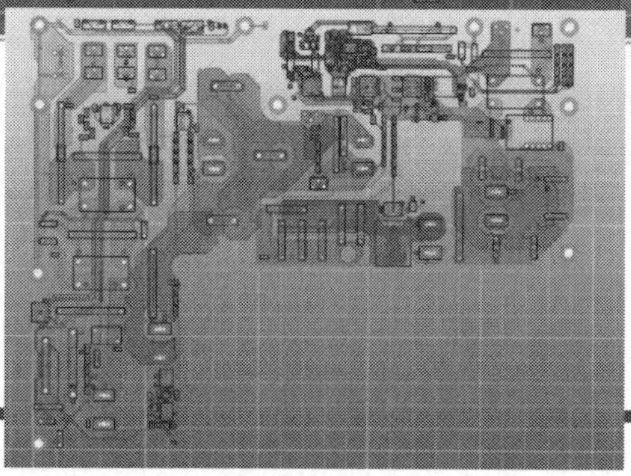

Power Board

MetaSystem

Pag. 8

IMS: Impedance Analysis of IMS

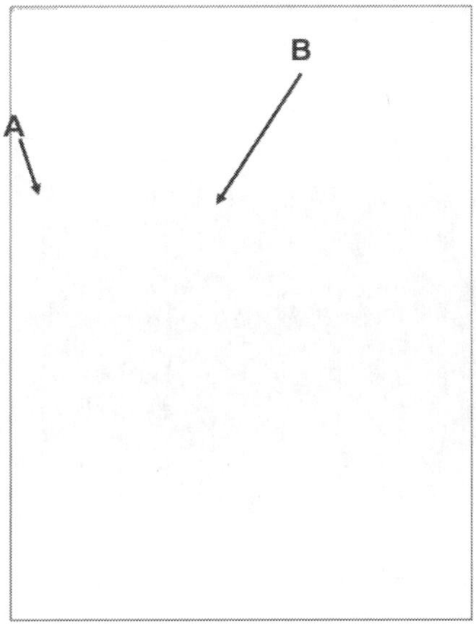

Impedence analysis in the frequency range 0.3-30MHz

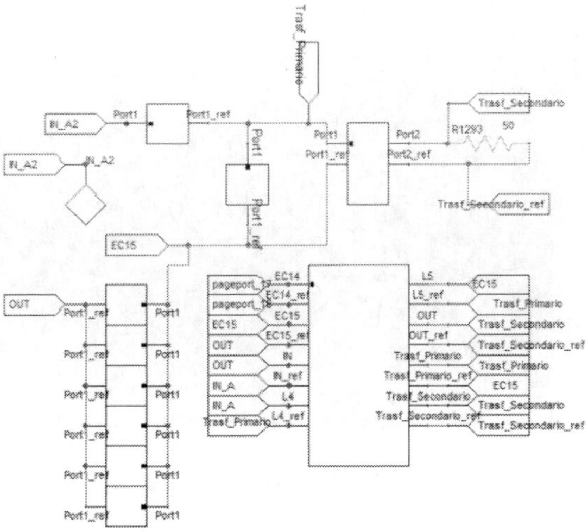

IMS Impedance

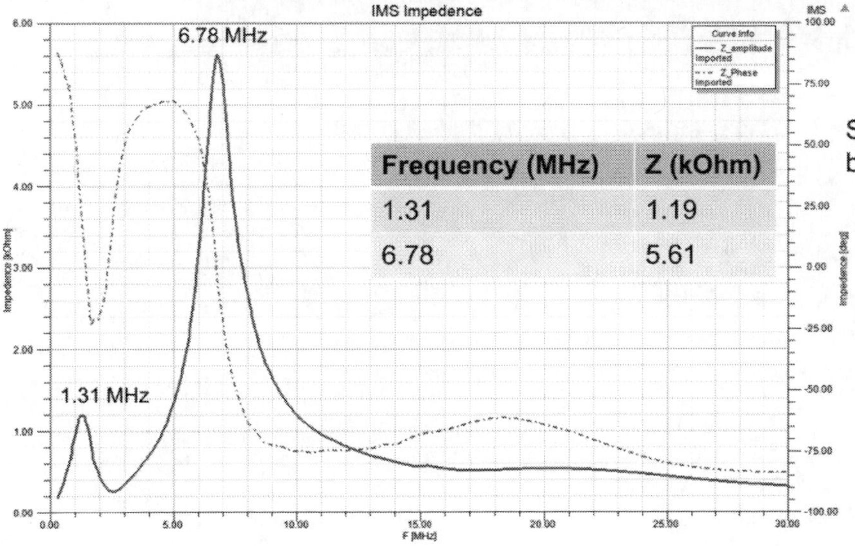

Frequency (MHz)	Z (kOhm)
1.31	1.19
6.78	5.61

Simulated Z between A and B

Conducted emission measure from battery charger

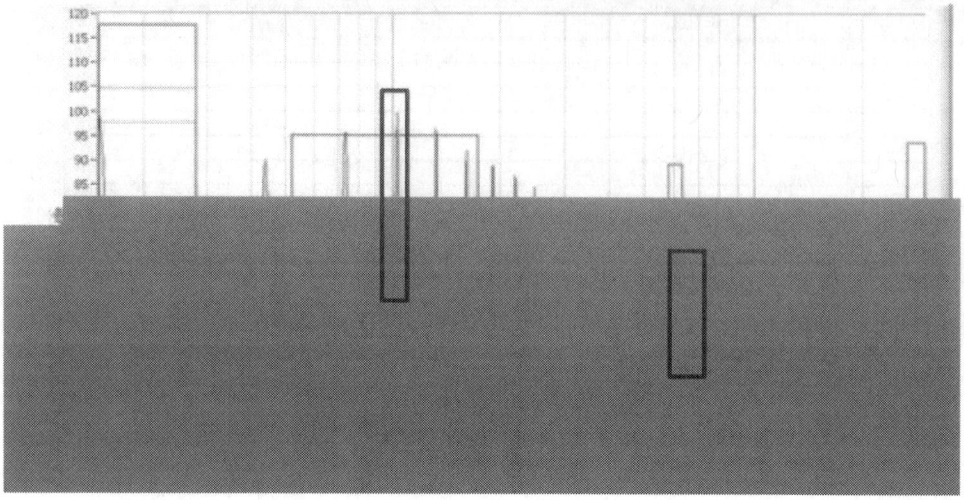

Emission spectrum is closely related to the impedance of IMS board

EMI Filter Design: Stand alone filter

Three Stages EMI filter designed with Mentor Pad and imported to SIwave

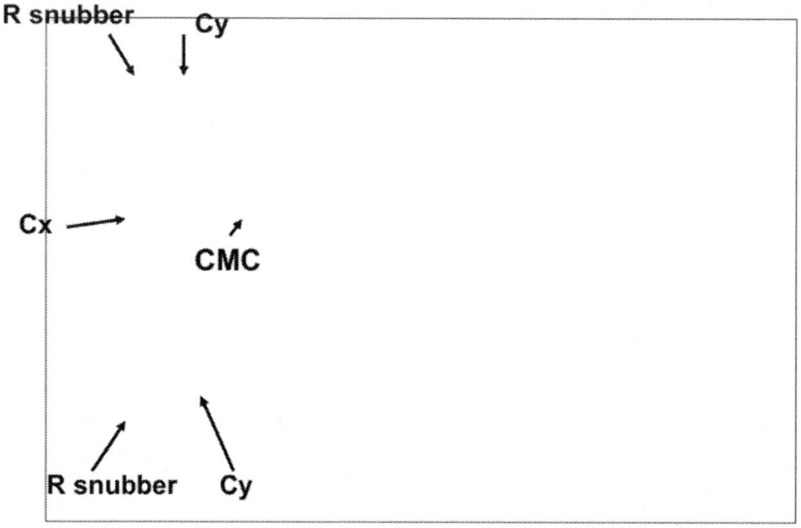

EMI Filter Model

THE GRAPHIC ON THIS PAGE WAS NOT AVAILABLE ON THE E-MEDIA VERSION OF THIS PAPER.

This study considers the **metallic enclosure**;
the connection points, between PCB and enclosure, are modelled as **4 Vias**

EMI Filter Model

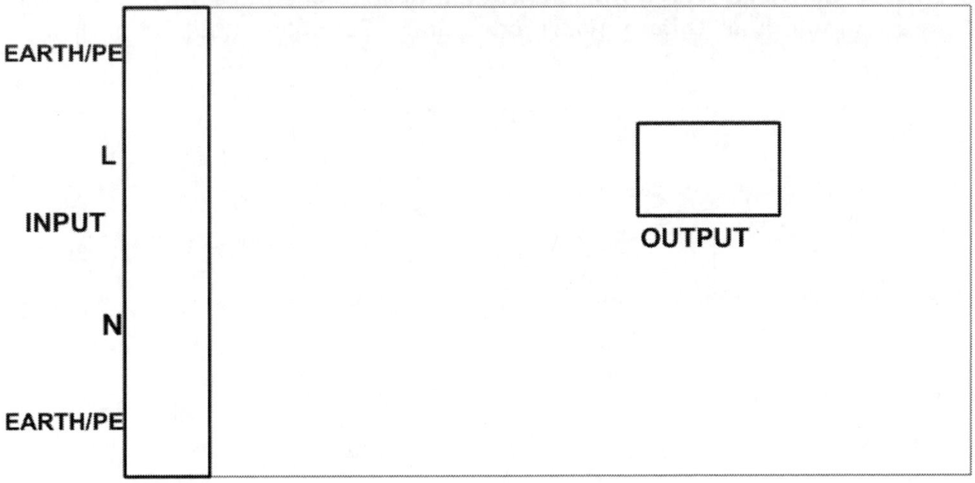

SIwave Model: ports created at the discrete components and input/output points

Discrete components: Capacitors

Discrete components (Capacitor Cx and Cy, Common mode choke and snubber resistor) are modeled with **Scattering parameters**; measures are carried out with VNA, in the frequency range 0,3-30MHz.

Capacitors

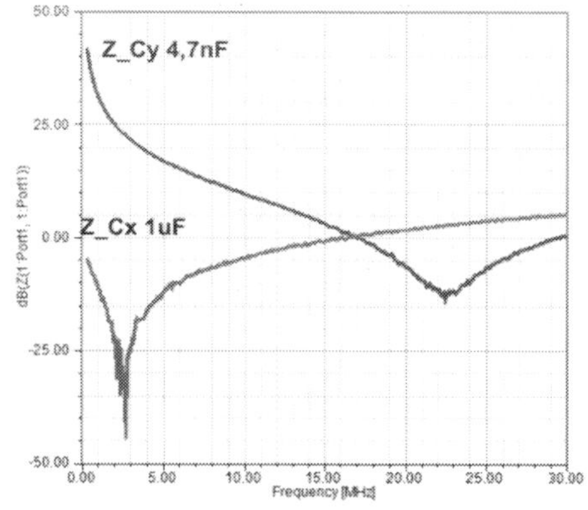

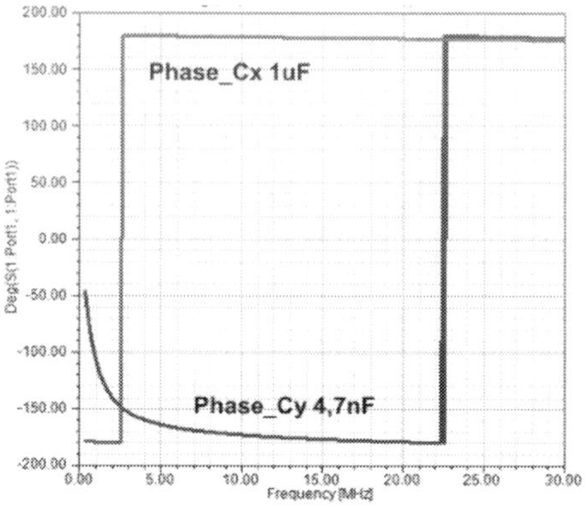

Discrete components: Common mode choke

Common mode choke has two different behaviours, linked to the type of noise: **Common mode and Differential mode**. Only Common mode behaviour is considerated.

Common mode choke

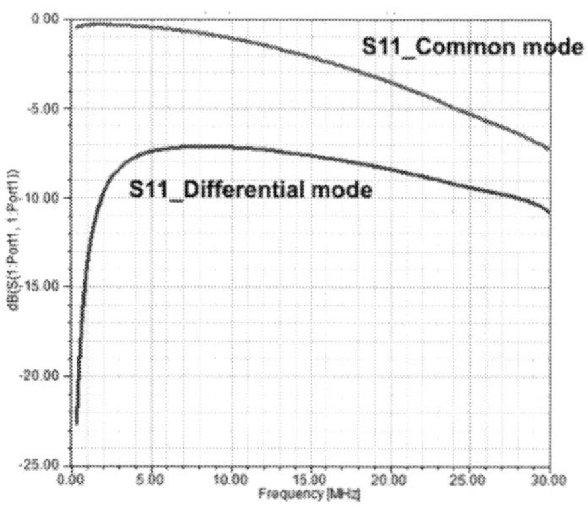

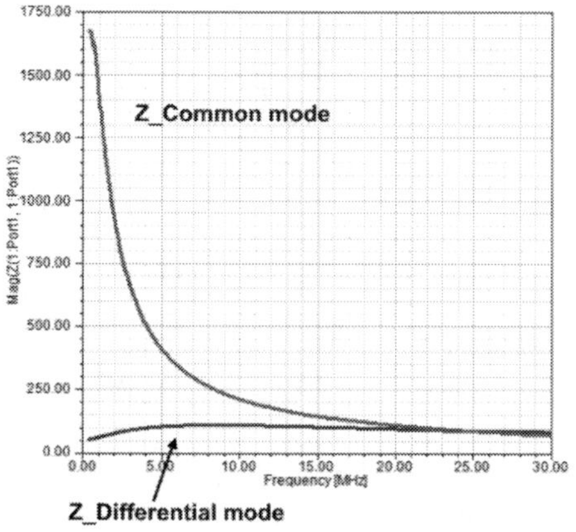

Schematic in Designer

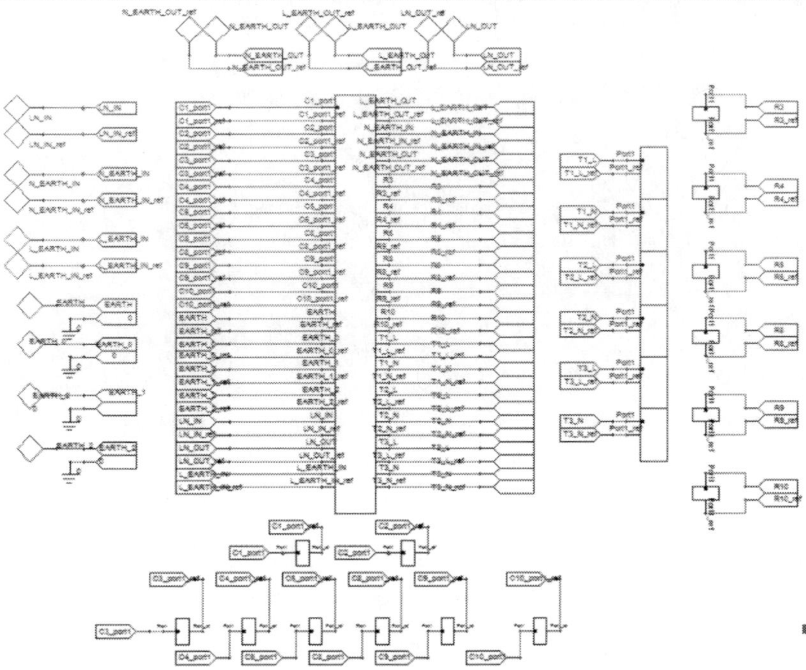

S parameter of **board** and S parameter of **discrete components**, are combined in Designer

Results: Stand alone filter Attenuation

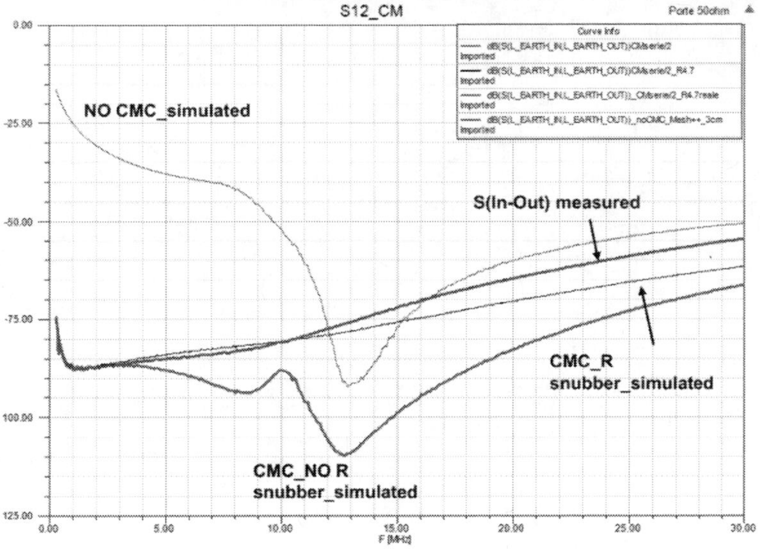

S(In-Out) simulated and measured between Phase and EARTH.

The snubber resistors degrade the attenuation but reduce resonances.

META SYSTEM

Integrated EMI filter

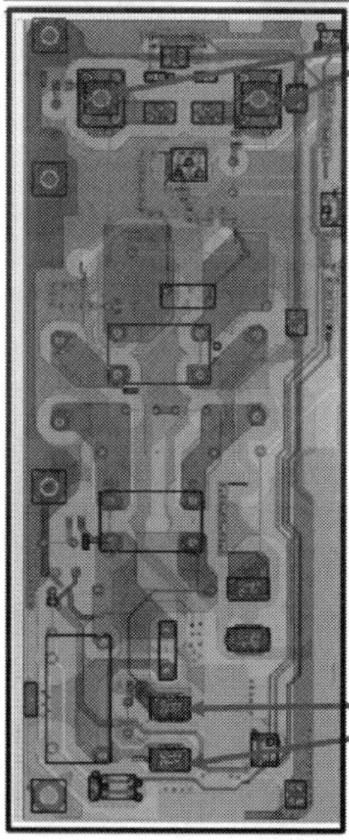

Connection to LISN
(L_out; N_out)

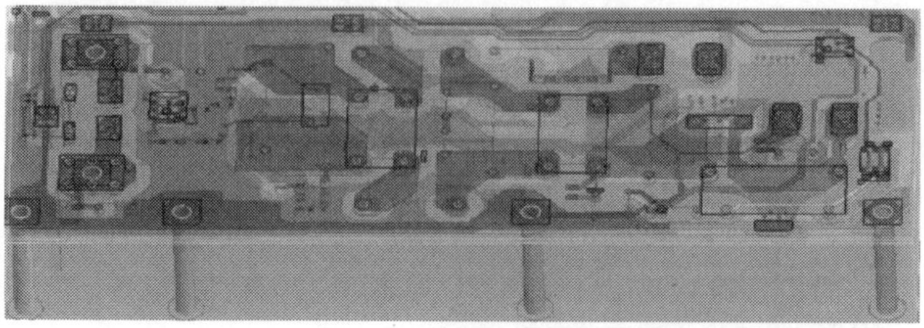

Connection with IMS
(L_in; N_in)

Filter components	
3	Cx 1uF
6	Cy 4,7nF
2	50uH Inductance
2	CMC
4	Snubber resistance

META SYSTEM

Schematic in Designer

S parameter from Filter+Power board

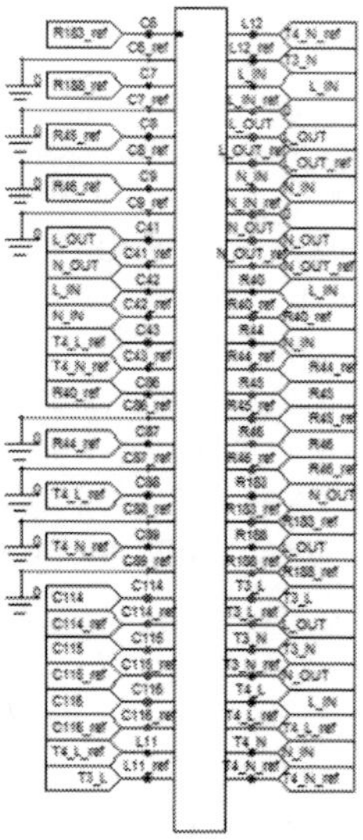

Discrete components

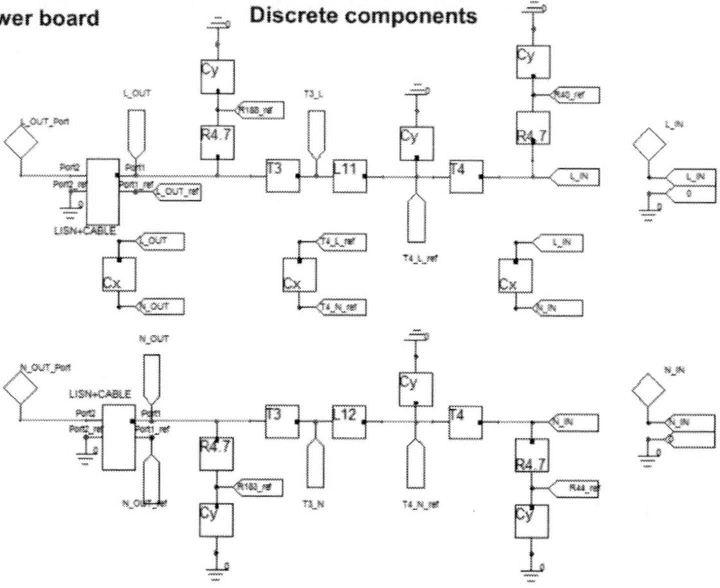

Result 2: Integrated EMI filter Attenuation

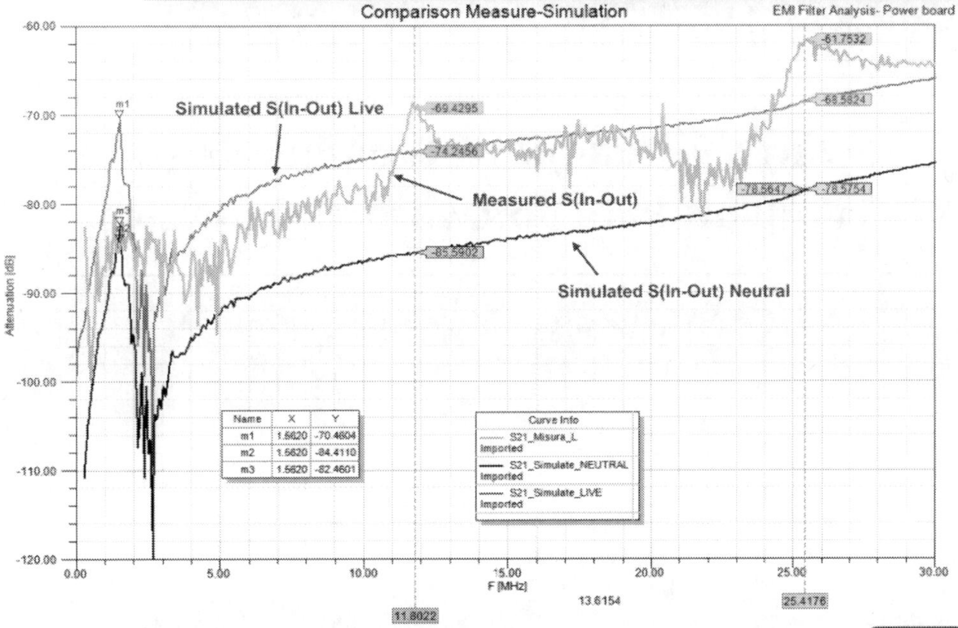

Name	X	Y
m1	1.5620	-70.4604
m2	1.5620	-84.4110
m3	1.5620	-82.4601

MetaSystem

Pag. 21

Conclusions

- **Impedence analysis and EMI filter design are investigated with Ansys Siwave and Designer software.**

- **The performance analysis of the EMI filter must consider the effects of the board and the real behavior of the components.**

- **Comparison between simulations and measurements of attenuation shows a good match in a wide frequency range for both filter.**

- **For the integrated filter, difference between LIVE and NEUTRAL are due to the different route of nets and positions of components.**

- **At points of mismatch between measurement and simulation, models of the CMC must be improved, to take account of more parasitic effects, such as the parasitic capacitances between turns.**

MetaSystem

Pag. 22

2013 Automotive Simulation World Congress

Towards virtual prototyping: Integrated simulation of software and physics

Christian Schrader, Esterel Technologies

When software & electronics become dominant in a product architecture

- Larger solution space – more design degrees of freedom
- Multiplicity of architectural interfaces have to be designed, tested and maintained
- Electronics & software are increasingly the primary cause of catastrophic failure modes

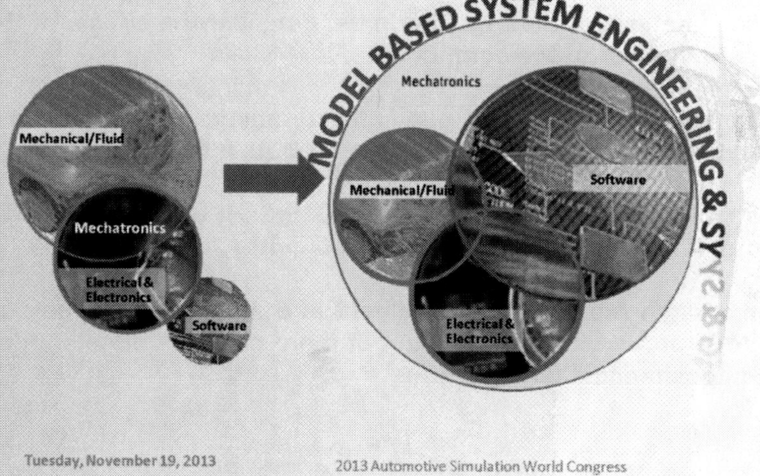

Manage Complexity
to design innovative, market leading products

Coordinate Interdisciplinary Engineering
to reduce design changes and development costs

Early & Reliable Verification
to deliver high quality, safe, and reliable products to the market faster

What is Embedded Software?

- Embedded software is the code embedded into cars, planes, industrial equipment and other modern products.
- It dynamically acquires data
- It monitors and controls the behavior of the physical product (electronics and hardware systems), as well as the human/machine interactions via GUIs.
- Software engineers need to simulate the embedded software behavior prior to its production.
- Products differentiate themselves increasingly by their software content, from a few thousand to several millions of lines of code.

SCADE Display integrates with SCADE Suite and ANSYS Simplorer for accurate plant modeling and co-simulation

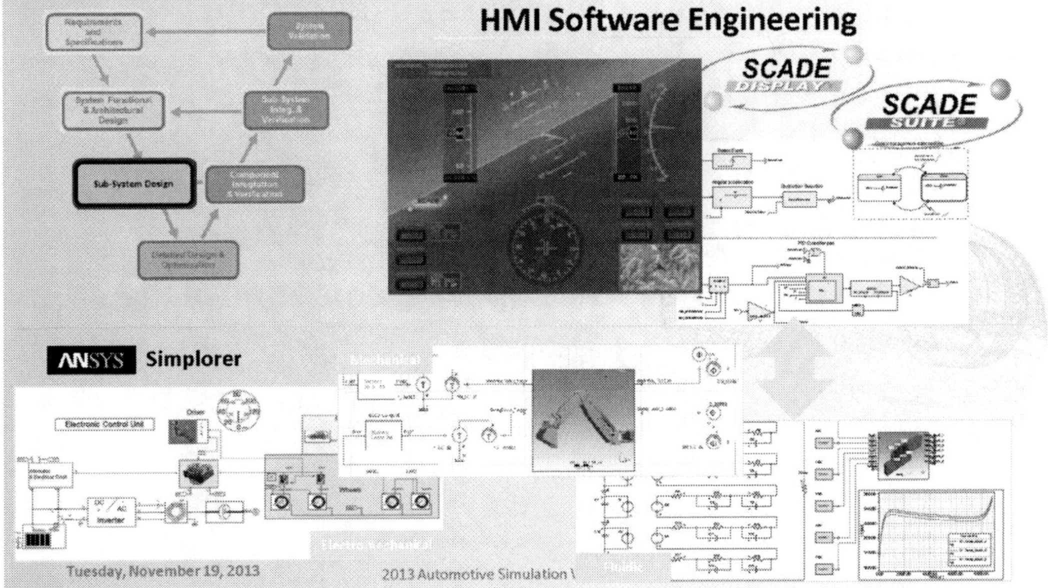

Example

- Electro-Mechanical System: **Automotive Entertainment Display**

- Rotational actuator inside console
- Mechanical setup trans-lates actuation into the effective movement of the screen
- Complex mechanics require controller for the actuator

Two scenarios for virtual prototype

- Opening and closing the display under normal conditions

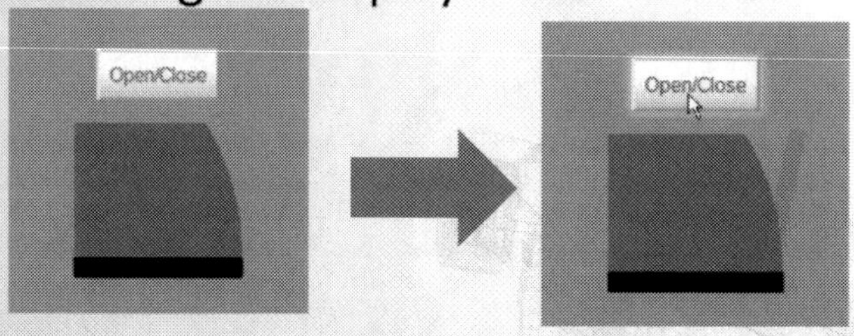

- Detect obstruction while opening and retrieve the display again

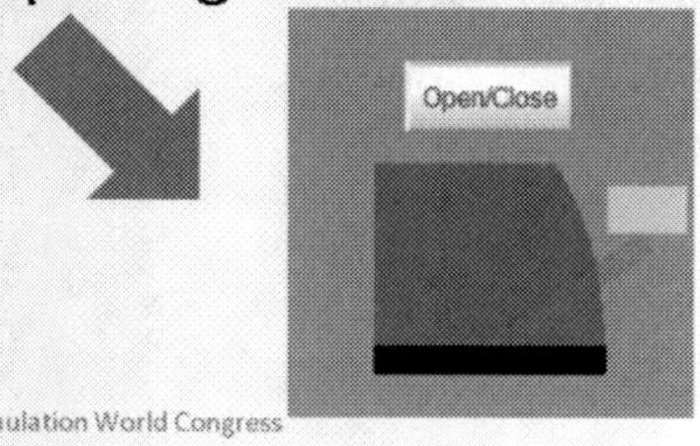

1st virtual prototype

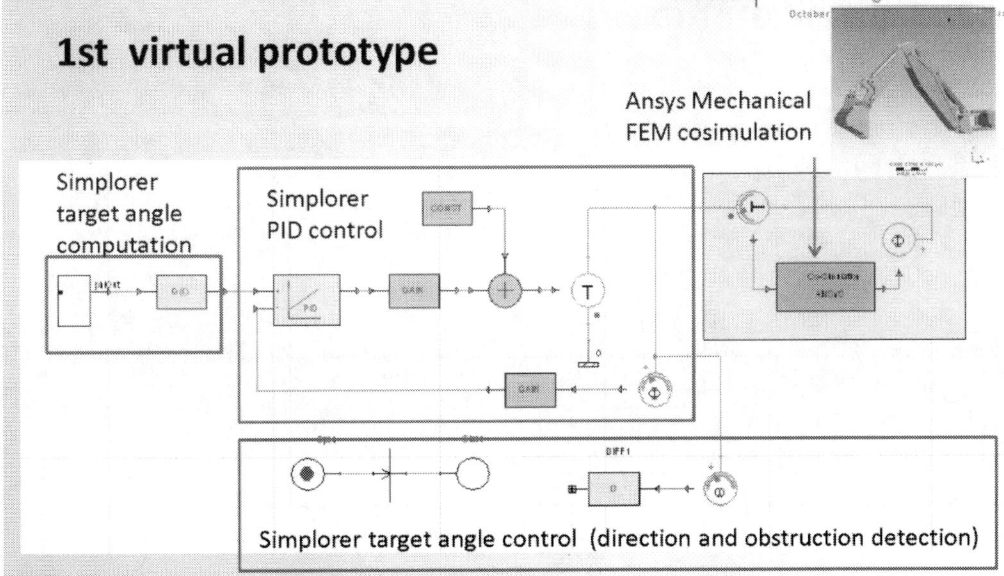

1st virtual prototype - Quick summary

- Conceptual development of controller part
- Validation with FEM simulation in the loop

Next step:

- Replacing the controller part with the real controller software

2nd virtual prototype

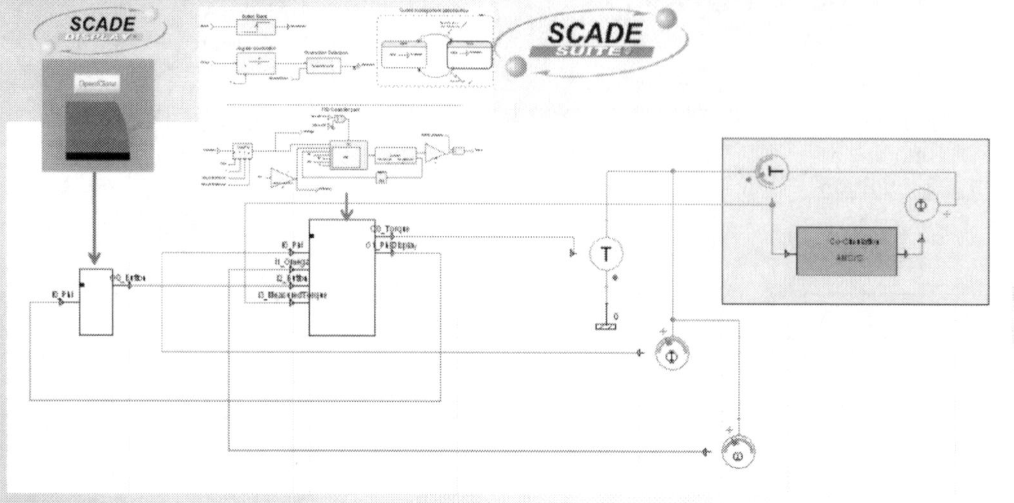

2013 Automotive Simulation World Congress
October 29 - 30 • Frankfurt am Main, Germany

A deeper look at the SCADE Suite controller model

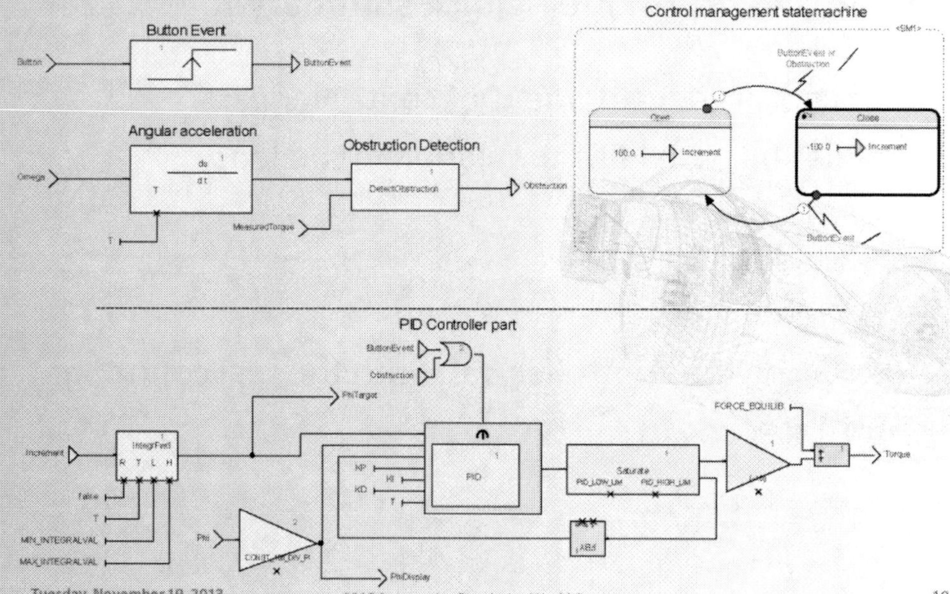

SCADE Suite

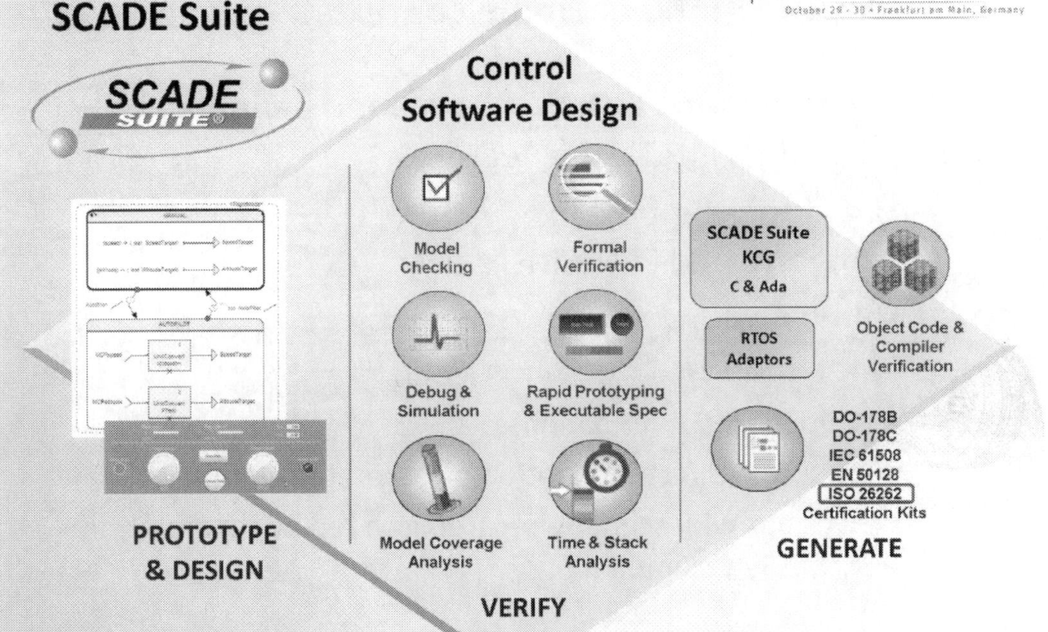

SCADE Suite at Subaru Electric Vehicles

- Subaru chose SCADE Suite for the design of its electric vehicle engine controls:

- Vehicle dynamics
- Engine functions
- Vehicle energy consumption
  - heating & air conditioning
  - breaking
  - body controls
- Battery load management

SCADE Suite in the example context

- In addition to the ISO 26262 oriented benefits the SCADE offers interesting features:
  - Direct tool support for Simplorer and FMI/FMU coupling
  - Three modes:
    - White box (co-simulation)
    - Black box (direct integration of generated code)
    - Black box with scenario recording to replay separately in SCADE Suite Simulator

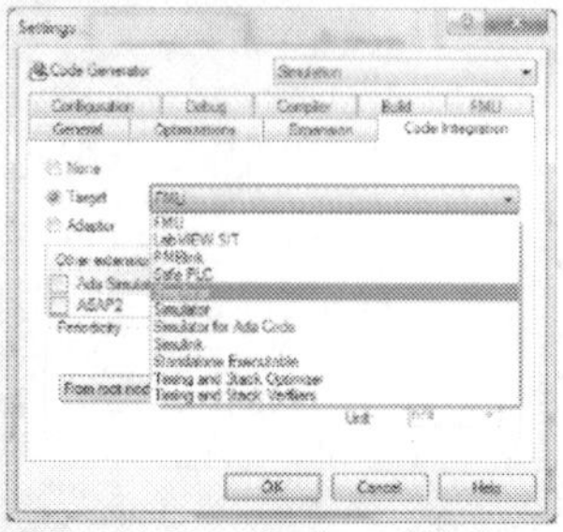

SCADE Display

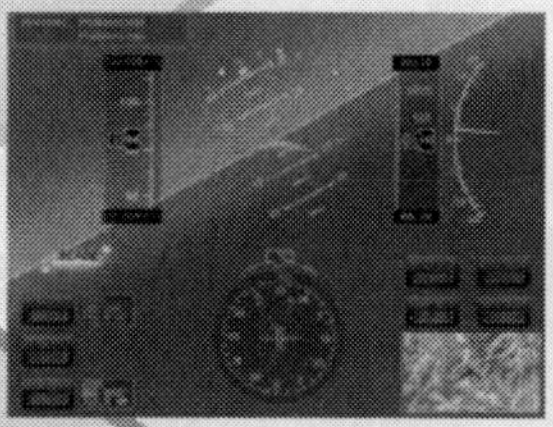

2013 Automotive Simulation World Congress
October 29 - 30 • Frankfurt am Main, Germany

SCADE Display in the example context

- Enables the user to interact with the simulation
- Enhanced flexibility is useful for prototyping
  - Revealed some problems that were not present in the static scenario of the first prototype ;-)
- Off topic: The HMI of the entertainment display itself could be designed using SCADE Display

2013 Automotive Simulation World Congress
October 29 - 30 • Frankfurt am Main, Germany

Next steps

- Simplorer serves as a simulation hub which can be enriched with other more detailed models to increase accuracy
  - Ideal torque sorce could be replaced by an ideal electric motor
  - ...or a more detailed motor model involving other FEM cosimulation (e.g. Maxwell for electro magnetics)
- To reduce simulation time FEM cosimulation blocks can be replaced by reduced order model (ROM) blocks

Conclusion

- Simulation Driven Product Development of Complete Systems enables efficient front loading in systems development

DEMO

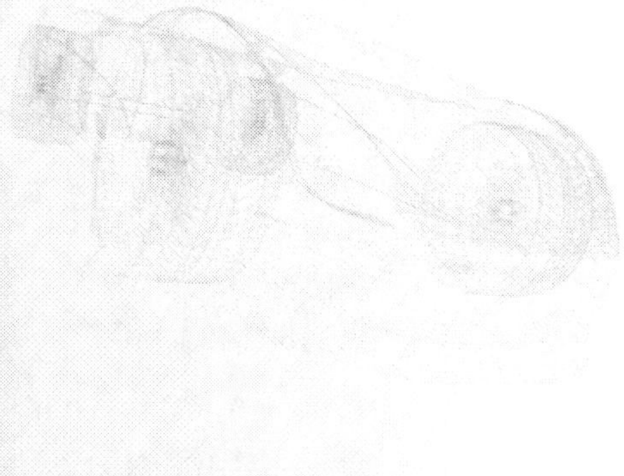

2013 **Automotive Simulation** World Congress

Effective Automotive System Design and Validation

Through the Adoption of Standardised Modelling Interfaces and Models Such as the VDA FAT-AK30 Library

Gerd Prillwitz

ANSYS, Inc.

Agenda

- **Motivation**
- **ANSYS' System Simulation Strategy**
- **VHDL-AMS**
  - **Overview**
  - **VHDL-AMS Libraries**
    - VHDL-AMS VDA AK30 Library (Data Driven Modeling)
    - VHDL-AMS HEV Library (Component Based Modeling)
- **FMI**
  - **Overview**
  - **FMI Support in ANSYS**
- **ROM and Co-simulation**
  - **State-space ROM**

Motivation

2013 Automotive Simulation World Congress
October 29 - 30 • Frankfurt am Main, Germany

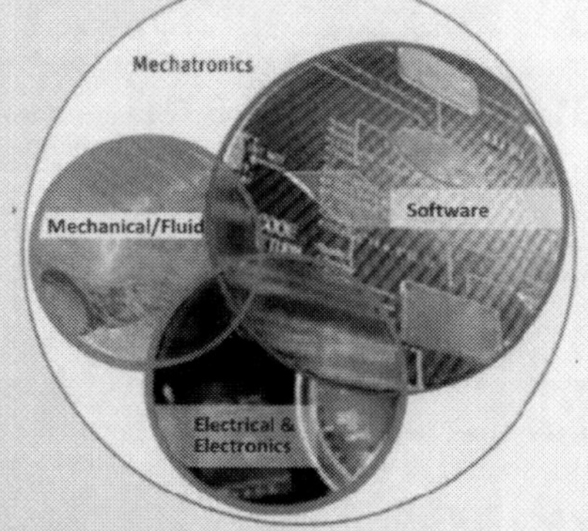

Manage Complexity
to design innovative, market leading products

Early & Reliable Verification
to deliver high quality products to the market faster

Coordinate Interdisciplinary Engineering
to reduce design changes and development costs

Agenda

2013 Automotive Simulation World Congress
October 29 - 30 • Frankfurt am Main, Germany

- Motivation
- **ANSYS' System Simulation Strategy**
- VHDL-AMS
    - Overview
    - VHDL-AMS Libraries
        - VHDL-AMS VDA AK30 Library (Data Driven Modeling)
        - VHDL-AMS HEV Library (Component Based Modeling)
- FMI
    - Overview
    - FMI Support in ANSYS
- ROM and Co-simulation
    - State-space ROM

Our Systems Engineering Vision

2013 Automotive Simulation World Congress
October 29 - 30 • Frankfurt am Main, Germany

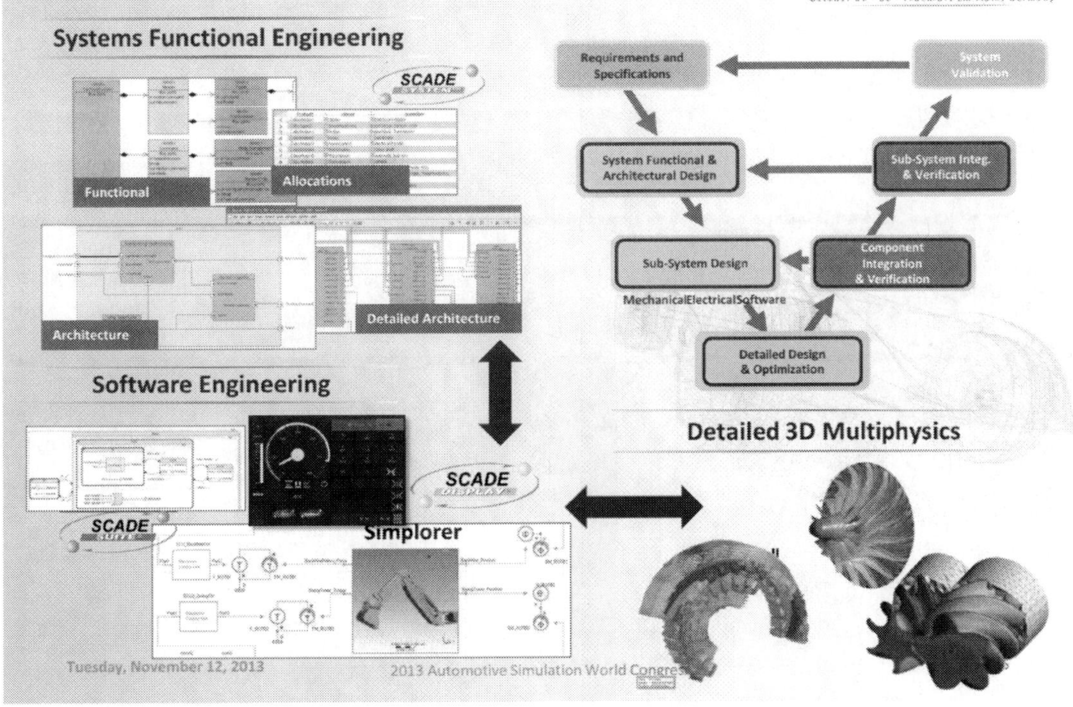

Tuesday, November 12, 2013 — 2013 Automotive Simulation World Congress

ANSYS System Strategy

2013 Automotive Simulation World Congress
October 29 - 30 • Frankfurt am Main, Germany

Enabling technology built on open standards
- SysML – VHDL-AMS – Modelica (in development)
- FMI/FMU standards for co-simulation
- Python scripting environment

Strong Links to Physics Based Solver
- ROM Technology – Cosimulation

Interoperability with 3rd Party Tools
- Matlab/Simulink
- LMS AMESim
- Dassault Systèmes Dymola
- NI LabView Veristand
- CAD & PLM neutral
- and many others via FMI/FMU ...

Tuesday, November 12, 2013 — 2013 Automotive Simulation World Congress — 6

Simulation Driven Product Development™

2013 Automotive Simulation World Congress
October 29 - 30 • Frankfurt am Main, Germany

	Model	Simulation Result	Tool
Requirement	Req 23: *On request, the valve should close in 500us*	X	MSWord, DOORS, ...
Functional			SCADE System
System – Behavioral Modeling	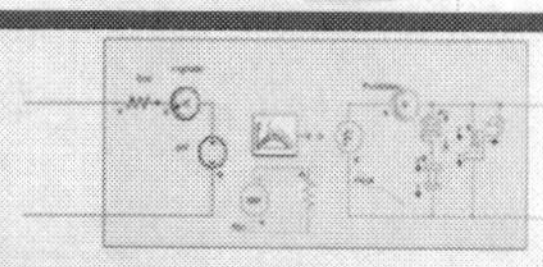	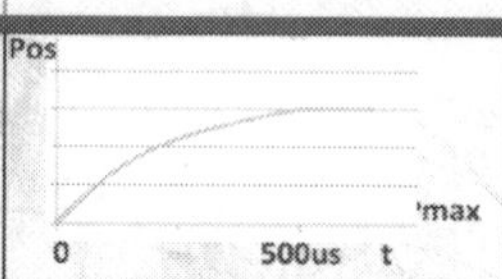	Simplorer VHDL-AMS Modelica FMI
Physics Based - Open loop validation	Actuator 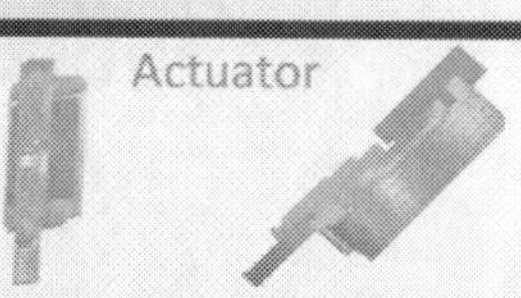	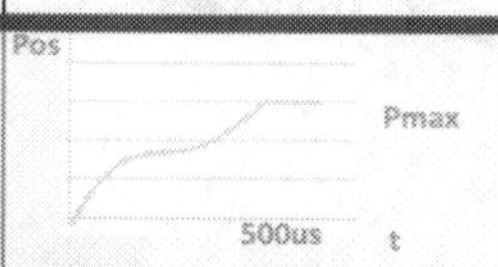	Fluent, Maxwell, Mechanics, HFSS
System - ROM Based - Close loop validation		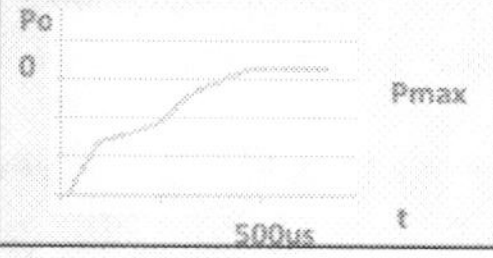	Simplorer with SCADE Suite

Agenda

2013 Automotive Simulation World Congress
October 29 - 30 • Frankfurt am Main, Germany

- **Motivation**
- **ANSYS' System Simulation Strategy**
- **VHDL-AMS**
  - **Overview**
  - **VHDL-AMS Libraries**
    - VHDL-AMS VDA AK30 Library (Data Driven Modeling)
    - VHDL-AMS HEV Library (Component Based Modeling)
- **FMI**
  - **Overview**
  - **FMI Support in ANSYS**
- **ROM and Co-simulation**
  - **State-space ROM**

What is VHDL-AMS

Very High Speed Integrated Circuit **H**ardware **D**escription **L**anguage – **A**nalog and **M**ixed-**S**ignal

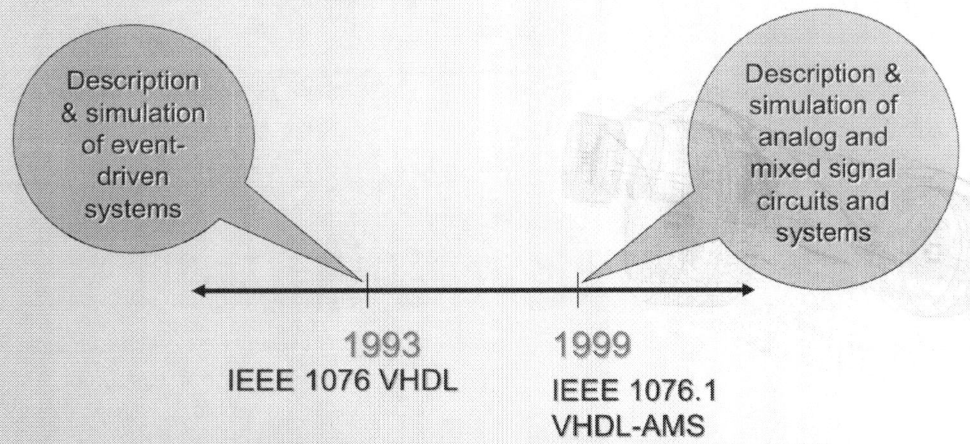

VHDL-AMS is a strict superset of IEEE Std. 1076

*Latest version of VHDL-AMS is 1076.1-2007

2013 Automotive Simulation World Congress
October 29 - 30 • Frankfurt am Main, Germany

VHDL-AMS Model Construct

- Entity
  - Interface description
  - Specifies input and output
- Architecture
  - Behavior description
  - Dataflow, structural, procedural, etc.
  - Analog (continuous) and digital (discrete) domains

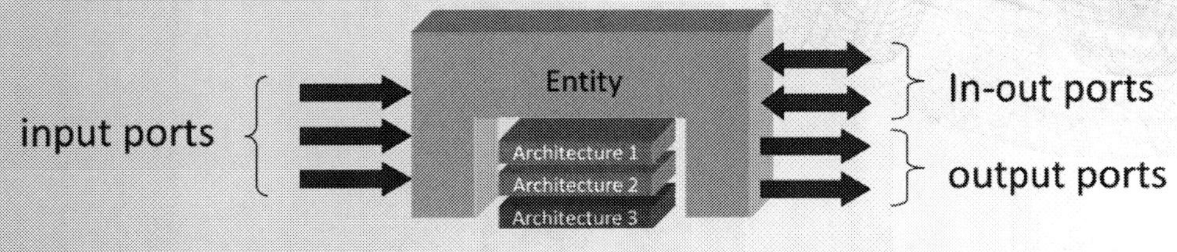

VHDL-AMS Model for a Capacitor

- **Entity**

```
LIBRARY Ieee;
use Ieee.electrical_systems.ALL;
ENTITY    cap IS
   GENERIC (
   capacitance : capacitance := 1.0e-3
   );
   PORT (
   QUANTITY init_v : IN voltage := 0.0;
   TERMINAL p :  electrical;
   TERMINAL n :  electrical
   );
END ENTITY cap;
```

The capacitor entity has one model constant, one input, and two terminals

- **Architecture**

```
ARCHITECTURE arch_cap OF cap IS
QUANTITY voltage ACROSS current THROUGH p TO n;
BEGIN

IF (domain = quiescent_domain) USE

 voltage == init_v;

ELSE

 current == capacitance * voltage'dot;

END USE;

END ARCHITECTURE arch_cap;
```

The model description essentially has two lines, one for initial condition and one for the equation!!

Agenda

2013 Automotive Simulation World Congress
October 29 - 30 • Frankfurt am Main, Germany

- **Motivation**
- **ANSYS' System Simulation Strategy**
- **VHDL-AMS**
  - **Overview**
  - **VHDL-AMS Libraries**
    - VHDL-AMS VDA AK30 Library (Data Driven Modeling)
    - VHDL-AMS HEV Library (Component Based Modeling)
- **FMI**
  - **Overview**
  - **FMI Support in ANSYS**
- **ROM and Co-simulation**
  - **State-space ROM**

VHDL-AMS Libraries for HEV

2013 Automotive Simulation World Congress
October 29 - 30 • Frankfurt am Main, Germany

- Two complementary approaches: Data Driven modeling (VDA AK30) and First Principles Modeling (HEV Library)

Data driven model (regressions and lookup):
- Easy to construct
- Ready for measurement data
- Working range limit to existing data, need proper data set for each different condition
- Hard to resize and adjust
- Example: VDA FAT-AK30

First principle/mathematical model:
- Need to develop from literature
- Not directly related to data
- Meaningful parameter settings
- Easier to modify and more physical understanding
- Better for design and testing
- Example: HEV Library

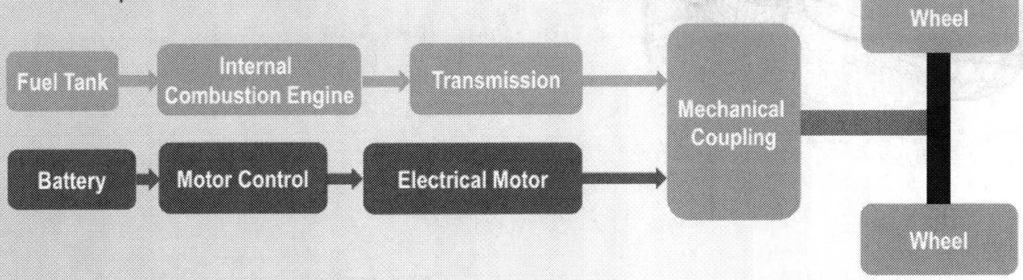

VDA FAT-AK30 Background

2013 Automotive Simulation World Congress
October 29 - 30 • Frankfurt am Main, Germany

- VDA: Federation of the Automotive Industry, consists of over 600 companies.
- They produce cars, Trailers, Buses, including spare parts and accessories. It includes AUDI AG, Volkswagen AG, Ford-Werke GmbH and many more.
- FAT-AK30: Research association automotive engineering – Working group 30
- FAT-AK30 started at July, 2003 with promotion of relationship between car manufacturers and their suppliers concerning simulation of mixed systems and model exchange.

More information: http://www.vda.de/de/verband/index.html
FAT-AK30 Library: http://fat-ak30.eas.iis.fhg.de/mb/index_en.html

14

Simplorer:
AK30 VHDL-AMS Libraries

- AK30 members developed generic *component libraries* frequently needed in automotive system design
  - FUNDAMENTALS_VDA – comprises basic models
    - Electrical, magnetic, mechanical, thermal, optical basic components
    - Transfer function blocks (continuous and discrete time)
    - Digital components, data conversion, sources, functions, LUT, ...
  - AUTOMOTIVE_VDA – special automotive components
    - Sources and loads
    - Powernet (battery, generator, fuses, lines, wires, lamps, ...)
    - Sensors, actuators, ...
  - SPICE2VHD – Spice3F5 comp's translated to VHDL-AMS
    - r, l, c, e, i, e(v, i), i(v, i), line, switches
    - D, BJT, MOSFET, JFET, ...
  - MEGMA – systems and controls

2013 Automotive Simulation World Congress
October 29 - 30 • Frankfurt am Main, Germany

Simplorer 2014 includes the full VDA-FAT AK30 VHDL-AMS Library

- 500+ Models for automotive/controls application
- Data driven, mostly lookup-table models with some dynamics
- Includes custom Simplorer symbols

~400 EV/HEV related models
in VHDL-AMS language
mixed signal, multidomain

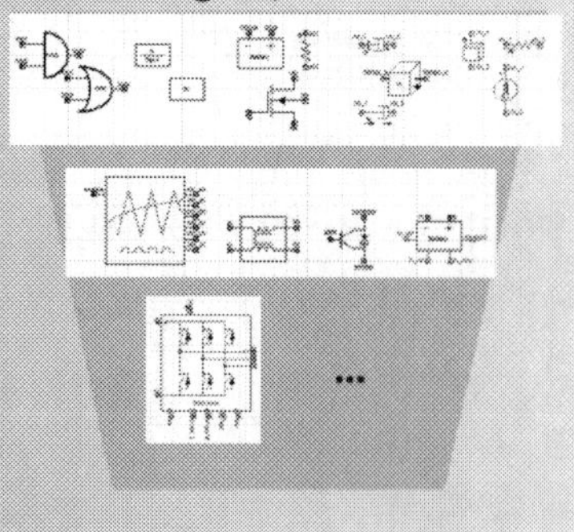

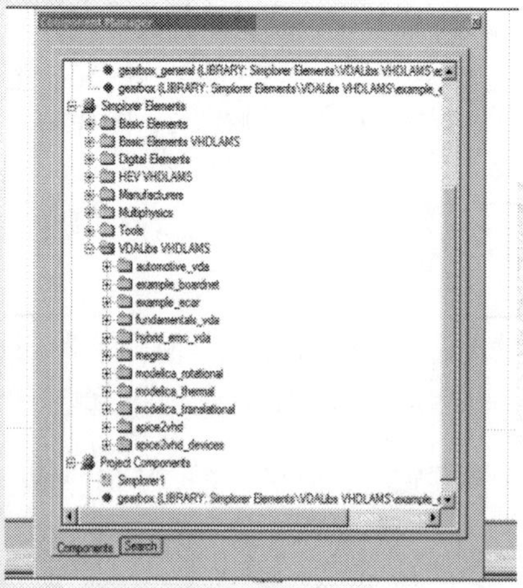

Example: Ecar

2013 Automotive Simulation World Congress
October 29 - 30 • Frankfurt am Main, Germany

- describe a 288V NiMH battery, 75 kW motor, a gearbox (ratio=7), and a car model similar to Ford Focus FCEV fuel cell vehicle
- The battery is connected to an electrical motor . The torque of the motor drives the gearbox, whose torque drives the vehicle.
- Profile_ecar includes drive cycle model and driver model, set demand torque for emotor and controls brake in vehicle

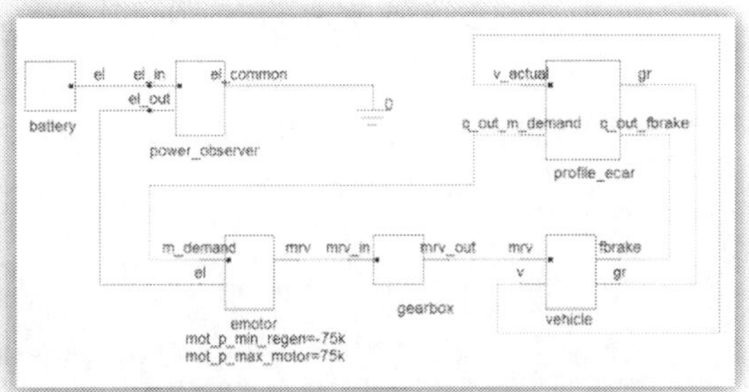

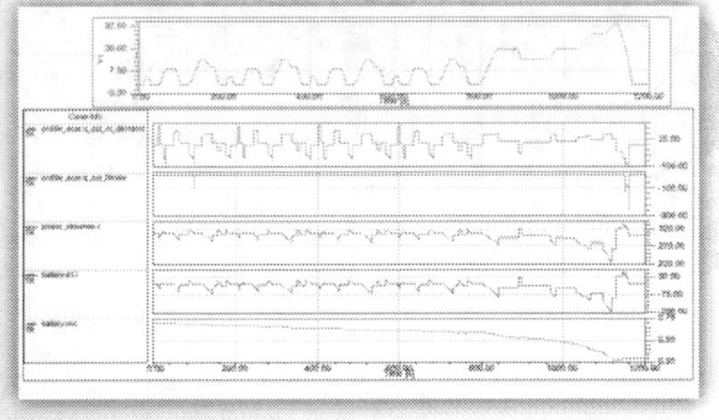

Example: Boardnet

2013 Automotive Simulation World Congress
October 29 - 30 • Frankfurt am Main, Germany

- The test-bench describes a simple boardnet that consists of a generator, a battery, a time-dependent load an some resistors
- The behavior of the generator is described by a curve family. The set of characteristic curves is defined by a file that contains in each row values of revolution, temperature, and generator current

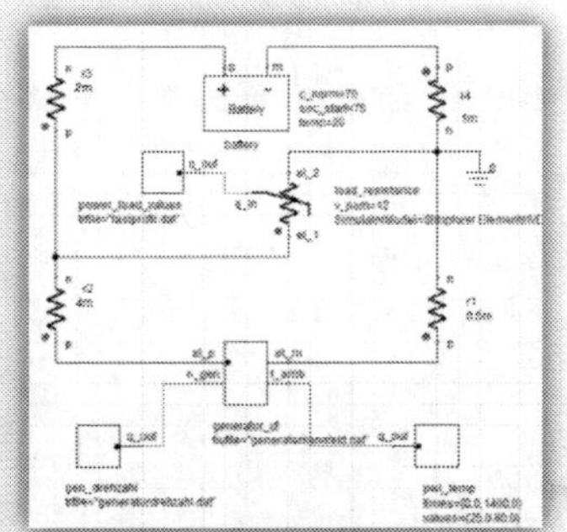

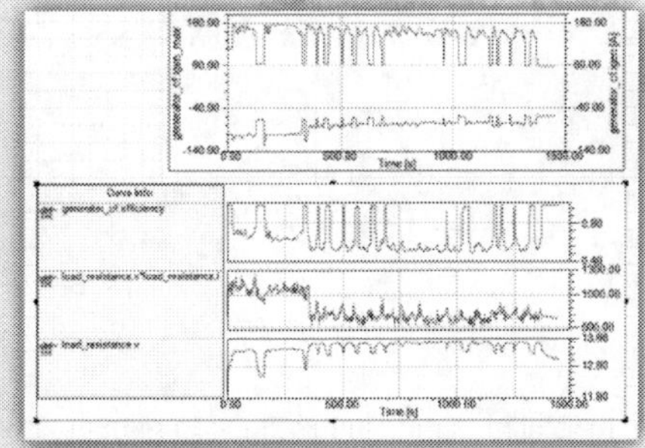

HEV Library in Simplorer R15

New "First Principles" based HEV Library is included in Simplorer R15

- Reusable and re-developable generic behavioral models of HEV components
- Framework for detailed component analysis (Battery, Motor, etc)

- HEV VHDLAMS
  - Data and Control
    - c_controller_cv: CV central controller model.
    - c_controller_ev: EV central controller model.
    - c_controller_hev: HEV central controller.
    - driver_hev: Driver model.
    - motor_controller_hev: Motor controller model.
  - Electrical Components
    - dc_em_hev: Permanent magnet DC motor (PMDC) model.
  - Mechanical Components
    - brake_hev: Single equivalent brake model.
    - chassis_hev: Vehicle body (chassis) model.
    - clutch_hev: Clutch model.
    - fuel_tank_hev: Fuel tank model.
    - ice_hev: So-called mean value internal combustion engine model.
    - mech_coupling_hev: Mechanical torque coupling model.
    - transmission_hev: Transmission model based on given data set.
    - wheel_hev: Wheel model.

Example: VHDL-AMS model of a PMSM

Input/Output: Electrical port (a,b,c). Mechanical port (mech_rv).

Park's Transformation: $V_{abc} = KV_{dq0}$, $I_{abc} = KI_{dq0}$ where

$$K = \begin{bmatrix} \cos(\theta_e) & -\sin(\theta_e) & 1 \\ \cos\left(\theta_e - \frac{2}{3}\pi\right) & -\sin\left(\theta_e - \frac{2}{3}\pi\right) & 1 \\ \cos\left(\theta_e + \frac{2}{3}\pi\right) & -\sin\left(\theta_e + \frac{2}{3}\pi\right) & 1 \end{bmatrix}$$

and

$$\begin{bmatrix} v_d \\ v_q \\ v_0 \end{bmatrix} = \begin{bmatrix} R_s i_d + \frac{d}{dt}\lambda_d - \lambda_q \omega_e \\ R_s i_q + \frac{d}{dt}\lambda_q + \lambda_d \omega_e \\ R_s i_0 + \frac{d}{dt}\lambda_0 \end{bmatrix}$$

where λ_d, λ_q are functions of (i_d, i_q)

$$\text{power} = \frac{3}{2}\left(v_d i_d + v_q i_q + 2 v_0 i_0\right)$$

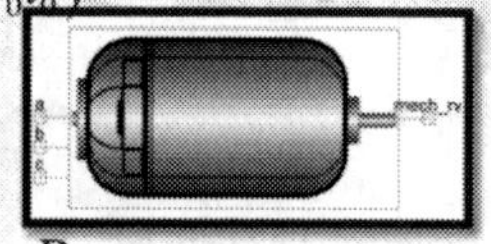

Torques and rotor dynamics: $\tau_e = \tau_{rotor} + \tau_m$

where $\tau_e = \frac{3}{2}\frac{P}{2}\left(\lambda_d i_q - \lambda_q i_d\right)$ and $\tau_{rotor} = J\frac{d\omega_m}{dt}$ where $\omega_e = \frac{P}{2}\omega_m$

Example: VHDL-AMS model of a PMSM

2013 Automotive Simulation World Congress

October 29 - 30 • Frankfurt am Main, Germany

```
Interface:
  quantity v_a across i_a through a to x;
  quantity v_b across i_b through b to x;
  quantity v_c across i_c through c to x;
```

```
function currents_lambda (current_d, current_q : real; current_d_vector,
current_q_vector, lambda_vector : real_vector)
return real is
        variable result  : real;
begin
        result := lookup_2d(current_d, current_q, current_d_vector,
current_q_vector, lambda_vector);
return result;
end function currents_lambda;
...
lambda_d == currents_lambda(i_d, i_q, id_vector, iq_vector, lambda_d_vector);
lambda_q == currents_lambda(i_d, i_q, id_vector, iq_vector, lambda_q_vector);
```

```
  tau == 3.0/2.0 * p_real/2.0 *(lambda_d * i_q - lambda_q * i_d);
-- rotor dynamics
  inertia * omega_m'dot == tau_r;

  phi_e == p_real/2.0 * phi_m;
  omega_e == p_real/2.0 * omega_m;
```

PMSM results compared with FEA

Compare with finite element analysis using Maxwell2D.
Differences are within 1%.

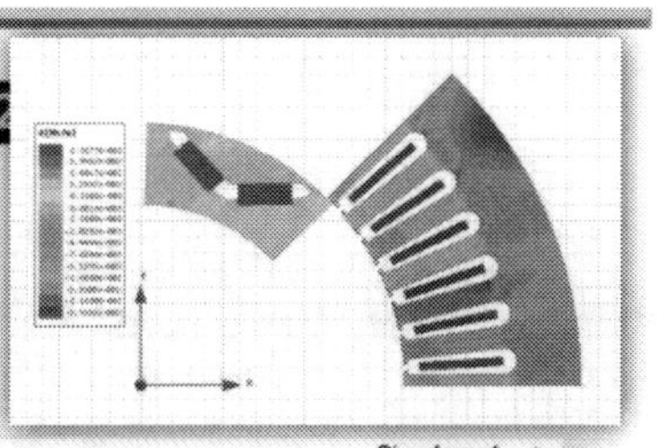

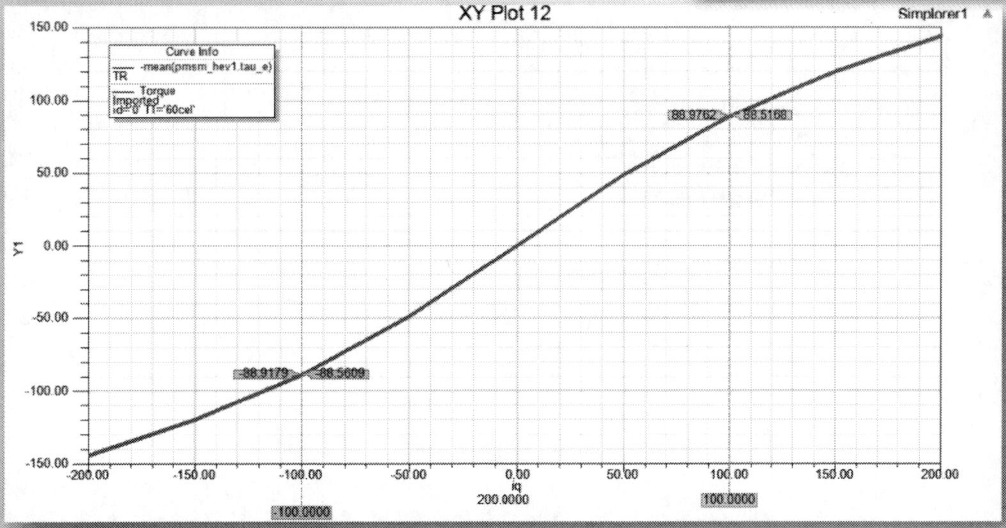

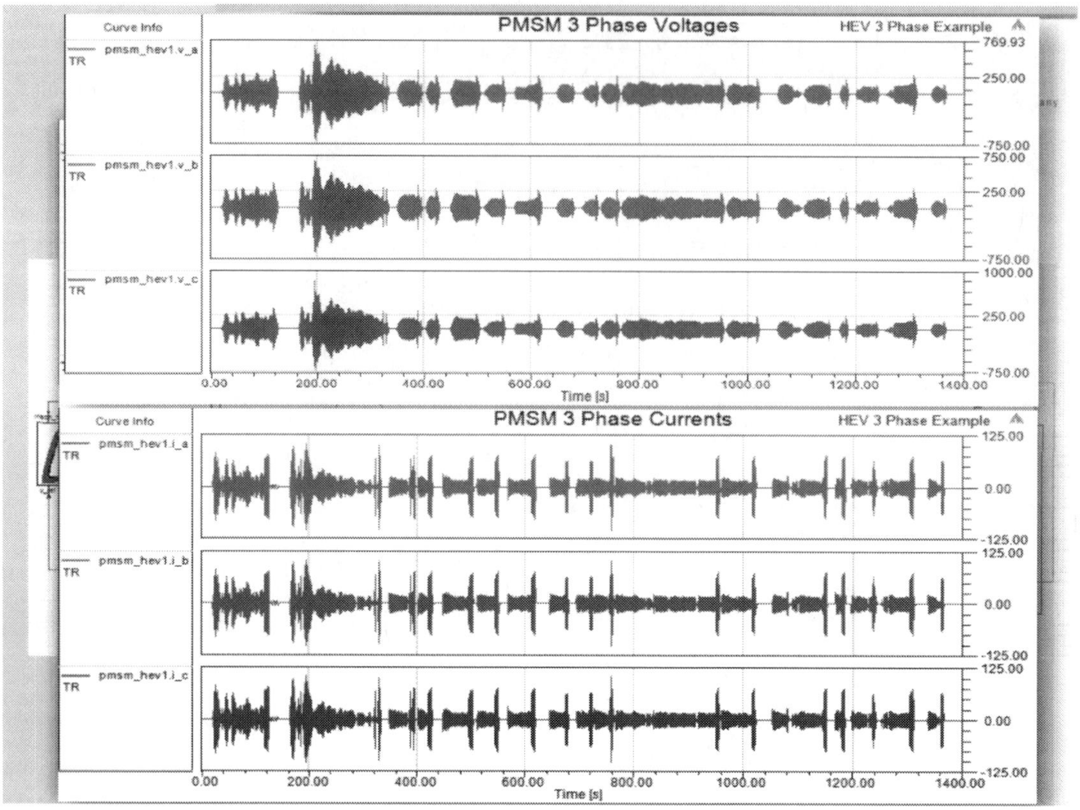

Agenda

2013 Automotive Simulation World Congress
October 29 - 30 • Frankfurt am Main, Germany

- Motivation
- ANSYS' System Simulation Strategy
- VHDL-AMS
  - Overview
  - VHDL-AMS Libraries
    - VHDL-AMS VDA AK30 Library (Data Driven Modeling)
    - VHDL-AMS HEV Library (Component Based Modeling)
- FMI
  - Overview
  - FMI Support in ANSYS
- ROM and Co-simulation
  - State-space ROM

Tuesday, November 12, 2013 2013 Automotive Simulation World Congress 24

Functional Mock-Up Interface (FMI): Motivation

2013 Automotive Simulation World Congress
October 29 - 30 • Frankfurt am Main, Germany

- Several types of Modeling Languages and Tools
- Need tool independent standard for model exchange and co-simulation
- Need to protect model IP of supplier
- Several tools offer model interfaces addressing above requirements

  Simplorer C interface, Simulink S-Functions, ADAMS, Silver, SIMPACK, SimulationX, etc.

  But all these interfaces are proprietary

Functional Mock-up Interface (FMI)

2013 Automotive Simulation World Congress
October 29 - 30 • Frankfurt am Main, Germany

- The Functional Mock-up Interface (FMI) is a tool independent standard to support model exchange and co-simulation of dynamic models
- Goal is to improve the exchange of simulation models between suppliers and OEMs

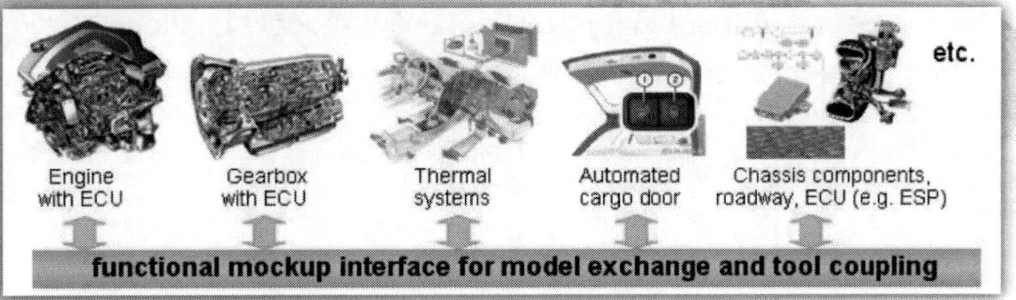

Supporting Tools

- Adams
- AMESim
- ANSYS Simplorer
- Asim – AUTOSAR
- Atego Ace
- @Source
- Building Controls Virtual Test Bed
- CATIA
- ControlBuild
- CosiMate
- Cybernetica CENIT
- Cybernetica ModelFit
- EnergyPlus
- FMI add-on for NI VeriStand
- FMI Blockset for Simulink
- FMI Library
- FMI Target for Simulink Coder
- FMI Toolbox for CarMaker
- FMI Toolbox for MATLAB
- FMU SDK
- ICOS "Independent Co-Simulation"
- JFMI
- Jmodelica.org
- LMS Virtual.Lab Motion
- MapleSim
- Mworks
- NI LabVIEW
- OpenModelica
- OPTIMICA Studio
- PyFMI
- RecurDyn
- SCADE Display
- SCADE Suite
- Silver
- SIMPACK
- SimulationX
- TLK FMI Suite
- TLK TISC Suite
- TWT Co-Simulation Framework
- TWT FMU Trust Centre
- xMOD

Functional Mock-Up Unit (FMU)

- A component which implements the FMI is called Functional Mockup Unit (FMU)
- Provided as one zip file (extension .fmu) containing:
  - Model description in XML – static information about the model such as definition of inputs, outputs, states and derivatives, unit information, etc.
  - Source (C) or binary files containing model equations
  - Model icon bitmap and documentations (optional)

FMI Interface Types

- Model-Exchange:

Defines the dynamics of the model by differential algebraic and discrete time equations. Depends on Solver for numerical integration.

- Co-Simulation:

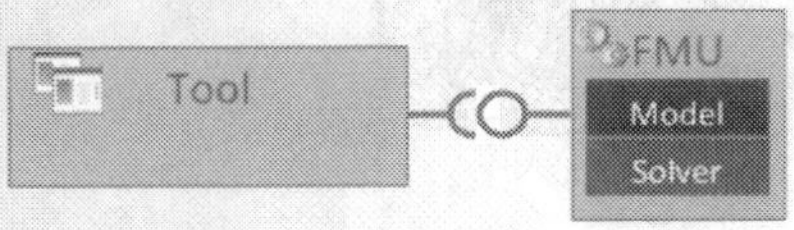

Defines coupling between simulation tools and subsystem models. Works master-slave manner with the target simulator.

2013 Automotive Simulation World Congress
October 29 - 30 • Frankfurt am Main, Germany

Cruise Control Example: Linking SCADE to Simplorer with FMI

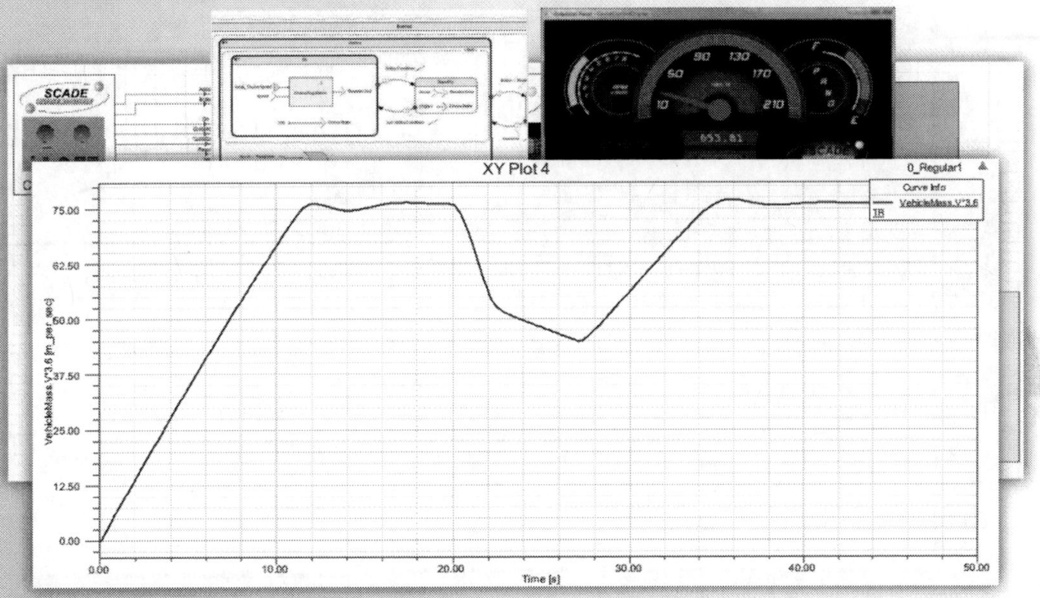

Agenda

- **Motivation**
- **ANSYS' System Simulation Strategy**
- **VHDL-AMS**
  - **Overview**
  - **VHDL-AMS Libraries**
    - VHDL-AMS VDA AK30 Library (Data Driven Modeling)
    - VHDL-AMS HEV Library (Component Based Modeling)
- **FMI**
  - **Overview**
  - **FMI Support in ANSYS**
- **ROM and Co-simulation**
  - **State-space ROM**

Reduced Order Models

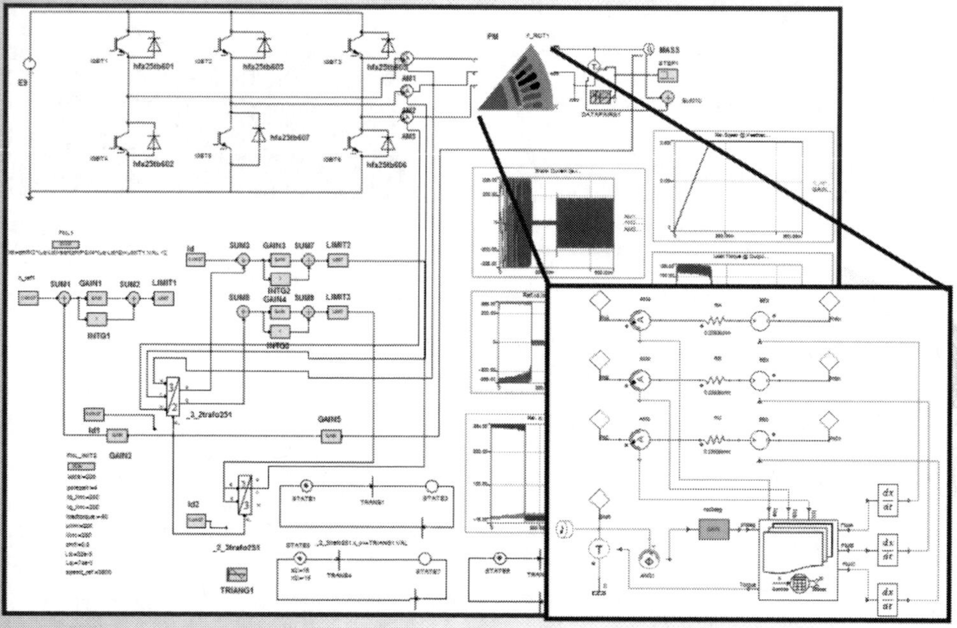

Thermal model

2013 Automotive Simulation World Congress
October 29 - 30 • Frankfurt am Main, Germany

ROM

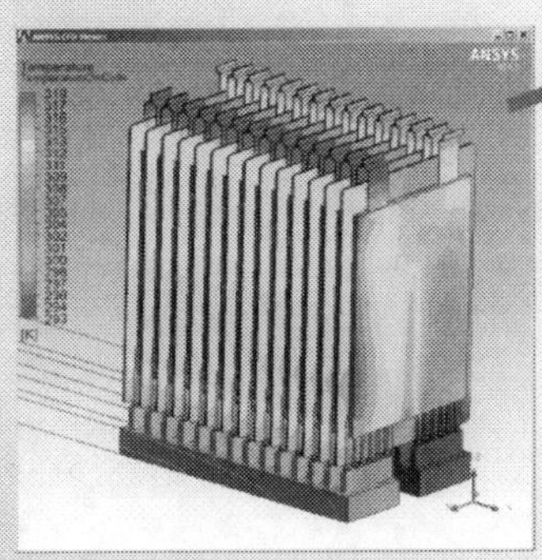

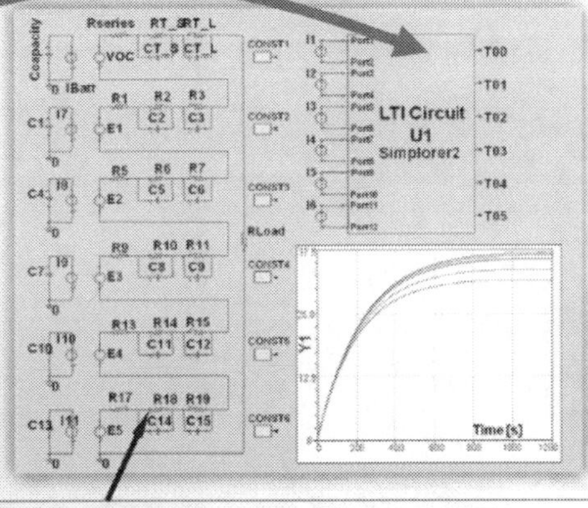

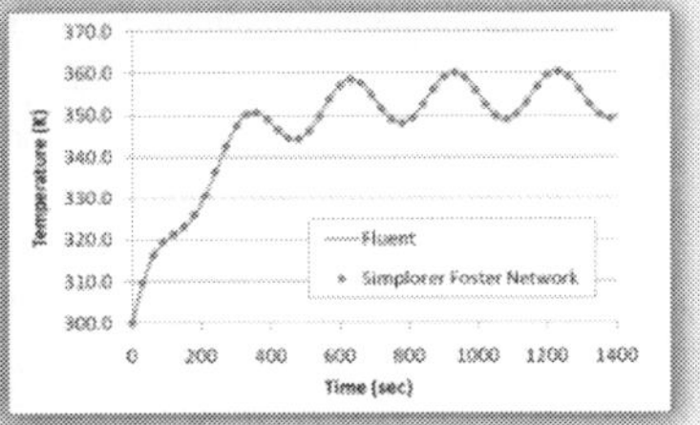

Voc = -1.031*exp(-35*(abs(IBatt.V/Vinit))) + 3.685 +
0.2156*(abs(IBatt.V/Vinit)) - 0.1178*(abs(IBatt.V/Vinit))^2 +
0.3201*(abs(IBatt.V/Vinit))^3 +
0.3/30.0*(U1.Temp_block_1-273)

Summary

2013 Automotive Simulation World Congress
October 29 - 30 • Frankfurt am Main, Germany

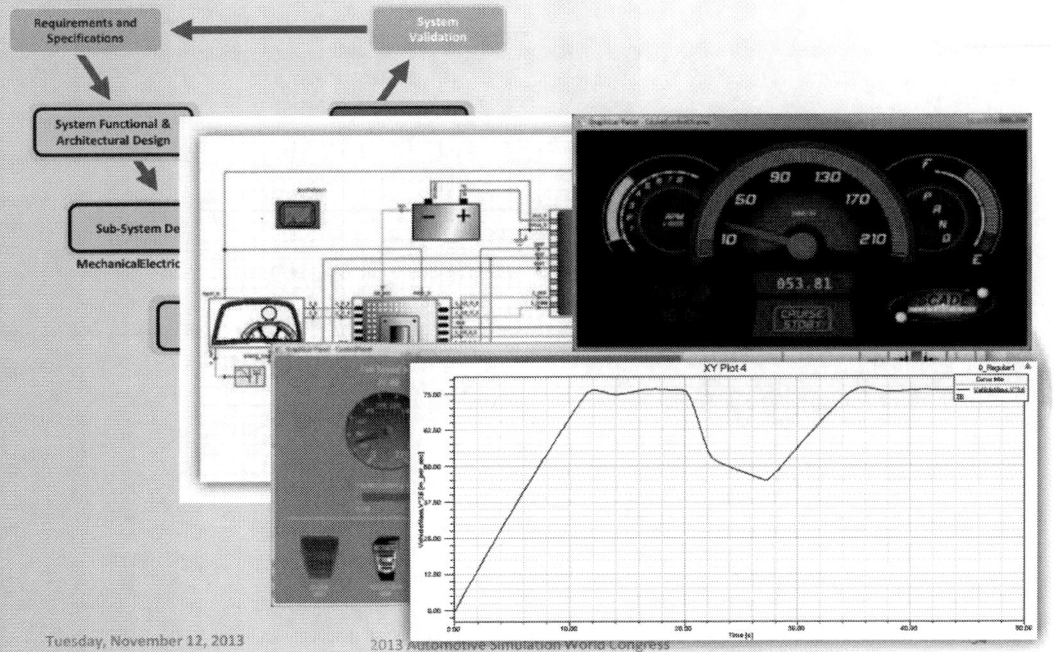

2013 Automotive Simulation World Congress
October 29 - 30 • Frankfurt am Main, Germany

Extra Slides

2013 Automotive Simulation World Congress
October 29 - 30 • Frankfurt am Main, Germany

Simplorer: an engine of innovation in EV/HEV design

Product capabilities that enables virtual prototyping to accelerate R&D procedure

Automotive industry faces the challenge of shorter innovation cycle

Multiphysics multitechnology simulation

Data driven simulation

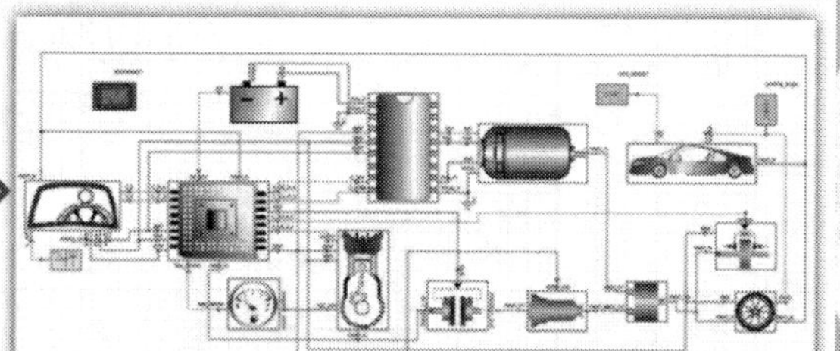

Support third party models

Capturing uncertainty in real world

ANSYS drives HEV Innovation through end-to-end Cell-Pack-System solution

- CAEBAT (Computer Aided Engineering for Batteries)
  - DOE program to develop software tools and models that enable automotive battery community to design and simulate batteries
  - GM-ANSYS team is one of CAEBAT industry partners
- Combine our product capabilities to provide end-to-end Cell/Pack/System solution
  - Include multiple physics and controls

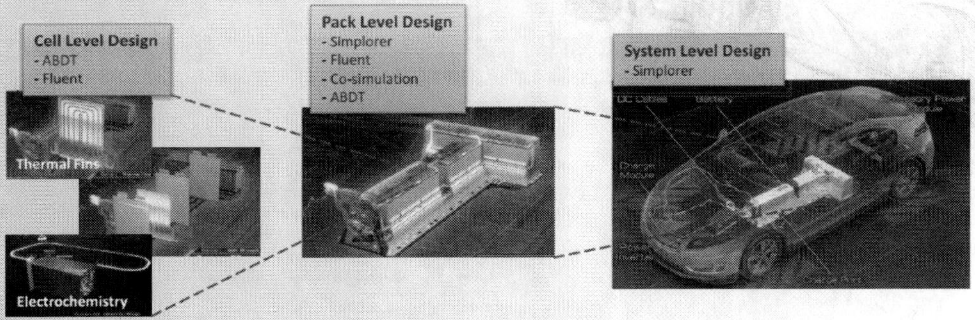

Newman Pseudo 2d Electrochemistry Model

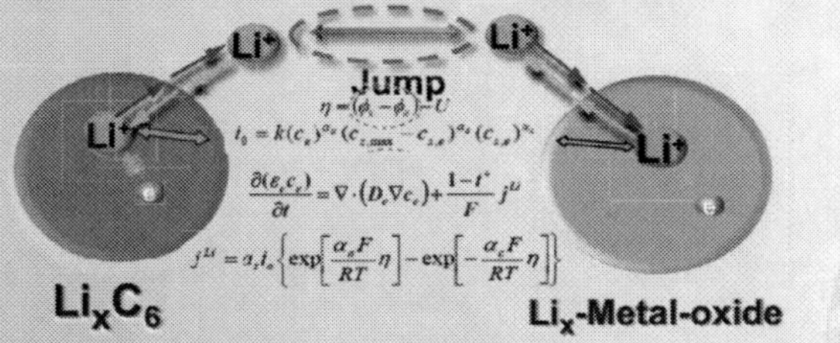

- Electrochemical Kinetics
- Solid-State Li Transport
- Electrolytic Li Transport
- Charge Conservation/Transport
- (Thermal) Energy Conservation

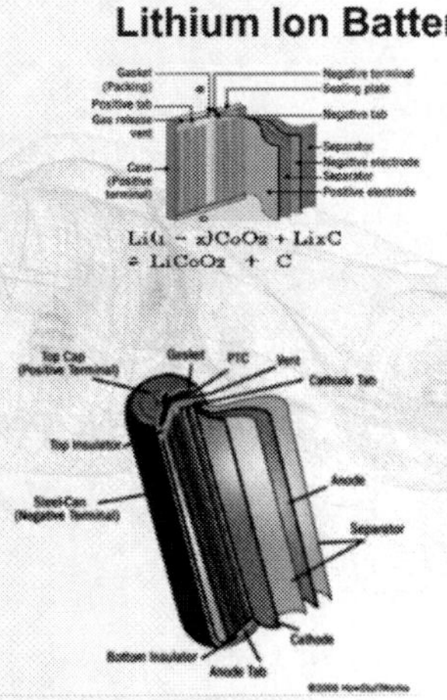

Lithium Ion Batteries

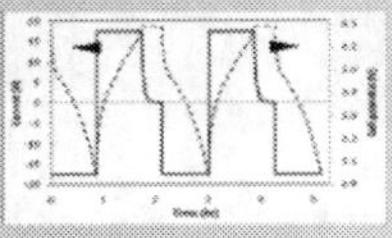

Results from Simplorer

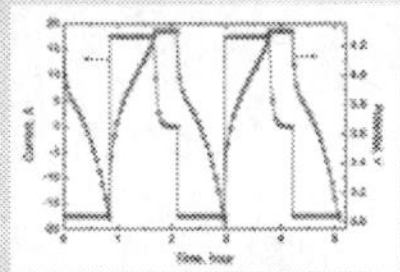

Results from Newman

2013 Automotive Simulation World Congress
October 29 - 30 • Frankfurt am Main, Germany

VHDL-AMS Model for Heat Conduction

Example:

$$\begin{cases} \dfrac{\rho C_p}{k}\dfrac{\partial T}{\partial t} = \dfrac{\partial^2 T}{\partial x^2} \\ T(x,0) = 20000x \quad x \in [0, 0.005] \\ T(x,0) = 20000(0.01 - x) \quad x \in [0, 0.01] \\ T(0.0, t) = 0.0 \\ T(0.01, t) = 0.0 \end{cases}$$

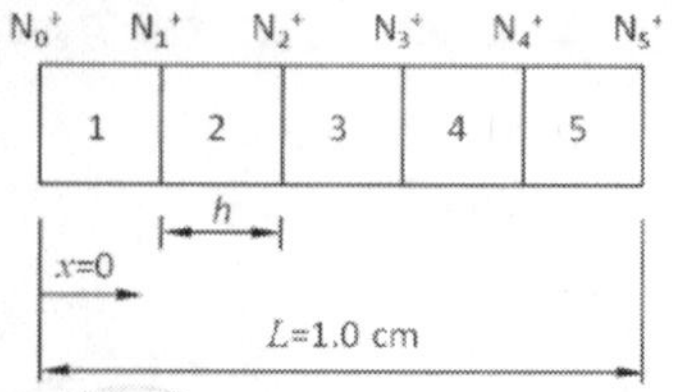

- **Entity**

```
ENTITY transient_diffusion IS
  generic (
        rho: real := 2000.0;
        k:   real := 2.0;
        Cp:  real := 1000.0);
END ENTITY transient_diffusion;
```

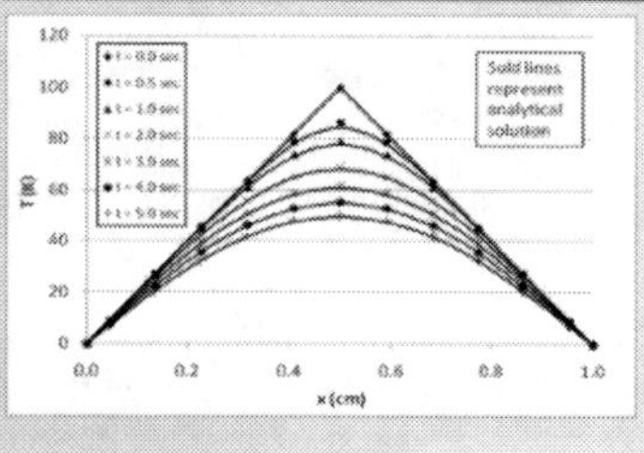

- **Architecture (main part)**

```
IF (domain = quiescent_domain) USE

     T1 == 20.0;
     T2 == 60.0;
     T3 == 100.0;
     T4 == 60.0;
     T5 == 20.0;

ELSE

     rho*Cp*T1'dot == -(N1p - N0p)/h;
     rho*Cp*T2'dot == -(N2p - N1p)/h;
     rho*Cp*T3'dot == -(N3p - N2p)/h;
     rho*Cp*T4'dot == -(N4p - N3p)/h;
     rho*Cp*T5'dot == -(N5p - N4p)/h;

END USE
```


2013 Automotive Simulation World Congress
October 29 - 30 • Frankfurt am Main, Germany

VHDL-AMS Model for Electrochemistry

The governing equations of porous electrode model of the lithium-ion battery (Electrochemical Systems, 3rd by John Newman)

$$\varepsilon \frac{\partial c}{\partial t} = \nabla \cdot (\varepsilon D \nabla c) - \frac{i_2 \cdot \nabla t_+^0}{z_+ \nu_+ F} + \frac{a j_n (1 - t_+^0)}{\nu_+}$$

PDE: initial value

$$i_2 = -\kappa \nabla \phi_2 + \frac{\kappa R T}{F}(1 - t_+^0)\nabla \ln c$$

$$I - i_2 = -\sigma \nabla \phi_1$$

$$a F j_n = \nabla \cdot i_2$$

$$j_n = i_0 \left\{ \exp\left(\frac{\alpha_a F}{RT}\eta_s\right) - \exp\left(-\frac{\alpha_c F}{RT}\eta_s\right) \right\}$$

Algebraic

A coupled set of PDEs. The way to solve them with VHDL-AMS uses the same technique as before.

Model Implementation

Ref: Newman et al (1993, 1995, 1996)

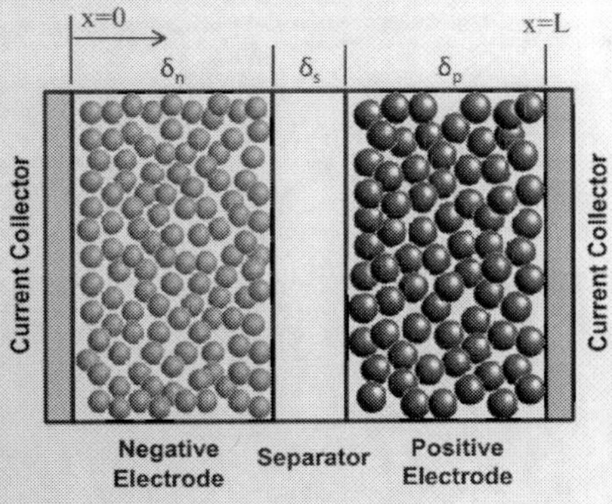

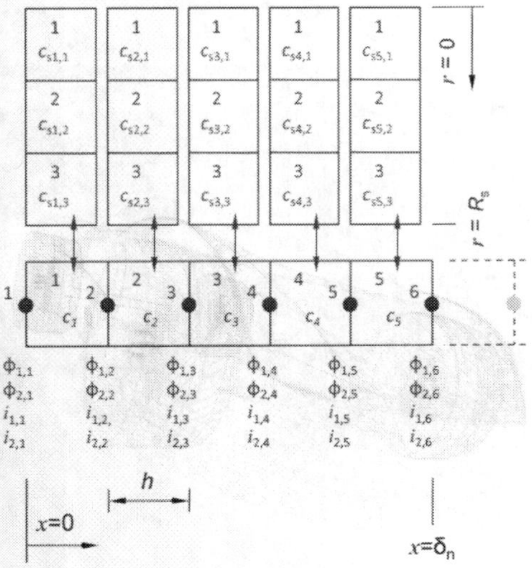

•Reference: M. Doyle, T.F. Fuller, and J. Newman, Journal of Electrochem. Soc., 140, 1526 (1993)
C. R. Pals and J. Newman Journal of Electrochem. Soc., 142 (10), 3274-3281 (1995)
M. Doyle, J. Newman, Journal of Electochem. Soc., 143, 1890 (1996)

2013 Automotive Simulation World Congress
October 29 - 30 • Frankfurt am Main, Germany

Newman Model – Quantitative Comparison

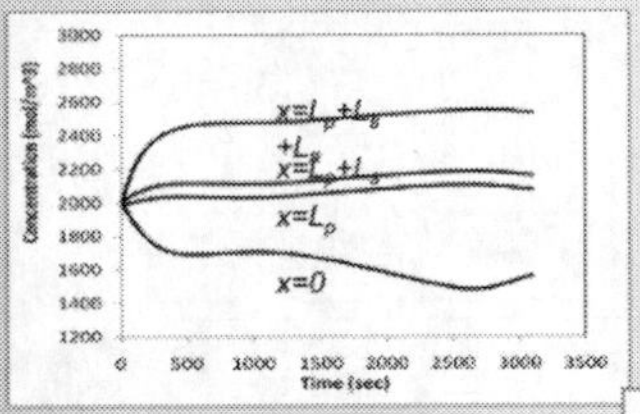

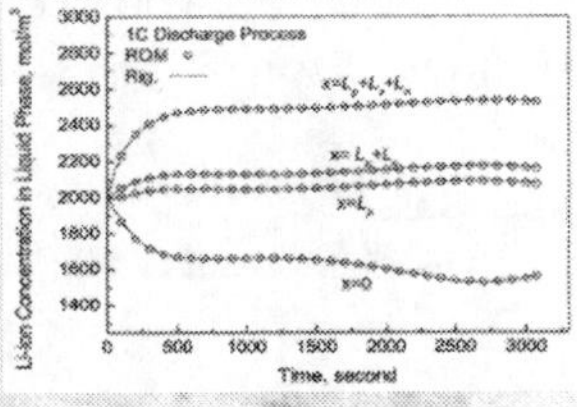

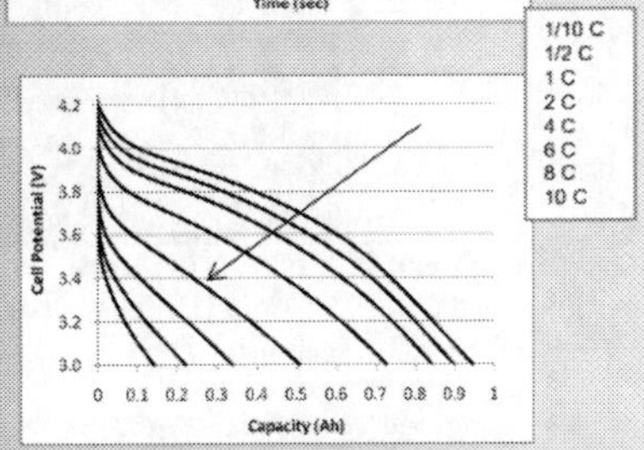

Simplorer's Results

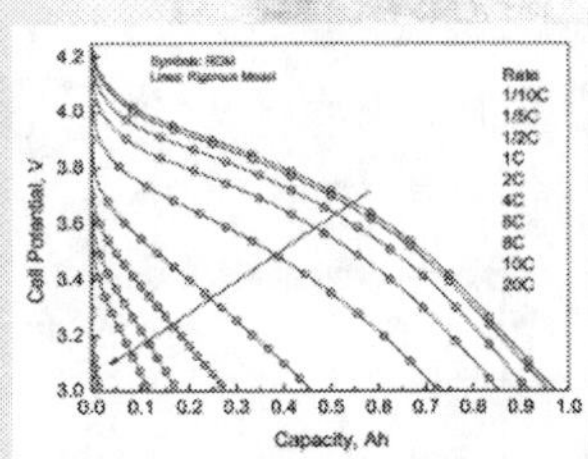

White's Results

•Reference: Long Cai, Ralph E. White, Journal of Electrochem. Soc., 156 (3), A154-A161 (2009)

2013 Automotive Simulation World Congress
October 29 - 30 • Frankfurt am Main, Germany

Results of a Prismatic Lithium-Ion Cell

Geometry & Mesh

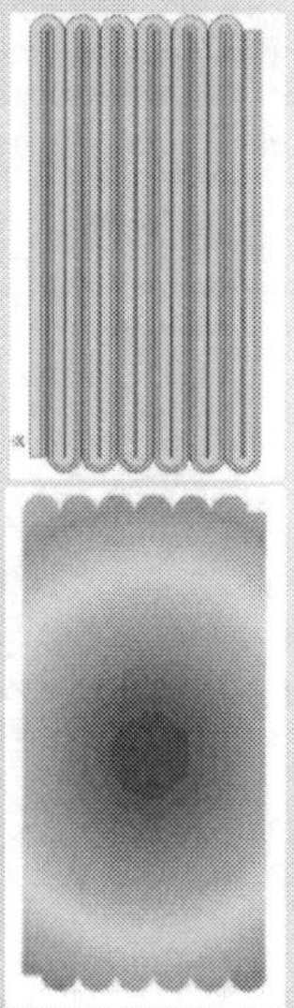

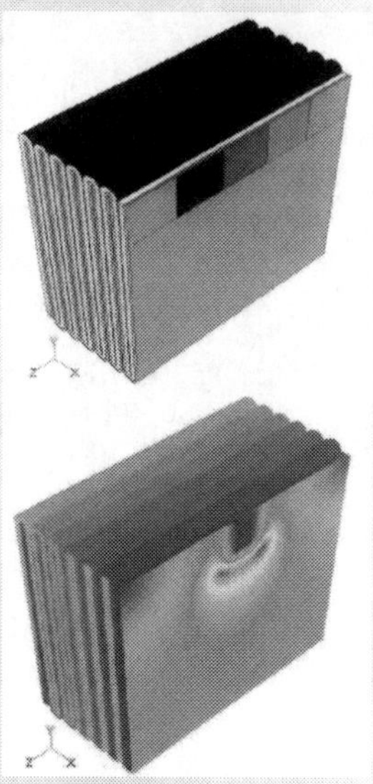

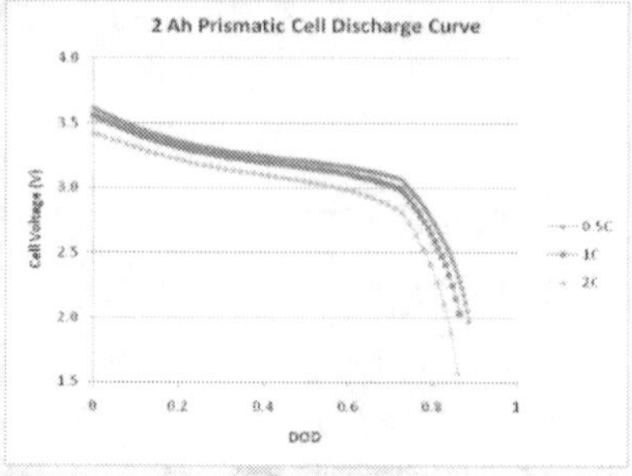

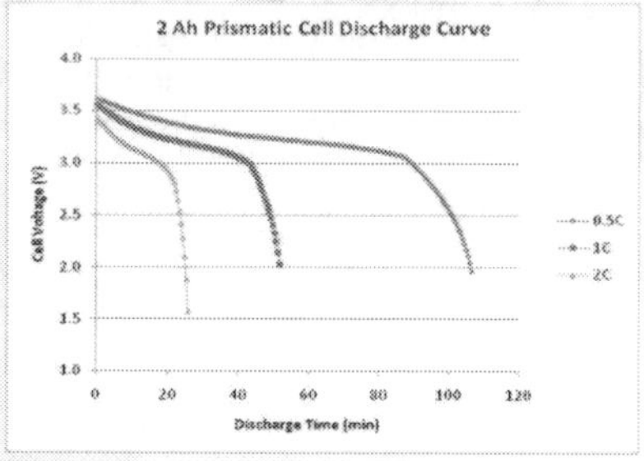

Temperature **Current Density**

2013 Automotive Simulation World Congress
October 29 - 30 • Frankfurt am Main, Germany

Multipack Simulation

- 60 Cells connected in matrix pack
- Packs are connected in matrix to final configuration

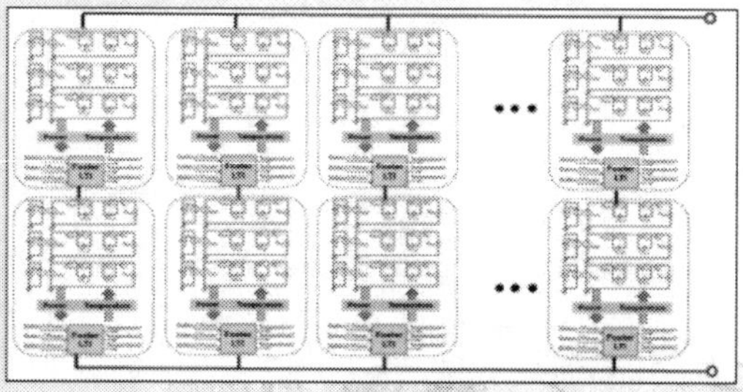

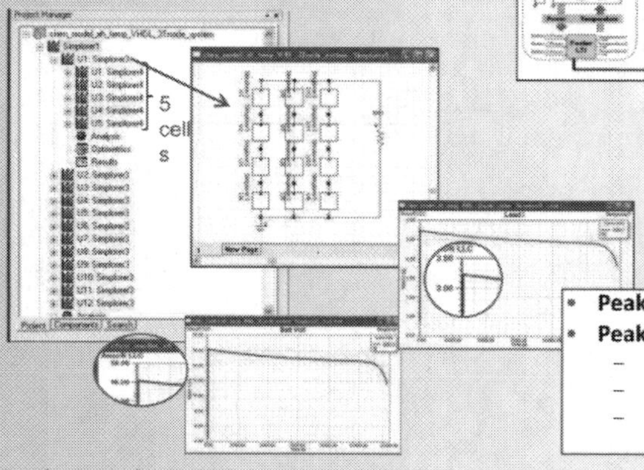

- **Peak voltage: 16 V (4 cells in series)**
- **Peak current: ~3.25 Amp (15 cells in parallel)**
  - 0.4 Amp for single battery case
  - And yet runtime is ~doubled
  - Estimated life: 0.4/(3.25/15)x8000 sec without rate factor consideration

Multidomain Simulation with Batteriepack

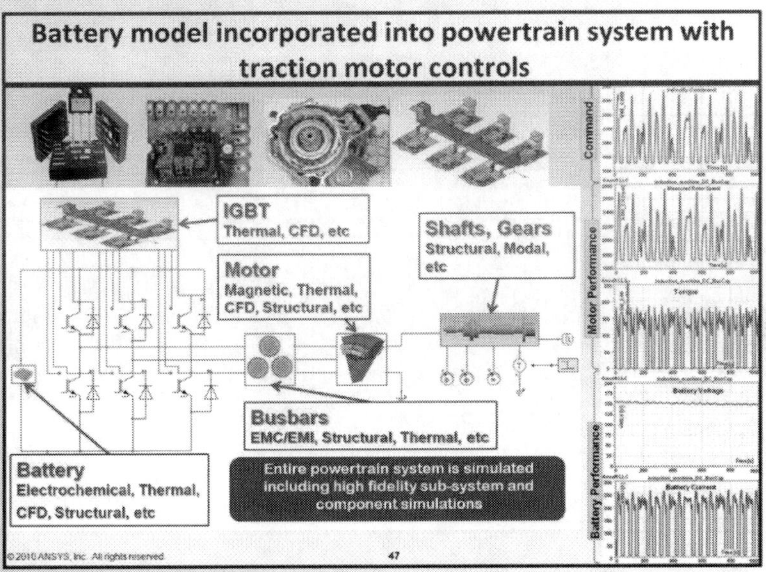

Supplier chain

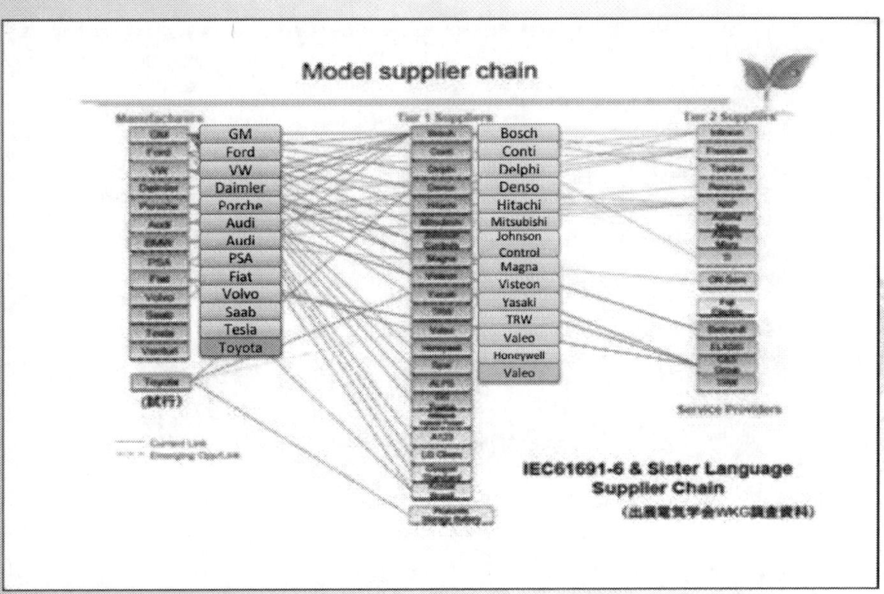

Simplorer

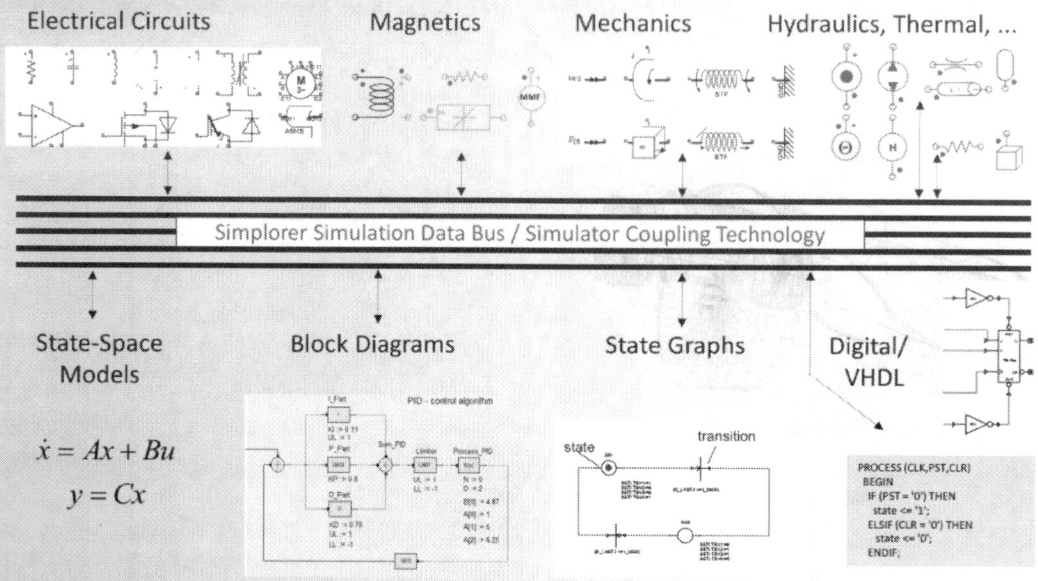

Simplorer Environment

- Windows-like
- Single project file
- Modeling
  - Graphical, SML, C/C++
  - VHDL-AMS, PSpice
- Hierarchical models
- All quantities accessible
- Postprocessing on sheet or in reports
- VB, Java, Python, ...
- Model libraries
- Wizards
- Physical units
- Optimetrics
- Distributed Solve
- WB integrated

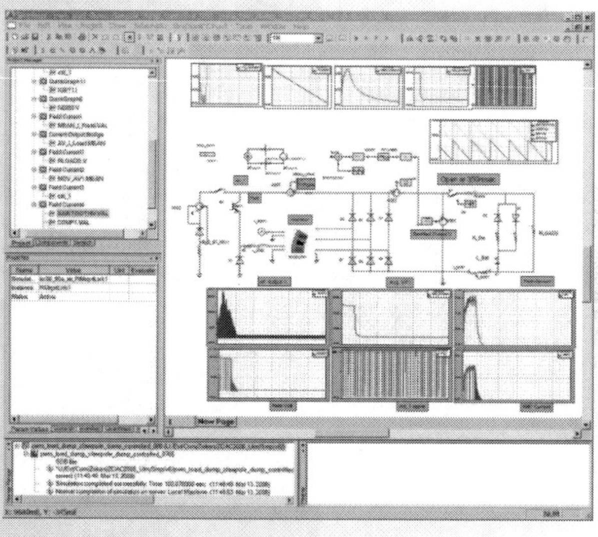

Simplorer: Component Characterization

In general:

Each parameter can be made dependent on time or any other quantity by expressions or tables.

- Characteristics
- Excitations
- Loads

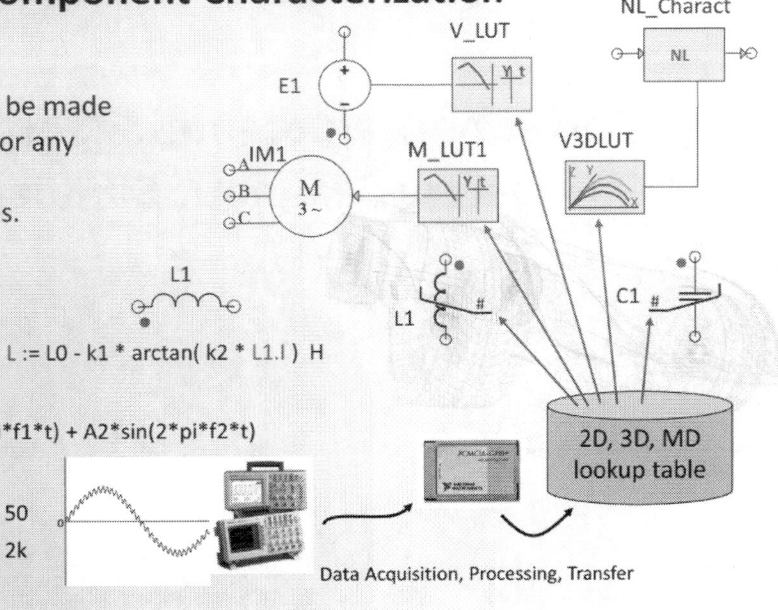

E1
EMF := A1*sin(2*pi*f1*t) + A2*sin(2*pi*f2*t)

EQU

FML1

A1 := 10   f1 := 50

A2 := A1/10   f2 := 2k

Links with Embedded through FMI extend our system simulation capabilities

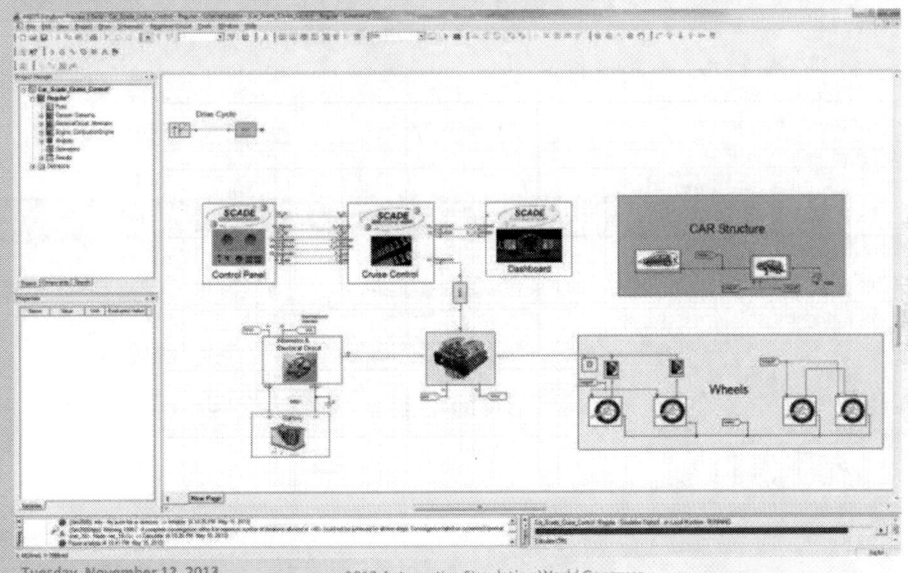

2013 Automotive Simulation World Congress

High-Performance Computing for a Full Vehicle Electro-Magnetic Simulation

Alain Michel, ANSYS France

2013 Automotive Simulation World Congress
October 29 - 30 • Frankfurt am Main, Germany

HPC Productivity Enhancements

- High Performance Computing (HPC)
  - HPC enables increased productivity and higher fidelity simulation - including more geometric detail and larger systems.
  - Using HPC you can make your engineering staff, and your product development process, more productive and efficient. Faster turnaround and larger models all mean better designs in less time.

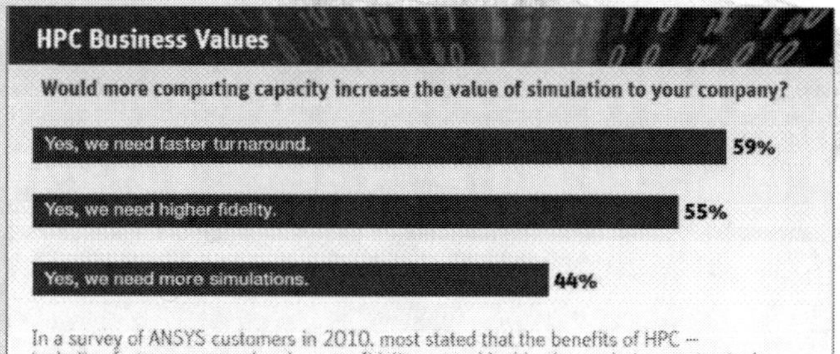

In a survey of ANSYS customers in 2010, most stated that the benefits of HPC – including faster turnaround and greater fidelity – would add value to their organization's use of engineering simulation.

2013 Automotive Simulation World Congress
October 29 - 30 • Frankfurt am Main, Germany

The Competitive Edge

Faster **- Solver technology targeted at utilizing multiple processor/cores to accelerate the solution process.**

Bigger **- Solver technology targeted at distributing the simulation memory across multiple computers. The distributed nature of the solution may also result in faster simulations, but it is primarily intended to increase capacity.**

Higher Fidelity **– Maintain true to life models because bigger is no longer a simulation limitation. Utilizing HPC means never having to compromise. Utilize faster solutions to do more.**

Monday, November 11, 2013 2013 Automotive Simulation World Congress 3

2013 Automotive Simulation World Congress
October 29 - 30 • Frankfurt am Main, Germany

Leveraging High Performance Computing Hardware

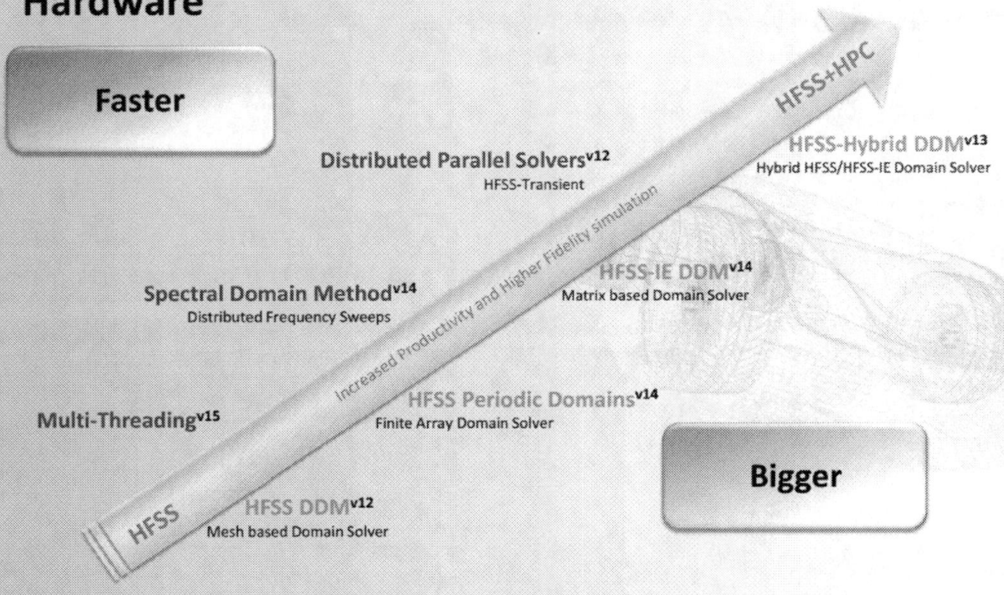

Monday, November 11, 2013 2013 Automotive Simulation World Congress 4

HFSS with HPC

Faster - Solver technology targeted at utilizing multiple processor/cores to accelerate the solution process.

Faster

Multi-Threading

Spectral Domain Method

Distributed HFSS-Transient

HPC: Multi-Threading (MT)

- **Multi-Threading (HPC-MT)**
  - Single workstation solution to increase the speed of the solver
  - **TAU Initial Mesh Generation**
    - Parallelized mesh generation
  - Direct Matrix Solver
    - Parallelized matrix solver
  - **Iterative Solver**
    - Parallelized matrix pre-conditioner
    - Parallelized excitations
  - **Field Recovery**
    - Parallelized field recovery for multiple excitations
  - Available with HFSS, HFSS-IE, and HFSS-Transient

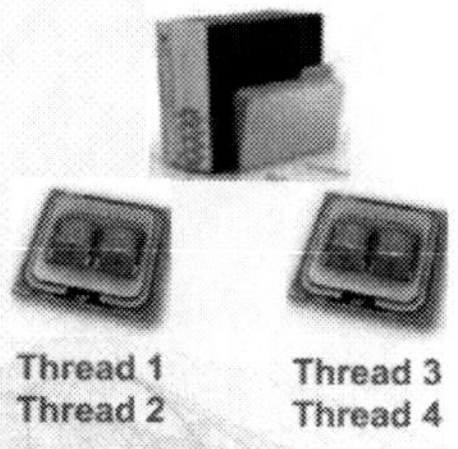

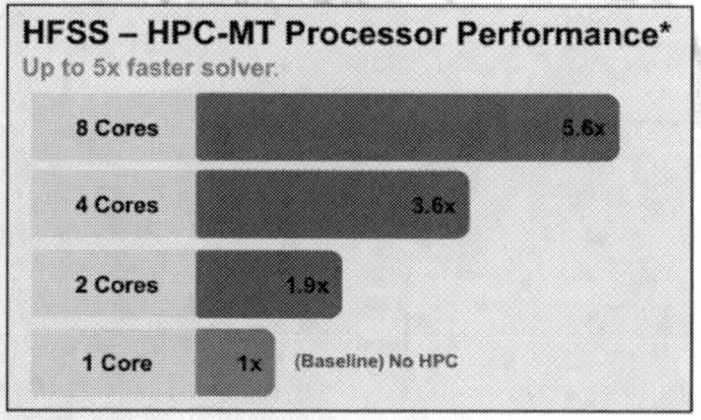

HFSS MP/HPC Comparison

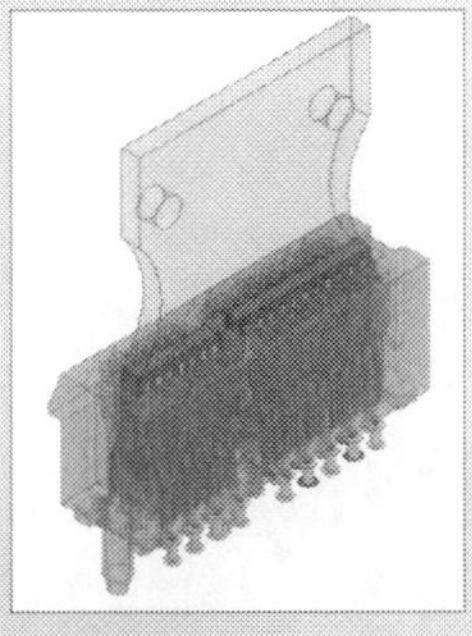

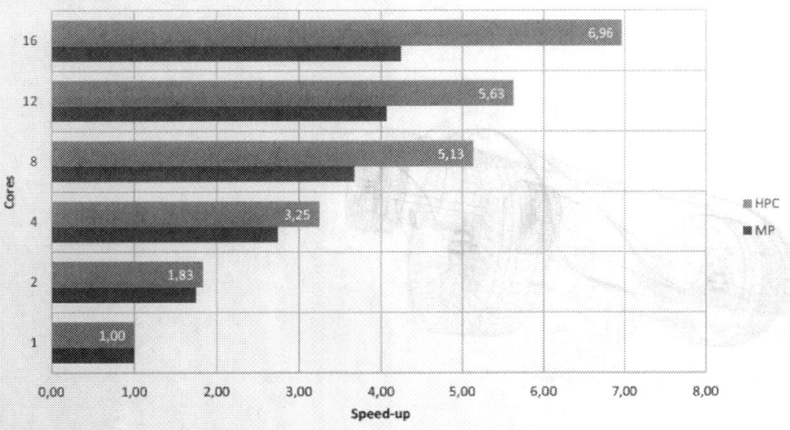

- HPC improves processor scaling (~40% faster with 16 cores)
- 1 HPC Pack out performs unlimited cores of MP license

HPC: Spectral Domain Method

- **Spectral Decomposition Method (HPC-SDM)**
  - Accelerates frequency sweeps by distributing the spectral content across a network of processors
  - Increases simulation speed
    - Combines with HPC-MT
  - Scalable to large numbers of cores
  - Available with HFSS, HFSS-IE, and PSI

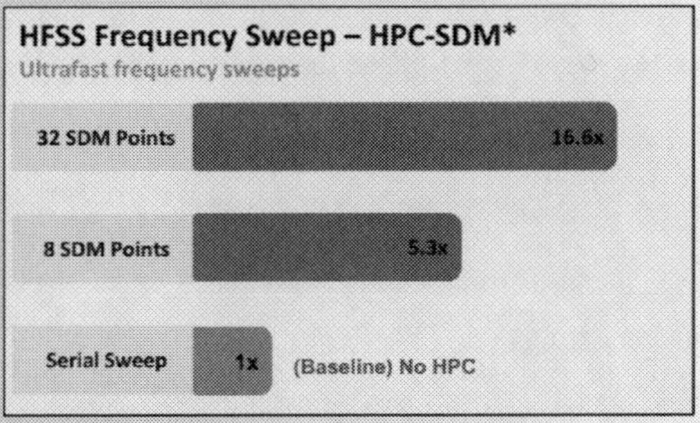

*Interpolating Sweep: 0-20GHz, 201 frequencies

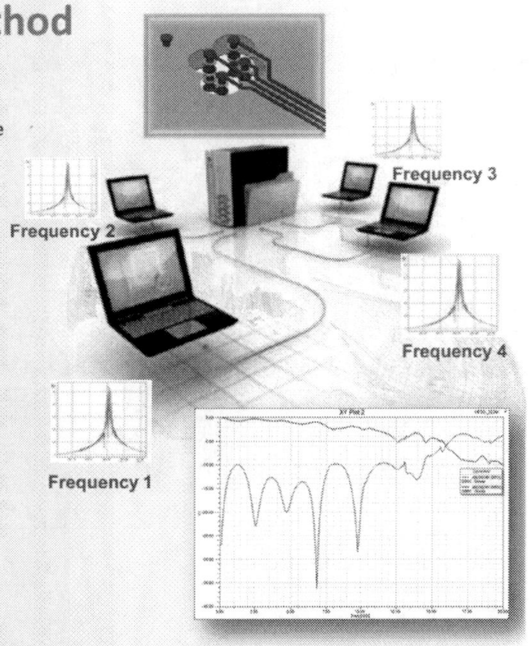

HFSS Frequency Sweeps

- **Discrete Sweep**
  - Runs a solution at each requested frequency point
- **Interpolating Sweep**
  - This method fits s-parameter data to a rational polynomial transfer function using a minimum number of discrete frequency solutions.
  - The interpolating sweep yields the poles and zeros of the transfer function. This information can be directly used in the Laplace Element from which a Full-Wave SPICE™ model can be generated (HSPICE, Spectre RF, PSPICE).

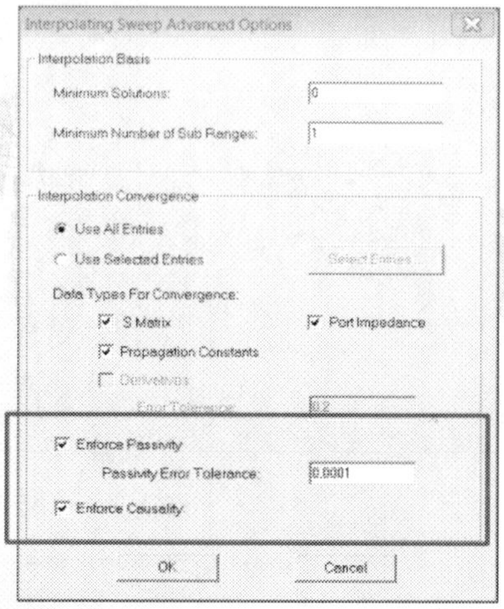

Rational Polynomial:

$$S=\frac{\beta_q(s-z_q)(s-z_{q-1})\dots(s-z_1)}{\alpha_q(s-p_q)(s-p_{q-1})\dots(s-p_1)}$$

See: IEEE Trans. Microwave Theory Tech., Vol. 46, No. 9, Sept. 1998

HFSS: HPC-SDM Interpolating Sweeps

- **Observations**
  - Great improvements for interpolating efficiency in HFSS v15
  - Total core count is only factor that offers improvement
    - Does not matter how you use cores: Frequency Points vs. Multi-Threading
      - On average same performance
      - Interpolating Efficiency increases as simultaneous frequency points decrease

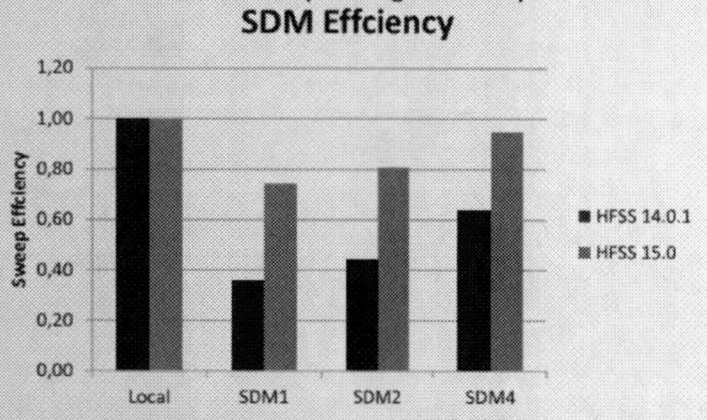

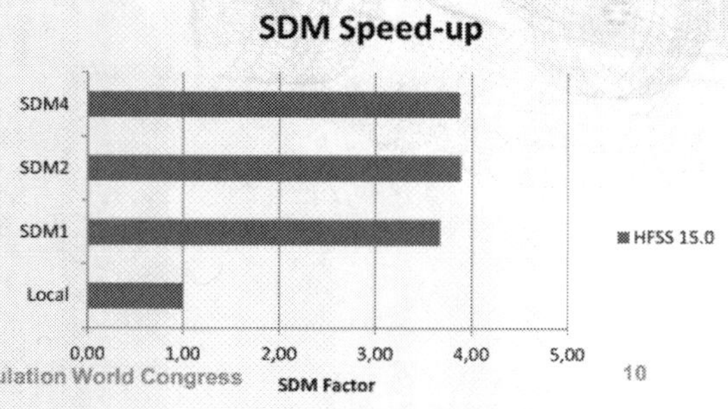

SDM1: 32 Freq
SDM2: 16 Freq/2 HPC-MT
SDM4: 8 Freq/4 HPC-MT

HFSS: HPC-SDM Discrete Sweeps

- **Observations**
  - Running more frequency points improves performance
    - Multi-Threading does not scale linearly with cores
  - Interpolating vs Discrete
    - Why do we have an interpolating sweep?
      - Minimize the number of solved points
    - With HPC-SDM it is compelling to run discrete sweeps
      - Passive/Causal – at least no interpolating noise
      - SI: Directly solve DC Point with HFSS Solver on Demand
      - RF: Save Fields at each frequency point
      - Single 32 Core Servers with large RAM footprint
      - HPC Packs
        - 2 Packs: 32 Cores
        - 3 Packs: 128 Cores

SDM1: 32 Freq
SDM2: 16 Freq/2 HPC-MT
SDM4: 8 Freq/4 HPC-MT

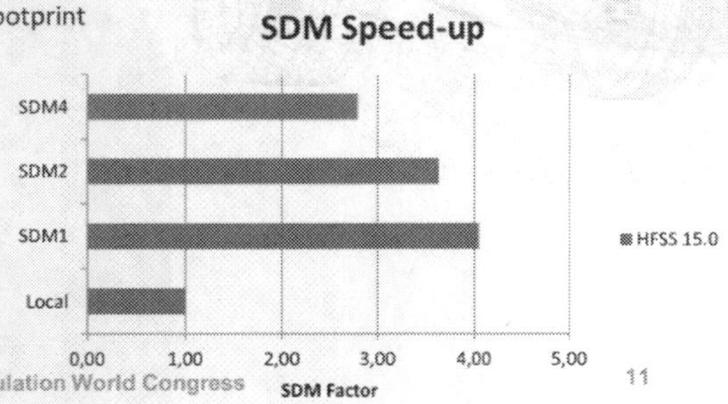

HPC: HFSS-Transient Distributed Parallel Solver

- **HFSS-Transient Distributed Parallel (HPC-DP)**
  - Accelerates HFSS-Transient solutions by distributing the excitations across a network of processors
  - Increases simulation speed
    - Combines with HPC-MT
  - Available with HFSS-Transient

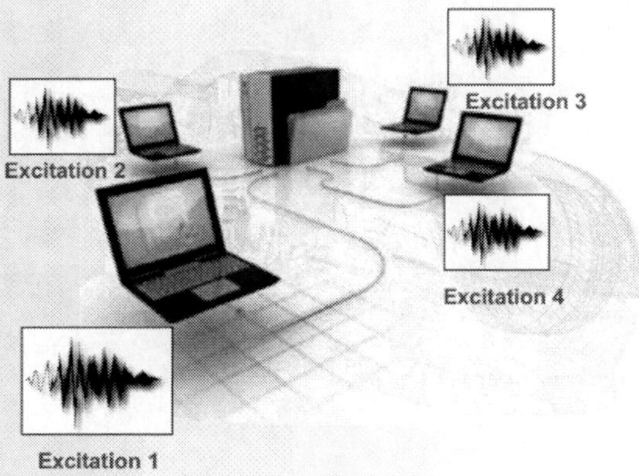

HFSS with HPC

Bigger **- Solver technology targeted at distributing the simulation memory across multiple computers. The distributed nature of the solution may also result in faster simulations, but it is primarily intended to increase capacity.**

Bigger

HFSS DDM (Mesh Based)

HFSS-IE DDM (Matrix Based)

HFSS-Hybrid DDM

HFSS Periodic Domains

HPC: HFSS-DDM (Mesh Based)

- **Domain Decomposition Method: Meshed Based**
  - Distributed memory parallel technique
    - Distributes mesh sub-domains to network of processors/RAM
  - Significantly increases simulation **capacity**
  - Highly scalable to large numbers of processors
    - Uses industry standard MPI
    - Combines with HPC-MT
  - Automatic generation of domains by mesh partitioning
    - User friendly
    - Load balance
  - Hybrid iterative & direct solver
    - Multi-frontal direct solver for each sub-domain
    - Sub-domains exchange information iteratively via Robin's transmission conditions (RTC)
  - Available with HFSS

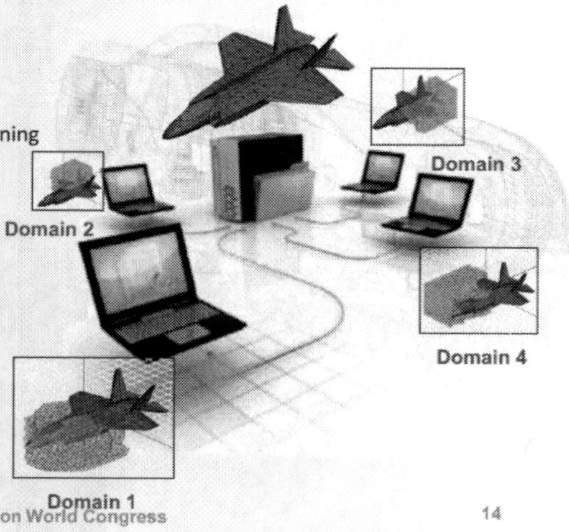

2013 Automotive Simulation World Congress
October 29 - 30 • Frankfurt am Main, Germany

HFSS DDM Example –car body with remote keyless entry (RKE) system

•Overall Project geometry

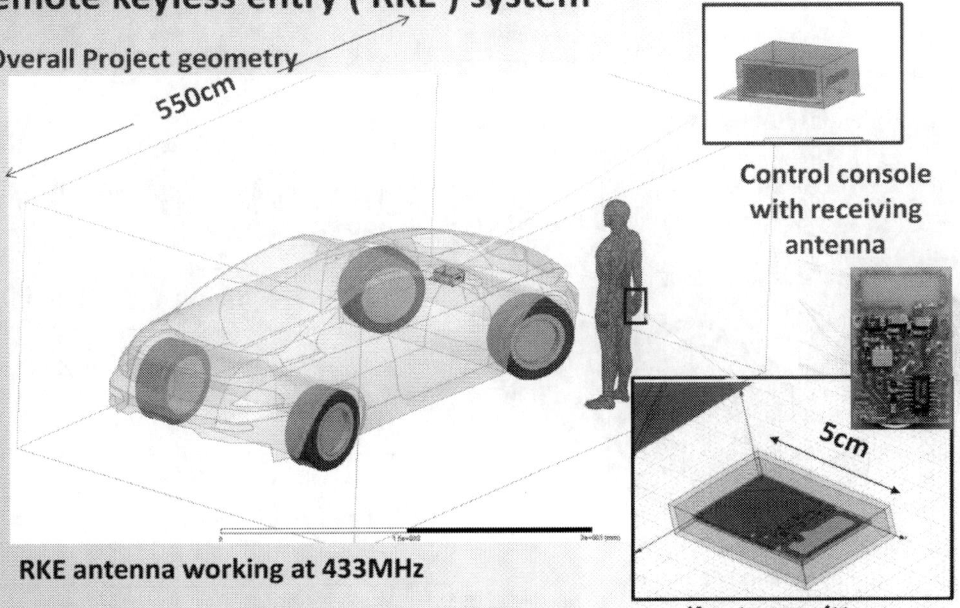

RKE antenna working at 433MHz

2013 Automotive Simulation World Congress
October 29 - 30 • Frankfurt am Main, Germany

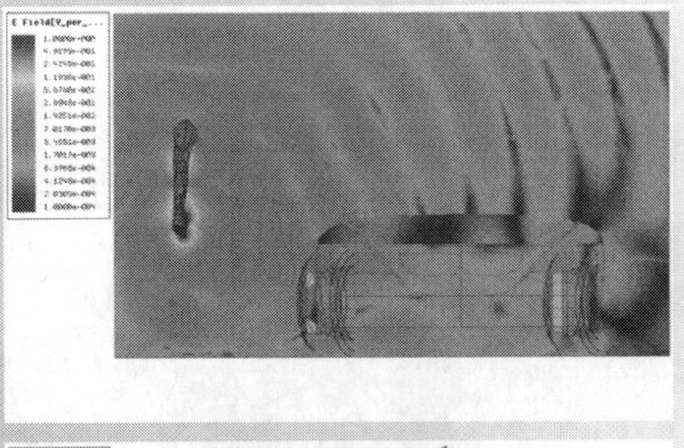

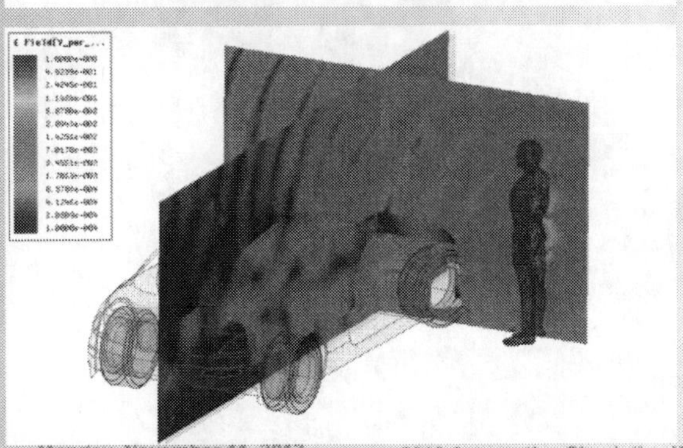

- **DDM**
  - 8 Domains, 14 cores

Domain 8 on winhpc9.2008HPC2.univ-brest.fr	
Matrix Assembly/Solver MCS14	

  - Total RAM9.59GB

Maximum domain memory: 1.893 GB
Average domain memory: 1.199 GB
Total memory for all domains: 9.588 GB

  - Simulation time 24 mn

Adaptive Process		
Total	00:24:52	01:47:10

2013 Automotive Simulation World Congress
October 29 - 30 • Frankfurt am Main, Germany

HPC: HFSS-IE DDM (Matrix Based)

- **Domain Decomposition Method: Matrix Based**
  - Distributed memory parallel technique
    - Distributes matrix solution to network of processors/RAM
  - Significantly increases simulation **capacity**
  - Highly scalable to large numbers of machines
    - Uses industry standard MPI
    - Combines with HPC-MT
  - Automatic generation and load balancing of matrix partitions
  - Available with HFSS-IE

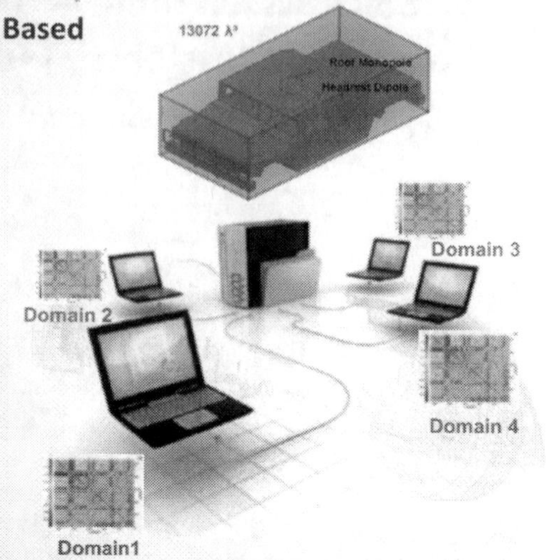

	HFSS	HFSS-IE
Time	6 hours	1:15 hour
RAM	98 GB	41 GB

2013 Automotive Simulation World Congress
October 29 - 30 • Frankfurt am Main, Germany

HPC: Hybrid HFSS - FEM DDM with IE Regions

- **Domain Decomposition Method**
  - Distributes mesh sub-domains to network of processors
    - FEM volume can be sub-divided into multiple domains
    - IE Domains that are discontinuous will be distributed to separate nodes when they become large
  - Significantly increases simulation **capacity**
  - Uses Industry Standard MPI
  - Extension of HFSS DDM to support the Hybrid FEM/IE solver
    - IE Regions & FE-BI
  - Available with HFSS+HFSS-IE Hybrid

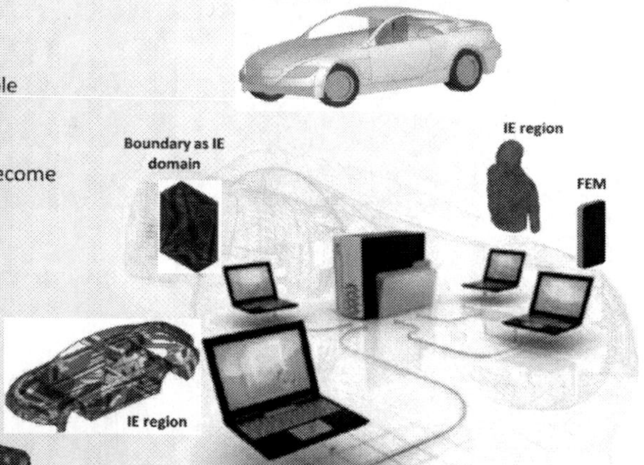

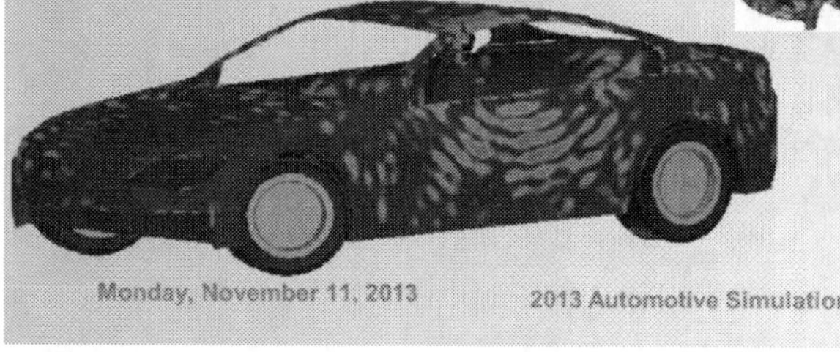

2013 Automotive Simulation World Congress
October 29 - 30 • Frankfurt am Main, Germany

Hummer Type Vehicle

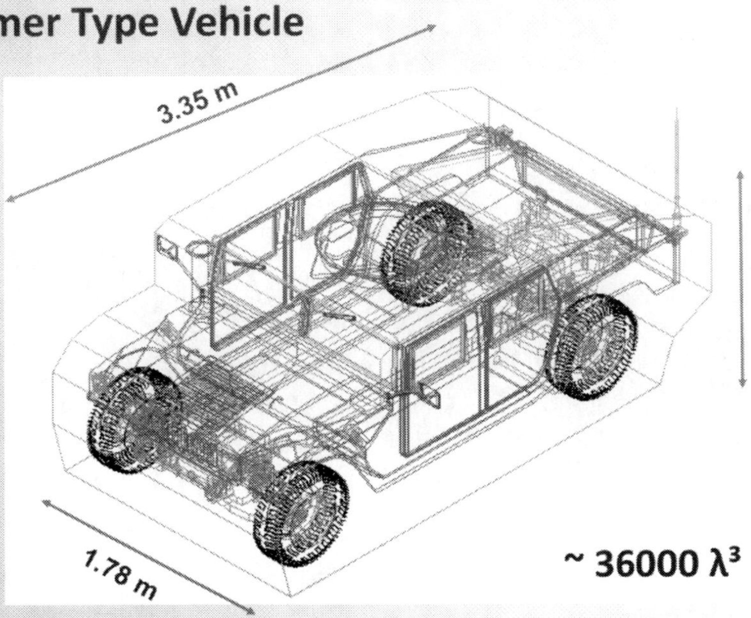

~ 36000 λ^3

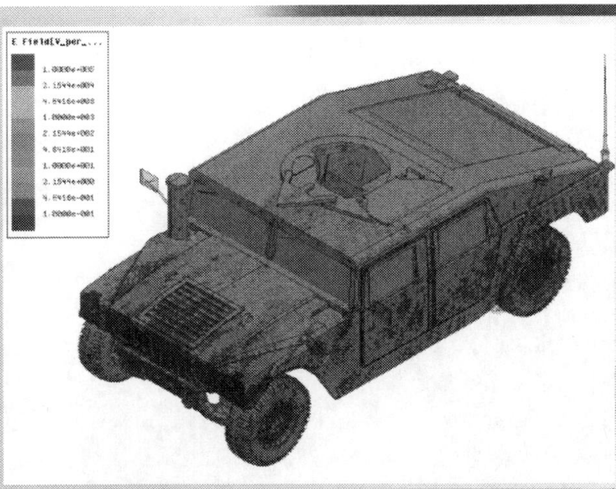

2013 Automotive Simulation World Congress
October 29 - 30 • Frankfurt am Main, Germany

13 IE Domain

Domain 1_13 on winhpc16.2008HPC2.univ-brest.fr	
Matrix Assembly/Solver DCS1, BI	4.3 G

151 FEM Domain

Matrix Assembly/Solver MCS1	1.6 G
Domain 151 on winhpc15.2008HPC2.univ-brest.fr	

Total Ram for all Domain: 288 GB

Maximum domain memory: 5.884 GB
Average domain memory: 1.768 GB
Total memory for all domains: 288.2 GB

Simulation done in 4h12mn

Elapsed time : 04:12:59 , I

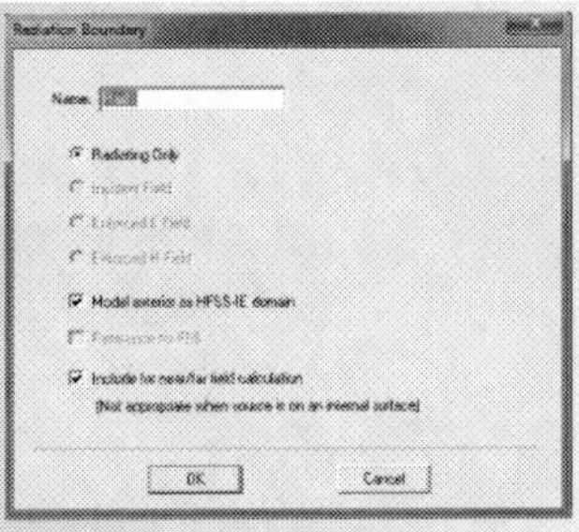

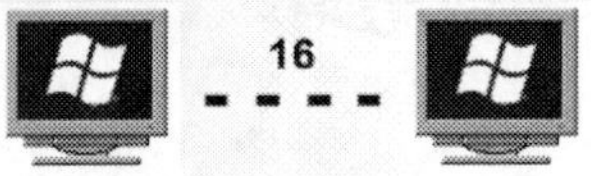

Intel(R) Xeon(R) CPU E5-2665 0 @ 2.40GHz 2.40 GHz (2 processeurs)
128 Go

HPC: HFSS-Periodic Domains (Finite Arrays)

- Periodic Domain Decomposition (HPC-PDM)
  - Distributed memory parallel technique for finite periodic geometries, such as finite antenna arrays
    - Distributes unit cell mesh sub-domains to network of processors/RAM
  - Significantly increases simulation capacity
  - Highly scalable to large numbers of processors
    - Uses industry standard MPI
    - Combines with HPC-MT
  - Automatic generation of domains
    - User friendly and easy to implement
    - Efficient simulation of only unique cells
  - Available with HFSS

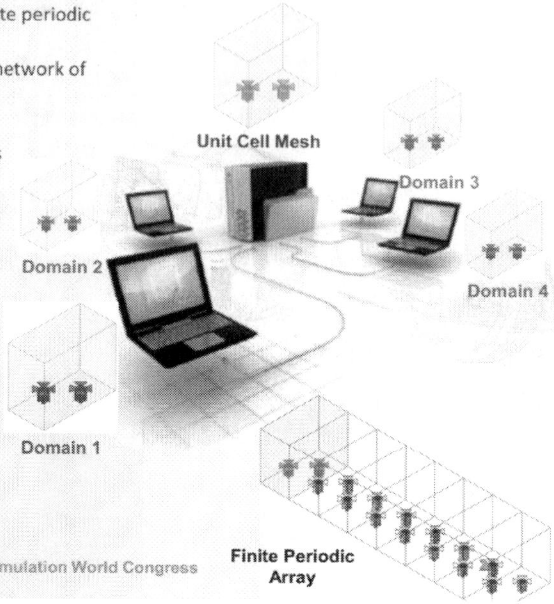

HFSS: HPC-PDM Snowflake Array*

E-field 5mm above aperture
Circularly polarized elements

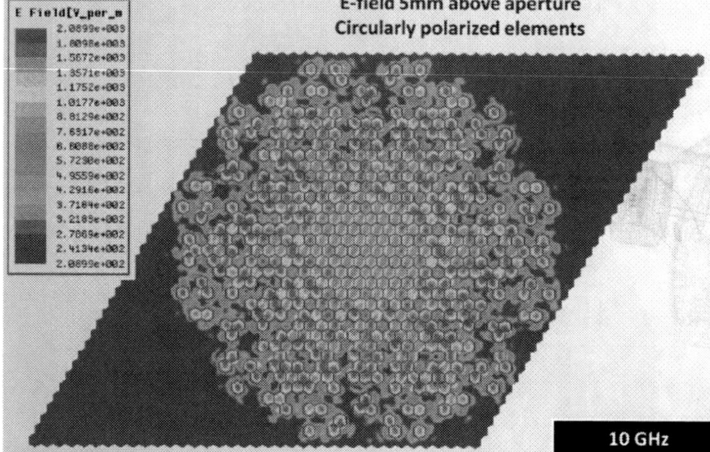

529 circular WG elements, 1058 modes

Array Mask^New
Composite Excitation^New

10 GHz	RAM	Elapsed Time
HFSS HPC-PDM	62G	27min

HPC: Fit for any application

High Performance Computing (HPC) License

Multi-Threaded Solvers (MT)	Domain Decomposition (DDM)	Distributed Parallel Solvers (DP)
HFSS	Mesh Based HFSS	HFSS-Transient Distributed Excitations
Designer	Matrix Based HFSS-IE	HFSS - Hybrid DDM with IE Domains
Q3D Extractor	Spectral Frequency Sweep	Q3DExtractor DC RL/AC RL/CG
SIwave	Periodic HFSS Phased Array	TPA Partitions
Maxwell		

Increase ANSYS Electronics solver performance with HPC

HFSS for Robust Design

Higher Fidelity – **Maintain true to life models because bigger is no longer a simulation limitation. Utilizing HPC means never having to compromise. Utilize faster solutions to do more.**

Higher Fidelity

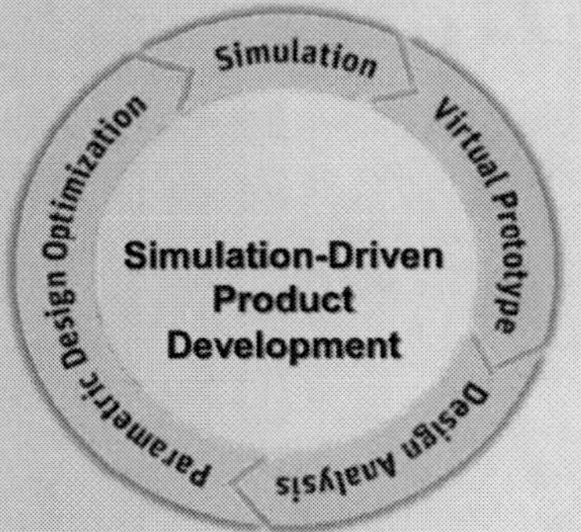

Simulation-Driven Product Development
- **Predictive Quality with Parametric Design**
- **Systematically test ideas**

Leverage Simulation for Robust Design

Simulate everything you think you know

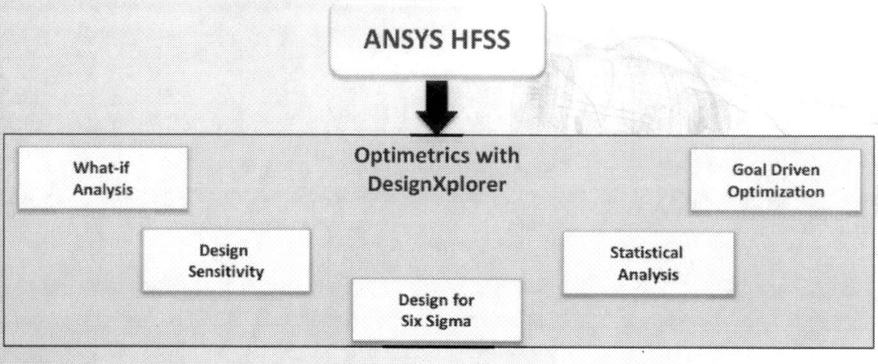

Accelerating Robust Design

- **Distributed Solve Option (DSO)**
  - Accelerates sweeps of design variations by distributing the design parameters across a network of processors
  - Increases simulation speed
    - Combines with HPC: MT, DDM
  - Scalable to large numbers of cores
    - Offers a near-linear speed up over conventional design sweeps
  - DesignXplorer Integration
  - Requires Optimetrics and a Distributed Solve license
    - Available with HFSS, Designer, Q3D Extractor, and Maxwell

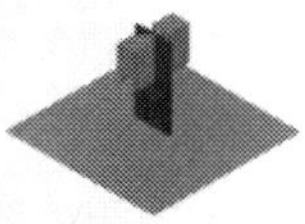

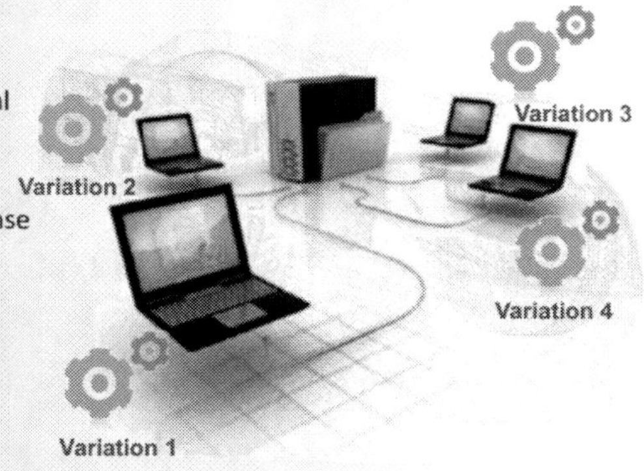

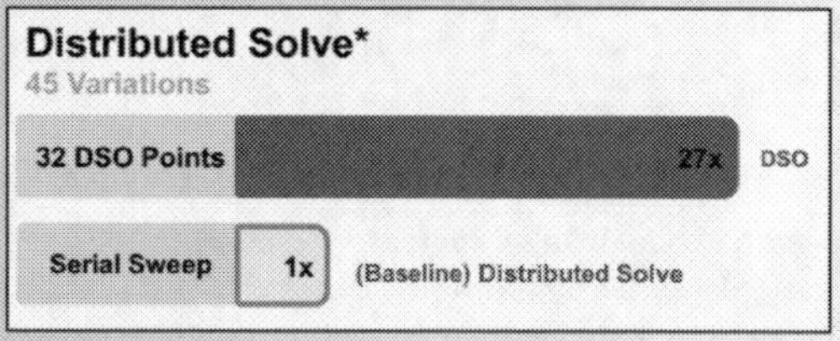

*Combined with HPC-MT

Hierarchical HPC

- **Additional level added to HPC**
  - Distribute parameters then...
    - DDM or new distributed Direct Matrix Solver
  - Distribute frequency points the...
    - DDM or new distributed Direct Matrix Solver

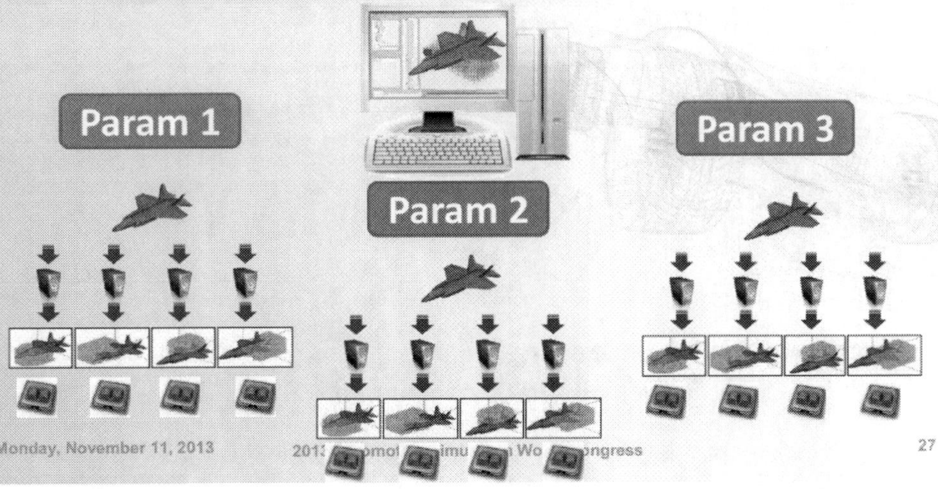

DSO + HPC on car body with RKE

- 9 variations from 0deg to 180deg step 22.5deg
- 9 DSO licenses used, 16 HPC used
- 9*16 = 144 cores used on 9 machines
- Simulation time 2hour 40mns

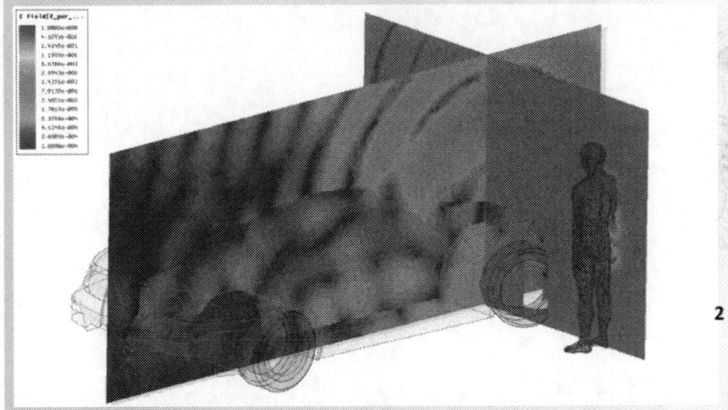

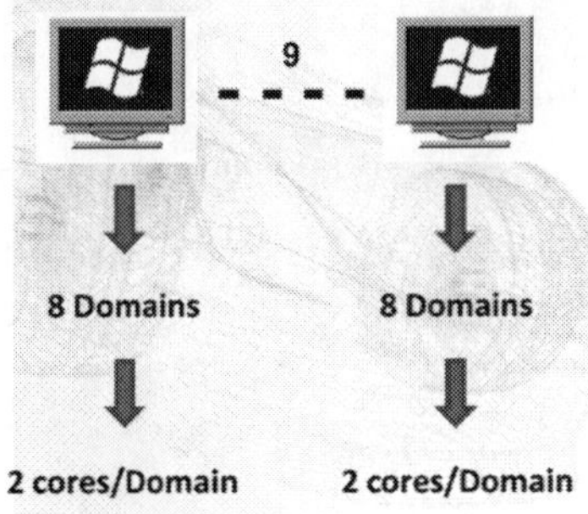

ANSYS HFSS with HPC

- **Faster simulations**
  - **Improved Multi-Threading**
    - Up to 5x faster solver with HPC enabled multi-threading
    - 20% faster compared to the previous generation multi-processing solver
  - **Spectral Domain Method**
    - Ultrafast frequency sweeps
    - Connect compute resources to a single frequency sweep for near-linear acceleration
- **Bigger simulations**
  - **New MPI based Domain Solutions**
    - Leverage the RAM across a network of computers to expand simulation capacity
    - Connect compute resources for an unparallel distributed RAM experience

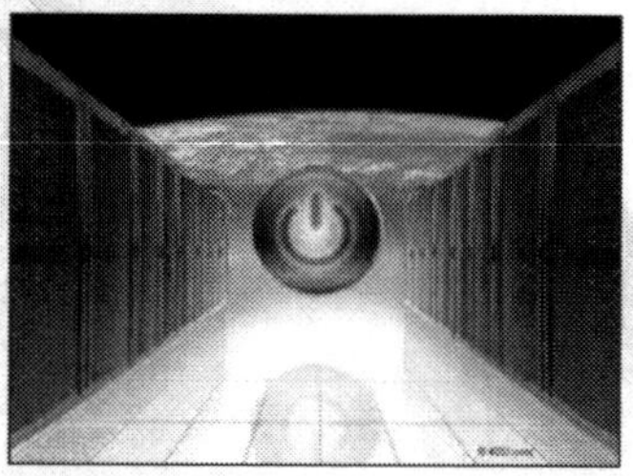

Modelling Inhomogeneous Heat Generation for Lithium-Ion Battery Cells

L. Kostetzer, E.Rudnyi : CADFEM GmbH, Grafing bei München, Germany
S. Stumpp, C. Günther, M.A. Danzer : Zentrum für Sonnenenergie- und Wasserstoff-ForschungBaden-Württemberg (ZSW), Ulm, Germany

ANSYS ANSYS Competence Center FEM CADFEM

Research project – Electro-thermal battery modeling

- Project partners:

ZSW
-Measurements
-Simulation application
-Validation

CADFEM
-Modeling development
-Simulation application

- Funding:
  - AiF ZIM project KF2613901WD0

ZIM impulse für wachstum Zentrales Innovationsprogramm Mittelstand

- 1 -

CADFEM

Outline

- Introduction & Motivation
- Model description – Multilevel approach
- Methodology for parameter ID
- Results
- FEM approach
- Summary

CADFEM

Introduction

- Li-ion batteries are temperature sensitive

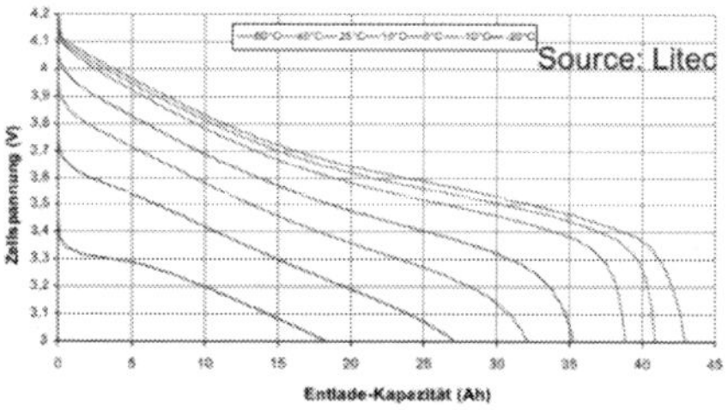

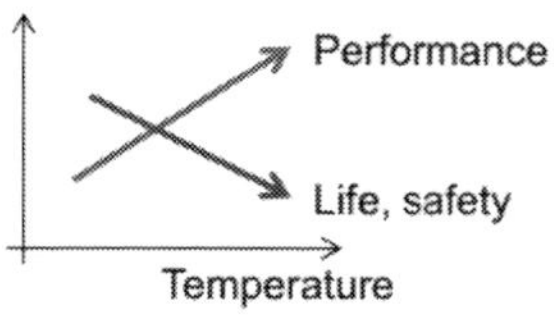

$$\text{Rate}_{\text{reaction}} \sim e^{\frac{E_A}{RT}}$$

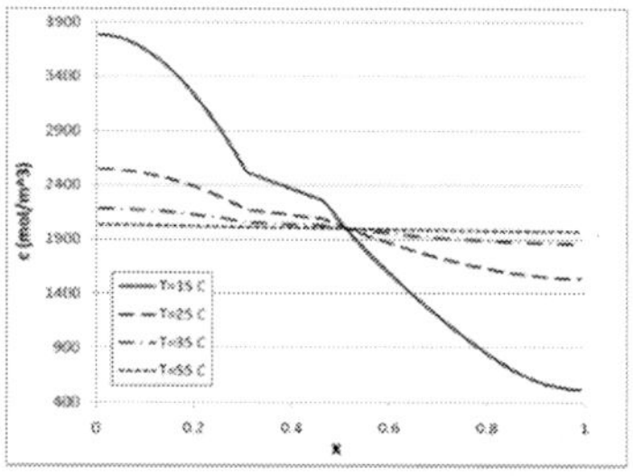

Mass transport effects [1]

CADFEM

Temperature influence in Lithium-ion cells

- Inhomogeneous degradation caused by cooling strategy
  - Accelerated aging in hotter regions[2]

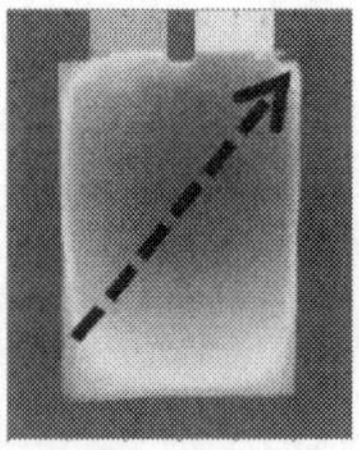

- Increased current flow through hotter cells [3,4]
  - This brings inhomogeneous in:
    - Current density
    - State of charge
    - Heat generation
- BMS is responsible to handle IN-homogeneity
  - Efficient cooling/heating systems
  - Control systems

- **We need first understand what happens at cell level**

CADFEM

Electro-thermal cell level modeling approach

Thermal

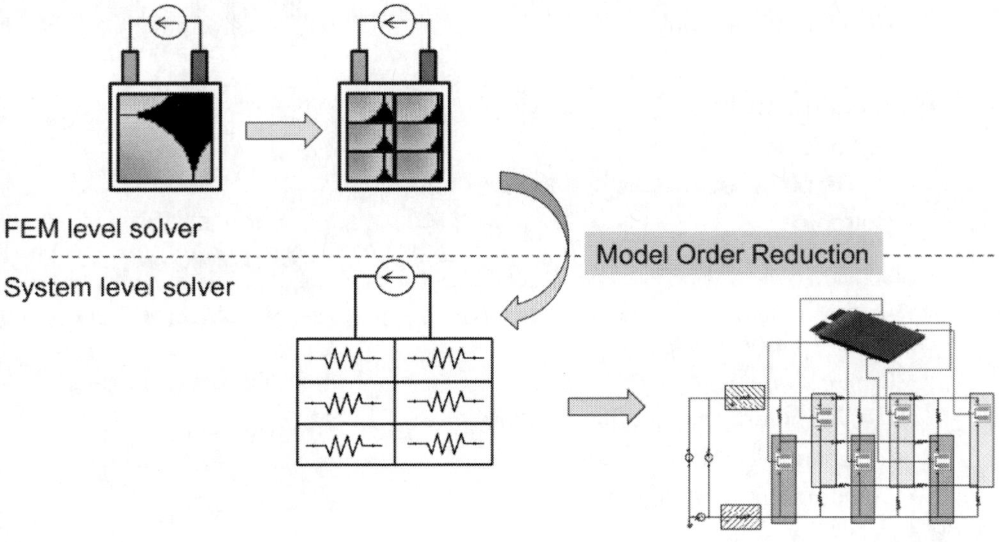

Electrical

CADFEM

Electrical model

- Sub- cells
  - Impedance RC models [5]
    - Temperature dependent
    - R,C,OCV
  - From experimental characterization

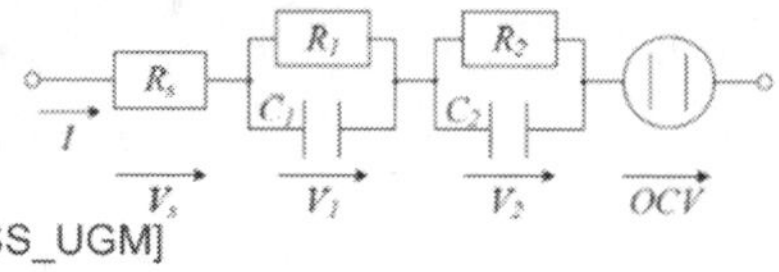

[SS_UGM]

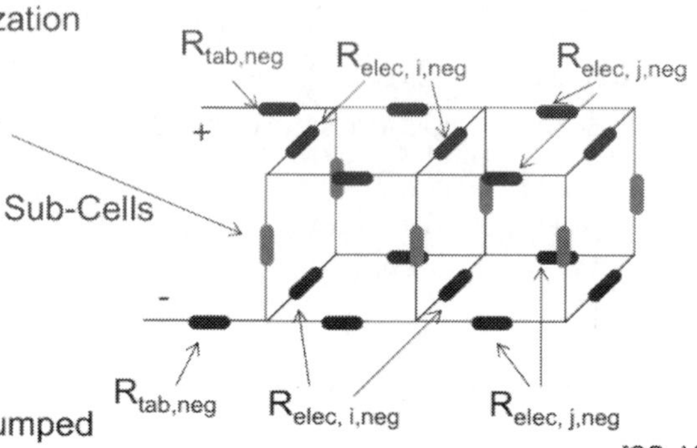

- Network of resistors
  - Conduction components are lumped

[SS_UGM]

[5] S. Stumpp, C. Günther, M.A. Danzer, L. Kostetzer, E.Rudnyi. ANSYS Conference & 31th CADFEM Users' Meeting 2013 | June 19-21, 2013 | Rosengarten Mannheim

- 6 -

CADFEM

Thermal Model

- FEM thermal model
  - ANSYS Workbench
  - Heat conduction only
- Heat conductivity
  - Orthotropic
    - Off plane low values due to stacking
  - Isotropic
    - All other material
- Boundary Conditions
  - Convection cooling
    - Measured Ambient temperature
    - Estimated Film confident
  - Temperature at tabs (measured)

Complete CAD [5]

Tabs [5]

Sub-Cells[5]

Pouch bag foil (external) [5]

- 7 -

CADFEM

Bring 3D thermal model to system level

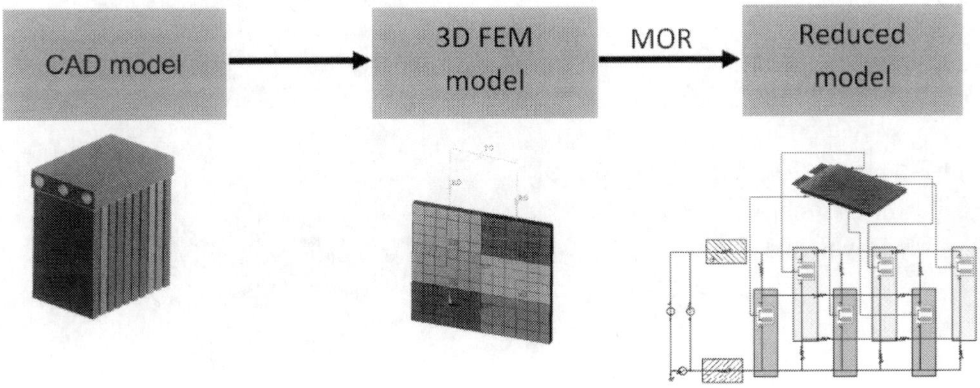

CADFEM

MOR for ANSYS: http://ModelReduction.com

$$E\dot{\mathbf{x}} + K\mathbf{x} = B\mathbf{u}$$

- Projection onto low-dimensional subspace[6]

$$\mathbf{x} = V\mathbf{z} + \varepsilon$$

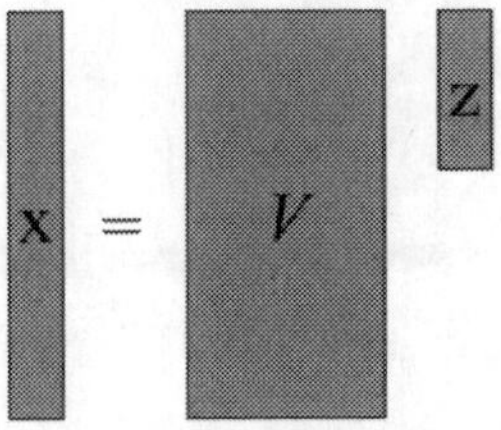

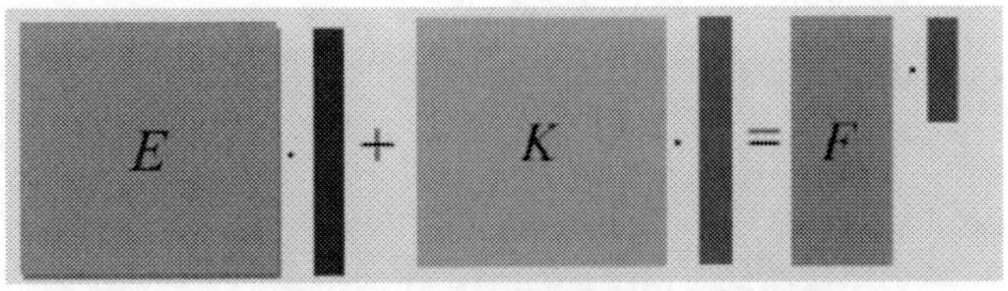

$$V^T E V\dot{\mathbf{z}} + V^T K V\mathbf{z} = V^T B\mathbf{u}$$

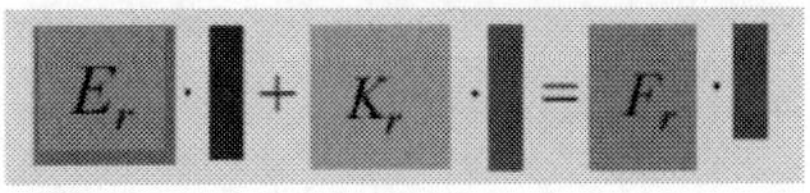

- Advantage:
  - NO transient solution with the original FEM model is necessary
  - Highly accurate, thermal models ~1% difference to Full FEM [7]
  - Speed up in the order to 1000x [7]

CADFEM

Electro-thermal coupling at system level

- System level = circuit solver (ODE's)
- Temperature from ROM
  - Middle of sub cell
- Electrical conductors
  - Joule heat
- Active material heat losses [5]
  - Irreversible
  - Reversible

$$q_{gen} = \sum I_i^2 R_i + \frac{dOCV}{dT} I\left(T - T_{ref}\right)$$

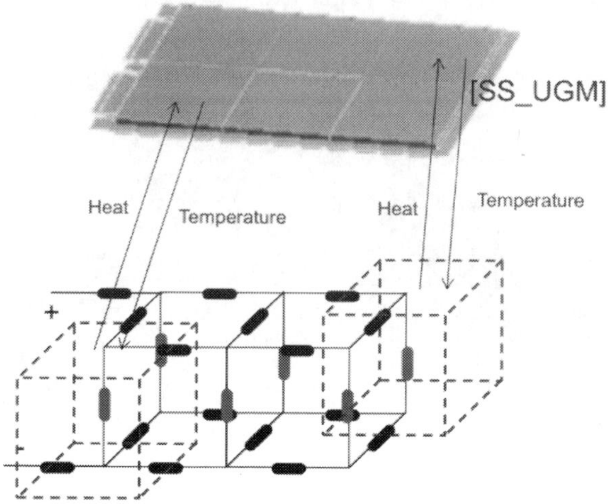

- 10 -

CADFEM

Discretization approach [5]

- Original impedance model is distributed
  - Parallel connection of sub-cells
- Volume fraction dependent [2]:

$$f_{\text{sub}} = \frac{V_{\text{sub}}}{V_T}$$

- Electrical parameters of sub-cells [2]:
  - Capacity : $\text{Cap}_{\text{sub}} = f_{\text{sub}} \text{Cap}_T$
  - Dynamics : $R_{\text{sub},1,2} = \frac{R_{1,2}}{f_{\text{sub}}}$ $\quad C_{\text{sub},1,2} = f_{\text{sub}} C_{1,2}$
  - Open Circuit Potential : $OCV_{\text{sub}} = OCV$

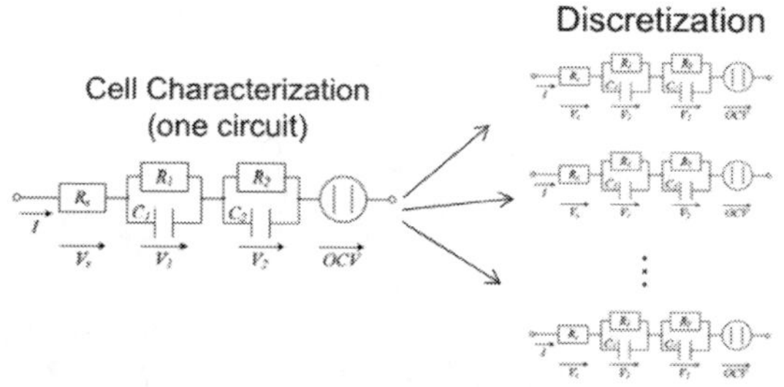

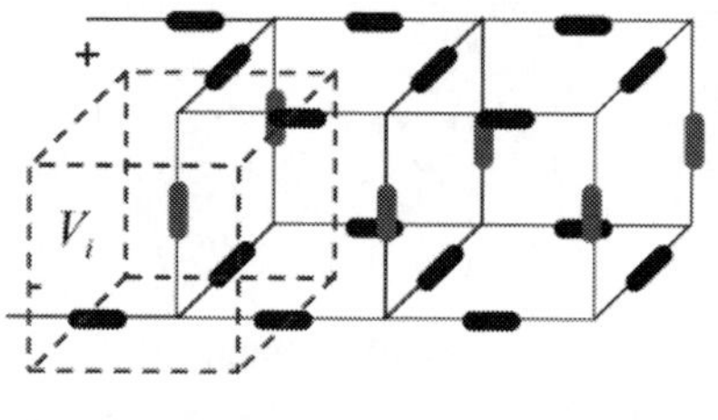

- 11 -

CADFEM

Parameterization [5]

- What is unknown?
  - Input parameterization
    - $R_{electrodes}$, R_{tabs}, $R_{s,sub}$
- What is know?
  - Constrains
    - Total DC resistance
  - Measurements
    - Voltage, Current
    - Temperature from **thermal camera**
- How the estimation is done?
  - Parameter Identification with optimization [5]:

$$f = \min \int \sum_{i=1}^{5} \left(\left(T_{sim,i} - T_{\exp,i} \right) \right)^2 dt$$

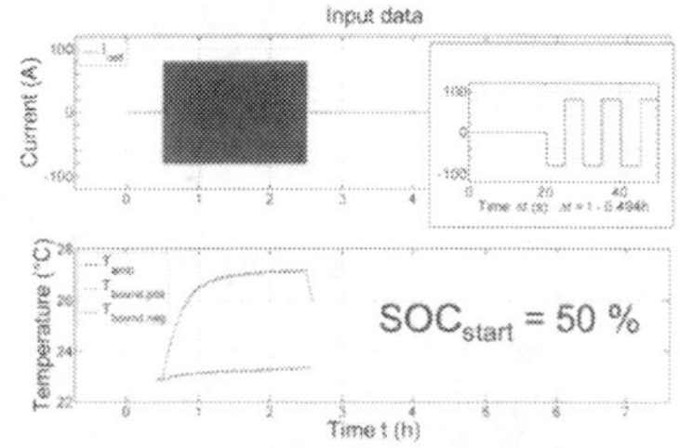

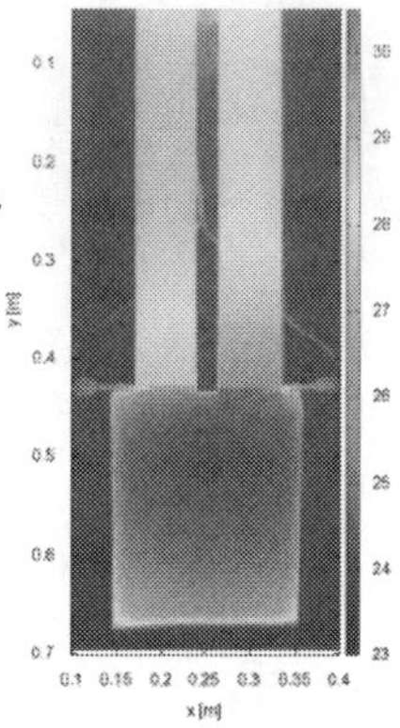

First results [8]

- The temperature profile is better reproduced
  - Inhomogeneous heat generation (G2)

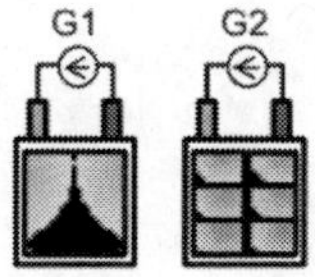

- Top of the cell could not be improved by G2 model

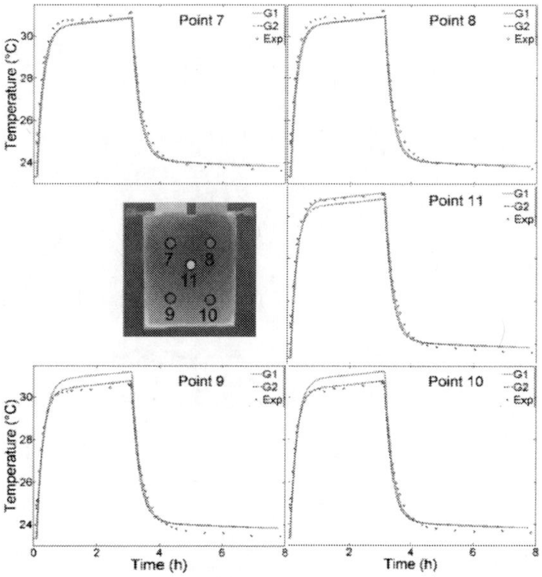

[8] S. Stumpp, L. Kostetzer, E. Rudnyi, M. Kellermeyer, R. Kuhn, und M. A. Danzer, „Evaluation of thermal modelling approaches for lithium-ion pouch cells", in *Advanced Battery Power 2012, Münster, 2012.*

Further investigations [5]

- Sensitivity of parameters
  - P1 region is sensitive to Resistance parameters

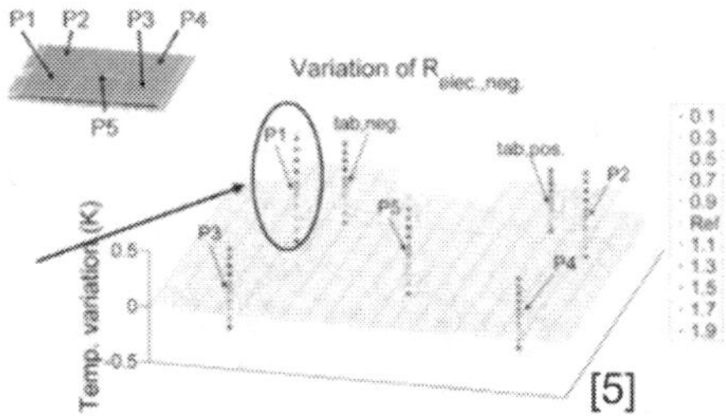

Improvements in experiments - > more numerical analysis

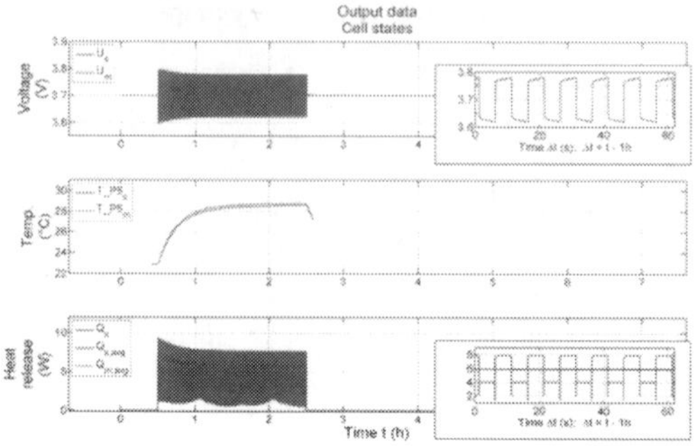

[5]

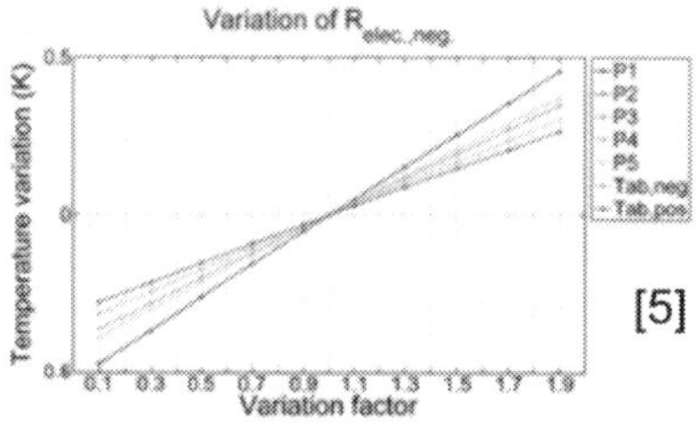

[5]

CADFEM

Results of parameterization [5]

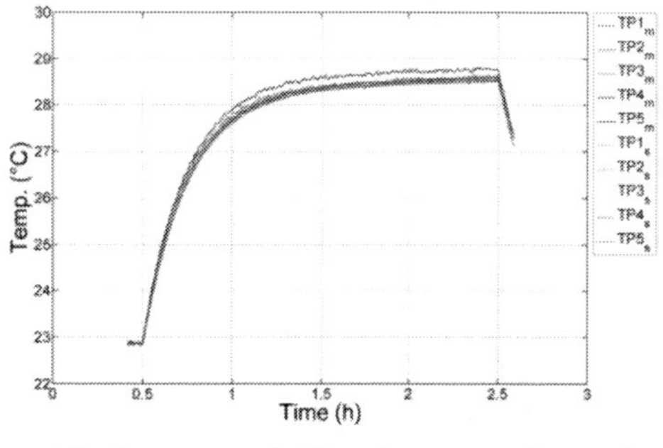

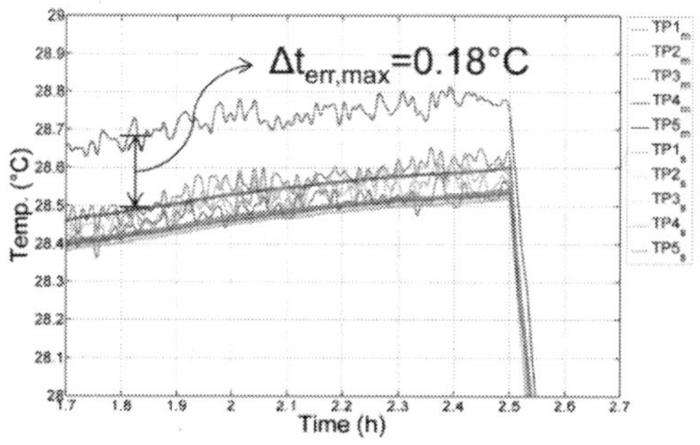

- Higher resolution in experiments
  - Parameter optimization improved

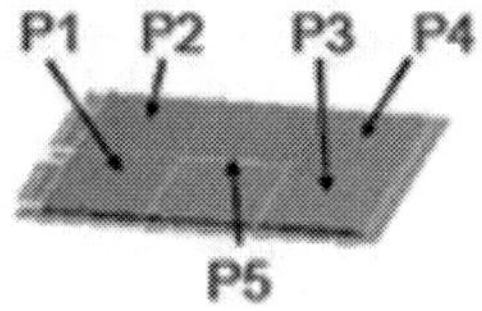

CADFEM

FEM approach [9]

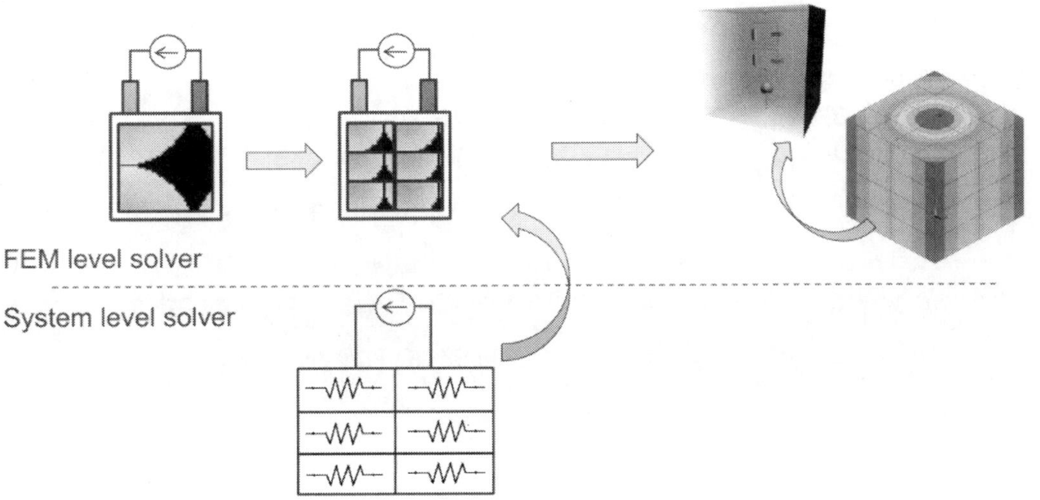

- 16 -

CADFEM

Battery element in ANSYS

- Impedance electrical model inside an ANSYS element

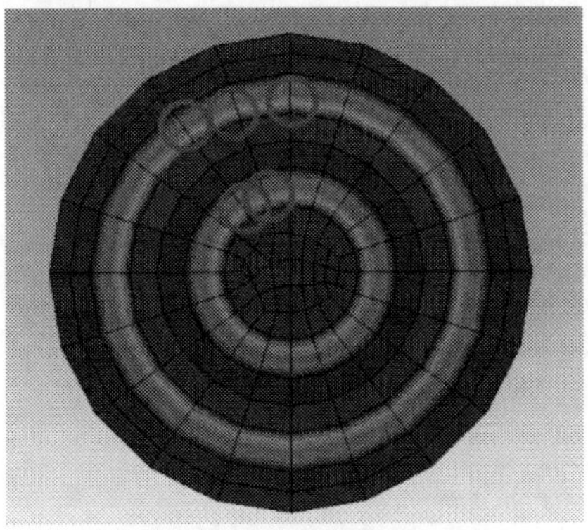

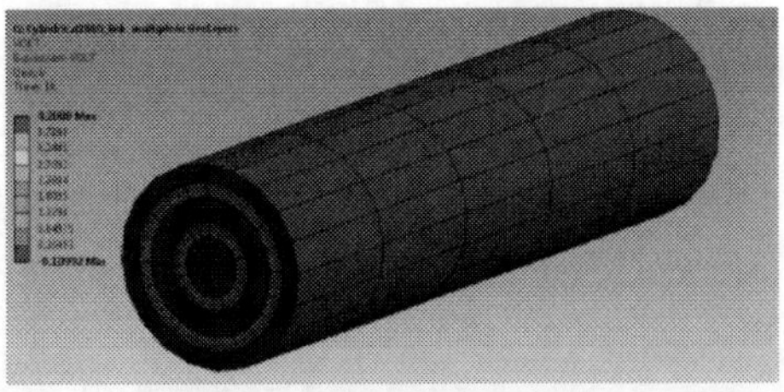

- More general method
  - Application in other geometries!

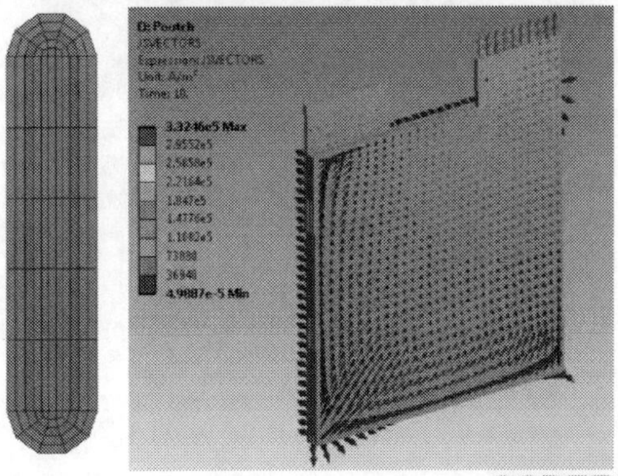

- 17 -

CADFEM

Summary

- Inhomogeneous heat generation modeling approach is shown
  - Distribution is done by sub-cell concept
- Electro-thermal coupling is done at the system level
  - MOR is essential to transform FEM to SYS level
- Parameters where identified based on experimental data
- FEM approach is a promising solution for higher resolution and more flexible geometric discretization
- Application in industry is necessary development

- 18 -

CADFEM

References

[1] Long Cai, Ralph E. White, Journal of Electrochem. Soc., 156 (3), A154-A161 (2009)

[2]J. B. Gerschler, F. N. Kirchhoff, H. Witzenhausen, F. E. Hust, und D. U. Sauer, „Spatially Resolved Model for Lithium-Ion Batteries for Identifying and Analyzing Influences of Inhomogeneous Stress Inside the Cells", in *5th International IEEE Vehicle Power and Propulsion Conference, Dearborn, Michigan, USA, 2009, S. 295–303.*

[3]M. Fleckenstein, O. Bohlen, M. A. Roscher, und B. Bäker, „Current density and state of charge inhomogeneities in Li-ion battery cells with LiFePO4 as cathode material due to temperature gradients", *Journal of Power Sources, Bd. 196, Nr. 10, S. 4769–4778, Mai 2011.*

[4]M. W. Verbrugge, „Three-dimensionai temperature and current distribution in a battery module", *AIChE Journal, Bd. 41, Nr. 6, S. 1550–1562, Juni 1995.*

[5] S. Stumpp, C. Günther, M.A. Danzer, L. Kostetzer, E.Rudnyi. ANSYS Conference & 31th CADFEM Users' Meeting 2013 | June 19-21, 2013 | Rosengarten Mannheim

[6]T. Bechtold, J.G. Korvink, E.B. Rudnyi, Fast simulation of electro-thermal MEMS efficient dynamic compact models, Springer, Berlin; New York, 2007.

[7]L. Kostetzer, „Battery Pack Electro-thermal Simulation", Master thesis, Landshut University of Applied Sciences, Ingolstadt University of Applied Sciences, 2011.

[8]S. Stumpp, L. Kostetzer, E. Rudnyi, M. Kellermeyer, R. Kuhn, und M. A. Danzer, „Evaluation of thermal modelling approaches for lithium-ion pouch cells", in *Advanced Battery Power 2012, Münster, 2012.*

[9] L. Kostetzer und S. Stumpp, „Electro-thermal battery modeling at the level of finite elements", in 16. Kongress mit Fachausstellung. SIMVEC Berechnung, Simulation und Erprobung Fahrzeugbau 2012, Baden-Baden, 2012.

- 19 -

CADFEM

Thank you for your attention

Contact:

Lucas Kostetzer

lkostetzer@cadfem.de

www.cadfem.de

- 20 -

CADFEM

The Future of Integrated Simulation: Where We Come From, Where We Have To Go

Dr. Peter Pirro, Manager Product Verification & Validation Global Mid Tractors

JOHN DEERE

Agenda

I. **John Deere**

II. **A Journey from Yesterday to Today**

- **Structural Analysis**
- **Fluid Dynamics**
- **Multi Body Simulation**
- **Dynamic System Simulation**

I. **Vision – Integrated Simulation / Multiphysics**

II. **Challenges**

- **Transfer Customer Requirements to Boundary Conditions and Acceptance Criteria**
- **The "three legged chair"**

JOHN DEERE

History of the John Deere Factory Mannheim

1837 – John Deere develops the first self-scouring steel plough

1918 – John Deere enters the tractor business

1931-1933 – John Deere didn't repossess machines from farmers that were unable to pay.

1859 – The year Heinrich Lanz AG, Mannheim was founded

1888 – Establishment of the factory in Lindenhof, Mannheim

1921 – First Bulldog

1956 – John Deere expands outside of the USA and takes over Heinrich Lanz AG Mannheim

John Deere worldwide

65 factories in 19 countries - 67,000 employees worldwide

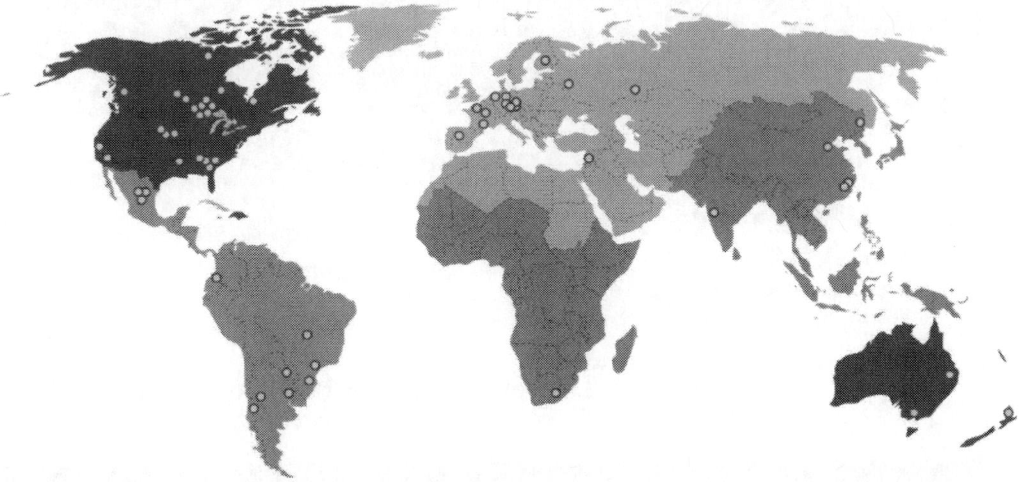

John Deere today

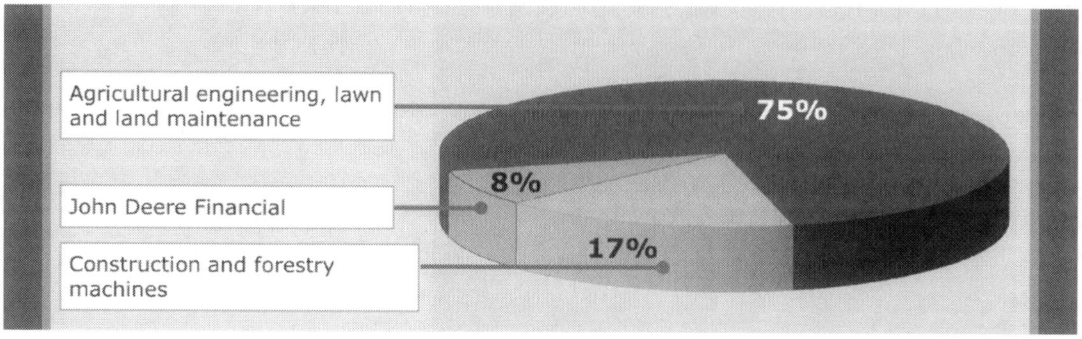

- **2012 turnover and other income:** $ 36 Billion
- **2012 profit:** $ 3,065 Million

Innovation through investment

2012 Investments: (8.4 % of the machine sales)

- **Capital investment:**
  $ 1,376 million (4.1 %)
- **Research and development:**
  $ 1,434 million (4.3 %)
- **John Deere invests more than**
  $ 5 million in research and development worldwide every day

Region 2 – Locations

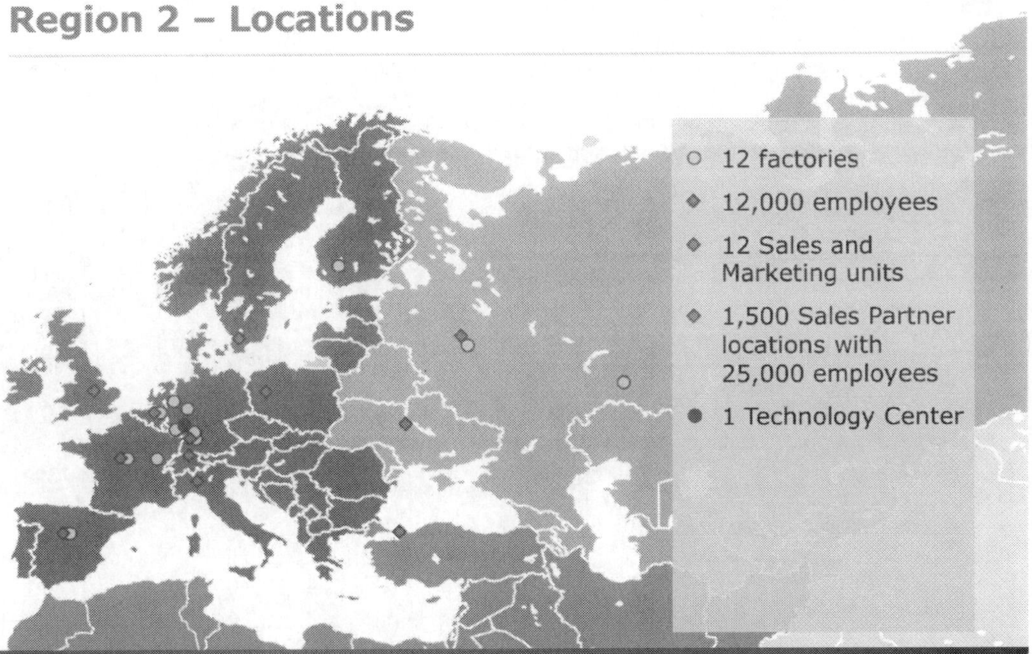

- ○ 12 factories
- ◆ 12,000 employees
- ◆ 12 Sales and Marketing units
- ◆ 1,500 Sales Partner locations with 25,000 employees
- ● 1 Technology Center

John Deere in Germany

- 6 locations
- 7,200 employees, (1 in 5 employees in the agricultural engineering industry)
- 3.29 billion turnover = 40 % of the turnover of the German agricultural engineering industry
- 96 Sales Partners with 4,500 employees in total

John Deere Factory Mannheim

John Deere Factory Mannheim and Bruchsal

Biggest John Deere factory outside of North America

- over 1.6 million tractors since 1921
- 1992 introduction of the full frame chassis construction
- 2/3 of all German tractor production
- 70 to 210 hp tractors (97/68/EC)
- 4,005 employees at the Mannheim site
- 1,375 employees at the Bruchsal site

The John Deere success story
Over 175 years of self-sustained commitment continuity and dynamic growth
Biggest agricultural machinery manufacturer worldwide
11 | Dr. Peter Pirro
2013 Automotive Simulation World Congress
October 29 – 30 Frankfurt am Main, Germany
JOHN DEERE

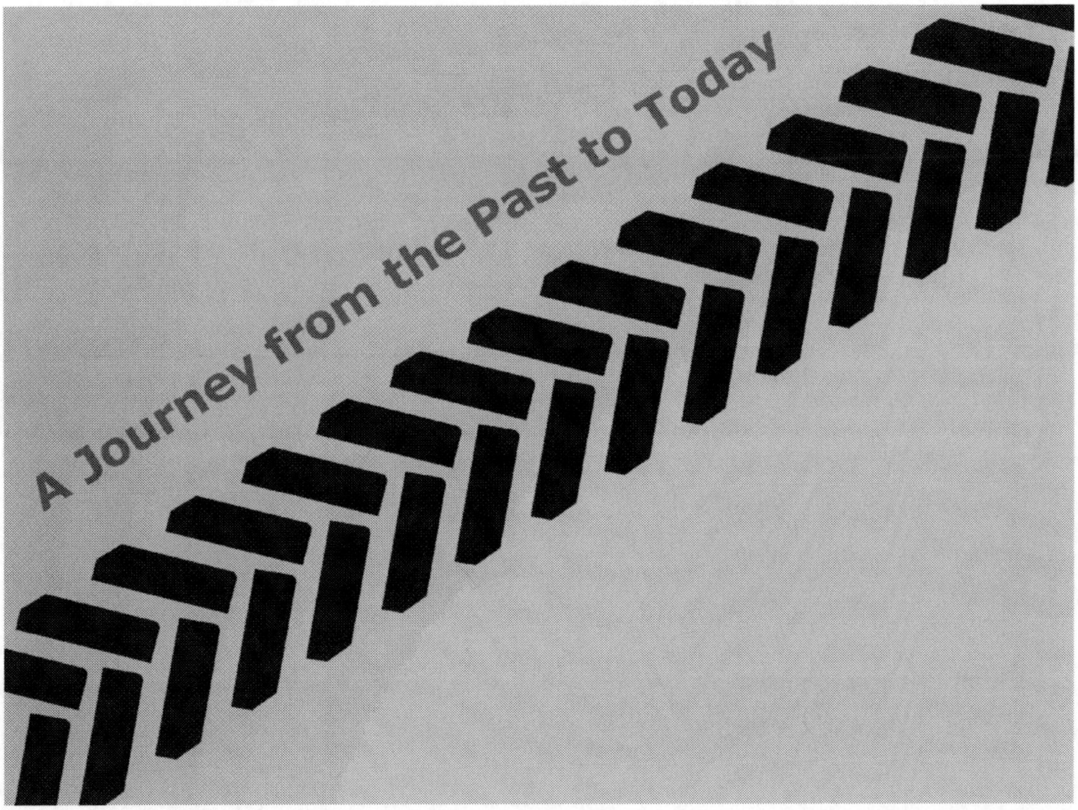
A Journey from the Past to Today

Roadmap

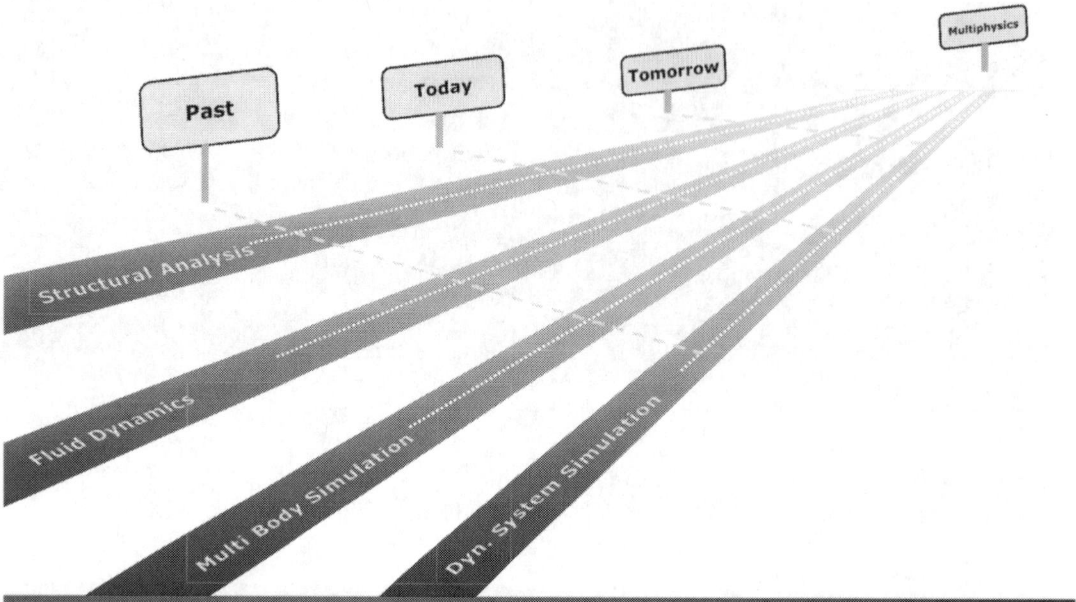

Structural Analysis

- Single part (planetary carrier)
- Bottom-up meshing, keypoints, lines, areas
- Section meshing
- Highly simplified geometry and boundary conditions
- Limited model size ->sub-models
- Scripting
- Calculations on centralized Main Frame Computers

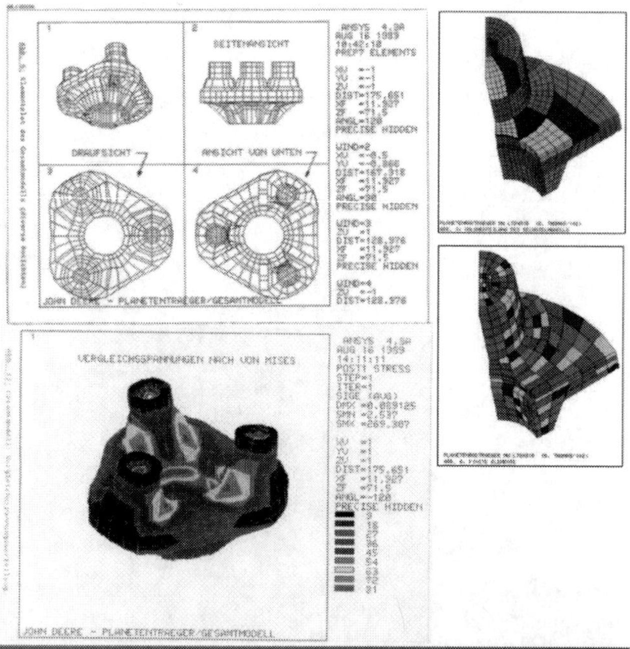

1990 1991 1992 1993 1994 1995 1996 1997 1998 1999 2000 2001 2002 2003 2004 2005 2006 2007 2008 2009 2010 2011 2012 2013

Structural Analysis

- 3D CAD Import
- Big time savings in modeling
- Complex casting geometry
- Assemblies
- Computation on workstation

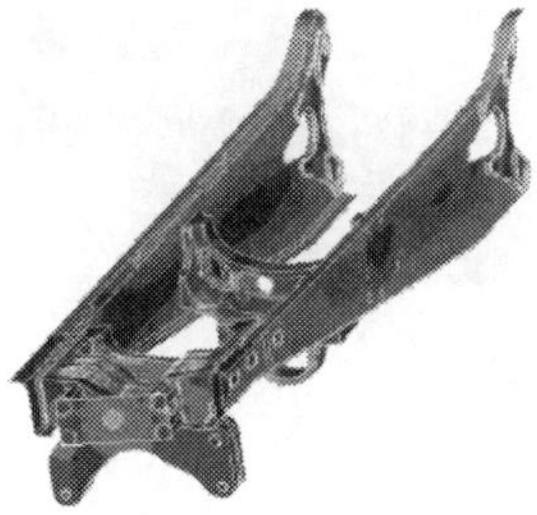

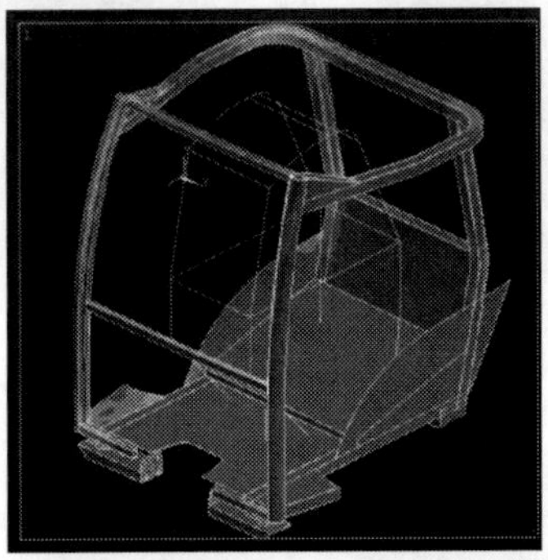

1990 1991 1992 1993 1994 1995 1996 1997 1998 1999 2000 2001 2002 2003 2004 2005 2006 2007 2008 2009 2010 2011 2012 2013

Structural Analysis

- Topology optimization / shape optimization
- Use numeric optimization methods for geometry finding
- Combination of load cases
- Interpretation of the results necessary -> new geometry

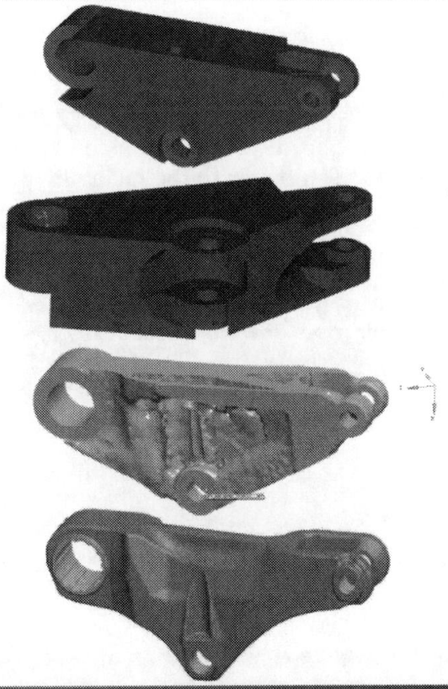

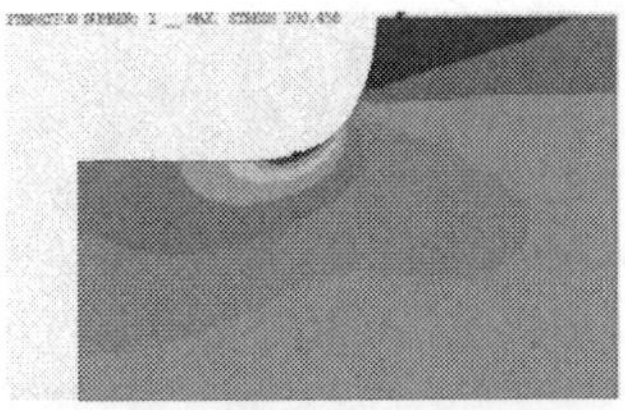

1990 1991 1992 1993 1994 1995 1996 1997 1998 1999 2000 2001 2002 2003 2004 2005 2006 2007 2008 2009 2010 2011 2012 2013

Structural Analysis

- Consideration of overall assembly
- Nonlinear mapping of the overall system
- Interactions included
- Use of computational clusters
- Computation of large models

17 | Dr. Peter Pirro 2013 Automotive Simulation World Congress October 29 – 30 Frankfurt am Main, Germany JOHN DEERE

1990 1991 1992 1993 1994 1995 1996 1997 1998 1999 2000 2001 2002 2003 2004 2005 2006 2007 2008 2009 2010 2011 2012 2013

Fluid Dynamics

- Cooling package and frontend
- Air distribution to the radiators
- Isothermal, incompressible, steady state
- Fan only as a boundary condition, air mass flow from test
- User had to simplify the geometry in favor of the model size

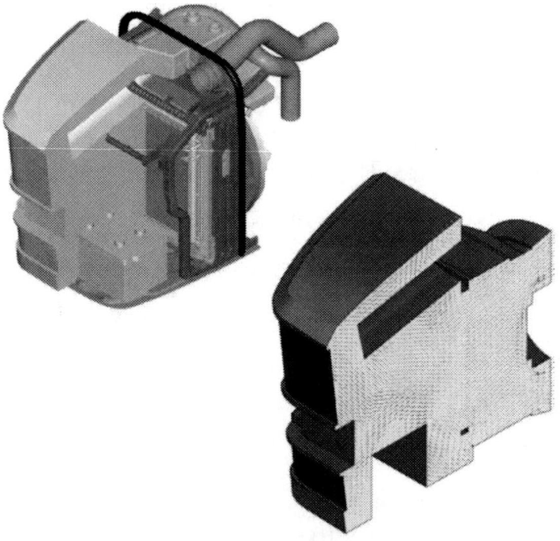

18 | Dr. Peter Pirro 2013 Automotive Simulation World Congress October 29 – 30 Frankfurt am Main, Germany JOHN DEERE

1990 1991 1992 1993 1994 1995 1996 1997 1998 1999 2000 2001 2002 2003 2004 2005 2006 2007 2008 2009 2010 2011 2012 2013

Fluid Dynamics

- Air flow to the windscreen
- No fan and heat exchanger
- Boundary conditions from test
- Simplified geometry
- Positioning the constraint has considerable influence on results => User experience

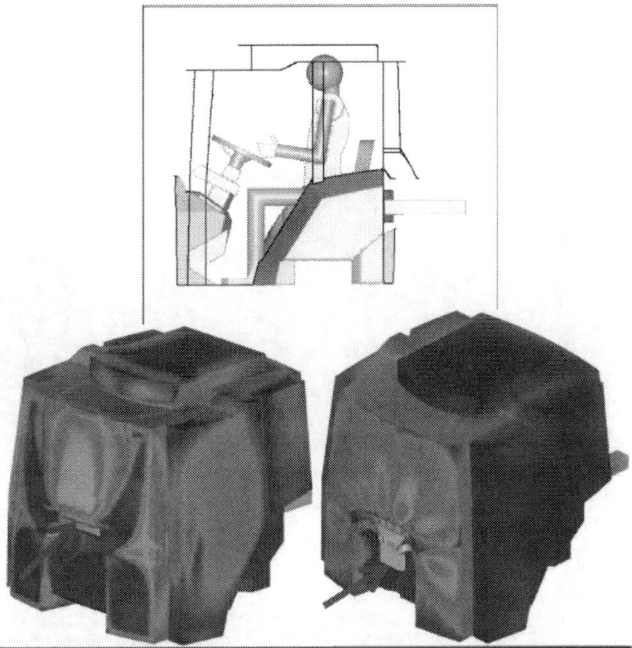

1990 1991 1992 1993 1994 1995 1996 1997 1998 1999 2000 2001 2002 2003 2004 2005 2006 2007 2008 2009 2010 2011 2012 2013

Fluid Dynamics

- Virtual Wind Tunnel
- Cooling air mass flow rate and thermal models (recirculation)
- Detailed vehicle model
- Fan in Moving Reference Frame
- Automation:
  Surface Wrapping
  Postprocessing Scripting
- Detailed system and application knowledge necessary

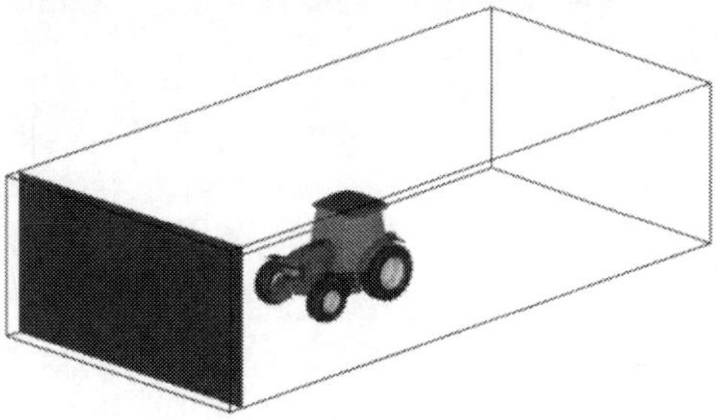

← 1990 1991 1992 1993 1994 1995 1996 1997 1998 1999 2000 2001 2002 2003 2004 2005 2006 2007 2008 2009 2010 2011 2012 2013 →

Fluid Dynamics

- Underhood temperature
- Transient Thermal Analysis
- Focus Hot Shutdown, only natural convection
- Underhood CFD can't provide absolute heat transfer coefficient yet
- User needs to make assumptions for each component for the convective heat transfer

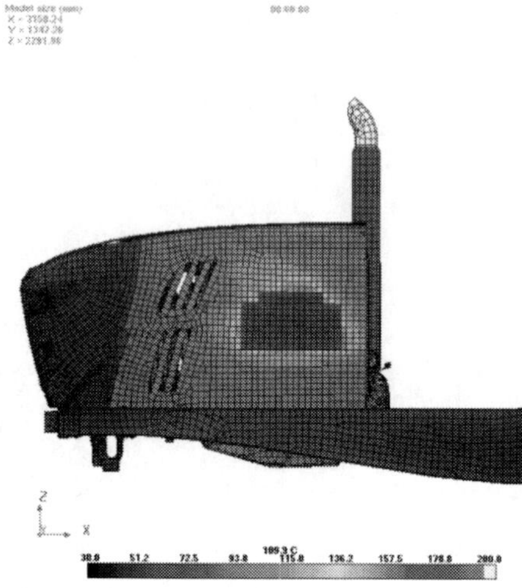

JOHN DEERE

← 1990 1991 1992 1993 1994 1995 1996 1997 1998 1999 2000 2001 2002 2003 2004 2005 2006 2007 2008 2009 2010 2011 2012 2013 →

Fluid Dynamics

- Topology Optimization
- "Ideas" for better shapes
- No automated results
- Transfer of optimization results to producible geometry by the user

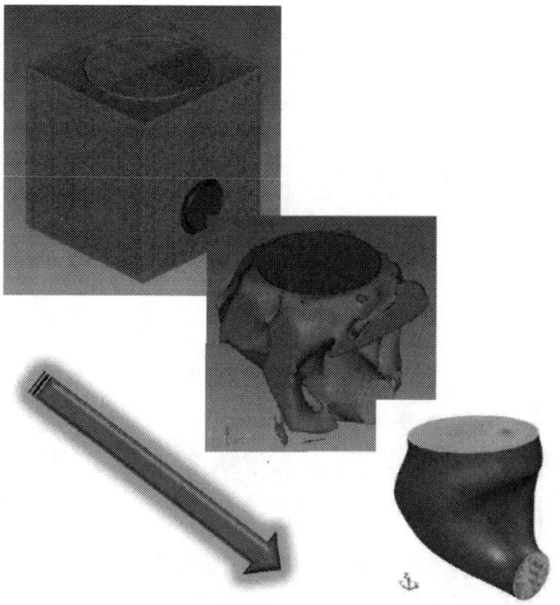

JOHN DEERE

1990 1991 1992 1993 1994 1995 1996 1997 1998 1999 2000 2001 2002 2003 2004 2005 2006 2007 2008 2009 2010 2011 2012 2013

Multi Body Simulation

- Single-Track model
- Equations of motion
- Numerical integration
- Results as time series
  - paths
  - velocities
  - accelerations
  - forces
  - moments
- Mainly vertical and longitudinal dynamics
- Front suspension
- Ride control

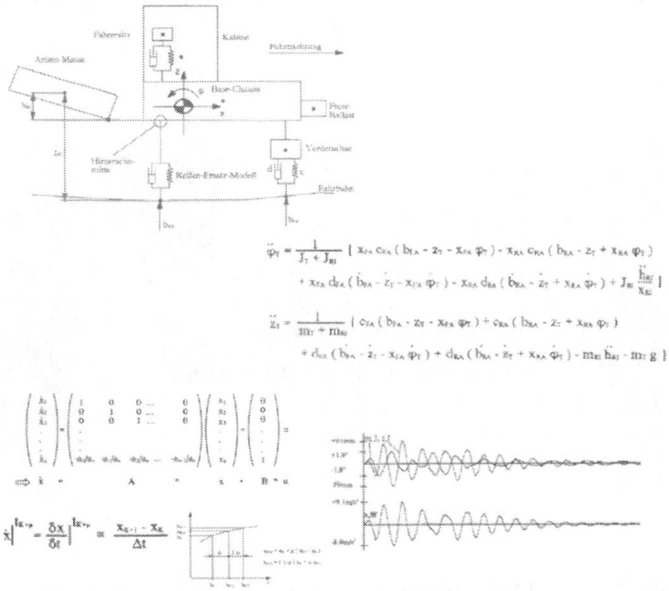

1990 1991 1992 1993 1994 1995 1996 1997 1998 1999 2000 2001 2002 2003 2004 2005 2006 2007 2008 2009 2010 2011 2012 2013

Multi Body Simulation

- Transition to two-track models
- Several software packages for 3D Multi-Body Simulation

  eg: DADS, Adams
- Integrated tire models
- 3D road profiles as excitation
- In addition, lateral dynamics
- Beginning of the vehicle dynamics simulation

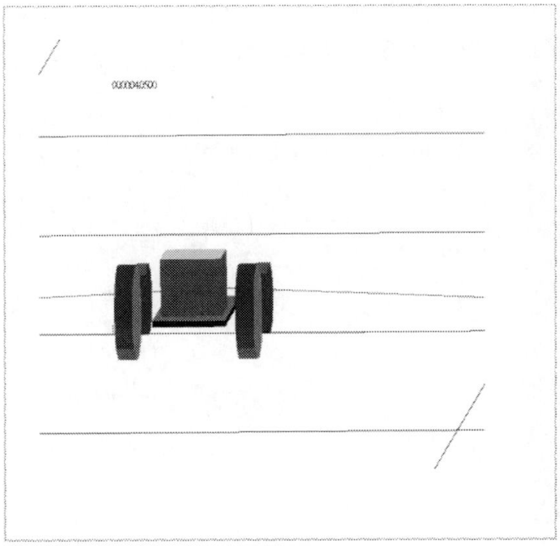

← 1990 1991 1992 1993 1994 1995 1996 1997 1998 1999 2000 2001 2002 2003 2004 2005 2006 2007 2008 2009 2010 2011 2012 2013 →

Multi Body Simulation

- Total vehicle model
- Dynamics studies
- Attachments
- Drawn equipment
- Comfort rating
- Optimization of suspension
- Load data derivation
- Co-simulation of Simulink / Virtual.Lab

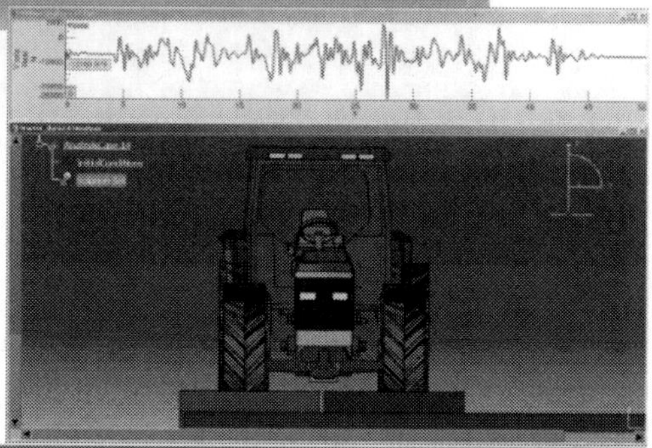

← 1990 1991 1992 1993 1994 1995 1996 1997 1998 1999 2000 2001 2002 2003 2004 2005 2006 2007 2008 2009 2010 2011 2012 2013 →

Dyn. System Simulation

- Stick diagram
- Steady-state calculations
- Gear ratios
- Rotational velocities (rpm)
- Forces
- Moments

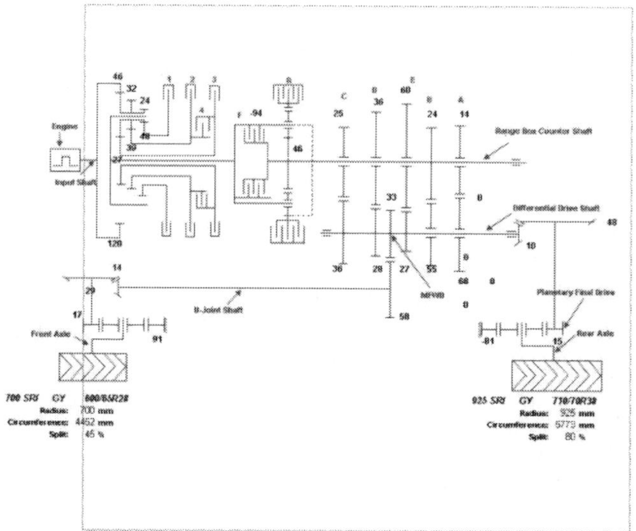

Dyn. System Simulation

- Dynamic calculations
  - Hydraulic
  - Mechanical systems
- Modal Analysis
- Gear Shifting
  - Hydraulic system
  - Mechanical systems
- Controller design
- SIL und HIL-Testing

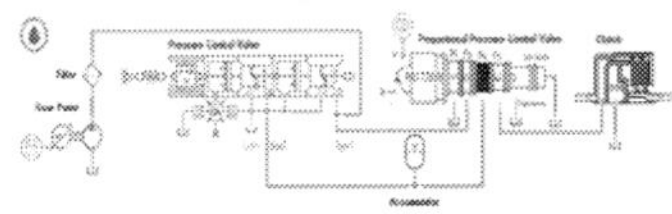

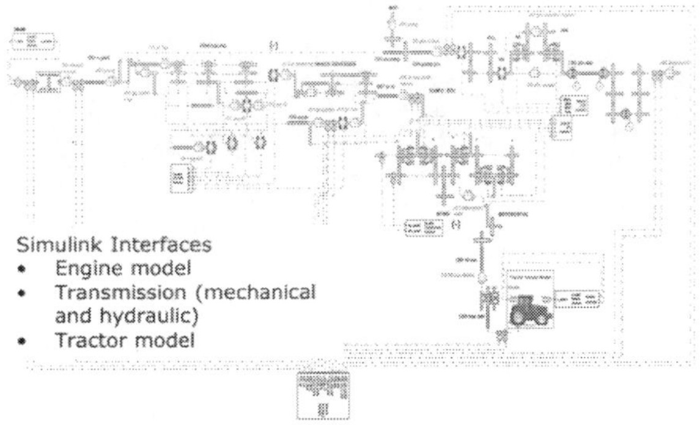

Simulink Interfaces
- Engine model
- Transmission (mechanical and hydraulic)
- Tractor model

Dyn. System Simulation

- Hydraulic system
- Coupling of mechanics
- Coupling with controllers
- Interaction of the systems
- Multi-domain approach
  - Co-simulation
  - AMESim
  - Simulink
  - Virtual.Lab

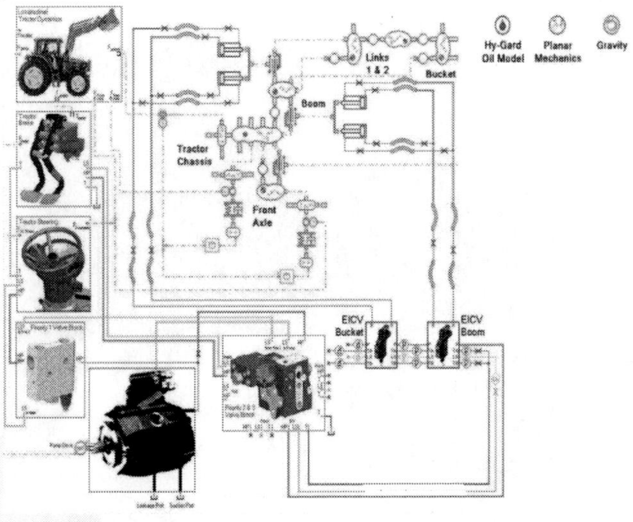

Multiphysics

Multiphysics

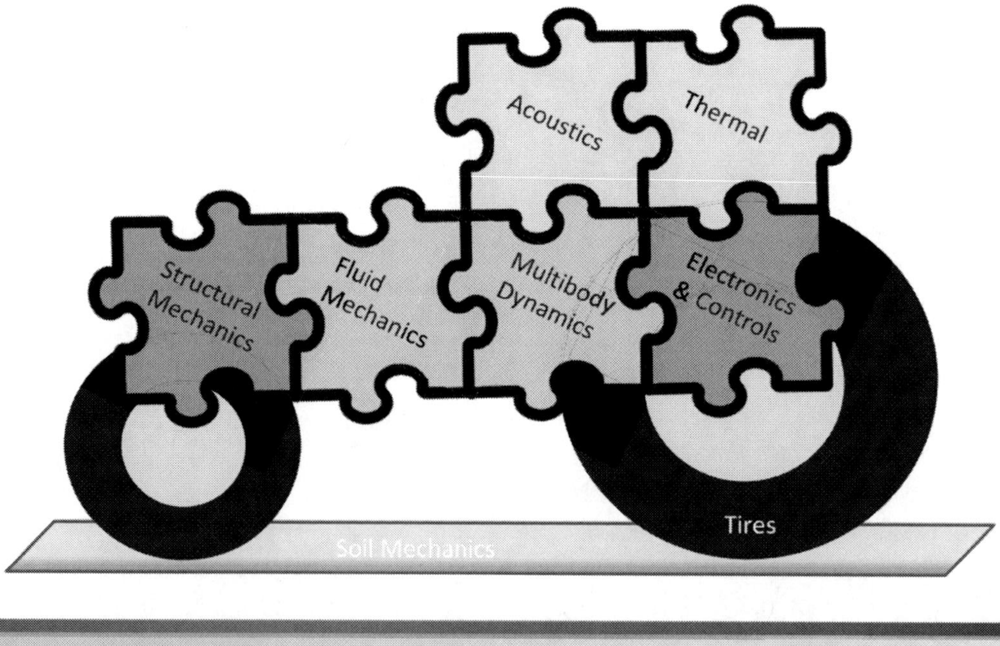

Multiphysics

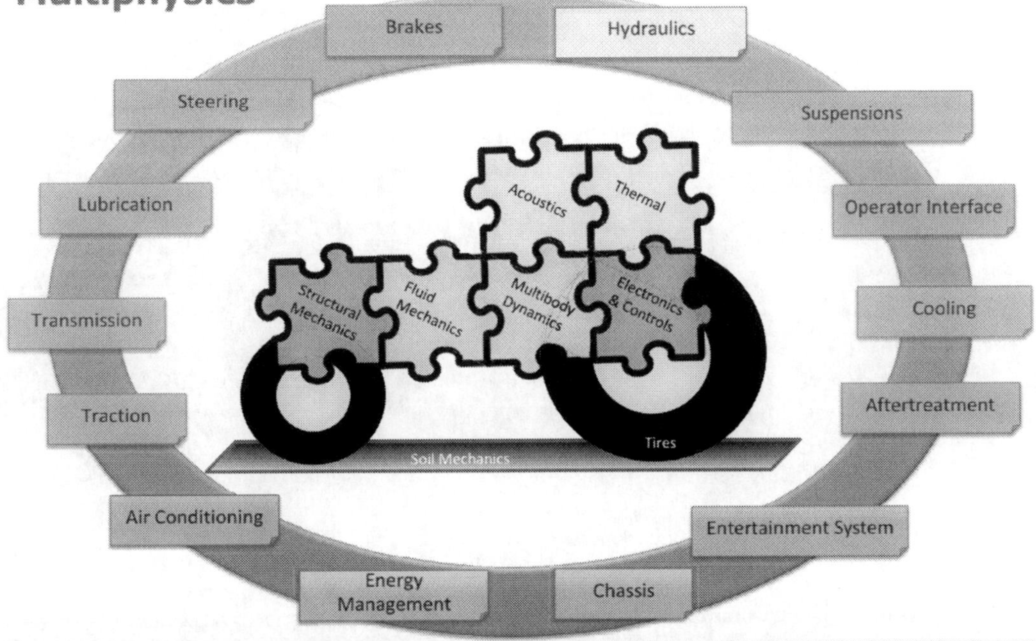

Multiphysics

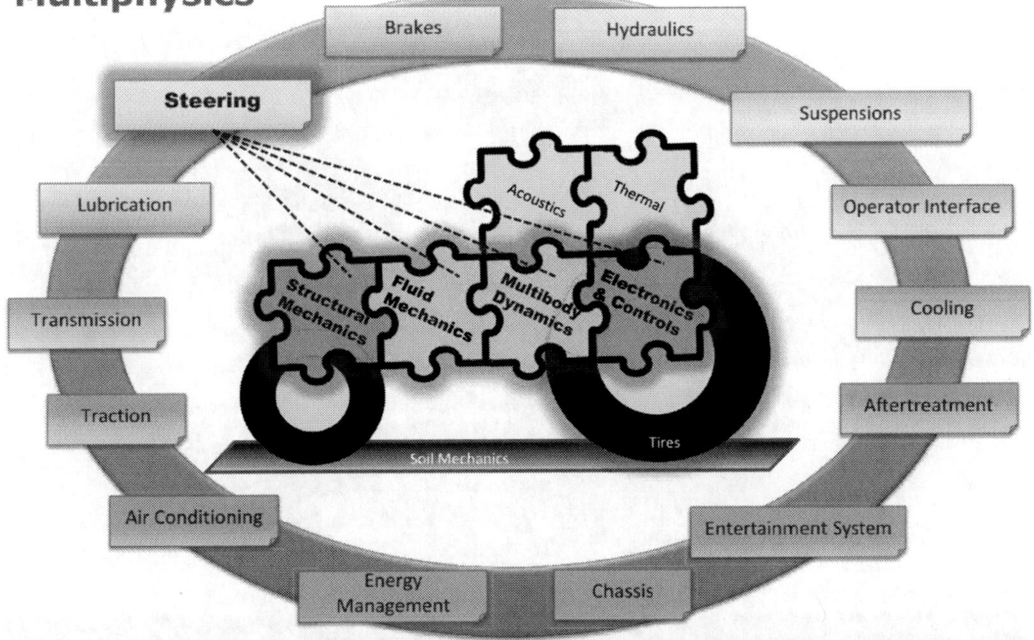

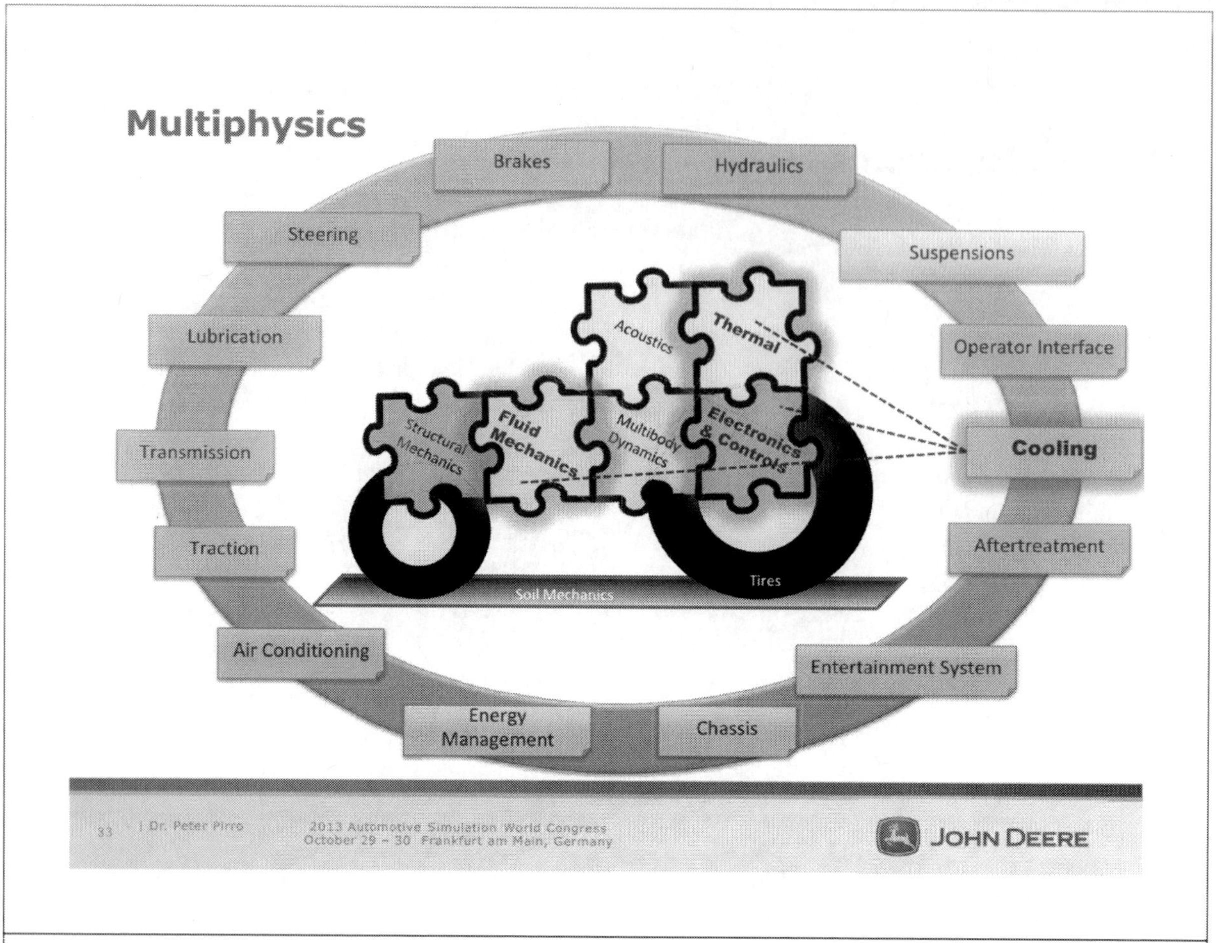
Multiphysics
Brakes
Hydraulics
Steering
Suspensions
Lubrication
Operator Interface
Acoustics
Thermal
Structural Mechanics
Fluid Mechanics
Multibody Dynamics
Electronics & Controls
Transmission
Cooling
Traction
Aftertreatment
Tires
Soil Mechanics
Air Conditioning
Entertainment System
Energy Management
Chassis
33 | Dr. Peter Pirro
2013 Automotive Simulation World Congress
October 29 – 30 Frankfurt am Main, Germany
JOHN DEERE

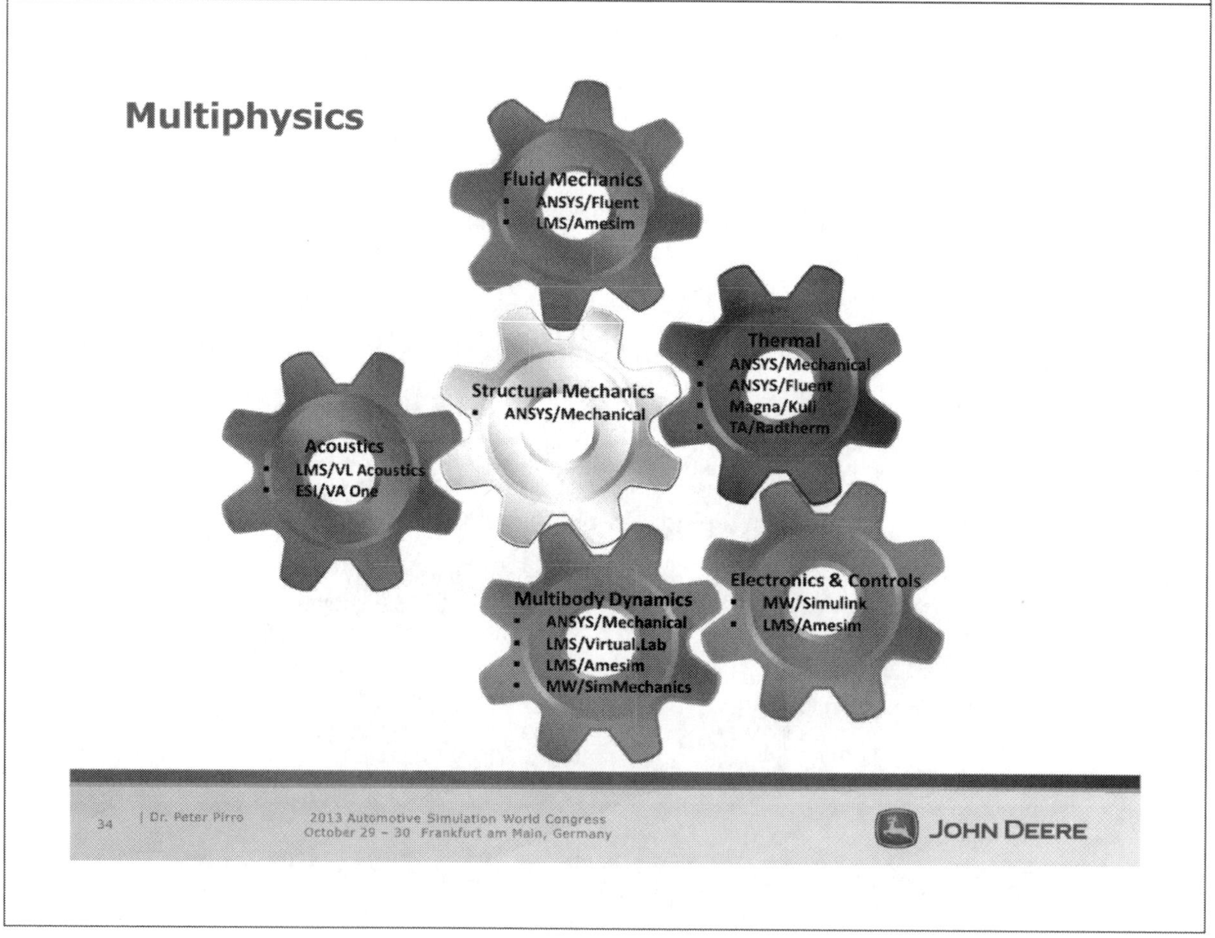
Multiphysics
Fluid Mechanics
ANSYS/Fluent
LMS/Amesim
Thermal
ANSYS/Mechanical
ANSYS/Fluent
Magna/Kuli
TA/Radtherm
Structural Mechanics
ANSYS/Mechanical
Acoustics
LMS/VL Acoustics
ESI/VA One
Electronics & Controls
MW/Simulink
LMS/Amesim
Multibody Dynamics
ANSYS/Mechanical
LMS/Virtual.Lab
LMS/Amesim
MW/SimMechanics
34 | Dr. Peter Pirro
2013 Automotive Simulation World Congress
October 29 – 30 Frankfurt am Main, Germany
JOHN DEERE

Multiphysics

35 | Dr. Peter Pirro 2013 Automotive Simulation World Congress
October 29 – 30 Frankfurt am Main, Germany
JOHN DEERE

Multiphysics

36 | Dr. Peter Pirro 2013 Automotive Simulation World Congress
October 29 – 30 Frankfurt am Main, Germany
JOHN DEERE

Challenges

Transfer Customer Requirements to Boundary Conditions and Acceptance Criteria

The Three-Legged Chair

Process:

Systematic integration of simulation in the development process early in the concept phase

Resources:

Availability of high qualified personal, High-Performance Computing, software-licenses ….

Competencies:

Building and maintaining the necessary core competencies / qualifications in the company, knowledge management

High Performance Computing

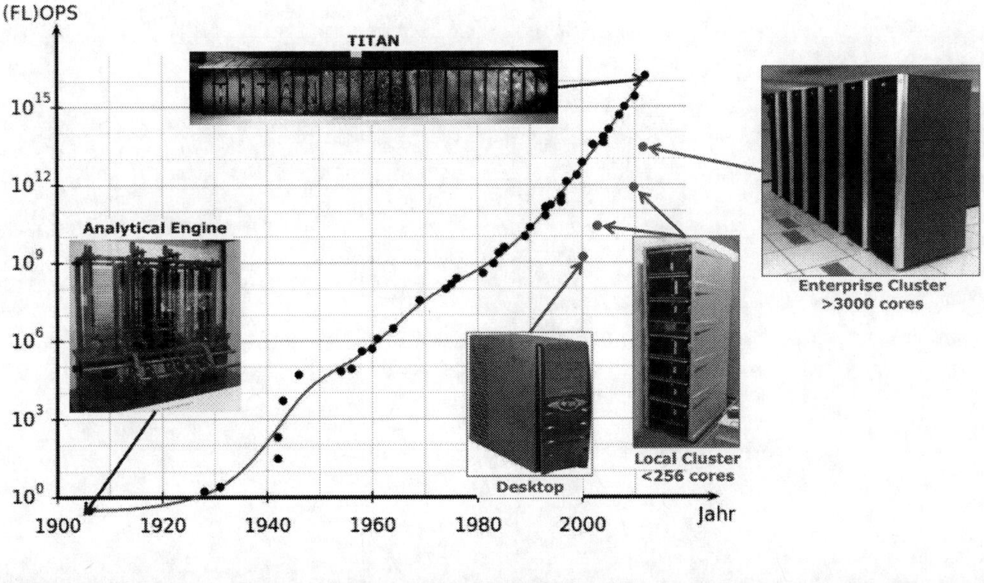

Required Competencies

Requirements for a Simulator/Simulant/CAE Engineer

- Basic Engineering Principles
- Special CAE Knowledge
- Continuous Training
- Product Knowledge
- ...

41 | Dr. Peter Pirro 2013 Automotive Simulation World Congress October 29 – 30 Frankfurt am Main, Germany JOHN DEERE

JOHN DEERE

„With the methods from yesterday
you
develop products from yesterday"

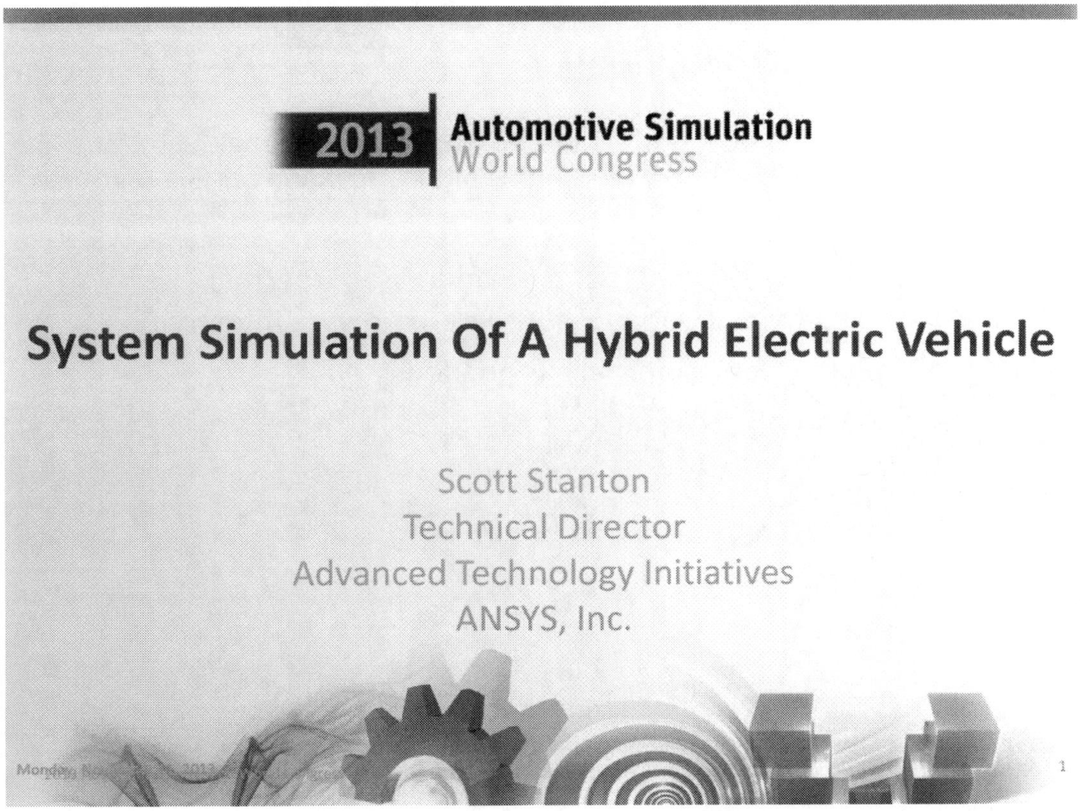
2013 Automotive Simulation World Congress
System Simulation Of A Hybrid Electric Vehicle
Scott Stanton
Technical Director
Advanced Technology Initiatives
ANSYS, Inc.
1

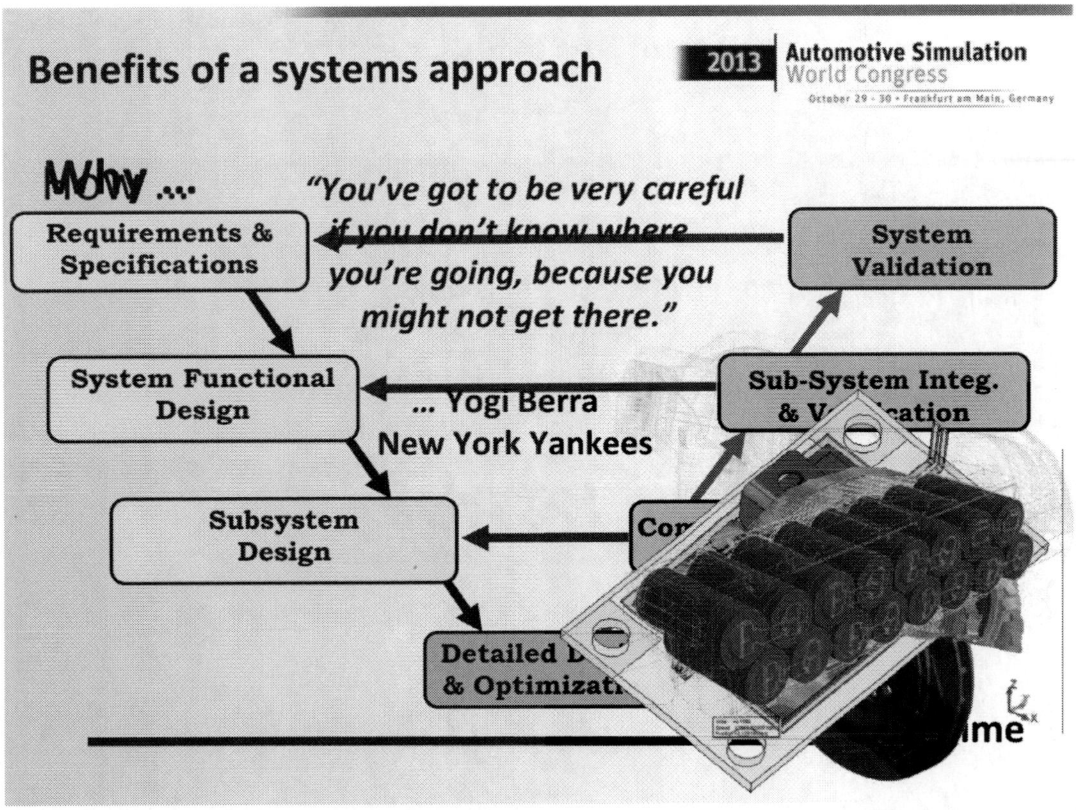
Benefits of a systems approach
2013 Automotive Simulation World Congress
October 29 - 30 • Frankfurt am Main, Germany
"You've got to be very careful if you don't know where you're going, because you might not get there."
... Yogi Berra
New York Yankees
Requirements & Specifications
System Validation
System Functional Design
Sub-System Integ.
Subsystem Design
Detailed
& Optimizati
ime

2013 Automotive Simulation World Congress
October 29 - 30 • Frankfurt am Main, Germany

Introduction

- For the next generation of drives, the challenge is to achieve smaller footprint, higher efficiency with less material usage.
- Multiples issues need to be addressed: thermal management, fatigue, software-hardware interaction.
- A full systems approach is needed

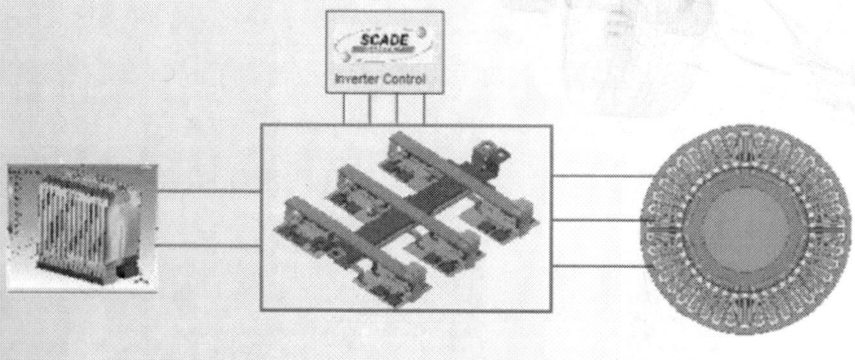

2013 Automotive Simulation World Congress
October 29 - 30 • Frankfurt am Main, Germany

System functional engineering

- System engineering process:
  - > Step 1: Define what the system shall do:
    - » the system function design
  - > Step 2: Define how the system is built:
    - » the system architecture
- Process supported by:

SCADE SYSTEM™

Step 1: Define what the system shall do

2013 Automotive Simulation World Congress
October 29 - 30 • Frankfurt am Main, Germany

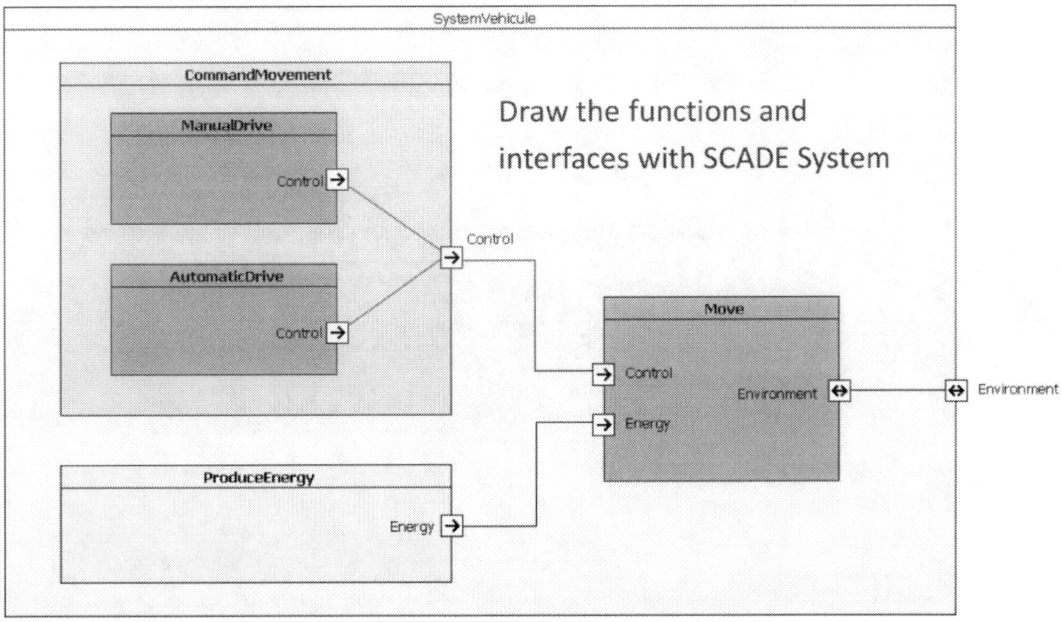

Step 1: Define what the system shall do

2013 Automotive Simulation World Congress
October 29 - 30 • Frankfurt am Main, Germany

Verify design completeness:
Requirement Traceability

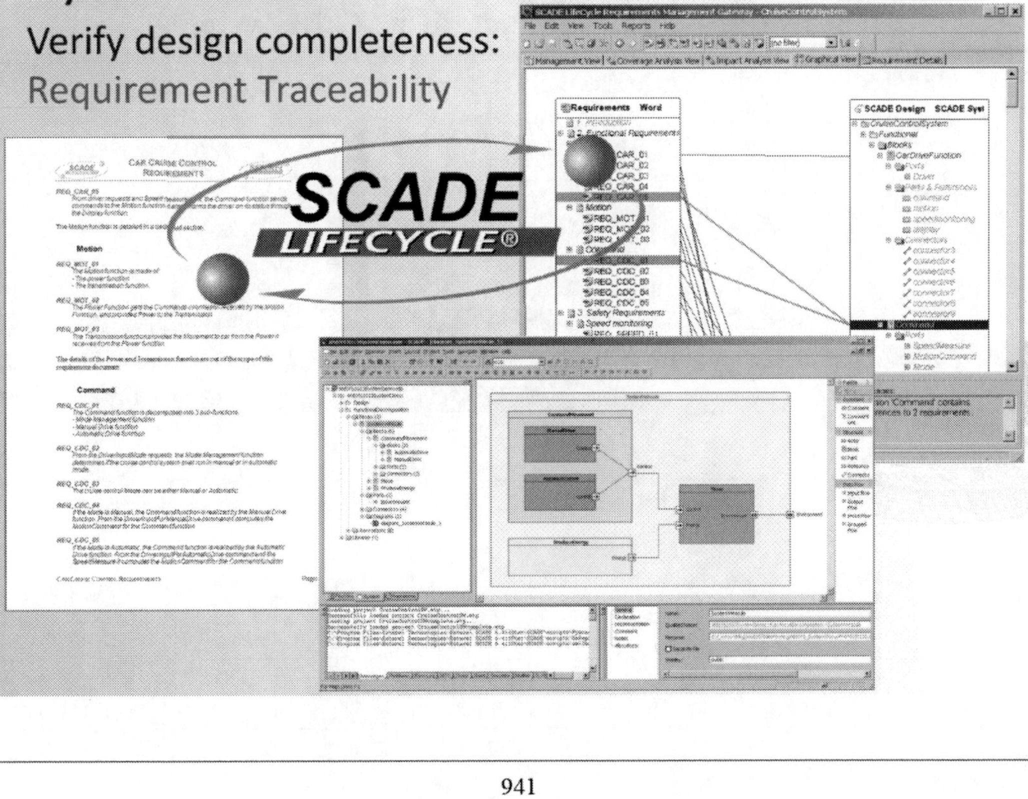

Step 1: Define what the system shall do

Verify design consistency: SCADE System Checker

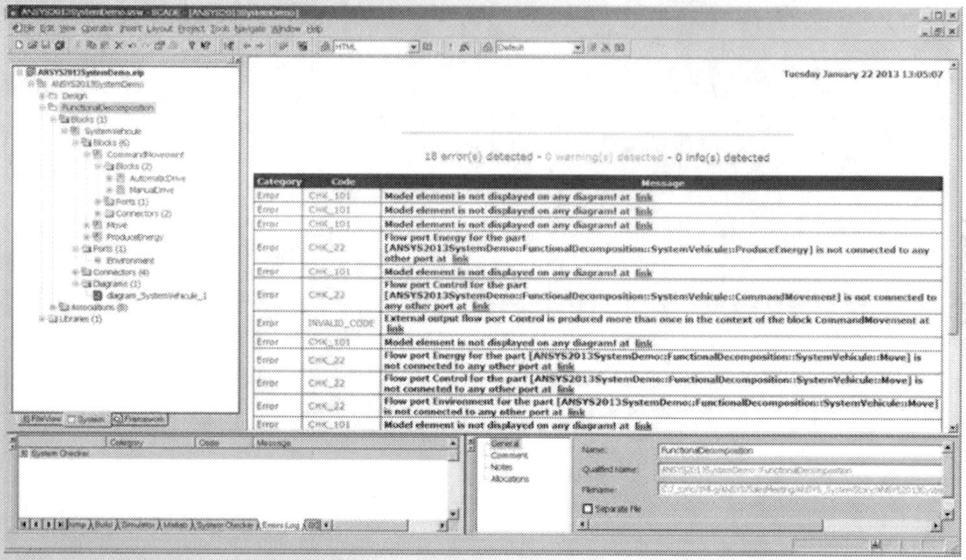

Step 2: Define how the system is built

2013 Automotive Simulation World Congress
October 29 - 30 • Frankfurt am Main, Germany

Draw the system architecture in SCADE System

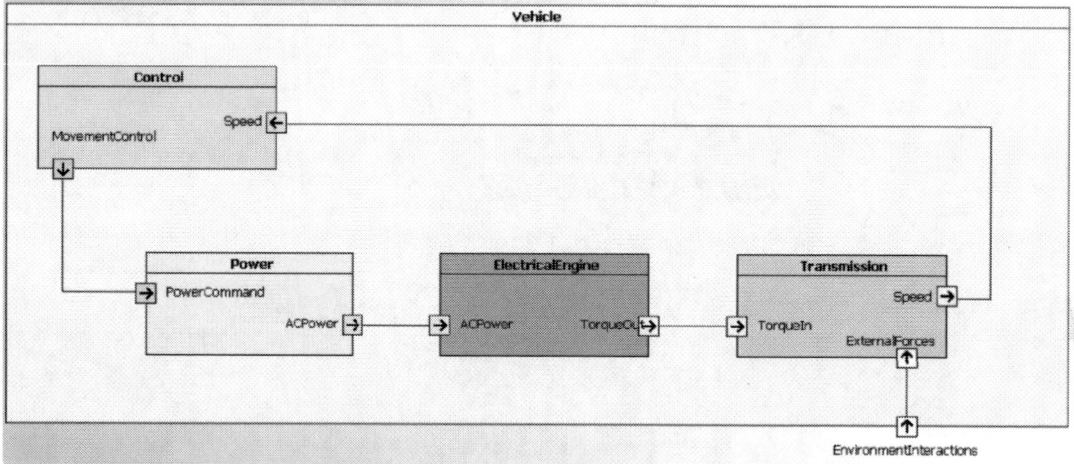

Step 2: Define how the system is built

2013 Automotive Simulation World Congress
October 29 - 30 • Frankfurt am Main, Germany

Draw the system architecture in SCADE System

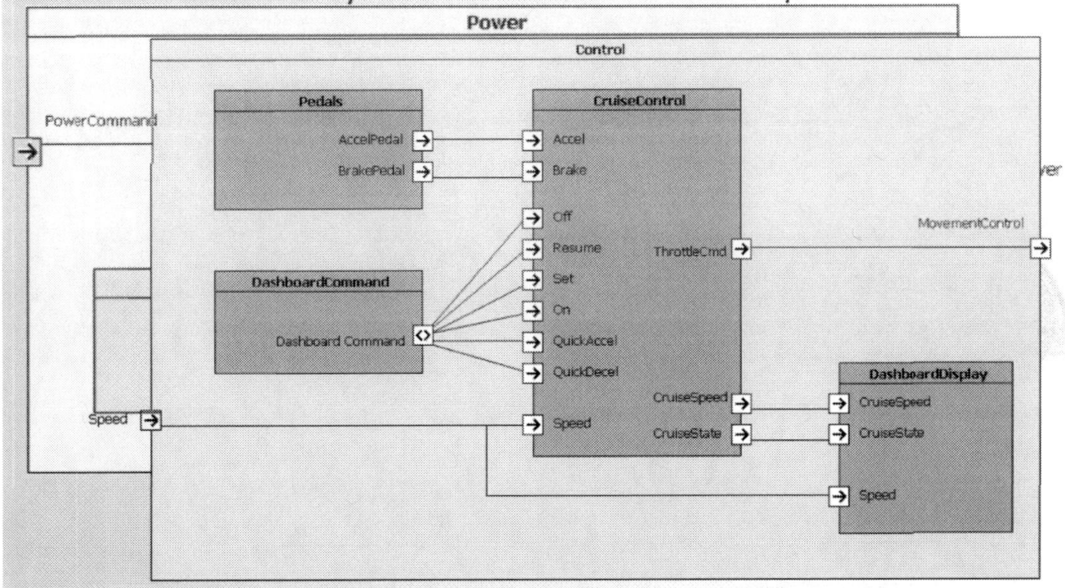

2013 Automotive Simulation World Congress
October 29 - 30 • Frankfurt am Main, Germany

2- Battery Pack Design

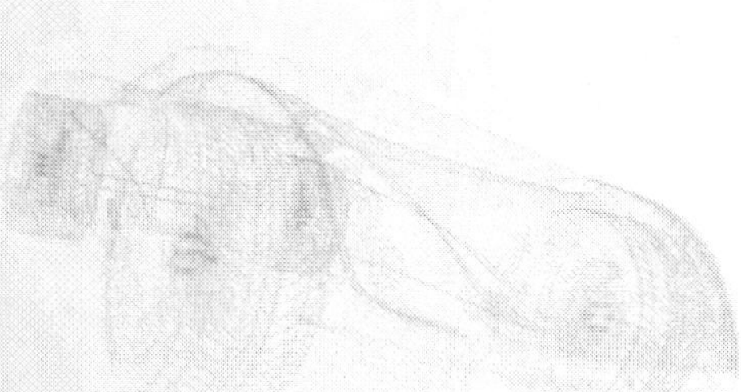

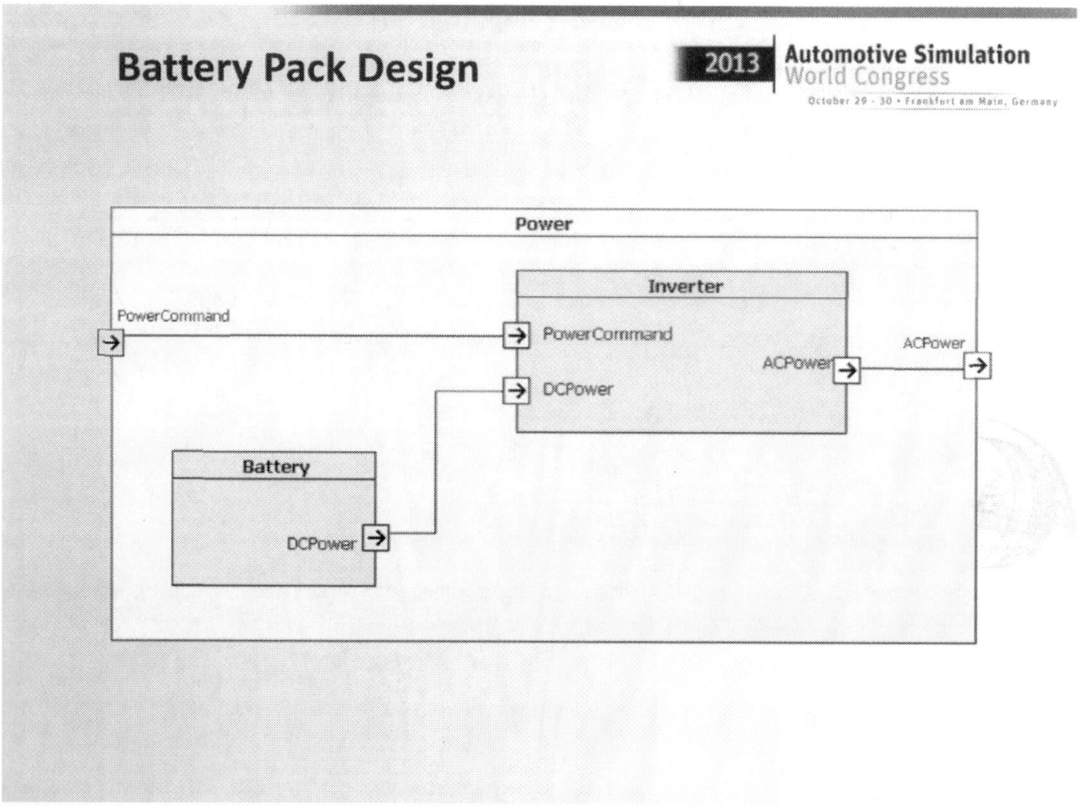
Battery Pack Design
2013
Automotive Simulation
World Congress
October 29 - 30 • Frankfurt am Main, Germany
Power
Inverter
PowerCommand
PowerCommand
DCPower
ACPower
ACPower
Battery
DCPower

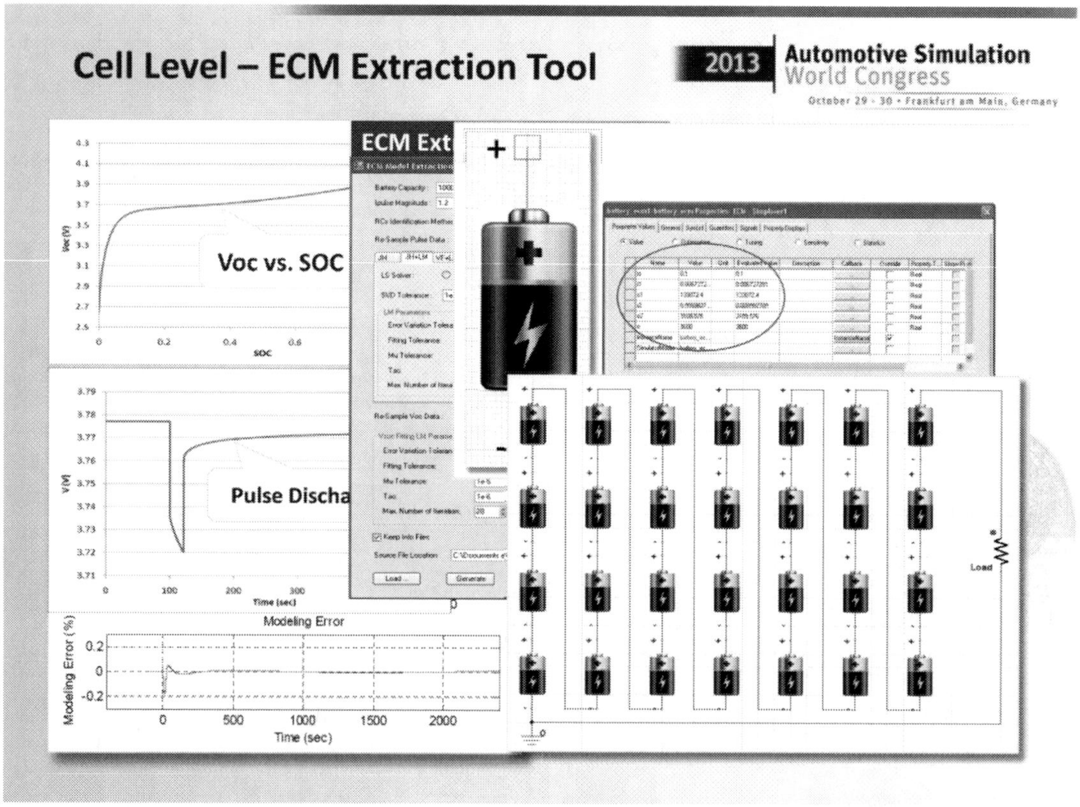
Cell Level – ECM Extraction Tool
2013
Automotive Simulation
World Congress
October 29 - 30 • Frankfurt am Main, Germany
Voc vs. SOC
Pulse Discha
ECM Ext
Modeling Error
Modeling Error (%)
Time (sec)
Load

Cell Level – ECM Extraction Tool

2013 Automotive Simulation World Congress
October 29 - 30 • Frankfurt am Main, Germany

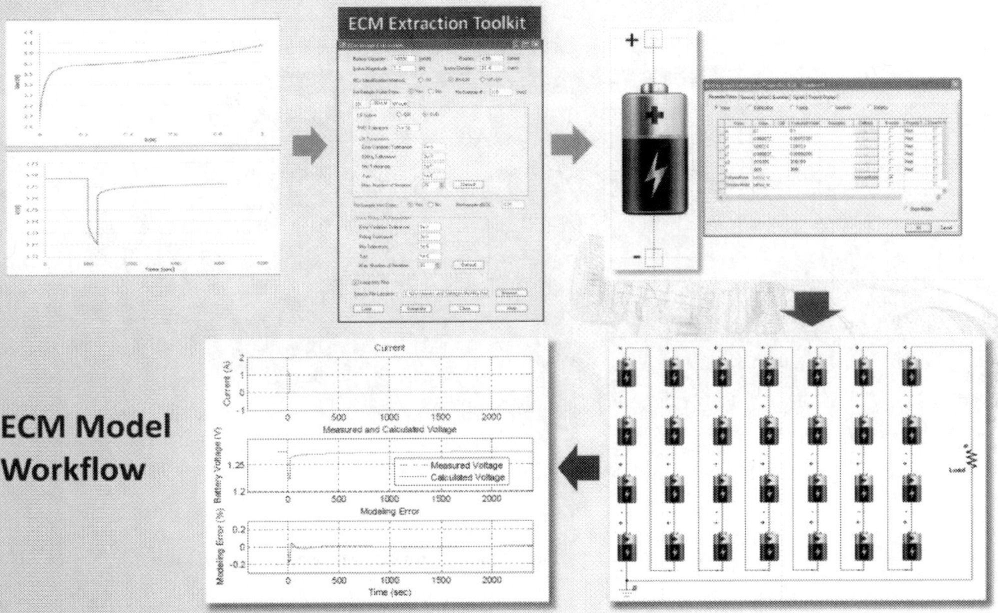

ECM Model Workflow

X. Hu, L. Collins, S. Stanton, S. Jiang, "A Model Parameter Identification Method for Battery Applications", SAE 2013-01-1529.

Complete IGBT Module

2013 Automotive Simulation World Congress
October 29 - 30 • Frankfurt am Main, Germany

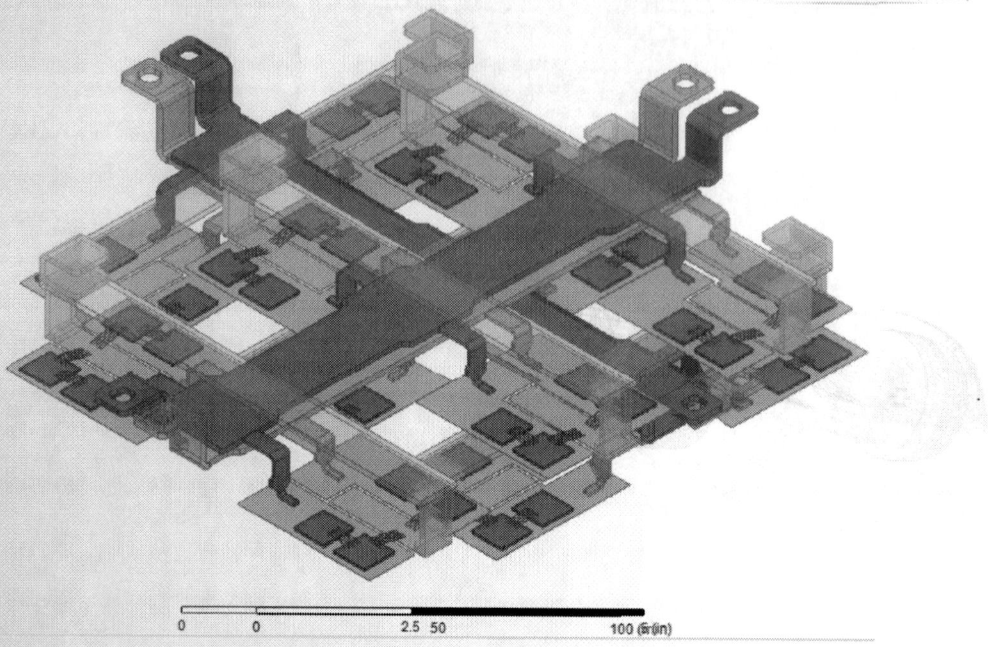

Complete IGBT Module

2013 Automotive Simulation World Congress
October 29 - 30 • Frankfurt am Main, Germany

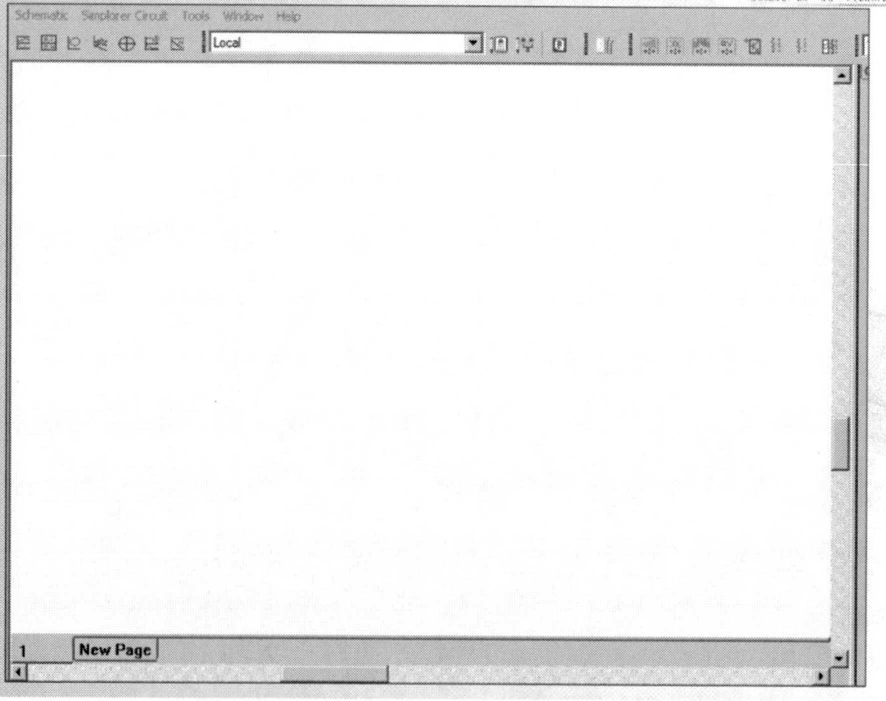

16

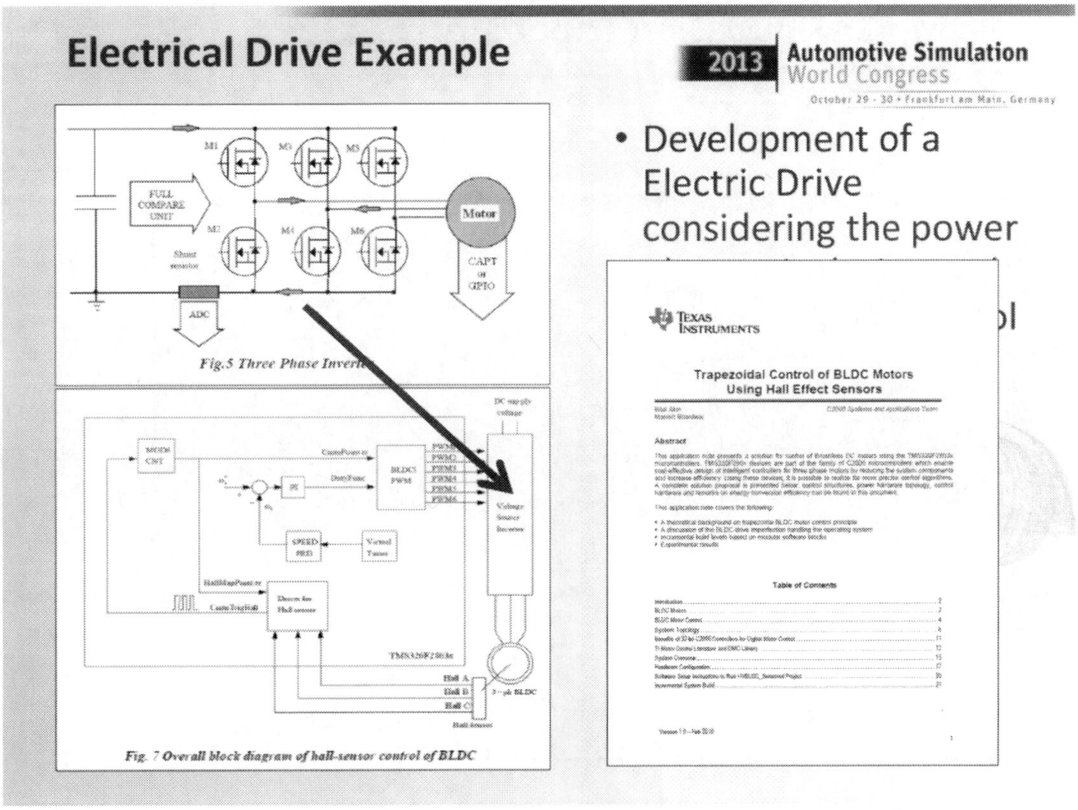

Electrical Drive Example
2013
Automotive Simulation
World Congress
October 29 - 30 • Frankfurt am Main, Germany
• Development of a
Electric Drive
considering the power
ol
Motor
CAPT
or
GPIO
ADC
Fig.5 Three Phase Inverter
Fig. 7 Overall block diagram of hall-sensor control of BLDC
Texas Instruments
Trapezoidal Control of BLDC Motors
Using Hall Effect Sensors
Abstract
Table of Contents

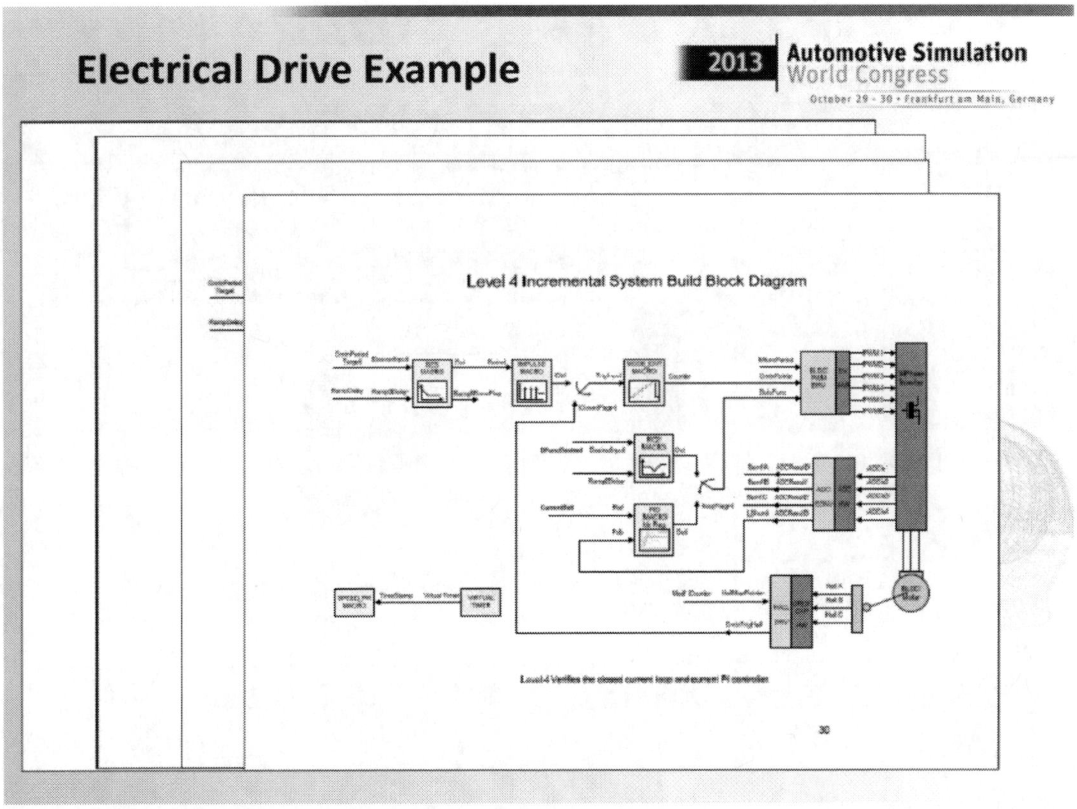

Electrical Drive Example
2013
Automotive Simulation
World Congress
October 29 - 30 • Frankfurt am Main, Germany
Level 4 Incremental System Build Block Diagram
30

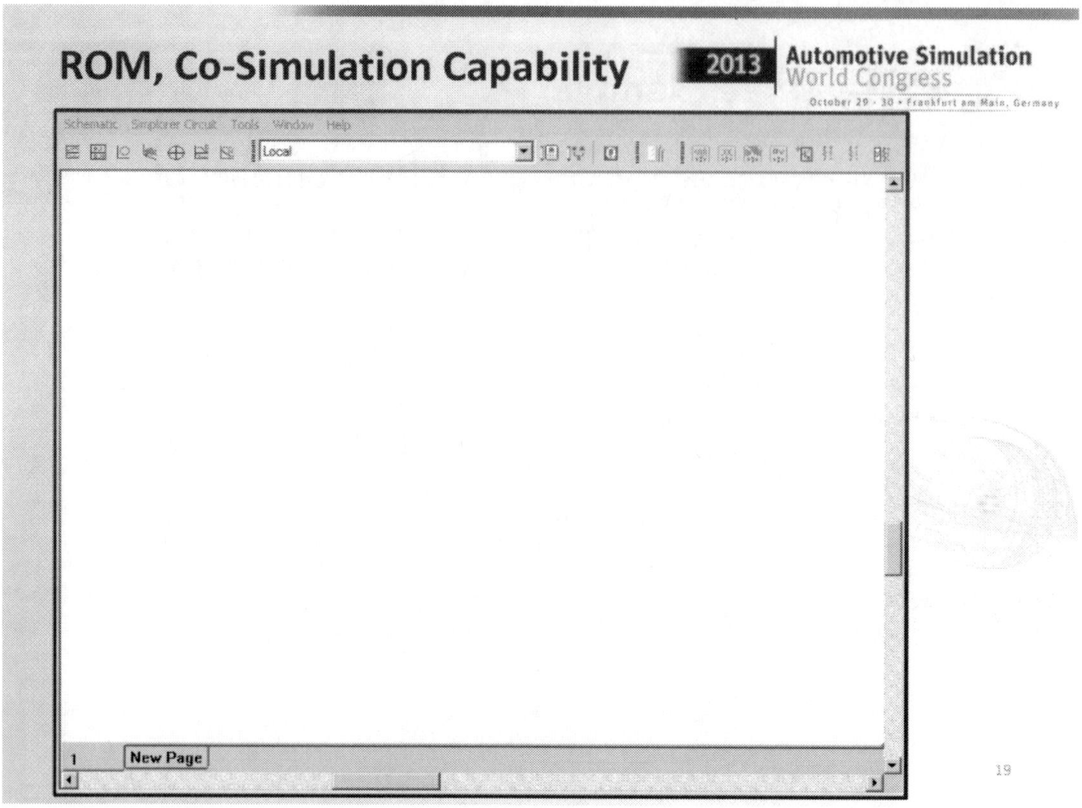
ROM, Co-Simulation Capability
2013
Automotive Simulation
World Congress
October 29 - 30 • Frankfurt am Main, Germany
Schematic Simplorer Circuit Tools Window Help
Local
1
New Page
19

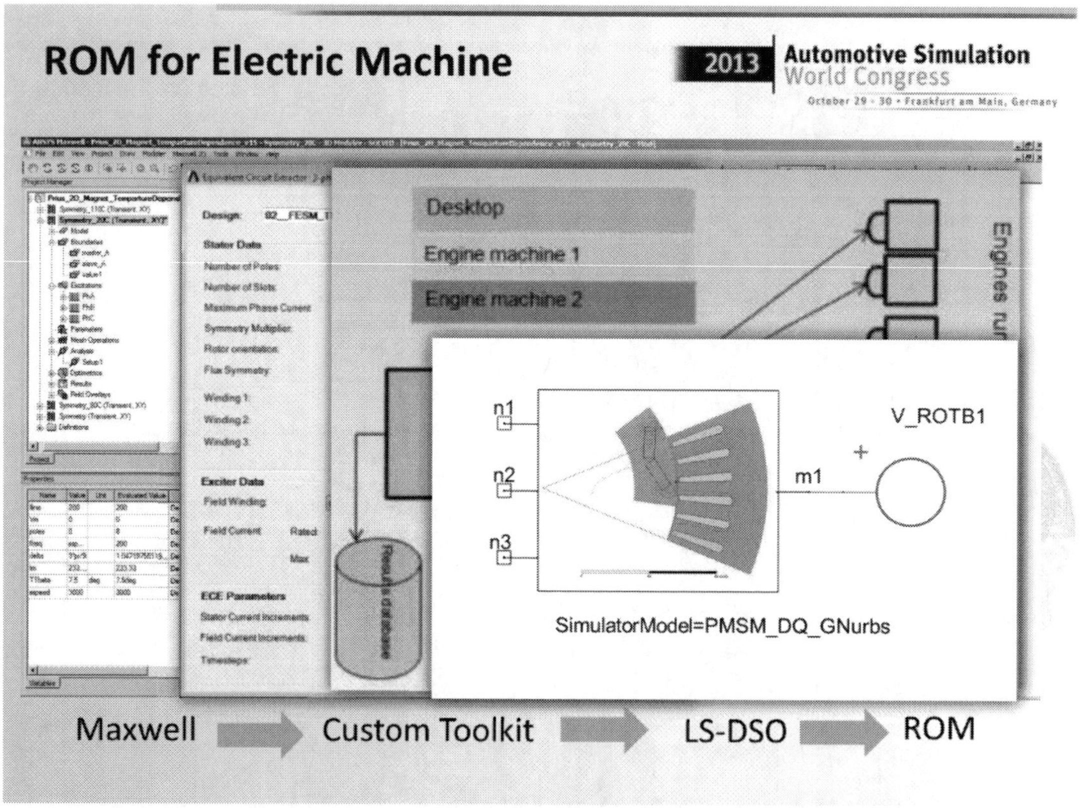
ROM for Electric Machine
2013
Automotive Simulation
World Congress
October 29 - 30 • Frankfurt am Main, Germany
Desktop
Engine machine 1
Engine machine 2
Stator Data
Exciter Data
ECE Parameters
n1
n2
n3
V_ROTB1
+
m1
SimulatorModel=PMSM_DQ_GNurbs
Maxwell
Custom Toolkit
LS-DSO
ROM

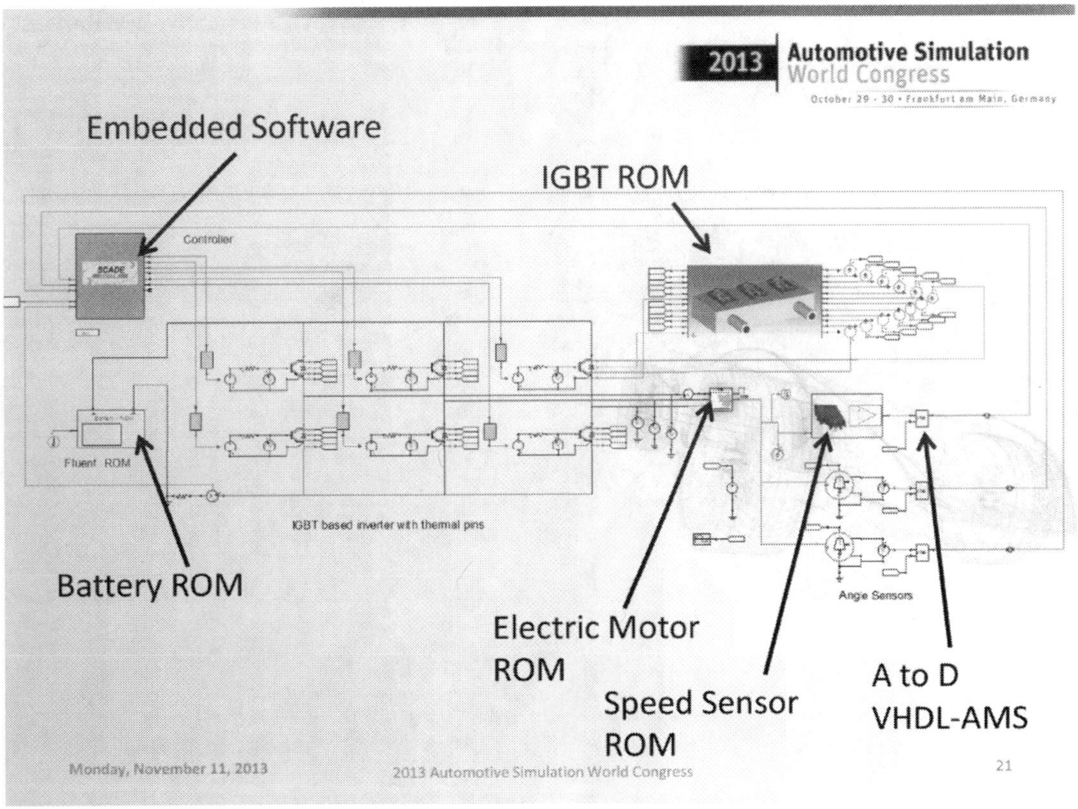
2013 Automotive Simulation World Congress
October 29 - 30 • Frankfurt am Main, Germany
Embedded Software
IGBT ROM
Controller
SCADE
Fluent ROM
IGBT based inverter with thermal pins
Battery ROM
Angle Sensors
Electric Motor ROM
Speed Sensor ROM
A to D VHDL-AMS
Monday, November 11, 2013
2013 Automotive Simulation World Congress
21

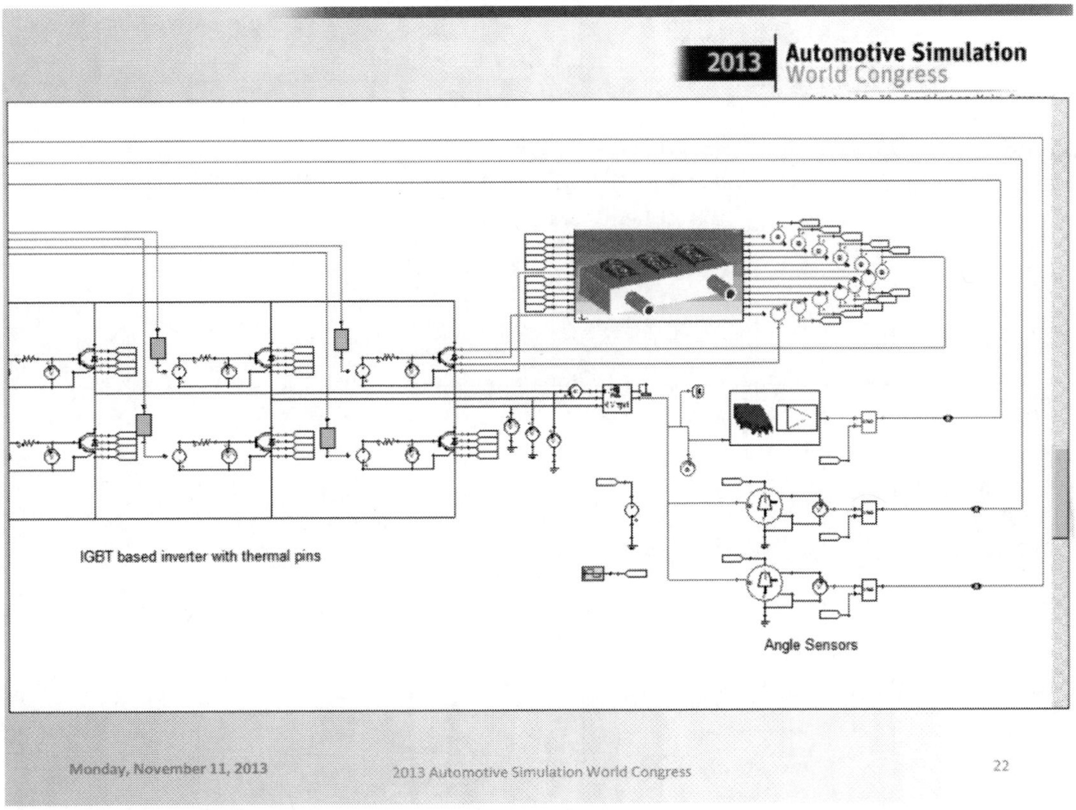
2013 Automotive Simulation World Congress
IGBT based inverter with thermal pins
Angle Sensors
Monday, November 11, 2013
2013 Automotive Simulation World Congress
22

Sensor Physics and Behavior via Reduced Order Model (ROM)

2013 Automotive Simulation World Congress
October 29 - 30 • Frankfurt am Main, Germany

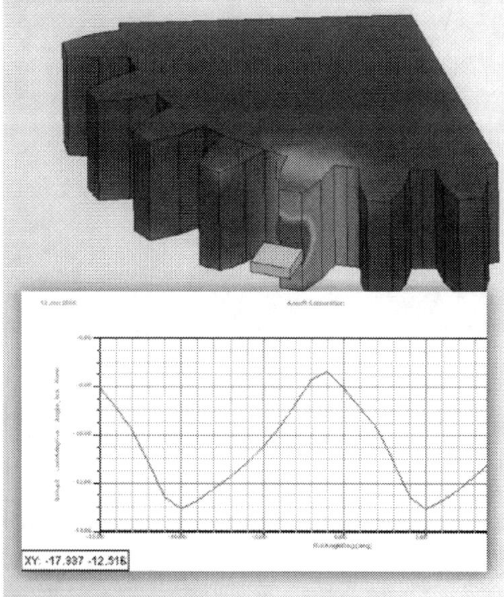

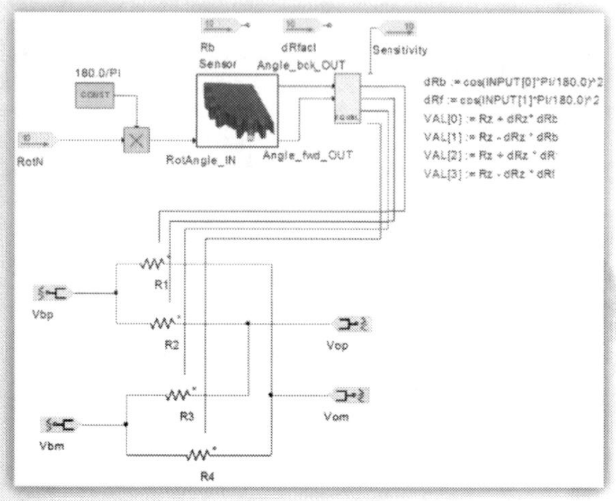

Simulation Results

2013 Automotive Simulation World Congress
October 29 - 30 • Frankfurt am Main, Germany

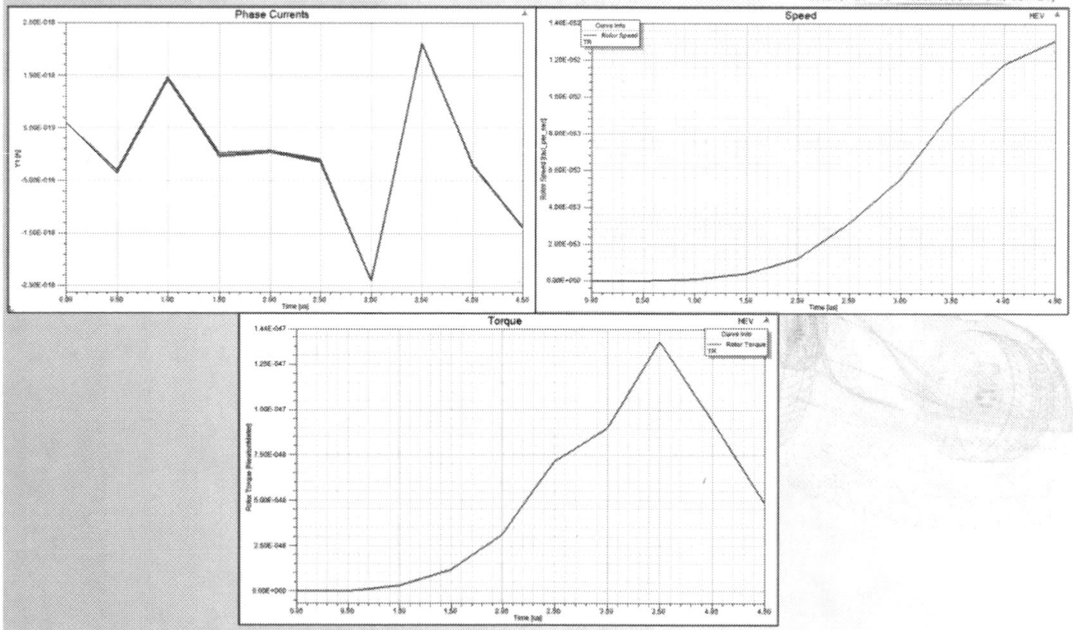

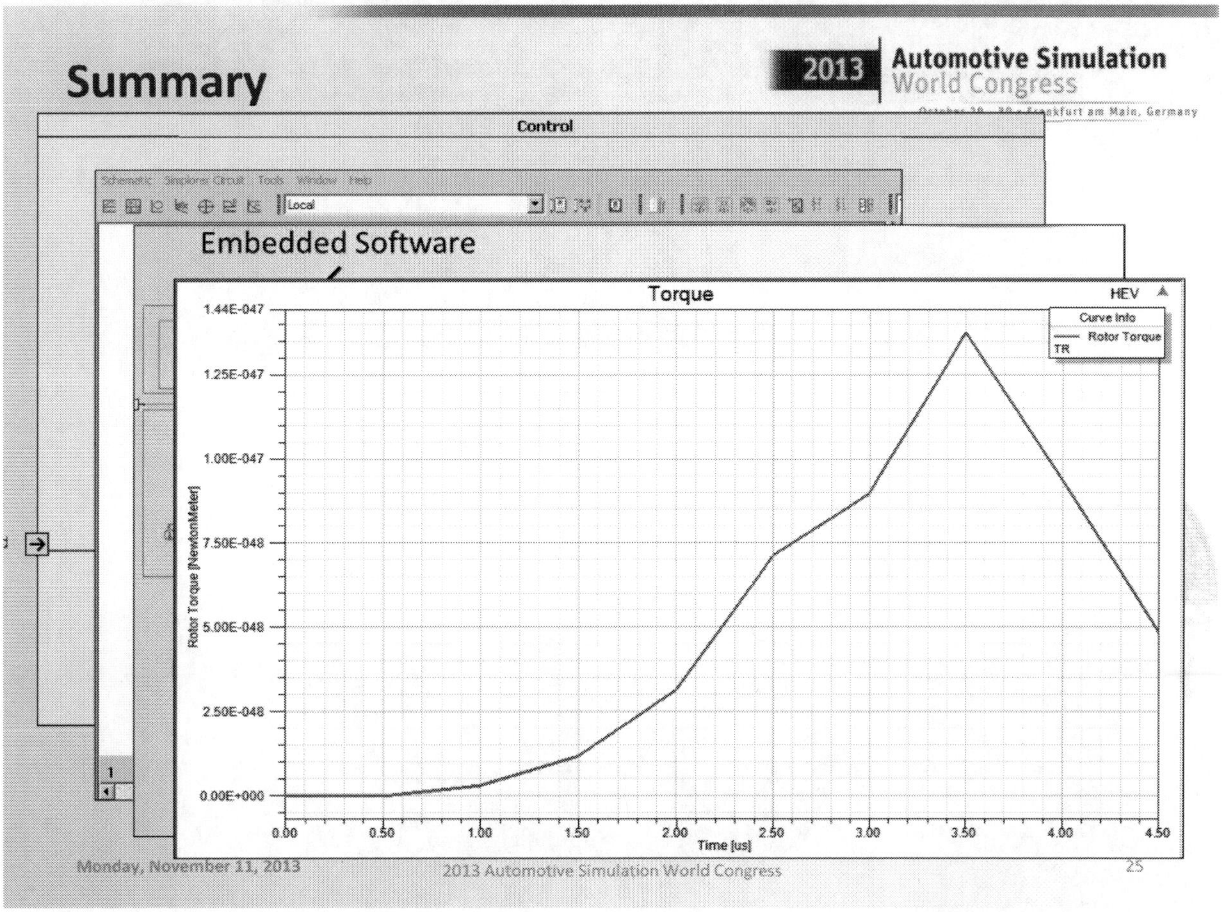

Summary
2013
Automotive Simulation
World Congress
Control
Embedded Software
Torque
HEV
Curve Info
Rotor Torque
TR
1.44E-047
1.25E-047
1.00E-047
7.50E-048
5.00E-048
2.50E-048
0.00E+000
Rotor Torque [NewtonMeter]
0.00
0.50
1.00
1.50
2.00
2.50
3.00
3.50
4.00
4.50
Time [us]
Monday, November 11, 2013
2013 Automotive Simulation World Congress
25

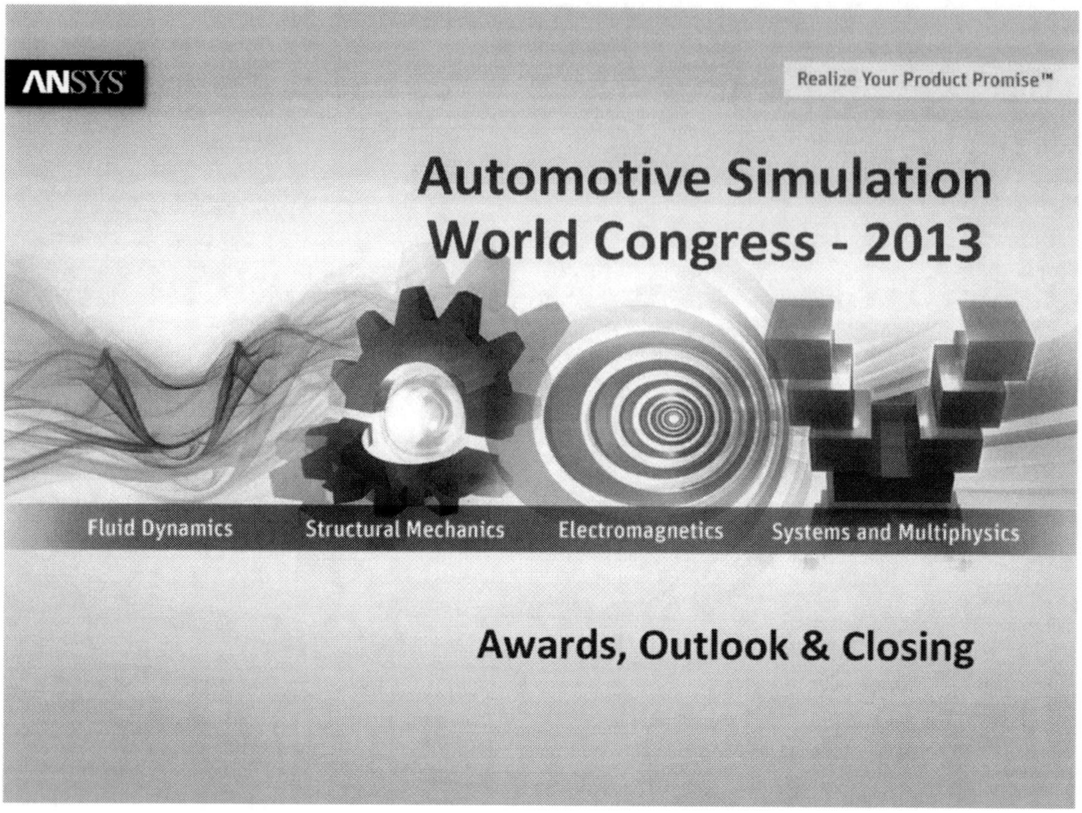
ANSYS
Realize Your Product Promise™
Automotive Simulation
World Congress - 2013
Fluid Dynamics
Structural Mechanics
Electromagnetics
Systems and Multiphysics
Awards, Outlook & Closing

ANSYS
Awards
2
© 2011 ANSYS, Inc.
November 11, 2013

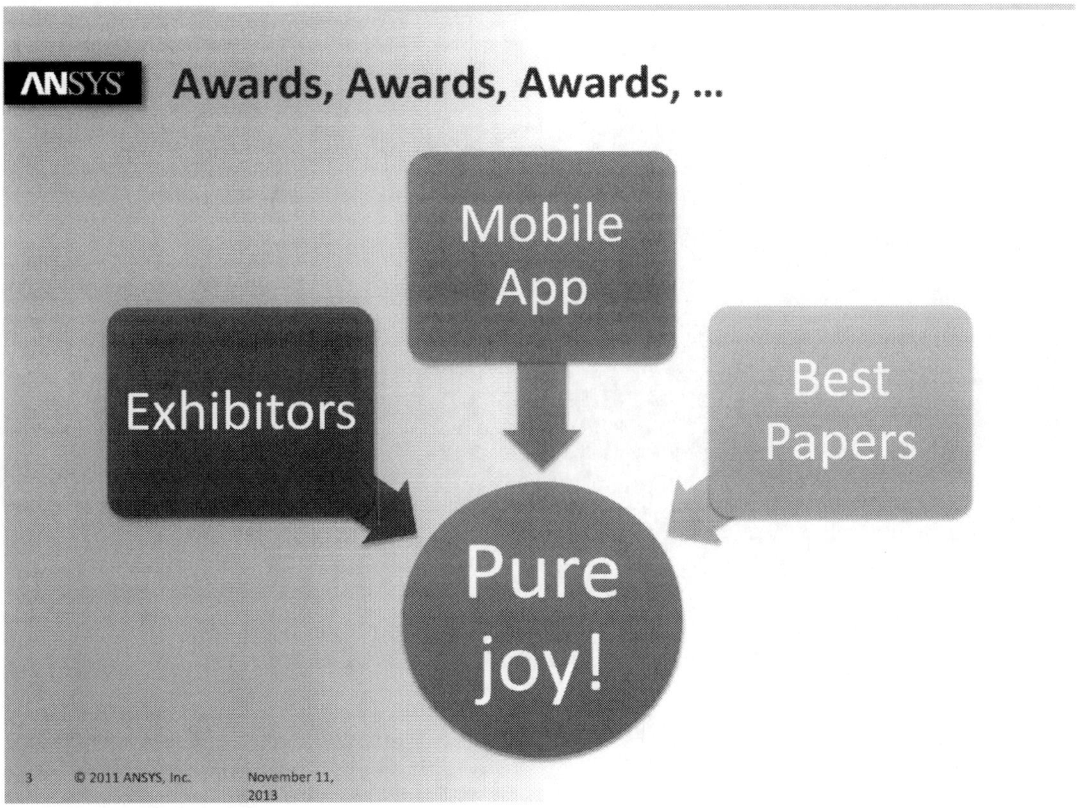
ANSYS
Awards, Awards, Awards, ...
Mobile App
Exhibitors
Best Papers
Pure joy!
3
© 2011 ANSYS, Inc.
November 11, 2013

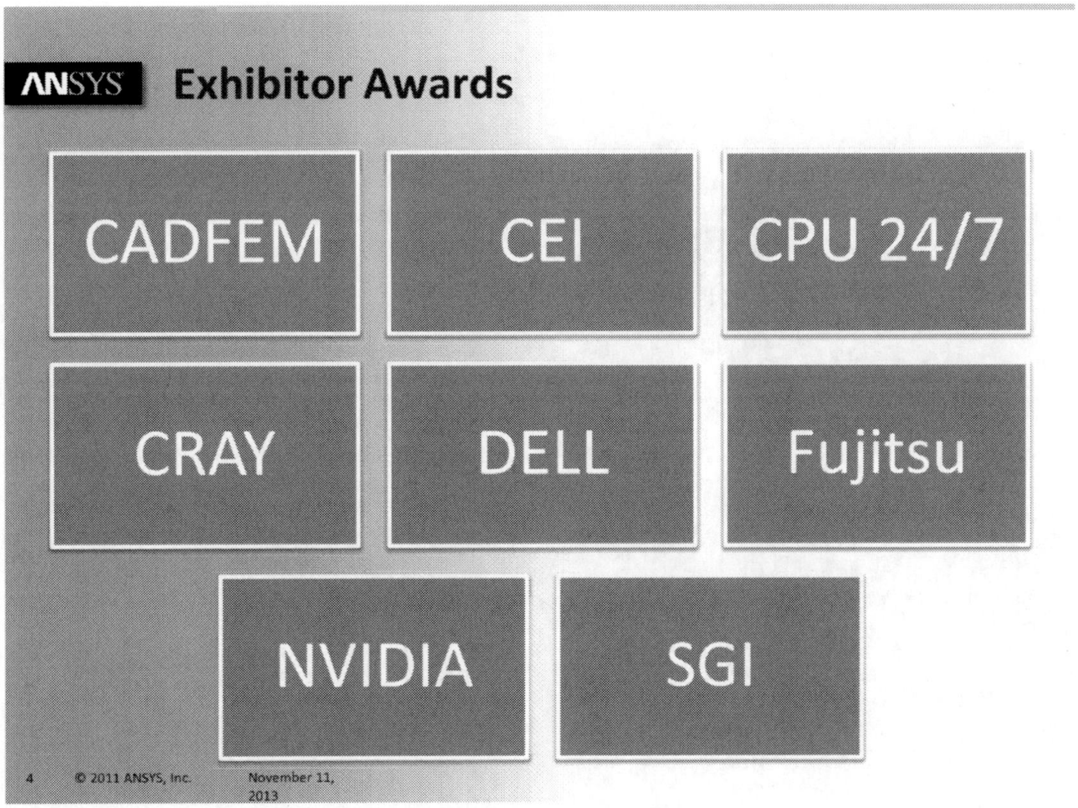
ANSYS
Exhibitor Awards
CADFEM
CEI
CPU 24/7
CRAY
DELL
Fujitsu
NVIDIA
SGI
4
© 2011 ANSYS, Inc.
November 11, 2013

ANSYS Mobile Apps Awards

- Badges
- Photographer
- Leader Board

ANSYS Mobile Apps Award

Mobile Apps Awards

Badges

- Rodrigo Coelho

Photographer

- Joel Sousa

Leader Board

- Bernd Thomas

Best Paper Awards

Best Paper

Best Application of Multi-Disciplinary or Multi-Scale Simulation

Most Innovative Use of Simulation

Best Use of HPC

ANSYS Award Criteria

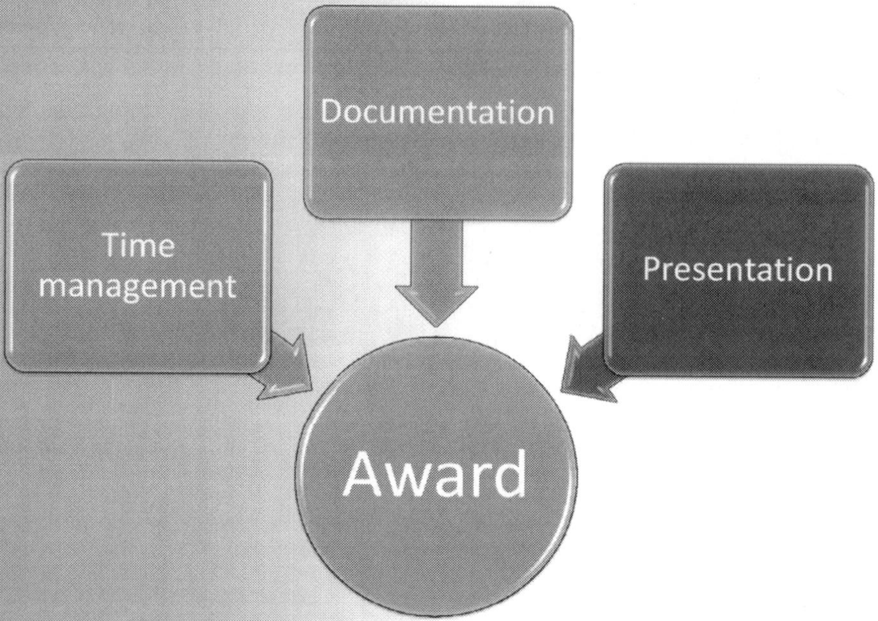

9 © 2011 ANSYS, Inc. November 11, 2013

ANSYS Best Paper Awards

10 © 2011 ANSYS, Inc. November 11, 2013

ANSYS Best Paper Awards

Best Paper
- Oleksiy Kurenkov, Continental Automotive, Regensburg

Best Application of Multi-Disciplinary or Multi-Scale Simulation
- Matteo Pelosi, Öhlins Racing, Väsby

Most Innovative Use of Simulation
- Simona Invernizzi, Dallara Engineering, Parma

Best Use of HPC
- Marco Biancolini, University of Rome, Rome

ANSYS Proceedings → EKM Download

ANSYS

Thanks

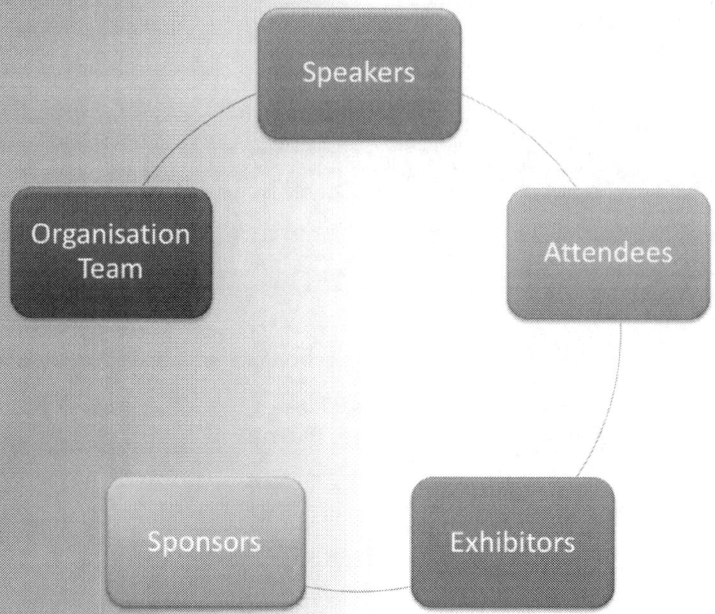

ANSYS

ASWC 2014 - Tokyo

Model Order Reduction for Electronics Cooling – Case Study of Lear Power Inverter model

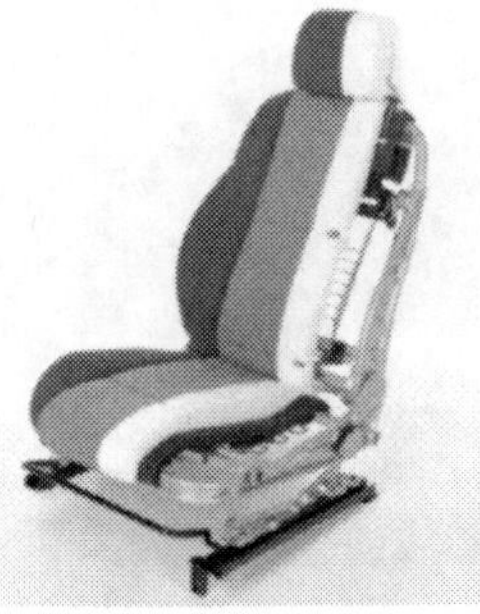

Rutunj Rai : Lear Corporation
Amit Shah and Xiao Hu : ANSYS Inc.

Lear Case Study Traction Inverter

- Inverter converts high-voltage DC power into multi-phase AC power required to drive three-phase induction or permanent magnet electric machines used in hybrid and electric vehicle propulsion.
- Depending on driving condition, power level in inverter changes with respect to time.
- Steady state analysis is not enough.
- Thermal design needs have to be verified verify in time domain.

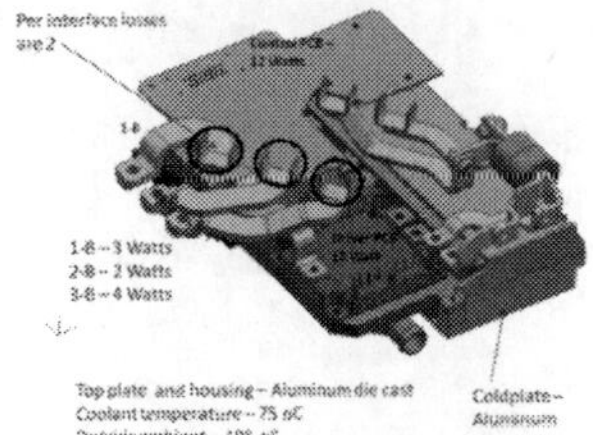

 3

Why we need MOR

- CFD as a general thermal analysis tool is accurate.
- Can be computationally expensive for large cases running transient CFD analysis
- MOR can significantly reduce the model size and simulation time.

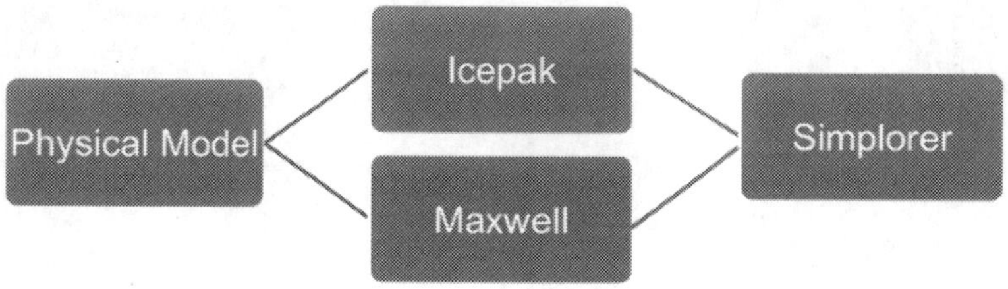

How ANSYS MOR Technology Works

- Model order reduction (MOR) for linear problems
  - **LTI based (aka transfer function based)**
  - System matrix based

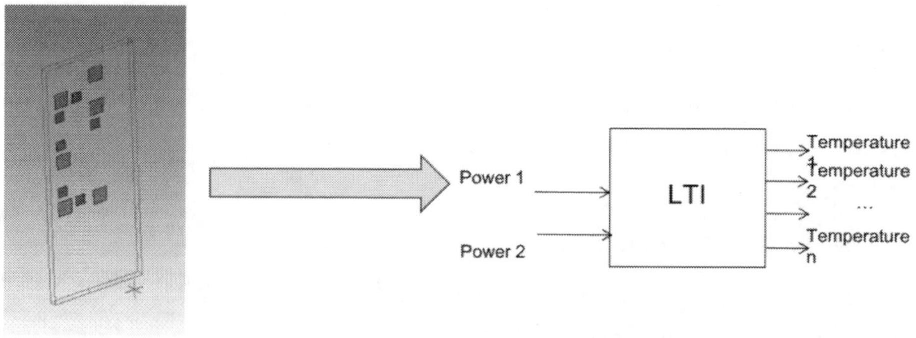

Outline

- Traction Inverter
- MOR
  - Need for Model order reduction
- Lear Case Study
  - Icepak model – Steady State, Transient
  - Icepak - Simplorer connection – LTI model extraction
  - Simplorer model
- Results
  - Simulation 1
  - Simulation 2
- Conclusion

MOR Using Icepak - Simplorer Coupling

1. Create the Icepak model
2. Generate step responses
   - Use Icepak *Parametrics* tool
3. Extract LTI model
   - Use Simplorer
4. Simulate inside Simplorer

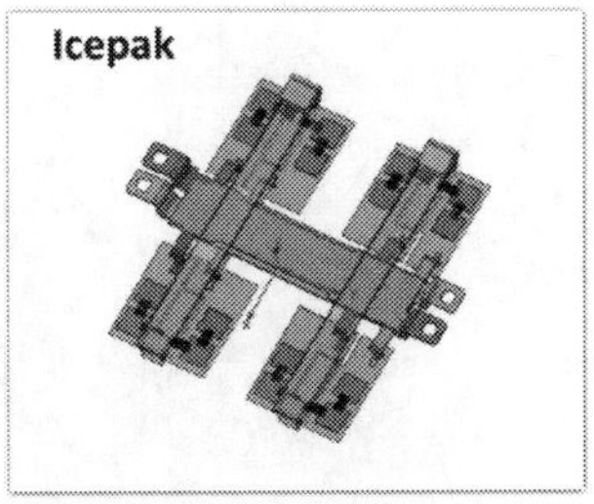

1

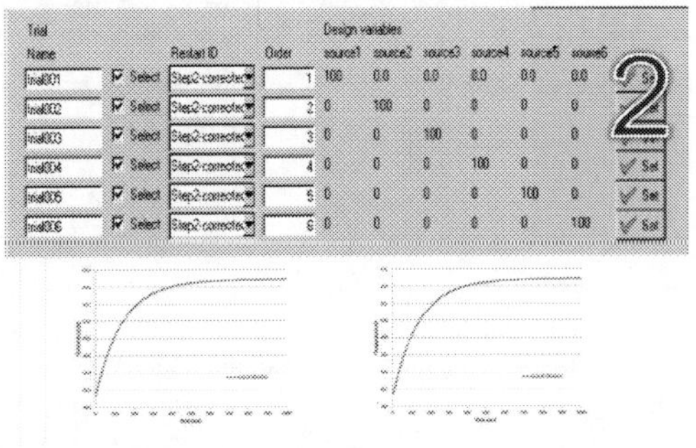

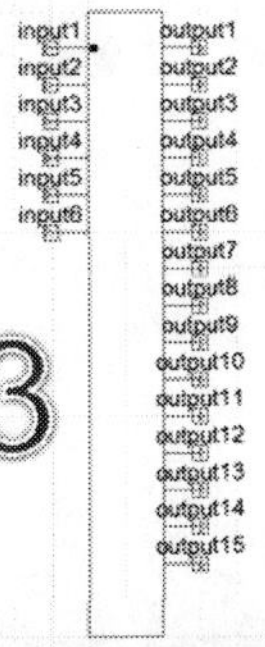

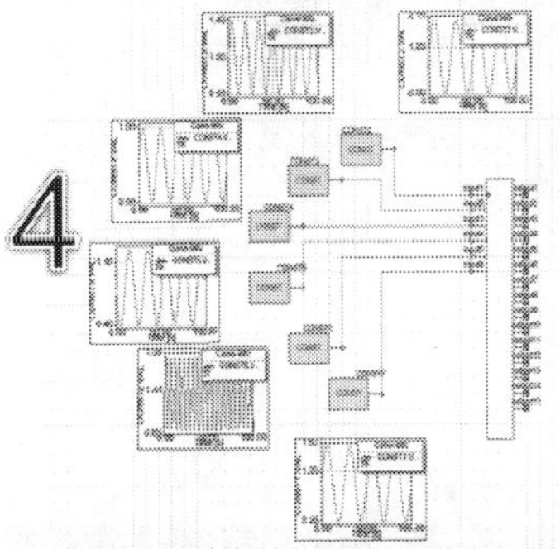

Icepak Model Details

- Forced convection cold plate model.
- Coolant flows through cold plate channel.
- IGBTs mounted on cold plate.
- Six transient heat sources are applied on the inverter model.
- Fifteen monitor points are set up throughout the model.

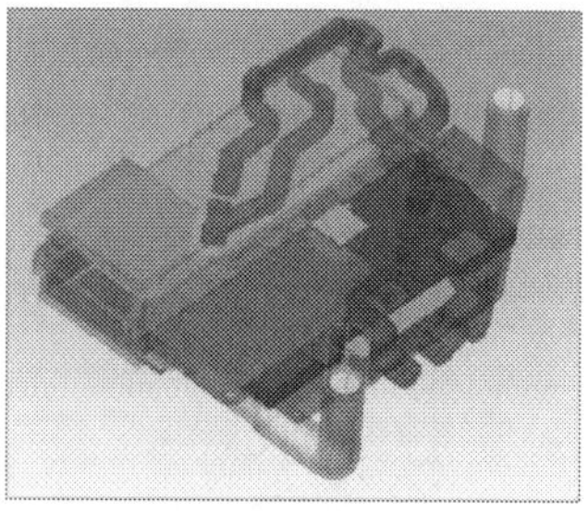

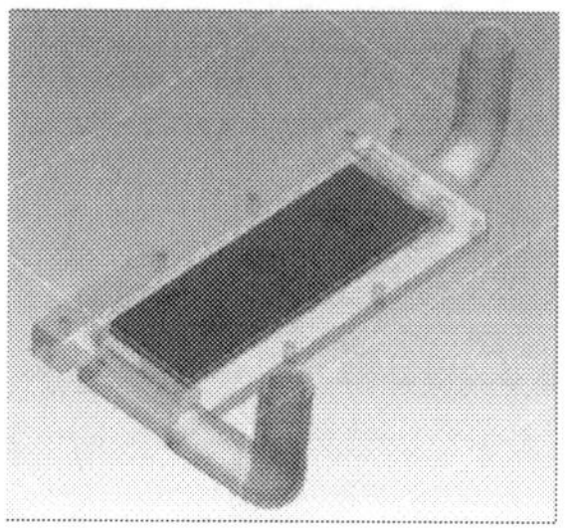

Step 1: Model Definition

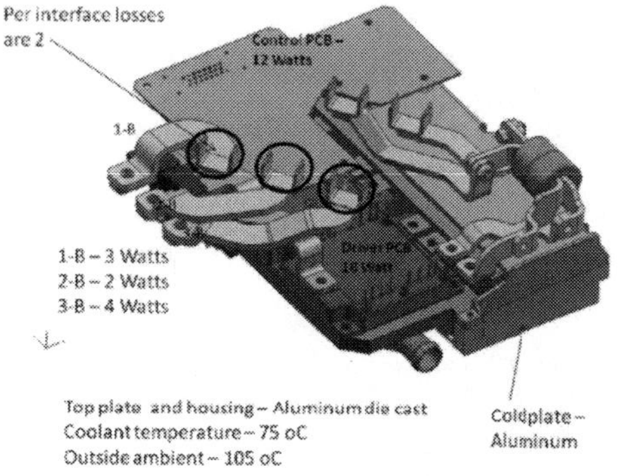

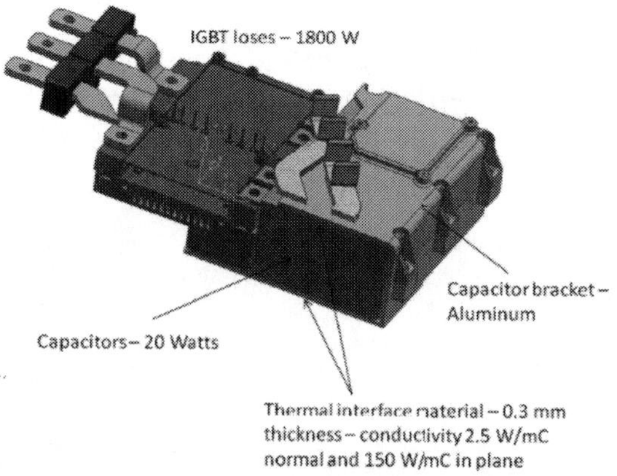

Step 1: Bus Bar (2B) Loss Calculation

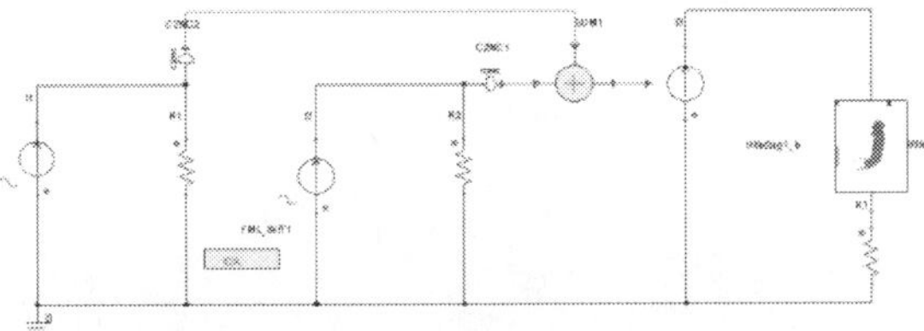

Simplorer Model

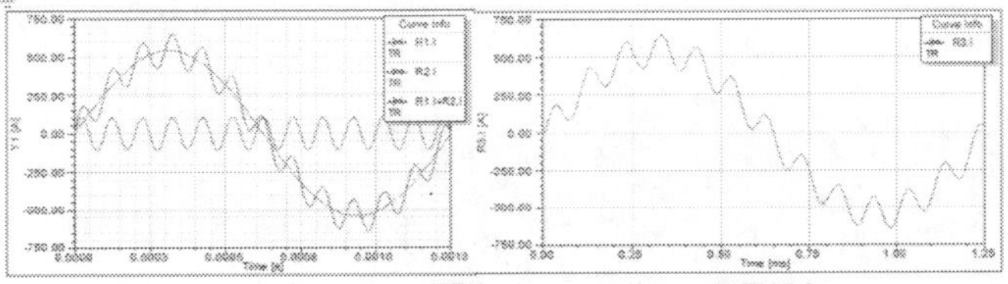

Input current waveform

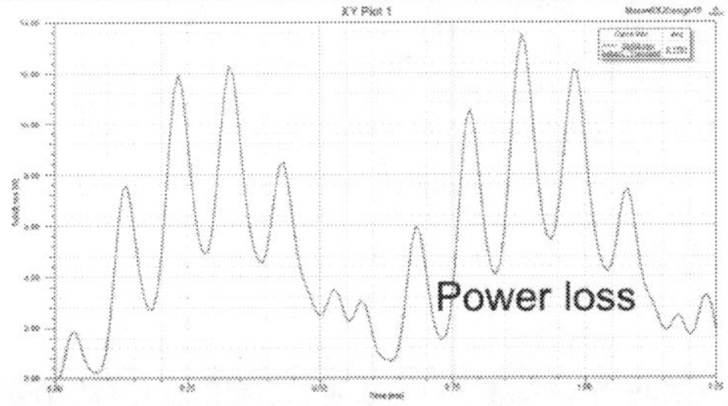

Area under curve - 5.1 Watts

Step 1: Icepak Mesh Model

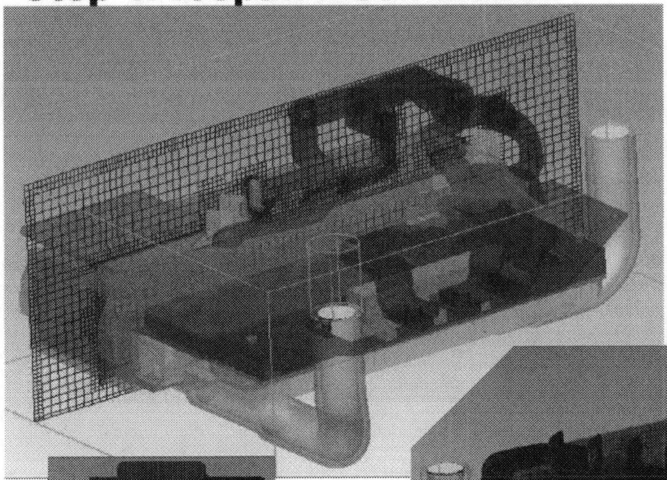

- Mesher-HD
- Mesh Size
  - 2.86 M
- Multi level Mesh
  - Levels: 3
- Non-Conformal Mesh

Step 1: Steady State Temperature Distribution

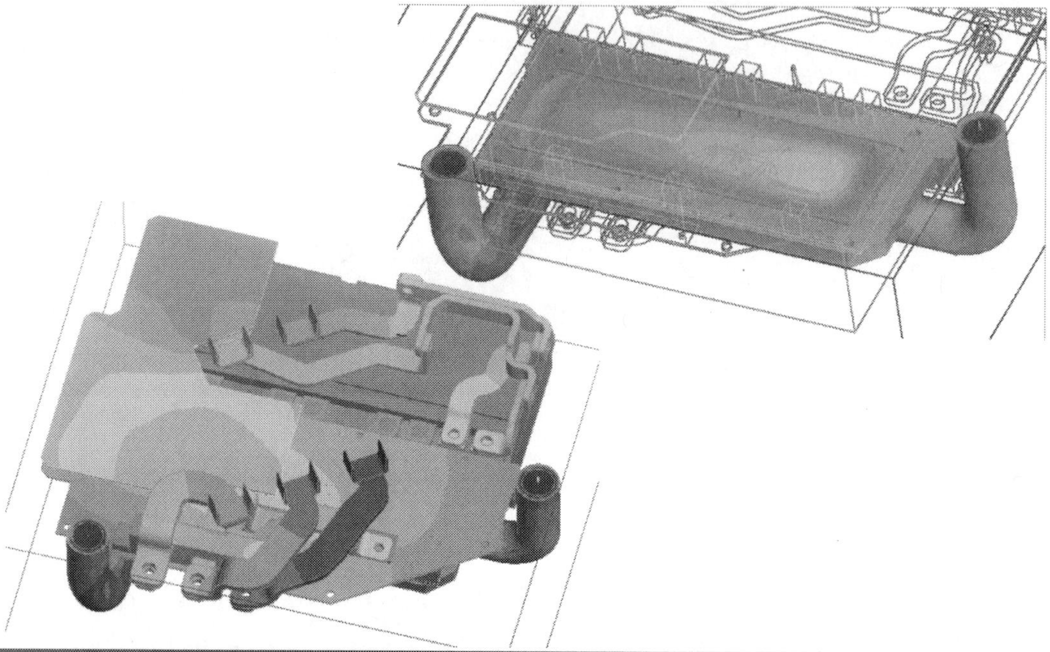

Step 2: Transient Icepak runs

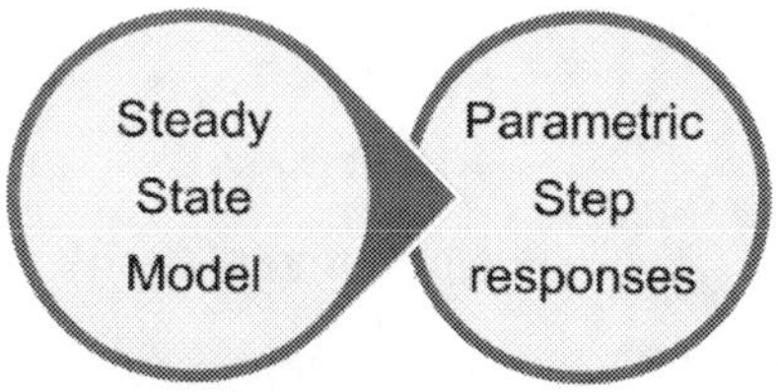

Trial Name		Restart ID	Order	Design variables source1	source2	source3	source4	source5	source6	
trial001	Select	Step2-correcte	1	100	0.0	0.0	0.0	0.0	0.0	Set
trial002	Select	Step2-correcte	2	0	100	0	0	0	0	Set
trial003	Select	Step2-correcte	3	0	0	100	0	0	0	Set
trial004	Select	Step2-correcte	4	0	0	0	100	0	0	Set
trial005	Select	Step2-correcte	5	0	0	0	0	100	0	Set
trial006	Select	Step2-correcte	6	0	0	0	0	0	100	Set

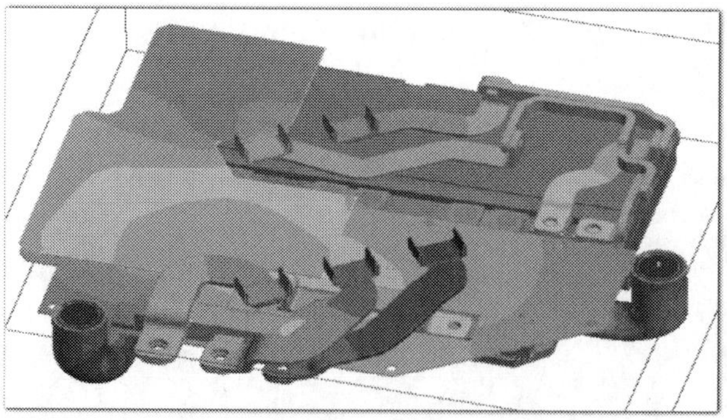

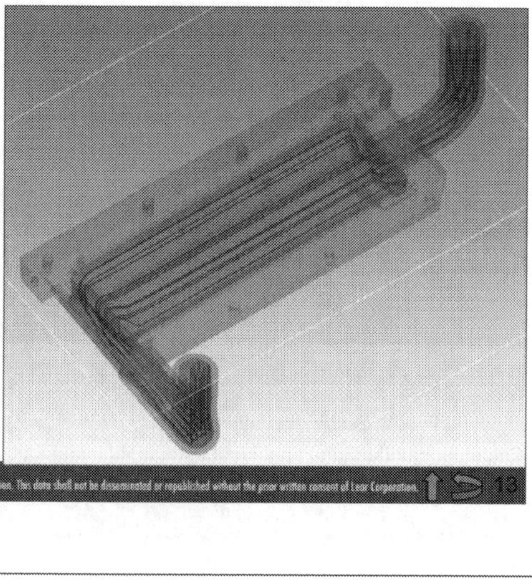

 13

Icepak - Simplorer connection

- Simply point Simplorer to locate the files from the Icepak project folder.

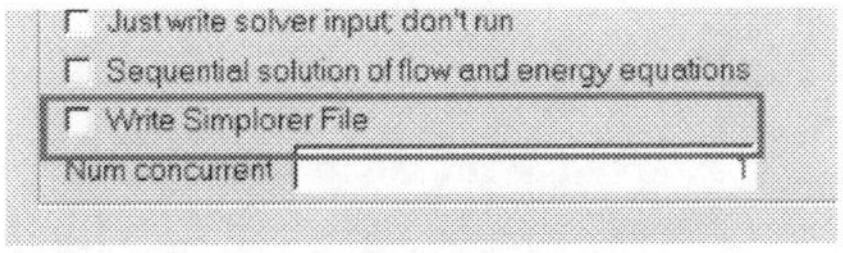

Icepak

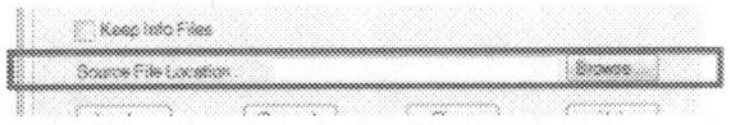

Simplorer

Step 3: MIMO - Simplorer Model

LTI Model generated from Icepak step responses

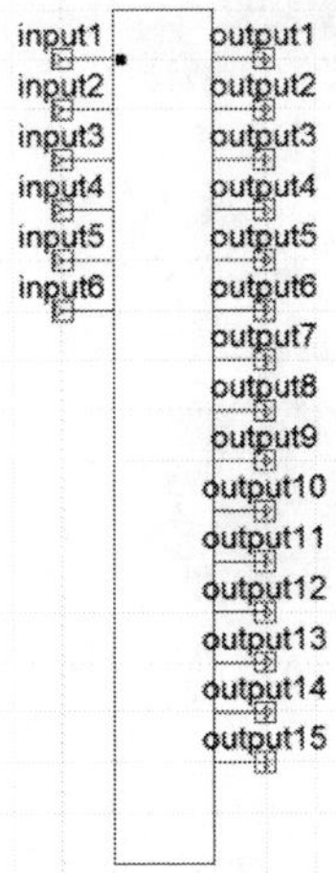

Model validation in Simplorer

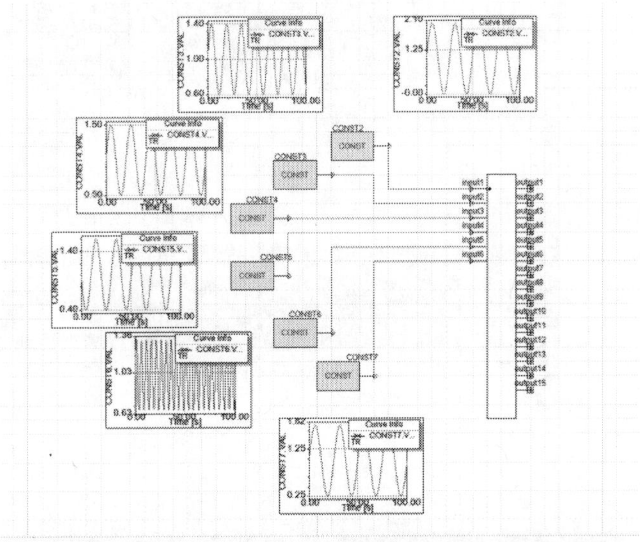

Results

Verification

Icepak

Transient verification simulation in Icepak for variable heat sources.

Simplorer

LTI model simulation with variable inputs to the model. Use the same transient heat sources as in Icepak.

Transient Heat Sources and monitor point locations

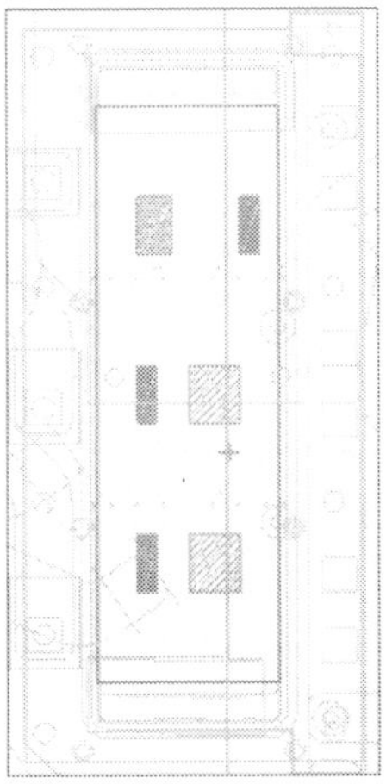

Heat Source Locations

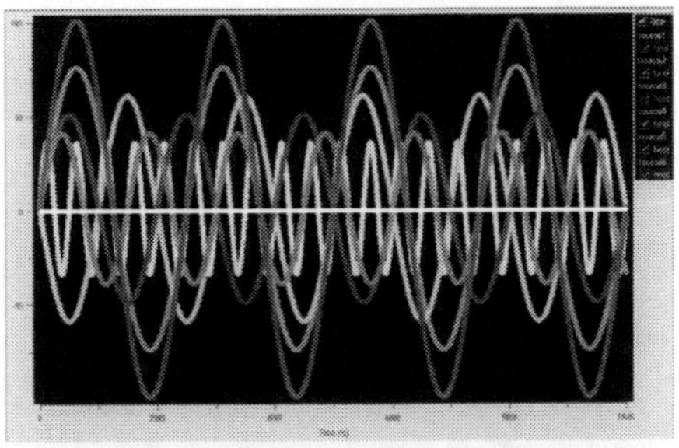

Transient Powers

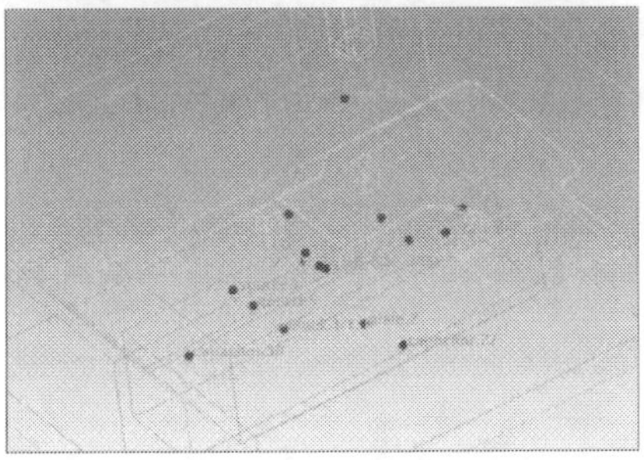

Monitor Point locations

 18

Results: Simulation 1 long range ~5000s

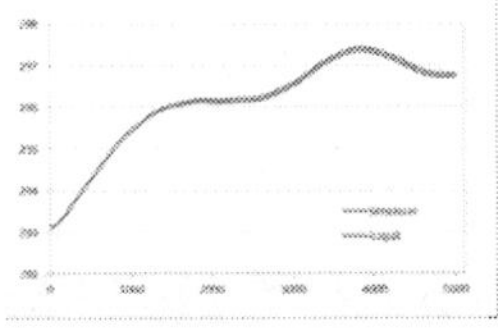

Monitor 1

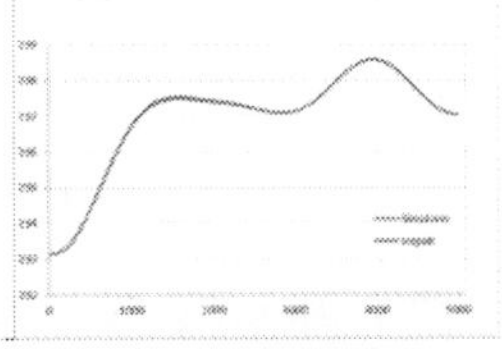

Monitor 2

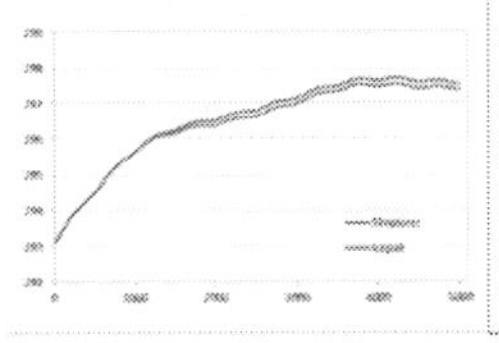

Monitor 3

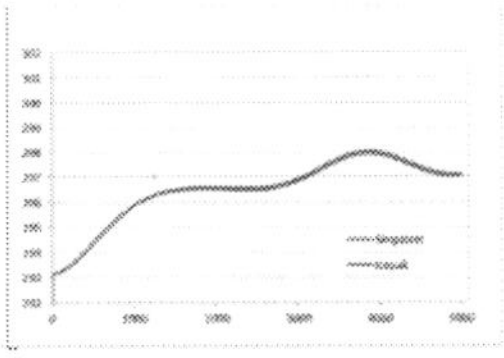

Monitor 4

Results: Simulation 2 Short range ~100s

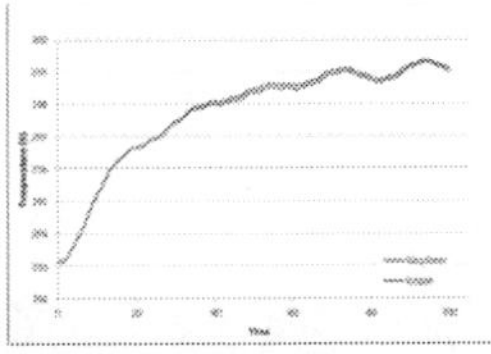

Monitor 1

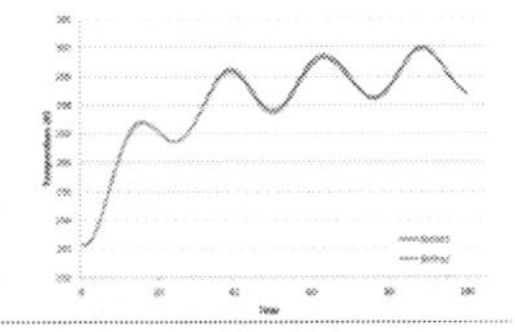

Monitor 2

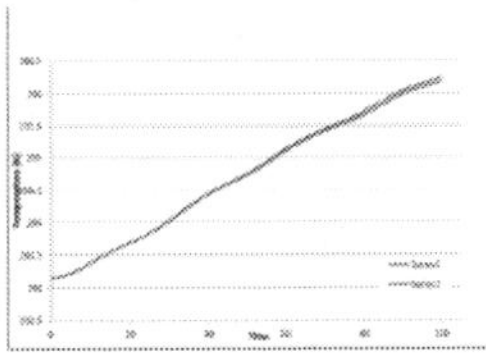

Monitor 3

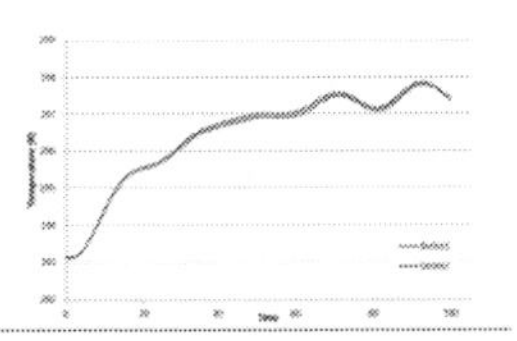

Monitor 4

 19

Conclusion

- LTI based MOR has shown to significantly reduce transient simulation time while maintaining great accuracy.

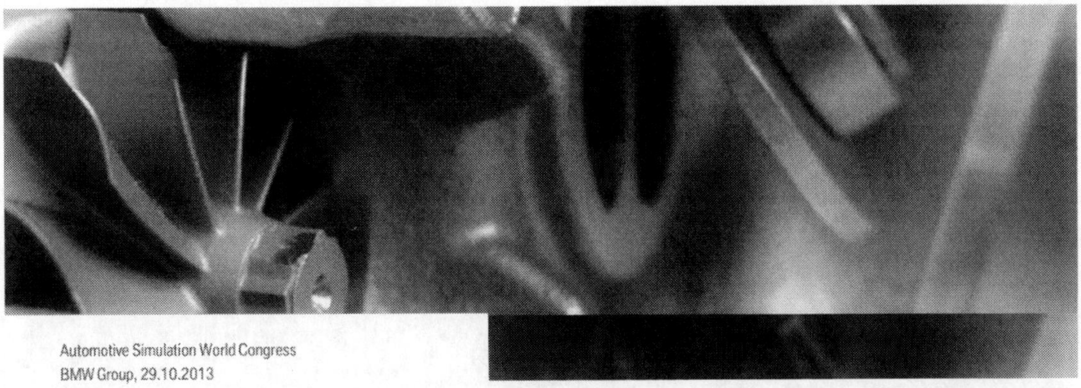

Automotive Simulation World Congress
BMW Group, 29.10.2013

CHALLENGES IN THE SIMULATION OF FLUID-STRUCTURE INTERACTIONS IN TURBOCHARGERS.

NICOLAS LACHENMAIER, THOMAS SAILER, DR. OLIVER HAUSNER, PROF. DR. MARC AVILA (UNIVERSITY OF ERLANGEN-NUREMBERG)

BMW
GROUP

FLUID-STRUCTURE INTERACTIONS. MOTIVATION.

– Compressor forces more air into the combustion chamber

FLUID-STRUCTURE INTERACTIONS. MOTIVATION.

- Compressor forces more air into the combustion chamber
- Compressor driven by the turbine which itself is driven by exhaust gases

FLUID-STRUCTURE INTERACTIONS. MOTIVATION.

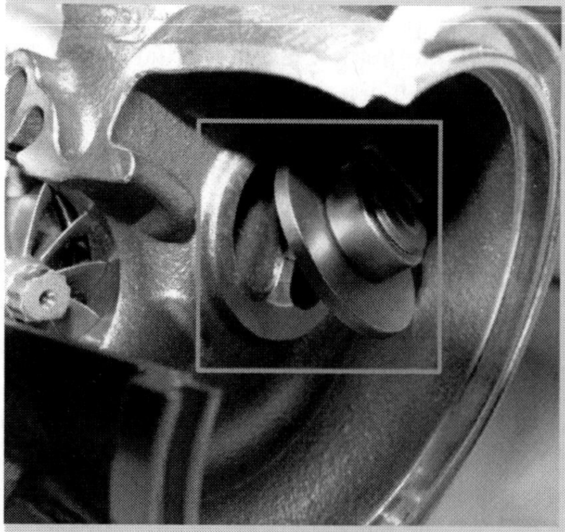

- Compressor forces more air into the combustion chamber
- Compressor driven by the turbine which itself is driven by exhaust gases
- Control unit for power of compressor: bypass called wastegate

FLUID-STRUCTURE INTERACTIONS. MOTIVATION.

- Compressor forces more air into the combustion chamber
- Compressor driven by the turbine which itself is driven by exhaust gases
- Control unit for power of compressor: bypass called wastegate
- High dynamic and thermal stresses lead to wear and rattling noise

FLUID-STRUCTURE INTERACTIONS. SIMULATION.

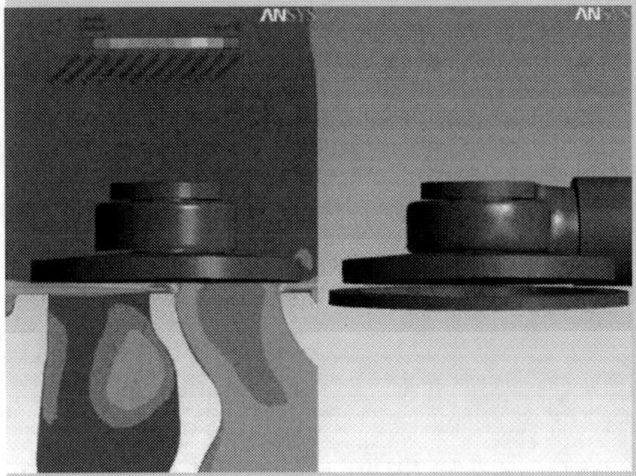

- Detailed unterstanding of the behaviour of the wastegate is crucial for optimisation
- Upcoming questions:
  - How good can we model the wastegate movement in the simulations?
  - Is the impact of the wastegate movement on the fluid stream negligible?

FLUID-STRUCTURE INTERACTIONS. THE ALGORITHM.

STRUCTURE

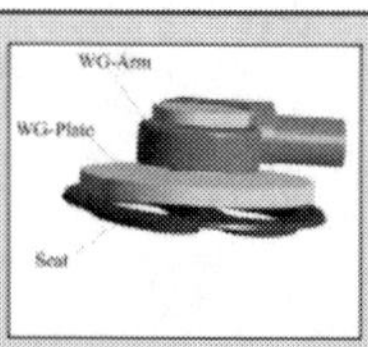

FLUID-STRUCTURE INTERACTIONS. THE ALGORITHM.

STRUCTURE

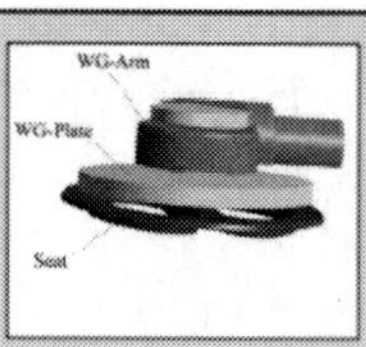

FLUID

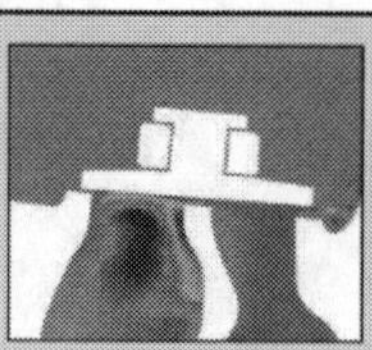

FLUID-STRUCTURE INTERACTIONS. THE ALGORITHM.

STRUCTURE

fluid-structure interactions (FSI)
in order to model
the coupling of these fields

FLUID

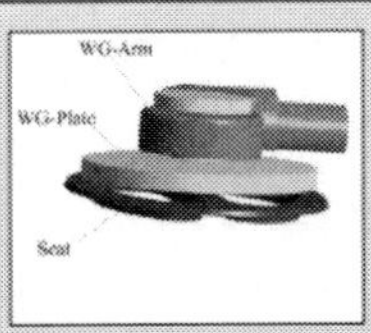

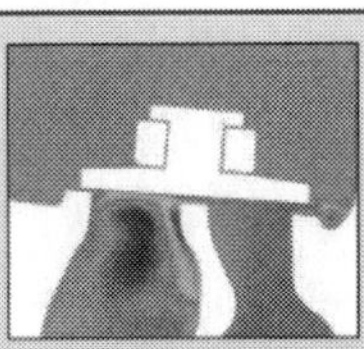

FLUID-STRUCTURE INTERACTIONS. THE ALGORITHM.

STRUCTURE
ANSYS Mechanical:

FEM-based solver

m inner iterations

FLUID
ANSYS CFX:

FVM-based solver

n inner iterations

Displacement diffusion equation for mesh movements

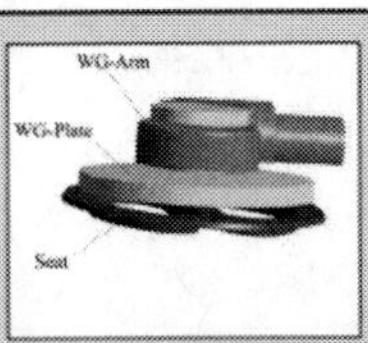

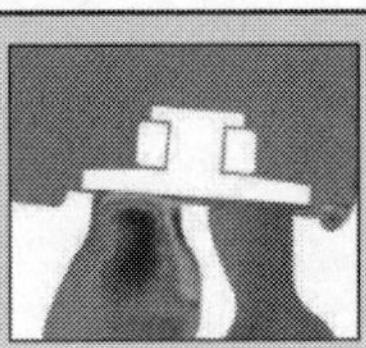

FLUID-STRUCTURE INTERACTIONS.
THE ALGORITHM.

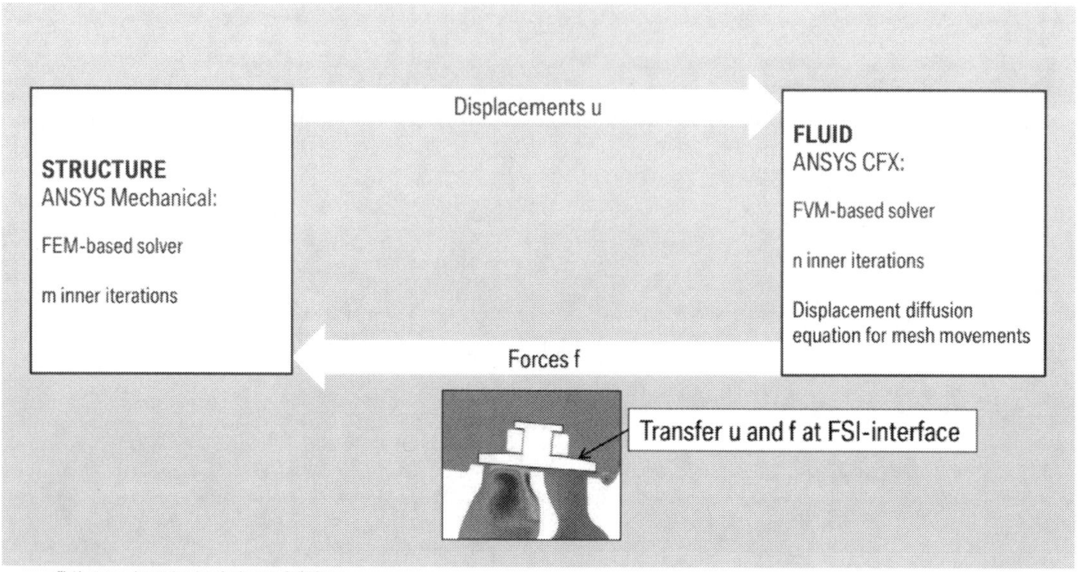

FLUID-STRUCTURE INTERACTIONS.
THE ALGORITHM.

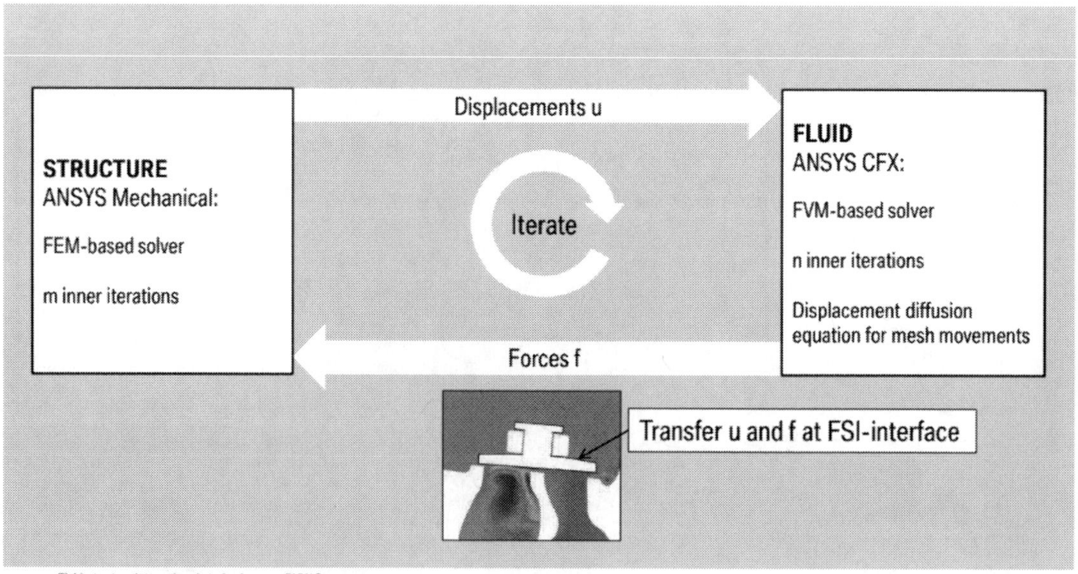

FLUID-STRUCTURE INTERACTIONS. THE ALGORITHM.

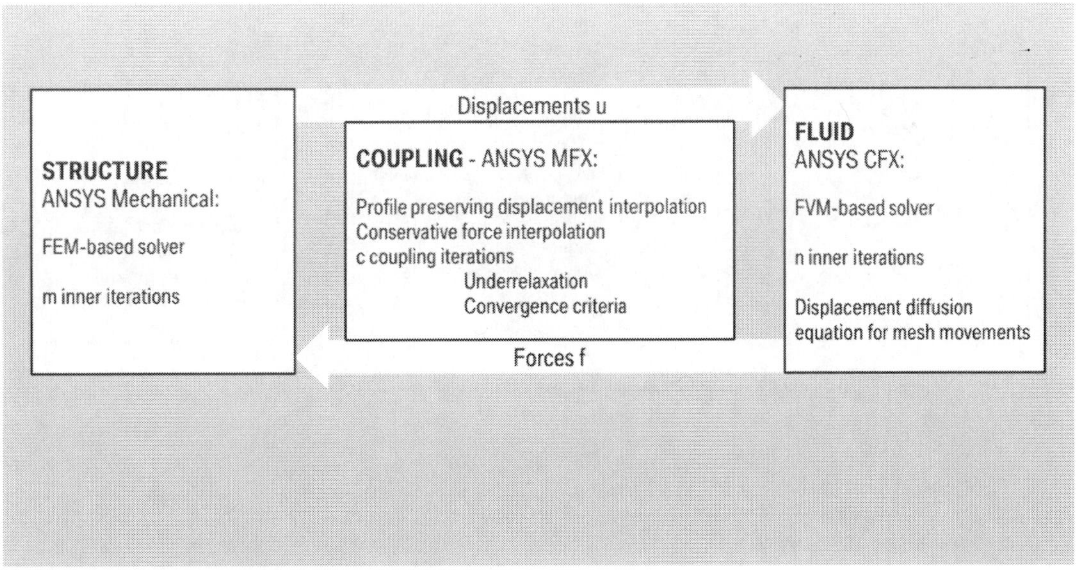

Fluid-structure interactions in turbochargers, BMW Group, 29.10.2013

Seite 13

FLUID-STRUCTURE INTERACTIONS. THE ALGORITHM.

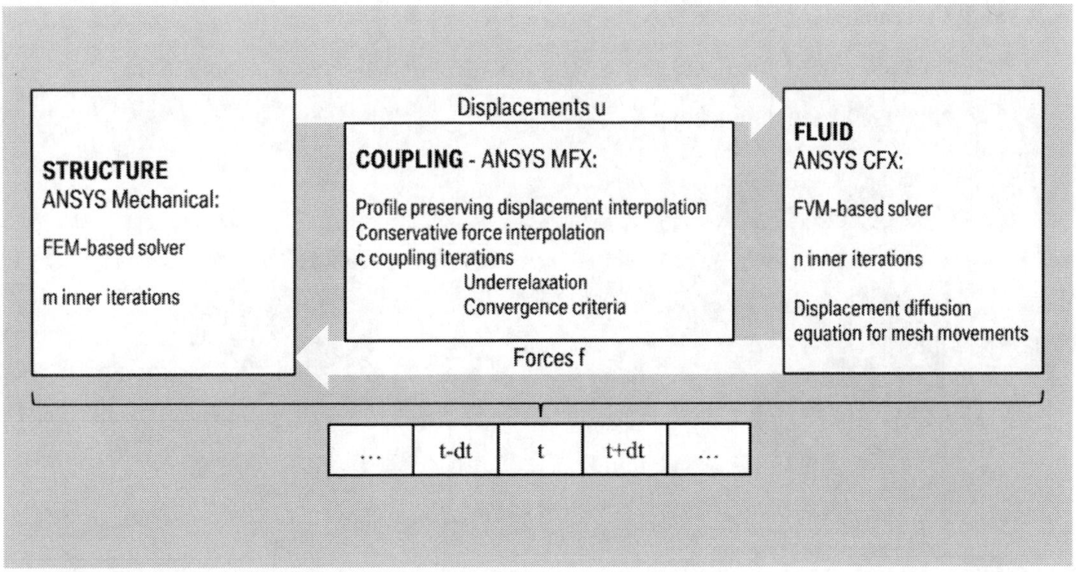

Fluid-structure interactions in turbochargers, BMW Group, 29.10.2013

Seite 14

FLUID-STRUCTURE INTERACTIONS. THE CHALLENGES.

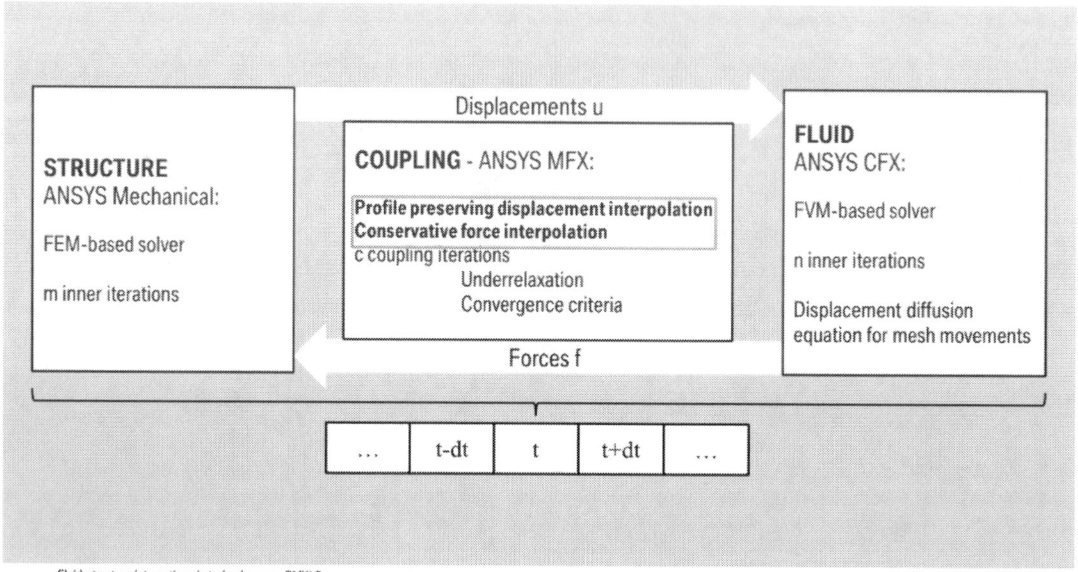

FLUID-STRUCTURE INTERACTIONS. INTERPOLATION AT THE INTERFACE.

- Different meshes at FSI-interface
- Interpolation algorithms necessary for transmission of displacements and forces
- Implemented interpolation algorithms work well

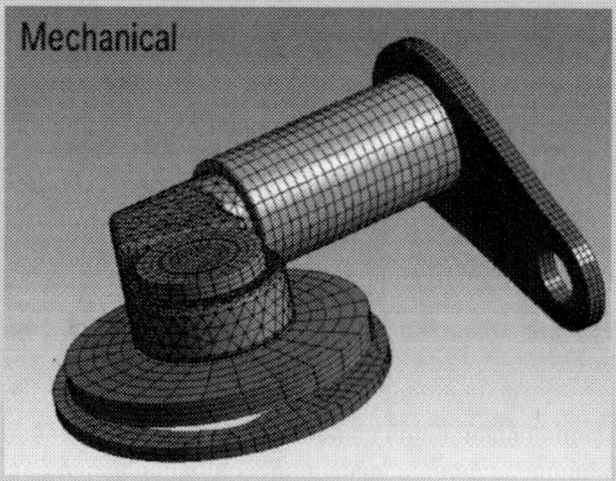

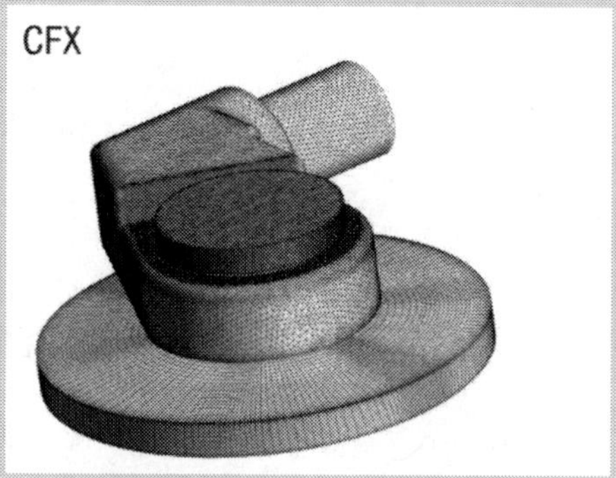

FLUID-STRUCTURE INTERACTIONS. THE CHALLENGES.

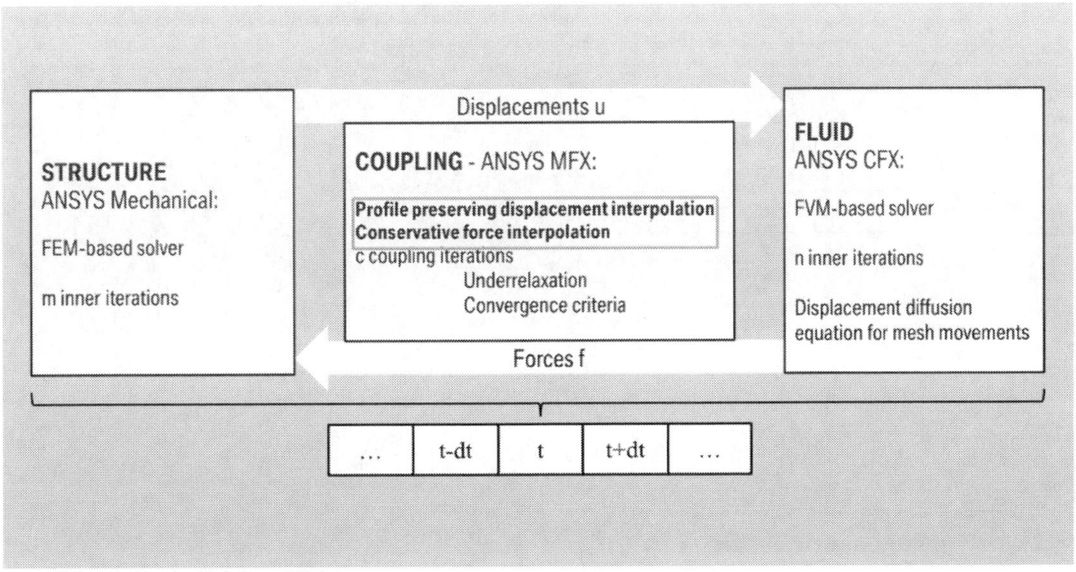

Fluid-structure interactions in turbochargers, BMW Group, 29.10.2013

Seite 17

FLUID-STRUCTURE INTERACTIONS. THE CHALLENGES.

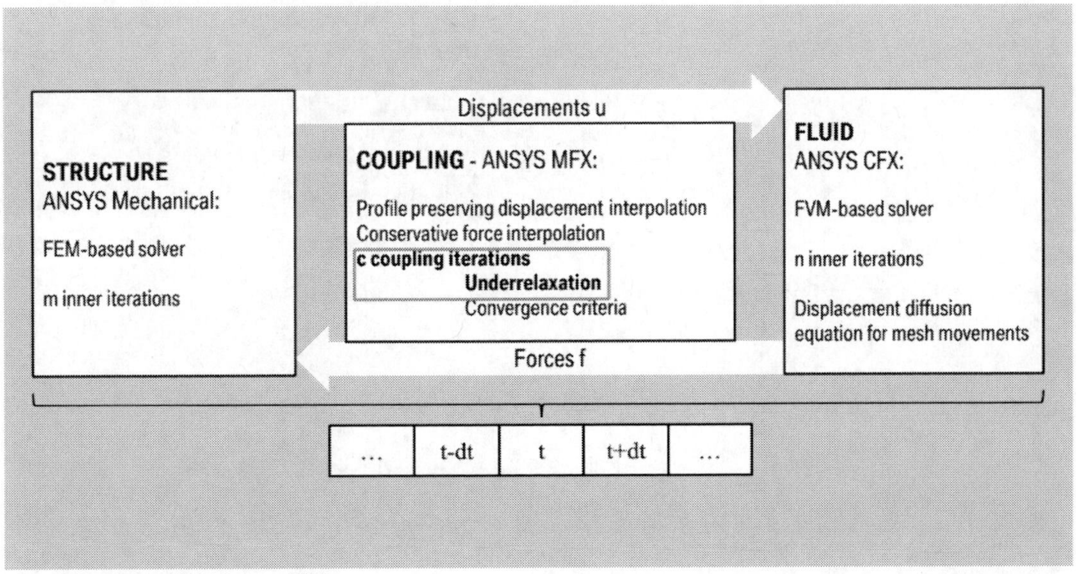

Fluid-structure interactions in turbochargers, BMW Group, 29.10.2013

Seite 18

FLUID-STRUCTURE INTERACTIONS.
THE ARTIFICIAL ADDED MASS EFFECT.

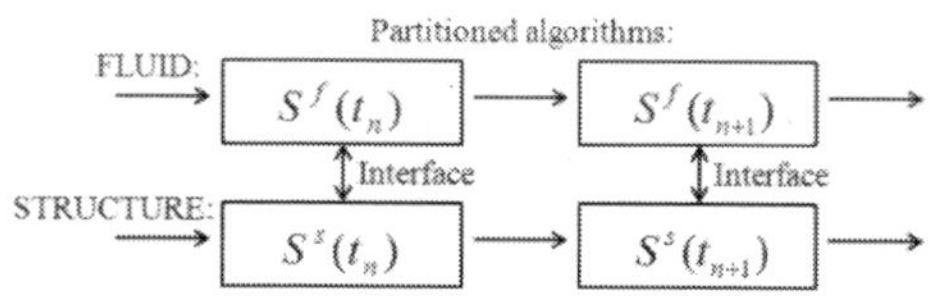

- Partitioned algorithms are naturally unstable
- Software independent

FLUID-STRUCTURE INTERACTIONS.
THE ARTIFICIAL ADDED MASS EFFECT.

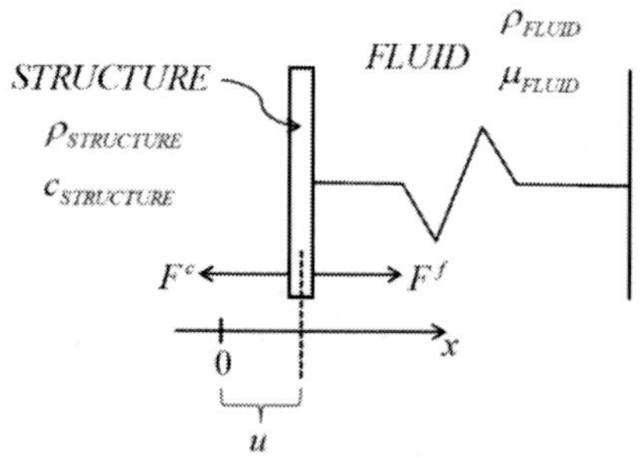

- Partitioned algorithms are naturally unstable
- Software independent
- Physical properties have a big influence on stability of algorithm
  - $c_{STRUCTURE}$
  - μ_{FLUID}
  - $\rho_{FLUID} / \rho_{STRUCTURE}$

FLUID-STRUCTURE INTERACTIONS. THE ARTIFICIAL ADDED MASS EFFECT.

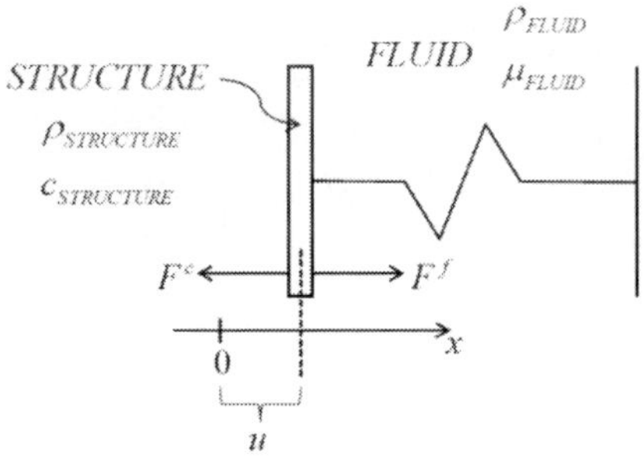

$$u_t^i = u_t^{i-1} + \alpha_u(u_t^{i,explizit} - u_t^{i-1})$$
$$f_t^i = f_t^{i-1} + \alpha_f(f_t^{i,explizit} - f_t^{i-1})$$

- Partitioned algorithms are naturally unstable
- Software independent
- Physical properties have a big influence on stability of algorithm
  - $c_{STRUCTURE}$
  - μ_{FLUID}
  - $\rho_{FLUID} / \rho_{STRUCTURE}$
- Underrelaxation techniques need to be applied with the underrelaxation factors α_f , α_u

FLUID-STRUCTURE INTERACTIONS. THE CHALLENGES.

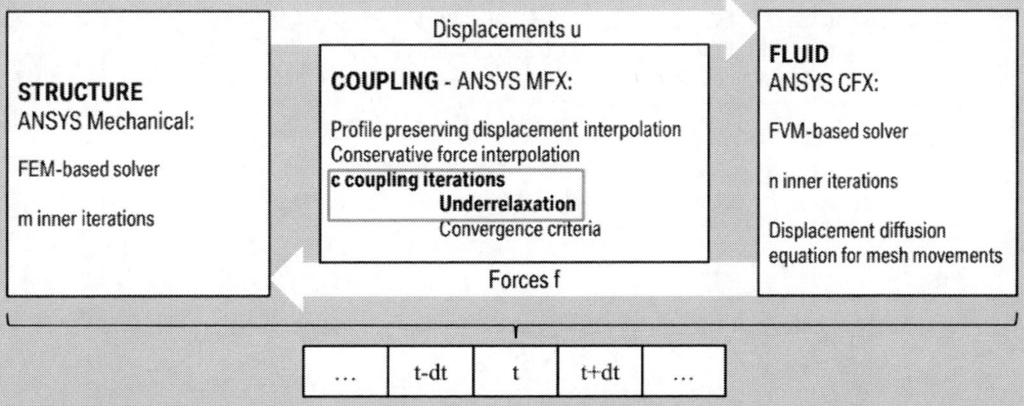

FLUID-STRUCTURE INTERACTIONS.
THE CHALLENGES.

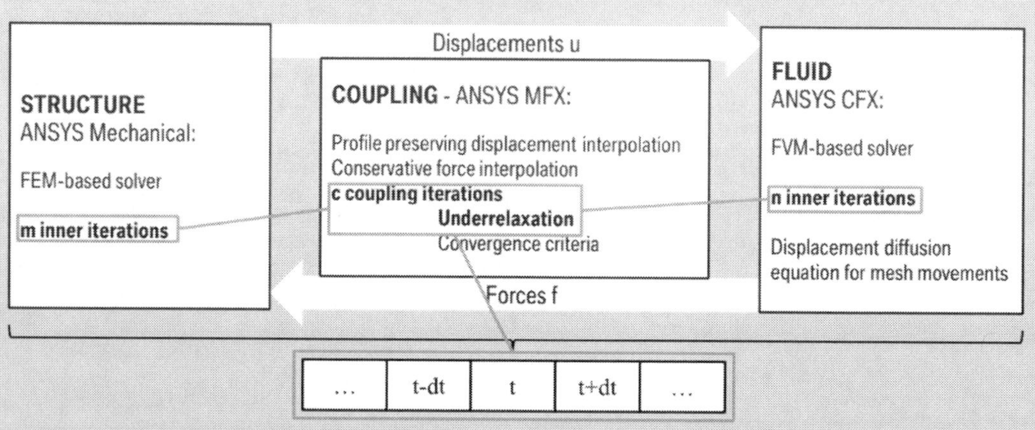

FLUID-STRUCTURE INTERACTIONS.
RUNTIME.

– Example:

t time steps
  c coupling iterations
    n inner CFX iterations
    m inner Mechanical iterations

– Runtime: $O((n+m)\cdot c\cdot t)$

– Number of iterations per time step might be quite big

FLUID-STRUCTURE INTERACTIONS.
RUNTIME.

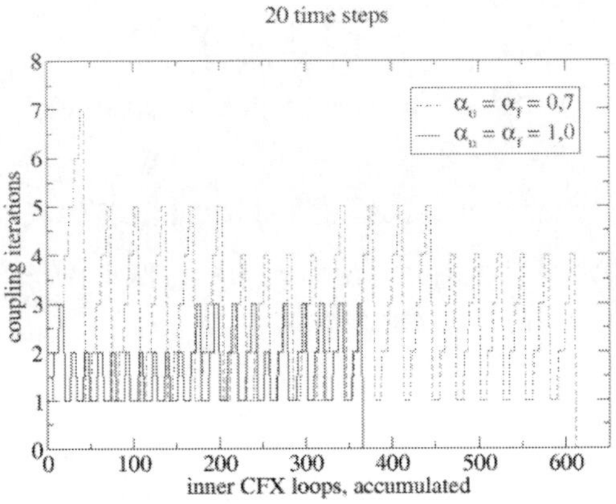

- Number of iterations per time step might be quite big
- Choice of underrelaxation factor is crucial for the runtime of the transient simulations

FLUID-STRUCTURE INTERACTIONS.
RUNTIME.

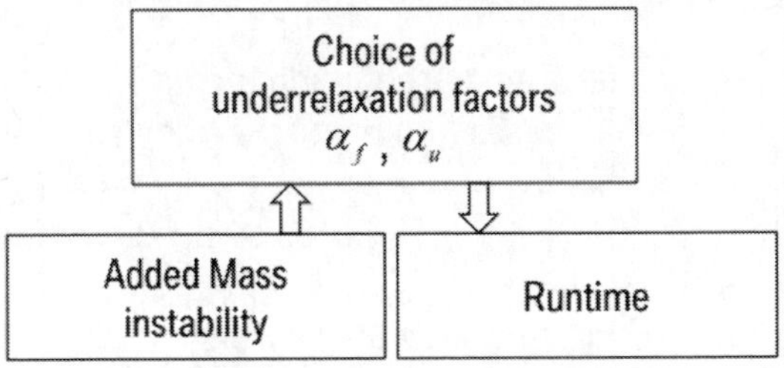

- Number of iterations per time step might be quite big
- Choice of underrelaxation factor is crucial for the runtime of the transient simulations
- What about the Added Mass Effect?

FLUID-STRUCTURE INTERACTIONS. RUNTIME.

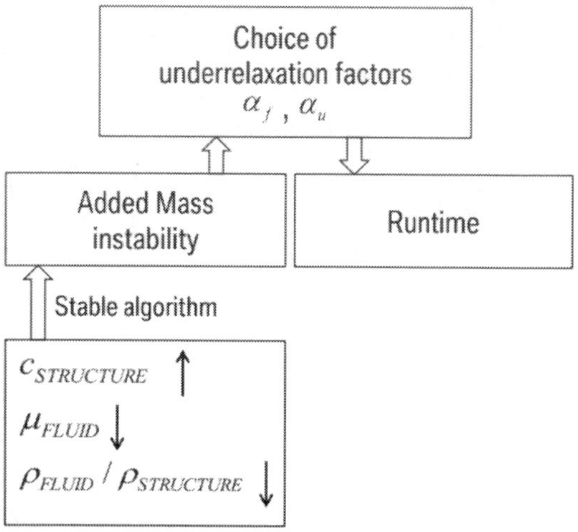

- Number of iterations per time step might be quite big
- Choice of underrelaxation factor is crucial for the runtime of the transient simulations
- What about the Added Mass Effect?
  - Properties in turbochargers fit very well to these conditions!

FLUID-STRUCTURE INTERACTIONS. RUNTIME.

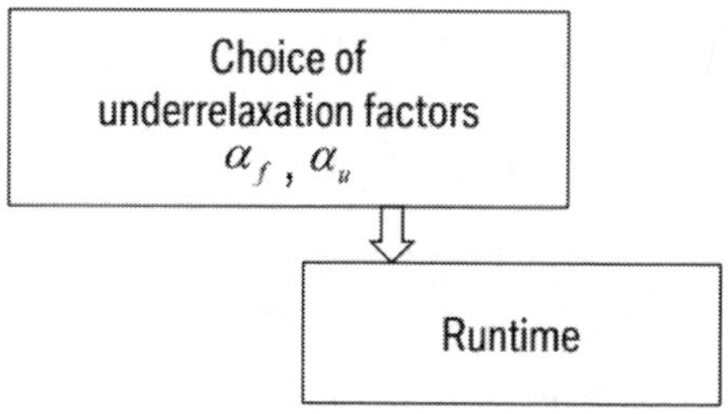

- Number of iterations per time step might be quite big
- Choice of underrelaxation factor is crucial for the runtime of the transient simulations
- What about the Added Mass Effect?
  - Properties in turbochargers fit very well to these conditions!
- Several parameter studies:
  - Default setting: + ~200% runtime
  - Improved setting: + ~40% runtime

FLUID-STRUCTURE INTERACTIONS. THE CHALLENGES.

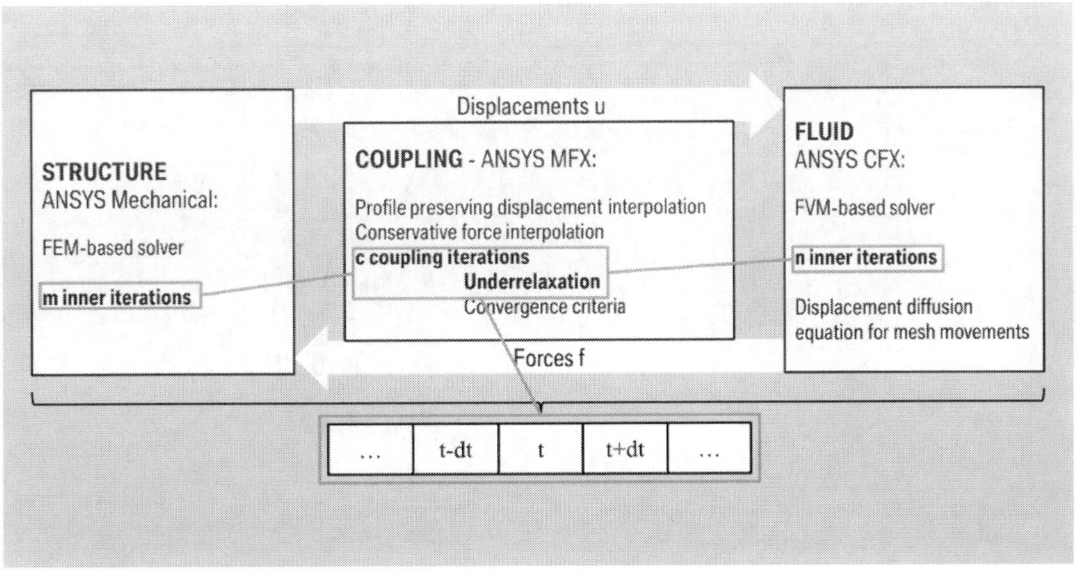

Fluid-structure interactions in turbochargers, BMW Group, 29.10.2013

Seite 29

FLUID-STRUCTURE INTERACTIONS. THE CHALLENGES.

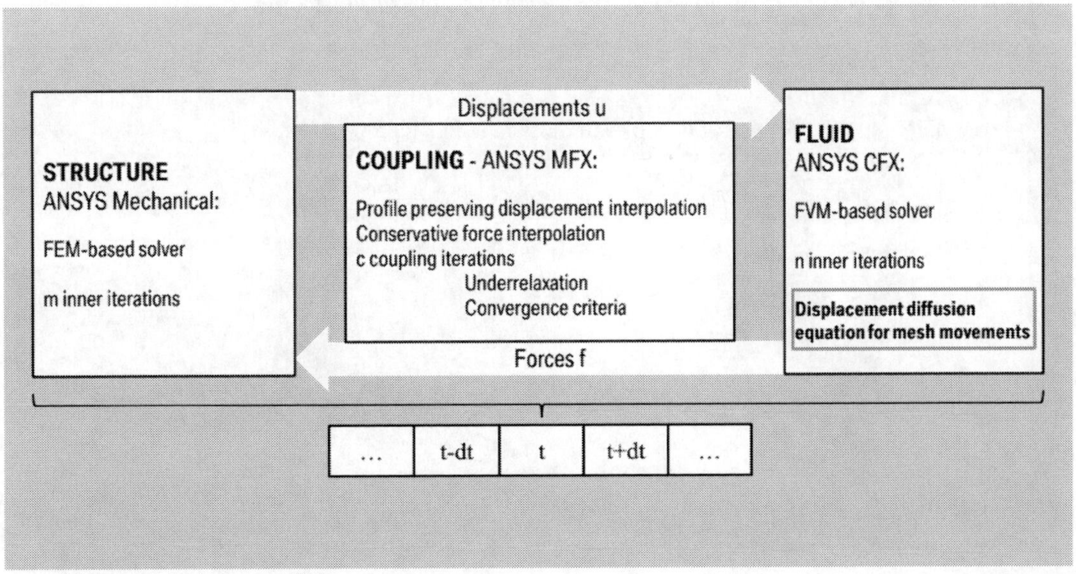

Fluid-structure interactions in turbochargers, BMW Group, 29.10.2013

Seite 30

FLUID-STRUCTURE INTERACTIONS. MESH MOVEMENT.

- Mesh movement problem is solved with displacement diffusion equation
- Leads to
  - small element angles
  - negative volumes
- Parameter studies necessary to choose stiffness function appropriately

$$\nabla \cdot (\Gamma_{disp} \nabla \delta) = 0$$

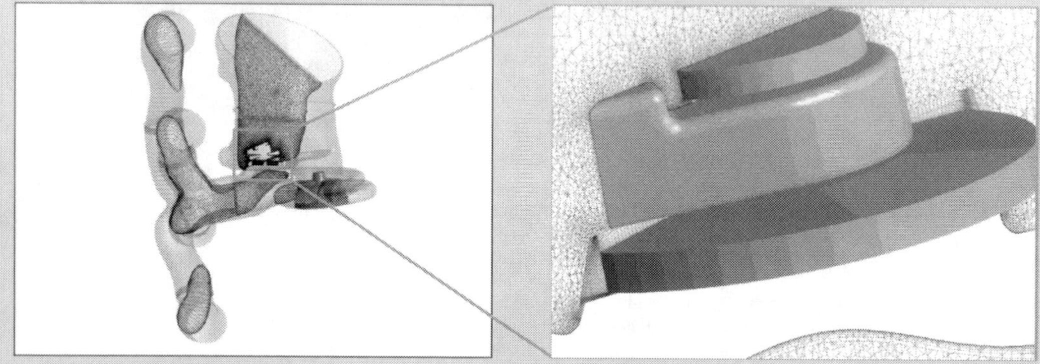

Fluid-structure interactions in turbochargers, BMW Group, 29.10.2013

Seite 31

FLUID-STRUCTURE INTERACTIONS. MESH MOVEMENT.

- Mesh movement problem is solved with displacement diffusion equation
- Leads to
  - small element angles
  - negative volumes
- Parameter studies necessary to choose stiffness function appropriately

$$\nabla \cdot (\Gamma_{disp} \nabla \delta) = 0$$

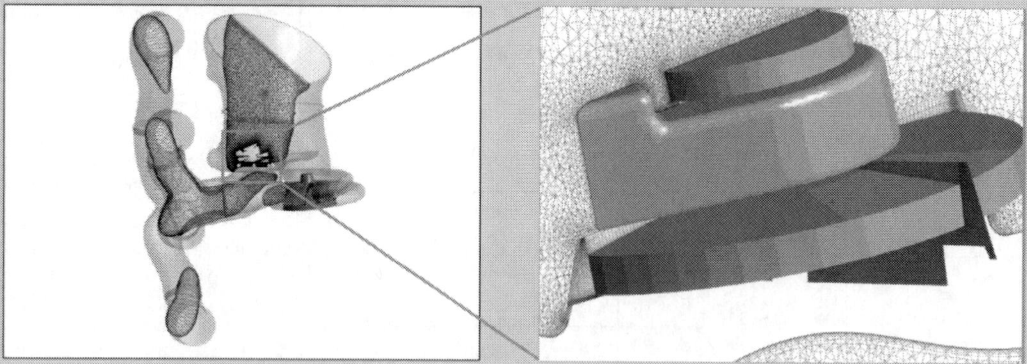

Fluid-structure interactions in turbochargers, BMW Group, 29.10.2013

Seite 32

FLUID-STRUCTURE INTERACTIONS. MESH MOVEMENT.

- Mesh movement problem is solved with displacement diffusion equation
- Leads to
  - small element angles
  - negative volumes
- Parameter studies necessary to choose stiffness function appropriately

$$\nabla \cdot (\Gamma_{disp} \nabla \delta) = 0$$

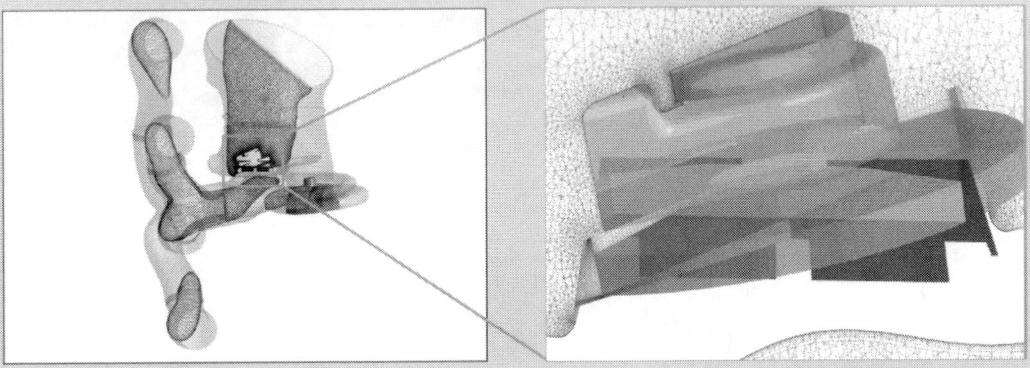

Fluid-structure interactions in turbochargers, BMW Group, 29.10.2013

Seite 33

FLUID-STRUCTURE INTERACTIONS. MESH MOVEMENT.

- Mesh movement problem is solved with displacement diffusion equation
- Leads to
  - small element angles
  - negative volumes
- Parameter studies necessary to choose stiffness function appropriately

$$\nabla \cdot (\Gamma_{disp} \nabla \delta) = 0$$

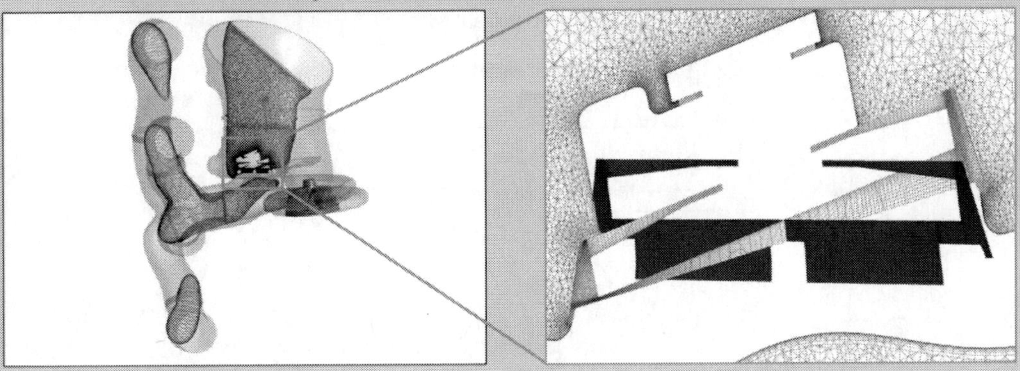

Fluid-structure interactions in turbochargers, BMW Group, 29.10.2013

Seite 34

FLUID-STRUCTURE INTERACTIONS. THE CHALLENGES.

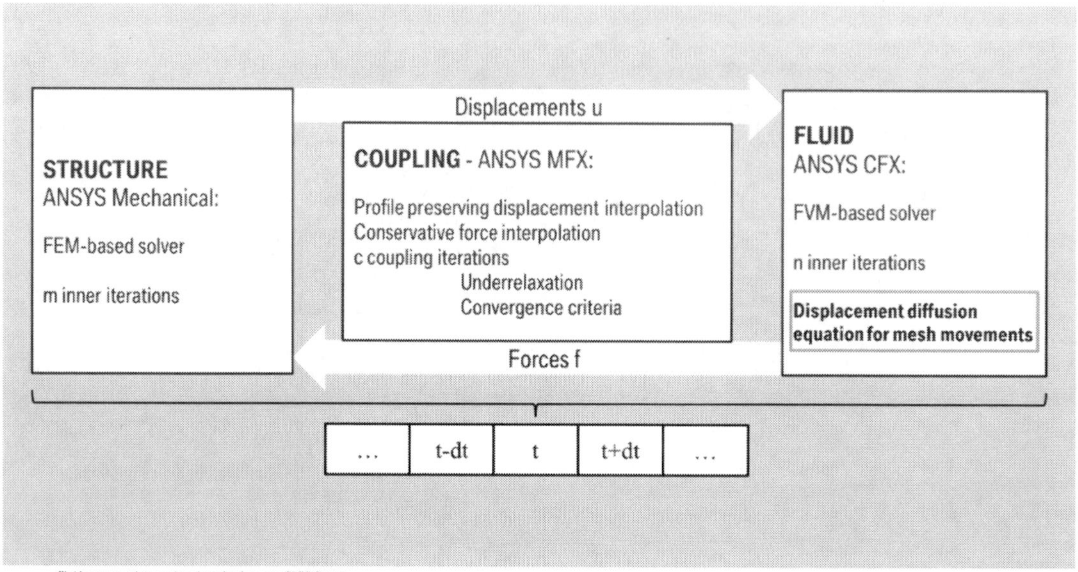

FLUID-STRUCTURE INTERACTIONS. THE CHALLENGES.

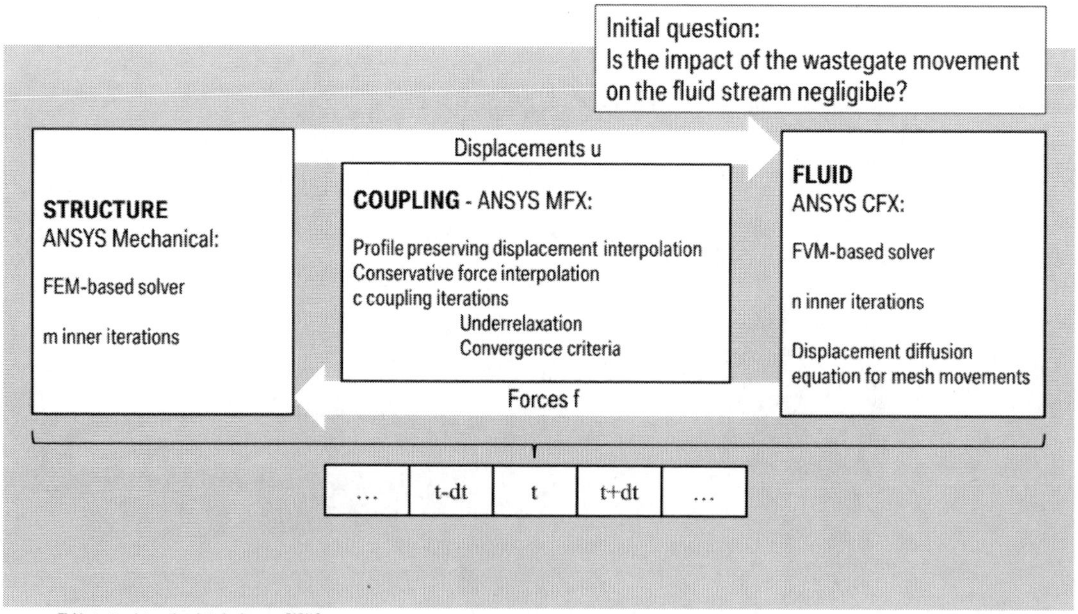

FLUID-STRUCTURE INTERACTIONS. ONE-WAY-COUPLED SIMULATIONS.

Initial question:
Is the impact of the wastegate movement on the fluid stream negligible?

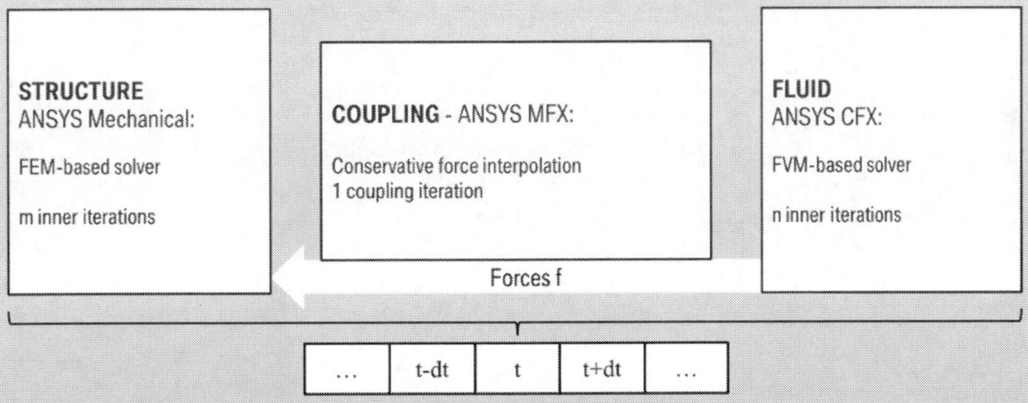

FLUID-STRUCTURE INTERACTIONS. ONE-WAY-COUPLED SIMULATIONS.

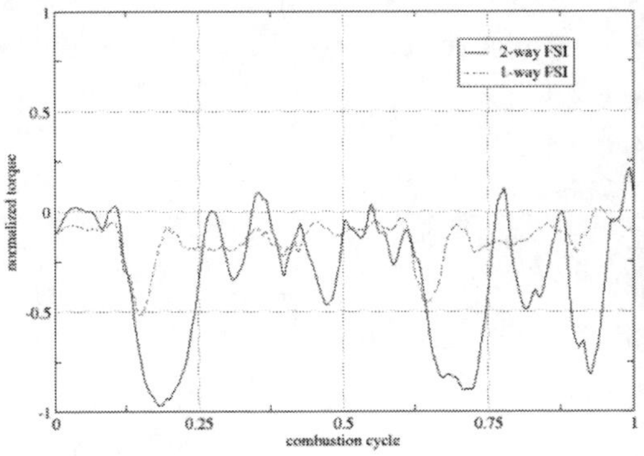

Initial question:
Is the impact of the wastegate movement on the fluid stream negligible?

Answer:
No, our studies have shown that there are severe differences in some of the calculated loads on the wastegate.

FLUID-STRUCTURE INTERACTIONS. ONE-WAY-COUPLED SIMULATIONS.

Second question:
How good can we model the wastegate movement in the simulations?

FLUID-STRUCTURE INTERACTIONS. RESULTS.

– The load on and behaviour of the wastegate can be described very detailed with the help of FSI.

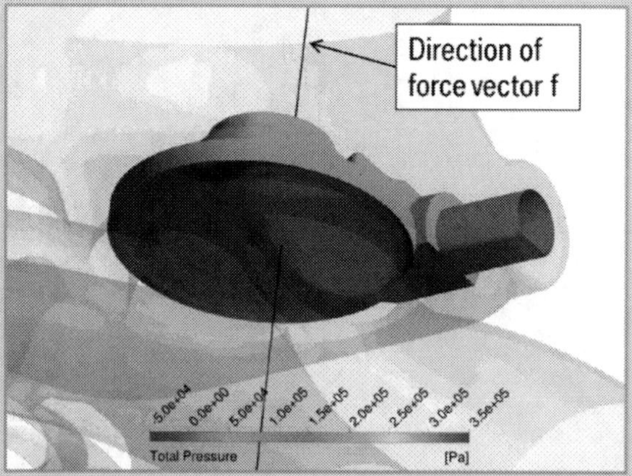

FLUID-STRUCTURE INTERACTIONS. RESULTS.

- The load on and behaviour of the wastegate can be described very detailed with the help of FSI.
- Qualitative comparisons between experiments at the test rig and the simulation are in good agreement.

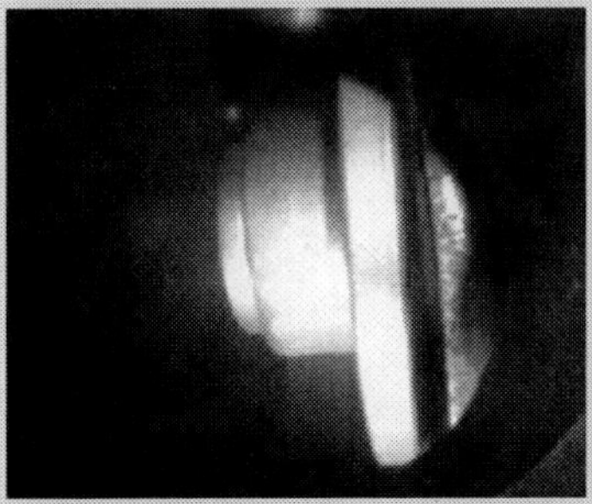

FLUID-STRUCTURE INTERACTIONS. RESULTS.

- The load on and behaviour of the wastegate can be described very detailed with the help of FSI.
- Qualitative comparisons between experiments at the test rig and the simulation are in good agreement.
- Analysis of frequency spectrum of the rattling plate in simulations fits to expectations.

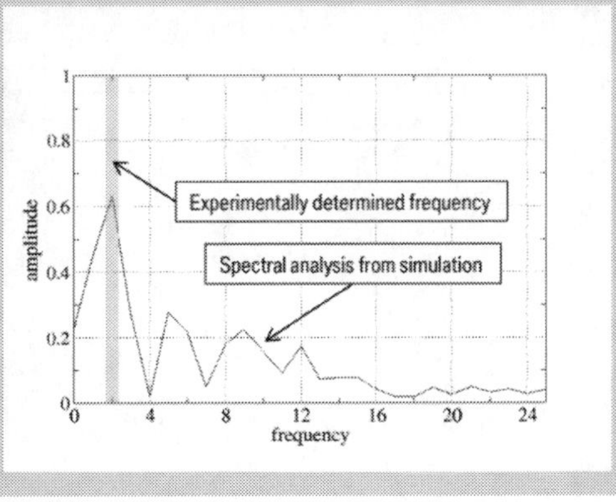

FLUID-STRUCTURE INTERACTIONS.
SUMMARY.

- The data interpolation at the FSI-interface works.
- The runtime can be reduced significantly by performing detailed parameter studies and choosing the coupling settings appropriately.
- The physical properties of the fluid flow in turbochargers fits nicely to the conditional Added Mass instability.
- The mesh movements turn out to be the biggest challenge as there are thin and angled structures around the wastegate that do no allow a lot of mesh morphing.
- There are severe differences in the calculated results of one-way- and two-way-coupled simulations.
- Results can be validated qualitatively with the help of test rig experiments.

FLUID-STRUCTURE INTERACTIONS.
FUTURE DIRECTIONS.

- Include thermal effects in coupling process.

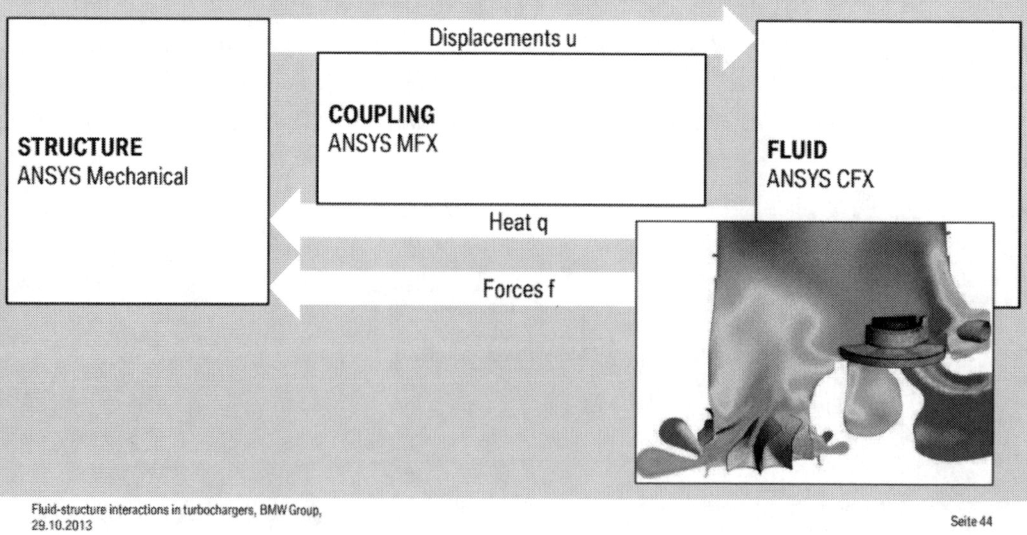

FLUID-STRUCTURE INTERACTIONS. FUTURE DIRECTIONS.

– Include thermal effects in coupling process.
– Parametrized automatic remeshing around wastegate to increase stability of solving process.

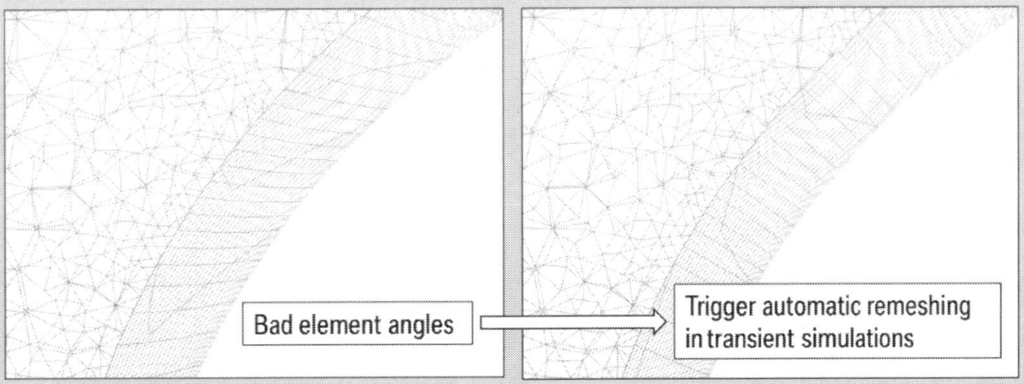

Fluid-structure interactions in turbochargers, BMW Group, 29.10.2013

Seite 45

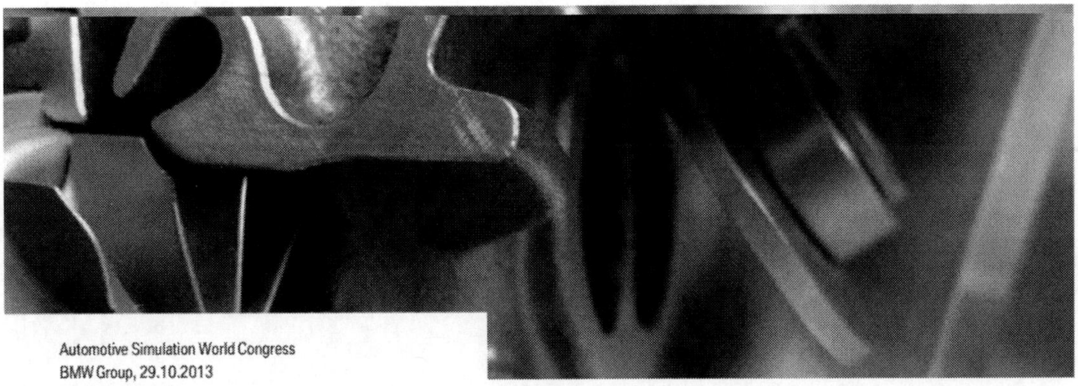

Automotive Simulation World Congress
BMW Group, 29.10.2013
Nicolas Lachenmaier

THANK YOU FOR YOUR ATTENTION!

ANY QUESTIONS?

BMW GROUP

ANSYS Inc.
Southpointe
2600 ANSYS Drive
Canonsburg, PA 15317
USA

CURRAN ASSOCIATES INC.
proceedings.com

ISBN 978-1-5108-4102-4
9 781510 841024